Physiological Reference Values

Parameter	Normal range	Units
Cellular elements of the blood		
Hematocrit	40 to 54	%
Hemoglobin	14 to 18	gm/dl
Red blood cells	4.6 to 6.2	Million/mm³
White blood cells	5000 to 10,000	Cell/mm³
Blood plasma		
Albumin	3.2 to 5.6	gm/dl
Bicarbonate	21 to 27	mEq/L
Calcium (total)	4.3 to 5.3	mEq/L
Calcium (ionized)	2.1 to 2.6	mEq/L
Chloride	95 to 103	mEq/L
Cholesterol	150 to 250	mg/dl
Immunoglobulins	2.3 to 3.5	gm/dl
Glucose	65 to 100	mg/dl
Iron	50 to 150	μg/dl
Lactic acid	5 to 20	mg/dl
Magnesium	1.5 to 2.5	mEq/L
Osmolarity	280 to 290	mOsm/L
pH	7.35 to 7.45	
Phosphate	1.8 to 2.6	mEq/L
Potassium	4.0 to 4.8	mEq/L
Sodium	136 to 142	mEq/L
Sulfate	0.2 to 1.3	mEq/L
Urea	8 to 20	mg/dl
Uric acid	2.1 to 7.6	mg/dl

Parameter	Average value	Units
The lung volumes		
Static lung volumes and capacities		
Alveolar ventilation ($\dot{V}_{Alv}$)	5000	ml/min
Anatomical dead space (VD)	150	ml
Expiratory reserve volume (ERV)	1200	ml
Functional residual capacity (FRC)	2500	ml
Inspiratory capacity (IC)	3500	ml
Minute volume ($\dot{V}$)	7500	ml/min
Residual volume (RV)	1200	ml
Tidal volume (TV)	500	ml
Total lung capacity (TLC)	6000	ml
Vital capacity (VC)	5000	ml
Dynamic lung volumes		
Forced expiratory volume ($FEV_{1.0}$)	4000	ml
Forced vital capacity (FVC)	5000	ml

Parameter		
The cardiovascular system		
Cardiac output	4.0 to 8.0	L/min
A-VO$_2$ difference	3.5 to 4.7	ml/dl
Systemic arterial		
Central venous pressure	2 to 7	mm Hg
Diastolic pressure	60 to 90	mm Hg
Mean pressure	70 to 105	mm Hg
Systolic pressure	90 to 140	mm Hg
Pulmonary arterial		
Diastolic pressure	4 to 13	mm Hg
Mean pressure	9 to 19	mm Hg
Systolic pressure	17 to 32	mm Hg
Left ventricle		
Diastolic pressure	5 to 12	mm Hg
Systolic pressure	90 to 140	mm Hg
The blood gases		
Arterial blood gases		
Carbon dioxide content	47 to 51	ml/dl
Carbon dioxide partial pressure (P_{CO_2})	35 to 45	mm Hg
Oxygen content	16.5 to 20.0	ml/dl
Oxygen partial pressure (P_{O_2})	93 to 97	mm Hg
Oxygen saturation	96 to 98	Percent
Venous blood gases		
Carbon dioxide content	51 to 55	ml/dl
Carbon dioxide partial pressure (P_{CO_2})	41 to 51	mm Hg
Oxygen partial pressure (P_{O_2})	35 to 45	mm Hg
Oxygen saturation	70 to 80	Percent
The kidney and urine		
Ammonia (normal)	20 to 70	mEq/24 hr
Ammonia (acid load)	70 to 160	mEq/24 hr
Calcium	100 to 250	mEq/24 hr
Chloride	110 to 250	mEq/24 hr
Creatinine clearance	100 to 135	ml/min
Osmolarity (normal)	500 to 800	mOsm/L
Osmolarity (extremes)	40 to 1400	mOsm/L
pH (normal)	6.0 to 6.6	
pH (extremes)	4.6 to 8.0	
Phosphate	0.9 to 1.3	gm/24 hr
Potassium	25 to 100	mEq/24 hr
Sodium	130 to 260	mEq/24 hr
Urea	6 to 17	gm/24 hr
Uric acid	250 to 750	mg/24 hr
Volume (normal)	600 to 1600	ml/24 hr

Human Physiology
Foundations & Frontiers

SECOND EDITION

*In the Möbius strip, the continuity of the loop is maintained
despite the continuous shifting of its plane. We chose the Möbius
strip as a physiological motif because in a homeostatic state,
the internal environment is not kept absolutely constant,
but rather is held within limits of fluctuations that the body
can tolerate without disruption of function.
The Möbius strip is constructed from a rectangular strip by
holding one end fixed, rotating the opposite end through 180°
and attaching it to the first end.*

Human Physiology
Foundations & Frontiers

DAVID F. MOFFETT

Associate Professor
Department of Zoology
Program in Zoophysiology
Washington State University

STACIA B. MOFFETT

Associate Professor
Department of Zoology
Program in Zoophysiology
Biological Sciences Program
Washington State University

CHARLES L. SCHAUF

 Mosby

St. Louis Baltimore Boston Chicago London Philadelphia Sydney Toronto

Editor-in-Chief: James M. Smith
Editor: Robert J. Callanan
Senior Developmental Editor: Jean Babrick
Project Manager: Mark Spann
Production Editors: Daniel J. Johnson, Kathy Teal, Teresa Breckwoldt
Book Designer: Susan Lane
Illustrators: Barbara DesNoyer Stackhouse
        Gail Morey Hudson
        Jack Tandy
        Graphic Works Inc.
Cover Photograph: William Leslie Photography

Credits for all materials used by permission appear after the glossary.

SECOND EDITION

Printed in the United States of America

Mosby-Year Book, Inc
11830 Westline Industrial Drive
St. Louis, Missouri 63146

Library of Congress Cataloging-in-Publication Data

Moffett, David F.
    Human physiology : foundations & frontiers / David F. Moffett,
Stacia B. Moffett, Charles L. Schauf. — 2nd ed.
      p.   cm.
    Schauf's name appears first on the earlier edition.
    Includes bibliographical references and index.
    ISBN 0-8016-6903-0
    1. Human physiology.   I. Moffett, Stacia B.   II. Shauf, Charles
L.   III. Title.
    [DNLM: 1. Physiology.   QT 104 M6953h]
QP36.S32   1993
612—dc20
DNLM/DLC
for Library of Congress   92-48327
                                     CIP

93 94 95 96 97 Cl/VH/VH 9 8 7 6 5 4 3 2 1

Stacia Moffett and **David Moffett** received their Ph.D. degrees in Biology from the University of Miami in Coral Gables. They did postdoctoral training in the Departments of Physiology, Pharmacology, and Biomedical Engineering at Duke University. Currently at Washington State University, they teach human, mammalian and comparative physiology in the Department of Zoology and teach in the Graduate Program in Zoophysiology. Their research, funded by the National Science Foundation and the National Institutes of Health, emphasizes the use of invertebrates for learning about basic physiological processes. Stacia studies neurobiology and endocrinology, especially regeneration in the central nervous system and muscle and, also, the control of reproduction by neuropeptides in molluscs. David, a membrane physiologist, studies the cellular mechanisms of transepithelial ion and nutrient transport in both vertebrate and invertebrate animals. He also serves as Coordinator of the premedical and predental curriculum at the university. In addition to teaching, research, and writing, they enjoy hiking, sailing, and listening to music with their three children.

Dr. Charles L. Schauf received a B.S. degree with honors from the University of Chicago in 1965 and a Ph.D. degree in physics in 1969. Following a 2-year postdoctoral fellowship at the University of Maryland, Dr. Schauf was appointed Assistant Professor of Physiology at Rush University, Chicago, in 1972. He was promoted to Associate Professor in 1975 and to Full Professor in 1979. At Rush, Dr. Schauf directed the program in physiology, organized a human physiology course for nursing students, and eventually assumed administrative responsibility as Associate Chairman of the Physiology Department. Dr. Schauf was chairman of the Biology Department at Indiana University—Purdue University at Indianapolis from 1986 to 1992, where he also taught a 2-semester sequence in human physiology and endocrinology to a broad range of undergraduates. Dr. Schauf's research in the general area of neurophysiology and membrane biophysics has been supported by the National Science Foundation and the National Multiple Sclerosis Society. Dr. Schauf has published over 80 research papers and has served on several national advisory review groups. In his leisure time, Dr. Schauf enjoys backpacking, cycling, and tennis.

Preface

*H*uman physiology is the science that explains the normal operation of the human machine. To understand that operation, the body must be studied at all levels of integration, from the submolecular to the whole organism. Each level is a web of complex relationships. Molecular biology has shown that the ultimate components of the machine—proteins—can be understood as machines in themselves. Cells, long considered the basic building blocks of organisms, are now understood as highly organized intercommunicating systems. At the highest level of complexity, the brain (probably the most complex structure known) cannot be understood by study of molecular interactions aone but requires a highly integrated approach.

Today's students of human physiology may be preparing for careers in health care delivery or health related research or education. Others simply are curious about how their own bodies work. For all of our students, our goal in preparing the second edition of *Human Physiology: Foundations & Frontiers,* as it was in the first edition, was to provide an up-to-date grasp of the subject as well as to confer the excitement of its current discoveries. We included a sense of the history of the science by giving credit to some of its important pioneers. Respecting the intelligence of today's students and the awesome complexity of the human organism, we endeavored to present clear, complete explanations without distorting oversimplification. Early in the text we distinguish between "how" and "why" explanations of body functions and related the latter to human evolution.

Elementary textbooks of science tend to present as revealed fact what is in actuality only current consensus. By including some of the experimental background for key insights into how the body works, we remind students that experimentation, including animal studies, is a primary source of information in physiology. Where important questions remain unanswered, we point this out and give current hypotheses.

Although this text focuses mainly on the normal function of the body, we introduce disease conditions where they illustrate fundamental physiological principles. This challenges the students to synthesize information and begin applying the course content to conditions they will encounter, whether in medical practice or in everyday life.

The working lives of the students who use the second edition of *Human Physiology: Foundations & Frontiers* will extend well into the next century. They will live and work in a time of enormous promise and challenge. During this time, the entire human genome may be described. Devastating diseases such as AIDS, Alzheimer's syndrome, cystic fibrosis, and multiple sclerosis can probably be cured or prevented, and other as yet unforeseen diseases may arise. A higher standard of health and well-being can be extended to all people through the future developments in human physiology and its allied sciences. We hope to endow our students with the tools and the inspiration to take up these tasks.

ORGANIZATION

Human Physiology: Foundations & Frontiers is divided into seven sections. Section I presents the Foundations of Physiology. The first chapter discusses the origins of the science of physiology and introduces the encompassing concept of homeostasis in physiological systems. It also introduces the tissue and organ systems of the body to provide a basis for the discussion of structure-function relationships found later in the text. Chapters 2, 3, and 4 encompass the highlighted *Foundations Unit,* which presents introductory concepts in cell biology, chemistry, physics, and mathematics with appropriate physiological examples. Chapter 5 discusses feedback mechanisms and control system loops involved in the maintenance of homeostasis. Local, neural, reflex, and hormonal control are introduced at this point in the text because they are fundamental to the integration and regulation of all body systems. Finally, Chapter 6 discusses transport across cell membranes and the establishment of an electrical resting potential in all living cells.

Section II begins with a discussion of action potentials in Chapter 7, followed by coverage of neurophysiological principles of transmission and integration in Chapter 8. This information is then used to show how stimulus-sensitive ion channels allow sensory receptors to convert external events to an internal neural code, and how the central nervous system processes sensory signals (Chapters 9 and

10). Chapter 11 presents the physiology of skeletal, cardiac, and smooth muscle. The body usually reacts to the environment by movements, changes in visceral muscle function, or secretion from glands, each mediated by components of the somatic and autonomic motor systems (Chapter 12).

Section III examines the functional interconnection of organ systems by the cardiovascular system. The properties of individual elements of the cardiovascular system—heart, arteries, capillaries, and veins—are covered first, followed by a description of neural and hormonal control of the cardiovascular system. Because the cardiovascular and respiratory systems combine to deliver oxygen to cells and remove carbon dioxide, the treatment of the respiratory system follows the cardiovascular system in Section IV. Renal physiology is covered in Section V because the kidney regulates salt and water balance and helps adjust blood pH. These functions relate directly to cardiovascular and respiratory functions.

Sections III through V of *Human Physiology: Foundations & Frontiers,* covering cardiovascular, respiratory, and renal physiology, compose almost a third of the text and, we believe, fittingly so. Not only do these systems include the "heart" and "breath" of modern physiology, but abnormalities in these three systems are responsible for the majority of human illnesses. Undergraduates who are planning careers as health professionals will benefit from the care we have taken to present this material at a level of detail that is both comfortable and complete.

Section VI covers gastrointestinal physiology. It begins with a description of the motility and secretion characteristics of each region, along with the neural and hormonal control of these processes. Next, students learn how the basic food groups are broken down by the gastrointestinal tract into nutrients that can be carried in the blood. The section concludes with the mechanisms and implications of nutrient absorption.

The first chapter of Section VII follows absorbed nutrients as they enter pathways of synthesis and storage during the absorptive state. It then shows how nutrients are released from storage during the postabsorptive state. The next two chapters of this section discuss reproductive physiology as well as development, birth, and the physiology of the newborn. The final chapter shows how the immune system protects the body from various potentially destructive invaders.

Endocrine Coverage

Most physiology texts cover the subject of endocrinology on a gland-by-gland basis. In contrast, we have chosen to describe general endocrine mechanisms early in the text and to reserve discussion of individual hormones until they are encountered in the context of other systems. Throughout Sections

III through VII of *Human Physiology: Foundations & Frontiers* there is an emphasis on how organ systems are regulated and how they interact. Neural and hormonal mechanisms are integral to this coverage. For this reason, we discuss each of the hormones and their effects as they are encountered in our presentation of systems physiology. We believe that this synthesis better enables students to understand the interactions among systems and the role each plays in the overall homeostasis of the body.

A supplement to this approach is provided by the *Focus Unit* on endocrine glands located at the end of Chapter 5. Expanded from the first edition, this *Focus Unit* presents, in a more traditional form, the location and function of endocrine glands. This overview allows students to grasp the entirety of the endocrine system and is also a valuable reference throughout the course.

TEACHING AND LEARNING AIDS

Each chapter begins with an outline of the contents and is prefaced with a list of learning **objectives.** The text of the chapter begins with a short, highly readable **overview** that presents the major concepts of the chapter without employing unfamiliar terms. Each major division of the chapter begins with a few major questions to refocus student's attention on the relevant chapter objectives. New and unfamiliar terms are given in boldface, fully defined when first used, and available for reference in the glossary. Each chapter concludes with a numbered summary, study questions, and suggested reading including both popular and authoritative sources. Answers to some study questions are provided in an appendix at the back of the book. This arrangement evolved with the help of reviewers who shared their years of teaching experience with the first edition of this text and other physiology textbooks.

We have included more than 60 essays. Focus on Physiology essays provide historical anecdotes, clinical applications or in-depth coverage of specific topics. Frontiers essays describe recent scientific advances. These have been very popular in the first edition and their number is increased in the second edition.

Focus Units located at the end of selected chapters are intended to provide students with optional, more detailed coverage of specific topics. These include: Endocrine Glands and Their Functions (Chapter 5), Diffusion, Osmosis, and Bioelectricity (Chapter 6) How Drugs Affect the Nervous System (Chapter 9), Exercise (Chapter 18) and Acids, Bases, and Buffer Systems (Chapter 20).

Tables present information in a logical and condensed form that is easy for students to compare, contrast, and retain.

Illustrations are central to communication of physiological relationships. All illustrations use color to enhance comprehension. Approximately one-third of the illustrations were prepared by Barbara Stackhouse, working closely with the authors. Many new illustrations have been added and old ones have been revised to accommodate revisions in the second edition. Each chapter includes a number of flow charts that allow students to visualize the relationships involved in the qualitative relationships of physiological processes. Graphs are included where necessary to show quantitative relationships.

The first edition of *Human Physiology: Foundations & Frontiers* was unique among physiology texts in its extensive use of photographic illustrations, many of which were made expressly for this text. The second edition incorporates more than 100 photographs and photomicrographs, many of which are new.

Reference information is provided on the inside covers of the book on such topics as measurements, standard physiological values, and conversion factors.

Development of the Second Edition

To meet the challenge of preparing a text that is thoroughly up-to-date, we drew on all possible resources. We combined our experiences in teaching undergraduate physiology with our own research backgrounds and extensive survey of the contemporary scientific literature. We have incorporated suggestions and corrections from students, colleagues, and a number of expert reviewers. Every chapter was revised for better continuity and organization and to reflect current developments in the field. The most current explanations of contraction of skeletal, cardiac, and smooth muscle are presented. In particular, the chapters on cardiovascular physiology, the nervous system, reproduction, and immunology, were extensively revised.

New boxed essays were chosen both for timeliness and to illustrate basic physiological principles. Topics include:

An explanation of how cystic fibrosis can be traced to a genetic defect in membrane CI-channels.

The importance of new kinds of microscopes for physiology

Lithotripsy

Cardiovascular "spare parts"

The use of "clot buster" enzymes in treating coronary artery disease

Endometriosis

The mechanism and use of RU486 as an abortion inducer.

The AIDS crisis

The role of cytokines in the immune response

SUPPLEMENTS

Human Physiology: Foundations & Frontiers is accompanied by a complete set of supplements to aid both students and instructors in dealing with what sometimes seems to be an immense amount of material. As with the text itself, we were concerned about producing supplements of extraordinary quality and utility. All supplements that are intended for use by students were thoroughly reviewed by panels of human physiology instructors. Often these were the same instructors who reviewed the text.

Study Guide

Written by Ann Vernon, an experienced community college instructor at St. Charles County Community College, in conjunction with Charles Schauf, the *Study Guide* far surpasses study guides that accompany other texts of this type. A direct companion to the text, this study guide provides the student with valuable study aids such as a brief focus on key material, word part and key term exercises, a series of varied activities such as matching and labeling, quick recall, a set of concept checks for review, and a multiple choice practice test. Answers to all questions are also provided. A unique feature of the study guide is the chapter on survival, which identifies and explains the study skills necessary for success in the course, including some special techniques for reading science texts effectively.

Laboratory Manual

The *Laboratory Manual* was written by B. J. Weddell and David Moffett, both instructors in human physiology at Washington State University. Each of the exercises focuses on a key physiological concept and has been successfully performed by beginning students. Selected exercises are available to qualified adopters as videotaped demonstrations. These can be used in place of actual laboratory sessions that require equipment or preparations that may not always be available. They can also be used as a preview for more complicated techniques or as a review of data collection and analysis.

In addition to the standard physiology exercises, the manual includes special **computer exercises** designed to accompany commercially available computer software. The *Preparator's Manual,* which accompanies the lab manual, provides a list of available software for using the computer laboratory exercises, instructions for preparing solutions and setting up equipment, and the answers to the questions in the laboratory manual.

Instructor's Resource Guide

Written by Charles Schauf, the *Instructor's Resource Guide* provides useful material for both the experienced and the novice instructor. Each chapter includes a comparison of the coverage in *Human Physiology: Foundations & Frontiers* with corresponding coverage in other leading textbooks. Information is presented in a tabular form designed to allow easy conversion of lecture notes. For each chapter, there is also a summary of chapter material, a detailed outline, key terms, study questions for use in class discussion or as essay items on tests, and up-to-date lists of related readings, audiovisual materials, and computer software. Answers to the study questions appear at the end of the text.

Test Bank

The **test bank,** also written by Charles Schauf, is another feature found in the *Instructor's Resource Guide.* The test bank includes about 1200 multiple choice or short-answer items that have been class-tested and carefully reviewed and edited. Each question is coded for content and level of difficulty. Text page references where answers may be found are also provided.

Computerized Test Bank

The test bank is also available in a **computerized version** for IBM PC or Macintosh computers. The computerized form of the test bank allows users to edit, delete, or add test items and to create and print tests and answer keys.

Tutorial Software

Mosby's interactive tutorial software on human physiology is available to qualified adopters. This consists of separate modules covering topics such as regulatory mechanisms, the cardiovascular system, and special topics such as exercise physiology.

Transparency Acetates and Masters

Key physiological processes as depicted in the text are emphasized by 84 color transparency acetates and 102 transparency masters. Graphs and flow charts are provided as transparency masters to supply a common vehicle for communication between the lecturer and the student. The transparency acetates are also available as slides to qualified adopters.

ACKNOWLEDGMENTS

We are particularly grateful to colleagues who have provided original micrographs and photographs for this text, including Dr. Brenda Eisenberg at the University of Illinois; Drs. Andrew Evan, William De Meyer, Ralph Jersild, Omar Markand, Glen Lehman, Ken Julian, and Nyla Heerma of the Indiana University School of Medicine; and Drs. Carolyn Coulan and John McIntyre of the Center for Reproduction and Transplantation Immunology of Methodist Hospital. We also wish to thank Ms. Patricia Kane and the Department of Radiology at the Indiana University School of Medicine for allowing us to use the original radiographs appearing in this text. Dr. Philip Hamann provided expert advice on advances in monoclonal antibody research, and Dr. Lucy A. Bradley-Springer of the University of New Mexico did the same for HIV and AIDS. Ms. Peggy Watson deserves much thanks for her invaluable secretarial assistance.

This book would not have been possible without the helpful feedback provided by the reviewer panels and from the many students we have taught over the years.

DAVID F. MOFFETT
STACIA B. MOFFETT
CHARLES L. SCHAUF

Contents in Brief

Contents

SECTION **II**

NERVE AND MUSCLE

Human Physiology
Foundations & Frontiers

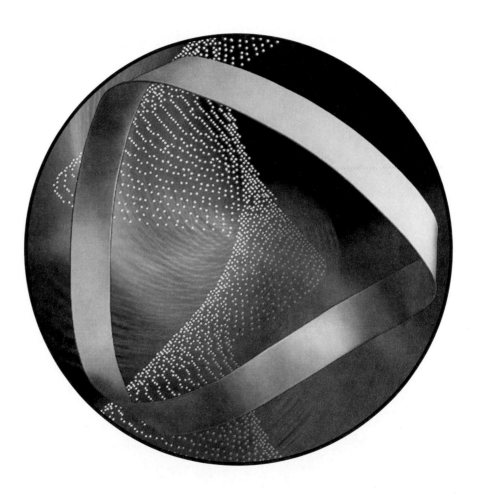

Human Physiology in Perspective

*P*hysiology is the science that describes and explains how the bodies of living organisms work. The name comes originally from the Greek word *physiologoi,* a name given to a school of Greek philosophers of the sixth and fifth centuries BC. These philosophers, including Thales, Heraclitus, and Democritus, studied all aspects of nature—mathematics, astronomy, physics, and medicine. They rejected supernatural explanations for the nature of things, arguing that the universe and all its parts are understandable and that nothing happens without cause.

In modern terms the physiologoi should be called scientists rather than philosophers. The inscription on their temple at Delphi was "Know yourself." What you will learn in your study of human physiology has practical application in any occupation related to human health and medicine—it forms an important part of the basis for medicine, dentistry, pharmacy, nursing, psychology, physical education, and public health. But even if you do not plan to become a health professional, you still experience fatigue, hunger, thirst, sexual arousal, pain, fear, and anger—the normal states of the human machine. Understanding the connections between one's life in the world and the inner life of one's body is an essential part of self-knowledge.

HUMAN PHYSIOLOGY: AN EXPERIMENTAL SCIENCE

> - *What is physiology?*
> - *What are valid and invalid uses of teleological explanations?*
> - *How do chance and natural selection operate to determine the characteristics of organisms?*
> - *Is there a genetic basis for the similarity of human and animal physiology?*
> - *How does a teleological explanation differ from a mechanistic explanation?*

The modern forms of physiology, like most other sciences, had their rebirth during the revolution in thinking called the Renaissance (fourteenth to sixteenth centuries AD). The first steps toward modern physiology were taken by physicians of the Renaissance who attempted to understand the workings of the human body by systematic study of its anatomy. It is a byword that the form of an organ or tissue follows its function, but historically, knowledge of form has typically come before understanding of function. The history of physiology is the record of attempts to understand body function by observing and experimenting on living structures while they are functioning.

One example of the importance of experimentation in physiology is the information we have about the heart and circulation. The heart and other major internal organs were well-described by anatomists by the middle of the sixteenth century. However, in spite of the fact that one-way valves were clearly visible in the heart and veins, most physicians of that time believed that contractions of the heart simply caused blood to wash back and forth in the blood vessels. They were following the teaching of the Greek physician Galen, whose conclusions were accepted without question for centuries. It was not until 1628 that the English scientist William Harvey correctly described the pumping action of the heart and the circulation of the blood. To reach this goal Harvey used not only his knowledge of the structure of the heart and blood vessels, gained by human dissection, but also observations and experiments on living subjects, both human and animal (Figure 1-1).

The stomach is another example of a system that was described long before the first real understanding of its function. The first direct observations of digestion of food in the stomach were not made until early in the nineteenth century, most notably by the American physician William Beaumont. Beaumont conducted his observations and experiments on a Canadian trapper named Alexis St. Martin, who had received a musket wound in 1822 that created a permanent opening (called a fistula) between his stomach and the outside of his body. Beaumont's observations started in 1825 and lasted for a

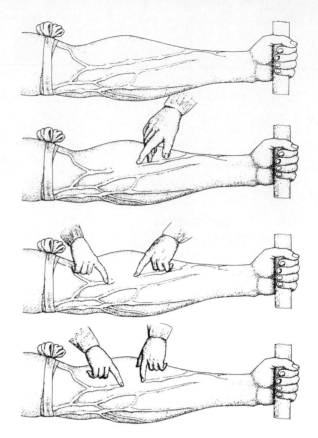

FIGURE 1-1
Illustration of an experiment proving the circulation of the blood, redrawn from William Harvey's book, *De Motu Cordis* (1628).

number of years (Figure 1-2). Beaumont observed that St. Martin's stomach responded to emotions. He removed some of the stomach's secretion and showed by chemical analysis that it contained hydrochloric acid. These studies were the first major contribution to physiology by an American. They laid the foundation of human physiology as an experimental science.

Clinical medicine provides many examples like that of St. Martin in which a disease or injury provides a window for understanding normal function. However, studies of human patients are limited by the overriding concern for patient welfare. Fortunately, the basic mechanisms of the bodies of other animal species usually correspond very closely to those of the human body. For example, in the muscles of all animal species studied thus far, there are two types of protein molecules, called actin and myosin, which interact to cause contraction. Furthermore, as explained in Chapter 11, the control of this interaction always seems to involve calcium. The physiological similarities are closest between humans and other mammals, so human physiology is to a very large extent equivalent to mammalian physiology. Therefore, experiments on nonhuman mammals can often provide essential information about human physiology.

An early example of direct contribution of ani-

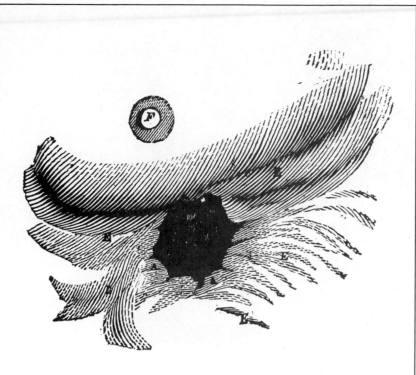

This engraving represents the appearance of the aperture with the valve depressed.

A A A Edges of the aperture through the integuments and intercostals, on the inside and around which is the union of the lacerated edges of the perforated coats of the stomach with the intercostals and skin.

B The cavity of the stomach, when the valve is depressed.

C Valve, depressed within the cavity of the stomach.

E E E E Cicatrice of the original wound.

F The nipple.

FIGURE 1-2
Page from William Beaumont's reports of his observations and experiments on the human stomach (1833).

mal studies to human welfare is the work of the Russian physiologist Ivan Pavlov (1849-1936) on control of the stomach by the nervous system. In the first half of the twentieth century, Pavlov created stomach fistulas in dogs, similar to that which William Beaumont had studied in Alexis St. Martin. Pavlov was able to demonstrate that acid secretion by the stomach occurred in response to feeding and other stimuli associated with food. He surgically disconnected the stomach from the rest of the diges-

tive system and showed that the stomach's activity was responsive to inputs from the nervous system (Figure 1-3). The surgical techniques developed by Pavlov and other investigators who followed him are still in use for studies of digestion and the physiological mechanisms that control hunger. These experiments, together with studies of human patients, have provided a basis for understanding how stress can cause stomach ulcers and why some people respond to stress by overeating. They also gave physi-

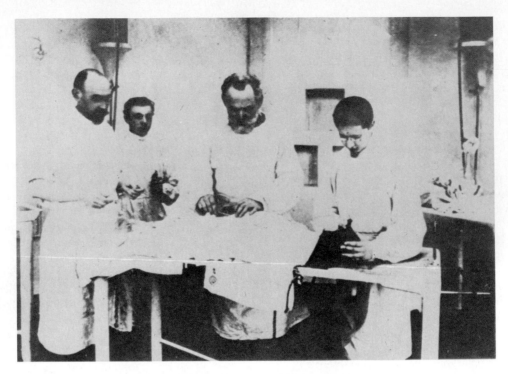

FIGURE 1-3
Ivan Pavlov (second from right) and some of his co-workers in his laboratory.

cians the knowledge and confidence needed to conduct successful surgeries on the stomachs of human patients and provided the basis for development of drugs for treatment of stomach ulcers.

The ethical acceptability of experimentation on animals has always been controversial. This is especially so for mammals. After all, they are treasured pet animals, the characters in nursery stories, the soft animal toys that comfort children, and the mascots of athletic teams. They have even entered our culture as heroes (such as Lassie) and myths (such as the Easter Bunny). Meanwhile, most people in industrialized nations eat meat and wear leather and wool, and many hunt for sport. It is unarguable that animal experimentation has revealed most of what we know about how the human body works. Had it not been possible to use nonhuman mammals as experimental subjects, today's medicine would be based to a far greater extent on speculation, trial and error, and anecdote.

The value of the knowledge obtained in animal experiments must be balanced with the highest concern for the welfare of the animals. The United States and most European nations require that animals used for research or education be kept under healthy conditions, receive veterinary care, and experience pain only when absolutely necessary for the goals of the research. In the United States, the law also requires investigators to use the minimum number of animals consistent with experimental aims, to refine the experiments to gain as much information as possible from each animal used, and to replace animals with nonanimal models where possible. It is sometimes possible to use cells and tissues raised in the laboratory in tests and experiments formerly conducted using whole animals. This is only a partial solution for the problem of animal use, because animals must usually be killed to obtain cells to start the cultures, and the cells usually will not grow in culture without growth factors and other essential substances extracted from blood or tissues.

HOW AND WHY: EXPLANATIONS IN PHYSIOLOGY

Until recently, the state of being alive seemed so remarkable that it was possible to believe that living organisms possessed some mysterious vital life force that shaped body parts to their function and guided the seemingly purposeful working of these parts. This way of explaining body form and function was called **vitalism.** Vitalism dominated physiological thinking until the middle of the nineteenth century, and only within this century has it become clear that living organisms are subject to the same physical and chemical laws as nonliving matter. Physiologists now view organisms as incredibly complex machines that are nevertheless understandable by study of their working parts. This is called the **mechanistic** approach.

The mechanistic approach concentrates on the process that leads to a final result. For any particular result, a sequence of causes may be traced from the immediate to the ultimate. For example, if the proper conditions exist, an avalanche may be triggered by a very small force, such as the sound of a twig snapping. A full explanation of the avalanche would consider both the immediate cause (the twig) and the ultimate causes: the geological processes of mountain uplift and erosion that placed the mass of loose rock on the slope where it could be acted upon by gravity. The chain of cause and effect that led from the ultimate causes to the final cause and effect is the mechanism of such events.

In contrast to mechanistic explanations, **teleological** explanations justify events in terms of some grand design or ultimate benefit. Teleological explanations have no place in the physical sciences and generally have had a bad reputation in science. To the geologist, it would be unscientific to say that an avalanche occurred *in order* to accomplish some worthwhile end, such as leveling the mountain. However, in physiology and other life sciences, valid teleological explanations may be based on **evolution,** the process by which new animal species arise from other species and become better suited to particular environments and lifestyles over time.

Evolution is the result of changes in the information encoded in thousands of small inherited units called **genes.** The mechanisms by which this genetic blueprint is passed on from parents to children are explained in Chapter 25. Each individual inherits a mixture of the genes carried by his or her parents. The mechanisms by which the genetic blueprint determines the structures and activities of cells are explained in Chapter 4. The many possible combinations of genes are mainly responsible for the range of variation seen in individuals from the same population. Ultimately, genetic diversity is provided by small changes in individual genes, called **mutations,** that occur from time to time. Animal species differ from one another to the extent that they possess different genetic blueprints, but quite different species still possess some genes in common. For example, there is about 37% of overlap between the human genetic code and the mouse genetic code.

Although evolution is a historical fact, the mechanisms involved continue to be investigated. In 1858 British scientists Charles Darwin and Alfred Russel Wallace introduced the theory of evolution by **natural selection.** In natural selection, individuals whose inherited characteristics are most favorable for survival and reproduction in their particular environment have a better chance of passing on to succeeding generations the genes that specify those characteristics. In the Darwin-Wallace theory of evolution, natural selection gradually leads to predominance in the population of the genes that favor survival and reproduction in that particular environment.

As far as any particular body part or physiological process is concerned, natural selection may result in evolutionary change or in relative stability over time. When a gene coding for a characteristic that improves reproductive success appears in a population, natural selection will favor the few individuals possessing that gene—this could be called **positive selection.** After the favorable characteristic has become established in the population, selection tends to eliminate mutations that affect that characteristic—this could be called **stabilizing selection.** In view of this, the contemporary American biologist Stephen Jay Gould has proposed that newly evolving species do not gradually diverge from existing species. In his modification of the Darwin-Wallace theory, evolutionary change could be compared to crossing a stream by hopping from one stepping-stone to another. Each successful hop to a new stone would correspond to relatively rapid evolution of a new form by positive selection, followed by a period during which stabilizing selection fine-tunes the new successful form. Human evolution has involved both types of selection, with the result that some body features have remained stable while others have undergone extraordinary adaptation.

Many features of human physiology existed long before there were humans. For example, the genes that code for the molecules that carry out the steps of **oxidative phosphorylation** are "old" genes that appear to have been stabilized by selection. Oxidative phosphorylation is the final stage of **oxidative metabolism,** the process in which cells convert sugar and other carbon-containing food substances to carbon dioxide and water and capture the chemical energy released for growth and activity. (The details of this process are explained in Chapter 4.) Early in evolutionary time, some cells evolved the ability to transfer electrons from carbon compounds to oxygen and capture the energy released by these reactions. In the beginning, these reactions probably took place at the cell surface, and they still occur there in bacterial cells. As **eucaryotic cells** (cells of animals and plants, in which the genetic material is contained in a central nucleus) evolved, the responsibility for oxidative phosphorylation was taken up by structures called mitochondria, located in the interior of the cells. An intriguing hypothesis holds that mitochondria are the evolutionary remnants of bacteria that originally lived as guests within the cells.

Well before the evolution of mammals, positive selection had almost maximized the efficiency of oxidative phosphorylation. After the effectiveness of the system was well established, mutations of the genes coding for the molecules that carry out oxidative phosphorylation continued to occur, but the only mutations that survived were "neutral" ones that did not affect the efficiency of energy capture. Cytochrome C, one of the molecules involved in oxidative phosphorylation, is an excellent

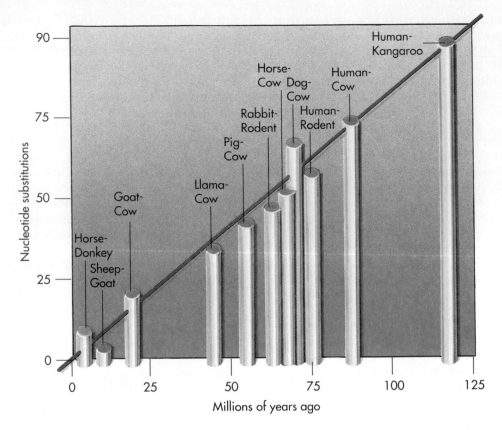

FIGURE 1-4

The estimated number of nucleotide substitutions in the gene for the protein cytochrome C over evolutionary time. As further explained in Chapter 3, a nucleotide is one of the "letters" of the genetic code, and each amino acid in the protein is coded for by a three letter "word" in the code. A change in one of the "letters" usually results in a change in the amino acid specified at that point in the sequence of the protein. The fact that the relationship can be plotted as a straight line indicates that the rate of neutral mutations has been approximately constant over tens of millions of years. These neutral mutations had essentially no effect on the function of the protein; unfavorable mutations can be assumed to have been eliminated.

example of how such neutral mutations accumulate slowly over evolutionary time. Proteins are large molecules that cells assemble by linking subunits (different types of amino acids) together in a genetically dictated sequence. (The mechanism of protein synthesis is described in Chapter 4.) A mutation of the gene for cytochrome C would cause substitution of a different amino acid at a single point in the sequence. The amino acid sequences of cytochrome C are known for a wide variety of animal species. Fossil evidence shows that the major groups of mammals diverged from one another about 90 million years ago, but the mean difference in cytochrome C sequences between modern mammal groups (a measure of the rate at which neutral mutations have occurred over the 90 million years) is less than 5% (Figure 1-4). A similar accumulation of neutral mutations occurs in genes for other cellular components involved in the fundamental processes of growth, regulation, and energy capture.

In contrast to the stabilizing selection at the cellular and subcellular level of integration illustrated by cytochrome C, the brain and body form of hominids (immediate ancestors of the present-day human species Homo sapiens) have undergone dramatic evolutionary change. If upright posture is taken as an indicator of humanity, the first hominids arose roughly 4 million years ago in Africa. Over this period (brief by evolutionary standards), the ratio of brain weight to body weight of hominids tripled (Figure 1-5). In the process, the capability of the brain to plan, remember, and communicate was greatly increased. The human brain's capabilities are probably still rapidly evolving as new selective pressures are placed on it by human culture. Remarkably, the number of gene changes that occurred was small; the human genetic code differs by only about 2% from that of the chimpanzee, our closest living relative. However, in contrast to the neutral mutations that accumulated in the code for cytochrome C, the few genetic changes that distin-

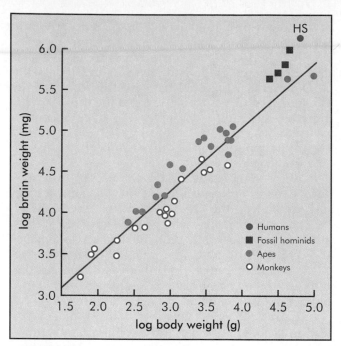

FIGURE 1-5

Plot of brain weight as a function of body weight for monkeys (open circles), apes (solid circles), some fossil hominids (solid squares) and present day humans *(Homo sapiens) (HS)*. Brain size does not increase in a 1:1 ratio with body size, but by a lower power, so the relationship is a straight line if the logarithms of the values are plotted rather than the values themselves. Fossil hominids have higher ratios of brain weight to body weight than nonhominids, and the ratio for *Homo sapiens* is the highest of all, being slightly more than 3 times greater than expected for a nonhuman primate of the same body size.

guished *Homo sapiens* from ancestral apes had dramatic anatomical effects.

Physiological mechanisms had to work to have survived positive selection—so it is natural and even convenient to talk about them in teleological terms. This is especially true for responses that involve the nervous system. For example, a painful injury to a finger or toe results in rapid withdrawal of the injured limb. In most cases, this withdrawal protects the limb from further injury. It would be a mistake to conclude (with the vitalists) that the body makes that response because it somehow "knows" the response is needed. It would be equally wrong to conclude that the nervous system evolved the organization that makes the withdrawal response possible because the response was needed. This would be an improper teleological explanation, because natural selection operates only on inherited characteristics that already exist (or are produced by mutations)—it cannot call new ones into existence. From the evolutionary point of view, humans (and all other vertebrate animals with jointed limbs) have the withdrawal response because at some time far back in our common ancestry it happened to be coded for and it worked. This is what we are really saying if we say that the purpose of the withdrawal reflex is to protect limbs from injury. This evolutionary teleological explanation gives no information about how the reflex works.

In contrast, an explanation of the mechanism of the withdrawal response would trace the step-by-step sequence of events that starts with the injury and ends with contraction of the muscle groups that flex the injured limb. Such an explanation could be outlined in the following way (Figure 1-6): the in-

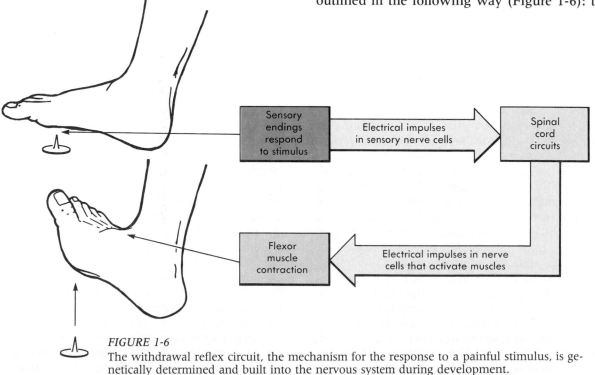

FIGURE 1-6

The withdrawal reflex circuit, the mechanism for the response to a painful stimulus, is genetically determined and built into the nervous system during development.

Human Physiology in Perspective

jury excites specific nerve endings in the skin, which relay a message by way of electrical impulses to nerve cells in the spinal cord. There are connections between the nerve cells that respond to the injury and other nerve cells in the spinal cord. These connections form a circuit, carrying electrical impulses that activate the population of nerve cells that controls contraction of the flexor muscles. Electrical impulses pass from the spinal cord to the flexor muscles, causing them to contract. (A more complete description of the withdrawal response and other such responses of the nervous system, which are called **reflexes,** is presented in Chapters 5 and 12.) A complete explanation of the mechanism of the withdrawal response would involve every level of integration of the body, including ultimately some processes that occur on the molecular scale. The science of physiology is overwhelmingly concerned with such mechanisms—"how" the body works.

FROM SINGLE CELLS TO ORGAN SYSTEMS

- *What is the role of differentiation in the development of multicellular animals?*
- *What cell types can be replaced by differentiation of stem cells after embryonic development has ended?*
- *What is a tissue? What are the four general categories of tissues?*
- *What is an organ? What are the major organ systems?*

Cells are the fundamental organizational units of organisms. Cells, in turn, are composed of the atoms and molecules that are the building blocks of the physical universe (see Chapter 2). Most animals that are visible without a microscope are **multicellular** animals, consisting of a number of cells. The number of cells in complex animals is typically very large; for example, the body of an adult person contains roughly 100 trillion (100,000,000,000,000 or 10^{14}) cells. The first organisms were probably **unicellular** (single cells), similar to some groups of modern bacteria. Unicellular organisms must possess all of the structures for activities such as feeding, locomotion, and reproduction. In evolutionary terms, unicellular organisms have been very successful; they are abundant in every one of the earth's environments. However, one advantage of being a multicellular organism is that individual cells can be highly specialized to carry out specific functions. The division of labor among specialized cells confers greater efficiency, which translates into greater reproductive success in some environments.

Differentiation is the developmental process by which relatively unspecialized cells become specialized to perform specific functions in a multicellular organism. Functional specialization of cells is reflected in their structures; sometimes the cellular commitment to a single function is quite obvious. For example, as shown in Chapter 11, muscle cells are filled with a highly organized array of the contractile proteins actin and myosin, together with other structures involved in regulating contraction. The mucus-secreting goblet cells of the lining of the digestive system (see Chapter 21) have a central cavity in which mucus made by the cell is stored. The cavity typically makes up most of the total volume of the cell.

Differentiated cells are usually unable to reproduce themselves by cell division. In some cases, such as skeletal muscle cells and the nerve cells of the central nervous system, all differentiation takes place from embryonic cells during development, and cells that die cannot be replaced during the life of the mature organism. In other cases, such as the skin and blood, dead cells are replaced from a reserve of undifferentiated cells called **stem cells.**

Tissues consist of cells that share similar structure and cooperate to perform a common basic function. There are four general tissue types. **Muscle** tissue is specialized for contraction and generation of force. The different types of muscle tissue are functional adaptations of the basic contractile apparatus of actin and myosin. **Skeletal muscle** is attached to the skeleton and moves parts of it relative to one another; **cardiac muscle** is responsible for the contractions of the heart that circulate the blood; **smooth muscle** surrounds hollow internal organs, such as the stomach, intestines, and blood vessels.

Nervous tissue consists of nerve cells, or **neurons,** which are specialized for generation and conduction of electrical impulses, and **glial cells,** which are believed to support neurons and maintain a favorable environment for their function.

Epithelial tissue makes up the membranes that cover body surfaces and line hollow internal organs, forming barriers between the interior of the body and the environment. Epithelial cells may be modified to serve as sensory receptors that detect stimuli from the environment, such as in the retina of the eye and the taste buds of the tongue. Epithelial cells also form the **endocrine glands** (pituitary, parathyroid, thyroid, adrenal, ovary, and testis), which secrete substances called **hormones** directly into the blood, and the **exocrine glands,** which secrete substances via ducts (for example, the pancreas and liver secrete into the gut, the contents of which are effectively outside the body).

Connective tissue joins body parts and contributes to the structural integrity of the other tissue types. Connective tissue characteristically has large amounts of extracellular material. Examples include **bone** and the elastic elements of the skeleton and other organs. **Adipose tissue** (fat) and **blood** are usually grouped with connective tissue.

Organs are structures composed of two or more tissue types. Organs that perform related functions are grouped into **organ systems.** The major organ

Metaphors in the History of Physiology: from the Hearth to the Computer

From the beginning of physiology as a science (about the fifth century BC) to the present day, scientists have compared the workings of the body to other natural and artificial processes that were characteristic of their time. During the pretechnological era fire was so dynamic, mysterious, and useful that it was regarded with a religious awe. Attempts to understand the equally mysterious chemistry of body metabolism led to the idea of the "fire of life." As people began to construct large dwellings and provide them with plumbing, heating, and ventilation systems, they saw that the body could be compared with a house, in which the lungs provided ventilation, the brain served to cool the blood, and the arteries provided a route for a spirit, the breath of life, or *pneuma*, to reach all of the body.

With the development of complex machines such as pumps, mills and engines, the machine became a metaphor for the body that has survived to this day. The most sophisticated machines of the eighteenth century were clocks, so it was natural to see the bodies of animals as little clocks. However, even the best clocks of the time were imperfect machines that had to be wound up regularly and adjusted frequently. Vitalists could still argue for a kind of biological clock-winder in each living organism. The nineteenth century brought the beginning of the era of physiology as an experimental science. The French scientist Claude Bernard (1813-1878) showed that the liver is a source of glucose in fasting animals and also that glucose could be released by the liver after the animal itself was dead. Such experiments showed that animals, like machines, could be studied by being taken apart. If kept under favorable conditions, the individual parts could continue to function.

A further revolution in physiology came with the availability of radioisotopes of common elements. With radioactive labels it was possible to trace the movements and fate of atoms and molecules in the body, and it quickly became clear that all structures of the body, even bones, are subject to a continuous turnover of their molecular constituents and are thus in dynamic steady state. Unlike machines that show wear with use, the healthy body maintains and repairs itself and, in principle, could last forever. In fact, many tissues actually increase their capabilities in response to use; for example, muscles get larger and stronger with regular use and bones get stronger with regular stress and strain. It is no longer possible to think of biological aging as the wearing-out of a machine, and it is clear that, in the absence of disease, aging and death are the result of a failure of homeostasis.

The newest physiological metaphor is the computer. Computer jargon has crept into physiology as it has other disciplines. Sensory signals and motor signals are the biological equivalents of input and output. The capabilities of the brain are frequently compared to those of the computer. Contemporary neurophysiologists have viewed the electrical circuits of the cerebellum and hippocampus of the brain as repetitive units, like the circuits of a microchip. The metaphor works both ways: experimenters at the Bell Laboratories are now using simple models of neural function derived from animals such as snails and slugs to improve the flexibility of information processing in computers.

TABLE 1-1   *Organ Systems: Components and Functions*

System	Components	Functions
Nervous	Brain, spinal cord, peripheral nerves, ganglia	Controls activities of other systems; receives information from the environment; stores memories; initiates and controls behavior
Skeletal	Bone, connective tissue	Support; mineral storage; production of blood cells
Muscular	Skeletal muscle	Movement
Integumentary	Skin, hair, associated glands, blood vessels	Protection against drying and infection; temperature regulation
Cardiovascular	Heart, blood vessels, red blood cells	Transports nutrients, gases, metabolic end products, and hormones between organ systems
Respiratory	Nose and throat, trachea, bronchi, bronchioles, lungs	Takes up oxygen and releases carbon dioxide to atmosphere; produces sounds; partly responsible for regulating blood acidity
Digestive	Mouth, esophagus, stomach, small and large intestines, salivary glands, liver, gall bladder, pancreas	Digestion, food storage, absorption of nutrients; protects against infection
Urinary	Kidneys, ureters, bladder, urethra	Homeostasis of extracellular fluid volume and composition; excretion of waste products
Endocrine	Pituitary, adrenals, thyroid, parathyroids, gonads, pancreas; many other organs secrete hormones in addition to their other functions	Regulation of reproduction, growth, metabolism, energy balance, extracellular fluid composition
Reproductive	Male: testes, associated glands and ducts, penis Female: ovaries, fallopian tubes, uterus, vagina, clitoris, breasts	Reproduction; sexual gratification
Lymphatic	Lymph vessels, lymph nodes	Fluid balance; transport of digested fat; cells of the immune system are also located within it
Immune	Lymphoid tissues, bone marrow, white blood cells, thymus	Resists infection, parasitization, and cancer

systems are the **nervous system,** the **skeletal system,** the **muscular system,** the **integumentary system,** the **cardiovascular system,** the **respiratory system,** the **digestive system,** the **urinary system,** the **endocrine system,** the **reproductive system,** the **lymphatic system,** and the **immune system** (Table 1-1). These functional designations include some overlap. For example, hormones are secreted by several organ systems besides the endocrine system, and many components of the immune system are located within vessels of the lymphatic system. The multiple functions of organ systems reflect functional integration of the body as a whole.

HOMEOSTASIS: A DELICATELY BALANCED STATE

- *How does diffusion control cell size?*
- *What are the selective advantages of size and multicellularity?*
- *What is homeostasis? What features of the internal environment are subject to homeostasis?*
- *How do interstitial fluid, plasma, extracellular fluid, and intracellular fluid differ?*
- *How does the rate of solute movement by diffusion necessitate circulatory, respiratory, and digestive systems?*

Unicellular organisms have little or no protection from the environment. As complex multicellular organisms evolved, the internal environment for cells on the interior of the body became more and more protected from the external environment. The importance of a stable internal environment was emphasized by the French physiologist Claude Bernard as early as 1859. By maintaining a relatively stable internal environment, complex multicellular animals are able to live freely in changing external environments. The American physiologist Walter Cannon (1871-1945) called this stable state of the internal environment **homeostasis,** from the Greek words *homeo* (same) and *stasis* (staying). Most control systems in the body utilize the principle of **negative feedback** (Figure 1-7), described in more detail in Chapter 5. In negative feedback, the control system continuously compares a controlled variable (such as body temperature or blood pressure) to a **setpoint** value that is established by the organism. Changes in the controlled variable initiate responses that oppose the change and restore the variable to its setpoint value.

What are the key features of the internal environment that must be kept constant? First of all, cells must be wet. Some small organisms can sur-

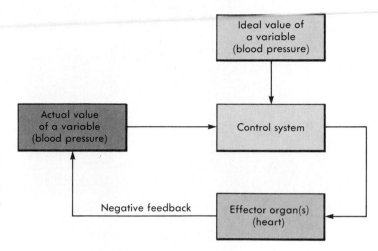

FIGURE 1-7
The principle of negative feedback control. The control system compares the actual value of a physiological variable such as blood pressure with an ideal value (set point). The control system then regulates effector organs such as the heart (and blood vessels) to correct any deviation that has occurred.

vive drying, but while they are dry they can be said to be only potentially alive. The fluid outside the cells of complex multicellular animals is called **extracellular fluid.** The two major volumes of extracellular fluid are **blood plasma** (the fluid component of blood exclusive of blood cells) and **interstitial fluid** (the fluid in the spaces between cells). The chemical composition of the extracellular fluid is also homeostatically controlled—it must contain particular dissolved substances (**solutes**) in specific concentrations. For example, the extracellular fluid of all vertebrate animals contains a high concentration of sodium and chloride, the constituents of table salt. It contains smaller concentrations of potassium, calcium, magnesium, bicarbonate, phosphate, and other solutes. The volume of extracellular fluid and the concentration of each of its solutes are regulated by homeostatic mechanisms that typically involve several organs or organ systems.

The **intracellular fluid,** the water inside cells, must also contain particular solutes in particular concentrations. Virtually all cells contain a high concentration of potassium and low concentrations of sodium, chloride, magnesium, and calcium. In addition, cells must have a ready supply of **nutrients,** substances that serve as molecular building blocks and sources of chemical energy.

Finally, temperature has a powerful effect on physiological processes. The range of temperatures in which life is possible is a tiny fraction of the known range of temperatures. Within the range of temperatures compatible with life, cooler temperatures favor preservation of cellular structure but slow the rates of the chemical reactions carried out by cells. Higher temperatures accelerate chemical reactions but also may disrupt the structure of the proteins and other large molecules within cells. Most complex multicellular animals can regulate their body temperatures to some extent using behavioral mechanisms. The most precise regulation is found in mammals, in which the temperature of the brain and body core is homeostatically regulated with a precision of better than 0.1° C. This precise

regulation is called **homeothermy** (same heat). Homeothermy in cold environments usually requires significant amounts of heat energy generated as a byproduct of oxidative metabolism. Generation of body heat by metabolism is called **endothermy** (heat from within).

Needless to say, the water content, chemical composition, and temperature of the external environment are seldom ideal for the molecular processes carried out in the interior of cells. Unicellular organisms are highly vulnerable to environmental stress because their surfaces are in direct contact with the external environment. The larger an organism is, the less vulnerable its interior is to disruption by uncontrolled movement of materials and energy between it and the environment. On the other hand, some substances must be exchanged with the environment. In particular, in most cell types the exploitation of chemical energy from food requires that oxygen and nutrients reach the interior of cells and that carbon dioxide and other chemical end products be transferred to the environment (Figure 1-8). Much exchange between cells and their immediate surroundings occurs by **diffusion** (see Chapter 6), a process in which the random movements of individual molecules result in net movement of a diffusing substance if there is a **concentration gradient**—a difference in the concentration of the substance over distance.

Net movement of dissolved nutrients, gases, and end products by diffusion is rapid enough to meet the needs of active cells if the distance is short compared to cellular dimensions, but not if the distance is long. For example, substances dissolved in water require about 5/100 of a second to diffuse 1/100,000 of a meter, about the distance between the center and the surface of a typical cell. If the distance is increased to 1 cm, the time increases to 13 hours, and for 10 cm the time is about 53 days. Diffusion of substances over distances of meters, the scale of the human body, would require decades. The circulatory, respiratory and digestive systems of multicellular organisms overcome the limited effec-

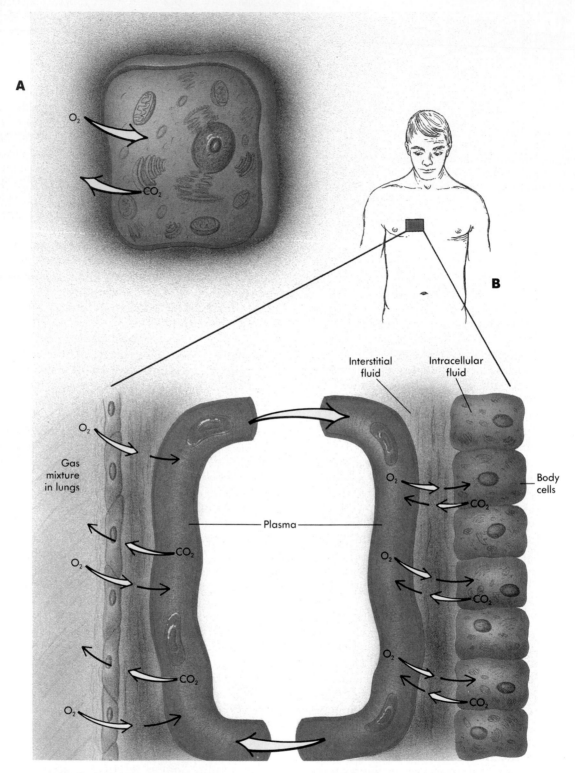

FIGURE 1-8
A Cells exchange materials such as oxygen and carbon dioxide with their immediate environment by diffusion across the cell membrane. The path for diffusion of materials between cells and the extracellular fluid is very short.
B In the body, cells do not have direct access to the external environment. The circulatory system carries dissolved gases and other materials over the long distance between the lungs and the cells much more rapidly than would be possible by diffusion.

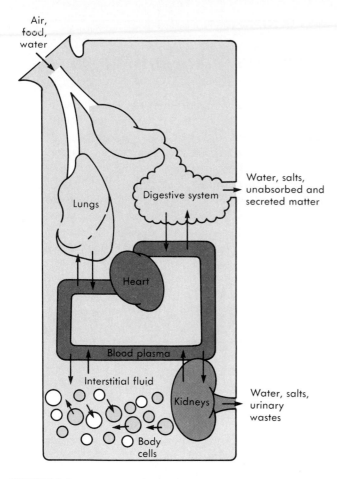

FIGURE 1-9

In the human body plan, the immediate environment of the cells is interstitial fluid. All materials exchanged between cells and the environment pass through this fluid. The circulatory system links interstitial fluid with the sources of oxygen (the lungs) and nutrients (the digestive system) and permits carbon dioxide and waste products to be transferred to the environment by the lungs, kidneys, and digestive system.

tiveness of simple diffusion for moving substances over long distances.

In the circulatory system, blood is rapidly moved among the respiratory system, where gases are exchanged with the environment; the kidney, where nongaseous wastes and excess fluid and solutes are removed; and the digestive system, where nutrients are absorbed (Figure 1-9). Rapid transport of the substances within the internal environment by blood flow overcomes the diffusional limit on large body size. The volume of extracellular fluid in the human body is only about one third the volume of intracellular fluid. The circulatory system allows this relatively small extracellular fluid volume to serve as an effective substitute for the large volume of fluid that typically surrounds unicellular organisms.

Homeostasis is a dynamic steady state—a delicate balance maintained by the many separate regulatory processes carried out by all organ systems (Table 1-1). This balance can be overwhelmed when

the external environment stresses homeostatic processes beyond their limits. For example, prolonged exposure to cold may lead to an intolerable reduction in the temperature of the body core; exercise in a hot environment may result in depletion of body fluid and an increase in the core body temperature, resulting in heat stroke. The cells of most mammals have adapted so thoroughly to a regulated temperature that even a few degrees of variation in body temperature may have fatal effects. *Homo sapiens* is a tropical species and has relatively weak ability to regulate its temperature. Unclothed and unprotected human beings can tolerate only a few tens of degrees centigrade of difference between body temperature and environmental temperature. Nevertheless, *Homo sapiens* has been able to live in hostile external environments by creating engineered microenvironments (such as clothes, houses, space vehicles, and diving suits) with appropriate conditions of temperature, atmosphere, and availability of food and water.

Almost all diseases are failures of homeostasis, as is the deterioration of the body in aging. One responsibility of physicians and other health professionals is to help maintain homeostasis. This responsibility is most obvious in critical care facilities for acutely ill patients (Figure 1-10). In such facilities a number of indicators of homeostasis are monitored, including heart rate and blood pressure, respiration, body temperature, blood chemistry, and gain and loss of body fluid. The mission of the critical care facility is to take over the responsibility for those aspects of homeostasis that the injured or diseased organ systems of the patient's own body are unable to perform. Knowledge of the functioning of organ systems gained through the work of physiologists enables the critical care facility to do this vital work.

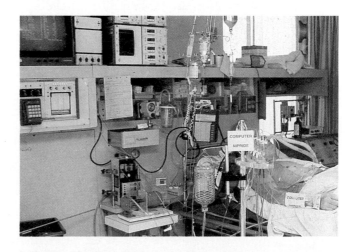

FIGURE 1-10

A patient's vital functions are being monitored in an intensive care unit. This patient's blood pressure is being regulated by a computer that monitors the pressure and injects a heart-stimulant drug when the pressure falls below a predetermined value.

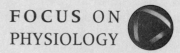

A Classic Experiment You Can Repeat

One piece of evidence William Harvey advanced for one-way flow of blood was his discovery of one-way valves in peripheral veins. These valves are apparent in dissected veins, but their effect can be seen without dissection. Find a clear unbranched stretch of vein on the inner surface of your forearm. Place and keep one finger firmly on the distal (handward) end of this stretch. Have someone else expel the blood from the vein in the proximal (shoulderward) direction by passing his or her finger along the vein. There will probably be a venous valve somewhere in this stretch of vein. If so, when your helper lifts his or her finger, blood will flow backward until it reaches the valve, but will not refill the portion of vein that lies between the valve and your finger. When you lift your own finger, you will see the collapsed portion of the vein refill with blood flowing in the "right" direction—toward the heart.

SUMMARY

1. Physiology is the science that explains how the bodies of living organisms work.

2. The development of human physiology as a science depended on integration of anatomy, with knowledge gained from experiments on living animals and parts of animals.

3. Vitalism attempts to explain life processes as consequences of a mysterious life principle or force. The mechanisms of body function are now believed to obey the basic principles of physics and chemistry.

4. In the course of natural selection, the stress placed on organisms by their environment causes those inherited characteristics that favor survival to predominate in succeeding generations.

5. A teleological explanation equates the cause of a process with its value to the organism. In biology, teleological explanations can be based on continuous refinement of body structures and mechanisms by natural selection. Mechanistic explanations trace the sequences of cause and effect that underlie body processes.

6. In multicellular organisms, cells become specialized for performance of specific functions by differentiation. Cells aggregate to form tissues. Tissues are classified into four basic functional categories: muscle tissue, nervous tissue, epithelial tissue, and connective tissue. Organs are composed of two or more tissue types; major organ systems include the nervous, muscular and skeletal, integumentary, cardiovascular, respiratory, digestive, urinary, endocrine, reproductive, and immune systems.

7. Diffusion is a process in which the random movements of individual molecules result in net movement down a concentration gradient. The slowness of diffusion over long distances limits the diameter of cells. Large complex organisms are composed of many cells bathed by an internal fluid environment. Extracellular fluid consists of the interstitial fluid, immediately surrounding the cells, and plasma, the fluid component of blood. Movement of materials between cells and the external environment is assisted by blood circulation.

8. Homeostasis is the dynamic steady state of the internal environment. Departures from the steady state are opposed by negative feedback regulation.

1. What are two important evolutionary advantages of being multicellular?

2. Describe the relationship between distance and the rate of transport of dissolved substances by diffusion.

3. What is environmental stress? Give some examples of environmental stress.

4. What is natural selection? Do you think that natural selection is still operating on human populations?

5. The following statements were overheard in a comparative anatomy class:
   Student 1: "I learned that in humans, the appendix is just an evolutionary vestige that has no function - in fact, mine was removed when I was nine and I'm getting along fine."
   Student 2: "Natural selection would have eliminated an organ without a function - therefore it must be that we just haven't discovered the function of the appendix yet."
   Instructor: "Monkeys don't have an appendix; apes and humans do, and the appendix of humans is the largest. Therefore the appendix is a relatively recent adaptation."
   Do any of these arguments have merit?

6. What is negative feedback? Name some aspects of the internal environment that are subject to homeostatic control.

Choose the MOST CORRECT Answer.

7. This type of selection "fine-tunes" the successful new form of an organism after establishment in the population:
   a. Natural selection
   b. Reverse mutation selection
   c. Stabilizing selection
   d. Natural selection

8. The American physician who made the first direct observation of digestion of food in the stomach:
   a. Alexis St. Martin
   b. William Harvey
   c. Walter Cannon
   d. William Beaumont

9. Which of the following is NOT an advantage of multicellularity in organisms?
   a. Greater specialization of cells
   b. Greater numbers of organisms
   c. Greater efficiency due to division of labor
   d. Greater reproductive success

10. Generation of body heat by metabolism is called:
    a. Exothermy
    b. Endothermy
    c. Homeothermy
    d. Diathermy

11. Which of the following tissues is specialized for generation and conduction of electrical impulses?
    a. Dense connective tissue
    b. Muscle tissue
    c. Simple epithelium
    d. Nervous tissue

● SUGGESTED READINGS

BUTTERFIELD H: *The study of the heart down to William Harvey*. In *The origins of modern science, 1300-1800,* MacMillan, New York, 1951. Places Harvey's work in the context of the development of other sciences during the Renaissance.

CANNON WB: Organization for physiological homeostatics, *Physiological Review* 9:399-431, 1929. Reprinted in *Homeostasis—origins of the concept,* Benchmark Papers. In Langley LL, editor, *Human physiology.* Dowden, Hutchinson & Ross, Stroudsburg, Pa., 1973. This historic paper introduces the concept of homeostasis and outlines what was known at the time about regulation of blood sugar, extracellular fluid volume, blood calcium, and body temperature.

GOULD SJ: The episodic nature of evolutionary change. In *The panda's thumb,* London, 1982, WW Norton. First published as an essay in the magazine *Natural History,* this explains the author's reinterpretation of the Darwin-Wallace theory of evolution.

MCCOURT R: Model patients; among the inhabitants of the animal world are some that hold the cures for many human afflictions. *Discover* 11 (8):36, 1990. Discusses a variety of animal models of human disease.

MORSELLI M: *Amedeo Avogadro.* D Reidel Publishing Co., Dordrecht 1984. The life and contributions of this theoretician and independent thinker who proposed the molecular hypothesis in 1811.

SALZBERG HW: *From caveman to chemist—circumstances and achievements.* American Chemical Society, Washington, DC, 1991. The history of chemistry from its primitive beginnings through its domination by alchemy to its blossoming in the modern world. Basic ideas are presented as the information unfolded within the historical context of the people's lives.

SCHOENHEIMER R: *The dynamic state of body constituents,* Cambridge, Mass, 1946, Harvard University Press. A pioneer report of the use of radioisotopes in physiology.

SCHULTZ PG: The interplay between chemistry and biology in the design of enzymatic catalysts, *Science* 240:426-433, 1988. Two strategies are described for the design of catalysts to promote reactions of medical and commercial importance. Information about molecular mechanisms of recognition by antibodies and enzymes is being provided by this research.

SMITH HW: The philosophic limitations of physiology. In Veith I, editor, *Perspectives in physiology,* Washington, DC, 1954, American Physiological Society. An overview of the history of physiological thought by one of the foremost American kidney physiologists.

WILSON AC: The molecular basis of evolution, *Scientific American* 256:164, October 1985. A short explanation of the genetic basis of evolution.

WOLF S: *The stomach,* New York, 1965, Oxford University Press. This book, now somewhat dated scientifically, gives a wonderfully readable account of the life and death of a patient with a gastric fistula whose stomach was studied by Dr. Wolf over many years.

CHAPTER 2

The Chemical and Physical Foundations of Physiology

On completing this chapter you should be able to:

- Describe the structure of atoms and explain the principles of chemical bonds as they relate to the structure and behavior of biologically important compounds.
- Understand why water is the universal solvent for biological systems, and define hydrophilic and hydrophobic.
- Describe solutions in terms of their molarity and osmolarity.
- Explain the basic chemistry of acids and bases and calculate the pH of solutions.
- Understand the difference between equilibrium and steady state.
- Describe the relationship among force, friction, and movement that will be applied in subsequent chapters to such processes as blood flow, muscle movement, and the behavior of respiratory structures.
- Define compliance.
- Understand the application of thermodynamics to biological systems, and define entropy and free energy.

*T*he science of physiology developed as the applicability of physical and chemical laws to biological systems became apparent. Physical aspects of human body function such as blood flow, ventilation of gases, and the transmission of electrical signals in the nervous system can best be described in mathematical terms. The ability of physiologists to predict the workings of the body by using mathematical equations helped to demystify the human body and paved the way to development of medicine as a science.

The human body is a living system in dynamic interaction with the organic and inorganic elements in its environment. Energy from the sun enters the earth's ecosystem and is trapped by photosynthesis. This provides the chemical energy that supports animal life. The chemical environment of both plants and animals has been shaped by the properties of water molecules, because water is the solvent in which biological reactions occur. The water-based solutions of the extracellular and intracellular fluids are separated by the water-repelling properties of lipid membranes. Flow of material across these membranes is determined by defined chemical and physical forces and the physical properties of the membranes themselves.

Physical laws also govern the work performed by cells in maintaining an internal environment different from the extracellular solution and synthesizing complex molecules. Living systems do not create order out of disorder without paying the price. Cells pay by sacrificing high-energy molecules that they have synthesized for this purpose. A by-product of work is heat production, which is wasted in most machines made by humans, but which can be viewed as a useful by-product in mammals because it provides the basis for maintaining a body temperature above environmental temperature. In fact, the more physiologists understand how the body works, the more they appreciate how efficiently it functions.

THE CHEMICAL BASIS OF PHYSIOLOGY

- *What is atomic number? Atomic weight?*
- *What force is most often operative in physiology?*
- *How strong are the different types of chemical bonds?*
- *How important are weak bonds in physiology?*

Atomic Structure

Atoms, the basic units of all matter, consist of three fundamental types of subatomic particles: **protons, neutrons,** and **electrons.** The central **nucleus** contains a variable number of positively charged protons and uncharged neutrons. The nucleus is surrounded by a cloud of negatively charged electrons.

The chemical identity of an atom is determined by its **atomic number,** the number of protons in its nucleus. **Elements** are fundamental substances; each element has a different atomic number. There are slightly more than 100 elements, but only 4 of them—hydrogen, oxygen, carbon, and nitrogen—make up over 95% of the substance of living organisms (Table 2-1). Elements can be converted from one to another only by removing or adding protons to the nucleus, a feat that does not occur in ordinary chemical reactions.

The **atomic weight** of an atom is the sum of the numbers of protons and neutrons in the nucleus (Figure 2-1, A). Most elements exist in two or more **isotopes,** variant forms whose atoms have the same atomic number but different atomic weights, due to different numbers of neutrons (Figure 2-1, B). Most chemical properties are the same for all of the isotopic forms of any particular element. The atoms of some isotopes are unstable—they tend to decay by emitting high-energy particles from their nuclei until a stable form is reached. Such isotopes are called **radioactive.** Radioactive isotopes provide the

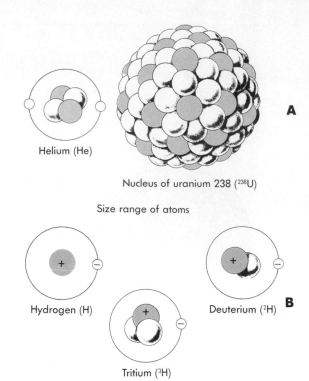

Size range of atoms

Helium (He)

Nucleus of uranium 238 (^{238}U)

A

Hydrogen (H)   Tritium (^{3}H)   Deuterium (^{2}H) **B**

Isotopes of hydrogen

FIGURE 2-1

A The helium nucleus contains two protons and has an atomic weight of 4. The largest naturally occurring atom is uranium (atomic weight 238), whose nucleus contains 92 protons and 146 neutrons.
B The smallest atom is hydrogen (atomic weight 1), whose nucleus consists of a single proton. Hydrogen has two naturally occurring isotopes, deuterium and tritium, with one and two neutrons, respectively.

means for tracing the routes and rates of movement of particular elements within the body, as described in the boxed essay on p. 21.

The electrical force is one of four basic types of forces present in and around matter. Gravity is another such force, and the remaining two are called simply the strong force and the weak force. Of the four forces, the electrical force is surely the most frequently met with in physiology. It operates at the subatomic level to hold atoms together, at the atomic level in interactions between atoms, and at the cellular level in the electrical impulses generated by nerves and muscles. A single **charge** is a discrete unit of electrical force that may be either positive or negative. Like charges repel each other whereas opposite charges attract each other. Protons, each having a single positive charge, give the nucleus a net positive electrical charge that attracts electrons, each of which has a single negative charge. In a **neutral atom** there is no excess of electrical charges; the number of electrons surrounding the nucleus equals the number of protons in the nucleus. Free atoms bearing an excess of positive or negative charge are

TABLE 2-1		Elements in the Human Body		
Element	Symbol	Atomic number	Valence	Body weight (%)
Hydrogen	H	1	1	9.5
Carbon	C	6	+4 or −4	18.5
Nitrogen	N	7	−3	3.3
Oxygen	O	8	−2	65.0
Sodium	Na	11	+1	0.2
Magnesium	Mg	12	+2	0.1
Phosphorus	P	15	+3	1.0
Sulfur	S	16	+4 or −4	0.3
Chlorine	Cl	17	−1	0.2
Potassium	K	19	+1	0.4
Calcium	Ca	20	+2	1.5

(Hydrogen, Carbon, Nitrogen, Oxygen: 96.3)

FOCUS ON PHYSIOLOGY

The Use of Isotopes in Physiology

The isotopes of an element are forms of that element that possess the same number of protons but different numbers of neutrons. Almost every element of physiological importance exists in two or more isotopic forms. Thus for hydrogen, which normally has only a proton as its nucleus, isotopes are deuterium with one neutron and tritium with two neutrons (see Figure 2-1). Some isotopes are relatively stable; others can decay to more stable atomic forms by releasing energy in the form of alpha particles, beta particles, or gamma rays and are thus called radioisotopes. With only a few exceptions, isotopes are indistinguishable to organisms. Stable isotopes can be separated by mass spectrometry; radioisotopes can be detected by their radioactivity.

Isotopes exist in nature in varying abundances; they became available as scientific and medical tools with the development of nuclear reactors, in which they can be produced. In physiology isotopes are sometimes called tracers or labels because they can be used to follow the fate of different substances in the body or in isolated cells, tissues, or cytoplasmic components. Tracer experiments are particularly useful for systems that are in steady state, since it might not be possible to measure the movements of substances in such systems by chemical methods. At first, isotopes were used to destroy cancerous tissue inside the body, but radioactive isotopes can also be effective diagnostic tools.

A classic experiment provides a measure of the rate at which the thyroid gland takes up iodine from the blood. The iodine content of the gland is in steady state because the secretion of iodine-containing hormone from the gland is matched to its rate of iodine uptake from the blood; no change in its iodine content over time can be detected chemically. Radioactive iodine is injected into the blood, and the animal is placed with its thyroid over a scintillation detector. The detector converts the emissions of the isotope to small flashes of light, which are counted by a sensitive phototube. In a short time the detector begins to register an accumulation of isotope in the thyroid. At first the rate of uptake of tracer is much greater than the rate of return of tracer to the blood since the blood contains much iodine isotope and the gland initially contains none. While this condition is satisfied, the rate of increase of isotope in the gland is an accurate measure of its normal rate of uptake of nonradioactive iodine.

In a nuclear scan a very small amount of a radionuclide with special affinity for a particular organ or part of the body (like the thyroid just discussed) is administered by mouth or injection. Instead of a gross measure of uptake, a scanning "camera" picks up the radiation being emitted and transforms it into an image. Nuclear scans lack clear definition, but can reveal areas of an organ that are not functioning normally. Some radioactive isotopes concentrate in tumors, appearing as "hot spots." Nuclear scans of bones can detect bone tumors long before they show up on X-rays, and aid in diagnosing bone injury, infection, and arthritis. Nuclear scans are also used to locate blood clots in the lungs, scans of the liver help diagnose cirrhosis and hepatitis, and brain scans can uncover tumors and strokes.

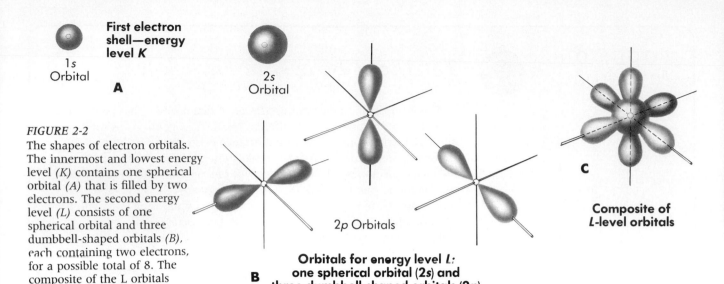

First electron shell—energy level K

1s Orbital

A

2s Orbital

2p Orbitals

C

Composite of L-level orbitals

FIGURE 2-2
The shapes of electron orbitals. The innermost and lowest energy level *(K)* contains one spherical orbital *(A)* that is filled by two electrons. The second energy level *(L)* consists of one spherical orbital and three dumbbell-shaped orbitals *(B),* each containing two electrons, for a possible total of 8. The composite of the L orbitals is shown in part C.

B Orbitals for energy level *L:* one spherical orbital (2s) and three dumbbell-shaped orbitals (2p)

called **ions;** if the net charge is positive the ion is a **cation,** if negative an **anion.**

Chemical Bonds

Atoms may attach themselves to one another by **chemical bonds,** forming stable associations called **molecules.** The resulting substances are **compounds,** consisting of two or more elements associated by chemical bonds. Any process in which chemical bonds are formed, broken, or rearranged is called a **chemical reaction;** in the course of a chemical reaction compounds may be formed, changed, or degraded to their elements. The rules for formation of chemical bonds apply to almost every aspect of physiology at the cellular and subcellular level, because at this level of organization all of the mechanisms involve chemical reactions. In growth and repair, cells promote, or **catalyze,** synthesis of specific types of molecules. In this process, chemical energy is used to increase the molecular order of the body. In metabolism, cells transform compounds by breaking some bonds and forming others. In the process, energy is transferred from one molecule to another, ultimately being made available to drive activity and growth.

While the nucleus determines the chemical identity of an atom, the electrons are directly responsible for the formation of chemical bonds. Each electron is in constant, unpredictable motion around the nucleus. Its location at any instant cannot be predicted but can be described by a statement of probability. The space within which an electron is likely to be 90% of the time is called its **orbital.** Each orbital is filled when it is occupied by 2 electrons. Orbitals are arranged around the nucleus in **energy levels** (Figure 2-2). The first energy level (the **K level**) is a single spherical orbital. The second, third, and fourth energy levels (the **L, M,**

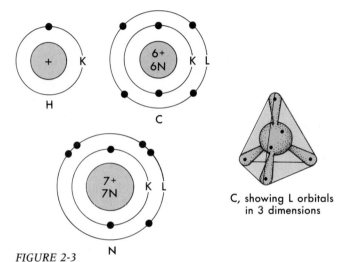

C, showing L orbitals in 3 dimensions

FIGURE 2-3
Energy levels may be represented as rings and electrons as dots. For example, a hydrogen atom with its one electron in the K energy level, and a carbon atom with a full K level and four unpaired electrons in the L level, one in each of the four p orbitals. Ring diagrams do not show which orbitals the electrons are in, so three-dimensional information is left out. For example, when represented in three dimensions, the four p orbitals of the L level of carbon project out from the nucleus to form a four-sided triangle (tetrahedron), as shown. The L energy level of nitrogen consists of two full orbitals with electrons shown as pairs, one orbital containing an unpaired electron, and one empty orbital.

and **N levels**) each contain four orbitals that may be spherical or dumbell-shaped. The higher levels thus are full when they contain eight electrons each—this rule holds for elements up to an atomic weight of 20. In elements of higher atomic number the third energy level may have up to five additional orbitals that allow it to hold up to 18 electrons. For simplicity, the energy levels can be represented as rings and the electrons in them as dots (Figure 2-3).

As the atomic number of the nucleus increases from one element to another, the additional elec-

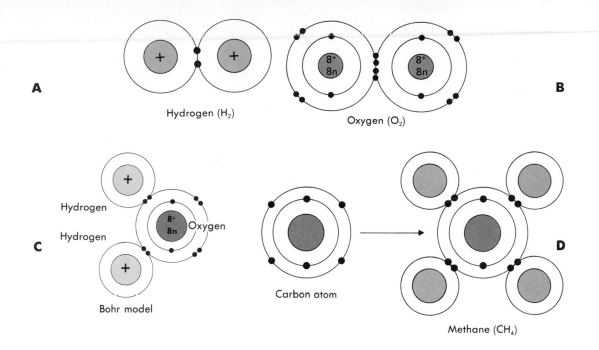

A Hydrogen (H₂)
B Oxygen (O₂)
C Hydrogen, Hydrogen, Oxygen — Bohr model
Carbon atom
D Methane (CH₄)

FIGURE 2-4
A Hydrogen gas is a diatomic molecule composed of two hydrogen atoms, each sharing its electron with the other.
B Oxygen gas is a diatomic molecule composed of two oxygen atoms, each sharing two electrons with the other.
C In water, two hydrogens each share one electron with one oxygen, completing the outer energy levels of all three atoms.
D In methane, a carbon atom shares each of its four unpaired L electrons with a hydrogen.

trons fill the energy levels from lowest to highest. For example, hydrogen (atomic number = 1) has one electron in the first energy level, helium (atomic number = 2) has two electrons in the first level whereas carbon (atomic number = 6) has two electrons in the first level and four in the second level. An atom is least reactive when its outermost energy level is completely filled; for example, helium (atomic number = 2) does not readily participate in chemical reactions whereas hydrogen and carbon do. Atoms with unfilled orbitals are reactive because the unfilled orbitals can be filled by sharing of one or more of the unpaired electrons with another atom that also has unpaired electrons. In forming chemical bonds, atoms with nearly full outer shells tend to acquire the extra electrons they need to complete their outer energy level and are called **electron acceptors.** Those with substantially unfilled outer levels tend to give up the extra electrons, acting as **electron donors.** The **valence** of an element is the number of electrons its atoms can donate or accept. For example, oxygen, an electron acceptor, lacks two electrons and has a valence of −2; sodium has a single electron in its outermost energy level and so acts as an electron donor with a valence of +1. Carbon has an outer energy level that is half full. It can act either as an electron acceptor or an electron donor (that is, its valence may be either +4 or −4).

Covalent Bonds

In **covalent bonds,** the nuclei of the bonded atoms share electrons, completing the outer electron shell of each nucleus. Covalent bonds are the strongest chemical bonds. Covalent bonds may be **single, double,** or **triple** depending on the number of electrons shared. For example, two hydrogen atoms (atomic number =1) can share an electron, forming a single bond that results in molecular hydrogen (H₂) (Figure 2-4, A). Two oxygen atoms (atomic number = 8) can each share two electrons, forming a double bond that results in molecular oxygen (Figure 2-4, *B*). Such molecules, formed from two atoms of the same element, are called **diatomic.** Or two hydrogen atoms can each share an unpaired electron with the two unpaired electrons of oxygen to form water (H₂O) (Figure 2-4, C). Nitrogen, with an atomic number of 7, has five electrons in the second energy level. It can share the unpaired electron with another nitrogen to form N₂ (the major constituent gas of the atmosphere), or accept a share in as many as three additional electrons provided by other atoms. Each of the four electrons in the second energy level of carbon is in its own orbital and so is unpaired, so a carbon atom can form single covalent bonds with up to four other atoms, as for example the methane molecule shown in Figure 2-4, D. Alternatively, it can form one double and two single

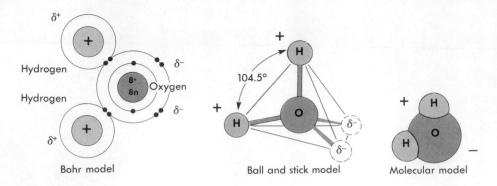

FIGURE 2-5
In water the hydrogen atoms have a slight excess positive charge, while the oxygen atom has a slight excess negative charge, indicated as $^+$ and $^-$ respectively. These charges form a dipole, and such compounds are therefore referred to as polar.

bonds or two double bonds, and some or all of these bonds may be with other carbons. As shown in the next chapter, the remarkable ability of carbon to form chemical bonds makes possible an immense variety of carbon-based compounds. Some carbon-based molecules synthesized by cells are so large and complex that they are best understood as molecular machines.

Covalent bonds fall into two classes, **nonpolar** and **polar,** depending on whether the two constituent nuclei have equal or unequal abilities to donate or accept an electron. If the two nuclei have about the same tendency to give up or to take on electrons, the shared electrons spend equal amounts of time orbiting each nucleus, and a nonpolar covalent bond results. An example of a nonpolar bond is the bond between carbon and hydrogen. One of the best examples of unequal sharing of electrons occurs in water, where electrons donated by the hydrogen atoms spend a slightly larger fraction of their time near the oxygen nucleus (Figure 2-5). Covalent bonds of this type are termed **polar.** The oxygen side of the water molecule acts as if it has a slight negative charge, while each hydrogen molecule has a slight positive charge. The unequal average charge distributions of polar bonds result in molecular **dipoles** with two oppositely charged regions, or "poles." The relative presence of polar bonds versus non-polar bonds in a molecule affects its ability to dissolve in water or **water solubility.** Molecules consisting largely or entirely of nonpolar bonds have low water solubility and tend to associate with other nonpolar molecules rather than water. This is because the positive and negative poles of the water do not find any charges to interact with on non-polar molecules. Such molecules are called **hydrophobic** (water-fearing). In contrast, **hydrophilic** (water-loving) molecules contain a large proportion of polar bonds.

Ionic Bonds

An element that readily gives up electrons may react with an element that readily accepts them, so that one or more electrons are donated outright from one atom to another. In the process the elements acquire opposite electrical charges, and the bond so formed between them is called an **ionic bond** (Figure 2-6). Ionic bonds are more easily broken than covalent bonds. In the presence of water, the ionically-bonded atoms may separate into charged ions, each having a complete outer electron shell that resulted from the transfer of electrons (Figure 2-6, *A*). In table salt (NaCl) the sodium atom donates its single unpaired electron to chlorine, and in the process empties its outer shell and becomes a cation. The electron donated by the sodium atom completes the chlorine's unfilled outer shell and becomes an anion (chloride). Solid sodium chloride is a cube-shaped molecular array held together by the electrical attractions between neighboring, oppositely charged sodium and chloride ions (Figure 2-6, *B, C*). When salt is placed in water, it dissolves due to the solvent properties of the polar water molecules, as described in the next section.

Weak Bonds

There are two other types of bonds between atoms that, though 10 to 20 times weaker than covalent and ionic bonds, are nevertheless very important for physiology. These are **hydrogen bonds** and **van der Waal's forces.** Hydrogen bonds are possible because the bond between hydrogen and oxygen or nitrogen is of the polar covalent type; the electrons are not distributed equally between the two bonded atoms. Instead, hydrogen retains a small amount of positive charge, and oxygen or nitrogen retains a small amount of negative charge. The positively-charged hydrogen can then form a hydrogen bond with negatively-charged atoms of other molecules.

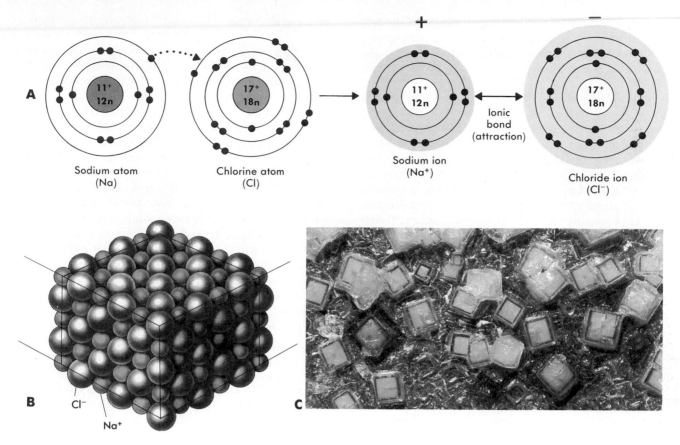

FIGURE 2-6
Ionic bonds involve outright donation and acceptance of electrons to form a crystalline array.
A   When a sodium atom donates an electron to a chlorine atom, the sodium atom becomes a positively charged sodium ion, while the chlorine atom becomes a negatively charged chloride ion.
B   Solid salt consists of highly ordered crystals of ionically bonded NaCl.
C   The highly ordered packing of ionically bonded atoms is repeated in this example of salt crystals.

FIGURE 2-7
An illustration of the hydrogen bonds that form between water molecules.

For example, hydrogen bonds are readily formed between the hydrogens and oxygens of adjacent water molecules (Figure 2-7).

Van der Waal's forces are another type of weak bond that results from interaction between the electrons around the atoms of two adjacent molecules. A nearby polar molecule can slightly shift the charge distribution in a nonpolar molecule. The nonpolar molecule develops a dipole-like character and the two molecules are attracted to one another. A similar interaction occurs between two nonpolar

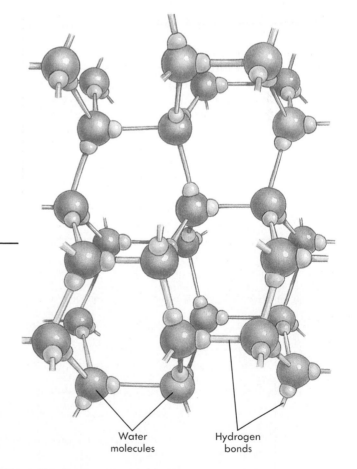

molecules because of random fluctuations in the position of the electrons around them. Each nonpolar molecule induces a redistribution of charge in the other, and a weak attraction between the two molecules occurs.

Although hydrogen bonds and van der Waal's forces are referred to as weak, these bonds play a major role in determining the three-dimensional structure of large molecules such as proteins, RNA, and DNA. These large biological molecules are linear (threadlike) when synthesized by cells. To function as molecular machines, they must assume a correct three-dimensional structure, or **conformation.** This usually involves complex folding, twisting, and curling of the molecules. Hydrogen bonding and van der Waal's forces between different parts of the molecule cause it to assume its correct conformation automatically after synthesis.

Properties of Water

> - *What are the properties of water?*
> - *What is: a solvent, an ion, an electrolyte, an acid, a base, pH?*
> - *How can the equilibrium concentration ratio of dissociated to undissociated weak acid or weak base be stated in mathematical terms?*
> - *What is the general relationship between molarity and osmolarity?*
> - *How can the osmolarity of solutions for which the molarity is given be estimated?*

Water makes up 70% to 80% of the mass of most body tissues (excluding bone), and its unique properties (Table 2-2) are critical for almost every mechanism of the human machine. Both its physical properties (its ability to accept and conduct heat and its behavior as a liquid and a gas) and its chemical properties (the ability of water molecules to interact with other atoms and molecules) are important.

The **specific heat** of a substance is the amount of energy required to increase the temperature of 1 gram of the substance by 1° C. The specific heat of water is very high, providing protection against changes in body temperature. Water readily evaporates from body surfaces into dry atmospheres. The **heat of vaporization** of a substance is the amount of energy absorbed by the substance as it passes from the liquid state to the gaseous state. Because water has a high heat of vaporization (586 cal/g), evaporation of moisture from the skin and the wet surfaces of the respiratory system results in a high rate of heat loss. Heat loss can be increased by reflexive sweating. Water conducts heat rapidly (that is, has a high **thermal conductivity**). As a result, when one's body is immersed in cool water, heat is lost much more rapidly than in air of the same temperature.

The hydrogen bonds that form between water molecules near the surface of liquid water result in a large **surface tension,** which draws the surface to the smallest possible area (causing water droplets to assume a spherical shape, for example). Surface tension makes a thin layer of water surprisingly resistant to stretching by forces acting along the plane of the layer. Surface tension has potential consequences for the structure of the lungs and the work of expanding them in breathing. The inner surfaces of the lungs are covered by a thin layer of fluid. If this fluid had the surface tension of pure water, the delicate air pockets that make up the gas-exchange surface of the lung would collapse and be extremely difficult to reinflate with each breath. This does not normally occur because the lungs secrete pulmonary surfactant, a substance that greatly reduces the ability of water molecules to bond to each other.

TABLE 2-2   *The Properties of Water*

Property	Explanation	Physiological effect
High polarity	Water molecules orient around ions and polar compounds	Makes a variety of compounds soluble in cells so they can react chemically
High specific heat	Hydrogen bonds absorb heat when they break and release heat when they form	Water stabilizes body temperature
High heat of vaporization	Many hydrogen bonds must be broken for water to evaporate	Evaporation of water cools the body and is an important way of regulating body temperature
High thermal conductivity	The high rate of molecular interaction in water allows heat energy to be transferred rapidly	Tissues composed largely of water are poor insulators compared with hair and fat
High surface tension	Hydrogen bonds are especially common near water surface	Layer of water on lung surfaces increases the work needed to breathe

Salt crystal

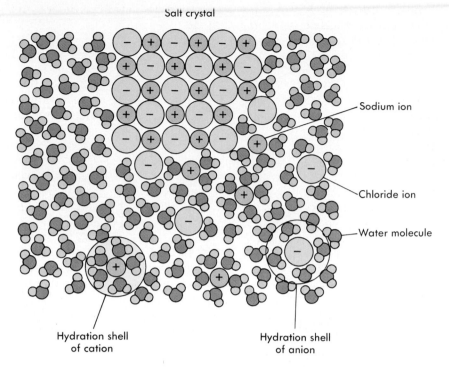

Sodium ion

Chloride ion

Water molecule

Hydration shell
of cation

Hydration shell
of anion

FIGURE 2-8

The molecular properties of water and solutions. When a crystal of table salt dissolves in water, individual Na^+ and Cl^- ions break away from the salt lattice and become surrounded by water molecules in hydration shells. Water molecules orient around Cl^- ions so that their positive ends face in toward the negative Cl^- ions; water molecules surrounding $Na+$ ions orient in the opposite way.

Aqueous Solutions

Solvents are liquids that are able to cause other substances to leave the solid or gas states and mingle with the solvent molecules as freely-moving ions, atoms, or molecules. The dissolved substance is called a **solute.** The **solubility** of a substance in a given solute is the amount of that substance that can be dissolved per unit volume of solvent under standard conditions of temperature and pressure.

Water has been called the universal solvent. Entry of a solid into water solution may involve breaking of ionic bonds and release of free ions. For example, when a salt crystal (such as NaCl) is placed in water, water molecules orient around the positively charged cations and negatively charged anions in the salt and pull apart the ionic bonds between them (Figure 2-8), mixing the ions among the water molecules. Solutes that dissociate into ions in water are called **electrolytes.** The electrical forces between charged ions and polar water molecules in solution cause each ion to be surrounded by a **hydration shell** of oriented water molecules (see circled structure in Figure 2-8).

Acids are hydrogen-containing compounds that tend to liberate protons (H^+) when dissolved in water. Equation 2-1 is the simplest way of representing this dissociation reaction.

$$HA -> A^- + H^+$$
Eq. 2-1

The proton released by the acid is conventionally called a hydrogen ion, but actually it is accepted by a water molecule, forming a **hydronium ion** (H_3O^+). Consequently, Equation 2-2 is a more correct representation of the reaction.

$$HA + H_2O -> A^- + H_3O^+$$
Eq. 2-2

Bases are compounds that can act as proton acceptors. In solution, bases bring about an increase in the concentration of OH^-. In the reaction shown above the product A^- formed by release of the proton is called the **conjugate base** of the acid HA. Acid-base reactions always result in the formation of an acid-conjugate base pair (Figure 2-9), so that Cl^- is the conjugate base of HCl and HCO_3^- (bicarbonate) is the conjugate base of H_2CO_3 (carbonic acid).

Strong acids and bases completely dissociate in solution. Hydrochloric acid, sulfuric acid, and sodium hydroxide are some examples of strong acids and bases. Acids and bases that do not dissociate completely are referred to as **weak.** In solution, weak acids and bases exist in equilibrium with

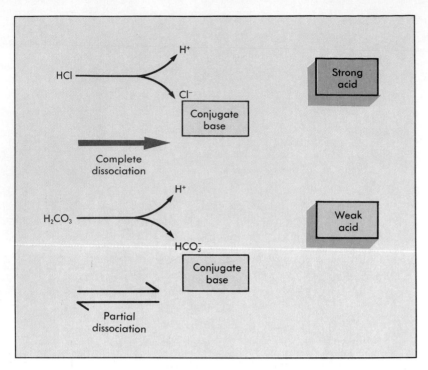

FIGURE 2-9
Strong acids dissociate completely in solution. Weak acids partly dissociate. For acids, one product is the H^+ ion; the other is called the conjugate base.

their conjugate forms. The **dissociation constant** is a measure of the extent to which a particular weak acid or base dissociates in solution. For example, the dissociation constant for weak acid dissociation (K_a in Equation 2-3) is the ratio of the concentrations of the products of the reaction to the concentration of undissociated acid (Equation 2-4). The smaller the value of K_a, the weaker the acid.

$$HA + H_2O \leftrightharpoons A^- + H_3O^+$$
Eq. 2-3

$$Ka = [H_3O^+][A^-]/[HA]$$
Eq. 2-4

Biologically important weak acids include many of the amino acids and carbonic acid (H_2CO_3), a compound formed when carbon dioxide reacts with water. Water can act both as a weak acid by dissociating into H^+ and OH^- ions, and as a weak base by accepting a proton to form H_3O^+. The K_a of water is quite low. Only one water molecule in 10 million (1 x 10^{-7}) dissociates to yield H^+ and OH^- (a **hydroxide** ion), so that the pH of pure water is 7.0.

Hydrogen ion concentration can vary by a factor of 10,000 trillion, so it is convenient to express it using a logarithmic scale of **pH units**, rather than molarity. The **pH** of a solution is defined as the negative logarithm of the hydrogen ion concentration (pH = -$\log_{10}$ [H^+]). The pH scale runs from 0 to 14 (Figure 2-10). Each change of one pH unit on the scale represents a tenfold step of H^+ concentration:

a solution with an H^+ ion concentration of 1.0 molar will have a pH of zero (the log of 1 is zero); a 1 mM (1 x 10^{-3} M) H^+ solution will have a pH of 3.0; and a H^+ concentration of 1 mM (1 x 10^{-6} M) corresponds to a pH of 6.0. Biological solutions usually have a pH near 7.0, so that they are in the middle of the range of possible H^+ ion concentrations.

A **buffer system** is a mixture of a weak acid or base and its salt. A buffer system accepts some protons if acid is added to the solution and gives up some if base is added, thus minimizing changes in hydrogen ion concentration. There are two major buffer systems in the extracellular fluid: the bicarbonate system, which is based on the equilibrium among carbon dioxide (a gas), the weak acid carbonic acid (H_2CO_3), and the conjugate base bicarbonate (HCO_3^-); and the phosphate system, which is based on the equilibrium between the weak acid form $H_2PO_4^-$ and the weak base form HPO_4^{2-} (Table 2-4). Proteins are able to act as buffers because they contain many weak organic acid groups. Proteins are relatively unimportant as buffers in the extracellular fluid because of their low concentration, but they are the major buffer system of intracellular fluid. A more detailed explanation of the regulation of the pH of extracellular fluid and of acids, bases and buffers is given in Chapter 20 and the accompanying "Focus Unit."

Molecules that do not dissociate into ions may nevertheless dissolve in water. Molecules that are po-

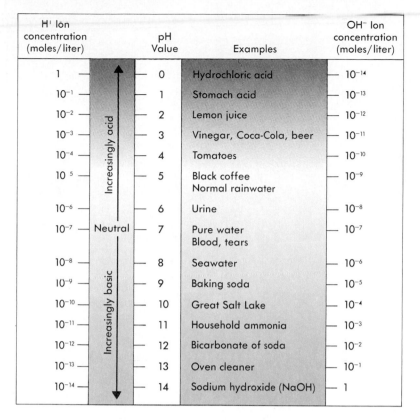

FIGURE 2-10
The pH scale. At pH = 7.0, the concentration of H⁺ and OH⁻ are equal; a solution with a
pH of 7.0 is said to be neutral. Solutions with pH below 7.0 are acidic, reflecting H⁺ concentrations greater than OH⁻ concentrations. Solutions with pH greater than 7.0 are basic
or alkaline; at these values OH⁻ is more abundant than H⁺. Some commonly encountered
example solutions and their approximate pH values are shown below the scale. Blood and
most other body fluids are slightly alkaline. A notable exception is the fluid secreted by the
epithelium of the stomach.

TABLE 2-3   Common Buffer Systems in the Body

Buffer system	Constituents	Physiological role
Bicarbonate	$NaHCO_3$ H_2CO_3	Main buffer system that maintains the normal plasma pH
Phosphate	Na_2HPO_4 NaH_2PO_4	Main buffer system in the urine
Proteins	$-COO^-$ groups $-NH_3^+$ groups	Important buffer system regulating the internal pH of cells; accessory system for plasma pH regulation

lar enough to mix with water form a category called
hydrophilic, or water-loving. Water is a good solvent for these compounds because the water molecules can attach themselves to the positive and negative parts of the solute molecules. For example, sugar
molecules have projecting polar hydroxyl (OH)

groups. Adjacent molecules in a sugar crystal orient
with oppositely charged regions, attracting one another. Water molecules can get inside the sugar crystal, orient around the polar groups of each sugar molecule, and thus disrupt the attractive forces holding
the compound together (Figure 2-11). The more po-

The Chemical and Physical Foundations of Physiology **29**

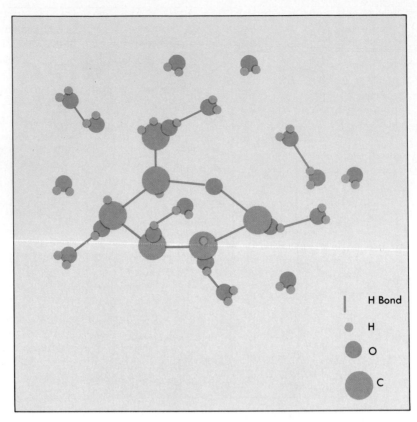

FIGURE 2-11
Solvation of polar compounds, like glucose (shown here), depends on interaction of polar water molecules with polar groups on the solute.

lar the solute molecules are, the stronger is the interaction between the solute and water molecules and the more soluble the compound.

Compounds that have too few polar groups to dissolve readily in water are called **hydrophobic,** or water-fearing. Most oils and fats fall into the second category. When water is present, **hydrophobic bonds** can form between hydrophobic molecules or between hydrophobic regions of large molecules. Such hydrophobic bonds hold cell membranes together and, along with hydrogen bonds and van der Waal's forces, help determine the three-dimensional structure of proteins (see Chapter 3).

Gases may dissolve in water. Oxygen (O_2) and carbon dioxide (CO_2) are consumed and produced, respectively, in oxidative metabolism and so are of the greatest importance in physiology. The route of movement of these gases between cells and the environment involves entry and exit from solution in blood plasma and diffusion as dissolved gas in interstitial fluid and cytoplasmic fluid. The factors that affect gas solubility and diffusion of dissolved gas will be discussed in greater detail in Chapter 17.

Molarity, Osmolarity, and Equivalency

A solution is described in terms of both the amount of solute in it and the total number of particles it contains. The **molecular weight** of a compound is defined as the sum of the atomic weights of its constituent atoms; this number, in grams, is 1 g molecular weight, or one **mole.** A mole of any compound always contains 6.023×10^{23} molecules **(Avogadro's number).** The number of moles of solute in a liter of solution is termed the **molarity** (Figure 2-12, *A*). For example, if the molecular weight of a compound is 100, a 1.0 molar or 1000 millimolar (mM) solution is prepared by adding 100 grams of the compound to slightly less than a liter of water, and then adding water to give a final volume of exactly 1 L. More generally, the molarity is equal to the number of grams of solute contained in a liter of solution divided by the solute's molecular weight. Water has a molecular weight of 18 and a density of 1.00 g/mL, so there are about 55 moles of water in 1 L (1000 g divided by 18). In biological solutions, solute concentrations are seldom more than a few tenths of a mole per liter, so there are always many more water molecules than solute molecules.

The **osmolarity** of a solution (Figure 2-12, *B*) is the number of dissolved particles in a liter. For a given compound the osmolarity is the product of the molarity and the number of particles into which each molecule dissociates in solution:

Osmolarity=(Molarity)x(Number of dissolved particles/molecule)

Eq. 2-5

Polar nonelectrolyte molecules do not dissociate when they dissolve in water. The number of particles in a solution of nonelectrolytes equals the number of molecules in the nonelectrolyte compound originally added to it, so the osmolarities and molarities are equal for such substances. For example, the osmolarity of a 100 mM glucose solution is 100 mOsM, and the osmolarity of a 50 mM glycine solution is 50 mOsM.

In electrolyte solutions the total number of dissolved particles that results when a given number of salt molecules are placed in solution depends on the number of ions each salt molecule dissociates into.

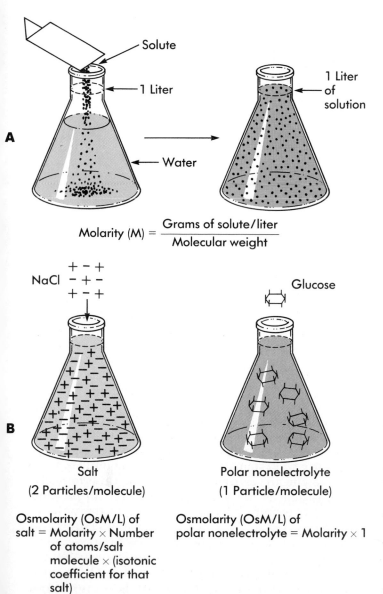

Molarity (M) = $\dfrac{\text{Grams of solute/liter}}{\text{Molecular weight}}$

NaCl
Salt
(2 Particles/molecule)

Glucose
Polar nonelectrolyte
(1 Particle/molecule)

Osmolarity (OsM/L) of salt = Molarity × Number of atoms/salt molecule × (isotonic coefficient for that salt)

Osmolarity (OsM/L) of polar nonelectrolyte = Molarity × 1

FIGURE 2-12

An illustration of the definitions of molarity (**A**) and osmolarity (**B**).

For example, one NaCl dissociates to give one Na^+ ion and one Cl^- ion, so the number of dissolved particles is twice as great as the number of NaCl molecules (Figure 2-12, *B*). The salt $CaCl_2$ dissociates into one Ca^{++} ion and two Cl^- ions, so the number of dissolved particles is three times the number of molecules. Solute ions interact with each other to a small extent, as well as with the water molecules. To the extent that this occurs, a small correcting factor has to be added to the equations for osmolarity and osmotic pressure. The magnitude of this factor, called the **isotonic coefficient**, depends on the particular ions involved and on their concentration. This correction is small for most ions in body fluids. If this correction is neglected, a 1 molar (1.0 M) solution of NaCl is 2 osmolar (2.0 OsM); a 100 mM solution of NaCl is 200 mOsM; and a 100 mM solution of $CaCl_2$ is 300 mOsM.

It is conventional to express the concentration of ions in terms of the number of electrical charges the ion contributes to the solution. The unit of ionic charge is the **equivalent** (Eq) that is equal to 1 Avogadro's number of charges. For example, in a 1M solution of NaCl, Na^+ and Cl^- each contribute 1 mole of charges, and the Na^+ and Cl^- concentrations are each 1 Eq/L. In a 1M solution of $CaCl_2$ each ion type contributes the equivalent of 2 moles of charges, and the Ca^{++} and Cl^- concentrations are each 2 Eq/L.

PHYSIOLOGICAL FORCES AND FLOWS
The Relationship Between Force, Resistance, and Movement

- *How does work differ from energy?*
- *What is the relationship among force, resistance, and movement?*
- *How do acceleration, friction, elasticity, and hydrostatic pressure operate in physiology?*
- *How can units of osmolarity be converted to units of osmotic pressure?*
- *What is the relationship among voltage, current, and resistance or conductance?*

In physics, **work** is done when material is moved, and is defined as the product of force and distance. **Energy** is the capability to do work. The living body is in constant motion. An arm swings when pitching a baseball; blood flows within an artery; air enters the lungs; plasma is filtered across the wall of a capillary into the kidney or the interstitial space between cells; and ions cross membranes. Different driving forces are present in each of these examples, and different materials are being moved. Nevertheless, there are three variables involved in all situations in which work is done: the driving force, the rate of movement, and the resistance to movement. **Driving forces** are forms of energy with the potential to cause movement. **Rate of move-**

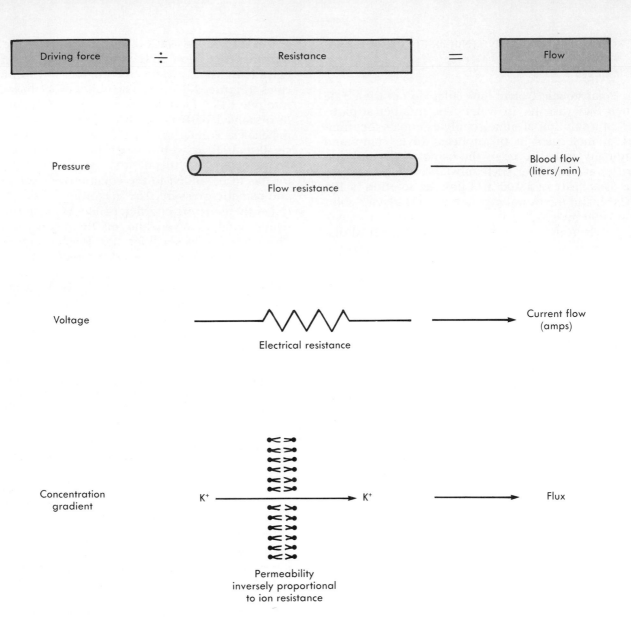

FIGURE 2-13
The force-flow relationship can be applied to many physical situations in which there is friction, including the flow of fluids such as blood (upper diagram), the flow of electrical current (middle diagram), and the flux of ions across membranes (lower diagram).

ment is the amount of material (or charge in the case of electric current) moved a given distance per unit of time. In all physiologically-relevant situations, moving materials meet with some form of opposing force or resistance. These three variables are related in an equation that takes the following general form:

Rate of movement = driving force/resistance

Eq. 2-6

Depending on whether what is moving is solid, liquid, gas, or electric current, different specific terms will be used for the three variables (Figure 2-13).

In calculations involving several measured

quantities, all must be expressed in mutually consistent units. The system of measurements generally employed in chemistry is the centimeter-gram-second (c.g.s.) system. In this system, the units of force or weight are **dynes** ($g \ cm \ sec^{-2}$). Unfortunately, this system is not yet universally used in physiology. For example, in the c.g.s. system the units of pressure are the **dyne cm^{-2} for** small pressures and the bar (1 bar = 10^6 dyne cm^{-2}) for large pressures. However, in physiology pressure is usually expressed in units of **mm Hg** or **mm H_2O** for small pressures and **atmospheres** for large pressures. Conversion factors for these units are given in Table 2-4. In the c.g.s. system, work or energy is expressed in **ergs** (dyne cm g cm^2 sec^{-2}) or **joules** (1

TABLE 2-4	c.g.s. Units of Measurement and Equivalent Customary Units	

Quantity	c.g.s. units	Some customary units
Length	cm	1 m = 100 cm
		1 mm = 0.001 m
		1 μm = 10^{-6} m
Mass	gm	1 kg = 1000 gm
Volume	cm^3	1 L = 1000 cm^3
Molecules		1 g molecular wt (mol) = 6.022×10^{23} molecules
Charges		1 Eq = 6.022×10^{23} charges
Concentration		moles/liter (M) or equivalents/liter (Eq/L)
Velocity	cm sec^{-1}	
Acceleration	cm sec^{-2}	
Force or Weight	dyne = gm cm sec^{-2}	1 gm = 980.7 dyne
Pressure	dyne cm^{-2}	1 mm Hg = 1 Torr = 1,333 dyne cm^{-2}
		1 mm Hg = 13.5 mm H_2O
		1 atm = 760 mm Hg
		1 bar = 10^6 dyne cm^{-2}
		1 Osmole/L = 22.4 atm
Work	erg = dyne cm g cm^2 sec^{-1}	1 J = 10^7 ergs
		1 calorie = 4.184 J
		1 volt equivalent = 96,487 joules

joule = 10^7 ergs). Traditionally, the unit of measure of thermal energy and chemical potential energy is the **calorie,** the amount of heat that will increase the temperature of 1 gram of water by 1° C. One advantage of the c.g.s. system is that a common unit is used for all forms of energy, whether mechanical, chemical, thermal, or electrical.

Mechanical Forces

Mechanical forces are those that affect the motion of objects or masses. First, let us consider a moving object, such as the arm of a baseball pitcher. Newton's second law of mechanics states that a mass subjected to an applied force will accelerate in the direction of the force with an acceleration that is proportional to the force. **Acceleration** is the rate of change of velocity with time. If there are no opposing forces, positive acceleration means the velocity of an object will continuously increase. The pitching arm is accelerated by contracting skeletal muscle, and in turn accelerates the baseball. This is an example of movement resulting from a mechanical driving force; other examples are the acceleration of blood by the contracting heart and the acceleration of air in airways of the respiratory system by volume changes of the chest.

Frictional force is resistance to motion that arises from interaction between the moving material and its surroundings. Friction occurs whenever mechanical work is performed, including muscular contraction. Whatever their origin, all frictional forces share two common characteristics: (1) the

magnitude of the force is proportional to the velocity of the object; and (2) frictional forces always act in a direction exactly opposite to the direction of the object's motion. If the driving force—the acceleration—is constant, the rate of movement of the material—called **velocity** for objects and **flow rate** for liquids and gases—will ultimately become constant. This is because frictional forces increase with velocity. The velocity becomes constant when the driving force and the frictional force come into balance. The magnitude of the constant velocity depends on magnitude of the driving force and the resistance or degree to which the moving mass is subject to friction.

Hydrostatic pressure is the form of potential energy that results when a driving force is applied to a confined liquid. Most children have inflated a rubber balloon with water under pressure from the tap and then watched the energy stored in the elastic walls of the balloon press the water out the nozzle of the balloon in a stream, perhaps directed at a little brother or sister. In this case the water is flowing down a difference of hydrostatic pressure between the contents of the balloon and the exterior. The rate of flow depends on the magnitude of the hydrostatic pressure and the resistance to flow. For example, the driving force for fluid movement through a tube is a pressure difference between the two ends of the tube, usually expressed in units of millimeters of mercury (mm Hg). If the pressure difference is 100 mm Hg and the flow resistance of the tube is 100 mm Hg/L/min, the flow rate will be 1

The Chemical and Physical Foundations of Physiology

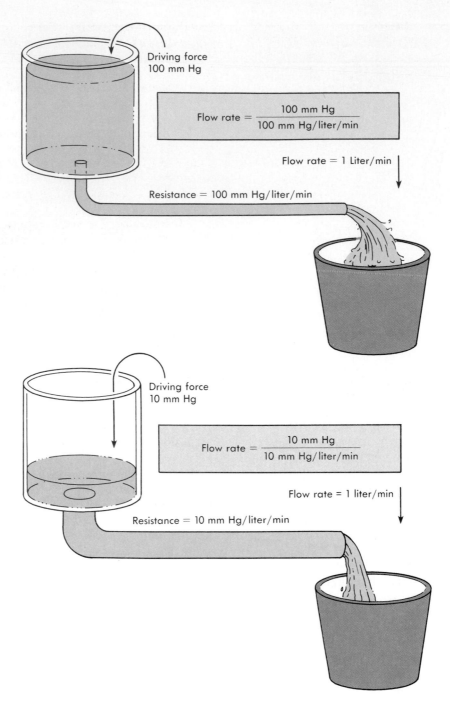

FIGURE 2-14
The flow rate is determined by dividing the driving force (in this case a gradient of hydrostatic pressure) by the resistance. If the resistance is reduced by tenfold, the driving force need be only one-tenth as large to achieve the same flow rate.

L/min. The same flow rate could be obtained with a pressure difference of only 10 mm Hg if the flow resistance were 10 mm Hg/L/min (Figure 2-14).

In a standing person, the hydrostatic pressure of blood in blood vessels (Figure 2-15), is the sum of the force that the heart imparts to the blood and the gravitational acceleration of the column of blood in the mainly vertical arteries and veins. The gravita-

tional component can be modeled by a column of fluid as tall as the distance between the heart and the soles of the feet. As blood descends from the heart toward the feet, the gravitational component of blood pressure increases, while the force imparted to the blood by the heart decreases slightly, due to friction with the vessel walls. The same gradient of gravitational pressure is experienced by

THE FOUNDATIONS OF PHYSIOLOGY

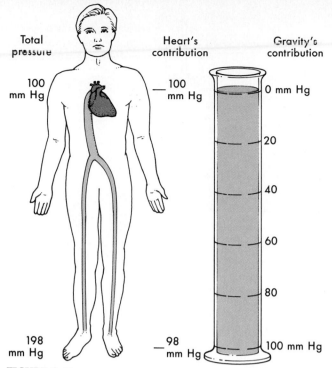

Total pressure

Heart's contribution

Gravity's contribution

100 mm Hg — 100 mm Hg — 0 mm Hg

— 20

— 40

— 60

— 80

198 mm Hg — 98 mm Hg — 100 mm Hg

FIGURE 2-15

The contractions of the heart impart a mean hydrostatic pressure of about 100 mm Hg to blood leaving the heart (center). The force imparted to the blood diminishes by about 2 mm Hg as blood travels over the distance from the heart to the feet. The gravitational component due to the weight of the column of blood in arteries corresponds to the pressure gradient in a column of fluid level with the heart (right). The sum of the heart's contribution and gravity's contribution is the total pressure applied to arterial walls, and rises from about 100 mmHg at the level of the heart to about 198 mmHg at the level of the feet (left).

blood returning uphill to the heart, so that the effect of gravity cancels out and does not contribute to the driving force for blood movement through the circulatory system. For this reason, blood pressure is conventionally measured with reference to the pressure at the level of the heart. However the gravitational component causes increased stress on the walls of the vessels of the legs and feet of standing people. It also increases the rate at which fluid is squeezed from the plasma into the interstitial spaces in the lower part of the body. This is the reason that people who stand for long periods tend to have swollen feet.

Elastic materials resist deformation and tend to return to their original dimensions after being stretched or deformed. For example, a spring loaded with a weight opposes stretching with an elastic restoring force that is greater the more the spring is stretched (Figure 2-16). The spring stops stretching at the length at which the restoring force becomes equal to the gravitational acceleration of the load. When the load is removed, energy stored in the spring is released, returning it to its rest length. Connective tissue is one of the most elastic of biological materials. When skeletal muscles contract under load, energy is stored in the elastic tendons that connect muscles to bones, to be released as the muscle relaxes. This temporary storage of energy smooths body movement and increases the efficiency of force generation.

Another example of an elastic restoring force at work is the balloon filled with water in the descrip-

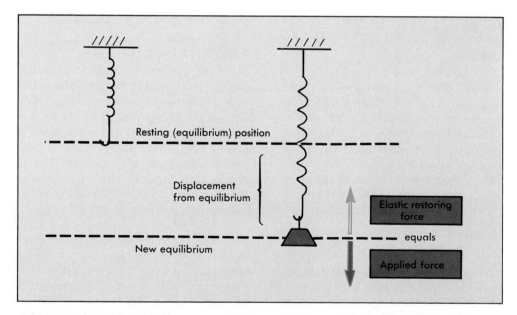

Resting (equilibrium) position

Displacement from equilibrium

Elastic restoring force

equals

New equilibrium

Applied force

FIGURE 2-16

When objects are displaced from their equilibrium positions in elastic systems, restoring forces oppose the applied force. Thus elastic restoring forces determine the equilibrium positions or conformations of objects. The effects of such restoring forces are important in several systems, including the chest wall/lung, skeletal muscles, and arteries.

tion of hydrostatic pressure. The balloon's diameter is increased by the entry of water under pressure. If the nozzle is open, the elastic restoring force returns the balloon to its original size. The transfer of energy from heart muscle to blood involves a similar elastic restoring force. Each heartbeat rapidly ejects a quantity of blood from the heart, stretching the walls of the large arteries that carry blood away from the heart. While the heart is resting between beats, the elastic restoring force of the stretched arterial wall maintains pressure on the blood, converting the intermittent spurts of blood from the heart into a steady flow in the blood vessels.

The **compliance** of a structure is a measure of how much distortion results when a given force is applied to the structure. In other words, the less stiff or rigid a structure, the greater its compliance. The appropriate dimension for measurement of compliance is determined by the structure and function of the tissue. The compliance of a skeletal muscle is the ratio of its change in length to the force applied to its ends; the compliance of a blood vessel is the ratio between its diameter and the pressure difference across its walls; and lung compliance is the ratio between lung volume and the pressure of air within it. Compliance is an important quality of the walls of blood vessels because it determines the relationship between the volume of fluid contained by the vessel and the resulting pressure. For example, the walls of veins possess less muscle and connective tissue and are much more compliant than the walls of arteries. As a consequence, veins contain a large fraction of the total blood volume under a relatively low pressure, while arteries contain a small fraction of the total blood volume under a much higher pressure.

Osmotic Pressure

The osmolarity of a solution was defined (p. 30) as the number of moles of dissolved particles in a liter. The solute particles dilute the solvent molecules, resulting in a lower water concentration than that of pure water. The **osmotic pressure** of a solution is that pressure that would be generated by diffusional movement of water across a partition separating the solution from pure water. The osmotic pressure (P_{osm}) corresponding to any given osmolarity of a particular solute can be calculated as:

$$P_{osm} = kRTC_{osm}$$
Eq. 2-7

where R is the universal gas constant, T is the temperature, C_{osm} is the osmolarity, and k is a correction for the extent to which that particular solute deviates from ideal solute behavior. If there is a difference in osmolarity between two solutions separated by a water-selective partition, water molecules will tend to move from the side of lower solute concentration (with a higher water concentration) to

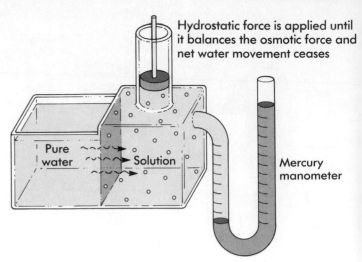

FIGURE 2-17

One form of osmometer measures the osmotic pressure of a solution as the hydrostatic pressure (delivered by the piston) that just stops osmotic movement across a water-selective membrane with the solution in question on one side and pure water on the other. In this diagram the hydrostatic pressure is measured by a mercury manometer; a U-shaped tube filled with mercury in which the pressure corresponds to the difference in heights of the two mercury columns.

the more concentrated solution (a lower water concentration), driven by the difference in the osmotic pressure of the two solutions. In contrast to hydrostatic pressure, osmotic pressure is an attractive force: water moves from lower osmotic pressures (where the water concentration is high) toward higher osmotic pressures (where the water concentration is low [Figure 2-17]). The difference in osmotic pressure between interstitial fluid and blood is one of the forces that affects movement of fluid between the circulatory system and the interstitial fluid.

Electrical Forces

Both the cytoplasm and extracellular fluid contain electrolytes. An electrolyte solution must be neutral as a whole because the total number of positive and negative ions produced by dissociation of an electrolyte are equal. However, in some situations there can be small local imbalances in the distribution of positive and negative charge. The most important charge imbalance occurs when ions move across cell membranes, a topic discussed in detail in Chapter 6. Charge imbalances produce electrical forces.

Voltage (V or E) is an electrical driving force that results from charge imbalance. The relationship is best seen by considering what happens in a wire when a battery is connected to it (Figure 2-18). Electrons will move away from the negatively charged pole of the battery and toward the positive pole. The voltage is a measure of the electrical driving force of the battery. The magnitude of the resulting **current** (I) of electrons obeys the general relationship between driving force and flow developed earlier. The relationship among electrical current,

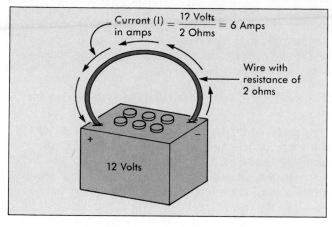

FIGURE 2-18
Ohm's Law for current flow in a conductor. The battery gives a voltage of 12 volts. If the wire has a resistance of 2 ohms, the current will be 6 amps.

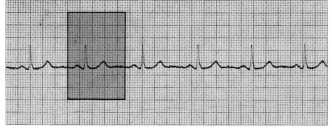

Electrical activity during a single heart beat

driving force, and electrical resistance (**R**) is referred to as **Ohm's law** and is written:

$$I = (V_2 - V_1)/R$$
Eq. 2-8

where V2 and V1 refer to the voltages at two different points. The unit of electrical resistance is the **ohm;** voltages are measured in **volts,** and current flow is measured in **amperes.** One ampere is a flow of 1 **Coulomb** or 1.6×10^{19} charges per second. The reciprocal of resistance is the **conductance (G),** measured in **Siemens** (1/ohm). According to the relationship given above, 1 volt of electrical driving force will drive a current of 1 ampere through a resistance of 1 ohm.

Compared with pure water, all body fluids are relatively good conductors of electricity because they contain a high concentration of ions. As a result, the body behaves as a **volume conductor** of electricity. Some organs, such as the heart, brain, and skeletal muscles, generate electrical currents within themselves when they are active. The flowing charges within such an organ in turn exert fields of electromagnetic forces on charges in the surrounding parts of the body. As a result, currents flow around as well as within these organs. The electromagnetic forces diminish with the square of the distance from the organ, but the resulting currents can be detected at some distance from the organ. With sensitive equipment they can even be detected at the body surface. This is the basis of records of the electrical activity of the heart, called **electrocardiograms** (Figure 2-19, *A*), and of the

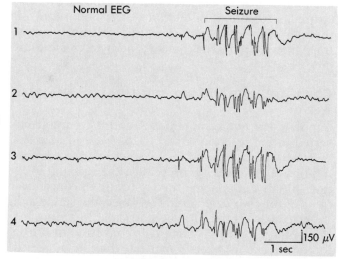

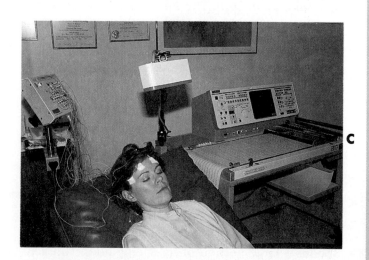

FIGURE 2-19
A A normal electrocardiogram (ECG). Wires are connected to the chest. The deflections seen in the electrocardiogram reflect current flow between different parts of the heart.
B An electroencephalogram (EEG). Here electrical currents in the brain are measured by electrodes placed at four different points on the scalp. Seizures are associated with a dramatic increase in the amplitude and frequency of the electroencephalogram at all four areas.
C A photograph of a patient with EEG electrodes connected to her head.

The Chemical and Physical Foundations of Physiology   **37**

brain, called **electroencephalograms** (Figure 2-19, *B* and *C*). The electromagnetic fields generated around large skeletal muscles when they contract can also be easily detected at the body surface (the records are **electromyograms**) and can be used for diagnosis of muscle disorders.

THERMODYNAMICS IN PHYSIOLOGY
The First Law: Energy Conservation

- *What are the First and Second Laws?*
- *How can equilibrium be defined in terms of the Second Law?*
- *How can the equilibrium ratio of reactants and products of a reversible reaction be expressed mathematically?*
- *How does the Law of Mass Action operate when the system is displaced from equilibrium?*
- *How can the bodies of living organisms maintain internal order and homeostasis without violating the Second Law?*

Thermodynamics is the science that deals with energy transformations. In this science, a **system** is an assemblage of matter and energy with definite boundaries in space. The **state** of the system is a description of the forms and amounts of matter and energy in it. A change in the state of a system is called a **process. Heat flow** is a process in which the energy content of the system changes as the result of a difference in temperature between the system and its surroundings, while **work** is an energy transfer between the system and the surroundings that occurs by some other means. A system can go from one state to another by doing work or having work done on it, by losing or gaining heat, or by a combination of both. The **First Law of Thermodynamics** states that the total quantity of energy in the universe must be conserved in any process—no energy disappears and no new energy is generated. For example, if there is flow of heat into a system and it also does work on its surroundings, the increase in internal energy of the system equals the difference between the heat input and work output.

Living organisms are continuously transforming energy from one form to another. The intestine converts the complex molecules of foods into simpler compounds that can be absorbed. Cells convert the chemical potential energy of these food molecules into forms that can be used to support the energy needs of cells. Muscles transform this chemical potential energy into kinetic energy. Nervous tissue spends a large portion of its energy budget in generating electrical signals. All cells expend some of their share of the chemical energy from food in growth and repair—processes in which energy expenditure results in an increase in the order of the system. In each of these transformations some of the energy is unavoidably lost as heat. This heat is an important resource for endotherms ("warm-blooded" animals). In these animals body temperature is regulated by feedback loops that affect both the rate of heat production and the rate of heat loss to the environment.

The Second Law: Disorder of the Universe Increases

Organisms could be thought of as islands of order in a more or less disorderly environment. Order in the body takes many forms. Order is incorporated in structures, such as protein molecules, and also in differences of chemical concentration, such as those between the body and the environment and between the cytoplasm of cells and extracellular fluid. The **Second Law of Thermodynamics** says that any spontaneous process results in a net increase in the amount of disorder (**entropy**) in the universe. However, the increase in order that occurs when organisms grow, maintain homeostasis and repair themselves does not violate the Second Law, because the law applies to the universe as a whole. Order may be increased locally by expenditure of energy, and the higher the degree of order, the more energy required. For example, joining individual amino acids together to form a protein requires energy to form the chemical bonds and results in an increase in order, compared to the less orderly free amino acids. Furthermore, making a protein with different types of amino acids joined in a specific sequence represents an even greater increase in order and requires 10 to 20 times as much energy as would be needed to simply join the amino acids together in random order. The Second Law simply requires that the gain in order represented by the completed protein be more than equalled by an increase in disorder elsewhere. The body can be thought of as an engine that uses chemical energy from the environment (that is, food) to create local order.

Chemical Equilibrium and the Law of Mass Action

Suppose that two compounds A and B are interconverted in the reaction $A \leftrightharpoons B$. This kind of reaction is described as **reversible.** No matter what concentrations of A and B are present in the reaction mixture initially, after some time the ratio of concentrations of the two compounds will become stable at a value characteristic of that particular reaction. At this point the rate at which A is converted to B (the forward reaction) exactly matches the rate at which B is converted to A (the backward reaction), and the system of reactants is said to be in equilibrium. The characteristic ratio of concentrations is called the **equilibrium constant.**

In a reversible reaction that has not yet reached equilibrium, the reaction that leads toward equilibrium will predominate. For example, if the equilibrium constant for the reaction A ⇌ B is 3, at equilibrium the mixture will contain one part of A to 3 parts of B. If the reaction is started with a mixture that contains equal parts of A and B, the forward reaction (A → B) will go at a greater rate than the reverse reaction. As equilibrium is approached, the rate of the forward reaction decreases and that of the reverse reaction increases, until the two rates match. If more of compound A is added to the reaction mixture at this point, the forward reaction will again predominate until the equilibrium ratio of A and B is restored. On the other hand, if more B is added, the reverse reaction (A ← B) will be driven until the equilibrium ratio of A and B is restored. This is an example of the **Law of Mass Action,** which states that the rate of a reaction is proportional to the concentrations of the reactants. The same principle applies to the dissociation of weak acids and bases described earlier in this chapter.

Free energy is defined as the portion of a system's total energy that is available to do work. The thermodynamic definition of equilibrium is the state in which the free energy of the system is minimal. The Second Law can be rephrased in terms of free energy: to occur spontaneously, a process must result in a decrease in the system's free energy. In the example of A and B, the free energy of the compound B must be three times less than that of compound A. The difference between the free energy of the initial mixture and that of the equilibrium state appears as heat released by the reaction. As noted above, for many reactions that take place in the body the free energy of the products is greater than that of the reactants. Remember that the Second Law demands only that a process result in a *net* negative free energy change. Reactions that result in an increase in free energy may be driven by being coupled to reactions that result in an even greater decrease in free energy, so that the net free energy change of the driving and driven reactions is negative.

The Concept of Dynamic Steady State Revisited

The thermodynamic laws that govern the equilibrium state are able to explain static **closed systems** in which only energy, and not matter, is exchanged with the surroundings. Living organisms are dynamic **open systems** that continuously take in and give off both matter and energy. For organisms, perfect equilibrium would correspond to a body chemical composition, temperature, and pressure identical to that of a like-sized slice of the environment; in other words, for organisms to equilibrate would be to cease to exist. The concept of chemical equilibrium is useful in understanding some chemical reactions that take place in the body, but the stability of the internal environment described in Chapter 1 is not due to the system having attained an equilibrium state. Instead, the features of the internal environment are maintained in dynamic **steady states,** in which homeostatic processes oppose the tendency of the open system to go towards equilibrium.

One example of such a steady state is the normal condition of water balance. Water is lost from the body by continuous evaporation from body surfaces and, from time to time, in feces and urine. These losses are normally balanced quite precisely by water intake, so the average water content of the body does not change much. Maintenance of steady states requires both expenditure of chemical potential energy (in the example, finding water, drinking, and water absorption from the gastrointestinal tract all require energy) and information (to control thirst the brain must have information about the state of water balance). In almost all homeostatic systems, the energy-requiring reactions are coupled to a reaction in which a high-energy molecule called adenosine trisphosphate (ATP) is converted to its low-energy form, adenosine disphosphate (ADP). Chapter 4 describes the mechanisms by which ATP is used and regenerated inside cells. Subsequent chapters will describe the ways in which information is gained and used to direct the homeostatic processes that maintain the dynamic steady states of body systems.

SUMMARY

1. Strong chemical bonds between atoms are stable interactions in which electron sharing (for **covalent bonds**) or electron donation (for **ionic bonds**) stabilizes the outer **electron shells** of the participating atoms. **Hydrogen bonds** and **van der Waal's forces** are less stable bonds between atoms; these weak bonds nevertheless are important because, for example, they stabilize the three-dimensional structure of proteins and give water its solvent properties.

2. The ability of water molecules to form hydrogen bonds accounts for many of its solvent properties, including its high specific heat and its liquidity at physiological temperatures. Water is an effective solvent for molecules with electrically charged regions (**polar** molecules) because water molecules are also polar. Nonpolar molecules are much less soluble in water and may form **hydrophobic interactions** with one another in the presence of water.

3. Solutions can be characterized by their **molarity** (the number of gram molecular weights per liter) and by their **osmolarity** (the number of moles of particles per liter).

4. **Acids** and **bases** are solutes whose dissociation in solution results in hydrogen ion donation or acceptance. Strong acids and bases dissociate almost completely in solution; weak acids or bases exist in solution in an equilibrium between the dissociated and undissociated forms.

5. All moving masses in the body experience **friction** that resists their movement. The general relation between forces and flows is: **velocity = driving force/frictional resistance.** If the driving force is steady, the velocity will be steady; **steady states** of variables result from steady expenditures of energy. In **equilibrium states,** forces are balanced and no energy is being expended.

6. In some body systems, physical distortions are countered by **elastic restoring forces** whose effect tends to restore the systems to mechanical equilibrium. The relationship between the degree of distortion and the force applied is the **compliance.**

7. The **First Law of Thermodynamics** asserts that **systems** (the body is a system) change their states by transfer of **work, heat,** or both. In such changes of state, all of the energy transferred between the system and its environment can be accounted for.

8. The **Second Law of Thermodynamics** shows that biological order is bought at the cost of energy captured from the environment. The capture of this energy results in a net increase in entropy. Following the Second Law, chemical reactions proceed spontaneously in the direction that results in a decrease in the **free energy** of their system (a net increase in disorder).

9. The **equilibrium state** of a system is the state of lowest free energy. In a reversible chemical reaction at equilibrium there is a fixed ratio of reactants to products and the rate of the forward reaction is equal to the rate of the reverse reaction. If the system is displaced from equilibrium, the **Law of Mass Action** dictates that the reaction that leads in the direction of equilibrium will predominate.

10. The **dynamic steady state** of the body represents a system that is kept far from equilibrium by expenditure of chemical energy. The stability of this system is due to feedback regulation.

1. Define the following terms:
   Valence
   Electrolyte
   Molecular weight
   Acid

2. Molecules may interact with one another by forming covalent, hydrogen, or ionic bonds, or by van der Waal's forces. Give an example of each type of bonding and describe its relative stability.

3. What factor dictates that an oxygen atom can form two covalent bonds whereas a carbon atom can form four?

4. What is a polar molecule? What important chemical and physical properties of water result from its being polar?

5. Under what circumstances is there a difference between one mole of a solute and one osmole of that solute?

6. Why are hydrogen bonds important in determining the structure of proteins?

7. Why is energy required for assembly of elements into complex molecules? How does this relate to the concept of entropy?

8. Define the term resistance. Resistance enters into descriptions of blood flow, electric currents across membranes, and the movement of air in and out of the lung. These processes seem physically different. What is the relationship between force and flow that they all obey?

Choose the MOST CORRECT Answer.

9. Select the strongest chemical bond:
   a. Ionic bond
   b. Hydrogen bond
   c. van der Waal's forces
   d. Covalent bond

10. The major buffer systems in intracellular fluid is (are) the:
    a. Phospholipids
    b. Proteins
    c. Bicarbonate system
    d. Phosphate system

11. Glucose ($C_6H_{12}O_6$) has a molecular weight of 180. How many grams of glucose are added to water to prepare a 1.0 molar solution?
    a. 1800 g/1000 ml
    b. 180 g/100 ml
    c. 180 g/1000 ml
    d. 1.8 g/100 ml

12. The number of dissolved particles in a liter is the:
    a. Osmolarity of a solution
    b. Molarity of a solution
    c. Molecular weight of a solution
    d. Isotonic coefficient of a solution

13. Which of the following is NOT a variable in situations in which work is done?
    a. Resistance
    b. Acceleration
    c. Driving force
    d. Rate of movement

● *SUGGESTED READINGS*

ATKINS PW: *Atoms, electrons, and change.* Scientific American Library, WH Freeman and Co, New York, 1990.

BENNETT CH: Demons, engines, and the second law, *Scientific American,* 257:108, November 1987. Discusses attempts to design perpetual motion machines in violation of the second law of thermodynamics.

VEGGEBERG S: New uses found for electricity in bone disease, *Nutrition Health Review,* Spring 1990, p. 8. Describes use of electromagnetic fields to stimulate bone growth.

CHAPTER 3

The Chemistry of Cells

On completing this chapter you should be able to:

- Describe the structure of lipids, including triacylglycerols, fatty acids, steroids, and phospholipids, and state the major storage form of lipids in humans.
- Describe the structure of carbohydrates, including monosaccharides, disaccharides, and the starches. State the major storage form of carbohydrates in humans.
- Understand peptide bond formation.
- Compare primary, secondary, tertiary, and quaternary structure of proteins.
- State the way the sequence of amino acids in a protein is coded for by a sequence of nucleotide codons.
- Understand how enzymes catalyze chemical reactions.
- Define the terms active site, saturation, competition, and allosteric site.
- Describe DNA and how DNA replicates by complementary base pairing.
- Understand the process of transcription of the nucleotide sequence of a gene into messenger RNA.
- Describe how messenger RNA is translated into protein.

*T*he internal processes of cells involve transformations of carbon compounds, carried out and controlled by other carbon compounds. Organic chemistry, the branch of chemistry that concerns itself with carbon compounds, reflects a vitalistic belief that such compounds could be made only in living cells. The modern science of biochemistry is devoted to the structures and functions of organic molecules that are of biological importance. In the last 20 to 30 years, the science of biochemistry has undergone a revolution. The revolution began with the study of proteins, the most abundant large molecules in the body. Proteins could be thought of as the working parts of cells. Some proteins form basic structural elements; others act as catalysts for biochemical processes, including synthesis of carbohydrates and lipids, the other two main types of organic compounds present in cells. Protein molecules are structurally complex and have the potential for great diversity. The structure and properties of a protein are determined by its amino acid sequence, the order in which its different amino acids are strung together.

The study of proteins led investigators inevitably to the molecules deoxyribonucleic acid (DNA) and ribonucleic acid (RNA). These giant molecules are now known to specify the amino acid sequences of proteins and regulate their synthesis. Essentially every aspect of the structure and function of the body is determined ultimately by information about the structure of proteins stored in DNA and communicated to working cells by RNA. Molecular biology is the name for the new science of the structure and function of nucleic acids. The development of this science is enabling investigators to tackle some of the most fundamental problems in physiology and medicine: How does a fertilized egg give rise to a multitude of different specialized cell types? How are hormones synthesized, and how do they exert their effects on the body's activity, growth, and development? What are the mechanisms of thought and memory? Why are some diseases inherited?

SOME MOLECULAR TERMINOLOGY

From one point of view, all molecules are small. However, it is useful to divide biologically important organic molecules roughly into two groups: "small" molecules with molecular weights of less than 1000, and "large" or macromolecules with molecular weights that may be several hundred thousand or more. Typically the mass of a large molecule whose chemical formula may not be exactly known (such as a protein) is given in units of **daltons (d)**, a unit of mass essentially equivalent to that of a hydrogen atom. So the mass of a protein with a molecular weight of 100,000 would be 100,000 d or 100 kd (kilodaltons).

Cells contain perhaps a thousand different small organic compounds. Most of these fall into one of four main families: sugars, fatty acids, amino acids, and nucleotides. The large molecules fall into four main categories: lipids, polysaccharides, proteins, and nucleotides. One role of the small molecules is to serve as material for construction of larger molecules. Some general terms are used in this relationship. If the small molecules are linked together to form a chain, but retain essential features of their structures, the chain is called a **polymer** ("many parts") and the constituent small molecules are called **subunits, residues** or **monomers** ("one part"). Small molecules that essentially lose their identity in the course of synthesis of larger ones are usually called **precursors** ("coming before"). Fatty acids may be thought of as subunits of some lipids and precursors of others; sugars are the monomers of polysaccharides; amino acids are the monomers of proteins; and nucleotides are the monomers of nucleic acids such as RNA and DNA.

Many physiologically important organic molecules exist in the form of **isomers,** multiple forms having the same molecular weight and chemical composition but different structures. Isomers are of two types: **structural isomers,** in which the locations of bonds differ; and **stereoisomers,** in which the relative positions of the atoms in space differ. For example, sugars and amino acids can exist in two forms called the **D** and **L isomers.** As a general rule, only the D isomers of sugars and only the L isomers of amino acids are used by organisms. Carbohydrates in particular may have a large number of stereoisomeric forms, and a detailed treatment is beyond the scope of this text. In subsequent parts of this chapter, specific examples of isomers will be presented in cases in which the difference has physiological importance.

In many molecular diagrams shown in this chapter, a shorthand is used in which lines represent bonds, intersections represent carbons, and each carbon is assumed to have hydrogen side groups unless otherwise indicated. The single bonds of carbon project in four directions, so organic molecules have a three-dimensional structure.

Bonds indicated by heavy lines are meant to project out of the page toward the reader.

LIPIDS
Types of Lipids

Lipids are a large group of poorly water-soluble organic compounds containing a high ratio of carbon and hydrogen to oxygen and other elements. The main forms of lipid that have physiological importance are (1) **acylglycerols;** (2) **phospholipids;** (3) **glycolipids,** (4) **steroids,** and (5) **eicosanoids** (Table 3-1). Lipids have three basic functional roles in the body: (1) as constituents of cellular membranes (phospholipids, glycolipids, and cholesterol), (2) as depot fat (triacylglycerols) in adipose cells they contribute to body form and serve as an energy reserve, and (3) as chemical regulatory signals (steroid hormones, eicosanoids, and diacylglycerols).

Fatty Acids

- *What is the general structure of fatty acids and acylglycerols?*
- *How do saturated and unsaturated fatty acids differ?*
- *What are the characteristics of amphipathic molecules?*
- *How can amphipathic molecules aggregate into bilayers and micelles in the presence of water?*
- *What is the structure of glucose, fructose, and galactose?*
- *How can the condensation reaction leading to disaccharides be diagrammed?*
- *How can alpha and beta glycosidic bonds be distinguished?*

Fatty acids, the major precursors for synthesis of lipids, consist of a chain of carbons linked together by single or double bonds, with a carboxyl (-COOH) group at one end (Figure 3-1, *A*). In humans, most fatty acids contain an even number of carbons (usually 16 or more). Each carbon in the chain is bonded to one or more hydrogens as well as to the adjacent carbons. **Saturated** fatty acids contain only single carbon-carbon bonds and thus the maximum ratio of H to C; **unsaturated** fatty acids contain some double carbon-carbon bonds and have a lower ratio of H to C (Figure 3-1, *A*).

Acylglycerols

Acylglycerols are compounds based on **glycerol** (Figure 3-1, *B*), a three-carbon alcohol (an organic compound having one or more -OH groups). To glycerol are coupled one, two, or three fatty acid chains to form **monoacylglycerols, diacylglycerols,** and **triacylglycerols,** respectively. Triacylglycerols (Figure 3-1, *B*) are also called **triglyceride** or **neutral fat.** Different fatty acid chains may be bound to each glycerol carbon. Animal fats such as butter tend to have higher proportions of saturated fatty acids than do vegetable fats such as linseed oil and corn oil.

TABLE 3-1   Major Categories of Organic Molecules in the Human Body

Category	Subclass	Subunits	Function
Lipids	Triglycerides	3 fatty acids + glycerol	Insoluble in water; major store of reserve fuel in the body
	Phospholipids	2 fatty acids + glycerol + phosphate + charged nitrogen molecule	Molecule has polar and nonpolar ends; major constituent of membranes
	Steroids		Cholesterol and steroid hormones
	Eicosanoids	Modified phospholipid	Important regulatory agent
Carbohydrates	Monosaccharides (sugars)		Major fuel used by cells for energy
	Polysaccharides	Monosaccharides	Storage form of carbohydrate
Proteins		Amino acids	Catalysis (enzymes)
			Cell structures
			Protein hormones
			Recognition (antibodies and receptors)
			Cell membrane transport and permeability
Nucleic acids	DNA	Nucleotides containing adenine, cytosine, guanine, and thymine	Storage of genetic information
	RNA	Nucleotides containing adenine, cytosine, guanine, and uracil	Translate genetic information into protein synthesis

Phospholipids

At first glance, **phospholipids** appear similar to triacylglycerols because they also have a glycerol backbone (Figure 3-2, *A*). However, only two of the three glycerol carbons are linked to fatty acid chains. The third carbon has a phosphate group ($-PO_4$) linked to a polar molecule. Phospholipids are named according to the type of molecule attached to the terminal carbon of the glycerol. **Phosphatidylcholines** (lecithins), have choline ($-CH_2-CH_2-N(CH_3)_3$) on the terminal carbon (Figure 3-2, *A*); phosphatidylinositol has inositol, a sugar; phosphatidylserine has an amino acid, serine (see Figure 3-8); and phosphatidylethanolamine has ethanolamine ($-CH_2-CH_2-NH_3$).

Phospholipids are important components of the membranes that surround cells and form extensive internal systems of tubes and vesicles within cells. These membranes assemble themselves and are able to reseal themselves if punctured. These abilities arise from the properties of the phospholipid molecules. The end of a phospholipid consisting of hydrocarbon chains is hydrophobic, whereas the end containing the polar group is hydrophilic (Figure 3-2, *B*). Such molecules are called **amphipathic**—the name means "feeling both." In the presence of water, the polar "heads" of phospholipid molecules tend to interact with the water molecules (Figure 3-3, *A*), while their hydrocarbon "tails" associate with the tails of other phospholipids or nonpolar regions of other molecules. The consequence is that at the surface of the cytoplasm, phospholipids aggregate to form a two-molecule-thick **bilayer** (Figure 3-3, *B*).

Lipids and phospholipids are too poorly soluble in water to be transported in blood in the free form, but they can be transported in the form of spherical **micelles** that form spontaneously when lipids are dispersed in water (Figure 3-3, *C*). Micelles are the basic form in which dietary lipids enter the blood stream from the intestine. Specific plasma proteins in the circulating blood attach themselves to the micelles by hydrophobic bonds, forming complexes of lipid and protein called **lipoproteins** (Figure 3-3, *D*). The attached protein serves to tag the lipoprotein for uptake and metabolism by particular cell types (see Chapter 23).

Glycolipids

Glycolipids are lipids with covalently bound carbohydrate (see next section) that may range from a single sugar to a straight or branched chain of seven or more sugars. Glycolipids are commonly found on the outer surfaces of cell membranes with the polar carbohydrate chains projecting into the surrounding water. In this position they may serve as molecular identity tags that allow the immune system to recognize foreign cells. For example, the ABO blood type system is based on genetically determined differences in the carbohydrate chains of particular glycolipids on the cell surface (see Chapter 24). Transfused blood cells may be attacked by the recip-

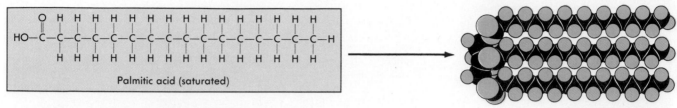

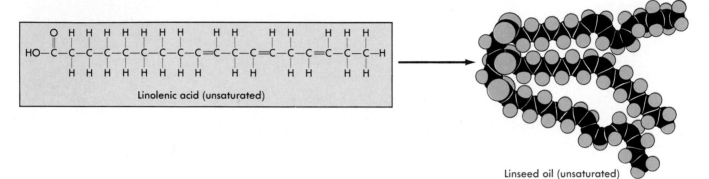

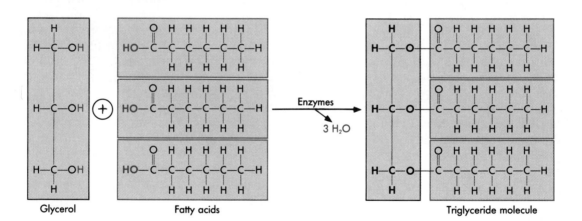

FIGURE 3-1

A   Fatty acids can be saturated (no double bonds in the hydrocarbon tail; palmitic acid) or unsaturated (one or more double bonds; linolenic acid). Many of the animal triacylglycerols are saturated. Because their fatty acid chains can fit closely together, these triacylglycerols form immobile arrays called hard fat. Vegetable oils, such as linseed oil, by contrast, are typically unsaturated, and the multiple double bonds prevent close association of the triacylglycerols.

B   Triacylglycerols are composite molecules made up of three fatty acids coupled to a single glycerol backbone.

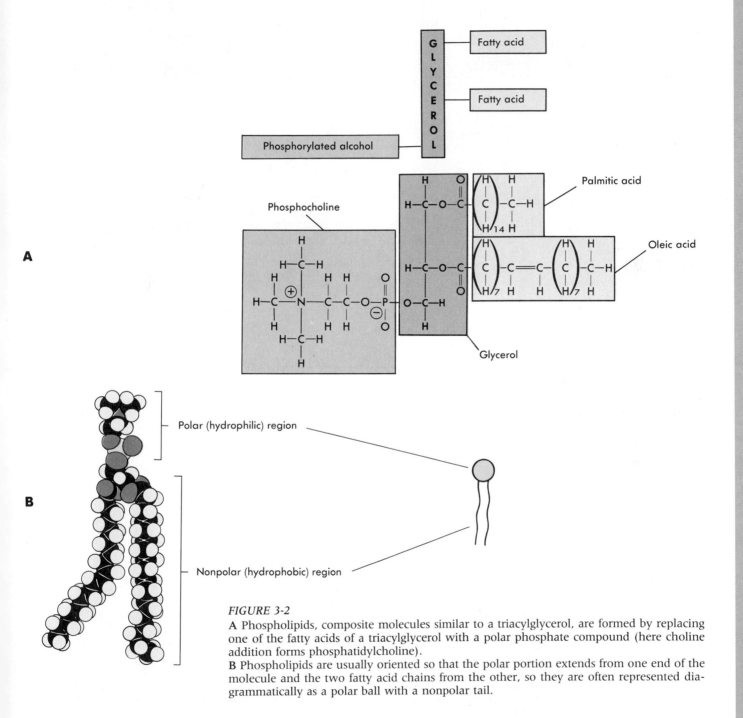

FIGURE 3-2

A Phospholipids, composite molecules similar to a triacylglycerol, are formed by replacing one of the fatty acids of a triacylglycerol with a polar phosphate compound (here choline addition forms phosphatidylcholine).

B Phospholipids are usually oriented so that the polar portion extends from one end of the molecule and the two fatty acid chains from the other, so they are often represented diagrammatically as a polar ball with a nonpolar tail.

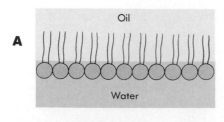

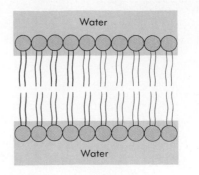

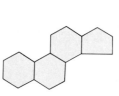

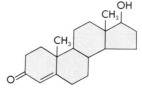

FIGURE 3-3

A At an oil-water interface, phospholipid molecules orient so that their polar heads are in water while the nonpolar tails extend into the oil phase. In an aqueous environment, phospholipids arrange themselves to form either extended bilayers (**B**) or spherical micelles (**C**). In the bloodstream, poorly-soluble lipids travel as lipoprotein: complexes of lipid surrounded by a phospholipid monolayer and associated with protein (**D**).

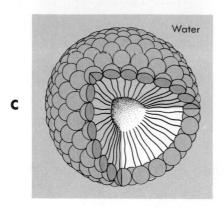

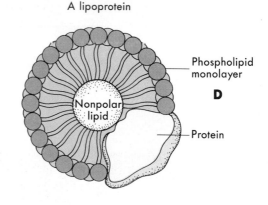

FIGURE 3-4

Steroids have a backbone of four carbon rings. Physiologically important steroids are synthesized from cholesterol; shown as examples are the sex hormone testosterone and the adrenal steroid cortisol.

ient's immune system if their surface glycolipids bear oligosaccharides different from those of the recipient's own blood cells.

Steroids

Steroids (Figure 3-4) have a basic structure composed of four interconnected carbon rings. Recall that in the shorthand used in such diagrams the lines represent C-C bonds; the intersections represent carbons, and if not otherwise indicated, each C is assumed to have H side groups. Because the four bonds of a carbon with all single bonds project to form a three-sided pyramid, the side groups of such carbons project either toward or away from the viewer in such diagrams. Steroids got their name from their great potential for formation of stereoisomers. The side groups at various positions on these carbon rings define specific steroid classes. **Choles-**

FRONTIERS IN PHYSIOLOGY

Dietary Fat and the Incidence of Cancer

Human diet has changed profoundly since prehistoric times. The fat intake of prehistoric peoples living in temperate climates was probably about 20% of their total caloric intake, as compared with about 40% in the modern United States. The consumption of complex carbohydrates, such as plant starches, was much greater than that of simple carbohydrates, such as sucrose and glucose. Epidemiological evidence suggests that there may be a relationship between specific dietary elements and the incidence of some types of cancer, although it is difficult to obtain positive proof because long periods of time may elapse between the initiating event that transforms a normal cell into a potential cancer cell and the actual development of a tumor. The best evidence comes from following the dietary habits of a large study population for many years and comparing populations of subjects with different diets. For example, the Japanese population might be compared with the U.S. population, or Japanese immigrants in the United States could be compared with Japanese who remained in Japan.

When such comparisons are made, the incidence of breast cancer is found to be high in natives of the United States and in Japanese immigrants who adopted typical American diets, but low in natives of Japan. Similar comparisons have been made for 39 countries and confirm a strong relationship between fat intake and mortality due to breast cancer. The rate is highest in countries where the average consumption of fat is 140 to 160 grams/day. It approaches zero in populations that consume less than 40 grams of fat per day. In contrast to breast cancer, stomach cancer is common in Japan and rare in the United States. The incidence of stomach cancer may be related to the consumption of smoked or pickled food. Some studies suggest that it is not the calories contained in fat that matters, but rather which fatty acids a diet includes. In experimental animals certain fatty acids act as cancer-promoting substances (carcinogens).

These laboratory and epidemiological studies cannot answer the question of whether an individual's decision to change her diet would affect her risk of cancer. But even a small beneficial effect of diet modification would have important implications for public health. For every nine women who live a normal lifespan, one will get breast cancer. Even a small reduction in the risk factor for the population would mean that a large number of women would not develop the disease.

---

terol is the precursor for biological synthesis of other steroids. Cholesterol and other steroids are important components of cell membranes. Steroid hormones include the sex hormones estrogens, progesterone, and testosterone, and the adrenal hormones cortisol and aldosterone.

Eicosanoids

Eicosanoids are derivatives of the unsaturated fatty acid **arachidonic acid** (Figure 3-5, *A*), one of the phospholipids normally present in cell membranes. The best-known of eicosanoids are the **prostaglandins**, 20-carbon fatty acids that contain a 5-carbon ring (Figure 3-5, *B*). Different prostaglandins are denoted by an abbreviation in which PG stands for prostaglandin, followed by a letter that designates the stereoisomeric configuration of the 5-carbon ring and a number subscript indicating the number of double bonds outside the ring. For example, PGE_1 (Figure 3-5, *B*) has a ring isomer denoted E and one double bond outside the ring.

Prostaglandins are used for chemical communication within tissues; they operate over shorter ranges than those typical of hormones. They have been implicated in a large number of regulatory

Arachidonic acid

PGE₁ (a representative prostaglandin)

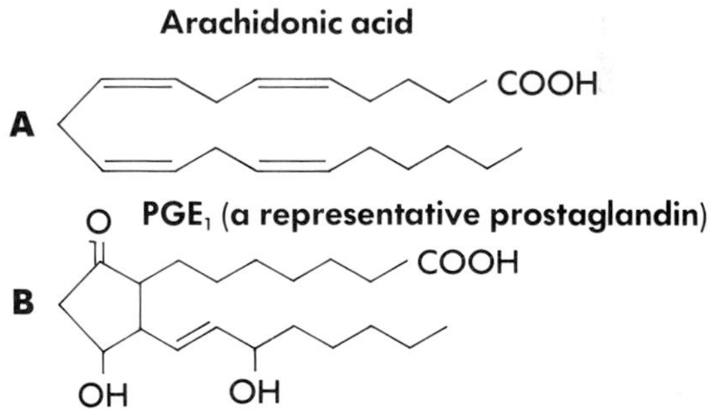

FIGURE 3-5

A Arachidonic acid, the unsaturated fatty acid precursor of eicosanoids, shown folded so the locations of the carbons correspond to those in the structure of a prostaglandin. **B** A representative prostaglandin, PGE_1, so named because of the stereoisomer form of its ring and the single double bond outside the ring.

functions, including inflammation and blood clotting; ovulation, menstruation, and labor; and secretion of acid by the stomach.

CARBOHYDRATES
Monosaccharides

Carbohydrates (Table 3-1) are composed of carbon, hydrogen, and oxygen in the ratio 1:2:1.

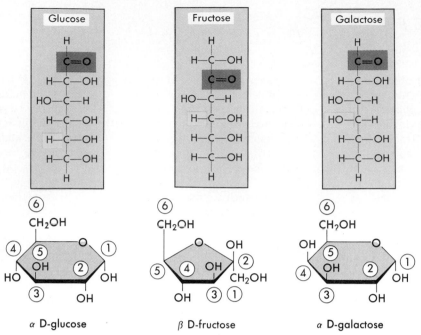

FIGURE 3-6

Structure of three monosaccharides: glucose, fructose, and galactose. In the solid form these are linear 6-carbon molecules, but in solution they form rings (lower diagrams). Fructose is a structural isomer of glucose with identical chemical groups bonded to different carbon atoms. Galactose is a stereoisomer of glucose with identical chemical groups bonded to the same carbon atoms, but in a different orientation. Note the numbers of the carbon atoms.

Among the simplest carbohydrates are the **monosaccharides** (Figure 3-6). All but one of the carbon atoms in a sugar molecule have a polar hydroxyl (-OH) group attached to them, making the carbohydrates freely water-soluble. The most common monosaccharides include **pentose** sugars (containing 5 carbons) and **hexose** sugars (containing 6 carbons). The pentose sugars ribose and deoxyribose are important constituents of nucleic acids (see Figure 3-18); the hexose sugars are mainly important as sources of energy and as the carbohydrate components of glycoproteins and glycolipids, discussed in other parts of this chapter.

The simplest representation of a monosaccharide molecule is its linear structure (upper diagrams in Figure 3-6). Conventionally, the carbon at the end of the molecule with a double bonded O in the linear structure is designated as C1 and the rest are numbered in sequence. In water solution almost all of the O that was double-bonded to C1 in the linear structure forms an additional single bond to another carbon; the result is a four or five carbon ring (lower diagrams in Figure 3-6). Conventionally this ring is represented as if it were at right angles to the plane of the page, with C1 on the right hand side of the molecule and the bonds shown by thicker lines projecting toward the viewer. Because the sugar ring can open and close from time to time, the -OH of C1

in glucose and galactose (C2 in fructose) can happen to be in either the up or down orientation when the ring recloses. In the alpha stereoisomer (as shown in Figure 3-6) the -OH is down when the molecule is viewed from the standard position. In the beta isomer, it is oriented up. This distinction becomes physiologically important when this -OH group is involved in formation of bonds between monosaccharides (see below).

The hexose **D-glucose** (also called **dextrose**; Figure 3-6) is the most abundant monosaccharide in nature and surely the most important in physiology. D-glucose is sometimes called blood sugar and is the major form in which carbohydrate is transported from place to place in the body. Monosaccharides characteristically exist in isomeric forms that sometimes have different physiological roles. For example, among the many isomers of glucose, there is a structural isomer, **fructose**, which has a 4-carbon rather than a 5-carbon ring structure, and a stereoisomer, **galactose**, which differs from glucose only in the orientation of the carboxyl group on the 4 carbon (Figure 3-6). These isomers are handled differently by the body. For example, glucose is actively absorbed by the intestine (Chapter 22) while fructose and galactose are not. Glucose and its isomers fructose and galactose are the major hexoses of physiological importance.

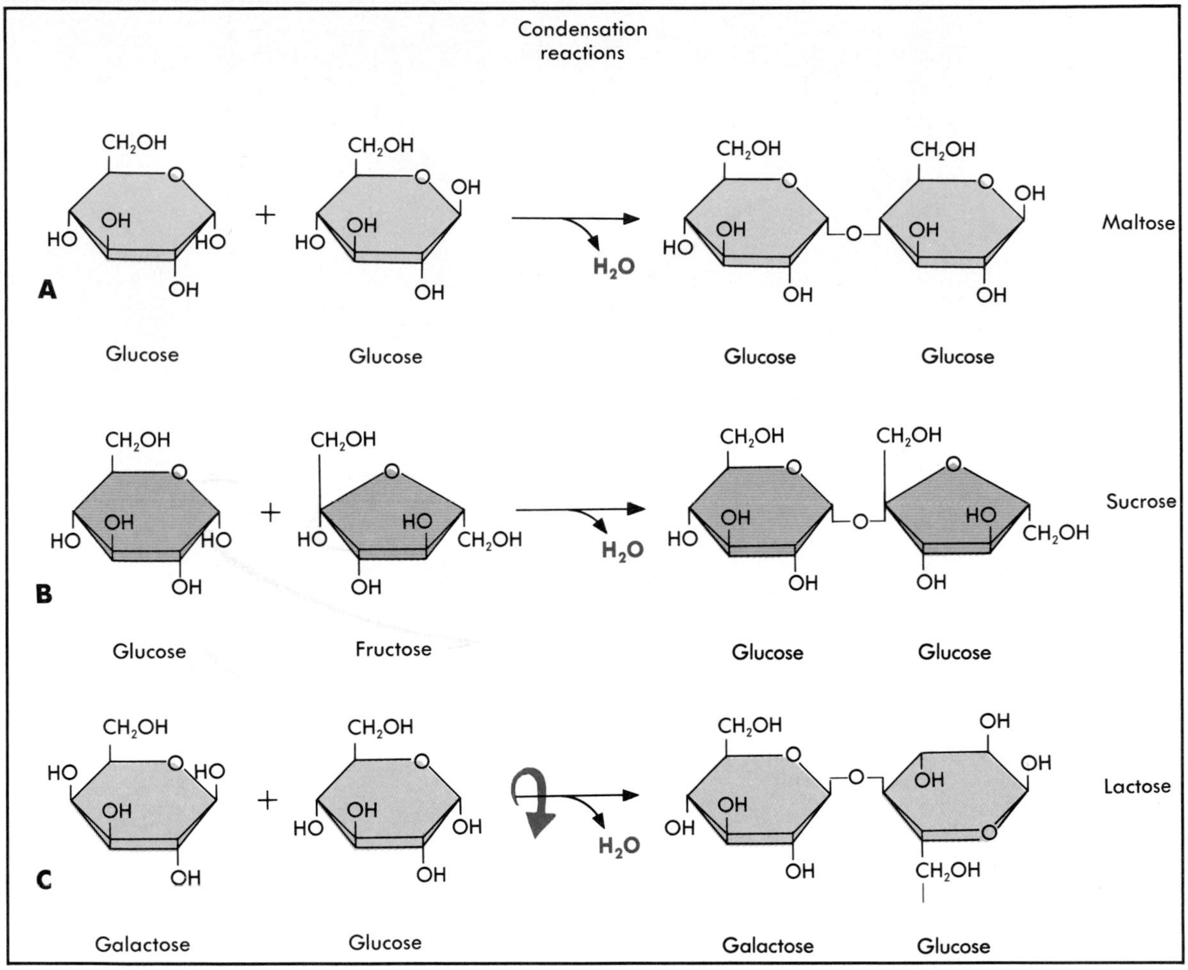

Condensation reactions

FIGURE 3-7
Some examples of disaccharides.
A Apples are rich in maltose, composed of two glucose molecules joined by an alpha glycosidic bond.
B In table sugar, glucose and fructose are joined by an alpha glycosidic bond to form sucrose.
C Milk contains a disaccharide of glucose and galactose termed lactose or milk sugar. Lactose contains a beta galactosidic bond, making lactose difficult to digest for individuals who lack a specific enzyme (lactase).

Disaccharides, Oligosaccharides, and Polysaccharides

Disaccharides are formed by the covalent linkage of two monosaccharides. As with acylglycerol synthesis, this is a dehydration reaction. The type of disaccharide that results depends on two factors: the identity of the two monosaccharides involved and which carbons of the two molecules become bonded. For example, maltose is a disaccharide of two alpha D glucose molecules in which the bond is between C1 of one glucose and C4 of the other and both monosaccharides are oriented right-side-up relative to one another; this is an **alpha 1,4 glycosidic bond** (Figure 3-7, A). Sucrose (table sugar) is the alpha 1,4 glycoside disaccharide of glucose and fructose (Figure 3-7, B). Lactose (milk sugar) is the disaccharide of glucose and galactose. The orienta-

tion of the -OH group of C1 of galactose is opposite to that of the C4 of glucose, so for the bond to form, the galactose must be flipped relative to the glucose (Figure 3-7, C). The result is a **beta 1,4 galactosidic bond.** Different enzymes are required to break the two types of bonds. People of all ages have digestive enzymes that can hydrolyze the alpha 1,4 glycosidic bond of sucrose. Some adults of European origin, and almost all those of Asiatic and African origin, lack enzymes capable of breaking beta galactosidic bonds. The resulting inability to digest lactose is called **lactose intolerance** (see box essay in Chapter 22).

Monosaccharides may be linked to form chains of from several monosaccharide subunits (an **oligosaccharide**) to several hundred or thousand subunits (a **polysaccharide).** The chains may be

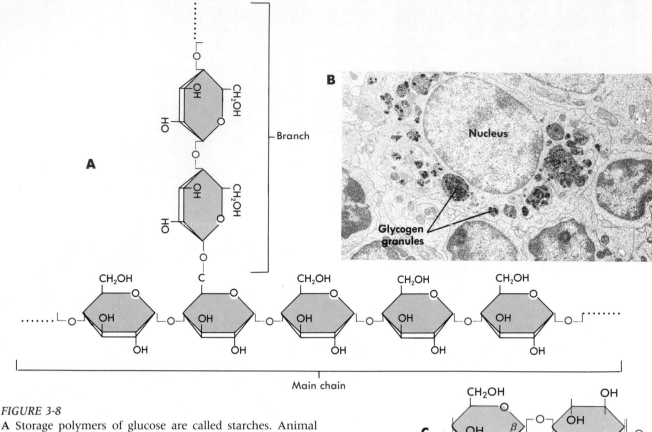

FIGURE 3-8

A Storage polymers of glucose are called starches. Animal starch is called glycogen and is highly branched.
B An electron micrograph of glycogen granules in a liver cell.
C The disaccharide unit of cellulose contains a beta glycosidic bond, so named because the left-hand glucose is the beta isomer. The beta glycosidic bond of cellulose cannot be broken by the types of enzymes that break alpha glycosidic bonds of glycogen. This bond makes the glucose units of cellulose unavailable to most animals.

branched or straight. Polysaccharides can have a wide range of molecular weights, and their physical properties depend on the nature of the internal chemical bonds and the degree of branching.

Glycogen, or animal starch (Figure 3-8, *A*), is the most highly branched polymer of glucose and the major storage form for carbohydrates in animals. The large size of glycogen molecules confines them to the cells in which they are synthesized. Most glycogen is in the form of insoluble granules scattered throughout the cytoplasm (Figure 3-8, *B*). In contrast to glycogen, plant starches are less branched. In starches the glucose monomers are linked by alpha glycosidic bonds. Enzymes that break these bonds are widely found in many cell types, allowing the stored glucose to be released when glucose is scarce. The rates of the two processes are regulated so that glycogen synthesis dominates when glucose is plentiful, and glycogen breakdown takes over when glucose is scarce.

Cellulose, the structural polysaccharide of plants, is a polymer of glucose linked by beta 1,4 glycosidic bonds (Figure 3-8, *C*). Very few animal species have enzymes that can break beta glycosidic bonds. Consequently, cellulose and other molecules that contain beta 1,4 linkages cannot be degraded into monosaccharide by digestive enzymes and pass intact through the small intestines of most animals as **dietary fiber** (see Chapter 22). Some animals, such as cattle and termites, contain in their digestive tracts microorganisms that can digest beta glycosidic bonds. These animals are able to exploit the glucose units of cellulose.

PROTEINS
Amino Acids

- *How do the primary, secondary, tertiary, and quaternary structures of a protein differ?*
- *What specifies the primary structure?*
- *What thermodynamic law determines the final tertiary structure?*
- *What kinds of chemical interactions determine the secondary, tertiary, and quaternary structures of proteins?*

lou zipper

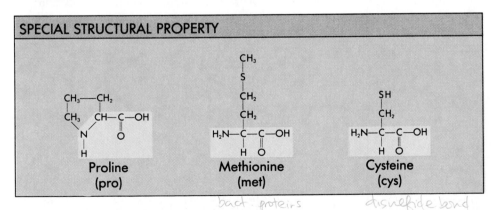

| NONAROMATIC | AROMATIC |

NONPOLAR

Alanine (ala) Valine (val) Leucine (leu) Isoleucine (ile) Phenylalanine (phe) Tryptophan (trp)

SMALLER ← → BULKIER

POLAR UNCHARGED

Glycine (gly) Serine (ser) Threonine (thr) Asparagine (asn) Glutamine (gln) Tyrosine (tyr)

LESS POLAR ← → MORE POLAR

IONIZABLE

Glutamic acid (glu) Aspartic acid (asp) Histidine (his) Lysine (lys) Arginine (arg)

ACIDIC ← histones → BASIC

SPECIAL STRUCTURAL PROPERTY

Proline (pro) Methionine (met) Cysteine (cys)

bact. proteins   disulfide bond

FIGURE 3-9

The 20 common amino acids. Amino acids consist of an amino group, a carboxylic acid group, and a side chain, which is different for each amino acid. Different side chains can be nonpolar, polar, or ionic. Some amino acids have special roles in forming links between proteins and sugars, between different parts of the same protein, or between different amino acid chains of protein complexes.

Amino acids, the subunits of proteins, are organic compounds including at least one carbon atom that is bound to (1) a carboxylic acid (-COOH) group, (2) an amino (-NH₂) group, (3) a side chain of variable length and complexity, and (4) a hydrogen atom. The side chain may be nonpolar, as in alanine and leucine; polar, as in serine and glutamine; acidic, as in glutamic and aspartic acids; basic, as in lysine and arginine; or aromatic, as in tryptophan and tyrosine (Figure 3-9).

Essential amino acids are those for which human cells cannot synthesize the carbon backbone (Table 3-2), including histidine, which is essential in infants (but not adults). There are amino acids that are not themselves regarded as essential but are derived from the essential amino acid methionine; therefore, the concept of essentiality is a relative rather than an absolute distinction.

The characteristic properties of individual amino acids contribute to the structure and function of proteins. A portion of a protein sequence that contains a disproportionate number of nonpolar amino acids will tend to form hydrophobic bonds with other such regions of the protein. In proteins that are associated with the cell membrane, such regions serve to attach the proteins to the lipids of the membrane. Sulfur-containing cysteines tend to form disulfide (-S-S-) bonds with other cysteine groups that may attach two protein chains to one another or join different regions of the same protein (Figure 3-10). Prolines form attachment sites for the sugar side chains in glycoproteins. Since the side chains of ionizable amino acids are weak acids or bases, both amino acids and proteins function as buffers. The histidine groups of proteins are good buffers in the near-neutral part of the pH scale and have an important role in buffering the cytoplasm against changes in pH.

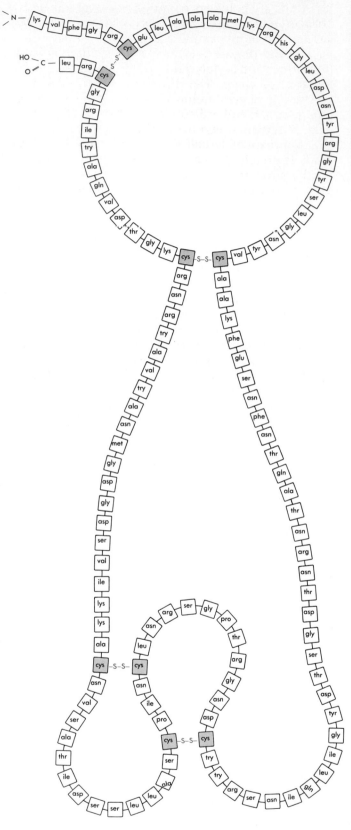

FIGURE 3-10
The amino acid sequence of the enzyme lysozyme. The four disulfide bridges are important in determining the three-dimensional shape of the enzyme.

TABLE 3-2	Essential and Nonessential Amino Acids

Essential	Nonessential
Histidine*	Alanine
Isoleucine	Arginine
Leucine	Asparagine
Lysine	Aspartic acid
Methionine	Cysteine
Phenylalanine	Glutamic acid
Threonine	Glutamine
Tryptophan	Glycine
Tyrosine†	Proline
Valine	Serine

*Histidine is essential only in infants.
†Tyrosine can be synthesized from phenylalanine.

Protein Structure and Function

Proteins are polymers of amino acids assembled by formation of **peptide bonds** between the carboxylic acid group of one amino acid and the amino group of a second (Figure 3-11). As in the synthesis of disaccharides and acylglycerols, this occurs through the removal of water. The dipeptide that results from the union of two amino acids has a free carboxyl group at one end, called its **C-terminal.** The opposite, **N-terminal** end has a free amino group.

Approximately half the organic material of the body is protein (Table 3-1). Proteins compose many of the structural elements of cells. They also function as molecular machines that develop the force that causes muscles to contract and cells to divide and change their shape. They form pathways by which ions and polar substances can enter and leave cells, catalyze intracellular chemical reactions, regulate gene action, transport oxygen, act as carriers of electrons, and serve as markers that allow cells to identify one another. These diverse functions depend on specific adaptations of protein structure.

The structure of a protein is organized on three levels. The simplest level of organization, or **primary structure,** is determined by the linear sequence of amino acids (Figure 3-12, A) and the location of disulfide (-S-S-) bonds between different parts of the chain (Figure 3-10). The primary structure is two dimensional; the functional protein is three-dimensional, but all of the properties of the protein arise ultimately from its genetically specified primary structure. The great diversity of protein structures and functions seen in cells is made possible by the very large number of possible primary structures. For example, there are 20^{100} different possible sequences for a protein containing 100 amino acids.

As the primary structure of a protein is assembled, thermal energy causes the molecule to bend and twist randomly. These motions automatically bring different parts of the molecule together and allow it to assume the secondary and tertiary levels of structural organization described in the following paragraphs. In this process of self-assembly, the molecule is following the laws of thermodynamics that favor changes in form that result in decreases in free energy. In the final, three-dimensional form of a protein, single amino acids located some distance apart in terms of the primary amino acid sequence may come to be spatially adjacent.

The **secondary structure** of a protein is a folding or coiling of the primary structure that results from formation of hydrogen bonds between the carboxyl portion of a peptide bond and the amino group of a peptide bond at another point on the chain (Figure 3-12, B). Two types of regular secondary structures are particularly common. The coiled secondary structure is called an **alpha helix;** the pleated one a **pleated sheet.**

The folded or coiled secondary structure undergoes further twisting, folding, and coiling into a compact three-dimensional **tertiary structure** (Figure 3-12, C). This is the result of bonds and interactions that do not include covalent bonds (that is, hydrophobic interactions and ionic, hydrogen, and van der Waal's bonds) between amino acids at different points in the primary structure. Regions of the primary structure that contain a preponderance of hydrophobic amino acids are forced to the center of the tertiary structure, and regions that contain mainly polar amino acids tend to lie on the outside of it. The weakness of the interactions that determine tertiary structure gives it a certain amount of flexibility, a fact that has important consequences.

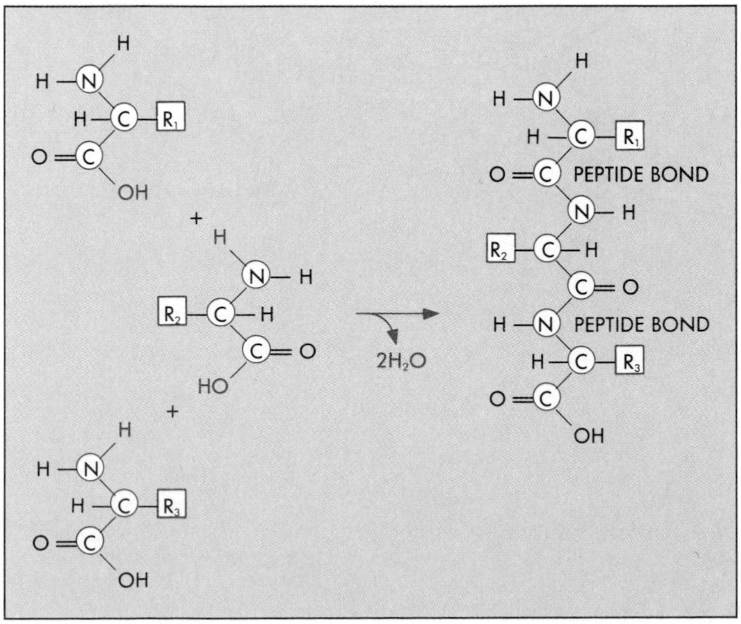

FIGURE 3-11
Amino acids combine by way of peptide bonds to form polypeptide chains. Peptide bonds are formed by the removal of one water molecule per peptide bond, referred to as a condensation reaction.

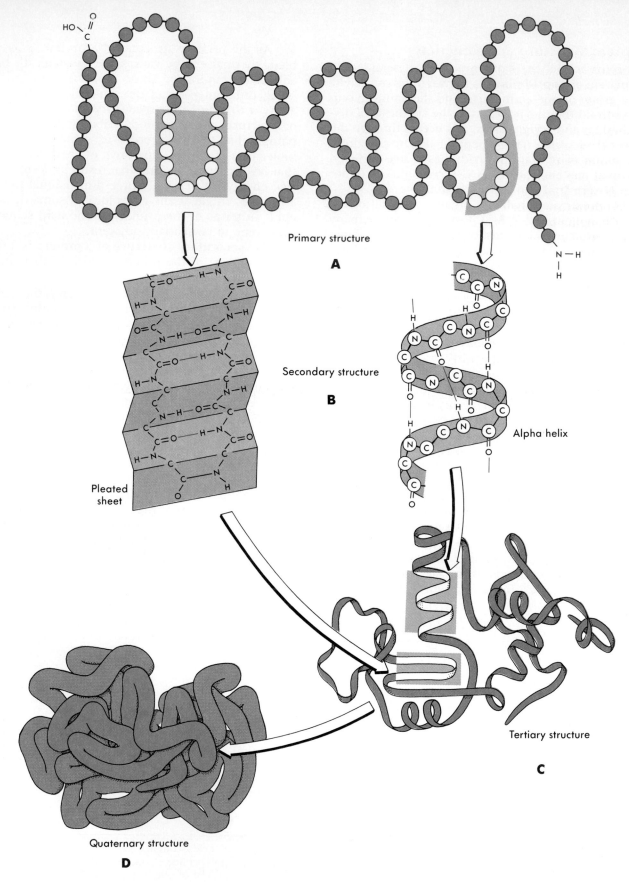

Primary structure

A

Secondary structure

B

Pleated sheet

Alpha helix

Tertiary structure

C

Quaternary structure

D

FIGURE 3-12

A The sequence of amino acids in a protein is termed its primary structure.

B Hydrogen bonds between peptide linkages may produce an alpha helical secondary structure or a pleated sheet.

C Formation of noncovalent bonds between different parts of the primary structure results in the tertiary structure of a protein, which represents the lowest free energy state of the molecule.

D In some proteins, separate polypeptide chains associate with one another to yield a quaternary, or subunit, structure.

Changes in the tertiary structure are sometimes involved in regulating the function of proteins. By the same token, factors that change the interactions between amino acids affect the tertiary structure of a protein and may affect its function. Physiological examples of such factors include temperature and pH. Temperatures in excess of about 45° C cause the parts of protein molecules to flex more strongly relative to one another, breaking some of the weak bonds and causing changes in tertiary structure called **denaturation.**

Changing only a few amino acids in the primary structure of a protein can profoundly alter its tertiary structure. An example of such an effect is provided by the genetic disease **sickle cell anemia,** which is the result of a substitution of valine for glutamic acid in position 6 in the beta chain of oxygen-carrying protein hemoglobin. The result of this substitution is that the hemoglobin can aggregate, forming strands. Red blood cells in which some of the hemoglobin has aggregated are distorted into a half-moon shape, referred to as sickled (Figure 3-13). Such cells do not pass readily into the smallest blood vessels.

At the highest level of protein organization, two or more protein chains may be assembled to form a protein complex; the relationship of the chains to one another composes the **quaternary structure** of the complex (Figure 3-12, *D*). The general terminology for such complexes is based on the number of subunits and whether the subunits are the same or different. For example, a complex of two structurally different proteins is a **heterodimer,** with three identical members a **homotrimer,** with four different ones a **heterotetramer,** and so on. The adaptive significance of such complexes lies in the fact that they allow the individual chains to interact functionally. Some enzymes (see next section of this chapter) are tetramers that, depending on the cell type, may be assembled as homotetramers or in various combinations to form heterotetramers. The variable mix of different protein subunits allows the functional characteristics of the enzyme to be adapted to the needs of the cell type. An example of such organization is the enzyme lactate dehydrogenase (see Chapter 4). Hemoglobin, the oxygen-transporting protein of red blood cells, is a homotetramer that can bind one O_2 per subunit. Interaction between the subunits affects the binding properties of the complex (see Chapter 17). Other examples of such protein complexes include ion channels of the cell membrane (Chapter 6) and antibodies produced by the immune system (Chapter 26).

ENZYME CATALYSIS
Enzymes and the Rates of Biochemical Reactions

> - *How does an enzyme catalyze a reaction?*
> - *How do free energy, activation energy, and an active site function in a reaction?*

Chemical reactions that result in net decreases in free energy can occur spontaneously. However, at normal body temperature the rate may be very slow. For example, even though it produces heat, the oxidation of sugars is so slow that there is no danger of the sugar bowl on the dining-room table spontaneously catching fire. This is because to complete their reaction the sugar and oxygen must pass through a **transition state,** whose free energy is much higher than the total free energy of either the reactants or products (Figure 3-14, *A*). The difference between the free energy of the reactants and the free energy of the transition state is called the **activation energy** of a reaction. The activation energy could be thought of as a barrier that the sugar oxidation reaction must cross to reach the lower free energy of the products. Such free-energy barriers lie in the way of most of the spontaneous or energy-releasing reactions carried out in the body.

The pace of life would be very slow if there were no way of overcoming the barriers of activation energy. In the body, specific proteins called **enzymes** greatly increase the rate of reactions while remaining unchanged themselves. This increase in reaction rate is called **catalysis.** To understand the mechanism of enzyme catalysis it is necessary to consider the energy changes reactants undergo as they become products. The rates of the forward and reverse reactions are determined by the magnitude of the activation energies. That is, a decrease in the energy of the transition state would increase both the forward and reverse reaction rates, leaving the equilibrium constant (K_{eq}) unchanged. Catalysts accelerate reactions by reducing their activation ener-

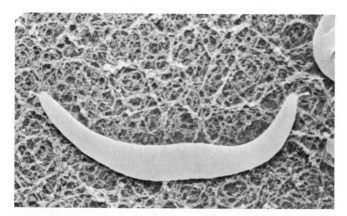

FIGURE 3-13
Sickle cell hemoglobin is produced by a mutated form of the gene that codes for the beta-chain of hemoglobin. Sickling results from aggregation of the hemoglobin into long strands, producing a change in the shape of the red blood cell, as seen in this scanning electron micrograph.

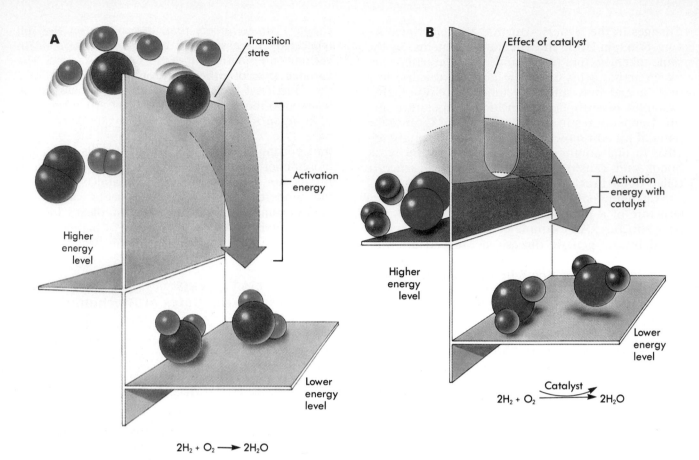

A $2H_2 + O_2 \longrightarrow 2H_2O$

B $2H_2 + O_2 \xrightarrow{\text{Catalyst}} 2H_2O$

FIGURE 3-14

A Chemical reactions involve intermediate transitional states that can be viewed as energy barriers somewhat higher than the mean thermal energy of reactants and products, illustrated as shelves of unequal height. Even though energy is given up in the reaction, the barrier must be surmounted for the reaction to occur. The height of the barrier is termed the activation energy, and it affects both the forward and reverse reaction rates. The difference in the energies of the reactants and products determines the ratio of reactants to products at equilibrium.

B Enzymes reduce the activation energy for chemical reactions and thus increase the probability that a reactant or product will have sufficient energy to surmount the barrier. As a result, enzymatic catalysis increases the rate of both the forward and the reverse reactions, and thus the overall reaction rate. Catalysis does not change the energy of reactants and products, so it does not change the equilibrium constant.

gies (Figure 3-14, *B*). For example, the invertase enzyme that catalyzes sucrose hydrolysis reduces the activation energy from 26 kcal/mole to about 9 kcal/mole and increases the rate of the forward reaction by 12 orders of magnitude. Since the change in activation energy does not affect the equilibrium constant for the reaction (K_{eq}), which is determined by the difference in free energy between reactants and products, the equilibrium ratio of reactants to products remains the same.

The Names of Enzymes

Enzyme-catalyzed reactions fall into a small number of general categories. The suffix *-ase* is usually used to designate an enzyme involved in a particular chemical reaction. For example, addition of a carboxyl group (-COOH) is catalyzed by carboxylases and removal of a carboxyl group by decarboxylases. Formation or degradation of a compound by the removal or addition of water, previously defined as condensation and hydrolysis, are carried out by synthetases and hydroxylases, respectively. If one molecule gives up an electron, becoming oxidized, another must accept the electron and become reduced. It is therefore usual to refer to such processes as paired **oxidation-reduction reactions,** catalyzed by **oxidases** and **reductases.** Oxidation-reduction reactions are the basis for energy production in cells (see Chapter 4). Some additional examples of physiologically important reactions include: removal of a hydrogen (**dehydrogenation**), catalyzed by **dehydrogenases,** the addition or removal of phosphate groups (**phosphorylation** or **dephosphorylation**) catalyzed by **kinases** and **phosphorylases** respec-

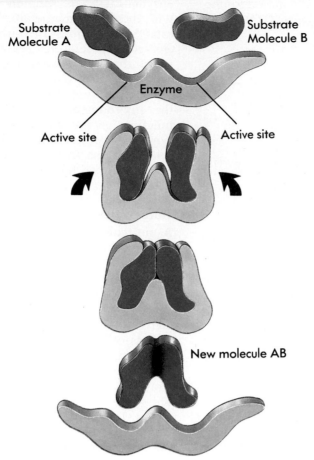

Substrate Molecule A

Substrate Molecule B

Enzyme

Active site   Active site

New molecule AB

FIGURE 3-15

Enzymes have active sites that bind substrates. The active sites must recognize some aspect of the molecular structure of their substrates. The enzyme molecule is not permanently altered in the reaction and can catalyze further reactions.

tively, and degradation of proteins into their constituent amino acids, catalyzed by **peptidases** or **proteases.** Another example of physiological importance is the reaction of carbon dioxide with water to form bicarbonate and the hydrogen ion. This reversible reaction is greatly accelerated by the enzyme **carbonic anhydrase.** This enzyme is present at various sites in the body where CO_2 must be rapidly converted to bicarbonate.

Many enzyme-catalyzed reactions involve **coenzymes,** proteins that are not enzymes themselves but that assist the enzyme-catalyzed reaction; or **cofactors,** compounds that must be present for an enzyme-catalyzed reaction to occur. Examples of substances that serve as cofactors include vitamins, trace elements, and ions such as Ca^{++}, Mg^{++}, Fe^{+++} and Zn^{++}. Like enzymes, coenzymes and cofactors are catalysts, because they are not used up in the reaction.

The Importance of Active Sites in Enzymatic Catalysis

A reactant in an enzyme-catalyzed reaction is referred to as a **substrate** for that enzyme. The first step in enzymatic catalysis is binding a substrate molecule to an **active site** on the surface of the enzyme (Figure 3-15). The active site is a pocket in the enzyme into which the appropriate part of the substrate just fits. When random motions of the enzyme and a substrate molecule bring the appropriate part of the substrate into contact with the active site, weak bonds form between the enzyme and the substrate. These bonds momentarily fasten the two molecules together, and may even improve the fit between the substrate and the active site. The tightness with which an enzyme binds a substrate is referred to as its **affinity** for the substrate.

The binding of substrate to the active site can be viewed as an act of molecular recognition; the appropriate substrates are recognized by the fact that they bind the enzyme and also induce an appropriate conformational change. When a substrate binds to an active site it is brought into contact with specific chemical groups on the enzyme that can assist the reaction, making the substrate more reactive. In cases in which an enzyme catalyzes the reaction of two substrates, the enzyme also plays the role of matchmaker, promoting the reaction by bringing the reactants close together in an orientation to one another that favors the reaction (Figure 3-15).

The Concentration Dependence of Enzyme-Catalyzed Reactions

- *What do V_{max} and K_m mean?*
- *How do competitive and noncompetitive effects on enzyme activity differ?*
- *How can the competitive and noncompetitive effects be explained in terms of the Michaelis-Menten plot?*
- *What is an "allosteric effect"?*
- *What is the physiological importance of an allosteric effect?*

In a mixture of enzyme and substrate the rate of formation of product molecules is directly proportional to the number of active sites that are occupied by substrate molecules. If only a little substrate is present at any instant some, but not all, of the active sites will be occupied. If more substrate is present, the number of sites in use increases, and the rate of product formation increases. Ultimately, enough substrate is present so that all the sites are constantly occupied. At this point the enzyme is said to be **saturated,** and further increases in substrate concentration do not increase the rate of formation of product.

The **Michaelis-Menten plot** (Figure 3-16, *A*) is a common way of graphing the relationship between the substrate concentration and the rate of product formation. The catalytic power of an enzyme can be described by two numbers taken from the Michaelis-Menten plot. The **maximal velocity (V_{max})** is the maximum rate of product formation in the presence of unlimited substrate, and is determined by the concentration of enzyme and the rate at which each active site can bind a substrate molecule, catalyze the reaction, and release the product. The **Michaelis Constant (K_m)** is the concentration of substrate that causes half of the active sites to be occupied at any instant, thus resulting in half the maximal rate of product formation. The K_m is a measure of the affinity of the enzyme's active site for the substrate, with lower values of K_m corresponding to higher affinities. The values of both V_{max} and K_m are different for different enzymes, but as a rule K_m is within the range of concentrations of the substrate that the enzyme normally encounters within cells. Under these conditions, the enzyme will tend to regulate the concentration of substrate, using up the substrate when the substrate concentration rises and not using it so fast when the substrate concentration falls.

The Effects of Temperature and pH

The activity of enzymes is sensitive to both temperature and pH. As for chemical reactions generally, enzyme-catalyzed reaction rates usually increase with increasing temperature. However, for enzyme-catalyzed reactions the increase in rate with increased temperature ceases at temperatures in which the enzyme begins to be denatured. Each enzyme also has a **pH optimum.** Usually the value of the pH optimum corresponds closely to the pH of the environment in which the enzyme normally functions. For example, the protein digesting enzyme in the stomach—pepsin—is active only in acid solutions (pH less than 3.0), whereas the enzymes secreted by the pancreas are active only at a neutral pH (about 7.0).

Competitive Inhibition

A molecule structurally similar to the normal substrate of an enzyme may be able to bind to the enzyme's active site. While it occupies the active site, a substrate cannot bind. In the presence of such molecules the fraction of enzyme molecules bound to the normal substrate will be reduced and the rate of conversion of substrate to product will decrease. This is called **competitive inhibition.** The definitive feature of competitive inhibition is that it can be overcome by increasing the concentration of substrate. In terms of the Michaelis-Menten plot, the K_m of the enzyme is increased by competitive inhibition, but the V_{max} of the enzyme is not affected (Figure 3-16, *B*). Many enzymes are competitively inhibited by their products; this is reasonable since the product is frequently similar in structure to the substrate. Competitive inhibition by the product may be of adaptive significance since it helps regulate levels of the product within cells. The more product present, the greater the inhibition and the lower the rate of formation of more product.

An example of clinical use of the principle of competition is in cases of poisoning with ethylene glycol, an ingredient of automobile antifreeze. Ethylene glycol is not very toxic, but it is converted to much more harmful oxalic acid by the liver. The first step in this process is catalyzed by the enzyme alcohol dehydrogenase. The toxic effects can be blocked by administration of ethanol, which competes with the ethylene glycol for the dehydrogenase, allowing the ethylene glycol to be excreted harmlessly.

Allosteric Regulation of Enzymes

The catalytic activity of many enzymes is sensitive to binding of molecules at sites other than the active site (Figure 3-17). Such binding sites are called **allosteric sites** (*allosteric* means "a different place"). An **allosteric activator** increases the activity of the active site; an **allosteric inhibitor** decreases it. These are forms of **noncompetitive activation** and **inhibition.** In contrast to competitive inhibition, noncompetitive inhibition cannot be overcome by increasing the concentration of substrate. In terms

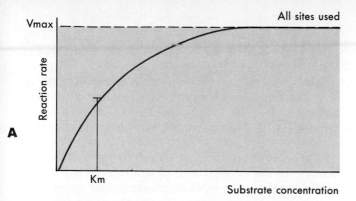

A

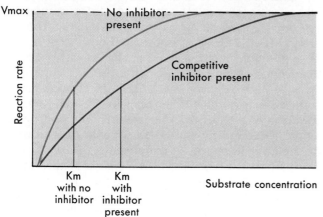

B

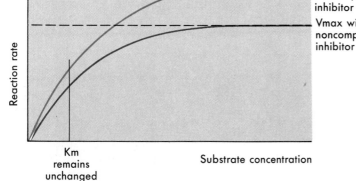

C

FIGURE 3-16

A The Michaelis-Menten plot shows the function that describes the relationship between the reaction rate and the substrate concentration for a given solution of enzyme. When all active sites of a population of enzyme molecules are occupied, the enzyme is said to be saturated.

B Substances whose molecular structure is similar to that of the natural substrates can sometimes bind to the active site and act as competitive inhibitors. In the presence of competitive inhibitor, the Michaelis-Menten plot shows an apparent increase in K_m, but the same V_{max} can ultimately be reached by adding a higher concentration of substrate.

C The active sites are unaffected by noncompetitive inhibitor, so the K_m is not affected. However the rate at which the enzyme catalyzes the reaction is decreased, so V_{max} is decreased.

of the Michaelis-Menten plot, allosteric activation or inhibition affects the maximum rate of conversion of substrate to product (V_{max}), but not the K_m (Figure 3-16, *C*).

Allosteric effects are a major mechanism by which the activities of enzymes are regulated. Two general classes of allosteric regulation are frequently seen: regulation by products and regulation by second messengers. If the product of an enzyme-catalyzed reaction is an allosteric inhibitor of the enzyme, the rate of conversion of substrate to product is made sensitive to the concentration of product and to the concentration of substrate. This is a mechanism of cellular self-regulation. In contrast, intracellular second messenger systems stimulate or inhibit major cellular activities in response to a first messenger's signal from outside the cell. In many cases, the second messenger system causes an enzyme called a **protein kinase** to attach phosphate

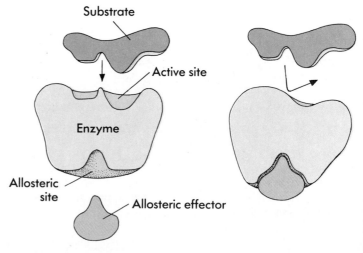

FIGURE 3-17

Many enzymes have allosteric sites that can influence enzyme activity.

groups to the allosteric control sites of all key enzymes involved in a particular cellular process. The specific mechanisms involved in enzyme regulation will be treated in greater detail in Chapter 5.

A Deoxyribose

B Ribose

C Phosphate group / Nitrogen base / 5-Carbon sugar

FIGURE 3-18

The structures of the nucleic acids. Nucleotides are composed of a sugar, deoxyribose in DNA (**A**) or ribose in RNA, (**B**) linked to a phosphate group and a base (**C**).

NUCLEIC ACIDS
Structure of Nucleic Acids

Nucleic acids are long polymers of subunit molecules called **nucleotides** joined by **phosphodiester bonds** (Figure 3-18). Each nucleotide is made up of a **pentose** (a 5-carbon sugar), a phosphate group, and either a **purine** or **pyrimidine** base (Figure 3-19). In the ribonucleic acids (**RNA**) the sugar is **ribose**, whereas deoxyribonucleic acids (**DNA**) contain **deoxyribose**. In DNA the base can be **adenine** (A), **guanine** (G), **cytosine** (C), or **thymine** (T). In RNA, **uracil** (U) replaces thymine as one of the pyrimidine bases. RNA consists of a single strand of nucleotides. Hydrogen bonds between the bases cause the molecule to assume a characteristic three-dimensional shape that is different for different forms of RNA (Figure 3-27). DNA has two parallel strands of nucleotides. The bases in the two parallel strands are bound to one another by hydrogen bonds. The bonding twists the strands into a **double helix** resembling a spiral staircase, with the base pairs corresponding to the treads of the staircase (Figure 3-20).

Most of the DNA in cells is in the form of units called **chromosomes** in the nuclei of cells; some is associated with specialized cytoplasmic organelles called mitochondria. These structures will be described in more detail in the next chapter. Each

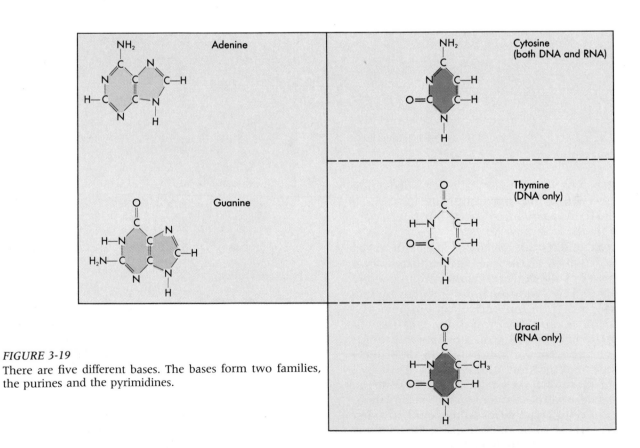

FIGURE 3-19

There are five different bases. The bases form two families, the purines and the pyrimidines.

PYRIMIDINES

THE FOUNDATIONS OF PHYSIOLOGY

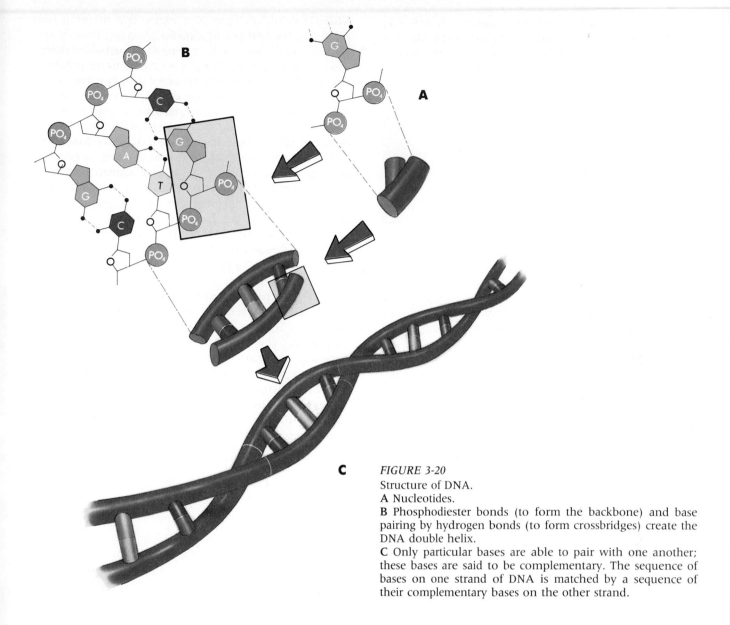

FIGURE 3-20
Structure of DNA.
A Nucleotides.
B Phosphodiester bonds (to form the backbone) and base pairing by hydrogen bonds (to form crossbridges) create the DNA double helix.
C Only particular bases are able to pair with one another; these bases are said to be complementary. The sequence of bases on one strand of DNA is matched by a sequence of their complementary bases on the other strand.

chromosome consists of two interconnected strands of DNA about 50 nm long and containing millions of nucleotide subunits, with some associated smaller molecules. Human somatic ("body" as opposed to reproductive) cells have 46 chromosomes in 2 sets of 23 **homologous pairs**; this is referred to as the **diploid number** of chromosomes.

Functions of DNA and RNA

The DNA has two major functions:

1. It **replicates**, or makes an exact copy of itself, in the course of division of somatic cells (**mitosis**) and in the process of cell division that leads to reproductive cells (**meiosis**). Because DNA molecules can copy, or **replicate** themselves, a dividing cell can pass on a copy of its genes to each of its daughter cells, and parents can endow their offspring with genetically determined characteristics.

2. It contains a code that specifies the amino acid sequences of all proteins that are ever synthesized in the body. The code is in the form of a sequence of nucleotide triplets. A unit of the DNA that contains the information for a single protein is called a **gene**.

Each of these two functions of DNA will be treated in detail in the next sections of this chapter.

Three forms of RNA are involved in expression of the genetic code of DNA. The code for a protein sequence is **transcribed** or "read" from the DNA into a corresponding code in molecules of **messenger ribonucleic acid (mRNA)**. The transcribed code in the form of the nucleotide sequence of mRNA is then **translated** into the corresponding amino acid sequence of a protein in the cytoplasm. In this process, **transfer RNA (tRNA)** molecules bind to specific amino acids and attach them to the protein being synthesized in accordance with the mRNA code sequence. **Ribosomal RNA (rRNA)** organizes the structure of **ribosomes,** cytoplasmic organelles that serve as frameworks for the translation process.

DNA Replication and Base Pairing

- *What events occur in the process of replication?*
- *What is the importance of complementary base pairing in DNA replication?*

The key to the ability of DNA to replicate itself is the fact that each nucleotide base in a single strand of DNA will bind to only one other **complementary base;** adenine pairs only with thymine, and guanine only with cytosine. The first step in replication is unwinding of the double helix and separation of the two strands. In the second step, a **complementary strand** develops along each single strand by the action of **DNA polymerases,** which catalyze base pairing (Figure 3-21). The result is the production of two identical double-stranded DNA molecules, each containing one of the original nucleotide strands and one entirely new complementary strand.

The Cell Cycle and Mitosis

- *What is the sequence of phases of the cell cycle?*
- *What are the characteristic events of each phase?*

Mitosis is the process of somatic cell division that increases the cell population in growing tissues and replaces cells lost in injury or by normal attrition. This process begins with a single parent cell possessing the **diploid number** of chromosomes (23 pairs in the case of human cells) and results in two diploid daughter cells. In the process, the chromosomes are replicated, and the cytoplasm and nuclear material of the parent cell are divided equally between the daughter cells, with each daughter cell receiving a complete set of chromosomes. (**Meiosis,** the process of cell division that occurs in reproductive organs and results in gametes (reproductive cells) containing the **haploid number** of chromosomes (23 chromosomes in the case of human cells), will be described in Chapter 24.)

The events of mitosis occur during defined periods of the **cell cycle** (Figures 3-22 and 3-23). During the **interphase** period of the cell cycle, the chromosomes are extended (or uncondensed) and not visible by light microscopy. During the first part of interphase, the G_1 **period,** the DNA directs the synthesis of proteins. During the **S period,** the chromosomes replicate in preparation for the next mitosis. Each chromosome now consists of two identical **sister chromatids.** The S period is followed by the G_2 **period.** The first visible sign of the onset of mitosis is condensation of the chromosomes into visible entities (**prophase**). The nuclear membrane breaks down. The chromosomes become attached to the **mitotic spindle** (**metaphase**) and line up be-

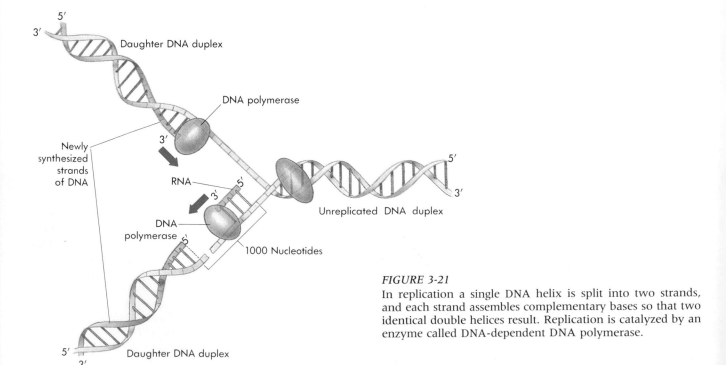

FIGURE 3-21

In replication a single DNA helix is split into two strands, and each strand assembles complementary bases so that two identical double helices result. Replication is catalyzed by an enzyme called DNA-dependent DNA polymerase.

THE FOUNDATIONS OF PHYSIOLOGY

fore separating as the spindle fibers draw sister chromatids in opposite directions (**anaphase**). The result is division of the replicated chromosomes into two equal groups. New nuclear membranes form (**telophase**), and the cytoplasm divides (termed **cytokinesis**) to generate two daughter cells, each with a single nucleus containing chromosomes identical to those of the parent cell. The chromosomes uncoil, and the daughter cells enter interphase.

The rate of mitosis is much higher during growth and development than during adulthood and is characteristically higher in some tissues than in others. The cells of some tissues, such as nervous tissue and skeletal muscle, have the potential to survive for the life of the individual and are not normally replaced if they die. Other tissues, such as epithelial tissue (including both the skin and the lining of the gastrointestinal tract), undergo continuous loss and replacement of cells during an individual's lifetime.

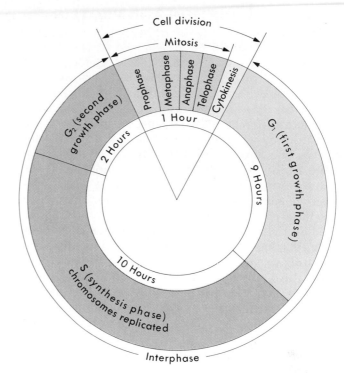

FIGURE 3-22
The cell cycle of liver cells grown in culture. During the S phase the chromosomes replicate.

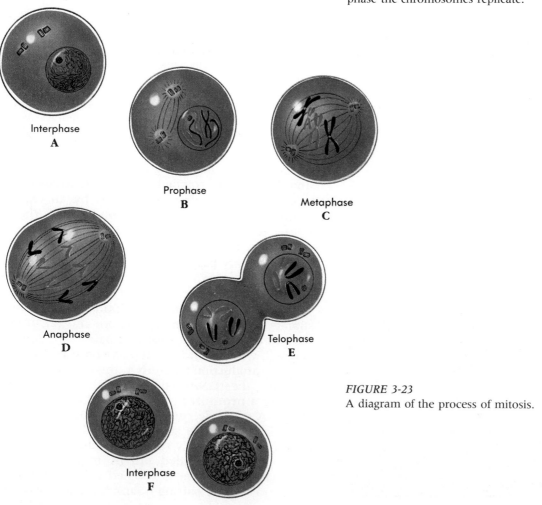

FIGURE 3-23
A diagram of the process of mitosis.

DNA TRANSCRIPTION AND PROTEIN SYNTHESIS
The Genetic Code

The functional units of nucleic acid molecules are sequences of three nucleotides called **codons,** each of which specifies a particular amino acid. For example, the base sequences UUU, UAA, and UGC represent different RNA codons. Because there are four nucleotides, there are 64 (4 ×4 ×4) different codons. Because there are only 20 amino acids, some amino acids are specified by more than one codon (termed **redundancy**). The first two nucleotides of a codon are the most critical for amino acid specification. Thus CCA, CCC, CCG, and CCU all code for the amino acid proline, whereas UUx codes for either leucine or phenylalanine, depending on the third nucleotide (Table 3-3). One codon (AUG, which also codes for the amino acid methionine) serves as a **start codon** signaling the beginning of a sequence to be transcribed. Three codons (UAA, UAG, and UGA) signal the end of a sequence to be read and are therefore known as **termination sequences** or "**stop**" **codons.**

DNA Transcription into mRNA

> - *What is the importance of complementary base pairing in mRNA synthesis?*
> - *How do the fates of the intron and exon sequences of a primary transcript differ?*

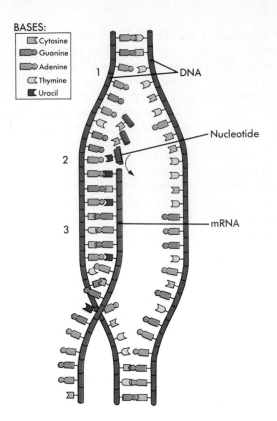

TABLE 3-3 *Amino Acid Codons*

Amino acid	Nucleotide code			
Alanine	GCU	GCC	GCA	GCG
Arginine	CGU	CGC	CGA	CGG
Asparagine	AAU	AAC		
Cysteine	UGU	UGC		
Glutamic acid			GAA	GAG
Glutamine			CAA	CAG
Glycine	GGU	GGC	GGA	GGG
Histidine	CAU	CAC		
Isoleucine	AUU	AUC	AUA	
Leucine	CUU	CUC	CUA	CUG
Lysine			AAA	AAG
Methionine				AUG
Phenylalanine	UUU	UUC		
Proline	CCU	CCC	CCA	CCG
Serine	UCU	UCC	UCA	UCG
Threonine	ACU	ACC	ACA	ACG
Tryptophan	UAU	UAC		
Tyrosine	UAU	AUC		
Valine	GUU	GUC	GUA	GUG
Stop codons			UAA	UAG
			UGA	

FIGURE 3-24

In transcription, DNA-dependent RNA polymerase opens the double helix for a short distance. The unpaired DNA bases of one strand of the open part of the helix are used as a template to assemble a primary transcript, a single-stranded RNA molecule whose base sequence is complementary to that of the DNA strand. The polymerase closes the helix after the base sequence has been transcribed, at the same time opening a space for transcription of the next few bases in the sequence. The polymerase continues in this fashion down the DNA helix, trailing a lengthening strand of RNA behind it.

In the nucleus of a cell, specific portions of one strand of double-stranded DNA are transcribed into single-stranded molecules of RNA by an enzyme called **DNA-dependent RNA polymerase.** The resulting mRNA has a nucleotide sequence complementary to that of the DNA. This is the result of complementary base pairing just as in DNA replication. There are two important differences between replication and transcription: (1) only one strand of the DNA is involved in transcription, whereas both strands are involved in replication, and (2) in production of mRNA the base that pairs with adenine is uracil, whereas in DNA replication it is thymine.

The production of mRNA involves two steps. Initially, the DNA-dependent RNA polymerase binds to a **promoter region** located on one strand of a DNA molecule, causing the two DNA strands to separate locally. The strand that will act as a template for the assembly of RNA produces a complementary single-stranded mRNA molecule until a termination sequence is reached (Figure 3-24). In this process, base pairing occurs spontaneously, but

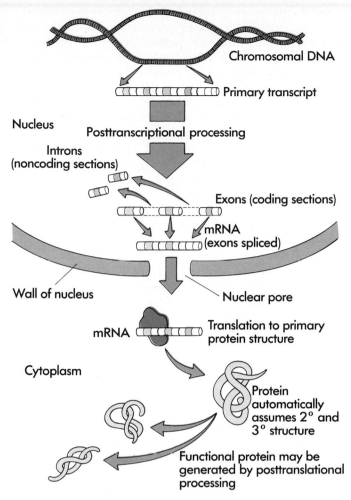

FIGURE 3-25
Messenger RNA synthesized from DNA, the primary transcript, may contain coding regions known as exons, and noncoding regions known as introns. In posttranscriptional processing, introns are removed from the primary transcript; the exons form the final messenger RNA (mRNA) that is exported to the cytoplasm to serve as a template for protein synthesis. Once in the cytoplasm, the mRNA is translated into a protein amino acid sequence. The protein automatically assembles its secondary and tertiary structure. It may then undergo posttranslational modification, usually by enzymatic removal of some part of its primary structure.

formation of bonds linking the nucleotides in linear sequence requires DNA-dependent RNA polymerase. This first portion of the process results in formation of the **primary transcript** (primary mRNA in Figure 3-25).

The primary transcript is subsequently modified in the nucleus of the cell, a procedure called **posttranscriptional processing.** As much as 75% to 90% of the primary transcript of mRNA does not specify for amino acids present in the protein; such noncoding sections are **introns.** The remaining regions of the primary transcript that together carry the code for the protein sequence are called **exons.** Specific nucleotide sequences designate the beginning and end of an intron. Before the mRNA leaves

the nucleus, the introns are deleted and the exons are spliced together to form the final version of the transcript. The exons in a particular primary transcript can sometimes be spliced together in different ways so that several distinct mRNAs (and therefore different proteins) can be produced from the same primary transcript. This versatility may explain how different antibodies are generated (see Chapter 26).

mRNA Translation into Proteins

- *What is the importance of complementary base pairing in mRNA translation?*
- *What distinguishes between the processes of posttranscriptional and posttranslational modification?*

Messenger RNA molecules pass from the interior of the nucleus into the cytoplasm by way of openings in the nuclear membrane called **nuclear pores** (Figure 3-25). In the cytoplasm the codes they carry are translated into the amino acid sequences of proteins. In the first step of translation, one end of a mRNA becomes attached to a cytoplasmic organelle called a **ribosome** (Figure 3-26). Ribosomes are protein-synthesizing molecular machines. Each ribosome is composed of several different ribosomal RNAs (rRNA) together with some ribosomal proteins. The ribosome travels along the length of the mRNA, "reading" its code one base triplet at a time and trailing behind it a growing protein chain.

As for DNA replication and mRNA synthesis, the key to the process of translation is base complementarity. There is a different transfer RNA molecule for each of the 20 amino acids, and in the process of translation each tRNA can be thought of as both a label and a handle for a single amino acid. At one end of the tRNA an amino acid can be bound, and at the other end there is a sequence of three nucleotides called an **anticodon** that is complementary to the codons of the mRNA molecules (Figure 3-27). At each codon it encounters as it travels along the length of the mRNA transcript, the ribosome accepts a tRNA that possesses a complementary anticodon (Figure 3-28, *A*). The associated amino acid is incorporated into the growing polypeptide chain (Figure 3-28, *B*), and the ribosome goes on to the next mRNA codon (Figure 3-28, *C*). Translation goes on in this way until a stop codon is encountered (Figure 3-28, *D*). At this point the completed protein is detached from the ribosome, and the ribosome accepts another mRNA and begins the process anew. Each molecule of mRNA can be translated repeatedly to yield a large number of protein molecules; typically at any moment each mRNA has a number of ribosomes lined up along its length, each with a protein chain at a different stage of completion (Figure 3-28, *E*).

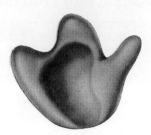

Small subunit

Large subunit

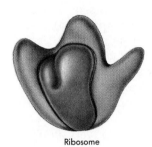

Ribosome

FIGURE 3-26
A ribosome is composed of a large and a small subunit, each containing both RNA and protein. The small subunit has the mRNA binding site.

There are several types of **posttranslational modifications** of proteins. Some proteins destined to serve outside the cell as enzymes or hormones are synthesized as inactive **proenzymes** or **prohormones** and become active only when they are later modified by enzymatic removal of parts of their primary structures. Proteins destined to be incorporated into membranes often have sugar residues added to them, forming glycoproteins, and in other cases particular amino acids of proteins are phosphorylated after the protein is synthesized. Large proteins may also be broken down into a series of smaller, related polypeptides, each of which may have a different function.

Selective Reading of the Genetic Code in Differentiated Cells

The four basic tissues of the body (epithelial, connective, nervous, and muscle) include at least 200 different cell types. **Differentiation** is the process by which cell types adapted for particular functions arise during development. It has been known for some time that each somatic cell has all the genetic information needed to become any of the types of differentiated cells, and that only a part of that information is transcribed by any particular cell type. Generally, differentiation is a one-way street; differentiated cells cannot dedifferentiate and reassume the wider developmental potential of undifferentiated cells. Therefore, in differentiated cells many genes must be permanently "turned off" so that they are not transcribed. The mechanisms that cause a cell to transcribe only a particular subset of its genes and thus differentiate into, for example, muscle cell instead of brain cell or liver cell, is one of the most interesting questions in modern biology, and the answers are only beginning to come in.

FIGURE 3-27
The structure of a transfer RNA (tRNA) molecule shown diagrammatically (left) and in its three-dimensional form.

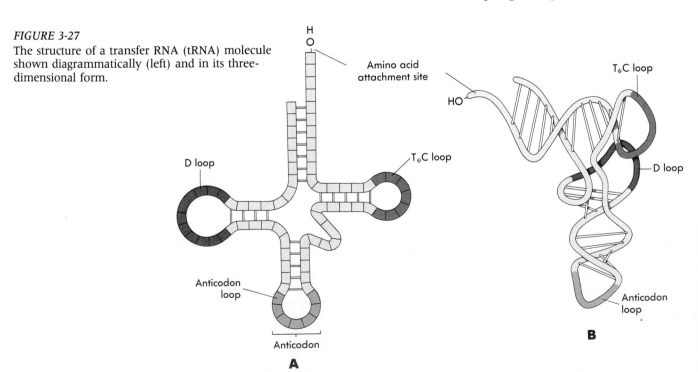

THE FOUNDATIONS OF PHYSIOLOGY

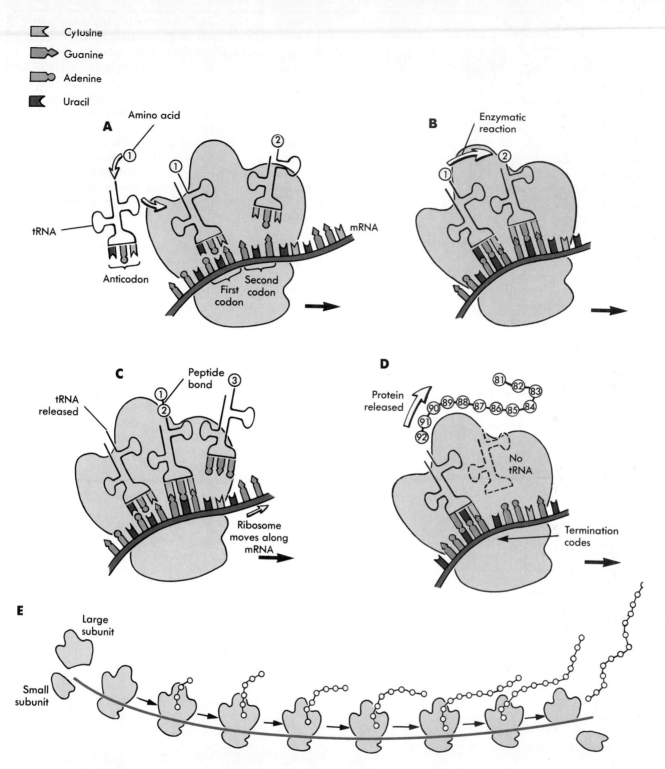

Cytosine

Guanine

Adenine

Uracil

A — Amino acid — ① ② — tRNA — Anticodon — First codon — Second codon — mRNA

B — Enzymatic reaction — ① ②

C — tRNA released — Peptide bond — ① ② ③ — Ribosome moves along mRNA

D — Protein released — 81 82 83 89 88 87 86 85 84 90 91 92 — No tRNA — Termination codes

E — Large subunit — Small subunit

FIGURE 3-28

Protein synthesis from messenger RNA by translation.

A The mRNA strand becomes associated with the ribosome. Amino acids attach to specific transfer RNA (tRNA) molecules that recognize (through their complementary anticodons) triplets of three bases on mRNA called codons. The other end of each transfer RNA binds the corresponding amino acid.

B An enzyme within the ribosome catalyzes the formation of peptide bonds between the amino acids that are delivered by tRNA.

C The formation of a dipeptide removes the amino acid from one tRNA, which is released from the ribosome as another tRNA enters the ribosome and its anticodon combines with the adjacent codon on the mRNA.

D The peptide chain is formed as amino acids from tRNAs are joined by peptide bonds until a termination codon on the mRNA stops chain elongation and the protein chain then separates from the ribosome. The ribosome then separates into its two subunits and separates from the mRNA.

E A schematic overview of the entire process.

Genetic engineering has many goals. One is to treat or cure victims of genetic diseases. Another is to speed up the process of selection of desirable traits in domestic animals and plants. Still another is to use the synthetic machinery of cells in culture to make large amounts of proteins that are important as drugs.

In many cases the genetic engineer's first step toward manipulating the gene that produces a particular protein is to determine the amino acid sequence of the protein. Once this has been done, a corresponding mRNA can be synthesized. The next step may be to synthesize the corresponding DNA from the mRNA, using an enzyme called reverse transcriptase. The engineer then possesses the gene coding for the protein of interest. Alternatively, a cell's own DNA can serve as the starting point. There are a variety of restriction endonucleases that will cut DNA wherever certain sequences of four to six base pairs occur. Resulting fragments of DNA

are incorporated into virus-like agents, termed vectors. The vectors transfer the fragments into bacteria in which the "new" DNA can be made to fuse with the bacterial DNA. Bacterial cells that receive a fragment containing the gene of interest will synthesize the corresponding protein and can be selected for large-scale culture. Foreign DNA that is incorporated into the genetic material of a cell is called recombinant DNA.

Bacteria can be genetically engineered to express recombinant DNA coding for human hormones. For example, growth hormone, insulin (a hormone essential for metabolism of glucose), interferon (a protein important in the immune response against viruses and cancer), and relaxin (a hormone used to relax the birth canal in delivery) are produced by this method. Clearly the potential benefits of recombinant DNA technology are just beginning to be realized, but at the same time genetic engineers are beginning to face ecological concerns and ethical questions.

Most objections to the current use of recombinant DNA technology focus on the risks of modifying the characteristics of presently existing organisms. In many cases, such as those of the rabbit in Australia and the starling in America, species introduced from other parts of the world have multiplied rapidly in the absence of predators and diseases, becoming pests and displacing native species. Release or escape of genetically engineered organisms may be assumed to present similar risks. Finally, some critics fear that it might be but a short step from treating human disease to "improving" the human species, and who would decide which traits need to be improved? Parents might want athletic, intelligent, and attractive children, whereas governments might want docile and obedient citizens. Clearly an ethical consensus must be reached before the technology can be applied to humans.

SUMMARY

1. Carbon forms bonds to other carbon atoms and to hydrogen and oxygen to form **lipids** and **carbohydrates.** Lipids and carbohydrates serve as elements in cellular structure and as sources of chemical energy. In animals, the major storage form of lipid is **triacylglycerol,** and the major storage form of carbohydrate is **glycogen,** a polymer of **glucose.**

2. **Proteins** are polymers of **amino acids** joined by **peptide bonds.** The sequence of amino acids in a protein is coded for by a sequence of nucleotide **codons** in DNA. The three-dimensional structure of a protein is the result of the formation of internal bonds between amino acids in different parts of the molecule—the structure that is formed is the most stable of the many structures possible.

3. **Enzymes** are proteins that catalyze specific chemical reactions. Catalysis involves a decrease in the **activation energy** that reactants must surmount to become products.

4. The double-stranded DNA molecule provides the sequence code for all proteins made in the body. In the process of **mitosis,** the DNA molecules separate into single nucleotide strands; each strand then **replicates** a **complementary** strand by base pairing.

5. The first step in synthesizing a specific protein is **transcription** of the nucleotide sequence of a gene into messenger RNA.

6. In the second step of protein synthesis, messenger RNA passes through nuclear membrane pores into the cytoplasm, where ribosomes **translate** its message into protein. In this process, the ribosome moves along the messenger RNA, attaching molecules of transfer RNA that carry the amino acids specified by the messenger RNA.

1. What are the four classes of organic compounds found in living cells? What are the unique characteristics of each class? How are their chemical structures related to their physiological function?

2. Define "enzyme." How do such molecules affect the rate of chemical reactions? What features do all enzyme-mediated reactions have in common? What is meant by the term "allosteric regulation"?

3. At what stage in the cycle of cell growth and mitosis do the following events occur:
   DNA replication
   Chromosome condensation
   Separation of sister chromatids

4. Describe the process by which a chromosome forms a replica of itself.

5. How is the base sequence of a strand of DNA transcribed into a strand of RNA? What are the differences between the DNA strand and its transcript RNA?

6. Describe the process by which a primary transcript is shortened to form messenger RNA. What is the name of this process?

7. What is a codon? The genetic code has been called redundant. Why?
   Choose the MOST Correct Answer.

8. These lipids act as regulatory agents:
   a. Acylglycerols
   b. Phospholipids
   c. Glycolipids
   d. Eicosanoids

9. This carbohydrate cannot be degraded into monosaccarides by digestive enzymes:
   a. Cellulose
   b. Glycogen
   c. Starch
   d. Sucrose

10. Proteins perform which of these functions:
    a. Regulate gene action
    b. Act as electron carriers
    c. Serve as identity markers
    d. All of the above

11. Select the FALSE statement:
    a. Enzymes increase the rate of reactions without being changed.
    b. Enzymes accelerate reactions by increasing the activation energy.
    c. Enzymes exhibit specificity.
    d. Enzymes are sensitive to both temperature and pH changes.

12. Meiosis:
    a. Results in two diploid daughter cells
    b. Results in the cytoplasm and nuclear material of the parent cell divided equally between the daughter cells.
    c. Occurs in the reproductive organs resulting in gamete production.
    d. Is the process of somatic cell division resulting in growth and repair.

13. The "start codon" for all proteins codes for the amino acid:
    a. Methionine
    b. Tyrosine
    c. Glycine
    d. Histidine

● SUGGESTED READINGS

AN 'INSULIN' FOR ANEMIA SUFFERERS, U.S. NEWS & WORLD REPORT 106(23):13, 1989. Describes the use of erythropoietin in kidney dialysis patients. The new drug was made by inserting human erythropoietin genes into laboratory grown hamster cells.

BROWNLEE S: The assurances of genes: is disease prediction a boon or a nightmare? U.S. News & World Report 109(4):57, 1990. Discusses the insurance, health, moral, and ethical aspects of genetic therapy.

BYLINSKY G: Coming: Star Wars medicine, Fortune 115:153, April 27, 1987. Describes how monoclonal antibodies from the body's own immune system can be used to speed medical diagnosis and precisely target drugs to kill cancer cells.

CARPENTER B: The body's master controls: unraveling proteins to tackle disease at its roots. U.S. News & World Report 106(18):57, 1989. Describes gene therapy strategies.

HOOD L: Biotechnology and the medicine of the future, JAMA 259:1837, March 25, 1988. Discusses how recombinant DNA technology can be applied to clinical medicine.

LERNER RA, TRAMONTANO A: Catalytic antibodies, Scientific American 258:58, March 1988. Nature has "designed" a limited number of enzymes. This article explores the possibility of designing antibodies to serve as catalysts in biochemical reactions.

LOCKSHIN RA, ZAKERI ZF: Programmed cell death: new thoughts and relevance to aging, J Gerontol 45:135, Sep 1990. Cell death is as tightly regulated as mitosis. Reviews the connections between cell death and aging.

MONTGOMERY G: Sperm truck, Discover 11(1):40, Jan 1990. New Italian method of gene transfer via sperm.

STEITZ JA: Snurps, Scientific American 259:56, June 1988. Shows how small nuclear ribonucleoproteins (snurps) delete introns from primary messenger RNA to produce the mRNA that actually directs protein synthesis.

CHAPTER 4

The Structure and Energy Metabolism of Cells

On completing this chapter you should be able to:

- Identify the major organelles of cells and provide brief descriptions of their physiological roles.
- Describe the structure and function of desmosomes, tight junctions, and gap junctions.
- Outline the biochemical pathways by which carbohydrates are metabolized, distinguishing between the processes that occur under anaerobic and aerobic conditions.
- Distinguish between substrate-level phosphorylation and oxidative phosphorylation.
- Describe the ways amino acids and fatty acids can enter the metabolic pathways of the cell and thus be used as energy sources.
- Describe the mechanism of oxidation-phosphorylation coupling according to the chemiosmotic hypothesis.
- Summarize the net yields of ATP obtained under anaerobic and aerobic conditions.
- Describe the mechanisms of respiratory control and control of glycolysis by phosphofructokinase.

*T*he structural organization of cells can be seen with microscopes of various kinds, and pictures (micrographs) made of what is seen. However, looking at pictures of cell structure is like looking into a large factory building during a business holiday. Machines, offices, raw materials, and finished goods are visible, but the processes of supply, repair, administration, shipping, and finance that carry out the function of the company are not visible. Life is, in the final analysis, a sum of processes. Cells may retain their structures if carefully frozen and begin to live again when thawed, but while they are frozen they cannot be said to be alive.

Living organisms are machines that use materials from the environment—foods—to obtain energy and increase order within themselves. The organic compounds that the body metabolizes for energy are characterized by a tendency to give up electrons in chemical reactions. Ultimately these electrons are transferred to oxygen in the powerhouses of the cell, its mitochondria. In the process of electron transfer, some of the energy contained in foods is captured in the form of high-energy phosphate bonds in ATP; the rest is released as heat. Cells use the captured energy to fuel energy-requiring processes, such as synthesis of proteins and other cellular constituents, contraction of muscles, and transport of materials across cell membranes.

BASIC CELL STRUCTURE

The complex organization of cells could not be fully appreciated until the development of the electron microscope. Each cell is surrounded by a **cell membrane** that mediates all interactions between the cell and its environment. The interior of cells contains a liquid, **cytoplasm**, filled with numerous small structures called **organelles**. Figure 4-1 shows an idealized cell and the major organelles found within cells. Although any particular cell may not have all the organelles illustrated, most contain a large number of them (Figure 4-2). Table 4-1 summarizes the components of cells and briefly outlines their physiological function.

The Cell Membrane

- *What are the molecular components of the cell membrane?*
- *What factors determine membrane structure?*

The cell membrane (Figure 4-3), is a thin film composed mainly of phospholipids, together with lipids and proteins. The orderly structure of the membrane is due to the amphipathic nature of phospholipid molecules. Because there is water on both sides of the membrane, the phospholipids all stand at right angles to the surface of the membrane, forming a double layer (a **bilayer**) with their hydrophilic po-

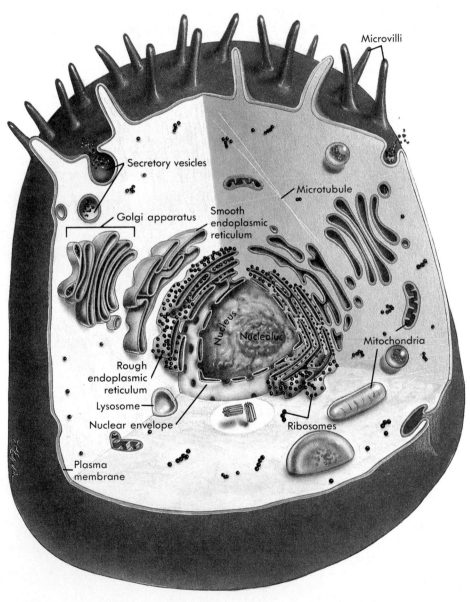

FIGURE 4-1

A diagram of an idealized cell showing the major organelles common to most cells. Microvilli are fingerlike projections that increase the cell surface area. Note that the intracellular organelles are not all drawn to the same scale.

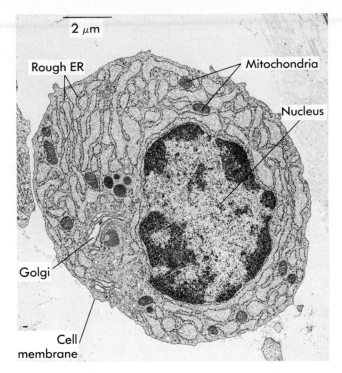

FIGURE 4-2
An electron micrograph of a plasma cell. The plasma membrane surrounds the cell; rough endoplasmic reticulum is studded with ribosomes that are synthesizing proteins for transport to the Golgi apparatus and eventual export by exocytosis, while mitochondria provide most of the energy needed by cells. The nucleus contains the diffuse form of DNA most characteristic of nondividing cells.

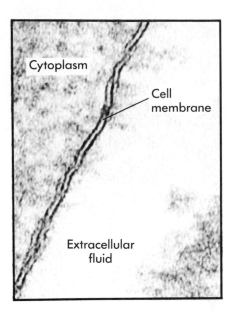

FIGURE 4-3
Electron micrograph of a human kidney cell showing the cell membrane. The membrane has a less dense central layer surrounded by two dense outer layers.

TABLE 4-1   Intracellular Organelles

Organelle	Characteristics	Functional role
Cell membrane	Lipid bilayer	Defines limit of cell Solute transport
Vesicle	Membrane-bound Sphere	Pinocytosis Phagocytosis Exocytosis (secretion)
Nucleus	Bounded by double membrane Contains DNA	Controls all cell processes
Rough endoplasmic reticulum	Sheets of membrane Ribosomes	Synthesis of protein to be secreted
Smooth endoplasmic reticulum	Sheets of membrane Lack of ribosomes	Membrane synthesis Fatty acid and steroid synthesis
Golgi apparatus	Series of flattened membranous sacs	Incorporation of materials into secretory vesicles
Lysosome	Specialized vesicles containing digestive enzymes	Intracellular digestive apparatus
Mitochondria	Double membrane Cristae	Oxidative phosphorylation
Cytoskeleton	Microfilaments Microtubules Intermediate filaments	Cell shape Cell movement Anchoring of some membrane proteins

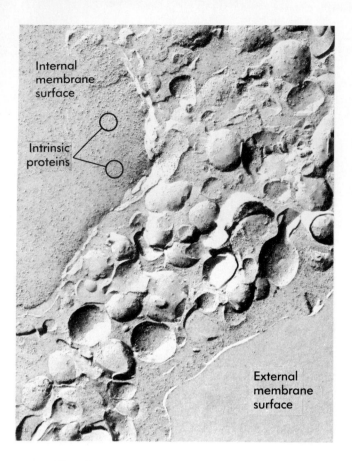

FIGURE 4-4

This scanning electron micrograph shows a white blood cell that has been frozen and then fractured. In the fracturing process, the outer leaflet of the lipid bilayer split away from the inner leaflet, revealing a population of intrinsic membrane proteins, the particles that project from the fracture face at the upper left. In contrast, the external face of the membrane, shown at the lower right, appears smooth. The area between these two layers consists of several fractures through different cell processes.

lar headgroups pointing outward and their hydrophobic fatty acid chains pointing inward. The forces that separate hydrophilic and hydrophobic groups impose a degree of order on the membrane; its constituent molecules can jostle about like people in a crowd, but flips or handstands are not allowed because they would bring polar and nonpolar groups together. Similarly, phospholipids in one leaflet of the bilayer cannot dive through to the other leaflet because their polar heads would have to pass through the double layer of nonpolar phospholipid tails.

All material that enters or leaves cells must cross the cell membrane, which is a highly selective barrier. Pure phospholipids form bilayers that are impermeable to electrolytes and large polar compounds; only nonpolar lipid-soluble compounds and small polar molecules such as water can penetrate them. The selective entry or exit of particular

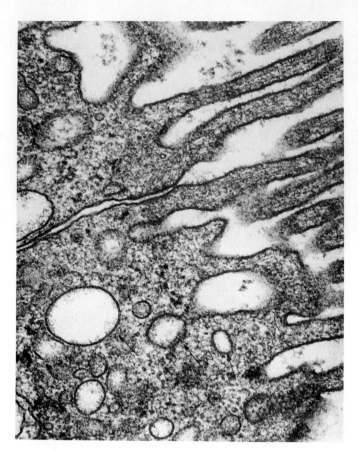

FIGURE 4-5

Pinocytosis is the ingestion by cells of water and dissolved solutes. The vesicle circled is in the process of budding off from the cell membrane. The arrows show pinocytotic vesicles within the cell.

types of ions, polar molecules, or large molecules is carried out by specific **transport proteins** of the cell membrane. In cells frozen and then fractured, transport proteins and other intrinsic membrane proteins appear as small particles (Figure 4-4). Chapter 6 discusses the aspects of membrane structure that have to do with transport of small molecules across cell membranes.

Endocytosis and Exocytosis

- *How do endocytosis, pinocytosis, and phagocytosis differ in function?*
- *What is the role of nuclear pores?*

In addition to transport of small molecules by way of specialized proteins in cell membranes, cells also carry out bulk exchange of materials with the extracellular space. This process is called **endocytosis** in the case of entry of material from outside the cell, and **exocytosis** in the case of exit of materials from the cell.

Extracellular
compartment

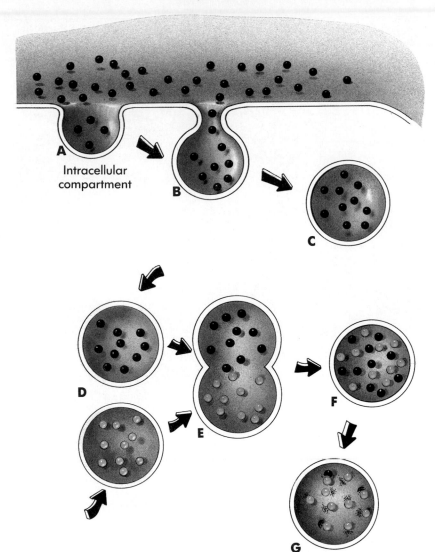

Intracellular
compartment

FIGURE 4-6
An illustration of endocytosis and the action
of lysosomes. In this diagram, the side of the
cell membrane that faces the extracellular
fluid is indicated by color; the cytoplasm surrounding the vesicles is white.
A Material is taken into the cell by endocytosis.
B, C The vesicle pinches off from the cell
membrane.
D A lysosomal vesicle approaches the endocytotic vesicle.
E, F, G The vesicles merge, allowing digestive enzymes to degrade the contents of the
endocytotic vesicle.

In endocytosis the membrane folds inward and
pinches off to form a tiny sphere of membrane,
called an **endocytotic vesicle,** whose interior contains extracellular material. There are several forms
of endocytosis. **Pinocytosis** or "cell drinking" (Figure 4-5) involves the uptake of bulk extracellular
fluid, along with whatever solutes are present. In
receptor-mediated endocytosis, specific molecules bind to receptor proteins in a region of the
membrane that subsequently is pinched off to form
a vesicle.

Endocytosis serves different functions depending on the cell type involved. In the endothelial
cells of capillaries and the epithelial cells of
the intestine, endocytotic vesicles carry proteins
from one side of the tissue to the other. Bacteria,
cell debris, and other foreign materials are engulfed
by specialized cells of the immune system in a form
of endocytosis called **phagocytosis** (discussed
further in Chapter 26). In phagocytosis, the endocy-

totic vesicles usually combine with digestive
organelles, termed **lysosomes** (see below), where
digestive enzymes act to break down their contents in an isolated compartment of the cell (Figure
4-6).

Exocytosis is a mechanism of cellular secretion.
Compounds made inside cells (in the ER and the
Golgi apparatus) are packaged into **secretory vesicles,** and these vesicles are transported to the cell
membrane (Figure 4-7, *A*). Here they fuse with the
membrane and discharge their contents into the extracellular space (Figure 4-7, *B*). In the process, the
phospholipids and proteins of the vesicular membranes are incorporated into the cell membrane.
Thus exocytosis is also a way by which cells add to
their surface areas or replace cell membrane lost in
the process of endocytosis.

The cell membrane is a selective barrier for information and substances. Cells must communicate
with and recognize one another, and these pro-

The Structure and Energy Metabolism of Cells

FIGURE 4-7

A Exocytosis involves the fusion of cytoplasmic vesicles with the cell membrane, resulting in the release of materials into the extracellular space.

B A transmission electron micrograph of exocytosis.

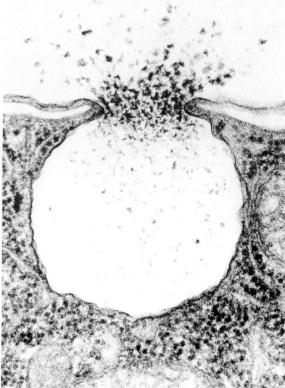

B

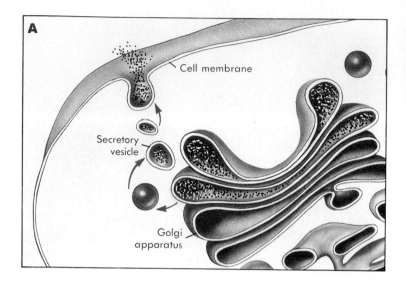

A

Cell membrane

Secretory vesicle

Golgi apparatus

cesses generally involve protein receptor molecules located in cell membranes. For example, electrically excitable cells communicate by exocytosis of a chemical, called a **neurotransmitter,** from a neuron; this chemical then interacts with a receptor on another neuron or muscle cell. Many hormones interact with specific membrane receptors, as do components of the immune system called antibodies (Chapter 26). During development, cells that must remain together to form tissues identify one another by virtue of markers in the cell membrane.

Nucleus

All mammalian cells (except mature red blood cells) possess a **nucleus** (see Table 4-1). The nucleus contains the DNA molecules that determine the structure and function of the cell, and it is bounded by a **nuclear envelope** consisting of a double phospholipid bilayer. The inner and outer bilayers of the nuclear envelope fuse in places to form **nuclear pores** 80 to 100 nm in diameter. In the nucleus, DNA is combined with certain specific protein molecules to form chromosomes, which are extended in a lace-like network except during cell division (Figure 4-8). Within the nucleus of many cells there is a dense **nucleolus,** composed of ribosomal RNA (rRNA) and associated ribosomal proteins. It is in the nucleolus that the basic structure of ribosomes is assembled.

Nuclear pores are selective passages. Transfer RNA (tRNA) and rRNA can pass through them, as can a variety of small molecules and ions. Nuclear pores prevent DNA from leaving the nucleus and only allow messenger RNA (mRNA) to enter the cytoplasm when it has been processed and is ready to guide protein synthesis (see Chapter 3). The nucleus itself has no ribosomes, so all of the nuclear proteins are synthesized in the cytoplasm and selectively imported into the nucleus through nuclear pores. Some hormones act by affecting the transcription of particular genes; these hormones bind to specific cytoplasmic proteins which allow them to be imported into the nucleus. The mechanisms that allow selective entry and exit of molecules through nuclear pores are not yet understood.

CYTOPLASMIC ORGANELLES
Endoplasmic Reticulum

- *How do rough endoplasmic reticulum and smooth endoplasmic reticulum differ?*
- *How are proteins that will be secreted by exocytosis synthesized and modified?*
- *What are the components of the cytoskeleton?*

The **endoplasmic reticulum (ER)** (Figure 4-9) forms an extensive intracellular membrane system. In some cells, this system ultimately connects with the cell membrane and the exterior of the cell. Two types of endoplasmic reticulum are found in most cells. **Rough ER** appears granular because it has ribosomes attached to its surface, while **smooth ER** lacks ribosomes (Figure 4-9). The smooth ER is the site of synthesis of new mem-

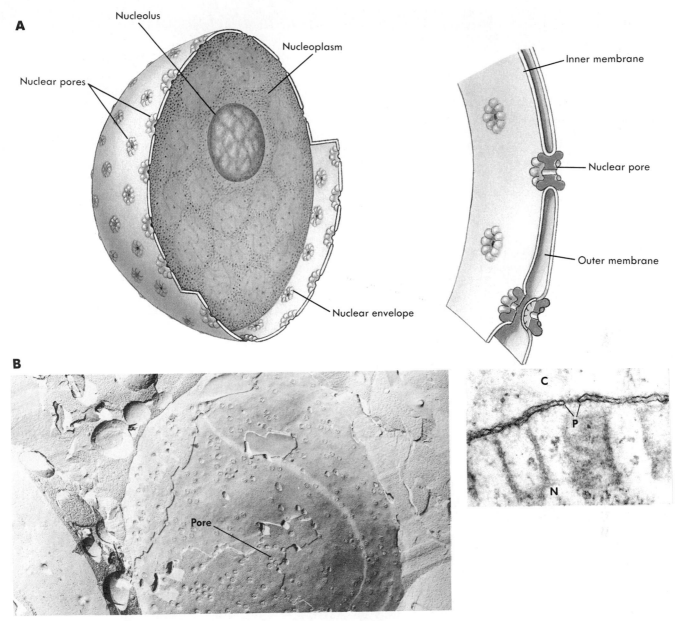

A

Nucleolus

Nucleoplasm

Nuclear pores

Nuclear envelope

Inner membrane

Nuclear pore

Outer membrane

B

Pore

C

P

N

FIGURE 4-8
The nucleus (**A**) is surrounded by a double membrane studded with pores clearly visible in freeze-etched (**B**) or transmission (**C**) electron micrographs (*C*, cytoplasm, *P*, pore, *N*, nucleus).

brane phospholipids, fatty acids, and steroids, but not proteins (Table 4-1). In muscle, the smooth endoplasmic reticulum is adapted for storage and release of calcium (Ca^{++}) ions and is called the **sarcoplasmic reticulum**.

Ribosomes are protein-synthesizing organelles (see Chapter 3). The ribosomes of the rough ER synthesize proteins destined to be secreted by cells or to be inserted into their cell membranes or the membranes of intracellular organelles (Table 4-1). The ribosomes of the rough ER do not appear to differ from free ribosomes. Rather, free ribosomes become attached to the endoplasmic reticulum when

they are translating mRNA that bears a **signal sequence**, an initial portion of the mRNA transcript that labels it as the code for a protein destined for secretion. Ribosomes attached to the rough ER release completed proteins directly into its lumen. The mechanism by which this occurs is not fully understood, but each ribosome bound to the rough ER is believed to be associated with a protein that forms a channel through the ER membrane. As it is synthesized by the ribosome, the protein passes through the channel into the lumen of the ER (Figure 4-9, *B*). After the protein enters the lumen of the ER, an enzyme cleaves off the signal sequence. Pro-

Inspecting Biological Molecules: New Frontiers

Microscopes, originally developed by physicists to visualize and manipulate atoms in metals, are giving biologists a new way to examine the molecules of living cells. The scanning-tunneling microscope takes advantage of electron tunneling to produce three-dimensional images of atoms and molecules. When the distance between two electrically conductive materials approaches atomic dimensions, electrons jump across, producing a minute electric current

that increases as the distance decreases. In the scanning-tunneling microscope a sharpened tungsten probe is moved along a sample a few billionths of an inch above its surface. The surface contour is seen as tiny fluctuations in the current flow from the probe. However, the scanning-tunneling microscope works only on samples that conduct electricity, and most organic molecules are nonconductors. A new instrument, the atomic force microscope, gets

around the limitations of the scanning-tunneling microscope by using a microscopic ceramic pyramid only one carbon atom wide at the tip as a probe to "feel" the solid surface, detecting surface features as small as single atoms. At the molecular level, the process is similar to the way a phonograph needle detects the slight irregularities in the grooves of a record that correspond to the recorded sounds.

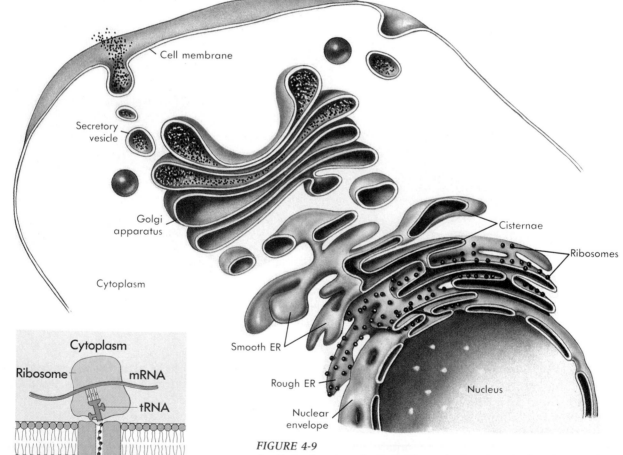

FIGURE 4-9

A Endoplasmic reticulum. The endoplasmic reticulum is continuous with the nuclear membrane and may appear as either rough or smooth endoplasmic reticulum. Ribosomes (*black dots*) are associated with only one side of the rough endoplasmic reticulum; the other side bounds a separate compartment within the cell into which the ribosomes extrude newly synthesized proteins.

B Diagrammatic view of a ribosome on rough ER membrane. The ribosome is associated with a transmembrane protein channel through which the newly synthesized protein passes into the ER lumen. The signal sequence that labeled this protein as one to be exported was on the leading end of the protein but was clipped off before the leading end entered the channel.

teins destined to be further modified before secretion are carried to the Golgi apparatus (see below) by **transfer vesicles** budded off by the rough ER.

Many cells are specialized for production of proteins that function outside the cell. Examples include the cells of the pancreas that synthesize digestive enzymes, those of the liver that produce plasma proteins, and connective tissue cells that synthesize collagen. The cytoplasm of these cells contains large amounts of rough ER (see Figure 4-2).

Golgi Apparatus

The **Golgi apparatus** is a stack of flattened membranous sacs surrounded by smaller spherical vesicles (Figure 4-10). Although this structure has generally been thought of as an organelle, it is best regarded as a specialized region of the ER. The vesicles surrounding the Golgi apparatus represent a two-way traffic of proteins. Proteins synthesized in rough ER are delivered by transfer vesicles (see above) to the Golgi apparatus. In the Golgi apparatus, the proteins are modified by glycosylation, becoming glycoproteins (see Chapter 3). These modified proteins are then repackaged into **secretory vesicles**. Secretory vesicles may be transported to the periphery of the cell and undergo exocytosis (see Figure 4-9), or they can fuse with other organelles, such as lysosomes (see below). In general, the Golgi apparatus seems to direct intracellular trafficking of proteins, probably by means of carbohydrate "destination labels" attached to the protein molecules (Table 4-1).

Lysosomes

Lysosomes (Figure 4-11) are vesicles manufactured in the Golgi apparatus that contain digestive (hydrolytic) enzymes capable of destroying a variety of substances. Lysosomes keep potentially harmful reactions isolated from the remainder of the cell interior. Within the cell, lysosomes encapsulate and break down damaged organelles and other structural components. When foreign substances are taken up by endocytosis or phagocytosis, the resulting vesicle often fuses with a lysosome that destroys the engulfed material.

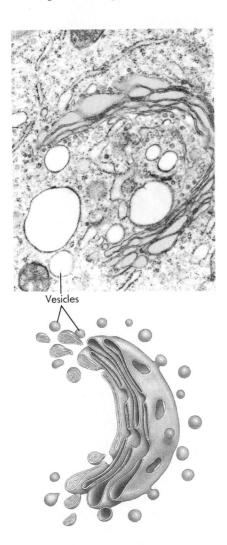

Vesicles

FIGURE 4-10

A Golgi apparatus in cross section. Within the Golgi complex the membrane is smooth and pinches together at its terminus to produce membrane-bound vesicles containing whatever enzymes or other substances have been transported from the endoplasmic reticulum.

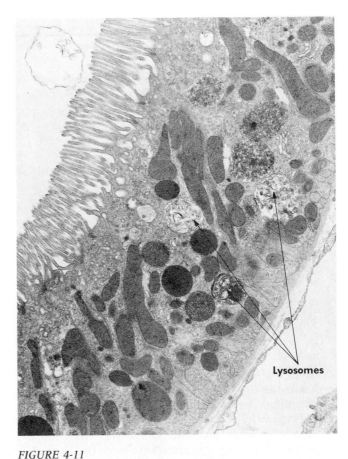

Lysosomes

FIGURE 4-11

An electron micrograph showing lysosomes (*arrows*) at various stages of intracellular digestion.

The Structure and Energy Metabolism of Cells

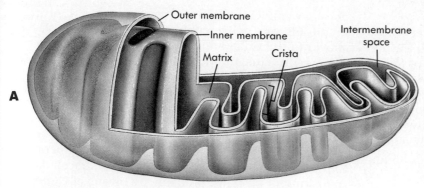

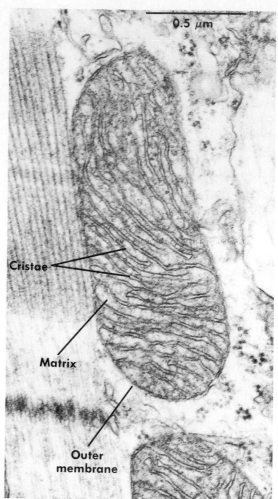

FIGURE 4-12

A schematic illustration (**A**) and electron micrograph (**B**) of a mitochondrion (the latter in a muscle cell). The mitochondrial outer membrane surrounds the inner membrane, which is folded into cristae. The enzymes of oxidative phosphorylation are located on the inner membrane, while the other enzymes for oxidative metabolism are located in the matrix.

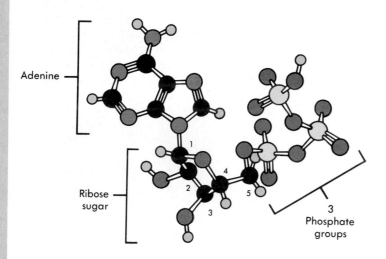

● = Oxygen

● = Carbon

● = Nitrogen

○ = Phosphorus

○ = Hydrogen

FIGURE 4-13

The structure of adenosine triphosphate (ATP).

Mitochondria

Mitochondria (Figure 4-12), like the nucleus, are surrounded by a double bilayer membrane. The outer membrane is smooth, while the inner is folded; the folds are called **cristae.** Mitochondria contain all of the enzymes involved in **oxidative metabolism,** a metabolic pathway that converts carbohydrate and fatty acid substrates to CO_2 and water, while capturing the released energy in the form of high-energy phosphate bonds in the compound **adenosine triphosphate (ATP)** (Figure 4-13). ATP is often referred to as the energy "currency" of cells. The enzymes and coenzymes needed for oxidative metabolism are located within the **inner mitochondrial membrane** and in the innermost space, called the **mitochondrial matrix.** The region between the outer and inner mitochondrial membranes is where H^+ ions are translocated by the electron transport chain (discussed on p. 91).

Cytoskeleton

All cells have a network of interconnecting and interlacing proteins collectively referred to as the **cytoskeleton** (Figure 4-14, *A*). The proteins of the cytoskeleton are organized into three types of strandlike organelles that together provide a framework that (1) determines the shapes of cells, (2) controls the distribution of the cell's internal organelles, (3) holds some proteins of the cell membrane in stable positions, and (4) allows some cells to move. **Microfilaments,** the smallest in diameter of the three types, are largely composed of **actin,** a protein

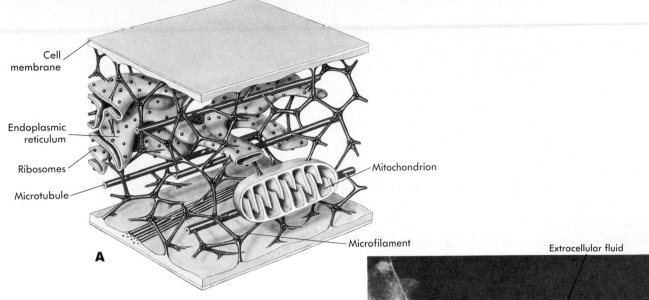

Cell membrane

Endoplasmic reticulum

Ribosomes

Microtubule

Mitochondrion

Microfilament

A

Extracellular fluid

B

Leading edge of a sheet of migrating cells

Cytoplasm

Actin filament

FIGURE 4-14
A   A diagram of the inner cytoskeleton of a cell.
B   An electron micrograph of the cytoskeleton of several cells in a migrating sheet of epithelial tissue. The dye rhodamine has been used to stain actin filaments in the cytoplasm. The boundaries of individual cells cannot be resolved.

that also plays a key role in muscle contraction (Figure 4-14, *B*). **Microtubules**, the largest in diameter of the three types, are hollow tubes composed of a protein called **tubulin**. **Intermediate filaments** can be formed by several types of cytoskeletal protein, and are the most stable component of the cytoskeleton. In general, intermediate filaments probably maintain the overall structures of cells, while microfilaments and microtubules cooperate to allow movement. Some examples of movement mediated by cytoskeletal elements are:

(1) The segregation of chromosomes and division of the cytoplasm of daughter cells in mitosis (see Chapter 3)

(2) Fast axoplasmic transport of intracellular vesicles and organelles along nerve axons (see Chapter 8)

(3) Muscle contraction (see Chapter 12),

(4) Locomotion and phagocytosis of bacteria by white blood cells (see Chapter 24).

JUNCTIONS BETWEEN CELLS

- *How do desmosomes, tight junctions, and gap junctions differ in structure?*
- *What is the function of each type of junction?*

Intercellular junctions are structures that attach adjacent cells to one another. Junctions are formed by aggregations of specific membrane proteins. Junctions may have three roles, depending on the

TABLE 4-2   *Junctions Between Cells*

Junction	Characteristics	Functional role
Desmosomes	Filaments attached to cytoskeleton	Binds cells together
Gap junction	Connexons	Impermeable Couples cytoplasm of adjacent cells
Tight junction	Fusing of cell membranes	Joins cells in epithelial sheets

function and structure of the tissue in which they are found (Table 4-2). Besides maintaining the structure of tissues, **junctions** may serve as lines of electrical and chemical communication between adjacent cells, and are important in determining the tightness of barriers between different body compartments. Three major types of junctions may be distinguished in photomicrographs: **desmosomes, gap junctions,** and **tight junctions.**

Desmosomes

The role of **desmosomes** (shown diagrammatically in Figure 4-15) is to maintain tissue integrity. For greater strength, desmosomal filaments extend across the junction and are attached to the cytoskeleton on each side of the junction. Desmosomes may be scattered widely on the cell surface like spot-welds (in this arrangement each spot is sometimes called a **macula adherens**), or clustered in a belt called a **zonula adherens** (Figure 4-16).

Gap Junctions

Gap junctions (Figure 4-17) couple the cytoplasm of a cell to that of its neighbors, spanning a "gap" between the cell membranes. Each gap junction consists of two doughnut-shaped halves, one in each membrane of the two joined cells, joined so that the holes of the doughnuts line up. Each doughnut consists of 6 wedge-shaped protein subunits called **connexons** arranged in a circle, with their tapered edges facing inward (Figure 4-17). A slight twisting of the connexons relative to one another opens or closes a central pore large enough to allow small molecules to pass from one cell to another. The opening and closing of gap junctions is believed to be regulated by a number of factors, including changes in the cytoplasmic concentrations of Ca^{++} and H^+.

Gap junctions allow passage of chemical messengers and ATP, so adjacent cells communicate with one another and provide metabolic assistance to one another. Epithelial cells are typically connected by gap junctions (see Figure 4-16). Electrical

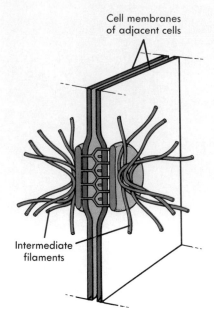

FIGURE 4-15
Desmosomes hold cells together, as shown in this schematic diagram.

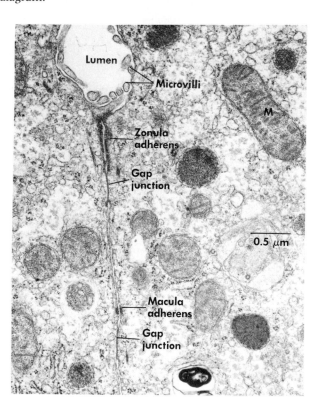

FIGURE 4-16
An electron micrograph of intestinal epithelial cells. Two types of desmosomes can be seen connecting intestinal epithelial cells—the zonula adherens and the macula adherens. In addition to desmosomes, these epithelial cells are coupled via gap junctions.

currents, carried by small ions, can also flow through gap junctions. Electrical coupling is important in heart muscle and in some types of smooth

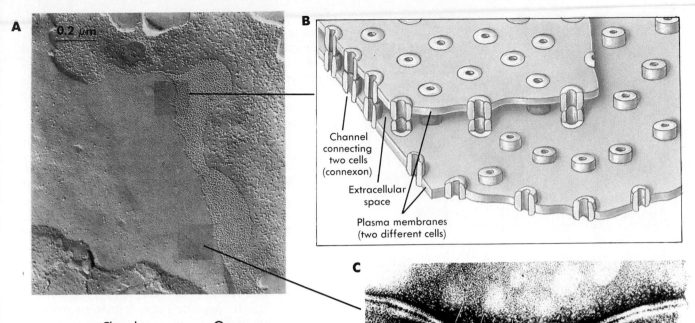

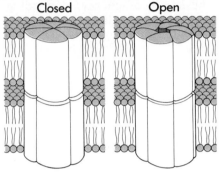

FIGURE 4-17

Gap junctions provide electrical and chemical communication between adjacent cells.

A A freeze-etch micrograph of a gap junction.

B A schematic illustration of a gap junction.

C A gap junction in cross section.

D Each cell has half of a gap junction; each half consists of six connexons. Twisting of the connexions relative to one another opens or closes the junction.

muscle. In muscle the effect of gap junctions between cells is to allow electrical excitation arising in one cell to spread throughout all of the cells coupled to it.

When cells are coupled, damage to the cell membrane of any one cell would threaten the cytoplasmic composition of all of them. A built-in damage control system, rather like the water-tight compartment doors of ships, is provided by the sensitivity of gap junctions to cytoplasmic Ca^{++}. The normal Ca^{++} concentration of cytoplasm is about three orders of magnitude lower than that of extracellular fluid. When the membrane of a cell is breached, Ca^{++} rushes in down its concentration gradient. The resulting increase in cytoplasmic Ca^{++} concentration of the damaged cell closes its gap junctions, disconnecting it from surrounding intact cells.

Tight Junctions

Tight junctions take the form of strips that completely encircle a cell, attaching it to similar strips on the surfaces of adjacent cells. At the point of attachment, the membranes of the two adjacent cells appear to be in contact with one another. Tight junctions are characteristically found in epithelia, the tissues that separate fluid compartments of the body. These tissues, including for example the linings of the stomach and intestine, the gallbladder, the nephrons of the kidney, and the walls of exocrine glands, must be selective in their permeability to solutes and water.

Substances that might diffuse through epithelia take one or both of two possible routes: through the cells (the **transcellular route**) or between them (the **paracellular route**). The transcellular route is possible for lipid-soluble substances and for substances for which there are specific membrane-transport proteins. The paracellular route is guarded by tight junctions (Figure 4-18). In spite of their name, the tight junctions can be arranged so that the paracellular pathway is either tight or leaky to small polar solutes and water. **Tight epithelia** are found where an osmotic gradient between fluid compartments must be maintained. An example is the urinary bladder, which stores urine that must remain at a different concentration from extracellular fluid. In tight epithelia, there are several bands of tight junctions without any gaps or discontinuities in the bands. **Leaky epithelia** are found where

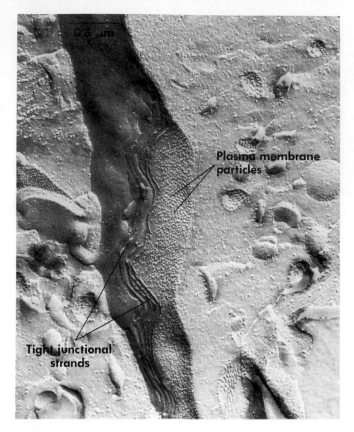

FIGURE 4-18
A freeze-fracture electron micrograph of a tight junction showing the strands that connect adjacent cells.

osmotic flow of water is important. An example is the small intestine, in which absorption of nutrients is accompanied by a proportional absorption of water. The tight junctions of leaky epithelia are more sparse and have occasional discontinuities. However, even leaky epithelia usually do not permit the free passage of large molecules such as proteins.

ENERGY METABOLISM

- *How do substrate-level phosphorylation and oxidative phosphorylation differ?*
- *What are the roles of the coenzymes NADH and FADH₂ in energy metabolism?*
- *What is the advantage of converting pyruvate to lactate in the terminal reaction of anaerobic glycolysis?*
- *In the chemiosmotic hypothesis, what forms does energy pass through before it is finally captured as ATP?*
- *What is the difference in terms of net ATP production between anaerobic glycolysis and oxidative phosphorylation?*

Metabolism, in its broadest sense, refers to the sum of chemical reactions carried out by the body. Within this great scheme, the atoms of any particular compound that enters the body may undergo a series of rearrangements, additions, and deletions in successive reactions of a **metabolic pathway.** The metabolic pathway followed by the compound could be traced experimentally by substituting radioactive isotopes for some of the atoms of the compound and chemically identifying the **metabolic intermediates,** or compounds, in which those "labeled" atoms subsequently appeared.

The sum of pathways in which organic compounds are broken down into more disorderly forms with lower free energies is referred to as **catabolism.** Catabolism releases chemical potential energy, some of which is captured in chemical bonds and some of which takes the form of heat. Synthesis of materials from simpler substrates is called **anabolism,** the predominant process in tissue replacement, growth, and development. In anabolic processes, chemical energy that originated outside the body is used to increase order and structure. For example, peptide bond formation requires 0.5 to 4.0 kcal/mole of energy, depending on the specific amino acids to be linked.

High Energy Phosphates: The Energy Currency of Cells

The energy needs of the body vary markedly with age, sex, and activity level. A healthy adult maintaining constant body weight takes in and expends about 2500 kcal (10455 kj) per day. This energy is used in muscular contraction, for transport of substrates and ions across cell membranes, for synthesis of molecules such as hormones and digestive enzymes that must be constantly renewed, for repair of damaged tissues, and for replacement of cells that die. In children and adolescents, a significant fraction of total energy intake is used for growth. If the net energy value of food intake exceeds the body's energy expenditure, the excess is stored in the form of fat and glycogen.

Carbohydrates, lipids, and proteins are metabolized for energy in varying proportions, depending on diet and activity level. Immediately after a carbohydrate meal, and during brief periods of exercise, much of the energy used by the body is derived from oxidation of glucose to carbon dioxide and water. The free energy change of this reaction can be determined by burning a known quantity of glucose in a special furnace called a bomb calorimeter in which the amount of heat released can be measured. If a mole of glucose (180 g) were burned in a bomb calorimeter, its complete oxidation would require only a few seconds and would yield 686 kcal of heat energy.

In cells, glucose is not broken down into carbon dioxide and water all at once, but in a series of biochemical reactions involving numerous intermediate compounds. This process releases the energy in manageable portions, and much of the theoretical yield of 686 kcal/mole is used to **phosphorylate** (add a phosphate group to) nucleotides, converting them to "high-energy" forms. The four high-energy nucleotide phosphates are **guanosine triphosphate (GTP)**, **cytosine triphosphate (CTP)**, **uracil triphosphate (UTP)**, and **adenosine triphosphate (ATP)**. Of these the most important is ATP (Figure 4-16); the other phosphorylated nucleotides are specifically required in some reactions but are interconvertible with ATP. Hydrolysis of the bond holding the terminal phosphate group of ATP yields **adenosine diphosphate (ADP)** and **inorganic phosphate** ($H_2PO_4^-$) and occurs with a free energy change of about 7.3 kcal/mole.

There is a relatively small reserve of ATP in the cytoplasm, so that as ATP is used it must constantly be regenerated by the energy-yielding reactions that phosphorylate ADP. There are two basic mechanisms for regeneration of ATP. The first is referred to as **substrate-level phosphorylation**. In substrate-level phosphorylation, a phosphate group (together with its bond energy) is transferred to ADP from a phosphorylated metabolic intermediate. Substrate-level phosphorylation takes place in the cytoplasm and is the only mechanism by which ATP can be regenerated in the absence of oxygen. The reactions that result in substrate-level phosphorylation of ADP in the absence of oxygen are collectively called **anaerobic metabolism**.

The second means of ATP production, **oxidative phosphorylation**, occurs in the inner mitochondrial membranes, requires oxygen, and produces most of the ATP used by the body. Oxidative phosphorylation and the reaction sequences that serve it are referred to collectively as **oxidative** or **aerobic metabolism**. However, one step in the pathway of oxidative metabolism also involves substrate-level phosphorylation.

Electron Carriers: NADH and FADH$_2$

If one molecule gives up an electron and becomes oxidized, another must accept the electron and be reduced; therefore, it is helpful to think of such processes as paired **oxidation-reduction reactions**. Most of the potential energy carried by glucose is first captured by oxidation-reduction reactions. A key element in such reactions is the participation of **coenzymes**. The most important electron-carrying coenzymes are **nicotinamide adenine dinucleotide** and **flavine adenine dinucleotide** (Figure 4-19). The oxidized forms of these coenzymes are abbreviated **NAD$^+$ and FAD**, respectively; the reduced forms are **NADH and FADH$_2$**.

The electron-carrying coenzymes are present in cells in only very small amounts. At some steps in the pathways of energy metabolism, the oxidized coenzymes are converted to NADH and FADH$_2$. If there were no means to regenerate NAD$^+$ and FAD, all of the oxidized coenzymes would quickly be converted to the reduced forms, and the reaction steps

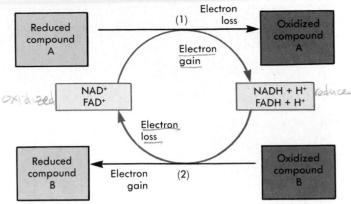

FIGURE 4-19
The role of the coenzymes NAD$^+$ and FAD$^+$ as electron carriers.

that require the oxidized forms would stop. The electron-carrying coenzymes can be reoxidized in two different ways, depending on whether oxygen is available. If oxygen is not available, some other metabolic intermediate can be reduced (Figure 4-19). Reoxidation of NADH and FADH$_2$ by this means has two disadvantages. First, it does not yield any energy, and second, it results in an accumulation of the reduced compound, either locally or in the body as a whole.

Under aerobic conditions a more effective solution to the problem of regenerating oxidized coenzymes is to couple their regeneration to reduction of oxygen to O^{2-}, a multistep process called **terminal oxidative metabolism**. The overall reactions (neglecting the intermediate steps) are:

$$NADH + H^+ + 1/2\ O_2 \rightarrow NAD^+ + H_2O$$
$$FADH_2 + 1/2 O_2 \rightarrow FAD + H_2O$$

In each of these reactions, two electrons are transferred to atomic oxygen ($1/2O_2$), resulting in O^{2-}. Each O^{2-} immediately reacts with two H^+ to form H_2O. The advantages of this alternative are that some of the free energy of reduced coenzyme can be used to phosphorylate ADP (oxidative phosphorylation) and the end product of the process is H_2O, which is more easily disposed of by cells than is lactate.

Glycolysis

The first pathway in energy metabolism of glucose is **glycolysis** (Figure 4-20). All of the reactions of the glycolytic pathway, except for the last, are indifferent to the presence or absence of oxygen. However, it is useful to distinguish between **anaerobic** and **aerobic glycolysis**. Anaerobic glycolysis is carried on in the absence of oxygen or by cells that lack mitochondria, and yields lactate as its end product. In **aerobic glycolysis**, the glycolytic pathway is the first step in complete oxidation of glucose, and the

The Structure and Energy Metabolism of Cells

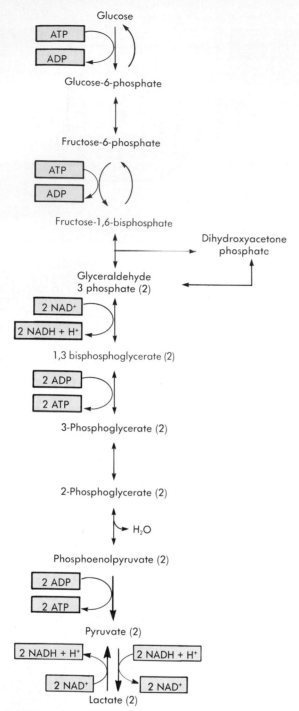

FIGURE 4-20
The reactions of glycolysis. The initial reactions involve the phosphorylation of glucose and consumption of ATP. Later reactions produce ATP with a net gain of two ATP molecules per molecule of glucose. The end product of glycolysis is pyruvate.

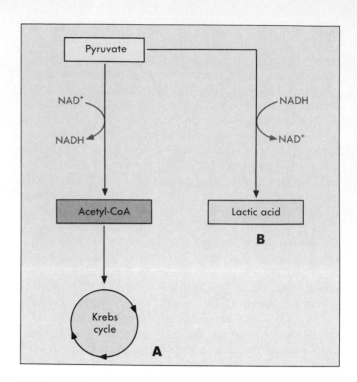

FIGURE 4-21
Alternative fates of pyruvate. Pyruvate can enter the Krebs cycle via acetyl-CoA (**A**) or be converted to lactic acid to regenerate NAD$^+$ (**B**).

specific carrier proteins. In most tissues the glucose carriers are active only in the presence of the pancreatic hormone **insulin.** Glucose is actively accumulated only by epithelial cells in the intestine and kidney. In other cell types, glucose entry is driven by a concentration gradient between extracellular fluid to cytoplasm. The difference in glucose concentration between the exterior and interior of cells is maintained by two processes. First, the extracellular glucose concentration is under homeostatic regulation, so that it normally does not drop below about 5 mM. Also, glucose molecules entering the cytoplasm react almost immediately to enter the pathway of glycolysis or a pathway leading to storage as glycogen. As a result, the intracellular glucose concentration remains much less than 5 mM, and a concentration gradient favorable for glucose to diffuse into cells is always maintained.

In the first step of glycolysis (see Figure 4-20), glucose is converted to **glucose-6-phosphate,** a reaction driven by ATP. The second reaction of glycolysis is conversion of glucose-6-phosphate to **fructose-6-phosphate** (with the use of another ATP molecule). In the next several steps in glycolysis, the 6-carbon ring of glucose is broken into two 3-carbon pieces, yielding two molecules of **glyceraldehyde-3-phosphate,** and ultimately two molecules of **pyruvate.** Conversion of two glyceraldehyde-3-phosphate molecules to two pyruvate molecules generates two ATP molecules by substrate-

pyruvate produced in the next-to-last step is not converted to lactate but enters a second pathway of **intermediary oxidative metabolism** that completes its oxidation to CO$_2$ and H$_2$O (Figure 4-21, A).

Because glucose is a highly polar molecule it can cross lipid membranes only with the assistance of

level phosphorylation, two molecules of NADH, and 56 kcal/mole of heat. Since two ATP molecules were used in the initial steps of glycolysis, the net yield is two ATP molecules per glucose molecule.

Conversion of one molecule of glucose to two molecules of pyruvate results in production of 2 moles of NADH. The NADH must be reoxidized if glycolysis is to continue. In the presence of oxygen and mitochondria, the pyruvate enters the pathway of intermediary oxidative metabolism, and the NADH is reoxidized by the terminal oxidative metabolism and results in a substantial additional phosphorylation of ADP. If a cell does not have the enzymes of the oxidative pathway, or if oxygen is not available, it regenerates the NAD^+ by converting pyruvate to lactate (Figure 4-21, B), with no energy gained from production of NADH.

The interconversion of lactate and pyruvate is catalyzed by an enzyme called **lactate dehydrogenase (LDH)**. Different cell types make different forms of the enzyme. One form favors production of lactate from pyruvate; this form predominates in skeletal muscle cells that depend exclusively on anaerobic glycolysis. Another form favors the conversion of lactate to pyruvate; this form predominates in heart muscle cells that depend on oxidative metabolism and even allows heart muscle cells to oxidize lactate produced by anaerobic skeletal muscle cells.

Intermediary Oxidative Metabolism: The Krebs Cycle

Pyruvate produced in aerobic glycolysis is prepared for entry into the pathway of intermediary oxidative metabolism by **decarboxylation**. This complex reaction (Figure 4-22, A) that releases CO_2, is coupled to reduction of NAD^+ to NADH, and requires the participation of **coenzyme A (CoA)** (Figure 4-22). The active site of coenzyme A is a sulfhydryl (-SH) group, so the free form of CoA is CoASH. When CO_2 is removed from pyruvate, the two remaining carbon molecules (in the form of an acetyl group) are substituted for the H of the sulfhydryl group of CoASH, forming **acetyl-CoA**. Acetyl-CoA is a carrier of acetyl groups in several important steps of energy metabolism (Figure 4-22, B).

The **Krebs cycle** (Figure 4-23) is a circular sequence of eight reactions that occurs in the interior (matrix) of the mitochondria. In the first step of the cycle, an acetyl-CoA combines with a molecule of oxaloacetate (a 4-carbon carboxylic acid) to form citrate (a 6-carbon acid). In subsequent steps, the 6-carbon backbone is decarboxylated twice, releasing two molecules of carbon dioxide; four pairs of electrons are transferred to electron-carrier coenzymes (three to NAD^+ and one to FAD); and one ATP molecule is formed by substrate phosphorylation. Since each glucose entering the glycolytic pathway forms two molecules of pyruvate, the cycle makes two turns for each glucose metabolized.

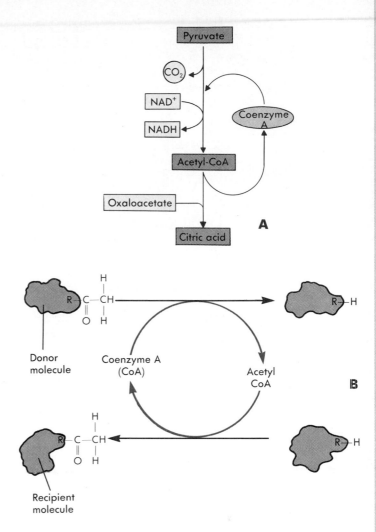

FIGURE 4-22

A, B The decarboxylation of pyruvate yields acetyl-CoA. Acetyl-CoA donates its acetyl group to oxaloacetate to form citric acid, the initial stage of the Krebs cycle.

Terminal Oxidative Metabolism and Oxidative Phosphorylation

At the point of pyruvate decarboxylation and at four points along the Krebs cycle, electrons are transferred to NAD^+ or FAD (see Figures 4-21 and 4-23). These electrons are passed to oxygen in the reactions of **terminal oxidative metabolism**. These reactions are carried out by the **electron transport chain**, a series of enzyme complexes in the inner mitochondrial membrane. Like NADH and $FADH_2$, each enzyme in the chain can exist in either an oxidized or reduced form. Some of the enzymes possess **heme** groups containing an iron atom that is oxidized or reduced, as in cytochrome C (Figure 4-24). Electrons from NADH are accepted by the first complex in the series, **NADH-CoQ reductase**, which becomes reduced. Subsequently it transfers electrons to succinate-CoQ reductase, reducing it. Succinate-CoQ reductase is also the acceptor of electrons from $FADH_2$. These transfers proceed in the same way through the chain (Figure 4-25). Each

FIGURE 4-23

In the Krebs cycle, pyruvate, amino acids, and fatty acids transfer 2-carbon segments to coenzyme A to form acetyl-CoA. The 2-carbon segments become part of a 6-carbon backbone (citrate). Successive reactions split off 2-carbon segments from the backbone to form CO_2, restoring a 4-carbon molecule that can accept another 2-carbon segment and start the cycle again. In the process, energy is captured in the form of NADH and FADH. There is also one substrate-level phosphorylation step (α ketoglutarate to succinate).

Pyruvate

Acetyl-CoA

NADH

Oxaloacetate (4) H_2O Citrate (6)

NAD⁺

Malate (4)   STAGE A PREPARATION   cis-Aconitate (6)

STAGE C REGENERATION H_2O

H_2O

Fumarate (4)   STAGE B ENERGY EXTRACTION   Isocitrate (6)

FADH₂   NAD⁺

FAD CO_2 NADH

Succinate (4) CO_2 H_2O α-Ketoglutarate (5)

ATP   NAD⁺

ADP   NADH

FIGURE 4-24

The structure of cytochrome C, one of the components of the electron transport chain. The electrons are carried by a Fe atom that is part of a heme group (box). The position of the heme group in the cytochrome is shown in red.

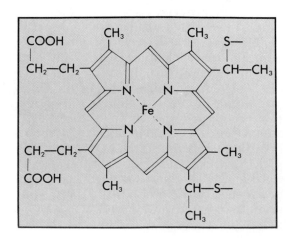

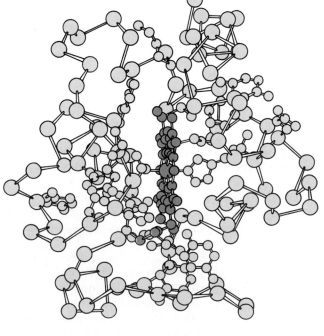

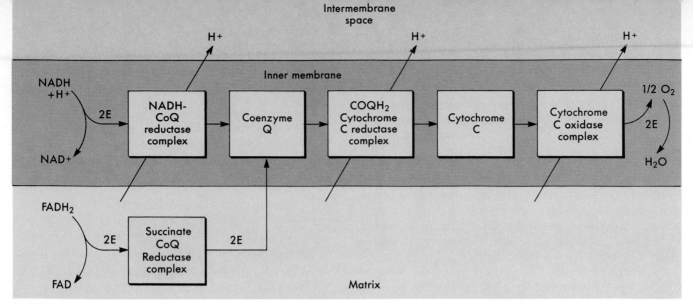

FIGURE 4-25

The complexes of the electron transport chain are located in the inner mitochondrial membrane. Electrons from NADH enter the chain at the NADH-CoQ reductase complex; those from $FADH_2$ are accepted by coenzyme Q. At three of the complexes, the energy released in transfer of electrons through the complex drives pumping of H^+ across the inner mitochondrial membrane. The last complex in the chain donates electrons to oxygen, reducing it to H_2O.

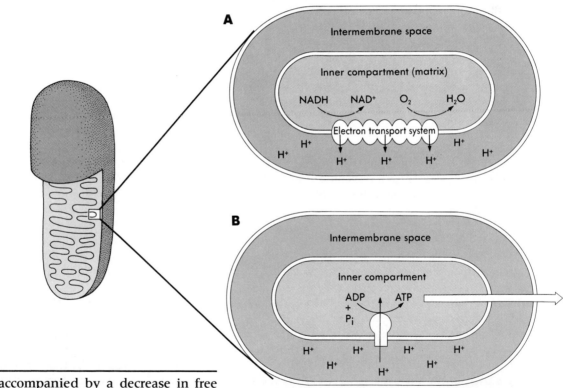

transfer step is accompanied by a decrease in free energy. The last carrier in the chain, **cytochrome C oxidase**, gives the electrons up to oxygen, reducing it to water.

Oxidation-phosphorylation coupling, the molecular mechanism by which the energy released in reoxidation of NADH and $FADH_2$ is made to drive phosphorylation of ADP, has been studied intensely for several decades. It is now clear that the reactions of terminal oxidative metabolism and phosphorylation are distinct processes carried out by separate protein complexes of the mitochondrial inner membrane (Figure 4-26). The coupling mechanism described

FIGURE 4-26

A The electron transport chain in the mitochondrial inner membrane. Electrons are transferred between a series of cytochromes and eventually combined with molecular oxygen to form water.

B Proton movement and ADP phosphorylation in mitochondria. Energy released in electron transport is used to force H^+ ions across the mitochondrial inner membrane into the intermembrane space. The H^+ (protons) may return to the mitochondrial matrix through lollipop-shaped F_0F_1 complexes that stud the inner surface of the inner membrane. In the process, the proton-motive force that drives the H^+ inward is used to phosphorylate ADP.

here, called the **chemiosmotic hypothesis**, is generally accepted even though some details are still unclear.

The key to the development of the chemiosmotic hypothesis was the discovery that reoxidation of NADH and $FADH_2$ by the electron transport chain energizes a process that pumps H^+ (protons) across the inner mitochondrial membrane from the matrix to the intermembrane space (Figures 4-25 and 4-26, *A*). As a result, the intermembrane space develops both an excess of H^+ and an excess of positive charges, compared to the matrix. The difference in H^+ concentration and the electrical difference add up to a substantial driving force that favors return of H^+ from the intermembrane space to the matrix (Figure 4-26, *B*). This driving force is called the **proton-motive force**.

Phosphorylation is carried out by F_0F_1 **complexes** that can be seen as lollipop-shaped structures on the inner surface of the mitochondrial inner membrane (Figure 4-26, *B*). The sticks of the lollipops (the F_0 units) form pores through the inner mitochondrial membrane. Hydrogen ions, driven by the proton-motive force, can return to the matrix from the intermembrane space through these pores. Each lollipop head (the F_1 unit) is analogous to a subway turnstile. A current of H^+ flows through it like a stream of commuters at rush hour. Every turn of the turnstile transfers energy from the proton-motive force to the F_1 unit, which uses it to phosphorylate ADP. Current experimental evidence suggests that at least four H^+ molecules are needed to phosphorylate one ADP molecule. In terms of the subway turnstile analogy, four commuters must push through a turnstile simultaneously to phosphorylate an ADP.

The Yield of ATP from Complete Oxidation of Glucose

Proton pumping is carried out at three sites in the electron transport chain where the transfer of electrons occurs with a large decrease in free energy: the NADH-CoQ reductase complex, the CoQ-cytochrome C reductase complex, and the cytochrome C oxidase complex (Figure 4-25). Current evidence indicates that 10 H^+ are forced across the mitochondrial inner membrane for each NADH oxidized and six for each $FADH_2$. The difference between the numbers of H^+ pumped in reoxidation of the two coenzymes is due to the fact that the electrons from NADH enter at the head of the chain, while electrons from $FADH_2$ enter the chain downstream from the first pumping site at NADH-CoQ reductase, and energize pumping only at the last two sites. Assuming that four H^+ return across the mitochondrial inner membrane per ADP molecule phosphorylated, 2.5 molecules of ADP would be phosphorylated for each molecule of NADH oxidized and 1.5 for each $FADH_2$.

Using these figures, the yield of ATP from complete oxidation of glucose to CO_2 and H_2O works out as follows:

glucose → 2 pyruvate: 2 ATP by substrate-level phosphorylation, 2 NADH

2 pyruvate → 2 acetyl-CoA: 2 NADH

2 Acetyl-CoA through Krebs Cycle: 2 ATP by substrate-level phosphorylation, 6 $NADH_2$ $FADH_2$

Totals from Glycolysis and Krebs Cycle: 4 ATP + 10 NADH + 2 $FADH_2$

The total yield of ATP per glucose should be:

4 from substrate-level phosphorylation

$10(2.5) = 15$ from reoxidation of the NADH

$2(1.5) = 3$ from reoxidation of the $FADH_2$

TOTAL 30 ATP

Actual measured yields of ATP are usually slightly less than this value. One reason is that the two NADH molecules produced by glycolysis have to cross from the cytoplasm into the mitochondria. Depending on the route, this process may require the equivalent of two ATP molecules.

The free energy of oxidation of one mole of glucose to CO_2 and H_2O is 686 kcal/mole, and the approximate free energy of hydrolysis of ATP is 7.3 kcal/mole. If cells packaged 30 kcal/mole of energy from glucose as ATP, the total energy captured as ATP would be 30 X 7.3 kcal = 219 kcal/mole of glucose, and the energy efficiency for complete carbohydrate metabolism would be about (219 kcal/686 kcal) × 100 = 32%. The First Law of Thermodynamics (Chapter 2) dictates that energy is conserved; therefore, in complete metabolism of glucose the 467 kcal of energy/mole that is *not* captured in the form of ATP is released as heat. Complete metabolism of glucose to CO_2 and H_2O yields much more energy as ATP (and also much more heat) than incomplete metabolism of glucose to lactate.

Energy Metabolism of Fatty Acids, Amino Acids, and Nucleic Acids

The Krebs cycle is the final common path for complete oxidation of all the major classes of energy-yielding molecules: lipids, proteins, nucleic acids, and carbohydrates. The major energy-yielding pathways are summarized in Figure 4-28. Dietary lipids and the lipids stored in adipose tissue represent particularly potent sources of energy; the yield of ATP per gram of fat is about six times greater than that of hydrated protein or glycogen. Catabolism of body proteins and stored lipid is under hormonal control, as described in Chapter 23.

The first step in energy metabolism of acylglycerols is hydrolysis in the cytoplasm to release fatty acids and glycerol. The glycerol enters the glycolytic pathway. The fatty acids enter mitochondria and are converted to multiple molecules of acetyl-CoA by a

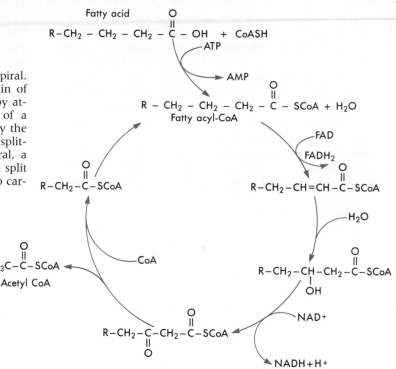

FIGURE 4-27
The pathway of beta oxidation of fatty acids forms a spiral. The R of the fatty acid stands for a hydrocarbon chain of variable length. The fatty acid must first be activated by attachment to coenzyme A—this involves formation of a -S-C- bond by dehydration; the process is energized by the splitting of an ATP to AMP (the energetic equivalent of splitting two ATPs to ADPs). With each turn of the spiral, a NADH and a FADH$_2$ are formed and an acetyl-CoA is split off, leaving the hydrocarbon chain of the fatty acid two carbons shorter.

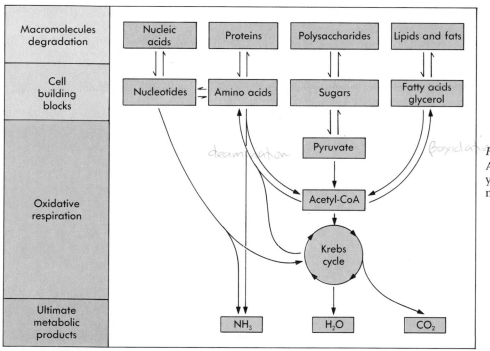

FIGURE 4-28
A summary of the energy-yielding pathways of cellular metabolism.

cyclical process called **beta oxidation** (Figure 4-27). First the fatty acid is primed for entry into the cycle by formation of fatty acyl-CoA, a reaction that requires a one-time-only investment of the equivalent of 2 ATPs. The cycle has four steps. In the first steps of the cycle, a FADH$_2$ and a NADH are formed and an oxygen is attached to the third carbon of the fatty acyl-CoA. Next an acetyl-CoA is split off from the original fatty acid. Finally, the fatty acid, now two carbons shorter, accepts another CoA, and the cycle begins again.

Every turn of the beta oxidation cycle yields an acetyl-CoA, a FADH$_2$ and a NADH. As an example, the net yield from an 18-carbon fatty acid is 9 NADH + 9 FADH$_2$ + 9 acetyl CoA − 2ATP. The acetyl-CoA enters the Krebs cycle, and the reduced coenzymes produced are reoxidized by the terminal oxidative metabolism. Using the same values for yields of ATP from reduced coenzymes that were given in the description of glucose metabolism above, the final yield of ATP from the 18-carbon fatty acid can be calculated to be 124.

Carbons from protein enter the pathways of energy metabolism under two conditions: (1) When

The Structure and Energy Metabolism of Cells

dietary protein intake exceeds the rate at which the body is synthesizing protein for growth and repair, the excess protein is catabolized for energy. (2) During fasting some body proteins are catabolized along with fat and glycogen. The first step in either case is conversion of protein to free amino acids; dietary proteins undergo this in the intestine before they can be absorbed into the bloodstream. The next step for both dietary and body proteins is **deamination**— removal of the NH_3 group to yield a carboxylic acid. The carboxylic acids released from most amino acids can be converted either to pyruvate or one of the intermediates of the Krebs cycle. The yield of ATP from different types of amino acids varies, depending on the level at which the resulting carboxylic acid enters the Krebs cycle. At most, a single amino acid could yield the 12.5 ATP that result from metabolism of 1 pyruvate through the Krebs cycle. Nucleotides from degradation of nucleic acids can undergo a similar deamination followed by conversion to Krebs cycle intermediates.

CONTROL OF ENERGY METABOLISM
Respiratory Control by ADP

In isolated mitochondria provided with adequate amounts of NADH and $FADH_2$, oxygen and phosphate, oxygen uptake and reoxidization of the reduced coenzymes quickly comes to a stop if ADP is not also added. The reason is that when the proton-motive force is not continuously drawn upon by phosphorylation, the H^+ pumps of the electron transport chain rapidly build up a proton-motive force so great that no more H^+ can be forced out into the intermembrane space. As a result, the levels of reduced intermediates increase and electron transport stops. If ADP is added to mitochondria under these conditions, there is a burst of oxygen uptake and phosphorylation that continues until the ADP concentration has returned to a low value. This experiment simulates what would happen in mitochondria of a cell undergoing a sudden transition from rest to activity; for example, in a resting muscle cell that was stimulated to contract. The control of terminal oxidative metabolism by ADP is called **respiratory control.**

Control of Enzymes by ADP, ATP, and Metabolic Intermediates

In an intact cell, a rise in ADP stimulates the reactions of glycolysis and the Krebs cycle, and the terminal oxidative metabolism. The increase in the intermediary reactions would increase the flow of reduced coenzymes to the terminal oxidative metabolism, ensuring that there would always be enough proton-motive force to meet the increased demand. This coordinated response to changes in cellular activity level is achieved by allosteric control of key enzymes of glycolysis and the Krebs cycle. The allosteric controls typically make the enzymes responsive both to the intracellular levels of ATP and ADP and to the rates of other key reactions in the pathways.

This control may be illustrated by an important example, phosphofructokinase. This enzyme catalyzes the third step in glycolysis, the conversion of fructose-6-phosphate to fructose-1,6-disphosphate. The activity of phosphofructokinase usually limits the rate of the entire glycolytic pathway. Phosphofructokinase is inhibited by NADH, citrate, and ATP, and is activated by ADP. In resting cells, NADH, citrate, and ATP are at high levels. An increase in cellular utilization of ATP stimulates phosphofructokinase both directly, and indirectly by way of the decreased levels of citrate and NADH that result from the stimulation of oxidative metabolism.

● STUDY QUESTIONS

1. Describe the function of each of the following organelles:
   Nucleus
   Cell membrane
   Golgi apparatus
   Mitochondria
   Ribosome
   Endoplasmic reticulum
   Lysosomes

2. What are the functional and structural differences between gap junctions, tight junctions, and desmosomes?

3. What is meant by the terms "tight" and "leaky" epithelia?

4. Red blood cells lack nuclei and mitochondria. How does this affect the functions they can perform?

5. What are the three major metabolic pathways that contribute to phosphorylation of ADP? What is the functional role of each pathway in the body?

6. What is the advantage of aerobic metabolism over anaerobic metabolism in energy production in living systems?

7. How is energy liberated in the electron transport chain? Where do the electrons come from and what is their final destination?

8. The end products of aerobic glycolysis are CO_2 and H_2O. Where do the oxygens in the CO_2 come from? Where do the oxygens in the water come from?

9. What are the routes by which proteins and fats can be metabolized? How does the energy yield compare with carbohydrate metabolism?

Choose the MOST Correct Answer.

10. Proteins are modified into glycoproteins in this cellular structure:
    a. Endoplasmic reticulum
    b. Ribosome
    c. Golgi apparatus
    d. Lysosome

11. Pyruvic acid is converted to CO_2 and H_2O in this cytoplasmic organelle:
    a. Golgi apparatus
    b. Mitochondria
    c. Endoplasmic reticulum
    d. Ribosome

SUMMARY

1. The major organelles of cells are:
   A. The **cell membrane,** which serves as a selective barrier between cytoplasm and extracellular fluid.
   B. The **nucleus** (actually an assembly of organelles), which contains the DNA.
   C. The **endoplasmic reticulum** and **Golgi apparatus,** which serve as the site of synthesis of cell membrane elements and of proteins destined for secretion by exocytosis.
   D. **Lysosomes,** which are the source of enzymes for intracellular digestive processes.
   E. **Mitochondria,** the sites of the **Krebs cycle** of intermediary oxidative metabolism, of **terminal oxidative metabolism,** and of **oxidative phosphorylation** of ADP.
   F. The **cytoskeleton,** which determines the forms of cells and is involved in cellular movement.
2. Three types of junctions couple cells together: **desmosomes,** which contribute to the physical integrity of tissues, **tight junctions,** which contribute to the barrier function of epithelial tissues, and **gap junctions,** which serve as routes of direct communication between adjacent cells.
3. Carbohydrates can be metabolized to lactate in the absence of oxygen, with the capture by **substrate-level phosphorylation** of a small fraction of the potential energy of the carbohydrate.
4. In the presence of oxygen, **pyruvate** formed by glycolysis is converted to **acetyl-CoA,** which enters the **Krebs cycle** (the pathway of **intermediary oxidative metabolism**). Beta-oxidation of fatty acids also produces acetyl-CoA. Amino acids can contribute to the Krebs cycle by deamination followed by conversion to pyruvate or to a Krebs cycle intermediate. Nucleic acids may be metabolized for energy by depolymerization to nucleotides, which are converted to Krebs cycle intermediates by deamination.
5. Complete metabolism of substrates to CO_2 and water (oxidative metabolism) involves the use of oxygen as an electron receptor. In the **chemiosmotic hypothesis** of **oxidative phosphorylation,** reducing power captured by **NADH** and **FADH$_2$ is converted to a proton-motive force** across the mitochondrial inner membrane by the **electron transport chain** in **terminal oxidative metabolism.** The proton-motive force is used to drive oxidative phosphorylation, yielding ATP with high efficiency.
6. The rate of reoxidation of NADH and FADH$_2$ by mitochondria is regulated by cellular levels of ADP (**respiratory control**). The rates of the reactions of glycolysis and intermediary oxidative metabolism are regulated at key steps by allosteric activators and inhibitors, including ADP, ATP, NADH, and other key metabolic intermediates. These controls coordinate the rates of reactions within the pathways, and allow the energy metabolism of a cell to respond almost immediately to an increase in demand for ATP.

12. Which of the following is NOT a function of the cytoskeleton?
    a. Determines the shapes of cells
    b. Holds some proteins of the cell membrane in stable positions
    c. Allows some cells to move
    d. Forms an extensive intracellular membrane system

13. The intercellular junction designed for greater strength to maintain tissue integrity:
    a. Desmosome
    b. Gap junction
    c. Tight junction
    d. Plasmadesmata

14. Which of the following pertains to substrate-level phosphorylation?
    a. Occurs in the inner mitochondrial membranes
    b. Requires oxygen
    c. Produces most of the ATP used by the body
    d. These reactions are collectively called anaerobic metabolism

15. This compound is common to the oxidative pathways of carbohydrates, fats, and proteins:
    a. Acetyl C$_o$A
    b. Lactate
    c. Pyruvate
    d. Glycerol

16. Respiratory control of the reoxidation rate of NADH and FADH$_2$ is dependent upon cellular levels of:
    a. ATP
    b. ADP
    c. Oxygen
    d. H$^+$

● *SUGGESTED READINGS*

BRETSCHER MS: How animal cells move. *Scientific American* December 1987, pp. 72-90. Explains the common basis for cell movement and endocytosis-exocytosis.

DARNELL J, LODISH H, BALTIMORE D: *Molecular cell biology,* ed 2, New York, 1990, Scientific American Books. An excellent high-level text with detailed coverage of the topics of this chapter.

DULBECCO R: *The design of life,* Yale University Press, New Haven, Conn., 1987. A noted cell biologist gives his view of cellular function.

PRESCOTT DM: *Cells: principles of molecular structure and function,* Boston, 1988, Jones & Bartlett. A readable text covering the content of this chapter.

Homeostatic Control: Neural and Endocrine Control Mechanisms

On completing this chapter you should be able to:

- Describe the elements of a negative feedback control system and show how such systems stabilize physiological variables.
- Understand how positive feedback leads to instability.
- Provide examples of intrinsic and extrinsic control processes in the body.
- Recognize several common mechanisms of intrinsic regulation.
- Appreciate the variety and importance of paracrine and autocrine agents.
- Distinguish between somatic, autonomic, and hormonal reflexes.
- State the characteristics of hormones and describe the three major chemical classes of hormones.
- Understand what is meant by receptor down-regulation and up-regulation.
- Describe the factors controlling hormone release from the posterior pituitary. Describe the role of releasing and release-inhibiting hormones in control of anterior pituitary hormone secretion.
- Provide examples of several second messenger systems.
- Compare the intracellular mechanisms of action of hormones with nuclear receptors and those with cell surface receptors.

*T*he human body is a self-controlling unit. Within the unit, regulated variables usually remain within a predetermined range of values, in spite of outside influences that tend to force departures from this range. This is not to imply that regulated variables do not change. The temperature of the body core is a regulated variable that changes slightly in exercise and cold or heat exposure, more so over the body's daily cycle, and more yet during fever. Control is also important during changes of state; in "voluntary" movements, the length and tension of skeletal muscle and the position of joints are under constant control. Even minor departures from the predicted path of a hand or foot are corrected in midcourse, so that the feet usually find the stairtreads and the cup usually reaches the lip without spilling.

The principles of control that are important to physiology were first worked out by engineers and physical scientists. Biological machines can be shown to follow the same control-system principles that were first put to use in such devices as the thermostat and the governor that regulated the speed of primitive industrial engines, and later in the electronic amplifier and the airplane autopilot.

What is most striking about biological control systems is their formidable complexity and the enormous range of size and time scales over which they operate. Intracellular control systems frequently involve a multistep cascade of chemical reactions that amplifies a single initiating event thousands of times. A similar complexity is seen in the central nervous system, in which millions of neurons may be involved in as "simple" an act as walking up stairs. Some control systems in the nervous system operate on the time scale of milliseconds; intracellular regulatory processes operate at the size scale of individual molecules or ions. On the other end of the time and size scales, the developmental plan of the human body, orchestrated on many levels of organization and involving many billions of cells, is fulfilled on a time scale of decades.

As physiologists have increased their understanding of body function, function and control have become inseparably linked. Several generations of physiologists had been accustomed to think of a single endocrine system consisting of several types of ductless glands secreting chemical messages into the blood. Recent developments have made such a view less realistic. It is now clear that essentially all tissues secrete chemical messengers; a rich traffic of chemical communication goes on between neighboring cells, between different tissues in the same organ, and between different organs. The same basic principles apply whether the chemical message reaches its target by diffusion in interstitial fluid, by way of the bloodstream, or by release from nerve endings, and the same chemical compounds frequently serve in more than one of these roles.

CONTROL THEORY
Feedback

- *Why is it that negative feedback cannot keep the regulated variable exactly at the setpoint in the presence of perturbing factors?*
- *How is the definition of gain of a negative feedback system defined?*
- *Why does the regulated variable in some negative feedback systems show a tendency to oscillate around the setpoint?*
- *What factors might affect the amplitude and frequency of the oscillation?*
- *Why is positive feedback characteristically destructive?*
- *In what situations is positive feedback useful?*

The term **feedback** is used for a variety of situations in which the rate of a process is affected by its end products or consequences. In **negative feedback** the end product or consequence diminishes the rate of the process, which tends to stabilize the system. Negative feedback could be summarized as "The more product or result you have, the less you get." In **positive feedback** the product or consequence accelerates the process, which drives systems to their extremes. Positive feedback could be summarized as "The more product or result you have, the more you get."

Negative feedback does not necessarily imply active control. For example, a weight suspended from a spring tends to return to its equilibrium position after it is displaced. Similarly, the Law of Mass Action tends to keep levels of metabolic intermediates constant because any given reaction in a pathway slows when the levels of its precursors fall or those of its products rise. These are examples of **passively controlled** systems in which negative feedback arises from the properties of the system itself rather than through the intervention of a separate control system. However, there are very few examples of purely passive control in physiology. Even the example above of enzyme-catalyzed reactions is not purely passive because, as noted in Chapter 3, key enzymes are typically under multiple allosteric controls. Greater stability and sensitivity are possible with active control systems that sense changes in the state of the system and intervene to apply a greater sensitivity than can be achieved with purely passive control.

Negative Feedback Loops

Homeostasis demands that important physiological parameters such as body temperature, blood composition, and blood pressure be maintained within appropriate limits. In the language of control theory, these are **regulated variables**. Negative feedback is the principle used by control systems that oppose the departure of a **regulated variable** from its appropriate value. Negative feedback is not restricted to keeping regulated variables constant. In skeletal muscle negative feedback is used for keeping muscle length constant, as when one maintains an erect position against the pull of gravity, but also for controlling body movements in which the lengths of muscles change. In the latter case the control system could be thought of as functioning like the autopilot of an airplane, making appropriate corrections when a moving appendage deviates from its predetermined path through space.

All of the many different physiological control systems incorporate the following features: The regulated variable is monitored by **sensors** or **receptors** that pass information to an **integrator.** The integrator compares the sensor's input with the setpoint. If there is a difference between the sensor's input and the setpoint, the integrator generates an **error signal** that is usually proportional to the magnitude of the difference. The error signal activates **effectors** that oppose the departure from the setpoint. The effector's response completes a **negative feedback loop** that runs from the regulated variable through the sensor to the integrator and back to the regulated variable by way of the effector (Figure 5-1, *A*). Such systems are sometimes called **closed loop systems.** In physiological control systems the sensors may be nerve or gland cells, the integrators may be at almost any place in the body. The error signal may take the form of nerve impulses or chemical substances carried in the blood over long distances (**hormones**) or diffusing over short distances between cells (**paracrine** and **autocrine agents**). The effectors may be muscles or **exocrine glands**—those whose secretions pass to the physiological exterior.

A Practical Example: Driving

A car and driver form a feedback control system (Figure 5-1, *B*). The regulated variable is the position of the car in its lane. The sensors are the eyes of the driver; the driver's brain (the integrator) constantly compares the car's position with the ideal or setpoint path down the exact center of the lane. Deviations from the ideal path that occur as a result of **perturbing factors,** such as bumps or curves in the road, are opposed by a system of effectors that includes the muscles of the driver and the car's steering wheel (car does not stay in the center of the lane, but wanders a little from side to side as errors are followed by corrections (Figure 5-2, *A*).

The error signal generated by the integrator is proportional to the difference between the setpoint and the value of the regulated variable. The body's effectors are usually capable of making larger or smaller efforts in response to larger or smaller error signals. In the driving analogy, the farther the car is from the center of the lane, the larger effort the driver makes to steer back toward the center.

The setpoint in physiological systems may be

THE FOUNDATIONS OF PHYSIOLOGY

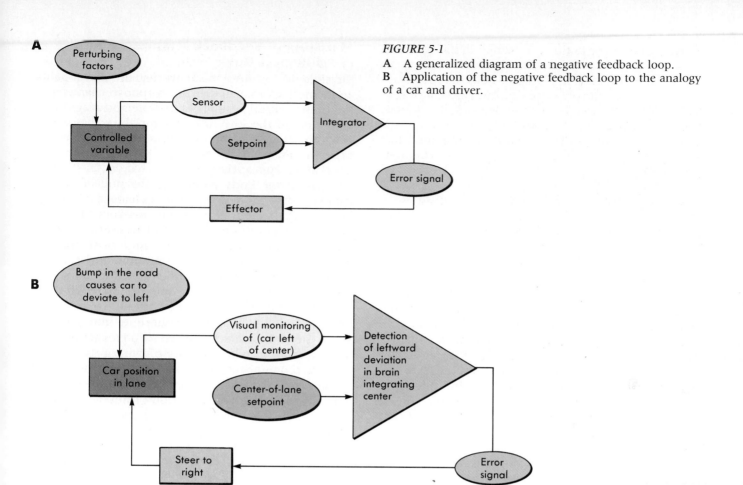

A

Perturbing factors

Sensor

Setpoint

Integrator

Controlled variable

Error signal

Effector

FIGURE 5-1
A A generalized diagram of a negative feedback loop.
B Application of the negative feedback loop to the analogy of a car and driver.

B

Bump in the road causes car to deviate to left

Visual monitoring of (car left of center)

Center-of-lane setpoint

Detection of leftward deviation in brain integrating center

Car position in lane

Steer to right

Error signal

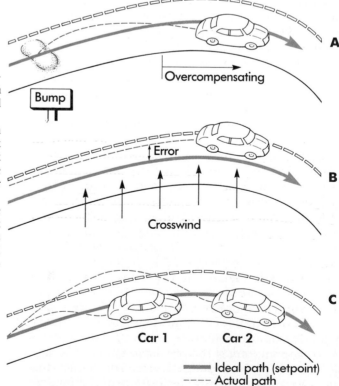

FIGURE 5-2

A The actual path followed by the car under negative feedback control deviates from the ideal or setpoint path down the center of the lane as perturbing factors are compensated by corrections.

B A steady perturbation, such as a cross-wind, causes a small, steady error between the actual path and the setpoint path. The error drives a continuous steering correction that is just enough to compensate for the effect of the wind. The higher the gain of the feedback loop, the smaller the error needs to be.

C The more rapid the response of the feedback loop, the shorter the time lag between a deviation from the setpoint and the beginning of the corrective response. Car 1 responds more rapidly than Car 2 to the same perturbing factor and also stops correcting more promptly when it has returned to the setpoint, so Car 1 stays closer to the ideal path.

Homeostatic Control: Neural and Endocrine Control Mechanisms

changed from time to time. A change in the setpoint might correspond to a lane change in the analogy of the car and driver. For example, body temperature is regulated at a lower value during sleep and at a higher value during fever. The setpoint for blood pressure varies with changes in levels of physical activity and arousal. In women, the setpoint for body temperature also varies predictably during the menstrual cycle.

Gain and Time Lag In Negative Feedback Systems

Two characteristics of the control system determine how closely the regulated variable will hover around the setpoint. The first characteristic, the **gain** of the system, is the degree to which the output of an effector is altered for a given amount of change in the regulated variable. Gain could be thought of as leverage built into the control system. The higher the gain of the system, the more difficult it is for perturbing factors to force the regulated variable away from the setpoint. In the driving analogy (Figure 5-2, *B*), a steady wind blowing across the path of the car from right to left would cause the car to deviate toward the left side of the lane. In the absence of negative feedback, the car would soon leave the road. With negative feedback operating, there is a small, constant error between the ideal path down the center of the lane and the actual path of the car. This difference between the actual and ideal paths corresponds to an error signal sufficient to stimulate a correction that just matches the perturbing force of the wind. The greater the gain of the feedback system, the smaller the deviation would be. In this analogy, power steering would increase the gain of the feedback system over manual steering. However, no matter how high the gain of the system, the actual path can never follow the setpoint exactly because no compensation can occur without a departure from the setpoint.

The second determining characteristic of the system's responsiveness is its **time lag**—the time that elapses between the departure of the regulated variable from the setpoint and the beginning of the response of effectors. Time lags may have several origins: the time needed to send sensory information to the integrator, the time for the signal to be processed in the integrator, the time needed for the command to travel to the effector, and the delay in the response of the effector itself. In the body, messages carried by nerve impulses may involve time lags of only a fraction of a second. However, if hormones are used to carry information, the delay due to the transit time of the messages may be much longer. Time lags may cause the system to oscillate around the setpoint if the regulated variable is rapidly altered by perturbing factors. The longer the time lag is, the greater the magnitude of the oscillation is (Figure 5-2, *C*).

Matching of Feedback Response to Physiological Role

Physiological control systems usually have gains and time lags that match the response characteristics of the control systems to the rate and magnitude of change of the physiological variables they regulate. For example, the system that controls blood pressure (Figure 5-3) has a large gain and little time lag. This is appropriate because changes in the orientation of the body relative to the pull of gravity could cause rapid and substantial changes in arterial blood pressure. The sensors for this control system are in the aorta and in the carotid arteries that serve the head. When a person stands up, the control system usually responds rapidly enough to keep the blood pressure in these arteries very close to the setpoint. If disease weakens the gain of the control system, blood pressure in the arteries serving the head cannot be maintained close to the setpoint when the body is erect; the condition sometimes results in fainting when the person stands up.

The refinements in feedback systems discussed before usually ensure that the error is small. These small deviations from the setpoint can be thought of as an acceptable cost of avoiding the much larger changes that would occur if the variable were not feedback regulated. However, in some cases even the small error can have significant consequences for health. If one begins to eat a lot of salt, the control systems that regulate extracellular fluid composition will react within a few days and cause the kidney to increase the rate of salt loss in urine. However, the gain of the renal control system is not infinite, and the consequences of a continued high salt intake will be an elevation of the salt content of the body by 1% to 2%. This small elevation constitutes the residual error, and it may be a contributing factor in certain types of high blood pressure (hypertension).

Most physiological control systems monitor several variables because of the complex interactions between different organ systems. The cardiovascular control center receives information on arterial blood pressure, blood volume, and oxygen and carbon dioxide content. In muscle control, both the absolute position and direction of movement of the limbs are constantly assessed. In both of these examples the relative importance of each regulated variable must be evaluated by the integrator to determine an optimum response.

In physiological systems, feedback may be applied by two sets of effectors that have opposite effects on the regulated variable (Figure 5-4). The most obvious examples are in the skeletal muscle system, in which the position of a limb is controlled by muscles that work to move it in opposite directions. Another example is the heart, which is inhibited by input from the parasympathetic nervous system and stimulated by input from the sympathetic

THE FOUNDATIONS OF PHYSIOLOGY

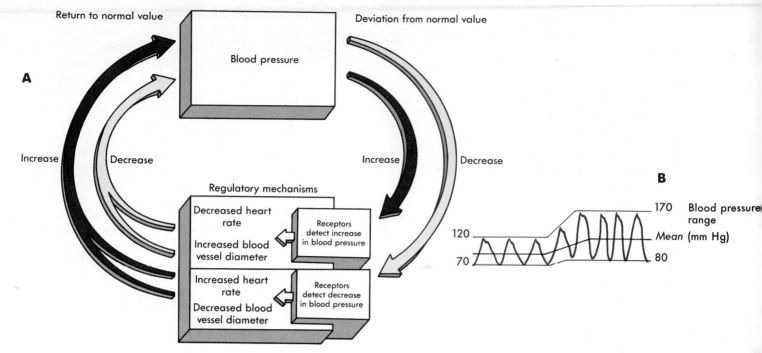

FIGURE 5-3

An example of negative feedback.

A Mechanisms involved in blood pressure regulation. Increases in blood pressure (*red arrow* on the right) cause changes in heart rate and vessel diameter that decrease the blood pressure toward its normal value (*yellow arrows* on left). Decreases in blood pressure (*yellow arrow* on the right) cause changes in heart rate and vessel diameter that increase the blood pressure toward its normal value (*red arrows* on left).

B Arterial blood pressure rises from a lower (diastolic) value to a higher (systolic) value with each heartbeat. The sensors of the blood pressure regulation loop average the pressure so that mean pressure is the regulated variable. In exercise the demand for oxygen by muscle tissues is increased. This is met in part by an increase in mean arterial blood pressure. The increased blood pressure is not an abnormal or nonhomeostatic condition but is simply a resetting of the normal homeostatic range.

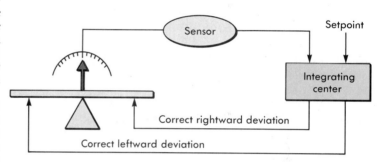

FIGURE 5-4

A control system with two sets of effectors; one increases the regulated variable and the other decreases the regulated variable. Opposing effectors are typical in physiological systems.

nervous system. Both branches are under the control of an integrating center in the brainstem; a reflexive increase in heart rate involves both an increase in sympathetic input and a decrease in parasympathetic input.

Positive Feedback in Disease and Normal Function

Negative feedback systems stabilize variables near their setpoints because the response of the effector minimizes the error signal. In positive feedback, a change in the regulated variable causes the effector to drive it even further away from the initial value. Systems in which there is positive feedback are highly unstable. The effect of the initial perturbation is analogous to that of a spark that

ignites an explosion. Note that positive feedback systems are distinct from open loop, uncontrolled systems.

Because it causes instability, positive feedback is usually an undesirable trait for physiological systems. Destructive positive feedback commonly occurs in disease, where it results in very rapid deterioration of homeostasis. For example, in certain types of heart disease, the heart becomes overloaded and cannot pump out all the blood returning to it. The volume of the heart increases, and this volume increase further diminishes the ability of the heart to pump blood. The end result is the conversion of a stabilizing negative feedback loop into an unstable positive feedback loop, with dangerous consequences for the patient.

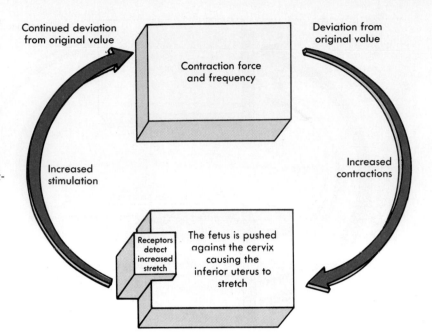

FIGURE 5-5
The positive feedback loop that increases contractions of the uterus in childbirth.

Positive feedback is not always abnormal. It may sometimes force a rapid buildup that can serve a signal function. For example, in the female reproductive cycle, increases in the production of estrogen by ovarian follicles is promoted by estrogen, and the large increases in estrogen then serve as a signal that leads to a surge of luteinizing hormone release, one of the hormones that support follicle development. These examples of positive feedback are self-limiting, in that the development of the follicle is brought to an end by release of the egg that the luteinizing hormone causes (Chapter 25). Another example is a nerve impulse, in which the opening of a few sodium channels causes the remainder to open (Chapter 6). Other examples include the normal response of the immune system to bacteria and viruses (Chapter 24) and the process of formation of a blood clot (Chapter 14). Positive feedback is useful in expulsive processes such as the contractions of the uterus in childbirth, in which a positive feedback loop increases uterine contractions in response to the pressure of the baby's head on the cervix (Figure 5-5). Typically responses driven to an extreme by positive feedback are either self terminating (like labor contractions) or are brought back into control by separate negative feedback systems.

Open Loop Systems

Open loop systems (Figure 5-6) have neither negative feedback control nor any positive feedback. In the analogy of car and driver, the negative feedback loop is opened if the driver shuts his eyes or takes

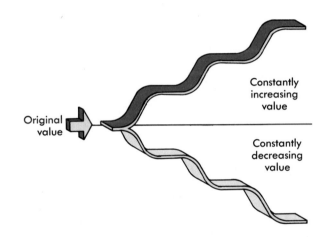

FIGURE 5-6
Open loop systems continue to change in one direction once perturbed from their original value.

his hands off the steering wheel. Deviations from the setpoint cannot be corrected, and the car will soon leave the road. Open loops can result from disease or damage to some part of the feedback loop. For example, damage to parts of the motor control system of the brain may result in uncontrolled body movements, as in Parkinson's disease (see Chapter 11). However, open loops are part of the normal function of some physiological systems. Very rapid body movements, such as the movements that the eyes make to follow an object when the head moves, the pitcher's pitch and the rapid movements of the pianist's fingers, must be carried out according to a learned pattern because they must be completed before feedback can be effective.

THE FOUNDATIONS OF PHYSIOLOGY

Levels of Physiological Regulation

- *How do intrinsic and extrinsic regulation differ?*
- *What are three possible mechanisms of intrinsic regulation?*
- *How do autocrine and paracrine agents differ?*
- *Why were many growth factors discovered only when attempts were made to grow cells in culture?*

Regulatory processes operate on all levels of integration, from single molecules to the whole body; all regulatory processes ultimately have their actions at the molecular level. Regulatory processes that involve tissues, organs, or organ systems can be divided into two great categories: (1) **intrinsic regulation,** or **autoregulation,** and (2) **extrinsic regulation.**

Intrinsic regulation, or autoregulation

Both names mean "self-regulation." This level of integration includes all regulatory processes that do not involve structures outside the organ or tissue being regulated. Intrinsic regulation integrates the growth and activities of cells or tissues within regions of an organ. This gives tissues or organs the ability to respond to changing conditions on a scale too small to be dealt with effectively by extrinsic nervous or hormonal controls. In some cases, auto-

regulation is the result of the properties or behavior of individual cells and there is no extracellular signal. In other cases, the signal between neighboring cells is an electrical one carried by gap junctions. However, in most cases the messages of intrinsic regulation are chemical substances released by one cell type that affect the activity of nearby cells, either of the same cell type (**autocrine agents**) or another cell type (**paracrine agents**).

Extrinsic regulation

"Extrinsic" means "from outside"; this is the level of regulation in which the activity of tissues and organs is controlled by signals that originate at a distance from the tissues or organs being controlled. The extrinsic regulatory signals to and from the controlled tissue may be carried by electrical impulses in nerves called action potentials (explained in detail in Chapter 6) or by **hormones**—chemical substances released into the blood that have their actions at a distance from the site of secretion. Extrinsic control by the nervous system and hormones could be characterized as "executive" or "top-down," although there are almost always pathways of negative feedback through which the effectors "talk back" to the executive.

Almost all tissues exhibit regulation on both

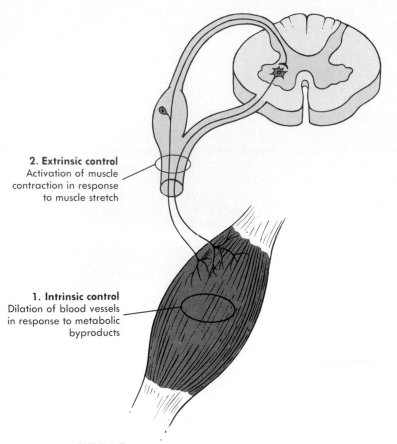

2. Extrinsic control
Activation of muscle
contraction in response
to muscle stretch

1. Intrinsic control
Dilation of blood vessels
in response to metabolic
byproducts

FIGURE 5-7
Intrinsic and extrinsic levels of regulation in an example tissue, skeletal muscle. Both levels of regulation ultimately act through events that take place at the molecular level of integration.

levels (Figure 5-7), but the relative importances of intrinsic and extrinsic regulation vary from tissue to tissue. The heart and intestine are examples of organs that show a large measure of intrinsic control. Skeletal muscle is perhaps the prime example of a tissue dominated by extrinsic control from the central nervous system.

MECHANISMS OF INTRINSIC REGULATION
Cellular Self-Regulation

In some cases, intrinsic regulation of whole tissues or organs is the result of the properties of the individual cells rather than any regulatory signals that pass between them. The heart is under extrinsic control but it is also a good example of intrinsic regulation that arises from the properties of the individual cells. In the absence of any extrinsic input, each contraction of the heart or heartbeat, tends to pump out a volume of blood equal to the volume that entered the heart during the preceding relaxation. This self-regulating negative feedback is the result of a direct relationship between stretching and force in heart muscle. As described in more de-

tail in Chapter 13, adjustments of contractile force in response to stretch are made independently by each heart muscle cell; the intrinsic regulation of the heart is the sum of these independent decisions. In hearts weakened by disease, the ability of the cells to generate additional force in response to stretch is impaired, and ultimately additional stretching results in less rather than more contractile force. This condition, called congestive heart failure, was presented earlier in the chapter as an example of destructive positive feedback.

Tissue Coordination by Gap Junctions

In many tissues, adjacent cells are coupled by gap junctions (see Chapter 4). These connections allow passage of both chemical substances and electrical current from cell to cell. The coordinating effect of gap junctions is particularly apparent in two types of muscle—cardiac (heart) muscle and single-unit smooth muscle (see Chapter 12). In muscle, contraction is initiated by electrical impulses called action potentials that are propagated across the cell surface. In cardiac and single-unit smooth muscle, an action potential initiated in any one of the cells spreads by flow of electric current through gap junctions to all cells coupled to it. As a result, all of the coupled cells contract in unison.

Chemical Regulation at the Tissue Level: Paracrine and Autocrine Agents

It is now clear that all tissues carry out a continuous internal chemical conversation between cells mediated by paracrine and autocrine agents that affect growth and differentiation as well as function. These agents include a large number of different compounds, and doubtless many more remain to be discovered. Some classes of molecules now recognized as important autocrine and paracrine agents

TABLE 5-1	*Effects of Prostaglandins*
Prostaglandin	*Physiological effects*
Tx$_{A2}$ (Thromboxane)	Constricts blood vessels
	Promotes blood clotting
PGI$_2$ (Prostacyclin)	Inhibits blood clotting
	Dilates blood vessels
PGE$_2$	Stimulates inflammation
	Dilates blood vessels
	Increases urine formation
	Contracts uterus
	Inhibits acid secretion in the stomach
PGF$_2$	Dilates airways
	Constricts blood vessels
	Contracts uterus

TABLE 5-2   Growth Factors

Name	Origin	Effect
Epidermal growth factor	Salivary and other glands	Growth of epidermal and nerve cells
Nerve growth factor	Tissues innervated by sympathetic neurons	Growth of sympathetic and sensory neurons
Insulin-like growth factors 1 & 2 (somatomedins 1 & 2)	Liver	Induced by growth hormone; stimulates growth and metabolism
Interleukins	T cells and macrophages of immune system	Autocrine agents for T cell growth
Granulocyte-macrophage colony-stimulating factor	T cells, endothelial cells and fibroblasts	Differentiation of stem cells for white blood cells
Transforming growth factors <alpha> and <beta>	Embryonic cells	Autocrine agent for embryonic growth, implicated in tumor growth
Platelet-derived growth factor	Blood platelets	Stimulates growth of connective tissue and smooth muscle in damaged blood vessels

include eicosanoids, growth factors, histamine, endothelins, and adenosine.

Eicosanoids are 20-carbon fatty acids derived from arachidonic acid. Eicosanoids include prostaglandins, prostacyclins, thromboxanes, and leucotrienes (see Chapter 3 for structures). Eicosanoids have been implicated in a large number of local regulatory functions (Table 5-1), including inflammation and blood clotting (see Chapter 14), ovulation, menstruation, and labor (see Chapters 25 and 26), and secretion of acid by the stomach (see Chapter 21). The initial step in prostaglandin production is the splitting off of arachidonic acid from a cell membrane phospholipid by a membrane-bound phospholipase enzyme. The next step, catalyzed by the enzyme **cyclooxygenase,** forms the 5-carbon ring characteristic of prostaglandins and thromboxanes. Cyclooxygenase is inhibited by the drugs aspirin and acetaminophen, which is why these drugs are effective against inflammation. At the same time, blockage of PGE_2 formation by aspirin and other aspirin-like drugs promotes increased acid secretion in the stomach, therefore large doses of these drugs can cause stomach ulcers.

Growth factors are polypeptides that regulate cell division and direct the growth of tissues in development and repair. The first growth factor to be discovered was **nerve growth factor (NGF)** (see box essay), which directs the neurons of the sympathetic branch of the autonomic nervous system (see Chapter 11) to grow toward and form connections with their target tissues. This factor is also important for normal development of the kidney. **Platelet-derived growth factor (PGDF)** is released by clot-forming blood cells (platelets) at the site of an injury to a blood vessel. PGDF stimulates growth of connective tissue and smooth muscle, promoting repair. Inappropriate secretion of PGDF may be a cause of the abnormal growth of arterial smooth muscle that causes narrowing of arteries in atherosclerosis. Production of **insulin-like growth factors (IGFs,** also called **somatomedins)** by the liver and other tissues is stimulated by **growth hormone** secreted by the anterior pituitary (see Chapter 23). IGFs stimulate growth and cell division and also uptake of glucose and amino acids by cells. These and additional growth factors are described in Table 5-2. Growth factors frequently act as autocrine agents, stimulating the growth and differentiation of the cells that secrete them.

Histamine is a derivative of the amino acid histidine. Mast cells scattered within tissues and basophils (one type of white blood cell) release histamine at sites of infection, as part of a general response to bacterial invasion (**inflammation**). The histamine causes blood vessels in the immediate vicinity to dilate and also to become leaky, so that they lose plasma to the surrounding interstitial space. These effects account for the redness and swelling of infection and allow circulating components of the immune system to gain easier access to the site of infection. Histamine also stimulates fluid secretion across airway surfaces of the respiratory system and acid secretion by the stomach.

Endothelins are small proteins containing 21 amino acids secreted by the endothelial cells that line all but the smallest blood vessels. The endothelium affects the activity of surrounding smooth muscle cells through its release of varying amounts of endothelins, which cause contraction, as well as prostacyclin, an eicosanoid that causes relaxation. Although recently discovered, endothelins have

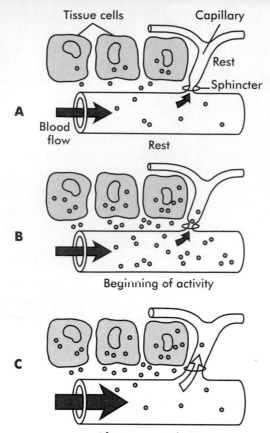

Tissue cells　　　　Capillary

A

Rest
Sphincter

Blood
flow

Rest

B

Beginning of activity

C

After autoregulation to
new higher flow setpoint

FIGURE 5-8
Blood flow autoregulation, an example of intrinsic regulation.

A The resting tissue releases low levels of local factors—metabolites and paracrine agents (indicated by colored particles), so that nearby arterioles are mostly constricted and the precapillary sphincters controlling blood flow into each capillary are mostly closed. The resulting low rate of blood flow is adequate to keep the concentrations of the local factors at the setpoint levels.

B When the tissue becomes active, its release of metabolites and paracrine agents increases, raising the levels of these local factors above their setpoint levels.

C The increase in local factors dilates precapillary sphincters and arterioles. Blood flow increases, returning the concentrations of local factors nearly to their setpoint levels.

been found in a number of different organs, including the brain, heart, kidney, and intestine. They may be involved in long-term regulation of blood pressure.

The purine **adenosine** and its nucleotides AMP, ADP, and ATP, are best known for their role in intracellular energy metabolism. However, purines have been found to relax smooth muscle and may be involved along with other metabolites in flow autoregulation (discussed below), especially in the case of the kidney.

An Example: Blood Flow Autoregulation in Tissues

In **blood flow autoregulation** (Figure 5-8), the flow of blood through each individual capillary bed is adjusted according to the changing needs of the small population of cells served by that bed. When the tissue is not active, the arterioles are mostly constricted, and precapillary sphincters, rings of smooth muscle that act as valves at the head of each capillary, are mostly closed. Increased metabolic activity, such as occurs in a muscle that is beginning to contract, causes an increase in the rate of production of carbon dioxide and lactate and a decrease in oxygen levels. In addition to end products of metabolism, active tissues increase their rate of release of a number of other substances, including potassium (released by nerves and muscles during activity) and paracrine agents such as prostaglandins and adenosine (Figure 5-8, *B*). The increase in the concentration of the local factors dilates arterioles serving that region of active tissue, and opens precapillary sphincters (Figure 5-8, *C*). The negative feedback loop is closed when the increased blood flow increases oxygen delivery to the active tissue and increases the rate at which local factors are flushed out. Thus, for any steady level of muscle activity, there is a corresponding setpoint for blood flow. In flow autoregulation the error signals are metabolites and paracrine agents released by the active tissue, and the effectors are the smooth muscle cells of arterioles and precapillary sphincters.

EXTRINSIC REGULATION: REFLEX CATEGORIES
Reflex Arcs

> • *What are the basic elements of a reflex arc?*
> • *What are the three general classes of reflexes?*
> • *How do somatic and autonomic reflexes differ?*

A **reflex arc** (Figure 5-9) is a feedback circuit that contains at least three elements: (1) an **afferent,** or sensory, component that detects variations in external or internal variables and relays information about the variable using neural or chemical signals; (2) an **integrator** (often called an **integrating center** or a collection of **interneurons** when it exists within the central nervous system) that determines the magnitude of the response that is appropriate; and (3) an **efferent** or **motor** component that sends neural or hormonal signals from the integrator to the effector organ.

Reflexive control can be used to stabilize physiological variables if the circuit is arranged in such a way as to provide negative feedback. Some examples of homeostatic reflexes are the **baroreceptor reflex,** which regulates arterial blood pressure, and the **respiratory reflexes,** which regulate ventilation of the lungs.

Neural and Endocrine Reflexes

In some reflex loops, integrator cells that are neurons (anatomically speaking) synthesize and release into the blood substances that act as hormones.

Nerve Growth Factor

Occasionally, a single discovery opens an unknown realm of scientific inquiry. Scientists are supposed to be open to the unexpected, but it takes an exceptional person to grasp the significance of an observation that does not fit into the accepted framework. The discovery of **nerve growth factor** has broad implications for cellular communication and control, and has resulted in a whole new field of research in growth factors and their effects.

The observation that started this new field was made by Rita Levi-Montalcini and for it she and her biochemist collaborator, Stanley Cohen, were awarded the Nobel Prize in 1986. Dr. Levi-Montalcini was trained as a physician in Turin, Italy, and completed her degree just as Fascist control of Italy closed the door to medical practice for Jewish women. Nevertheless, she had received training in research, and she decided to pursue questions of interest to her in a tiny laboratory that she set up in her bedroom. Working with a crude microscope

and performing operations with a sharpened sewing needle, she repeated experiments on chick embryos published earlier by an American scientist, Victor Hamburger. She published her interpretations in a Belgian journal because Italy was still not a place where a Jewish writer could publish. This led to an invitation from Dr. Hamburger to join him at Washington University in St. Louis. Dr. Levi-Montalcini's interpretations proved correct, and with the encouragement of Dr. Hamburger, she remained to collaborate with him and enjoy the more open atmosphere for scientific inquiry in the United States.

The line of observations that led to the unexpected discovery began in 1949 as the result of culturing a chick embryo together with a mouse tumor. What happened was that growth of several classes of nerve cells was stimulated enormously in response to the presence of the tumor. The nerve cells penetrated the tumor and grew with such wild abandon that they overran it. This experiment

had been performed before, but its significance had not been grasped. Levi-Montalcini proposed an explanation that had no foundation in the theoretical framework of the day: she thought that the sprouting was due to the release of a chemical factor, a trophic factor, from the tumor. To test this hypothesis, she cultured tumor cells with isolated sympathetic nerve cells. Within hours the nerve cells grew outward in a thick halo, indicating the presence of a powerful stimulant of nerve growth released by the tumor. After years of patient effort and Cohen's discovery that normal salivary glands contain large quantities of the same factor, they determined the chemical composition of nerve growth factor. Since then, many other growth factors have been isolated and studied, including **epidermal growth factor** discovered by Cohen. Nerve growth factor is believed to be released normally during development by target cells that receive connections from sympathetic nerves.

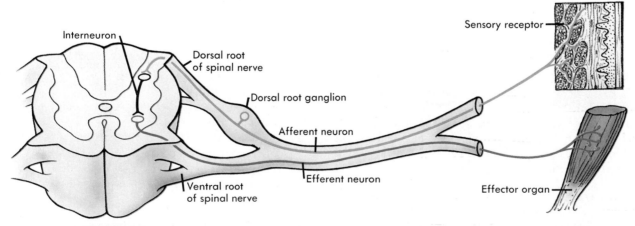

Interneuron

Dorsal root
of spinal nerve

Dorsal root ganglion

Afferent neuron

Ventral root
of spinal nerve

Efferent neuron

Sensory receptor

Effector organ

FIGURE 5-9

Basic diagram of a spinal reflex arc. Information from sensory receptors enters the spinal cord via the dorsal root. After synapsing with one or more association neurons, the efferent fiber transmits nerve impulses to effector organs.

These are **neurosecretory** or **neuroendocrine** cells. Endocrine glands that are not of neural origin may serve as lines of communication between the nervous system and effectors if their hormonal secretions are controlled by nervous input.

In some cases, endocrine glands combine the functions of sensor and integrator and respond to changes in the regulated variable by increasing or decreasing their rates of secretion. Although the term reflex has traditionally stood for a process involving the central nervous system, it is possible to refer to such a feedback loop as a **hormonal reflex** or **endocrine reflex.** Several organs not readily recognizable as glands serve an endocrine function. This is the case for the kidney and the heart, both of which secrete hormones important in regulating the salt and water economy of the body.

Reflexes can be divided into three classes:

1. **Somatic motor reflexes** that control skeletal muscle. A familiar example is the **withdrawal reflex,** in which a painful stimulus to an arm or leg results in rapid flexing of the appendage, removing it from threat of further injury.
2. **Autonomic reflexes** that modulate the activities of smooth muscle, exocrine glands, and the heart.
3. **Endocrine reflexes** in which the feedback loops may or may not involve the nervous system.

The basic principles of reflex control are the same for somatic motor, autonomic, and endocrine reflexes.

Somatic Motor Control

One function of somatic motor reflexes is to preserve constancy of body position with respect to the surroundings. Another function is to protect the body from dangerous stimuli delivered to the body surface. "Voluntary" movements of the body have a substantial reflexive component.

Afferent input to the central nervous system arrives in the spinal cord by way of nerve fibers from the special sense organs, muscles, and joints. Skeletal muscles are controlled by fibers from nerve cells called **motor neurons.** The pathway of information from sensory neurons to motor neurons almost always includes **interneurons.** These are cells that do not project out of the central nervous system. The specific connections that interneurons make between sensory neurons and motor neurons determine the reflex response that occurs. These connections are established during development so that sensory information is relayed to effectors that can make an appropriate response. The interneurons in a reflex pathway typically increase the possibility for control and modification of the response. Many somatic motor reflexes are so constant in their response pattern that they are referred to as **stereotyped.** If a muscle's tendon is lightly tapped, as by a physician testing the knee jerk response, only the muscle that was stretched by tapping on the tendon will contract. The contraction results from a **stretch reflex** (sometimes called a **tendon reflex** or **myotatic reflex**) that resists or opposes stretching of the muscle (Figure 5-10, *A*). Such reflex circuits are important in maintenance of posture because their negative feedback loop tends to return limbs to their original position.

The skeletal muscles of the human body are, for the most part, arranged as opposing muscle groups, or **antagonists.** Activity in the antagonistic muscles at a joint sets the position of the skeleton relative to a joint. If one muscle is to alter the limb position, it is necessary for the antagonistic muscle or muscle group not to oppose that action by contracting simultaneously. Typically, interneurons in the spinal

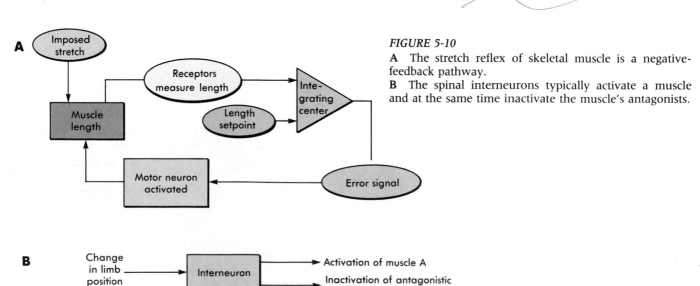

FIGURE 5-10
A The stretch reflex of skeletal muscle is a negative-feedback pathway.
B The spinal interneurons typically activate a muscle and at the same time inactivate the muscle's antagonists.

THE FOUNDATIONS OF PHYSIOLOGY

cord connect the motor neurons of antagonists in such a way that activation of a muscle is automatically accompanied by deactivation of its antagonists (Figure 5-10, B). Stereotyped spinal reflexes are the building blocks of far more complicated motor activities, as discussed in Chapter 11.

Autonomic Reflexes

The internal variables of the body are monitored by receptors that have functional names, such as baroreceptors (blood pressure detection), chemoreceptors (sensitivity to chemicals such as oxygen and carbon dioxide), osmoreceptors (measurement of osmolarity), and thermoreceptors (which monitor the temperature of the skin and body core). As is the case in the somatic motor system, the afferent limb for visceral reflexes is neural. Sensory information is evaluated by integrating centers within the central nervous system. Commands are sent out over efferent nerves and may activate or relax visceral smooth muscle, cause glandular secretion, or alter intracellular metabolism. These neural signals correct deviations from the setpoints programmed within the central nervous system.

There is no difference between the reflex mechanisms for visceral reflexes and those that control skeletal muscles. However, there are anatomical differences in the pathway between the central nervous system and visceral and somatic (skeletal) effectors, and it was formerly believed that voluntary control of visceral functions was not possible. Conventionally, the neurons that constitute the efferent limb of visceral reflexes are classified as the **autonomic,** or **involuntary,** division of the central nervous system.

ENDOCRINE REFLEXES
Mechanisms of Chemical Communication Between Cells

One central fact unites the nervous, endocrine, and paracrine systems: the final pathway of command or communication takes the form of arrival of a specific chemical substance at the effector cells. This substance, whether it is a hormone that arrived by the bloodstream or a paracrine or autocrine agent that diffused from neighboring cells or a neurotransmitter chemical released by a nearby neuron, is conventionally called the **first messenger.** The basic principles of communication by chemical first messengers are similar no matter what the origin of the messenger. In this section, these principles will be presented in the context of hormonal messages.

There are two important facts about the mechanism of information transfer by chemical first messengers. First, the reception of the message occurs as a result of binding of the first messenger by **receptors** on or in the **target cells,** those cells that possess receptors for the hormone (Figure 5-11). This occurs by a process similar to the interaction

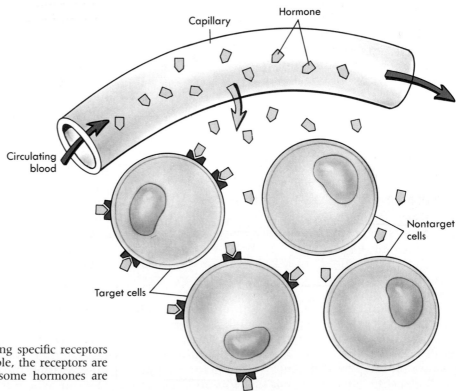

FIGURE 5-11
Hormones act only on target cells having specific receptors for a particular hormone. In this example, the receptors are on the cell surface, but receptors for some hormones are within the target cell.

between enzymes and substrates—such a bound molecule is generally called a **ligand** (in biochemical terminology). In this case, the complex of ligand and receptor causes changes in specific activities of the target cells, without a change in the chemical structure of the ligand.

The second important fact about chemical communication between cells is that the response of the target cells to the message depends on what happens inside the cell as a result of the binding of first messenger to receptor. The binding of first messenger and receptor causes formation or activation of one or more intracellular **second messengers** that in turn activate or inhibit specific enzymes. The meaning of the message thus depends on differing capabilities of the target cells; the same first messenger could, for example, simulate a target gland to increase its secretion and cause a target muscle to relax.

Chemical Classes of Hormones

Three basic chemical classes of hormones are listed here:

1. Derivatives of the amino acid tyrosine, including **norepinephrine, epinephrine,** and **dopamine** (see Figure 11-5 for structures), and the thyroid hormones **triiodothyronine** (T_3) and **tetraiodothyronine** (T_4 or **thyroxine**) (see Figure 23-20 for structures).
2. Peptides and proteins. These range in size from **thyrotropin-releasing hormone**, which has only three amino acids, to **growth hormone**, containing 191 amino acids.
3. Steroids and their derivatives. Steroid hormones are synthesized from cholesterol (see Figure 3-4 for structures) and include the sex hormones **estrogen, progesterone,** and **testosterone** produced by the gonads, as well as **cortisol** and **aldosterone** produced by the **adrenal cortex. "Vitamin D" (1,25 dihydroxycholecalciferol**), also derived from cholesterol, acts by mechanisms similar to the steroid hormones and is now considered a hormone by most endocrinologists.

Hormonal Signal Amplification

Hormones are typically present in endocrine glands in amounts so small that purification of an amount large enough for chemical analysis may require starting with hundreds of kilograms of tissue. The typical blood concentrations of hormones are very low indeed (typically in the range of 10^{-7} to 10^{-12} molar), and only a few molecules of hormone delivered to each target cell may be enough to cause a response. The target cells must greatly amplify this small message. The details of the amplification process are different for different types of hormone, but in all cases the effect is due to multiplication of the hormonal message within the cell.

Action of Steroid Hormones and Thyroid Hormones

- *Why must steroids be complexed with protein to travel in the blood?*
- *How is amplification of the hormonal message obtained in the case of steroid hormones? How about protein hormones?*

The effects of steroid hormones on their target tissues are largely due to the ability of these hormones to induce the production of specific proteins in the target cells. This is the result of "turning on" transcription of the specific portions of the target cells' chromosomes that code for the proteins. As a result, the structure of the target tissue may be changed, as happens when sex hormones induce the development of sex-specific body structures during early development and at puberty (Chapter 25). If the new proteins are enzymes, the tissue function served by those enzymes will be stimulated; this is what happens when aldosterone, the hormone secreted by the adrenal cortex, stimulates sodium transport in the distal portion of the nephrons of the kidney (Chapter 20).

Steroids are not water-soluble enough to travel as free molecules in blood plasma, so they travel in the blood in the form of complexes with protein. Although steroids can bind to a variety of blood proteins, specific carrier proteins exist for many steroid hormones. Steroid hormones are lipid-soluble and can therefore dissolve in cell membranes. When molecules detach from the protein carrier and dissolve in the cell membrane, they can enter a cell. If the cell is not a target cell for that hormone, the process of information transfer goes no further, and the hormone is ultimately metabolized as if it were any other membrane lipid.

Target tissues possess steroid hormone receptors in their nuclei (Figure 5-12). These nuclear receptors are probably proteins, but not all of them have been chemically characterized. In some cases, the steroid hormone must undergo enzymatic conversion to a different steroid before it can bind effectively with its receptor. For example, many tissues convert the male sex hormone testosterone to dihydrotestosterone (Chapter 25). This is particularly important in the case of embryonic external genitalia, which follow the female pattern of development unless dihydrotestosterone is present. Genetic males who lack the DNA coding for the enzyme that catalyzes the conversion of testosterone to dihydrotestosterone (called **5-α reductase**) are born with feminized genitalia.

Once formed, the complex of steroid hormone and nuclear receptor attaches itself to specific portions of particular genes, promoting transcription of mRNAs from those genes. The mRNAs enter the cytoplasm, where their message is translated into protein by ribosomes. Each steroid-receptor complex

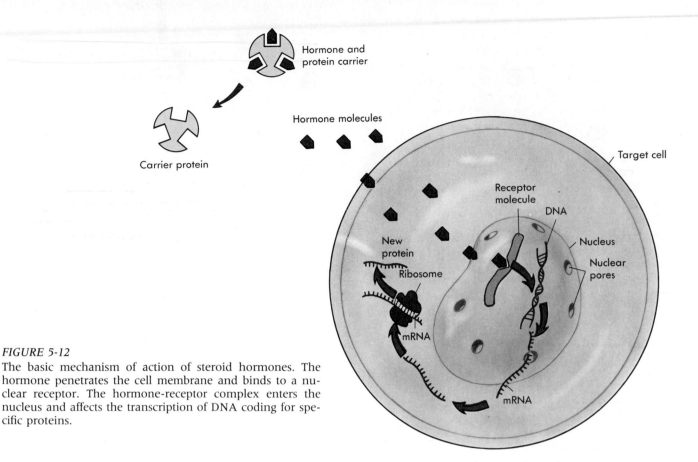

Hormone and
protein carrier

Carrier protein

Hormone molecules

Target cell

Receptor
molecule

DNA

New
protein

Nucleus

Nuclear
pores

Ribosome

mRNA

mRNA

FIGURE 5-12

The basic mechanism of action of steroid hormones. The hormone penetrates the cell membrane and binds to a nuclear receptor. The hormone-receptor complex enters the nucleus and affects the transcription of DNA coding for specific proteins.

can initiate the production of a large number of mRNAs, and each mRNA can result in the synthesis of a large number of copies of the protein. Thus a single hormone molecule might easily result in production of several thousand protein molecules, giving the feedback loop served by the hormone a gain of several thousandfold.

Hormones with Membrane-Bound Receptors: The Fixed Receptor Hypothesis

Most hormone classes (with the exception of steroids, thyroid hormones, and eicosanoids) are lipid-insoluble polar molecules that do not readily cross cell membranes. The responses to lipid-insoluble hormones are initiated by binding with receptors present on the surface of target cells (Figure 5-13). Although some of the effects that hormones of this type produce may be mediated by movement of the hormone into the cell (possibly by endocytosis of both hormone and receptor) with subsequent effects on mRNA synthesis, most of the rapid, obvious effects cannot be accounted for by hormone entry into the cell. Instead, interaction of the first messenger with receptors on the outside of the target cell surface causes one or both of the following: (1) formation within the cell of chemical second messengers that carry the message to enzymes in the cell interior (Figure 5-13), or (2) opening of Ca^{++} channels

in the membrane that allow the entry of Ca^{++}, which activates responses within the cell (Figure 5-13).

One common second messenger system (Figure 5-14, A) involves a compound called **cyclic AMP** (a form of adenosine monophosphate, abbreviated **cAMP**). Binding of the hormone with its receptor results in activation of a class of membrane protein now referred to as **G-proteins**. Different types of G-proteins in turn activate (**Gs**) or inhibit (**Gi**) an enzyme called **adenylate cyclase** located on the cytoplasm-facing surface of the cell membrane. Adenylate cyclase catalyzes the conversion of a small amount of the cell's stock of ATP into cAMP, the second messenger. The duration of the hormone effect is determined by the activity of a second enzyme called **phosphodiesterase** that converts cAMP into plain AMP, which is not active as a messenger. A similar mechanism utilizes **cyclic guanosine monophosphate (cGMP)** as a second messenger.

An equally important but more recently discovered second messenger system (Figure 5-14, B) involves phospholipid compounds called **phosphoinositides** that are normal constituents of cell membranes. In this case the hormone receptor is coupled by a membrane-bound G-protein (structurally similar to Gi) to **PIP$_2$ phosphodiesterase**, an enzyme

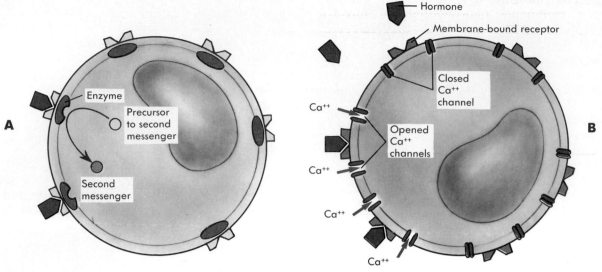

FIGURE 5-13

Basic mechanisms of action of hormones with receptors on the cell surface. The hormone binds with receptors on the exterior of the cell membrane, setting in motion second messengers.

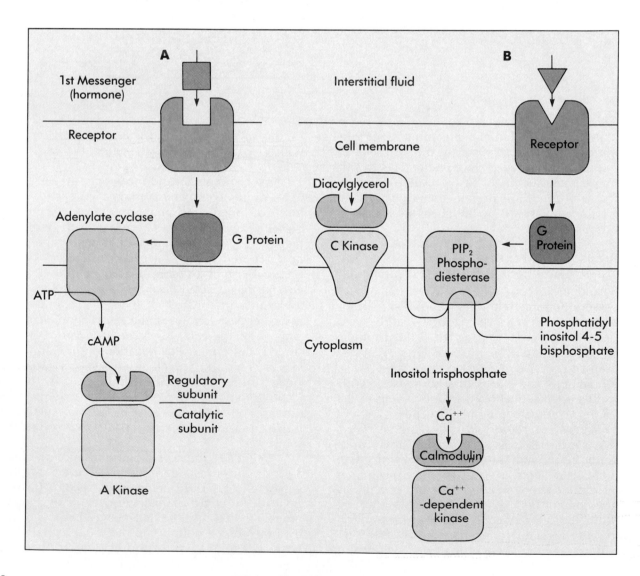

that converts **phosphatidyl inositol 4,5 bisphosphate,** a membrane phospholipid, to **inositol trisphosphate (ITP)** and **1-2 diacylglycerol (DAG**, a diglyceride). Hormone binding activates this enzyme, and both ITP and DAG act as internal second messengers at different sites. Specifically, DAG activates membrane-bound **protein kinase C (C kinase),** while ITP causes release of Ca^{++} from intracellular sites.

Intracellular Ca^{++} is involved in control of many cellular functions. It is important to realize that the use of Ca^{++} to control cell activities depends on maintaining a very low concentration of Ca^{++} in the cytoplasm (approximately 10^{-6} M). Hormonal signals can then be relayed by relatively few Ca^{++} ions. Depending on the particular cell type, messenger Ca^{++} may be allowed to enter the cell from extracellular fluid as a result of hormone binding, or it may be released from intracellular stores located in such organelles as mitochondria and the endoplasmic reticulum. The latter mechanism is activated by the inositol trisphosphate second messenger system described before (Figure 5-14, *B*). The most important general mechanism for translating a Ca^{++} second message into effects on cellular function involves an intracellular Ca^{++} receptor protein called **calmodulin** (Figure 5-15). Calmodulin is frequently attached to enzymes, where it serves as a regulatory subunit (Figure 5-14, *B*).

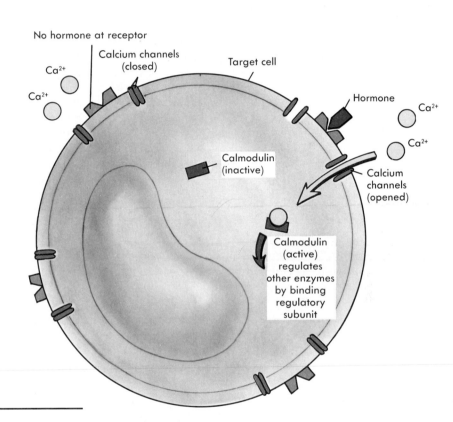

FIGURE 5-14

Two specific second messenger systems.

A The sequence of events that leads to the generation of the cAMP second messenger. Hormone binding can either increase or decrease adenylate cyclase activity via intermediate, membrane-bound G-proteins. The effects of cAMP within cells are mediated by protein kinases termed *A kinases.* The cAMP acts by binding to an allosteric site on a regulatory subunit of the enzyme.

B The sequence of events collectively termed the *phosphoinositol system.* Hormone binding increases phosphodiesterase activity via an intermediate G-protein. Activation of the phosphodiesterase results in the formation of two messengers, diacylglycerol and inositol trisphosphate. Diacylglycerol can bind to the regulatory subunit of a family of C kinases. Inositol trisphosphate can bring about entry of Ca^{++} into the cell. The Ca^{++} messenger binds to calmodulin, the regulatory subunit of Ca^{++}-dependent protein kinases.

FIGURE 5-15

Calcium-calmodulin as an intracellular second messenger. Certain hormones bind to receptors on the cell membrane, often causing an increase in the membrane permeability to calcium. Calcium entering the cell binds to calmodulin, and the complex regulates the activity of a variety of intracellular enzymes.

Second Messenger Regulation of Protein Kinases

A great many enzymes that catalyze or control key steps in cell metabolism, or important cell functions such as muscle contraction, can be activated or inhibited by attachment of phosphate groups to allosteric control sites. This process is referred to as enzyme phosphorylation. Phosphorylation of specific enzymes is catalyzed by a family of control enzymes called **protein kinases** (Figure 5-14). Protein kinases are, in turn, under the control of second messengers. Specifically, the family of A kinases is typically controlled by the cAMP system, while the family of C kinases is controlled by the phosphoinositide second messenger system. Because phosphate forms a covalent bond to the protein regulatory site, a second enzyme, **protein phosphatase,** serves to remove the phosphate groups so that the regulatory signals are not permanent. However, at the same time that protein kinase A is activated by cAMP, protein phosphatase is inhibited by cAMP, so that as long as cAMP levels remain high, there is no interference with the regulatory signal.

Duration of Hormone Effects

- *What factors determine the level of a hormone in the blood?*
- *What factors determine the duration of hormonal effects?*

The duration of the effects of a hormone is affected by three factors: the time pattern of hormone secretion, how long the hormonal signal persists in the circulation after secretion falls off, and how long the effects last after the hormone has been cleared from the blood. All three of these factors vary enormously from one hormone to another.

Many endocrine glands synthesize and release their hormones only in response to other hormones, various metabolites, or external signals. Other glands constantly secrete hormones at a fixed (basal) rate, which is then increased or decreased by the control systems that affect the particular gland. Basal hormone release may be steady or periodic; in the latter case there is often a **diurnal** (24-hour) pattern. For example, in the case of **growth hormone** (also called **somatotropin**) secreted by the pituitary, a large fraction of the daily secretion of the hormone (perhaps as much as half of the daily total) occurs in a pulse during the hour or two following the onset of deep sleep (Figure 5-16). Only during this pulse do the levels of hormone in the plasma rise high enough to have a significant stimulatory effect on growth.

Different hormone molecules have different life spans in the blood. The level of hormone in the blood is determined by the secretion rate and the rate at which the hormone is removed by metabolism or excretion. Usually the **half-life,** the time required for half of a given quantity of hormone in the plasma to be metabolized or excreted, is used to measure how rapidly a particular hormone is turned over by the body. The shorter the half-life, the higher the secretion rate must be to maintain a given level of hormone in the plasma.

The effects of hormones that act through cell surface receptors usually involve activating preexist-

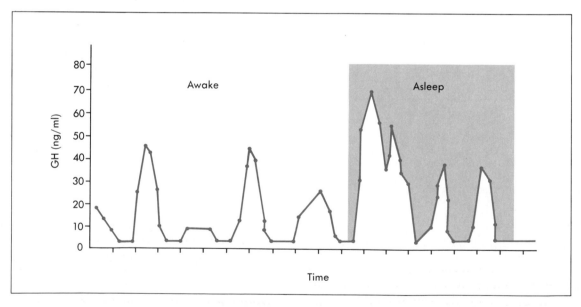

FIGURE 5-16
Secretion of growth hormone in a normal male adolescent occurs in small pulses during waking, with a large pulse just after the onset of sleep.

THE FOUNDATIONS OF PHYSIOLOGY

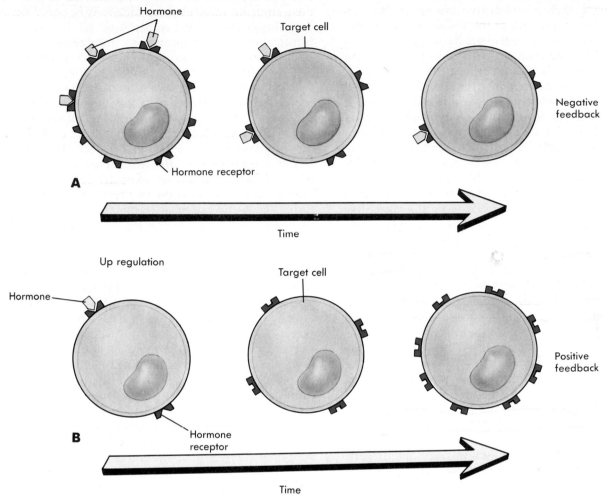

FIGURE 5-17
Down-regulation (A) and up-regulation (B) of hormone receptors.

ing cell systems. Therefore the effects of such hormones are typically both rapid and brief. For example, epinephrine has a plasma half-life of a few minutes and does not produce a prolonged response because the response of the target cells is also brief. Growth hormone, steroid hormones, and thyroid hormones may have plasma half-lives as short as 20 minutes, but their effects persist because they initiate long-lasting changes in their target cells.

Hormone Receptor Regulation

Hormone receptors, like all other cellular elements, are subject to turnover. The rate of synthesis of the receptor proteins is influenced by the circulating levels of the hormone. In some cases, continued exposure of cells to a high level of hormone results in a decrease in the number of receptors, called **down-regulation** (Figure 5-17, *A*). Down-regulation can be seen as a process of negative feedback. The opposite process, **up-regulation**, can also occur (Figure 5-17, *B*). This process results in positive feed-

back. A good example of up-regulation is found in the hormonal control of maturation of follicles in the female ovary (Chapter 25). Maturing follicles secrete the female sex hormone estrogen; as a follicle matures, it augments its secretion of estrogen by increasing the number of binding sites for a pituitary hormone that stimulates estrogen production. Sometimes this is referred to as a **priming effect.** This positive feedback cycle is brought to an end by ovulation (release) of the follicle.

Functional Classes of Hormones

• *How do the direct effects, trophic effects, and permissive effects of hormones differ?*

Hormones can have either **direct** or **permissive** actions. In a direct effect, the binding of hormone with receptor initiates a change in the activity of the target tissue. In some cases a second hormone must also be present in small quantity to permit the primary hormone to exert its full effect; the second hormone is called a **permissive hormone.** For example, many of the direct metabolic effects of the hormone epinephrine, secreted by the adrenal medulla, are dependent on the presence of an adequate concentration of the hormone cortisol, secreted by the adrenal cortex. The cortisol cannot substitute for epinephrine, but epinephrine is much less effective in its absence. The nature of the permissive effect is not known in all cases, but one mechanism by which permissive hormones can exert their effects is by stimulating the target cells to synthesize receptors for the other hormone.

In some cases, two or more chemically distinct hormones may have similar effects on the target tissue. If the effect of both hormones applied together is greater than the sum of their individual effects, the hormones are said to be **synergists.** If, when two hormones are present, their effects on the target tissue are merely additive, such hormones are called **agonists. Antagonistic** hormones are ones whose effects on the target tissue tend to cancel each other out. Antagonistic hormones exemplify the principle that regulation is often achieved by balancing opposing forces. In general, the sensitivity and versatility of control of a target tissue are enhanced by multiple lines of hormonal communication.

The term **trophic hormone** is applied to hormones that control the secretion of other hormones. The regulatory pathways that run from the brain to endocrine systems by way of the anterior pituitary are excellent examples of control of final hormone secretion by trophic hormones. Frequently, in addition to their immediate effect on hormone secretion, trophic hormones also have a long-term effect on growth and maintenance of the target endocrine tissue. For example, **thyroid-stimulating hormone (or thyrotropin)** secreted by the anterior pituitary gland affects both the size of the thyroid gland and the rate of secretion of thyroid hormones.

The Posterior Pituitary and Its Hormones

> • *What are the posterior pituitary hormones and their targets?*
> • *What are the anterior pituitary hormones and their targets?*
> • *How do the mechanisms of secretion of anterior pituitary hormones differ from those of the posterior pituitary?*

The **pituitary gland (hypophysis)** is a small but complex structure that is attached to the ventral side of the hypothalamus of the brain (Figure 5-18). From an embryological and functional view, its anterior and posterior parts are distinct. The **posterior pituitary,** or **neurohypophysis,** is embryologically

a part of the brain. Branches of neurons run from the hypothalamus into the posterior pituitary. These posterior pituitary hormones are synthesized in hypothalamic nuclei and transported to the posterior pituitary via the hypothalamic-hypophyseal nerve tract where they are stored in vesicles. Electrical activity in these hypothalamic neurons results in the release of two hormones: **antidiuretic hormone** (abbreviated **ADH,** also called **vasopressin**) and **oxytocin.** This is an example of secretion of hormones by neurons, or **neurosecretion.** Secretion of the hormones represents the output of neural reflexes whose inputs come from sensory projections that run to the brain, or from sensors within the brain itself (Figure 5-19, *A*). ADH is involved in fluid and electrolyte balance (Section V), and oxytocin is important in the processes of labor and lactation (Chapter 25).

The Anterior Pituitary and Its Hormones

A more complex process of endocrine communication between the brain and the body is seen in the **anterior pituitary** (also called the **adenohypophysis**). Unlike the posterior pituitary, the anterior pituitary does not arise from the nervous system in development, but from a cleft of epithelial tissue (Rathke's pouch) that folds and separates from the pharynx (throat) in early development. The anterior pituitary contains five recognizable populations of cells that, together, secrete at least seven hormones (Table 5-3). Six of these hormones have well-characterized roles in metabolism, growth, and reproduction: **adrenocorticotropic hormone (ACTH); thyroid-stimulating hormone (TSH,** also called **thyrotropin); growth hormone (GH,** also called **somatotropin);** the two **gonadotropins, luteinizing hormone (LH)** and **follicle-stimulating hormone (FSH);** and **prolactin (Pr).** The role of the seventh posterior pituitary hormone, **melanocyte-stimulating hormone (MSH)** is less well-understood, at least in humans.

Of the six well-characterized anterior pituitary hormones, four are trophic hormones, by means of which the brain controls the thyroid gland (TSH), the adrenal cortex (ACTH), and the gonads (FSH and LH). The other two hormones are released into the blood from storage sites in the anterior pituitary and have direct effects on growth (GH) and lactation (Pr). To control secretion of one of the anterior pituitary hormones, the brain secretes the appropriate **releasing hormone** (or in some cases a **release-inhibiting hormone**) into a branch of the circulatory system that carries blood from the brain to the anterior pituitary (Figure 5-19, *B*). Such circulatory systems linking the capillary networks of two organ systems are called **portal circulations;** this one is called the **hypothalamic-anterior pituitary portal circulation.** With one exception, all of the

THE FOUNDATIONS OF PHYSIOLOGY

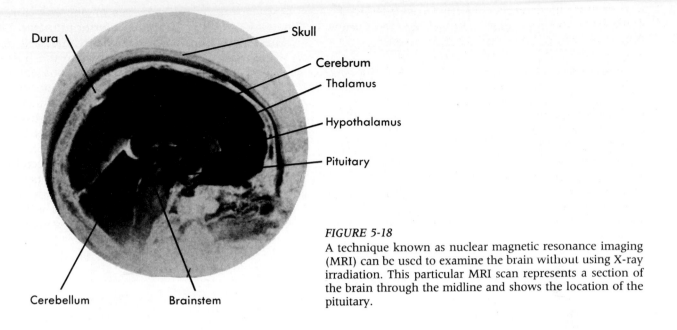

Dura
Skull
Cerebrum
Thalamus
Hypothalamus
Pituitary
Cerebellum
Brainstem

FIGURE 5-18
A technique known as nuclear magnetic resonance imaging (MRI) can be used to examine the brain without using X-ray irradiation. This particular MRI scan represents a section of the brain through the midline and shows the location of the pituitary.

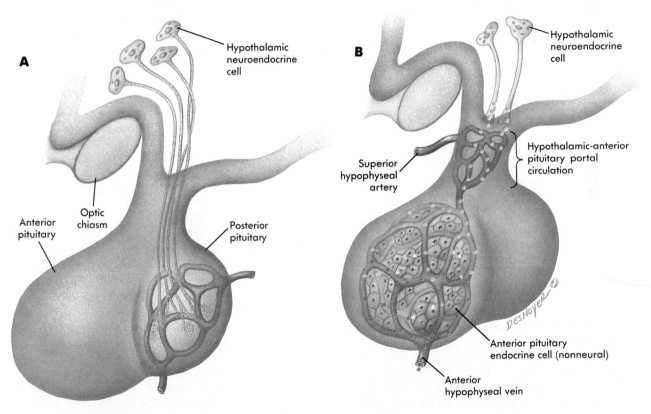

A

Hypothalamic neuroendocrine cell

Anterior pituitary

Optic chiasm

Posterior pituitary

B

Hypothalamic neuroendocrine cell

Superior hypophyseal artery

Hypothalamic-anterior pituitary portal circulation

Anterior pituitary endocrine cell (nonneural)

Anterior hypophyseal vein

FIGURE 5-19
A The control of hormone secretion by the posterior pituitary. Some hypothalamic neurons send processes down through the stalk of the pituitary into the posterior pituitary. The peptide hormones oxytocin and ADH are synthesized in the neuron cell bodies and transported down the processes into the pituitary, where they are stored. Electrical activity in the neurons results from sensory inputs and leads to release of the hormone.
B The control of hormone secretion by the anterior pituitary. In this case, hypothalamic neurons secrete releasing hormones that travel along the portal circulation to nonneuronal endocrine cells in the anterior pituitary. Binding of the releasing hormone *(aqua)* by the target cell stimulates release of the anterior pituitary hormone *(red)*.

TABLE 5-3   Hormones of the Anterior Pituitary

Hormone name, abbreviation, and structure	Releasing hormones (and release-inhibiting hormones)	Target/effect
Adrenocorticotropic H. (ACTH; 39 a.a.)	ACTHrh (41 a.a)	Adrenal cortex: cortisol release
Thyroid-stimulating H. (TSH; 204 a.a.)	TSHrh (3.a.a)	Thyroid; thyroxine release
Growth H. (Somatotropin, hGH; human form has 191 a.a.)	GHrh (40 a.a) stimulates; somatostatin (several forms) inhibits	Whole body growth; mediated by insulin-like growth factors.
Prolactin (Pr; several forms)	Prh?; dopamine inhibits	Breasts: lactation
Gonadotropins		
Luteinizing H. (LH; 207 a.a.)	Gnrh (10 a.a)	Gonads: development of reproductive cells and secretion of gonadal
Folliclc-stimulating H. (FSH; 210 a.a.)		sex hormones

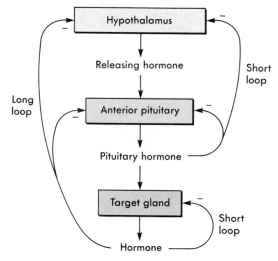

FIGURE 5-20
Feedback may act at several levels in endocrine systems controlled by the hypothalamus and anterior pituitary.

known releasing and release-inhibiting hormones are small peptides (Table 5-2).

Feedback control of hormones controlled by the anterior pituitary (Figure 5-20) is potentially quite complex because feedback can be applied at several levels, often referred to as **short-loop** or **long-loop** feedback. The controlled hormone may inhibit its own secretion, the secretion of the trophic hormone, or the secretion of the releasing hormone. The trophic hormone may also inhibit secretion of the releasing hormone. Recent evidence suggests that in addition to the portal circulation from hypothalamus to pituitary, there is a second portal circulation running from anterior pituitary to hypothalamus, which could mediate feedback of anterior pituitary hormones on the hypothalamus.

Regulation of Blood Glucose: An Example of the Principles of Endocrine Control

• *What are diabetogenic hormones?*

Blood glucose levels are typically within the range of 80 to 90 mg/100 ml in fasting adults. The level may rise above 100 mg/100 ml for no more than an hour or so after a carbohydrate meal. If blood glucose levels fall below about 50 mg/100 ml, brain function is immediately affected. If blood glucose levels remain above normal for months or years, damage to blood vessels, nerves, and eyes will occur. Regulation of blood glucose levels is part of the larger story of metabolic homeostasis treated in Chapter 23. This section will concentrate on some aspects of the system that illustrate the principles of intracellular second messengers, negative feedback, receptor regulation, and antagonistic hormones.

The major hormones involved in regulation of glucose metabolism over the time scale of hours are: (1) **insulin**, a protein hormone secreted by **beta cells** of the **islets of Langerhans**, which compose the endocrine portion of the pancreas; (2) **glucagon**, a protein secreted by **alpha cells** of the islets of Langerhans; (3) **epinephrine**, a catecholamine secreted by the adrenal medulla; and (4) **cortisol**, a steroid secreted by the adrenal cortex.

The rate of secretion of insulin is controlled primarily by plasma levels of glucose and amino acids acting directly on the beta cells, which serve as both sensors and integrators. Insulin is secreted at a very low rate during fasting, and blood insulin levels are almost undetectable (Figure 5-21). An increase in blood glucose above the setpoint level stimulates in-

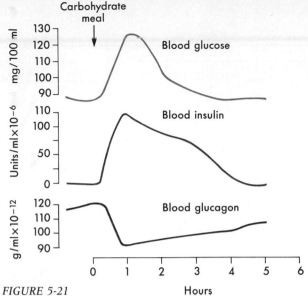

FIGURE 5-21

Hours

Changes in blood levels of glucose, insulin, and glucagon in response to a pure carbohydrate meal in a fasted subject reflect the responses of the regulatory system. Before the meal, insulin levels are very low and glucagon levels are high. This hormonal picture causes glycogen breakdown to predominate, sustaining blood glucose. The rise in glucose after the meal is followed closely by a rise in insulin and a fall in glucagon levels. These changes in the hormonal picture tip the balance in favor of glycogen synthesis. As blood glucose levels return toward the setpoint, insulin levels fall again and glucagon levels rise.

sulin secretion. Circulating insulin binds with its receptors, which are present on almost every cell type of the body. An increase in the rate of transfer of glucose and amino acids from blood into the cells results. Some of the glucose and amino acids are catabolized to meet the immediate energy requirements of the cells; the excess enters anabolic pathways, leading in the case of glucose to storage as fat and glycogen and in the case of amino acids to synthesis of protein and conversion of amino acids to carboxylic acids.

As a result of the insulin-stimulated movement of glucose into cells, blood levels of glucose fall back in the direction of the setpoint (Figure 5-21), turning off insulin secretion. Notice that, as in all negative feedback loops, there is a time lag between the rise in blood glucose and the rise in insulin, and also an overshoot as glucose levels return to the fasting baseline. When glucose is no longer entering the body from the intestine, insulin secretion drops and secretion of glucagon increases (Figure 5-21). If glucose levels drop below the setpoint of about 90 mg/100 ml, or during fasting that is accompanied by stress or exercise, secretion of epinephrine increases. Glucagon and epinephrine favor catabolism of fat and glycogen; their cellular actions are potentiated and reinforced by the permissive effect of cortisol.

Muscle and liver cells are the sites of most of the stored glycogen in the body. The action of epineph-

rine on glycogen breakdown in these two cell types (Figure 5-22) was one of the first intracellular hormonal effects to be worked out. The same second-messenger system is activated by glucagon in liver cells, but not in muscle cells, which lack glucagon receptors. The catecholamine (and glucagon) receptors are coupled to adenylate cyclase, so activation of these receptors results in production of the intracellular second messenger cAMP. In turn, cAMP activates protein kinase A molecules. The activated protein kinase A directly phosphorylates one of the key enzymes involved in glycogen metabolism: **glycogen synthetase** which catalyzes glycogen polymerization. The active kinase A also phosphorylates a second kinase that in turn phosphorylates a second key enzyme, **glycogen phosphorylase**. Glycogen phosphorylase catalyzes glycogen depolymerization, releasing glucose units in the form of glucose-1-phosphate. Phosphorylation has opposite effects on the two enzymes: it reduces the activity of glycogen synthetase but increases the activity of glycogen phosphorylase (Figure 5-22). In this way the cAMP second messenger tilts the balance decisively in favor of glycogen catabolism to release glucose units, and against glycogen synthesis.

Why are there so many control steps? Just as in the case of steroid hormones, almost every step gives amplification. The effects just described result from blood concentrations of epinephrine of the order of 10^{-10} M. Perhaps as many as a few thousand molecules of hormone arrive at the cell, each resulting in the formation of about 100 molecules of cAMP and raising the intracellular cAMP concentration from almost indetectible levels to the range of 10^{-6} M. Each cAMP molecule activates a single protein kinase molecule, but each protein kinase molecule activates many phosphorylase kinase molecules. Finally, each active molecule of phosphorylase releases many glucose units from glycogen. The final result is that each molecule of hormone bound to receptor results in the production of roughly 10^8 glucose units from glycogen, so the amplification factor, or gain, of the process is about 100 million.

Details of the pathway leading to activation of glycogen phosphorylase are shown in Figure 5-23. Note that both phosphorylation and Ca^{++}-calmodulin can activate glycogen phosphorylase kinase at different allosteric sites (Figure 5-23); for this enzyme they are agonistic second messengers. This is particularly important in muscle cells because, as explained in Chapter 12, Ca^{++} also serves as the second messenger that triggers contraction. There are other sites on the enzyme that can be phosphorylated by other protein kinases; some of these yield inhibitory effects. Why are there so many control sites on this molecule? The answer is that the key step of glycogen breakdown controlled by glycogen phosphorylase kinase must be responsive to many

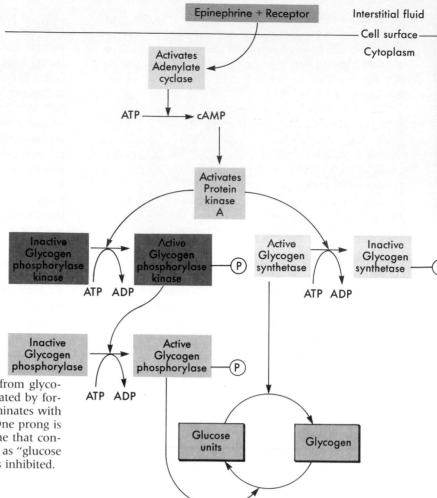

FIGURE 5-22

Epinephrine stimulates release of glucose units from glycogen by a regulatory cascade. The cascade is initiated by formation of the second messenger cAMP, and terminates with a two-pronged effect on glycogen metabolism. One prong is activation of glycogen phosphorylase, the enzyme that converts glycogen to glucose-1-phosphate (indicated as "glucose units"). At the same time, glycogen synthetase is inhibited.

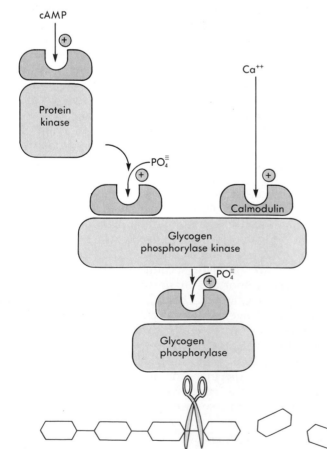

FIGURE 5-23

Activation of glycogen phosphorylase by cAMP and Ca^{++} is an example of enzyme control by multiple second messengers. Both second messengers stimulate glycogen breakdown. The protein kinase activated by cAMP is an A kinase. Calmodulin is a regulatory subunit of glycogen phosphorylase kinase, so this enzyme is an example of a Ca^{++}-dependent kinase. The actual conversion of glycogen to glucose-6-phosphate is catalyzed by glycogen phosphorylase, the final step in the regulatory cascade. The + symbols indicate an allosteric activating effect.

different homeostatic signals from different hormones and from other elements of the cell's energy metabolism. The enzyme molecule serves as an integrator, summing the inputs to arrive at a level of activity appropriate to serve the needs of the cell and the overall needs of the body.

The fate of the released glucose-1-phosphate depends on whether the cell is a muscle or liver cell. In the case of muscle, the glucose-1-phosphate is converted to glucose-6-phosphate and enters the glycolytic pathway. For muscle cells, the meaning of the hormonal message could be translated as, "Stop using blood glucose and start drawing on your reserves of glycogen." In liver cells, the glucose-1-phosphate is converted to glucose and released into the blood. In this case, the meaning of the message could be translated as, "Start using your reserves of glycogen to maintain blood glucose levels."

The binding of insulin to its receptor is essentially irreversible, so hormone-receptor complexes must be continuously removed from the cell membrane and replaced with fresh receptors. As detailed in Chapter 4, membrane proteins such as the insulin receptor are synthesized by rough endoplasmic reticulum and inserted in membrane vesicles that fuse with the cell membrane. The hormone-receptor complexes are removed by the reverse process of endocytosis; cell membrane is pinched off to form vesicles that subsequently fuse with lysosomes. The contents of the lysosomes digest the insulin. Some receptors may survive for reuse, but usually most are destroyed along with the hormone.

The number of insulin receptors present on the cell surface is not constant, but is regulated by the target cells. This regulation is responsive to blood levels of insulin averaged over time. In the case of insulin, excessive levels of hormone cause the target cells to down-regulate their insulin receptors. This is accomplished by endocytosis of receptors at a higher rate than they are replaced. In most cases this process is a normal part of feedback regulation of the target cell's responsiveness to extrinsic regulation. However, in some individuals it may lead to impaired glucose regulation, as described in the next section.

Failure of Glucose Regulation in Endocrine Diseases

> - *What are the differing mechanisms of type I and type II diabetes mellitus?*

The most common disorders of glucose regulation involve failure of glucose levels to return to the fasting range within the normal time period after a meal. This failure, called **diabetes mellitus,** is diagnosed by the **glucose tolerance test** in which the patient is given a meal of glucose after an overnight fast. Blood glucose levels are measured before the meal and at intervals afterward. Diabetes mellitus could result from failure of the feedback loop at any of a number of points. The first recognized cause of the condition was a failure of the pancreatic cells to secrete adequate insulin. This condition becomes apparent in adolescence and is now called **juvenile-onset** or **type I** diabetes mellitus.

A second form of diabetes mellitus, called **adult-onset** or **type II** diabetes, is associated with a patient history of excess carbohydrate consumption and obesity. In genetically-susceptible individuals who overeat, the repeated delivery of large doses of insulin to target cells results in such severe down-regulation of the insulin receptors that blood glucose can no longer be well-controlled with normal or even above-normal rates of insulin secretion. In most cases, adult-onset diabetes can be successfully treated by modifying the patient's diet.

Abnormally high levels of glucagon or epinephrine can also elevate blood glucose levels, as can excessive secretion of cortisol (a disease called **Cushing's syndrome**). Cortisol production is regulated by the hypothalamus of the brain (by way of the anterior pituitary), and normal levels of cortisol are maintained by negative feedback of cortisol at three levels (Figure 5-20): on the adrenal cortex itself (a "short" loop and an example of intrinsic regulation), on the anterior pituitary's secretion of ACTH, and on the secretion of ACTH-releasing hormone by the hypothalamus ("long" loops). The excessive secretion of cortisol in Cushing's syndrome could be the result of failure of any one of these three feedback loops. Successful treatment of the syndrome depends on determining which loop has failed.

Like hormones, growth factors control a variety of bodily functions. Some growth factors regulate normal processes, whereas others come into play in disease. Unlike "classical" hormones that typically travel in the blood plasma from the endocrine gland that produces them to distant areas of the body, most growth factors act locally. More than 30 growth factors have been identified, each with a specific ability to make certain cells divide more quickly. Growth factors get their names from the type of cell they stimulate. Because the body produces growth factors in such tiny amounts, scientists relying on nature did not have much to work with. It wasn't until the advent of biotechnology, that scientists were able to mass produce growth factors in the laboratory.

One category of growth factors are chemicals also called colony-stimulating factors (CSFs). Red blood cell growth factor, erythropoietin, boosts red cell synthesis and has been considered a hormone for some time. Erythropoietin is used to treat people with chronic anemia caused by kidney disease, who otherwise would require frequent blood transfusions. Three of the principal types of white cells are granulocytes, monocytes and lymphocytes. Granulocyte stimulating factor (G-CSF) stimulates formation of granulocytes; M-CSF stimulates monocyte formation; and GM-CSF stimulates both granulocytes and macrophages. Many leukemia patients have low levels of granulocytes. Although chemotherapy destroys cancer cells, it kills other rapidly dividing cells, including the disease-fighting white blood cells of the immune system. Treatment with GM-CSF or G-CSF increases the number of granulocytes, overcoming the reduction in white blood cells that usually follows cancer chemotherapy. The result is fewer

infections and less need for antibiotics. GM-CSF may also enable patients to tolerate higher doses of chemotherapy. As promising as current research seems, questions still remain, including the possibility of stimulating tumor cell growth itself.

Some 3 million Americans, including many elderly individuals and diabetics, suffer from wounds that heal poorly, such as bedsores. Epidermal growth factor (EGF) stimulates skin cells to multiply and has been reported to promote the healing of deep bedsores in half the normal time. EGF was first discovered in the salivary glands, which may explain why many animals instinctively lick their wounds (a response not entirely eliminated in humans). EGF has the potential to solve a major problem in treating severe burns, that of finding enough undamaged skin to cover the wound. When a small number of normal skin cells from burn victims are cultured together with EGF they rapidly proliferate. Sheets of laboratory-grown skin are applied to severely burned areas. Because the skin is the burn victim's own, these transplants aren't rejected by the immune system. Epidermal growth factor has also been used to treat eye injuries and surgical incisions made in the course of corneal transplants. Ointments containing the substance are now being tried on stubborn wounds that plague diabetics and other poor healers. Eventually, growth factors may go into bandages to treat everyday wounds.

Combinations of growth factors may be more effective than any one alone because they come closer to simulating nature. Growth factors act in groups during development. One growth factor, usually platelet-derived growth factor (PDGF) or fibroblast growth factor (FGF) will "prime" cells to enter a cycle of cell division. Then other growth factors such as EGF, the insulin-like growth factors, or

transforming growth factors (TGF) act to carry the cell the rest of the way through the division process. Growth factors act not only on the mitotic capability of cells, but also on the differentiation of cells - development of the ability of cells to assume their full physiological function. For example, EGF not only stimulates the division of epidermal cells but causes them to differentiate into an almost impermeable barrier on the outside of the skin.

During wound repair, growth factors like PDGF, FGF, and TGF act together to stimulate and regulate the regrowth of blood vessels into the wound site. This process is termed angiogenesis. When tumors grow they take unfair advantage of the process by producing the same growth factors to attract a blood supply into the tumor to continue tumor growth.

Bones, too, stand to benefit from growth factors. Fibroblast growth factor, important in maintaining healthy bone and connective tissue, can restore the blood supply to damaged bone. One of the most promising, though preliminary, uses of growth factors is to treat Alzheimer's disease and other nervous system problems. Once damaged, brain cells usually atrophy and die, as happens in Alzheimer's patients. But when researchers treated the brain cells of injured or old laboratory rats with nerve growth factor (NGF), the damaged cells thrived. Because NGF has been so effective in restoring nerve tissue in some experiments, Swiss researchers were puzzled when they were unable to induce brain and spinal cord tissue to grow in the laboratory. This setback led to the discovery of another piece of the growth-factor puzzle. Many growth factors coexist with substances now called growth inhibitors. One of these prevents nerve tissue from regenerating. Chemicals that neutralize these inhibitors would be important tools in the approach to nerve regeneration.

SUMMARY

1. In a **negative feedback loop,** information flows from the **sensors** to the **integrator,** where it is compared with a **setpoint.** A difference between the sensor's input and the setpoint results in generation of an **error signal** that is passed to effectors. The effect of the loop is to minimize the difference.

2. Positive feedback loops result in dramatic increases or decreases in effector activity, driving the regulated variable to extreme values.

3. In the body, homeostatic processes occur on all levels of integration. **Intrinsic control** operates on the tissue level and frequently involves the feedback effects of local chemical factors called **paracrine** and **autocrine agents. Extrinsic control** operates on the organ-system level in which the feedback loops are neural or hormonal reflexes.

4. Extrinsic feedback loops can be divided into three categories of reflexes: **somatic, autonomic, and hormonal.**

5. Hormone effects can be categorized as **direct, permissive, or trophic.**

6. The number of hormone receptors is under cellular control and may demonstrate either negative feedback **(down-regulation)** or positive feedback **(up-regulation).**

7. The brain releases two neurohormones, **antidiuretic hormone** and **oxytocin,** by way of the **posterior pituitary.** Using **releasing** and **release-inhibiting hormones,** it controls the release of at least seven **anterior pituitary** hormones. Four of the anterior pituitary hormones (ACTH, GH, FSH, and LH) are trophic hormones for endocrine glands.

8. The three major chemical classes of hormones are: **steroids, peptides/proteins,** and **amino acid derivatives.**

9. Steroid hormones are recognized by **mobile cytoplasmic receptors;** typically their action involves increases in the synthesis of specific proteins.

10. Peptide/protein hormones and amino acid derivatives are recognized by **fixed receptors** on the cell membrane; binding of hormone to receptor results in activation of an intracellular **second messenger.** Common second messengers are **cyclic adenosine monophosphate (cAMP), Ca^{++}, inositol trisphosphate,** and **1-2 diacylglycerol.** The effects of second messengers typically involve **phosphorylation** of an **allosteric control site** on key enzymes of the target cells by activated **protein kinase** enzymes.

● STUDY QUESTIONS

1. What are the basic elements of a feedback loop?

2. Why is the gain of a feedback loop important?

3. What happens when there is a delay between the origin of an error signal and the response of the effector?

4. At what two general levels of integration is negative feedback exerted?

5. Name some reflexes that act as negative feedback loops.

6. What are the major chemical classes of hormones?

7. What determines whether a given hormone will affect a particular target organ?

8. What are the two main intracellular second messenger systems? For each, what are the steps that lead from arrival of the hormone at the target cell to the response of the target cell?

9. What membrane constituents are the starting material for prostaglandin synthesis? At what level of regulation do prostaglandins typically operate?

10. What are the two senses of meaning of the term *trophic hormone*? Name some examples of trophic hormones and state the target tissue that each controls.

11. What are the functional differences between the two lobes of the pituitary?

12. Why might it be a mistake to attempt to evaluate the rate of secretion of a hormone on the basis of a single blood sample?

Choose the MOST CORRECT Answer.

13. This substance is released during an inflammatory response:
    a. Growth factor
    b. Melatonin
    c. Endothelin
    d. Histamine

14. This class of reflexes control activities of smooth muscle, exocrine glands and the heart:
    a. Somatic motor reflexes
    b. Autonomic reflexes
    c. Neurosecretory reflexes
    d. Endocrine reflexes

15. These receptors monitor changes in blood pressure:
    a. Chemoreceptors
    b. Baroreceptors
    c. Thermoreceptors
    d. Osmoreceptors

16. Which of the following is NOT a chemical class of hormone?
    a. Lipopolysaccharides and their derivatives
    b. Derivative of tyrosine
    c. Peptides and proteins
    d. Steroids and their derivatives

17. The "first messenger":
    a. Binds by receptors on or in the target cells
    b. Biochemically can be referred to as a ligand
    c. When bound causes activation of one or more intercellular second messenger
    d. All of the above are true

18. This hormone is a derivation of the amino acid tyrosine
    a. Growth hormone
    b. Testosterone
    c. Vitamin D
    d. Dopamine

19. Steroid hormones:
    a. Form complexes with protein to travel through blood plasma
    b. Are not water-soluble
    c. Dissolve in cell membranes
    d. All of the above are true

20. A common "second messenger system" that functions in the human body involves the compound:
    a. Estradiol
    b. cAMP
    c. Cycloxygenase
    d. Norepinephrine

21. Which of the following does NOT function as a "second messenger system"?
    a. cAMP
    b. cGMP
    c. Phosphoinositides
    d. Lipopolysaccharides

22. Pancreatic hormone that increases uptake and use of glucose and amino acids:
    a. Insulin
    b. Glucagon
    c. Somatostatin
    d. Cortisol

● SUGGESTED READINGS

DECHERNEY AH, NAFTOLIN F: Recombinant DNA-derived human luteinizing hormone, *JAMA* 259:3313, June 10, 1988. Discusses one case in which recombinant DNA technology has been rapidly applied to an important problem in clinical medicine.

HARDIE DG: *Biochemical messengers: Hormones, neurotransmitters, and growth factors.* London, 1991, Chapman & Hall. An overview of endocrinology from the molecular viewpoint.

Growing up with the endocrine system, *Current Health* 1:3, February 1987. Discussion of the structure and function of the endocrine system, including neuroendocrine mechanisms.

HOUK JC: Control strategies in physiological systems, *FASEB Journal* 2:97, February 1988. An article written for a very general audience that discusses electrical and chemical messages and control systems from both a cellular and integrative perspective.

McEWEN BS, et al.: Neuroendocrine aspects of cerebral aging. *Int J Clin Pharmacol Res* 10:7, 1990. Shows how age-related degeneration of neural tissue is the complex result of multiple factors that synergize to cause neural destruction, including endogenous excitatory amino acids, calcium ions, endogenous proteolytic enzymes, free radicals, and circulating glucocorticoids.

RUBANYI GM, PARKER BOTELHO LH: Endothelins. *FASEB Journal* 5:2713-2720, 1990. Summarizes what is currently known about the structure and function of endothelins and their possible role in high blood pressure and cardiovascular disease.

THOMPSON CC, WEINBERGER C, LEBO R, EVANS RM: Identification of a novel thyroid hormone receptor in the central nervous system. *Science* 237:1610, September 25, 1987. Shows how thyroid hormone influences the development of the central nervous system.

CHAPTER *5 FOCUS UNIT*

ENDOCRINE ORGANS

The term "hormone", from a Greek word meaning "to excite or arouse," was introduced by the pioneer British physiologists W.M. Bayliss and E.H. Starling in 1905. The fundamental concept of classical endocrinology is that of a ductless gland, a specialized secretory organ, that liberates small amounts of a highly specific chemical substance—a hormone—into the bloodstream to act on a distant target organ or tissue. The classically-recognized endocrine glands (Figure 5-*A*) are the pituitary, thyroid, parathyroid, thymus, adrenal, pancreas, gonads (testes in males, ovaries in females), and placenta (present in pregnant women and not shown in Figure 5-*A*). The pituitary plays a central role in classical endocrinology through its secretion of four trophic hormones and at least four primary hormones (Figure 5-*B*). With time, the classical list of endocrine organs was expanded to include the brain, heart, kidney, liver, and the gastrointestinal tract. Ironically, secretin, the substance for which Bayliss and Starling coined the name "hormone," is a distinctly nonclassical hormone. It was first shown to be produced by modified epithelial cells in the intestinal lining and to stimulate pancreatic secretion. It is now known to serve also as a neurotransmitter—a messenger substance released by neurons at points of contact with other neurons or with effectors such as muscles or glands.

It is now apparent that almost every cell type both sends and receives chemical messages. Furthermore, as in the example of secretin, the same substance may be used for communication between a number of different cell types, carrying a different message in each case. It may sometimes be carried to its targets by the blood, sometimes diffuse to nearby targets as a paracrine agent, be released by neurons, or serve as an autocrine agent or growth factor. These advances in understanding the nature and complexity of chemical communication between cells make the concept of a single integrated endocrine system obsolete. Furthermore, many hormones regulate more than one physiological process and, conversely, an individual organ system may be regulated by several hormones. For example, antidiuretic hormone, aldosterone, and atrial natriuretic hormone are all released from different sites and interact to affect cardiovascular performance and fluid and electrolyte balance. The cardiovascular response to exercise and stress involves the autonomic nervous system, cortisol, and epinephrine. Such considerations preclude the use of a gland-by-gland approach to endocrinology. Instead, this text introduces the general principles of control and communication between cells and integrates endocrine and autonomic control systems with the basic physiology of each of the organ systems.

This Focus Unit concentrates on some general features of endocrine organs that cannot be treated conveniently in the context of specific organ systems, explains characteristic mechanisms of endocrine disease, and gives a tabular overview of each of the recognized endocrine systems with page references to the text.

GENES AND HORMONES: SYNTHESIS OF PEPTIDE AND PROTEIN HORMONES

As with all proteins, the amino acid sequences of peptide and protein hormones are dictated by the nucleotide sequences of particular genes. The general features of this process were described in Chapter 3. The primary transcript of mRNA from each gene contains both exons and introns. Before the mRNA transcript leaves the nucleus, the introns are removed by gene splicing. In almost all cases that have been thoroughly studied, the resulting mRNA codes for a protein larger than the final hormone. In some cases at least part of the extra size may be accounted for by the presence of a signal sequence on the amino end of the protein that targets it to the endoplasmic reticulum for secretion. The form of hormone containing the signal sequence is called a prehormone (Figure 5-*C*); the signal sequence is clipped off as the protein enters the ER, yielding the final hormone. In some cases, the original protein is a preprohormone (Figure 5-*C*) that undergoes additional posttranslational modification in the endoplasmic reticulum. In this process, the sequence of the prohormone is enzymatically clipped into two or more pieces. In some cases, such as that of insulin (Figure 5-*D*), posttranslational modification results in a functional hormone and one or more nonfunctional polypeptide fragments. In other cases, more than one hormone sequence may be contained in the sequence of the prohormone.

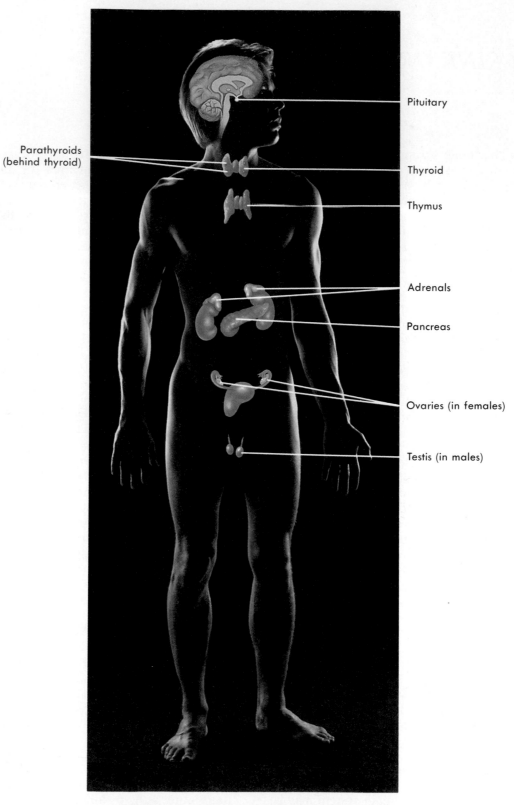

Parathyroids
(behind thyroid)

Pituitary

Thyroid

Thymus

Adrenals

Pancreas

Ovaries (in females)

Testis (in males)

FIGURE 5-A
Location of the major endocrine glands of the body.

THE FOUNDATIONS OF PHYSIOLOGY

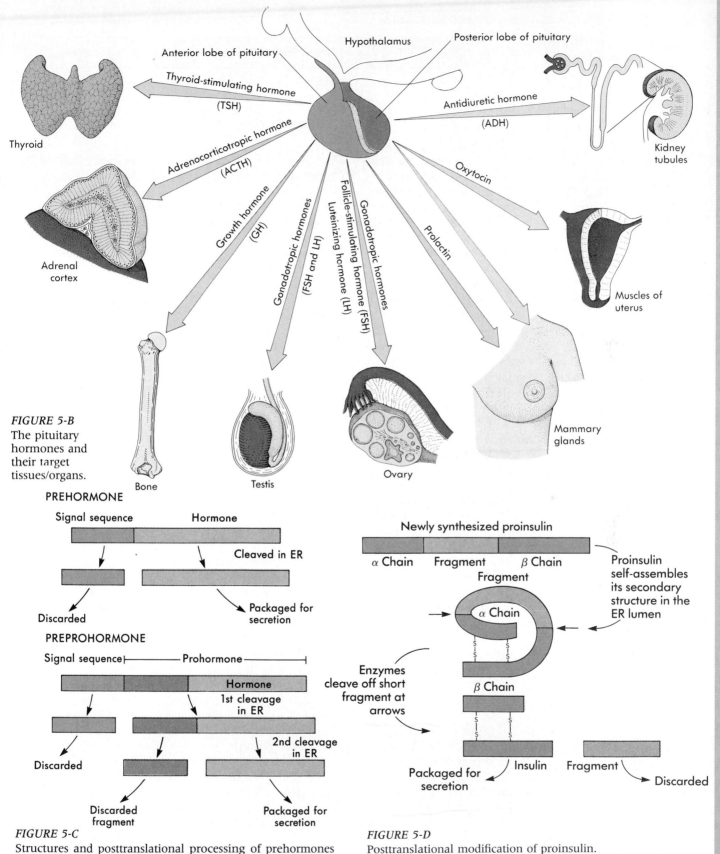

FIGURE 5-B
The pituitary hormones and their target tissues/organs.

PREHORMONE

Signal sequence   Hormone

Cleaved in ER

Discarded   Packaged for secretion

PREPROHORMONE

Signal sequence —————— Prohormone ——————

Hormone
1st cleavage in ER

Discarded

2nd cleavage in ER

Discarded fragment   Packaged for secretion

FIGURE 5-C
Structures and posttranslational processing of prehormones and preprohormones.

Newly synthesized proinsulin

α Chain   Fragment   β Chain

Fragment

Proinsulin self-assembles its secondary structure in the ER lumen

α Chain

Enzymes cleave off short fragment at arrows

β Chain

Packaged for secretion   Insulin   Fragment   Discarded

FIGURE 5-D
Posttranslational modification of proinsulin.

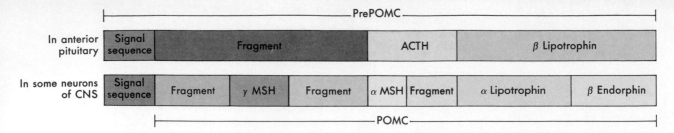

		PrePOMC		
In anterior pituitary	Signal sequence	Fragment	ACTH	β Lipotrophin

In some neurons of CNS	Signal sequence	Fragment	γ MSH	Fragment	α MSH	Fragment	α Lipotrophin	β Endorphin

POMC

FIGURE 5-E
Posttranslational processing of proopiomelanocortin (POMC) yields different messengers in different cell types.

The best studied example of a prohormone that contains multiple messengers is proopiomelanocortin (POMC), a protein synthesized in both the anterior pituitary and the nervous system. The POMC molecule can be clipped in various ways to yield six different chemical messengers (Figure 5, E). In the anterior pituitary, the cells responsible for secretion of adrenocorticotrophic hormone (ACTH) cleave POMC into ACTH; further enzymatic processing of the remaining fragments yields β lipotrophin and β endorphin, which are secreted along with ACTH. β Endorphin is one of a family of polypeptides that serves as modulators in nervous pathways involved in pain, appetite, and reward; its release into the blood along with ACTH may contribute to a decrease in pain sensitivity during stress. Little is known of the function of β lipotrophin. A different pattern of posttranslational modification would yield three forms of melanocyte stimulating hormone (MSH). Melanocyte stimulating hormone may be important in animals that undergo changes in body color but is not believed to be secreted into the blood in significant amounts in adult humans. The nervous system contains defined sets of neurons that use ACTH, β endorphin or MSH as neurotransmitters. In these neurons processing of the POMC protein is assumed to yield mainly the corresponding messenger.

MECHANISMS OF ENDOCRINE DISEASE
Endocrine diseases can be divided into three classes:
1. Hormone deficiency or hyposecretion diseases
2. Hormone excess or hypersecretion diseases
3. Transduction failure diseases, in which adequate hormone is secreted but target cells fail to respond appropriately.

All three classes cause failure of the control system served by that particular hormone, with consequences for all the physiological functions regulated by that system. Understanding of the normal function of the feedback loops that regulate hormone secretion is essential for diagnosis of the causes of endocrine diseases. In particular, deficiency or excess of a particular hormone may be caused by malfunction of the tissue that secretes that hormone (a primary hypo- or hypersecretion), or by inappropriately high or low levels of trophic hormone (a secondary hypo- or hypersecretion). Measurement of the level of trophic hormone provides essential information about the seat of the problem.

Hormone-Deficiency Diseases
Hormone deficiency can result from damage to the gland or tissue that secretes the deficient hormone. In juvenile-onset diabetes mellitus the damage is believed to be the result of an attack on the pancreatic beta cells by the body's own immune system. Such erroneous attacks are called autoimmune responses. In people with genetic susceptibility to develop diabetes mellitus, the autoimmune response may be triggered by a viral infection early in life.

Hormone deficiency can also result from inheritance of defective genetic code. As discussed in detail in Chapter 25, each somatic or nonreproductive body cell has two copies of each chromosome, one inherited from each parent. Genes may undergo a variety of accidents. The simplest accidents are point mutations that alter a single base of the DNA sequence and result in substitution of a single amino acid in the primary sequence of the protein coded for by the affected gene. In some cases the effect of point mutation is trivial or rarely even beneficial; in most cases it makes synthesis of a functional protein impossible. Occasionally a chromosome becomes twisted across itself, forming a loop. The chromosome may break at the point of contact and then rejoin in such a way that the loop is detached from the chromosome. The loss of DNA from the chromosome is called a deletion. Usually a defect in one chromosome is "backed up" by normal code in the other, but this backup fails if the code is defective or missing at the same site, or allele, on both chromosomes. For example, the gene for growth hormone is carried on chromosome 17; one form of growth retardation is caused by inheritance of copies of chromosome 17, both of which lack the gene for growth hormone. Both parents of the affected person may have grown normally; in that case each parent probably carried one copy of the defective allele which was backed up by a normal allele.

Similarly, a steroid hormone deficiency may arise from inheritance of defective code for one of the enzymes in the hormone's synthetic pathway. The adrenal cortical hormones provide a number of examples of such defects. The adrenal cortex normally produces three types of steroid hormones: the

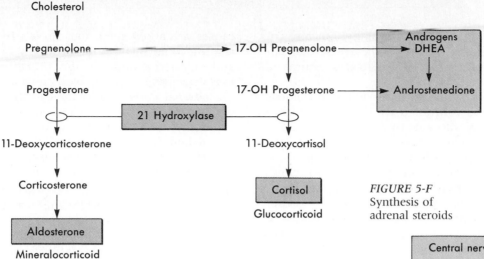

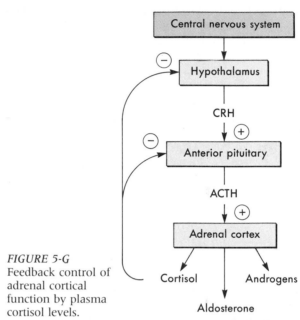

FIGURE 5-F
Synthesis of
adrenal steroids

FIGURE 5-G
Feedback control of
adrenal cortical
function by plasma
cortisol levels.

glucocorticoid cortisol, involved in control of blood glucose (see Chapter 5) and in resistance to stress and infection; the mineralocorticoid aldosterone, which stimulates the kidneys to conserve Na^+ (see Chapter 19), and small amounts of the androgens dehydroepiandrosterone (DHEA) and androstenedione. Androgens are steroids that, like the male sex hormone testosterone, cause the development of male secondary sexual structures in early development, as well as the male-specific traits that appear at sexual maturation or puberty. A simplified diagram of the pathways for synthesis of the adrenal cortical hormones is shown in Figure 5, *F*.

The most common inherited defect of adrenal steroid synthesis involves 21-hydroxylase, an enzyme that catalyzes two steps: the conversion of 17-hydroxyprogesterone to 11-deoxycortisol, a precursor of cortisol, and of 17-hydroxyprogesterone to 11-deoxycorticosterone, a precursor of aldosterone (Figure 5, *F*). In about a quarter of the cases, most or all of the gene for the enzyme is lost; in the remaining cases, the gene appears to have undergone mutation so that a defective enzyme is produced. In either case, neither cortisol nor aldosterone can be synthesized. The resulting condition of adrenal cortical hyposecretion causes one variety of Addison's syndrome. The lack of cortisol increases susceptibility to infection, and the lack of aldosterone interrupts one of the endocrine feedback loops involved in control of extracellular fluid volume, causing chronic low blood pressure.

The blockage of cortisol synthesis by 21-hydroxylase deficiency opens the feedback loop between the anterior pituitary and adrenal cortex that normally controls cortisol production (Figure 5, *G*; see also Figure 5-20), allowing ACTH secretion to reach very high levels. Indeed, the finding of a high level of ACTH in such cases confirms the diagnosis that the cortisol deficiency is due to a failure of the adrenal cortex rather than the anterior pituitary. The resulting high levels of ACTH secretion stimulate the adrenal cortex to produce maximal levels of all of the precursors formed before the 21-hydroxyla-

tion steps. However, since the branches of the synthetic pathway that lead to cortisol and aldosterone are blocked, all of the precursors are converted to DHE and androstenedione.

The excess androgen of 21-hydroxylase defect causes some of the most distressing symptoms of the condition, since high levels of androgen during a critical period before birth may interfere with differentiation of the external genitalia. In normal prenatal sexual differentiation, the external genitalia of female fetuses automatically take the female pattern; androgen secretion by the testes of male fetuses directs development of male external genitalia (see Chapter 25). Normally adrenal androgen production is not high enough to interfere with this process, but female fetuses with the 21-hydroxylase defect usually have masculinized external genitalia. Sometimes the masculinization is so complete that it causes the sex of the baby to be misassigned at birth. In males, the defect is not apparent at birth, but causes a premature puberty with appearance of adult male sexual characteristics between 3 and 10 years of age. Treatment of the 21-hydroxylase deficiency includes regular administration of cortisol. This not only replaces the cortisol that the adrenal cortex would have produced normally, but also inhibits secretion of ACTH, with the result that androgen production by the adrenal cortex returns to approximately normal levels.

Deficiencies of one or more of the hormones whose secretion is controlled by the anterior pituitary are frequently traceable to a failure of the anterior pituitary to secrete adequate amounts of the corresponding trophic hormones. For example, one form of depressed secretion of thyroid hormones (secondary hypothyroidism) is due to damage to the anterior pituitary, to the hypothalamus, or to impaired circulation through the portal blood vessels that link the hypothalamus and anterior pituitary, while the thyroid remains intact. Secondary hypothyroidism is characterized by depressed blood levels of both the thyroid hormones and of TSH; high levels of TSH would reflect the normal response of the hypothalamus-anterior pituitary in primary hypothyroidism, caused by failure of the thyroid gland itself.

Hormone Excess Diseases

As with deficiency diseases, hormone excess may be caused by a disorder of the endocrine tissue that secretes the final hormone (a primary hypersecretion) or by excessive secretion of the corresponding trophic hormone (a secondary hypersecretion). Cancers of the pituitary frequently secrete large amounts of one or more anterior pituitary hormones; these cells apparently do not respond to the feedback signals that regulate secretion in normal anterior pituitary.

An example of hormone excess disease is Cushing's syndrome, in which there is hypersecretion of both cortisol and aldosterone by the adrenal cortex. The cortisol causes elevated blood glucose levels as in diabetes mellitus, along with a distinctive concentration of fat in the upper body. The aldosterone causes a marked retention of Na^+ by the kidneys, leading to an increase in extracellular fluid volume and elevated blood pressure. Cushing's syndrome may arise from a spontaneous increase in activity of the adrenal cortex (primary Cushing's syndrome), or from an excess of ACTH (secondary Cushing's syndrome). The two causes of the syndrome are distinguished by measurement of plasma levels of ACTH. If they are low, the disease is primary; if high, the disease is secondary. In the secondary disease, the ACTH usually comes from a pituitary tumor, but it may be secreted by a tumor in some other part of the body.

Transduction Failure Diseases

Any defect in the receptor for a hormone or in the second messenger system utilized by the target cells can result in a disease that has all the symptoms of a deficiency of the hormone. Type II diabetes mellitus, discussed in Chapter 5, is one example of such a transduction failure disease. Another striking example is androgen-insensitivity or testicular feminization, in which genetic males whose testes secrete normal amounts of testosterone, nevertheless have female external genitalia at birth and undergo a feminizing puberty. This disorder is caused by mutation or loss of the gene for the androgen receptor, one of the nuclear steroid receptors described in Chapter 5. As described before and in Chapter 25, female external genitalia develop automatically unless affected by androgen. The feminizing puberty is the result of small amounts of the female sex hormone estrogen that are normally produced by the testes in the course of their synthesis of testosterone. Individuals with this syndrome typically receive medical attention because they are infertile and sexually unresponsive; otherwise they appear and behave as normal women.

THE FOUNDATIONS OF PHYSIOLOGY

Pituitary

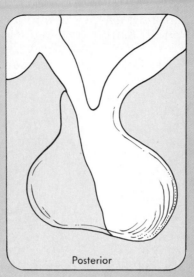

Posterior

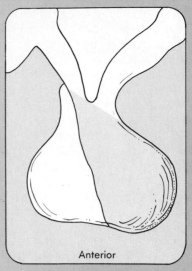

Anterior

Hormone	Source	Target	Action/Reference
Adrenocorticotropic hormone	Anterior lobe	Adrenal cortex	Secretion of adrenal cortex hormones p. 669
Follicle-stimulating hormone	Anterior lobe	Ovaries, testes	Oogenesis p. 694 Spermatogenesis p. 702 Menstrual cycle p. 707
Growth hormone	Anterior lobe	Most tissues	Anabolism, growth and development p. 671
Luteinizing hormone	Anterior lobe	Ovaries, testes	Oogenesis p. 695 Spermatogenesis p. 702 Menstrual cycle p. 707
Melanocyte-stimulating hormone	Anterior lobe	Varied	Believed insignificant in humans after birth p. 116
Prolactin	Anterior lobe	Mammary glands	Milk production p. 745
Thyroid-stimulating hormone	Anterior lobe	Thyroid	Production and secretion of thyroid hormone p. 678
Antidiuretic hormone	Posterior lobe	Kidney	Water reabsorption p. 569, 426-427, 547-550
Oxytocin	Posterior lobe	Mammary glands, uterus	Milk letdown p. 745 Uterine contraction p. 742

Thyroid and Parathyroid

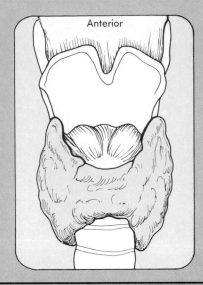
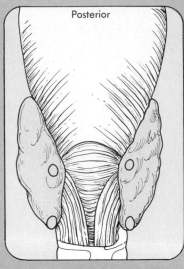

Hormone	Source	Target	Action/Reference
Calcitonin	Thyroid	Bone	Bone deposition p. 577 Plasma calcium concentration p. 578
Thyroxine	Thyroid	Most tissues	Growth and development p. 675 Response to stress and cold p. 678
Parathormone	Parathyroid	Bone, kidneys	Calcium and phosphate homeostatis p. 577

Hypothalamus

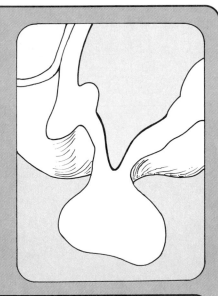

Hormone	Source	Target	Action/Reference
Corticotropin-releasing hormone	Hypothalamus	Anterior pituitary	Increased ACTH secretion p. 670
Gonadotropin-releasing hormone	Hypothalamus	Anterior pituitary	Increased FSH, LH secretion p. 695
Growth-hormone–releasing hormone	Hypothalamus	Anterior pituitary	Increased GH secretion p. 671
Prolactin-inhibiting hormone	Hypothalamus	Anterior pituitary	Decreased Pr secretion p. 745
Prolactin-releasing hormone	Hypothalamus	Anterior pituitary	Increased PR secretion p. 745
Somatostatin	Hypothalamus	Anterior pituitary	Decreased GH secretion p. 671
Thyrotropin-releasing hormone	Hypothalamus	Anterior pituitary	Increased TSH secretion p. 678

Adrenals and Kidneys

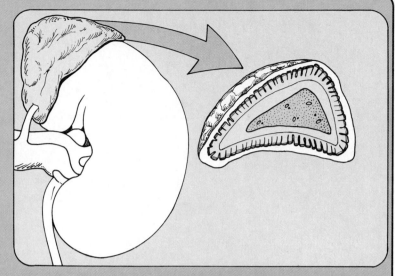

Hormone	Source	Target	Action/Reference
Aldosterone	Adrenal cortex	Kidney, most tissues	Na retention, K excretion p. 572
Androgens	Adrenal cortex	Most tissues	In females, sexual drive, hair growth p. 700
Cortisol	Adrenal cortex	Most tissues	Catabolism p. 667
			Inhibition of inflammation p. 755
Epinephrine	Adrenal medulla	Most tissues	Response to stress p. 669
Cholecalciferol (DOHCC)	Kidney	Intestine	Calcium and phosphate homeostasis p. 577
Erythropoietin	Kidney	Bone marrow	Erythrocyte synthesis p. 393

Reproductive Hormones

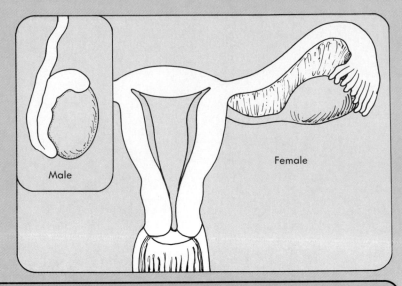

Male

Female

Hormone	Source	Target	Action/Reference
Testosterone	Testes	Most tissues	Spermatogenesis p. 690 Secondary sex characteristics p. 697
Inhibin	Gonads	Pituitary	Feedback from gonads to the pituitary p. 702
Estrogen (Estradiol)	Ovaries	Most tissues	Menstrual cycle p. 711 Development of mammary glands and uterus p. 698 Secondary sex characteristics p. 700
Progesterone	Ovaries	Most tissues	Menstrual cycle p. 711 Development of mammary glands and uterus p. 698 Secondary sex characteristics p. 700
Human chorionic gonadotropin (HCG)	Placenta	Corpus luteum	Maintain corpus luteum in pregnancy p. 730
Human placental lactogen	Placenta	Mammary glands, fetus	Preparation for lactation p. 732 Fetal bone growth p. 732
Relaxin	Placenta	Bones and ligaments of pelvis, cervix	Preparation for delivery p. 732
Estrogen	Placenta	Breasts, uterus	Maintains uterine lining p. 711 Stimulates breasts and menstrual cycle p. 744
Progesterone	Placenta	Breasts, uterus	Inhibits uterine contraction p. 697 Stimulates breasts and menstrual cycle p. 744

Gastrointestinal Tract

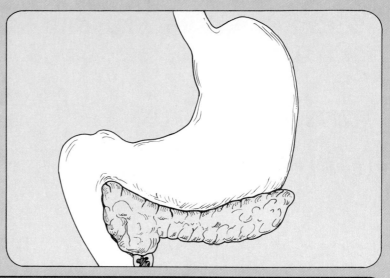

Hormone	Source	Target	Action/Reference
Cholecystokinin (CCK)	Duodenal mucosa	Gallbladder, pancreas	Contraction of gallbladder p. 648 Pancreatic enzyme secretion p. 648
Gastric inhibitory peptide (GIP)	Duodenal mucosa	Stomach, pancreas	Inhibits stomach motility p. 647 Increases insulin and glucagon secretion p. 658 At higher concentrations, motility and secretion p. 648
Gastrin	Stomach mucosa	Stomach	Acid secretion p. 643 Motility p. 646
Motilin	Stomach mucosa	Stomach, intestine	Motility p. 647
Secretin	Duodenal mucosa	Stomach, pancreas	Inhibits acid secretion and gastrin release p. 647 Pancreas enzyme and fluid secretion p. 648
Glucagon	Pancreas (alpha cells)	Liver	Increases breakdown of glycogen p. 662
Insulin	Pancreas (beta cells)	Liver, muscle, fat	Increases uptake and use of glucose and amino acids p. 656
Somatostatin	Pancreas (delta cells)	Alpha and beta cells	Inhibits insulin and glucagon secretion p. 662

Other Structures

Hormone	Source	Target	Action/Reference
Atrial natriuretic hormone (ANH)	Atrium	Kidney	Regulates solute and water loss p. 572
Thymosin	Thymus	Immune tissues	Development and function of the immune system p. 764
Prostaglandins	Most tissues	Most tissues	Inflammation p. 755 Uterine contraction p. 742
Melatonin	Pineal	Gland	Brain p. 223 Regulation of circadian rhythm p. 241

Transport Across Cell Membranes

On completing this chapter you should be able to:

- Describe the bilayer structure of cell membranes, identifying intrinsic and extrinsic proteins.
- Understand why pure lipid bilayers are impermeable to polar solutes and ions.
- Understand the factors that determine the rate of diffusion.
- Define the terms isotonic, hypotonic, and hypertonic, and be able to use them to determine whether cells will change their volume.
- Understand why cell volume regulation requires that Na^+ be effectively impermeant.
- Describe the characteristics of mediated transport systems.
- Distinguish among facilitated diffusion, primary active transport, and secondary or gradient-driven active transport.
- Describe the origin of the membrane potential, and be able to describe the effect on the potential of changes in the permeabilities and concentration gradients of Na^+ and K^+.

*T*he cell membrane is a thin film of lipid, studded with proteins, that defines the boundaries of each cell. It maintains an appropriate chemical environment for the metabolic processes of the cell, regulates the volume of cytoplasm, and mediates information transfer in the form of chemical and electrical signals. This chapter is about the cell membrane and how it permits or even promotes entry and exit of specific solutes and water. These **membrane transport processes** are critical to the life of individual cells, and they play an important part in the function of organs that form boundaries between different compartments of the body, such as the epithelial lining of the gastrointestinal and respiratory systems, the kidney, and exocrine glands.

The cell membrane is not an absolute barrier between the interstitial fluid and the cytoplasm. While some substances are excluded, there is a continuous molecular traffic across the membranes of living cells. Substances that cross the membrane include water, nutrients such as glucose and amino acids, and ions such as Na^+, K^+ and Cl^-. The membrane is said to be **permeable** to those substances that are able to cross it and **impermeable** to those that do not cross. (The corresponding adjectives for the substances are **permeant** and **impermeant**.) The permeability of a membrane to a particular substance is expressed as the rate of solute movement across a unit of membrane surface area per unit of driving force.

If the cell membrane was composed of pure phospholipid, it would be permeable only to nonpolar solutes. The fact that the membrane is selectively permeable to specific polar solutes and impermeable to others is due to the presence of specialized proteins that span the membrane. The selective permeability of the membrane to particular polar solutes is thus ultimately determined by the cell's genes. However, the permeability to particular solutes can be changed by intracellular chemical messages or by changes in the cell's environment.

THE MOLECULAR STRUCTURE OF CELL MEMBRANES
Properties of Phospholipids in Membranes

- *What is a "lipid bilayer"?*
- *What are the forces that cause the phospholipids of cell membranes to take the form of a bilayer?*
- *What are the features of the fluid mosaic model of a cell membrane?*
- *How do extrinsic and intrinsic membrane proteins differ?*
- *What are the general functional classes of membrane proteins?*

Membranes contain mainly phospholipids, together with cholesterol and smaller amounts of a few other lipids. When they are surrounded by water molecules, the phospholipid molecules behave in a fashion reminiscent of a herd of musk oxen surrounded by wolves. The hydrophilic "heads" of the phospholipids point outward toward the water while the hydrophobic hydrocarbon chain "tails" face away from the water and interact with one another. This is the state with the lowest free energy and thus the highest stability. A small amount of phospholipid in water can form a spherical **micelle** with all the tails in contact with one another (see Figure 3-3, *B*). When the phospholipid is arranged as a film between two aqueous environments, as is the case of the cell membrane and the membranes of intracellular organelles, the state with the lowest free energy consists of two monolayers of phospholipids with the tails in each monolayer oriented toward the inside of the membrane. This structure is called a **lipid bilayer** (see Figure 3-3, *C*). The hydrocarbon interior of the membrane forms a barrier that is effectively impassable to polar solutes, such as amino acids, sugars, and ions. Water molecules are an exception to the rule. In spite of their polarity, they can pass through lipid bilayers with relative ease because they are small enough to slip between the hydrocarbon tails.

The Fluid Mosaic Model of Proteins and Phospholipids in the Membrane

Cellular membranes consist of a mixture of lipid and protein. The current understanding of roles of the two constituents in the structure of a cell membrane as shown in Figure 6-1 is called the **fluid mosaic model**. The phospholipids can move freely in the plane of the membrane, so the membrane is fluid rather than solid. Many (but not all, as explained below) membrane proteins are also free to wander about within the plane of the membrane. One piece of evidence for fluidity of cell membranes was provided by experiments in which certain proteins were labelled by application of a fluorescent label that bound to the extracellular portions of the

proteins, allowing their location on the cell surface to be tracked. Initially, the marked proteins were scattered over the cell surface, but when an antibody that crosslinked membrane proteins was added, clusters of membrane proteins were formed. These experiments showed that proteins may move freely within the plane of the membrane. However, the freedom of movement of proteins does not include rotation; the protein-phospholipid interactions are strong, and therefore the energy required to turn a protein around so that its external and internal ends are interchanged would be very large. Membrane proteins are thus like giants in a crowd; they can shuffle along, but they cannot turn handsprings.

Membrane proteins can be divided into two classes, **extrinsic** and **intrinsic**, based on how tightly they interact with membrane phospholipids. **Extrinsic** proteins are loosely bound to the external or internal surfaces by electrostatic forces and can be removed by mild chemical procedures, such as changes in salt concentration, that disrupt weak electrostatic forces (see Chapter 3).

Intrinsic or **integral** membrane proteins are difficult to dissociate from membrane lipids; drastic treatments are required, such as addition of strong detergents or organic solvents. The primary sequence of intrinsic proteins consists of specialized regions, or **domains**. Some domains contain mainly amino acids with hydrophobic side groups, which lie within the lipid bilayer because they associate strongly with the hydrophobic tails of the membrane phospholipids. Other domains contain mainly hydrophilic amino acids; these domains prefer to associate with water and thus extend into either the cytoplasm or the extracellular fluid. Some intrinsic membrane proteins extend completely across the membrane, while others only partially penetrate it (Figure 6-1 shows space-filling representations of membrane proteins). In membrane proteins that extend across the membrane, the membrane-spanning domains generally have an alpha-helical secondary structure (in Figure 6-2 the secondary structure of intramembrane domains is shown; also see Figure 3-12). In many cases, portions of a membrane protein that project into the extracellular fluid have carbohydrate chains attached to them, forming a glycoprotein (Figure 6-1).

Functions of Membrane Proteins

The functions of membrane proteins fall into six general categories (Table 6-1; Figure 6-3), with some overlap between categories. As described in Chapter 5, **receptor proteins** are involved in converting chemical messages arriving at the membrane surface into intracellular responses. Others, called **recognition proteins**, serve as identity tags that allow transplanted cells or cancer cells to be

THE FOUNDATIONS OF PHYSIOLOGY

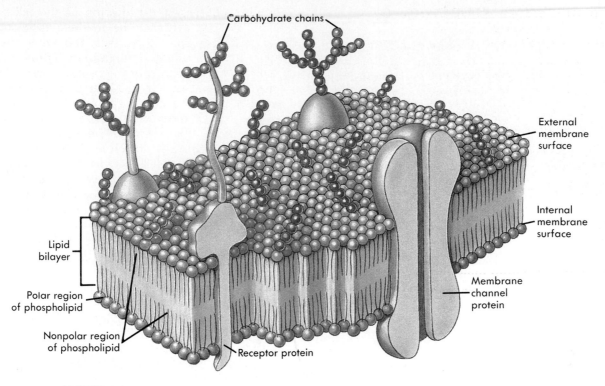

Carbohydrate chains

External
membrane
surface

Internal
membrane
surface

Membrane
channel
protein

Lipid
bilayer

Polar region
of phospholipid

Nonpolar region
of phospholipid

Receptor protein

FIGURE 6-1
Fluid mosaic model of a cell membrane composed of a lipid bilayer with proteins "float-ing" in the membrane. The nonpolar hydrophobic portion of each phospholipid molecule is directed toward the center of the membrane, and the polar hydrophilic portion is di-rected toward the water environment either outside or inside the cell. Intrinsic proteins partially or fully penetrate the membrane and may serve as transport or receptor mole-cules.

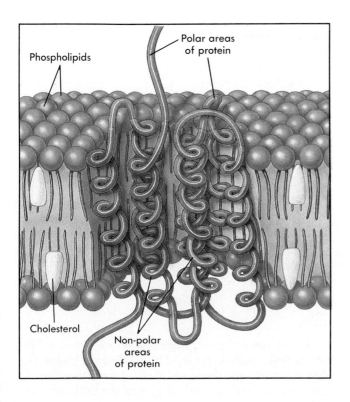

Phospholipids

Polar areas
of protein

Cholesterol

Non-polar
areas
of protein

FIGURE 6-2
Membrane proteins are anchored within the lipid bilayer by nonpolar segments having a helical secondary structure. Many membrane proteins consist of several hydrophobic segments within the membrane connected by hydrophilic segments projecting into the aqueous environment on either side of the membrane.

recognized by the immune system (see Chapter 26). **Transport proteins** confer permeability to specific polar solutes. Some transport proteins are also re-ceptors; in those cases a change in permeability to specific solutes is the way in which the receptor re-sponds to the message. As described in Chapter 4 (see Figure 4-17), **junctional proteins** form adhe-sions that hold adjacent cells together and, in the case of gap junctions, also link the cytoplasm of the joined cells.

Membrane-bound enzymes may be attached to the interior or exterior surface of membranes, cat-alyzing reactions that take place near the cell surface

TABLE 6-1   Functional Classes of Cell Membrane Proteins

Class	Function	Examples
1. Receptors	Bind molecules with specific properties	Acetylcholine receptor (Chapter 8)
2. Recognition proteins	Allow the immune system to distinguish cancer cells and invading organisms from normal body cells	Histocompatibility antigens (Chapter 26)
3. Transport proteins	Permit specific polar solutes and ions to cross the cell membrane; in some cases these proteins are also enzymes that split ATP	Na$^+$-K$^+$ ATPase (This chapter)
4. Junctional proteins	Form links between adjacent cells	Gap junctional protein (Chapter 4)
5. Enzymes	Catalyze specific reactions of substrates in intracellular or extracellular fluid	Acetylcholine esterase (Chapter 8)
6. Cytoskeletal anchors	To attach cytoskeleton to cell surface	Myofibril attachments and smooth and cardiac muscle (chapter 11)

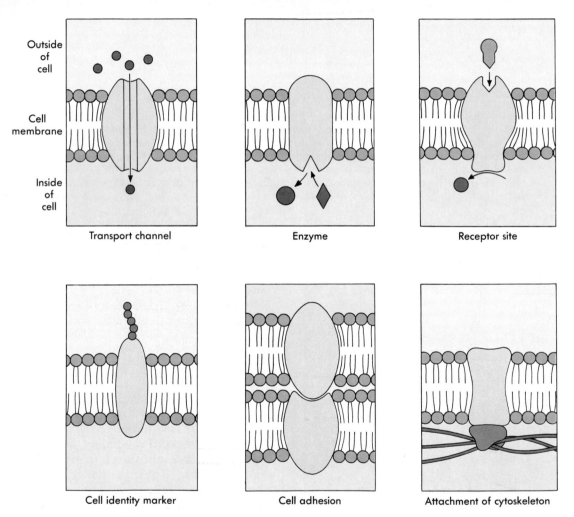

FIGURE 6-3
An illustration of the various functions membrane proteins may serve.

or generating intracellular second messengers in response to receptor binding. As will be shown later in this chapter, some membrane-spanning transport proteins are also ATPase enzymes.

Some membrane proteins are anchored to the intracellular cytoskeleton. There are two major reasons for such attachments: first, they allow the cytoskeleton to change the shape of the cell, and second, they allow the cytoskeleton to hold particular proteins in particular places on the cell surface. This protein anchoring allows some regions of cell membrane to be specialized for particular functions. For example, membrane proteins are segregated in myelinated nerve axons (Chapter 7) in which the transport proteins that allow the ion movements responsible for nerve impulses are confined to regions, called nodes, free of insulating myelin. Spatial segregation also occurs in muscle fibers (Chapter 12), in which the receptors that must be activated for contraction to occur are limited to regions called **endplates.** Typically, epithelial cells have lumen-facing and blood-facing surfaces with different functional specializations. The proteins associated with specific functions are confined to one face of such cells, as in epithelial cells of the gastrointestinal tract (Chapter 21) and renal tubules (Chapter 19). In some cases the proteins do not move laterally because they are attached to the cytoskeleton (Figure 6-3; see also Figure 4-14). The question of how different proteins get to the right membrane sites and stay there is a major issue in current research.

DIFFUSION
Random Motion and Diffusional Equilibrium

- *What is diffusion?*
- *What is the relationship between the two unidirectional fluxes of a solute that is in diffusional equilibrium?*

Diffusion is the movement of atoms, molecules, or small particles through space by random thermal motion. Molecules have greater freedom to diffuse in liquids than in solids, because there is less hindrance from surrounding molecules. The freedom to move is greatest in gases, where there is little or no contact between individual gas molecules. In solutions, the water molecules and solute particles collide continually with one another, so their pathways include frequent, random changes in direction (Figure 6-4). Diffusion is a spontaneous process that increases the disorder or entropy of the system (see Chapter 2 for the characteristics of spontaneous processes). Thus the equilibrium state of a system containing randomly moving molecules is one in which the disorder (entropy) is maximal, the free energy is minimal, and the solute molecules are, on the average, uniformly distributed throughout the system. If a **concentration gradient** is initially present in the system, the random motion of individual solute molecules causes net movement from areas of high concentration to those of low concentration, until the equilibrium state of uniform distribution is reached. For this reason, net movement of a substance from a region of higher concentration to one

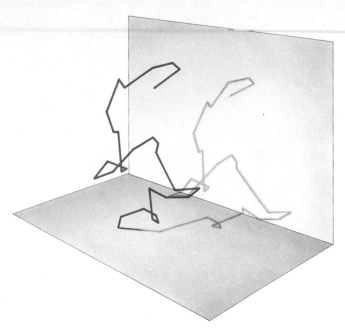

FIGURE 6-4
An illustration of the random path a molecule may take due to its inherent thermal energy. This random movement is the basis of diffusion.

of lower concentration is sometimes referred to as "passive" or "downhill."

The dispersion of concentration gradients by diffusion can be illustrated with an example of some crystals of glucose that are poured into an vessel containing unstirred water (Figure 6-5, *A*). In the first stage of equilibration, the water that is in the immediate vicinity of the glucose dissolves it, but all of the glucose is near the bottom of the vessel (Figure 6-5, *B*). The result is a gradient of glucose concentration in the vessel.

At any instant, some glucose molecules are moving toward the top of the vessel, others are moving toward the bottom, and some are moving sideways. Since the glucose concentration is initially high near the bottom, at any instant there are many more molecules near the bottom but moving toward the top than there are molecules near the top but moving toward the bottom. As a result, there is a large unidirectional **flux,** or flow of molecules toward the top. Toward the top of the vessel, the concentration is low, so the unidirectional flux from the top toward the bottom is low. The **net flux** is the algebraic sum of these two unidirectional fluxes. As more molecules approach the top, the concentration gradient diminishes (Figure 6-5, *C*), and ultimately the glucose will be uniformly distributed throughout the water and the system will have come into **equilibrium,** with equal concentrations at top and bottom and equal fluxes of glucose molecules moving in all directions (Figure 6-5, *D*).

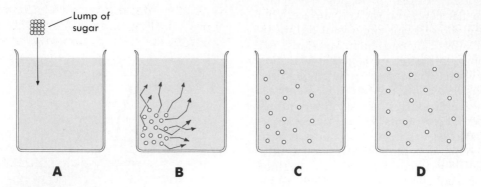

FIGURE 6-5
Diffusion. If a lump of glucose is dropped into a beaker of water, its molecules dissolve (A) and diffuse (B, C). At equilibrium the concentration of glucose is the same throughout the vessel (D), and the average number of glucose molecules moving in any given direction is approximately the same at any instant.

Determinants of the Rate of Diffusion

> • *What is Fick's Law of Diffusion?*
> • *What two factors determine the diffusion coefficient of a particular solute?*

In the presence of a concentration gradient, three factors determine the rate of net flux by diffusion of a substance between two points in space: (1) the velocity of the individual particles, (2) the magnitude of the concentration gradient, and (3) the dimensions of the path through which diffusion occurs. The first factor, the velocity of individual molecules, is determined by the temperature and the mass of the particles. This factor is expressed as a **diffusion coefficient** that is characteristic of each chemical species. Diffusion coefficients have units of cm^2/sec.

The second factor, the concentration gradient, represents stored energy because work must have been done to concentrate the solute. Although the movements of the individual particles are not affected by the presence or absence of a concentration gradient, it is useful to think of the stored energy of a concentration gradient as providing a driving force for net movement of the solute. Thus the greater the magnitude of the concentration gradient, the greater the rate of net movement by diffusion.

The third factor, the dimensions of the path available to the diffusing molecules, includes both the cross-sectional area and the path length. The greater the cross-sectional area through which the molecules are diffusing and the shorter the path length, the greater the net flux. As noted in Chapter 1, the rate of net movement of substances by diffusion is strongly affected by the length of the path. The accompanying Focus Unit gives an explicit mathematical treatment of this relationship. A substance with a diffusion coefficient of 1×10^{-5} cm^2/sec can approach diffusional equilibrium in 50 msec when the path length is 10μ (roughly the radius of a cell); for a distance of 1 cm the same substance would require 13 hours to approach diffusional equilibrium. This is the reason that most cells are small. Where diffusion is important for exchange of substances between the body and the environment, as in the lung and intestine, it is advantageous for the path length to be short and the area available for diffusion to be large. This advantage is reflected in structural adaptations of these organs that greatly increase the ratio of surface area to path length.

These three factors are incorporated in **Fick's Law of Diffusion** (see also the Focus Unit), which says that the net flux of a substance by diffusion between two points in space is equal to the product of the cross-sectional area, the diffusion coefficient for the substance at that temperature and the difference between concentrations of the substance at the two points.

Diffusion through Cell Membranes

The same principles apply to passive movement of substances through a cell membranes as to diffusion in free solution, but with one difference. This difference is the need to substitute for the diffusion coefficient the permeability of the membrane for the substance. This is done by substitution of a **permeability coefficient** for the diffusion coefficient in Fick's Law. For almost all solutes, the membrane permeability coefficients are much smaller (typically about a million times smaller) than diffusion coefficients. Since permeability is a property of the membrane, the membranes of different cell types may have different permeability constants for the same substance. Because in these calculations the membrane is taken as infinitely thin, the units of the permeability coefficient are cm/sec rather than cm^2/sec as in the diffusion coefficient.

If biological membranes were composed of pure lipid, the permeability of the membrane to a given

THE FOUNDATIONS OF PHYSIOLOGY

solute would be determined solely by the solute's molecular weight and lipid solubility. In this case, membrane permeability would be high to lipid-soluble hydrocarbons (such as fatty acids) and very small polar solutes (such as water). Membranes would be much less permeable to larger polar solutes. For example, the permeability of a pure lipid membrane to glucose is five orders of magnitude less than that to water.

Passage of ions directly through pure lipid membranes is particularly difficult because ions are surrounded by a **hydration shell** of water molecules oriented toward the charge of the ion (see Chapter 2). As a result, pure lipid membranes have permeability coefficients to ionic solutes (such as Na^+ and K^+) that are some 10 orders of magnitude less than the diffusion coefficients of the same ions in water. However, in contrast to pure lipid membranes, cell membranes may be quite permeable to glucose, Na^+, K^+, and other solutes that cross pure lipid membranes with great difficulty. Transport proteins, specific for each solute, allow these substances to bypass the energetic barrier posed by the membrane lipids.

Osmosis

- *What is osmosis?*
- *What is the relationship between molarity and osmolarity for any particular ideal solute?*

Osmosis is diffusion of water from a region of higher water concentration to one of lower concentration. Differences in water concentration arise because solute particles dilute the water (solvent) in which they are dissolved. For ideal solutes, the diluting effect of the solute on the water depends only on the concentration of solute particles and not on any particular features of the particles themselves.

The **osmolarity** of a solution is the number of moles of solute particles per liter of solution (see Chapter 2). Solutions with the same concentration of solute particles have the same osmolarity (note that some solutes require a correction for nonideal behavior), and are therefore termed **isosmotic**. A solution having a greater concentration of solute particles (hence a lower water concentration) than another is **hyperosmotic** to the comparison solution, while a solution with a lower concentration of solute particles (hence a higher water concentration) is **hyposmotic**.

The **osmotic pressure** of a solution is the driving force for water movement developed by the difference in water concentration between a particular solution in question and pure water (see also the Focus Unit). The osmotic pressure of any solution can be measured experimentally by placing the solution into a bag made of a membrane permeable to water (but not to the solute), connecting a vertical tube to the bag, and measuring the height of the solution in the tube (Figure 6-6). At equilibrium, the osmotic pressure is equal, but opposite in direction, to the hydrostatic pressure (see Chapter 2) exerted by the column of solution in the tube. Osmotic pressure can be expressed in centimeters of water (cm H_2O) but is usually given in millimeters of mercury (mm Hg) (see Chapter 2 for conversion between these units).

The greater the concentration of solute particles in a solution, the greater the difference between the water concentration of the solution and pure water, and thus the greater the osmotic pressure of the solution. At body temperature, a solution with an osmolarity of 1 milliosmole per liter (1 mOsm/L) produces an osmotic pressure difference of 19.3 mm Hg, compared with pure water. Two solutions differing by 1 mOsm (for example, 300 mOsm/L and 301 mOsm) would also have an osmotic pressure difference of 19.3 mm Hg between them.

Although water molecules are polar, their small size allows them to move rapidly through lipid bilayers. As a result of their high water permeability, lipid membranes are unable to oppose significant osmotic forces. At even a small osmotic pressure difference between the cytoplasm and the extracellular fluid there would be net water movement across the cell membrane, causing cells to swell or shrink. Although mechanisms for primary active transport of solutes are present in all cell membranes, there are no cellular mechanisms for direct active transport of water molecules. Consequently, osmotic force is responsible for all net movement of water between the interior of cells and interstitial fluid around them. Primary active transport of solutes by cells allows them to regulate their own volume, as described in greater detail in the following paragraphs. It also drives osmotic movement of water in epithelial tissues that absorb or secrete fluid. For example, solute transport drives fluid secretion in sweat and salivary glands, is responsible for fluid absorption across the wall of the intestine and for water reabsorption from urine in renal tubules.

Two concepts related to solute diffusion and osmosis are **dialysis** and **ultrafiltration.** Dialysis refers to the separation of smaller molecule from larger molecules by diffusion through a membrane permeable to only small molecules (see the boxed essay in Chapter 19). Ultrafiltration refers to the use of hydrostatic pressure to "push" water through a membrane permeable to water and some solutes against its concentration gradient, increasing the concentration of those impermeable solutes that are left behind. This process occurs to some extent across capillary walls in almost all parts of the body, and is important for generation of interstitial fluid. It is the first step in formation of urine by the kidney.

FIGURE 6-6

Osmotic pressure is the result of a difference in the water concentration between pure water and a solution. The osmotic pressure of a solution is defined as the driving force developed by the difference in water concentration between a particular solution and pure water. Osmotic pressure can be measured by placing the solution into a bag made of a membrane permeable to water but not to the solute (**A**), connecting a vertical tube to the bag, and measuring the height of the solution in the tube (**B**). The osmotic pressure is equal, but opposite in direction, to the hydrostatic pressure exerted by the column of solution in the tube. The more concentrated the solute, the less concentrated the water and thus the greater the osmotic pressure of the solution.

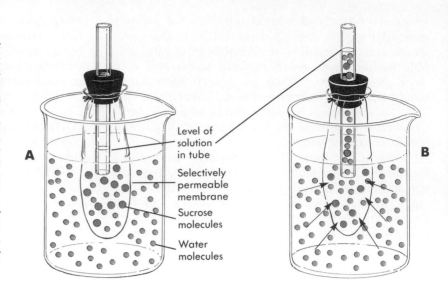

Level of solution in tube

Selectively permeable membrane

Sucrose molecules

Water molecules

Determinants of Cell Volume

- *What is tonicity?*
- *Why is the concentration of impermeant solute what matters in determining tonicity?*
- *How does the Na^+/K^+ pump protect cells from swelling?*

The **tonicity** of a solution refers to the tendency of cells to shrink or swell when placed in it. A solution is defined as **isotonic** if it produces no change in the volume of a cell placed in it, **hypertonic** if cells shrink, and **hypotonic** if cells swell (Table 6-2, Figure 6-7). A distinction between tonicity and osmolarity is necessary because the effect of a given solute on osmotic water movement across a cell membrane depends on the membrane permeability to the solute. This can be illustrated by two contrasting examples. If a cell is bathed in extracellular solution that is initially isotonic, addition of some solute to the extracellular solution would have opposite effects on the cell's volume at equilibrium, depending on whether or not the cell membrane is permeable to the solute.

If the cell is permeable to the added solute, some of the solute will diffuse across the membrane and enter the cytoplasm, increasing the total number of solute particles in the cell. After diffusional and osmotic equilibrium have been reached, the osmolarity of the cytoplasm and extracellular fluid will be equal, but the volume of water inside the cell will be greater, and the cell will have swelled. Thus in this case, addition of solute makes the extracellular solution hypotonic, even though the new extracellular solution is hyperosmotic to the initial extracellular solution.

If the cell is impermeable to the added solute, the effect will be opposite. The increase in the con-

TABLE 6-2	Osmolarity versus Tonicity

Term	Definition
Isosmotic	Having the same concentration of solute particles per unit volume as a comparison solution, regardless of the identity of the solutes.
Isotonic	Containing sufficient nonpenetrating solute to exactly match the osmotic effect of intracellular solute.
Hyperosmotic Hyposmotic	Having a higher (hyperosmotic) or lower (hyposmotic) concentration of solute particles per unit volume than a comparison solution, regardless of the identity of the solutes. Having an osmolarity of >300 mOsm/L (hyperosmotic) or <300 mOsm/L (hyposmotic if extracellular fluid is the comparison solution.
Hypertonic Hypotonic	A solution in which cells shrink (hypertonic) because of the presence of excess nonpenetrating solute that causes an osmotic net efflux of water or swell (hypotonic) because of insufficient nonpenetrating solute to match the osmotic effect of intacellular solutes.

centration of solute particles outside the cell creates a gradient of osmotic pressure that will cause water to move from the cytoplasm to the extracellular solution, shrinking the cell (see Figure 6-7, *A*). Thus it is the concentration of impermeant molecules that determines the tonicity of a solution.

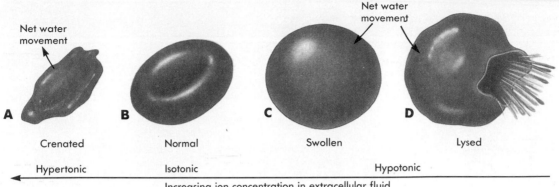

FIGURE 6-7

Effects of hypertonic, isotonic, and hypotonic solutions on red blood cells.

A   Hypertonic solutions result in crenation (shrinkage) of the cell.

B   Isotonic solutions result in normal-shaped cells.

C   Hypotonic solutions result in swelling and, eventually, lysis of cells (D).

TABLE 6-3	Factors that Affect the Rate of Solute Movement by Diffusion

Factor	Rationale
The magnitude of the concentration gradient	Provides the driving force for net movement
The diffusion coefficient	Incorporates the factors of temperature and the size of the solute molecule. The diffusion coefficient is higher if the molecules are smaller and the temperature is high.
The dimensions of the path	The greater the area available to the diffusing solute and the shorter the distance that must be covered, the greater the rate of diffusion.

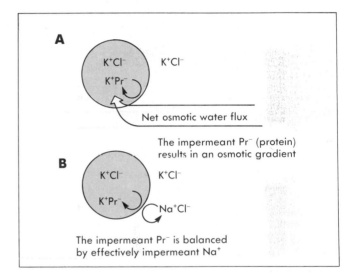

FIGURE 6-8

A   A cell bathed in isosmotic KCl swells as water moves into the cell along the osmotic pressure gradient caused by the impermeant intracellular protein anions (Pr⁻).

B   Swelling does not occur in normal extracellular fluid because the osmotic pressure of the internal impermeant solutes is matched by the osmotic pressure of the effectively impermeant extracellular Na^+.

The normal osmolarity of cytoplasm and extracellular fluid is 300 mosm/L. Extracellular fluid must be isotonic as well as isosmotic to cytoplasm because even small changes in the tonicity of extracellular fluid can cause cells to experience dangerous volume changes. To be isotonic, the extracellular fluid must contain a concentration of impermeant solute particles outside the cell equal to whatever impermeant solute particles are present in the cytoplasm. Protein molecules, which are important intracellular solutes, are large and normally membrane-impermeant. If the extracellular solution contained only permeant solute, the impermeant protein would attract an inward osmotic flux of water that would ultimately burst the cell (Figure 6-8,

A). To be isotonic, the extracellular fluid must therefore contain a concentration of impermeant solute particles equal to the protein concentration of the cytoplasm.

Cells are permeable to Na^+, but by expending energy to keep the intracellular Na^+ concentration low, they make the Na^+ in the extracellular fluid act as if it were impermeant. Thus it is the Na^+ of extracellular fluid that balances the osmotic effect of negatively charged cytoplasmic proteins and prevents swelling (Figure 6-8, B). The $Na^+ - K^+$ **pump,** an example of primary active transport (discussed in the next section) is an intrinsic membrane

Table 6-4   Classes of Mediated Transport Systems

Class	Characteristics	Example
1. Channel	A pore through the membrane that mediates a high rate of solute transfer (up to 10^8 ions/sec); driving force is the solute gradient	Na^+ channels of nerve and muscle (Chapter 7)
2. Carrier	An intrinisic protein binds the transported solute(s) and undergoes a conformational change that moves the solute(s) to the other side of the membrane; rates of transfer slower than channels (up to 10^4 molecules/sec)	Transport of glucose into cells (This chapter)
a. "Facilitated diffusion"	Driving force is the transported solute's gradient	Glucose uptake in most cells (Chapter 23)
b. Active transport	Driven directly by ATP	NA^+-K^+ pump (This chapter)
c. Gradient-driven cotransport	Gradient of one transported substance drives transport of other(s) in same direction	Na^+-coupled glucose and amino acid absoption in intestine and kidney (Chapters 20 and 22)
d. Gradient-driven countertransport	Gradient of one transported substance drives other(s) in opposite direction	HCO_3^--Cl^- exchange in intestinal cells (Chapter 22)

protein that returns Na^+ to the extracellular fluid in exchange for K^+ from the extracellular fluid. In this process, ATP is dephosphorylated to provide the energy needed to move both the Na^+ and the K^+ against their concentration gradients. Consequently, the pump is also an ATPase enzyme and is sometimes called the Na^+-K^+ ATPase.

The Na^+-K^+ pump probably evolved in primitive animals as an adaptation that protected cells from swelling. However, it has come to play a central role in two important physiological processes. One of these is the electrical excitability of nerve and muscle cells, discussed in Chapter 7. The second is in transport of solutes such as Na^+, glucose and amino acids across epithelial tissues. Such transport processes occur in almost every epithelial tissue, including the intestinal lining, the tubules of the kidney, and in exocrine glands such as the sweat glands, salivary glands and the exocrine portion of the pancreas. The role of the Na^+-K^+ pump in each of these processes will be described in the appropriate chapters.

MEMBRANE TRANSPORT PROTEINS
Mediated Transport

- *How do active and passive transport differ?*
- *What is facilitated diffusion?*
- *Explain the differences in function of channels and carriers.*
- *Distinguish between primary active transport and secondary or gradient-driven active transport.*

Because cells are in dynamic steady state, exchange of substances with the extracellular fluid is highly regulated. Except for bulk exchanges of material with the cellular environment such as pinocytosis, exocytosis, and endocytosis (described in Chapter 4), this homeostatic role is played largely by transport proteins. In **mediated transport** (Table 6-4), polar solutes and ions cross the cell membrane with the help, or mediation, of intrinsic membrane proteins. Like enzymes, mediated transport processes are specific for particular chemical structures and can be saturated by high concentrations of transported solute (see Figure 3-15, *A*). In some cases, solutes with similar structures can compete with each other to be transported.

Active Versus Passive Transport

Membrane transport processes are classified as either **active** or **passive**. This is a thermodynamic distinction. Passive processes resemble diffusion in free solution in that the driving force for net movement of a solute is the concentration gradient of that solute. By definition, passive processes do not require metabolic energy and cannot move substances "uphill" against a concentration gradient. In fact, such processes have been called **facilitated diffusion**, because the protein-mediated transport process allows the transported substance to pass through the membrane much more rapidly than would be expected from its lipid solubility. Facilitated diffusion is an especially effective mechanism for cellular uptake of substances that are metabolized, because conversion to a chemically different

THE FOUNDATIONS OF PHYSIOLOGY

form inside the cell sustains the concentration gradient. For example, glucose enters most cells by facilitated diffusion. Once in the cell, glucose does not accumulate but is rapidly converted to glucose-6-phosphate, so the intracellular concentration of glucose is always about an order of magnitude lower than that in extracellular fluid.

By definition, an **active transport** process is one that moves solute "uphill" against a concentration gradient. An active transport process has a positive free energy change and must be coupled to a second energetically-favorable process so that the sum of the free energy changes of the two coupled processes is negative.

There are several distinct mechanisms of active and passive membrane transport. Transport proteins can be divided into two major groups: **channels** and **carriers**. The basic characteristics of each group are explained in the following paragraphs.

Channels

The simplest way for ions to cross the cell membrane is by way of **channels** (Figure 6-9, *A*). A channel is a tube or pore through the membrane formed by one or more intrinsic membrane proteins. On the outside of the tube, facing the phospholipid, are many hydrophobic groups. The inside of the tube is lined with hydrophilic groups, allow-

FIGURE 6-9
Comparison of the operations of membrane channels and membrane carriers.
A   A channel is a tube through the membrane formed by one or more intrinsic membrane proteins, usually containing a local region acting as a "gate" to open or close the channel.
B   In contrast to channels, carriers must undergo a cycle of binding and conformational change.

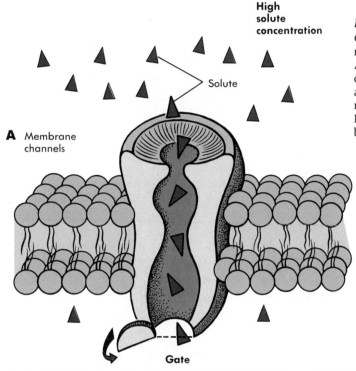

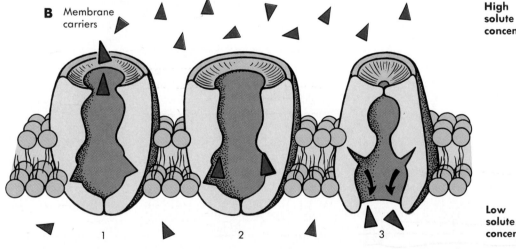

ing the channel to be filled with water. Thus ions can diffuse through the membrane in aqueous solution. Channel-mediated transport occurs at a high rate—from several million to several hundred million ions can pass through a typical channel each second. Since the ions are diffusing through the channel as if in free solution, channels mediate only passive movement.

Channels typically contain a region called the **selectivity filter** that determines which solutes may enter the channel. For example, the permeability of the cell membrane to Na^+ and K^+ is due to the presence of structurally different Na^+ and K^+ channels. Because most channels prefer either cations or anions, ionic flow through channels results in net charge movement, so channels are conductors of electric current. Channels typically open and close spontaneously; some are **gated channels,** whose opening is regulated by an external factor. Both of these features of channel function are especially important in generation of electrical signals in nerve and muscle cells, described in Chapter 7.

Carriers

In contrast to channels, **carriers** bind the transported solute at one face of the membrane and undergo a conformational change that transfers the bound solute to the other side of the membrane (Figure 6-9, *B*). Each cycle transfers at best only a few solute molecules or ions, so each carrier mediates a much lower rate of solute transfer than a channel—perhaps at most a few hundred thousand molecules per second. Depending on the particular system, carrier-mediated transport may be active or passive. As noted above, glucose entry into most cells is by carrier-mediated facilitated diffusion. There are two basic energy sources for carrier-mediated active transport in animal cell membranes. One is ATP; transport energized directly by ATP is termed **primary active transport.** The second is a transmembrane gradient of a second solute; such processes are called **secondary** or **gradient-driven active transport.**

The Na^+-K^+ pump (Na^+-K^+ ATPase) is an example of primary active transport. The pump protein has several membrane-spanning domains. The elements of the protein that are within the membrane can assume several different conformations (Figure 6-10). In the conformation labelled *A*, the pump is open to the cytoplasm, where it can bind Na^+, and closed to the external solution. The pump hydrolyzes ATP to phosphorylate a specific site. As a result of this phosphorylation, the pump undergoes a conformational change (*B*) that has three effects: (1) the affinity for Na^+ is decreased; (2) the binding sites for K^+ are uncovered; and (3) the pump cavity opens to the external solution and closes to the cytoplasm (*C*). When external K^+ ion is bound, the phosphate dissociates and reverses the conformational change, returning the pump to its initial state (*D*). The net result is that Na^+ is bound on the inside and released into the external solu-

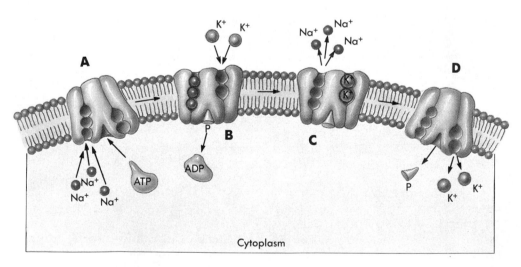

FIGURE 6-10
Schematic diagram of the Na^+-K^+ pump cycle.
A   Three Na^+ ions bind from the cytoplasmic side.
B   The pump protein is phosphorylated by ATP.
C   Phosphorylation causes a conformational change of the protein that also involves a decrease in the affinity of Na^+ binding sites and an increase in affinity of K^+ binding sites. The three Na^+ ions are then released to the extracellular side.
D   Two K^+ ions occupy the K^+ binding sites, and the pump protein releases the phosphate and returns to its original conformation. The affinity of the K^+ binding sites decreases and that of the Na^+ increases. The K^+ ions are released to the cytoplasmic side of the membrane, the pump is ready to bind three Na^+, and the cycle starts again.

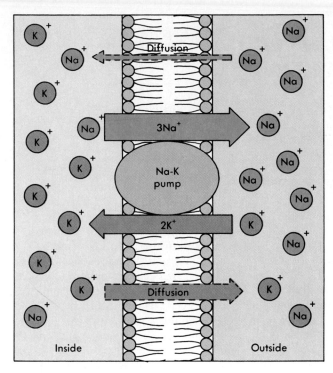

FIGURE 6-11
The $Na^+ - K^+$ pump maintains Na^+ and K^+ gradients, countering diffusional entry of Na^+ and exit of K^+.

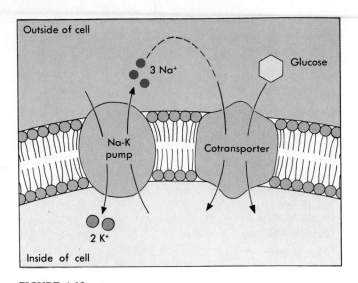

FIGURE 6-12
The Na^+ gradient produced by the $Na^+ - K^+$ pump can be used to drive absorption of glucose against its concentration gradient by coupling glucose absorption to the diffusion of Na^+ down its concentration gradient.

tion, while at the same time K^+ is taken up at the external surface and discharged into the cytoplasm. Each cycle of the pump is believed to dephosphorylate one ATP and exchange two K^+ for three Na^+. The rate of cycling of the pump is regulated by the intracellular Na^+ concentration. Accumulated Na^+ is removed before its concentration gradient can change significantly (Figure 6-11). The $Na^+ - K^+$ pump is an important membrane protein in every cell of the body; use of ATP by the pump is the largest single item in the body's energy budget, accounting for perhaps as much as one third of the total energy expended by the body.

Other important primary active transport proteins include several types of **H^+ ATPases** that function in acidification of vesicles of the endoplasmic reticulum and in H^+ secretion in the kidney (see Chapter 19), the gastric **$H^+ - K^+$ ATPase** that is responsible for active H^+ secretion by the gastric mucosa (see Chapter 21) and the **Ca^{++} ATPase** that pumps Ca^{++} into the sarcoplasmic reticulum of muscle cells, bringing about relaxation (see Chapter 12).

Secondary active transport is not directly energized by dephosphorylation of ATP, but instead by transmembrane chemical or electrical gradients. One important example is **Na^+-linked cotransport.** This is the mechanism by which the intestine

absorbs glucose (Figure 6-12). In this process, the carrier, located in the surface of the cell facing the intestinal lumen binds a molecule of glucose and also a Na^+ ion. A conformational change follows, in which both the Na^+ and the glucose molecule are translocated to the interior of the cell. The glucose may be moved against a concentration gradient, so thermodynamically this is an active process. The energy for this is provided by the Na^+, which is going downhill. Since the Na^+ concentration is high outside and is maintained low inside the cells by the $Na^+ - K^+$ pump, the energy for this process really comes ultimately from the primary active transport of Na^+ by the $Na^+ - K^+$ pump, driven by ATP.

Similar cotransport systems exist for many small molecules, including amino acids. One interesting example is the Na^+, K^+, 2 Cl^- cotransporter, which is important for solute reabsorption in renal tubules (Chapter 19). In this case the Na^+ gradient drives both K^+ and Cl^- against their electrochemical gradients. There are also gradient-driven transport systems in which the driver and the driven solutes move in opposite directions: these are referred to as **counter-transport** systems (Table 6-4). Examples are exchange of Cl^- for HCO_3^- in the kidney (Chapter 20) and the gastrointestinal tract (Chapter 21).

Artificial Membranes as Valuable Experimental Tools

Cell membranes are complex structures that carry out many functions simultaneously. Simpler membrane systems can be created in the laboratory for study of the basic principles of transport and permeation. An artificial hydrocarbon membrane can be made by placing a drop of hydrocarbon on a small hole drilled in a partition that separates two reservoirs of solution. The hydrocarbon forms a film two molecules thick across the hole (Figure 6, *A*).

Some substances that previously had been known for their antibiotic activity were the first models for the behavior of transport proteins. These substances, called ionophores (the name means "ion carriers"), are molecules with a highly polar interior that can dissolve in artificial lipid membranes and confer permeability to specific ions. One example is valinomycin, a small doughnut-shaped peptide with six polar oxygen atoms facing the interior of the molecule. A K^+ ion just fits in the hole of the doughnut.

The polar oxygen molecules replace its hydration shell. Valinomycin can pass from one side of the membrane to the other, shuttling K^+ ions. Biological carrier proteins were once thought to function like this, but intrinsic proteins are now known to be much too bulky to shuttle back and forth within the membrane.

Some ionophores form relatively stable pores in lipid membranes. One example is a helical peptide antibiotic called gramicidin. Polar groups within the helix allow ions with the correct diameter to discard their hydration shells and move through the pore. Gramicidin is a useful model of the behavior of biological channels.

Recently it has been proved possible to extract transport proteins from biological membranes and reconstitute them in artificial membranes. In some cases the mRNA coding for transport proteins has been injected into frog oocytes, where the oocyte's translational machinery is able to synthesize the transport proteins and insert them into its cell membrane. Several important channels and pumps have been studied in reconstituted form with electrical measurement techniques similar to those employed for any living cell (for example, the patch-clamp described in the boxed essay on stretch-activated channels in Chapter 8).

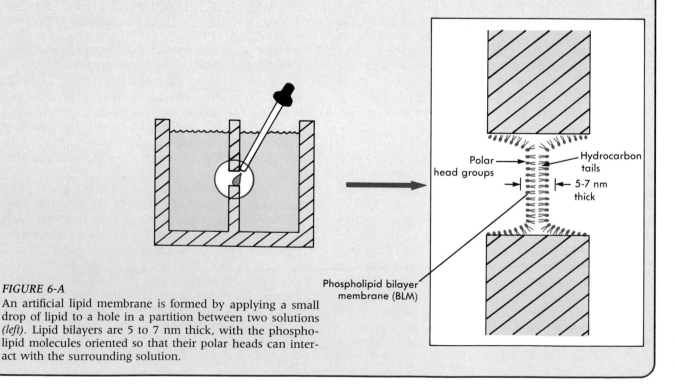

FIGURE 6-A

An artificial lipid membrane is formed by applying a small drop of lipid to a hole in a partition between two solutions *(left)*. Lipid bilayers are 5 to 7 nm thick, with the phospholipid molecules oriented so that their polar heads can interact with the surrounding solution.

Polar head groups — Hydrocarbon tails — 5-7 nm thick — Phospholipid bilayer membrane (BLM)

THE MEMBRANE POTENTIAL
Electrical Gradients Across Cell Membranes

> - What is the relationship between the chemical and electrical forces acting on an ion that is in electrochemical equilibrium?
> - What factors determine the value and sign of the resting potential?
> - What four changes could cause a cell to become hyperpolarized or depolarized?

All living cells maintain an imbalance of electrical charge, called the **resting membrane potential**, or simply **resting potential** (V_{rest}), across their cell membranes. Rapid, reversible changes in the membrane potential are the basis for rapid communication between different parts of the nervous system, and are important for such responses as secretion by glands, response of sensors to environmental changes, and contraction of muscle cells.

Membrane potentials can be measured by first inserting a fine glass capillary, or micropipette, filled with conductive solution into the cell and then connecting a voltmeter between the fluid-filled capillary, which functions as a **microelectrode,** and the extracellular solution (Figure 6-13) which serves as the reference for comparison. The resting potential is typically between 40 and 80 mV, depending on the cell type, with the inside of the cell negative compared to the outside.

The amount of charge imbalance needed to produce resting potentials of the size typical of those across biological membranes is very small in comparison with the total number of ions in the intracellular or extracellular solutions (Figure 6-14). That is, even though there will be a slight excess of negative charge near the interior of the cell membrane and the same amount of excess positive charge near the exterior, there will be no measurable change in the overall concentrations of any ion. For example, for a membrane 8 nm thick, a voltage of 100 mV (inside negative) could be produced if only those K^+ ions located within one membrane thickness were to move from the inside to the outside.

Ionic Gradients and Diffusion Potentials

The separation of electrical charges across the cell membrane represents stored or potential energy. The energy comes ultimately from the activity of the Na^+-K^+ pump, which contributes to the membrane potential in two ways. One way is by direct separation of charge. Recall that the pump exchanges only two K^+ for each three Na^+, so that each cycle results in net transfer of one positive charge to the outside of the membrane. However, in most cells this mechanism is responsible for only a few mV of the resting potential.

The second and main contribution of the Na^+-K^+ pump to the resting potential is by way of the diffusion of Na^+ and K^+ down the transmembrane gradients of Na^+ and K^+ that the pump maintains (Table 6-5). Such potentials caused by ionic diffusion are called **diffusion potentials.** To understand how an ionic concentration gradient can cause a diffusion potential, imagine a cell that contains 150 mEq K^+/L, together with other solutes to total 300 mOsm/L, as in Figure 6-16. Imagine further that this cell is bathed in an isotonic solution that contains 5 mEq K^+/L (Figure 6-15). If the cell membrane is permeable only to K^+, the transmembrane K^+ concentration gradient will drive a net flux of K^+ from inside to outside. However, each K^+ that

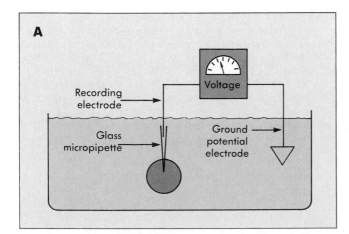
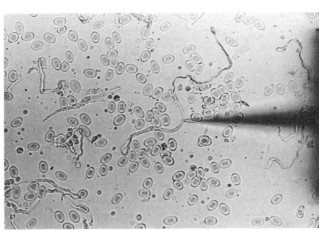

FIGURE 6-13
A   Cell membrane potentials can be recorded using glass capillary microelectrodes.
B   A patch-clamp micropipette is shown penetrating a tissue-cultured muscle cell. This is a freshly-isolated muscle cell, as can be seen from the presence of red blood cells in the field of view.

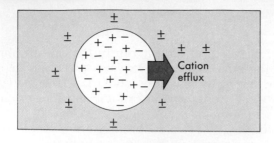

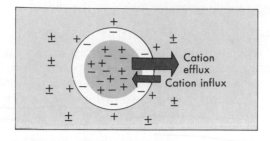

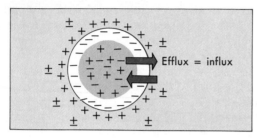

FIGURE 6-14

This diagram emphasizes that membrane potentials involve a very small number of the ions present in the intracellular and extracellular solutions.

TABLE 6-5	The Ionic Composition of Cytoplasm and Extracellular Fluid (Concentration in mMoles)

	Cytoplasm	Extracellular fluid	Ratio	Equilibrium potential
Na^+	15	150	10:1	+60 mV
K^+	150	5	1:30	−90 mV
Cl^-	7	110	15:1	−70 mV

leaves the cell leaves behind an unpaired negative charge, which accumulates at the inner membrane surface. As negative charges accumulate, an inside-negative membrane potential is generated. Within a few milliseconds, the electrical "pull" of the left-behind negative charges exactly equals the "push" of the K^+ concentration gradient. At this point the net flux of K^+ becomes zero (that is, the unidirectional fluxes are equal). The imaginary cell has come into **electrochemical equilibrium**. The diffusion potential that exists under these conditions is called a K^+ **equilibrium potential** (E_K).

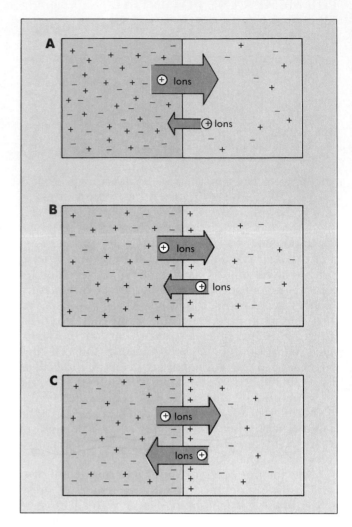

FIGURE 6-15

Development of electrochemical equilibrium in an imaginary cell in which the membrane is permeable only to the positive ion (K^+, indicated simply as + signs) of the salt in solution.

A   Initially, the concentration gradient and lack of opposing charge result in a large net efflux across the membrane.

B   As ion movement results in charge separation, the size of the efflux driven by the concentration gradient declines, and the influx increases, driven by the electrical attraction of the charge on the membrane.

C   At equilibrium, the electrical and concentration forces are equal and opposite, the influx and efflux are equal, and there is no net flux across the membrane. So few ions have moved relative to the total number in solution that the change in the concentrations of the two compartments has been negligible.

The relationship between the concentration ratio and the equilibrium potential, given by the **Nernst Equation** (see Focus Unit), has a logarithmic relation to the ratio of concentrations. This means that at body temperature, for every tenfold change in the transmembrane concentration ratio, the equilibrium potential increases by 60 mV. For the thirtyfold ratio of K+ concentrations in this ex-

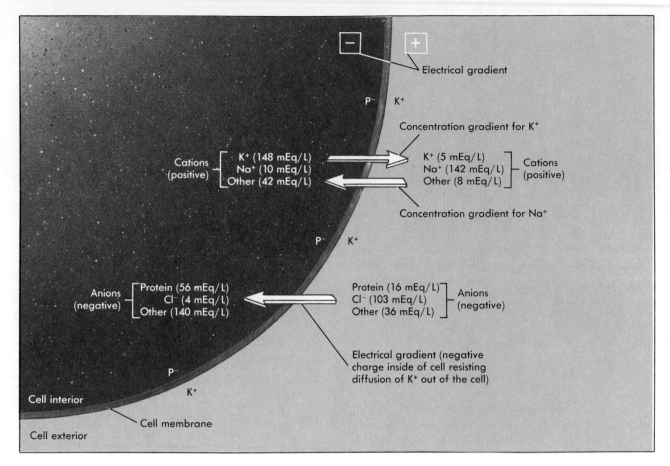

FIGURE 6-16
A diagram of a cell showing the intracellular and extracellular concentrations of ions and molecules and the gradients for Na^+, K^+, and voltage.

ample, the corresponding E_K given by the Nernst Equation is -90 mV.

Determinants of the Resting Potential

At first glance it would seem that the resting potential of real cells might be accounted for by the transmembrane K^+ gradient, as in this imaginary cell. However, careful measurements of the resting potentials and chemical composition of real cells revealed that the resting potential, while inside-negative, is significantly less inside-negative than E_K, about -70 mV for typical nerve and muscle cells. Furthermore, permeability measurements showed that cell membranes, though permeable to K^+, have a small but significant permeability to Na^+. The Na^+ concentration of normal extracellular solution is 142 mEq/L, while a typical cytoplasmic Na^+ concentration is 15 mEq/L (Table 6-5; Figure 6-16). The **Na^+ equilibrium potential (E Na)**, calculated using the Nernst equation and taking the ratio of Na^+ concentrations as 0.10, is 60 mV *inside-positive*. Thus the resting potential is not an equilibrium potential for either K^+ or Na^+.

The resting potential could be thought of as the product of a continuous tug of war between the transmembrane Na^+ and K^+ gradients. In this tug of war, the strength of each gradient is determined by two factors: (1) the concentration ratios for the two ions, and (2) the relative permeabilities of the membrane to the two ions. Considering both factors, the K^+ gradient is clearly stronger. The K^+ concentration ratio is thirtyfold, compared with the Na^+ concentration ratio of tenfold, and the membrane of resting nerve and muscle cells is 50 to 75 times more permeable to K^+ than Na^+. As a result, the K^+ almost wins: V_{rest} is close to E_K, but the effect of the Na^+ gradient prevents K^+ from carrying V_{rest} all the way to E_K. The **Goldman Equation**, a modification of the Nernst equation, allows calculation of the membrane potential for any combination of Na^+ and K^+ concentrations and permeabilities (see the Focus Unit).

The fact that both K^+ and Na^+ are not in electrochemical equilibrium implies that transmembrane driving forces exist for both ions. The driving force for K^+ is equal to $E_K - V_{rest} = -30$ mV and is directed outward. The driving force for Na^+ is equal to $E_{Na} - V_{rest} = +130$ mV and is directed inward.

Transport Across Cell Membranes

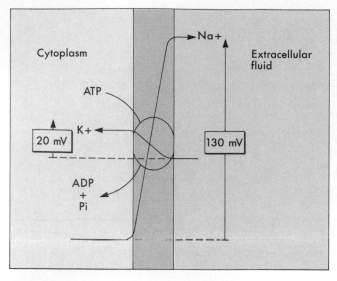

FIGURE 6-17

The work done by the Na^+-K^+ pump. Both Na^+ and K^+ are moved uphill against electrochemical gradients. In typical nerve and muscle cells the gradient for Na^+ is equivalent to about 130 mV extracellular fluid to cytoplasm; the gradient for K^+ is about 20 mV, cytoplasm to extracellular fluid.

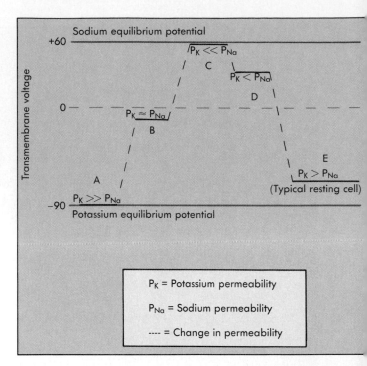

P_K = Potassium permeability

P_{Na} = Sodium permeability

---- = Change in permeability

FIGURE 6-18

Effect of different Na^+-K^+ permeability ratios on the membrane potential. At **A**, the potassium permeability (P_K) is much greater than the sodium permeability (P_{Na}). At position **B**, P_K and P_{Na} are nearly equal, while at **C**, P_K is much less than P_{Na}. **D** and **E** represent intermediate cases. Changes in the relative magnitude of P_{Na} and P_K cause shifts in the membrane potential, as illustrated by the dashed lines.

Because the membrane is permeable to both ions, a continuous outward diffusion of K^+ and inward diffusion of Na^+ occurs. This leakage would dissipate the concentration gradients if the Na^+-K^+ pump did not also operate continuously to keep the intracellular K^+ and Na^+ concentrations in steady state (Figure 6-17).

In contrast to K^+ and Na^+, Cl^- is in electrochemical equilibrium with V_{rest} in many cell types, indicating that in those cells it is not actively transported. However in some cell types, including some nerve cells, Cl^- is actively transported. In these cells the effect of the nonequilibrium distribution of Cl^- must be taken into account in computing the resting potential using the Goldman Equation.

Note that a change in the permeability of the membrane to either Na^+ or K^+ would change the balance of forces and bring about a rapid change in membrane potential (Figure 6-18). For example, an increase in K^+ permeability would carry V_{rest} closer to E_K; such a change in V_{rest} is called **hyperpolarization**. In contrast, an increase in Na^+ permeability would carry V_{rest} in the direction of E_{Na}. Depending on the size of the change, V_{rest} might be changed in the direction of zero voltage or even made inside-positive. By convention, such changes in V_{rest} are called **depolarization**, even if the potential crosses zero and becomes inside-positive. As will be shown in Chapter 7, rapid, reversible changes in the relative permeability of the membrane to Na^+, K^+ and other ions are the basis for the electrical excitability of nerve and muscle cells and of information processing in the nervous system.

Changes in the intracellular or extracellular concentrations of Na^+ or K^+ would also change the balance of forces that determines the resting potential. For example, increasing the extracellular concentration of K^+ would decrease the gradient for K^+ (E_K would be less inside-negative), allowing a greater influence of Na^+ and making the resting potential less inside-negative (Figure 6-18). Interestingly, an increase in extracellular Na^+ concentration would cause a similar change in V_{rest}, since such a change would increase the magnitude of the Na^+ gradient and make E_{Na} more inside-positive. When E_{Na} becomes more inside-positive, the transmembrane driving force for Na^+ also increases. This would increase the influence of Na^+, causing V_{rest} to become less inside-negative. Changes in extracellular concentrations of Na^+ or K^+ would change the resting potentials of all body cells. Such a change would affect electrical communication between excitable cells of the nervous system, the rhythm of the heart, and the ability of the nervous system to control skeletal muscles. Regulation of the Na^+ and K^+ concentrations of the extracellular fluid by the kidney (Chapter 20) is therefore important for these organ systems.

SUMMARY

1. Cell membranes consist of phospholipid molecules arranged as a **bilayer** in which **intrinsic proteins** are embedded. Extrinsic proteins are attached to the surface.

2. Pure lipid membranes are impermeable to polar solutes and ions; specialized **carriers** or **channels** are needed for such substances to cross cell membranes and are the basis of **mediated transport**. Mediated transport systems are characterized by specificity and saturability.

3. **Diffusion** is a process in which net solute movement down concentration gradients is the result of random thermal motion of solute molecules. The rate of solute diffusion is affected by the magnitude of the concentration gradient, the cross-sectional area and length of the diffusion path, the temperature, and the size of the diffusing solute.

4. Lipid membranes are highly permeable to water. The **osmotic pressure** of a solution is defined as the driving force developed by the difference in water concentration between a particular solution in question and pure water. Gradients of **osmotic pressure** across cell membranes can result when there is an imbalance of impermeant solutes across the membrane.

5. **Isotonic** solutions contain impermeant solutes sufficient to balance the osmotic pressure of impermeant solutes in the cytoplasm, so that cells bathed in isotonic solution neither swell nor shrink. **Hypotonic** solutions contain insufficient impermeant solute, and cells bathed in such solutions swell. **Hypertonic** solutions contain excessive impermeant solute, resulting in cellular shrinkage.

6. Cell volume stability requires there to be a high concentration of Na^+ in the extracellular fluid and for Na^+ to be impermeant. Even though it slowly leaks into the cell, Na^+ is made to be effectively impermeant by the $Na^+ - K^+$ **pump** of the cell membrane.

7. The **membrane potential,** an inside-negative gradient of electrical charge across the membrane, results from the chemical energy of Na^+ and K^+ concentration gradients and the differential permeability of the membrane to the two ions. If only one of the two ions were permeant, the resulting potential would be an **equilibrium potential** whose magnitude would be predicted by the **Nernst equation.** Because membranes are typically permeable to both Na^+ and K^+, the membrane potential is not an equilibrium potential for either ion.

8. The magnitude and sign of the membrane potential are determined by the relative magnitudes of the permeabilities and concentration gradients of Na^+ and K^+. Since typical membranes are more permeable to K^+ than to Na^+ and the K^+ concentration gradient is larger than the Na^+ gradient, the membrane potential is close to, but not equal to, the K^+ equilibrium potential and far from the Na^+ equilibrium potential.

9. There are three basic mechanisms of mediated transport of solutes across cell membranes: **facilitated diffusion,** in which the solute's own concentration gradient drives the **net flux; active transport,** in which the solute is driven against its concentration gradient by energy released from ATP; and **secondary** or **gradient-driven counter-transport** or **cotransport,** in which a net flux of one solute against its concentration gradient is driven by a net flux of another solute (frequently Na^+) down its concentration gradient.

1. Describe the fluid mosaic model of membrane structure. What features of this model would allow the internal surface of a membrane to have properties different from its external face?

2. What features of membrane and solute would determine the permeability coefficient of a particular solute through a particular membrane?

3. Specify the concentration in millimoles of three solutions with an osmolarity of 300 mOsm. One of these solutions should contain only a nonelectrolyte such as glucose, another should contain only univalent ions, and the third should contain divalent cations and univalent anions.

4. Plasma has an osmolarity of 300 mOsm. What must the intracellular osmolarity be? Why?

5. What is the role of the Na^+-K^+ pump in regulating cell volume?

6. What factors make the membrane potential inside-negative?

7. If a cell were bathed in a solution like normal extracellular fluid except containing 15 mM K^+ instead of 5 mM, what effect would this have on the membrane potential?

8. What are the two common features of all mediated transport systems?

Choose the MOST CORRECT Answer.

9. These cell membrane proteins function as identity tags, as in the case of histocompatibility antigens in the immune system:
   a. Transport proteins
   b. Recognition proteins
   c. Functional proteins
   d. Receptor proteins

10. Which of these ions is found in greater concentration in cytoplasm?
   a. Na^+
   b. K^+
   c. Cl^-
   d. Ca^{++}

11. Rate of net flux of diffusion of a substance between two points can be increased by:
   a. Decreasing the magnitude in concentration gradient
   b. Decreasing the temperature
   c. Increasing the cross-sectional area
   d. Increasing the path length of diffusion

12. A solution with a lower concentration of solute particles (a higher water concentration) is termed:
   a. Hyperosmotic
   b. Hyposmotic
   c. Isosmotic
   d. None of these terms applies

13. _____ is an example of secondary or gradient-driven active transport:
   a. Na^+-linked cotransport
   b. Na^+-K^+ pump
   c. H^+ ATPases
   d. Ca^{++}ATPase

14. _____ uses hydrostatic pressure to "push" water through a membrane permeable to water and some solutes against its concentration gradient, increasing the concentration of impermeable solutes left behind:
   a. Diffusion
   b. Osmosis
   c. Dialysis
   d. Ultrafiltration

● SUGGESTED READINGS

BERG HC: *Random walks in biology*, Princeton, NJ, 1983, Princeton University Press, A classic text on diffusion and other varieties of random movement.

HILLE B: *Ionic channels of excitable membranes*, ed 2. Sinauer Associates, Inc. Sunderland, Mass. This is the best contemporary monograph on ion channels.

QUINTON PM: Cystic fibrosis: a disease in electrolyte transport. In Diseases of receptors and channels, *FASEB Journal* 4:2709, 1990. This and other papers in this issue examine recent findings on defective ion channels as the cause for cystic fibrosis.

RAJAN AS, ET. AL: Ion channels and insulin secretion, *Diabetes Care* 13:340, March 1990. A review of the role of ion channels in regulating insulin secretion from pancreatic beta cells. In particular it describes how the closing of ATP- and glucose-sensitive K+ channels results in cell depolarization, activation of voltage-gated Ca^{++} channels, and exocytosis.

YEAGLE P: *The membranes of cells*, Orlando, FL, 1987, Academic Press. A more advanced, but still readable, summary of transport across membranes. Covers the fluid mosaic model, lipid-protein interactions in membranes, the various types of mediated diffusion, and the intracellular synthesis of cell membrane components.

CHAPTER 6 FOCUS UNIT

Diffusion, Osmosis and Bioelectricity

The purpose of this unit is to provide additional information on the physics of diffusion, membrane potentials and osmosis for students with some background in chemistry.

DIFFUSION

Diffusion is migration of atoms, ions, molecules, or even small particles that arises from random motion due to thermal energy. For consistency, in this discussion the diffusing entities will be called particles. A particle at any particular absolute temperature T has, on the average, a kinetic energy of $3kT/2$, where k is Bolzmann's Constant. The energy is the same no matter how large or small the particle is. The value of kT at 300° K (27° C) is 4.14×10^{-14} g cm^2/sec^2. Since particles of different species all have the same mean energy, the mean velocity of any particular single ion or molecule depends on its mass, so that each species has its own diffusion coefficient (D_s), which varies with the temperature.

Although each diffusing particle is in rapid motion, this motion is interrupted every picosecond or so by a collision with a water molecule or another solute particle. As a result, each particle takes a highly random path in three dimensions. However, for most situations encountered in physiology, diffusion can be considered as a one-dimensional process. At any instant, any particle moving in one direction along the x axis has an equal probability of going forward or reversing direction. When this is applied to a large number of particles traveling in one dimension, at any instant approximately half the particles will be moving in one direction and half in the other. If there is a concentration gradient, the number of particles moving down the gradient will be higher than the number moving up, simply because of the larger number of particles at the high end of the gradient. There will be a net flux (J_{net}) from the higher to the lower concentration. As noted in Chapter 6, the net flux of a diffusing solute S in one dimension x is the product of the concentration gradient dC_s/dx and the diffusion coefficient D_s for that solute. This is **Fick's Law of Diffusion:**

Jnet = Ds(dCs/dx)

The units of J_{net} are moles/sec · cm^2/sec and of the concentration gradient, moles/cm_3 cm.

The motion of each individual solute particle is equally random and erratic no matter whether there is a concentration gradient or not. However, it helps to think of a concentration gradient as a chemical force that drives the system in the direction of its equilibrium state. This is the state in which solute particles are randomly distributed throughout the system, J_{1-2} equals J_{2-1}, and J_{net} is zero.

Einstein solved the Fick equation to show that in an interval of time t, an average diffusing particle will travel a distance of $(2D_s t)^{1/2}$ away from its starting point. This solution says that the distance gained by diffusional motion increases as the square root of time, rather than in direct proportion to time as in straight-line motion. For example, a particle with D_s of 2×10^{-5} cm^2/sec (in the range for water molecules and small ions) travel at a velocity of about 566 m/sec when it is moving in a straight line. It can diffuse an average of 1 micron in 250 microsec, but requires 25 msec to travel 10 microns and 2.5 sec to travel 100 microns. The slowness with which diffusing particles travel long distances, in spite of their high linear volecity, is due to the cumulative effect of the setbacks that occur with each reverse in direction. This is why diffusion is an effective means of mass transport over very short distances, but very ineffective if the distance to be covered is more than a few microns.

ELECTROCHEMICAL EQUILBRIUM AND THE NERNST EQUATION

Figure 6-16 shows that in the simple case of a membrane permeable to a single ion, establishment of a concentration gradient leads immediately to generation of a diffusion potential across the membrane. When the system comes into electrochemical equilibrium, the electrical potential (voltage) across the membrane exactly balances the diffusional potential energy of the concentration gradient. The relationship between chemical potential energy and electrical potential energy for k^+ at equilibrium in the situation shown in Figure 6-16 is described by the **Nernst equation:**

EK = RT/zF ln ([K+]/out/[K+]in)

where R is the universal gas constant (0.082 L. atm/mol °K), T the absolute temperature, z the valence of the ion (+1 in this case) and F is Faraday's number (9,648 Coulombs/mole; Coulombs are units of electrical charge). Notice that in the Nernst equation the concentration gradient takes the form of a ratio rather than a difference, as in Fick's Law of Diffusion. Incorporating the numerical values of the constants and changing to base 10 logarithms gives

$$EK = 60 \text{ mV log } ([K+] \text{ out}/[K+]\text{in})$$

With adjustments, the Nernst equation can be written for any transmembrane ion gradient. Note that there is a logarithmic relation between E and the concentration ratio, so that the absolute value of E doubles with every order-of-magnitude change in the ratio. In this case where the K^+ concentration to one hundredfold greater on the inside than the outside would increase E_K to -120 mV.

THE GOLDMAN EQUATION

By definition, the Nernst equation describes states of electrochemical equilibrium. Since neither K_+ nor Na^+ is in electrochemical equilibrium across the cell membrane, it is necessary to modify the Nernst Equation to express the relationship between the Na^+ and K^+ concentration gradients and the resting potential. Since there is net diffusion of both ions across the membrane, the modified form of the equation must take into account not only the gradients but also the permeabilities of the membrane to the two ions. The result is the **Goldman equation.**

$$V_{rest} = RT/F \text{ ln } (P_K[K_+]\text{out}_{+P}na_{[na+]}\text{out})/(P_K[K+]\text{in}^{+P}Na[^{Na^+}]\text{in})$$

in which P_K is the K^+ permeability, P_{Na} is the Na_+ permeability, and the rest of the terms are as in the Nernst equation.

The Goldman equation says that the resting potential is determined by the relative magnitudes of the concentration gradients of the two ions, weighted by their relative permeabilities. The equation should include all ions that are actively transported. For cells that actively transport Cl^-, a term for Cl_- can be added to the equation. Changes in the relative permeability of the membrane to either ion would change the membrane potential. The ability to undergo rapid, reversible changes in permeability is the mechanism for generation of electrical signals in neurons and muscle cells. Note that if the Na_+ permeability, for example, were made infinitely small, the Goldman equation would turn back into the Nernst equation.

OSMOSIS AND OSMOTIC PRESSURE

Osmosis is a process whereby water spontaneously moves across a semipermeable membrane from a region of higher water concentration to one of lower water concentration. The osmotic pressure of a solution is defined as the pressure that would have to be applied to prevent osmotic flow into the solution from pure water.

The relationship between osmolarity and osmotic pressure (P_{osm}) in ideal solutions is given by the **Van't Hoff equation:**

$$Posm = MRT$$

Where the units of P
osm are atmospheres, M is the osmolarity of the solution, R is the universal gas constant, and T is the absolute temperature. Frequently the term "osmotic pressure" is used as if it were a synonym for osmolarity, although strictly it is meaningless to speak of an isolated solution as if it had a pressure like that of a gas.

Osmotic pressure is one of the three **colligative properties** of solutions, the other two being the freezing point and the vapor pressure. "Colligative" means "bound together," because the three are related. The higher to osmotic pressure, the lower the freezing point and the lower the vapor pressure compared to pure water. In practice, osmotic pressure is seldom measured directly. Until recently, most osmometers used in clinical and research laboratories measured freezing point depression; currently, vapor pressure osmometers have become more popular because they can handle smaller sample volumes and are less subject to error.

NERVE AND MUSCLE

CHAPTER 7

Action Potentials

On completing this chapter you should be able to:

- Compare action potentials, receptor potentials, and synaptic potentials.
- Describe the behavior of voltage-gated Na^+ and K^+ channels in excitable membranes as a function of voltage and time.
- Understand how an action potential is initiated using the concept of threshold.
- Understand how inactivation of Na^+ channels and opening of K^+ channels combine to terminate the action potential.
- State the factors that govern the velocity of conduction of action potentials in axons.
- Understand the role of myelin as an insulator and characterize saltatory conduction.
- Understand the origin of the compound action potential of a nerve trunk and account for the origin of its various components.

The most beautiful thoughts, the greatest human performances, and the simplest everyday acts depend on a rapid, complex flow of information within the nervous system and between the nervous system and the rest of the body. The information is encoded in electrical signals. The ability to send information rapidly over long distances makes high-speed control of the body possible. In the nervous system and muscles, signals are carried over long distances by rapid temporary changes in membrane potential called action potentials.

The ability to generate and transmit action potentials is a cellular specialization; cells that can produce action potentials are said to be excitable. Excitability is a property of the cell membrane and is due to the presence of proteins called ion channels that span the membrane. The membrane potential, which is a characteristic of all living cells, provides the stored energy that allows changes in membrane potential to serve as the basis of information transfer in excitable cells.

Channels that can open and close rapidly in response to environmental influences respond to stimuli that give sensations such as taste, sight, hearing, or touch. An open channel permits a flow of ions across the membrane—an electrical current. These tiny individual currents, multiplied many times, provide the information flow that links the 10 billion brain cells with one another. Currents flowing in a group of motor neurons selected by the brain excite the muscles in the surgeon's hand or carry the gymnast through her routine.

ELECTRICAL EVENTS IN MEMBRANES
Categories of Electrical Events

> - *What is "driving force"?*
> - *What are the comparative signs of the driving forces for Na^+ and K^+ at the resting potential?*
> - *How do the terms decremental and nondecremental conduction relate to action potentials, receptors, and synaptic potentials?*

The plasma membranes of some cells are able to undergo rapid, reversible changes in ion permeability that displace the membrane potential from the normal level. The ability to undergo these changes is utilized in signaling in the nervous system. The electrical changes register sensory stimulation or the presence of incoming signals from other nerve cells. Such information may act locally or be relayed for relatively great distances before it is ultimately transformed into a mechanical or chemical change in the cell. The transformation can result in contraction in muscles, secretion from glands, and in initiating dramatic chemical cascades in a wide variety of cell types, including immune system cells and human sperm and egg.

Changes in the membrane potential that correspond to a message can occur (1) as local events focused on a small region of membrane or (2) as simple electrical waves that are transmitted unchanged for great distances along the cell's surface. The first category, the local events, represent the response to a **stimulus** or **input** impinging on the membrane. These localized **receptor potentials** and **synaptic potentials** are important in acquisition and analysis of information and in communication between nerves and muscles. The second category, **action potentials,** are used for rapid long-distance signalling in nerve and muscle cells. Differences between these basic types of electrical events are summarized in Table 7-1.

Resting Potential

The membrane potential of cells is determined primarily by (1) the relative permeability of the membrane to specific ions, and (2) the concentration gradients of those ions across the membrane. Permeability to Na^+, K^+ and other ions such as Cl^- is a property of the number and kind of channels built into the membrane structure; collectively, channels that are commonly open are referred to as **leak channels.** There are many more leak channels that permit K^+ to cross the membrane than allow Na^+ to cross. Ion concentration gradients responsible for the resting potential rarely change very much because the body's homeostatic systems maintain the composition of the extracellular fluid, while the Na^+-K^+ pump (see Chapter 6) maintains constant intracellular concentrations.

In cells that utilize membrane shifts as signals, the **resting potential** is the term used to describe the relatively stable membrane potential that these cells exhibit when unstimulated. The resting potential is much closer to E_k than E_{Na} and is therefore inside-negative, typically around -70 mV for nerve cells and less internally negative for cells in smooth muscle and tissues other than nerve and skeletal muscle. When stimulated, the membrane potential shifts that cells can exhibit are limited by the physiological extremes set by the K^+ equilibrium potential (E_K, about -90 mV) and the Na^+ equilibrium potential (E_{Na}, around +60 mV).

Another way of saying that a membrane is charged is to say that it is **polarized,** with a membrane potential not equal to zero. If the resting potential is the starting point against which membrane potential changes are measured, a shift to a more polarized membrane potential (internally more negatively charged and moving further away from 0 mV) is called a **hyperpolarization** (Figure 7-1). Changes that make the membrane potential less inside-negative (moving closer to 0 mV) are **de-**

TABLE 7-1 *Comparison of Electrical Events in Membranes*

Characteristic	*Receptor and synaptic potentials*	*Action potentials*
Location	Sensory endings and synapses	Axons of nerve cells, muscle
Characteristic of initiating event	Physical or chemical	Electrical
Nature of response	Graded	All-or-nothing
Size	Variable but usually less than 10 mV	Usually greater than 100 mV
Propagated	No	Yes
Mechanism of current spread	Decremental conduction	Nondecremental conduction
Threshold	No	Yes
Summation	Yes	No
Direction of potential change	Depolarizing or hyperpolarizing	Depolarizing
Membrane type	Electically inexcitable	Electrically excitable
Type of ion channel present	Stimulus-gated	Voltage-gated

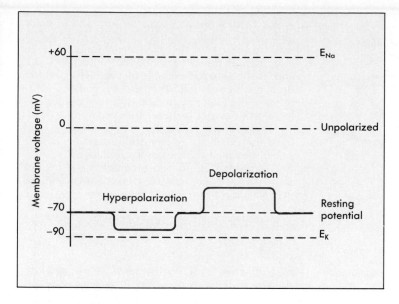

FIGURE 7-1

Illustration of hyperpolarizing and depolarizing voltage shifts. The Na⁺ and K⁺ equilibrium potentials (E_{Na} = +60 mV; E_K = -90 mV) are shown in relation to the resting potential (-70 mV) and the unpolarized state of the membrane (0 mV).

polarizations (Figure 7-1). Receptor potentials typically involve small depolarizations, and synaptic potentials may involve depolarizations or hyperpolarizations. In contrast to these signals in inexcitable membrane regions, the action potentials that occur in excitable membrane regions are much larger and they depolarize to the point of overshooting 0 mV and briefly make the membrane inside-positive (Table 7-1).

Shifts in membrane potential can be brought about in excitable cells by temporary alteration of the permeability to specific ions. As more channels of a particular type open, the membrane potential shifts away from the resting potential and moves in the direction of the equilibrium potential of the ion that becomes able to move more freely across the membrane. A measurable electrical current flows across the membrane when the permeability for an ion increases. The current carried by any particular ion is a function of the ease with which that ion crosses the membrane (the number of open channels) and the **driving force** for that ion. Driving force is the difference (in millivolts) between that ion's equilibrium potential and the membrane potential (see Chapter 6). For example, at the normal resting potential of -70 mV, the driving force for K⁺ is low, only 20 mV (the difference between -90 mV and -70 mV), while the driving force for Na⁺ is much larger, 130 mV (the difference between +60 mV and -70 mV). The large driving force tending to cause Na⁺ to cross the membrane is counteracted by the low permeability of the membrane to Na⁺. In contrast, the small driving force for K⁺ is assisted by high K⁺ permeability in the resting state. The consequence is that the net movement of these two ions across the resting membrane is roughly balanced.

In most cases, the terms permeability (a property relating to the presence of open membrane channels) and conductance (the movement of ions through those channels, creating a current) can be used interchangeably. However, under artificial circumstances, channels could open under conditions in which the relevant ion was absent from the intracellular or extracellular solutions, so no conductance could reflect the altered permeability.

Receptor and Synaptic Potentials

Receptor potentials and synaptic potentials have three common features: (1) their amplitudes are **graded**, increasing with the magnitude of the chemical or physical stimulus that activates them; (2) they cannot be transmitted over long distances because the voltage change diminishes with distance from the stimulus; and (3) they can add together, or **summate.**

Different classes of sensory (receptor) cells are specialized to respond to different sensory modalities, such as touch, odor, and sounds. The "inputs" resulting from the responses to the stimulus energy determine the activity of the sensory cell. The activity of cells that receive their inputs from other neurons is determined primarily by the synaptic potentials initiated following activation of neurons that are connected to them (Figure 7-2).

The graded response of a receptor such as a touch receptor is illustrated in Figure 7-3. If the skin receives a large displacement, many channels will be opened and a large membrane potential shift will occur in the membrane of a pressure-sensitive receptor. When the skin receives a moderate or small displacement, correspondingly fewer channels will be opened. Synaptic potentials and receptor potentials immediately begin to diminish after the channels opened by the stimulus begin to close.

Synaptic and receptor potentials are primarily local changes in potential because the current spreads to surrounding regions of the cell by **decre-**

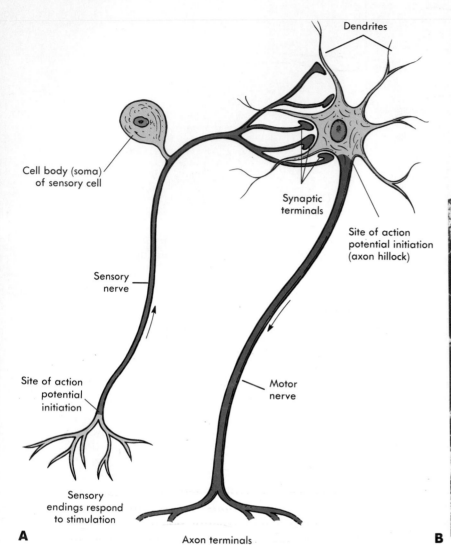

Dendrites

Cell body (soma)
of sensory cell

Synaptic
terminals

Site of action
potential initiation
(axon hillock)

Sensory
nerve

Motor
nerve

Site of action
potential
initiation

Sensory
endings respond
to stimulation

Axon terminals

A

FIGURE 7-2
A A schematic drawing of a sensory and a motor neuron. Electrically excitable regions *(indicated in red)* are present in the axon membranes of these two cells, at the synapses of the sensory neuron, and at the synaptic endings of the motor neuron onto the muscle cell. The sensory endings, dendrites of the motor neuron, and the cell bodies (somata) are electrically inexcitable *(indicated in blue)*. The region where the motor neuron's axon projects from the cell body is electrically excitable and is called the initial segment or axon hillock.
B A scanning electron micrograph of a neuron showing the soma, dendrites, and a small portion of the axon.

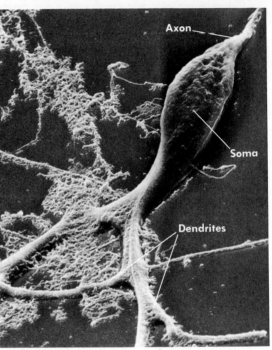

Axon

Soma

Dendrites

B

FIGURE 7-3
Graded responses of a touch receptor. The lower portion of the figure shows four stimuli of increasing size; the upper portion shows the corresponding receptor potentials on the same type of graph. These responses would typically last several milliseconds.

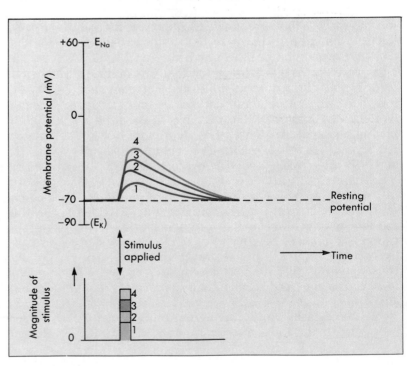

$+60$ — E_{Na}

0

-70

-90 — (E_K)

Membrane potential (mV)

4
3
2
1

Resting
potential

Stimulus
applied

Time

Magnitude of
stimulus

4
3
2
1

0

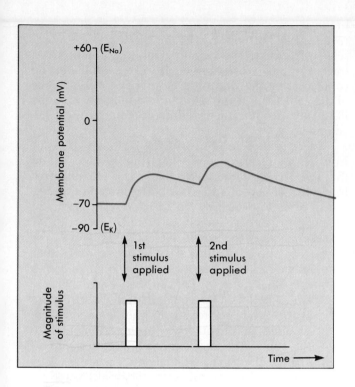

FIGURE 7-4

Summation. The lower portion of the figure shows two stimuli occurring close together in time. The upper portion shows how the two receptor potentials would add together when a second stimulus occurs before the first receptor potential has faded away.

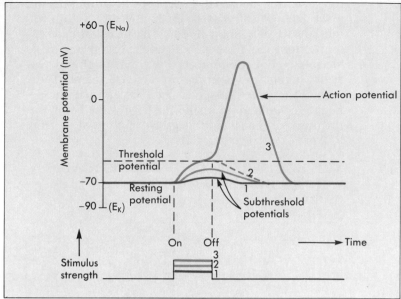

FIGURE 7-5

An illustration of the all-or-nothing property of action potentials. The responses of the excitable membrane to the stimulus, applied at *(ON)* and removed at *(OFF)*. When a stimulus is applied, the membrane potential depolarizes; the magnitude of the depolarization increases with larger stimuli. The responses to subthreshold stimuli fade away when the stimuli are removed. At threshold, an all-or-nothing action potential is generated 50% of the time and fails the other 50% of the time (dotted line returning to resting potential).

mental conduction. Although neither receptor potentials nor synaptic potentials spread far along the membrane before they become insignificantly small, individual signals typically fade slowly enough (within 4 msec or much longer, up to many seconds and even minutes) that other signals can add to them, a process called **summation** (Figure 7-4). The ability of neurons to combine information arriving from several sources or arriving from the same repeatedly activated source within a short time is the basis of **integration,** the decision-making capability of the nervous system.

Action Potentials

Action potentials differ from receptor potentials and synaptic potentials in two important respects (Table 7-1): (1) the amplitude of an action potential is nearly constant and is not related to the size of the stimulus, so action potentials are **all-or-nothing** events; and (2) action potentials do not fade away with distance from the site of initiation. The process by which action potentials spread along cell membranes without loss of strength is called **propagation** or **nondecremental conduction,** to distinguish it from the decremental spread of receptor and synaptic potentials that diminish with distance from the site of stimulation. Regions of the cell membrane that can undergo action potentials are

called **electrically excitable** regions (indicated in red in Figure 7-2); portions of a cell that exhibit only graded receptor or synaptic potentials are **electrically inexcitable** membrane regions (blue in Figure 7-2).

When the terminal portion (electrically inexcitable region) of a sensory cell responds with a large graded potential to a large stimulus, an action potential is initiated in the nearby portion of the long axon (electrically excitable) which links the sensory cell ending with the central nervous system. An action potential either occurs or fails to occur (Figure 7-5). Because it is an all-or-nothing type of event, there are no small action potentials to reflect weak stimulation or large action potentials to correspond to the more dramatic events that the nervous system communicates. The information about the size of the stimulus is relayed by the number of action potentials that the graded sensory signal is able to initiate in the axon; this information is transmitted unchanged down the axon to the output portion of the cell.

Action potentials are a response of standard size that moves along the length of the axon, rather like the burning of a fuse, but unlike the burnt region left behind as a fuse burns, the region behind the moving action potential recovers its ability to transmit an action potential. Except for heart muscle and some smooth muscles (see Chapter 12), action potentials are very short-lived shifts in membrane potential that last from 1 to 5 msec. Action potentials are relatively large compared with receptor and synaptic potentials. During an action potential the membrane potential goes from its normal inside-negative value to an inside-positive value before returning to the inside negative potential (Figure 7-5).

MOLECULAR BASIS OF ELECTRICAL EVENTS IN MEMBRANES
Membrane Repolarization

> - *How is the membrane potential restored to its resting state following a hyperpolarizing or a subthreshold depolarizing shift in membrane potential?*
> - *What opposing forces are balanced at the threshold potential?*

Electrical events can be initiated in nerve and muscle membranes because these regions possess special channels. Such membrane regions also contain the leak channels present in all cells. To appreciate what occurs in sensory and synaptic potentials and in the generation of an action potential, it is first helpful to understand the powerful forces that tend to keep the membrane potential at its resting level.

Figure 7-6, *A* shows the resting potential and the equilibrium potentials for K^+ and Na^+ ions and how these relate to the driving forces on each ion. At rest, the membrane potential is much closer to the equilibrium potential for K^+ because the mem-brane is much more permeable to K^+ than to Na^+ (Figure 7-6, B). This makes the driving force for K^+ relatively small. The negative internal potential attracts the positively charged K^+ ions, opposing the concentration gradient that otherwise forces K^+ out of the cell. Thus at the resting potential few K^+ ions flow out, despite the presence of many open K^+ channels.

The situation is different for Na^+, which is far from its equilibrium potential (Figure 7-6, *A*), so that the driving force for Na^+ is large. The negative internal charge tends to draw Na^+ ions into the cell, but in a resting cell few Na^+ channels are open, so the net entry of Na^+ ions into the cell at rest approximately matches the small net exit of K^+ ions. For simplicity, the contribution of chloride and other ions will be ignored, but these can also move across the membrane.

When a cell is hyperpolarized slightly (Figure 7-7, *A*), the driving force on K^+ ions is decreased, with the effect that even fewer of these ions tend to flow out across the membrane, but rather remain behind with the additional negative charges. The hyperpolarization has the opposite effect on Na^+, increasing its rate of entry as its driving force is increased. As a consequence, the temporary hyperpolarization will be "defeated" by Na^+ entry until the membrane potential is driven back to the steady state—the point at which Na^+ and K^+ entry are matched, otherwise known as the resting potential. An imposed depolarization will decrease the Na^+ driving force and increase the K^+ driving force, so the excess of K^+ exit over Na^+ entry will return the cell to its resting potential (Figure 7-7, *B*).

Just as in the examples of hyperpolarization and

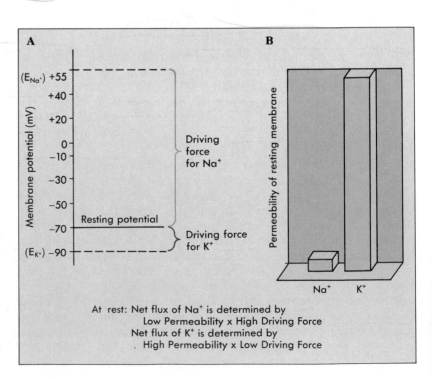

FIGURE 7-6
Comparison of the driving forces (A) and permeabilities (B) for Na^+ and K^+ at the resting membrane potential.

NERVE AND MUSCLE

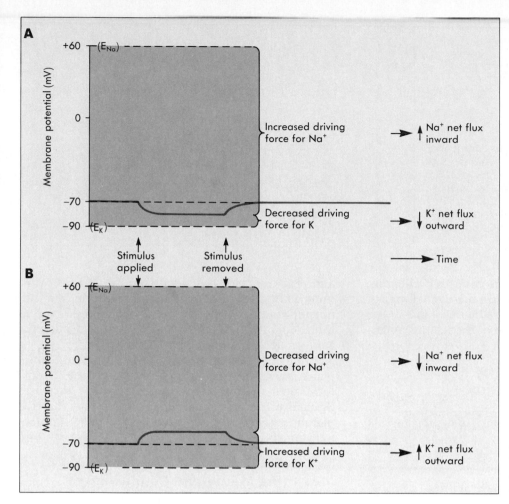

FIGURE 7-7
A The effect of a hyperpolarizing stimulus is to increase the driving force for Na⁺ and decrease the driving force for K⁺ (compare the size of the bracketed voltages indicating driving forces). When the imposed hyperpolarization is removed, the combination of an increased net entry of Na⁺ and a decreased net exit of K⁺ returns the membrane potential to the resting potential.
B The effect of a depolarizing stimulus is to decrease the driving force for Na⁺ and increase the driving force for K⁺. When the stimulus is removed, the combination of a decreased net entry of Na⁺ and an increased net exit of K⁺ returns the membrane potential to the resting potential.

depolarization given in Figure 7-7, restoration of the membrane potential to the resting potential following a sensory stimulus or a synaptic event is accomplished by movement of ions through leak channels. Consequently, following an *alteration* of the membrane potential, the return to the resting value depends only on forces that are responsible for establishing the resting potential—the resting (leak) permeabilities and the concentration gradients for K⁺ and Na⁺.

Action Potential Threshold

For an action potential to be initiated: (1) the ions flowing across the leak channels must fail to overwhelm the externally-imposed depolarization and (2) the depolarization must open voltage-gated Na⁺ channels that exhibit positive-feedback properties. When a large enough depolarization spreads from inexcitable membrane regions, the entry of Na⁺ takes over and the permeability properties of the membrane are radically altered for a brief period of time.

Threshold is defined as the value of the mem-

brane potential at which an action potential will occur 50% of the time. Levels of membrane potential below the value needed to produce an action potential are **subthreshold** and such depolarizations disappear without sending any signal down the axon. To understand what is happening around threshold, it is helpful to picture the competition between opposing forces (Figure 7-8). As the membrane potential becomes depolarized, the driving force for K⁺ increases and there is an increased movement of K⁺ ions out of the cell. This force, which tends to repolarize the membrane, gets stronger as the cell becomes more depolarized. However, some of the gates on voltage-gated Na⁺ channels are opening in response to the depolarization. Because the driving force for Na⁺ entry is very large, opening of even a few additional Na⁺ channels allows a large increase in Na⁺ entry, which further depolarizes the cell and causes more voltage-gated Na⁺ channels to open. Na⁺ entry opens more Na⁺ gates that allow more Na⁺ in—an example of positive feedback.

Threshold is exceeded if the voltage-gated Na⁺ channels let just a few more Na⁺ ions into the cell

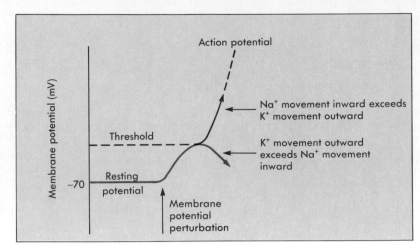

FIGURE 7-8
The positive-feedback and negative-feedback forces that determine membrane responses at threshold. At threshold, if the net entry of Na$^+$ exceeds the net exit of K$^+$, an additional depolarization will cause the opening of still more voltage-gated Na$^+$ channels, allowing the positive feedback mechanism of the action potential to take over.

than the number of K$^+$ ions flowing out, allowing the positive-feedback system to take over. Entry of slightly fewer Na$^+$ ions will result in failure of the action potential and restoration of the resting potential (Figure 7-8).

VOLTAGE CLAMP STUDIES
Voltage Clamp of Giant Axons

> • What differences in Na$^+$ and K$^+$ conductances were revealed by voltage clamping?
> • What limits the positive feedback effect on membrane potential that results when the membrane is depolarized above threshold?

Progress from the earliest demonstrations that nerve and muscle cells possess "animal electricity" to actually recording such events was slow because the science of electronics was in its infancy. Two groups of researchers, the American scientists Kenneth Cole and Howard Curtis and the British scientists Alan Hodgkin, Andrew Huxley, and Bernard Katz, began a landmark series of studies of action potentials in the 1930s and (after the interruption for World War II) continued into the early 1950s.

The unwitting but invaluable collaborator in these experiments was an ideal experimental animal, the squid. This animal, like many invertebrates, has a few nerve cells that mediate escape reflexes and conduct action potentials very rapidly. In the invertebrates, the major adaptation for speed of conduction is large diameter axons. The giant axon of the squid is so large that it was initially described as a blood vessel; it can be seen through the mantle in species with more transparent body surfaces (Figure 7-9). The pair of giant axons are part of a giant conducting system that activates the mantle muscle contractions, sending the animal jetting backwards to escape a threat.

The giant axons can be removed from the animal, placed in artificial solutions of known composition, and the axoplasm can even be squeezed out

and replaced with a defined solution of ions. Most importantly, the axons, which measure up to a millimeter in diameter, can have one or more wires threaded into them so that the electrical potential along a length of the membrane cylinder can be controlled. In the early recordings employing squid giant axons, researchers found that the overall movement of ions across the membrane increases dramatically during action potentials and that the membrane potential does not just go to zero potential but briefly becomes internally positive (approaching E_{Na}).

Measuring Membrane Conductances

The electronic circuitry of the **voltage clamp** was devised so that the membrane potential of the axon could be shifted from its resting value to some chosen value and held there while the current flow resulting from movement of ions across the membrane was measured. The principle underlying the voltage clamp circuit (Figure 7-10) is that of negative feedback: when the flow of current across the membrane starts to alter the membrane voltage, the electronic circuitry forces current of opposite polarity into the cell to keep the membrane potential from changing. Thus the currents injected by the voltage clamp negative-feedback circuit (the "bucking currents") are equal in magnitude and opposite in charge to the currents flowing across the membrane.

A major aim of the investigators was to determine which ions were flowing across the membrane during the early and late periods of the brief action potential. They clamped the membrane potential at values above the resting potential and held it at that command level for longer than the action potential duration. The experiments were run with the axon maintained in two conditions: in normal seawater, which is very similar to the blood of a squid, and in an artificial solution in which a large impermeant positive ion was substituted for Na$^+$. The contribution of K$^+$ ions alone was apparent from the re-

A

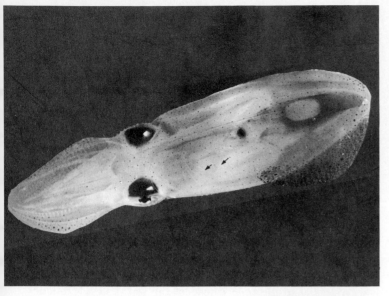

B

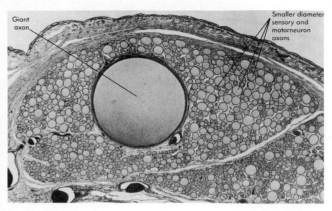

FIGURE 7-9
A Photograph of a squid. The arrows indicate the path followed by the giant axon, which extends posteriorly, seen as the darker streak.
B Cross section of the mantle of a squid showing one of the giant axons.

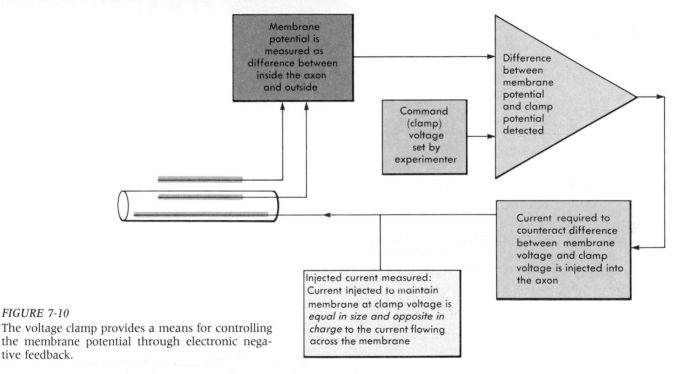

Membrane potential is measured as difference between inside the axon and outside

Command (clamp) voltage set by experimenter

Difference between membrane potential and clamp potential detected

Current required to counteract difference between membrane voltage and clamp voltage is injected into the axon

Injected current measured: Current injected to maintain membrane at clamp voltage is *equal in size and opposite in charge* to the current flowing across the membrane

FIGURE 7-10
The voltage clamp provides a means for controlling the membrane potential through electronic negative feedback.

sponses of the membrane when no Na$^+$ ions were present, and the pure contribution of Na$^+$ could be calculated by subtracting the response in Na$^+$-free seawater from the response recorded from the combined action of both ions in normal seawater. Luckily, only these two ions contribute significantly to the action potential in the squid axon.

The Na$^+$ and K$^+$ conductances were found to differ in how soon they responded to a membrane potential change and in their rate of change in permeability (Figure 7-11). At a membrane potential just above threshold (-40 mV), Na$^+$ conductance is elevated, but even under constant voltage clamp

conditions this is not maintained, but declines within a few milliseconds (Figure 7-11, *A*). In other words, the opening of Na$^+$ gates depends on the voltage, but it is terminated by an intrinsic time-dependent process. In contrast, the K$^+$ ions respond to a clamping current just above threshold with a much slower increase in K$^+$ ion flux that is maintained at a constant level as long as the clamp current is on (Figure 7-11, *B*). Thus the opening of the K$^+$ gates is dependent only on voltage and unaffected by the passage of time.

Responses to a series of clamp currents were collected, a few of which are illustrated in Figure

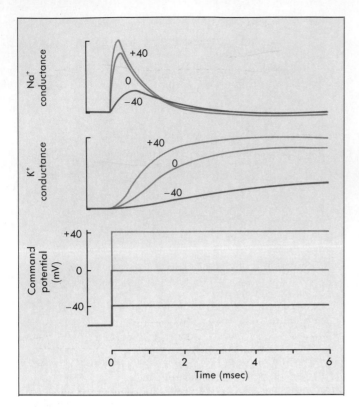

FIGURE 7-11

Responses under voltage clamp for three command potentials show that the Na$^+$ conductance (A) is low around threshold (-40 mV), intermediate at 0 mV, and very high at +40 mV. The Na$^+$ conductance is initiated rapidly (activation) and is subsequently inactivated. The K$^+$ conductance (B) has a slower onset and does not inactivate as long as the command potential is held.

7-11. In all cases, the Na$^+$ response involves a rapid onset (less than 1 msec) and a self-determined turn-off, whereas the K$^+$ response has a slower onset (1 to 2 msec) and is maintained.

If the membrane potential is allowed to change rather than being clamped, the underlying currents change with time after the action potential is initiated. Figure 7-12 illustrates the action potential on the same time scale as the calculated Na$^+$ and K$^+$ conductances. The rising phase of the action potential is driven by the Na$^+$ influx, and the falling phase occurs as the Na$^+$ current turns off and the K$^+$ efflux is turning on. Even though K$^+$ conductance is maintained under voltage clamp conditions, it is normally turned off by the repolarization of the membrane that results from K$^+$ flow in the unclamped condition.

Data gathered with the voltage clamp allowed Hodgkin and Huxley to describe with a set of equations the ionic currents flowing across the membrane at each point in time during the action potential. Their equations employed three components: Na+ conductance (both activation and inactivation features), K+ conductance, and leak channel con-

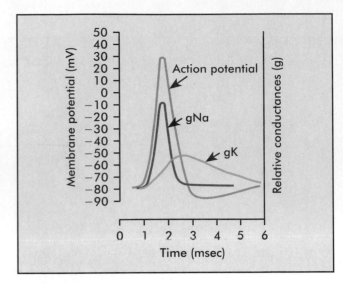

FIGURE 7-12

An illustration relating the voltage changes during the action potential to the time course of the Na$^+$ and K$^+$ conductances (abbreviated gNa and gK).

ductances. The significance of their model (for which they shared the Nobel Prize in Physiology or Medicine with John C. Eccles in 1963) was not only that their equations fit the shape of the action potential curves but that the model was consistent with molecular mechanisms and stimulated further work on the nature of the channels.

VOLTAGE-GATED CHANNELS
Channels in Electrically Excitable Membranes

The understanding of the action potential gained from voltage clamp recordings reflected the measurement of the behavior of large populations of what could only, at that time, be hypothetical channels. It is now possible to interpret the events that underlie the action potential in the light of what is known about proteins in excitable cell membranes. The **gated ion channels** are proteins that are thought to possess a region that acts as a selectivity filter (determining which ions pass through) and a second region of the molecule that acts as a gate (Figure 7-13). The gate is probably a highly mobile portion of the channel structure that opens or closes the ion pathway when the molecule's shape is altered by electrical or chemical interactions. Changes in the membrane voltage will alter the electrical environment of **voltage-gated** channels and trigger conformational rearrangements of the entire molecule that have the effect of converting it from a closed to an open (activated) pore.

Activation of a population of gated channels does not occur in an all-or-nothing manner. The behavior of individual channels and the average re-

FIGURE 7-13

A hypothetical model of the Na$^+$ channel illustrating the three states: resting, activated, and inactivated.

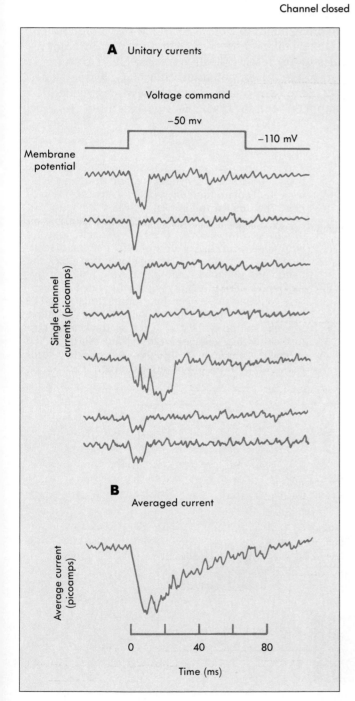

A Unitary currents

Voltage command

−50 mv

−110 mV

Membrane potential

Single channel currents (picoamps)

B Averaged current

Average current (picoamps)

0   40   80

Time (ms)

FIGURE 7-14

A Patch-clamp data showing the currents associated with individual Na$^+$ channels in response to a depolarizing voltage step. The individual channels can be seen to open and close independently. Currents associated with individual channels are on the order of 1 picoampere.

B The averaged currents for individual channel openings resemble the activation-inactivation time-course characteristic of the whole cell.

sponses of many channels can be seen using a technique called patch-clamping, which permits electrical isolation of a small region of cell membrane (a patch) containing only a few channels (Figure 7-14). The overall permeability of the membrane patch is determined by the number of Na$^+$ and K$^+$ channels and their opening behavior.

It has become possible to biochemically isolate various types of gated ion channels, determine their amino acid sequence and three-dimensional structure, and even clone them using the recombinant DNA techniques described in Chapter 3 (see boxed essay, p. 70). In addition to voltage-gated channels, **stimulus-gated** channels open and close in response to physical or chemical stimuli, and the **ligand-gated** channels of synaptic potentials are gated by chemicals released by nerve cells (Table 7-1).

Properties of Na$^+$ and K$^+$ Channels

Opening of the voltage-sensitive Na$^+$ channels occurs rapidly (within a few tenths of a millisecond) and is known as **Na$^+$ activation.** Once a Na$^+$ channel has opened, it closes spontaneously after a time, even though the membrane potential is still depolarized. Closing of the opened channels, called **Na$^+$ inactivation,** takes about 10 times longer than Na$^+$

activation. An inactivated channel is not simply a closed channel but a closed and latched channel (Figure 7-13). The inactivated state lasts as long as the membrane remains depolarized. When the membrane repolarizes, the channels revert to the initial resting state and can again be reopened by depolarization. On the average, potassium channels open more slowly than Na$^+$ channels. Potassium channels do not inactivate spontaneously; they remain open as long as the membrane is depolarized and can close only if the membrane potential returns to its original resting level.

Changes in Permeability During Action Potentials

When some Na$^+$ gates open in response to depolarization, the conductance to Na$^+$ increases by the positive-feedback effect of depolarization of the voltage-gated channels, quickly leading to a very high Na$^+$ conductance (Figure 7-15). The limits to the increase in the Na$^+$ conductance are (1) the decrease in driving force for Na$^+$ movement as the membrane potential approaches the Na$^+$ equilibrium potential, and (2) the spontaneous closing (inactivation) of Na$^+$ gates within a set time after their opening. However, throughout the action potential,

the resting permeability channels are present, and ions flowing through them contribute to the shape of the action potential. The action potential never reaches E$_{Na}$ because of K$^+$ exit during the rising phase.

Inactivation of Na$^+$ channels reduces the Na$^+$ permeability to the low value characteristic of the resting membrane. Inactivation of Na$^+$ channels, by itself, results in membrane repolarization, but repolarization under these circumstances is relatively slow. K$^+$ channels activate about the same rate that Na$^+$ channels inactivate (Figure 7-15). Activation of K$^+$ channels adds to the K$^+$ leak channels and greatly accelerates repolarization, limiting the recovery phase of the action potential to 1 to 2 msec or less. As the membrane repolarizes, voltage-gated K$^+$ channels begin to close. Closing of the population of channels takes time, so K$^+$ conductance does not return to its resting level immediately. During this part of the action potential, after Na$^+$ channels have inactivated and while many of the voltage-gated K$^+$ channels remain open, K$^+$ permeability is even

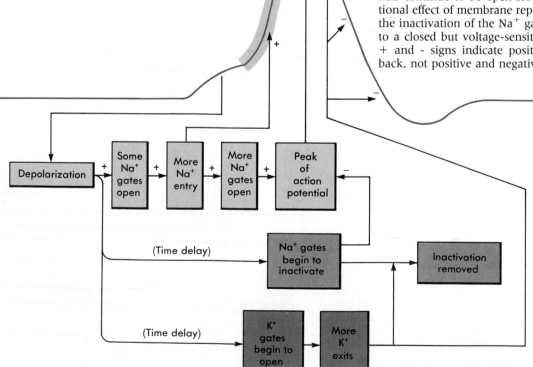

FIGURE 7-15

The positive-feedback contribution to the rising phase of the action potential is made through the effect of depolarization on voltage-gated Na$^+$ channels. The depolarization also activates voltage-sensitive K$^+$ channels after a time delay. The contribution of current flow through the K$^+$ channels, together with coincident inactivation of Na$^+$ channels, repolarizes the membrane. The repolarization takes the membrane potential below the resting potential (after-hyperpolarization) because some gated K channels continue to be open for a short time. An additional effect of membrane repolarization is to remove the inactivation of the Na$^+$ gates, so that they return to a closed but voltage-sensitive state. Note that the + and - signs indicate positive and negative feedback, not positive and negative charge.

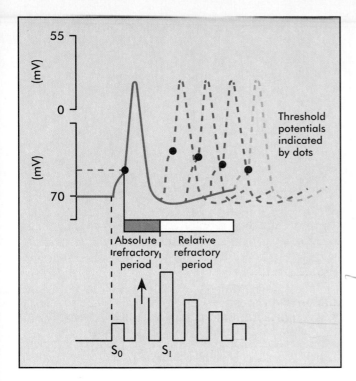

FIGURE 7-16

No stimulus, however great, will produce a second action potential immediately following a stimulus (S_o) that initiates an action potential; this is the absolute refractory period. The magnitude of a second stimulus (S_1) needed to elicit a second action potential during the relative refractory period is greater (that is, the threshold is higher) than for the first action potential, but decreases with time until the end of the relative refractory period when it has returned to the initial value.

greater than during rest, and the membrane potential hyperpolarizes past the resting value and approaches the K^+ equilibrium potential (Figure 7-16). This is called either a **hyperpolarizing afterpotential** or an **after-hyperpolarization**.

Role of the Na^+-K^+ Pump in Excitability

The membrane potential is maintained in the face of continual leakage of Na^+ and K^+ across the membrane occurs in all body cells. The action potentials of excitable cells allow additional K^+ out and an equal amount of Na^+ into the cell. All cells possess Na^+-K^+ pumps (Na^+-K^+ ATPases) that counteract this movement of ions down their concentration gradients. The Na^+-K^+ pump operates steadily to prevent significant reduction of the concentration gradients, which in excitable cells would result in loss of excitability. It is easy to appreciate the contribution of the pump by observing the effect of metabolic poisons that block production of ATP. Depending on the size of the cell, the membrane potential will decline sooner or later.

The amounts of Na^+ and K^+ that move in the course of a single action potential are typically small, compared with the amounts of these ions that are responsible for the transmembrane concentration gradients. In an extreme example, a single action potential in a squid giant axon results in loss of about only one ten-millionth of the cytoplasmic K^+. The Na^+-K^+ pumps of an axon can be blocked pharmacologically without interfering with the steady generation of action potentials for many hours in cells with such large ion stores. This shows that the production of action potentials is not an energy-dependent event, but rather is driven by the stored energy of the concentration gradient. Smaller cells will have their concentration gradients reduced more significantly by high levels of activity because their surface areas are larger relative to their cytoplasmic volumes. A measurable stimulation of the Na^+-K^+ pump activity occurs after a high level of repetitive firing of small axons. However, ion pumps operate on a much slower time scale than individual action potentials and thus have no direct role in the recovery phase of action potentials.

REFRACTORY PERIOD AND FIRING FREQUENCY
Absolute and Relative Refractory Periods

- *What two factors are responsible for refractoriness of an axon?*
- *If action potentials are of uniform size, how do they relay information about the size of the depolarization in the nonexcitable portion of the cell?*
- *How does extracellular Ca^{++} affect membrane excitability?*

The action potential of excitable cells is a discrete event of uniform size and highly repeatable nature. Historically, two terms have been used to quantify the observation that action potentials cannot be forced to "pile up" on one another in the way that sensory and synaptic potentials are able to summate. These terms are the **absolute refractory period** and the **relative refractory period** of the action potential. When an excitable cell is refractory, it resists attempts to initiate an action potential.

The basis of this refractoriness is now understood in terms of the changes in excitable channels that underlie the action potential. First, consider the rising phase of the action potential: the membrane possesses just so many Na gates, and when they are activated, there is no way additional stimulation can increase the height of the rising phase of the action potential because no more gates can be called into service.

During the falling phase, the incapacity of the membrane to respond to depolarizing stimulation can be explained by the two changes in channels that bring an end to the action potential: Na^+ inactivation and opening of K^+ channels. Both of these changes were set to occur with their characteristic time lags after the depolarization exceeded threshold, and both reduce the effectiveness of a depolarizing stimulus. First, a depolarizing stimulus will drive even more K⁺ outward through the combination of leak and activated K⁺ channels. Second, inactivated Na⁺ channels cannot be opened by depolarizing current and therefore a regenerative positive feed-back response cannot be initiated. The consequence is that, during the rising phase and most of the falling phase of an action potential, it is impossible to do anything to the membrane that will bring about activation of another action potential, and this is the absolute refractory period.

The action potential exists as a discrete, unitary event because the Na inactivation can be removed only by return of the membrane potential to near the resting level. The molecules that comprise the Na channels are rearranged from the inactivated state to the closed and voltage-sensitive state when the membrane around the channels bears an internally negative charge. The transition from inactivated to the merely closed state occurs above the resting potential for some channels, and others are still making this transition during the after-hyperpolarization. Somewhere in this range of recovery, enough Na channels return to the closed and voltage-sensitive state that another action potential can be forced by application of high levels of depolarization. At this point, the absolute refractory period is over and the relative refractory period has begun (Figure 7-16).

In the relative refractory period, more depolarization is required to bring the cell to threshold and initiate an action potential than was required when that cell had been previously inactive. The requirement for higher than normal amounts of depolarization during the relative refractory period is due to (1) some activated K gates that remain open, and (2) some Na gates that are still inactivated. This shows that the threshold is not a fixed value of membrane potential but varies according to the recent activity in an excitable cell (Figure 7-16).

Repetitive Firing During Prolonged Depolarization

Although absolute and relative refractory period are defined by experimental manipulations, a physiological consequence of the refractoriness of excitable membranes is that it sets an upper limit to the rate at which a cell can fire action potentials. Over a range of depolarization, a cell will fire action potentials at an increasing rate if the level of depolarization increases above threshold (Figure 7-17). Under extreme depolarization, (and for brief periods of time) some mammalian neurons can approach a firing rate of 1000 action potentials per second. This requires that the action potential depolarization and recovery occupy no more than a millisecond. Most excitable cells have a lower maximum rate of action

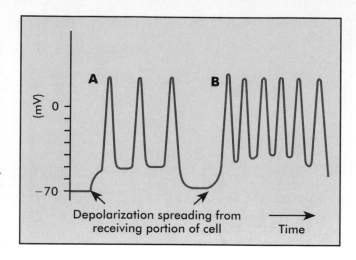

FIGURE 7-17
The rate of repetitive firing is a function of the membrane depolarization level.
A The maintained depolarization was slightly above threshold, and three action potentials were generated.
B The depolarization was much higher, and action potentials were generated during the relative refractory period, producing a higher frequency of firing.

potential discharge, due to factors such as cell geometry and the density and characteristics of each type of channel found in the cell.

Extracellular Ca⁺⁺ and Excitability

Changes in extracellular Ca^{++} concentration can affect membrane excitability by altering the amount of depolarization required to initiate an action potential (Figure 7-18). This is because Ca^{++} exerts a stabilizing effect on the membrane by interacting with negative charges on the polar heads of membrane phospholipids and extensions of membrane proteins in the extracellular region. By themselves, these negative charges on the outside of the membrane make a small but measurable depolarizing contribution to the total membrane potential. If many of these negative charges are neutralized by extracellular Ca^{++}, less fixed negative charge is left on the outside to reduce the membrane polarization. The interactions between these charges and Ca^{++} means that the membrane potential is very sensitive to extracellular Ca^{++} concentration.

Disorders of calcium homeostasis produce large fluctuations in heart rate and are thought to contribute to some cases of hypertension (elevated blood pressure). In conditions that elevate Ca^{++} in the plasma, the action potential threshold is farther from resting potential. Because a larger stimulus is needed to produce an action potential, the excitable membranes of nerves and muscles become refractory to stimulation when Ca^{++} levels are elevated. In conditions that cause low plasma Ca^{++} (**hypocalcemia**) the threshold is closer to the resting potential, and the membranes of nerve cells and mus-

NERVE AND MUSCLE

FIGURE 7-18
The effect of extracellular Ca^{++} on electrically excitable membranes. The normal condition (**A**) is contrasted with the hyperpolarizing effect of high Ca^{++} on the resting potential (**B**) and the increased excitability that results from the depolarizing effect of low Ca^{++} levels in the membrane potential (**C**). In the latter case, spontaneous generation of action potentials can occur if the resting potential rises to threshold potential.

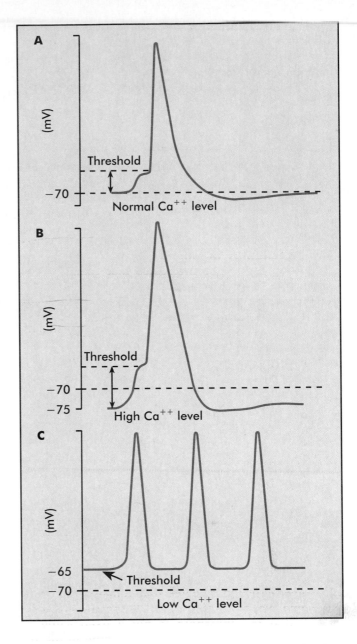

cle cells become **hyperexcitable.** They may even discharge action potentials continuously, just as if they were being stimulated. Motor nerve cells and skeletal muscle fibers may discharge spontaneously, causing muscle spasms that are called **hypocalcemic tetany.**

PROPAGATION OF ACTION POTENTIALS
Spread of Excitation

> • *What two factors determine action potential conduction velocity?*
> • *What is the immediate source of the energy that drives an action potential?*
> • *What is a compound action potential? Under what circumstances might one be recorded?*

In neurons, the propagation of action potentials along the axon transmits the signal in the cable-like extension of the cell, which may link a brain region with the spinal cord or the spinal cord with the body extremities. Under normal conditions, action potentials are initiated in membrane regions close to the electrically inexcitable portion of the cell where stimuli are received. The site of initiation is called the **axon hillock** or **initial segment** (Figure 7-2). The axon hillock possesses voltage-gated Na^+ channels that can respond to depolarizing current spreading from the input sites. If the depolarization is above threshold, an action potential will be initiated. Once an action potential is initiated, it flows from the axon hillock down the axon because the depolarization associated with the action potential activates next-in-line voltage-gated Na^+ channels in adjacent excitable membrane (Figure 7-19).

A path of current flow is set up between the region experiencing the peak of the action potential (where positive charge is drawn across the membrane most strongly) and adjacent regions on the outside of the membrane, where positive charge is readily drawn off. This removal of positive charge from the outside of the membrane in front of the action potential depolarizes that region, adding to the depolarizing effect of the influx of Na^+ across the membrane. The combined depolarizing forces cause the region ahead of the action potential to become the next area to have Na^+ gates opening. Paths of current flow associated with the repolarization be-

hind the moving action potential move positive charge out across the membrane to complete the circuit (Figure 7-19).

In this way an action potential, once initiated, can spread along the axon. The non-decremental conduction that characterizes action potentials is necessary to relay signals over relatively long distances. The distance between a sensory nerve ending in the foot and its cell body near the spinal cord is about 1 m. Without some way of recharging the signal as it spreads along the membrane, an electrical signal would travel only a few millimeters before it died out. The cycle by which Na^+ channels open, allow current stored in the membrane potential battery to briefly flow, and then inactivate after the current has opened adjacent Na^+ channels, provides continuous regeneration of the signal so that it does not die away with distance.

Toxins That Block Channels: The History of TTX and STX

Until the mid-1960s biophysicists had no clear idea of how ions actually moved across cell membranes. Were there transient water-filled pores? Did ions dissolve in a homogeneous lipid membrane? Were lipid-soluble protein carriers involved? Answers to these questions were provided in part by the discovery of chemical "magic bullets" that aim unerringly at membrane channel proteins. Tetrodotoxin (TTX) is a poison found in the sex organs and liver of marine puffer fish and other species of the order Tetraodontiformes, in the blue-ringed octopus and some salamanders. Saxitoxin (STX) is a related molecule found in single-celled marine protozoa called dinoflagellates. Both toxins are small, water-soluble compounds that block voltage-gated Na^+ channels in nerve and muscle fibers at nanomolar concentrations, making conduction of action potentials in these cells difficult or impossible. Their high affinity for Na^+ channels ranks them among the most potent poisons.

The earliest reference to TTX appeared in the first Chinese pharmacopoeia, written approximately 200 BC, where it was recommended as an effective means of arresting convulsions. A more detailed account appeared in a pharmacopoeia published in 1600 AD, describing its effects as: "In the mouth they (the liver and eggs of puffer fish) rot the tongue; internally they rot the guts"; "the poisoning no remedy can relieve"; and "to throw away life, eat puffer fish." In Japan the puffer fish, called *fugu,* is regarded as both a cult object and a delicacy. If the flesh is to be eaten safely, the fish must be prepared without puncturing either the liver or the reproductive organs. Restaurants that serve fugu are strictly licensed, but fascination with the consumption of fugu accounts for some 200 to 400 cases of fugu poisoning annually. Most poisonings occur when the fish is prepared by amateurs, but there is reason to believe that risk-taking behavior, in the form of deliberate addition of some of the toxin-containing parts of the animal to the flesh, plays at least as large a role as carelessness or ignorance.

TTX was a new experience for Europeans who began to visit the Orient in the seventeenth century. A detailed account of the symptoms of fugu poisoning appears in the log of Captain Cook's second circumnavigation of the globe in 1670. Mild cases involve limb numbness, flushing of the skin, muscle weakness, and a tingling sensation in the mouth and tongue. (This tingling is regarded by some Japanese as part of the sought-after experience of consuming fugu.) Larger doses produce severe disturbances of the heart rhythm. Lethal doses paralyze the diaphragm, leading to asphyxiation.

STX-containing dinoflagellates "bloom" at certain times of the year in certain regions of the ocean. The concentration of organisms sometimes reaches 20 million per liter, giving the water a reddish tint referred to as a "red tide." The dinoflagellates are consumed by clams, mussels, and scallops which themselves have Na+ channels that are relatively resistant to the toxins. The tissues of these edible shellfish accumulate the toxins, which account for several accidental deaths each year despite the fact that the waters are constantly monitored for the appearance of the dinoflagellates.

Some dinoflagellates in the waters around Florida and the Caribbean contain the marine toxin Ciguatoxin. Unlike TTX and STX which block Na^+ channels, CTX holds Na^+ channels open by interacting with a different binding site. The dinoflagellates are ingested by small fish which, in turn, are eaten by larger fish. Levels of CTX gradually accumulate. Barracuda, a fish high on the food chain, cannot be legally imported into this country because the meat contains significant levels of CTX. CTX is also found in Pacific red snapper and grouper, where it causes some 10,000 cases of CTX poisoning annually. Mild CTX poisoning produces intestinal and cardiovascular disturbances. With increasing interest in seafood as a healthy alternative to red meat, there has been more attention to seafood inspection. Very recently, specific antibodies to various marine toxins have been developed to aid in identifying low levels of these toxins in contaminated seafood.

Toxins have been used by neurophysiologists to show that sodium channels are physically distinct from other membrane channels, such as the K^+ channels, and have helped in characterizing the entryway of the Na^+ channel and in separating and counting the channel molecules. Other useful toxins have been extracted from plants, the skin of amphibians, and the poison glands of scorpions, spiders, and sea anemones. A great diversity of toxins has recently been discovered in marine predatory snails. Because of their specificity for ion channels, natural toxins may also be useful as pharmacological agents. For example, some marine toxins have antitumor activity, whereas others suppress the immune response, an effect that could aid recipients of organ transplants.

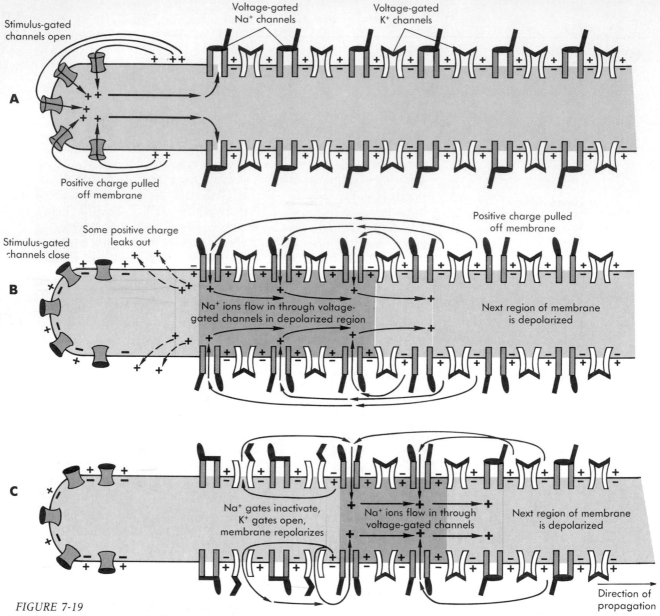

FIGURE 7-19

A The decremental spread of current entering receptor channels in the receiving portion of an excitable cell may be great enough to cause opening of voltage-gated Na$^+$ channels in the electrically excitable portion of the cell.

B Current flow through the Na$^+$ channels spreads in one direction, down the axon, because voltage-gated channels are not present in the electrically inexcitable, receiving portion of the cell.

C The path of current flow along an axon when an action potential is propagating down it includes depolarizing, repolarizing, and as yet unexcited regions of the membrane.

Differences in Propagation Velocity

Some responses made by the body require rapid conduction, whereas others can be transmitted much more slowly without reducing the effectiveness of the response. Rapid conduction of information on development of the internal state associated with hunger is unnecessary, but rapid reflex withdrawal from a hot object or to avoid the path of a moving projectile is crucial. Preferential survival of individuals with quick reflexes has resulted in the evolution of giant fiber systems in the neural pathways that mediate escape in many invertebrates, such as the squid.

Effect of Diameter and Myelin on Propagation Velocity

The velocity with which an action potential can propagate in an excitable membrane is determined by how much adjacent membrane the action potential can bring to threshold, or to put it another way, how far ahead of themselves the open Na$^+$ channels can open more Na$^+$ channels. Two factors determine this distance. One is the diameter of the fiber. The larger the diameter, the better a conductor of electricity it will be and the further current will spread. In vertebrate animals, however, another factor contributes much more to the increase in conduction velocity than diameter.

NERVE AND MUSCLE

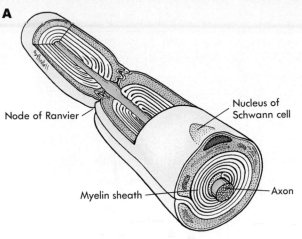

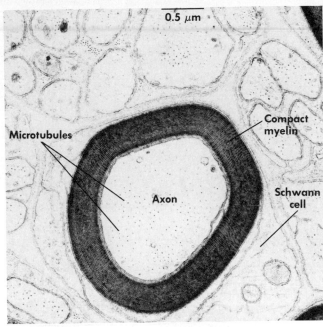

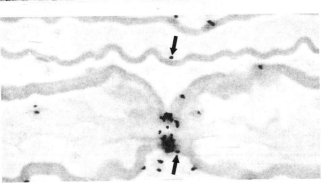

FIGURE 7-20

A A longitudinal section of a myelinated axon from the central nervous system, showing a node of Ranvier and the terminal loops of the myelin sheath.
B A cross section of a myelinated axon showing the layers of membrane forming compact myelin and their relationship to the surrounding Schwann cell.
C An autoradiograph of a node of Ranvier. The black dots indicated by the arrows represent monoclonal antibodies directed at sodium channels and are thus concentrated at the nodes.

The effectiveness of current spread from regions where channels have been opened to adjacent regions of membrane is reduced by the current that flows out across the membrane rather than down the length of the nerve fiber. In many axons of the vertebrate nervous system, insulation against current leakage is provided by supporting cells (glia) of the nervous system called **Schwann cells** or **oligodendrocytes,** depending on where in the nervous system they are located.

In the course of development, these cells send out tonguelike processes that wrap themselves around nearby axons, forming **myelin sheaths.** Each myelin sheath is composed of many layers of phospholipid cell membrane. The myelin sheaths are periodically interrupted, leaving small gaps of uncovered membrane 1 to 2 μm wide called **nodes of Ranvier.** Figure 7-20, *A* illustrates a longitudinal section of a myelinated fiber including a node of Ranvier, while Figure 7-20, *B* is a micrograph of the myelin sheath in cross section. In a fiber with a diameter of 10 to 20 μm, the nodes are about 1 mm apart. The membrane under the myelin does not have voltage-gated channels; these are clustered in the membrane at the nodes (identified by the black dots in Figure 7-20, *C*).

Propagation in Myelinated Nerve Fibers

The effect of the myelin on the pattern of current flow is illustrated in Figure 7-21. Node 1 is producing an action potential. During the action potential,

Na$^+$ ions enter the fiber, carrying an inward current that is pulled off the membrane at the next node. Because of the insulation, most of the current crossing the membrane flows down the axon, with only a small leakage out through the myelin. As a result of the insulation, the current is adequate to depolarize node 2 to threshold, even though it is relatively far away.

When node 2 is depolarized to threshold, an action potential will be produced at that point. The inward current spreads forward to node 3, but also backward to node 1. An action potential will result at node 3, but not at node 1 because it is in its refractory period (see Figure 7-21, *B*). Thus the action potential is transmitted in only one direction. As this process is repeated at each node, the action potential appears to jump from one node to another. This is called **saltatory conduction.** Speed is gained because the electrical forces are maximized to depolarize the membrane at the next node rather that involving ion flow through each region of membrane. The jumps between nodes are almost instantaneous, and then the progress of the action potential is slowed at each node as the recharging takes place.

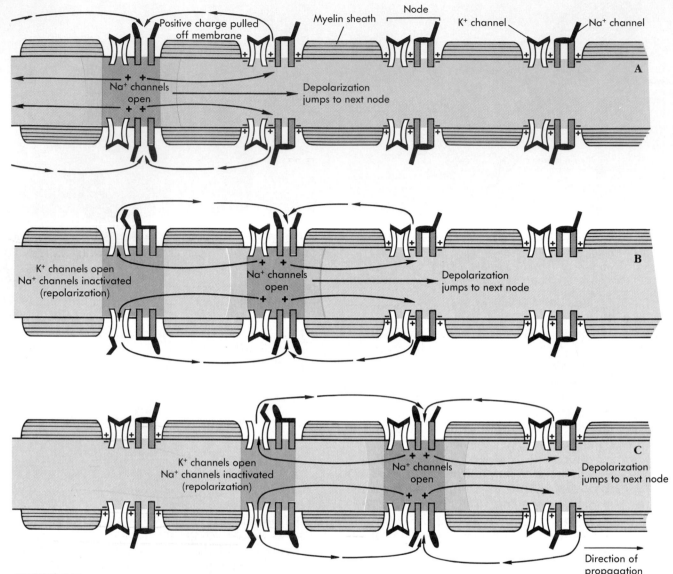

FIGURE 7-21
A Saltatory conduction along a myelinated axon involves spread of depolarization along the core of the axon and regeneration of the signal at the nodes.
B and **C** Current flow is unidirectional because the channels behind the moving wave of depolarization are inactivated. (Note that the distance between nodes is not drawn to scale.)

The spacing of nodes is an important feature of the development of the nervous system. If the distance were too great, the current at one node could not depolarize the next node, and conduction would fail. The developing nervous system seems to err on the safe side—the nodes are close enough that current flow between them is several times the minimum needed to bring a node to threshold. The proportion of extra current over the minimum needed to bring the node to threshold is referred to as the **safety factor.** The current actually flows so effectively that early in the onset of an action potential at one node, the next node is already depolarized above threshold, so more than one node is involved in the same action potential at any one time.

Comparison of Nonmyelinated and Myelinated Nerves

The relative advantages of size and myelination can be seen in the following examples. An unmyelinated axon 10 μm in diameter conducts action potentials at a velocity of about 0.5 m/sec. If such axons were present in the pathway that withdraws an appendage from a painful stimulus, the response would take about 4 seconds. If the diameter of these axons were increased to the size of a squid giant axon, about 500 μm, the conduction velocity would increase to approximately 25 m/sec (Table 7-2), and the response would take about 80 msec. If, on the other hand, the 10 μm axon were myelinated, the myelin would add some to the diameter, perhaps

NERVE AND MUSCLE

TABLE 7-2	Conduction Velocities of Axon Fibers		
Nerve	**Diameter**	**Myelin**	**Conduction velocity**
Squid giant axon	500μm	No	25m/sec
Large motor axon to leg muscle	20μm	Yes	120m/sec
Axon from skin pressure receptor	10μm	Yes	50m/sec
Axon from temperature receptor in skin	5μm	Yes	30m/sec
Axon from pain receptor	1μm	No	2m/sec

doubling it, but the conduction velocity would rise from 0.5 m/sec to about 50 m/sec and the response would require 40 msec. Clearly myelination results in a saving of space compared with the strategy of increasing the axon diameter.

Myelination and relatively large axonal diameter are used selectively in vertebrate nervous systems in pathways where speed is essential. In sensory nerve fibers from pressure receptors in the skin and motor nerves that run to skeletal muscles in arms and legs, conduction velocities can exceed 100 m/sec. Sensory fibers that carry information from temperature receptors are smaller but still myelinated, whereas many motor nerves leading to internal organs (where response time is not so critical) are not myelinated and conduct at less than 1 m/sec (see Table 7-2).

Compound Action Potentials in Nerve Trunks

The structures called nerves are actually bundles containing many axons surrounded and supported by glia. Electrodes placed on or near these nerve trunks can record electrical signals that represent the summation of a large number of individual action potentials. If a nerve is stimulated with sufficient current to initiate an action potential in all the axons, the synchronous activation of all the axons is a **compound action potential.** The action potentials will propagate in both directions from the point of stimulation. This is an unphysiological situation because the normal site of initiation of an action potential (the axon hillock) is adjacent to nonexcitable membrane, therefore the action potential can flow in only one direction. Normally, some of the axons in a nerve carry information toward the central nervous system (sensory cell axons) and others carry commands away from the central nervous system to muscles and glands.

The axons in nerves are typically distributed into several size groups (Figures 7-22 and 7-23), each having a different conduction velocity. If a pair of recording electrodes is placed some distance away from the point of stimulation, action potentials traveling in the different size groups will arrive at the recording electrodes at different times after the stimulus has been delivered. The compound action potential thus consists of several bumps occurring over time, each bump corresponding to a family of axons of about the same diameter (see Figure 7-23).

The axons of different sizes in a nerve trunk have action potentials with different sensitivity to the electrical stimulation; larger fibers are depolarized to fire action potentials by smaller currents. This does not represent a difference in their thresholds in response to natural stimulation, but is simply a result of the way the stimulating current spreads in the nerve trunk. As a result, the compound action potential is not all-or-none, even though the action potentials of the individual fibers are such. As the stimulus current is increased, more and more fibers are brought to threshold. Each contributes its individual increment to the composite response.

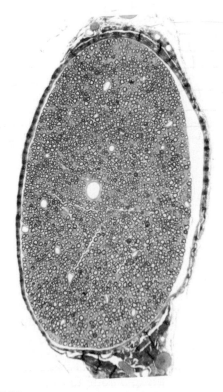

FIGURE 7-22

An electron micrograph of a cross section of a spinal nerve. Each small circle is a myelinated axon. Note that the axons fall into several size classes.

The Consequences of Nerve Demyelination

Axons conduct reliably even when the firing rate is as high as several hundred action potentials per second. In a myelinated axon the amount of current generated during an impulse at one node is five to seven times the amount needed to excite the next node; this is the safety factor.

The situation is quite different in diseases that result in loss of myelin. Myelin loss, or demyelination, occurs in diabetes and can also be a consequence of alcoholism. However the most common demyelinating disease is multiple sclerosis. This disease is currently understood to be an autoimmune disorder in which an immune attack is mounted against myelinated parts of the central nervous system.

Depending on the site of the demyelination, symptoms may involve muscle weakness, paralysis, or abnormal sensations in any part of the body; the symptoms may come and go. The more extensive the myelin loss and the more nerve tracts that are affected, the worse the symptoms become.

In myelinated axons, the voltage-dependent Na^+ channels are confined to the membrane of the nodes of Ranvier, held in place by the cytoskeleton. Myelin loss increases the amount of current that leaks out of the fiber (Figure 7-A). The first effect is an increase in the time needed to bring the next node to its threshold, so that conduction slows. As demyelination progresses, there may not be enough current to bring the next node to threshold, and conduction is blocked.

Conduction block in a group of motor axons results in paralysis of the muscles they control; in sensory nerves the effect is a loss of sensation. Even slowing of conduction has profound effects, because information processing within the central nervous system often depends on the timing of incoming signals.

The effects of demyelination reveal the crucial importance of transmission success and timing of signal flow within the nervous system. There is hope that a better understanding of factors that normally prevent the immune system from attacking normal body tissues can be applied to produce a therapy for those suffering from demyelinating diseases.

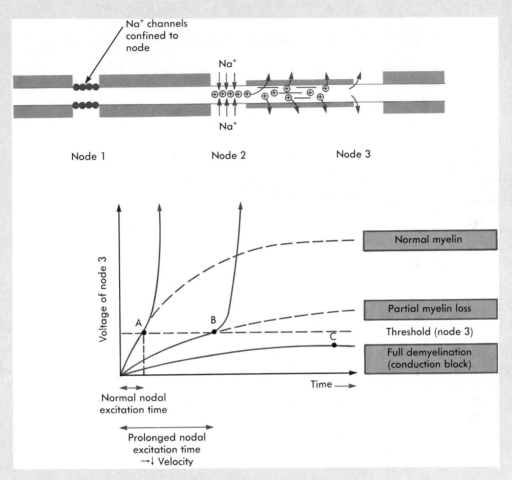

FIGURE 7-A

The upper part of the figure shows the Na⁺ channels localized at the nodes of Ranvier. Activating channels at one node normally results in sufficient current flow down the core of the axon to bring the next node rapidly to threshold. Demyelination is occurring between nodes 2 and 3. The lower part of the figure shows the consequences for node 3 of progressing demyelination. Before demyelination starts *(A)*, the node is rapidly brought to threshold by currents that reach it from node 2. As demyelination progresses, more and more current is lost through the myelin, and node 3 reaches threshold more slowly; the conduction velocity is reduced. At last, the current leakage is so great that node 3 is not brought to threshold at all; the demyelination has blocked conduction, and the axon no longer functions.

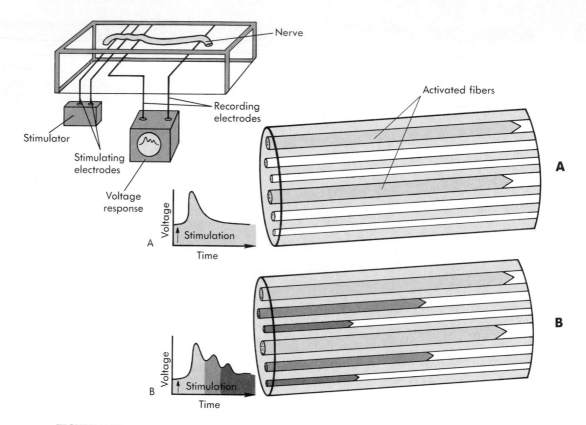

FIGURE 7-23

Stimulation of a whole nerve causes a compound action potential.

A Small stimulating currents activate only the largest class of axons. These have similar conduction velocities and produce a single peak at the recording electrode.

B Larger stimulating currents can activate all classes of axons in the nerve. The different conduction velocity of the large, intermediate, and small axons results in multiple peaks in the compound action potential as the action potential of the different axon classes arrive at the recording electrode at different times after the stimulus.

SUMMARY

1. **Action potentials** differ from receptor potentials and synaptic potentials in that they have a **threshold,** are all-or-none, cannot summate, are propagated regeneratively across excitable membranes, and leave the membrane temporarily refractory to a second stimulus.

2. Action potentials are explained on the basis of the behavior of **voltage-gated ion channels** in the excitable membranes: Na^+ channels are activated rapidly by depolarization and inactivate spontaneously; K^+ channels are activated more slowly by depolarization and do not inactivate, but rather close in response to repolarization.

3. An action potential is initiated when a depolarizing stimulus opens enough voltage-gated Na^+ channels that the Na^+ conductance exceeds the K^+ conductance, setting up a positive-feedback cycle in which depolarization rapidly opens the remaining Na^+ channels.

4. Repolarization is the result of two factors: (1) spontaneous inactivation of Na^+ channels, and (2) the opening of K^+ channels. These two factors make the membrane **absolutely refractory** to a second stimulus for a few milliseconds after an action potential is initiated. When a sufficient number of inactivated Na^+ channels have reverted to the closed but potentially responsive state, the membrane is still **relatively refractory** until the K^+ channels have closed.

5. The velocity of conduction of an action potential down an axon or muscle cell is determined by the diameter of the fiber and is also greatly increased by myelination in the case of axons. In myelinated axons, conduction is **saltatory:** a regenerative action potential occurs at intervals along the fiber at nodes of Ranvier, and the signal covers the intervening distance rapidly in the form of a current between the depolarized node and the next polarized node.

6. In a nerve trunk containing many axons, stimulation evokes many simultaneous action potentials that propagate in both directions from the site of stimulation and whose currents can be recorded as a **compound action potential.** Since different fibers conduct at different rates, the compound action potential separates with distance into several peaks that correspond to the action potentials of axons of different size populations in the nerve trunk.

1. The venoms of most scorpions and many insecticides act by slowing the inactivation of Na^+ channels. What effects would such venoms have on the duration of an action potential? Would its amplitude be changed significantly? What would happen to the refractory period?

2. How does decremental conduction differ from nondecremental conduction?

3. Relate the membrane potential changes that occur during an action potential to the ion permeability changes that produce them.

4. What determines the height of the action potential? Why is it that a small stimulus does not produces a small action potential and a large one produce a large action potential?

5. What are the negative feedback forces that return the membrane potential to the resting level after a hyperpolarizing shift?

6. What is Na^+ channel inactivation? What removes the inactivation and what state of the Na^+ channel results?

7. What effect does myelin have? What are nodes of Ranvier and what function is associated with them?

8. Why does a compound action potential appear to violate the all-or-none law?

9. Local anesthetics used by dentists to deaden pain are blockers of Na^+ channels. If half of the Na^+ channels in a sensory nerve were blocked, how would this affect the threshold? Could it affect the ability of the nerve to propagate an action potential?

Choose the MOST CORRECT Answer.

10. Which of these characteristics does NOT describe action potentials in membranes?
   a. Membranes are electrically excitable.
   b. A threshold stimulus is required for initiation.
   c. These are all-or-none events.
   d. The current is spread by decremental conduction.

11. The membrane potential that an stimulated cell exhibits:
   a. Resting potential
   b. Action potential
   c. Receptor potential
   d. Synaptic potential

12. If the membrane potential moves further away from O mV the membrane is said to be:
   a. Unpolarized
   b. Depolarized
   c. Hyperpolarized
   d. Polarized

13. Choose the FALSE statement.
   a. Receptor and synaptic potential increase with the size of the stimulus that activates them.
   b. Receptor and synaptic potentials decrease in size with distance.
   c. Receptor and synaptic potentials can summate.
   d. Receptor and synaptic potentials in comparison to action potentials involve large changes in the membrane potential.

14. Choose the FALSE statement:
   a. Na^+ activation depends on the voltage.
   b. Na^+ inactivation is an intrinsic time dependent process.
   c. Na^+ activation is ten times slower than K^+ activation.
   d. K^+ activation greatly accelerates repolarization.

15. Action potential velocity is greatest in an (a):
   a. Unmyelinated axon; 10 μ m in diameter
   b. Unmyelinated axon; 5 μ m in diameter
   c. Myelinated axon; 50 μ m in diameter
   d. Myelinated axon; 10 μ m in diameter

● *SUGGESTED READINGS*

CATTERALL WA: Structure and function of voltage-sensitive ion channels. *Science* volume 242, 1988, p. 50-61. This review examines structural models of ion channel function and their responses to membrane voltage changes.

EDWARDS DD: Still stalking MS, *Science News* 132:234, Oct 10, 1987. Article summarizes progress in understanding the cause of multiple sclerosis and reviews the efficacy of several current therapies.

HILLE B: *Ionic channels of excitable membranes,* ed 2, Sunderland, 1991, Sinauer Associates. An excellent treatment of the structure and molecular mechanisms of channels, including second-messenger pathways and channel modulation.

KANDEL ER, SCHWATZ JH, JESSELL TM: *Principles of neural science,* ed 3, New York, 1991, Elsevier. A comprehensive treatment of the neural basis of behavior, from molecular mechanisms to cognitive processes.

CHAPTER 8

Initiation, Transmission, and Integration of Neural Signals

On completing this chapter you should be able to:

- Understand the functions of the input and output segments of neurons.
- Explain how intensity of a stimulus is coded by the receptor potential and how this is translated into action potentials in the afferent neuron.
- Describe the two types of specialized cells that transduce the energy provided by stimuli into membrane potential changes.
- Define a modality and discuss categories of sensory information that are not included in modalities because they are not consciously experienced.
- Summarize the types of receptor adaptation.
- Understand the functional differences between electrical and chemical synapses.
- Know major classes into which neurotransmitters can be categorized.
- Distinguish between excitatory postsynaptic potentials and inhibitory postsynaptic potentials.
- Understand the role of Ca^{++} in transmitter release and how synaptic efficacy can be modulated.
- Understand how painful sensations can be modulated in the periphery and in the central nervous system (CNS).

*T*he nervous system detects events in the external world by means of receptors sensitive to physical and chemical stimuli in the environment. It also monitors changes in the body interior using additional specialized receptors. The inputs from sensory receptors are not the only initiators of neural processes; these neural processes also arise from spontaneously active neural elements and circuits within the central nervous system. We are beginning to understand the neural circuits regulating some processes, such as breathing patterns or wake-sleep cycles. We can follow other neural functions, such as the complex mental processes underlying the waking and dreaming human mind, in terms of what portions of the brain are active. At this point, however, we still do not know how these electrical paths are related to consciousness.

The process by which information from external and internal sensory receptors and ongoing events in the active brain are coordinated to initiate appropriate responses is called integration. Integration involves the weighing of many inputs, some tending to activate the neurons of a particular circuit and others tending to inhibit them. A similar but less complicated process of decision-making is seen in sensory cells, which usually deal only with relative degrees of activation. For that reason, we will present the simpler coding of sensory stimulation into action potentials before we approach the weighing of the positive and negative elements that occurs during integration within the central nervous system.

The action potentials that transmit signals within the peripheral nerves and axon fiber tracts of the central nervous system are all-or-nothing events. In contrast, the degree of activation of a sensory ending is reflected in graded responses of nonexcitable membrane regions. Similarly, integration within the central nervous system occurs on nonexcitable membrane regions. The membrane potential changes flowing to the site of action potential initiation from the input regions are reflected in the constantly changing pattern of activity of the cell: silence (no action potentials), an occasional action potential, or many action potentials fired in rapid succession.

A large portion of the neurons within the brain and spinal cord are local circuit neurons that lack long axonal processes. These neurons can receive inputs from many other neurons, weigh those inputs, and deliver an output to their follower cells without ever initiating an action potential. In such cells the membrane potential fluctuations caused by the inputs affect the communication of that cell with its follower cells. The presence of vast numbers of such local circuit neurons underlies the parallel processing of information that characterizes the brain of higher animals and particularly the human brain. This chapter gives some insights into how the human brain functions. This will require an understanding of the design of the neuron and its specialized input and output regions.

NEURON STRUCTURE AND INFORMATION TRANSFER
Input and Output Segments of Neurons

> • *What are the functional subdivisions of a neuron's structure?*
> • *What role does each subdivision play in communication?*
> • *What is a stimulus modality?*
> • *What is the relationship between receptor specificity and stimulus modality?*

All neurons communicate. They respond to information from other nerve cells or energy in the environment and convey a modified form of the information they receive to other nerve cells or effector cells. The **input** and **output** elements of neural function are typically located in separate, specialized regions of the neurons. In neurons that relay information over considerable distances, a specialized region called an **axon** transmits the cell's response to its inputs all the way to the output region without loss of signal strength (Figure 8-1). The axon is a single cablelike process that relays the action potentials to the axon's **synaptic terminals**, the output region where the cell's level of activation can be communicated to other cells by **synaptic transmission.** At the output region the neuron's depolarization will influence the next cell in the neuronal circuit.

The anatomy of neurons differs with their anatomical location and function, but each neuron has a **cell body**, or **soma**, that provides metabolic support for other parts of the neuron. The neuron shown in Figure 8-1, *A* is typical of some neurons in the **central nervous system** (**CNS**), including **motor neurons** and some types of **interneurons.** In such neurons the cell body (soma, perikaryon) is covered with fine branching processes called **dendrites.** The dendrites and cell body receive many synapses from other neurons, and thus constitute the **input segment** in these cells. Inputs (**synaptic boutons**) on the neurons in sympathetic ganglia are shown in Figure 8-1, *C*.

Sensory neurons have a quite different mor-

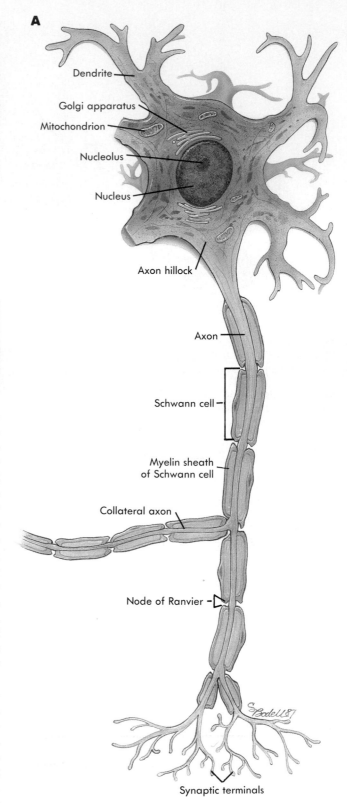

A

FIGURE 8-1
A A generalized neuron. The input segment of the neuron includes the dendrites and cell body. The axon connects the input segment to the output segment, the synaptic terminals. Action potentials are initiated at the axon hillock, where the axon joins the cell body, and are conducted to the synaptic terminals. The cell body contains the usual intracellular organelles.

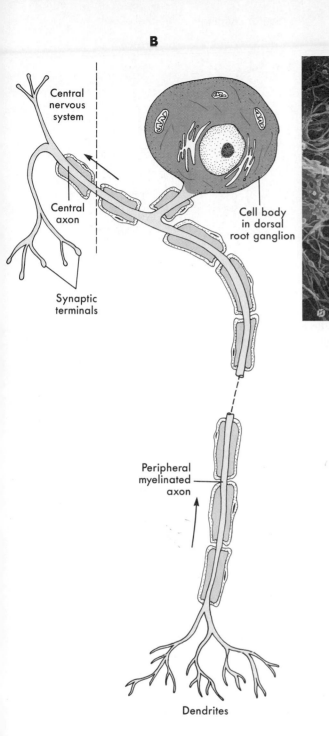

B Most sensory neurons have the cell body connected to the axon by a short segment; the axon extends centrally, where it branches to form synaptic connections and distally to receive the sensory signals, which flow in the direction indicated by the arrow. The dotted line represents the division between the central part (left) of a spinal sensory neuron and the peripheral part (right). The cell body would be in the dorsal root ganglion.

C A scanning electron micrograph of a sympathetic ganglion neuron showing numerous synaptic boutons on the soma, dendrites of other cells, and the axon hillock and axon.

D A scanning electron micrograph of a sensory neuron soma in the dorsal root ganglion showing the axon hillock (AK), and a node of Ranvier.

phology. The input segment of sensory neurons is confined to the specialized membrane region in the periphery that receives stimuli or inputs from epithelial sensory cells. The sensory cell body is located in the CNS or in **dorsal root ganglia** immediately adjacent to the spinal cord (Figure 8-1, *B,* and *D*). Cell bodies in the dorsal root ganglia are separated from the axon by a short process and are electrically inexcitable; they do not participate in the conduction of the action potential or the integration of sensory inputs. In the CNS the sensory cell's axon branches to form the output segment, the **synapses** with other neurons. The long process that transmits action potentials is called an axon, no matter where it is located relative to the cell body. This is a functional rather than an anatomical classification that is preferred by **electrophysiologists,** the physiologists that record electrical changes in nerve cells and other excitable cells.

We can make some generalizations about the properties of input and output segments of neurons. The input segment responds to stimulus energy or the synaptic signals from another cell because it possesses ion channels that can be opened or closed by a specific kind of stimulation, such as odor molecules, pressure stimulation, or the chemicals released by other neurons. The input region possesses few if any voltage-sensitive channels. It therefore has a high threshold and usually is incapable of producing an action potential. Signals in the input segment are **local currents** that spread decrementally throughout the input region. If the distance to the output region approaches a millimeter or more, the signal is converted into action potentials at a region of the membrane where the axon, characterized by voltage sensitive channels, joins the input region (see Chapter 7). At this site, action potentials are initiated if the currents depolarize the membrane beyond threshold. The first node of Ranvier may be the site of action potential initiation in myelinated sensory cells (Figure 8-1, *B*); the axon hillock (Figure 8-1, *A*) typically has the lowest threshold in neurons in which the soma or cell body is part of the input segment.

Axoplasmic Transport

In the cell body, proteins and membranes used for packaging the neurotransmitter molecules are synthesized. The output or input segment may be located far from this center of synthesis. For instance, the terminals of a motor neuron that innervates muscles that move the big toe may be a meter from the cell body, located in the spinal cord. Material moves between the cell body and the axon terminals of motor neurons or the dendrites of sensory neurons by **axoplasmic transport**. Material travels from the cell body down the axon (**anterograde transport**) at three rates: fast, intermediate, and slow axoplasmic transport. Material also travels in the reverse direction (**retrograde transport**) at a rate that equals the fast transport in the anterograde direction (Figure 8-2).

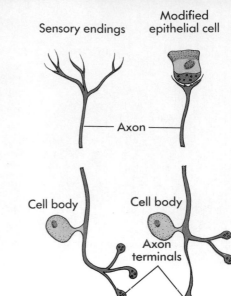

FIGURE 8-3

Two ways that sensory neurons can receive stimulation. In **A** the energy, such as pressure or heat, is detected by the endings of the neuron. In **B**, modified epithelial cells, such as those in taste buds, respond to the energy with a change in membrane potential that influences the rate at which they release synaptic transmitter. Input and output regions of the cells are indicated by orange and green, and the axon of the sensory neuron, which relays the information by action potentials, is blue.

The structural basis for the fastest form of axoplasmic transport is provided by tracks of microtubules arranged lengthwise within the axon (Figure 8-2). Computer-enhanced videos of axons show synaptic vesicles and small granules riding along these tubules as if on a monorail. This railway allows traffic in both directions: full synaptic vesicles run in the direction of axon terminals and the membrane remnants of empty vesicles pass them going toward the cell body for degradation and recycling. Mitochondria and some associated proteins move at intermediate rates of transport; structural proteins used in growth and repair move at the slowest rates of transport (see Figure 8-18).

OVERVIEW OF SENSORY SYSTEM ORGANIZATION

How Sensory Neurons Receive Information

Figure 8-3 shows two ways sensory information may reach afferent neurons. In the simplest arrangement, illustrated by touch and pressure sensors, the afferent neuron is directly responsive to the stimulus (Figure 8-3, *A*). In the taste, visual, and auditory systems and the vestibular system (which mediates balance and equilibrium), specialized epithelial cells are the primary receivers of the stimulus (Figure 8-3, *B*). Sensory epithelial cells do not have axons or generate action potentials, but their changes in permeability alter the release of a chemical at synapses onto the sensory neurons (Figure 8-4). In this way, these specialized epithelial receptor cells function as part of the nervous system.

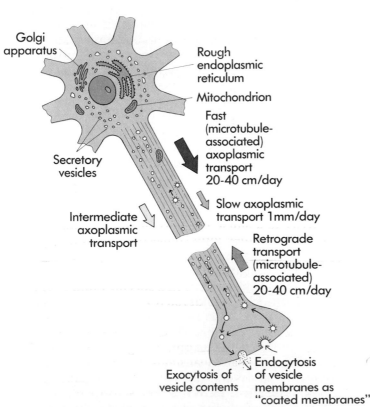

FIGURE 8-2

Material synthesized in the cell body moves down the axon (the anterograde direction) at three different rates; material is also transported toward the cell body by retrograde transport. Fast transport in both directions involves the microtubules that extend along the axon.

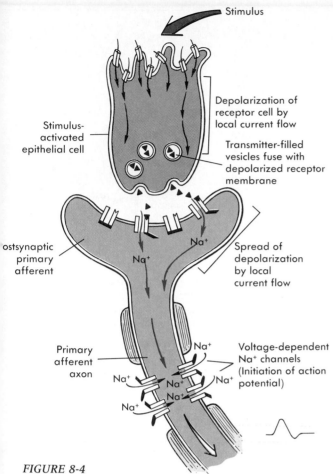

Stimulus

Stimulus-activated epithelial cell

Depolarization of receptor cell by local current flow

Transmitter-filled vesicles fuse with depolarized receptor membrane

Postsynaptic primary afferent

Na⁺

Na⁺

Spread of depolarization by local current flow

Primary afferent axon

Na⁺

Na⁺

Na⁺

Na⁺

Na⁺

Na⁺

Voltage-dependent Na⁺ channels (Initiation of action potential)

FIGURE 8-4

Initiation of action potentials in a sensory system in which the input segment is a modified epithelial cell. A stimulus activates the epithelial cell, which undergoes a receptor potential but cannot generate an action potential. The receptor potential modulates the rate of transmitter release. The transmitter depolarizes the postsynaptic primary afferent ending. A local current, shown by the red arrows, flows between the depolarized nerve ending and the axon (containing voltage-gated Na⁺ channels), initiating action potentials in the axon that leads to the CNS.

Sensory Modalities

Different categories of sensory receptors are able to detect different types of stimulation. Sensations are the conscious experiences that reflect changes in the types of energy our sensory receptors can detect. They result from the relaying and subsequent processing of information from each category of receptor to the appropriate regions of the brain. Classical sensory categories or **modalities** are the senses of touch, taste, hearing, vision, and olfaction. The modalities do not cover all the sensory afferent categories, but only those that are consciously evaluated. For example, information from receptors that sense blood pressure is not analyzed at the conscious level. In addition, information that lacks novelty or significance is filtered from our consciousness. For instance, the contact of our clothes with our bodies ceases to be noticed, although conscious effort can facilitate access to some of that sensory information.

Light does not produce a sensation of sweetness in the mouth, and a light directed into the ear does not give either a visual or an auditory perception. Artificially activating the pathway that mediates a particular modality with a different form of energy, such as applying pressure on the eyeball, results in the false perception of light because activity in visual pathways cannot be interpreted as any other sensation. Each category of receptor cells is so much more sensitive to its own modality than to other forms of energy that the information sent to the brain normally provides the correct answer to the question "What kind?".

An apparent exception to this generalization is seen in the case of pain, which is really not a single sensory modality at all, but rather the presence of excess stimulation in several sensory pathways, such as touch, temperature, sound, and even light. The perception of pain is subject to personal differences in threshold and modulation that depend on the circumstances (see pain modulation below).

Receptive Fields and Their Central Representations

The location of the input segment of an afferent neuron determines its **receptive field,** the region over which an appropriate stimulus of adequate intensity can produce a change in membrane permeability (Figure 8-5). For instance, the endings of a receptor for skin temperature have a specific distribution in the skin that corresponds to their receptive field, and the position of a photoreceptor cell in the retina determines the small portion of the visual input that corresponds to its receptive field.

During neural development, each afferent neuron sends an axon out to establish a receptive field in an appropriate place within the body or on its surface. The endings of the sensory neuron develop sensitivity to a single modality of stimulation, such as pressure, or they form connections with the epithelial receptor cells that respond to a particular modality, such as taste. The central processes establish axon terminals on neurons in a specific brain or spinal cord region that is devoted to processing information of that modality. As a consequence, lines of communication (sometimes called **labelled lines**) are established between central regions specialized for processing a particular modality and the sensory inputs from receptors specific for that modality.

The projections formed by the sensory elements of a particular modality do not converge on higher sensory processing centers in an unorganized jumble. One of the breakthroughs in understanding sensory processing came with the recognition that the location of the receptive fields of afferent neurons determines the space they occupy within their

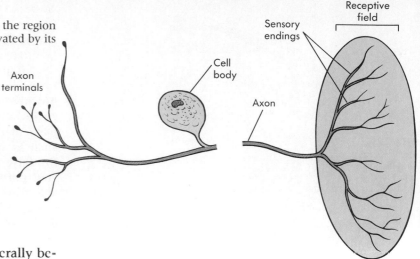

Axon terminals

Cell body

Axon

Sensory endings

Receptive field

central connections. The central regions literally become maps of the layout of individual receptive fields (see Figure 9-17). In children, injuries that reduce the body surface, such as loss of a finger, have been found to result in a disappearance of the finger from the brain map, with the representation of other fingers expanded to fill the vacated space. Not only are the body surface and sites of internal receptors mapped in an orderly fashion, but visual space and auditory space are also mapped. Sounds, for instance, are mapped in a way that reflects whether the sound is of a high, low, or intermediate pitch, and at higher levels of analysis the information from the two ears is compared to provide a map of the origin of the sound in space (see box, Chapter 9).

PHYSIOLOGICAL PROPERTIES OF SENSORY RECEPTORS
Sensory Receptors as Transducers

> - *In what two senses is the term receptor used?*
> - *What is the effect of receptor adaptation on transmission of information about stimulus intensity?*
> - *What types of stimulus modalities require a logarithmic response of receptors? Why?*

Sensory receptors are specialized to transform one of the several categories of energy into electrical energy by producing a membrane voltage change. They are thus transducers (Chapter 5). The categories of sensory perception, or modalities, are determined by the physical nature of the energy and its source. We are aware of the information that comes to us from the external environment and generally unaware of the large proportion of receptors that are directed at internal stimuli. For instance, some chemicals in the external environment can be tasted or smelled, but we are unaware of many chemicals (such as the sugar glucose) that are detected by internal chemoreceptors. Proprioceptors provide information about the position of body segments relative to one another (joint angles, for example) and the position of the body in space. Other internal receptors measure oxygen and carbon dioxide levels,

osmolarity, and acidity of body fluids.

The exquisite sensitivity to one particular stimulus modality that marks sensory receptors is apparent in the rods and cones of the visual system, which can detect single photons, and the hair cells of the auditory system, which can detect displacement equivalent to the diameter of a single atom. The remarkable sensitivity to some stimuli is obtained by limiting the range of effective stimuli. For example, the human eye is sensitive to only a tiny fraction of the entire spectrum of electromagnetic energy, which includes infrared and ultraviolet light, microwaves, radio waves, x-rays, and gamma rays. The visual range of other kinds of animals presents them with a very different picture of the world. The human ear is quite insensitive to sound energy of frequencies below about 10 cycles/sec and above about 20,000 cycles/sec, although other mammals are able to respond to higher or lower portions of the sound spectrum. The selectivity of our sensory receptors limits our experience of the world. The addition of machines that extend our ability to experience forms of energy beyond those our receptors detect is an interesting example of cultural evolution that supplements our biological limitations.

Receptor Potentials
Sensory cells respond to specific forms of energy because they possess the appropriate **stimulus-sensitive receptor proteins** in their membranes. (Note that the term *receptor* is now used to describe the protein that responds to a form of energy rather than to refer to an entire sensory cell.) The receptor protein may have an ion channel in its structure that allows cations such as Na^+ and K^+ to cross (Figure 8-6, *A*). Alternatively the receptor proteins may be coupled to ion channels through other membrane-bound proteins or second messenger systems (Figure 8-6, *B* and *C*).

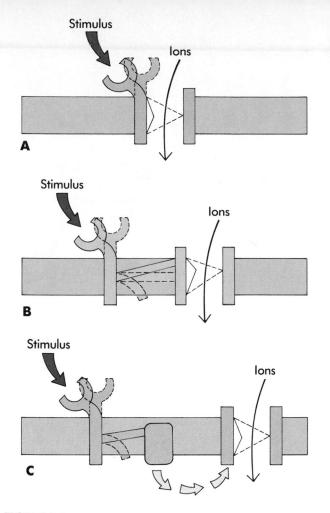

FIGURE 8-6

A Diagram of sensory receptor proteins that are ion channels. The stimulus reacts with a transducing element that is part of the channel molecule and the interaction causes the gates to open, primarily admitting Na^+.
B Diagram of receptors coupled to ion channels: the stimulus activates a receptor protein that physically interacts with the channel molecule by movement within the membrane.
C Diagram of sensory protein interactions that involve a second messenger such as a G protein in the cytoplasm (see Chapter 5).

If a cation channel is opened by the stimulus, the driving force for Na^+ is much larger than that for K^+. Consequently, more Na^+ enters than K^+ exits and the dominant effect is depolarization. A stimulus usually causes channels to open, but in the visual system, detection of light causes Na^+ channels to close, producing a hyperpolarization (Chapter 10). The voltage changes that occur in sensory receptors upon stimulation are called **generator potentials** or **receptor potentials**.

Intensity Coding by the Receptor Potential

The receptor endings (the input regions) are unable to generate action potentials because they are electrically inexcitable. In a sensory neuron, the membrane region is depolarized when energy is trans-

duced into an electrical charge, and the difference between the resting potential of the axon and the depolarized endings results in a flow of current between endings and axon (Figure 8-7, recording site *A*). Chapter 7 described how a steady electrical current elicits a series of action potentials. The response of sensory neurons to depolarization by a stimulus is similar. The voltage change in the endings where the channels open is larger than the voltage change that reaches the spike initiation site, but even after diminishing in size due to decremental conduction, it can still be well above threshold. This will result in the generation of many action potentials before the membrane potential in that region again falls below threshold (Figure 8-7, *B*).

Spatial and Temporal Summation

The greater the stimulus intensity, the larger will be the receptor potential (Figure 8-8, compare stimulus 1 and 2). This proportional response makes it possible for sensory receptors to register information about **stimulus intensity** in the size of the receptor potential. They then encode the information about intensity in terms of the number of action potentials generated per unit time. The larger the depolarization, the faster the action potentials will be initiated.

Because the receptor depolarization is not transformed into action potentials at the point where the stimulus-sensitive channels open, summation of inputs is possible. Two types of summation can be distinguished: (1) **spatial summation** and (2) **temporal summation**. Spatial summation results from the addition of inputs arising at more than one site on the receptor input region. Only sensory cells with branching or distributed endings display this type of summation. The typical example is a cutaneous (skin) touch receptor. Inputs can add in "space" on the surface of the receptor membrane and together determine the overall receptor potential. Stimulation of more of the receptive field (Figure 8-8, stimulus condition 3) produces correspondingly higher frequencies of action potentials in the axon.

Temporal summation is the result of stimulation of the same input region by inputs arriving in rapid succession (Figure 8-9). As with spatial summation, the effect is to build up the size of the receptor potential because additional channel-opening events occur while the membrane potential is still depolarized by the preceding input.

Linear and Logarithmic Coding

The frequency of action potentials reflects important features of the receptor potential. Sensory cells can respond to the energy present in the stimulus according to two basic patterns. If the receptor potential is directly proportional to the stimulus intensity throughout the detection range, it displays a **linear response pattern** (Figure 8-10). If the receptor is

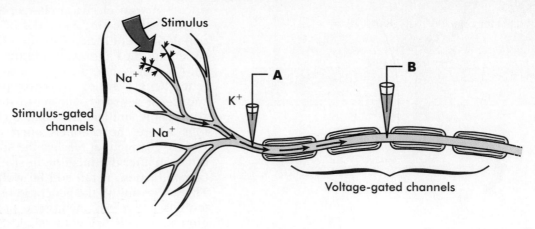

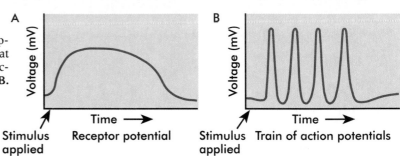

FIGURE 8 7

Depolarization of a free nerve ending leads to a receptor potential (see *A*), recorded by electrode **A**, that spreads by local current flow to the axon, where action potentials (see *B*) are initiated at recording site **B**.

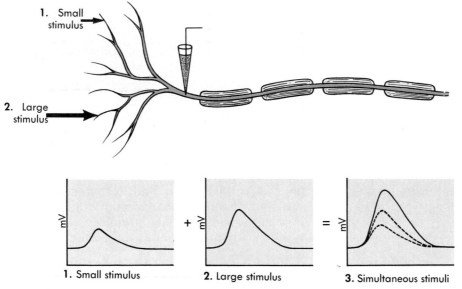

FIGURE 8-8

Spatial summation occurs in the input segment of a sensory receptor. Stimuli applied to different regions of the receptive field spread to the recording site, indicated by the electrode. If the inputs are applied simultaneously, they summate, producing a larger receptor potential.

quite sensitive to differences in stimulus intensity when the stimuli are weak (compare stimuli 1 and 2 in Figure 8-10) but becomes progressively less sensitive to stimulus intensity as the stimulus intensity increases, this is a **logarithmic response pattern.**

The logarithmic response pattern is typically seen in light and sound receptors, which must register the intensity of stimulation over many orders of magnitude. Such a range cannot be effectively reflected by a linear conversion because the refractory period limits the frequency of action potentials to several hundred per second, which would cover approximately two orders of magnitude. Rather than responding with a linear conversion, the logarithmic response of visual and auditory receptors results in action potential frequencies that change only sevenfold (less than one order of magnitude) in response to a 10 millionfold change in stimulus intensity. This sevenfold difference might correspond to the difference between a barely audible whisper and a jackhammer.

FIGURE 8-9

Temporal summation in the input segment of a sensory receptor. The receptor potential generated at a single input site by a single stimulus is compared with the summated receptor potential resulting from repeated stimulation.

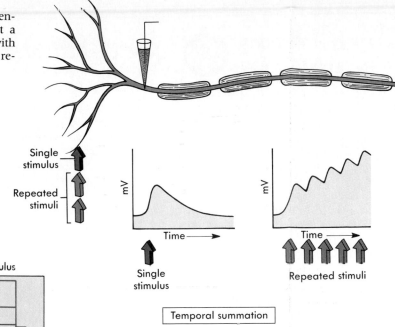

Temporal summation

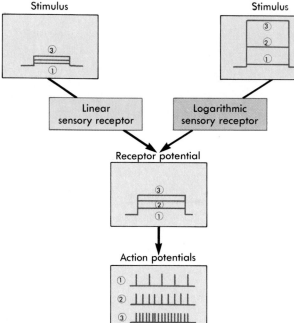

FIGURE 8-10

Intensity coding by sensory receptors. In systems where the intensity does not vary over a wide range, the relationship between stimulus intensity and receptor potential is linear. For systems that must represent a wide range of stimulus intensity, a logarithmic relationship is seen. In both cases, action potential frequency is directly proportional to receptor potential amplitude.

Sensory Adaptation

In sensory physiology, adaptation refers to the pattern of decline in responsiveness when a stimulus of constant magnitude is maintained a long time. Sensory responses to maintained stimulation fall into three general classes. The first class of receptors continues to discharge at the same rate no matter how long the stimulus lasts—these are called **nonadapting receptors (tonic receptors)**. Nonadapting receptors monitor things such as temperature, limb position, blood pressure, and blood oxygen content. In these receptors, a receptor potential is

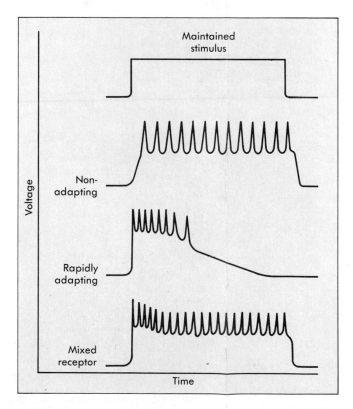

FIGURE 8-11

Responses of a nonadaptating receptor, a rapidly adapting receptor, and a mixed receptor to a maintained stimulus.

produced as long as the stimulus is present, and the magnitude of the receptor potential is proportional to the stimulus intensity (Figure 8-11).

The second category of receptors exhibits a burst of action potentials at the onset of stimulation, but as the stimulus continues, the rate of action po-

Stretch-Activated Ion Channels

A technique called patch-clamping, in which a patch of cell membrane small enough to contain a single channel is isolated, has provided detailed information on the nature of ion channels. The Nobel Prize in Physiology or Medicine was awarded in 1991 to cell physiologists Erwin Neher and Bert Sakmann for their development of this technique, which has revealed that the number of different types of channels that neurons possess is much larger than previously suspected. The patch-clamp technique uses blunt glass pipettes with tip diameters of about 1 μm. The tip of the pipette is pressed against the membrane of a cell, and a small amount of suction is applied. An electrically tight seal forms between the cell membrane and the glass tip (Figure 8-A). If an ion channel is in this patch, its opening and closing can be seen as a steplike change in the membrane voltage (Figure 8-B).

If desired, the patch pipette can be pulled away from the cell, taking with it the attached membrane (Figure 8-C). Both sides of these excised patches then can be exposed to solutions of controlled composition. For example, it is possible to add cAMP, the catalytic subunit of adenyl cyclase, Ca^{++}-calmodulin, or protein kinase C to determine the effect of these second messengers on ion channels of various types.

Stretch-activated channels are present in sensory cells that detect distortion of their membrane. These include hair cells, touch cells, muscle stretch receptors, and receptors of blood pressure and osmolarity. The sensitivity of these cells to stretch can now be tested by applying controlled suction to a patch of membrane. An increase in membrane area of less than 2% can maximally activate stretch-sensitive channels.

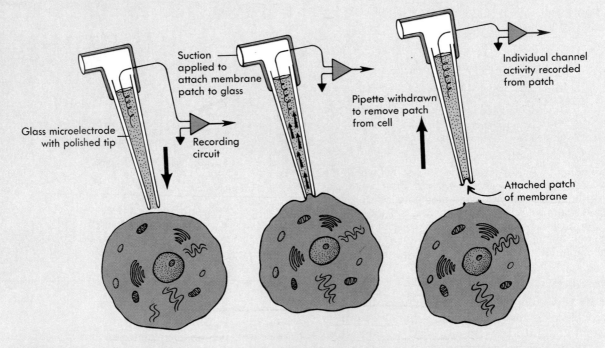

FIGURE 8-A
Steps in the formation of an inside-out membrane patch. Contact with the cell membrane *(left)*. Application of suction to attach the membrane to the electrode *(middle)*. The withdrawal of the electrode to remove the membrane patch from the cell. The size of the patch is about 1 μm²—the relationship of a membrane ion channel to this patch would be about the size of a thumbtack in a football field *(right)*.

The ions that flow through the stretch-activated channels are typically cations, with the relative selectivity to K$^+$, Na$^+$, and Ca^{++} varying in the cell types that have been examined.

The mechanism by which stretch opens channels has not yet been elucidated, but cytoskeletal fibers may focus membrane distortion from a relatively large area onto the channel site. Evidence for stretch-activated channels has been found in bacterial membranes and the membranes of protozoa, as well as in many cell types in multicellular organisms. Stretch applied to the membrane patch in the pipette can cause distortion and lead to ion entry even in cells that are not thought to detect stretch, a fact that has produced some erroneous experimental results and also provided evidence that stretch-activated channels may represent a very primitive mechanism for responding to external stimulation from which voltage and stimulus-gated channels could have evolved.

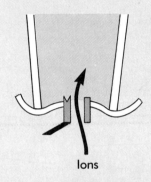

Ions

FIGURE 8-B
Drawing illustrates the presumed presence of a single channel (open state) in the membrane in the patch.

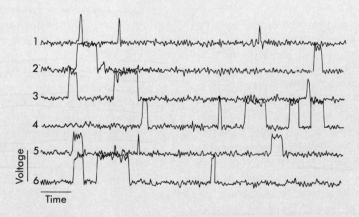

FIGURE 8-C
Sample of currents recorded from a single ion channel under patch clamp. Each steplike change in the level of the trace represents an opening or closing of the ion channel.

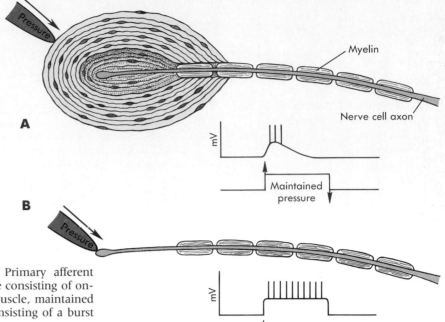

FIGURE 8-12

A   Illustration of a Pacinian corpuscle. Primary afferent nerve endings are surrounded by a capsule consisting of onion-like layers of tissue. In the intact corpuscle, maintained pressure produces a transient response consisting of a burst of action potentials.

B   Removal of the capsule alters the response properties of the cell so that maintained pressure is reflected by a constant level of action potential firing.

tentials diminishes, and ultimately no more action potentials occur. These are **rapidly adapting receptors (phasic receptors)**. The response of rapidly adapting receptors generally reflects the rate of change of a stimulus rather than its final magnitude (Figure 8-10). In these receptors the receptor potential disappears after a time, even if the stimulus continues at a steady level. Rapidly adapting receptors are specialized to register novel stimuli or changes in stimulus intensity and are found in systems where change is especially significant. Examples include receptors that detect the speed of muscle length changes.

The mechanism of the adaptation is not known for all receptors. In some systems the stimulus-sensitive channels include an inactivation feature that is responsible for adaptation. In others, the excitation of the cell slowly activates additional channels not present in the resting cell, which alter its threshold. The responsiveness of sensory receptors can also be modified by nonneural structures associated with them. For example, **Pacinian corpuscles** (Figure 8-12) are vibration-sensitive receptors found in the deeper skin layers and other sites in the body. In the Pacinian corpuscle, the neuron's input region is surrounded by many layers of connective tissue. When pressure is applied, it is initially transmitted to the ending through the layers of membrane, but when the pressure is maintained, the layers slip and slide over one another, relieving the distortion of the receptor ending. When the pressure is removed, the elasticity of the layers

causes them to resume their former shape, and their sliding back into place temporarily distorts the sensory ending. The information transmitted by these receptors is the time of onset of the stimulus and its termination, but nothing reflecting the maintained pressure (Figure 8-12). Clearly, vibration would be a very effective stimulus for these receptors. The Pacinian corpuscle is one example of how associated structures can tune sensory receptors to selected features of the stimulus; another example will be encountered in the auditory system.

The third category of receptors, the **mixed receptors**, incorporate some features of both rapidly adapting and nonadapting receptors. The receptor potential in these mixed receptors displays an initial rapid rise that is proportional to the rate of change of the stimulus (Figure 8-11), followed by a steady-state response of smaller amplitude, which is proportional to the intensity of the stimulus. This receptor potential pattern results in an initial burst of action potentials at a rate that reflects the rate of onset of the stimulus, followed by a decline in the frequency of action potentials to a steady rate, proportional to the final magnitude of the stimulus.

SYNAPTIC TRANSMISSION
Electrical and Chemical Synapses

Communication between cells of the nervous system and between neurons and effectors occurs at cell-to-cell junctions called **synapses. Electrical synapses** are regions where the membranes are

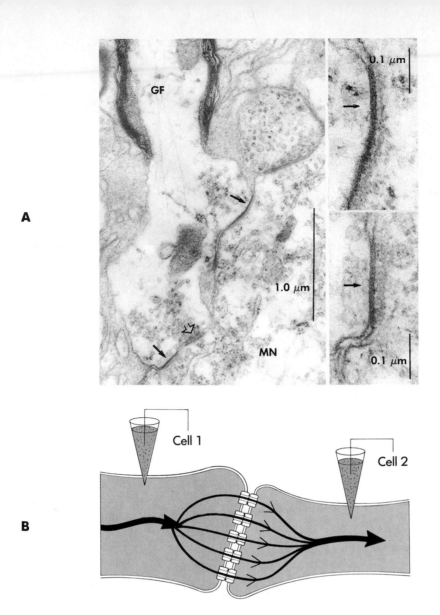

GF

1.0 μm

MN

0.1 μm

B

Cell 1

Cell 2

FIGURE 8-13

A   An electron micrograph of an electrical synapse made by a giant fiber on a motor neuron in a hatchetfish. *Left,* A giant fiber *(GF)* loses its myelin sheath near the top of the figure and forms gap junctions, the substrate of electrical transmission, with a motor neuron *(MN)* near the center and at the lower left *(solid arrows). Right,* Enlargement of two gap junctions. **B**   Diagram of the spread of current through an electrical synapse.

very close together (Figure 8-13, *A*) and joined by intramembrane-spanning channels characteristic of gap junctions (see Figure 4-17). These channels allow current to flow between cells. When an action potential arrives at such an electrical synapse, depolarizing current flows across the synapse into the next cell with no delay (Figure 8-13, *B*). The greater the area of electrical synaptic contact, the more current will be shared between the cells. Electrical synapses typically allow current to flow between the connected cells in either direction.

In the nervous system of invertebrates, electrical synapses are well described and fairly common, and one obvious function that they serve is to synchronize neurons that are normally activated together. Typically, current can spread equally well in both directions between electrically coupled cells (see Figures 8-C and 8-D). Likewise, among the vertebrates, the electrical synapses between the motor

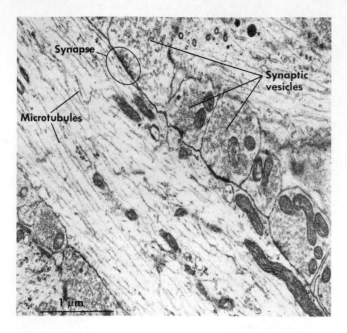

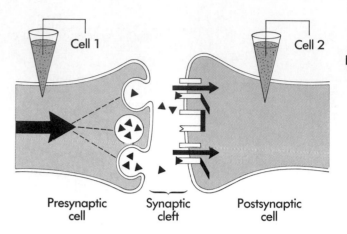

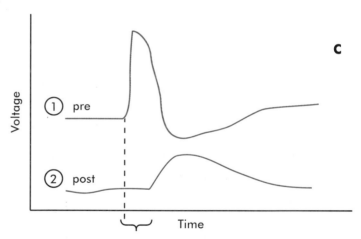

FIGURE 8-14

A   Example of several chemical synapses onto a single postsynaptic cell. Note that the synaptic terminals are full of vesicles containing transmitter chemical, and no vesicles are observed on the postsynaptic side of the membrane. Densely staining material is present on the interior surface of both presynaptic and postsynaptic membranes. The postsynaptic cell contains a number of microtubules.

B   The steps between arrival of an action potential in the presynaptic cell and the fusion of vesicles and release of transmitter are indicated by the dotted lines. The initiation of a change in membrane potential in the postsynaptic cell by receptor-mediated channel opening occurs after a delay. No current flows directly from the presynaptic cell to the postsynaptic cell.

C   Illustration of the action potential that might be recorded in the presynaptic cell and the delayed onset of a membrane potential change in the postsynaptic cell (compare with Figure 8-13, C).

neurons that activate structures such as the pectoral fins of flying fish promote synchronous movements. Although electrical synapses are prevalent in some areas of the mammalian CNS, they are relatively difficult to study in brain tissue, so their potential contributions to neural function have received little attention.

In the **chemical synapse** (Figure 8-14, A), there is a narrow space called a **synaptic cleft** 30 to 50 nm wide between the membranes of the two cells. This cleft prevents the flow of electrical currents from cell to cell; instead, chemicals called **neurotransmitters** or simply **transmitters** are released as a result of depolarization in one of the cells (Figure 8-14, B). The cell that releases transmitters is the **presynaptic** cell; the cell that receives the message is the **postsynaptic** cell. The transmitter molecules diffuse across the synaptic cleft and interact with chemical-sensitive receptors on the postsynaptic cell. This mechanism of information transfer is called **chemical synaptic transmission.** In contrast to the situation in electrical synapses, information transfer in chemical synapses is typically polarized: it crosses the synapse in only one direction, from presynaptic to postsynaptic cell. The process of transmitter release involves a slight delay, known as the **synaptic delay,** between the presynaptic membrane electrical change and the initiation of the postsynaptic electrical response (Figure 8-14, C).

Anatomical Specializations of Chemical Synapses

A synapse is a structure that belongs to both cells: the presynaptic terminal is one element of the output segment of the presynaptic cell, and the corresponding portion of the postsynaptic cell is one element of its input segment. The most obvious characteristic of the presynaptic terminal is the presence

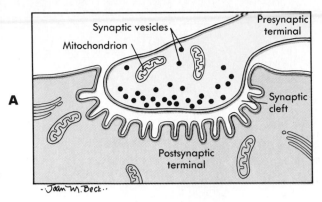

A

Synaptic vesicles
Mitochondrion
Presynaptic terminal
Synaptic cleft
Postsynaptic terminal

·Joan M. Beck··

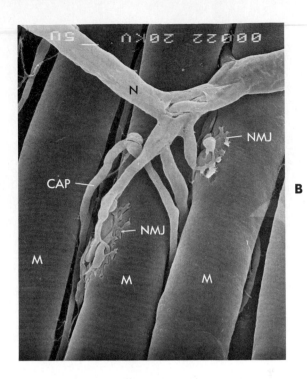

B

FIGURE 8-15

A A diagram illustrating the characteristic features of the neuromuscular junction in skeletal muscle.

B A scanning electron micrograph of three muscle fibers (M) showing the external morphology of two neuromuscular junctions (NMJ) formed by the branches of a single motor nerve fiber (N). Note also the presence of a capillary supplying these muscle fibers.

of intracellular **synaptic vesicles,** which are small, membrane-bound spheres that contain neurotransmitters (Figure 8-14). The electron microscope reveals several recognizable types of transmitter vesicles with discrete size categories that range in diameter from 30 to 1600 nm. A given synaptic terminal may have one or more of these morphological types, and each type may contain a single type of neurotransmitter molecule or two or more types present together.

A chemical synapse is characterized in electron micrographs not only by the aggregation of vesicles on one side of a pair of closely spaced membranes but also by the **densities**—darkly staining material along both the presynaptic and postsynaptic membranes (Figure 8-14, *C*). Special staining techniques reveal that the presynaptic membrane possesses a gridwork that may channel the vesicles to **active sites** on the membrane, where they can fuse with the presynaptic membrane. The postsynaptic membrane is more variable in its associated protein elements, which probably reflects the variety of responses that the postsynaptic cell can make to the transmitter (described below). Postsynaptic densities sometimes include cytoskeletal elements that can change the area of contact with the presynaptic cell. This is one of the alterations known to occur in some synapses as a result of synaptic activity.

Important functional differences in CNS synapses, synapses on skeletal muscle, and the synapses made by neurons of the autonomic nervous system onto their effectors are reflected in their different structures. In the CNS, the amount of cell surface taken up by each individual synapse is relatively small. These individual synapses typically provide a very small portion of the total number of inputs to the postsynaptic cell, and each one alone

is insufficient to bring the postsynaptic cell to threshold.

In synapses on skeletal muscle, the axon terminal occupies a relatively large membrane area and is buried within folds of the postsynaptic membrane (Figure 8-15). This structure restricts the loss of transmitter by diffusion and provides extensive postsynaptic membrane for the transmitter to act upon. Skeletal muscle cells receive a single synapse, and each action potential arriving at that synapse provides the muscle cell with sufficient depolarization to reach its threshold.

In the autonomic nervous system, efferent neurons penetrate the organs they innervate and release transmitter molecules from a series of enlargements, or **axon varicosities** (Figure 8-16). The distance between the axon varicosities and the membranes of effector cells is quite large compared with synaptic clefts in the CNS, so transmitter is distributed over a wide membrane area rather than a restricted postsynaptic region. The postsynaptic receptor molecules likewise are distributed over much of the membrane of the postsynaptic cell. Such synapses typically mediate slower, more prolonged responses from the postsynaptic cell. The distribution of receptors over the postsynaptic cell membrane allows the postsynaptic cell to be responsive both to synaptically released transmitters and to hormones (which can be identical or very similar molecules) delivered by the blood.

Neurotransmitters

The number of chemicals that are recognized or strongly suspected of being neurotransmitters would make a very long list, and the number con-

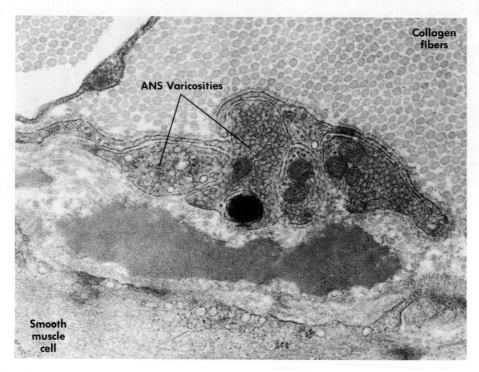

Collagen fibers

ANS Varicosities

Smooth muscle cell

FIGURE 8-16
A photomicrograph showing synapses in the autonomic nervous system. Synaptic vesicles are present in swellings, called axon varicosities, near a smooth muscle cell, but there are no membrane specializations comparable to those at synapses in the CNS.

tinues to grow, particularly in the peptide category, as more chemicals are found to function as neurotransmitters. We are only now beginning to appreciate the elaborate chemical language of the nervous system, and along with knowledge of the effects of these chemicals has come a greater appreciation for the subtleties of synaptic function.

Acetylcholine was the first neurotransmitter to be identified, and its action has been intensively studied. Unlike other transmitters, which belong to families of related chemicals, acetylcholine is in a chemical category all by itself. It can be synthesized in axon terminals and packaged into vesicles there. It is released at neuromuscular junctions, at synapses in the autonomic nervous system, and in the CNS. In the brain, destruction of neurons that release acetylcholine has been associated with Alzheimer's disease.

Norepinephrine, epinephrine, dopamine, and **serotonin** are examples of a class of compounds known as **monoamines.** Norepinephrine (noradrenalin) and the closely related chemical epinephrine (adrenalin) are present in the sympathetic nervous system as well as the brain, and appear to have a role in regulating emotions and arousal responses in the body. Dopamine is involved in emotional behavior, with excesses causing certain types of schizophrenia. Dopamine is also important in

motor control; lack of dopamine leads to the shaking and uncontrolled movements seen in Parkinson's disease. Serotonin is located in discrete groups of neurons in the brainstem (lower portion of the brain) that have projections all over the brain; its functions are many, including focusing of attention and modulation of pain perception.

Another important class of transmitters is the amino acids and their derivatives, such as **glutamate, glycine,** and **gamma aminobutyric acid (GABA).** Glutamate is probably responsible for the majority of the excitatory synaptic actions in the CNS. Glycine and GABA are associated with inhibitory synaptic effects. A deficit of GABA appears to be responsible for certain types of anxiety, because antianxiety drugs such as the benzodiazapines Valium and Librium act by binding to the chloride channel and increasing its response to GABA.

Peptides are the largest group of recognized transmitters. Their role as hormones was described first, and more recently many of the same molecules have been shown to be released in synaptic transmission. For instance, the hormones **cholecystokinin** and **vasoactive intestinal peptide** are released from secretory cells in the digestive system. These peptides are also released from neurons to act on other neurons in the brain. The endogenous opiates, **enkephalins** and **endorphins,** are peptides that were discovered after investigations of the actions of opium and morphine strongly suggested that these derivatives of the poppy plant must act by mimicking substances normally present in the brain.

Peptide transmitters are typically released at synapses that also release another, smaller type of transmitter molecule. In some synapses it has been established that the peptides are preferentially released when the neuron is strongly activated, causing it to fire action potentials in rapid succession. Sometimes the effect of peptides on the postsynaptic cell is slow or "silent" and may only be revealed when the postsynaptic cell is activated by other pathways. For these reasons, peptides have sometimes been classified as **neuromodulators** rather than neurotransmitters. However, any possible distinction between neurotransmitters and neuromodulators has become blurred by the recognition that some "fast-acting" transmitters such as acetylcholine and monoamines may also initiate slower, more modulatory effects on the postsynaptic cell. Neural function has components that last over a continuous range of timecourses, from milliseconds to days or longer. Moreover, the speed of action of any transmitter is determined not by its chemical class but by the channels and other pathways it activates in the postsynaptic cell.

Control of Neurotransmitter Release by Calcium

Ca^{++} concentration is maintained at very low levels in resting neurons by pumps that move free Ca^{++} from the cytoplasm into the extracellular fluid. Axon terminals contain voltage-dependent Na^+ and K^+ channels and also **voltage-dependent Ca^{++} channels.** When an action potential depolarizes a presynaptic terminal (event 1 in Figure 8-17), these Ca^{++} channels open briefly and Ca^{++} ions enter from the extracellular fluid (event 2). The elevated level of cytoplasmic Ca^{++} brings about a cascade of events involving proteins in the cytoplasm, vesicle membrane, and presynaptic membrane that culminate in fusion of the synaptic vesicles at active sites on the presynaptic membrane. Fusion results in exocytosis of the transmitter molecules (event 3). Then the Ca^{++} channels close (event 4) and Ca^{++} levels are returned to the resting level by the ATP-driven Ca^{++} pumps (event 5). The membrane added to the presynaptic membrane by the fusion of vesicles is removed by the process of endocytosis (event 6) and the membrane in these **coated vesicles** is recycled either in the neuron ending or after traveling to the nucleus by retrograde axoplasmic transport (event 7), carrying trophic messages from the extracellular environment.

The **synaptic delay** of about 0.5 msec between the arrival of the action potential at the axon terminal and the beginning of the voltage response of the

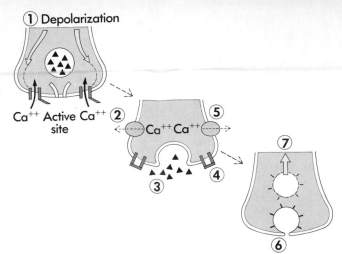

FIGURE 8-17

Diagram of the presynaptic events, shown in three stages. The depolarization (1) leads to opening of Ca^{++} channels (2) that are located in the presynaptic terminal in the vicinity of the active sites. The Ca^{++} leads, in a complex pathway, to fusion of the vesicle and release of the transmitter molecules into the cleft (3). The Ca^{++} channels close (4) and the cytoplasmic Ca^{++} is pumped out of the cell (5). The membrane that enclosed the transmitter is removed from the terminal ending by endocytosis (6) and may either have more transmitter added, to make a new synaptic vesicle, or it may travel to the cell body by axoplasmic retrograde transport (7).

postsynaptic cell is attributable to events that lead to fusion of the vesicle membrane with the axon terminal membrane; diffusion of transmitter across the synaptic cleft and activation of postsynaptic channels only occupy a small fraction of the delay.

Although the synaptic vesicles in axon terminals are within a few μm of the presynaptic membrane, they rarely fuse with it as long as the concentration of Ca^{++} in the cytoplasm remains at its resting value of less than 0.1 μM. Occasionally, a single vesicle fusion occurs in the absence of stimulation, and the effect on the postsynaptic cell of occasional "miniature" synaptic events corresponding to a single vesicle fusion is negligible. The vesicles contain so nearly the same number of transmitter molecules that they produce changes of the same size in the membrane potential of the postsynaptic cell. The change in voltage brought about by release of one vesicle is called a **quantum.**

When an action potential invades the presynaptic terminals, the miniature potential change, or quantum, resulting from one vesicle fusion is multiplied by some large integer, producing the much larger **postsynaptic potential.** The **quantal content** of the postsynaptic potential is therefore determined by a discrete number of quanta that are thought to correspond to the number of vesicles that fuse with the presynaptic membrane.

Synaptic Efficacy

The number of synaptic vesicles that fuse with the membrane when a single action potential invades

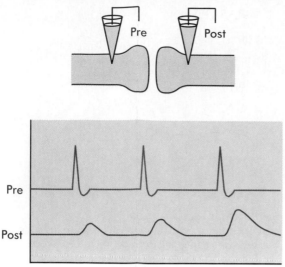

FIGURE 8-18

Time

Recording from the presynaptic and postsynaptic elements in a synapse that exhibits facilitation when the action potentials arrive in rapid succession. Note that the postsynaptic potentials increase in size; this is thought to be due to enhanced release of transmitter due to residual Ca^{++} in the terminal.

the terminal is determined by how much Ca^{++} enters the terminal. Varying this factor can therefore change the **synaptic efficacy,** or how big an effect a synapse has on the postsynaptic cell.

Synaptic efficacy can be altered in many ways. Some alterations represent very short-term changes and others last much longer. For instance, when a neuron fires action potentials at a high rate, some Ca^{++} that enters its terminals during the invasion of one action potential will still be there when the next action potential arrives. Residual Ca^{++} in the presynaptic terminal increases synaptic efficacy by causing more vesicles to be released. This could be viewed as a summation of intracellular Ca^{++} concentration rather than membrane depolarization, and it results in larger and larger postsynaptic potentials (Figure 8-18). In other cells, rapid firing of action potentials could eventually deplete the number of vesicles, and this would reduce the synaptic efficacy.

The size and shape of the action potential as it enters the terminal is another factor that determines how much Ca^{++} comes in. This is because Ca^{++} channels open relatively slowly in response to depolarization. Special channels can be opened or closed by synaptic inputs close to the terminals. The opening or closing of these channels can reshape the action potential (Figure 8-19). An example of the anatomical arrangement that is associated with reduction in synaptic efficacy by **presynaptic inhibition** is given in Figure 8-20. A similar anatomical arrangement could also produce **presynaptic facilitation** if conductances that prolong the action potential are activated. Small increases or decreases in

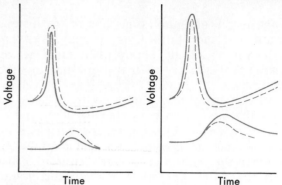

FIGURE 8-19

Presynaptic inputs on the axon terminals open channels that alter the shape of the action potential and the release of transmitter.

A The effect of a presynaptic inhibitory input on the shape of the action potential and the postsynaptic response (from dotted line condition, before, to continuous line condition, after the input).

B Effect of a presynaptic facilitatory input changes the response from dotted line, before, to continuous line condition after the input. The resulting synaptic potential is much larger.

the amount of Ca^{++} that enters can have profound effects on synaptic transmission because the sequence of events that results in vesicle release is very sensitive to small changes in Ca^{++} level.

Presynaptic inhibition (Figure 8-19, A) can result from the opening of additional Cl^- channels, which keep the membrane potential close to the resting potential and reduce the size of the action potential. If the presynaptic input opened K^+ channels before the action potential arrived at the terminal, the additional leakiness of the membrane to K^+ would oppose the depolarization by Na^+ ions during the action potential's rising phase and add to the repolarizing brought about by the opening of voltage-sensitive K^+ channels in the falling phase.

On the other hand, presynaptic facilitation could result from closing of special K^+ or Cl^- channels in the presynaptic membrane. Under either condition, the height and duration of the action potential would increase (Figure 8-19, B). The Ca^{++} channels of the presynaptic terminal may also be modulated, causing a direct effect on the influx of Ca^{++}. Timing is crucial in the presynaptic modulation of transmitter release. The fact that associations involving reward and punishment are also dependent upon the timing of events suggests that such alterations in neural function may contribute to learning (see box on p. 212).

Fates of Released Neurotransmitter

At synapses in the CNS, the exposure of the postsynaptic cell (and presynaptic cell) to the neurotransmitter is typically very brief. This brief exposure is due to removal of the transmitter from the synaptic cleft by one or more of several mechanisms. After the molecules have diffused into the

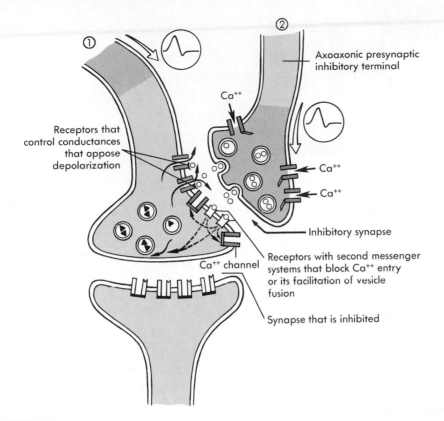

FIGURE 8-20

In presynaptic inhibition a particular presynaptic terminal *(1)* receives synaptic input from another neuron *(2,* an axoaxonic synapse). Presynaptic inhibition can change the K^+ or Cl^- conductance of the terminal and reduce the amplitude of the invading potential and the quantal content. In some cases presynaptic inhibition involves modulation of voltage-sensitive Ca^{++} channels by intracellular second messenger systems.

cleft and some have interacted with receptor molecules, they may (1) be actively recovered by the presynaptic cell for recycling, (2) diffuse away from the area, or (3) be degraded by enzymes on the postsynaptic and/or presynaptic membranes (Figure 8-21). The reuptake, diffusion, or degradation ensure that the neurotransmitter concentration remains high for only a few milliseconds. This allows for high rates of information flow to the postsynaptic cell.

Neurotransmitter Receptor Families

The neurotransmitter is the first messenger in a chain of events that alter channel activity and therefore the membrane potential of the postsynaptic cell. Just as was the case for sensory receptor proteins, **neurotransmitter-sensitive receptors** fall into two categories, based on mechanism of action. In the first category, they may themselves be part of the channel molecule. (These are called the **ligand gated channels.**) In the second category, they may be coupled to channels by other molecules (see Figure 8-5).

The great variety of receptor molecules found in the human nervous system belong to a few families, as can be deduced by similarities in their amino

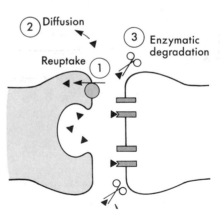

FIGURE 8-21

Representation of the ways transmitter concentration can be reduced following its release: (1) reuptake, (2) diffusion, (3) enzymatic degradation. Typically only one mechanism is utilized in a given synapse.

acid sequences. The family in which the receptor element and the channel are part of the same molecule includes the acetylcholine nicotinic receptor, one of the receptors for GABA (called the $GABA_A$ receptor), and the glycine receptor. The channel that is opened by GABA or glycine allows Cl^- to pass through, whereas the nicotinic acetylcholine recep-

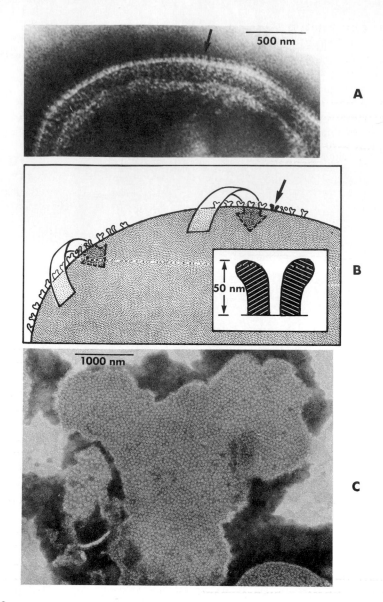

FIGURE 8-22
A A view of a membrane with acetylcholine receptor/channels *(arrow).*
B A drawing of the surface of a membrane such as that shown in A with the projecting part of the receptors indicated and an enlargement of one of the receptors *(arrow)* providing the dimensions of the receptor. The binding site for acetylcholine is located on the part of the receptor that projects from the membrane.
C Acetylcholine receptors seen end-on; the pore in the center of each receptor is the ion channel.

tor channel admits cations, primarily Na$^+$.

Each acetylcholine nicotinic receptor consists of a transmitter-binding site that projects from the membrane surface and controls a transmembrane protein. Figure 8-22, *A,* shows a high-magnification electron micrograph of the nicotinic acetylcholine receptor. The projecting portion of the molecule is illustrated in 8-22, *B,* and the end-on view of the channels are shown in 8-22, *C.* This protein is made up of five subunits that form a gated channel; opening of the channel may involve the swiveling of the subunits brought about by binding of the receptor with acetylcholine.

The second family of receptor molecules changes permeability of the cell through coupling to G proteins. Many transmitters have receptors in this category, including the muscarinic acetylcholine receptor, epinephrine and norepinephrine, monoamines such as serotonin, and the neuropeptides. Coupling to a G protein in the membrane can activate a variety of secondary mechanisms, depending on small differences in the G protein structure. The G protein action on a channel can occur by movement within the membrane, as in the heart cell response to acetylcholine. Alternatively, additional intracellular steps can involve the messengers **cyclic adenosine mono-**

phosphate (cAMP) or phosphatidyl inositol 4,5 bisphosphate (see Figure 5-13).

Individual neurotransmitters can be detected by more than one type of receptor, and the different receptors can be coupled to more than one type of channel. Consequently, a particular neurotransmitter can affect some postsynaptic cells one way and perhaps have the opposite effect on other cells. Sometimes the transmitter causes both a short-term opening or closing of existing channels and also a longer-term regulation of gene expression that results in additional channels or receptors being inserted into the membrane.

> - *What are the important differences between electrical and chemical synapses?*
> - *What ion is most intimately associated with the alteration of synaptic efficacy, and how can the level of this ion be modulated?*
> - *What determines the effect a neurotransmitter will have on the postsynaptic cell?*

PRINCIPLES OF INFORMATION PROCESSING IN THE NERVOUS SYSTEM
Categorizing Postsynaptic Potentials

> - *What conductance changes underlie the two types of synaptic potentials, and why is this classification of synaptic interactions an oversimplification?*
> - *How do temporal and spatial summation of postsynaptic potentials differ?*
> - *How can integration of synapses activated by descending pathways block some or all of the nociceptive inputs?*

Until this point, we have dealt only with excitatory inputs, and with their modification by presynaptic inputs, which can either facilitate or inhibit the release of neurotransmitters. Postsynaptic potentials have classically been divided into two categories, excitatory and inhibitory. **Excitatory postsynaptic potentials (EPSPs)** are depolarizing potentials that increase the probability that the postsynaptic cell will initiate an action potential. **Inhibitory postsynaptic potentials (IPSPs)** reduce the chance that the postsynaptic cell will initiate an action potential. The EPSPs increase in the permeability of the postsynaptic membrane to both Na^+ and K^+, causing a net depolarization because the driving force for Na^+ is so much larger than that for K^+. The IPSPs can result from increasing K^+ permeability or increasing Cl^- permeability or a combination of both. Increasing K^+ permeability hyperpolarizes the postsynaptic membrane because the equilibrium potential for K^+ is usually more internally negative than the resting potential (see Chapter 7). In a hyperpolarized cell, the membrane potential is farther from threshold, and the probability that depolarizing synaptic potentials will result in action potentials is reduced. The inhibitory effect of increasing Cl^- permeability is based on the fact that additional open Cl^- channels allow Cl^- to more readily flow into the cell to neutralize a depolarizing inward flow of Na^+.

Although categorizing a synapse between two neurons as excitatory and inhibitory is a useful tool, it is now recognized that in some cases this is an oversimplification. A transmitter can have a short-term excitatory effect but over the long term the response can be converted to inhibition, or vice versa. This might be due to activation of genes or alteration of the responsiveness of receptors or channels. Added to this complexity is the fact that coreleased peptide transmitters may have actions that are quite different from the quicker-acting transmitter.

Integration of Postsynaptic Potentials

The neuromuscular junction is not a site of integration of inputs. This is because there is only one synaptic site on each muscle cell and it acts as a simple relay, passing information on, one-for-one, from neuron to muscle cell. All decisions about whether or not to activate the muscle are made in the CNS before the action potential was initiated in the motor neuron. At the neuromuscular junction the quantal content is about 100 vesicles per presynaptic impulse and the resulting EPSP is about 50 mV, far greater than the minimum needed to bring the muscle cell to threshold.

In contrast to skeletal muscle cells, neurons in the CNS receive inputs from many neurons, a condition called **convergence** (Figure 8-23, *A*), and in turn they distribute their terminals to provide inputs to many other neurons, a condition called **divergence** (Figure 8-23, *B*). The quantal content of each input to a central neuron is much lower than at the neuromuscular junction and the neurons would never be activated unless inputs summated.

As was true for the sensory cells, summation can occur with rapid repetition of inputs from one source (**temporal summation**) or can result from inputs arriving at synapses onto the cell from different sources (**spatial summation**). Temporal summation of postsynaptic potentials occurs if additional channels in the postsynaptic cell are opened while the membrane potential is still altered by the currents activated by the preceding PSP; when this occurs the resulting PSPs summate.

Spatial summation results if two or more PSPs are generated at nearly the same time at different synapses on the cell's input segment (Figure 8-24). In the spatial summation of sensory inputs, all of the inputs are of the same polarity (typically all depolarizing) but synaptic summation is algebraic: it calls for adding positive and negative numbers. If all the inputs to a neuron happen to be of the same polarity at a particular time (all EPSPs or all IPSPs),

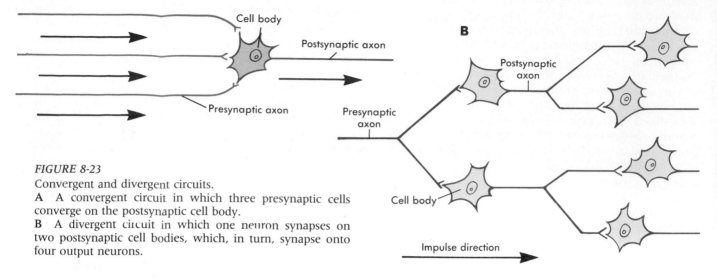

A

Cell body

Postsynaptic axon

Presynaptic axon

B

Postsynaptic axon

Presynaptic axon

Cell body

Impulse direction

FIGURE 8-23
Convergent and divergent circuits.
A A convergent circuit in which three presynaptic cells converge on the postsynaptic cell body.
B A divergent circuit in which one neuron synapses on two postsynaptic cell bodies, which, in turn, synapse onto four output neurons.

they summate constructively and the change in membrane voltage gets larger and larger (Figure 8-24, *A* and *B*). However, if they are of opposite polarity (a mix of EPSPs and IPSPs), the IPSPs tend to cancel the EPSPs and the resulting membrane change can be very small (Figure 8-24, *C*).

The process of integration in central neurons might be likened to the deliberation of a wise statesman who will listen to all sides of an argument before making a decision. A special feature that makes the neuron good at this type of deliberation is the ability to "listen" simultaneously to the pros (EPSPs) and cons (IPSPs). The weight of the evidence at any one moment may call for action (an action potential) or no action (no action potential).

The integration of EPSPs and IPSPs is accomplished on the inexcitable membrane of the cell, where the excitatory effects of net cation entry are summed with the inhibitory effects of K$^+$ exit or the stabilizing effect of opening additional Cl$^-$ channels. A large depolarizing potential may result from very active excitatory inputs on one dendrite, but another dendrite may receive many inhibitory inputs around the same time; both of these signals will diminish in size by the time they are conducted decrementally to the spike initiating site. A "decision" will be made at that site, based on the size of the summed inputs relative to the threshold of the cell, and an action potential will or will not be initiated. In the next instant, the decision will be reconsidered and another outcome can result, depending on the balance between the excitatory and inhibitory inputs impinging on the cell. This process is illustrated in Figure 8-25.

Not all inputs to a neuron have equal ability to influence the cell's activity. The strength of a synapse is influenced by: (1) the quantal content and (2) how close the synapse is to the spike initiating site. The closer an input is, the less it will diminish in strength before it reaches the decision-making region of the cell. Inhibitory synapses from neurons in higher brain centers are frequently located close to the spike initiating site of spinal neurons, and in that position, they can exert "veto power" over the activity of the neuron.

Processing of Pain Signals

This chapter began with a description of how energy can be detected and transformed into membrane potential changes by sensory transduction. Features of the stimulus are then conveyed to the appropriate regions of the nervous system, where they interact with ongoing neural processes. Pain perception is a relatively complex phenomenon that illustrates many features of neural processing. Survival can depend on rapid and appropriate reflex responses to pain. Pain can also preserve the individual by mediating the transition to rest from painful effort, to allow for recuperation. On the other hand, a person in a life-threatening situation exhibits altered responses to pain, and painful sensory signals may be blocked.

The catch-all category of receptors known as nociceptors includes those heat receptors which begin to respond at temperatures over 45° F, cold receptors, receptors that respond to deep depression of the skin or sharp, penetrating stimuli, and cells that apparently can transduce more than one of these types of stimulation. Many of the receptors in the skin share the characteristic of releasing the neuropeptide called **substance P.**

The responses of sensory neurons to stimulation is modulated in the periphery by the release of substances such as **serotonin, prostaglandins,** and **bradykinin** from injured cells. These substances in-

axon
hillock

NERVE AND MUSCLE

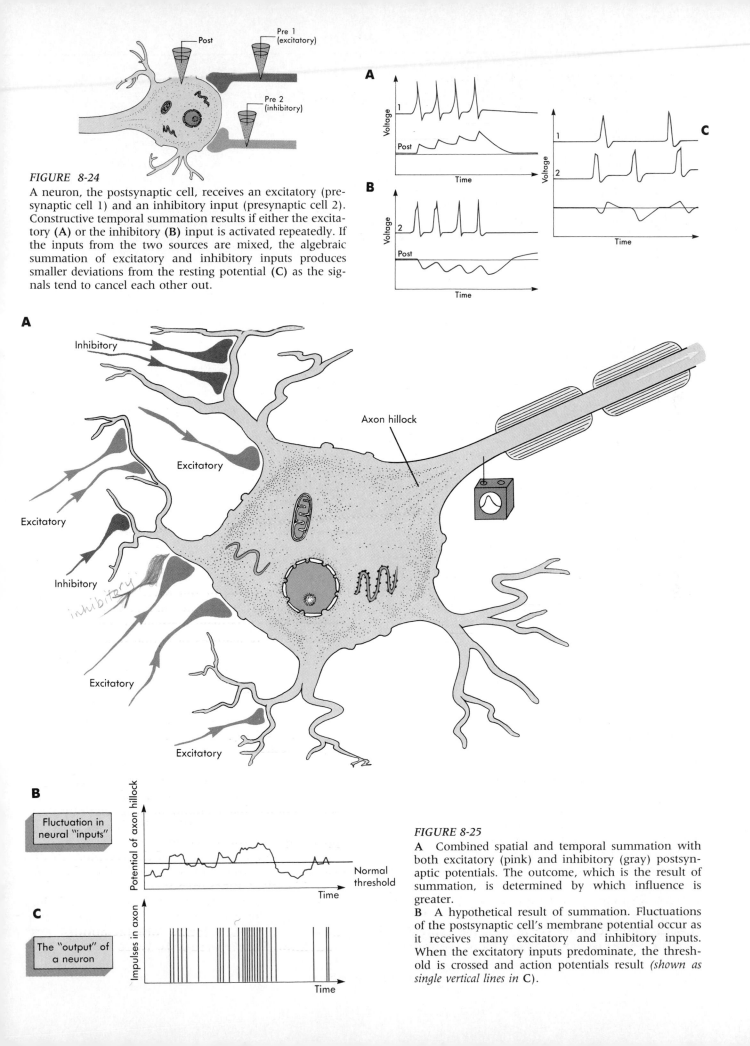

FIGURE 8-24

A neuron, the postsynaptic cell, receives an excitatory (presynaptic cell 1) and an inhibitory input (presynaptic cell 2). Constructive temporal summation results if either the excitatory **(A)** or the inhibitory **(B)** input is activated repeatedly. If the inputs from the two sources are mixed, the algebraic summation of excitatory and inhibitory inputs produces smaller deviations from the resting potential **(C)** as the signals tend to cancel each other out.

FIGURE 8-25

A Combined spatial and temporal summation with both excitatory (pink) and inhibitory (gray) postsynaptic potentials. The outcome, which is the result of summation, is determined by which influence is greater.

B A hypothetical result of summation. Fluctuations of the postsynaptic cell's membrane potential occur as it receives many excitatory and inhibitory inputs. When the excitatory inputs predominate, the threshold is crossed and action potentials result *(shown as single vertical lines in* **C***).*

Information Storage at the Synapse

The hippocampus is a region of the brain that has long served as a model for learning in the mammalian nervous system. This region is crucial for rats learning a maze or students learning to locate all the proper classrooms in the proper buildings at the beginning of a new school year.

The brain processes sensory modalities separately at one level, and then brings the separate parts of a sensory experience together to integrate the "whole picture." The hippocampus is one of the sites of multisensory integration; the neurons there often respond to more than one modality because projections from several sensory areas converge onto them. The pattern of projections that was experienced at one moment may be mentally recreated at a later time by the reassembling of sounds, shapes, colors, and textures and their relative positions in space. The initial association of these inputs to neurons in the hippocampus depended upon coincidence in time. This temporal association of otherwise unrelated stimuli underlies a type of learning known as classical conditioning: two stimuli arising from different sources must converge on the response pathway and must overlap in time. In the hippocampus, strong activation of a single input pathway or pairing of inputs to a cell can cause an increase in the efficacy of the active synapses, and the changes persist for days to weeks. Such lasting changes in synaptic function are called **long-term potentiation** and probably correspond to stored associations. Recollection may be promoted by the ease with which such synaptic pathways can be reactivated.

The mechanism of long-term potentiation has been studied in thin slices of hippocampal tissue. Slicing the tissue shortens the distance oxygen must travel by diffusion, allowing the cells to survive when placed in oxygenated physiological solution. Fortunately, the physical layout of neuronal circuits in the hippocampus is largely two-dimensional, so large parts of the circuitry remain intact in thin slices. In such slices, repeated simultaneous stimulation of two synaptic inputs to a single cell was found to potentiate both synapses. The potentiation occurs when two conditions are met: (1) the postsynaptic cell must be depolarized, and (2) the transmitter glutamate must be released and bind to a particular class of glutamate receptors, the NMDA (N-methyl-D-aspartate) receptors. Unless the cell is depolarized, the NMDA channel opened by glutamate is blocked by Mg^{++}, and so very little Ca^{++} enters. Depolarization drives the Mg^{++} out, and under that condition, glutamate binding opens the channel for a relatively long time, allowing a large surge of Ca^{++} ions to enter (Figure 8-D). This makes the NMDA receptor for glutamate a coincidence detector: opening of the channels requires that the cell already be excited by another input.

The Ca^{++} enters the postsynaptic cell and initiates changes that strengthen the efficacy of the active synapses. Possible mechanisms of synapse alteration acting within the postsynaptic cell include Ca^{++}-activated enzymes that alter the cytoskeleton and mechanically reshape the postsynaptic part of the synapse so that less current is lost in decremental conduction (Figure 8-D). Such physical changes are known to occur in some regions of the rat brain in response to enriched environments. However, most changes in synaptic efficacy have been shown to involve alteration in release of transmitter by the presynaptic cell, and the hippocampal potentiated synapses are no exception. For this to occur, the altered state of the postsynaptic cell resulting from Ca^{++} entry must be communicated to the presynaptic cell. Recent findings implicate nitric oxide as this retrograde messenger. Nitric oxide is a gas with a half-life of 5 seconds and the ability to move across membranes without having to interact with receptors. Its production in response to Ca^{++} entry in the postsynaptic cell would allow it to diffuse rapidly to the presynaptic cell. Just how it might interact with targets in that cell to bring about changes in synaptic efficacy remains to be explored. Although inhibitors of nitric oxide production block long-term potentiation, the story is far from complete. However, the findings suggest memory formation involves some very elusive elements.

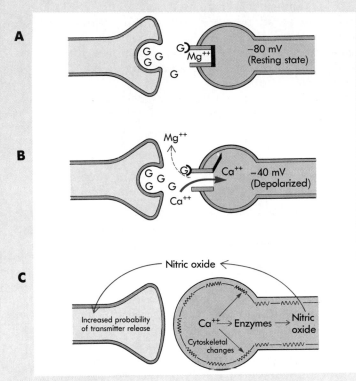

FIGURE 8-D

Simplified illustration of a synapse in which glutamate release from the presynaptic cell is represented by a single vesicle and the postsynaptic cell's responses are represented by a single NMDA-type glutamate receptor channel. In **A** the Ca^{++}-dependent events in presynaptic transmitter release are not shown, but the interaction of glutamate with the postsynaptic receptor does not admit Ca^{++} because the channel is blocked by Mg^{++} when the cell is at its resting potential of -80 mV. In **B** the channel can open because some other input onto the postsynaptic cell has depolarized it, driving out the Mg^{++} and allowing Ca^{++} to enter. In **C** the hypothetical changes brought about by Ca^{++} in the postsynaptic cell include alteration of the cytoskeleton and synthesis of nitric oxide, which diffuses out of the cell and acts as a retrograde messenger to the presynaptic cell, where it may increase the probability of transmitter release.

crease the sensitivity of nociceptors and other receptors by increasing the probability of channel openings, so that a slightly painful stimulus becomes more painful or a stimulus that was perceived as tickling or mild contact is converted to a pain sensation after cell injury has caused the release of modulators. An example is the increased sensitivity of sunburned skin to all types of stimulation. With high levels of injury, the generator potential is greatly enhanced and large numbers of impulses pass into the spinal cord or brain, depending on the receptive field location.

The responses to the incoming pain message are highly variable. Reflexes causing withdrawal responses result from connections onto neurons that activate motor neurons. Such responses can occur before a person is conscious of pain, because the conscious response is dependent upon activation of processing areas in the brain, which are several synapses away from the reflex response pathway. Pain that enters consciousness includes "neutral" information such as where the pain originates, and an emotionally laden aspect that includes suffering and visceral responses. The intensity of the perceived pain depends very strongly on other events taking place at the same time. The "high" experienced in the midst of a competitive sport can cause an athlete to be completely unaware of personal injury.

Variations in pain perception are mediated in large part by a family of peptides called **endogenous opioid peptides,** the **enkephalins** and **endorphins.** These substances are released by neurons that synapse on the pain afferents and onto the neurons to which they relay information. Neurons projecting from the brain that release serotonin or norepinephrine activate these enkephalin-containing neurons (Figure 8-26). The enkephalins act at two sites: presynaptic inhibition reduces the duration of the sensory cell action potentials, and postsynaptic hyperpolarizing IPSPs on the relay neurons reduce the effectiveness of the depolarizing sensory inputs. The net effect of these descending pathways can be so strong that the message fails to be transmitted to higher brain centers and even the reflexes can be blocked. In this way, enkephalins are thought to suppress the sense of pain during natural childbirth. Opium derivatives, such as morphine or heroin, have an analgesic, or pain-reducing, effect because they are similar enough in chemical structure to bind to the receptors normally utilized by enkephalins.

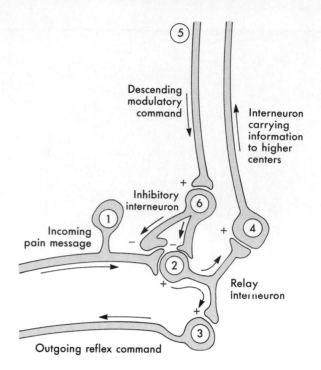

FIGURE 8-26
A simplified illustration of interactions thought to underlie pain pathways and their modification. The incoming pain afferent (1) synapses onto an interneuron (2), which relays the information to a motor neuron (3), which carries an outgoing motor command to a muscle that mediates a withdrawal reflex. The interneuron also relays the message to an ascending interneuron (4), which will carry the pain signal to higher brain centers required for the experience of pain and localization of the pain stimulus. These relays can be blocked if the descending modulatory pathway (5) is active. It activates an inhibitory interneuron (6), which can block the incoming signal presynaptically and inhibit the relay neuron postsynaptically.

SUMMARY

1. The **input regions** of neurons and sensory cells are specialized to respond to a particular type of energy (such as mechanical or chemical) with a change in conductance. The membrane of the input region does not possess voltage-sensitive channels, so inputs throughout the input region **summate,** jointly contributing to the voltage change that is decrementally conducted. If the sensory cell, or neuron, has an axon, the voltage change can be transformed into action potentials, but if there is no axon, the voltage change is communicated to a neuron by an altered rate of neurotransmitter vesicle release.

2. Sensory experiences are related to sensory **modalities,** the classes of stimuli to which the human body is sensitive. The classical modalities of touch, taste, hearing, vision, and olfaction do not include the monitoring of internal states, such as muscle position, blood glucose level, and blood pressure, but these also are detected by sensory cells and feed into neural pathways that control the behavior of the body. The location of the sensory endings corresponds to the receptive field of the cell. Centrally, the sensory information is mapped in a precise way that corresponds to the peripheral origin of the inputs.

3. The intensity of stimuli is encoded by the frequency of action potentials in afferent nerves. The initial event in intensity coding is a graded potential change (usually a depolarization), called a **receptor** or **generator potential,** that results from activation of **stimulus-sensitive ion channels.** The magnitude of the local current that results from this event determines the frequency of action potentials that arise in adjacent excitable parts of the receptor.

4. **Adaptation,** a decrease in receptor responsiveness with stimulation, occurs in some receptors that are more sensitive to changes in stimulus intensity than to maintained, constant conditions. Receptors that must encode a large range of stimulus intensities display a logarithmic rather than a linear relationship between stimulus intensity and action potential frequency.

5. Action potentials in a neuron can affect the electrical activity of another neuron or an effector cell if the two are connected by a **synapse.** The synapse corresponds to part of the **output region** of one cell and the **input region** of the other cell. In **electrical synapses,** some of the current in one cell flows across the synapse through gap junctions. In chemical synapses, depolarization of the synaptic terminal leads to an influx of Ca^{++}, through special channels present in the output region of the cell. The Ca^{++} entry brings about fusion of synaptic vesicles with the presynaptic membrane. Some of the transmitter diffuses to the postsynaptic cell and is bound by receptor molecules on its surface. Receptor binding activates ion channels directly or indirectly, leading to a change in the membrane conductance and a **postsynaptic potential.**

6. The postsynaptic potential may **summate** with other synaptic inputs to increase or decrease the probability of an action potential in the postsynaptic cell. **Excitatory** (depolarizing) **postsynaptic potentials** result from an increase in conductance to all small cations; **inhibitory postsynaptic potentials** may be hyperpolarizing (resulting from an increase in K^+ conductance) or silent (resulting from an increase in Cl^- conductance that stabilizes the membrane potential near its resting value).

7. Synaptic efficacy can be influenced by the activity level of the neuron itself and by presynaptic inputs from other neurons that affect the probability of vesicle release, either increasing it **(presynaptic facilitation)** or decreasing it **(presynaptic inhibition).**

1. Describe the distinctive features of the input and output regions of a neuron.

2. How could the receptive field of a skin touch receptor be mapped experimentally? What stimuli could cause spatial summation in such a receptor?

3. Both spatial and temporal summation require that the stimulation arrive at the input region of a cell within a brief period. What special feature of temporal summation distinguishes it from spatial summation?

4. What ions flow through channels that produce depolarizing receptor potentials? Hyperpolarizing receptor potentials?

5. How does the neural code give information about stimulus intensity? How does the fact that the initial response is in an electrically inexcitable part of the cell make this possible?

6. Compare the response properties of fast-adapting and slowly adapting receptors. How might the two types of information they provide be used in integrating information about a single stimulus modality into an appropriate response?

7. How is the release of neurotransmitter regulated and why is it such an important element in neuronal function?

Choose the MOST CORRECT Answer.

8. The input segment of a neuron includes all EXCEPT:
   a. Dendrites
   b. Axon
   c. Axon hillock
   d. Cell body

9. Spatial summation:
   a. Results from stimulation of the same input region by inputs arriving in rapid succession
   b. Decreases the size of the receptor potential
   c. Occurs only in motor neurons
   d. None of the above is true

10. This category of sensory receptors monitor such things as temperature, blood pressure, blood oxygen, content and limb position:
    a. Epithelial receptors
    b. Nonadapting receptors (tonic receptors)
    c. Rapidly-adapting receptors (phasic receptors)
    d. Mixed receptors

11. This neurotransmitter plays a role in regulating emotions and arousal responses in the body:
    a. Acetylcholine
    b. Dopamine
    c. Epinephrine
    d. Serotonin

12. _____ is an example of an inhibitory neurotransmitter:
    a. Acetylcholine
    b. Dopamine
    c. Cholecystokinin
    d. Gamma aminobutyric acid

13. This ion brings about fusion of synaptic vesicles with the presynaptic membrane:
    a. Ca^{++}
    b. K^+
    c. Na^+
    d. Cl^-

SUGGESTED READINGS

BARINAGA M: Is nitric oxide the retrograde messenger? *Science* 254:1296, Nov 29, 1991.

CHURCHLAND PM, CHURCHLAND PS: Could a machine think? *Scientific American* 262(1):32, 1990. Describes the debate concerning whether computer-based artificial intelligence can ever rival the brain.

CORSI P, editor: *The enchanted loom. Chapters in the history of neuroscience*, London, 1991, Oxford University Press. A fascinating trip through the discoveries that have shaped the present-day understanding of brain function.

MILLER RJ: Receptor-mediated regulation of calcium channels and neurotransmitter release. *FASEB Journal* 4:3291, 1990.

NAZIF FA, BYRNE JH, CLEARY LJ: Cyclic AMP induces long-term morphological changes in sensory neurons of *Aplysia, Brain Research* 539:324, 1991. Cellular mechanisms that underlie short-term and long-term memory have been extensively investigated in the marine snail *Aplysia,* which has been used as a "model system" for investigations of neural function.

SHEPPARD GM: *Neurobiology,* ed 2, New York, 1988, Oxford University Press. Available in paperback, it is one of the most readable texts in the field. Contains a particularly good discussion of the higher functions of the nervous system such as perception, learning, and memory.

SYED NY, BULLOCH AGM, LUKOWIAK K: *In vitro* reconstruction of the respiratory central pattern generator of the mollusk *Lymnaea, Science* 250:282, Oct 12, 1990. The identified neurons that control a simple behavior have been taken out of the animal and when the neurons form synaptic connections in culture, they produce the same pattern of activity .

VISI ES, KISS J, ELENKOV IJ: Presynaptic modulation of cholinergic and noradrenergic neurotransmission: interaction between them, *News in the Physiological Sciences* 6:119, 1991. Receptors for transmitters are found not only on the postsynaptic cell but also on the neuron which releases the transmitter; this is another way that activity can regulate transmitter release.

CHAPTER 9

The Somatosensory System and an Introduction to Brain Function

On completing this chapter you should be able to:

- Distinguish between the peripheral and central nervous system, and describe the location of the major subdivisions of the brain: the brainstem, cerebellum, thalamus, hypothalamus, and cerebral hemispheres.
- Understand what is meant by a somatotopic representation and why such representations are important in the central processing of sensory information.
- Outline the elements of the somatosensory afferent pathways, identifying the different routes taken by touch, vibration, temperature, and pain projections.
- State the function of the reticular activating system.
- Understand the difference between short-term memory and long-term memory.
- Appreciate the significance of the term *dominant cerebral hemisphere* and identify the location of the language areas of the brain.
- Identify the characteristics of the stages of sleep.

*T*he brain and spinal cord make up the central nervous system (CNS). The peripheral nervous system (PNS) consists of nerves and peripheral ganglia which link the CNS, our "onboard computer," with the external and internal world. The CNS receives and integrates sensory information in order to initiate appropriate responses. It mediates reflexes and generates behavior, the complex patterns of activity in the body's effector systems. The nervous system also controls many endocrine systems, including those that regulate growth, metabolism, and reproduction. The output of the nervous system is modified by experience; that is, behavior patterns are altered on the basis of stored information. Finally, the nervous system creates a sense of self, built upon the unique organization of each individual's nerve cells and the modifications of their connectivity by life's experiences.

Acquisition of sensory information is a key element in most of these functions. Sensory information is collected and analyzed in discrete regions of the brain. Information provided by different sensory pathways is integrated by reference to memories. For example, when we hear a friend's voice, we recognize it, recall his face, and remember his name. The voice, the face, and the name are experienced as part of one reality. In principle, everything that the CNS does can be understood as arising from the capabilities of individual neurons. Nevertheless, the ability of the nervous system to sort through billions of bits of information and recognize a kindred face is one of the most remarkable feats of living systems and requires a vast underpinning of neuronal activity.

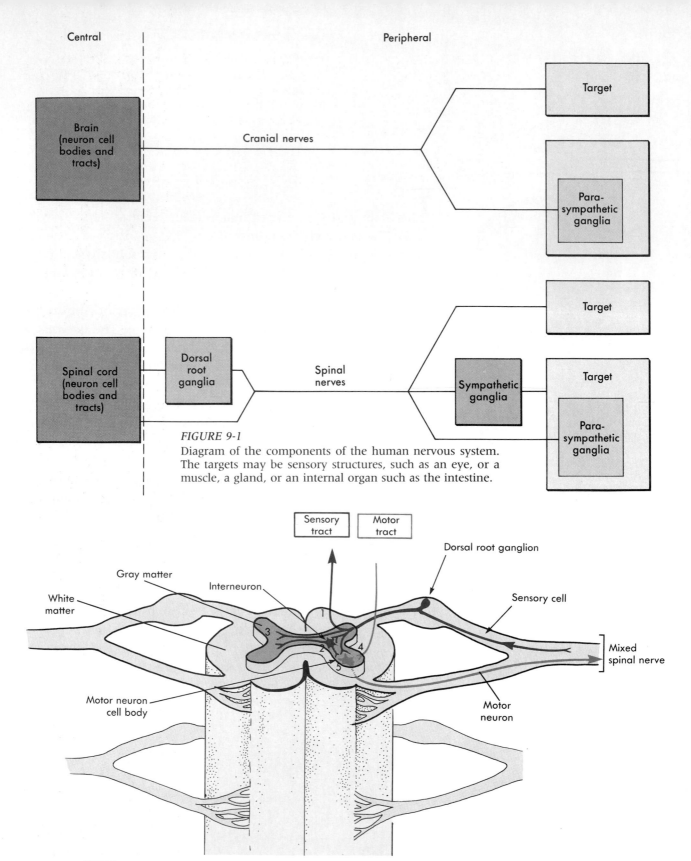

FIGURE 9-1

Diagram of the components of the human nervous system. The targets may be sensory structures, such as an eye, or a muscle, a gland, or an internal organ such as the intestine.

FIGURE 9-2

The spinal cord. Gray matter consists of neuron cell bodies and is located centrally in the spinal cord. White matter consists of ascending and descending axon tracts that run outside the central gray matter. Sensory neurons have their cell bodies in the dorsal root ganglion and project into the spinal cord. Branches of sensory axon terminals synapse on interneurons or directly on motor neurons. Motor neurons (and also interneurons and sensory endings) receive descending inputs from the brain. Motor neuron axons leave the spinal cord by way of the ventral root and join the sensory axons. The separation of motor and sensory portions of the spinal nerves just outside the spinal cord has allowed selective stimulation of sensory or motor components of the mixed spinal nerves. The sensory cells that contribute to the cranial nerves do not have cell bodies in the periphery (see Fig. 9-1).

220

ANATOMY OF THE NERVOUS SYSTEM
Segregation of Axons and Cell Bodies

- *What are the major divisions of the CNS?*
- *What is the function of the limbic system? The reticular activating system?*
- *What is a cortical column?*
- *What is the functional relationship between neurons that make up a cortical column?*

The nervous system has many distinct components (Figure 9-1), but each is made up of neurons and/or their processes (axons and dendrites), along with several types of supporting glial cells. Within the nervous system, input segments of neurons, their dendrites and cell bodies, are often segregated from the output segments, the axons. Cell bodies and dendrites are the **gray matter** of the nervous system; the **white matter** is composed of axons.

In the spinal cord and brainstem, the gray matter is in the center, surrounded by an outer layer of white matter, the axon tracts (Figure 9-2). The axons in the spinal cord are segregated into ascending **sensory tracts** and descending **motor tracts**. A narrow central canal extends down the spinal cord. This canal is continuous with the fluid-filled spaces in the brain, called the *cerebral ventricles*. Both the central canal of the spinal cord and the ventricles of

the brain are filled with **cerebrospinal fluid,** which provides a highway for the circulation of hormones, nutrients, and white blood cells within the CNS. The cerebrospinal fluid cushions the brain and the spinal cord by absorbing shocks that might otherwise harm neural tissue.

Sensory neurons, or **afferent neurons,** transmit action potentials to the dorsal roots of the spinal cord or to the brain (see Figure 9-2). The axons of **motor neurons,** or **efferent neurons,** transmit action potentials from the brain or the ventral root of the spinal cord to skeletal muscles, smooth muscles, glands, and visceral organs. Most efferent neurons control or modulate the activities of effectors but some efferents modulate the sensitivity of sensory receptors.

Central and Peripheral Components of the Nervous System

The nervous system can be divided into two components, the CNS and the PNS. The CNS consists of the spinal cord and the brain with its many subdivisions (Figure 9-3). Estimates of the total number of neurons in the CNS vary, but about 10^{10} cells is a currently accepted figure. Most neurons are neither sensory neurons nor motor (efferent) neurons but are **interneurons (association neurons).** These neurons form links between the input (sensory) and output (efferent) neurons. The large number of pos-

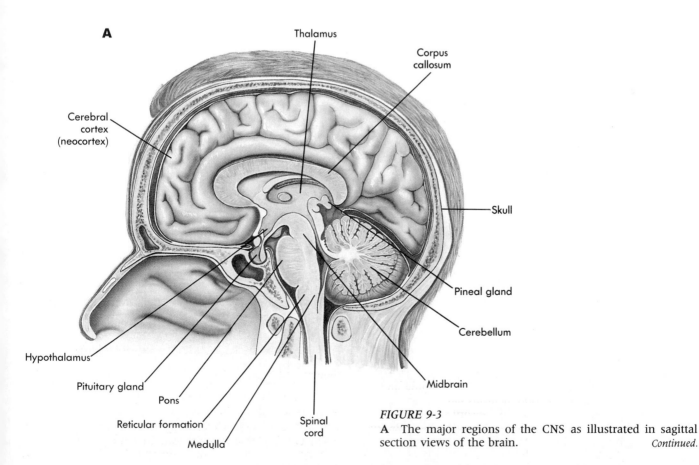

A

Thalamus

Corpus callosum

Cerebral cortex (neocortex)

Skull

Pineal gland

Cerebellum

Hypothalamus

Pituitary gland

Pons

Reticular formation

Medulla

Spinal cord

Midbrain

FIGURE 9-3
A The major regions of the CNS as illustrated in sagittal section views of the brain.   *Continued.*

The Somatosensory System and an Introduction to Brain Function

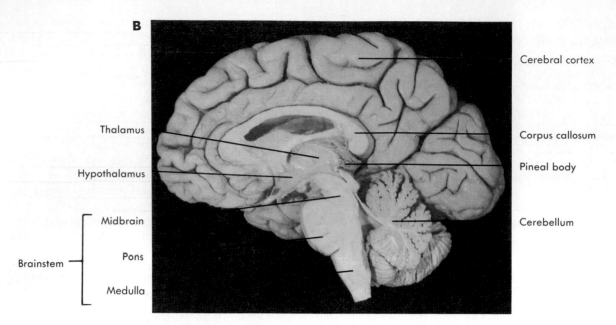

B

Cerebral cortex

Thalamus

Corpus callosum

Hypothalamus

Pineal body

Midbrain

Pons

Cerebellum

Medulla

Brainstem

FIGURE 9-3, cont'd
B   A photograph of a sectioned, preserved brain.

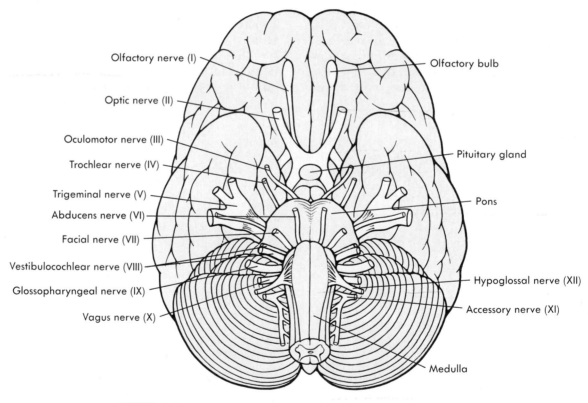

Olfactory nerve (I)

Olfactory bulb

Optic nerve (II)

Oculomotor nerve (III)

Pituitary gland

Trochlear nerve (IV)

Trigeminal nerve (V)

Pons

Abducens nerve (VI)

Facial nerve (VII)

Vestibulocochlear nerve (VIII)

Hypoglossal nerve (XII)

Glossopharyngeal nerve (IX)

Accessory nerve (XI)

Vagus nerve (X)

Medulla

FIGURE 9-4
Inferior surface of the brain showing the origin of the cranial nerves.

sible connections provided by the many interneurons provides enormous integrating power. There are also many neurons in the periphery. Groups of neuron cell bodies within the CNS that are surrounded by axon tracts are called **nuclei** but groups of neuronal cell bodies in the PNS are called **ganglia.**

The PNS includes all the neural elements outside the brain and spinal cord. It consists of peripheral nerves and the neuronal cell bodies located in the dorsal root ganglia (see Figure 9-2) and the ganglia of the autonomic nervous system (see Chapter 5 and Chapter 12). **Cranial nerves** extend from the brain (Figure 9-4), and **spinal nerves** extend from

NERVE AND MUSCLE

TABLE 9-1   The Cranial Nerves

Nerve	Pathway and function
I Olfactory	Sensory—from olfactory epithelium to olfactory bulb
II Optic	Sensory—from retina to lateral geniculate nucleus of thalamus
III Oculomotor	Carries motor and muscle stretch receptor pathways between midbrain and eye that mediate movement of eyeball and eyelid, focusing, and control of pupillary diameter
IV Trochlear	Carries motor and muscle stretch receptor pathways that mediate eye movement
V Trigeminal	Carries motor pathways from pons to muscles involved in chewing; carries sensory pathways from cornea, face, lips, tongue, and teeth to pons
VI Abducens	Carries motor and muscle stretch receptor pathways that mediate eye movement
VII Facial	Carries pathways that mediate facial expression, secretion of saliva and tears, and taste sensation
VIII Vestibulocochlear	Sensory—carries auditory and vestibular afferents
IX Glossopharyngeal	Carries taste afferents and pathways involved in swallowing and salivary secretion
X Vagus	Carries sensory and autonomic pathways running between medulla and thoracic and upper abdominal viscera
XI Accessory	Carries pathways between medulla and throat and neck muscles involved in swallowing and head movements
XII Hypoglossal	Carries pathways between medulla and tongue involved in speech and swallowing

the spinal cord. Twelve pairs of cranial nerves (Table 9-1) arise from the undersurface of the brain. Spinal nerves are formed by the unification of dorsal (sensory) and ventral (motor) roots on each side of each spinal segment (see Figure 9-2) and are thus mixed nerves. There are 31 pairs of spinal nerves (Figure 9-5): 8 cervical, 12 thoracic, 5 lumbar, 5 sacral, and 1 coccygeal.

The output of the CNS can consist of a single relay (one neuron in the pathway) or a more complicated relay (at least two neurons in the pathway). The axons of **somatic** motor neurons, those that control skeletal muscles, run directly from the CNS to the muscle fibers. Motor axons of **autonomic** motor neurons run from the CNS to peripheral autonomic ganglia, where incoming signals may be integrated in circuits involving peripheral neurons before being relayed to visceral effectors by other neurons (see Chapter 11).

Subdivisions of the Central Nervous System

The major anatomical subdivisions of the CNS are the spinal cord, the brainstem, the cerebellum, the diencephalon and the cerebral hemispheres (Table 9-2 and Figure 9-3). The spinal cord carries information to and from the brain via well-defined sensory and motor tracts. The medulla, pons, and midbrain constitute the **brainstem,** an extension of the spinal cord devoted primarily to coordinating the internal organs (cardiovascular control centers, respiratory control centers, and so on). The brainstem links the rest of the brain with the spinal cord and contains motor neurons of cranial nerves. The brainstem also contains the **red nucleus,** which is in-

volved in motor function, and the **reticular formation.**

The **diencephalon** ("between brain") is located between the midbrain and the **cerebral hemispheres (telencephalon).** Its two components are stacked one on the other: the **thalamus** and the **hypothalamus** (Figure 9-6). The thalamus is a primary site of sensory integration. The thalamus is also "informed" by the cortex about voluntary motor movements, and gets information about posture and body orientation from the cerebellum. The thalamus is like a major telephone exchange, processing sensory and motor information and channeling it to the appropriate destinations. The hypothalamus controls body temperature, influences respiration and heartbeat, and directs the secretions of the **pituitary gland** (Chapter 5, Figure 9-6). Posterior to the thalamus is the pineal gland, which has structural features that reveal its evolutionary derivation from the pineal eye. The pineal has a daily cycle of secretion of the hormone **melatonin,** which may coordinate activity cycles and play a role in reproduction (see boxed essay p. 240).

The hippocampus and olfactory lobes are two components of the **limbic system** (Figure 9-7), an evolutionarily old group of structures that link the hypothalamus with the cerebral cortex. Together, the activity of the hypothalamus and limbic system are responsible for emotional responses. The ability of odors to evoke memories and associated emotional or sexual responses reflects the importance of such associations for survival. The hippocampus is folded under the neocortex (see below) and is important in the formation and recall of memories.

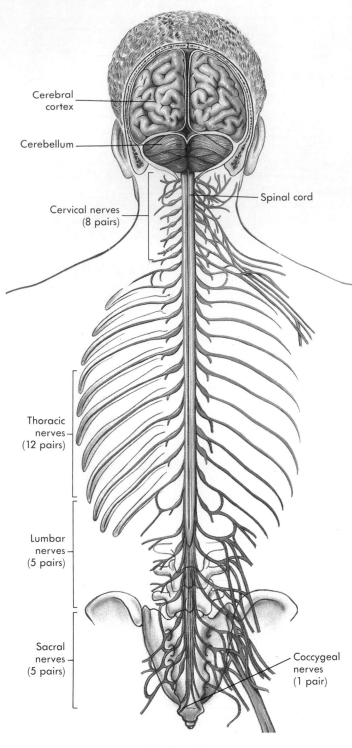

Cerebral cortex

Cerebellum

Cervical nerves
(8 pairs)

Spinal cord

Thoracic
nerves
(12 pairs)

Lumbar
nerves
(5 pairs)

Sacral
nerves
(5 pairs)

Coccygeal
nerves
(1 pair)

FIGURE 9-5
The spinal cord and spinal nerves.

The two hemispheres of the cerebellum are located dorsal to the brainstem, and are sometimes classified as part of the brainstem. The primary function of the cerebellum is motor coordination. The cerebellum receives information about the current position and movement of each limb and the state of relaxation or contraction of the muscles af-

fecting that limb. The cerebellum is crucial to the learning of complex motor tasks.

The cerebral hemispheres are so highly developed in human beings and other primates they almost cover the evolutionarily older parts of the brain. Deep within each cerebral hemisphere there is a complex structure called the **corpus striatum,** which contains several large groups of neurons, the **basal ganglia** (the globus pallidus, caudate nucleus, and putamen in Figure 9-8). (This designation of "ganglia" is an exception to the general rule that collections of CNS neurons are called nuclei.) The basal ganglia, together with the cerebellum and the red nucleus of the midbrain, are involved in generating patterns of activity that drive complex movements (see Chapter 11).

Neurons in the cerebral hemispheres communicate by two major pathways. Communication within different regions in a hemisphere involves pathways into and out of the corpus striatum. The two hemispheres communicate with one another by way of the **corpus callosum** (see Figures 9-3 and 9-8), a thick bundle of axons (white matter) that connects the two hemispheres. One of the anatomical differences between male and female brains is the larger number of axons in the corpus callosum of the female brain.

The Cortex of the Cerebral Hemisphere

The **cerebral cortex,** a convoluted cap of gray matter (cell bodies) 2 to 4 mm thick, has three subdivisions: the **neocortex,** the **hippocampus,** and the **olfactory lobes.** The neocortex is largest in area and is what one sees when one looks at the human brain. The neocortex is, from an evolutionary point of view, the newest part of the mammalian brain, and is often simply called the cortex. In primates the neocortex has a dominant role in processing sensations (other than odors) and in generating behavior.

The cerebral cortex of some mammals has fissures, called **sulci** (singular, sulcus), and ridges, called **gyri** (singular, gyrus), that increase its surface area. The folding is most pronounced in the human brain, where it approximately triples the surface area. The folding creates important anatomical landmarks. The **longitudinal fissure** divides the cerebrum into its two hemispheres (see Figure 9-8). The fissure of Rolando (central sulcus) separates **frontal** and **parietal lobes;** the fissure of Sylvius divides the **temporal lobe** from the frontal and parietal lobes; and the parieto-occipital fissure (not as easily identified) separates the **occipital lobe** from the parietal lobe (Figure 9-9).

Cortical Organization

The functional subdivisions of the cortex have been mapped using two kinds of information. Injuries to

Text continued on p. 227.

TABLE 9-2	Subdivisions of the Central Nervous System	

Major subdivision	Components	Function
Spinal cord		Spinal reflexes; relay sensory information
Brainstem	Medulla	Sensory afferent nuclei; reticular activating system; visceral control centers
	Pons	Reticular activating system; visceral control centers
	Midbrain	Similar to pons
Cerebellum		Coordination of movements; balance
Diencephalon	Thalamus	Relay station for ascending sensory and descending motor neurons; control of visceral function
	Hypothalamus	Visceral function; neuroendocrine control
Cerebral hemispheres	Basal ganglia	Motor control
	Red nucleus	Motor control
	Corpus callosum	Connects the two hemispheres
	Hippocampus (limbic system)	Memory; emotion
	Olfactory lobes	Smell
	Neocortex	Higher functions

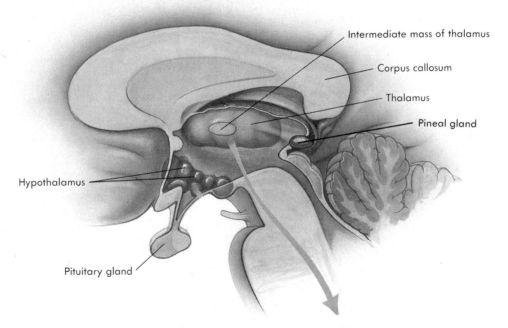

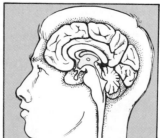

FIGURE 9-6

The diencephalon showing the thalamus, hypothalamus, and the location of the pituitary and pineal glands.

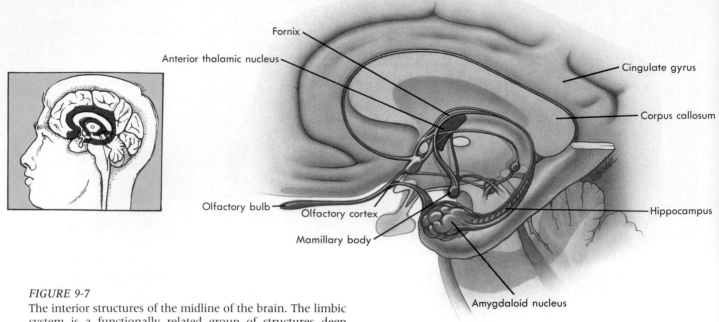

FIGURE 9-7
The interior structures of the midline of the brain. The limbic system is a functionally related group of structures deep within the cerebral hemispheres. These structures mediate emotional responses and are connected to autonomic pathways through the hypothalamus.

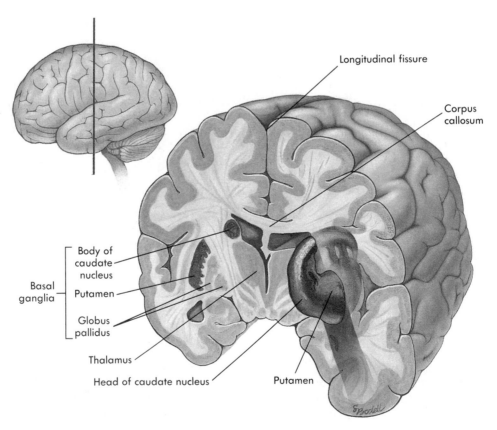

FIGURE 9-8
A section through the middle of the brain showing the corpus callosum and basal ganglia. The caudate nucleus and the putamen together form the corpus striatum. White myelinated axon tracts run between the areas of gray matter (the basal ganglia), producing the striped appearance of the corpus striatum.

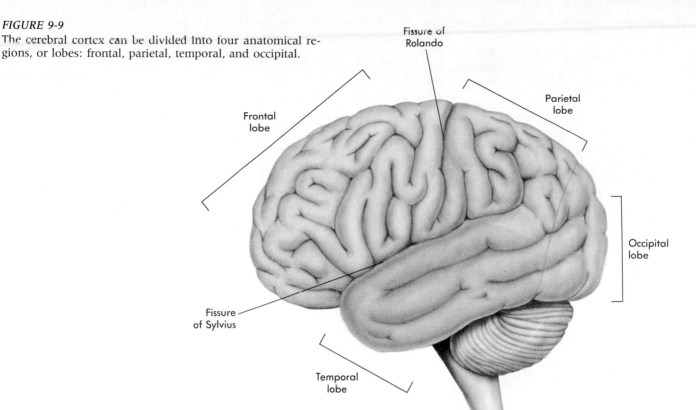

Fissure of
Rolando

Frontal
lobe

Parietal
lobe

Occipital
lobe

Fissure
of Sylvius

Temporal
lobe

particular sites on the cerebrum have been associated with functional deficits. Additional clues to the location of function were provided by electrical stimulation of the cortical surface in conscious patients undergoing brain surgery. The primary motor cortex is the most posterior portion of the frontal lobe, and each point on its surface is associated with movement of part of the body. The primary somatosensory cortex lies just behind the motor cortex, on the leading edge of the parietal lobe. Each patch of the surface of the somatosensory cortex receives inputs from sensory receptors in a discrete region of the body, and the patches form a map of the periphery. In the temporal lobe, or auditory cortex, adjacent regions respond to adjacent sound frequencies, producing a tone map. In the visual cortex of the occipital lobe, adjacent sites correspond to adjacent points in the retina.

The motor cortex and sensory cortex occupy only a small portion of the cerebral cortex. The remainder is called the association cortex, the site of higher mental activities. The association cortex is a larger portion of the total cortex in primates than in other mammals. In a mouse, for example, 95% of the surface of the cerebral cortex is occupied by motor and sensory areas, and communication between the two occurs through lower brain regions. In humans, only 5% of the surface is devoted to primary motor and sensory functions, and the remainder is devoted to the associations between these two regions.

The cortex is composed of six layers (Figure 9-10, A and B), each of which has a characteristic pattern of synaptic connections with other parts of the cortex and other parts of the nervous system. The two innermost layers (V and VI) contain axons of output cortical cells that pass to the spinal cord and thalamus. Layer IV receives sensory input via the thalamus.

Some regions of the cortex, particularly the primary sensory and motor areas, are organized as columns. A **cortical column** is a small (about 1 mm diameter) cylinder of cortex containing neurons that process information about a single discrete part of the body. For example, in the somatosensory system, a single column may represent information arising from a small area on the fingertip. In the motor system, a single column may concern itself with movement around a single joint.

SENSORY PROCESSING
Projection of Sensory Inputs

Conscious sensations arise from activity in sensory pathways that project to the cerebral cortex. A process called **feature extraction** results from the detailed analysis of information provided by a given sensory modality. For instance, features of the visual world include shape analysis, which is used for recognition of familiar objects, and movement analysis, which is used in recognizing changes that may call for a response. Information arising from a single population of receptors can follow **divergent path-**

Imaging the Brain

Any new diagnostic tool must satisfy certain requirements before it can be useful to the medical community. First, the risks of the procedure must be determined in animal experiments. If the risks and side effects are acceptably low, then tests are conducted on healthy adult volunteers. Extension of a technique to all segments of the population, including young children, pregnant women, and other high-risk groups requires additional careful assessment.

Another criterion a new tool must satisfy is that the information it provides must help in the choices between treatment options. It is unacceptable to subject a patient to a test that provides a wealth of data that cannot be used by health care professionals in treating the patient. One way the utility of a new diagnostic tool can be established is for volunteer patients to submit to evaluation by both a familiar technique and the new technique, so that the information provided by the new technique can be interpreted in the light of experience with the established technique. This method was used in studies of two noninvasive methods of imaging the soft tissues of the body, including the brain: (1) computed tomography (CT scans), and (2) a newer technique, magnetic resonance imaging (MRI).

In a CT scan (Figure 9, *A*) an x-ray source travels in a circle around a patient's head or body, while on the other side sensors

FIGURE 9-A
Apparatus for computed tomography (CT scan).

FIGURE 9-B
A sagittal magnetic resonance image (MRI scan) of a normal brain.

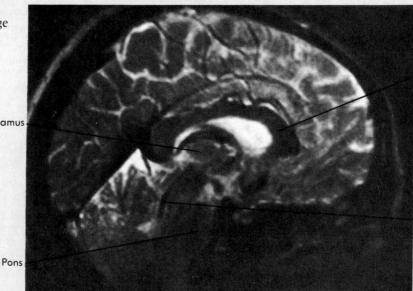

Thalamus

Corpus callosum

Fourth ventricle

Pons

detect x-rays emerging from the body. The information is fed to a computer that constructs an image on a computer monitor based upon the differences in the absorbance of tissues to x-rays.

The single photon emission computed tomography (SPECT) scan resembles a CT scan, except that radiation from injected radionucleotides is detected and processed to form an image. One new injectable imaging agent, called SPECTamine, is designed to pass through the blood-brain barrier to give a quick, accurate assessment following a stroke.

The radiation doses used in CT scans are much lower than occurs with conventional x-ray imaging, but patients are still exposed to radiation. In the MRI, the movements of molecules in the body are detected using a magnetic field thousands of times stronger than the Earth's. The magnetic field

acts on atomic nuclei in all body tissues, lining them up like compass needles. A brief pulse of electromagnetic energy is used to "bump" the aligned nuclei out of position. When the pulse of energy stops, the displaced nuclei return to their original position. As they return, the nuclei emit a faint signal that can be picked up by detector coils. The nucleus of each element has a distinct "signature" that also depends on the local environment. Since 75% of the body is composed of water, it is hydrogen atom vibrations that are most often detected in MRIs. Hemorrhage is detected easily in an MRI because the iron atoms in blood and blood breakdown products can be distinguished from normal brain tissues. Hydrogen nuclei of water molecules in normal tissue behave differently from those in cancerous tissue. By

changing the settings on the detector coils, the image of a tumor can be outlined and then examined to determine if it is solid or fluid-filled, thus offering clues to whether it is benign or malignant.

All these techniques involve computer-assisted reconstruction of data. For example, Figure 9, *B* shows an MRI scan equivalent to the anatomical drawing in Figure 9-3. Neither CT scans nor MRIs would be feasible without the recent advances in computer speed and memory capacity. Both CT and MRI scans reveal the location of tumors, sites of hemorrhage, and other abnormalities. In Figure 9, *C*, bleeding following a concussion is revealed by both the MRI and CT scans, but more information about the site and extent of intracranial bleeding is available in the MRI. This is because the MRI technique has greater contrast resolution than the CT scan, and bony structures do not obscure MRI images. MRI was used first to image the brain and spinal cord but is now used to diagnose disorders of bones, joints, and muscle; heart and blood vessel disease; and cancer in reproductive organs, liver, kidneys, lymph nodes, bladder, and pancreas. Although MRI does not require that a patient receive x-rays, it is not the method of choice for determining the relationship between bony structures and a site of injury or abnormality.

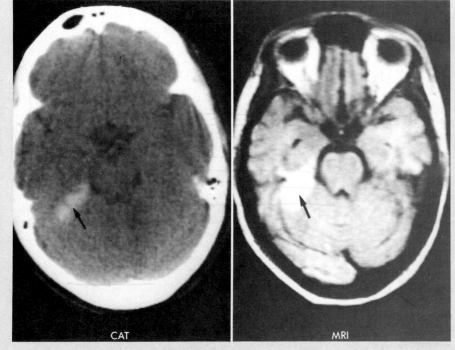

FIGURE 9-C
Hemorrhage resulting from a concussion is imaged by an MRI and a CT scan. Site of injury is indicated by the arrows.

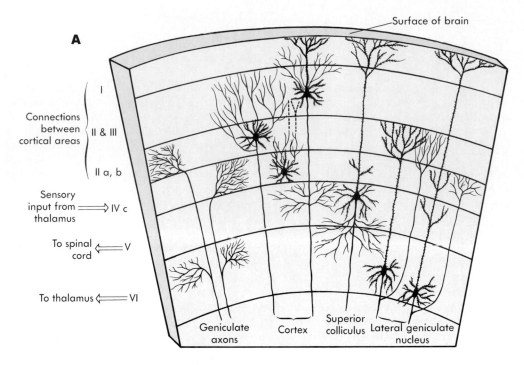

A

Surface of brain

I

Connections between cortical areas { II & III

II a, b

Sensory input from ⟹ IV c
thalamus

To spinal ⟸ V
cord

To thalamus ⟸ VI

Geniculate axons   Cortex   Superior colliculus   Lateral geniculate nucleus

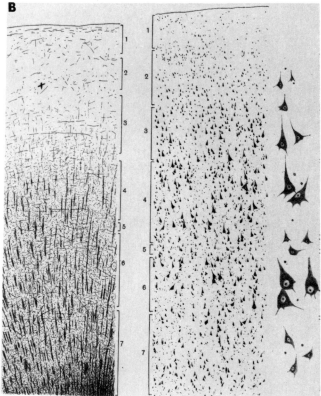

FIGURE 9-10
A A diagram of the six-layer structure of the cortex with representative types of neurons.
B Micrographs of the cortex illustrating its six-layer structure. The left micrograph has been stained to show axons and the right side has been stained to show the location of cell bodies. The shapes of the cell bodies in each layer are indicated in the enlargement on the far right.

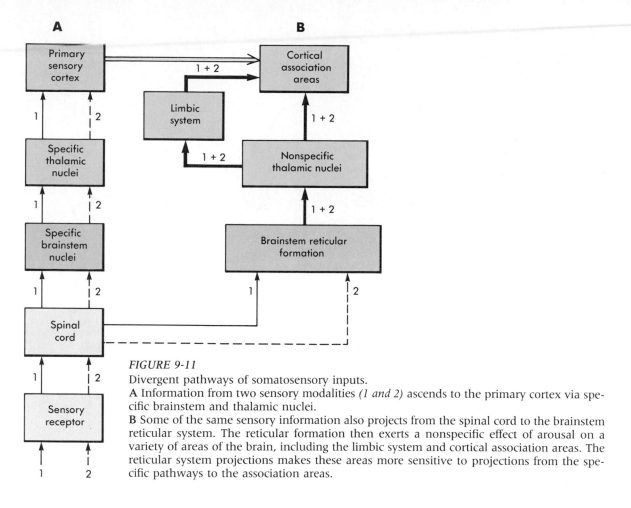

A

B

FIGURE 9-11

Divergent pathways of somatosensory inputs.

A Information from two sensory modalities *(1 and 2)* ascends to the primary cortex via specific brainstem and thalamic nuclei.

B Some of the same sensory information also projects from the spinal cord to the brainstem reticular system. The reticular formation then exerts a nonspecific effect of arousal on a variety of areas of the brain, including the limbic system and cortical association areas. The reticular system projections makes these areas more sensitive to projections from the specific pathways to the association areas.

ways (Figure 9-11) and be mapped in several regions of the brain, in addition to the cortical representations, so that different relevant features can be extracted and integrated for appropriate responses. This is the principle of **parallel processing.** Within the visual system, the field of view is analyzed for the chance event in which an object moves rapidly toward the eye. The protective eyeblink is activated long before knowledge of the object's movement could register at the level of conscious visual experience.

THE SOMATOSENSORY SYSTEM
Overview of Somatosensory Receptors

Nerve trunks contain many axons, but all sensory axons fall into one of four axon size classes, three that are myelinated and one that is unmyelinated. As a consequence of the different diameters and the presence or absence of myelination, different categories of sensory information travel to the CNS at different speeds. Two classification schemes are used: the Group system applies to sensory axons in nerves that innervate muscles, and the A_α, A_β, A_Δ and C system is used for nerves that innervate skin (Figure 9-13 and Table 9-3).

Each spinal nerve innervates a restricted region of the body surface, called a **dermatome** (Figure

9-12). In each dermatome, deep receptors and surface, or cutaneous, receptors are present. Deep receptors include pressure receptors (Group II axons) and proprioceptors. Proprioceptors detect joint position and muscle length and have rapidly conducting axons (Group I) that evoke rapid reflex responses. Proprioceptors are described together with their function in motor control in Chapter 12. Cutaneous receptors conduct action potentials over a range of velocities (Groups II, III, and IV axons) and detect simple contact, complex mechanical stimuli such as vibration, temperature (hot and cold are separate sensations), or surface injury.

Touch Receptors

Several types of mechanically sensitive receptors are present in the dermis and subcutaneous tissue (Figure 9-14). In most areas of the body surface, mechanical contact causes activity in diffusely distributed **free nerve endings** in the dermis and subcutaneous tissue. The ability to localize stimuli delivered to these receptors is not very acute. The resulting sensation is referred to as "coarse" touch. Morphologically specialized receptors mediate fine touch and vibration and are densely concentrated on areas such as the fingertips and face. They localize cutaneous stimuli precisely, and include rapidly

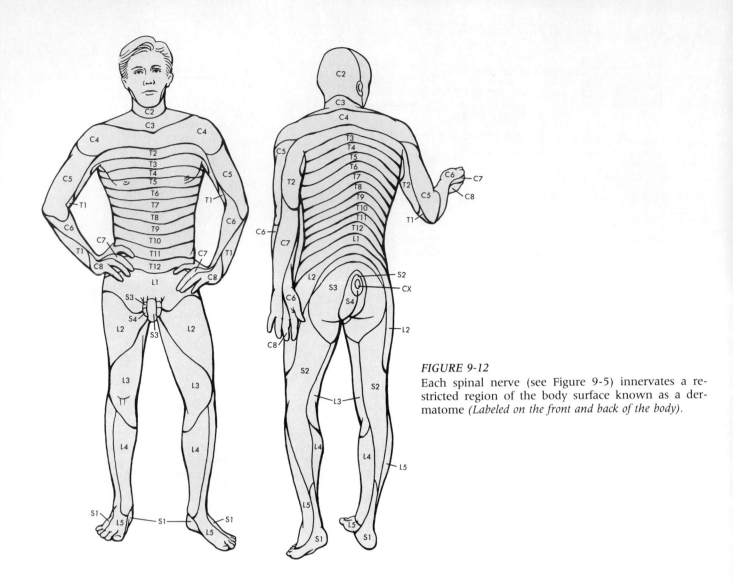

FIGURE 9-12
Each spinal nerve (see Figure 9-5) innervates a restricted region of the body surface known as a dermatome *(Labeled on the front and back of the body).*

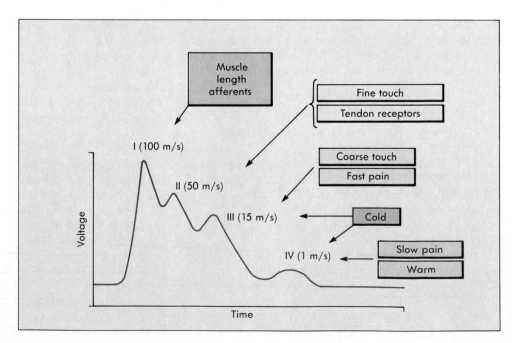

FIGURE 9-13
Axon groups are separated by their rates of axonal conduction in the compound action potential of a spinal nerve. Sensory axons fall into four size classes.

TABLE 9-3   Classes of Sensory Fibers

Classifications	Diameter	Conduction velocity	Myelination	Function
Group I (A$_\alpha$)	13-20 μm	70-110 m/s	Yes	Muscle length
Group II (A$_\beta$)	6-12 μm	25-70 m/s	Yes	Tendon receptor; rapidly adapting touch receptors; Pacinian corpuscle
Group III (A$_\Delta$)	1-5 μm	3.5-20 m/s	Yes	Touch; fast pain; cold
Group IV(c)	1 μm or less	Less than 1 m/s	No	Slow pain; temperature; itch; tickle

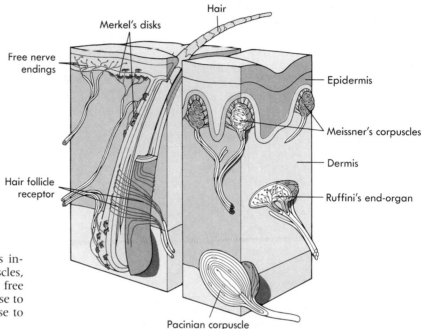

FIGURE 9-14

Mechanically sensitive receptors in the dermis include hair follicle receptors, Meissner's corpuscles, Pacinian corpuscles, Ruffini's endings, and free nerve endings. The specialized receptors give rise to Group II axons and free nerve endings give rise to Group III axons.

adapting or slowly adapting types. Rapidly adapting receptors include **hair follicle receptors** and in hairless areas, the **Meissner's corpuscles.** The most effective stimulus for both is periodic displacement at a frequency of 30 Hz to 40 Hz (cycles per second), providing a sensation of flutter. The **Pacinian corpuscle** (see Chapter 8) is also rapidly adapting but is found deeper in the subcutaneous tissue, and its most effective stimulus is a high-frequency periodic displacement (300 Hz), sensed as vibration. There are two types of slowly adapting receptors. **Touch dome endings (Merkel's disks)** are located near the surface of the skin, and **Ruffini's endings** are in the dermis.

Temperature Receptors

There is no such thing as absolute cold, only varying degrees of warmth. However, separate **warm** and **cold receptors** are identified by the way they respond to changes in temperature. Temperature receptors are free nerve endings (specialized in an un-

known way) that can detect changes of a few tenths of a degree. "Cold" receptors typically begin to discharge at 35° C, and their firing rate increases as the temperature decreases to 20° C. Warm receptors are activated above 30° C, and their discharge rate increases with increasing temperature to 45° C. The combined response of warm and cold receptors can be compared to the shape of an inverted bell, with minimum activity at temperatures slightly lower than normal body temperature and increases in either direction of temperature. Temperature sensations are significantly influenced by the adaptation temperature of the skin, as can be shown by holding one hand in cold water and the other in hot water and then plunging both into tepid water. This is just one example of the fact that receptor systems do not give an absolute measure of the variable they detect, but only a relative one.

Cold receptors are five to ten times more abundant than warm receptors. Both are most dense on the face and hands. Axons of warm receptors are

FRONTIERS
IN PHYSIOLOGY

Selective Stimulation and Anesthetic Block Demonstrate Distinct Receptor Categories

Sensory cell axons fall into several classes on the basis of axon diameters and the presence or absence of myelination. For example, when a large population of receptors is simultaneously stimulated in the toe region the different categories of information will reach the CNS at different times. If the compound activity in the nerve is recorded with electrodes and displayed on an oscilloscope, the action potentials in the most rapidly conducting axons will be seen as the first peak, followed by the distinct peaks that correspond to the other two size groups of myelinated axons and finally the summed spikes of the unmyelinated axon group (Figure 9-13).

Somatosensory modalities will be segregated to different components of the compound action potential because modalities are confined to specific axon diameters. In addition to the separation of the modalities on the basis of conduction velocity, it is possible to selectively activate the largest-diameter group present in the skin (Group II). When a nerve is stimulated electrically, the largest axons are more effectively depolarized by the current and are activated at the lowest voltages. If electrodes are attached to a subject's finger and the subject is asked to describe what she feels as the current is increased, she will

first report the illusion of mechanical contacts as Group II axons are activated. When Group III axons are stimulated, the subject will feel fast pain, but when Group IV (unmyelinated) axons are activated, the subjective experience is slow pain. At the lowest concentrations, a local anesthetic blocks only the smaller axons, so that slow pain is lost. Increasing the concentration eliminates first temperature, then fast pain, and lastly touch. Pressure blocks large axons first, so that touch is lost before pain (think about what happens when your arm "falls asleep").

unmyelinated, as are some of the axons of cold receptors; other cold receptors belong to the smallest myelinated axon category (Group III).

Pain Perception

There are two types of pain experiences. **Fast pain** is evoked by a needle jab, for example. Such pain is typically described as sharp, prickling, or bright. It is well localized but rapidly disappears without residual effects. **Slow pain** is the unpleasant, dull, burning sensation that follows tissue injury such as a burn. Slow pain is difficult to localize, persists after the painful stimulus is removed, and evokes autonomic reflexes and emotional reactions. Fast and slow pain are both detected by free nerve endings, but a significant difference in pain relief is related to the fact that fast pain is relayed by Group III and slow pain is relayed by the unmyelinated Group IV axons. Aspirin crosses into the unmyelinated axons and prevents the formation of prostaglandins in these cells, interfering with their activity. Luckily, aspirin does not block transmission in the other types of receptors, so that aspirin does not cause the numbness that local anesthetics such as xylocaine produce by blocking all types of sensory cells. Fast

pain is insensitive to both narcotics and analgesics but slow pain is blocked centrally by narcotics (see Chapter 8).

In internal organs, pain sensations can arise from stretching, chemical stimuli, and pressure. Internal pain is more difficult to localize than cutaneous pain because the axons that carry visceral pain information enter the spinal cord along with axons from skin receptors, and the nervous system does not distinguish the exact origin of these stimuli. As a result, visceral pain is often **referred pain**, because it is referred to the cutaneous region, which receives sensory inputs that join the inputs from a particular region of the viscera. An example is heart attack pain, which is felt in the chest and left arm, or menstrual cramps, which may seem to be localized in the back or thighs.

After amputation, the amputee may experience **phantom limb** sensations—an illusion that the amputated limb is still present. This effect is produced by action potentials arising in axons of sensory cells that formerly projected into the limb. If only the limb nerve had been severed, the axons might have regenerated to proper receptive field regions, but after amputation, they have no normal destination to regenerate into. Instead, their projections may grow into a tangled ball that can be easily activated. Activity in these sensory cells is perceived as coming from the absent limb. Other types of pain illusions may arise from rearrangements of central neural connections following injury.

Integration in the Somatosensory System

Lateral inhibition is a process of neural integration that sharpens the distinction between the signal relayed by the most activated sensory cell and other sensory cells that are also activated by the same stimulus. The process of lateral inhibition depends upon the ability of the most active cell to suppress its neighbors by activating inhibitory pathways that lower their output. In essence, the most activated cell "gets the jump" on its neighbors by firing first, and in addition to sending signals to higher levels, it activates inhibitory inputs onto pathways that carry information from other sensory cells with close-by receptive fields.

The somatosensory system is one of several sensory systems that utilize lateral inhibition. Pinpointing the site of stimulation of the skin by employing lateral inhibition is one of the kinds of distortion of the nature of stimulus that sensory systems utilize in the process of feature extraction. Contrast is enhanced if the signal from mildly activated receptors can be suppressed so that only the signal from the most active cells is relayed centrally. For instance, if the skin on a finger is depressed by the tip of a pencil, the center of a stimulated area and the surrounding less strongly stimulated area may be represented centrally as if only the center were getting stimulated. Thus the shades of gray are erased and only a clarified picture of the actual pattern of stimulation is received by the brain.

Somatosensory Projections to the Thalamus

Spinal cord circuits mediate **segmental reflexes** in which the reflex loop is completed within one spinal segment. These are especially important in control of posture (see Chapter 12). For coordinated activities of the body, sensory information must also be passed to higher centers. There are two basic routes for sensory information to reach the brain: (1) a direct system carrying fine touch, proprioceptive, and kinesthetic information; and (2) an indirect and less specific route involved in pain and temperature sensation.

Axons mediating proprioception (position and movement of body appendages) and fine touch reach higher centers by joining a specific, anatomically ordered route that is called the **dorsal column pathway** (or **lemniscal system**) (Figure 9-15, *A*). The axons do not cross the midline, but rather project ipsilaterally (same side as entry) in the dorsal column pathway up to the medulla. There, axons

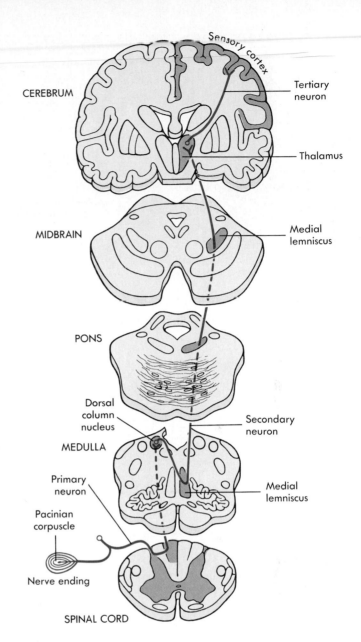

FIGURE 9-15

A The dorsal column (medial lemniscal) pathway is the most direct route for somatosensory information travelling to the cortex. Sensory axons enter the spinal cord (represented here by only one segment) and branches ascend ipsilaterally within the spinal cord to the dorsal column nucleus (nucleus gracilis) in the medulla. The postsynaptic fibers of secondary neurons decussate (cross) in the medulla and ascend through the midbrain into the somatosensory nuclei of the thalamus, where they synapse on tertiary neurons whose axons pass to the sensory cortex. The primary somatosensory cortex gets information only from the opposite side of the body. *(Figure continued on p. 236).*

of the postsynaptic neurons **decussate** (cross to the opposite side of the nervous system) in the upper brainstem and then ascend to synapse in a specific sensory area of the thalamus. From the thalamus, axons of relay interneurons extend to the primary somatosensory region of the cortex.

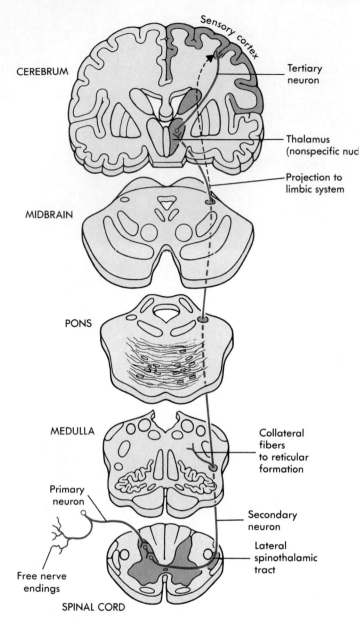

CEREBRUM

Sensory cortex

Tertiary
neuron

MIDBRAIN

Thalamus
(nonspecific nuclei)

Projection to
limbic system

PONS

MEDULLA

Collateral
fibers
to reticular
formation

Primary
neuron

Secondary
neuron

Lateral
spinothalamic
tract

Free nerve
endings

SPINAL CORD

FIGURE 9-15, cont'd

B The nonspecific (anterolateral) ascending pathways. Group III and IV axons synapse on interneurons that cross the cord in the segment of entry and ascend in different tracts from the specific projections. All Group IV axons project to nonspecific thalamic nuclei or to the reticular formation.

The receptors for sensations of heat, cold, and pain do not travel in the dorsal column system. Instead, they cross (decussate) in the spinal segment where they enter and ascend in the **ante-rolateral** tract (Figure 9-15, *B*). Their projections terminate within nonspecific nuclei in the thalamus; sensations from information that follows this pathway are not so easily localized on the body surface as the modalities that ascend in the dorsal column.

Point to ponder: Because dorsal column nuclei include ipsilateral projections that only cross the spinal cord after they reach the medulla, whereas the spinothalamic tract is formed by axons that cross at the level of entry, injuries that sever half of the spinal cord cause a loss of sensitivity to pain and temperature on the opposite side of the body, and a loss of proprioceptive input and fine touch from the same side of the body as the injury.

Somatosensory Maps

- *What is meant by somatotopic representation?*
- *How do specific sensory pathways and nonspecific pathways differ in function?*
- *What are the roles of the reticular activating system? What inputs affect it?*

Projections from the thalamus produce several maps of the body surface. For example, in the primary somatosensory cortex, the somatotopic representation of the body surface on the primary somatosensory cortex (Figure 9-16) is called the **somatosensory homunculus.** This map is exaggerated for those areas of the body surface that have the greatest density of receptors. Because the ascending axons that project to the sensory cortex decussate, the map in the left cortex represents the right side of the body and the map in the right cortex represents the left side of the body.

The parietal lobe (Figure 9-9) contains the primary somatosensory processing area. Lesions in this area cause sensory deficits. Adjacent brain areas are sensory integrating sites that synthesize an overall body image. A lesion in these areas will cause an individual to neglect the part of the body represented in that region, almost as if it did not exist. The neglected body part will be on the side opposite the brain lesion, due to the decussation of sensory projections.

Body surface receptors produce two maps of the body on the cerebellar surface (Figure 9-17). The cerebellar maps are not restricted solely to a somatosensory representation of the body. For instance, the head region also receives information from the auditory, visual, and vestibular systems. Information coming from body receptors overlaps in the cerebellar map with descending information from corresponding areas of the sensory and motor regions of the cortex. The convergence of this information in the cerebellum allows the intended position in space to be compared with the actual position. This convergence is important for the continuous modulation of movement commands, which are necessary to achieve coordination in difficult circumstances, such as walking over uneven ground.

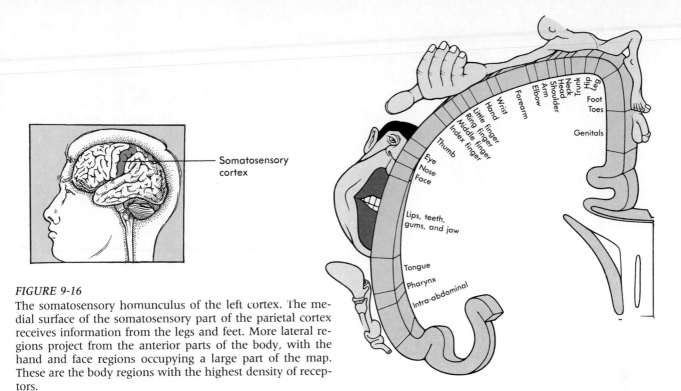

FIGURE 9-16

The somatosensory homunculus of the left cortex. The medial surface of the somatosensory part of the parietal cortex receives information from the legs and feet. More lateral regions project from the anterior parts of the body, with the hand and face regions occupying a large part of the map. These are the body regions with the highest density of receptors.

Two-Point Discrimination

One measure of sensitivity of the somatosensory system is the **two-point discrimination test.** This is performed by stimulating the skin with a divider and measuring the minimum separation that can be perceived by the subject as two separate points. One determinant of the two-point resolution limit is the relative density of receptors. More afferents come from the fingertips than from regions where tactile discrimination is poor, so the ability to resolve nearby stimuli ranges from 1 to 2 mm on the face and fingers to as much as 40 mm on the back (Figure 9-18). Another determinant of discrimination is the degree of **convergence** of afferent pathways. Convergence results when a central neuron receives inputs from all the receptors in a given area of skin. It will be difficult, if not impossible, for the subject to detect two points of stimulation within that region. Thus if the level of convergence is high for a given body region, two-point resolution will be correspondingly poor. In the somatotopic map, the region represented by convergent inputs will have a very small representation, whereas the region with many direct (non-converging) inputs will have a large representation on the map.

The Reticular Activating System

The **reticular activating system** is part of the reticular formation, which is located in the brainstem and thalamus (Figure 9-19). The activating system controls the level of consciousness and the transitions between sleep and the different waking states (revealed by the electroencephalogram: see below). Behavioral arousal is only one of the important functions of the reticular formation, which also controls muscle tone, cardiovascular and respiratory function, and pain threshold. It receives inputs from all sensory modalities and also a share of the activity that arises in the cortex. Inputs from cutaneous sensory receptors or from muscle and joint receptors increase alertness through reticular projections to almost all areas of the diencephalon and cortex. Pain or proprioceptive inputs that signify a change in body position (such as falling) require immediate responses and significantly alter the output of the reticular activating system. This is why shaking is an effective means of waking someone.

In addition to its outputs to higher brain regions, the reticular system projects into the spinal cord, where its inputs modulate the strength of reflex responses. These inputs contribute to the difference between the size of the knee jerk response exhibited by a comfortably seated person deep in a problem of mental arithmetic and the response of a person terrified that his precarious seating situation is going to collapse momentarily.

LEARNING AND MEMORY
Distinctions Between Learning and Memory

Learning is the process whereby experience modifies centrally controlled behavior. **Memory** is the storage and retrieval of experience. The two concepts are related in that any relatively long-lasting

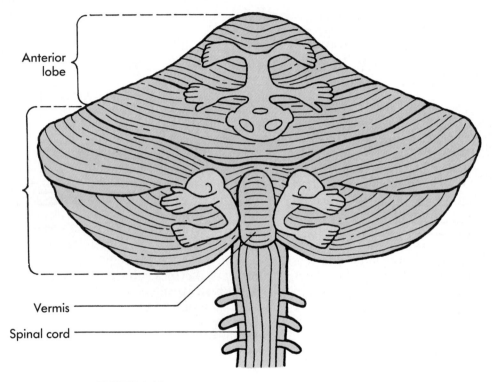

Anterior
lobe

Vermis

Spinal cord

FIGURE 9-17
Maps of the body are found in two regions of the cerebellum, where somatotopic representations are supplemented by inputs from the auditory, vestibular and visual systems.

FIGURE 9-18

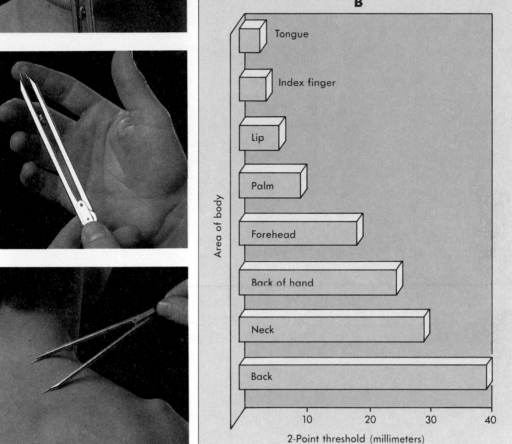

A Two-point discrimination can be demonstrated by touching a person's skin with the two points of a compass. When the two points are close together, the individual perceives only one point. When the points of the compass are opened wider, the person becomes aware of two points.

B A bar graph showing typical two-point discrimination thresholds for different parts of the body. Below the threshold, two points are perceived as one. Above the threshold, two separate points can be discriminated. Compare the size of the thresholds with the relative size of the sensory cortex for each region as shown in Figure 9-16.

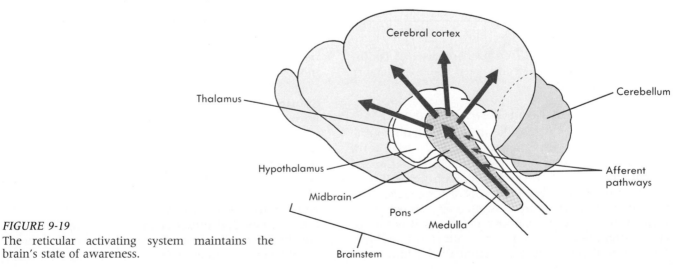

FIGURE 9-19
The reticular activating system maintains the brain's state of awareness.

Jet-lag and Shift-work: The Human Biological Clock

The body has daily (circadian) rhythms. Although these rhythms seem to be coupled to the dark-light cycle, they persist in the absence of environmental cues, as has been demonstrated when experimental subjects have lived isolated in deep bunkers or caves, far from cues about natural changes in light or the activity patterns of society. In the test subjects, who are free to set their own activity and sleep-wake cycles, the rhythms continue, but typically drift from a 24-hour cycle to a cycle of approximately 25 hours. In addition to the sleep-wakefulness cycle, the cycle of body temperature varies a degree or two over the daily cycle, peaking in late afternoon and dropping to its lowest point in the early morning. Pulse rate and blood pressure also peak around the same time as body temperature. Many hormones exhibit a circadian rhythm of secretion. Important life events are also controlled by circadian rhythms. Women go into labor most often between 1:30 and 2:30 AM and least frequently about midday. Heart attacks are twice as likely to occur between 8 and 10 AM as in the early morning or the evening.

What controls these rhythms? Neurons in the **superchiasmatic nucleus** of the hypothalamus are an important part of the circadian clock. These neurons maintain a daily rhythm of activity and keep the internal rhythm coordinated with the external world by adjusting their pattern to correspond to cues related to the light cycle. Much of the control over body rhythms exerted by the superchiasmatic nucleus is achieved by neural connections to critical brain regions. Coordination with internal clock patterns in other tissues may be mediated by the hormone melatonin, whose circadian pattern of secretion from the pineal gland (Figure 9-6) is driven by the superchiasmatic nucleus.

The human circadian rhythm is disrupted in people who work night shifts or experience "jet lag" after they rapidly cross time zones. The attempt to be wakeful and function normally during the period of the day when the body is in its "rest" phase results in impaired concentration and poor judgment and may sometimes produce mild depression. Enterprises that are open 24 hours a day, such as hospitals

change in the response to a particular set of conditions requires storage and retrieval of sensory events or ideas.

Short-Term and Long-Term Memory

Memory operates in at least two different time scales: (1) short-term memory and (2) long-term memory. Short-term memory stores information for seconds to minutes. For example, looking up a telephone number and remembering it long enough to dial correctly uses short-term memory. Information resides in short-term memory as long as it is the focus of immediate attention. In contrast, a telephone number that has been transferred to the long-term memory store may be retrieved hours, days, months, or even years later. Long-term memory involves relatively permanent changes in the brain. Shifting events and ideas from short-term to long-term memory is **memory consolidation.**

Long-term memory is not disrupted by temporarily blocking the conduction of nerve impulses, so it cannot be dependent upon reverberating neural circuits. (Such circuits underlie the repetitions of a phone number that keep it in mind long enough for you to find a pencil and write it down). Neuroscientists have evidence that changes in synaptic efficacy are the basis of long-term memory (see chapter 8). A relatively permanent change in synaptic connec-

and supermarkets, often require that people work in one of three 8-hour shifts. Individuals who rotate between these shifts work the night shift at a time of lowered mental alertness. Human error is responsible for the majority of accidents on the job, and there is a much higher incidence of accidents at night, when chemical plants and nuclear power stations are maintained by individuals who are not as alert as their counterparts on the day shift.

People working a temporary night shift not only are attempting to function at a time when their internal clock reduces their mental efficiency but also they have difficulty getting sufficient sleep in their off hours. When night shift workers go to bed in the morning, they may fall asleep rapidly but are prematurely awakened by "get active" aspects of their circadian rhythm. The ability of people permanently on night shift duty to shift their circadian rhythm varies among individuals, but a complete shift is rarely achieved.

Jet lag results from moving to a different time zone. It is more severe when people travel from west to east, which shortens the day, than from east to west, which lengthens the day. This may be because the body's clock is reset by "light on" signals and the human clock's duration is closer to 25 hours than 24. Travelers take from days to weeks to adjust to the different time zone, and the different parts of their rhythm may become desynchronized in the process.

Secretion of the hormone melatonin is thought by some to play a part in coordinating the biological clock, so there is an interest in studying whether melatonin will alleviate jet lag. Women who suffer from premenstrual syndrome appear to have an abnormal pattern of melatonin secretion in the days before menstruation. This may be treatable with exposure to bright lights in the evening, which act on the superchiasmatic nucleus to lengthen the period of melatonin release and reduce the premenstrual symptoms. Melatonin secretion patterns are also being investigated in the hope of treating a type of depression known as seasonal affective disorder.

tions will change the likelihood that information will flow through a particular pathway. Consolidation of short-term into long-term memory can be blocked by agents that inhibit protein synthesis, but the sites that require protein synthesis are widespread, because relatively extensive cortical damage does not selectively remove memories.

Regions of the temporal lobes, the hippocampus, and the **amygdala** (part of the limbic system) contribute to memory consolidation because lesions in these regions affect the ability to process recent events into long-term memories. Such lesions do not affect previously stored long-term memories but they prevent formation of new long-term memories.

SPECIALIZATIONS OF THE HUMAN BRAIN
Hemispheric Specializations

- *What functions does the dominant hemisphere dominate?*
- *How does the metabolic activity of the brain in sleep differ from its activity in a comatose state?*
- *How does slow-wave sleep differ from REM sleep?*

The two hemispheres of the human brain are roughly symmetrical, but different functions can be identified with the right and left hemispheres. The best studied hemispheric specialization is language.

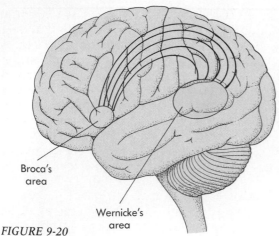

Broca's
area

Wernicke's
area

FIGURE 9-20

Lateral surface of the left hemisphere, showing the location of cortical areas specialized for language and the major interconnecting tract.

The hemisphere in which language ability develops is called the **dominant hemisphere.** Cerebral dominance is not the same as handedness because the left side of the cortex is dominant in about 95% of all right-handed people and 70% of all left-handed people.

There are two major language areas in the dominant hemisphere: (1) **Wernicke's area** and (2) **Broca's area** (Figure 9-20). Wernicke's area, located between the primary auditory and visual areas, is important for interpretation of language and formulation of thoughts into speech. Broca's area, found near that part of the motor cortex that controls the face, is responsible for generation of the patterns of motor output that result in meaningful speech. Broca's area is connected with Wernicke's area by a thick bundle of nerve axons. Regions of the cortex surrounding these two areas are also important for language. Development of these areas of the human brain is greatly influenced by early language experience. Recent studies have revealed that the precise locations of the language areas differ between individuals and also between the sexes.

Evidence for Language Specialization

One type of evidence of hemispheric specialization is provided by studies of individuals who suffer strokes that affect their ability to communicate. Damage in the vicinity of Broca's area leads to a condition called **expressive aphasia,** in which concepts are communicated accurately, but speech does not flow smoothly and consists of a series of logically related nouns, verbs, or short phrases with the interconnecting parts of speech omitted. Similar defects occur in writing. Individuals with expressive aphasia function normally in other tasks that are of comparable complexity to language construction, such as singing melodies or drawing. They comprehend verbal messages, which suggests that auditory

information is decoded in Wernicke's area before being relayed to Broca's area.

Damage to Wernicke's area is more serious because it impairs the ability to express and/or comprehend language, either written or verbal. The patients exhibit **receptive aphasia.** The speech of receptive aphasics has been described as "word salad." They may produce words fluently and generate grammatically correct sentences, but these are illogical and unrelated to actual events or intentions.

If the neural connection between Broca's area and Wernicke's area is disrupted, individuals can understand words, but they cannot repeat what has just been said. If there are lesions in the pathways connecting Wernicke's area with the primary auditory cortex, comprehension is impaired only for spoken language, and speaking, reading, and writing remain unaffected.

Split-Brain Individuals

An important source of information about hemispheric specialization was provided as a result of surgeries in which the corpus callosum was severed. The corpus callosum is a large bundle of axons that connect the two hemispheres. The surgeries were performed in hopes of limiting the spread of epileptic seizures from one hemisphere to the other. In some cases, the desired goal was achieved by the surgery, but another consequence was that it produced patients with a **split-brain** condition. Under special testing conditions, the different response capabilities of the two hemispheres can be revealed. The left cortex receives sensory input from the right half of the body, including the right visual field. A picture viewed by only the right eye could be correctly described because the information would project to the left cortex, which contains the language centers. Presenting the picture to the other eye would result in no apparent recognition, or the denial of any object. Yet if the picture had an emotional content, presenting it to the nonverbal hemisphere may result in denial but the test subject may smile or reveal by other body language that the picture was interpreted at some level.

Specializations of the Right Hemisphere

The right (nonverbal) cortex is specialized to recognize individuals by their facial features alone. Trauma to the ventral portion of the occipital lobe anterior to the primary visual cortex in the nondominant hemisphere eliminates the capacity to recall faces. Reading, writing, and oral comprehension remain normal, and patients with this disability can still recognize acquaintances by their voices. Trauma to other parts of the right hemisphere may lead to an inability to appreciate spatial relationships and may impair musical activities such as singing, but does not usually result in aphasia. The

NERVE AND MUSCLE

left hemisphere is required for consolidation of verbal memories whereas the right hemisphere is important for remembering nonverbal experiences.

All of these observations suggest that the two hemispheres handle information differently. The dominant hemisphere is adept at sequential processing, such as that needed to formulate a sentence. The nondominant hemisphere is adept at spatial reasoning, such as that needed to assemble a puzzle or draw a picture.

Cortical Regions Involved in Memory, Emotion, and Self-Image

Each cortical lobe is associated with functions that contribute importantly to personality. Lesions in the frontal lobe eliminate rage and decrease anxiety. Damage to the frontal lobe also impairs the ability to plan, to set priorities, or to anticipate the consequences of actions.

During brain surgeries performed to relieve epilepsy, the patient can remain conscious under local anesthesia because the brain itself is insensitive to pain. In fact, the cooperation of the patient is important, because frequently it is necessary to map the cortex to determine exactly what functions will be disrupted by the surgery. In the course of many such surgeries, the American scientist Wilder Penfield explored most of the cortical surface with stimulating electrodes. He found that stimulating the primary visual, auditory, and somatosensory areas did not elicit well-defined sensory experiences, but only rudimentary hallucinations such as flashes of color, simple sounds, or vague contacts. On the other hand, stimulating the temporal lobe often evoked coherent, detailed memories or fantasies of sensory experiences along with associated emotions. Initially, these experiments were interpreted as indicating that the temporal lobe had access to a large store of memories of past experiences, even those that had been "forgotten." More recent evidence suggests that in such experiments the temporal lobe stimulation spreads to the limbic system, a portion of the brain known to be involved with both memory and emotion. However, patients with damage to the right temporal lobe do have deficits in processing spatial and visual information and in remembering such information.

SLEEP AND THE WAKEFUL STATE
Sleep as an Active Function

During the waking hours of the daily cycle, the pattern of output from the reticular formation stimulates the cortex and increases the ability of the brain to focus on information provided by specific sensory pathways.

An early hypothesis about sleep held that falling asleep involved a decrease in activation of the cortex by the reticular formation. Sleep would then involve a transient loss of consciousness resembling the comatose state that follows severe head injury. It is now known that the reticular activating system controls both sleep and the waking state. Sleep is an active multistate process that involves output from the reticular activating system, which is driven by inputs from many regions of the brain. Although the pattern of brain activity during sleep differs from that of the waking brain, overall the brain consumes as much energy during the sleep states as during the waking states. In contrast, energy use by the brain is depressed in comatose persons.

Sleep States and Waking States

Electrical activity of the cortex is measured in an **electroencephalogram (EEG)** (Fig. 9-21). An EEG represents the external current generated by neuronal activity in the cortex. The more synchronized this activity is, the larger the extracellularly recorded signal.

Three criteria define the stages of sleep: (1) the EEG pattern (2) the ease with which the sleeping individual can be aroused, and (3) muscular tone. The first change seen in the EEG with the onset of drowsiness is slowing of the frequency and reduction in the overall amplitude of the brain waves (Figure 9-22, A).

In a relaxed individual whose eyes are shut, the pattern of activity in the EEG consists of large slow waves that occur at a frequency of 8 to 13 Hz; these are **alpha waves** (Figure 9-22). Alpha waves are most easily recorded from the occipital lobe. In an alert subject whose eyes are open, faster, lower amplitude (18 to 20 Hz) **beta waves** are recorded. These beta waves in an alert individual are less synchronized than alpha waves because multiple sensory inputs are being received, processed, and translated into motor activities.

The sleep states consist of four stages of slow wave sleep and the dream state, characterized by rapid eye movements (REM). In the first stage of slow wave sleep, **theta waves** appear (Figure 9-21). In the second stage, bursts of high-frequency activity known as sleep spindles occur, along with occasional large-amplitude, slow **delta waves.** Delta waves become more frequent in the third stage of sleep. In the fourth stage, **deep sleep,** the EEG is dominated by delta waves. Slow-wave sleep involves a decrease in arousability, skeletal muscle tone, heart rate, blood pressure, and respiratory frequency.

During the REM phase of sleep, the EEG resembles that seen in a relaxed, awake individual. The heart rate, blood pressure, and respiration are all increased. Although eye movements occur, skeletal muscle tone is decreased. Individuals in REM sleep are difficult to arouse but are more likely to awaken spontaneously than those in the deeper phases of slow-wave sleep. Experimental studies in which subjects were awakened during various stages dem-

A

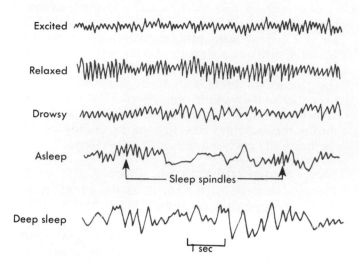

Excited

Relaxed

Drowsy

Asleep
↳ Sleep spindles ↰

Deep sleep

⊢ 1 sec ⊣

FIGURE 9-21
A Changes in the EEG in a relaxed individual, a drowsy individual, and an individual in successively deeper stages of sleep.
B The sequence of sleep cycles in a typical night. Each cycle is terminated by an episode of dreaming in REM sleep *(shading)*. Spontaneous waking can occur most readily from the REM sleep stage.
C The overall amount of sleeping time *(solid curve)*, the proportion of this that is REM sleep *(boxes)*, and the amount of time spent in slow-wave sleep *(broken curve)* change with age.

B

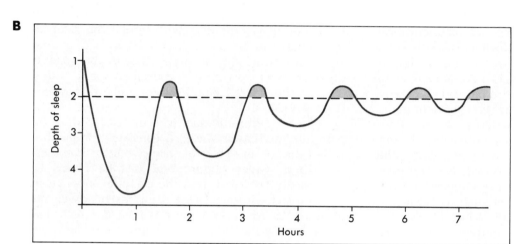

C

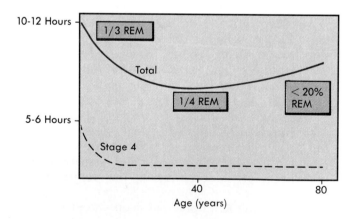

10-12 Hours

1/3 REM

Total

1/4 REM

< 20% REM

5-6 Hours

Stage 4

40   80
Age (years)

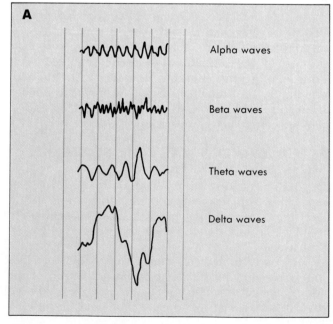

A

Alpha waves

Beta waves

Theta waves

Delta waves

FIGURE 9-22
A The EEG is a reflection of the type of activity present in the cortex. There are four basic rhythms of the brain: alpha, beta, theta, and delta waves. *Continued.*

The Ear Can Learn from the Eye

In several places in the brain, topographic maps of space are formed by overlapping projections from two or more sensory systems. For instance, pit vipers, snakes that can locate prey on the basis of infrared radiation from warm bodies, have maps of the world in which inputs from visual and pit organ receptors provide information from the same point in space. Thus a mouse can be recognized on the basis of both its appearance and its pattern of heat radiation above background level.

Sounds are located in space by comparing the inputs from the two ears. This information is mapped in a separate place in the brain from the tonotopic map that functions in recognition and comparison of sounds of different frequencies. The formation of this auditory maps has not been extensively studied in human beings, but auditory maps

of space have been studied in other species, such as barn owls. These birds, which are nocturnal predators, are able to catch prey on the basis of a single rustling sound, and blind owls perform this task as efficiently as ones with normal vision.

Studies in which young owls had one ear plugged for a period of time during development have revealed that visual and auditory information are used together in the formation of the map. The auditory inputs from the owls' plugged ears were greatly reduced, but the owls still learned to locate prey by comparing the sounds arriving at their two ears. If the plug was then removed, tests performed in the dark revealed that the owls misguessed the location of the prey in a predictable way: they assumed that a sound originating close to the midline was closer to the side with the formerly plugged

ear, because they were used to receiving little information from that ear unless the sound was very close to that side of the head.

If allowed visual experience, the young owls could adjust to the removal of the plug and again learn to accurately locate auditory stimuli in space. Only through visually locating the source of a sound in space and associating it with the pattern of sounds received from the two ears can the owls learn to properly interpret the auditory information. Thus comparison of sights with sounds is necessary for the formation of an auditory map of space. Although experience can influence the strength of synaptic connections in young owls, this adaptability is lost after a critical period, and owls that did not have the ear plug removed until they were adults never learned to locate prey accurately in the dark.

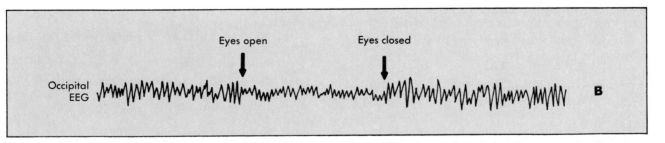

FIGURE 9-22, cont'd
B Recordings from the occipital cortex change when a relaxed subject with eyes closed is asked to open his or her eyes.

onstrated that dreaming occurs during REM stage sleep. The rapid eye movements that occur during this stage are similar to the tracking movements made by the eyes during waking, suggesting that dreamers "watch" their dreams.

The Sleep Cycle

Figure 9-21, *B* shows that when an individual starts to fall asleep the initial progression through successively deeper levels of slow-wave sleep (stages 1 to 4) typically requires a half-hour or so. After a relatively brief period in stage 4 sleep the process reverses, only to be interrupted by an abrupt transition to REM sleep. After a variable amount of time the REM phase ends, and there is a return to progressively deeper levels of slow-wave sleep. This cycle is repeated several times and is terminated by spontaneous awakening, usually during the final period of REM sleep.

There are changes in the sleep cycle as the night progresses. Slow-wave sleep is reached very early, with more time being spent in stages 2 and 3 as morning approaches. In contrast, REM duration increases from only a few minutes initially to 30 to 45 minutes as successive REM cycles are entered throughout the night.

The human sleep pattern is also age-dependent (Figure 9-21, *C*). The total amount of sleep time and the amount of stage 4 sleep decrease to a minimum in middle age. The relative amount of REM sleep decreases from 50% in infants to 25% in middle age and to less than 20% in the elderly. Thus as age advances, more time is spent in transitional stages of sleep and less in REM sleep and slow-wave sleep.

● *STUDY QUESTIONS*

1. Students found the two-point threshold to be 1 mm on the lip and 5 mm on the shoulder. What predictions can they make about each of the following:
   a. The relative size of the lip and shoulder representations in the somatosensory cortex
   b. The relative sizes of the receptive fields on the lip and shoulder
   c. The density of receptors on the lip and shoulder

2. What is the role of the reticular activating system during wakefulness? During sleep? What evidence indicates that sleep is an active state of the brain?

3. What experimental evidence suggests that memories are distributed over wide areas in the brain?

4. A patient speaks very slowly and with long pauses between words, but can sing quite well. What neurological deficit might he have?

5. What is a mixed nerve? Give some specific examples.

6. Which kinds of sensory information are carried by myelinated axons and which kinds by unmyelinated axons?

7. What structures are present in white matter? How are gray matter and white matter distributed in the brain? How does the brain differ from the spinal cord in this regard?

8. What are the anatomical names of the two main ascending pathways for sensory information in the spinal cord? What types of information does each characteristically carry?

9. Name the types of cutaneous receptors. Which of these are fast-adapting and which are slowly adapting receptors?

Choose the MOST CORRECT Answer

10. What component of the limbic system is important in the formation and recall of memory?
    a. Thalamus
    b. Medulla
    c. Hypothalamus
    d. Hippocampus

11. The hypothalamus:
    a. Controls body temperature
    b. Directs secretions of the pituitary gland
    c. Influences respiration and heartbeat
    d. All of the above

12. Which of the following is NOT part of the brainstem?
    a. Basal ganglia
    b. Medulla
    c. Pons
    d. Red nucleus

13. The primary somatosensory cortex in the human brain is found in the:
    a. Frontal lobe   c. Temporal lobe
    b. Parietal lobe   d. Occipital lobe

14. Group IV sensory fibers:
    a. Conduct an impulse faster than the other classes of sensory fibers.
    b. Conduct impulses for touch and fast pain.
    c. Are unmyelinated.
    d. Are the largest diameter fibers.

15. These somatosensory receptors sense vibration:
    a. Ruffini endings
    b. Touch dome endings
    c. Pacinian corpuscles
    d. Free nerve endings

16. Choose the FALSE statement:
    a. If convergence is high for a body region the two-point resolution will be more acute.
    b. Two-point resolution is proportional to the density of receptors.
    c. Two-point resolution is more acute in the fingers than on the back.
    d. The two-point discrimination test measures sensitivity of the somatosensory system.

17. The level of consciousness and the transitions between sleep and different waking states are controlled by the:
    a. Limbic system
    b. Reticular activating system
    c. Basal ganglia
    d. Association cortex

18. If after a stroke, an individual can understand words but cannot repeat what has been said, damage has probably occurred in:
    a. Broca's area
    b. Wernicke's area
    c. The connection between Broca's area and Wernicke's area
    d. The reticular formation

SUMMARY

1. The nervous system is divisible into a **central** component, consisting of the brain and spinal cord, and a **peripheral** component, consisting of nerves, dorsal root ganglia, and autonomic ganglia. The brain is divisible into the **brainstem** (made up of the pons, medulla, and midbrain); the cerebellum; the **diencephalon** (made up of the thalamus and hypothalamus); and **telencephalon** (cerebral hemispheres). In the CNS, regions containing cell bodies are **gray matter**; those consisting of axons are **white matter**.

2. Sensory pathways commonly **diverge** to carry signals to more than one brain region for **parallel processing**. Central processing areas **filter** the incoming signals, **extracting** novel or important **features** for further analysis. Central processing areas are usually organized **topically** (somatotopic, tonotopic, and so on), so that receptors from a spot on the body surface project to a corresponding spot in the relevant brain area.

3. Somatosensory receptors are morphologically specialized for different submodalities of touch, vibration, temperature, and pain. **Ruffini endings** and **touch dome** endings mediate touch, and **Pacinian corpuscles** sense vibration. Temperature and pain are sensed by free nerve endings.

4. The **dorsal columns** of the spinal cord carry sensory pathways to specific cortical processing regions. The nonspecific pathways exercise a general arousal effect on the brain by way of the reticular activating system.

5. **Learning** occurs when sensory inputs result in altered behavior. Information may be stored in **verbal** and **nonverbal** forms, and it is not possible to identify a single discrete brain region in which memory stores are located. **Short-term memory** depends on electrical activity in neuronal circuits.

6. In the course of development, the **dominant hemisphere** specializes in control of voluntary motor activities and language. Usually the left hemisphere becomes dominant. The **nondominant hemisphere** specializes in nonverbal intellectual functions. Damage to language regions of the dominant hemisphere results in **aphasias** (language disabilities) and impairments of verbal memory. Damage to corresponding regions of the nondominant hemisphere results in impairment of spatial and nonverbal functions.

7. The waking states and the sleep states represent activity patterns of the brain. Waking states are characterized by the EEG as alpha, beta, theta, and delta waves. Sleep can be divided into **slow-wave sleep** and **REM sleep** on the basis of differences in the EEG, muscle tone, and arousability.

● SUGGESTED READING

BLACK IB, et al: Biochemistry of information storage in the nervous system, *Science* 236:1263, June 5, 1987. Discusses short- and long-term changes in synaptic efficacy and the appearance of specific intracellular proteins correlated with learning and memory in invertebrate model systems.

DICHTER MA: Cellular mechanisms of epilepsy: a status report, *Science* 237:137, July 10, 1987. Good explanation of the factors that may act to increase the excitability of neurons. Includes a description of the mechanisms of action of antiepileptic drugs.

EICHENBAUM H, OTTO P: The hippocampus: what does it do? *J Behav Neural Biology* 57:2-36, 1992. The involvement of the hippocampus in declarative memory begins with generation of the theta waves characteristic of an attentive state. This time-locks multiple channels of incoming sensory input, and maximally excites the activated cells in a pattern that can be recreated during the process of memory retrieval.

GORMAN C: Sizing up the sexes, *Time Magazine* January 20, 1992, p. 42. Scientists are discovering that gender differences have as much to do with the brain as with the way we are raised.

GIBBONS A: The anatomy of autism, *Discover* 10(1):54, 1989. Discusses possible CNS mechanisms related to the autistic child.

FINDLAY S, BROWNLEE S: The delicate dance of body and mind, *U.S. News & World Report*, 109(1):54, July 2, 1990. Discusses the current state of mind-body research including immune system research and theories of disease causation.

MISHKIN M, APPENZELLER T: The anatomy of memory, *Scientific American*, 260:80, June 1987. Discusses how memories may be stored in the CNS and the role of different regions of the brain in memory storage and recall.

THOMAS P: Pain syndrome begs for attention, *Medical World News* 31(15):18, September 1990. Discusses intractable pain care and treatment.

WINSON J: The meaning of dreams, *Scientific American*, 263(5):86, November 1990. Discusses the brain mechanisms in dreaming.

How Drugs Affect the Nervous System

This focus unit will cover the major types of drugs that are abused: drugs in some way harmful, but that individuals nevertheless consume. Since this is a physiology text and most abused substances affect the brain, we will limit ourselves to trying to understand the mechanism of action of abused drugs on the brain.

DRUGS AND DRUG ABUSE

What is a drug? Any compound not normally present in the body or in a usual diet that causes a demonstrable change in body function can be considered a drug. To this we might add compounds that the body manufactures or that are ordinarily consumed, if they are taken in amounts large enough to be harmful. Examples are consumption of testosterone by body-builders and the ingestion of megadoses of vitamins. Many prescription drugs, including tranquilizers and sedatives, are used compulsively in the absence of a medical justification. We are also besieged with a variety of over-the-counter medications aimed at relieving headaches, insomnia, fatigue, tension, heartburn, constipation, and about every ill known to man.

Extent of Drug Use

With only 5% of the world's population, Americans consume 50% to 60% of the world's legal and illegal drugs. A major inner-city crime laboratory will typically handle more than 50,000 drug cases each year, with 60% to 80% cocaine-related at the present time. Almost 66 million people have tried marijuana at least once, and some 20 million have tried cocaine. There are about 1 million heroin and 2.9 million cocaine addicts in the United States, and over 50,000 drug-related emergency room admissions. In inner-city hospitals some 10% to 15% of newborn babies test positive for cocaine as a result of prenatal drug abuse by their mothers. The cost of drug-related productivity losses and increased medical expenses to American business is estimated at $7000 per drug-abusing employee.

More and more medicines are now approved for over-the-counter (OTC) sales, including antacids, sleep aids, pain relievers, cold and allergy medica-

tions, laxatives, stimulants, and a variety of "cosmetic" products. The OTC market is enormous, with annual sales in excess of $20 billion, comparable to the level of prescription sales. In one study, 75% of all individuals reported experiencing "symptoms" of some illness on the average of once a month. Of these, two of three seek relief in OTC drugs. Americans spend an estimated $110 million each year on diet aids alone. **Phenylpropanolamine** (PPA), found in most diet drugs, is one of the less-potent members of the amphetamine family. It is commonly agreed among medical experts that one can become addicted to many over-the-counter medications. Indeed, supposedly harmless OTC drugs can prove far more debilitating than the ailment for which they were taken.

Addiction

Addiction refers to a behavioral pattern of drug use characterized by the compulsive use of a drug, preoccupation with its supply, and a high tendency to relapse. Addiction can be psychological or physical or a combination of both. The common thread is that the individual's whole life becomes focused on the drug, despite adverse health effects, and the possible loss of family, friends, and jobs. With repeated drug use, individuals may become **tolerant**, with repeated doses producing less of an effect. Physical **dependence** may also develop. If so, removal of the drug produces **withdrawal symptoms.** Tolerance can be due to the induction of enzymes that metabolize the drug, or a decrease in the response of the drug's target tissue or both. The decrease in response is frequently due to down-regulation of receptors for the drug.

Behaviors associated with a reward tend to be repeated (**positive reinforcement**). The euphoric effects of drugs are examples of positive reinforcement. A model of reinforcement-driven drug addiction is seen in brain stimulation reward studies. In these studies animals are able to control the delivery of low levels of electrical current to specific areas of the limbic system of the brain thought to be "pleasure centers." Opiates and stimulants like the amphetamines and PCP reduce the ani-

TABLE 9-A	Functional Classes of Drugs Affecting the CNS

Classification	Examples
Stimulants	Amphetamine
	Methamphetamine
	Cocaine
	Phenylcyclidine
	Ephedrine
	Phenylpropanolamine
	Nicotine
	Caffeine
General depressants	Alcohol
Tranquilizers	
Antianxiety	Diazepam (Valium)
	Librium
Antipsychotics	Chlorpromazine
	Thorazine
	Haloperidol
Sedative-Hypnotics	
Sedatives	Phenobarbital
	Secobarbital
Hypnotics	Flurazepam (Dalmane)
	Temazepam (Restoril)
	Triazolam (Halcion)
Narcotics	Morphine
	Codeine
	Heroin
	Methadone
	Phenothiazines

mal's rate of self-stimulation. Low doses of alcohol and other CNS depressants increase the rate of self-stimulation.

Drugs acting on the CNS fall into five major functional classes: (1) **stimulants,** (2) **tranquilizers,** (3) **sedative-hypnotics,** (4) **narcotics,** and (5) **antidepressants** (Table 9A). While all of these classes have proven medical utility, many of the CNS-active drugs in the first four categories can be abused. Most CNS-active drugs are either **agonists,** acting like normal neurotransmitters; or **antagonists,** blocking presynaptic or postsynaptic receptors for the various neurotransmitters.

CNS Stimulants

The prototype CNS stimulant is **amphetamine** and its many derivatives. Stimulants are widely used to increase the level of alertness and as dieting aids. Stimulants tend to increase the level of catecholamine-driven activity in the brain by interfering with the re-uptake of dopamine and norepinephrine and by acting as an agonist on particular subclasses of catecholamine receptors (D_2 receptors). Activation of these receptors opens Ca^{++} channels in the postsynaptic cell membrane. Amphetamines are also widely used in the management of hyperactive

children, in whom they have a paradoxical calming effect.

CNS Depressants

Drugs that depress the function of the CNS are the most widely used pharmacological agents in medicine. **Nonselective depressants** include alcohol and the general anesthetics used in surgery. Both are lipid-soluble and readily cross the blood-brain barrier. Selective depressants include tranquilizers and the sedative-hypnotics.

Tranquilizers

Tranquilizers are subdivided into **antianxiety** drugs (minor tranquilizers) and **antipsychotics** (major tranquilizers). The most widely used antianxiety drugs are the **benzodiazepines** such as diazepam (Valium). Benzodiazepines bind to membrane receptors that are normally occupied by the neurotransmitter GABA. This receptor is coupled to a Cl^- channel that inhibits neuronal excitability. The GABA receptor has separate binding sites for GABA, benzodiazepines, and the sedatives and convulsants.

The antipsychotic drugs are widely used to treat patients with schizophrenia and are also called **neuroleptics.** They include chlorpromazine, thorazine, and haloperidol. The antipsychotics inhibit a subclass of catecholamine receptors, labelled D_1, that when activated open K^+ channels. The antipsychotics also block serotonin re-uptake.

Sedative-Hypnotics

The term **sedative** is used to describe drugs that mainly relieve anxiety and produce relaxation without causing sleep. Drugs that primarily induce sleep are called **hypnotics.** In practice there are often overlapping effects, so it is convenient to consider them as a single category. The sedative-hypnotics depress the activity of all excitable tissue, particularly nerve cells. In large doses these drugs suppress brainstem autonomic control centers. The **barbiturates,** including phenobarbital and pentobarbital, are the oldest hypnotics. Fast-acting barbiturates include pentobarbital (Nembutal) and secobarbital (Seconol). Phenobarbital is a long-acting barbiturate. Non-barbituate sedative hypnotics include many benzodiazepines. Barbiturates activate GABA receptors, but they do so at a different binding site than the benzodiazepines.

While barbiturate use by physicians has declined, the benzodiazepines are probably the most overprescribed group of drugs in use today. Some benzodiazepines, like diazepam (Valium), were developed to treat anxiety, while others are sleeping medications, like flurazepam (Dalmane), temazepam (Restoril), and triazolam (Halcion). The main difference is in the duration of their effect. The benzodiazepines all have the potential to disrupt short-term memory and occasionally to produce a period of amnesia lasting up to a day after a dose is taken. Benzodiazepines are considered "safe" in the sense that, when used alone, it is difficult to overdose on them. However, the nature of the symptoms of these conditions and the fact that the drugs act directly on the brain can lead to abuse and addiction. In combination with alcohol, librium and valium actually lead heroin in overdose-related deaths.

Narcotics

Narcotics include **opium** and all its derivatives—**morphine, codeine,** and **heroin**—as well as **methadone, phenothiazines** such as chlorpromazine (also used as an antipsychotic), and nalaxone. All narcotics interact to some degree with receptors for the endorphins and enkaphalins, the brain's **endogenous opiates.** Opiate binding sites are concentrated in areas controlling pain perception and in areas involved in pleasure and reward.

LEGAL DRUGS
Alcohol

Excessive alcohol consumption produces temporary intoxication, an emotional state characterized by behavioral and cognitive changes. As blood ethanol concentration increases from 5 to 50 mM (25 to 250 mg/dl), there is a progressive impairment of motor coordination and mental ability. As blood ethanol increases from 50 to 100 mM (250 to 500 mg/dl), there is progressive CNS depression ranging from sedation to stupor and coma. Alcohol is lipid soluble and mixes with the normal lipid and protein constituents of the cell membrane, altering membrane fluidity. This may explain the general anesthetic properties of alcohol.

Glutamate binding to the N-methyl-D-aspartate (NMDA) receptor activates Ca^{++} channels that have a significant role in learning and memory (see the boxed essay, p. 212). There is evidence that alcohol concentrations that produce acute intoxication selectively inhibit the NMDA-activated Ca^{++} channel. This would explain the fact that moderate levels of alcohol can impair learning and memory. Because the inhibition of NMDA receptors at high ethanol levels is not much greater than at lower levels, the general anesthetic effects of ethanol do not appear to result from an inhibition of NMDA receptors. However, ethanol significantly inhibits two subtypes of glutamate receptors at higher concentrations. GABA receptors have binding sites for barbiturates and for the benzodiazepine tranquilizers, drugs with many effects similar to alcohol. The actions of alcohol are potentiated by barbiturates and vice versa. New evidence suggests that alcohol acts as a partial agonist on $GABA_A$ receptors, again increasing the level of inhibitory activity in the brain.

Alcohol abuse is usually defined as a pattern of drinking that impairs social or occupational function. **Alcohol dependence** involves the development of tolerance or withdrawal symptoms. Estimates of the number of alcohol-dependent individuals in the United States range from 10 to 15 million. Altered sensitivity to alcohol at the cellular level may explain some aspects of tolerance, either from the adaptation of cellular membranes to the changes in fluidity produced by alcohol, from alterations in the function of specific membrane proteins, or from both. Over the longer term, alcohol may damage dopamine-secreting neurons in the substantia nigra, one of several brain structures that control body movement. Depleted levels of dopamine in the substantia nigra are the cause of Parkinson's disease, and many alcoholics develop Parkinson-like tremors.

Tobacco

Cigarette smoking may be responsible for as many as 350,000 deaths annually in the United States, representing 18% of all deaths. Total costs of smoking-related health care and lost productivity exceed $65 billion each year. Consumption peaked in the early 1960s, when 40% of adults smoked (50% of men and 35% of women), and dropped after the Surgeon General reported tobacco use to be a major health hazard. Currently, about 30% of adults smoke (30% of men and 25% of women). Passive inhalation of tobacco smoke also leads to significant mortality. The substances in cigarette smoke can be divided into two categories: **nicotine** (an **alkaloid**) and products of combustion such as carbon dioxide, carbon monoxide, and tar. The latter contain the carcinogens responsible for elevated rates of cancer of the lung and other organs, as well as the toxic chemicals that are associated with coronary artery disease.

Nicotine is a psychoactive agent whose continued use usually leads to addiction. When nicotine has been given intravenously it has proved to be more addicting than cocaine, so smoking is not merely a "habit," but a true drug addiction. The most common form of nicotine dependence is associated with the inhalation of cigarette smoke, although tobacco chewing and the use of snuff can also lead to nicotine dependence. The average cigarette contains about 10 mg of nicotine, and 1 to 2 mg is delivered to the lungs when the cigarette is smoked. One puff of smoke results in a measurable nicotine level in the brain within seconds. Nicotine readily crosses the blood-brain barrier, where it acts as an agonist on muscarinic cholinergic receptors in the CNS. Nicotine is metabolized in the liver but, with regular use, nicotine accumulates in the body. Withdrawal from nicotine resembles withdrawal from CNS stimulants, such as cocaine and amphetamines.

Caffeine

A class of stimulant compounds called **methylated xanthines** (from a Greek term meaning yellow), slightly water-soluble alkaloids, are some of the oldest drugs known to man and one can find references to them in the earliest writings and cave drawings. **Caffeine,** named because it was isolated from coffee beans in the early nineteenth century, is the most common xanthine. Caffeine absorption is rapid with peak plasma levels seen within a half-hour. Since caffeine is an alkaloid, it readily crosses the blood-brain barrier, with maximal effects on the CNS in 1 to 2 hours. Caffeine is quickly broken down in the liver and the plasma half-life is 3 to 4 hours.

Caffeine blocks receptors for **adenosine,** a neuromodulator that acts in several areas of the brain to decrease the level of consciousness and alertness. By blocking adenosine receptors, caffeine removes the inhibitory effects of adenosine, stimulating the brain and the brainstem cardiovascular and respiratory centers, increasing heart rate and respiration. Caffeine also stimulates acid secretion in the stomach. In addition to coffee, which may contain as much as 100 mg of caffeine per cup, depending on brand and strength, caffeine is found in a variety of soft drinks, tea, many over-the-counter pain-killers and cold remedies, and stay-awake pills. On the average, both men and women consume 400 to 500 mg of caffeine per day.

ILLICIT DRUGS AND THEIR EFFECTS

Illegal drugs that are widely abused can be classified as stimulants (amphetamines and cocaine), narcotics (opium, morphine, and heroin), and sedative hypnotics, or hallucinogens (marijuana, LSD and PCP). Most of these drugs interact with components of the limbic system and some areas of the frontal lobe, the sites of our deepest emotions and drives. The active ingredients of some illicit drugs are plant alkaloids.

Amphetamines

The ancient Chinese made medicinal teas from herbs of the genus *Ephedra,* whose active ingredient is **ephedrine,** a mild amphetamine used as a bronchiodilator to treat asthma. Another mild amphetamine, **phenylpropanolamine,** is used as a nasal decongestant. The amphetamines are **sympathomimetic** drugs, turning on the "fight-or-flight" response. Amphetamines (mostly benzedrine or dexedrine) were widely used as "stay-awake" pills during the Second World War, and as appetite suppressants. There is a considerable structural similarity between the catecholamine neurotransmitters and the amphetamines. The difference is that ephedrine, phenylpropanolamine, and the catecholamine neurotransmitters have an $-OH$ group that decreases their ability to cross the blood-brain barrier. The methyl groups on the amphetamines increase their lipid solubility, allowing them to rapidly enter the brain. With two methyl groups, **methamphetamine** is the most potent.

Methamphetamine is the major form of amphetamine abused as a street drug, with 90% of all clandestine drug laboratories producing this substance. Estimates of occasional amphetamine users are in the range of 7 to 8 million individuals, including 4% of high-school seniors. The term **speed** is used to describe tablet, liquid, or fine-crystal forms of d,l-methamphetamine, while **ice** refers to large crystals of d-methamphetamine. Methamphetamine can be injected, snorted, smoked, or taken orally. When smoked it delivers a highly concentrated dose instantaneously to the brain via the lungs. The effects are even more intense than with crack cocaine (discussed next). Amphetamines are at least psychologically addictive, and perhaps physically. Abrupt termination is associated with a profound drop in energy level accompanied by depression. At high doses, amphetamines produce feelings of panic, create hallucinations, and in some individuals lead to a full-blown paranoid reaction with often violent behavior.

Amphetamines promote dopamine and norepinephrine release from presynaptic neurons and block their re-uptake, increasing the level of catecholamine-dependent neural activity. Amphetamines stimulate the adrenergic arousal and antifatiguing paths arising from the locus cerulus region of the brainstem. Stimulation of dopaminergic pathways within the limbic system are probably responsible for the euphoria and increased motor activity produced by amphetamines, as well as the paranoia and psychosis seen with high doses.

Cocaine

Cocaine is an alkaloid drug found in the leaves of the coca plant, *Erythroxylon coca.* References to it can be found in Peruvian grave sites dating from 500 AD. Cocaine was first used medically as a local anesthetic for the nose, throat, and cornea. It has been replaced by synthetic **local anesthetics** such as procaine (novocaine, a common local anesthetic used in dentistry) and dibucaine. Cocaine, like amphetamines, is a powerful stimulant to the CNS, producing feelings of euphoria, self-confidence, increased mental and physical abilities, and lessening fatigue.

Cocaine is available as a water-soluble hydrochloride salt or a water-insoluble free base, called **crack.** The hydrochloride salt of cocaine, a white powder, is generally inhaled or "snorted" (because heat destroys the drug) and is rapidly absorbed into the brain. Street cocaine is usually impure, often diluted with substances such as lactose or mannitol, procaine, amphetamine, or PCP. Some of these substances, especially PCP, enhance the toxic effects of cocaine. In contrast to the hydrochloride salt, crack is heat-stable and often nearly pure. Crack is usually smoked. Again, cocaine rapidly enters the brain.

Cocaine is psychologically habit-forming, and addiction develops quickly because of the self-reinforcing stimulation of the brain's "pleasure" centers that it produces. Like the amphetamines, large doses of cocaine can lead to irritability, anxiety, and auditory hallucinations (often threatening voices). A chronic cocaine user typically lacks appetite, suffers from attention disturbances, and avoids social activities. Chronic cocaine users can develop a full-blown paranoid psychosis, leading to violent anti-social behavior. When cocaine use is stopped, the user often experiences symptoms such as depression, muscle tremors, headache, and change in sleep patterns.

Although it is structurally different from amphetamines, cocaine blocks the reuptake of norepinephrine, dopamine, and also serotonin. This results in a general increase in the level of adrenergic and dopaminergic activities. Cocaine activates the brain's "pleasure centers" because these areas use dopamine as a neurotransmitter. Peripherally, cocaine acts like the amphetamines to enhance sympathetic activity, increasing the heart rate, blood pressure, and body temperature.

Opium Derivatives

Opium, used as a sedative since ancient Greek times, is obtained from the dried sap of the opium poppy plant, *Papaver somniferum.* Opium and all its derivatives are termed **narcotics.** Opium use can be traced back to the early Greek and Egyptian cultures, with reference to it in ancient papyrus scrolls and in Homer's *Odyssey.* Raw opium contains about **10% morphine** and a small amount of **codeine.** Crude opium was refined into pure morphine in the early nineteenth century, and slightly later into pure codeine. Morphine is useful clinically as an analgesic in cases of extreme pain (particularly the aching type of pain associated with tissue damage). Codeine has been used both as an analgesic and as a cough suppressant, the latter because of its specific actions on the medullary "cough center." **Heroin** (diacetylmorphine) is produced by the addition of two acetyl groups to morphine and is more powerful and far more addictive than morphine. Illicit heroin is 5% to 6% pure by the time it reaches the streets. Heroin is dissolved in water and injected directly into a vein, sniffed up the nose, or smoked (as the base).

All narcotics stimulate receptors for the endorphins and enkephalins in those areas of the brain that regulate the perception of pain and in neurons that interconnect the hypothalamus, the limbic system, and the locus ceruleus. Enkephalin receptors are also abundant in the amygdala and hypothala-

mus. All these areas contribute to the emotional state of an individual. Opiates are physically and psychologically addictive. When the opiate is removed, the opiate receptors become hypersensitive, resulting in withdrawal symptoms including shivering, profuse sweating, and profound restlessness. All narcotics depress the respiratory centers in the brainstem, so that overdoses of the opiates can easily become fatal, a risk accentuated by the fact that the opiates all cross-react with each other and with CNS-depressant drugs such as the barbituates and alcohol.

Marijuana

Marijuana contains a combination of as many as 60 alkaloids (called cannabinoids) from the plant *Cannabis sativa.* The chief mood-altering drug in cannabinoids is called THC (see Figure 2). The dried leaves, stems, and flowers have a concentration of THC which can vary from 0.5% to 1.0% in domestic plants, to 4% to 6% in Mexican and Central American. The actual content is strongly affected by the horticultural methods used. **Hashish,** the resin of the plant, is far more concentrated. In the brain THC may induce intoxication, depression, or hallucinations, while decreasing reaction time and motor skills. About 55,000 tons of marijuana are consumed annually in the United States, with 25% of that grown domestically.

Adverse short-term effects of marijuana use include a reduction in attention span, deficits in short-term memory, inability to concentrate, apathy or lethargy, and sometimes a feeling of anxiety or panic instead of the expected "high." Several studies suggest that smoking pot may do long-term damage, combining some of the worst effects of alcohol and tobacco. Marijuana cigarettes release five times as much carbon monoxide into the bloodstream and three times as much tar into the lungs of smokers as tobacco cigarettes. One study showed that three or four joints a day could do as much bronchial damage as 20 regular cigarettes. Unlike alcohol or nicotine, THC accumulates in the body. Habitual users may never be rid of it, and the long-term health effects of this accumulation include variable degrees of memory impairment that may continue for at least 6 weeks after discontinuing the drug.

LSD

Psilocin and **psilocybin** are naturally occurring hallucinogens present in the *Psilocybe* genus of mushrooms. These fungi grow in the wild: *Psilocybe semilanceata* is found on well-cultivated grasslands such as parks and golf courses in the northern parts of the United States, and *P. cubensis* and *P. mexicana* in the South. They have an action similar to that of LSD but are less active. **Lysergic acid diethylamide** (LSD) is a synthetic substance and is one of the most powerful hallucinogens known. The action of LSD mixes up the senses, causing vivid visual hallucinations, such as seeing music as colors. Although it is thought to be impossible to overdose on LSD, deaths attributed to accidents under the influence of the drug are relatively common.

LSD is thought to act by stimulating serotonin receptors in the brain because the basic indole ring of LSD resembles that of serotonin, and LSD binds to those areas of the brain with the greatest density of serotonin receptors. There are several types of serotonin receptors, and LSD appears to be an agonist for some and an antagonist for others. Flashbacks, recurrence of hallucinogenic effects, are also a symptom associated with LSD use. They are probably due to a very small amount of the drug's being absorbed by the fatty tissue in the brain and later released, causing a second "trip."

PCP

Phencyclidine (PCP) was first developed as an veterinary anesthetic, but had unacceptable side-effects. Known on the street as **angel dust** (also as horse and several other terms), PCP is one of today's most powerful and abused drugs. Euphoria, agitation, violence, hallucinations, catatonia, para-noia, panic, and suicidal tendencies are common effects. PCP users have increased strength and a decreased sense of pain, making them extremely dangerous and difficult to control. In extreme cases, PCP-intoxicated individuals become capable of terrifying, almost superhuman, acts of violence. Hypertension and tachycardia are also associated with the use of PCP, which may result in cardiac failure.

In the CNS, PCP interacts with the glutamate NMDA receptors. When activated by glutamate, NDMA receptors open a Ca^{++} channel. Once the Ca^{++} channel has been opened, PCP binds to a newly exposed site inside the Ca^{++} channel and blocks Ca^{++} entry. This renders the particular synapse involved nonfunctional. Why does the brain have receptor sites for PCP? The NMDA receptor is abundant in areas of the **limbic system** and **hippocampus,** regions known to be involved in learning and memory, and emotional behavior. What if there were a chemical similar to PCP in the brain that modulated the behavior of NMDA receptors? Disturbing the chemical balance between glutamate and this compound might cause the psychotic symptoms of patients with schizophrenia, symptoms often seen in PCP intoxication.

CHAPTER 10

The Special Senses

On completing this chapter you will be able to:

- Describe how the tympanic membrane and auditory ossicles transmit sound energy to the cochlea
- Understand how the resonance properties of the cochlea translate sound energy into displacements of hair cells on the basilar membrane
- Understand how sound sources are localized in space
- Appreciate how the structure of the vestibular system translates gravitational or inertial forces into hair cell displacement
- Understand the process of image formation in the eye and describe accommodation
- Appreciate how lateral inhibition makes retinal ganglion cells respond to spots on contrasting backgrounds. Distinguish between ON center/OFF surround and OFF center/ON surround cells in terms of the synaptic connections in the retina
- Summarize the principles of feature extraction in the visual system
- Describe the organization of cortical columns in the visual cortex
- Describe the neural pathways for the modalities of taste and olfaction

*T*he senses of hearing, equilibrium, vision, taste, and smell are mediated by specialized multicellular receptor systems. The inputs from these receptors are rich with information that is extensively analyzed by the nervous system. In a multicellular sense organ such as the eye, each sensory cell provides a piece of the puzzle that is reassembled by the brain into a picture of the external world. The eye projects an image on photoreceptors, each of which provides a small portion of the picture to the visual cortex. Similarly, the ear transforms different sound frequencies to distinct receptors in the inner ear. The brain's comparison of the activity of different cells in the receptors provides frequency discrimination. In the vestibular system, a group of sensory structures in the inner ear, the structure of the system causes different receptors to be stimulated by different orientation or movement of the head.

Taste and olfaction (smell) are complicated sensory systems that are difficult to understand because there appear to be only a few basic tastes and no fundamental odors. Instead, a large variety of stimulus molecules interact with single olfactory or taste receptors, somehow giving rise to fairly specific sensations.

The entire range of information provided by the inputs of receptor cells in a multicellular receptor system is much greater than the information provided by individual receptors. In the brain's reconstruction of the sensory world, the whole is greater than the sum of the individual parts.

THE AUDITORY SYSTEM
Physics of Sound

- *What is the physical nature of sound?*
- *Why is it useful for the sound intensity scale of decibels to be logarithmic rather than linear?*
- *What is the function of the ossicles of the middle ear?*
- *How are the receptor hair cells protected from intense sounds?*
- *What is the path of sound energy from the tympanic membrane to the receptors on the basilar membrane?*
- *Define resonant frequency.*
- *Explain how the gradient of resonant frequencies along the basilar membrane helps the auditory system resolve sounds into pure tones.*
- *Compare the tonotopic organization of the auditory cortex with the organization of the somatosensory cortex.*

The auditory system converts sound energy into electrical signals that are relayed to the central nervous system for interpretation. Sound waves consist of periodic increases and decreases in density (**compression** and **rarefaction**) of the air (Figure 10-1). A **pure tone** is a sound with only one frequency, expressed in cycles per second or **Hertz** (Hz). Laboratory sound sources are capable of generating pure tones, whereas musical instruments, voices, and other sound sources usually generate sounds containing a mixture of frequencies. A sound is said to have **pitch** if there is a dominant frequency in the mixture. The human auditory sys-

tem is normally able to discriminate pitches in the range of frequencies between 16 and 20,000 Hz.

The intensity or loudness of a sound is a reflection of the amplitude of the sound pressure wave (see Figure 10-1). Since audible sounds may vary in intensity by many levels of magnitude, a logarithmic scale is used to measure sound intensity. A change in intensity of 10 **decibels** (dB) represents a change in sound-pressure amplitude of a factor of 10. The threshold for detection of sound is defined as 0 dB. Intensity differences as small as 0.1 dB can be detected. Normal conversation generates about 50 to 60 dB, a city street has a sound level of around 75 dB, and an auto horn at close range is about 90 dB. Sounds become painful (and potentially damaging to auditory receptors) at 130 dB, a level approached in some rock music concerts. There is a 10 millionfold difference in intensity between conversation (60 dB) and painful sounds (130 dB).

The Outer Ear and Tympanic Membrane

The structure of the ear can be divided into three regions: (1) the **outer ear**, (2) the **middle ear**, and (3) the **inner ear**, or **cochlea** (Figure 10-2). The outer ear consists of the **pinna, auditory canal,** and **tympanic membrane** (eardrum). The pinna is the visible outer ear. The tympanic membrane spans the auditory canal and separates the outer and middle ear. Sound waves traveling into the auditory canal strike the tympanic membrane, causing it to vibrate.

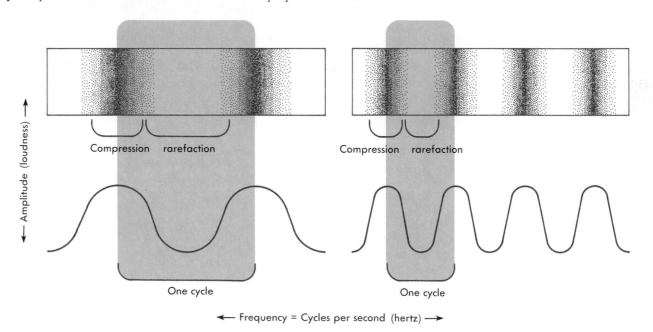

FIGURE 10-1

Sound waves consist of alternating regions of compression and rarefaction of air molecules. Two pure tones are shown; most sounds are mixtures of pure tones and have a much more complex waveform. The two properties of a pure tone are its frequency (the number of wave peaks per second) and its amplitude (the difference between the pressure in the peaks of compression and in the valleys of rarefaction). The frequency is experienced as pitch (Hertz) and the amplitude as loudness (decibels).

NERVE AND MUSCLE

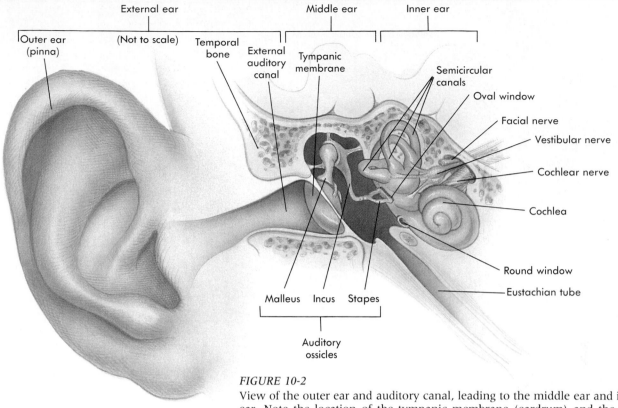

External ear | Middle ear | Inner ear

Outer ear (pinna) | (Not to scale) | Temporal bone | External auditory canal | Tympanic membrane | Semicircular canals | Oval window | Facial nerve | Vestibular nerve | Cochlear nerve | Cochlea | Round window | Eustachian tube

Malleus Incus Stapes

Auditory ossicles

FIGURE 10-2

View of the outer ear and auditory canal, leading to the middle ear and inner ear. Note the location of the tympanic membrane (eardrum) and the relationship between the bones of the middle ear and the cochlea. The middle ear is connected to the pharynx by the eustachian tube. The structures shown are not drawn to scale.

The tympanic membrane vibrates freely only if the pressures are equal on both sides. The **Eustachian tube** (Figure 10-2) connects the middle ear with the atmosphere by way of the throat and normally prevents the development of pressure differences between the middle ear and the atmosphere. However, in colds or illnesses such as **otitis media** (an infection of the middle ear), inflammation can close the Eustachian tube or otherwise increase the pressure in the middle ear and dampen vibrations of the tympanic membrane, causing temporary hearing impairment.

Sound Transmission in the Middle Ear

The middle ear contains three small bones, or **auditory ossicles:** (1) the **malleus,** (2) the **incus,** and (3) the **stapes** (see Figure 10-2). These bones relay vibration from the tympanic membrane to the membrane of the **oval window** of the cochlea. The outer and middle ear amplify the sound wave pressure delivered to receptors in the inner ear by about eightyfold. This increase in sound wave pressure is the result of three features of the relay system. First, the outer ear acts as a funnel to concentrate sound on the tympanic membrane. Second, the ossicles form a lever system that applies a mechanical ad-

vantage in transmitting sounds to the oval window. Finally, the oval window has only about 1/16 the area of the tympanic membrane. These factors constitute an **impedance matching system** that increases the effectiveness of transmission of sound energy from the low-resistance medium of air into the higher-resistance medium of the fluid in the inner ear.

Two muscles, the **tensor tympani,** attached to the tympanic membrane, and the **stapedius** (Figure 10-2), attached to the stapes, control sound transmission through the ossicles. Loud sounds initiate a reflex that activates these muscles, stiffening the chain of ossicles and decreasing sound transmission. This reflex has a latency of about 30 milliseconds, so it cannot protect the ears from sudden loud sounds such as gunshots. It also only partly protects the receptor cells from prolonged intense sounds such as those associated with jackhammers or rock music. Overexposure to sounds such as these is responsible for hearing loss in an ever-increasing number of individuals in industrialized societies. The cumulative damage of loud sounds on hearing is not seen in societies isolated from artificial sources of sounds, and individuals in such societies show much less hearing loss as they grow older.

The Special Senses **257**

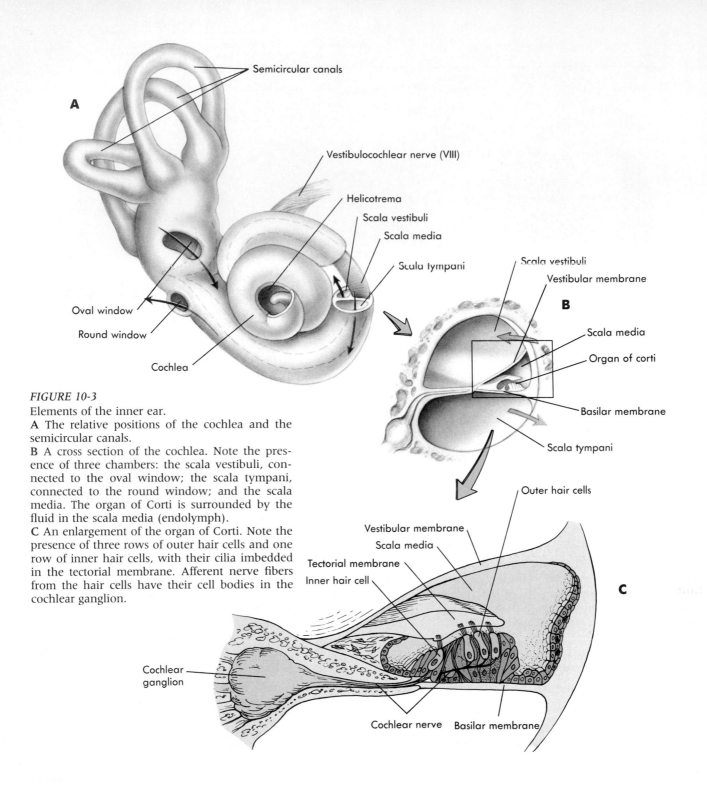

Semicircular canals

A

Vestibulocochlear nerve (VIII)

Helicotrema
Scala vestibuli
Scala media

Scala tympani

Oval window
Round window

Cochlea

Scala vestibuli
Vestibular membrane

B

Scala media
Organ of corti

Basilar membrane

Scala tympani

Outer hair cells

Vestibular membrane
Scala media
Tectorial membrane
Inner hair cell

C

Cochlear
ganglion

Cochlear nerve   Basilar membrane

FIGURE 10-3
Elements of the inner ear.
A The relative positions of the cochlea and the semicircular canals.
B A cross section of the cochlea. Note the presence of three chambers: the scala vestibuli, connected to the oval window; the scala tympani, connected to the round window; and the scala media. The organ of Corti is surrounded by the fluid in the scala media (endolymph).
C An enlargement of the organ of Corti. Note the presence of three rows of outer hair cells and one row of inner hair cells, with their cilia imbedded in the tectorial membrane. Afferent nerve fibers from the hair cells have their cell bodies in the cochlear ganglion.

Structure of the Cochlea

The actual conversion of sound energy into electrical impulses occurs in the cochlea. The cochlea is a 3.5-cm long, spiral-shaped chamber that looks like a snail's shell (Figure 10-3, *A*). The end nearest the middle ear is called the **base** and the opposite end the **apex.** Figure 10-3, *B* shows a cross-sectional view of the cochlea, with the **organ of Corti** en-

larged in Figure 10-3, *C.* The organ of Corti contains the auditory receptor cells, called **hair cells** (Figures 10-4 and 10-5).

The cochlea is divided into an upper chamber, the **scala vestibuli,** and a lower chamber, the **scala tympani,** by a middle chamber, the **scala media,** or **cochlear duct** (Figure 10-3, *B*). The stapes acts on the oval window to set up waves in the fluid within

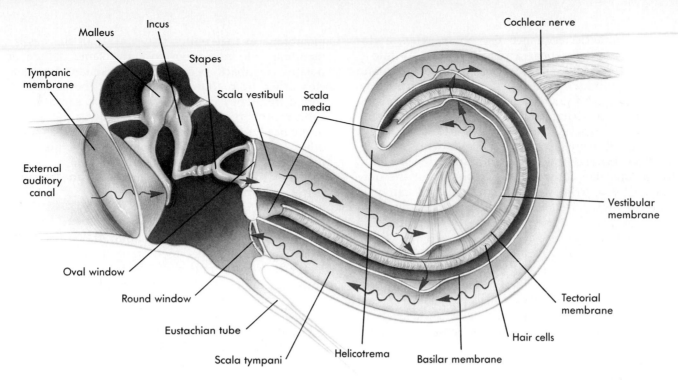

Malleus
Incus
Stapes
Tympanic membrane
Scala vestibuli
Scala media
Cochlear nerve
External auditory canal
Oval window
Round window
Eustachian tube
Scala tympani
Helicotrema
Basilar membrane
Hair cells
Tectorial membrane
Vestibular membrane

FIGURE 10-4

Effect of sound waves on the cochlear structures. Sound waves *(arrows)* strike the tympanic membrane and cause it to vibrate. This vibration is transmitted to the oval window by the ossicles of the middle ear. Vibration of the oval window causes the perilymph in the scala vestibuli to vibrate. As a consequence the basilar membrane is deflected (effect greatly exaggerated).

the scala vestibuli (Figure 10-4). At the apical end of the cochlea, the scala vestibuli is connected by a narrow passageway (the **helicotrema**) to the scala tympani. The scala tympani is separated from the middle ear by the **round window** (Figure 10-3, *A*).

The fluid in the scala vestibuli and the scala tympani is called **perilymph.** Between the scala vestibuli and the scala tympani there is a closed chamber, called the **scala media,** bounded by the **vestibular membrane** and the **basilar membrane** (see Figure 10-3, *C*). The **endolymph** within the scala media differs in composition from perilymph, and this difference in ionic composition contributes to the magnitude of the membrane potential changes that occur in the auditory receptor cells.

Sympathetic Resonance and Frequency Coding in the Cochlea

When the stapes vibrates, it bows the oval window in and out with each arriving sound wave (see Figure 10-4). The oval window vibrates readily although the cochlea is filled with fluid, because as the oval window is deflected inward, the round window deflects outward and vice versa. Vibrations of the oval window are transferred to the perilymph in the scala vestibuli, which transfers them to the basilar membrane, exciting the auditory receptors.

One could imagine that the auditory receptors of the cochlea might generate a burst of action potentials corresponding to each arriving sound wave. In fact, such bursts are seen in response to low-frequency sound. However, the maximum rate at which action potentials can be generated in an axon is no greater than about 1000 Hz, while the auditory system can discriminate the pitch of sounds up to frequencies of about 20,000 Hz. Frequency coding throughout most of the range of audible sound could not be accomplished by one-for-one following of sound waves by action potentials of the individual neurons. The complex structure of the cochlea makes it possible for the brain to distinguish the pitch of sounds in the middle- to high-frequency range. To understand how the cochlea accomplishes this, it is necessary to know something about the phenomenon of **resonance.**

When struck, a tuning fork vibrates at a characteristic **resonant frequency,** as do the strings of musical instruments. A swing also shows resonance. It tends to oscillate at a resonant frequency that depends on its length and the total weight of swing and swinger. When the swinger pumps her legs at the resonant frequency of the swing, the amplitude increases because the force always occurs at the same point in time relative to the natural oscillation of the swing. This phenomenon is a type of positive feedback and is referred to as **sympathetic resonance.** While it is possible to cause a structure with a resonant frequency to move at other frequencies by applying a force to it (someone pushing the swing, for example), positive feedback will not occur and therefore the amplitude of the oscillation will be smaller.

Sympathetic resonance also occurs when a violin note is played near a piano. Thus a middle C (512 Hz) on a violin causes the corresponding piano string to vibrate. In stringed musical instruments, the resonant frequency of a string depends on the length of the string, its thickness, and its tautness (which corresponds to the length and total weight of a swing). Thus a string on a cello or bass is normally of lower frequency than a string on a violin, but a musician can raise the frequency of the note played by a string by tightening it and can play notes of higher frequency by placing her fingers on the fingerboard to shorten the portion of the string free to vibrate.

The ability of the ear to analyze the frequency components of sounds depends on the resonance of the basilar membrane. The basilar membrane consists of elastic fibers of varying length and stiffness imbedded in a gelatinous matrix. It could be compared with a piano into which someone has poured a large quantity of Jell-O. At its base, the fibers of the basilar membrane are short and stiff, but at its apex the fibers are 5 times longer (the basilar membrane is wider) and 100 times more flexible. Like a piano, the resonant frequency of the basilar membrane is highest at the base and lowest at the apex.

Because the elastic fibers in the basilar membrane are imbedded in a gelatinous matrix, they behave as if they are loosely coupled to one another. When a wave of sound energy enters the scala vestibuli from the oval window, it does not just cause one fiber to vibrate but initiates a traveling up-and-down motion of the basilar membrane. This wave imparts most of its energy to that part of the basilar membrane that has resonant frequencies near the particular frequency of the sound wave (Figure 10-5), resulting in a maximum deflection of the basilar membrane at that point. Other areas of the basilar membrane do not experience the same degree of positive feedback because their resonant frequencies differ from that of the sound. The coding of sound frequency is therefore based on selective response properties (**place coding**) along the length of the basilar membrane. Intensity of sounds, on the other hand, is reflected in the degree of displacement of the basilar membrane and is therefore coded by the rate of firing of neurons and the number of cells that are affected.

Hair Cells: The Auditory Receptors

The organ of Corti includes about 20,000 hair cells attached to the basilar membrane in three rows of **outer hair cells** and a single row of **inner hair cells** (see Figure 10-3, C). Hair cells in the adult auditory system have a surface covered with several rows of **stereocilia** and also a cytoplasmic structure, called the **basal body,** which provides polarity to the hair cells (Figures 10-6 and 10-7).

Deflections of the stereocilia toward the basal body depolarize hair cells, while movements away from the basal body hyperpolarize them. Thus the receptor potential can be either depolarizing or hyperpolarizing, and during the movements of the basilar membrane caused by passage of a sound wave, both changes typically occur in succession. Like presynaptic nerve terminals (Chapter 8), the bases of hair cells contain transmitter-filled vesicles (Figure 10-7). Resting (unstimulated) hair cells release some transmitter continuously, causing a steady low level of action potentials in auditory afferent neurons. Depolarization increases transmitter release and elevates afferent activity above this basal level, whereas hyperpolarization inhibits release.

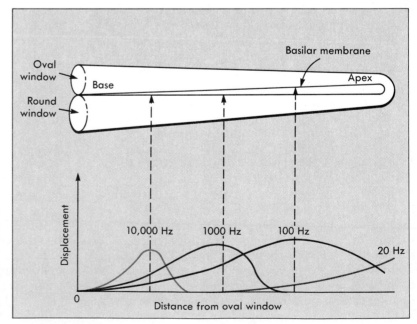

FIGURE 10-5
Frequency localization in the cochlea. The cochlea is shown unwound at the top, while the resonance of the basilar membrane in response to different sound frequencies (tones) is illustrated below. The parts of the basilar membrane nearest the oval window resonate preferentially to sounds of high frequency; at increasing distances from the oval window, progressively lower resonant frequencies are encountered.

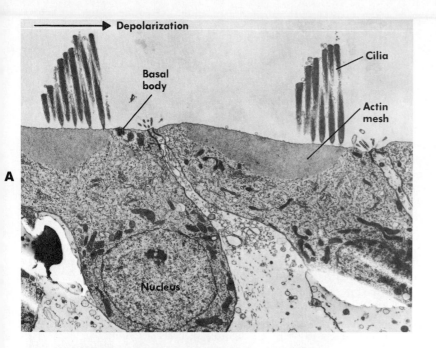

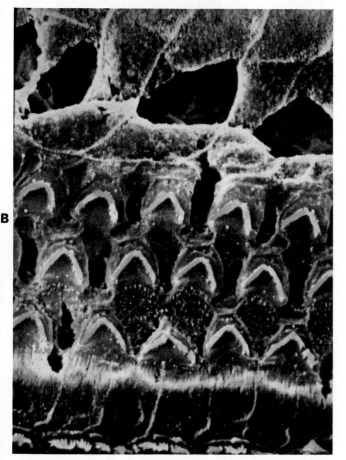

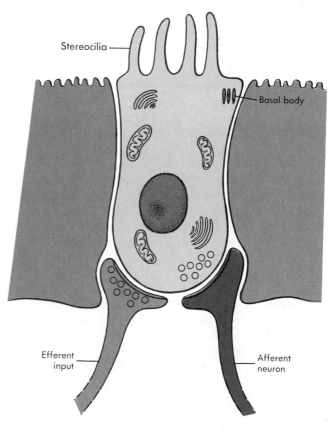

FIGURE 10-6
Transmission (A) and scanning (B) electron micrographs of cochlear hair cells.

FIGURE 10-7
Diagrammatic representation of an auditory hair cell. Auditory hairs cells in the adult possess stereocilia and are morphologically polarized. They are depolarized when the stereocilia are deflected toward the basal body. The hair cell communicates with the afferent neuron by release of transmitter and also receives efferent synaptic inputs that modulate its sensitivity.

The mechanical responses of the cochlea to sound provide better localization than can be explained by the traveling wave mechanism alone. The displacement of the tectorial membrane, which would only amount to a millimeter trip if the basilar membrane were hundreds of miles long, is greater than might be predicted on hydrodynamic grounds. The outer hair cells were suspected of somehow boosting the response at the peak of the basilar membrane displacement, but proof of their role was only obtained when the cells were isolated and studied with whole cell–patch clamp techniques (pp. 198). The outer hair cells were found to respond to changes in membrane voltage with changes in the dimensions of the cell: when the cells are hyperpolarized, they lengthen, and when they are depolarized, they shorten.

The changes in dimensions of the isolated cells is in the order of microns, up to 5% of the cell length. The significance for the wave modification is that the attachment of the outer hair cells to the tectorial membrane transforms their tendency to undergo length changes into a force that acts on the basilar membrane to amplify the wave at the points where it passes through the basilar membrane. Thus a clearer signal is picked up by the inner hair cells. This action by the outer hair cells has been confirmed in recordings from the cells in their normal positions in the organ of Corti.

The really remarkable part of this story is the rapidity of the shape changes in the outer hair cells. The response to voltage changes is so rapid that it defies explanation by any of the known movement-producing mechanisms. Hair cells are capable of following auditory frequencies with a mechanical response to each cycle of the waveform. This is much faster than the movement-generating mechanism in muscle or the subcellular systems that involve microtubules and microfilaments. Inhibitors that block the known force-generating mechanisms do not affect the hair cell responses, so a new mechanism awaits exploration.

The selective attention to a particular portion of the auditory spectrum also involves modification of the capacity of the outer hair cells to amplify the response to sounds. The major efferent innervation of the cochlea ends on the outer hair cells, and neurotransmitter-activated membrane currents that reduce the voltage changes that they make in response to distortion of the basilar membrane would reduce the amount of active force generation that the cells add. This would turn down the amplifier in part of the organ of Corti, allowing the inner hair cells in the more amplified portions of the auditory spectrum to dominate the afferent signal.

The stereocilia of outer hair cells are embedded in the **tectorial membrane,** which forms a roof over the basilar membrane. In this position, they can be excited in proportion to the degree of displacement of the basilar membrane. The stereocilia of the inner hair cells extend into the gelatinous matrix (see Figure 10-3, *C*). They also respond to the movement of the basilar membrane, but the pattern of their response differs from that of the outer hair cells. The inner hair cells are more densely innervated by afferent fibers than the outer hair cells and constitute the major output of the auditory receptor system.

Hair cells, particularly the outer ones, are also innervated by efferent fibers from the central nervous system (Figure 10-7). Increases in activity in the efferent fibers can reduce the active response of the outer hair cells to tectorial membrane movements, reducing displacement in some regions (see box above). This example of central control of receptor sensitivity demonstrates the increased ability of individuals to concentrate on one part of the auditory range (for example, a conversation) in the midst of background noise, which is effectively "tuned out" by the efferent control (see box above).

Auditory Afferent Pathways

Primary auditory afferents join fibers from the vestibular system to form the eighth cranial nerve, which passes into the brainstem (Figure 10-8). The auditory fibers synapse on interneurons in two auditory nuclei of the medulla, the **cochlear nuclei** and **superior olives.** Interneurons of these nuclei send axons into the thalamus. Most of these axons cross in the brainstem and ascend into the thalamus on the opposite side of the body (Figure 10-8). From the thalamus, pathways project to the **auditory cortex** in the temporal lobe.

Tonotopic Organization of the Auditory System

Sound frequencies are mapped onto specific regions of the basilar membrane. As a result, frequency discrimination in the auditory system resembles position discrimination in the somatosensory system. The first requirement for discrimination in each case

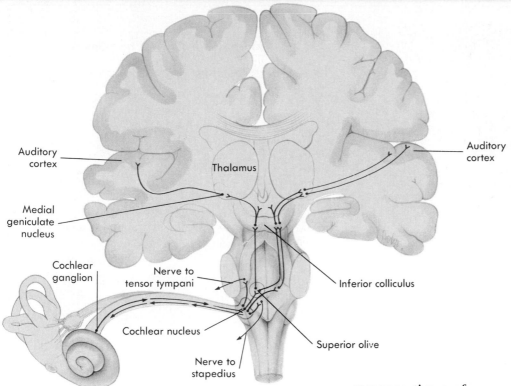

FIGURE 10-8
Central nervous system pathways for hearing. Afferent fibers *(blue arrows)* from the cochlear (spiral) ganglion terminate in the cochlear nucleus in the brainstem. Axons from neurons in the cochlear nucleus project to the superior olive, inferior colliculus, medial geniculate nucleus, and the auditory cortex. Note that auditory input projects to both sides of the cortex and that neurons in the superior olive send efferent axons back to the cochlea and to the muscles of the middle ear *(red arrows)*.

is the existence of a map of the sensory surface. In the auditory system, cochlear afferents enter the nervous system in an orderly way. Ascending auditory pathways are mapped in terms of frequencies, so that different pure tones are represented in a sequence on the cortical surface, called a **tonotopic** map. Lateral inhibition (see Chapter 9) in the auditory system aids in the discrimination of tones that differ only slightly in frequency. When a tone excites one population of neurons, neurons that project from adjacent regions of the organ of Corti are inhibited by pathways from the more excited neurons.

Localization of Sounds in Space
In the lower frequency range, **phase differences** (time of arrival of a particular point on the sinusoidal sound wave) in the sounds that strike the two ears are most important for sound localization. At higher frequencies (above 3000 Hz), the difference in the time of arrival is small in comparison to the

response times of neurons, and **intensity differences** of the sound seem to be most important for sound localization (Figure 10-9). Superior olivary neurons that receive **binaural inputs** (inputs from both ears) and are sensitive to small differences in phase and intensity seem to be most instrumental in localizing sounds in space.

THE VESTIBULAR SYSTEM
Functions of the Vestibular System

- *What is the importance of inertia for the functions of the vestibular system?*
- *How do angular acceleration and linear acceleration differ?*
- *How do the structures of the macular organs and the semicircular canals make the two systems sensitive to different types of acceleration?*

The vestibular system is a part of the inner ear adjacent to the cochlea. Both the cochlea and the canals that house the vestibular system are located in a labyrinth within the temporal bones on the right and left sides of the skull (see Figure 10-2). The vestibular system provides information about the **orientation** of the head with respect to gravity, **linear acceleration** (or change in velocity) of the head when it is moving in a straight line, and **angular acceleration**, or change in the direction or velocity of rotation of the head. This information is integrated with information from the visual system, inputs from receptors signaling the angle of joints, and signals from muscle length detectors to give a sense of position or movement in space. The vestib-

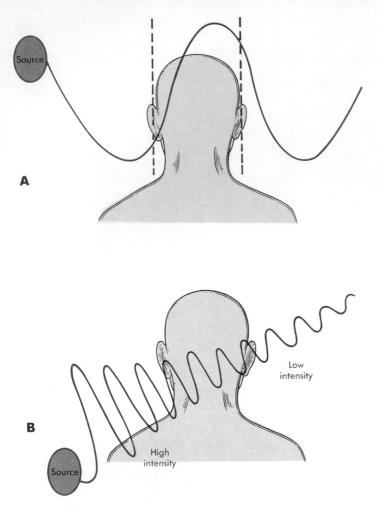

A

B

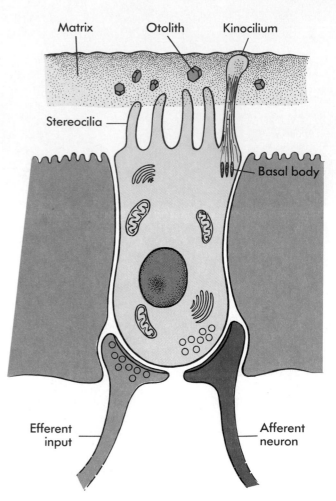

Matrix   Otolith   Kinocilium

Stereocilia

Basal body

Efferent input

Afferent neuron

FIGURE 10-9

A Sound localization clues are provided by phase differences for sounds of low frequency, in which each pressure wave strikes the right ear and the left ear at slightly different times. **B** For sounds of high frequency, phase differences would give ambiguous information because there is time for more than one wave cycle while the sound travels the width of the head. Thus the intensity difference must be used as a directional clue. The head absorbs some of the sound, emphasizing the intensity difference.

ular inputs provide information critical to those somatic reflexes that maintain the position of the head and body with respect to gravity and stabilize images on the retinas of the eyes during head movements.

Vestibular Receptors

As in the auditory system, hair cells are the primary sensory receptors of the vestibular system. In the vestibular system, the same basic receptor type responds to displacement from changes in the orientation and acceleration of the head. The fact that the auditory system responds to sound, whereas the vestibular system detects head orientation and movement, is accounted for by the physical coupling of the two sets of receptors to different types of mechanical stimulation.

FIGURE 10-10

Hair cells of the human vestibular system possess both stereocilia and a knoblike kinocilium projecting from a basal body. Like auditory hair cells, vestibular hair cells transmit to an afferent neuron and receive modulation from vestibular efferents.

Hair cells in the vestibular system have a single knoblike process called a **kinocilium,** as well as a number of smaller stereocilia (Figure 10-10). The structures of the vestibular system position the hair cells so that they can detect the position of the head with respect to gravity and acceleration of the head. Head position and linear acceleration are detected by two structures in each inner ear, the **utricles and saccules** (Figure 10-11, *A*). Rotation is sensed by the **semicircular canals,** of which there is one set of three canals in each inner ear (Figure 10-12, *A*).

The structures of the vestibular system are filled with endolymph. The **macular organs**—the utricle and saccule—are sacs containing gelatinous **otolithic membranes** weighted with small crystals of calcium carbonate called **otoliths** (Figure 10-11, *B*). The stereocilia of hair cells are embedded in the otolithic membrane. Each vestibular apparatus has

A

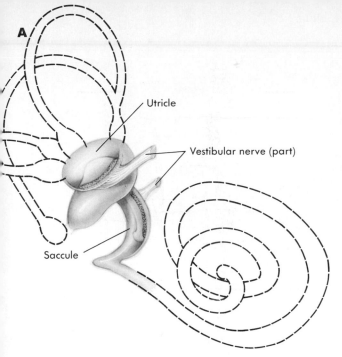

Utricle

Vestibular nerve (part)

Saccule

B

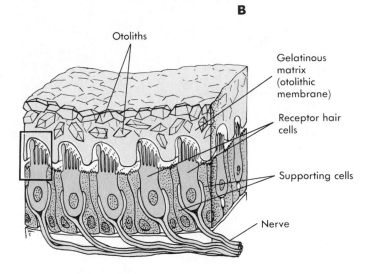

Otoliths

Gelatinous matrix (otolithic membrane)

Receptor hair cells

Supporting cells

Nerve

FIGURE 10-11
A The position of the saccular macula and utricular macula in relation to the cochlea and semicircular canals.
B Enlargement of a section of the utricle showing the otoliths embedded in the gelatinous matrix that covers the hair-cell receptors.

A

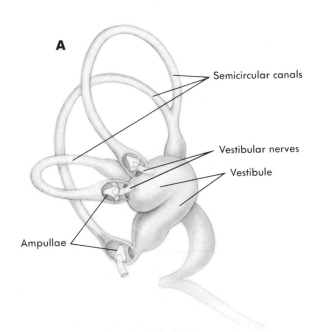

Semicircular canals

Vestibular nerves

Vestibule

Ampullae

B

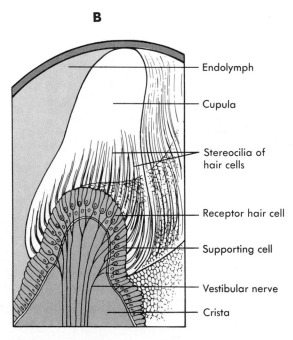

Endolymph

Cupula

Stereocilia of hair cells

Receptor hair cell

Supporting cell

Vestibular nerve

Crista

FIGURE 10-12
A The position of the semicircular canals in relation to the rest of the inner ear showing the location of the ampulla.
B Enlargement of a section of the ampulla showing how hair cell stereocilia insert into the cupula.

three semicircular canals positioned at right angles to one another (Figure 10-12, *A*). Each semicircular canal has an enlargement, the **ampulla**, at the end closest to the utricle. The ampulla contains a mound of tissue, the **crista**, supporting a tuft of vestibular hair cells. The vestibular kinocilia and stereocilia extend into a sheet of gelatinous material, called the **cupula**, that occludes the semicircular canal (Figure 10-12, *B*). For many years the cupula was thought to move freely in the endolymph (because of destruction of its delicate connection with the opposite

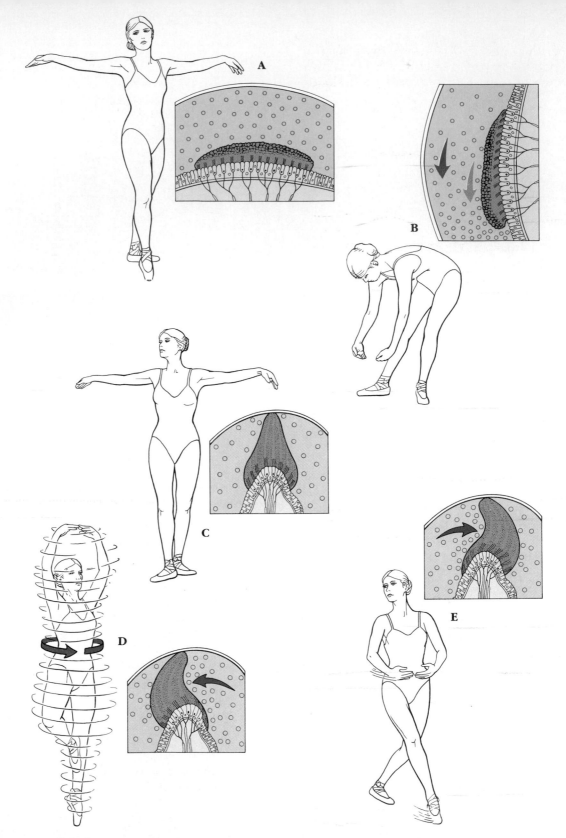

FIGURE 10-13
Function of the otoliths and the semicircular canals in maintaining balance.
A The utricle responds to changes in the position of the head relative to the direction of gravity. When the head is upright the utricular macula exerts a uniform downward force on the vestibular hair cells.
B Tilting the head forward causes a shearing force. Similar deflections can also be caused by linear accelerations, such as are experienced in a car or elevator.
C to **E** The semicircular canals respond to rotations. As a person begins to spin, the cupula is displaced by the endolymph in a direction opposite to the direction of spin. This displacement exerts a shearing force on the vestibular hair cells. When the person stops spinning, the inertia of the endolymph causes the cupula to be displaced in the same direction as the original spin.

wall of the canal during dissection), but now that the structural attachment is established, the known sensitivity of the system is more easily understood.

Detection of Head Position and Linear Acceleration

The macular organs detect head position and linear acceleration. The utricle is oriented so that when the head is vertical, gravity pulls the utricular otolithic membrane straight down on the stereocilia (Figure 10-13, A). When the head is tilted sideward, gravity shifts the utricular otolith relative to the hair cells, bending the stereocilia. The orientations of hair cells (defined by the location of their kinocilia) vary over the utricular surface. Any particular head tilt excites one population of hair cells and inhibits another, generating a unique pattern of afferent activity (Figure 10-13, B).

Forward linear accelerations (such as experienced in an automobile when the driver steps on the gas) are similar to backward head tilts because inertia causes the dense utricular otolithic membrane to be left behind. The hair cells in the saccule extend horizontally when the head is in the upright position. Some hair cells are excited more by gravity than others when the head is in this normal position. Head tilts will alter the normal discharge pattern from the saccule, with an opposite pattern of excitation generated for the right and left saccule sensory cells. When the head is upside down, gravity pulls the otolithic membrane away from the hair cells. Because they are oriented at right angles, the saccule and utricle of each ear provide information on all head positions and linear accelerations.

Detection of Head Rotations

The semicircular canals detect angular acceleration. Angular acceleration causes the endolymph to move relative to the cupula (Figure 10-14). For example, if the head begins to rotate in the horizontal plane, inertia causes the endolymph to lag behind the canal rotation. Relative to the canal, the endolymph is flowing in a direction opposite to the head rotation. When the head rotates to the right, the relative endolymph flow in the right horizontal canal will be toward the left, while that in the left canal will be to the right (see Figure 10-13, C and D). This flow deflects the cupula.

Recall that deflections of the stereocilia either excite or inhibit hair cells, depending on whether the stereocilia are bent toward or away from the kinocilium. In the horizontal canals, all of the hair cells are oriented so that their kinocilia face the utricle, and movements of endolymph toward the utricle therefore excite all of them. Head rotations to the right excite hair cells in the right horizontal canals and inhibit those in the left canals. The reverse occurs when rotations stop because inertia causes the endolymph to continue to flow, deflecting the cup-

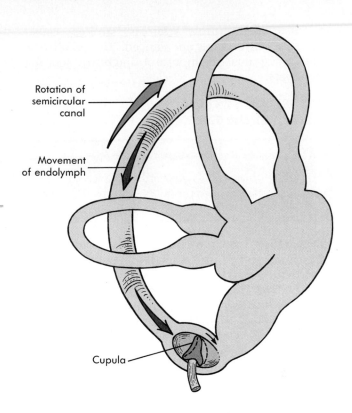

FIGURE 10-14
The effect of inertia on the movement of endolymph in an idealized representation of a semicircular canal. Rotation results in acceleration of the semicircular canal relative to the endolymph. The inertia of the endolymph causes it to lag behind, displacing the cupula and stimulating the hair cells.

ula in the direction opposite to that occurring when the rotation began. In the superior and posterior canals, the same physical responses occur, and the only difference is the plane of rotation that maximally stimulates each canal. Each semicircular canal is excited (or inhibited) to the extent that there is a component of rotation in its plane of orientation. Together, the canals can provide information about rotation in any plane.

Rotational Nystagmus

If a subject is rotated for some time in a swivel chair, the motion of endolymph "catches up" with that of the walls of the horizontal semicircular canals. When rotation is stopped, the endolymph continues to move in the direction of the rotation for a few seconds (see Figure 10-13, E). During this time, the subject finds it difficult to maintain his balance if asked to walk, and his eyes rotate slowly in the direction of endolymph movement, snap back rapidly, and move slowly in the original direction again, as if to keep up with the perceived head rotation. This eye movement is called **rotational nystagmus.** In the absence of rotation, the inputs from the left and right semicircular canals are the same and nystagmus does not occur. However, some diseases (multiple sclerosis, strokes) cause these inputs to become unbalanced. The result is a resting nystagmus, an important diagnostic tool for neurologists.

THE VISUAL SYSTEM
Structure of the Eye

- *What are the roles of the cornea and the lens in image formation?*
- *How does the eye accommodate for distance?*
- *What developmental defects in the structure of the eye result in myopia and hyperopia?*

Light from the large area that constitutes the visual field is concentrated into a small image projected on a dense array of light-sensitive receptors in the **retina** at the back of the eye. The gross structure of the eye reflects its adaptation for forming images on the retina.

The eyeball is covered with a protective fibrous coat called the **sclera** (Figure 10-15). Light passes through a transparent anterior portion of the sclera called the **cornea.** Interior to the sclera is the **choroid layer,** which contains blood vessels in the posterior part of the eye. In the anterior region of the eye the choroid layer is elaborated into the **lens** and associated structures responsible for regulating light entry and for image focusing. The **iris** is a pigmented disc of smooth muscle with a central aperture called the **pupil.** Reflexive changes in pupillary diameter regulate the amount of light that reaches the lens. The lens is attached to a structure called the **ciliary body** via a series of elastic **zonular fibers** (Figure 10-16) that are continuous with similar fibers in the lens.

The lens is normally transparent, but **cataracts** (regions of the lens that become opaque) (Figure 10-17) can impair vision or even cause blindness. Cataracts are caused by slow accumulation of products of chemical reactions within the lens of the eye that make the lens opaque. These reactions are accelerated by the ultraviolet rays of sunlight.

The **anterior cavity,** the part of the eye between the lens and cornea, is divided into two compartments—the **anterior** and **posterior chambers** (Figure 10-16). The anterior cavity is filled with a watery fluid, the **aqueous humor.** The **posterior cavity** behind the lens contains a much thicker fluid, the **vitreous humor** (Figure 10-15). Aqueous humor is secreted from the ciliary body into the posterior chamber and drains into the venous blood from the anterior chamber via the **canals of Schlemm.**

The Eye as an Image-Generating Device

In a camera the lens forms an inverted image of an object on film by bending incoming rays of light. The total amount of light entering a camera is controlled by adjusting its **aperture,** the diameter of the opening through which light reaches the lens. The inside of a camera is painted black to keep light from being reflected inside. In many ways the eye resembles a camera. For example, both the eye and the camera form an inverted image (Figure 10-18). The retina is analogous to the film in a camera. The **pigment layer** behind the retina, like the black-painted interior of the camera, absorbs the light not captured by photoreceptors. The amount of light entering the eye depends on the diameter of the **pupil,** or opening of the iris. Pupillary diameter is controlled by a reflex that monitors the intensity of light reaching the retina.

Both the eye and the camera contain lenses that

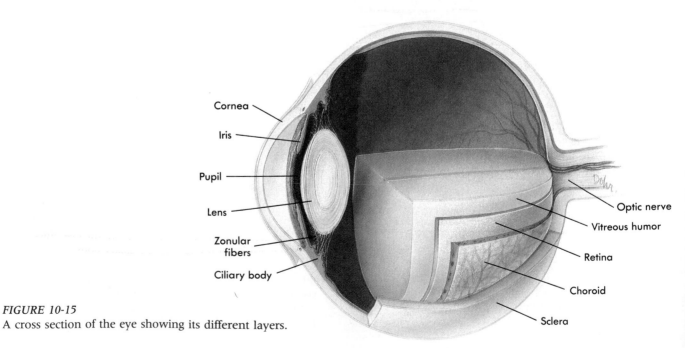

FIGURE 10-15
A cross section of the eye showing its different layers.

Causes and Treatment of Glaucoma

If the canals of Schlemm are plugged, the pressure in the eye increases, resulting in a condition called **glaucoma.** Glaucoma is divided into two types, **open-angle** and **closed-angle.** In closed-angle glaucoma the drainage system is acutely blocked by injury, abnormal structure, certain drugs, or diseases like diabetes. It is usually painful. In open-angle glaucoma there is no pain or visual symptoms to alert the person that something is wrong, and the underlying cause of the increased pressure is unknown.

The increased intraocular pressure of glaucoma closes off the blood vessels that supply the retina, making the retina appear paler and causing starvation and death of the retinal cells (see Figure 10-20, *C*). A new device called the Molteno implant can be placed in the eye where it acts like a tiny faucet, allowing fluid to flow out whenever the intraocular pressure increases. The implant consists of a small anchoring plate sutured to the conjunctiva and attached to a silicone drainage tube that connects the anterior chamber of the eye to the posterior part of the eye where any fluid that drains out can be absorbed. The procedure takes less than an hour and vision returns to normal in a week or two.

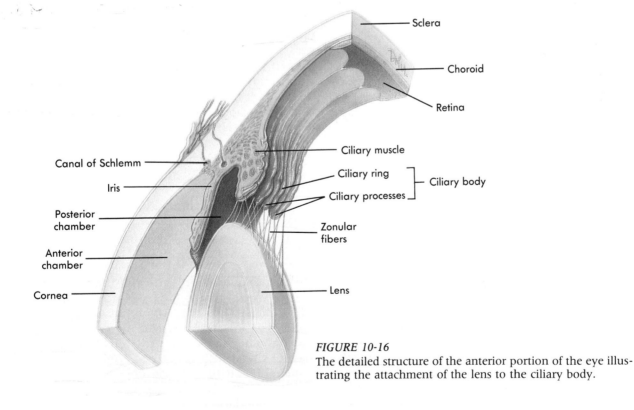

FIGURE 10-16
The detailed structure of the anterior portion of the eye illustrating the attachment of the lens to the ciliary body.

The Special Senses

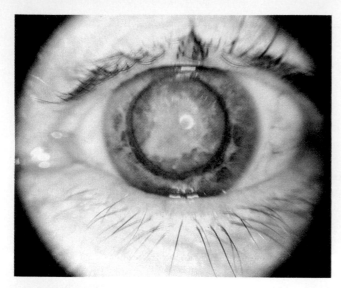

FIGURE 10-17

Photograph of a cataract. The cataract is the white spot approximately in the center of the cornea.

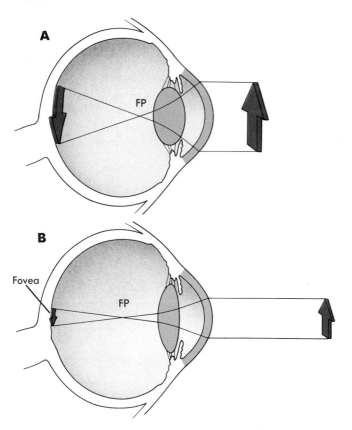

FIGURE 10-18

Image formation on the retina. Note the location of the fovea, the area with the greatest photoreceptor density. *FP* indicates the focal point.

A For near vision the ciliary muscles contract, decreasing the force exerted by the suspensory ligaments and causing the lens to become less flattened (more convex). A more convex lens bends the entering light rays more.

B Distant objects are focused by a flattened (less convex) lens. The lens is flattened because the ciliary muscle is relaxed, and the suspensory ligaments are exerting their greatest force on the lens.

focus the image by bending light. In a camera the amount of light bending depends on the curvature of the lens and on the difference between the refractive index of the air and the glass of the lens. The same principles apply to the eye, except that in the eye most of the of light bending is carried out not by the lens but by the cornea, the transparent extension of the sclera over the front of the eye. The cornea bends light more than the lens because the difference between the refractive index (a measure of the light-bending power of a material) of the cornea and air is about five times greater than the difference between the lens and cornea.

Accommodation for Distance

Accommodation is the process that adjusts the focus of the eyes to compensate for distance. In a camera, both the curvature of the lens and the refractive index are constant for any particular lens, so that the camera is focused by changing the distance between the lens and film. In contrast, accommodation of the eye does not involve a change in the distance between the lens and retina, but instead, a change in the curvature of the lens.

The lens has a natural elasticity. If it is removed from the eye, the lens assumes a round shape appropriate for focusing at close range (see Figure 10-18, *A*). In the relaxed eye, the lens is subject to forces exerted by the zonular fibers of the suspensory ligament attached to the lens. These fibers act against the intrinsic elasticity of the lens to pull it more nearly flat, a condition that is appropriate for focusing on distant objects (see Figure 10-18, *B*). The zonular fibers suspend the lens in the center of a doughnut-shaped structure called the **ciliary body.** When the ciliary body thickens because of contraction of the ciliary muscles, the doughnut ring becomes smaller and the tension in the zonular fibers diminishes, allowing the elastic lens to assume the more rounded shape it would have if removed from the eye.

Accommodation is a reflex mediated by the autonomic nervous system. Active contraction of the ciliary muscles is required for focusing on close objects, whereas distance vision is performed with a relaxed eye, which explains why focusing on near objects can result in eyestrain.

Disorders of Image Formation

The natural elastic properties of the lens permit the eyes of young people to focus on objects as close as a few inches. The closest point at which a clear image can be formed is termed the **near point** for vision. In older individuals the lens loses some of its elasticity, increasing the near point so that it becomes difficult to focus on objects close to the eye. This condition is called **presbyopia,** or restriction of accommodation.

In some individuals the eyeball is too long or too short relative to the power of the lens and cornea. If the eyeball is too long (Figure 10-19, *A*), the

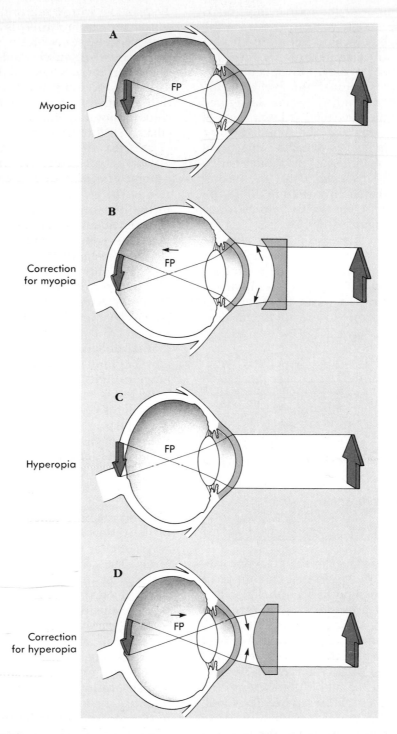

FIGURE 10-19

Visual disorders and their correction by lenses. *FP* is the focal point.

A Myopia (nearsightedness) is a defect in which the cornea and lens are too powerful, or the eyeball is too long, causing the focal point to be too near the lens. The image *(red arrow)* is formed in front of the retina.

B The effect of placing a concave lens in front of the myopic eyeball is to spread out the light rays *(black arrows)*, moving the focal point away from the lens so that there is a focused image on the retina.

C Hyperopia (farsightedness) is a disorder in which the cornea and lens are too weak or the eyeball is too short. The image is formed behind the retina.

D The effect of placing a convex lens in front of the hyperopic eyeball is to bend the incoming light more, moving the focal point toward the lens.

image is focused in front of the retina; this condition is **myopia, or nearsightedness.** If the eyeball is too short, the image is focused behind the retina and the condition is called **hyperopia, or farsightedness** (Figure 10-19, C). Whereas myopia often occurs with a normal lens because the eyeball is too long, normal individuals tend to become farsighted as they get older. As the lens ages, it loses some of its elasticity and cannot assume the rounded shape needed to focus on near objects. Thus the changes of aging tend to correct myopia. Myopia can also be corrected by corrective lenses that cause the incoming light rays to diverge slightly (Figure 10-19, B). Hyperopia is treated with lenses that help converge parallel rays of light (Figure 10-19, D).

Recently, lasers have been used to reshape the cornea, offering a way to correct visual disorders without eyeglasses or contact lenses. The remodeled cornea acts like a corrective lens. The laser involved is capable of removing surface tissue layers to a precisely predetermined depth to create the desired curved surface.

The Retina

> - *What factors determine the ability of different parts of the retina to resolve the details of an image?*
> - *What is the sequence of events between the arrival of light at the outer segment of a photoreceptor and hyperpolarization of the cell's membrane potential?*
> - *What are the roles of the iris and retina in adaptation for changes in light level?*
> - *What sequence of events occurs between illumination of a patch of the retinal surface and the response of ganglion cells with receptive fields in the vicinity of the illuminated spot?*

The retina, the innermost layer of the posterior cavity, is itself organized into layers that contain **photoreceptors** (which are modified epithelial cells), several types of interneurons and their synaptic connections, and blood vessels (Figure 10-20, A). The embryonic development of the retina results in an inside-out design, so that the photoreceptors are nearest the sclera at the back of the eye, and light must pass through the retinal interneurons and blood vessels to reach the photoreceptors. The blood vessels serving the retina are nearest the interior of the eye and can be seen easily when the retina is viewed with an ophthalmoscope (Figure 10-20, B). There are four basic types of retinal interneurons: **Ganglion cells** are the output cells of the retina; their axons join to form the optic nerve of each eye. The signals of photoreceptors are relayed to ganglion cells by way of **bipolar cells, horizontal cells** and **amacrine cells.**

Cone and Rod Photoreceptors

The two types of photoreceptors, **cones** and **rods,** have some common structural features (Figure 10-21). The **outer segment** (nearest to the sclera) consists of stacked **membrane discs** that contain a high concentration of the **visual photopigment.** The **inner segment** contains mitochondria, the photoreceptor nucleus, and transmitter-filled vesicles. The outer and inner segments are joined by a thin connecting zone containing the remnant of a cilium, showing that photoreceptors evolved from ciliated epithelial cells.

Visible light is electromagnetic energy with wavelengths between 400 (the color violet) and 700 nanometers (nm) (red) (Figure 10-22, A). The portion of the electromagnetic energy spectrum visible to the human eye is only a small part of the total electromagnetic spectrum, which includes wavelengths longer than visible light (infrared radiation, microwaves, and radio waves) and shorter than visible light (ultraviolet rays, x-rays, and gamma rays). The portion of the electromagnetic spectrum that can be detected by the eye is determined by the response characteristics of the photoreceptors, which in turn are determined by the wavelengths of light that are absorbed by the retinal photopigments.

The human eye functions over a remarkable range of light intensity; the difference between the lowest light levels that can be detected and those of the brightest sunny day is about 10^{10}-fold. Only a small part of this range can be accounted for by changes in the diameter of the pupil, which can be changed by only about sixfold, resulting in a thirtyfold difference in light entry. The wide range of light intensities over which the eyes can operate is therefore largely due to the photoreceptors and associated neural pathways of the retina. One feature of the retina is the presence of two sensory pathways, one for high light levels and one for low light levels. The cone or **photopic** system, which functions only in bright light, provides information about color and detail. The rod or **scotopic** system is more sensitive to light and functions in low light levels but provides poor resolution of detail and no information about color.

There are four basic types of retinal photopigments; any photoreceptor possesses only one type. The absorption spectrum of each photopigment determines what portion of the visible range that photoreceptor can respond to. The information provided by the cone receptors is used to mediate color vision. There are three types of cones, each having a different wavelength sensitivity. **Blue-sensitive cones** absorb maximally at 420 nm, **green-sensitive cones** at 560 nm, and **red-sensitive cones** at 630 nm (see Figure 10-22, B). Rods absorb best at about 540 nm (Figure 10-22, B).

Photopigment Bleaching—The First Step in Photoreception

The process of photoreception begins when light activates visual pigments. Visual pigment molecules

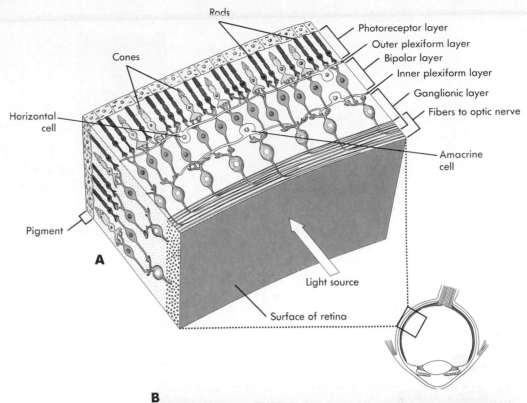

Rods
Cones
Horizontal
cell
Pigment
A

Photoreceptor layer
Outer plexiform layer
Bipolar layer
Inner plexiform layer
Ganglionic layer
Fibers to optic nerve
Amacrine cell
Light source
Surface of retina

FIGURE 10-20

A A diagram of the retina. The afferent nerve fibers from the ganglion cells are closest to the light source. The bipolar cell layer is between the ganglion cell layer and the photoreceptors. Synaptic contacts are made in the inner and outer plexiform layers.

B A photograph of a normal retina. The blood vessels enter and leave the retina at the optic disk.

C The retina in an individual with glaucoma. In glaucoma, the blood supply to the retina is reduced by the pressure in the eye, causing the retina to appear pale.

B

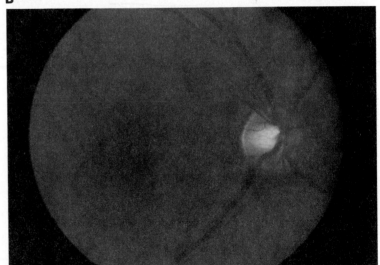

C

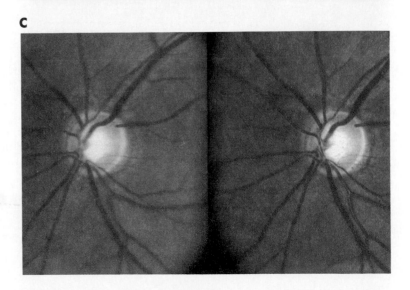

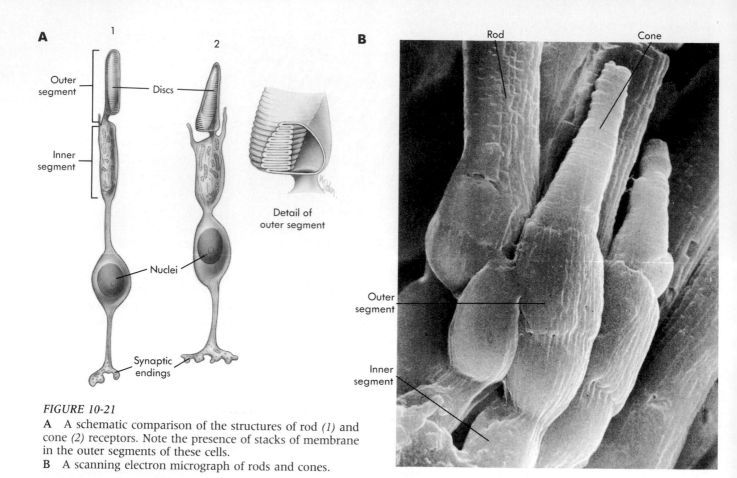

A

Outer segment

Discs

Inner segment

Nuclei

Synaptic endings

Detail of outer segment

B

Rod

Cone

Outer segment

Inner segment

FIGURE 10-21

A A schematic comparison of the structures of rod *(1)* and cone *(2)* receptors. Note the presence of stacks of membrane in the outer segments of these cells.
B A scanning electron micrograph of rods and cones.

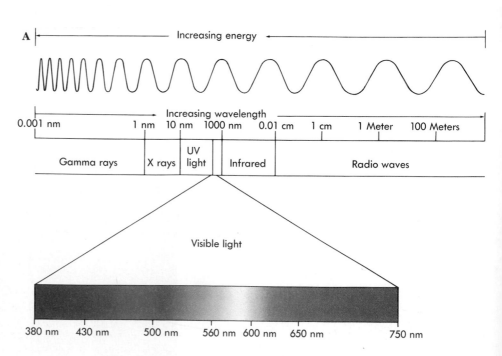

A Increasing energy

Increasing wavelength

0.001 nm   1 nm   10 nm   1000 nm   0.01 cm   1 cm   1 Meter   100 Meters

| Gamma rays | X rays | UV light | Infrared | Radio waves |

Visible light

380 nm   430 nm   500 nm   560 nm   600 nm   650 nm   750 nm

FIGURE 10-22

A The electromagnetic spectrum and the small segment of it that is visible, with the corresponding color spectrum.

274

NERVE AND MUSCLE

Color Sensation Is Genetically Determined

The ability to distinguish subtle variations in color is the result of integration of signals from photoreceptor classes that are selective for overlapping ranges of wavelengths. For example, the sensation of yellow is obtained when the retina is stimulated with **monochromatic** (single wavelength) light of 580 nm, which stimulates red cones about twice as effectively as green cones. The same sensation can be obtained if two beams of light of 560 nm (yellow-green) and 600 nm (orange) are mixed. The net effect of the mixed light is to stimulate red and green cones in the same proportion as did the monochromatic light. Thus relative responses of different cone classes are integrated by the brain to confer the sensation of color.

Normal individuals with three cone pigments are referred to as **trichromats. Color blindness** is the absence of one or more of the cone pigments. It may be complete, involving the total absence of a particular pigment, or partial, with reduced levels of a pigment. **Dichromats** have only two cone pigments. If the red pigment is lacking **(protanopia)**, discrimination of red and green is impossible and the visual system is insensitive to deep red colors. If the green pigment is lacking **(deuteranopia)**, it is still impossible to distinguish red and green, but the visual system is nevertheless sensitive to light in the range normally served by the green pigment because the responsiveness of the red and blue pigments overlaps into this range. **Tritanopes** lack the blue pigment and cannot discriminate shades of blue and green. **Monochromats** lack all three cone pigments and cannot distinguish colors at all. Because monochromats must use the scotopic system for seeing in bright, as well as dim, light, they find bright light very unpleasant.

In humans, most forms of hereditary color blindness are recessive X-linked traits (i.e, the gene is on the X chromosome and does not have a corresponding allele on the Y chromosome). As a result, the incidence of color blindness is higher in males than females. In Western Europe, approximately 8% of all males have some form of color blindness, but only about 1% of females are color blind. Color blindness is often assessed using charts in which normal individuals see one number, while color-blind individuals see a different number.

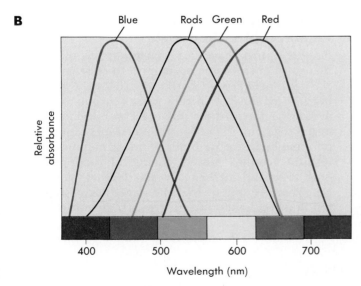

FIGURE 10-22, cont'd
B The sensitivity of rods and the three categories of cones to light of different wavelengths. Each receptor responds most vigorously to an optimum wavelength and less vigorously to higher and lower wavelengths.

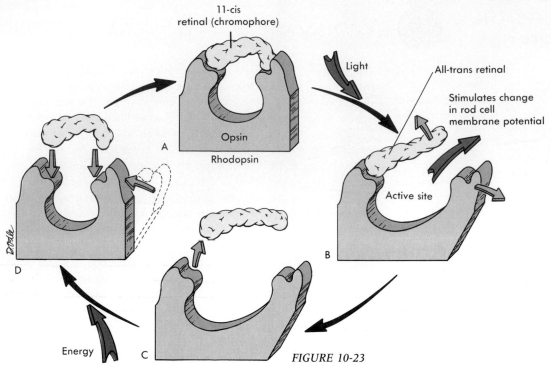

FIGURE 10-23

The cycle of bleaching and restoration of rhodopsin.
A Retinal (the chromophore of rhodopsin) is attached to opsin.
B Light causes retinal and opsin to change shape *(red arrows)*. This begins a second messenger cascade that causes a change in the rod cell membrane potential.
C Retinal separates from opsin (bleaching).
D Metabolic energy is required to bring opsin back to its original form and to attach retinal to it. Until this is done, the photopigment is unable to respond to light.

are made up of two parts: a **chromophore** that absorbs light, coupled to one of four different proteins called **opsins.** For both rods and cones, the chromophore is **retinal,** a modified form of vitamin A. The opsins confer wavelength specificity to the photopigment so each of the three cone types possesses a different opsin. All rods contain a single type of opsin; the resulting visual pigment is **rhodopsin,** or **visual purple.**

Rhodopsin was the first photopigment to be studied and serves as an example of photopigment reactions in general. In the dark, retinal is in a conformation called **11-cis-retinal** (Figure 10-23, *A*). When a photon of the appropriate wavelength is absorbed by a rhodopsin molecule, 11-**cis**-retinal undergoes a conformational change to **all-trans-retinal;** this process is called **bleaching** (Figure 10-23, *B* and *C*). Bleaching initiates a sequence of events leading ultimately to a change in the rod cell membrane potential, as will be described.

Maintenance of Photoreceptor Sensitivity and Dark Adaptation

Bleached rhodopsin cannot respond to light, so that all-**trans**-retinal must be restored to the 11-**cis**-retinal conformation if the photosensitivity of the retina is to be maintained (Figure 10-23, *D*). This process uses ATP and occurs at a rate much lower than the rate of bleaching in bright light. After the retina has been exposed to dim light for some minutes, almost all of the rhodopsin is in the unbleached form and the retina is **dark adapted.** When the retina is exposed to bright light of appropriate wavelengths, much of the rhodopsin is converted to the bleached

form, and the scotopic pathway becomes much less sensitive. This is one way in which the retina adjusts its sensitivity to the light level. Since rods are not responsive to red light (see Figure 10-22, *B*), red-lit instruments are used in airplane cockpits and other situations where it is important to preserve the high sensitivity of the dark-adapted retina.

From Photopigment Bleaching to Photoreceptor Potentials

Photoreceptors have Na^+ channels in the membranes of their outer segments. In the dark these Na^+ channels are open (Figure 10-24). The resulting inward flow of current passes through the connecting segment of the photoreceptor and keeps the inner segment depolarized. When photons bleach photopigment in the outer segment membrane, the resulting conformational change of rhodopsin begins a second-messenger cascade that closes the Na^+ channels opened in the dark, decreasing current flow and allowing the inner segment to hyperpolarize.

Rhodopsin is chemically similar to membrane

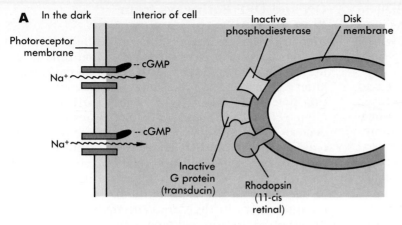

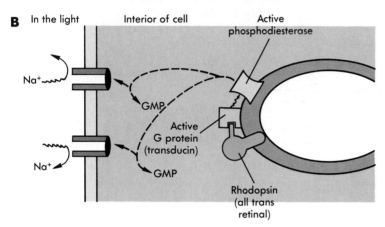

A In the dark

Photoreceptor membrane

Interior of cell

Inactive phosphodiesterase

Disk membrane

Na⁺ ·· cGMP

Na⁺ ·· cGMP

Inactive G protein (transducin)

Rhodopsin (11-cis retinal)

B In the light

Interior of cell

Active phosphodiesterase

Na⁺

GMP

Active G protein (transducin)

GMP

Na⁺

Rhodopsin (all trans retinal)

FIGURE 10-24

The molecular events of transduction of light energy to membrane permeability changes in photoreceptors.

A In the dark, relatively high levels of cGMP keep Na⁺ channels in the cell membrane of the outer photoreceptor segment open.

B When photons interact with retina in photoreceptor disk membrane, formation of all-trans rhodopsin activates G proteins, which activate phosphodiesterase. Phosphodiesterase degrades cGMP, closing Na⁺ channels. The decrease in Na⁺ conductance hyperpolarizes the receptor and reduces transmitter release.

receptors for neurochemical transmitters described in Chapter 7. The second messenger cascade set in motion by rhodopsin bleaching involves a G protein called **transducin** that activates phosphodiesterase. Phosphodiesterase degrades **cyclic guanosine monophosphate (cGMP),** an intracellular second messenger. cGMP keeps the Na⁺ channels open in the dark, and the effect of light acting on rhodopsin is to reduce the levels of cGMP, closing the channels (Figure 10-24). This second messenger system is much like those used by some hormones (Chapter 5). However, cGMP seems to act directly to gate ion-channel proteins, rather than phosphorylating them.

Intracellular Ca⁺⁺ seems not to play a direct role in the response of photoreceptors to light. Instead, recent findings suggest that it modulates the recovery of photoreceptors from light stimulation. In the dark, cGMP opens membrane Ca⁺⁺ channels, as well as Na⁺ channels, keeping intracellular Ca⁺⁺ relatively high. As explained (Figure 10-24), photoexcitation decreases intracellular cGMP levels. The drop in cGMP blocks Ca⁺⁺ influx, so intracellular Ca⁺⁺ levels decrease when photoreceptors are illuminated. The enzyme **guanylate cyclase** is responsible for regenerating cGMP after a light stimulus. Guanylate cyclase is activated by a drop in intracellular Ca⁺⁺. The effect of decreased Ca⁺⁺ on guany-

late cyclase is mediated by a regulatory protein, **recoverin.** Recoverin is a Ca⁺⁺-binding protein structurally similar to calmodulin. The Ca⁺⁺-free form of recoverin binds to guanylate cyclase and activates it allosterically. The active form of the enzyme rapidly restores the levels of cGMP to their prestimulus levels, reopening Na⁺ channels that were closed by the light stimulus. By this mechanism, stimulation not only activates photoreceptors but also enhances their ability to recover their previous resting potential when illumination is terminated.

Distribution of Photoreceptors on the Retina

Visual acuity is the ability to resolve details of the visual world that are at a distance from the eye. This is what is tested by eye charts. The resolving power of the eye is affected by the power of its lens system and by the sensitivity of the retina. The distribution of rods and cones on the retina and the connections to retinal interneurons are not uniform across the retinal surface. Instead, a small part of the retina, the **central fovea** (see Figure 10-18) is specialized for high acuity and color vision, while surrounding regions of the retina are weaker at providing information about color and detail. The fovea has the highest photoreceptor density. Each fovea contains about 4000 receptor cells, all of which are cones.

Furthermore, the cones of the fovea are wired to their ganglion cells with little convergence. As a result, the fovea provides a fine-grained image with a resolution 100,000 times greater than that of a standard color TV set.

The relative density of cones diminishes and that of rods increases as the distance from the fovea increases, so in the majority of the retinal area outside the fovea, rods greatly outnumber cones. Moreover, in the regions outside the fovea there is a much higher degree of convergence of photoreceptor inputs on retinal interneurons. This convergence increases the probability that a single photon will excite an interneuron and thus improves the sensitivity of dim light vision at the expense of a loss in resolving power.

Ganglion cell axons come together to form the **optic nerve** (see Figure 10-15) at a location called the **optic disc**. The optic disc has no photoreceptors and therefore is a "blind spot" in the retina. Ordinarily, this blind spot is not a handicap because the part of the visual field that falls in the blind spot of one retina does not fall in the blind spot of the other. Also, the brain tends to continue lines and patterns from surrounding parts of the retinal image and thus "fills in" missing parts in the picture of the visual field that is presented to consciousness.

ON and OFF Pathways from Photoreceptors to Ganglion Cells

Bipolar cells are pathways for vertical flow of information from photoreceptors to ganglion cells (Figure 10-25; see also Figure 10-20). In the dark, ganglion cells produce a low, steady baseline rate of action potentials. The activity of any particular ganglion cell either increases (an ON response) or decreases (an OFF response) in response to illumination. The ON and OFF pathways from receptor cells to ganglion cells are mediated by separate types of bipolar cells. These differ in their response to the transmitter that is continuously released by photoreceptors in the dark. The transmitter causes hyperpolarization in one type of bipolar cell and depolarization in the other (Figure 10-25). Bipolar cells that are hyperpolarized by the photoreceptor transmitter in the dark become relatively depolarized when light excites the receptors, these constitute the ON pathway (Figure 10-25). Those bipolar cells that are depolarized by the photoreceptor transmitter in the dark become relatively hyperpolarized when light excites the receptors, and constitute the OFF pathway.

Lateral Inhibition in the Retina

The receptive field of a ganglion cell is that area on the retina, consisting of a population of photorecep-

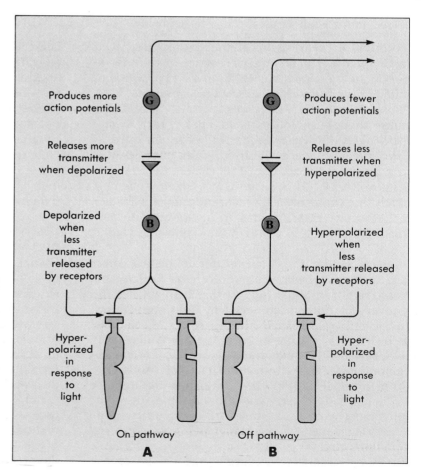

FIGURE 10-25
The response of the photoreceptors (both rods and cones) to light is always hyperpolarization.
A The bipolar cell *(B)* in this pathway is excited by the decrease in transmitter release from the photoreceptors, that is, the transmitter had an inhibitory effect that was turned off. The bipolar excitation increases the number of action potentials in the ganglion cell *(G)* and therefore this is known as the ON pathway.
B Hyperpolarization of the photoreceptors decreases transmitter release, which inhibits the bipolar cell. This effect indicates that receptor cell transmitter has an excitatory effect on this class of bipolar cells. The inhibited bipolar cell releases less transmitter at its terminals on the ganglion cell, and this results in a decreased number of action potentials from ganglion cells in this pathway. This is therefore the OFF pathway.

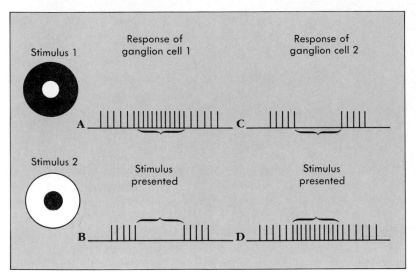

FIGURE 10-26
Electrical recordings from two categories of retinal ganglion cells that are responding to two stimuli. The traces show the rate of action potentials (indicated by *vertical spikes*) over time.
A Ganglion cell 1 is turned on (the frequency of action potentials increases) by a stimulus *(1)* consisting of a bright spot (the "center") surrounded by a dark background (the "surround").
B Ganglion cell 1 is turned off by a stimulus *(2)* consisting of a dark center surrounded by a bright background. It is therefore called an "OFF center/ ON surround" cell. C Ganglion cell 2 is turned off by stimulus 1 and turned on *(D)* by stimulus 2. It is therefore called an "OFF center/ON surround" cell.

tors, in which light stimulation can affect the activity of the ganglion cell. In sensory systems that show lateral inhibition, stimulated receptors not only activate pathways leading from them but also pathways that modulate the flow of information from surrounding receptors. An especially sophisticated form of lateral inhibition is seen when ganglion receptive fields are mapped experimentally. This is done by stimulating the retina with a small spot of light while recording action potentials from the ganglion cell's axon in the optic nerve. Some ganglion cells respond to the general level of illumination in their receptive fields, but typically ganglion cells have receptive fields with central regions in which a stimulus results either in a burst of action potentials or a decrease in the steady, low rate of action potentials (Figure 10-26). These are called **ON center** and **OFF center** ganglion cells, respectively.

Most receptive fields combine excitation and inhibition, so that if light in the center of the receptive field excites a ganglion cell, the same cell will be inhibited by light in a circular area around the center (the **surround**); these are **ON center/OFF surround** cells (*ganglion cell 1* in Figure 10-26). In other ganglion cells, a spot of light falling on the center is inhibitory, while in the surrounding region it is excitatory; these are **OFF center/ON surround** cells (*ganglion cell 2* in Figure 10-26). The opposite effects of stimulating the center and surround of a ganglion cell's receptive field are probably due to the receptors of the surround and center being wired to the ganglion cell through different types of bipolar cells. However, horizontal cells (see Figure 10-20) are also believed to be involved in lateral inhibition. These cells have horizontal processes that reach for some distance across the plane of the retina, making them pathways for lateral flow of information in the retina.

What the Retina Tells the Brain

Ganglion cells are the retinal cells whose axons form the optic nerve, so their output is the final product of the information processing that occurs in the retina. The optimum stimulus for an ON center/ OFF surround cell is a spot of light of the right size on a dark background. The optimum stimulus for an OFF center/ON surround cell is a dark spot on a white background. In some cases, the basic receptive field organization incorporates selectivity for colors. For example, a ganglion cell may be excited by a spot of green light on a red background but inhibited by a spot of red on a green background.

Each ganglion cell may send three different messages to the brain (Figure 10-27). A burst of action potentials constitutes a signal to the brain that most of the light falling on the cell's receptive field is on the excitatory part of the field. A decrease in the rate of action potentials means that most of the light falling on the receptive field is on the inhibitory part of the field. No change in the rate of action potentials means that light, if present, does not vary in intensity over its receptive field. In sum, the effect of lateral inhibition in the retina is to favor response to contrast in the visual field and to suppress response to uniformity, so the retina informs the brain of the locations of spots in the image where there is contrast, either of light intensity or of color.

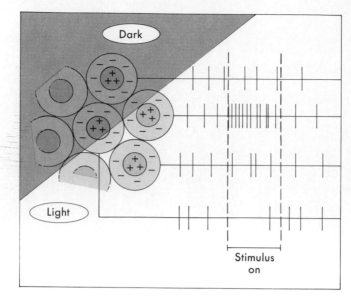

FIGURE 10-27

A border between light and dark parts of an image is shown falling on the receptive fields of several ganglion cells. One receptive field *(right center)* receives light only on part of the inhibitory surround, resulting in inhibition. An adjacent field on the other side of the border *(right center)* receives light on the excitatory center but on only part of the surround; the net result is excitation. These responses make these cells accurate indicators of the position of the border. In contrast, cells *(upper left* and *lower right)* whose fields lie entirely within the dark or light parts of the retinal image, do not show changes in activity.

Central Visual Pathways

- *How is point-to-point correspondence in the visual fields of the two eyes established during early development? How is it maintained?*
- *What features of the image are extracted by simple cells of the primary visual cortex? Complex cells? Hypercomplex cells?*
- *What is a macrocolumn of the visual cortex? What information about an image can a single macrocolumn provide?*

Ganglion cell axons enter the optic nerve in an orderly fashion, so that adjacent axons in the nerve correspond to adjacent receptive fields on the retinal surface. The pathway ascends to the lateral geniculate nucleus of the thalamus and then projects to the primary visual cortex in the occipital lobes of the brain (Figure 10-28, *A*).

The visual fields of human eyes overlap. Objects in the right visual field form images on the lateral (temporal) half of the retina of the left eye (Figure 10-28, *B*) and on the medial (nasal) half of the retina of the right eye. To avoid double vision, the images of objects that lie in the overlapping part of the visual field must be projected onto **corresponding points** on both retinas.

The correspondence between the left and right retinal maps of the visual field is maintained by oculomotor reflexes that move the eyes simultaneously in the same direction, fix the gaze of both eyes on an object of interest in the visual field, and maintain this fixation even if both the object and the body of the viewer are moving. The vestibular system, the brainstem, and the visual cortex participate in oculomotor reflexes. Developmental failure of some aspect of these reflexes causes a condition called **strabismus** in which it is impossible for both eyes to fixate on a common point. This usually causes loss of vision in one or both eyes.

Receptors located at corresponding points on each retina must be connected to a single point on the brain's map of the visual field. The merging of two retinal maps into one visual cortical map is accomplished by formation of precise connections between the eyes and the brain. These connections are made during the development of the nervous sys-

A

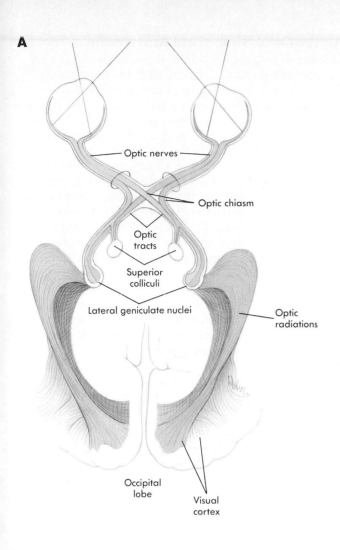

Optic nerves

Optic chiasm

Optic tracts

Superior colliculi

Lateral geniculate nuclei

Optic radiations

Occipital lobe

Visual cortex

FIGURE 10-28

A Central visual pathways. The axons of ganglion cells of the retina synapse with cells of the lateral geniculate nucleus of the thalamus. They also send collaterals to the superior colliculi where visual information is used to control eye movements. Axons from the lateral geniculate project to the visual cortex (the optic radiations). Note the partial crossing (decussation) of optic nerve fibers at the optic chiasm.

B The axons of ganglion cells that project from both eyes to the left visual cortex have receptive fields that collect information from the right visual field *(blue circle)*. This includes ganglion cells on the temporal (lateral) side of the left retina and on the nasal (medial) side of the right retina. The axons of ganglion cells that project from both eyes to the right visual cortex have receptive fields that collect information from the left visual field *(green circle)*. This includes ganglion cells on the nasal side of the left retina and on the temporal side of the right retina. The nasal part of each visual field overlaps to produce a central area of binocular vision.

C An MRI through the optic chiasm.

B

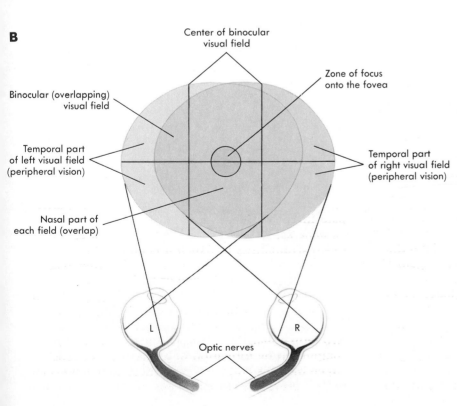

Center of binocular visual field

Zone of focus onto the fovea

Binocular (overlapping) visual field

Temporal part of left visual field (peripheral vision)

Temporal part of right visual field (peripheral vision)

Nasal part of each field (overlap)

L   R

Optic nerves

C

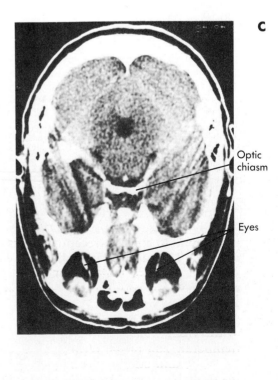

Optic chiasm

Eyes

tem before birth and refined during the first months of visual experience after birth. Ganglion cell axons from the temporal half of the left eye project to the lateral geniculate nucleus without crossing (decussating). Afferent fibers from the nasal half of the right eye decussate at the optic chiasm (shown in an MRI of the brain in Figure 10-28, C) and go to the opposite lateral geniculate. The left lateral geniculate receives binocular input from the right visual field, while the right lateral geniculate receives input from the left visual field. Fibers from the lateral geniculate go only to the corresponding primary visual areas (see Figure 10-28, A).

As an example of how this system works, consider a point in the part of the right half of the visual field that is visible by both eyes. The corresponding points in the retinal maps are on the left sides of both retinas, one lies on the temporal side of the left retina and the other on the nasal side of the right retina. The pathway from the point on the temporal side of the left retina passes to the left visual cortex without decussating; the pathway from the corresponding point on the nasal side of the right retina decussates and also passes to the same spot in the left visual cortex.

As in the cortical maps of other special senses, the visual map is not uniform. The number of visual cortex neurons surveying a particular region of the visual field is inversely related to the degree of convergence in the corresponding area of the retina. As a result, the fovea, which occupies a small portion of the retinal surface, takes up a large part of the visual cortical map.

Detection of Shape and Movement

In contrast to ganglion cells, visual cortical cells are inactive without stimulation and are most responsive to lines or edges rather than spots. One type of cortical neuron, the **simple cell,** is most sensitive to lines oriented in a particular angle to the vertical. Simple cells probably receive a convergence of inputs from lateral geniculate cells whose receptive fields partly overlap (Figure 10-29), so that the most effective stimulus would be a line covering all the excitatory areas. Other cortical neurons, called **complex cells,** are sensitive to both orientation and movement. For example, a complex cell might be stimulated by a line or edge with a particular tilt only if it was also moving in the right direction. Complex cells are less sensitive to the exact position of lines in the visual field than simple cells. **Hypercomplex cells** require not only that the line or edge be moving but also that it be a certain length or have a corner. In sum, the image is analyzed into information about spots in the retina; this information is analyzed in the primary visual cortex for information about contrasting and moving edges and borders.

The columnar organization of the cortex is the basis for analysis of the visual image for the feature of orientation. The visual cortex consists of **macrocolumns,** three of which are shown schematically in Figure 10-30. Each macrocolumn takes up about 1 mm^2 of the cortical surface. All the cells in the various layers of a macrocolumn have receptive fields in the same area of the visual field, and adjacent macrocolumns have adjacent receptive fields. Mac-

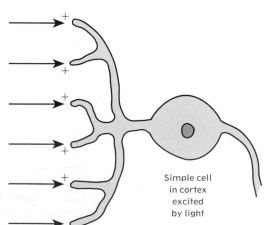

Simple cell in cortex excited by light

FIGURE 10-29
Convergence on a cortical simple cell of inputs from ganglion cells having overlapping fields. The arrangement of the receptive fields with respect to the vertical determines the orientation to which the cortical cell will respond.

NERVE AND MUSCLE

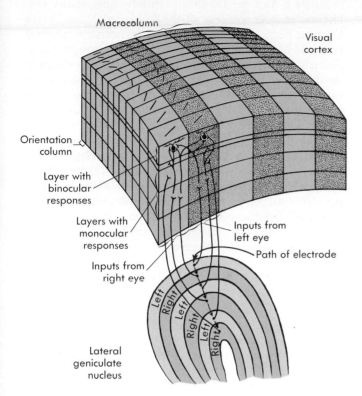

Macrocolumn

Visual cortex

Orientation column

Layer with binocular responses

Layers with monocular responses

Inputs from left eye

Inputs from right eye

Path of electrode

Lateral geniculate nucleus

Left / Right / Left / Right / Left / Right / Left / Right

FIGURE 10-30

Projections from the lateral geniculate nucleus (LGN) to the cortex relay information in a very orderly fashion. The path of an electrode that penetrates the LGN perpendicular to the surface penetrates a series of neurons that process information from one area of the visual field. As the electrode passes through the successive layers, it encounters cells driven first by one eye and then the other. The projections from each eye are segregated as the information is relayed to the visual cortex, producing macrocolumns consisting of the two adjacent regions dominated by corresponding right and left eye monocular responses. The monocular dominance macrocolumns (*stippled* for the left eye projections) run at right angles to the orientation columns. In the more superficial layers of the cortex, interneuronal connections have crossed the boundary between the left and right eye projections and provided neurons with binocular inputs. The projections shown here relate to only one of the many orientation columns detected in that part of the visual field, and such projections are repeated over and over for each part of the visual field. Adjacent macrocolumns represent the bar orientations for all the other corresponding regions of the two retinas.

rocolumns consist of **orientation microcolumns,** each of which contains simple, complex, and hypercomplex cells that all respond maximally to a single orientation.

Within a macrocolumn, the preferred orientations of adjacent microcolumns shift systematically in increments of about 10° of arc, which is shown as the lines oriented in different directions on the surface of the microcolumns in Figure 10-30. Within a single macrocolumn is a band of microcolumns dominated by input from the left eye and a band dominated by input from the right eye. Each cortical macrocolumn seems to contain all of the circuitry

needed to analyze a small portion of the visual world.

THE CHEMICAL SENSES— TASTE AND SMELL

- *What are the functional differences between taste and olfactory receptors?*
- *What are the differences between the central projections of taste and olfactory information?*

Two types of chemical sensory systems use different receptors and process information at different locations in the brain: (1) *taste,* in which the receptors are specialized sensory cells, and (2) *smell,* or *olfaction,* in which the receptors are neurons. Taste and smell are complex, subtle sensations. While we all see a specific wavelength of light as "red," subjective descriptions of the same combination of chemicals in food typically differ. What is pleasing to one person, another may find repulsive. An individual's taste preferences may change with age or experience, and different cultures and subcultures have characteristic food preferences.

Submodalities of Taste

About 10,000 **taste buds** are located in the crevices of small, domelike papillae that cover the tongue, as well as portions of the palate, pharynx, larynx, and upper third of the esophagus. Each taste bud resembles an orange in which individual receptor cells are arranged like the segments (Figure 10-31). The primary receptor cells, taste cells, are modified epithelial cells that extend short, microvillus-like **taste (gustatory) hairs** through a pore to the surface of the tongue. The portion of the receptor cell exposed to sensory stimulation is the apical membrane surface; the rest of the cell membrane (the basolateral surface) is sealed off from the stimuli by a ring of tight junctions between sensory cells and supporting epithelial cells.

Stimulation of taste receptor cells by a large variety of solute molecules can give rise to relatively few sensations. The ones classically recognized as **primary flavors** are sour, sweet, salty, and bitter (Figure 10-32). Whereas an individual taste cell responds most strongly to one of these basic tastes, it also may respond weakly to others. Specific taste buds are concentrated on different parts of the tongue, with sweet and salty on the front, sour on the sides, and bitter at the back. Current research suggests that there might be additional primary taste sensations, such as metallic, but in any case the number of primary taste sensations is small compared with the variety of different stimulus molecules.

Afferent information from the taste buds is relayed directly to specific locations near the mouth region of the somatosensory cortex by way of the

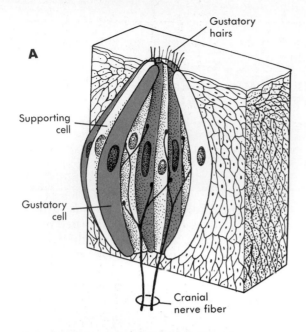

brainstem and the thalamus. In addition, many collaterals project to areas of the brain concerned with feeding behavior, including the hypothalamus and several brainstem structures. Even though taste transduction mechanisms are more or less discrete at the level of the receptors, gustatory input to higher centers is not in the form of "labeled lines" for sweetness, saltiness, and so on. The result of the interactions within the taste buds and the relaying of inputs from several types of receptors to individual afferent neurons is that the information carried by any one axon is ambiguous and only the central comparison of the inputs from many afferent fibers provides sufficient information for decoding the nature and intensity of the stimulus. The subjective sensations of taste are strongly influenced by input from the olfactory receptors.

Mechanisms of Transduction in Taste Buds

The chemistry of the initial events in taste reception differs for each basic taste. The sensation of saltiness is produced by NaCl, KCl, and a few other inorganic compounds. The mechanism of transduction of saltiness, at least in animals, does not involve a specific receptor. Instead Na^+ or K^+ in food enters the taste cell through an open cation-selective channel in the membrane, causing a depolarization. This channel is similar to channels in the apical surface of the epithelial cells of the kidney. The sour taste also does not involve a specific receptor. The

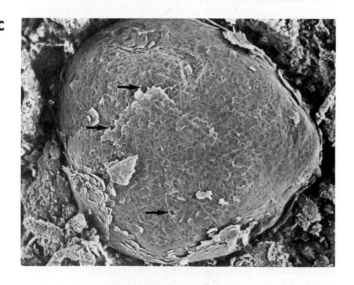

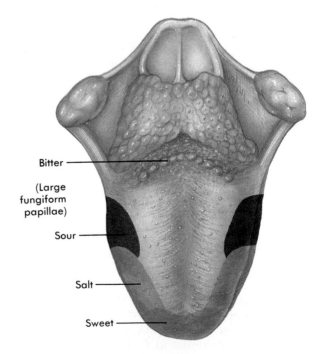

FIGURE 10-31
A Taste bud showing supportive and gustatory cells with gustatory hairs projecting through taste pore.
B Filiform papilla (non-gustatory, contact-sensitive).
C Fungiform papilla (arrows indicate taste bud pores).

FIGURE 10-32
Regions of the tongue sensitive to particular primary flavors.

NERVE AND MUSCLE

Stimulation Is Critical for the Developing Visual System

The responsiveness of the nervous system is easily influenced during its development by changes in the environment; this quality is referred to as **plasticity.** Plasticity of the visual system has been demonstrated by experiments in which experimental animals were deprived of normal visual stimulation. These experiments have shown that there are times of development—**critical periods**—during which even brief periods of deprivation have profound effects on the visual system. In the most extreme case, if one eye is prevented from receiving any light during the critical period, binocular vision will be permanently impaired.

The visual cortex contains bands of microcolumns that are dominated by input from one eye, alternating with bands that are dominated by input from the other eye. This can be demonstrated experimentally by injecting the eye of an animal with a radioactive labeled amino acid. The cells of the retina pass the amino acid to successive cells in the visual pathway, and ultimately some of the labeled molecules are taken up by visual cortical cells. Autoradiograms of the visual cortex of animals that have been visually deprived in one eye show that the portion of the visual cortex dominated by input from the undeprived eye increases. The synapses from the stimulated eye literally take over the cortical neurons that would normally have been dominated by inputs from the unstimulated eye.

In other experiments, visual stimulation during the critical period was restricted to lines of a single orientation. In this case, most of the animal's behavior after the critical period seems normal, but behavioral tests show that the animal is blind to visual stimuli of all orientations except the one that it was exposed to during the critical period. When the visual cortex of such an animal is explored with microelectrodes, all of the orientation microcolumns are found to have become sensitive to the single orientation that the animal had been exposed to during the critical period. The effect was not reversed by exposure to a rich visual environment after the critical period was over.

These experiments have important implications for humans, whose critical period for visual development may continue until about 6 years of age. First, they suggest that medical treatments that restrict visual input, for example, surgery that would necessitate bandaging the eye for a long period, be undertaken with care to minimize developmental effects if the patient is a child. More broadly, these experiments emphasize the importance to children's mental development of providing an environment rich in all kinds of sensory stimulation.

H^+ ions produced from the dissociation of acids in foods block a H^+-gated K^+ channel that is concentrated in the apical portion of the cell. When this population of K^+ channels is blocked, the membrane potential drifts toward the Na^+ equilibrium potential and the depolarized cell releases synaptic vesicles.

Mammals are some 1000 times more sensitive to bitter molecules than to sweetness or salt. This may be an adaptation that reflects the fact that plant alkaloids, which are generally bitter, are often poisonous. Chemicals with quite different structures taste bitter, and more than one receptor may be involved. The most bitter substance known, denatonium, was developed as a result of the search for a better local anesthetic and has been used on fingernails of children in attempts to control thumb-sucking and nail-biting. In isolated mouse taste cells, denatonium blocked K^+ channels, causing a brief depolarization. Denatonium also increased internal Ca^{++} concentration because of release from internal stores, an effect involving inositol triphosphate as a second messenger. This also tends to depolarize receptor cells.

The nature of sweetness has been the topic of much research in the food industry. Sweetness appears to result from the binding of sucrose or other molecules to a specific receptor. However, the mammalian sweetness receptor has been difficult to examine because it is destroyed by the proteolytic enzymes used to dissociate taste cells. The binding of the sweet stimulus molecule to a membrane protein activates a second-messenger system. At least two second messenger systems, one involving adenylate cyclase closure of K^+ channels and the other involving Na^+ channel activation, may mediate sweet sensations.

Olfactory Receptors

Olfactory receptors are located in a small patch of **olfactory mucosa** on the dorsal surface of the nasal cavity (Figure 10-33). Each receptor is a neuron with a long, thin dendrite that terminates in a small knob bearing several cilia (Figure 10-34). Supporting cells of the olfactory epithelium secrete mucus; to be smelled, odor molecules must first dissolve in the layer of mucus that covers all surfaces of the nasal cavity. A protein called the **olfactory binding protein** is secreted by nasal glands and mixed with the incoming air. This protein is similar to a family of carrier proteins that transport lipophilic molecules within aqueous media. The binding affinity of odorant molecules with the olfactory binding protein predicts how easily they can be detected, suggesting that this molecule, which is not specific enough to be a receptor, may play a role in the delivery of volatile molecules to the sensory cell membrane. Transduction of olfactory stimuli is not well understood at the level of the receptor protein, but the second messenger system is mediated by G proteins that activate cyclic nucleotide-gated ion channels. In addition to the types of G proteins typically found in neural tissue, a unique G protein has been isolated from olfactory cell cilia. Some failures of olfactory function (**anosmias**) have been correlated with congenital defects in G proteins.

Olfactory neurons are not specialized to detect single fundamental odors. A given receptor may respond vigorously to some stimulus molecules, weakly to several others, and not at all to others, so that the concept of a few primary submodalities is not useful for understanding the olfactory system. A dozen specific anosmias for particular odor categories have been shown to be inheritable traits, but this information would set only the lower limit to the number of submodalities, which is probably much larger and could be as diverse and individualized as the molecules generated by the immune system. Sensations of odor must therefore arise in the brain from the integration of many different receptor responses. Consequently, the olfactory system can mediate a large number of different odor sensations. The number may be very large indeed. Each person possesses a unique odor signature that trained dogs can distinguish from those of a practically unlimited number of other odors.

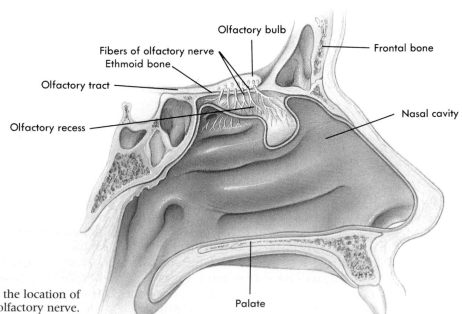

FIGURE 10-33
Lateral wall of the nasal cavity showing the location of the olfactory bulb and the fibers of the olfactory nerve.

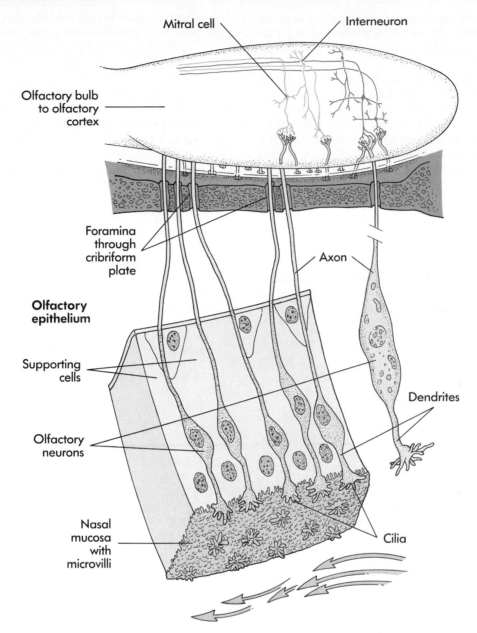

Mitral cell Interneuron

Olfactory bulb to olfactory cortex

Foramina through cribriform plate

Olfactory epithelium

Axon

Supporting cells

Olfactory neurons

Dendrites

Nasal mucosa with microvilli

Cilia

FIGURE 10-34
Schematic diagram of receptors and supporting cells in the olfactory mucosa. The cytoplasm of the supporting cells is filled with material that forms mucus when secreted. The olfactory receptors send a long projection tipped with cilia to the surface of the epithelium. The cilia of the olfactory receptors form a shaggy coating on the surface of the epithelium. The axons of the receptor cells pass through the cribriform plate and enter the olfactory bulb of the brain, where they synapse with interneurons.

Central Olfactory Pathways

Axons from olfactory receptors pass through perforations in the cribriform plate of the ethmoid bone (see Figure 10-34) and synapse with **mitral cells** and **tufted cells** in the **olfactory bulb** of the brain. The subsequent pathways followed in the processing of olfactory information do not parallel those described for the other special senses. A major pathway leads into the limbic system, the seat of emotion and memory. This route can affect visceral responses and influence the activities of the hypothalamus and pituitary. The importance of olfaction in mammalian evolution is reflected in the direct line that the olfactory receptors have to relatively primitive parts of the brain responsible for sex, aggression, feeding, and visceral homeostasis. It is possi-

ble that olfactory signals processed along this route influence moods and behavior without entering conscious awareness. The second route leads to the olfactory cortex. Projections from there to the frontal cortex can lead to the recognition of odors. Connections between the olfactory cortex and other sensory regions of the cortex allow integration of olfactory sensations with those arising from other sensory modalities. Signals that are processed along this route are more likely to lead to conscious sensations.

The sensitivity of the olfactory receptors for odor stimulants is several orders of magnitude greater than the sensitivity of taste receptors. The sensitivity of the olfactory system is modified by the rapid rate of adaptation exhibited by the central neural pathways processing the odor inputs. Adaptation is responsible for the loss of perception of an odor soon after it is first encountered and serves to filter non-novel stimuli from the stream of olfactory information presented to higher processing areas. The central nervous system also alters odor perceptions through efferent control of sensory synapses in the olfactory bulb. This control reflects the overall state of the individual and is altered, for example, by hunger and reproductive status. As with other sensory systems, the efferent control modulates the gain and selectivity of the filtration process.

● *STUDY QUESTIONS*

1. Under what conditions might the mechanisms that operate to decrease the effect of loud noises on the ear fail to provide adequate protection?

2. What are the principles that determine how pure tones are resolved in the organ of Corti?

3. How is it that excitation of a relatively large fraction of the hair cells can result in activation of only a narrow band of cells in the auditory cortex?

4. Describe factors that determine whether the hair cells in the ampulla will depolarize or hyperpolarize in response to a motion of the head.

5. Why is the nystagmus that occurs following rotation opposite in direction to the nystagmus occurring at the start of a rotation?

6. Trace the pathways of light detection from receptor cells to ganglion cells in the retina. What factor distinguishes the ON pathway from the OFF pathway?

7. What is the sequence of events that leads from photopigment bleaching to a membrane potential change in a photoreceptor?

8. What pattern of connections in the projections from the lateral geniculate nucleus could account for the response pattern of a simple cell? A complex cell?

9. Compare the developmental origin of the sensory cells that are responsible for the two chemical senses. Describe the ambiguous nature of the information provided by any single receptor's responses to stimulation. How does the brain process this type of information?

10. Why is the concept of primary odors relatively useless in explaining the responses of the olfactory system?

Choose the MOST CORRECT Answer.

11. The vestibular system provides information about all EXCEPT:
    a. Orientation of the head with respect to gravity
    b. The frequency components of a sound
    c. Linear acceleration of the head when it is moving in a straight line
    d. Angular acceleration when changing direction or velocity of rotation of the head

12. These structures detect angular acceleration:
    a. Semicircular canals
    b. Hair cells in organ of Corti
    c. Macular organs
    d. Elastic fibers of basilar membrane

13. The ability of the ear to analyze the frequency components of sounds depends on the resonance of the:
    a. Organ of Corti
    b. Tectorial membrane
    c. Basiliar membrane
    d. Cochlea

14. The transparent anterior portion of the sclera is the:
    a. Choroid layer
    b. Vitreous humor
    c. Ciliary body
    d. Cornea

15. These retinal interneurons function as the "output cells" of the retina, with their axons joining to form the optic nerve of each eye:
    a. Ganglion cells
    b. Bipolar cell
    c. Horizontal cells
    d. Amacrine cells

16. Choose the TRUE statement about the cones or photopic system:
    a. Cones provide poor resolution of detail
    b. Cones provide no information about color.
    c. Cones function only in bright light.
    d. Cones are more sensitive to light than are rods.

17. Choose the FALSE statement:
    a. Taste receptors are specialized sensory cells.
    b. Olfaction receptors are neurons.
    c. Taste receptors respond to the primary flavors: sour, salty, sweet, and bitter.
    d. Olfactory neurons are specialized to detect single fundamental odors.

SUMMARY

1. The **tympanic membrane** and auditory **ossicles** transmit sound energy to the **cochlea.** The sensitivity of this transmission is regulated by small muscles in the middle ear.
2. The resonance properties of the cochlea translate sound energy into displacement in such a way that specific hair cells are stimulated according to the pure tone components of the sound.
3. Sound localization in space depends on the detection of phase and intensity differences between the two ears.
4. The structure of the vestibular system translates gravitational or inertial forces into hair-cell displacement. Each **semicircular canal** is oriented in a different plane and is preferentially stimulated by rotations of the head in that plane. The **saccule** and **utricle** are oriented at right angles to one another.
5. The saccule and utricle are responsive to orientation and linear acceleration of the head; the semicircular canals are responsive to angular acceleration.
6. The **cornea** and **lens** of the eye focus images on the **retina**, which contains **photoreceptors** and several types of **interneurons.**
7. In the retina, lateral inhibition filters the image to increase contrast. Retinal **ganglion cells,** whose axons project to the thalamus, are tuned to respond to spots on contrasting backgrounds.
8. The central processing of visual images involves feature extraction at increasing levels of analysis. Each **orientation microcolumn** of the visual cortex contains cells that respond to lines of a single orientation, with additional conditions of motion and length that depend on the cell type. A complete analysis of the part of the image that falls on any two corresponding points of the retina is carried out by a **macrocolumn** containing orientation microcolumns, which analyze the input from each retina for all possible orientations.
9. The taste modality is characterized by a small number of primary taste perceptions; in contrast, the olfactory modality gives rise to a wide variety of olfactory perceptions. The taste receptors are epithelial cells; the olfactory receptors are neurons. Taste information is processed in a specific region of the neocortex.
10. The olfactory system is unusual in that processing of olfactory information goes on in the limbic system in parallel with processing in the olfactory cortex.

● SUGGESTED READINGS

FINGER TE, SILVER WL: *Neurobiology of taste and smell,* Somerset, NJ, 1987. John Wiley & Sons. A summary of what is and is not known about these special senses.

FREEMAN WJ: *The physiology of perception,* Scientific American, February, 1991, p. 78. Description of analysis of chaos as it relates to perception, particularly in the olfactory system.

KINNAMON SC: *Taste transduction: a diversity of mechanisms,* Trends in Neuroscience, 11:491, 1988. Readable summary of vertebrate taste transduction mechanisms.

NICHOLLS JG, MARTIN RA, WALLACE BG: *From neuron to brain,* ed 3, Sunderland, Mass, 1992, Sinauer Associates. A comprehensive textbook in the neurosciences.

PIERCE JR: *The science of musical sound,* New York, 1983, W.H. Freeman. Nice description of the principles of sound localization in the cochlea.

POGGIO T, KOCH C: *Synapses that compute motion,* Scientific American, May, 1987, p. 46. Describes how certain cells in the visual system become differentially sensitive to moving objects.

SCHNAPF JL, BAYLOR DA: *How photoreceptor cells respond to light,* Scientific American, April, 1987, p. 40. Discussion of the role of second messengers in visual transduction.

TROTTER DM: *How the human eye focuses,* Scientific American, July 1988, p. 92. Explores the biochemical and geometrical factors that cause the eye to gradually lose its ability to focus on nearby objects.

Skeletal, Cardiac, and Smooth Muscle

On completing this chapter you will be able to:

- Identify the components of the contractile machinery of muscle
- Understand the origin of force generation in muscle
- Describe the structure of a sarcomere
- State the sliding filament model of muscle contraction
- Describe the sequence of events mediating excitation-contraction coupling in muscle
- Distinguish between isotonic and isometric contractions
- State the functional differences between cardiac and skeletal muscle
- Understand the initiation and control of contraction in smooth muscle
- Appreciate the different roles of calmodulin and caldesmon in smooth muscle contraction
- Distinguish between multi-unit and unitary smooth muscle

Muscle accounts for approximately 40% to 50% of the body weight of an adult. Muscles consume 25% of the total oxygen used at rest; during strenuous exercise, the oxygen consumption of muscle can increase 10 to 20 times. Thus muscle metabolism accounts for a major part of the body's energy budget. The three basic muscle types are skeletal, cardiac (heart), and smooth muscle. Both skeletal and cardiac muscle have alternating light and dark bands called striations, which result from the ordered way in which contractile proteins (actin and myosin) are arranged within the muscle cells. Thus skeletal and cardiac muscle are referred to collectively as striated muscle. The basic contractile machinery is similar in both skeletal and cardiac muscle.

Skeletal muscles are connected to the bones of the skeleton by tendons and cause movement about joints. Each skeletal muscle contains many muscle cells, called muscle fibers because of their elongated shape. Motor neurons innervate small groups of fibers, the motor units, which are the smallest functional elements of skeletal muscle. Skeletal muscles are subject to fatigue with continued use, although the rate of fatigue development is much higher in muscle fibers using glycolysis than in fibers using oxidative pathways for metabolism.

Cardiac muscle resembles all skeletal muscle in that it is striated, and in its resistance to fatigue it corresponds to the fatigue-resistant oxidative skeletal muscle fibers. However, cardiac muscle also differs from skeletal muscle in several important respects. First, individual cardiac muscle fibers are in electrical communication with each other. Also, muscle fibers from some parts of the heart are able to generate rhythmic spontaneous action potentials in the absence of any nervous stimulation.

Smooth muscle contains the same contractile elements as striated muscle. However, because the contractile proteins are not organized in a regular array, smooth muscle lacks striations. Smooth muscle is a major constituent of the internal muscular organs of the body, including the arteries and veins, the urinary bladder, and the gastrointestinal tract. In many cases, smooth muscle exhibits spontaneous activity, which can be modified by neural and hormonal factors.

THE STRUCTURE OF SKELETAL MUSCLE
Organization of Skeletal Muscle Tissue

> - *What structural feature gives striated muscle its banded appearance?*
> - *What is the "sliding-filament" model?*
> - *What are the roles of the muscle proteins actin, myosin, troponin, and tropomyosin?*

Most skeletal muscles are attached to the bones of the skeleton by **tendons composed of collagen** (Figure 11-1). Tendons are various lengths. For example, the muscles that control finger movements are located in the arm and their tendons are quite long, whereas the tendons of many postural muscles are fairly short. The two places a muscle attaches to bone, the origin and insertion, are defined by the result of shortening the muscle. The **origin is the less movable** of the two points, and contraction pulls the **insertion** toward the origin.

Limb muscles that cause the angle of a joint to decrease are classified as **flexors; extensors** have the opposite effect. Typically, the angle of a joint is affected by several flexors and extensors. Muscles that cooperate to produce movement in the same direction are synergistic muscles; those that oppose each other are **antagonistic.** For example, the biceps and brachialis are synergists in that they both flex the forearm. The triceps extends the forearm and is therefore an antagonist to the biceps (and vice versa).

In skeletal muscle, the fibers frequently extend the entire length of the muscle. A muscle may contain a few hundred fibers to several thousand fibers. The diameter of an individual fiber ranges from 10 to 100 μ. Skeletal muscle fibers are formed by the fusion of cells in development, so each fiber has many nuclei. Muscle fibers are present for life and are not replaced.

Each muscle is surrounded by a connective tissue sheath (the **epimysium**) that is continuous with the muscle's tendons (Figure 11-2, *A*). Within the muscle, bundles of fibers are bound together by sheaths of **perimysium** to form **fascicles.** Fascicles contain many muscle fibers, each enclosed in its own sheath of **endomysium.** These connective tissue elements transfer the force generated by muscle contraction to the tendons.

The cell membrane of a skeletal muscle fiber is called the **sarcolemma** (Figure 11-2, *B*). Muscle fibers contain long, parallel **myofibrils** 1 to 2 μ in diameter, which make up 75% of a muscle's total volume. Myofibrils are made up of repeating subunits called **sarcomeres.** Adjacent sarcomeres are stacked "in register" within the muscle fibers, to form the characteristic striation patterns of repeating bands and lines seen in muscle tissue viewed with a light microscope.

A sarcomere is a cylindrical structure formed from the overlapping of **thin** and **thick filaments,** which contain the contractile proteins **actin** and **myosin,** respectively (Figure 11-2, *C*). The thin filaments appear smooth, whereas the thick filaments have numerous protrusions referred to as **cross-bridges.** A relaxed sarcomere is 1.5 to 2.0 μ long. Seen from the side, it has a central **H zone** in which the thick filaments are not overlapped by thin filaments. On either side of the H zone is a region of overlapping of thick and thin filaments.

The entire region of the sarcomere occupied by thick myosin filaments is the **A band** (for anisotropic). Toward the ends of the sarcomere the thin filaments are not overlapped by thick filaments; these two regions of nonoverlap in each sarcomere are the **I bands** (for isotropic) that continue into the adjacent sarcomeres. The ends of the sarcomere and beginnings of adjacent sarcomeres are marked by disklike **Z structures,** to which the thin filaments are attached. The Z structures appear as lines in longitudinally sectioned muscle. The sarcomeres of all the myofibrils within a muscle fiber are typically lined up so that all the Z lines form continuous, thin bands (Figure 11-3).

Sliding Filament Model of Muscle Contraction

Theories about the mechanism of muscle contraction began in ancient Greece with the physiologist Galen, who held that muscles contracted when they were inflated with a fluid delivered to them by nerves. A later theory explained contraction as a form of shrinkage. This theory was disproved by the finding that muscles do not change their volume as they contract.

A key piece of evidence about the nature of muscle contraction was provided by the discovery

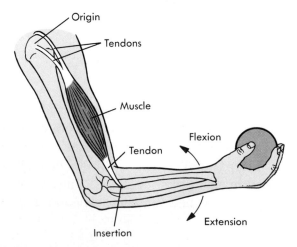

FIGURE 11-1

The gross anatomy of a skeletal muscle, showing its attachment to the skeleton by tendons.

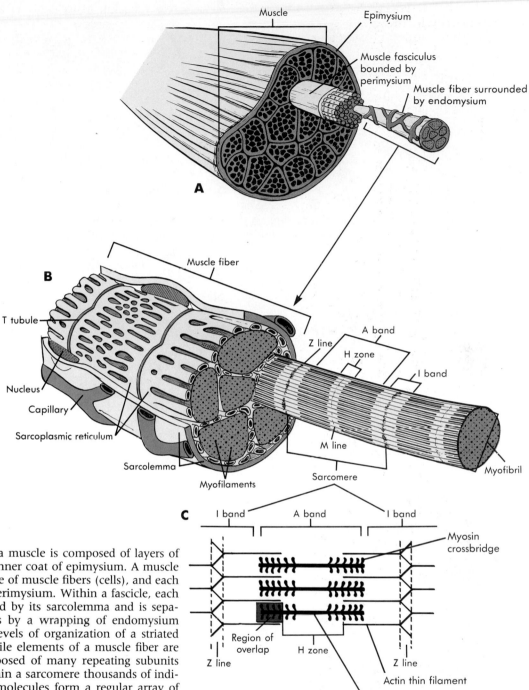

FIGURE 11-2

A, The outer covering of a muscle is composed of layers of connective tissue and an inner coat of epimysium. A muscle fascicle consists of a bundle of muscle fibers (cells), and each fasicle is surrounded by perimysium. Within a fascicle, each individual fiber is bounded by its sarcolemma and is separated from adjacent fibers by a wrapping of endomysium (not illustrated). **B,** The levels of organization of a striated muscle fiber. The contractile elements of a muscle fiber are the myofibrils, each composed of many repeating subunits called sarcomeres. **C,** Within a sarcomere thousands of individual actin and myosin molecules form a regular array of thick and thin filaments.

that the degree of overlap of thick and thin filaments was greater and sarcomere length was shorter in muscles that were contracting when they were fixed for microscopy (Figure 11-4). On the basis of such evidence, British scientists Huxley and Hanson proposed the **sliding filament theory** of contraction in 1955.

If thick and thin filaments slide across one another, physical links that generate force must form between the two filament types during contraction. Evidence for such links was provided by electron micrographs of contracting muscle, which revealed numerous crossbridges—connections between thick and thin filaments—in muscles fixed during activity. These crossbridges are best seen in cross section (Figure 11-5). Sections through the I band (Figure 11-5, *A*) show only thin filaments. In the region of filament overlap (Figure 11-5, *B*), there is a characteristic pattern of thick filaments, with crossbridges located 60 degrees apart and surrounded by hexagonal arrays of thin filaments. Each thick filament interacts with six thin filaments, whereas each

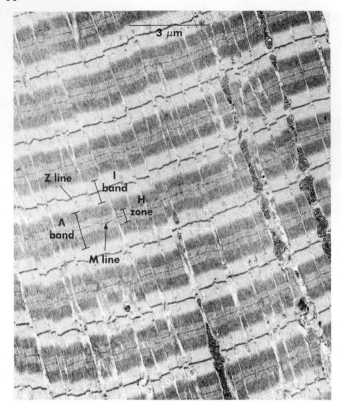

A

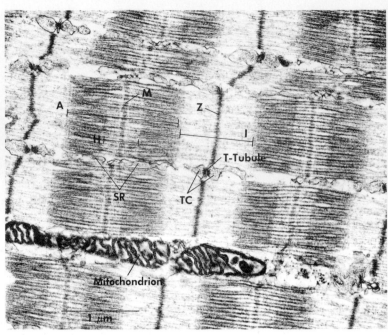

B

FIGURE 11-3
A Longitudinal view of a portion of a muscle fiber including many myofibrils. The sarcomeres of the myofibrils are lined up, giving the whole muscle fiber a banded appearance. Mitochondria are scattered among the myofibrils.
B Several sarcomeres from the muscle shown in **A**, viewed at higher magnification. *TC,* Terminal cisternae; *SR,* sarcoplasmic reticulum; *M,* mitochondria.

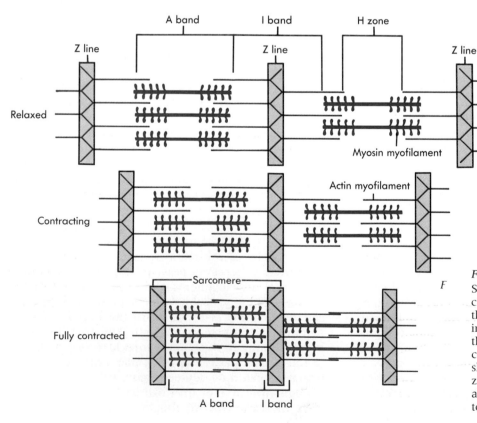

FIGURE 11-4
Sarcomere shortening in response to crossbridge formation. Shortening of the whole myofibril is achieved by an increase in the degree of overlap of thick and thin filaments in each sarcomere. During shortening, the I bands shorten, but the A bands do not. The H zone narrows and may even disappear as the actin filaments meet at the center of the sarcomere.

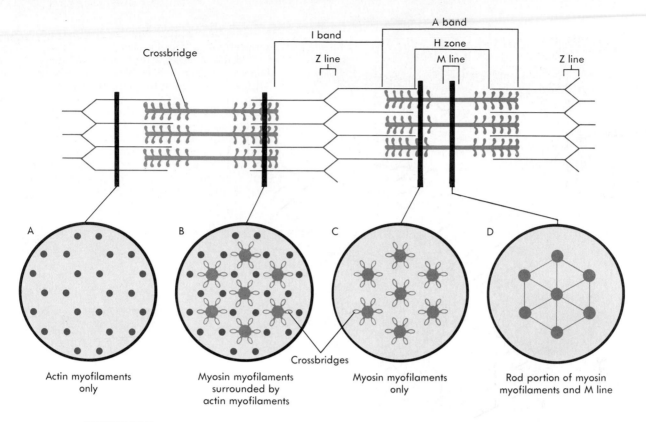

Crossbridge

I band

A band

Z line

H zone

M line

Z line

A
Actin myofilaments
only

B
Myosin myofilaments
surrounded by
actin myofilaments

Crossbridges

C
Myosin myofilaments
only

D
Rod portion of myosin
myofilaments and M line

FIGURE 11-5

Cross sections at various points in a sarcomere. **A** In the I band there are only actin filaments. **B** Cross sections in the region of overlap of thin and thick filaments show that each thick filament is surrounded by a hexagonal array of thin filaments. **C** In the region of non-overlap only myosin molecules are present. **D** In the middle of a sarcomere, the myosin molecules lack crossbridges.

thin filament interacts with three thick filaments.

The Hungarian scientist A. Szent-Gyorgyi extracted actin and myosin from muscle and showed that when the actin and myosin were mixed together, they formed actomyosin. This interaction is the basis of crossbridge formation. Two questions are posed by this discovery. How do crossbridges generate force? What regulates formation of crossbridges? These questions have largely been answered by studies of the protein molecules that compose the thick filaments and thin filaments.

MUSCLE CONTRACTION AT THE MOLECULAR LEVEL
The Structure of Thin Filaments

Thin filaments are composed of a double helical strand of identical globular actin molecules that resembles two strands of pearls twisted around each other (Figure 11-6, *A*). The single globular molecules of actin are called G-actin (4 nm in diameter; molecular weight about 50,000 daltons). Each strand of G-actin subunits is about 1.0 μ long and is called **F-actin**. A single twist of the F-actin helix is 70 nm long and contains 13.5 G-actin subunits. Each G-actin subunit has a single site where a myosin head can bind to it. Because the thin filament

is formed by the twisting of two F-actin strands, the binding sites of some of the G-actin subunits face the inside of the helix and are not accessible to the myosin heads. G-actin subunits that have myosin-binding sites projecting outward and available for myosin binding occur every 2.7 nm along the thin filament.

In addition to actin, thin filaments contain two regulatory proteins (Figure 11-6, *A*). Strands of the protein **tropomyosin** lie on the actin filament. Each tropomyosin strand extends along about 7 G-actin subunits, and the strands are joined end to end. The second regulatory protein of thin filaments, **troponin**, has three globular subunits. One subunit binds to a strand of tropomyosin, the second attaches to actin, and the third has a binding site for Ca^{++}.

When Ca^{++} is not present, troponin is bound to both actin and tropomyosin in such a way that the tropomyosin strands block the myosin-binding sites present on each G-actin subunit of F-actin (Figure 11-6, *B*). The presence of Ca^{++} and its binding to troponin change the shape of the troponin molecule so that the tropomyosin strand shifts relative to the F-actin filament and thus exposes the myosin-binding sites on the G-actin subunits, allowing actin and myosin to interact (Figure 11-6, *B, right*). Thus

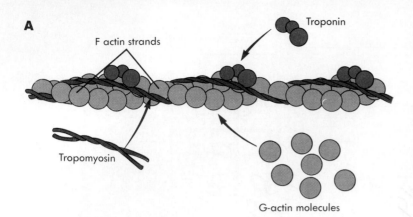

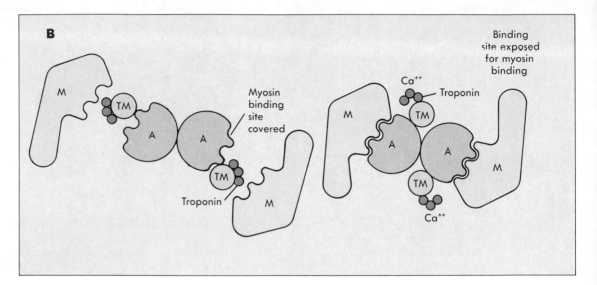

FIGURE 11-6

A The components of thin filaments. G-actin is assembled into single-stranded F-actin. F-actin strands are twisted together to form the backbone of a thin filament. Double helixes of tropomyosin lie on the F-actin strands. Each tropomyosin has a molecule of troponin positioned at one end.
B A view of a thin filament showing the relationship of troponin, tropomyosin (TM), and F-actin. Troponin consists of three subunits. Binding of myosin heads (M) is impossible when tropomyosin covers the binding sites (left), but is possible when Ca^{++} binding to troponin causes the tropomyosin to move away from the binding sites (right).

part of the question of how contraction is controlled in striated muscle is answered by an understanding of the roles of troponin, tropomyosin, and Ca^{++}.

The Myosin Subunit Structure of Thick Filaments

Each thick filament is 1.5 μ long and 15 nm in diameter and is composed of several hundred myosin molecules, each with a molecular weight of about 500,000 daltons. The backbone of the thick filament is a bundle of the long chains of myosin molecules, the heads of which extend out to the sides (Figure 11-7). The myosin molecules are oriented with their heads extending away from the center of the filament, so that the two ends of the filament have heads oriented oppositely and the center region has no heads (see Figure 11-5).

Myosin molecules have a polarity that affects how they associate with each other to form the thick filaments. Each myosin unit is a double structure composed of two heavy polypeptide chains that form a rodlike tail and a double head (Figures 11-7 and 11-8). The heads, which bind actin and engage in ATPase activity, are joined to the rest of the myosin molecule by peptide regions that act as hinges. Three or four smaller (light) polypeptide chains are associated with the head and play a regulatory role.

Myosin molecules in solution do not generate force when they interact with actin because they are randomly oriented and not anchored. However, when the myosin molecules associate to form the thick filaments of the sarcomeres, binding to actin allows the energy to produce a force directed toward the center of the thick filament.

Powerstroke of Muscle Contraction

In a muscle at resting length there is some overlap between the thick and thin filaments. During contraction, the thick filaments pull the populations of thin filaments on each side of the sarcomere toward the center of the sarcomere (see Figure 11-4). This pulling force is caused by the combined action of all the crossbridges formed when myosin heads are able to contact actin binding sites. In a contracting muscle, each myosin head undergoes a cycle of (1) attachment to an adjacent thin filament, followed by (2) a **powerstroke** that moves the head about 10 nm relative to the site of attachment, followed by (3) detachment from the thin filament, and followed by (4) the beginning of another cycle. An individual myosin head can perform this cycle several times a second.

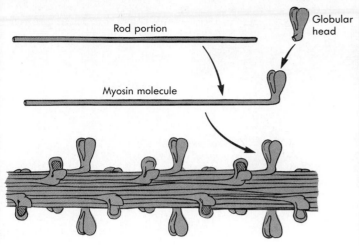

FIGURE 11-7
Single myosin units have a double globular head at the end of a long rod. The long rod is composed of two heavy polypeptide chains; the ends of the heavy chains, together with the light polypeptide chains of myosin, form the heads. If myosin units are mixed together, they self-assemble into thick filaments. The myosin heads project from the thick filament at the proper angles to interact with the hexagonal array of thin filaments.

A description of the myosin cycle should start at the point just before the head attaches to the thin filament. This is when the myosin is in the state that it would be in at the instant of activation of a previously resting muscle (Figure 11-9, *A*). A myosin head that is ready to bind to actin has hydrolyzed a molecule of ATP. ADP and inorganic phosphate remain bound to the myosin, forming a **myosin:ADP complex.** Myosin:ADP represents a "high-energy" form of myosin that can bind to exposed actin sites in the presence of Ca^{++}.

Myosin heads are thought to have several sites that can interact with several sites on actin. After the first site on the myosin head has bound to the first actin site, the energy of the myosin:ADP complex is gradually transformed into mechanical energy by sequential binding with sites of greater and greater affinity. Sequential binding rotates the myosin head, generating tension between the thick and thin filaments (Figure 11-9, *B*). This is the powerstroke of the crossbridge cycle. During the powerstroke, ADP and phosphate are released from the head.

At the end of the powerstroke, the myosin head is tightly bound to the thin filament in a low-energy form of actomyosin. This state is called **rigor complex** because the thick and thin filaments cannot slide past each other and the muscle is stiff. In living muscle, the rigor complex is quickly broken by the rebinding of a molecule of ATP to the head (Figure 11-9, *C*). When blood circulation and breathing stop at death, the energy metabolism of muscle

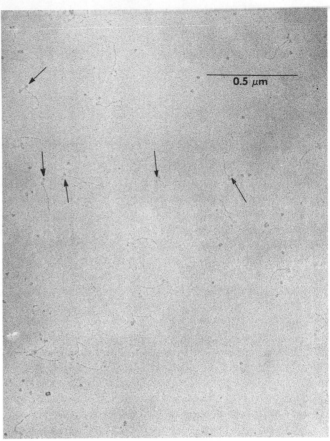

FIGURE 11-8
An electron micrograph of free myosin units. The arrows indicate the myosin heads.

halts. The small supply of ATP is quickly used up and the sarcoplasmic reticulum loses its ability to sequester Ca^{++}. As a result, actin and myosin remain locked together in the rigor complex. This complex is responsible for the rigor mortis of muscle that occurs after death.

In biochemical terms (Figure 11-10), when the rigor complex is broken by the binding of a molecule of ATP to the head, the low-energy actin:myosin complex dissociates to yield actin and a **myosin:ATP complex.** As soon as ATP can bind to myosin, it is hydrolyzed to ADP and inorganic phosphate to regenerate the high-energy state of myosin, which can begin another crossbridge cycle. The rate at which ATP can be hydrolyzed is a measure of the cycling rate and is termed the **ATPase activity** of the myosin.

Muscle Contractility

The contractility of a muscle fiber, the extent to which it is able to generate force and to shorten rapidly, is determined by the molecular properties of the thick and thin filaments. For example, different forms of myosin, or **myosin isozymes,** have different ATPase activities. If the rate of ATP hydrolysis is rapid, force development and shortening occur

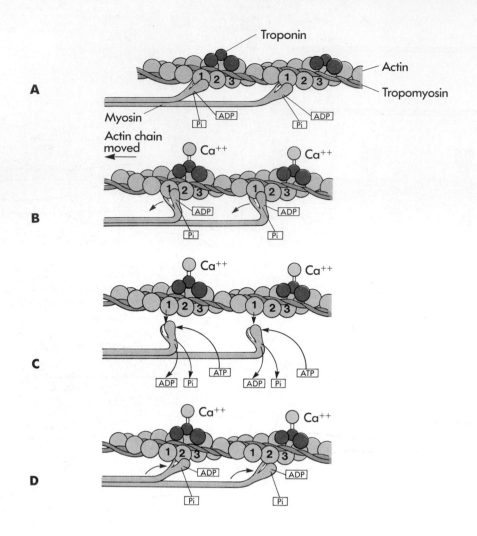

FIGURE 11-9

The crossbridge cycle. In the relaxed state (Ca^{++} concentration less than 10^{-7} M), the myosin-binding site on the actin filament is blocked by tropomyosin, and myosin and actin are unable to interact. **A** The myosin heads are in the "high energy" state, with the resulting ADP and (P) still attached. In the presence of elevated Ca^{++}, troponin complexes form, moving tropomyosin away from actin-binding sites. Myosin then binds to exposed active sites *(1)* on the actin molecules. **B** The myosin head rotates causing the thin filament to move relative to the thick filament. This is the powerstroke. **C** After the powerstroke, ATP binding releases the head from the rigor complex. The ATP is split, returning the myosin head to the high energy state. **D** The cycle is repeated with myosin binding to the next site *(2)* on the actin filament.

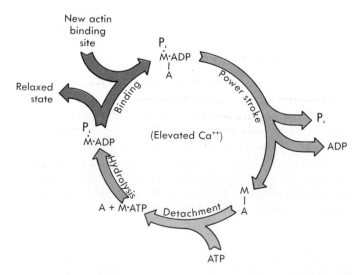

FIGURE 11-10

Summary of the biochemical reactions associated with the crossbridge cycle. *A* Actin; *M* Myosin; *P* Phosphate.

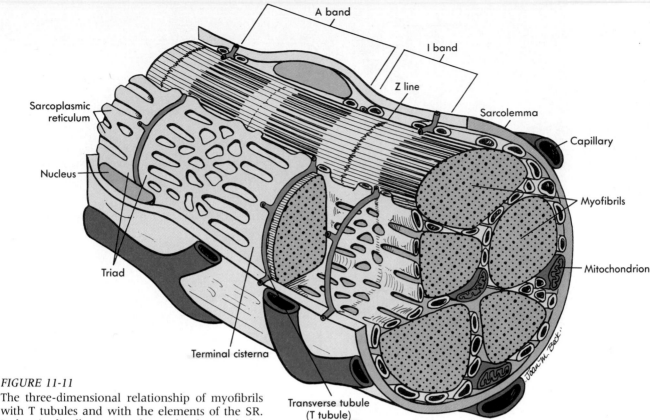

Labels on figure:
A band
I band
Z line
Sarcolemma
Capillary
Myofibrils
Mitochondrion
Transverse tubule (T tubule)
Terminal cisterna
Triad
Nucleus
Sarcoplasmic reticulum

FIGURE 11-11

The three-dimensional relationship of myofibrils with T tubules and with the elements of the SR. In longitudinally sectioned muscle, the T tubules and terminal cisternae appear as three-part structures called muscle triads.

- *What is excitation-contraction coupling?*
- *At what stage in crossbridge cycling is energy transferred from ATP to myosin?*
- *What is a twitch? What process determines the duration of a twitch? What factors determine the ability of a muscle to undergo tetanic contraction?*

rapidly and the force developed is relatively great. Each myosin head repeats the crossbridge cycle five times per second in fibers that have a myosin isozyme with high ATPase activity. Such muscles are called **fast-twitch muscles** and are found in the eyes and fingers. A slower rate of ATP hydrolysis and smaller contractile force are characteristic of **slow-twitch muscles**, which possess a different isozyme and are seen in the back muscles.

EXCITATION-CONTRACTION COUPLING
T-Tubules of Striated Muscle

Every few micrometers along the length of a striated muscle fiber, the sarcolemma invaginates to form **transverse tubules,** (T tubules) fingers of membrane that penetrate the muscle fiber interior (Figure 11-11). Tubules branch extensively to come into close proximity to each myofibril. The lumina of the T tubules are near the junctions between sarcom-

ercs. Because T tubules are continuous with the muscle membrane, the extracellular space extends into the interior of the muscle fiber.

The membrane of the T tubule system is part of the plasma membrane, and when the membrane is depolarized during a muscle action potential, the action potential spreads deeply into the fiber interior. The T tubule system is an adaptation for rapid activation of all parts of a muscle fiber. Each fiber is 10 to 100 μm in diameter, a long distance for a diffusing messenger to travel. The passage of an action potential across the sarcolemma, an event termed **excitation**, must be communicated rapidly to the innermost parts of the muscle if all parts of the fiber are to contract simultaneously. The T tubules carry the electrical signal into the cellular interior, minimizing the distance that must be covered by a diffusing intracellular second messenger.

Sarcoplasmic Reticulum

The second tubule system, which is greatly elaborated in skeletal muscle fibers and less elaborated in cardiac muscle fibers, is the smooth endoplasmic reticulum. It forms an intracellular system of interconnected tubules known in these fibers as the **sarcoplasmic reticulum (SR)** (Figure 11-11). The interior of the SR surrounds all the myofibrils, but, although it comes close, it never unites with the T tubules. The SR enlarges into structures known as **terminal cisternae (lateral sacs)** located next to

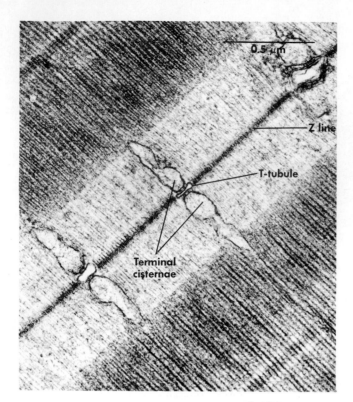

TABLE 11-1	The Events of Muscle Contraction

Sequence	Event
1	Muscle action potential
2	Spread of depolarization into the transverse tubule system
3	Communication of transverse tubule depolarization to the terminal cisternae of the sarcoplasmic reticulum
4	Release of Ca^{++} from the terminal cisternae
5	Binding of Ca^{++} to troponin
6	Exposure of myosin binding sites on actin by tropomyosin
7	The powerstroke of muscle contraction, a process repeated so long as Ca^{++} and ATP are present (crossbridge cycling)
8	Uptake of Ca^{++} into the sarcoplasmic reticulum, resulting in the actin binding sites for myosin becoming covered again by tropomyosin

FIGURE 11-12
A electron micrograph illustrating the structure of a muscle triad.

the T tubules. The association of a T tubule with two adjacent terminal cisternae is termed a **muscle triad** (Figure 11-12). The triad is the site where muscle action potentials are coupled with the internal mechanisms responsible for force generation, a process known as **excitation-contraction coupling.**

Calcium and Excitation-Contraction Coupling

When a muscle fiber is at rest, an active transport system in the sarcoplasmic reticulum pumps Ca^{++} out of the cytoplasm into the sarcoplasmic reticulum and maintains the intracellular Ca^{++} concentration at less than 0.1 μmol/L, well below the threshold for significant binding to troponin. In a resting muscle, the amount of Ca^{++} that is **sequestered** (taken up as a result of Ca^{++} pumping) inside the sarcoplasmic reticulum is so great that the pumps must work against a concentration gradient of about six orders of magnitude. This gradient would be even greater (about eight orders of magnitude), except that inside the sarcoplasmic reticulum are large quantities of a protein called **calsequestrin.** Calsequestrin binds Ca^{++} reversibly, taking some of it out of solution.

The events of excitation-contraction coupling in skeletal muscle are shown in Figure 11-13 and summarized in Table 11-1. The first event is the production of an action potential at the neuromuscular junction, a specialized synapse that faithfully relays every impulse from the motor nerve fiber to the muscle (see Chapter 8 and Figure 8-17). At the neuromuscular junction the axon terminal is buried within folds of the postsynaptic membrane that prevent the loss of the neurotransmitter acetylcholine by diffusion. The action of acetylcholine is terminated by the enzyme **acetylcholinesterase,** which is present at a very high density in the postsynaptic membrane at the neuromuscular junction. Once initiated, the muscle action potential rapidly propagates over the sarcolemma and also spreads into the T tubules. The passage of the action potential past the lateral cisternae of the SR causes the SR to release its store of Ca^{++}. The released Ca^{++} acts as an intracellular messenger to turn on the contractile apparatus of the sarcomeres.

The mechanism by which the T tubule communicates its signal to the lateral cisternal membrane has not been completely determined, but Figure 11-13 shows the presently accepted hypothesis. The T-tubular membrane contains many units of an intrinsic protein called the **dihydropyridine (DHP) receptor.** These molecules are related to the voltage-sensitive Na^+ channel of nerve and muscle membranes and are believed to act as sensors of T-tubular depolarization. The lateral cisternal membrane contains **SR endfeet (ryanodine receptors).** These proteins are believed to form the Ca^{++} release channels of the SR. Scanning electron micrographs of the junction between the T tubule and lateral cisternal membrane show a very orderly one-to-

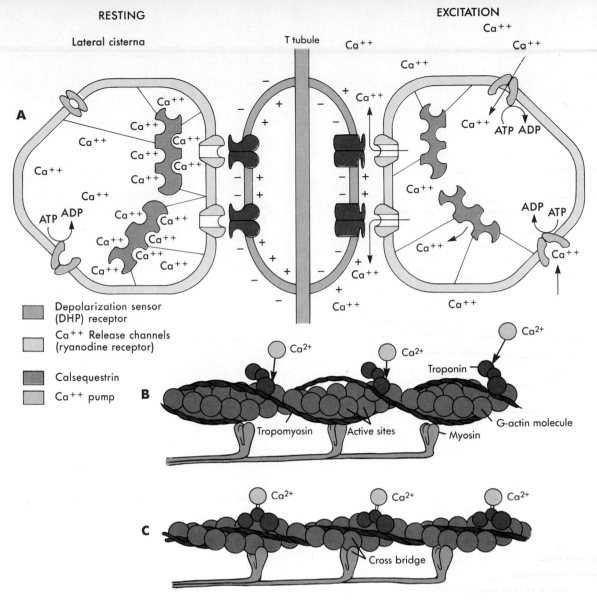

Lateral cisterna   T tubule

A

■ Depolarization sensor
(DHP) receptor

□ Ca⁺⁺ Release channels
(ryanodine receptor)

■ Calsequestrin

□ Ca⁺⁺ pump

B

Troponin

Tropomyosin   Active sites   Myosin

G-actin molecule

C

Cross bridge

FIGURE 11-13

The sequence of events in excitation-contraction coupling. The resting condition of the triad is shown at left in **A.** In resting muscle the Ca^{++} ATPase of the SR membrane has sequestered most of the Ca^{++} inside the SR. Much of this sequestered Ca^{++} is bound by calsequestrin molecules. Filaments attach the molecules of calsequestrin to the walls of the terminal cisternae, keeping them near the junctions with the T tubules.
The initial event of excitation-contraction coupling is an action potential that propagates across the sarcolemma and into the T tubule system. The events in the right side of part **A** follow. The depolarization of the T-tubule membrane causes T-tubule voltage detectors (DHP receptors) to undergo a conformational change. This opens the Ca^{++} release channels of the SR endfeet (ryanodine receptors). (A single Ca^{++}-release channel is seen from the point of view of a DHP receptor.) Calcium ions rush out into the cytoplasm. As the concentration of Ca^{++} in the SR drops, additional Ca^{++} is released from calsequestrin. **B** The released Ca^{++} binds to troponin molecules on the thin filament. **C** This binding causes the tropomyosin molecule to move deeper into the groove along the actin myofilament, exposing active sites on actin and initiating the crossbridge cycle.

one relationship in which a DHP receptor sits atop each of the four subunits of each Ca^{++} release channel (Figure 11-13). The four subunits of each endfoot fit together to form a central channel that opens into four slots that form a cross on the side of the endfoot that faces the DHP receptor (Figure 11-13, *inset*). It may well be that the cytoplasmic ends of the DHP receptors literally act as stoppers for the

Ca^{++} release channel when the T-tubular membrane is at its resting potential. In this hypothesis, depolarization would then rearrange the configuration of the DHP receptors, lift the stoppers and allow the Ca^{++} release channel to open.

The result is that Ca^{++} channels in the SR are opened, allowing Ca^{++} to leave the SR and enter the cytoplasm. As the Ca^{++} levels in the SR drop,

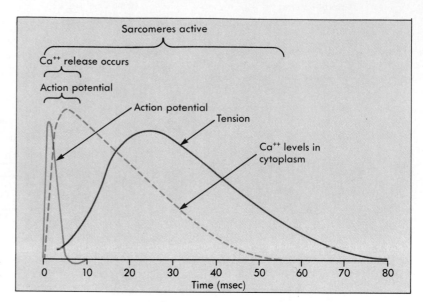

FIGURE 11-14
The time relationships of the muscle action potential, rise and fall of cytoplasmic Ca^{++}, and force development by sarcomeres during a twitch.

calsequestrin gives up its bound Ca^{++}, making additional Ca^{++} available. The cytoplasmic Ca^{++} level increases rapidly from less than 0.1 μM to more than 10 μM, a concentration sufficient to saturate binding sites on troponin. Binding of Ca^{++} to troponin shifts the tropomyosin molecules, exposing the myosin-binding sites on the G-actin subunits so that myosin:ADP and actin interact with each other (see Figure 11-13, B and C). The combination of troponin and tropomyosin is a **Ca^{++}-activated switch** that turns on the contractile machinery.

In mice with muscular dysgenesis the skeletal muscles are paralyzed. Electron microscopy of muscle fibers from dysgenic mice reveal an absence of junctional feet, suggesting that the dihydropyridine receptor that triggers intracellular Ca^{++} release is not present. The disease has now been traced to an autosomal recessive lethal mutation in the gene coding for the α1 subunit of the dihydropyridine receptor. It is possible that a similar defect may occur in some human muscle diseases.

The contraction that results from a single action potential is a **twitch.** After an action potential, the contractile machinery will remain active as long as cytoplasmic Ca^{++} levels remain elevated (Figure 11-14). The period of activity depends on how quickly the released Ca^{++} can be returned to the SR. The active transport system of the SR is stimulated by an increase in the intracellular free Ca^{++} concentration. In fast-twitch muscles, the Ca^{++} pump is so powerful that after only 10 to 20 msec, all the released Ca^{++} is pumped into the SR and the intracellular Ca^{++} concentration drops below the threshold for significant crossbridge activation. In slow-twitch fibers, the SR contains fewer Ca^{++} pump proteins and the duration of twitch contractions may be as long as 40 to 50 msec.

The states of the contractile machinery of striated muscle during relaxation and activity are summarized in Table 11-2.

TABLE 11-2	The States of the Contractile Machinery of Striated Muscle

	Resting	Contracting
Cytoplasmic Ca^{++} concentration	Less than 10^{-7} M	About 10^{-5} M
Troponin	Ca^{++} binding sites free	Ca^{++} binding sites occupied
Tropomyosin	Blocks actin binding sites	Exposes actin binding sites
State of myosin heads	High-energy myosin:ADP	Cycling

Summation of Contractions in Single Muscle Fibers

The total amount of force developed by skeletal muscles depends on two factors: (1) **summation** of twitches in individual muscle fibers, and (2) **recruitment,** increases in the number of fibers active in a given muscle (see Chapter 12).

Repeated activation of the motor neuron innervating a given muscle fiber results in summation when the firing frequency is sufficiently high. If the muscle action potentials occur infrequently, the Ca^{++} concentration returns to rest levels before the arrival of the next action potential, and the muscle's sarcomeres return to rest length (Figure 11-15, A). When two or more muscle action potentials occur in rapid succession, not all of the Ca^{++} released in response to the first activation can be resequestered before more Ca^{++} is released. As a result, the contractions summate (Figure 11-15, B).

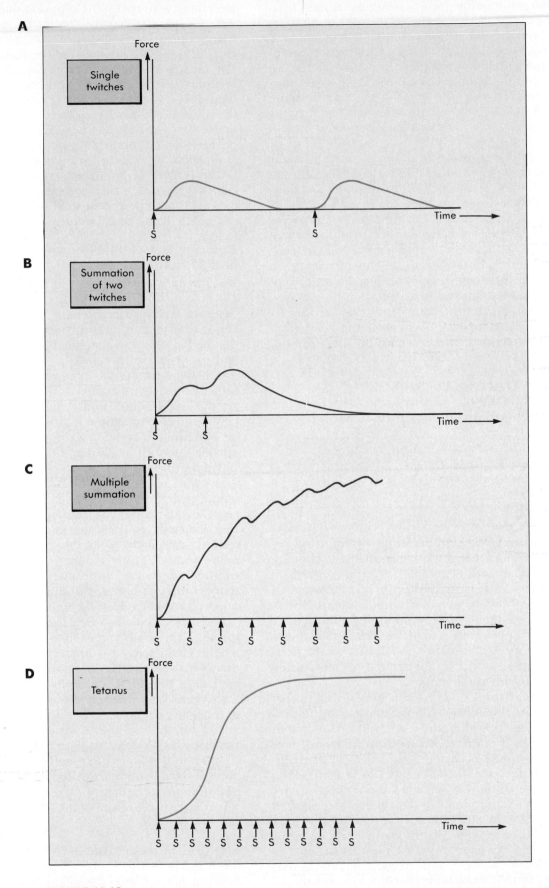

FIGURE 11-15
A Single twitches in a muscle fiber initiated by stimuli *(S arrows)* well separated in time. **B** and **C** Individual twitches in a skeletal muscle can summate with each another when stimuli are closely spaced. **D** At high rates of stimulation, the contractile force rises smoothly in a tetanus.

What summation means in terms of the cross-bridge activity is that some of the sliding that resulted from the first action potential is retained, and the second surge of Ca^{++} allows additional shortening of the sarcomere. As the stimulation rate is increased (multiple summation), the Ca^{++} concentration increases and the total force rises (Figure 11-15, C). At high stimulation frequencies, **tetanus** occurs, and force development becomes smooth (Figure 11-15, D).

The ability of a muscle fiber to undergo tetanic contraction depends on two factors: (1) the contractile machinery must be able to respond to maintained high levels of internal Ca^{++} with continual crossbridge activity, and (2) the muscle action potential must not last as long as the resulting twitch contraction. This second requirement is met in skeletal muscle, where tension development lasts much longer than the action potential, but the action potential in cardiac muscle is so long that the sarcomeres return to precontraction length before the action potential is over. This protects the heart from tetanic contractions, which would prevent filling of the heart.

THE MECHANICS OF MUSCLE CONTRACTION
Properties of Isometric Contractions

> • *How do isometric and isotonic contractions differ?*
> • *What makes a preloaded contraction different from an afterloaded contraction? Which has the shorter latent period?*

An **isometric** (constant length) contraction occurs when an activated muscle is unable to shorten. In this kind of contraction, the effect of crossbridge formation is to increase the **tension**, or tautness, of the muscle. Force generation without shortening is a normal activity for postural muscles; this also occurs when a person tries to lift an immovable object. In laboratory experiments, isometric contractions can be studied by fixing the tendons of an isolated muscle between rigid supports.

If a muscle is loaded by attaching a weight that stretches it, the tension of the muscle is proportional to the weight applied, regardless of whether crossbridges are active. An **isotonic** (constant tension) contraction occurs if the loaded muscle contracts and lifts the load. In practice, a contraction can change from isometric to isotonic and vice versa. For instance, when a bucket of water is lifted from the ground, the contraction is isotonic as long as the load is moving. If the load is held at arm's length, the force exerted by the muscles is equal to the bucket's weight, and muscle length is not changing, so the contraction becomes isometric.

The relationship between the length of a resting muscle and its tension can be measured by attaching a muscle to two rigid supports that can be set at different distances apart. One of the supports has a **force transducer** —a device that measures the tension transmitted to the support from the muscle. The muscle assumes some minimal length when it is removed from the body and has no load. Passive tension is zero at this length (Figure 11-16, A—note the transducer shows zero force). Stimulation of the motor nerve to the muscle results in an active force on the lower support (Figure 11-16, B). If an unstimulated muscle is stretched beyond its rest length (Figure 11-16, C), the increase in tension becomes greater with each increase in length. Stimulation will then produce a total force that is the sum of the active and passive forces (Figure 11-16, D).

Relation between Maximum Force and Muscle Length

The **passive-tension** curve of a muscle (Figure 11-17) is a measure of the muscle's **elasticity**, or tendency to return to rest length when stretched. The **series elastic elements** include all elastic structures that lie between the attachments of the crossbridges and the ends of the tendons. When a muscle contracts isometrically, the active tension generated by the crossbridges is added to the passive tension to get the total tension (Figure 11-17). Conversely, the length-tension relationship for the sarcomeres can be determined by subtracting the passive length-tension curve from the length-tension curve of the active muscle.

Active tension is maximal near the resting length of a muscle in the body attached to the skeleton; at this length the overlap between thick and thin filaments is such that all myosin heads may form crossbridges. When the muscle is stretched by more than a few percentages of resting length, the overlap is decreased and some myosin heads can no longer form crossbridges (Figure 11-17). The consequence is a sharp decrease in active tension.

Active tension also decreases if the muscle is allowed to shorten by more than a few percent of rest length. This decrease is probably caused by the hindrance to crossbridge formation that results from overlap of thin filaments from opposite ends of the sarcomere. With further shortening, active tension declines to zero because both ends of the thick filaments bump into the Z structures. In summary, skeletal muscle delivers its maximum force at rest length and is adapted to give optimum force development over a very narrow length range. This range of optimum performance corresponds closely to the actual **operating range** of muscles in the body (Figure 11-17).

Skeletal Muscle and Skeletal Lever Co-adaptation

When a muscle lifts a load directly (Figure 11-18, A), the muscle tension must equal the pull of gravity on the load. In many cases, muscles are attached to

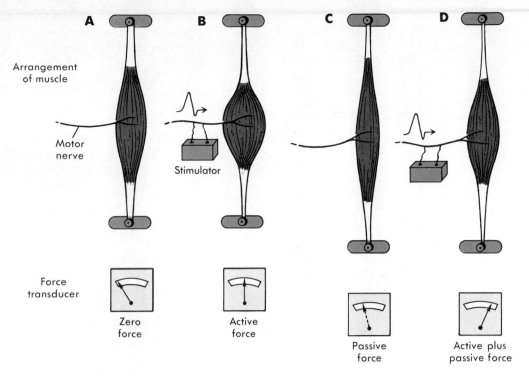

Arrangement of muscle

Motor nerve

Stimulator

Force transducer

Zero force

Active force

Passive force

Active plus passive force

FIGURE 11-16

The method of stimulating a muscle and recording the resulting force during an isometric contraction. **A** An unstimulated muscle is fixed between two supports at its unloaded length. **B** The muscle is stimulated, producing an active force that can be measured by a transducer. **C** The muscle has been stretched slightly, resulting in an increase in passive force. **D** The stretched muscle is stimulated. Note that the total force in an isometric contraction is the sum of the active force produced by actin-myosin interaction and the passive force that is continuously present if the muscle is stretched beyond its rest length.

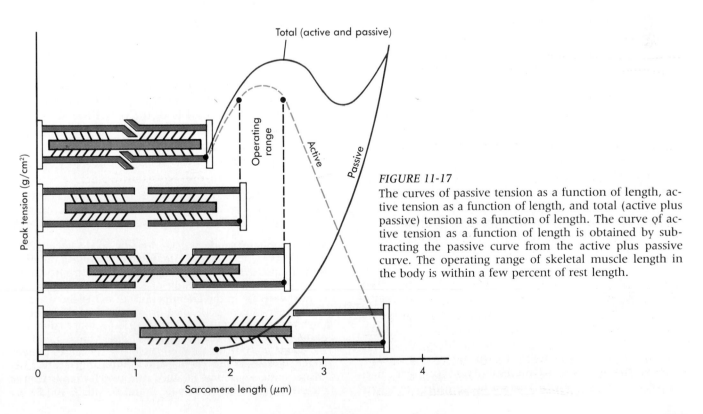

Peak tension (g/cm²)

Total (active and passive)

Operating range

Active

Passive

Sarcomere length (μm)

FIGURE 11-17

The curves of passive tension as a function of length, active tension as a function of length, and total (active plus passive) tension as a function of length. The curve of active tension as a function of length is obtained by subtracting the passive curve from the active plus passive curve. The operating range of skeletal muscle length in the body is within a few percent of rest length.

Skeletal, Cardiac, and Smooth Muscle

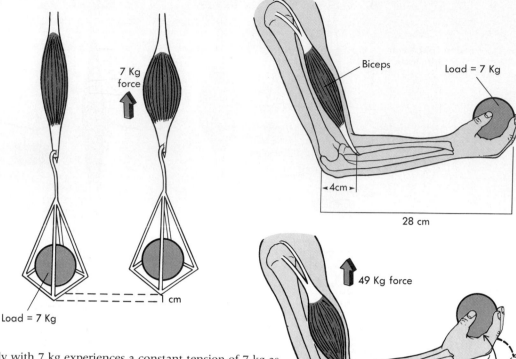

FIGURE 11-18
A A muscle loaded directly with 7 kg experiences a constant tension of 7 kg as it lifts the load. The load is lifted a distance equal to the amount of muscle shortening. **B** When the arm and hand are supporting a 7 kg load, the tension in the biceps is multiplied by the leverage of the arm. If the distance between elbow and biceps insertion is ¹⁄₇ of the total length of the arm, the tension in the biceps will be 7×7 kg, or 49 kg. If the biceps shortens 1 cm, the load will be moved 7 cm.

parts of the skeleton that act as levers. Muscle insertions are usually close to joints, placing skeletal muscles at a mechanical disadvantage. To lift a loaded appendage, they must exert a force many times greater than that of the load. The **leverage factor** is the factor by which the muscle force must exceed the load. It is determined by dividing the total length of the lever by the fraction of its length that lies between the joint and the muscle insertion. For example, to lift a weight held in the hand, the biceps muscle must exert a force about seven times that of the load (Figure 11-18, **B**). However, because the insertion is close to the elbow, the biceps can lift the loaded hand through a wide range of movement with a relatively small amount of shortening. The construction of sarcomeres, which maximizes force generation over a narrow operating range, allows skeletal muscles to function well within the lever system of the skeleton.

Latent Period in Afterloaded Contractions

A delay, or **latent period**, occurs between excitation of a muscle and the beginning of shortening (Figure 11-19, **A**). Part of this delay involves excitation-contraction coupling. Another phase of the latent period is attributable to the necessary stretching of the series elastic elements in proportion to the load. A **preloaded muscle** supports the load before it is stimulated to lift it. The latent period of a preloaded muscle does not change much with increased load (Figure 11-19, **B**).

An isolated muscle is **afterloaded** if it does not support the load until it begins to shorten. An afterloaded muscle is under no tension until its sarcomeres begin to contract. Before the muscle can begin to shorten and lift the load, its series elastic elements must be stretched in proportion to the weight to be lifted. If the load is light, very little stretching is necessary. If the load is greater, more stretching is necessary, just as a rubber band will stretch more if it is supporting a heavier load. The latent period of

FIGURE 11-19
The effect of preloading and afterloading on the latent period and the shortening attained in a twitch. The effect on a preloaded muscle of an increased preload (compare **B** with **A**) is a slightly decreased amount of shortening without a substantial decrease in the latent period. **C** and **D** show a comparable experiment performed on an afterloaded muscle. In the afterloaded muscle, the latent period is longer than that of the preloaded muscle (compare **C** with **A** and **D** with **B**). Increasing the load in an afterloaded muscle causes a much larger increase in the latent period and a much larger decrease in the shortening attained in a twitch than is seen in the preloaded muscle (compare **C** and **D** with **A** and **B**).

NERVE AND MUSCLE

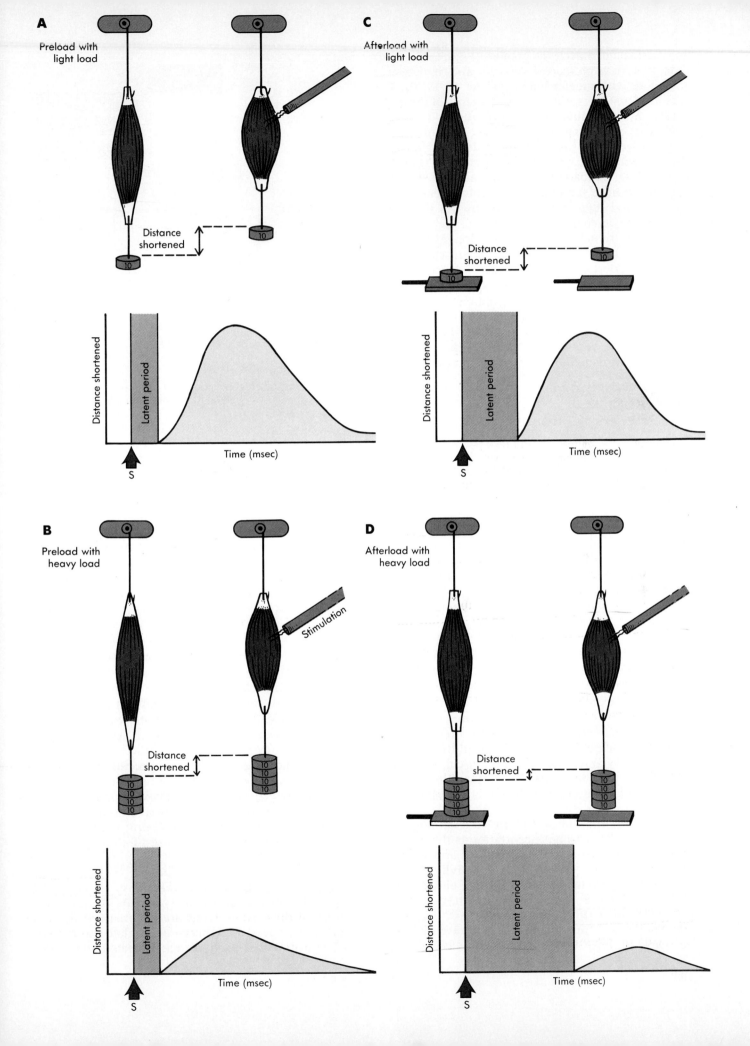

A

Preload with light load

Distance shortened

Distance shortened

Latent period

Time (msec)

S

B

Preload with heavy load

Stimulation

Distance shortened

Distance shortened

Latent period

Time (msec)

S

C

Afterload with light load

Distance shortened

Distance shortened

Latent period

Time (msec)

S

D

Afterload with heavy load

Distance shortened

Distance shortened

Latent period

Time (msec)

S

an afterloaded muscle is greater (Figure 11-19, *C*) than that in a preloaded muscle, and it increases still more with greater loads (Figure 11-19, *D*). The reason is that the series elastic elements of the afterloaded muscle must stretch in proportion to the load. This stretching requires time. In contrast, the latent period of a preloaded muscle increases only slightly with increased load. The series elastic elements of the preloaded muscle are already stretched in proportion to the load; the slight increase in the latent period reflects the greater inertia the active force must overcome.

Because twitches are brief, the longer the latent period, the less time remaining for muscle shortening. The period during which Ca^{++} levels are elevated is the same for all four examples in Figure 11-19 because each response is the result of a single action potential. The afterloaded muscle is prevented from shortening as much as does the preloaded muscle because the Ca^{++} levels decline before a similar degree of shortening can occur.

MUSCLE ENERGETICS AND METABOLISM
Muscle Energy Consumption and Activity

- *What processes might contribute to elevated oxygen consumption for a time after exercise (oxygen debt)?*
- *How does hypertrophy in oxidative muscle and in glycolytic muscle differ? Which muscle type is trained selectively in bodybuilding?*
- *What is the general definition of muscle fatigue?*

Most of the ATP in a resting muscle is used by the sodium-potassium (Na^+-K^+) pump, which maintains the concentration gradients across the sarcolemma. In a contracting muscle, the Na^+-K^+ pump uses ATP at a higher rate because it must counteract the additional ion movement resulting from the action potentials. Active muscle also uses more ATP for resequestration of Ca^{++} than does resting muscle. However, the active crossbridges account for most of the extra ATP use in contracting muscles. Because each crossbridge cycle requires the hydrolysis of one ATP molecule, the energy requirements of active muscle are greatly influenced by the rate of crossbridge cycling. The rate of energy use is higher in isotonic than in isometric contractions because crossbridges cycle more rapidly when filaments are in motion.

The concentration of ATP in skeletal muscles is about 5 mM. This ATP concentration is the immediate source of energy for contraction, and by itself can support only a few seconds of activity. Muscles contain a second immediate reserve of energy in the form of **creatine phosphate**, which can donate phosphate to ADP, becoming **creatine** (Figure 11-20). The creatine phosphate reserve prevents a large decrease in ATP levels during the few seconds

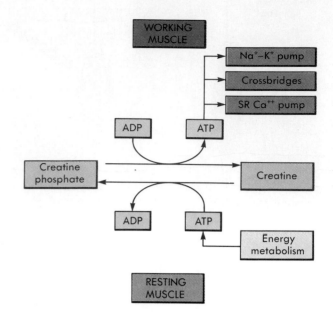

FIGURE 11-20
The reserve of creatine phosphate is drawn on in working muscle and is recharged by energy metabolism after the muscle returns to rest.

needed for energy metabolism to respond to the increased rate of ATP use that occurs when contraction begins. The reaction is reversed to recharge the creatine phosphate reserve in resting muscle. Both forward and reverse reactions are catalyzed by the same enzyme, **creatine phosphokinase.** The reaction of creatine phosphate with ADP to yield creatine and ATP is favored by a decrease in ATP concentration and an increase in ADP concentration. When the muscle is not contracting the abundance of ATP drives the reaction toward creatine phosphate synthesis.

Skeletal Muscle Fiber Types and Energy Metabolism

Different muscle groups serve different functions and display corresponding contractile properties. For example, finger and eye muscles contract rapidly and become fatigued with continued use, whereas postural muscles contract slowly and do not become fatigued as easily. Functional specialization of muscle is possible because three types of muscle fibers exist, and the relative numbers of each fiber type are different in different muscles. **Fiber types** are distinguished by several factors, including the metabolic pathways used for energy and resistance to fatigue.

The three muscle fiber types are (1) **slow oxidative, or type I,** (2) **fast oxidative,** or **type IIA,** and, (3) **fast glycolytic,** or **type IIB.** The characteristics of these three types are summarized in Table 11-3. The speed of contraction is related to the type of myosin a fiber contains. Fast-twitch fibers contain

TABLE 11-3   Comparison of Skeletal Muscle Fiber Types

Characteristics	Type I	Type IIA	Type IIB
Contractile			
Contraction velocity	Slow	Fast	Fast
Myosin ATPase	Slow	Fast	Fast
Twitch duration	Long	Short	Short
Ca^{++} sequestration	Slow	Rapid	Rapid
Metabolic			
Capillaries	Abundant	Intermediate	Sparse
Glycolytic capacity	Low	Intermediate	High
Oxidative capacity	High	High	Low
Myoglobin content	High	Intermediate	Low
Glycogen content	Low	Intermediate	High
Fiber diameter	Small	Intermediate	Large
Motor unit size	Small	Intermediate	Large
Recruitment order	Early	Intermediate	Late

myosin with a high ATPase activity, whereas slow twitch fibers have myosin with low ATPase activity. The higher the rate of ATP hydrolysis, the faster the crossbridges can cycle and the more rapidly a muscle can shorten. The price of faster shortening is a higher rate of ATP use.

Type I muscle fibers are found in the greatest numbers in **postural muscles** (for example, the long muscles in the back) and generate some 10 grams of force per square centimeter of muscle cross-sectional area. Oxidative metabolism is appropriate for muscles that must maintain contraction for long periods because it produces ATP with high efficiency. The commitment of type I fibers to oxidative energy metabolism is reflected in their enzyme profiles, high mitochondrial density, and rich capillary blood supply. Their relatively small diameter facilitates gas exchange with the circulatory system. Type I fibers are red in color because they contain an oxygen carrier known as **myoglobin.** Myoglobin facilitates oxygen diffusion within the fiber and provides a reserve oxygen supply when dissolved oxygen levels in the cytoplasm fall very low.

The two types of fast muscle (types IIA and IIB) produce more than 100 grams of force per square centimeter of muscle cross-sectional area. They are distinguished from each other by their susceptibility to fatigue. Type IIA fibers (found, for example, in the soleus muscle of the calf) primarily use oxidative processes and are moderately resistant to fatigue but not as resistant as type I fibers. They have many mitochondria, considerable myoglobin, and a moderate amount of **glycogen,** which can be broken down to provide glucose and thus ATP via glycolysis (see Chapter 4). Like type I fibers, type IIA fibers are red.

Type IIB fibers rely on anaerobic glycolysis almost exclusively and thus become fatigued quite easily. They have an extremely high glycolytic capacity, large stores of glycogen, and few mitochondria. They possess little myoglobin and thus are sometimes called white fibers. They are mixed with the other muscle fiber types in the limb muscles, particularly in the flexors, and the eye and finger muscles have a high proportion of this fiber type.

Anaerobic glycolysis is advantageous for muscle fibers that are specialized for brief, powerful contraction because the maximum force can be much greater if it is not limited by the rate at which glucose and oxygen can be delivered by the circulatory system. The duration of use of type IIB fibers is limited by their stores of glycogen. Although type IIB fibers contain substantial stores of glycogen, anaerobic glycolysis is relatively inefficient in converting the stored energy to ATP. The rapid fatiguing of these fibers during sustained, submaximal exercise is highly correlated with the depletion of their glycogen stores.

Type I fibers are innervated by small motor neurons that activate only a small number of fibers; the size principle of recruitment (see Chapter 12) causes them to be recruited when small forces are required. Type IIB fibers are supplied by large motor neurons that activate large numbers of fibers; these fibers are recruited for heavier loads. Type IIA fibers have intermediate properties. Thus the size principle provides for smooth increases in force and also activates first and most frequently those fibers (type I) which are most resistant to fatigue.

The types of individual fibers can be identified under the microscope in tissue stained for particular metabolic enzymes. For example, the fast fibers of the muscle shown in Figure 11-21 are darkened by a stain specific for myosin ATPase.

The Cori Cycle

After strenuous exercise, oxygen consumption does not fall to its resting level immediately. Full return to the resting level of oxygen consumption may require up to several hours. The extra oxygen consumption occurring after exercise is referred to as **oxygen debt.** Some of the oxygen debt is the result of accumulated lactic acid, which must be metabolized to CO_2, and of a corresponding decrease in the glycogen stores of type IIB muscle fibers. Repayment of the debt occurs when the lactate produced during exercise is returned to the Type IIB fibers in the form of glucose. The process by which some of the lactic acid can be converted to glucose in the liver and returned to muscle is called the Cori cycle (Figure 11-22).

During heavy exercise and for a time after exercise is stopped, lactic acid is released by muscle fibers into the bloodstream. Some lactic acid is taken up by the liver and resynthesized to glucose by an energy-requiring pathway (see Chapter 4) supported by the oxidative metabolism of the liver. The resulting glucose is released into the bloodstream and can be taken up by any fibers that need it, including muscles that have depleted their glycogen reserves. In principle, the extra oxygen used by the liver for converting lactic acid to glucose could account for the oxygen debt.

In fact, however, the oxygen debt cannot be accounted for by the Cori cycle alone. First, lactate metabolism goes on during, as well as after, exercise. Also, much of the lactate produced during exercise does not enter the Cori cycle, but is metabolized to CO_2 by the oxidative metabolism of the heart, liver, brain, and type I fibers of skeletal muscle (Figure 11-22). Finally, much of the increased oxygen consumption occurring during recovery apparently does not result from lactic acid metabolism.

FIGURE 11-21
A micrograph of a muscle containing both fast and slow fiber types. Myosin ATPase reacts with ATP in a procedure that stains the reaction product. The darkly stained fibers have high levels of myosin ATPase and thus are Type II fibers. The unstained fibers are Type I fibers.

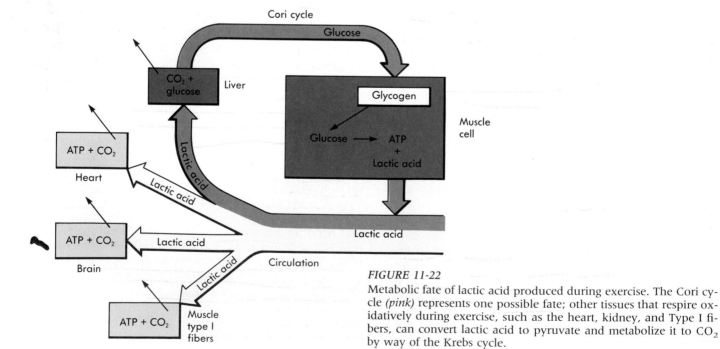

FIGURE 11-22
Metabolic fate of lactic acid produced during exercise. The Cori cycle *(pink)* represents one possible fate; other tissues that respire oxidatively during exercise, such as the heart, kidney, and Type I fibers, can convert lactic acid to pyruvate and metabolize it to CO_2 by way of the Krebs cycle.

NERVE AND MUSCLE

The differences among the three types of muscle fiber arise during development. Mammalian muscle fibers are all slow-twitch at birth, and those fibers destined to be fast (Types IIA and IIB) differentiate by two processes: (1) the "switching off" of the synthesis of myosin with low ATPase activity in some fibers, and (2) the dying off of some fibers to produce the proportion of Types I, IIA, and IIB appropriate for each particular muscle. The destiny of muscle fibers can be altered by changing the signals the fibers receive from the nervous system.

A pair of muscles, one fast and the other slow, can have their innervation cut, and the cut ends of the nerves redirected so that the nerves regenerate connections with the wrong muscles.

The result of this cross innervation is that the properties of the two muscles are dramatically altered: the former slow muscle will contract faster and the fast muscle will contract slower (Figure 11-A). The biochemical and structural properties of the two muscles will alter accordingly. Thus the motor neurons exercise a trophic effect—they determine the specific functional properties of their own motor units.

The influence is exerted both by chemical messengers that travel between the muscle and the nervous system via axoplasmic flow and by the frequency and pattern of recruitment of the muscles by the motor neurons. Although the results of cross innervation are more dramatic when they are performed on young animals, similar effects can be demonstrated in adult animals. These studies demonstrate the capacity of muscle fibers to adjust their properties in response to the signals they receive from the nervous system.

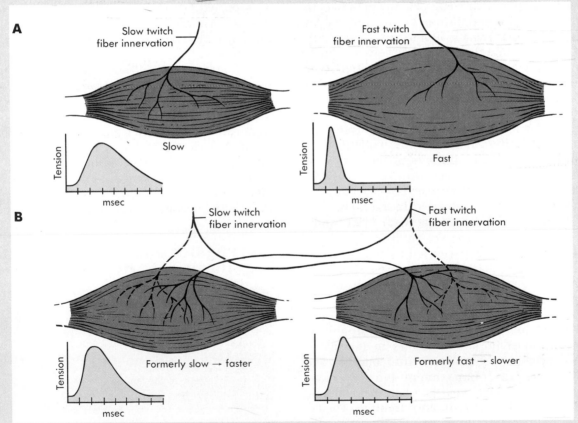

FIGURE 11-A

The effect of changes in innervation on the contractile properties of muscle. The nerve running to a muscle composed predominantly of slow-twitch fibers is surgically switched with a nerve innervating a muscle composed predominantly of fast-twitch fibers. The original twitch properties of the muscles are shown in **A**; the effects of the crossed innervation on twitch properties are shown in **B**.

A large part of the postexercise metabolic rate can be explained by physiological changes occurring during exercise that persist into the recovery phase. These include hormonal responses, circulatory changes, and elevated body core temperature.

Effects of Training on Specific Fiber Types

Muscle fibers that are frequently active respond by increasing their performance capabilities. The changes that occur differ for different fiber types. Oxidative fibers respond with increases in the myoglobin content, the number of mitochondria, and the number of capillaries serving the muscle (see Table 11-3). These changes involve little change in the mass of the muscle. Glycolytic fibers respond by increasing the number of myofibrils and the size of the glycogen stores; these changes necessitate an increase in fiber diameter and thus an increase in muscle mass (see Table 11-3). Tissue growth resulting from an increase in fiber size, rather than in the number of fibers, is called **hypertrophy.**

Hypertrophy of skeletal muscle in human athletes is not believed to involve an increase in the number of fibers, although some animal experiments suggest that such increases are possible. In any individual, the proportion of the three fiber types in a particular muscle is not altered by training and appears to be determined largely by heredity. However, selective use of fibers of a single type results in hypertrophy of those fibers. Exercise that calls for sustained effort, such as distance swimming, running, or skiing, results in development of endurance. This result causes an increase in the circulatory supply, mitochondria, and oxidative enzymes and myoglobin in all fibers. Athletes trained for endurance increase oxygen uptake capacity by up to 20% and produce lower levels of lactic acid during exercise than do untrained individuals.

The situation is quite different for brief, high-intensity training such as weight lifting. The increase in muscle mass experienced by bodybuilders is the result of two changes: (1) increase in the diameters of glycolytic fibers, and (2) addition of collagen and other connective tissues required to sustain the passive tension of heavy loads. The rate and amount of tension developed during training exercises are the most important factors in increasing contractile protein incorporation into muscle. The short-term demands of maximum force development do not cause the circulatory and metabolic changes associated with endurance training. The effects of both types of training are rapidly reversed if the regimen is not maintained; this process is called **disuse atrophy.**

Muscle Fatigue

Fatigue is a use-dependent decrease in the ability of muscle to generate force. The **rapid fatigue** seen in high-intensity exercise apparently results from operating muscles under anaerobic conditions. High-intensity exercise causes an increase in the lactic acid in muscle and blood. There is evidence that the resulting acidic conditions alter several aspects of muscle physiology, including the activity of key glycolytic enzymes and possibly the contractile machinery. Use of anaerobic pathways also depletes high-energy phosphate compounds and alters the ionic gradients across the fiber membrane and T tubules. Changes in the energy status of the fibers may impair the ability of the sarcoplasmic reticulum to control intracellular Ca^{++} levels.

The **slow fatigue** that results from prolonged submaximal exercise coincides with depletion of the glycogen stores of the liver and working muscle fibers. When this occurs, the body must use fat as the sole energy source. Energy production from fat occurs at about half the rate of energy production from glucose, so the depletion of glycogen stores is marked by a substantial decrease in muscle performance. Distance runners refer to the onset of slow fatigue as "hitting the wall." With adequate amounts of oxygen, fatigue of glycolytic motor units is partly compensated for at first by increased reliance on oxidative units. However, repeated activation of oxidative units results in failure of neuromuscular transmission and a further decline in performance.

Endurance in athletic events requiring sustained exercise can be improved by **glycogen loading,** a dietary regimen in which consumption of a low-carbohydrate diet is followed by a switch to a high-carbohydrate diet just before an important event. Muscle fibers adapt to the low-carbohydrate diet by increasing their ability to take up glucose from the blood; the switch to the high-carbohydrate diet results in a temporary but substantial increase in muscle glycogen levels that confers greater endurance.

THE PHYSIOLOGY OF CARDIAC MUSCLE
The Structure of Cardiac Muscle

- *What is the contractility of muscle?*
- *What intracellular processes affect contractility?*
- *What process normally prevents tetanic contractions of cardiac muscle?*
- *How does the length-tension relationship of cardiac muscle compare with that of skeletal muscle?*
- *In what part of the length-tension curve is the cardiac operating range?*

Cardiac muscle fibers (Figure 11-23), like skeletal muscle fibers, contain myofibrils and a network of T tubules and SR. The molecular mechanism of force generation and its control by Ca^{++} are similar in cardiac and skeletal muscle. Cardiac muscle fibers resemble type I skeletal muscle fibers in having abundant myoglobin, many mitochondria, and a

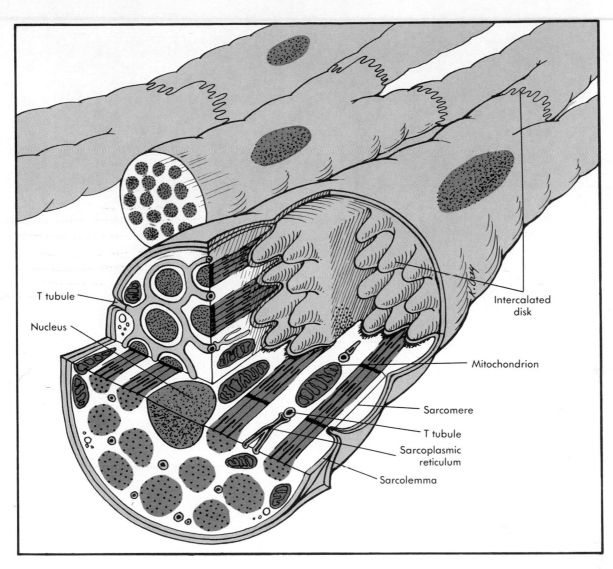

FIGURE 11-23
The major structural features of cardiac muscle fibers. Note the presence of intercalated disks connecting successive sarcomeres.

rich supply of capillaries. These characteristics make heart muscle fatigue resistant but also very dependent on a continuous oxygen supply. Cardiac muscle fibers are much shorter than skeletal fibers. The fibers are linked together by strong connections between their ends that form a mesh; the connections are called **intercalated disks** (Figure 11-23).

The Heart as an Electrical Unit
One distinction between cardiac and skeletal muscle is that the fibers of cardiac muscle are electrically coupled to each other. When one fiber is excited, the action potential spreads throughout the entire muscle, allowing the cardiac muscle to contract as a unit. In this respect, cardiac muscle resembles single-unit smooth muscle, described later in this chapter. The electrical communication is mediated by gap junctions (see Chapter 4) in the intercalated disks (Figure 11-24).

Lack of Summation in Cardiac Muscle
The characteristics of action potentials in cardiac fibers are described in Chapter 13, but the important points to understand now are that (1) some cardiac fibers initiate action potentials spontaneously in a regular rhythm and (2) action potentials last much longer in the heart than in nerve axons and skeletal muscle fibers. A typical nerve or skeletal muscle action potential lasts several milliseconds; some cardiac fibers remain depolarized for several hundred milliseconds (Figure 11-25). Because of the prolonged depolarization, the membrane of cardiac muscle fibers does not recover from its refractory period until the active state of the myofibrils is almost over; thus individual contractions of cardiac muscle cannot summate. The advantage of the prolonged action potential is that it prevents tetanic contractions, which would interfere with the heart's pumping cycle of contraction and relaxation. The

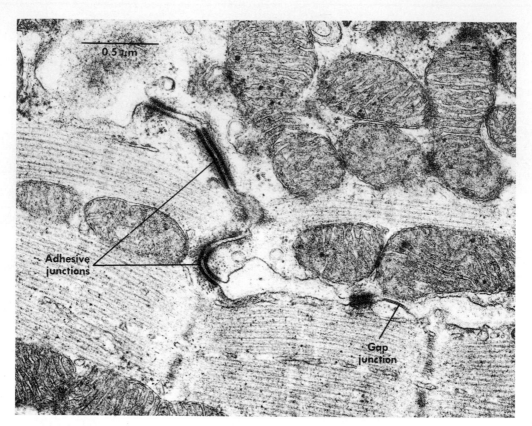

FIGURE 11-24

A longitudinal section of an intercalated disk between two cardiac muscle fibers. Desmosomes (adhesive junctions) within the intercalated disk weld the two fibers together. Gap junctions line the disk, forming electrical connections between the fibers.

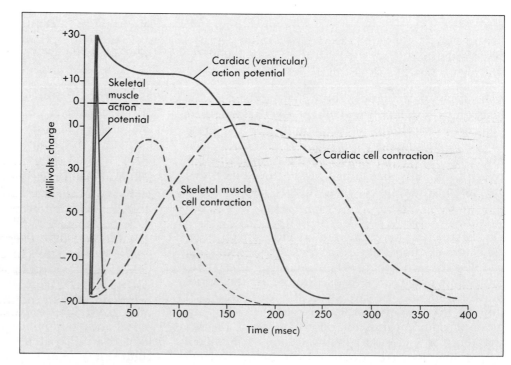

FIGURE 11-25
Comparison between the action potentials of a skeletal muscle fiber and a fiber of the heart ventricle. Note that the contractions of the two fibers are similar, but the heart action potential lasts much longer.

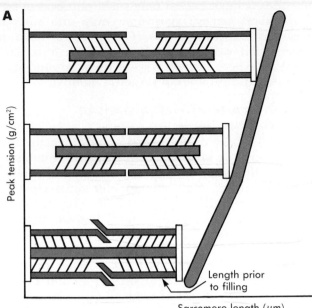

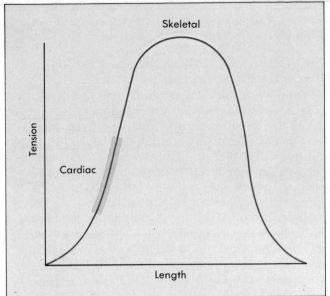

FIGURE 11-26
A The relationship between sarcomere length and active tension in a cardiac muscle fiber. If the curve were continued to greater lengths, the tension would decrease, as it does in the length-tension curve of skeletal muscle shown in Figure 11-17.
B The general relationship between length and tension in muscle. The normal range of lengths for skeletal muscle is near the peak of the curve; length of cardiac muscle is on the rising part of the curve.

long refractory period of cardiac muscle also prevents "circus movements" around the wall of the heart that would otherwise tend to continuously re-excite it.

Excitation-Contraction Coupling in Cardiac Muscle

In one respect, the mechanism of excitation-contraction coupling in cardiac muscle is fundamentally different from that in skeletal muscle. In cardiac muscle the surface membrane contains Ca^{++} channels and a significant fraction of the Ca^{++} that triggers contraction enters the fiber through voltage-dependent Ca^{++} channels in the membrane activated by the action potential. This rise in Ca^{++} inside the fiber triggers further release of Ca^{++} from the SR. This process is called Ca^{++}-induced Ca^{++} release. Thus cardiac muscle must have extracellular Ca^{++} to contract normally, whereas in skeletal muscle all of the Ca^{++} needed for excitation-contraction coupling comes from the internal stores of the SR. As a result, cardiac muscle can be affected by drugs that block Ca^{++} channels without such drugs affecting skeletal muscles.

Determinants of Cardiac Muscle Force

Just as in skeletal muscle, the force delivered by cardiac muscle is affected by stretch. In skeletal muscle, the leverage factors of the skeleton ensure that the muscle is always near the sarcomere length optimal for force production. The heart is not constrained to operate within a narrow range of lengths by origins and insertions. In its general shape, the length-tension curve of cardiac muscles (Figure 11-26, *A*) resembles that of skeletal muscle. The difference is that the operating range is in a region of the length-tension curve that is considerably shorter than the length at which maximal force is generated (Figure 11-26, *B*). Thus, in the healthy heart, additional stretch results in additional force. This intrinsic ability to regulate itself is referred to as **Starling's Law of the Heart;** it is discussed in Chapter 12.

Historically, the increase in force generation with stretch in cardiac muscle was regarded as corresponding exactly to the ascending limb of the length-tension curve of skeletal muscle and was explained on the same basis. Recent evidence suggests that the increase in force with stretching is too great in cardiac muscle to be explained solely on the basis of more favorable overlap between thick and thin filaments. At least part of the effect is caused by an increase in the binding affinity of troponin for Ca^{++} with stretch. The mechanism by which stretch is communicated to the thin filaments to result in an increase in Ca^{++} binding is not understood. In any case, Starling's Law is not unlimited—stretching the heart beyond its normal limits (as occurs in some types of heart failure) will eventually result in a decrease in the force of contraction with further stretching. This decrease corresponds to the descending limb of the length-tension curve of skeletal muscle.

Modulation of Contractility in Cardiac Muscle by External Inputs

The contractility of cardiac fibers may be altered by nervous or hormonal inputs. Contractility is increased by norepinephrine released from sympathetic nerve fibers innervating the heart and epinephrine from the adrenal glands. Several mechanisms are involved; one is based on a positive relationship between the intracellular Ca^{++} concentration achieved in activation and the force developed. In cardiac muscle, the plasma membrane has both voltage-dependent Ca^{++} channels and a Ca^{++} ion pump. The intracellular Ca^{++} level, and thus contractility, depends on the balance between two factors: (1) Ca^{++} influx and active extrusion across the surface membrane, and (2) the release and reuptake of Ca^{++} by the sarcoplasmic reticulum.

A second mechanism that modulates contractility affects the cardiac contractile machinery. Both the thick and thin filaments of cardiac muscle have regulatory sites for second messengers. The ATPase activity of cardiac-muscle myosin can be modified by allosteric effects to alter the crossbridge cycling rate; the Ca^{++} binding properties of troponin may also be affected by intracellular messengers.

THE PHYSIOLOGY OF SMOOTH MUSCLE
Classes of Smooth Muscle

- *How do single-unit and multi-unit smooth muscle differ?*
- *What two major systems regulate contractility of smooth muscle?*
- *What is the role of calmodulin in each of the control systems?*
- *What conditions favor the formation of latchbridges?*

Smooth muscle surrounds hollow internal organs including the gastrointestinal tract and all blood vessels except capillaries. Many aspects of smooth muscle contraction are particularly relevant to the functioning of organ systems and are described in Chapter 15 (smooth muscle of blood vessels) and Chapter 23 (gastrointestinal smooth muscle).

Two types of smooth muscle are distinguishable on the basis of features associated with electrical coupling between fibers. The fibers of **single-unit (unitary)** smooth muscle are electrically and mechanically coupled to each other. This type of smooth muscle is typical of visceral organs. Gastrointestinal smooth muscle is largely single-unit, as is the muscle around small blood vessels. **Multi-unit** smooth muscle contains relatively few gap junctions, so that at best only small numbers of adjacent fibers can act as a unit. Multi-unit smooth muscle is found in the iris of the eye, the walls of larger blood vessels, the airways of the lung, and the skin, where smooth muscle fibers around hair follicles cause **piloerection** (goose bumps).

Length-Tension Relationships in Smooth Muscle

Smooth muscle contains actin and myosin, but these contractile proteins are not organized into sarcomeres. Instead, contractile units consist of parallel arrangements of thick and thin filaments crossing diagonally from one side of the fiber to the other. Within these units, myosin molecules are attached either to structures called **dense bodies** (the functional equivalents of Z structures) or to the sarcolemma. Actin filaments have no constant relationship to myosin filaments (Figures 11-27 and 11-28). The ratio of actin filaments to myosin filaments is much larger in smooth muscle than in striated muscle; most smooth muscles contain 10 to 15 actin filaments per myosin filament, as compared with the 3 thin filaments per thick filament of striated muscle.

In smooth muscle there is no practical limit to the distance that individual thick and thin filaments can slide relative to each other. In this respect smooth muscle differs radically from striated muscle, in which the thick and thin filaments can slide only a short distance before interfering with each other. The organization of smooth muscle is an adaptation for force development over a broad range of operating lengths. This adaptation is especially appropriate for muscle fibers in the walls of organs that experience large volume changes, such as the stomach and intestine.

Excitation-Contraction Coupling in Smooth Muscle

Excitation of smooth muscle fibers occurs by means of intrinsic and extrinsic mechanisms. In single-unit smooth muscles, spontaneous depolarizations of the muscle membrane occur at regular intervals, resulting in a **basic electrical rhythm**, or **BER** (Figure 11-29). The depolarizing waves of the BER may or may not result in action potentials, depending on the smooth muscle involved. This intrinsic **pacemaker activity** can be modified by extrinsic inputs from the sympathetic and parasympathetic divisions of the autonomic nervous system. The effects of autonomic input differ in different smooth muscle systems (Table 11-4). Hormones may also activate or inhibit contraction. Single-unit smooth muscle resembles cardiac muscle in its intrinsic excitability and possession of gap junctions.

In multi-unit smooth muscle, changes in contractile activity usually result from extrinsic inputs from the autonomic nervous system rather than from an intrinsic BER. Because there are few gap

A Relaxed

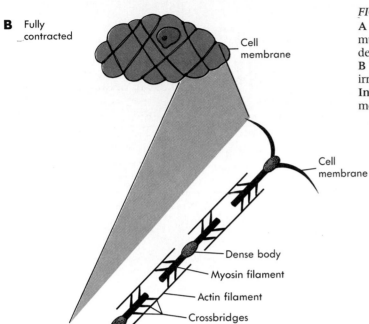

B Fully contracted

Cell membrane

Cell membrane

Dense body

Myosin filament

Actin filament

Crossbridges

FIGURE 11-27
A Contractile units span the diameter of a relaxed smooth muscle fiber and are connected to the fiber membrane by dense bodies.
B When the fiber contracts, its surface becomes rounded and irregular.
Inset The relationship between thick filaments, thin filaments, and dense bodies in a single active contractile unit.

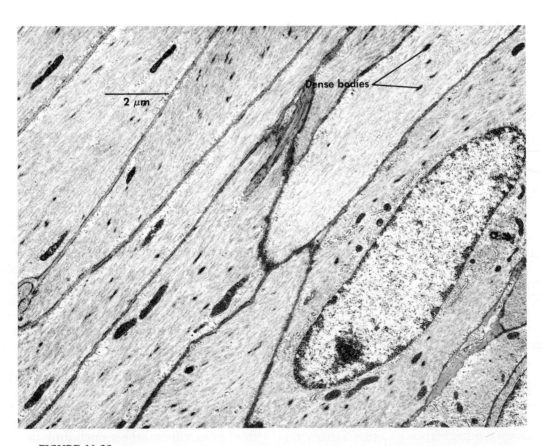

Dense bodies

2 μm

FIGURE 11-28
Micrograph of longitudinally sectioned smooth muscle fibers. The cytoplasm is filled with thick and thin filaments. Note the dense bodies scattered throughout the cytoplasm. The sarcoplasmic reticulum is represented by the small vesicles just next to the sarcolemma in each fiber.

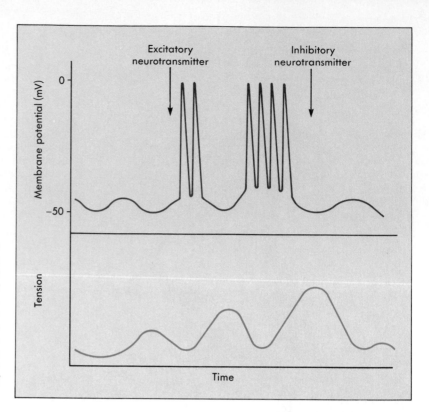

FIGURE 11-29

The basic electrical rhythm of a smooth muscle fiber consists of slow waves of depolarization and repolarization. Depending on the specific smooth muscle, the waves themselves may or may not cause changes in force production. Application of excitatory transmitter or hormone increases the probability that each depolarizing wave will trigger action potentials. Action potentials cause larger force increases than does slow depolarization alone. Inhibitory neurotransmitter decreases the probability of action potentials.

TABLE 11-4 Contractile States of Smooth Muscle

	Relaxed	Actively shortening	Maintaining tension
Cytoplasmic Ca^{++} concentration	Less than 10^{-7} M	3 to 5 $\times$ 10^{-6}M	10^{-7}M to (3 $\times$ 10^{-6}M)
Calmodulin	Binding sites open	Binding sites occupied	Binding sites open
MLCK	Inactive	Active	Inactive
Myosin	Dephosphorylated, no Ca^{++} bound	Phosphorylated, Ca^{++} bound	Dephosphorylated, Ca^{++} bound
Myosin ATPase activity	Very low	High	Low
Caldesmon	Bound to actin	Free	Bound to actin
Crossbridge state	No crossbridges	Crossbridges cycling	Latchbridges

junctions, each fiber must typically receive direct motor input. In these respects, multi-unit smooth muscle is like skeletal muscle.

Smooth muscle fibers do not have the well-developed SR characteristic of striated muscle fibers; instead, the SR takes the form of small vesicles near the cell membrane. Smooth muscle fibers are much smaller than skeletal muscle fibers and contract much more slowly, so they can depend primarily on Ca^{++} influx from the extracellular space for the activation of contraction. Electrical events are difficult to study in fibers that are electrically coupled to each other. The solution has been to dissociate them. The membrane properties of single muscle fibers can be studied using the **patch clamp technique** (see Chapter 8). Smooth muscle fibers iso-

lated from the stomach have spontaneous action potentials and can also be activated by neural inputs. Like cardiac muscle, smooth muscle fibers have Ca^{++} channels in the surface membrane. The depolarization results from Ca^{++} rather than Na^+ entry, and the rate of repolarization in smooth muscle caused by opening K^+ channels is itself sensitive to the rise in the internal Ca^{++} level. In many types of smooth muscle, the concentration of Ca^{++} in the cytoplasm is strongly affected by hormonal and neurochemical inputs that act through the ITP second messenger system (see Chapter 5). This second messenger system may bring about release of Ca^{++} from intracellular stores, including the SR.

As in skeletal muscle, Ca^{++} is the link between excitation and contraction. Control of contraction by

A Relaxed state

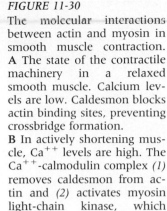

Myosin

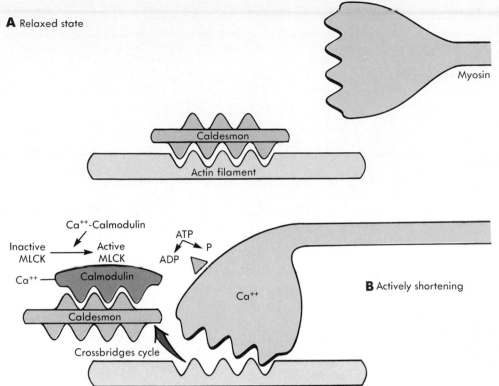

Caldesmon

Actin filament

Ca⁺⁺-Calmodulin

Inactive MLCK → Active MLCK

Ca⁺⁺

Calmodulin

Caldesmon

Crossbridges cycle

ATP ADP → P

Ca⁺⁺

B Actively shortening

C Maintaining tension

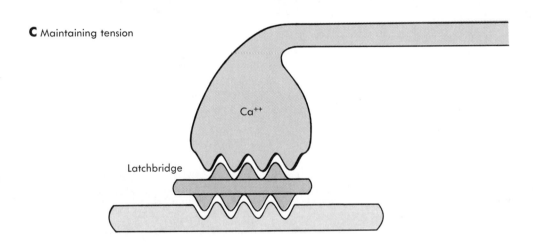

Ca⁺⁺

Latchbridge

FIGURE 11-30
The molecular interactions between actin and myosin in smooth muscle contraction.
A The state of the contractile machinery in a relaxed smooth muscle. Calcium levels are low. Caldesmon blocks actin binding sites, preventing crossbridge formation.
B In actively shortening muscle, Ca^{++} levels are high. The Ca^{++}-calmodulin complex *(1)* removes caldesmon from actin and *(2)* activates myosin light-chain kinase, which phosphorylates myosin, promoting cycling. Ca^{++} is bound to myosin heads *(3)*.
C Tension is maintained in smooth muscle when Ca^{++} levels are intermediate. At these levels, the calmodulin system is not activated, so caldesmon occupies actin sites. However, the myosin heads retain their bound Ca^{++} *(3)*, allowing non-cycling crossbridges to form.

Ca^{++} in smooth muscle is less well understood than in skeletal muscle, and the mechanisms may differ somewhat in smooth muscle from different organs. What follows is a general picture derived from several types of smooth muscle. Inside the smooth muscle fiber the Ca^{++} combines with at least two target proteins: the light chain of myosin and a protein called **calmodulin.** Both of these molecules are structurally similar to one of the subunits of troponin. Initiation of contraction by Ca^{++} may involve both thick and thin filaments. In this aspect, smooth muscle differs from striated muscle, in which the control of contraction primarily involves the thin filament.

The activity of the thick filaments is turned on by the following sequence of events: (1) the Ca^{++}-calmodulin complex activates the enzyme myosin light-chain kinase (MLCK), then (2) MLCK phosphorylates one of the light chains of the myosin heads, and the myosin phosphorylation exposes a binding site for actin. The myosin ATPase can now be activated by myosin binding to actin, and the interaction generates force. The actomyosin complex formed is broken down by a second enzyme called **myosin phosphatase.**

In some types of smooth muscle, the activity of the thin filaments is believed to be regulated by a molecule called **caldesmon** (Figure 11-30). This control system is believed to be responsible for the ability of smooth muscle to maintain tension for

Feature	Skeletal	Cardiac	Smooth
Calcium receptor in excitation-contraction coupling	Troponin	Troponin	Calmodulin, myosin
Ca^{++} sources	Sarcoplasmic reticulum	Sarcoplasmic reticulum and extracelular space	Primarily extracellular space
Force regulation	Summation, recruitment	Stretch, contractility modulation	Contractility modulation
Organization	Many motor units	Single-unit	Single-unit or multiunit

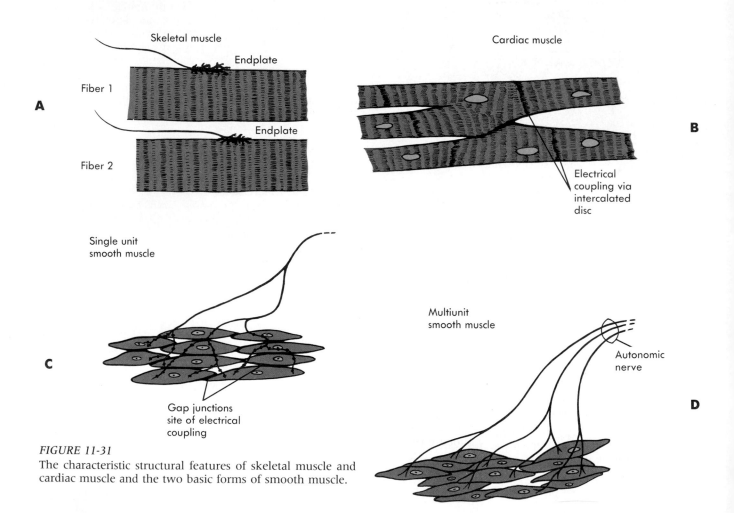

FIGURE 11-31

The characteristic structural features of skeletal muscle and cardiac muscle and the two basic forms of smooth muscle.

long periods with relatively little expenditure of energy. When cytoplasmic Ca^{++} levels are insufficient to maintain myosin phosphorylation but are still high enough to allow myosin to bind Ca^{++}, crossbridges that are formed do not cycle but are maintained. These are called **latchbridges.** Caldesmon is believed to form a link between actin and myosin heads that keeps the latchbridges from cycling. At higher Ca^{++} levels, latchbridges do not form because caldesmon is bound by the Ca^{++}-calmodulin complex and is not available to link actin and myosin. The ability to form latchbridges that do not hy-

Anabolic Steroids: Muscle in a Bottle?

The male gonadal hormone, testosterone, is one of the hormones responsible for the differences in skeletal mass, muscle mass, and fat distribution of men and women that begin to appear at the onset of sexual maturation. The effects of both male and female hormones on growth have been known for centuries: originally castration and more recently female hormones have been used in animal production to increase fattening and make male animals more tractable.

In sports competition, the opposite effects are desired. As methods of strength training have been perfected, it has become clear that the upper limit of performance is set by intrinsic body capabilities, such as the number of fibers in muscles and their fiber type distribution. These quantities vary from individual to individual, and the variations are believed to be genetically determined. Frustration with this upper limit has led many athletes to resort to using testosterone, or derivatives of it (anabolic steroids), that retain the growth-promoting effect. Using such drugs without a prescription is illegal, has been condemned as unethical by such organizations as the American College of Sports Medicine, and disqualifies competitors in most major international competitions, such as the Olympics and the Pan American Games. Nevertheless, steroid use is widespread among both professional and amateur athletes.

Are anabolic steroids effective? Are there significant risks involved in their use? To answer the first question, it is necessary to compare athletes who are using optimal training practices and diet with those who are doing the same but also using steroids. A number of such studies confirmed that anabolic steroids can enhance muscle mass and strength in well-trained male subjects. Furthermore, many athletes are convinced that anabolic steroids have markedly enhanced their performance.

The answer to the second question is unequivocal: at levels characteristic of those used by many athletes, testosterone and other anabolic steroids present a significant risk of liver disease and cardiovascular diseases, such as atherosclerosis. If used before the end of the growth period, anabolic steroids may terminate bone lengthening and cause the user to be shorter than he or she would otherwise have been if growth had proceeded normally. Furthermore, all anabolic steroids retain some ability to masculinize secondary sexual characteristics. If they are used by women, the normal growth pattern may be disrupted, abnormal hair growth may result, and the female external genitalia may be partly masculinized. Males who use testosterone or its analogues may suffer a reduction in the ability of their own testes to produce testosterone. This is a result of the feedback control of

hormonal levels described in Chapter 5. The depression of testicular endocrine function may last quite long; a study of 5 male power athletes showed that when steroid use was stopped after 26 weeks, testosterone levels remained at less than half the presteroid levels for at least 12 to 16 weeks. There is some overlap between the feedback loop that regulates testosterone production and the one that regulates sperm production, so testosterone abuse can also cause infertility. Case histories suggest that heavy users of anabolic steroids may experience mood changes and increased irritability, with many users displaying outbreaks of "roid rage." Recognition of these effects led to reclassification of anabolic steroids as high-risk controlled substances under federal law in 1990.

Sensitive chemical techniques can be used to detect the presence of anabolic steroids in the urine or blood of athletes and thus exclude abusers from serious competition. Some competitors have tried to avoid detection by switching from anabolic steroids to testosterone for a time before competition. However, abuse of testosterone can be detected by comparing testosterone levels with those of related metabolites; a relatively high level of testosterone indicates that some of it came from a source other than the competitor's own body.

---

drolyze ATP makes tension maintenance in smooth muscle considerably more economical than in skeletal muscle, in which force maintenance requires continuous crossbridge cycling and ATP hydrolysis.

Table 11-4 summarizes the features of the contractile machinery of smooth muscle in three basic states of activity: relaxed, shortening, and maintaining tension.

Similarities in and Differences Among the Three Muscle Types

Muscle fibers of all three types are highly specialized cells in which almost every cellular process and organelle is specifically adapted to provide particular contractile properties. Some of the important functional and structural similarities and differences between the three basic types of muscle are summarized in Table 11-5 and Figure 11-31.

SUMMARY

1. The contractile machinery of muscle consists of thick filaments of **myosin** and thin filaments containing **actin.**

2. Force generation occurs when **crossbridges** on myosin interact with actin binding sites.

3. **Striated** (skeletal and cardiac) muscle contains **myofibrils** in which thick and thin **filaments** are organized into **sarcomeres,** an adaptation for rapid force development.

4. The thin filaments of striated muscle contain the regulatory proteins, **troponin,** which binds Ca^{++} and **tropomyosin,** which controls myosin binding sites.

5. Contraction of striated muscle is triggered when a muscle action potential causes the release of Ca^{++} from the **terminal cisternae** of the SR in a process referred to as excitation-contraction coupling. Calcium binds to troponin and causes tropomyosin to shift away from myosin binding sites.

6. In **isotonic** contractions, muscle tension is constant, and shortening results from force development. In **isometric** contractions, muscle length is constant, and sarcomere shortening is accompanied by stretching of series elastic elements and an increase in muscle tension.

7. Cardiac muscle differs from skeletal muscle in that the spontaneously generated cardiac action potential outlasts contraction so that contractions cannot summate. Cardiac muscle fibers are connected by intercalated disks containing **gap junctions** so that the heart contracts as a **single unit.**

8. The thick and thin filaments of smooth muscle are not organized into sarcomeres but form a network that is able to develop force over a broad operating range.

9. Contraction in smooth muscle is controlled by a second messenger system in which the primary step is Ca^{++} entry. Ca^{++} binds to **calmodulin** and also to myosin. The Ca^{++}-calmodulin complex activates myosin light-chain kinase, which phosphorylates myosin, causing it to form crossbridges that cycle. In smooth muscle, tension can be maintained with little expenditure of ATP. A second level of control of smooth muscle contractility occurs when **caldesmon** links actin and myosin leads to produce crossbridges that do not cycle.

10. Each fiber is innervated in **multi-unit** smooth muscle. The fibers of **single-unit** smooth muscle are electrically coupled, and the tissue is usually spontaneously active as the result of a basic electrical rhythm.

● *STUDY QUESTIONS*

1. An action potential in a motor nerve fiber produces a single muscle contraction. Describe the sequence of events that occurs between the time that a nerve impulse arrives at a muscle and the onset of the muscle twitch.

2. What is the role of each of the following in excitation-contraction coupling in skeletal muscle?
   Transverse tubules   Sarcoplasmic reticulum
   Inositol triphosphate   Troponin
   Tropomyosin

3. What is the physical state of the crossbridges in a skeletal muscle when its internal ATP stores are exhausted? What is their state when ATP is present, but there is no free Ca^{++}?

4. What is the difference between an isometric and an isotonic contraction? Give a real-life example of each.

5. If you were holding a 1-kg weight in your right hand and a 100-gm weight in your left, how would the pattern of motor neuron activity differ for the muscles in your right and left arms?

6. Large amounts of stored glycogen are present in which type of skeletal muscle fiber? What is the functional role of this stored glycogen?

7. What specific processes contribute to rapid fatigue and slow fatigue?

8. How does the process of contraction in smooth muscle differ from that in skeletal muscle? What other features distinguish these two types of muscle?

9. Why are latchbridges particularly useful in the typical functions of smooth muscle?

10. Define single-unit and multi-unit smooth muscle. How can the heart be viewed as a single-unit muscle?

11. Summarize the differences in the mechanisms of excitation-contraction coupling between the three basic muscle types.

Choose the MOST CORRECT Answer

12. The site of excitation-contraction coupling in skeletal muscle is termed
    a. Transverse tubules
    b. Muscle triad
    c. Terminal cisternae
    d. Sarcolemma

13. The cytoplasmic ends of this intrinsic protein acts as "stoppers" for the Ca^{++} release channel when T-tubular membrane is at its resting potential:
    a. Ryanodine receptors
    b. Calsequestrin
    c. Dihydropyridine receptor
    d. G-actin

14. Which characteristic pertains to type II B (white) fibers?
    a. Large amount of myoglobin
    b. High mitochondrial density
    c. Moderately resistant to fatigue
    d. Large stores of glycogen

15. Which of the following is true of isotonic contraction?
    a. Muscle tension remains constant?
    b. Muscle length is constant?
    c. Elastic elements stretch without sarcomere shortening
    d. Muscle tension increases

16. This characteristic is exclusive to cardiac muscle tissue:
    a. Intercalated disks
    b. Abundant myoglobin
    c. Many mitochondria
    d. Myofibrils and t-tubules

17. Single-unit smooth muscle
    a. Causes piloerection
    b. Is found in the iris of the eye
    c. Is typical of visceral organs
    d. Contains relatively few gap junctions

● *SUGGESTED READING*

ADAMS BA, BEAM KG: Muscular dysgenesis in mice: a model system for studying excitation-contraction coupling, *FASEB Journal* 4(1):2809, July 1990. Muscular dysgenesis is a lethal autosomal, recessive mutation of skeletal muscle involving the failure of excitation-contraction coupling. This article is an up-to-date review of the molecular mechanisms involved.

DE-VORE S: Big muscles, big problems. *Current Health* 17(3):11, November, 1990. Discussion of the consequences of anabolic steroid abuse by athletes and bodybuilders.

KIRKENDALL DT: Mechanisms of peripheral fatigue. *Med Sci Sports Exerc* 22(4):444, August 1990. A discussion of the variety of mechanisms that can lead to fatigue, including the motor neuron, excitation-contraction coupling, accumulation of metabolites, and depletion of fuels.

MURPHY RA: Muscle cells of hollow organs, *News in Physiological Sciences,* 3:124-128, 1988. Explains the molecular basis for the efficiency of smooth muscle.

RASMUSSEN H, TAKUWA H, PARK S: Protein kinase C in the regulation of smooth muscle contraction, *FASEB Journal,* 1:177, September 1987. Comprehensive discussion of the mechanisms of smooth muscle contraction and excitation and of contraction coupling in smooth muscle.

WARSHAW DM, MCBRIDE WJ, WORK SS: Corkscrew-like shortening in single smooth muscle cells, *Science* 236:1457, June 12, 1987. Describes the use of digital video microscopy to track the movement of marker beads on the surface of smooth muscle cells and show how these cells contract.

CHAPTER 12

Somatic and Autonomic Motor Systems

On completing this chapter you will be able to:

- Compare innervation of skeletal muscle fibers by somatic motor neurons with autonomic innervation of visceral smooth muscle and glands
- State the anatomical differences between the pathways followed by parasympathetic and sympathetic efferents to their effectors
- Understand the role of dual innervation in the autonomic nervous system
- Describe the neurotransmitters used in the autonomic nervous system
- Outline the spinal circuitry and functional role of the stretch reflex, withdrawal reflex, and the Golgi tendon reflex
- Appreciate the role of subcortical structures in the initiation and control of movements
- Distinguish between the lateral and proximal motor control systems
- Understand the pharmacological differences between postsynaptic receptors in the autonomic nervous system

*T*his chapter describes the output side of the nervous system: the motor pathways that control effectors of the body. Historically, the motor nerves have been subdivided into a somatic motor system, which controls skeletal muscle, and an autonomic motor system, which controls visceral tissues such as the heart, exocrine glands (those that secrete via ducts in contrast to ductless endocrine glands), blood vessels, and the gastrointestinal system. The somatic motor system has also been called the voluntary nervous system because it is involved in consciously initiated activities; the autonomic nervous system was named because visceral activities, such as blood circulation, secretion of saliva, and digestion, seemed to go on by themselves.

The two motor systems operate according to the principles of negative feedback control explained in Chapter 5. With some slight differences, the same principles of action potential propagation and synaptic transmission apply to both systems. They differ most importantly in the anatomy of their peripheral components and in the nature of the effectors they serve. From a functional point of view, the role of the somatic nervous system generally is to command, while the role of the autonomic system generally is to modulate or influence. Skeletal muscles do not contract unless stimulated. Many of the effectors served by the autonomic nervous system, such as the heart and intestines, are active in the absence of any nervous input, and their activity can be increased or decreased by the autonomic nervous system.

The distinction between purely voluntary and purely involuntary activities became less sharp with advances in understanding of both systems. For instance, although the somatic motor neurons respond to voluntary commands that originate in the brain cortex, they also are responding continually to signals from muscle receptors whose inputs are ordinarily not consciously perceived. On the other hand, certain individuals are able to voluntarily alter some autonomic functions normally not consciously controlled, such as blood pressure, if appropriate operant conditioning methods are used.

MOTOR PATHWAYS OF THE PERIPHERAL NERVOUS SYSTEM
Motor Units

> • *What are the paths from the central nervous system to effectors in the autonomic nervous system? How do they differ from the path of the somatic nervous system?*
> • *What is the definition of a motor unit in the somatic nervous system?*
> • *Why is dual innervation an important organizational feature of the autonomic nervous system?*

Each skeletal muscle fiber, the cellular unit of a muscle, is innervated by only one motor neuron. However, most motor nerve axons innervate and therefore control more than one muscle fiber. The set of muscle fibers innervated by all branches of the axon of a single motor neuron is called a **motor unit** (Figure 12-1, *A* and Figure 12-2). Every time the motor neuron produces an action potential, all muscle fibers in the motor unit contract together.

Thus a motor unit, rather than a muscle fiber, is the smallest functional element of a skeletal muscle.

Motor units differ in size within any given muscle, and the average number varies greatly from one muscle to another. Muscles that can perform delicate movements, such as those of the fingers or those that position the eyes, have motor units consisting of as few as two or three muscle fibers. At the other extreme, the lower leg muscles that produce the force for powerful, rapid movements, such as jumping, may have motor units consisting of several hundred muscle fibers. Individual fibers in a motor unit are not necessarily adjacent to one another but may be distributed throughout the entire cross-sectional area of the muscle (Figure 12-1, *A*).

Control of Contractile Force

How forcefully a skeletal muscle contracts depends on two factors: (1) the number and size of the motor units activated, or **recruited,** by excitatory synaptic inputs, and (2) the frequency of action potentials in axons serving each active motor unit. Not all motor units need be active at the same time, and the total tension of a whole muscle is an average over all the motor units in it (Figure 12-1, *B*). At the level of the

A

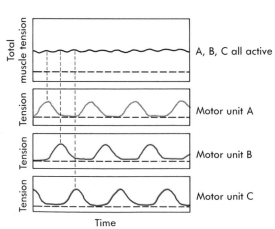

B

FIGURE 12-1
A motor unit is a group of skeletal muscle fibers controlled by a single motor neuron. **A** A schematic diagram of three different motor units in a single muscle, consisting of 3 to 5 individual muscle fibers. **B** The total tension in a muscle is given by summating the activity of all the active motor units.

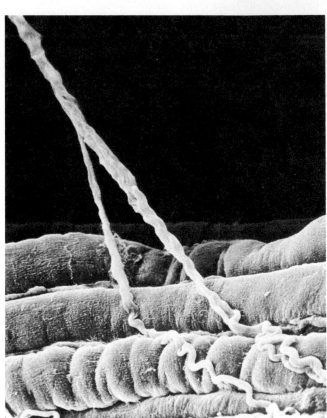

FIGURE 12-2
A scanning electron micrograph of a nerve fiber innervating several skeletal muscle fibers.

motor neurons there is a high degree of convergence; each spinal motor neuron may receive as many as 15,000 synaptic inputs. Most of these inputs come from spinal interneurons, but a small fraction come directly from brain motor centers. The integration of all these inputs at the level of each motor neuron is what causes particular muscle units to be activated at any given instant. Pioneer neurophysiologist Sherrington referred to spinal motor neurons as the "final common path" from neurons in motor areas of the brain to muscle cells.

Innervation of Skeletal Muscle

The cell bodies of somatic motor neurons that control the skeletal muscles of all parts of the body except the head are located in the ventral horns of spinal cord segments (Figure 12-3). The corresponding motor neurons for the face and head are in brainstem nuclei and project through the cranial nerves. In the spinal cord, motor neurons innervating a single muscle or group of muscles are typically found within two to four adjacent spinal segments. The axons of somatic motor nerves leave the ventral horns and pass into spinal nerves.

Spinal nerves innervate the muscles of a single muscle segment of the body trunk, or **myotome**. Myotomes are functionally similar to the dermatomes of the somatosensory system. There is an approximate anatomical correspondence as well, with each dermatome and its underlying myotome being served by the same spinal nerve (see Figure 9-13).

Four major **plexuses**, or networks of axons, result from the intermingling of axons from more than one spinal nerve. These are the **cervical, brachial, lumbar,** and **sacral plexuses**; most of the nerve branches that emerge from these plexuses bear the names of the regions that they innervate, such as the splanchnic and mesenteric nerves. The largest nerve in the body, the **sciatic nerve**, emerges from the lumbar plexus and innervates the posterior thigh, leg, and foot muscles.

Pathways to the Visceral Effectors

Autonomic motor pathways to visceral organs are composed of two neurons. **Preganglionic neurons** arise in the central nervous system and project axons to **postganglionic neurons,** whose cell bodies

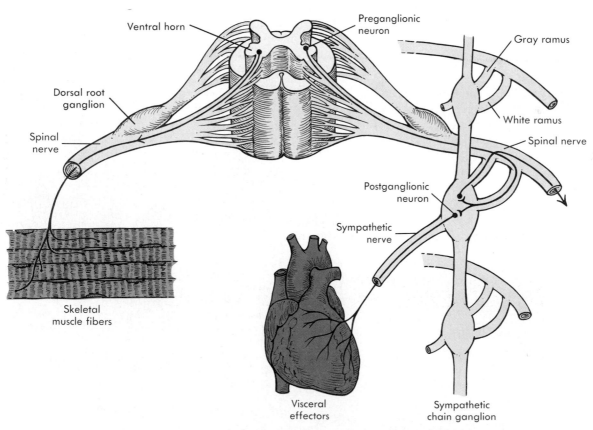

FIGURE 12-3

The somatic pathway *(left)* and autonomic sympathetic pathway *(right)*. For clarity, the two pathways are shown on opposite sides of the spinal segment; sympathetic and somatic motor neurons are actually present on both sides of the cord. Most sympathetic postganglionic axons run to effectors along sympathetic nerves, but some are found in spinal nerves along with the axons of somatic motor neurons.

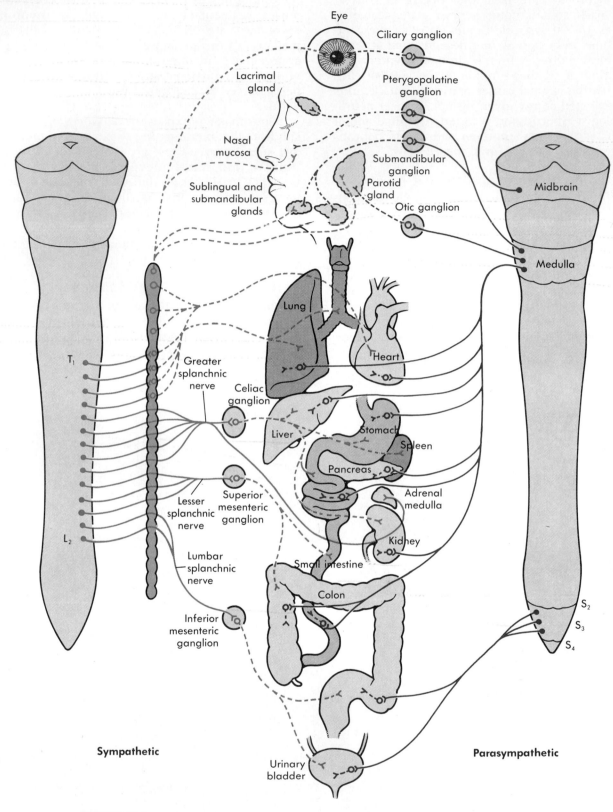

Sympathetic

Parasympathetic

FIGURE 12-4

Innervation of the major target organs by the autonomic nervous system. The right side shows connections of the parasympathetic branch *(red);* the left side those of the sympathetic branch *(blue).* Preganglionic fibers are indicated by the solid lines; postganglionic fibers by the broken lines.

are in the autonomic ganglia. The preganglionic neurons of the **sympathetic branch** are in the intermediate, lateral gray matter (intermediolateral horn) of the thoracic and lumbar spinal cord and are indistinguishable in microscopic appearance from the somatic motor neurons.

The **paravertebral sympathetic ganglia** form a chain on each side of the spinal cord (see Figure 12-3). Preganglionic fibers of the sympathetic branch run out of the ventral horn and pass into sympathetic ganglia through short connectives called **white rami** (*solid blue lines* in Figure 12-4). Within the ganglia, preganglionic axons synapse on postganglionic cells. The axons of most sympathetic postganglionic cells pass through **gray rami** (*broken blue lines* in Figure 12-4) into sympathetic nerves that run to blood vessels and internal organs. The axons of some sympathetic preganglionic neurons pass through the **gray rami** into spinal nerves and run to ganglia located near the effectors themselves (for example, the inferior mesenteric ganglion in Figure 12-4).

The **adrenal medulla** is derived during embryonic development from a sympathetic ganglion and is directly innervated by preganglionic fibers. It is a neurosecretory organ because the postganglionic cells of the adrenal medulla, called **chromaffin cells**, are modified neurons that do not possess axons. Instead, they release epinephrine and norepinephrine directly into the blood as hormones.

The preganglionic neurons of the **parasympathetic branch** are located in two regions of the CNS: the motor nuclei of some cranial nerves and the sacral segments of the spinal cord (Figure 12-4). An important difference between the sympathetic and parasympathetic branches of the autonomic nervous system is the location of their ganglia. In the parasympathetic system, preganglionic fibers are long (*solid red lines* in Figure 12-4), leading to ganglia located in or on the organs they serve (except for those that serve the eye, tear glands, and salivary glands). Conversely, in the sympathetic system the ganglia are either next to the spinal cord or in the abdomen but relatively remote from the organs they serve. Figure 12-5, *A* and Table 12-1 summarize the similarities and differences between the somatic nervous system and the two branches of the autonomic system.

Dual Autonomic Innervation of Visceral Effectors

Effectors innervated by the autonomic system typically receive inputs from both the parasympathetic and sympathetic branches; this is referred to as **dual innervation** (see Figure 12-4). Unlike the somatic nervous system, which can only excite its effectors, the autonomic nervous system generally can mediate either excitation or inhibition. This is possible because visceral effectors typically have some degree of intrinsic activity; the sum of autonomic inputs may either increase the activity or suppress it. Organs that receive dual autonomic innervation are generally excited by one of the branches and inhibited by the other; which branch is excitatory and which inhibitory varies from one organ system to another. Some tissues receive **single innervation** from only one branch of the autonomic nervous system. For example, most blood vessels receive only sympathetic innervation.

The morphology of synapses, described in Chapter 8, differs between the somatic and autonomic systems. Somatic motor neurons have elaborate motor end plates on each of the muscle fibers of their motor units (see Figure 8-17). The receptors for their neurotransmitter, acetylcholine, are confined to the immediate vicinity of the motor end plate. Autonomic synapses are less elaborate and may take the form of enlargements in autonomic axons, termed axon **varicosities,** which spread transmitter over a large area and do not have a one-to-one relationship with effector cells (see Figure 8-18). In such cases there is no discrete, morphologically specialized postsynaptic membrane and the effector cells possess receptors scattered over their entire surfaces.

Acetylcholine and Norepinephrine as Transmitters in the Motor Pathways

The transmitter chemical released at the synapse between somatic motor neurons and skeletal muscle fibers is acetylcholine, which is also the transmitter used between preganglionic and postganglionic cells in ganglia of both branches of the autonomic system (Figure 12-5, *B*). Thus the rule for all the motor pathways of the body is that the first synapse in the periphery (which is also the only synapse in the somatic system) is cholinergic. The postganglionic transmitters differ for the two autonomic branches: as a rule, acetylcholine is released on effectors by postganglionic axons of the parasympathetic nervous system, while **norepinephrine** is released by most postganglionic axons of the sympathetic system (Figure 12-5).

The discovery that **peptide neuromodulators** (see Chapter 8) are coreleased with acetylcholine or norepinephrine has greatly expanded appreciation of the integrative functions of autonomic peripheral synapses. For instance, in sympathetic pathways, enkephalin is often found in the preganglionic axons and somatostatin is often found in the postganglionic neurons. Peptide neuromodulators are also believed to be particularly important in the parasympathetic plexuses of the gastrointestinal tract.

The chromaffin cells of the adrenal medulla are an exception to the rule about postganglionic transmitters because chromaffin cells release primarily epinephrine, along with a small amount of norepi-

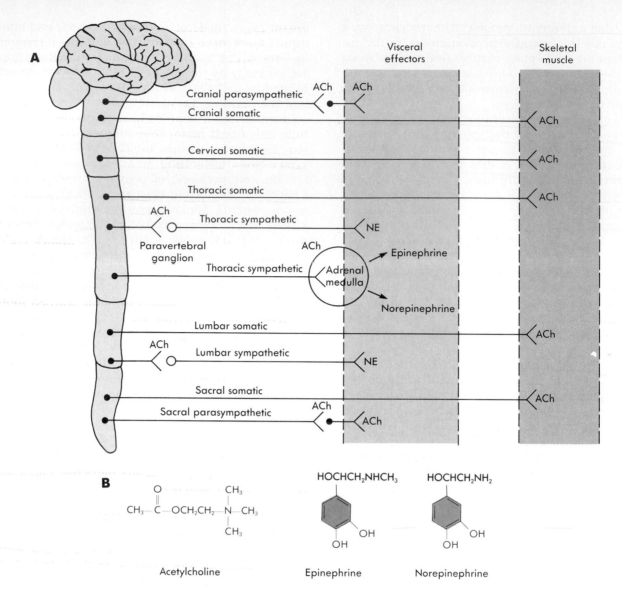

FIGURE 12-5

A A comparison of the central origins of somatic and autonomic motor pathways. Somatic motor neurons are present in cranial nuclei and all spinal segments; they synapse directly on their effectors after leaving the spinal cord and release acetylcholine. Parasympathetic preganglionic neurons are present at cranial and sacral levels only and acetylcholine is the transmitter at both the ganglionic and the postganglionic synapses. Sympathetic preganglionic neurons are present in thoracic and lumbar segments of the spinal cord; acetylcholine is the transmitter in paravertebral and prevertebral ganglia, but the postganglionic transmitter is usually norepinephrine (exceptions are noted in the text).

B The chemical structures of acetylcholine (ACh), epinephrine, and norepinephrine (NE).

nephrine. Epinephrine differs from norepinephrine only by the presence of a methyl (-CH₃) group; both are **catecholamines** (see Figure 12-5, *B*). Both epinephrine and norepinephrine are regarded as neurohormones as well as neurotransmitters. Adrenal release of epinephrine is an important component in many responses involving a general increase in the level of sympathetic activity, such as the classical "fight-or-flight" reaction. Because the adrenal medulla and sympathetic ganglia arise from the same tissues in the developing embryo, the adrenal

medulla could be regarded as a sympathetic ganglion modified for neurosecretion.

Sympathetic cholinergic fibers, which release acetylcholine, are another exception to the rule that sympathetic postganglionic fibers usually release norepinephrine. The best-known sympathetic cholinergic pathway serves sweat glands. This pathway is responsible for increased sweating during exercise and emotional stress. Other sympathetic cholinergic pathways have been proposed but evidence for them is incomplete.

TABLE 12-1 Comparison of Somatic and Autonomic Motor Systems

Characteristic	Somatic	Autonomic
Effectors	Skeletal muscle	Cardiac muscle Heart Smooth muscle Gastrointestinal tract Blood vessels Airways Exocrine glands
Effect of motor nerves	Excitation	Excitation or inhibition
Innervation of effector cells	Always single	Typically dual
Number of neurons in path to effector	One	Two
Peripheral ganglia	No	Yes
Transmitters	Acetylcholine	Acetylcholine Norepinephrine
Receptor types	Nicotinic	1. Cholinergic Nicotinic (ganglia) or muscarinic (effectors) 2. Adrenergic α or β (depending on the effector)

Somatic motor neurons resemble autonomic preganglionic neurons in two ways: both originate embryonically in similar spinal regions, and both release acetylcholine. Skeletal muscle fibers, while different in their embryonic origin, functionally resemble autonomic postganglionic neurons in that both receive input from central motor neurons and in both cases the transmitter is acetylcholine.

SOMATIC MOTOR CONTROL AT THE SPINAL LEVEL
The Size Principle of Motor Unit Recruitment

- *What features of the muscle spindle cause it to act as a detector of muscle length?*
- *What regulates the sensitivity of a muscle spindle?*
- *What is reciprocal inhibition? Why is it necessary?*

The total force exerted by an entire skeletal muscle can be regulated by selective activation of its motor units. The motor units activated at the lowest levels of effort are those with the smallest number of muscle fibers. The initial increments to the total force generated by a muscle are therefore relatively small. If greater force is required, larger and larger motor units are recruited and the force increments become larger. Recruitment of motor units in the order of increasing size results in smooth increases in force.

The basis for the size principle of recruitment is that spinal motor neurons have cell bodies of different sizes. Small motor units are innervated by spinal motor neurons whose cell bodies are small; larger motor units have motor neurons with correspondingly larger cell bodies. Synaptic inputs more readily depolarize a small neuron to threshold than a large one. That is, a given amount of depolarization provided by one excitatory synaptic event will more effectively depolarize the axon hillock region of a small neuron than a large one (Figure 12-6). Recruitment according to size is seen in both reflexive and voluntary contractions of a muscle. One exception to this rule occurs in some learned motor patterns that can recruit the large motor units earlier, thereby favoring very rapid shortening velocity.

When a muscle acts to lift a weight, individual motor units are recruited asynchronously so that an individual fiber is not active continuously but rather cycles on and off repeatedly. At any instant during the contraction a constant number of fibers may be active, but the composition of this population changes as different individual units turn on and off. Continuously changing the population of active motor units helps prevent any particular motor unit from becoming fatigued.

Spinal Reflexes and Postural Stability

Spinal reflexes are negative feedback systems (see Chapter 5). The most familiar example of a spinal reflex is the knee jerk a physician elicits in a physical examination; it is referred to as the stretch, or myotatic, reflex. When the physician taps the tendon of a muscle, the muscle is stretched momentarily, activating receptors, referred to as muscle spindles (described in the following section), in the muscle. The reflex response is a rapid contraction of the particular muscle that was stretched (Figure 12-7, A). The stretch reflex is a feedback loop that maintains muscle length and opposes passive increases in length.

Muscles that cooperate to move a joint in one direction are called synergists; opposing muscles are referred to as antagonists. Muscles that decrease the angle of a joint are called flexors, while those increasing the angle of a joint are termed extensors. In the knee-jerk reflex, while the stretched extensor muscle is contracting, its antagonist flexors are inhibited (Figure 12-7, B). This is the result of reciprocal inhibition in the pathways that activate the spinal neurons of antagonistic muscles. Reciprocal inhibition is a common feature of motor pathways and is advantageous for two reasons: (1) energy is wasted if a muscle has to overcome the force

Mind Over Matter? The Case for Biofeedback

Biofeedback uses instruments that measure internal variables, such as blood pressure, heart rate and muscle tone, to provide subjects with information they can use to influence organ systems usually regarded as "involuntary." The signal from whatever device is used is displayed as a graph on a computer monitor, or used to drive a loudspeaker. Visceral functions are thus fed back to the subject. The goal is for the subject to learn how to influence the output and thereby to control the "involuntary" function being measured. In the early days, biofeedback was fairly indiscriminately applied to just about everything. Currently the applications of biofeedback have narrowed, but its usefulness is well established for some conditions.

In electromyographic (EMG) biofeedback, an electrode picks up muscle action potentials, directly related to the level of muscle contraction, translating these action potentials into an audible tone. By learning to lower the tone's volume or frequency, the subject learns to reduce muscle tension. The EMG technique is used to treat tension headaches that stem from contraction of muscles in the forehead or neck. It is also used to treat chronic pain arising from muscle spasms following accidents or sports injuries. Temperature biofeedback uses a probe to measure skin temperature, a sensitive indicator of skin blood flow. Because skin blood flow is sympathetically controlled, learning to control blood flow in the hands or feet seems to give subjects some ability to affect the overall level of sympathetic tone. Galvanic skin-resistance biofeedback uses a probe that responds to sweat, again a sympathetic response to states of arousal and stress.

Progressive relaxation may be the main benefit of biofeedback. All these techniques help some individuals relax by lowering muscle tension, increasing blood flow, and reducing sweat gland activity. Successfully trained subjects seem to learn to reduce the activity in both the somatic motor system and the autonomic nervous system. Biofeedback has been used to treat some sleep disorders and several stomach and intestinal conditions associated with hyper-secretion. Sometimes individuals with partial paralysis of a muscle can regain some control through biofeedback by learning to contract weak muscles more effectively. Illnesses that are not caused by stress, such as asthma and diabetes, may be aggravated by the stress they induce. To the extent that biofeedback can help control stress, it can play a role in treating these conditions. There is also some recent evidence that biofeedback techniques may boost the body's immune system by increasing levels of helper T cells, cells that regulate the extent of an immune response.

Biofeedback therapy typically involves 6 to 18 weeks of 1 hour sessions, supplemented by daily home exercises. A key point about biofeedback is that it is a way of teaching skills in a clinical setting that must be applied throughout the day. The National Institutes of Health now recommends relaxation training in combination with losing weight and restricting salt intake as the treatment of choice for mild hypertension. Biofeedback is not a natural tranquilizer nor a cure for all ills, but recent research has shown that it can be an effective, painless, and noninvasive treatment for many psychological and physical disorders.

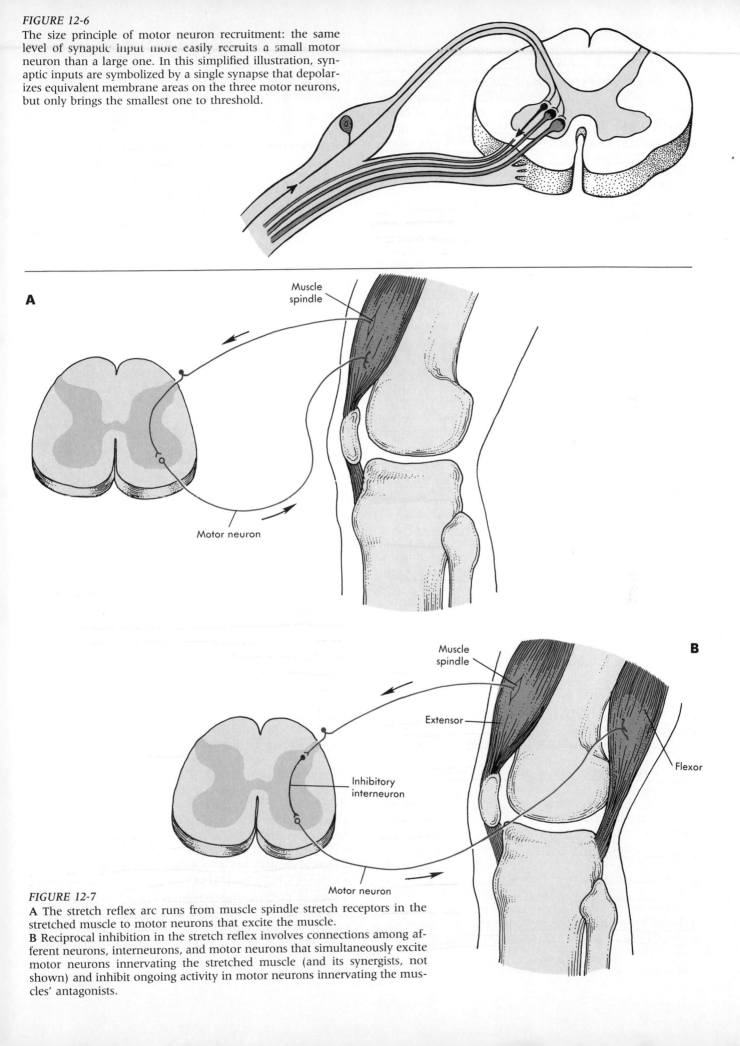

FIGURE 12-6
The size principle of motor neuron recruitment: the same level of synaptic input more easily recruits a small motor neuron than a large one. In this simplified illustration, synaptic inputs are symbolized by a single synapse that depolarizes equivalent membrane areas on the three motor neurons, but only brings the smallest one to threshold.

A

Muscle spindle

Motor neuron

B

Muscle spindle

Extensor

Flexor

Inhibitory interneuron

Motor neuron

FIGURE 12-7
A The stretch reflex arc runs from muscle spindle stretch receptors in the stretched muscle to motor neurons that excite the muscle.
B Reciprocal inhibition in the stretch reflex involves connections among afferent neurons, interneurons, and motor neurons that simultaneously excite motor neurons innervating the stretched muscle (and its synergists, not shown) and inhibit ongoing activity in motor neurons innervating the muscles' antagonists.

of its antagonists and (2) movements are smoother when opposing muscles do not interfere with one another.

Three examples illustrate the importance of the stretch reflex in both extensors and flexors. The muscle involved in the knee jerk extends the lower leg. In a standing person, gravity tends to buckle the knee, stretching the knee joint extensor. Gravitational stretch activates the stretch reflex, maintaining sufficient steady contraction in extensor muscles to prevent collapse of the knee joint.

The second example illustrates the ability of the stretch reflex to increase muscle force to compensate for an unexpected additional load that tends to lengthen an actively contracting muscle. Suppose several individuals are working together to lift a heavy object. If one of them suddenly slips, the others experience an increased load, and this load stretches their flexors. Stretch of the flexors produces a feedback signal from muscle receptors that increases the activity level of motor units already in use and recruits additional, larger motor units that had been inactive. The result is a rapid increase in muscle force that compensates for the increased load.

The third example involves an abrupt decrease in the load on a muscle, such as would occur if the rope broke during a game of tug-of-war. This is a case of rapid unloading of active muscles, the opposite of the previous example. The muscles that had been stretched would shorten because the force that had been opposing their contraction is removed. As the stretched muscle suddenly shortens, activity in the stretch reflex pathway rapidly diminishes, decreasing the excitatory drive on the motor neurons. In this way the spinal reflex, which works faster than conscious adjustment of muscle activity by the brain, reduces the amount of shortening the pulling muscles exhibit when the rope breaks.

Structure of the Muscle Spindle

Scattered throughout the mass of a muscle are tiny stretch receptors, the **muscle spindles**, that provide the central nervous system with information about muscle length (Figure 12-8 and Table 12-2). A muscle spindle consists of a small bundle of modified muscle fibers, called **intrafusal fibers**, to which the endings of several sensory nerves are attached. The nerve endings and the intrafusal muscle fibers are enclosed within a capsule. The muscle fibers that make up by far the greater part of the mass of the muscle and are responsible for generating all of its force are called **extrafusal fibers**. Although intrafusal muscle fibers are too weak to develop enough force to move body parts, their contractions are the means by which the body regulates the sensitivity of the muscle spindles. In order to differentiate between them, the motor neurons that innervate extrafusal fibers are termed α-**motor neurons**; intrafusal fibers are controlled by γ-**motor neurons**.

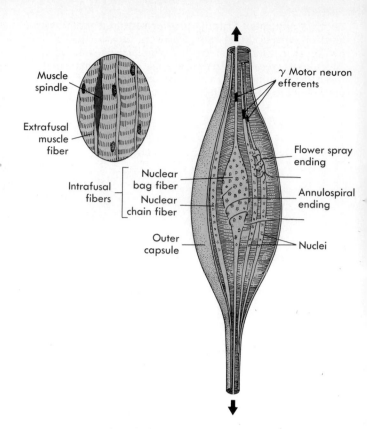

FIGURE 12-8

A diagram of the muscle spindle. The spindle consists of modified intrafusal muscle fibers. There are two types of intrafusal fibers: nuclear bag fibers are innervated by annulospiral endings, and nuclear chain fibers are innervated by flower-spray endings. The ends of the muscle spindles are also innervated by efferent fibers from γ-motor neurons. The muscle spindle is in parallel with the normal (extrafusal) muscle fibers.

There are actually two different types of intrafusal fibers (see Figure 12-8 and Table 12-2): (1) a **nuclear bag fiber** that has a swollen central region and (2) a **nuclear chain fiber** that is smaller and uniform in diameter throughout its length. The central regions of bag and chain fibers contain no myofibrils and are the most elastic parts of the fibers, so stretching the muscle spindle stretches the central regions of the intrafusal fibers preferentially. These elastic central regions are wrapped by endings of **group I** afferent neurons (sometimes termed annulospiral endings; see Chapter 9 for a discussion of nerve fiber groups). The lateral, contractile regions of chain fibers are where the endings of smaller **group II** afferent neurons are located (flower spray endings). There are no endings of group II neurons on nuclear bag fibers.

Muscle Spindles as Length and Length Change Detectors

When a muscle is stretched, the responses of its spindles include a phasic component in which action potential frequency is a measure of the rate of stretching and a tonic component in which action

TABLE 12-2   *Muscles and Their Receptors*

Component	Innervation	Functional rule
Extrafusal fibers	α-Motor neurons	Generation of muscle force
Intrafusal nuclear bag fibers	Group IA (annulospiral) afferents	Phasic stretch receptors
	γ-Motor neurons	Regulate spindle sensitivity
Intrafusal nuclear chain fibers	Group IB (flower spray) afferents	Tonic length detectors
	γ-Motor neurons	Regulate spindle sensitivity
Golgi tendon organ	Group II afferents	Monitor muscle tension

potential frequency is a measure of relative muscle length. Bag fibers are affected primarily by stretch and chain fibers primarily by length. Both bag and chain fibers have group I afferent endings in their central regions, so the group I response contains both a phasic and a tonic component. Group II afferents have endings only on chain fibers, so their activity relays only information on muscle length (see Table 11-2).

The density and the response characteristics of the stretch receptors present in a particular muscle are related to its function. Muscles typically involved in fine, highly controlled movements, such as finger muscles, have a high density of spindles containing both bag and chain fibers. They therefore provide information on both stretch and length. Muscles that are not involved in delicate movements but rather support the body and move large body parts contain relatively few spindles and those spindles consist in large part of chain fibers, so they primarily signal length.

Group IA afferents from bag and chain fibers branch within the spinal cord. Some branches ascend to higher centers; others synapse with interneurons or with the α-motor neurons. The branches that synapse on α-motor neurons are responsible for the reflexive contraction of a stretched muscle. The effect of spindle afferents on α-motor neurons is excitatory. The efficacy of these synapses must be high, because impulses resulting from spindle stretch are sufficient to recruit motor neurons, even though the number of synapses coming from the spindles is a small fraction of the total number of synapses on these neurons. The time interval, or **latency**, between activation of spindle afferents and the reflexive response is short because there is only one synapse in the reflex arc.

Most muscles of the body are not completely relaxed but show a low level of tension referred to as **muscle tone.** The importance of the stretch reflex for maintenance of muscle tone can be demonstrated by **denervation:** cutting the dorsal roots of a spinal segment, which interrupts afferent signals from the muscle spindles of that segment. When this is done, the tone of the muscles innervated by the segment decreases dramatically, even though the ventral roots containing the motor pathways to those muscles are still intact.

Control of the Stretch Reflex by Motor Neurons

The stretch reflex opposes departures from a set point of muscle length. This setpoint is achieved by the central excitatory drive onto the γ-efferents, which determines how much the ends of the spindle fiber are contracted and therefore how much the center region that contains the sensory endings is stretched. Muscle length is at its setpoint when spindles are "loaded," that is, taut enough to provide a small amount of stimulation to the receptor endings. Under these circumstances, feedback from spindle afferents contributes to muscle tension (Figure 12-9, *A*).

If the muscle is exhibiting the tone characteristic of a relaxed muscle, the spindle afferent may be almost the only drive on the motor neurons; if the muscle is maintaining tension during a voluntary contraction, spindle afferent feedback adds significantly to the excitatory drive from higher motor centers in determining which motor units are recruited (Figure 12-9, *B*).

Coactivation of α- and γ-Motor Neurons

Figure 12-10, *A*, illustrates the situation in a muscle before the initiation of a voluntary contraction. When voluntary changes in muscle length are initiated by motor control regions of the brain, the motor commands include changes of the set point of the muscle spindle system. These are initiated by a volley of impulses arriving at the α- and γ-motor neurons by way of interneurons that descend from higher motor centers (Figure 12-10, *B*). The simultaneous activation of extrafusal fibers (by way of α-motor neurons) and intrafusal fibers (by way of γ-motor neurons) is called **α- γ coactivation.** The functional role of coactivation can be illustrated by an example. If the extrafusal fibers shortened without a compensatory shortening of intrafusal fibers (this would be the case if only α motor neurons were activated), the spindles would become un-

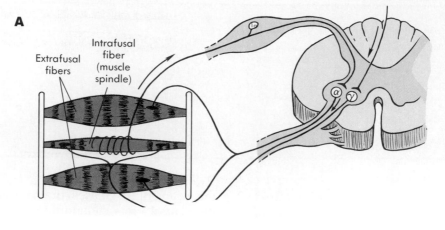

Muscle tone resulting from maintenance of
setpoint by central drive onto γ motor neurons

A

Extrafusal
fibers

Intrafusal
fiber
(muscle
spindle)

Central drive

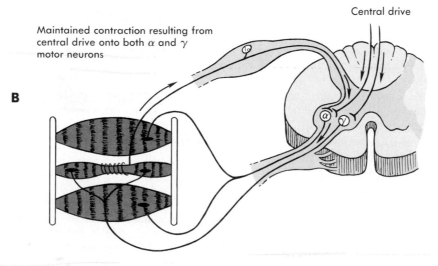

Maintained contraction resulting from
central drive onto both α and γ
motor neurons

B

Central drive

FIGURE 12-9

A In a relaxed muscle, some tone is maintained by central drive onto γ-motor neurons, which sets the activity level of the spindle afferents and maintains a low level of input onto α-motor neurons. **B** During a maintained contraction, central drive onto α-motor neurons summates with the feedback from IA afferents.

loaded, that is, so slack that they would no longer function as length sensors (Figure 12-10, *C*). By also activating the γ motor neurons innervating the intrafusal fibers, the sensitivity of spindles is readjusted over the entire range of muscle lengths. This allows the spindles to be functional at all times during a contraction (Figure 12-10, *D*).

The important benefit of incorporation of length-sensitivity into voluntary activities is illustrated by the following example. When a suitcase must be lifted, the motor centers of the brain estimate on the basis of experience approximately how many motor units must be activated to lift it. If the weight is assessed correctly, the amount of force called for will be adequate to the job. Extrafusal fibers will be able to shorten as rapidly as intrafusal fibers, so the activity of spindle afferents will not change much as the muscles shorten.

But if the suitcase is full of books instead of clothes, the amount of force initially called for by the motor centers may not be adequate to lift the load, and certainly the rate of shortening will be slower than expected. In this case, extrafusal fibers will not shorten much, but intrafusal fibers, which bear none of the load, will shorten. The central region of the spindles will be stretched and a burst of excitatory postsynaptic potentials (EPSPs) from the spindle afferents will summate with those already reaching the α-motor neurons from the descending pathways. Additional motor units will be recruited by this reflexive input, and the force of the muscle will increase until the load is lifted. In this example, the stretch reflex is activated by a difference between the desired muscle length and its actual length. Thus there are few "voluntary" motor actions that do not incorporate a substantial unconscious reflexive component.

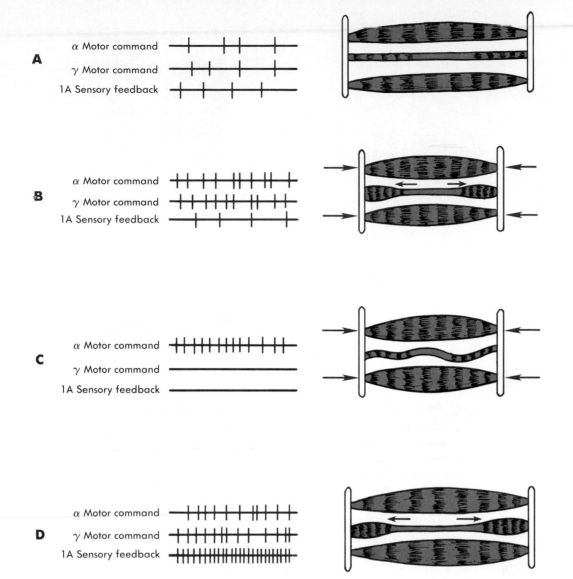

FIGURE 12-10

A Condition of the muscle and sensory system prior to initiation of a voluntary contraction. **B** Result of α-γ coactivation is shortening of the extrafusal fibers and shortening of the contractile ends of the intrafusal fibers to maintain sensitivity of the spindle afferents. Note that the rate of action potentials in the spindle afferent is similar in **A** and **B**. **C** The loss of sensory feedback information that would result if α-motor neurons were activated in the absence of γ coactivation. **D** α-γ coactivation in a situation in which the muscle contracting against an excessive load produces elevated levels of action potentials in IA afferent neurons. The effect of this feedback is to supplement the excitatory drive on α-motor neurons.

Golgi Tendon Organs: Monitors of Muscle Tension

> - *The reflex arc of the stretch reflex is a negative feedback loop. What structures correspond to the sensors, integrating center, and effectors?*
> - *Why do Golgi tendon organs provide information on muscle tension?*
> - *What is the significance of crossed extension?*

Tendons that connect muscles with the skeleton contain stretch receptors called **Golgi tendon or-** gans (Figure 12-11). Golgi tendon organs detect changes in the tension in a tendon and therefore monitor the force exerted by a muscle. Golgi tendon organs are sensitive to small changes in tension and discharge during muscle contractions, as well as during imposed stretch. Unlike primary spindle afferents, fibers from Golgi tendon afferents do not synapse directly on α-motor neurons but instead on spinal interneurons projecting to the motor neurons that control the relevant muscle, its synergists, and its antagonists. The effect of activity in Golgi tendon afferents is to inhibit the motor neurons of the mus-

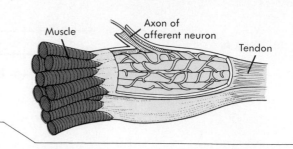

Muscle / Axon of afferent neuron / Tendon

FIGURE 12-11
The Golgi tendon organ is located in tendons and is thus in series with the muscle.

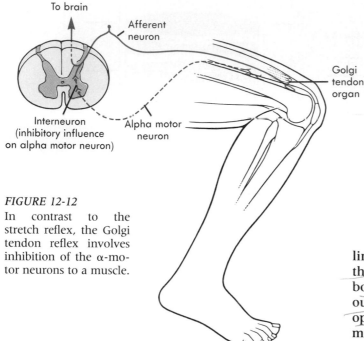

To brain

Afferent neuron

Golgi tendon organ

Interneuron (inhibitory influence on alpha motor neuron)

Alpha motor neuron

FIGURE 12-12
In contrast to the stretch reflex, the Golgi tendon reflex involves inhibition of the α-motor neurons to a muscle.

cle whose tendon was stretched and its synergists, while exciting the motor neurons of the antagonists (Figure 12-12). One possible function of Golgi tendon organs is to provide information to motor centers about the force of muscle contraction. Golgi tendon organs may also provide overload protection to relax muscles that otherwise would experience a dangerously high level of tension, preventing injury to the muscle or tendons.

The Withdrawal Reflex and Crossed Extension

Nociceptive cutaneous afferents synapse on spinal interneurons, some of which eventually relay information interpreted as pain to the brain. On the ipsilateral (same side) of the spinal cord they immediately excite the flexor muscles of the stimulated limb and inhibit extensors (Figure 12-13). Injury to an extremity of the body results in reflexive flexing of the injured arm or leg, withdrawing it from the source of the injury.

Branches of cutaneous nociceptor afferents also cross the spinal cord to excite extensors and inhibit flexors on the opposite, or contralateral side of the body (Figure 12-14). The effect is crossed extension, that is, bracing or extension of the opposite

limb. The significance of crossed extension lies in the fact that when a limb that is supporting the body must suddenly be flexed, the weight previously carried by the flexed limb is transferred to the opposite limb. The spinal reflex connections thus simultaneously bring about flexion of the injured limb and preparation of the muscles of the opposite limb to support more weight, all without the participation of the brain.

MOTOR CONTROL CENTERS OF THE BRAIN
Brainstem Motor Areas

The degree to which spinal reflexes contribute to actual movements can be regulated by the motor areas of the brain by way of descending pathways that increase or decrease the overall excitability of spinal interneurons. The control of reflex responsiveness by the brain was revealed by experiments in which the motor pathways were cut at different points. A cut below the brainstem (Figure 12-15, A) suppressed, for a time, all spinal reflexes beyond the point at which the spinal cord was sectioned (Table 12-3). Before the cut was made, spinal interneurons and motor neurons must have been receiving a steady stream of excitatory postsynaptic potentials from areas within or above the brainstem. Steady excitation that moves the spinal neurons closer to threshold and makes them more likely to respond to other excitatory inputs, such as those from stretch receptors, is termed facilitation.

The reticular formation of the brainstem includes several nuclei that are sources of descending pathways that facilitate spinal motor neurons (Figure 12-15). Of particular importance are the vestib-

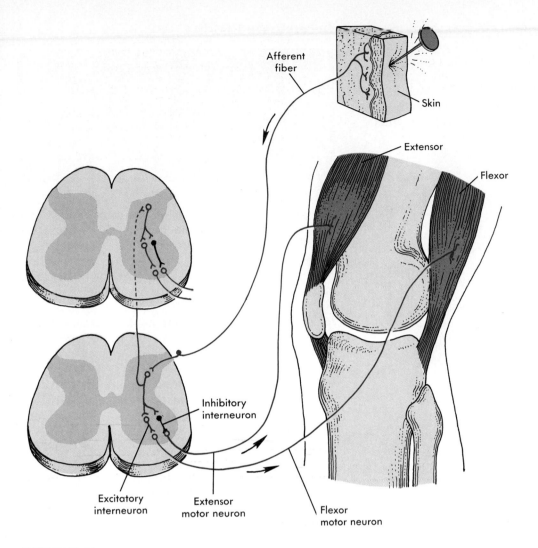

FIGURE 12-13
The spinal pathways that mediate the withdrawal reflex of a
limb on the same side of the body as the noxious stimulus.

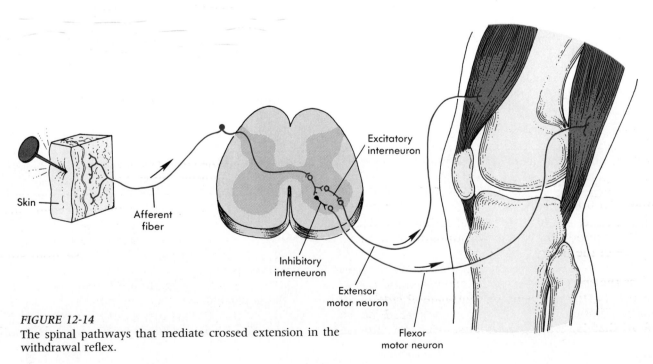

FIGURE 12-14
The spinal pathways that mediate crossed extension in the
withdrawal reflex.

vestib **ular nuclei,** which coordinate antigravity reflexes. Sudden accelerations or changes of orientation elicit vestibular reflexes that help maintain equilibrium. For example, a sudden tilt to one side, such as might occur from catching one's toe in a crack in the pavement, results in a powerful extension of the limb on the side toward which the body is tilted. This is called a **righting reflex.**

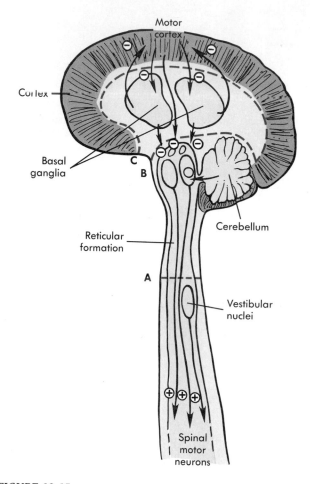

FIGURE 12-15
Brainstem areas that modulate the sensitivity of stretch reflexes. *Cut A* results in a spinal animal. *Cut B* results in a decerebrate animal. *Cut C* results in a decorticate animal.

Decerebrate Rigidity

- *What are the major sources of the synaptic inputs integrated by spinal motor neurons?*
- *How does the nervous system normally prevent decerebrate rigidity?*
- *What is a motor program? What is the role of the motor cortex in the execution of motor programs?*
- *How can the cerebellum affect the execution of motor programs when it has no direct connections to spinal neurons?*

The general effect of the vestibular nuclei on spinal motor neurons can be demonstrated by severing the brainstem above the vestibular nuclei. This enhances stretch reflexes, particularly in those muscles that oppose the pull of gravity. The result is that the limbs are rigidly extended. If the brainstem cut is made still higher (Figure 12-15, *B*) so as to leave all of the reticular nuclei attached to the spinal cord **(decerebration)** and the cerebellum is also removed, the rigidity is even greater (see Table 12-3).

The fact that decerebrate rigidity also involves sensory input from the muscle spindles can be demonstrated by cutting the dorsal roots of a decerebrate animal. The effect is an immediate loss of rigidity in the myotomes served by the cut roots. Decerebrate rigidity can also be reduced by destruction of the vestibular nucleus (or the nerve fibers that descend from it to the spinal cord) or by stimulation of the cerebellum. These experiments show that the major role of the brainstem is continuous facilitation of spinal neurons, particularly of those that serve "antigravity" muscles, which must operate against the pull of gravity. Higher motor centers in the cerebrum and the cerebellum suppress the output of the reticular formation to a greater or lesser degree, regulating it so that the responsiveness of stretch reflexes is sufficient to maintain posture. Turning off the brainstem's facilitation of specific spinal neurons allows the higher centers to "unlock" for use in voluntary motion those muscles that normally are involved in resisting the pull of gravity.

TABLE 12-3	Results of Interruption of Motor Pathways		
Level of section		**Terminology**	**Functional consequences**
Below brainstem		Spinal animal	Temporary loss of spinal reflexes below level of section
Above vestibular nuclei in brainstem		—	Enhanced spinal stretch reflexes in antigravity muscles
Above brainstem reticular formation		Decerebrate	Severe rigidity in all limbs
Removal of cortex		Decorticate	Severe handicaps in primates, but relatively minor deficits in nonprimates

NERVE AND MUSCLE

Role of the Basal Ganglia in Stereotyped Movements

The **basal ganglia** (see Figure 12-15 and Chapter 9) are subcortical forebrain nuclei including the **caudate nucleus, putamen,** and **globus pallidus.** Neurons in the basal ganglia receive input from wide areas of the cortex and send output to the reticular formation and the cortex. Damage to the basal ganglia or to connections between them and other motor centers results in a family of movement disorders called **dyskinesias.** Typically these disorders involve spontaneous, repeated, inappropriate movements of entire muscle groups. Patients may fling their limbs around or exhibit writhing movements or tremors.

The basal ganglia are believed to be organizing centers for motor programs. A **motor program** is a set of commands that is assembled in the brain before a movement begins and is sent to the motor units with the proper timing and sequence to result in the desired movement. A motor program thus is analogous to a script that specifies the parts that individual motor units will play in a complex movement. It still is not possible to describe the physical form in which motor programs are stored, or how they are translated into a pattern of activity in the appropriate neurons of the motor system. In diseases of the basal ganglia, movement disorders are thought to be the result of inappropriate execution of motor programs that are activated spontaneously by the injured nuclei.

The Cerebellum and Coordination of Rapid Movements

The cerebellum is an important destination for sensory information from muscle spindles, Golgi tendon organs, the vestibular system, and the visual system. All of the output of the cerebellum is inhibitory (see Figure 12-15); it is one source of the inhibition of the reticular formation that prevents decerebrate rigidity in intact animals. The structure of the cerebellum is highly ordered and this organization is thought to be a critical feature of its ability to coordinate movements (Figure 12-16).

Damage to the cerebellum results in motor activity that is rough, jerky, and uncoordinated, especially for rapid movements. Movements tend to be **decomposed;** for example, an individual trying to move a limb might do so by activating muscles around each joint separately, and the result is that the limb does not move smoothly. There are obvious disturbances in walking and postural equilibrium, and starting or terminating movements is difficult. Unlike damage to the basal ganglia, cerebellar damage does not result in inappropriate movements; therefore the cerebellum does not seem to be an organizing center for motor programs. Rather, the role of the cerebellum is to coordinate and smooth the execution of programs generated by the

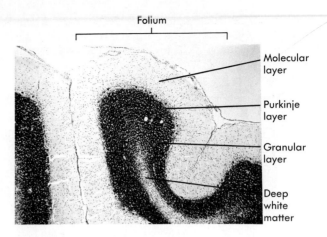

Folium

Molecular layer

Purkinje layer

Granular layer

Deep white matter

FIGURE 12-16
A micrograph showing the precisely layered structure of the cerebellum. This highly ordered structure is thought to be a critical feature of the cerebellum's ability to coordinate movements.

basal ganglia. Another important role of the cerebellum in motor control is prediction of the future position of moving body parts so that, for example, the hand does not knock over an object when reaching for it. When athletes and musicians practice, the effect is to train the cerebellum to achieve perfect timing and coordination of particular motor programs.

Regions of the Motor Cortex

There are at least three motor regions in the cortex: the **primary motor cortex** is located in the frontal lobe of the brain anterior to the primary somatosensory region of the parietal lobe, the **premotor cortex** is in front of the primary motor cortex, and a **supplementary motor area** is located on the medial surface of the cerebral hemispheres. Like the somatosensory cortex, the primary and premotor cortex are topographically organized. In the resulting motor cortical map or **motor homunculus** (Figure 12-17), the representation of the fingers, lips, tongue, and vocal cords is large in comparison to that of the trunk, legs, and upper arms. This is because many cortical neurons are needed to control the large number of small motor units involved in finely controlled movements of these body regions. Like sensory areas of the cortex, the motor areas are layered and contain distinctive neurons called Betz cells (Figure 12-18).

Neurons in the motor cortex that activate particular muscles receive three types of sensory information from the somatosensory cortex: (1) muscle spindles signal static and dynamic muscle length and tension, (2) joint capsule receptors provide information on the position and movement of all of the joints affected by a muscle or group of muscles, and (3) cutaneous receptors provide data on stimuli on the body surface in those areas affected by activation of a muscle or muscle group. Thus the motor

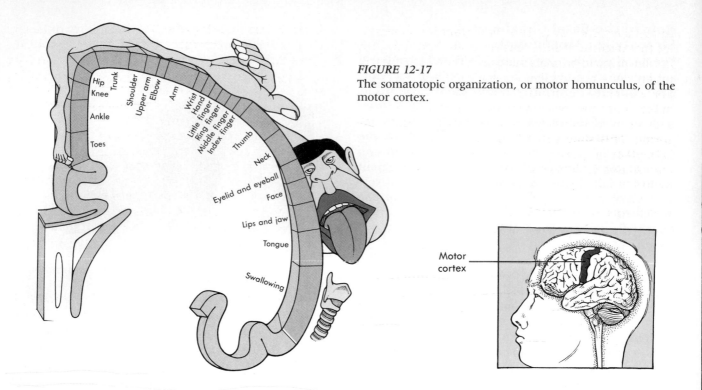

Motor cortex

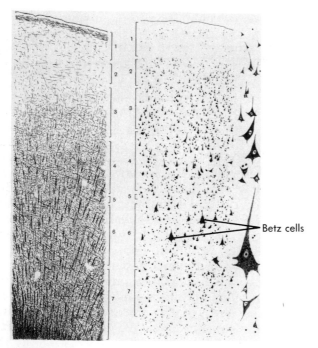

Betz cells

FIGURE 12-18
Sections through the motor cortex. The section on the left has been stained to show nerve fibers, while that on the right shows nerve cell bodies (further enlarged, *far right*). A characteristic feature of the motor cortex is the presence of large Betz cells in layer 6.

cortex is informed about all of the important factors needed to determine the rate, range, and force of movement.

The supplementary motor area differs from the primary motor and premotor areas in two ways: (1) its efferents do not reach the spinal cord by the same pathway, and (2) neurons in the supplementary motor area activate functionally-related groups of postural muscles.

Motor Programs in the Central Nervous System

The most thoroughly studied motor programs are those involved in limb movements. Several levels of programing can be identified by lesion experiments. For example, if the spinal cord segments that innervate an animal's leg are isolated by cutting the spinal cord above and below the region innervating the leg, the animal's leg will exhibit movements that resemble walking. These movements are crude compared with those seen in a normal animal, but the experiment demonstrates that some aspects of locomotion are programmed in the spinal segments. In decerebrate animals the brainstem is intact, and relatively normal movements of all four limbs are seen. When the **mesencephalic locomotor area** is active and thus overcomes rigidity, animals walk normally even if inputs from the muscle spindles and skin are blocked. Increasing the intensity of the stimuli eventually results in a transition from walking to running. Because co-activation of α- and γ-motor neurons occurs in these experiments, there is reason to think the movements are typical of those seen during voluntary locomotion.

If the cortex is surgically separated from the subcortical structures, a **decorticate** animal results (see Table 12-3). Cuts made in decortication are shown diagrammatically by line *C* in Figure 12-15.

In primates, the cortex is so important for all behavior that decortication results in severe handicap, but in non-primate animals such as cats or rats, decortication does not abolish all goal-directed behavior. For example, decorticate nonprimate animals can make exploratory movements directed toward sources of food and can care for their newborn offspring. This shows that in these animals the cortex is neither necessary for initiation of motor programs, nor is it solely responsible for generating them.

Readiness Potentials

How primate brain centers interact to initiate motor programs is not entirely clear, but some insight into the process has been gained by recording electrical signals called **evoked potentials** from different regions of the cortex. An evoked potential represents a change in the activity of a population of brain neurons in response to stimulation of an afferent pathway. In human subjects who are carrying out a task on command, there is a burst of neuronal activity (a form of evoked potential called a **readiness potential**) in several areas of association cortex and in the premotor area nearly a second before actual movements occur. Activity in the association cortex is followed by a nearly simultaneous increase in activity in both the basal ganglia and cerebellum; this takes place about 200 to 300 msec before movement. Activity in the primary motor cortex follows the onset of activity in the basal ganglia.

These experiments suggest that motor activity associated with voluntary tasks is planned in regions of the cortex other than the primary motor area. This planning involves the integration of sensory information with experience, selection of appropriate motor programs, and transmission of preliminary commands to the basal ganglia and the cerebellum, which organize an appropriate motor program. The results are communicated to the primary motor cortex, which in turn activates descending motor fibers running directly to the spinal motor neurons about 20 msec before muscles contract. If the subject is carrying out a program that requires postural changes, the pathways to postural muscle will also be active. In addition, the basal ganglia use a route to spinal motor neurons that does not involve the motor cortex. During movements, continuous sensory feedback to the cerebellum and subcortical motor areas allows for the moment-to-moment coordination of motor activity. Figure 12-19 summarizes current ideas concerning the interactions between the motor control regions of the brain.

CONNECTIONS BETWEEN THE BRAIN AND SPINAL MOTOR NEURONS
Comparison of the Medial and Lateral Motor Systems

- *How do the functions of distal and proximal muscles differ?*
- *Where are the motor neurons of proximal and distal muscles located in the spinal cord?*
- *How do the activities that demand the use of the corticospinal pathway differ from those that are controlled by the polysynaptic pathways?*
- *What deficits would be expected to occur if the corticospinal tract were severed?*

Skeletal muscles can be divided into two general categories on the basis of their location in the body and their functions: (1) **proximal muscles** are the muscles of the trunk, involved mainly in gross body movements and the maintenance of postural stability and (2) **distal muscles** are the muscles of the arms and fingers, important for controlled **manipulatory** movements, and the leg muscles. Of course, control of the two categories must be integrated, because fine movements require postural stability. For example, it is hard to perform delicate tasks with one's fingers unless the arm is steady.

In most cases, proximal muscles are innervated by motor neurons in the central, or medial, region of the ventral horn, whereas motor neurons that control distal muscles are located laterally (Figure 12-20, *A*). Similarly, spinal interneurons in the intermediate zone of the spinal cord can be subdivided into those located medially and those located laterally.

Descending motor tracts that control refined movements project to both lateral motor neurons and lateral interneurons (Figure 12-20, *B*). Lateral interneurons project to lateral motor neurons but only on one side of the spinal cord. Medial interneurons synapse onto medial motor neurons on both sides of the spinal cord and thus can control

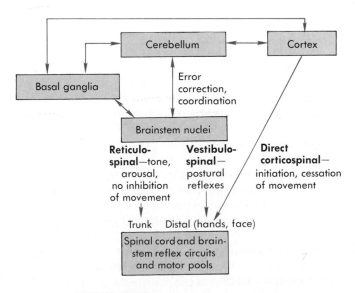

FIGURE 12-19
Interactions of motor areas in the brain.

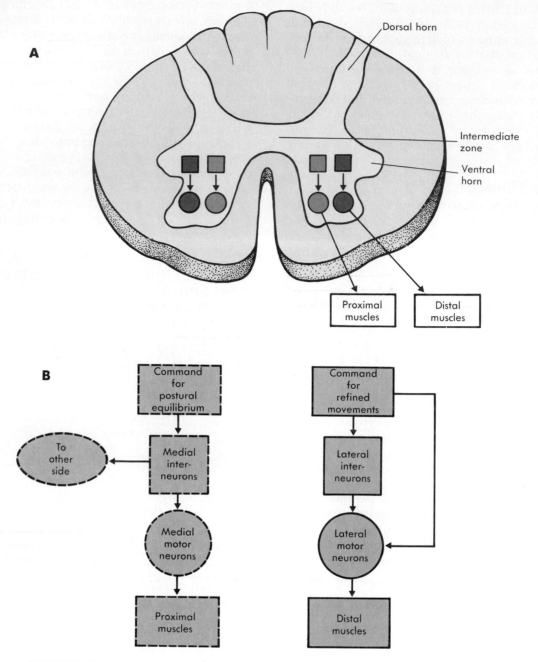

FIGURE 12-20
A Organization of the motor components of the spinal cord. Note the distinction between lateral and medial elements and the muscles they control. Squares represent interneurons; circles represent motor neurons.
B The pathways for proximal and distal muscles are contrasted.

muscles on both sides of the body. Descending motor tracts that regulate postural equilibrium synapse on medial spinal interneurons. They do not terminate directly on medial motor neurons.

Some spinal interneurons send their axons up and down the spinal cord where they synapse in other segments. These are **propriospinal interneurons,** and they also fall into two classes. Axons from lateral propriospinal interneurons extend for only a few segments and terminate on lateral motor neu-

rons or lateral interneurons. They interconnect motor neurons that innervate small groups of distal muscles, such as those acting across a single joint. Medial propriospinal interneurons affect segments located at long distances up or down the spinal cord and coordinate the variety of proximal muscles needed to maintain postural stability. Motor control centers coordinate medial and lateral propriospinal interneurons in those cases in which both manipulation and postural stability are simultaneously important.

Parkinsonism: Without Dopamine We Would All Freeze Up

One of the inevitable consequences of aging is the loss of neurons through a normal process of cell death. Parkinson's disease is the direct result of loss of neurons in the substantia nigra, a brainstem nucleus whose fibers project to the caudate nucleus. The incidence of Parkinson's disease is approaching 1 million in the United States, and the probability of developing clinical symptoms is over 1 in 40 for those living a normal lifespan. The functional disorders of Parkinson's disease appear very gradually, beginning with slight tremors in extremities. These tremors are followed by stiffness of the trunk and slowed movements. The characteristic tremors at rest, involuntary movements, and ultimately debilitating rigidity, result from loss of the transmitter dopamine, which is released from the nerve endings from the substantia nigra that project to cells of the caudate nucleus.

Dopamine is unable to cross the blood-brain barrier. However, Parkinson's patients have been treated with L-dopa, a precursor of dopamine that can enter the brain and be converted to dopamine. The problem with L-dopa therapy is that most patients progressively fail to respond, and at higher doses L-dopa causes hallucinations and convulsions.

Because Parkinson's disease does not appear to develop with age in laboratory animals, there have been no natural animal models in which to develop treatments of the disease. However, the tragic story of drug-induced Parkinson's disease has revealed a potent chemical toxin that specifically targets the substantia nigra neurons. Overnight transformation of young, healthy individuals into patients with incapacitating paralysis resulted from abuse of a synthetic narcotic that differed slightly from the intended chemical structure. Synthesis of the parkinsonism-inducing drug occurred in the course of attempts of street-drug chemists to synthesize meperidine (MPPP), a "designer drug" similar to heroin. The drug that actually resulted, 1-methyl-4-phenyl-1,2,3,6-tetrahydropyridine (MPTP), is converted into a toxic form by a naturally occurring enzyme, monoamine oxidase. The resulting compound is identical to the herbicide, cyperquat, which is chemically similar to the more commonly used herbicide paraquat. Once this fact was recognized, the incidence of Parkinson's disease was studied in relation to herbicide use, and a positive correlation was found. Additional evidence strongly supports the hypothesis that Parkinson's disease is an environmentally provoked disease of the industrial age, with exposure to a variety of chemicals probably contributing to the accelerated rate of neuronal death in the substantia nigra.

Researchers have tried to use this information to produce animal models in which to improve on the treatment of Parkinson's disease. The common laboratory rat does not develop Parkinson-like symptoms when given MPTP, but monkeys do. Thus primate relatives of humans provide an experimental model of the disease in which new therapies can be tested. One such experimental treatment is the grafting of cells from the patient's own adrenal medulla into the brain. These cells release some dopamine because they produce it as an intermediate in synthesis of norepinephrine and epinephrine.

Descending Motor Pathways from the Brain

A pathway, the corticospinal tract (Figure 12-21), running directly from the motor cortex to spinal motor neurons, has evolved in parallel with coordinated use of the fingers and reaches its highest evolutionary development in primates. Sometimes this pathway is called the **pyramidal tract.** The axons of the corticospinal tract cross to the opposite side of the body below the pons so that, just like the sensory systems, the left side of the brain controls the right side of the body and vice-versa. The corticospinal tract coordinates fine movements of the extremities, like the fingers of a violinist, and is part of the lateral motor control system.

Several fiber tracts connect the basal ganglia and brainstem motor areas with the spinal cord. These pathways are sometimes called the extrapyramidal tracts (Figure 12-22). They are all slower pathways because they are polysynaptic, that is, they include several synapses. The **rubrospinal tract** arises in the red nucleus of the brainstem, crosses to the opposite side of the body in the midbrain, and runs parallel to the lateral corticospinal tract. Like the corticospinal tract, the rubrospinal tract is part of the lateral motor control system. It participates in the control of fine movements, synapsing on lateral motor neurons and serving the contralateral side of the body.

The other polysynaptic routes (reticulospinal tract, tectospinal tract, tectobulbar tract in Figure 12-22) serve medial motor neurons that control proximal musculature and are the routes by which the brainstem facilitates postural reflexes. These routes include some fibers that decussate and others

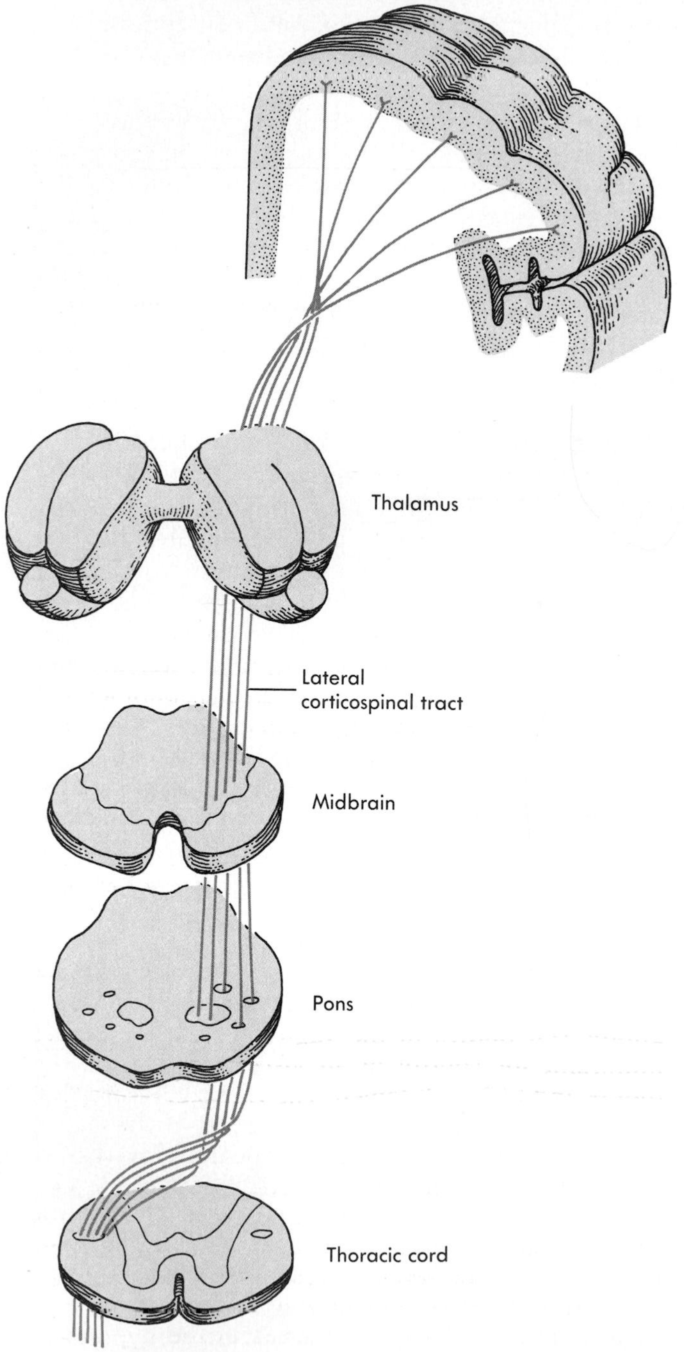

FIGURE 12-21
The corticospinal tract.

Thalamus

Lateral corticospinal tract

Midbrain

Pons

Thoracic cord

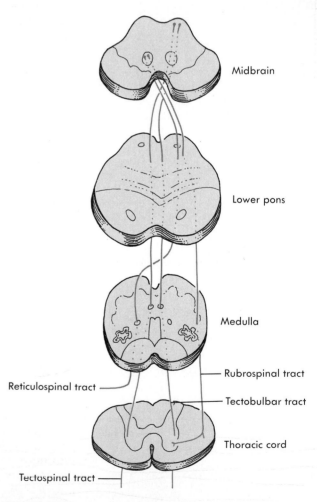

Midbrain

Lower pons

Medulla

Reticulospinal tract

Rubrospinal tract

Tectobulbar tract

Thoracic cord

Tectospinal tract

FIGURE 12-22
Extrapyramidal motor pathways.

NERVE AND MUSCLE

that do not, so that these pathways project to both sides of the body.

When the corticospinal tracts are cut, fine movement control is lost. Individuals with corticospinal lesions can walk normally because the medial motor pathways are unaffected. In contrast, if the medial motor pathways are interrupted, control of overall postural stability is lost and individuals wobble when trying to walk and cannot use their fingers for complex manipulatory tasks. However, if external support is provided, for example by holding an arm, the individual's ability to carry out fine movements is actually fairly normal.

THE AUTONOMIC NERVOUS SYSTEM
Brainstem Autonomic Control Centers

- *What are the different sympathetic and parasympathetic receptor types?*
- *What are receptor agonists and antagonists?*
- *Why can nicotinic antagonists block both sympathetic and parasympathetic actions?*
- *How does innervation by both branches of the autonomic motor system help in the control of spontaneously active visceral effectors?*

Afferents from visceral receptors enter the spinal cord along with axons from muscle spindles and cutaneous sensory endings. This internal sensory information is integrated within the central nervous system, either at the level of spinal segments or within the brain. The reticular formation of the pons, medulla, and midbrain (Figure 12-23) contains several nuclei associated with the autonomic nervous system.

The **cardiovascular centers** of the pons are involved in regulation of blood pressure. The **dorsal vagal nuclei** contain the parasympathetic preganglionic cell bodies of the vagus nerves, which are important for control of all visceral organs of the thorax and upper abdomen. The **Edinger-Westphal nuclei** of the midbrain mediate reflexive changes in pupillary diameter and accommodation by way of sympathetic and parasympathetic fibers to the eye. The **lacrimal** and **salivary nuclei** control secretion of tears and saliva. These nuclei receive inputs from other parts of the reticular system that are involved in control of the overall state of arousal of the nervous system, and also from areas of the hypothalamus, one of the most important areas of the brain coordinating autonomic, somatic, and emotional re-

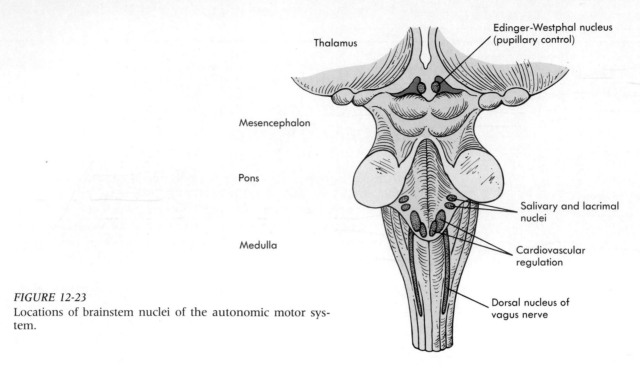

Thalamus

Edinger-Westphal nucleus (pupillary control)

Mesencephalon

Pons

Salivary and lacrimal nuclei

Cardiovascular regulation

Medulla

Dorsal nucleus of vagus nerve

FIGURE 12-23
Locations of brainstem nuclei of the autonomic motor system.

sponses. These connections are responsible for changes in visceral function that accompany changes in arousal or emotion.

Spinal Autonomic Reflexes

Some autonomic reflexes are mediated within the spinal cord. A relatively simple example of an autonomic reflex is emptying of the urinary bladder in infants. When the bladder is empty, stretch receptors in the bladder wall are inactive, but when it fills their activity increases. These receptors synapse on interneurons in the sacral part of the cord. The interneurons synapse on parasympathetic motor neurons whose effect is to increase the activity of the smooth muscle of the bladder. As the bladder fills, parasympathetic activity increases until the pressure in the bladder is sufficient to overcome the resistance of the internal urethral sphincter, a ring of smooth muscle that guards the junction between the bladder and urethra. At this point, reflexive urination occurs.

Voluntary control of urination, learned in childhood, is largely the result of learning to maintain contraction of the skeletal muscles of the pelvic diaphragm, which closes the neck of the bladder. Relaxation of the pelvic diaphragm permits urine to flow through the urethra, but this process is assisted by the tone of the smooth muscle of the bladder. When voluntary pathways from the cortex to the sacral cord are disrupted by injury in adulthood, bladder emptying reverts to being a local spinal reflex as it was in infancy. A similar autonomic reflex initiates reflexive defecation. Children learn to override reflexive defecation by voluntary control of the external anal sphincter, a skeletal muscle. These are examples of functional interaction of autonomic and somatic nervous systems.

The Role of Dual Autonomic Inputs

The importance of spontaneous activity of effectors and the regulatory role of tonic inputs from the autonomic nervous system can be illustrated by several examples. Smooth muscles around small airways in the lungs are activated by parasympathetic stimulation, causing the airways to **constrict,** or become smaller, while sympathetic stimulation causes **dilation** of airways.

The effect of parasympathetic input to the heart is to reduce the heart rate; one of the effects of sympathetic input is to increase it. The heart is spontaneously active, and in the absence of any autonomic input it would beat at about 100 beats per minute The resting heart rate is approximately 70 beats per minute. This is because, at mild to moderate activity levels, the heart is continuously inhibited by the parasympathetic nervous system via the vagus nerve. This steady inhibitory input is referred to as parasympathetic **tone** and is responsible for moderate changes in heart rate. However, at high levels of physical activity observed increases in heart rate are produced primarily by its sympathetic input. Sympathetic activity also increases the force of contraction of the heart and dilates the coronary arterioles (see Chapter 15).

Some organs are dominated by a single branch of the autonomic nervous system, either because they receive no other innervation, or because inputs from one of the two are more abundant. For example, the blood vessels of most organs are innervated only by the sympathetic nervous system. The normal functions of the gastrointestinal tract are largely controlled by the parasympathetic nervous system, which (1) stimulates smooth muscle in the esophagus, stomach, and intestines, (2) increases secretion, and (3) relaxes sphincters to allow food to pass from

Organ	Sympathetic stimulation	Parasympathetic stimulation
Eye		
Ciliary muscle	Relaxed	Contracted
Pupil	Dilated	Constricted
Glands		
Salivary	Vasoconstriction Slight secretions	Vasodilation Copious secretion
Gastric	Inhibition of secretion	Stimulation of secretion
Pancreas	Inhibition of secretion	Stimulation of secretion
Lacrimal	None	Secretion
Sweat	Sweating	None
GI tract		
Sphincters	Increased tone	Decreased tone
Wall	Decreased tone	Increased motility
Liver	Glucose released	None
Gallbladder	Relaxed	Contracted
Bladder		
Muscle	Relaxed	Contracted
Sphincter	Contracted	Relaxed
Heart		
Muscle	Increased rate and force	Slowed rate
Coronary arterioles	Dilated	Dilated
Lungs	Bronchioles dilated	Bronchioles constricted
Blood vessels		
Abdominal	Constricted	None
Skin	Constricted	None
Muscle	Constricted (α receptors)	None
	Dilated (β receptors)	None
Sex organs	Ejaculation	Erection
Metabolism	Increased	None

TABLE 12-4 *Autonomic Innervation of Target Tissues*

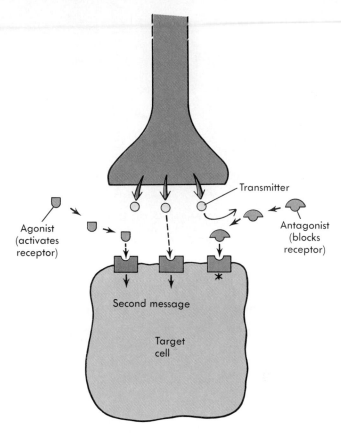

FIGURE 12-24
Agonists bind to postsynaptic receptors and have an effect similar to that of neurotransmitters. Antagonists block the receptor, interfering with transmission.

one part of the tract to another. The gastrointestinal tract also receives sympathetic fibers, but their influence is less important for normal function.

Membrane Receptors for Acetylcholine

Table 12-4 summarizes the effects of autonomic activation on each of the major organ systems. Subsequent chapters will describe some of these autonomic effects in more detail. These effects are mediated through specific receptors. The receptors can be classified according to their responses to drugs and toxins, so the distinctions are based on pharmacology. A drug or toxin that activates a particular receptor type is termed an **agonist**; agents that interfere with the function of a receptor type are **antagonists** (Figure 12-24).

The two types of **cholinergic** receptors (for acetylcholine) are **nicotinic**, named for the agonist nicotine, and **muscarinic**, named for the agonist muscarine (Figure 12-25; Table 12-5; see also Table 12-1). Nicotinic receptors are found on the postganglionic cells of both branches of the autonomic motor system and also on the postsynaptic membranes of skeletal muscle cells. Muscarinic receptors are found on effectors innervated by the parasympathetic branch. Thus the rule for cholinergic receptor types is that nicotinic receptors are found on the postsynaptic cells of the first synapse outside the central nervous system; muscarinic receptors at the second synapse if it is cholinergic.

Membrane Receptors for Norepinephrine

The situation with **adrenergic receptors** (for epinephrine and norepinephrine) is more complex. Historically two different types, α and β, were identified. As new antagonist and agonist drugs were

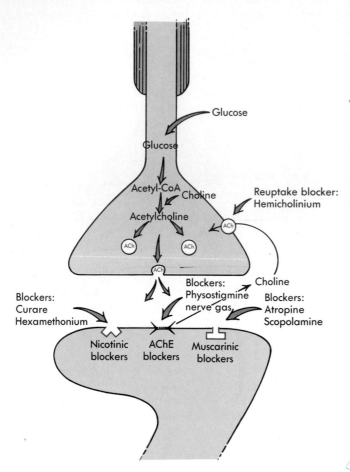

FIGURE 12-25

Neurotransmission at cholinergic synapses. Both nicotinic and muscarinic receptor types are shown together for simplicity, although actually both are not found together at the same synapse. Cholinergic transmission can be blocked by receptor antagonists and by inhibitors of choline absorption. Transmission can be enhanced by inhibitors of acetylcholinesterase.

TABLE 12-5	Location of Cholinergic Receptor Classes

Nicotinic	Muscarinic
Neuromuscular junction of skeletal muscle	Visceral effectors; heart pacemakers, gastrointestinal tract, arterioles of genitalia, iris of eye, salivary and sweat glands, lung airways, urinary bladder
Parasympathetic and sympathetic postganglionic neurons	
Cholinergic synapses of central nervous system	

agonists and antagonists that form the basis of medical therapies that result in selective activation or differential block of visceral responses. For any given receptor, the effects of transmitter and agonists are determined by the nature of the second messenger systems set in motion by receptor binding in that particular tissue. Thus, for example, acetylcholine binding to muscarinic receptors results in inhibition of heart pacemaker fibers and activation of intestinal smooth muscle.

Autonomic Receptor Antagonists

Transmission across autonomic ganglionic synapses (and across the neuromuscular junction) is specifically blocked by the nicotinic antagonists **curare** and **hexamethonium** (Figure 12-25). Curare-like compounds are useful in surgery because an anesthetist can achieve complete skeletal muscle relaxation without resorting to high levels of general anesthetic that would inhibit brainstem visceral control systems.

Muscarinic antagonists such as **atropine** and **scopolamine** (Figure 12-25) decrease gastrointestinal smooth muscle activity and gastrointestinal secretion. Such drugs are valuable in the treatment of gastrointestinal spasms, as well as disorders of hypersecretion such as ulcers. Atropine-like drugs are common ingredients of cold tablets and pills for motion sickness because they inhibit respiratory secretions and suppress the parasympathetically mediated vomiting reflex.

Distribution of Autonomic Receptor Types

Different subclasses of adrenergic receptors are found in different parts of the circulatory system (Figure 12-26 and Table 11-6). Heart muscle fibers, for example, possess β_1 receptors. A β_1 agonist such as **isoproterenol** would be appropriate therapy in situations of impaired pumping ability of the heart. Adrenergic β_2 receptors are present on the smooth muscle cells of arterioles that control the size of coronary arterioles. Activation of β_2 receptors relaxes these cells, causing dilation and an increase in blood flow to the heart. Thus β_2 agonists are important in the treatment of diseases in which coronary blood flow is inadequate.

In most tissues, the smooth muscle that surrounds arterioles and veins contracts when norepinephrine activates α_1 receptors. This contraction can be inhibited by α_1 antagonists such as **phentolamine.** Drugs that selectively stimulate α_1 receptors cause constriction of arterioles and veins. An example is **phenylephrine,** a common ingredient of cold medications and nasal sprays. The vasoconstriction it produces relieves nasal congestion. Similar drugs are useful in the treatment of low blood pressure because they increase the total flow resistance of the systemic vessels.

In the blood vessels of skeletal muscles, both α_1 and β_2 receptors are present. The vessels of most

identified, each of these types was subdivided into two subtypes designated by number subscripts (Figure 12-26; Table 12-6; see also Table 12-1).

Each receptor type initiates characteristic responses in effector organs. Each has a unique set of

TABLE 12-6   *Major Locations of Adrenergic Receptor Classes*

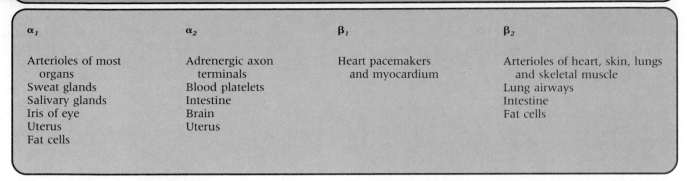

α_1	α_2	β_1	β_2
Arterioles of most organs	Adrenergic axon terminals	Heart pacemakers and myocardium	Arterioles of heart, skin, lungs and skeletal muscle
Sweat glands	Blood platelets		Lung airways
Salivary glands	Intestine		Intestine
Iris of eye	Brain		Fat cells
Uterus	Uterus		
Fat cells			

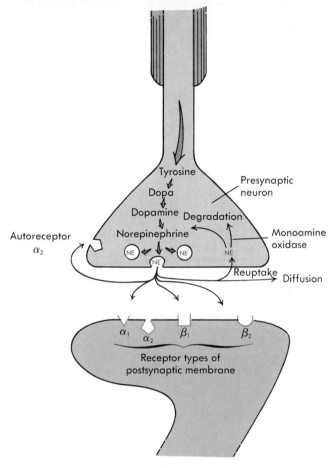

FIGURE 12-26

Neurotransmission at adrenergic synapses. Some antagonists and inhibitors are shown. Adrenergic transmission is blocked by antagonists specific for receptor types, such as propranolol for β receptors and phentolamine for α receptors. Norepinephrine is reabsorbed and subsequently degraded by monoamine oxidase.

visceral organs contain predominantly α_1 receptors. Activation of the α_1 receptors leads to vasoconstriction; activation of the β_2 receptors leads to vasodilation. This differential distribution of adrenergic receptors is important in the redistribution of blood flow into skeletal muscles during exercise.

The β_2 receptors differ from the other adrenergic receptors in being less sensitive to activation by norepinephrine than epinephrine. In tissues, such as the skeletal muscle vasculature, that possess both α_1 and β_2 receptors, a low dose of epinephrine can activate predominantly β_2 receptors, while either norepinephrine or a high dose of epinephrine elicits predominantly an α_1 effect. Thus the sympathetic nervous system can mediate either vasodilation or vasoconstriction in such tissues, depending on the level of input and whether the input is primarily neural (norepinephrine) or hormonal (epinephrine).

In the respiratory system, airways dilate in response to stimulation of β_2 receptors, so β_2 agonists are useful in asthma, a disease in which respiratory distress is caused by an abnormal constriction of airways. These examples emphasize the importance of autonomic receptor specificity.

Transmitter Synthesis in Adrenergic and Cholinergic Synapses

Synaptic transmission depends on a sequence of events that starts with transmitter synthesis and packaging; continues with transmitter release and inactivation, recycling, or loss by diffusion; and ends with a postsynaptic potential. The steps of this process differ between adrenergic and cholinergic synapses, and the differences are important for physiology and medicine.

Cholinergic transmission was illustrated in Figure 12-25. The precursors of acetylcholine are acetyl-CoA (derived from glucose in glycolysis; see Chapter 4) and choline. Choline, which results when acetylcholine is inactivated by postsynaptic acetylcholinesterase (AChE), is recycled into cholinergic terminals (see Chapter 8). Blockers of AChE such as **physostigmine** can be used to increase the synaptic efficacy of cholinergic synapses; some nerve gases and pesticides also block this enzyme. Recycling of acetylcholine can be blocked by the choline analog **hemicholinium**; the result is a decrease in the quantal content of the cholinergic synapses.

Figure 12-26 shows the processes that support adrenergic transmission. The immediate precursor of norepinephrine is dopamine. Dopamine may be synthesized from tyrosine that has been taken up by the terminal. Norepinephrine is reabsorbed from the synaptic cleft and then degraded within the presynaptic terminal by the enzyme **monoamine oxidase.**

Adrenergic terminals possess α_2 receptors for their own transmitter. The α_2 receptors are believed to be part of a negative feedback loop (see Figure 12-26) that regulates the excitability of the terminal. Some effectors possess α_2 receptors also (see Table 12-6).

● STUDY QUESTIONS

1. Movements can be broadly categorized as finely coordinated or postural. Where are the respective motor neurons located in the spinal cord? Are any motor neurons activated directly by the motor cortex? Which ones?

2. Describe the structure of a muscle spindle. What does the muscle spindle measure? How can the sensitivity of the spindle be changed?

3. The stretch reflex has been described as the "power steering" of voluntary movements. How does this reflex contribute to the ability to carry out voluntary movements? Why must the α- and γ-motor neurons be coactivated if this "power steering" is to occur?

4. What are the Golgi tendon organs? What information do they provide?

5. What is meant by the term reciprocal innervation? In which spinal reflexes is this organization most important? Why?

6. What is meant by the term facilitation? What is the physiological role of connections between the vestibular system and the spinal motor neurons?

7. What is the physiological role of the cerebellum? What are the general consequences of cerebellar damage?

8. State the major nuclei that make up the basal ganglia. Can the basal ganglia initiate movements by themselves?

9. What is the evidence for the existence of central motor programs for locomotion?

10. State the two different classes of receptors for acetylcholine and the four classes for norepinephrine and epinephrine. Where is each found? Give two examples of the physiological importance of differential distribution of receptor types.

11. What differences are there between the anatomy of the sympathetic and parasympathetic pathways? In what ways are both distinct from the somatic system?

12. What organs receive single innervation from the autonomic nervous system? Give three examples of organs that receive dual innervation. What characteristics of the target cells determine whether the effect of innervation by a branch of the autonomic system will be inhibitory or excitatory?

Choose the MOST CORRECT Answer.

13. The smallest functional element of a skeletal muscle
    a. Muscle fiber   c. Motor unit
    b. Sarcomere   d. Muscle spindle

14. The neurotransmitter for all the motor pathways of the body released at the first synapse in the periphery is:
    a. Acetylcholine   c. Epinephrine
    b. Norepinephrine   d. Somatostatin

15. The cell bodies of the postganglionic fibers of the sympathetic system lie in
    a. The dorsal root ganglia
    b. The gray matter of the brain
    c. A chain of ganglia running along each side of the spinal cord
    d. The ventral horns of the gray matter of the spinal cord

16. This spinal segmental reflex may protect muscles and their tendons from extremes of tension
    a. Stretch reflex
    b. Withdrawal reflex
    c. Golgi tendon reflex
    d. Crossed extension reflex

17. This area of the brain plays a role in the execution of stereotyped movements:
    a. Cerebellum
    b. Basal ganglia
    c. Vestibular nuclei of reticular formation
    d. Cerebral cortex supplementary motor area

18. Distal muscles
    a. Are muscles of the trunk
    b. Are mainly involved in gross body movements
    c. Are involved in maintenance of postural stability
    d. Are important for controlled manipulatory movements

19. The dorsal vagal nuclei
    a. Mediate reflexive changes in the diameter of the pupils and in accomodation of the eye
    b. Control secretion of saliva and tears
    c. Control all visceral organs of the thorax and upper abdomen
    d. Regulate blood pressure

● SUGGESTED READINGS

Brain Saver: new help for severe stroke, *Prevention,* 41(11):20. Describes new techniques to treat stroke patients. Borrowing a technique used on trauma victims, the doctors drain fluid from spaces in the brain, reducing pressure. Then the broken blood vessel is repaired to prevent further bleeding.

CLARKE E, JACYNA B: *Nineteenth century origins of neuroscientific concepts* Berkeley, 1987, University of California Press. Survey of the history of neurophysiology and the mind-body problem.

GEVINS AS, ET AL: Human neuroelectric patterns predict performance accuracy, *Science* 235:580, 1987. Discusses cortical-evoked potentials associated with activity in the motor system.

HIGGINS LC: Doctors use new drug for Parkinson's disease, *Medical World News* 30(16):13, August 28, 1989. Describes the effects of deprenyl, a monoamine oxidase inhibitor, in delaying the need for treatment with levodopa.

SUMMARY

1. Somatic **motor neurons** project directly to skeletal muscle fibers, innervating each muscle fiber with a single excitatory synapse. Autonomic **preganglionic** neurons whose cell bodies are in the central nervous system synapse in peripheral ganglia with postganglionic fibers that innervate visceral effectors.

2. The **parasympathetic** and **sympathetic** branches of the autonomic motor system differ anatomically in that the ganglia in the sympathetic branch are anatomically remote from the effector organs, while the ganglia of the parasympathetic branch are located in or on the effector organs.

3. Visceral effectors typically have **dual innervation** by autonomic fibers from both the sympathetic and parasympathetic branches and may possess more than one motor synapse on each effector cell.

4. As a general rule of organization of the motor systems, the transmitter chemical at the first synapse outside the nervous system is **acetylcholine.**

5. Spinal segmental reflexes that contribute to posture, protection, and voluntary movement include (1) the **stretch reflex,** which maintains muscle length constant; (2) the **withdrawal reflex,** which mediates rapid flexion of an injured limb; (3) the **crossed extension reflex,** which extends the limb contralateral to a flexed limb; and (4) the **Golgi tendon reflex,** which may protect muscles and their tendons from extremes of tension.

6. Motor programs may be generated at several levels in the central nervous system. Simple stepping can be carried out by the spinal cord. More complicated motor programs require the participation of the **basal ganglia, motor cortex,** and **cerebellum.**

7. The cell bodies of motor neurons in spinal cord segments are arranged so that **distal muscles** are served by lateral motor neurons, and medial motor neurons control **proximal muscles.**

8. Pathways from the motor centers of the brain to spinal motor neurons can be divided roughly into two categories: one class that serves primarily proximal muscles and mainly mediates adjustments of posture, and one that serves distal muscles and is important for fine movements. Control of distal muscles is exerted mainly by the direct **corticospinal** (pyramidal) tract; control of proximal muscles involves polysynaptic (extrapyramidal) tracts that pass through brainstem nuclei.

9. The responses of visceral effectors to autonomic inputs are mediated by pharmacologically distinguishable types of receptor molecules. **Nicotinic** cholinergic receptors are located on autonomic postganglionic cells of both branches, but **muscarinic** receptors are found on effectors innervated by cholinergic postganglionic fibers. Adrenergic receptors are divided into α and β types, and into α_1, α_2, β_1, and β_2 subtypes. The subtypes are typically associated with particular effectors, although some effectors have more than one subtype.

10. For any given tissue, the effect of autonomic inputs depends entirely on the second messenger systems activated by transmitter-receptor binding; the same receptor subtypes can frequently mediate opposite responses in different tissues.

HIGGINS LC: Fetal cell implants garner support, *Medical World News,* 31(5):39, March 12, 1990. Describes use of fetal tissue from the substantia nigra to alleviate symptoms of Parkinson's disease.

MASSION J, DUFOSSE M: Coordination between posture and movement: why and how? *News in Physiological Sciences* 3:88-93, 1988. Presents a model of normal interaction between brain centers that control posture and movement, and explains effects of disease and injury.

MAUGH TH: Medicine to change your mind, *Discover,* January 1988, p. 39. Description of how implants of adrenal tissue into the brain have been used to treat Parkinson's disease.

OLIWENSTEIN L: Monkey think, monkey do, *Discover,* 10(6):20, 1989. Describes how neuron activity in the monkey motor cortex shows pre-movement plans.

Stress can make you sick; but can managing stress make you well? *Consumer Reports Health Letter,* 2(1):1, 1990. Deals with the health aspects of stress and stress management as it concerns the immune system.

Ultramarathon man, *Discover,* February 1988, p. 16. Describes changes in pain perception and in the muscle composition of the legs of an individual who ran 2200 miles.

THE CARDIOVASCULAR SYSTEM

CHAPTER 13

The Heart

On completing this chapter you will be able to:

- Describe the components of the circulatory system: the heart, arteries, capillaries, and veins

- Understand the role of the valves in the heart

- Distinguish between the three types of cardiac muscle cells: pacemakers, conducting fibers, and contracting fibers, which generate the force of systole

- Describe the ionic basis of the action potentials of cardiac cells and the automatic generation of action potentials by cardiac pacemakers

- Identify the five phases of the cardiac cycle: early diastole, atrial contraction, isovolumetric ventricular contraction, ejection, and isovolumetric ventricular relaxation

- Understand how the Frank-Starling law of the heart can be derived from the length-tension relationship of cardiac muscle fibers

- Understand how pacemaker discharge rate is affected by sympathetic and parasympathetic input

- Identify the major factors that determine cardiac output

*E*ach minute, the heart of a resting adult pumps about 5 L of blood, or approximately the person's total blood volume. This works out to at least 7200 L/day; this volume of blood weighs about 100 times more than the body. Moreover, the rate of pumping may increase to 20 to 30 L/min during strenuous exercise. The relentless work of the heart uses 5% to 10% of the body's total energy budget.

Regulation of the heart to meet the demands of the body is exceedingly complex; thus it has been difficult to construct mechanical hearts with anything approaching the flexibility of the human heart. The heart has a large measure of intrinsic regulation—its basic rhythm is established by its own pacemaker cells, and the force of its contractions is largely self-adjusting. Inputs to the heart from both branches of the autonomic nervous system cause the changes in heart activity that accompany physical exertion, excitement, and emotion.

AN OVERVIEW OF THE CIRCULATION

The cardiovascular system consists of two pumps—the right and left sides of the heart—and three main types of vessels: **arteries,** which carry blood away from the heart, **capillaries,** the major sites of exchange of materials between blood and tissues, and **veins,** which carry blood back to the heart. The arteries branch successively, finally giving rise to **arterioles,** small vessels that branch to form capillaries. The capillaries join to form the smallest veins, called **venules.**

The circulatory system is arranged as a circuit in which the blood passes from right heart to lungs to left heart to body and then back to the right heart for another trip through the circuit (Figure 13-1). The **systemic loop** is the part of the circuit that passes from the left heart through the **aorta** to arteries that lead ultimately to capillaries of various body organs and then back to the right heart through the **superior** and **inferior venae cava**e. In the systemic capillaries the blood gives up some of its oxygen (O_2) in exchange for carbon dioxide (CO_2) produced by tissue metabolism. The **pulmonary loop** is the part of the circuit that passes from right heart through the **pulmonary arteries** to the pulmonary capillaries in the lungs and back to the left heart through **pulmonary veins.** In the pulmonary capillaries, the burden of carbon dioxide (CO_2) acquired by the blood in the systemic capillaries is transferred to pulmonary gas, and the O_2 that was unloaded in the systemic loop is replaced.

The central fact of circulatory physiology is that flow of blood through the two circulatory loops is driven by a gradient of pressure within each loop; the arteries contain blood under relatively high pressure (as much as 100 mm Hg or more in the systemic loop; about 10 times less in the pulmonary) whereas the veins contain blood under low pressure (usually no more than a few mm Hg). The heart maintains this driving force. The heart must fill with blood under the low pressure from veins and must eject blood into the arteries against a much higher pressure. The more effectively the heart pumps, the more it keeps the arterial pressure high and the venous pressure low. If the heart slows or weakens, the mean arterial pressure drops rapidly while at the same time the venous pressure rises.

The systemic arterial system is frequently compared to a tree with the aorta as the trunk. The first branches of the tree are **coronary arteries** (Figures 13-1 and 13-2) that serve the capillaries of the heart itself. Other major branches of the aorta serve other individual organs and major body regions. The smallest branches of arteries are **arterioles,** each of which serves a small network of capillaries, or **capillary bed.** As a rule, blood passes through only one set of capillaries before returning to the heart, so each organ or regional circulation is a pathway in parallel with the others of the systemic loop.

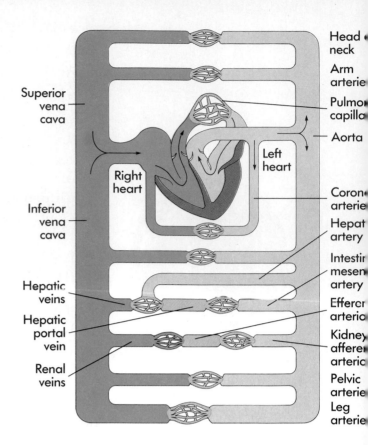

FIGURE 13-1

The circulatory system can be divided into the pulmonary loop and systemic loop, which together form a complete circuit. All of the blood pumped by the left heart passes through the systemic loop, returning to the right heart as systemic venous blood, which is enriched in CO_2 and partly depleted of O_2. The right heart pumps the blood into the pulmonary loop, where some CO_2 is lost and O_2 is gained by gas exchange with the atmosphere. This blood returns to the left heart for another trip through the circuit.

There are three important exceptions to the rule that blood passes through only one set of capillaries in its flow through the systemic loop. The first exception is the **hepatic portal circulation** (Figure 13-1) by which the liver receives blood that has first passed through the gastrointestinal tract's circulation. This allows the liver direct access to nutrients before they enter the general circulation. The second exception, the **hypothalamic-anterior pituitary portal system** (not shown; see Figure 5-19,*B*) carries chemical factors that affect the release of anterior pituitary hormones from the hypothalamus to the pituitary. The third exception to the rule is found in the kidney, where blood passes first through **glomerular capillaries** (Figure 13-1), which are involved in the initial stage of urine formation, and then through **peritubular capillaries,** which support regions of the kidney that are involved in modification of urine composition.

THE CARDIOVASCULAR SYSTEM

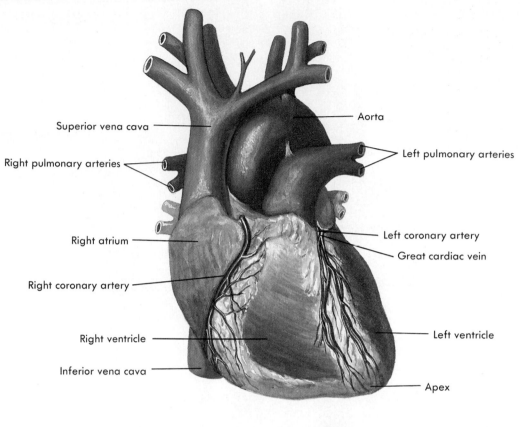

Superior vena cava

Right pulmonary arteries

Right atrium

Right coronary artery

Right ventricle

Inferior vena cava

Aorta

Left pulmonary arteries

Left coronary artery

Great cardiac vein

Left ventricle

Apex

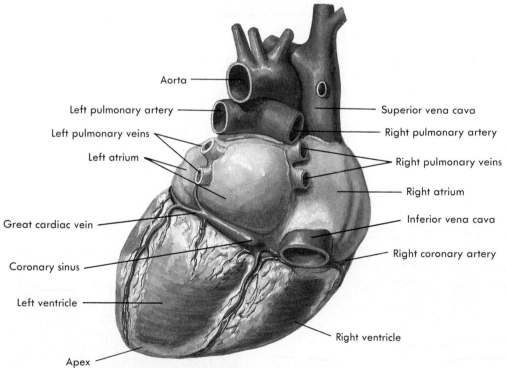

Aorta

Left pulmonary artery

Left pulmonary veins

Left atrium

Great cardiac vein

Coronary sinus

Left ventricle

Apex

Superior vena cava

Right pulmonary artery

Right pulmonary veins

Right atrium

Inferior vena cava

Right coronary artery

Right ventricle

FIGURE 13-2
Two diagrammatic external views of the heart and its major vessels.

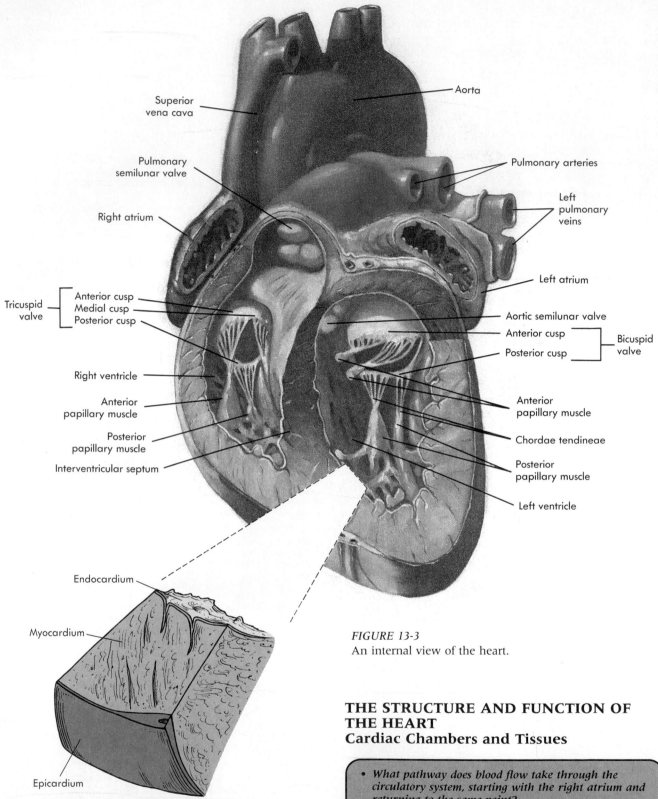

Superior
vena cava

Aorta

Pulmonary
semilunar valve

Pulmonary arteries

Left
pulmonary
veins

Right atrium

Left atrium

Tricuspid
valve
{ Anterior cusp
Medial cusp
Posterior cusp

Aortic semilunar valve

Anterior cusp

Posterior cusp
} Bicuspid
valve

Right ventricle

Anterior
papillary muscle

Anterior
papillary muscle

Posterior
papillary muscle

Chordae tendineae

Interventricular septum

Posterior
papillary muscle

Left ventricle

Endocardium

Myocardium

Epicardium

FIGURE 13-3
An internal view of the heart.

THE STRUCTURE AND FUNCTION OF THE HEART
Cardiac Chambers and Tissues

- *What pathway does blood flow take through the circulatory system, starting with the right atrium and returning to the same point?*
- *What are the locations and functions of the two sets of heart valves?*
- *What are the locations and functions of the three types of cardiac muscle cells?*
- *What do the terms* systole *and* diastole *mean?*
- *What are the origins of the first and second heart sounds?*

If the systemic arterial system can be likened to a tree, the proper analogy for the systemic venous system is a river with its tributaries. The smallest streams are **venules,** which empty into larger veins; finally, the mixed venous blood of the systemic loop drains into the venae cavae that fill the right heart.

The heart is located within a fluid-filled sac of membranous **pericardium** between the lungs in the center of the thorax. The inner lining of the pericardium is continuous with the covering of the heart itself, the **epicardium.** The pericardial sac does not contribute directly to the pumping action of the heart. However, in **pericarditis,** an inflammation of the pericardium, excessive pericardial fluid may accumulate, compressing the heart and impairing pumping. This condition is called **cardiac tamponade.**

Each half of the heart consists of two chambers: an **atrium** and a **ventricle** (Figures 13-2 and 13-3). A tough partition, the **septum,** separates the two halves of the heart, dividing the atria and ventricles longitudinally. The entire inner surface of the heart, including the valves, is covered with a delicate membrane, the **endocardium,** which forms a barrier between the blood and the heart tissue. In terms of force generation, the ventricles are the main pumps of the heart and have thick muscular walls. The atria are thin-walled and can serve as reservoirs of blood as well as boosters that aid in filling the ventricles.

Cardiac cells can be divided into three functional classes (Table 13-1) (in order of abundance): myocardial or **contractile cells, conducting cells,** and **pacemakers** or **nodal cells.** The myocardial cells make up about 99% of the heart's mass and are responsible for contraction and force generation.

Conducting cells form a **conducting system** (Figure 13-4), specialized for conducting action potentials rapidly from one part of the heart to another (Table 13-2). The conducting system consists of (1) the **sinoatrial (SA) node,** (2) one or more **internodal pathways,** (3) the **atrioventricular (AV) node,** (4) the **atrioventricular bundle** or **bundle of His** (described by the Swiss physician Wilhelm His in 1893), which divides to form (5) the **right** and **left bundle branches.** The bundle branches extend down the interventricular septum to the tip,

TABLE 13-1	Types of Cardiac Cells	

Cell type	Location	Physiological function
Myocardial (ordinary) muscle	Atria and ventricles	Generate force
Conducting	Bundle of His and branches; Purkinje fibers; some in atria	Coordinate contraction
Pacemaker	SA node and AV node	Initiate heartbeat; control heart rate

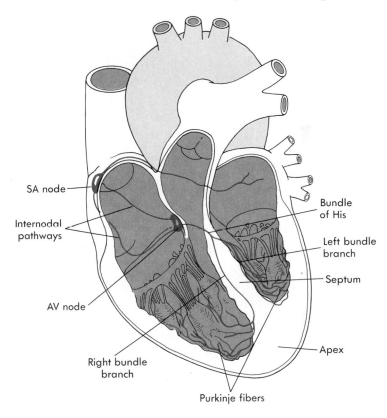

FIGURE 13-4
The conducting fiber system of the heart. The importance of the internodal pathways is not fully established.

TABLE 13-2	Electrical Characteristics and Innervation of Cardiac Cells		

Type	Conduction velocity (m/sec)	Pacemaker activity	Autonomic innervation
SA node	0.05	Normal (100/min)	Dual
Atrium	0.8 to 1.0	None	Minor
AV node	0.05	Latent	Dual
Purkinje fibers	2.0 to 4.0	Latent	Sympathetic
Ventricle	0.4 to 0.8	None	Sympathetic

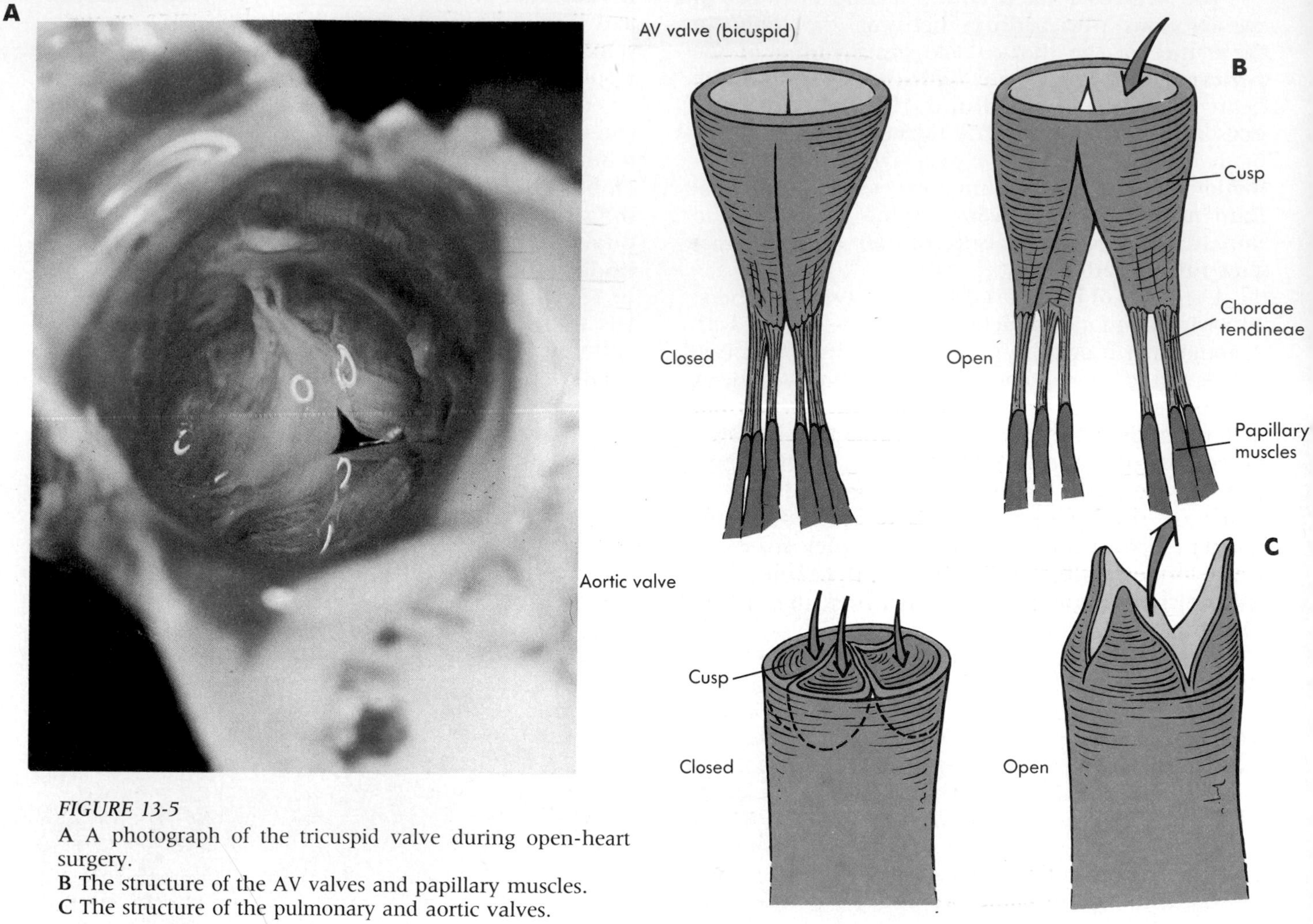

AV valve (bicuspid)

Cusp

Chordae tendineae

Papillary muscles

Closed

Open

Aortic valve

Cusp

Closed

Open

FIGURE 13-5
A A photograph of the tricuspid valve during open-heart surgery.
B The structure of the AV valves and papillary muscles.
C The structure of the pulmonary and aortic valves.

or **apex,** of the ventricles and then project upward toward the **base** or most superior part of the ventricles, subdividing into (6) a complex network of **Purkinje fibers** beneath the endocardium of both ventricles.*

The third functional class of cardiac cells are pacemaker cells found within the SA and AV nodes that are able to generate regular, spontaneous action potentials in the absence of any external stimulus. Normally, the SA node serves as the pacemaker that determines the frequency of heartbeats, or **heart rate.**

The Cardiac Cycle

The **cardiac cycle** consists of two major phases: **diastole,** during which the ventricles are relaxed and fill with blood from veins, and **systole,** during which the ventricles contract. During the first part

of diastole, all parts of the heart are relaxed, and blood flows from the great veins into the atria and ventricles. This process is called **passive filling.** At first this flow is rapid, but it slows as the ventricles become more and more stretched. **Atrial contraction** (sometimes called **atrial systole**) is the final event of diastole.

The ability of the ventricles to fill under low pressure from veins and to maintain high arterial pressures is critically dependent on two sets of cardiac valves (Figures 13-3 and 13-5). The first set are the **atrioventricular (AV) valves** (also called the **tricuspid valve** in the right heart and the **mitral valve** in the left heart) that separate the atria from the ventricles. The second set are the **pulmonary valve** and **aortic valve** (also called **semilunar valves**) that separate the right and left ventricles from the pulmonary artery and aorta respectively. The valves open and close in response to pressure differences. The role of the AV valves is to allow blood to pass from the atria into the ventricles during diastole and to prevent backflow of blood toward the atria during systole. The role of the pul-

---

*Jan Evangelista Purkinje was a Bohemian physiologist and anatomist who described the Purkinje fibers in 1845 and made a number of other important discoveries, including the Purkinje cells of the brain cortex.

THE CARDIOVASCULAR SYSTEM

monary and aortic valves is to allow blood to leave the ventricles into the great arteries during the peak of force generation by the ventricles in systole and to prevent backflow of blood from the great arteries into the ventricles during the rest of the cycle.

Diastole

The first event of diastole is opening of the AV valves. Recent studies suggest that the rapid passive filling of the ventricles that occurs immediately after the AV valves open is aided by suction generated by rebound of the ventricle from the distortion of its structure that occurred during systole. As the ventricle contracts during systole, energy is stored in stretched collagen strands that lace the myocardial cells together. As the ventricle begins to relax, this stored energy is released. The relaxing ventricle acts like the rubber bulb of a medicine dropper— squeezing the bulb corresponds to systole and releasing the rubber bulb allows the energy stored in the deformed bulb to refill the dropper. This effect may be most important when the heart is beating rapidly.

When the atria contract in the final phases of diastole, additional blood is driven through the AV valves into the ventricles. There are no valves at the openings of the venae cavae and pulmonary veins, so it would seem that atrial contraction would cause backflow of blood from the atria into the great veins. Some backflow does occur and can be felt as a pulse in the large veins of the neck, but it is minimized because atrial contraction partly closes the mouths of the great veins. Atrial contraction is responsible for roughly one-fifth of ventricular filling in resting people, but it becomes more important when the heart rate is elevated, as during exercise. For any single cardiac cycle, the volume of blood in each ventricle reaches a maximum at the end of atrial contraction; this volume is called the **end-diastolic volume.**

Systole

After atrial contraction, there is an **AV delay** of about 0.1 sec before ventricular contraction begins. This delay allows atrial contraction to be complete before ventricular contraction begins. At the beginning of ventricular contraction, the pressure of blood in the ventricles is slighter higher than venous pressure, or only several mm Hg greater than atmospheric pressure. The contracting ventricles immediately raise the ventricular pressure above that of the atria, closing the AV valves. The closure of the AV valves causes the **first heart sound ("lub").** The increase in ventricular pressure that continues after the AV valves close does not force the valves back into the atria because the valve leaflets are attached by threads of connective tissue called **chordae tendinae** to **papillary muscles**

(small muscles that project from the inside of the ventricular wall) (Figures 13-3 and 13-5). The contraction of the papillary muscles during systole takes up the slack in the chordae that otherwise would develop as the ventricles contract.

With the closure of the AV valves, the heart enters the phase of systole called **isovolumetric ventricular contraction.** During this phase, each ventricle is a completely closed chamber, because both the AV valves and the pulmonary and aortic valves are closed. Before blood can be **ejected** through the aortic valve into the aorta or through the pulmonary valve into the pulmonary artery, each ventricle must raise the pressure of the ventricular blood above the **diastolic pressure** of the blood in the corresponding artery.

When the pressure of blood in the ventricles exceeds the corresponding arterial diastolic pressure, the pulmonary and aortic valves open and each ventricle ejects blood into the corresponding artery. The volume of blood that is ejected is called the **stroke volume.** The volume of blood remaining in the ventricle at the end of the ejection phase is called the **end-systolic volume,** this is the minimal volume of the ventricle reached during any particular cardiac cycle (note the difference in ventricular volumes between systole and diastole visible in Figure 13-7).

The volume and momentum of the ejected blood cause the arterial pressure to rise rapidly. The pressure of blood reached in the artery and ventricle during the peak of the ejection phase is called the **systolic pressure.** It would not be an exaggeration to say that the stroke volume of blood is literally flung from the ventricle into the artery. This is especially true for the pumping action of the left ventricle, the walls of which are considerably thicker and more powerful than those of the right ventricle. Upon entering the aorta, the blood generates a shock wave that travels outward through all of the large arteries and can be felt as a **pulse** in the arteries of the wrist, ankle, and neck.

Two processes bring the ejection phase to an end. First, the ejected blood rebounds from the elastic arterial walls like a gymnast landing on a trampoline. The rebounding blood closes the pulmonary and aortic valves with a snap that is heard as the **second heart sound.** The second factor is that, meanwhile, the ventricle has begun to relax. With the closure of the pulmonary and aortic valves, systole is ended and the heart enters the first phase of diastole, called **isovolumetric ventricular relaxation.** During this phase, both the AV valves and the pulmonary and aortic valves are closed, and ventricular pressure falls toward atrial pressure. When ventricular pressure falls below atrial pressure, the AV valves open and the cardiac cycle begins again.

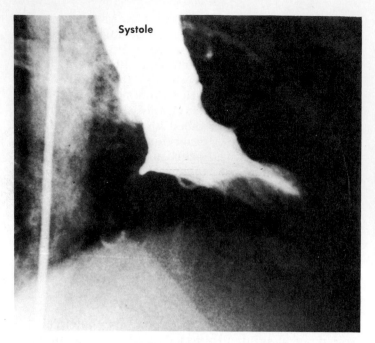

Systole

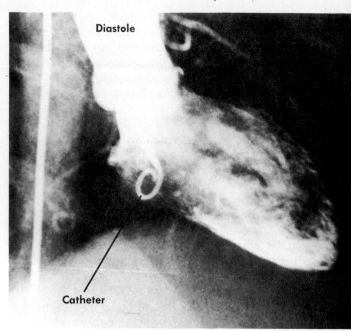

Diastole

Catheter

FIGURE 13-6
Two chest radiographs taken at the end of systole (**left**) and diastole (**right**) during cardiac catheterization. The probe used to measure pressures in the heart, seen clearly during diastole, was inserted into a vein and carefully guided into the heart.

THE ELECTRICAL ACTIVITY OF THE HEART

- *What factors cause pacemaker cells to undergo rhythmic action potentials?*
- *What is the ionic basis of the upstroke, the plateau and the repolarization of the action potential of nonpacemaker cells?*
- *How do the two major mechanisms that release Ca⁺⁺ into the cytoplasm of cardiac muscle cells during excitation-contraction coupling work?*
- *How does Ca⁺⁺ control force development in cardiac muscle?*
- *How is Ca⁺⁺ removed from the cytoplasm after each cardiac action potential?*

For the heart to be an effective pump, the contractions of the myocardial cells of the atria and ventricles must be coordinated. In skeletal muscle such coordination would require simultaneous activation of many motor neurons, but the heart excites itself and coordinates its own contractions. This section describes the mechanisms that generate the cardiac rhythm and coordinate the beating of atria and ventricles.

The Heart's Pacemaker

During a single heartbeat, each myocardial cell undergoes a single action potential that causes it to contract. The frequency of the heartbeat, or heart rate, is affected by inputs from the autonomic nervous system, but hearts that are deprived of all nervous and endocrine inputs continue to beat. The heart can excite itself at regular intervals because nodal cells undergo a repeated cycle in which repolarization after each action potential leads not to a stable resting potential but instead to a spontaneous depolarization called a **prepotential.** Each prepotential ultimately returns the nodal cell to its threshold, resulting in another action potential (Figure 13-8). The capability of generating rhythmic spontaneous action potentials is most highly developed in a small number of cells in the center of the SA node.

In the absence of any autonomic input, the intrinsic rate of production of action potentials by the SA node is normally about 100 per minute. As explained later in this chapter, parasympathetic input slows the pace set by the SA node while sympathetic input tends to accelerate it. Other cells in the conducting system, such as AV nodal cells and cells in the Bundle of His, have lower intrinsic rates (Table 13-2). As a result, the entire conducting system is normally driven by the SA node, which determines the heart rate. However, if the SA node is depressed, damaged, or electrically isolated from the rest of the heart, another portion of the conducting system typically is able to take over the function of pacemaker. The heart rate under these conditions is always lower than that normally set by the SA node.

The cellular mechanism of the SA nodal rhythm has been intensively studied, but still is not fully understood. The hypothesis presented here is based on the following experimental findings:

1. Unlike other excitable cells, including nerve axons, skeletal muscle cells, and myocardial cells, the membrane of pacemakers lacks a substantial steady permeability to K^+ but does have a significant steady background Na^+ permeability. By themselves these two factors would tend to keep the cells depolarized. In fact, the most negative voltages reached in pacemaker cells at the end of repolarization and the beginning of the prepotential are no more than about -65 mV, in contrast to the -80 mV to -90 mV resting potentials of myocardial cells.

2. Unlike skeletal muscle and neurons, voltage-gated Na^+ channels play no part in the upstroke of the action potential. Instead, the inward current of the action potential is carried by Ca^{++}. Two classes of Ca^{++} channels are present in pacemaker cells: **T channels** and **L channels.** The T channels open as the prepotential nears the threshold and are important in the last stage of the prepotential that carries the pacemaker to its threshold. The L channels begin to open around the threshold and are responsible for the upstroke of the pacemaker action potential.

3. As in other excitable cells, repolarization is caused by opening of voltage-gated K^+ channels.

The prepotential (Figure 13-7) begins immediately after the previous action potential, when the voltage-gated K^+ channels activated by the action potential are still open. These channels permit K^+ to diffuse outward and make the cell relatively hyperpolarized. As the K^+ channels close, the membrane potential drifts in the direction of depolarization, driven by the large "background" inward Na^+ leak. When the prepotential reaches about -60 mV, T Ca^{++} channels begin to open, allowing an inward flow of Ca^{++} that also contributes to further depolarization. When the threshold is reached at about -50 mV, L Ca^{++} channels open and an action potential results. The K^+ channels are reactivated by the action potential, repolarizing the cell and resetting the cycle.

In this hypothesis, the rate of closing of the K^+ channels determines the rate of depolarization during the early part of the prepotential, whereas the opening of T channels determines the rate of depolarization during the late part of the prepotential. These two types of channels determine the amount of time that elapses before the next pacemaker action potential and thus they control the heart rate. As will be shown later in this chapter, autonomic inputs modulate the heart rate by changing the rate of closing of the K^+ channels.

In addition to the levels of autonomic inputs received by the SA node, other factors may affect the

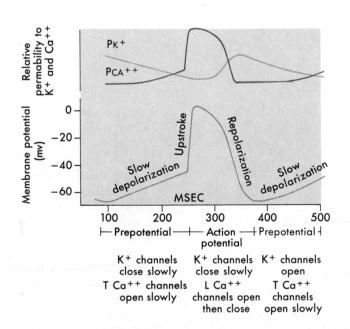

FIGURE 13-7
The ionic permeability changes responsible for the prepotential in cardiac pacemakers.

heart rate. Like almost every biological process, the rate of the SA nodal rhythm is affected by temperature, so the heart rate is increased by fever and decreased if core body temperature drops below normal. For open heart surgery or transplantation, the heart may be stopped by chilling; during such procedures the patient is maintained by connecting the systemic loop to a heart-lung machine. Also, like all other excitable cells, the pacemakers are affected by the ionic composition of the extracellular fluid, so departures from ionic homeostasis typically affect the heart rhythm.

The Action Potentials of Nonpacemaker Cells

The action potentials of all cardiac cells except nodal cells are characterized by three important features (Figure 13-8):

1. The upstroke is rapid, unlike that of pacemaker cells (and like that of nerve axons and skeletal muscle cells). This part of the action potential is due to "fast" voltage-gated Na^+ channels, as in axons and skeletal muscle cells.
2. The upstroke is followed by a **plateau phase** during which the membrane potential remains

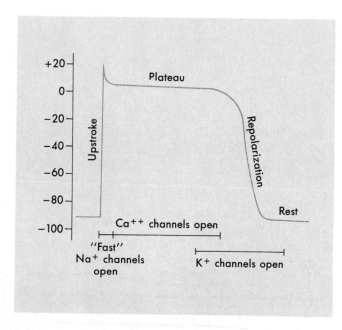

FIGURE 13-8

The permeability changes that occur during an action potential in a myocardial fiber. The pattern is similar to that in Purkinje cells. Rapid depolarization results from the activation of fast Na^+ channels, whereas rapid partial repolarization is caused by Na^+ inactivation. The plateau is produced by activation of Ca^{++} channels, whereas repolarization is the combined result of K^+ activation and Ca^{++} inactivation. During diastole, the membrane potential is stable at the resting potential.

depolarized for up to about 200 msec. Two types of channels are open during the plateau: voltage-gated K^+ channels that carry an outward current, and voltage-gated Ca^{++} channels that carry an inward current. These two currents approximately balance, keeping the cell depolarized.

3. The end of the plateau is marked by closure of the Ca^{++} channels, while the K^+ channels remain open for a time. The outward current carried by the K^+ channels is now not balanced by an inward current as in the plateau phase, resulting in repolarization.

Excitation-Contraction Coupling in Cardiac Muscle

Cardiac muscle has a number of features in common with skeletal muscle—these include the sarcomere organization of the contractile machinery and the presence of T tubules and a well-organized sarcoplasmic reticulum. In both muscle types there is a similar relationship between T tubular endfeet and Ca^{++} release channels of the terminal cisternae (shown in Figure 11-13 for skeletal muscle).

The process of excitation-contraction coupling in cardiac muscle (Figure 13-10) is believed to differ from that in skeletal muscle in two important respects—the mechanism for release of the Ca^{++} that activates the contractile machinery, and the fraction of the crossbridges that are typically activated. Recall from Chapter 11 that in skeletal muscle, all of the Ca^{++} needed to activate the contractile machinery comes from the SR. With each skeletal muscle action potential, the SR releases enough Ca^{++} to activate all of the crossbridges, and after excitation all of the Ca^{++} released into the cytoplasm is pumped back into the SR.

In cardiac muscle, much, but not all, of the Ca^{++} found in the cytoplasm during excitation comes from the SR. Release of Ca^{++} from the SR involves at least two mechanisms. The first is the T-tubular coupling to the lateral cisternae of the SR, which in cardiac muscle is believed to work essentially the same as in skeletal muscle (Figure 11-13).

The second mechanism for Ca^{++} release from the SR is set in motion by Ca^{++} that enters across the cell membrane. As shown in Figure 13-9, much of the transmembrane current that keeps the myocardial cell depolarized during the plateau of the cardiac action potential is carried by inward movement of Ca^{++} through voltage-gated channels. This Ca^{++} current is very significant for the action potential, but the amount of Ca^{++} that can be brought into the cell by this route is small in comparison to the amount needed to activate the contractile machinery. Nevertheless, the small amount of Ca^{++} that enters across the cell membrane during the cardiac action potential is important because it serves as "trigger Ca^{++}" that induces the SR to release a

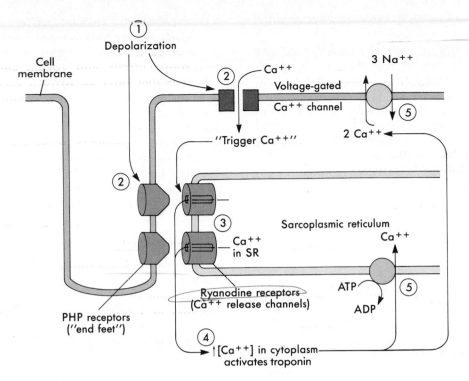

FIGURE 13-9
Pathways of excitation-contraction coupling in myocardial cells. Depolarization (1) opens (2) both membrane voltage-gated Ca++ channels and SR Ca++ channels controlled by T tubule DHP receptors. Entry of trigger Ca++ potentiates release of (3) Ca++ from the SR (Ca++-induced Ca++ release). The number of crossbridges that become active is determined by the amount of Ca++ released into the cytoplasm (4). Relaxation occurs as Ca++ is removed from the cytoplasm (5) by the Ca++ pump of the SR and by Na+/Ca++ exchange across the cell membrane.

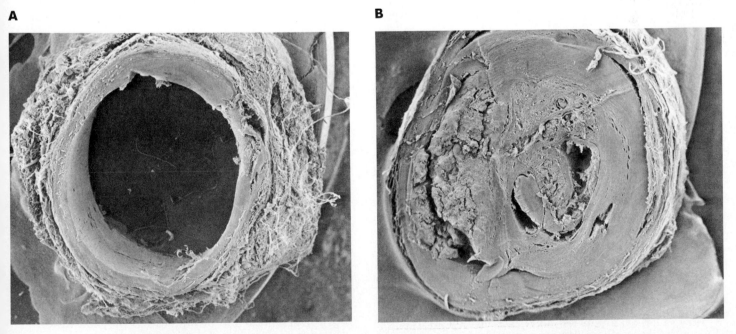

FIGURE 13-10
Scanning electron micrographs of a normal coronary artery (A) and an artery occluded by a thrombus (B).

much larger amount of Ca^{++}. This process is called **Ca^{++}-induced Ca^{++} release.**

The total amount of Ca^{++} released in cardiac muscle during excitation is typically enough to activate only about half of the myosin heads. Accordingly, the force developed by the ventricles during systole may be increased by increasing the concentration of Ca^{++} attained during excitation. The amount of trigger Ca^{++} exercises a strong influence on the final Ca^{++} levels attained in the cytoplasm during systole, so changes in the amount of trigger Ca^{++} will be important in extrinsic control of the heart.

Relaxation of the myocardium at the end of each systole is the result of removal of Ca^{++} from the cytoplasm. This is an active process energized by Ca^{++} pumps located both in the SR and the cell membrane (Figure 13-10). In the myocardial SR, as in skeletal muscle, the Ca^{++} pump is a primary active process driven by a transport ATPase. Extension of Ca^{++} across the cell membrane is mainly by an exchange of Ca^{++} out for Na^+ in. This exchange is energized by the transmembrane Na^+ gradient, which in turn is maintained by the cell's Na^+-K^+ pumps. When the heart is beating slowly, the two Ca^{++} removal mechanisms can reduce the cytoplasmic Ca^{++} concentration from its systolic value of about $10^{-5}M$ to about $10^{-7}M$ at the end of isovolumetric relaxation, and even further during the subsequent diastole. However, when the heart is stimulated to beat more rapidly, there is less time for Ca^{++} removal, and the diastolic level of Ca^{++} rises somewhat. The increase in the diastolic value causes the subsequent systolic value to be higher as well. Thus, increases in heart rate tend to increase systolic force, even in the absence of any other specific intracellular signal.

Spread of Excitation in the Heart

> - *What is the pathway of electrical excitation through the heart's conducting system?*
> - *What is arrhythmia?*
> - *What happens in each of the following arrhythmias?*
> - *Why are the consequences of atrial and ventricular fibrillation usually different?*

Cardiac muscle cells form a mesh in which the individual cells are connected to one another by intercalated disks. These structures make a durable mechanical connection between the cells and also contain gap junctions that permit passage of electrical current from cell to cell (see Figure 11-24). Action potentials can travel in the heart by two kinds of pathways: rapid conduction along the system of conducting cells and more slowly from one myocardial cell to another by way of the gap junctions between adjacent cells.

Recall that the atria are activated first in the cardiac cycle, and the ventricles are activated after a delay of about 0.1 sec (the AV delay). The action potential generated first by a few cells in the SA node spreads from cell to cell of the atrial wall until both atria are depolarized. An interatrial conducting system has been suggested, but simple cell-to-cell conduction by gap junctions seems adequate to account for the spread of excitation from the right to the left atrium. At the same time, the action potential is traveling toward the AV node along the internodal conduction pathways (Figure 13-4).

A tough ring of connective tissue surrounds each AV valve, forming a barrier to conduction of electric current between the atria and the ventricles. This barrier is penetrated at only one point—the AV node and bundle of His. The AV node is composed of small cells that conduct action potentials slowly. This is believed to account for the AV delay. Once the action potential has passed through the AV node, it travels rapidly through the bundle of His and the branch bundles that run down the septum.

The branch bundles feed into the system of Purkinje fibers that spread upward from the tip of the ventricle. This network rapidly carries the action potential across the inner surface of each ventricle, triggering a wave of ventricular contraction that starts at the apex of the heart and travels toward the base. As a result, the ventricles squeeze blood toward the pulmonary and aortic valves. Complete excitation of the ventricles is rapid, requiring only about 75 msec from the time that the action potential enters the bundle of His. This is because most of the distance is covered by way of the conducting cells. Only the last step of the route—conduction of the action potential from the Purkinje fiber branches underneath the endocardium outward through the thickness of the myocardium—involves slower conduction from one myocardial cell to another.

Abnormalities of Cardiac Rhythm

Abnormal or irregular heart rhythms are called **arrhythmias** (or **dysrhythmias**). These can be caused by (1) an abnormal rhythm of the SA nodal pacemaker itself, (2) blockage at some point in the normal conduction pathway for action potentials through the heart, (3) the presence of an abnormal pathway for action potential conduction through the heart, or (4) spontaneous generation of action potentials in a part of the heart other than the SA node—this region then is called an **ectopic pacemaker.** Several examples of arrhythmias will be described in this section; see also the Focus Unit on electrocardiograms.

The AV node and bundle of His may fail to conduct action potentials to the ventricle in some disease states or when the excitability of the AV node is depressed by drugs or parasympathetic input.

This condition is called **AV block** or **heart block.** There are several degrees of AV block. In the mildest case, **1st degree block,** the AV delay is simply abnormally long. In **2nd degree block,** there is a skipped ventricular systole during every second or third heartbeat. In the most severe case, **3rd degree block,** conduction through the AV node is completely blocked and atrial excitation never reaches the ventricles. In the last case, spontaneously active cells of the bundle of His begin to drive the ventricle at their own intrinsic rate of 25 to 40 beats/min, while the atria continue to be driven by the SA node. In 3rd degree heart block, atrial contractions are not synchronized with ventricular contractions and so do not assist ventricular filling.

An irritable or damaged region in the ventricular wall may occasionally generate a spontaneous action potential. This action potential then spreads throughout the ventricles, causing a **premature ventricular contraction** or **PVC.** This is an example of an ectopic pacemaker. A PVC typically makes the ventricles refractory when the next normal action potential generated by the SA node arrives. The result is a dropped beat or **compensatory pause,** followed by resumption of the normal rhythm. Anything that increases the excitability of the myocardium can cause an occasional PVC in a normal heart; for example, caffeine and sympathetic input can have this effect.

When a blood clot, or **thrombus,** blocks a coronary artery (Figure 13-10), the region of the myocardium served by that artery is damaged. This damage is called a **myocardial infarct;** and it is a common cause of "heart attack." The probability of clot formation is increased when coronary arteries are narrowed by the fatty deposits of **atherosclerosis** (discussed further in Chapter 15). If the victim survives the immediate damage, the cells in and around the infarct may be hyperexcitable and act as an ectopic pacemaker. Frequently, however, the infarcted cells create an abnormal conduction path that allows the wave of cardiac excitation to circle around and reexcite some parts of the heart, a phenomenon called **reentry.** Reentry can lead to a continual asynchronous contraction called **fibrillation,** which prevents coordinated pumping.

The consequences for the patient depend on whether the atria or the ventricles are fibrillating. In atrial fibrillation, the loss of atrial pumping is not in itself life-threatening because atrial contraction is not necessary for ventricular pumping. Action potentials arrive at the AV node from the fibrillating atria irregularly and more rapidly than they can be conducted into the ventricles. An irregular ventricular contraction rate results, but this is also typically not immediately life-threatening. However, over time it causes damage to cerebral blood vessels that may culminate in a stroke. In contrast, ventricular fibrillation is immediately life threatening because it halts blood flow to essential organs, including the heart itself. About 1/4 of all deaths involve ventricular fibrillation. Ventricular fibrillation very seldom reverses itself, but **defibrillation** can sometimes be achieved by giving the entire heart a brief, strong electrical shock that makes the entire myocardium refractory at the same time. After this refractory period has worn off, the SA node or some other pacemaker may be able to reestablish an effective cardiac rhythm.

The Electrocardiogram

- *What are the major components of the ECG?*
- *What is a cardiac vector?*
- *How is the cardiac vector measured using the standard limb leads?*

The **electrocardiogram** (**ECG** or **EKG**) is a record of the electrical activity of the heart. The ECG gives information about the timing of electrical events in the heart and also about the pattern by which excitation spreads in the heart. These processes are directly related to heart pumping. Usually the ECG is recorded by placing electrodes at different points on the body surface and measuring voltage differences between the different points. Because the body is a conductor of electricity, the arms, legs, and chest surface function as if they were wires connected to the heart.

To understand the principles of electrocardiography, it helps to consider first the ECG of a single cardiac muscle cell (Figure 13-11). This can be recorded by placing electrodes at two points along the length of the cell and connecting them to the amplifier of a pen recorder that measures voltage differences between the two electrodes. In this system, the pen deflects upward when the right-hand electrode is positive to the left-hand electrode.

Initially there is no voltage difference between the electrodes and the pen remains at the baseline. The cell is excited by stimulation at one end (Figure 13-11, *A*). In Figure 13-11, *B*, the leading edge of the action potential (a wave of depolarization) has reached the left-hand electrode. The inward movement of positive charge across the depolarized membrane has left a net negative charge outside the cell, so the left-hand electrode becomes negative with respect to the right-hand electrode and the pen deflects upward. When the action potential has spread as far as the right-hand electrode, both electrodes detect the same voltage and the pen returns to the baseline (Figure 13-11, *C*). When the action potential has passed the left-hand electrode but is still passing under the right-hand electrode, the amplifier again detects a voltage, but this time it is of the opposite polarity and the pen deflects downward (Figure 13-11, *D*). Finally, when the action potential has passed both electrodes, there is no

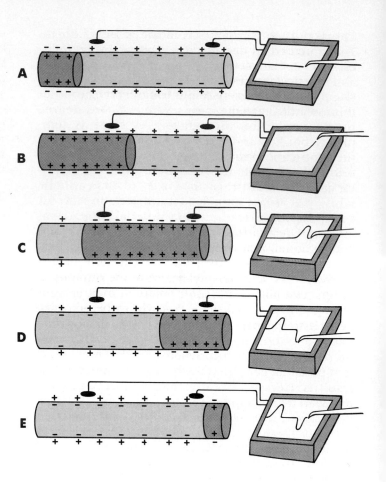

FIGURE 13-11

Extracellular recording of an action potential in a single cardiac fiber. The electrocardiograph records the voltage difference between the two electrodes. An upward deflection of the pen corresponds to an excess of negative charge at the first electrode. In **A** the action potential has not yet reached the first electrode, so the voltage difference is zero. In **B** the first electrode senses an excess of negative charge as compared with the second electrode, so the pen is deflected. In **C** the action potential occupies all of the fiber between the two electrodes, so no voltage difference exists between them and the pen returns to baseline. A deflection in the reverse direction is recorded as the tail of the action potential passes under the second electrode (**D**). The trace returns to baseline after the action potential has passed under the second electrode (**E**).

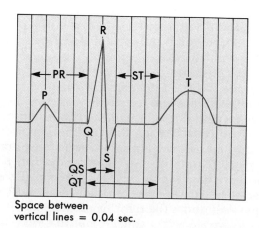

Space between
vertical lines = 0.04 sec.

FIGURE 13-12

An idealized ECG of a single heartbeat. The P-R interval is normally measured at the beginning of the QRS complex.

voltage difference between the electrodes and the pen returns to baseline (Figure 13-11, *E*).

The initial event of a heartbeat, the action potential of the SA node, is not seen in the ECG (Figure 13-12) because the node is so small that its current is too tiny to detect at any distance from the heart. However, the resulting wave of depolarization that spreads through the atria is seen as the **P wave** (Figure 13-12); this phase of the ECG corresponds to Figure 13-11, *B*. When the atrial myocardial cells are all in the plateau phase of their action potential, the ECG trace returns to its baseline; this phase of the ECG corresponds to Figure 13-11, *C*. In the meantime, the action potential has passed through the AV node and entered the bundle of His. As with the SA node, the depolarization of the AV node and bundle of His results in currents too small to be detected at any distance from the heart.

The next event visible in the ECG is the **QRS complex,** the signal of ventricular depolarization. As with atrial depolarization, the trace deflects only during the brief time that excitation is spreading through the ventricular myocardium. When all cells of the ventricular myocardium are in the plateau phase of their action potentials, the trace returns to baseline and remains there until ventricular repolarization begins. During this same time, the atria are repolarizing, but the current generated by atrial re-

370   *THE CARDIOVASCULAR SYSTEM*

polarization is usually not seen in the ECG because it is obscured by the much larger currents of ventricular depolarization.

Ventricular repolarization results in the **T wave**. The T wave characteristically has a lower peak deflection and lasts longer than the QRS complex because ventricular repolarization is less synchronous than depolarization. Contrary to what might be expected on the basis of Figure 13-11, the T wave usually has the same polarity as the R wave, the major component of the QRS complex.

In other words, the net current flow through the ventricle during repolarization has the same polarity as that of depolarization. This indicates the wave of repolarization follows a path through the heart that is exactly opposite to the path of depolarization. The reason is that there is a difference in the duration of the plateau phase in different parts of the ventricle, with the first cells in the ventricle to be activated being also the last to repolarize.

The **intervals** between the events of the ECG (Figure 13-12) reflect the timing of electrical events in the heart. The **P-R interval** corresponds to the time between the onset of atrial depolarization and the onset of ventricular depolarization. Most of this time reflects the AV delay. Normally the P-R interval lasts about 0.2 sec; this interval gets longer in 1st degree AV block. Normal ventricular activation requires about 0.075 sec; if the QRS complex lasts significantly longer than this, some part of the ventricular conducting system has failed.

Because heart is a complex, three-dimensional structure, the spread of excitation through it is also complex and three-dimensional. Any particular placement of electrodes on the body surface detects a particular portion of the current flow during excitation. A complete picture of the spread of excitation through the heart can be obtained by recording the ECG from several electrode placements. In electrocardiography, the electrode placements are called **leads.**

The three **standard limb leads**, numbered I, II, and III, are shown in Figure 13-13, *A*. In **Lead I** the positive terminal of the amplifier is connected to the left arm and the negative terminal to the right arm. In this lead the amplifier detects the component of excitation that is moving along an axis that passes from the left side of the heart to the right side. In **Lead II**, the positive terminal is connected to the left leg and the negative terminal to the right arm, and the amplifier detects the component of excitation moving along an axis that passes from the right upper corner of the heart toward the tip of the ventricles. In **Lead III**, the positive terminal of the amplifier is connected to the left leg and the negative terminal to the left arm, and the component of excitation moving along an axis that passes from the left atrium toward the ventricular tip is detected. These three electrode configurations define **Einthoven's**

triangle (Figure 13-13, *B*), named after the pioneer Dutch electrocardiologist W. Einthoven who worked out in the first two decades of the 20th century most of the basic methods of electrocardiography used today.

As shown in Figure 13-13, *B*, Einthoven's triangle can be used to determine the heart's **electrical axis** or **cardiac vector**—the magnitude and direction of the bulk of the excitation current. For example, the R wave results from the rapid spread of excitation downward from the AV node through the septum and into the tip of the ventricles. The normal axis of the R wave points downward and to the left, corresponding to the orientation of the septum.

The standard limb leads give the orientation of the cardiac vector in the two-dimensional frontal plane only. Additional information about the orientation of the vector in the three-dimensional volume of the chest is obtained using a set of **precordial leads** in which an exploring electrode is placed at seven positions ranging from the left side of the chest to the front of the chest. There is considerable individual variation in the axes of normal hearts. Some possible clinical implications of deviations of the axis from its normal range are explained in the accompanying Focus Unit.

SUMMARY OF EVENTS OF THE HEART CYCLE

> - *How do the peaks of the ECG relate to the sequence of electrical excitation of the heart and the mechanical events of the heart cycle?*
> - *How do arterial pressure, ventricular pressure, and atrial pressure change over the heart cycle?*
> - *How do these changes relate to the opening and closing of the two sets of cardiac valves?*

Having explained the electrical and mechanical events of the cardiac cycle and the resulting pressure changes in the atria, ventricles and arteries can now be summarized (Figures 13-14, 13-15, and Table 13-3). For simplicity, the cycle of the left heart will be described first, and afterward the important differences between the cycles of the left and right hearts will be noted.

Phase 1: Passive filling

All parts of the heart are relaxed; the ECG trace is at the baseline; the AV valve is open and the left atrium and ventricle are filling with blood from the pulmonary veins as shown by the rising trace of ventricular volume; the atrial and ventricular pressures are essentially the same as the central venous pressure. The pressure of blood in the aorta is much higher than that of the left ventricle, so the aortic valve is closed. The aortic pressure is falling as blood that was ejected into the aorta in the previous cardiac cycle runs off into the rest of the systemic loop.

The Heart

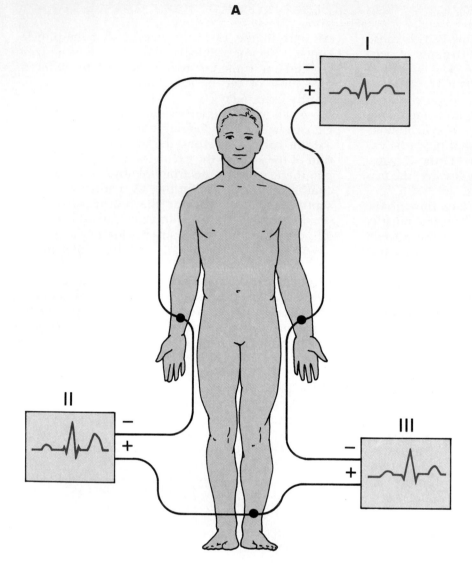

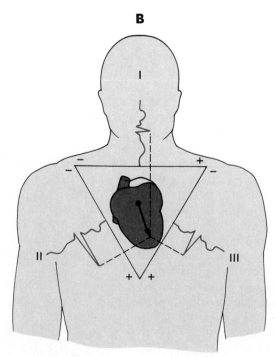

FIGURE 13-13

A The connections of the three standard limb leads.

B Einthoven's triangle is formed by the three standard limb leads. The cardiac vector for the R wave is determined by the relative magnitudes of the R wave in the three leads, plotted on Einthoven's triangle. For the R wave (shown) it usually points downward and to the left, lying along the axis of the septum and cardiac apex, since these parts are depolarizing during the R wave.

THE CARDIOVASCULAR SYSTEM

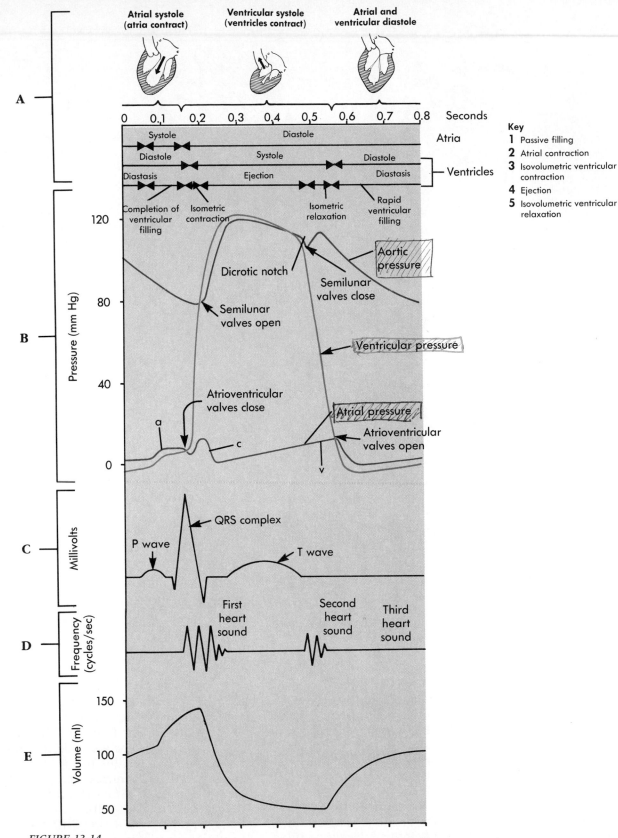

FIGURE 13-14

The cardiac cycle. Phases and approximate durations are indicated at the top of the figure. The blue, green, and red curves show the pressures in the aorta, left ventricle, and atrium, respectively. Below this are shown the ECG, the heart sounds, and the ventricular volume. The phases of the cardiac cycle occur in the following order: (1) passive filling (early diastole), (2) atrial contraction (late diastole), (3) isovolumetric ventricular contraction, (4) ejection, and (5) isovolumetric ventricular relaxation.

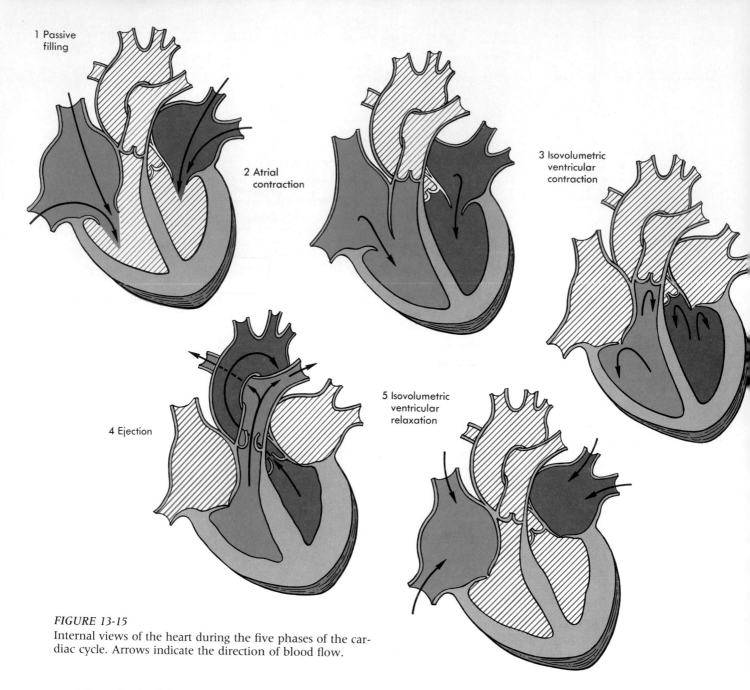

1 Passive filling

2 Atrial contraction

3 Isovolumetric ventricular contraction

4 Ejection

5 Isovolumetric ventricular relaxation

FIGURE 13-15
Internal views of the heart during the five phases of the cardiac cycle. Arrows indicate the direction of blood flow.

Phase 2: Atrial contraction

An action potential arises in the pacemakers of the SA node and spreads out through the atria, generating the P wave of the ECG. When the atria are fully depolarized, the ECG trace returns to the baseline. Atrial contraction adds perhaps 20% to the volume of blood that entered the ventricle passively, increasing the pressure of blood in the ventricle by several mmHg. The delayed conduction of the action potential through the AV node allows atrial contraction to be complete before ventricular activation occurs.

Phase 3: Ventricular excitation and isovolumetric ventricular contraction

The action potential passes through the AV node and is conducted rapidly through the bundle of His and the branch bundles to the Purkinje cells of the ventricle. The Purkinje cells spread the action potential through the ventricle, starting at the tip of the heart and spreading toward the atria. The depolarization of the ventricle generates the QRS complex of the ECG. When the ventricle is completely depolarized, the ECG returns to its baseline; during the interval between the QRS complex and the T wave, all ventricular myocardial cells are in the plateau phase of their action potentials. Excitation of the ventricle is followed within a few msec by the beginning of force development. The resulting rise in ventricular pressure closes the AV valve, causing the first heart sound. After the AV valve closes, making the ventricle a closed chamber, pressure in

TABLE 13-3   *Phases of the Cardiac Cycle*

Phase	Approximate duration	Atrial state	Ventricular state	AV valves	Pulmonary and Aortic valves
1	250 msec	Relaxed	Relaxed: filling	Open	Closed
2	100 msec	Contracting	Relaxed: filling	Open	Closed
				First heart sound	
3	20-30 msec	Relaxed	Isovolumetric contraction	Closed	Closed
4	250 msec	Relaxed	Contraction: ejection	Closed	Open
					Second heart sound
5	30-60 msec	Relaxed	Isovolumetric relaxation	Closed	Closed

the ventricle continues to rise, while aortic pressure is still slowly falling toward its minimum or diastolic value.

Phase 4: Ejection

When ventricular pressure exceeds aortic pressure, the aortic valve opens and blood is ejected into the aorta, as indicated by the rapid drop in ventricular volume and the rise in aortic pressure to the systolic value. Much of the energy imparted to the blood by ventricular contraction is stored in the elastic walls of the aorta and its large branches. This stored energy is released during diastole, keeping the arterial pressure high and maintaining blood flow from the aorta into the arterial tree while the heart is not ejecting blood.

Phase 5: Isovolumetric relaxation

The closing of the aortic valve marks the end of the ejection phase and the beginning of isovolumetric relaxation. Aortic valve closing is heard as the second heart sound and seen as a notch (called the **dicrotic notch**) in the aortic pressure trace. During the isovolumetric relaxation phase the pressure in the ventricle has fallen below that of the aorta but is still above that of the atrium, so both sets of valves are closed and no blood is leaving or entering the ventricle. Repolarization of the ventricular myocardium is seen as the T wave of the ECG. Repolarization is accompanied closely by decay of the active state of myocardial cells. As the ventricular myocardium continues to relax, ventricular pressure falls below that of the atrium. At this point, the AV valve opens and the heart returns to Phase 1.

Pressures in the Pulmonary and Systemic Loops

During all parts of the cardiac cycle, the pressures in the right heart and the pulmonary artery are considerably less than the corresponding pressures in the

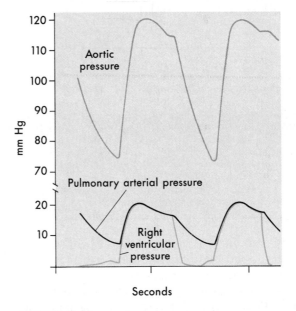

FIGURE 13-16
Comparison of pressure changes over the cardiac cycle in the right ventricle and pulmonary artery with the change in aortic pressure.

left heart and aorta (Figure 13-16). However, the amount of blood pumped per unit time by the right and left hearts is equal when measured over intervals of more than a few beats. Thus the right heart is able to drive blood through the pulmonary loop with approximately one-fifth the driving force required to drive the same amount of blood through the systemic loop. This difference is due to the fact that the resistance of the pulmonary loop to blood flow is much lower than that of the systemic loop. Reasons for this difference will be discussed further in Chapter 14. The easier task of the right heart is reflected in its thinner, less massive ventricular walls.

INTRINSIC AND EXTRINSIC REGULATION OF CARDIAC PERFORMANCE $= HR \times SV = CO$

The **cardiac output (CO)**—the amount of blood pumped by the heart per unit of time—is the product of the heart rate and the stroke volume. The heart rate varies from under 50 beats/min in resting, athletic adults to over 200 beats/min in maximal exercise. The stroke volume is normally between 70 and 140 ml. The cardiac output of resting adults is approximately 3 to 5 L/min during rest but can rapidly increase by four-to-seven fold or more during exercise. The highest values recorded—42 L/min—are from elite endurance athletes during maximal exercise. This section describes the mechanisms by which the heart supports such large changes in cardiac output.

Intrinsic Regulation of Stroke Volume: The Frank-Starling Law of the Heart

> - *What is the shape of the Frank-Starling curve of stroke volume as a function of end-diastolic volume?*
> - *How does this curve relate to the length-tension relationship of ventricular myocardium?*
> - *What is cardiac output?*
> - *How can the Frank-Starling curve be translated into a relationship between cardiac output and central venous pressure?*
> - *What effect does congestive heart failure have on the Frank-Starling curve?*
> - *Why does end-systolic volume increase as congestive heart failure progresses?*

As will be shown in Chapter 15, the cardiac output is not determined by the heart alone, but is the outcome of interaction between the heart and the blood vessels. An important factor in this interaction is the **venous return**—the rate at which blood returns to the heart. The venous return is strongly affected by the local responses of tissues to changes in metabolic rate, and by changes in autonomic inputs to blood vessels, so venous return may vary widely from second to second. Even in the absence of any extrinsic regulation, the heart is able to regulate its stroke volume to meet changes in the load placed on it by venous return.

The basis for the heart's intrinsic regulation of stroke volume was first studied in the amphibian heart by the German physiologist Otto Frank in the late 1800s. During the early 1900s the English physiologist E. H. Starling and his associates developed an isolated **heart-lung preparation** for studies of intrinsic regulation in the mammalian heart. In this preparation the heart of a dog was disconnected from the systemic loop but remained connected to the pulmonary loop. Reservoirs were attached to the right atrium and aorta so that the right atrial pressure and aortic pressure could be manipulated as experimental variables.

As described in Chapter 11, the performance of isolated skeletal muscle is affected by the preload applied to the muscle before it contracts as well as the afterload experienced by the contracting muscle. It was natural for Frank and Starling to apply to the isolated heart the concepts and terms already developed in studies of skeletal muscle. In the case of the heart, the **preload** is the right atrial pressure and the **afterload** is to the aortic pressure. These terms are still in use, but for clarity we will use the descriptive terms "central venous pressure" or "filling pressure" for preload and "mean arterial pressure" for afterload.

Starling and Frank found that the stroke volume of the isolated heart was increased when the filling pressure was increased. All other things remaining constant, an increase in filling pressure resulted in a proportionate increase in end-diastolic ventricular volume. Within the physiological range, each increase of end-diastolic volume was matched by an increase in the subsequent stroke volume (Figure 13-17, *A*). In other words, with increased filling the heart increases its contractile force so that end-systolic volume remains about the same. This intrinsic ability of the heart to adjust its active force to the load is called the **Frank-Starling Law of the Heart**. The law, simply stated, is: "In the subsequent systole, the ventricle pumps out the volume of blood that came to it during diastole."

The Frank-Starling Law is really a statement about the length-tension relationship of cardiac muscle. Recall from Chapter 11 that skeletal muscle develops its maximum active tension over a narrow range of lengths that correspond closely to the normal operating length of the muscle in the body. Stretching a skeletal muscle to a length greater than its normal operating length results in a decrease in contractile force because some myosin heads no longer can contact the thin filaments. However, as active skeletal muscle is allowed to shorten to lengths shorter than it normally attains in the body, it develops less and less force, because of interference between the overlapping thick and thin filaments. In this shorter range of lengths, stretching a skeletal muscle would increase its force.

The basic sarcomere organization of the contractile machinery of cardiac muscle is similar to that of skeletal muscle. Originally, the heart was regarded as simply a skeletal muscle that normally operated on the left side of its length-tension curve rather than at the peak. However, recent studies shows that cardiac muscle is much more sensitive to

THE CARDIOVASCULAR SYSTEM

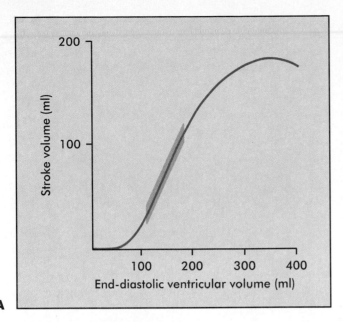
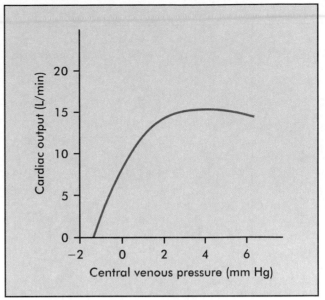

FIGURE 13-17
A The relationship between end-diastolic volume and stroke volume in the heart-lung preparation (Frank-Starling curve). The range of values typical of the normal heart is shown by the shaded portion of the curve.
B Corresponding curve of cardiac output versus central venous pressure.

stretch than can be explained on the basis of overlap between the sliding filaments. For example, within the operating range, a 15% stretch can result in a five-fold increase in contractile force. In some way, increases in sarcomere length are sensed by troponin, the regulatory protein of the thin filaments, which responds by becoming more sensitive to Ca^{++}. The result is that in cardiac muscle, stretch within the normal operating range increases the active tension by increasing the number of active crossbridges.

The Frank-Starling relationship between end-diastolic volume and stroke volume can be translated into a relationship between cardiac output and central venous pressure. This is possible because at any fixed heart rate the end-diastolic volume is proportional to the central venous pressure and the cardiac output is proportional to the stroke volume. The resulting plot of cardiac output versus central venous pressure (Figure 13-17, *B*) resembles that of the Frank-Starling curve (Figure 13-17, *A*). This relationship could be translated into words as: "The heart automatically adjusts its cardiac output to match its venous return."

Stimulation of the myocardium by adrenergic input or by cardiostimulatory drugs with positive inotropic effects raises the Frank-Starling curve and shift it leftward, increasing the stroke volume that is obtained for any given value of end-diastolic volume. This means that the stimulated heart pumps out blood that would have remained part of the end-systolic volume in the absence of the stimula-

tion. In contrast, damage to the myocardium or reduction of its blood supply depresses the Frank-Starling curve (Figure 13-18, *A* and *B*), reducing the stroke volume and cardiac output that result from any given end-diastolic volume. Disease processes that progressively weaken the myocardium result in a steady increase in the end-diastolic volume, called **congestive heart failure.** Early in the course of the disease, adequate cardiac output can still be maintained because stretch still results in a compensatory increase in force. Ultimately, however, the ventricle may be stretched into a range of end-diastolic volumes in which stretching results in no further increase in contractile force. At this point, further stretching results in positive feedback that causes the patient's cardiac output to deteriorate rapidly.

Extrinsic Regulation of the Heart by the Autonomic Nervous System

- *How is the autonomic innervation of the heart arranged?*
- *How do* **chronotropic** *and* **inotropic** *effects relate to the effect of autonomic innervation on cardiac performance?*
- *What is the cellular basis for chronotropic and inotropic effects?*

The heart is innervated by both branches of the autonomic nervous system (Table 13-4). The effect of parasympathetic input is to decrease the heart rate (**bradycardia**). The effects of sympathetic input are (1) to increase the heart rate (**tachycardia**), (2)

The Heart

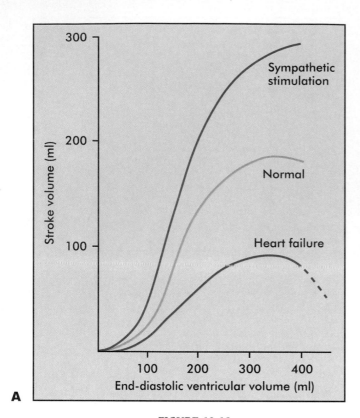

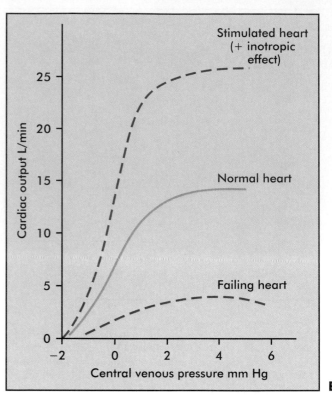

FIGURE 13-18
A Changes in the Frank-Starling curve with stimulation and heart failure.
B Corresponding curves of cardiac output versus central venous pressure.

TABLE 13-4   The Effects of Autonomic Transmitters on the Heart

Transmitter	Receptors	Cell type	Membrane effect	Heart effect
Acetylcholine	Muscarinic	Pacemaker	Slower closing of K^+ channels	Decreased heart rate
		Myocardial	Not innervated	Not innervated
Norepinephrine	Beta$_1$ adrenergic	Pacemaker	Faster closing of K^+ channels	Increased heart rate
Epinephrine		Myocardial	Increased Ca^{++} entry	Increased force of contraction
			Increased Ca^{++} pumping	Decreased duration of systole

to increase the force of ventricular contraction, and (3) to make each systole briefer by increasing the speed of myocardial relaxation.

The parasympathetic innervation of the heart runs in the paired vagus (10th cranial) nerves. The vagus contains preganglionic axons that synapse on postganglionic neurons in a ganglionic plexus that lies among the cardiac muscle cells. The transmitter released by both the preganglionic and postganglionic neurons is acetylcholine, but the postgangli-

onic cells possess nicotinic receptors while the cardiac cells possess muscarinic receptors. Most of the parasympathetic postganglionic neurons synapse on cells of the atria and conducting system (including the SA and AV nodes), with essentially no inputs to the ventricular myocardium. This is the reason that the parasympathetic branch can affect only the heart rate and not the contractility of the ventricles.

Sympathetic postganglionic cells run to the heart from the thoracic sympathetic chain ganglia.

THE CARDIOVASCULAR SYSTEM

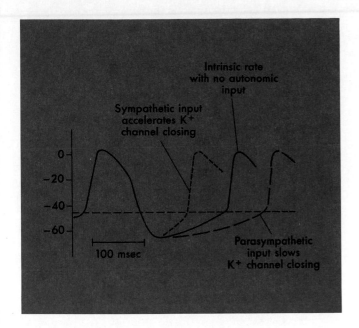

Intrinsic rate
with no autonomic
input

Sympathetic input
accelerates K⁺
channel closing

0
-20
-40
-60

100 msec

Parasympathetic
input slows
K⁺ channel closing

FIGURE 13-19
Factors influencing the rate of depolarization in pacemaker cells and thus the heart rate. The time between successive action potentials is determined by two factors: the magnitude of depolarization needed to reach threshold and the rate at which depolarization occurs. The rate of depolarization is increased by sympathetic stimulation, which hastens closing of K⁺ channels after an action potential, and is decreased by parasympathetic stimulation, which retards K⁺ channel closing.

In contrast to the parasympathetic innervation (Table 13-2), the sympathetic neurons make adrenergic synapses both on the pacemakers in the nodes and on the ventricular myocardium. The sympathetic branch also can deliver the adrenal medullary hormone epinephrine to all parts of the heart by way of the coronary circulation. As a result, the sympathetic branch can influence both the heart rate and the strength of ventricular contraction in systole. The adrenergic receptors of cardiac muscle are mainly of the beta₁ type.

A **chronotropic effect** is an alteration of heart rate; the bradycardia caused by parasympathetic input is a negative chronotropic effect, while the tachycardia caused sympathetic input is a positive chronotropic effect. Chronotrophic effects are the result of changes in the rate of **diastolic depolarization,** the process by which the pacemakers of the SA node automatically return to threshold after an action potential (Figure 13-7). The rate of diastolic depolarization is controlled by the rate at which K⁺ channels close after each pacemaker action potential. Binding of the parasympathetic transmitter acetylcholine to cardiac muscarinic receptors activates a G protein (G_i) that acts directly to slow the rate of K⁺ channel closing. This slows the rate of diastolic depolarization and extends the interval between heartbeats (Figure 13-19). Binding of norepinephrine or epinephrine to the beta₁ receptors of pacemaker cells activates a second type of G protein message (called G_s) that accelerates the rate of K⁺ channel closing and shortens the interval between beats (Figure 13-19).

An **inotropic effect** is an alteration of the contractile properties of myocardial cells. Activation of the beta₁ adrenergic receptors of myocardial cells by epinephrine or norepinephrine results in a **positive inotropic effect,** or increase in the force of contraction during systole. As for pacemakers, the first step in the process is the release of a G_s intramembrane message (Figure 13-20, *A*). The G_s activates phosphodiesterase, resulting in a cAMP second message. Both the G_s and the cAMP increase the amount of "trigger Ca⁺⁺" that enters through voltage-gated Ca⁺⁺ channels during the plateau phase of the myocardial action potential. The extra Ca⁺⁺ entering the cell causes the plateau potential to be more inside positive (Figure 13-20, *B*). Recall that the peak concentration of Ca⁺⁺ attained in the cytoplasm during excitation is affected by the amount of "trigger Ca⁺⁺". The ultimate effect of increasing the Ca⁺⁺ concentration is to increase the number of active crossbridges and thus the systolic force. The greater force generated by the ventricles in response to adrenergic input can result in ejection of a larger fraction of the end-diastolic volume and thus an increase in stroke volume and cardiac output (Figure 13-18).

In addition to its effect on contractile force, activation of the beta₁ receptors makes each systole briefer. It does this by increasing the rate at which Ca⁺⁺ is removed from the cytoplasm after excitation. The more rapid removal of Ca⁺⁺ from the cytoplasm is probably due to stimulation of Ca⁺⁺ pumps of both the SR and cell membrane (Figure 13-20, *A*) by the cAMP second message. This effect causes the cardiac action potential to be briefer (Figure 13-20, *B*). The shortening of systole helps pro-

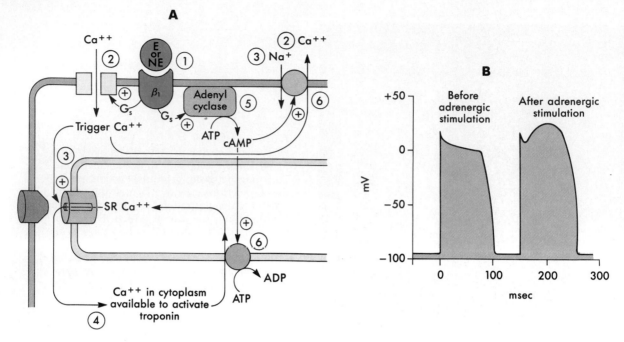

FIGURE 13-20

A Mechanisms of sympathetic inotropic effect. Binding of epinephrine or norepinephrine (1) activates the beta$_1$ receptor, sending a G$_s$ signal to (2) Ca^{++} channels, resulting in an increase in the amount of trigger Ca^{++} that enters during the plateau phase. This induces a greater release of Ca^{++} from the SR when (3) Ca^{++} release channels are activated by depolarization. The ultimate result is that (4) cytoplasmic Ca^{++} rises higher and more cross-bridges are activated, making systole more forceful. The cAMP second message (5) set in motion by the receptor stimulates Ca^{++} pumping (6), making systole briefer.

B Two action potentials from the same myocardial cell before (left) and after (right) adrenergic stimulation. The effect of activating beta$_1$ receptors increases the Ca^{++} current that flows during the plateau, heightening it.

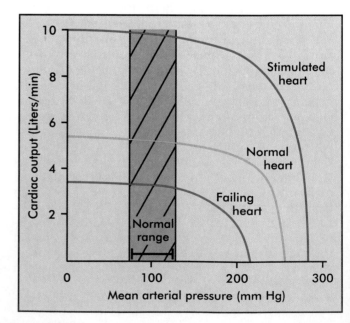

FIGURE 13-21

The relationship between cardiac output and mean arterial pressure for three states of contractility of an isolated heart: an unstimulated normal heart, a heart made hypereffective by sympathetic stimulation, and a heart made hypoeffective by disease. An important feature of these curves is that cardiac output is stable over a range which includes the normal values of mean arterial pressure (shaded area).

FOCUS ON PHYSIOLOGY

Valvular Heart Disease

There are two basic types of heart valve defects, **stenosis** and **insufficiency.** In stenosis, a valve is narrowed, restricting blood flow through it. In insufficiency, the valve leaflets do not fit closely together, allowing blood to leak backward through the valve, or **regurgitate.** Valve defects are typically accompanied by **murmurs** (abnormal heart sounds).

In **mitral stenosis,** blood has difficulty entering the left ventricle from the left atrium. Left atrial pressure increases, whereas the left ventricular end-diastolic volume and cardiac output decrease. During atrial contraction, the turbulent flow of blood through the narrowed mitral valve produces a distinctive blowing sound. This is an example of a **presystolic murmur.**

In **mitral insufficiency,** blood is regurgitated from the left ventricle into the left atrium during systole, increasing the left atrial pressure and reducing the amount of blood ejected into the aorta. Regurgitation through the mitral valve produces one form of **systolic murmur.** As a result of the regurgitation the left ventricle is less effective as a pump, and the output of the right heart continually threatens to become greater than that of the left heart. The outputs are rebalanced when the left atrial pressure and left ventricular end-diastolic volume rise, stretching the left ventricle and increasing its systolic force by the Frank-Starling law. As a result of this, cardiac output is maintained but the left ventricle is forced to do more work during systole than a ventricle with intact valves. Over time, this condition, called **left-ventricular volume overload,** causes **left-ventricular hypertrophy,** an adaptive increase in the mass and strength of the left ventricle in response to the increased load. Left ventricular hypertrophy causes a left-deviation of the cardiac vector (see Focus Unit on clinical use of the ECG). Ultimately the left ventricle cannot keep up with the work demand. The blood dammed on the venous side of the pulmonary loop causes pressures throughout the pulmonary loop to rise, causing interstitial fluid to accumulate in lung tissues, a condition called **pulmonary edema.**

Aortic valve stenosis results in an increased pressure gradient across the aortic valve during the ejection phase of systole. Turbulent flow through the narrowed valve causes a loud, high-pitched systolic murmur with vibrations that cannot only be heard but also felt with the hand on the upper chest and neck. Ejection velocity decreases, so aortic pressure rises more slowly during ejection, reducing the pulse pressure. At the same time, ventricular pressure rises higher during systole. This condition is called **pressure overload,** and as in mitral insufficiency, the increased work load on the ventricle causes left-ventricular hypertrophy.

Ventricular hypertrophy is a natural adjustment of the heart muscle mass to the increased workload imposed on it. As with skeletal muscle (see Chapter 11), the myofibrillar content of the muscle cells increases without an increase in the number of myocardial cells. Hypertrophy is initially beneficial because it enables the ventricle to generate higher pressures. However, the enlarged cells are more difficult to supply with adequate oxygen because the distances to be covered by gas diffusion are greater. In long-term overloads of the heart, some muscle cells die and are replaced by less compliant connective tissue. Furthermore, the internal work of the heart—the work it must do on itself to pump a given volume of blood—increases with the increasing thickness of the ventricle. Ultimately, overloads may result in heart failure.

Coronary Blood Flow and Cardiac Metabolism

Blood flow through the coronary arteries (Figure 13-A) varies greatly during the heart cycle, particularly in the left ventricle. When isovolumetric contraction begins, the tension in the walls of the ventricle is very high, compressing the coronary blood vessels and preventing blood flow. When ejection begins, the coronary blood flow increases and decreases with the rise and fall of the aortic pressure. Early in diastole, the combination of a high aortic pressure with the relaxation of the ventricle allows coronary blood flow to reach its maximum for the cycle, decreasing as the aortic pressure approaches the diastolic value. The flow pattern in the vessels of the right ventricle is similar, but the changes are less exaggerated because the changes in wall tension of the right ventricle are less dramatic.

A normal heart consumes 8 to 10 ml of oxygen/min/100 g of cardiac tissue under resting conditions. The oxygen consumption increases with increases in myocardial work. Myocardial work is affected by aortic blood pressure, ventricular diastolic filling, and changes in the heart rate and contractility. The resting heart extracts about 75% of the oxygen carried by the coronary arterial blood, so the heart cannot accommodate increased metabolic need by extracting more oxygen. Instead, the coronary blood flow must increase when the heart's workload increases. How does this occur?

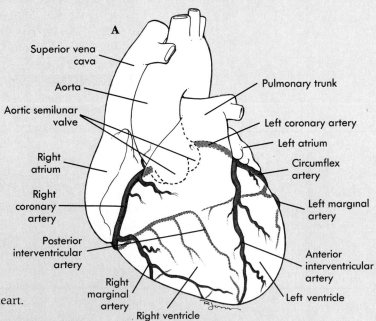

FIGURE 13-A
The coronary blood supply to the heart.

The most important determinant of coronary blood flow is **autoregulation,** a process in which local tissue metabolism affects the diameter of arterioles that serve that organ or region. Autoregulation of coronary arterioles is believed to result from changes in the rate of release of adenosine and other local metabolic factors in response to changes in the oxygen content of the blood in the capillaries of the heart. When the oxygen consumption of the cardiac muscle increases, the capillary oxygen content decreases, and the local concentration of adenosine increases. Adenosine dilates the coronary arterioles, increasing coronary blood flow. Sympathetic stimulation is a second and less important means of increasing coronary blood flow during increased demand for pumping. Coronary arterioles have mainly β_2 adrenergic receptors, which mediate dilation.

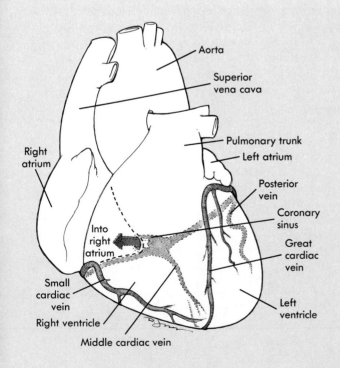

Cardiovascular Spare Parts

There are a number of ways to replace components of the cardiovascular system. Artificial heart valves have been in use for three decades while artificial blood and blood vessels are being developed. These "spare-parts" are combinations of synthetic materials and biological substances designed to mimic the body closely enough to prevent rejection by the immune system. A significant problem is that an artificial product must be ultra-smooth and behave like a living vessel. The slightest irregularity or the wrong surface charge can rupture platelets, triggering clot formation and obstructing blood flow.

Blood is a complex mixture of red cells, white cells, and platelets, suspended in plasma. The rise in blood-borne infections such as hepatitis and AIDS has made blood substitutes appealing. Of the oxygen-carrying chemicals that have been evaluated for use in artificial blood, the most promising are perfluorochemicals (PFCs), organic compounds containing fluorine. The first red cell replacement using PFCs was approved by the Food and Drug Administration in 1990. The product, called Fluosol, also contains a mild detergent and lipid molecules from egg yolk. Its use is currently restricted to angioplasty, a procedure in which the inflated balloon used to open up blocked coronary arteries temporarily blocks blood flow in nearby tissues.

An obvious red cell substitute is hemoglobin itself. Recently, human hemoglobin has been mass-produced by biotechnology. Unfortunately, free hemoglobin is filtered in the kidney and rapidly excreted. One way of overcoming this difficulty is to link individual hemoglobin molecules together. Without co-factors like 2,3 DPG and CO_2 that are normally present in red blood cells, hemoglobin cannot efficiently bind and release oxygen. Hemoglobin has been modified by genetic engineering so that its behavior as a free molecule more nearly resembles that of hemoglobin in red cells. A compromise between free hemoglobin molecule and nature's packaging is to enclose hemoglobin in liposomes to form neohemocytes. Unfortunately, neohemocytes are susceptible to attack by the immune system.

None of the methods of unclogging coronary vessels keeps them smooth and plaque-free indefinitely. What about synthetic blood vessels? One new type of vessel is constructed from cadaver-derived cells grown in tissue culture. It has an inner layer of endothelium, a middle layer of smooth muscle, and an outer layer of connective tissue strengthened by a Dacron mesh. Another approach to blood vessel replacement uses Dacron tubes coated with the patient's own endothelial cells. With the addition of growth factors, endothelial cells proliferate to form a one-cell-thick endothelial lining, smooth enough to prevent clotting, on the interior of the Dacron tubules. Because the immune system recognizes these cells as "self," it does not reject the replacement vessel.

About 75,000 people in the United States receive heart valves each year. Artificial valves are composites of a ceramic and a light, strong metal such as titanium. Replacement valves are taken from pig or cow hearts. Artificial valves are durable, but they are not as gentle in their operation as natural heart valves. Patients with such valves must take anticlotting drugs to prevent dangerous clot formation. Animal valves are less likely be blocked by clots, but carry the risk of rejection inherent in any transplant.

Unlike other spare parts for the cardiovascular system, artificial hearts have not been very successful. The Jarvik-7 device was used for temporary maintenance of patients awaiting heart transplant from a matched donor, but even this use was halted in January 1990. Another approach is the implantable ventricular assist device, a pump used to support only the patient's left ventricle. One implantable ventricular assist device is a blood pump about the size of a fist, inserted beneath the heart. Blood from the left ventricle is diverted to the pump, which then sends it to the aorta. The pump is timed using the electrocardiogram. The pump's battery is recharged by a battery pack worn by the patient using an external induction coil coupled to a second coil implanted under the skin.

Although many scientists hope a fully implantable electrically driven artificial heart can be developed, there are serious limitations in any mechanical device. One reason is that the length-tension properties of living cardiac muscle (Frank-Starling Law, see p. 376) allow the human heart to respond to a wide range of demands without neural or hormonal input. The degree of internal autoregulation of the living heart has been dramatically illustrated by experiments on racing greyhounds. When the heart is denervated by cutting the sympathetic and parasympathetic nerves, greyhounds still can perform at 80% to 90% of their maximum capacity before the surgery.

tect the time available for diastolic filling when the heart rate increases.

The contractile force of the ventricles is also affected by the heart rate as described previously (see *Excitation-Contraction Coupling in Cardiac Muscle*). It may be influenced as well by a number of other factors, such as coronary artery disease, reduced oxygen delivery, levels of thyroid hormones, cardioactive drugs, anesthetics, and alterations of plasma electrolytes and pH.

Overview of Interacting Factors that Affect Cardiac Performance

- *What changes result in enhanced cardiac performance during exercise?*
- *What major factors determine stroke volume?*
- *Why is arterial pressure usually relatively unimportant for determining stroke volume?*
- *What factors normally protect stroke volume when heart rate is increased?*

As work load is increased in exercise, cardiac output also increases, ultimately reaching values as much as 5 to 7 times that of rest. Sustained exercise is supported largely by oxidative metabolism. The maximum cardiac output determines the maximum rate at which O_2 and nutrients can be delivered to the working muscles and CO_2 removed from the muscles to the atmosphere. The maximum power output of fit individuals in sustained exercise is probably limited by the heart's ability to pump blood. How does the heart respond to the increased need for pumping during exercise, and what factors set an upper limit to its performance? Obviously, any factor that alters the stroke volume, the heart rate, or both will affect cardiac output.

The heart rate is determined by the pacemaker cells' integration of the inputs they receive from parasympathetic and sympathetic postganglionic neurons, with additional stimulatory input from circulating epinephrine (Table 13-3). At rest, the heart is believed to receive primarily parasympathetic input, because the resting heart rate is lower than the intrinsic heart rate. The increases in heart rate that accompany exertion and arousal are the result of a simultaneous decrease in parasympathetic input and increase in sympathetic input.

The stroke volume has two major determinants: (1) sympathetic inputs to the heart, and (2) the end-diastolic ventricular volume (Figure 13-17, *A*). The end-diastolic volume is determined in turn by two variables: (1) the time available for filling and (2) the central venous pressure.

Stroke volume is also potentially affected by the force that opposes blood ejection—the pressure of arterial blood. However, this effect is relatively minor under normal conditions, as is shown by data from heart-lung preparations. In experiments in which the mean arterial pressure was progressively increased while the central venous pressure and heart rate were held constant, the cardiac output declined relatively little until mean arterial pressures considerably higher than the normal range were reached (Figure 13-21). In other words, when there is an increase in the mean arterial pressure, the heart increases its systolic force, keeping cardiac output almost unchanged. This regulation of cardiac output is a negative feedback process that can be understood by imagining an increase in mean arterial pressure that initially causes stroke volume to drop. The resulting increase in end-diastolic volume over the next few beats, acting by the Frank-Starling law, would increase the heart's systolic force and return the stroke volume almost to its previous value. The same process operates in stimulated hearts, except that the systolic force is greater at all values of arterial pressure, so there is always a higher cardiac output. In failing hearts, systolic force is less than normal, so that the cardiac output is lower at all values of arterial pressure.

The positive effect of increased heart rate on cardiac output might seem to be offset by the resulting decrease in the time available for filling, and simple changes in heart rate do have small or even counterproductive effects on cardiac output. A person with tachycardia, caused by an overrapid atrial pacemaker, for example, may well have a reduced cardiac output even though the heart rate is increased. However, tachycardia is normally caused by sympathetic activation of the heart, and one of the effects of sympathetic activation on the myocardium is to make each systole briefer. This effect protects diastolic filling time when the heart rate is increased by sympathetic input. When combined with the increased force of contraction of the rapidly beating heart, this effect allows stroke volume to remain almost independent of heart rate until rates as high as 160 to 180 beats/min are reached (Figure 13-22). At rates this high, filling does become compromised, and further increases in heart rate do not increase cardiac output.

The large increases in cardiac output that normally occur with strenuous exercise could not occur as a result of cardiac stimulation alone; an increase in the ease with which blood flows through the circulatory system and returns to the heart must also occur. These changes are largely the result of intrinsic regulation of blood vessels by tissues and changes in the activity of sympathetic inputs to the the heart and vessels. The control of blood vessels by intrinsic and extrinsic factors will be described in Chapter 14. The interaction of the heart and vasculature that ultimately determines the mean arterial pressure and cardiac output will be examined in detail in Chapter 15.

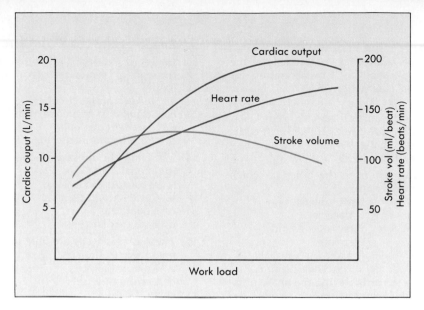

FIGURE 13-22
Changes in heart rate and stroke volume (right vertical scale) and their product, cardiac output (left vertical scale), as work load (horizontal axis) is increased during exercise. Note that heart rate increases continuously as the work load is increased, but stroke volume first becomes stable, then decreases as the heart rate becomes very high. At the highest work loads, cardiac output is limited by the decreasing stroke volume.

SUMMARY

1. In a complete circuit of the circulatory system, blood passes alternately through the **pulmonary loop** and the **systemic loop.** Blood pumped into the pulmonary loop by the right heart returns from the lungs to the left heart, which pumps it into the systemic loop.

2. The heart consists of right and left **atria** and right and left **ventricles.** The **atrioventricular valves** prevent backflow of blood from ventricles to atria; the **pulmonary** and aortic valves prevent backflow of arterial blood into the ventricles.

3. The three functional types of cardiac fibers are:
   a. **Contracting,** or myocardial, fibers, which generate the force of systole.
   b. **Pacemakers** located in the SA and AV nodes, which discharge action potentials at regular intervals.
   c. **Conducting** fibers specialized for impulse conduction, which carry excitation between nodes and from the AV node to the ventricle.

4. The rate at which pacemakers discharge action potentials is increased by sympathetic (adrenergic) input and decreased by parasympathetic (cholinergic) input. These effects are caused by increasing and decreasing, respectively, the rate at which K^+ channels close following each action potential.

5. Myocardial fibers possess "slow" channels for inward current of Na^+ and Ca^{++}, which are responsible for the **plateau phase** of the cardiac action potential. Ca^{++} that enters during the plateau phase triggers release of Ca^{++} from the sarcoplasmic reticulum. In contractile fibers, the cy toplasmic Ca^{++} concentration attained during the plateau phase determines the amount of force the contractile machinery can develop (contractility). Adrenergic (sympathetic) inputs increase the contractility.

6. The pump cycle of the heart consists of five phases:
   a. **Passive filling.**
   b. **Atrial contraction** (late diastole).
   c. **Isovolumetric ventricular contraction.**
   d. **Ejection.**
   e. **Isovolumetric ventricular relaxation.**

7. The length-tension relation of heart muscle is such that increased stretch by increased filling (increased end-diastolic volume) increases the force of systole and thus the stroke volume. This mechanism allows the heart to autoregulate its stroke volume to make cardiac output equal to venous return over a wide range of end-diastolic volumes (the **Frank-Starling Law** of the Heart).

8. The **cardiac output** is equal to the **stroke volume** multiplied by the **heart rate.** The major determinants of cardiac output are:
   a. The effects of adrenergic and cholinergic inputs to pacemakers.
   b. The **contractility** of the heart, determined largely by sympathetic input to contractile fibers.
   c. The **end-diastolic volume,** determined by the central venous pressure and the filling time between beats.

1. How is the force of contraction controlled in the heart? Increases in plasma Ca^{++} concentration increase the strength of contraction of cardiac muscle but not of skeletal muscle. Why is there a difference?

2. Describe the changes in aortic pressure, ventricular pressure, and atrial pressure over a single heart cycle. How do these pressure changes relate to the opening and closing of the cardiac valves?

3. Arrange the following events in the order in which they occur (some may overlap).

   QRS complex
   Atrial contraction
   Atrial repolarization
   T wave
   Isovolumetric relaxation

   Isovolumetric contraction
   First heart sound
   Ejection
   P wave
   Second heart sound

4. What is happening in the atria during the phase of isovolumetric ventricular contraction? Is ventricular pressure greater than or less than atrial pressure during this phase? Is it greater than or less than aortic pressure during this phase?

5. What properties of cardiac muscle membrane give rise to the plateau of the cardiac action potential? What is the functional significance of the plateau?

6. What events or processes in the electrical activation of the heart correspond to each of the following events of the ECG?

   P wave
   QRS complex
   T wave

   S-T interval
   P-R interval

7. What determines which population of spontaneously active cells will serve as the actual pacemakers for the heart cycle?

8. Define each of the following terms:

   Arrhythmia
   AV block
   Tachycardia

   Bradycardia
   Ectopic pacemaker

9. Diagram the pathway of electrical conduction of the heart. What is the role of gap junctions in spreading excitation? What is reentry? What property of heart cells normally prevents reentry?

10. What sequence of events occurs within myocardial cells as a result of activation of the β_1 receptors? What changes in heart performance result?

11. State the Frank-Starling Law of the Heart. What property of cardiac muscle is responsible for this law?

12. What factors affect the cardiac output?

Choose the MOST CORRECT Answer

13. In which of the following does the blood flow through only one set of capillaries in the systemic loop?
    a. Hepatic portal circulation
    b. Coronary portal circulation
    c. Renal circulation
    d. Hypothalamic-anterior pituitary circulation

14. _____ are the most abundant cardiac cells based on functional class:
    a. Myocardial cells
    b. Conducting cells
    c. Nodal cells
    d. Endocardial cells

15. These cardiac cells are able to generate regular action potentials in the absence of external stimuli:
    a. Myocardial cells
    b. Conducting cells
    c. Nodal cells
    d. Endocardial cells

16. Which of the following is TRUE of diastole?
    a. AV valves are closed.
    b. Semilunar valves are open.
    c. Passive filling of chambers occurs
    d. Ventricles contract at the end of this phase.

17. The upstroke of the action potential in SA node cells is in response to:Opening of voltage-gated Na^+ channels
    a. Opening of T Ca^{++} channels
    b. Opening of L Ca^{++} channels
    c. Opening of voltage-gated K^+ channels

18. Choose the correct conduction sequence:
    a. SA node → AV node → Bundle of His → Bundle branches → Purkinje fibers
    b. SA node → Bundle of His → Bundle branches → AV node → Purkinje fibers
    c. AV node → SA node → Bundle branches → Bundle of His → Purkinje fibers
    d. SA node → AV node → Bundle of His → Purkinje fibers → Bundle branches

19. In an ECG recording, the time between atrial depolarization and ventricular depolarization is represented by the :
    a. P wave
    b. QRS complex
    c. S-T interval
    d. P-R interval

20. This transmitter substance influences ONLY heart rate:
    a. Acetylcholine
    b. Epinephrine
    c. Norepinephrine
    d. G protein

● SUGGESTED READINGS

DAVID G, HIRST S, EDWARDS FR, BRAMICH N, and KLEMM MF: Neural control of cardiac pacemaker potentials, *News in Physiological Sciences* 6:185, August 1991. Describes what is currently known about how autonomic neurons affect cardiac pacemakers.

ERON C: Take heart: Ventricular tachycardia cure, *Science News*, August 1988, p. 133. Describes mechanisms of action of new antiarrhythmic drugs.

FRANCIS GS and COHN JN: Heart failure: mechanisms of cardiac and vascular dysfunction and the rationale for pharmacologic intervention, *FASEB Journal* 4(13):3068, October 1990. Describes remodeling of the ventricular myocardium in progressive heart failure and accompanying changes in blood vessels.

KATONA PG, MCLEAN M, DIGHTON DH, and GUZ A: Sympathetic and parasympathetic cardiac control in athletes and nonathletes at rest, *Journal of Applied Physiology* 52:1652, 1986. Discusses how cardiac performance is regulated and the effect of athletic training on the cardiovascular system.

LAKATTA EG: Excitation-contraction coupling in heart failure, *Hospital Practice* 26(7):85, 1991. A review of recent discoveries on the basis of the Frank-Starling law and its implications for heart failure.

OPIE LH: *The Heart—Physiology and Metabolism* (2nd edition). Raven Press, New York, 1991. One of the most recent comprehensive reference works on cardiac physiology.

ROBINSON TF, FACTOR SM, and SONNENBLICK EH: The heart as a suction pump, *Scientific American,* June 1986, p. 84. Presents evidence that suction contributes to ventricular filling.

VOGEL S: *Vital Circuits*, Oxford Univ. Press, New York, 1992. A very readable lay person's guide to the heart and circulation. The author is an expert in fluid dynamics of biological systems and wrote the book after his own recovery from a heart attack.

YATANI A and BROWN AM: Rapid beta-adrenergic modulation of cardiac calcium channel currents by a fast G protein pathway, *Science* 245:71, July 7, 1989. This study used the patch clamp method to discover one of the molecular mechanisms of sympathetic stimulation of myocardial cells.

CHAPTER 14

Blood and the Vascular System

On completing this chapter you should be able to:

- Identify the components of blood and define hematocrit
- Describe the process of blood clotting (hemostasis), noting the differences between the intrinsic and extrinsic pathways
- Appreciate the role of the arteries in maintaining blood flow into the vasculature during diastole
- Understand why veins are referred to as capacitance elements of the circulation
- Distinguish between linear velocity, average flow velocity, and flow rate
- Describe how vessel length, fluid viscosity, and vessel radius affect vessel flow resistance
- Understand the determinants of total peripheral resistance
- Describe the effects of extrinsic autonomic control on vascular smooth muscle
- Describe the factors that determine capillary fluid exchange
- Understand the anatomy and circulation of lymph

*T*he blood vessels are sometimes described as a "vascular tree." The metaphor is apt. The left heart pours its output into the aorta, the trunk of the arterial tree. Major arteries branch from the trunk of the aorta to course into each major body region—the head, arms, abdomen, and legs. Further branching forms successively smaller arteries. The final branches are arterioles, the smallest arteries, from which spring a profusion of capillaries, the twigs of the tree. At each successive branch the blood flow is further subdivided. Ultimately, a tiny fraction of the cardiac output passes through each capillary, bringing oxygen, nutrients, and hormones to each cell and removing carbon dioxide and metabolic end products. The veins form a second, reversed tree as capillaries merge to form first venules and then veins, finally joining the trunk of the vena cava. Thus the vasculature can be divided into a distributing system (the arteries and arterioles), an exchange system (the capillaries), and a collecting system (the venules and veins). Altogether, there are perhaps 100,000 kilometers of blood vessels in an adult.

The idea that blood circulates is relatively modern. Physicians in earlier times held ideas about the function of the cardiovascular system that seem quite bizarre now. At first, arteries were thought to contain air rather than blood because the arteries of cadavers sometimes filled with air when cut during dissection. By the time of the Greek physician Galen (200 AD), blood was thought to ebb and flow from the heart and to pass through imagined pores in the septum of the heart. The English physician William Harvey introduced the hypothesis of blood circulation in 1628 on the basis of careful observation of the heart, arteries, and veins. Harvey's ideas revolutionized physiology and were all the more remarkable because in his time the smallest blood vessels, capillaries, had not yet been seen. When the microscope was invented and used by Marcello Malpighi in 1661 to examine a frog's lung, it was finally possible for physiologists to observe the movement of blood through capillaries.

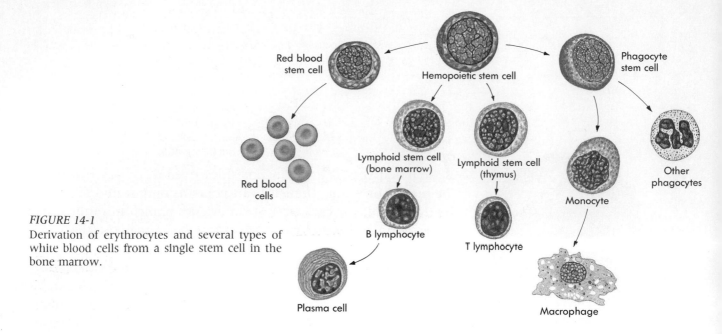

FIGURE 14-1
Derivation of erythrocytes and several types of white blood cells from a single stem cell in the bone marrow.

BLOOD: A LIQUID TISSUE
Blood Composition

- *What is the hematocrit?*
- *What are the respective roles of red cells, white cells, and platelets?*
- *What are the three stages of hemostasis?*

Blood consists of several types of cells suspended in a liquid called **plasma.** All of the cellular elements of blood have a common ancestral cell type, the hemopoietic stem cell, found in the bone marrow (Figure 14-1).

If a small, blood-filled tube is centrifuged, the plasma and cells are separated; the cells are packed in the bottom, whereas the less dense plasma occupies the top of the tube (Figure 14-2). A thin **buffy coat** of white cells (**leukocytes**) and **platelets** (**thrombocytes**) separates the packed **red cells** (**erythrocytes**) from the clear, straw-colored plasma. The **hematocrit,** the percentage of the total blood volume that is red cells, can be calculated by dividing the length of the column filled with red cells by the total length of the column of blood and multiplying the resulting fraction by 100. Normal hematocrit values range from 40% to 50% in men and 35% to 45% in women. These values correspond to cell densities of 5.1 to 5.8 million cells/mm^3 in men and 4.3 to 5.2 million cells/mm^3 in women (Table 14-1). Conditions in which the number of red blood cells is low are called **anemias,** whereas **polycythemia** refers to an abnormally high concentration of red blood cells.

Plasma components consist of water, inorganic ions, many organic compounds produced or consumed in metabolism, and **plasma proteins** (Table 14-1). Plasma proteins include **albumins,** which

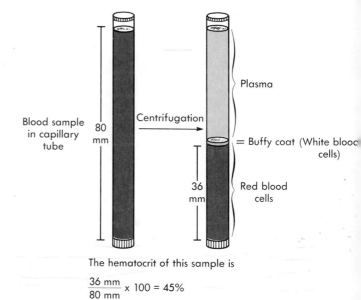

The hematocrit of this sample is

$$\frac{36 \text{ mm}}{80 \text{ mm}} \times 100 = 45\%$$

FIGURE 14-2
Separation of blood into cellular and fluid (plasma) components by centrifugation. A blood sample is placed in a glass capillary tube treated to prevent clotting. After centrifugation, the red cells are packed at the bottom of the tube with the white cells above them; the plasma remains at the top of the tube. The hematocrit level of this sample is calculated as shown.

serve as transport proteins for lipids and steroid hormones and are important in body fluid balance; **immunoglobulins,** which mediate specific immunity; and **fibrinogen** and several other proteins involved in the formation of blood clots.

Red cells (Figure 14-3, *A*) are disks that are 8.1 μm in diameter and about 2.7 μm thick with a concave center. About 25% of their cytoplasmic volume is taken up by hemoglobin, a protein that accounts

TABLE 14-1	Elements of the Blood	

Parameter	Normal range	Units
Cellular elements		
Hematocrit	40-54	
Hemoglobin	14-18	gm/dl
Red blood cells (erythrocytes)	4.6-6.2	Million/mm^3
White blood cells (leukocytes)	5000-10,000	Cells/mm^3
Platelets	0.2-0.4	Million/mm^3
Plasma components		
Water	91.5% of plasma volume	
Proteins	7.0% of plasma volume	
Albumin	3.2-5.6	gm/dl
Globulins	2.3-3.5	gm/dl
Fibrinogen	0.2-0.4	gm/dl
Ions		
Bicarbonate	21-27	mEq/L
Ca^{++}	2.1-2.6	mEq/L
Chloride	95-103	mEq/L
Iron	60-150	μg/dl
Magnesium	1.5-2.6	mEq/L
Phosphate	1.8-2.6	mEq/L
Potassium	4.0-4.8	mEq/L
Sodium	136-142	mEq/L
Sulfate	0.2-1.3	mEq/L
Cholesterol	150-250	mg/dl
Glucose	65-100	mg/dl
Other		
Urea	8-20	mg/dl
Uric acid	2.1-7.6	mg/dl

for the red color of blood (see Chapter 17). In hemoglobin the iron-containing **heme groups** that bind O_2 are attached to each of four subunit polypeptide chains. Hemoglobin is almost entirely responsible for O_2 transport in blood and plays an important role in CO_2 transport and regulation of blood acidity. Red cells lack a nucleus and cannot undergo mitosis. They have a lifetime of about 120 days in the circulation. Aging red cells are removed and destroyed by macrophages in the spleen and must be continuously replaced by formation of new ones from red blood stem cells (erythroblasts) in the bone marrow. This process, called **erythropoiesis**, requires adequate supplies of iron and also vitamin B_{12} (Figure 14-1).

An adult produces about 200 billion red blood cells each day, equivalent to the number of red cells in about 100 ml of whole blood, so that the 500 ml of blood lost in a typical blood donation can be replaced in less than a week. Erythropoiesis is regulated by the hormone, **erythropoietin**, which is released by the kidney and liver in response to decreases in the arterial partial pressure of oxygen. A deficiency in red blood cell formation can result in several types of anemia. **Pernicious anemia** is the result of a lack of vitamin B_{12}; **iron deficiency anemia** can be caused by inadequate iron intake or abnormal iron absorption and recycling; and **sickle-cell anemia** is the result of a genetic defect in hemoglobin formation. In kidney disease erythropoietin production can be impaired, again causing anemia.

The products of red blood cell disintegration in the spleen are processed in the spleen and liver cells. These macrophages remove heme groups from hemoglobin. The resulting globin is degraded into

A

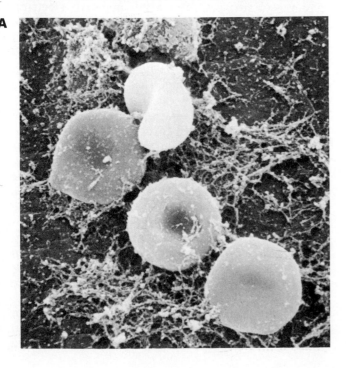

B

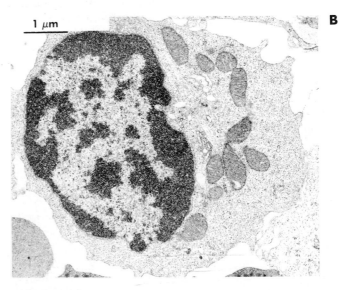

FIGURE 14-3
A A scanning electron micrograph of red blood cells trapped in a meshwork of fibrin threads.
B A transmission electron micrograph of a white blood cell.

TABLE 14-2

TABLE 14-2	Sites of Blood Cell Production During Different Life Stages

Life stage	Site of blood cell production
Fetus	Liver and spleen
Adolescent	Marrow of sternum, ribs, vertebrae, skull, pelvis, femur, and tibia
Adult	Marrow of sternum, ribs, skull, vertebrae, and pelvis

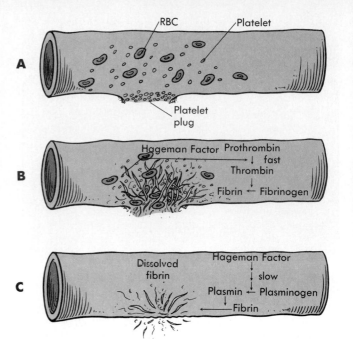

FIGURE 14-4

The sequence of events in formation and subsequent dissolution of a blood clot.
A A platelet plug is formed.
B Damage activates Hageman factor, which initiates a cascade of reactions (arrows) that ends with formation of a blood clot. Red and white blood cells and platelets are trapped among the fibrin strands of the clot.
C The clot is eventually dissolved by plasmin which is produced by a much slower reaction sequence also triggered by Hageman factor.

its constituent amino acids, while heme is converted to a compound called biliverdin. Biliverdin is then degraded to yield iron (bound to the plasma protein transferrin), which can be reused to form new hemoglobin, and **bilirubin.** Bilirubin combines with a plasma protein and is transported to the liver. Here bilirubin is conjugated with glucuronic acid and secreted as a component of the bile (see Chapter 22 for a further discussion of bile and iron recycling in the body). Bilirubin, also called a bile pigment, can accumulate in certain types of liver disease. In these conditions, bile pigment causes the body fluids and tissues to develop a yellowish discoloration referred to as **jaundice.**

Platelets are disks 2 to 5 μm in diameter, formed by the fragmentation of **megakaryocytes** in bone marrow. Like red cells, platelets lack nuclei and cannot reproduce. Their lifetime in the circulation is perhaps as little as 10 days.

In adults, red blood cells, platelets, and some types of white blood cells are formed in bone marrow (Table 14-2). Other types of white blood cells involved in the specific immune response develop in the lymph nodes. White blood cells (Figure 14-3, B) are agents of the immune system; their origin and function are discussed in detail in Chapter 26.

Hemostasis

After a person suffers a cut or scrape, blood may at first flow readily from damaged blood vessels, but after several minutes the flow slows or stops. Further blood loss is prevented by several processes, collectively called **hemostasis.** Hemostasis involves the action of substances produced by injured blood vessels, blood platelets, and a family of plasma proteins called clotting factors.

Hemostasis has three stages: (1) the initial sealing of a damaged vessel by a temporary **platelet plug,** (2) the formation of a **clot**—a network of threadlike **fibrin** molecules that forms a more lasting patch over the break in the blood vessel, and (3) the dissolution of the clot after vessel repair.

Formation of a platelet plug (Figure 14-4, A) requires a change in the behavior of platelets so that they are able to bind to the damaged surface of the vessel and to each other. The transformed platelets are called **sticky platelets.** Platelet sticking is triggered by exposure to an abnormal surface. Undamaged vessel surfaces carry a net negative electrical charge on their surfaces. Because platelets also have a negative surface charge, they are repelled by intact vessels. Injured vessel surfaces and the structural protein collagen have a net positive charge, which causes platelets that contact an injured surface to stick to it.

Binding to the damaged surface causes the platelets to release **arachidonic acid,** a fatty acid. In the plasma, some of the arachidonic acid is converted to **thromboxane A$_2$** (a prostaglandin). Thromboxane A$_2$ attracts additional platelets to the site of damage and causes the damaged vessel to constrict. It also causes platelets to release ADP. The ADP is a platelet-to-platelet signal that causes platelets to flatten, send out processes, and expose receptors for fibrin on their cell surfaces, making them sticky. This process is an example of positive feedback because a few sticking platelets can cause many more to become sticky.

The next event in the formation of a blood clot is the activation of the plasma protein **Hageman factor (factor XII)** by contact with a damaged area

FIGURE 14-5
A scanning electron micrograph of a blood clot, showing trapped blood cells.

of a blood vessel (Figure 14-4, *B*). Active Hageman factor triggers a series of reactions involving clotting factors (numbered I to XII) and, in many cases, Ca^{++} as an essential cofactor. Sticky platelets participate in what is termed the intrinsic pathway for blood clotting by releasing **phospholipid PF3**, a cofactor necessary for the activation of factor X. The fibrin formed in the clotting process glues the platelets to each other and traps them within the clot. This stage of the clotting cascade ends with the formation of prothrombin activator, which causes the conversion of **prothrombin** to **thrombin** (Figure 14-4, *B*). In the next step thrombin mediates a process in which many molecules of the soluble plasma protein, **fibrinogen**, interact (polymerize) to form long, insoluble strands of **fibrin**. Fibrin strands form a tight, meshlike lattice that binds the edges of the injured vessel together and traps platelets, erythrocytes, and leukocytes (Figure 14-5). Because several of the clotting factors require Ca^{++}, adequate calcium levels are required for the clotting process to be normal. The details of the clotting cascade are illustrated in Figure 14-6. There are two clotting pathways. The **extrinsic pathway** is initiated by chemical factors released from damaged tissue (Figure 14-6, *A*). Damaged tissues initiate the extrinsic pathway by release of **tissue thromboplastin**, or **factor III**, a complex mixture of lipoproteins. Factor III in the presence of Ca^{++} and activated factor V forms a complex with **factor VII** called **extrinsic thromboplastin**, which activates **factor X**. Stage 1 of the extrinsic pathway ends with the production of prothrombin activator.

The second, **intrinsic pathway**, requires only components present in blood (Figure 14-6, *B*). The intrinsic pathway is so named because it is initiated by the activation and aggregation of platelets together with the activation of a plasma coagulation factor, rather than by tissue factors. In addition to damaged blood vessel walls, various artificial surfaces such as glass can trigger the intrinsic pathway. The intrinsic and extrinsic pathways have a common end point, the production of prothrombin activator (Figure 14-6).

As in other regulatory system cascades (Chapter 5), the cascade of reactions that leads to clotting greatly amplifies the signal set in motion by tissue damage or sticking platelets. This is an example of the importance of positive feedback in hemostasis.

The clotting mechanism must be reliable because either failure of clot formation at an injury or spontaneous clot formation at an inappropriate site can result in death. The great number of reaction steps allows for much control over the clotting process, but it also provides many points at which the process can fail. Failure of clotting can result from a lack of or defect in any one of the factors in the cascade. For example, vitamin K is necessary for synthesis of some clotting factors, and a deficiency of this vitamin results in prolonged bleeding after injuries. **Hemophilia** is a genetic disease in which a single clotting factor is lacking or insufficient. Before modern methods of treating the disease were developed, most hemophiliacs died in childhood, typically from uncontrolled bleeding from minor internal or external injuries.

A clot that forms at an inappropriate site within vessels is called a thrombus, and can block the vessel in which it forms. Sometimes clots break loose from their sites of formation and travel within the circulatory system to block a smaller vessel. The general name for such a block is an **embolism.** If a clot forms in a coronary artery or a thrombus lodges in a coronary artery, the result is a **heart attack. If** either occurs in the brain, the result is a **stroke.** Clots that form in the systemic vasculature frequently lodge in the lungs, forming **pulmonary emboli.**

As shown in preceding paragraphs, clotting is strongly promoted by exposure of blood to an abnormal or injured tissue surface. In contrast, the normal endothelial cell actively protects undamaged vessels from inappropriate clot formation by synthesis of several natural **anticoagulants**—substances that interrupt or oppose clot formation. One important anticoagulant secreted by endothelial cells is **prostacyclin,** an antagonist of the clot-promoting prostaglandin thromboxane A_2.

A second anticoagulant mechanism of normal endothelial cells is the presence of a negatively-charged proteoglycan, **heparin**, bound to the cell surface. Because of its electrical charge, heparin op-

Hemophilias: Inherited Defects in Clotting Factors

Several inherited diseases of blood clotting involve defective genes for clotting factors. Hemophilia A, the most common form, results from the presence of a defective factor VIII protein; a less common form, hemophilia B, results from a defective factor IX protein. In severe cases of the disease, spontaneous bleeding into the joints, skin, and soft tissues is frequent. Without effective treatment, death usually occurs at an early age from complications resulting from internal bleeding. Bleeding episodes can be treated by transfusion of plasma fractions rich in clotting factors. This treatment only lessens the bleeding, and patients ultimately develop an immune response to the proteins of foreign plasma, so its use has to be reserved for the severest episodes.

Hemophilia almost always affects males but is inherited through maternal lines; such diseases are said to be sex linked. The basis for sex linkage lies in the genetics of sex determination. Human cells possess 23 pairs of chromosomes. One member of each pair is inherited from each parent. Generally the pairs are similar in appearance, but in the case of the sex chromosomes the pair is similar only in females, who carry two X chromosomes. In males the pair consists of an X chromosome and a much smaller Y chromosome, which carries few genes. The genes for factors VIII and IX are carried on the X chromosome. Defective genes for these factors are rare in the population. Possession of only one normal gene is enough to supply the needs of the body for normal factors, and a woman rarely inherits a defective gene on both the X chromosome from her mother and that from her father. Males have no such backup. If a male's single X chromosome, inherited from his

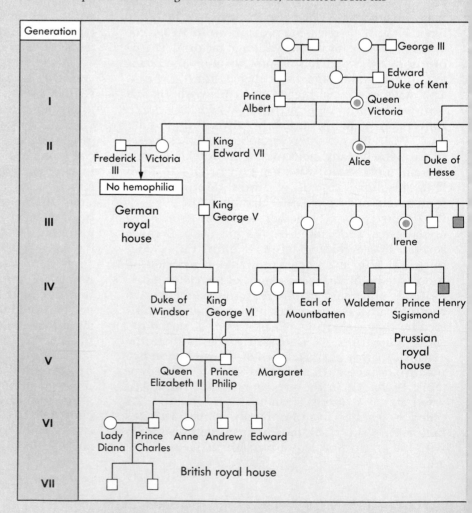

mother, bears a defective gene, only the defective amount of factor is made. A mother who carries one X chromosome with the defective gene has a 50:50 chance of transmitting the disease to a male child and a 50:50 chance of passing her carrier status to a female child.

The descendants of Queen Victoria present an interesting example of inheritance of hemophilia A. The defective gene in this family apparently arose as a mutation in one of Queen Victoria's parents or at an early embryonic stage in the queen herself. Relatively complete pedigrees, termed family histories, have been kept for royal families, and royal medical problems receive close attention. Consequently, we know that of Queen Victoria's five female children two were carriers of the gene. Their marriages to other European royalty transmitted the gene to several royal families (Figure 14-A), including that of Czar Nicholas II of Russia. Nicholas' son, Alexis, inherited the defective gene; his illness was a significant factor in the events leading to the Russian revolution. Of Victoria's four sons, one affected son survived long enough to pass the trait to descendants; an unaffected son became King Edward VII, establishing the current line of succession, which does not carry the defective genes.

Women who have a family history of hemophilia may want to know whether they carry the defective gene. Usually this can be determined by testing the woman's plasma for the defective clotting factor. Genetic counseling can help those who carry the defective gene to decide whether to bear children. Cures for such inherited errors of protein structure are a major promise of genetic engineering.

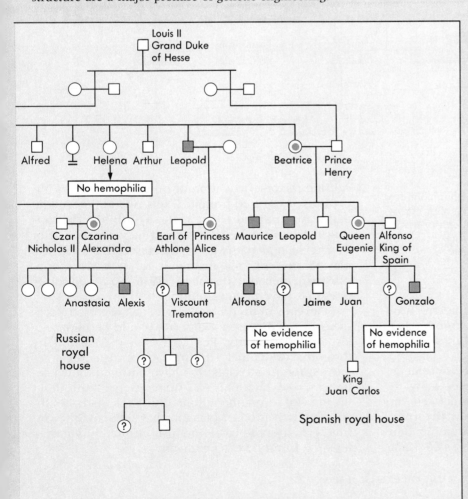

FIGURE 14-A

Family tree (pedigree) of descendants of Queen Victoria showing the sex-linked inheritance of hemophilia A. The circles represent female members of the family; squares represent males. Each branch in the family tree represents one generation of offspring. Filled squares indicate males that have hemophilia. Circles with a dot indicate women assumed to be carriers of the defective factor.

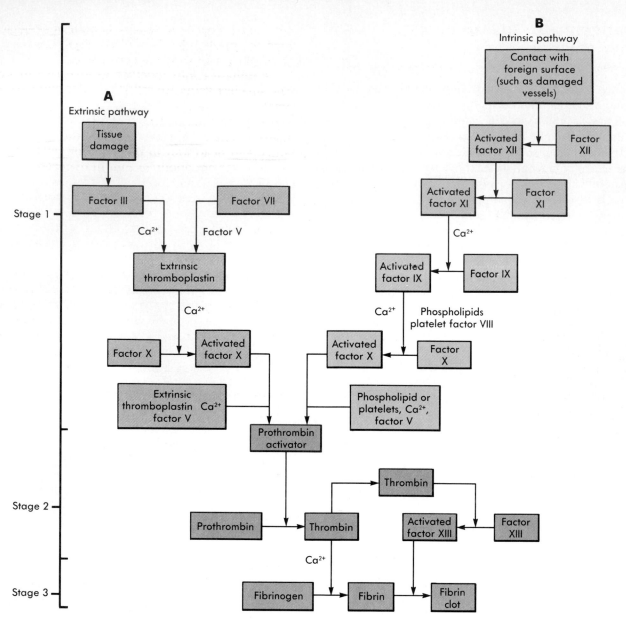

FIGURE 14-6

A The extrinsic pathway, which is stimulated by tissue damage.

B The intrinsic pathway for clot formation within blood vessels. Both these pathways cause the production of prothrombin activator at the end of stage 1 of coagulation. Stage 2 involves the production of thrombin, which then causes the polymerization of fibrinogen into fibrin in stage 3.

poses platelet sticking. Heparin is abundant in organs that are especially treatened by thrombi, such as the lung, and can be extracted from these tissues for use as an anticoagulant drug.

A third mechanism that protects the normal endothelial surface against clots is triggered by the clot-promoting enzyme thrombin. Normal cell surfaces carry a thrombin-binding protein, **thrombomodulin.** When thrombin binds to thrombomodulin, the complex activates **protein C.** As for some

clotting factors, activation of protein C requires Vitamin K. Activated protein C sets two anticoagulant pathways in motion. One pathway inhibits the clotting mechanism by destroying a key factor, Factor V. The second pathway stimulates conversion of a plasma protein, **plasminogen** to its active form, **plasmin.** Plasmin dissolves fibrin, so it dissolves clots that have begun to form.

Once an injured blood vessel has repaired itself, the continued presence of a clot would be both useless and dangerous. The activation of plasmin by thrombin and other clotting factors also helps remove appropriate clots that have outlived their usefulness (see Figure 14-4). Generation of thrombin during clot formation automatically results in a subsequent activation of plasmin that prevents the clot from growing too large and also ultimately removes it as the injured vessel recovers.

STRUCTURE OF BLOOD VESSELS
Types of Blood Vessels

- *What are the distinguishing features of each of the following vessel types?*
  - *Arteries*
  - *Capillaries*
  - *Arterioles*
  - *Veins*
  - *Metarterioles*
- *What is compliance?*
- *What is the significance of the difference in compliance between arteries and veins?*
- *Why can the veins be termed capacitance elements?*

The largest artery of the systemic circulation is the aorta, with a diameter of 20 to 30 mm (Table 14-3). The aorta branches to form arteries, which carry blood to individual organs and body regions. Further branching of the major arteries gives rise to small arteries, which are barely visible to the naked eye. Still further branching gives rise to **arterioles** (Figure 14-7, *A*), whose diameter of about 70 μm places them on the microscopic scale (Table 14-3). Each arteriole usually branches into several **metarterioles** of 10 to 20 μm in diameter, the walls of which are wrapped with smooth muscle at intervals along their length. Metarterioles branch to form **capillaries** (Figure 14-7, *B*), thin-walled vessels that are 5 to 10 μm in diameter, just wide enough for red cells to pass through in single file. At the origin of each capillary is a cuff-like **precapillary sphincter** consisting of a single smooth muscle fiber. Precapillary sphincters open and close their capillaries at intervals in response to changes in their immediate environment.

The capillaries are the sites of exchange of nutrients and wastes between tissues and blood. The capillaries arising from a single arteriole typically form a network called a **capillary bed** that serves a discrete area of tissue. Capillary density varies greatly from one tissue to another and is usually related to the maximum metabolic needs of the tissue concerned. Capillary density is high in skeletal and cardiac muscle, glands, and the CNS, but very low in cartilage and subcutaneous tissue.

Capillaries join together to give rise to **veins.** The smallest veins are called **venules** and have a diameter of about 20 to 30 μm. The veins draining the lower thorax, abdomen, and legs join to form the inferior vena cava; those draining the upper thorax, arms, and head join in the superior vena cava. The venae cavae are about 13 mm in diameter.

Generally, blood must pass through capillaries to go from arteries to veins, but two types of vessels allow blood to bypass capillary beds. In some cases, metarterioles make connections directly to venules (Figure 14-8, *A*). Such connections are particularly common in skeletal muscle. In some capillary beds there are even more direct pathways for blood to bypass the capillaries. These direct pathways are called **arteriolar-venular shunts** (Figure 14-8, *B*). Shunts can be found between arterioles and venules and also between small arteries and veins. Arteriolar-venular shunts are a prominent feature of the cutaneous microcirculation.

Walls of Arteries and Veins

Arteries and veins, viewed in cross section in Figure 14-9, *A*, *B*, consist of a coat of longitudinal smooth muscle, the **tunica externa,** surrounding an inner coat of circular smooth muscle, the **tunica media.** The inner surface is lined with a single layer of endothelial cells, the **tunica intima.** The arteries and veins differ in that at intervals along the length of veins, the tunica intima forms one-way valves. This favors blood flow to the heart in regions where the veins are subject to periodic compression, most notably the skeletal muscles. Also, the thickness of the smooth muscle of arteries is considerably greater, and the lumen is smaller than that of corresponding veins.

Compliance is the term physicists use to describe how much strain (the change in shape of an object) a given force causes when it is applied to the

---

TABLE 14-3   Scale of the Systemic Blood Vessels

Vessel type	Pressure* (mm Hg)	Length (mm)	Typical diameter	Number in parallel	Area (cm²)	Relative resistance (%)
Aorta	100	500	25 mm	1†	2.5	4
Arteries	98	10	2 mm	600	5	5
Arterioles	90	3	70 μm	5×10^7	40	40
Capillaries	35	1	5 μm	10^9	1700	25
Venules	15	10	30 μm	10^8	375	5
Venae cavae	5	300	13 mm	2	10	2

*The pressure is given at the start of each vessel type. †The aorta branches after leaving the heart to give rise to the carotid arteries, subclavian arteries, and thoracic aorta.

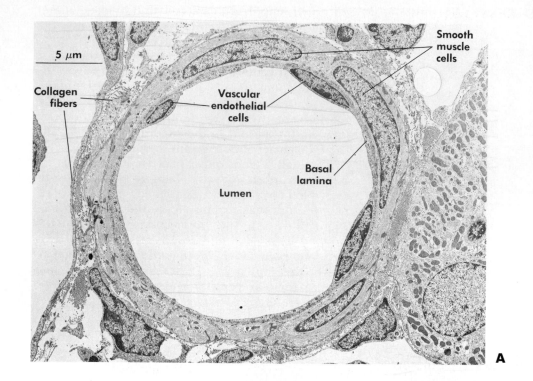

5 μm

Collagen
fibers

Smooth
muscle
cells

Vascular
endothelial
cells

Basal
lamina

Lumen

A

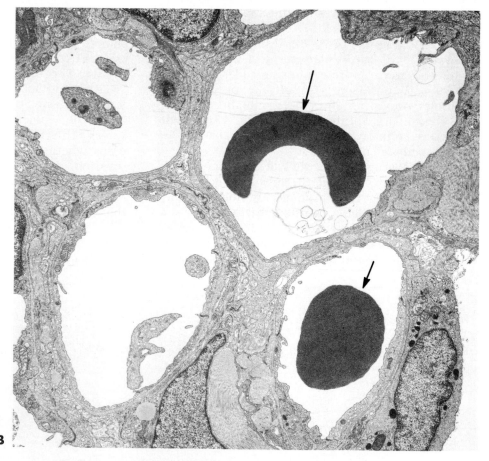

B

FIGURE 14-7
A Transmission electron micrograph of an arteriole.
B Transmission electron micrograph of several capillaries, two of which contain red blood cells *(arrows).*

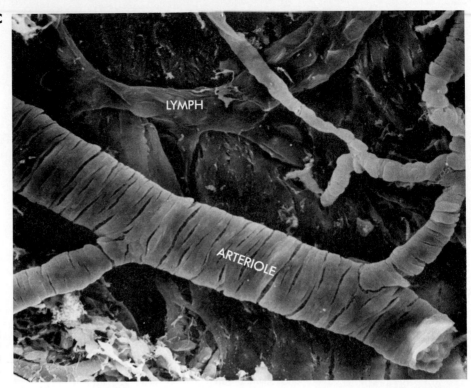

LYMPH

ARTERIOLE

FIGURE 14-7, cont'd
C A scanning electron micrograph of an arteriole and a lymphatic vessel. Note the presence of numerous individual smooth muscle cells surrounding the arteriole.

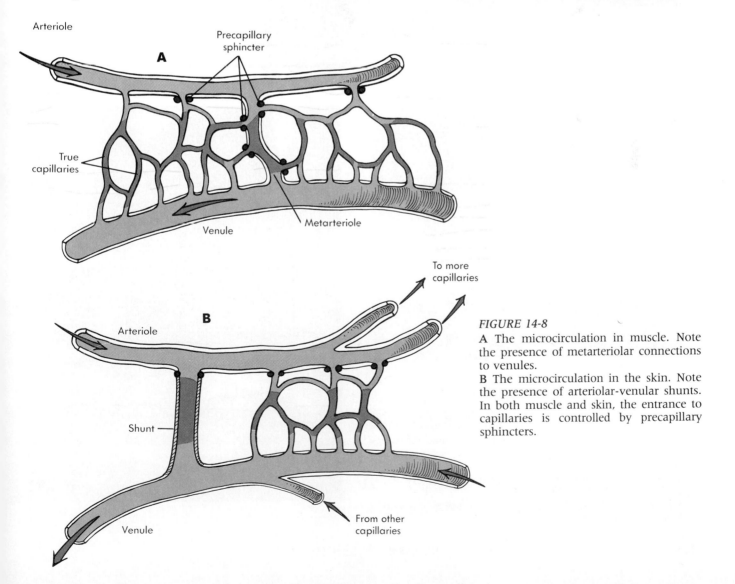

Arteriole

Precapillary sphincter

A

True capillaries

Venule

Metarteriole

To more capillaries

B

Arteriole

Shunt

Venule

From other capillaries

FIGURE 14-8
A The microcirculation in muscle. Note the presence of metarteriolar connections to venules.
B The microcirculation in the skin. Note the presence of arteriolar-venular shunts. In both muscle and skin, the entrance to capillaries is controlled by precapillary sphincters.

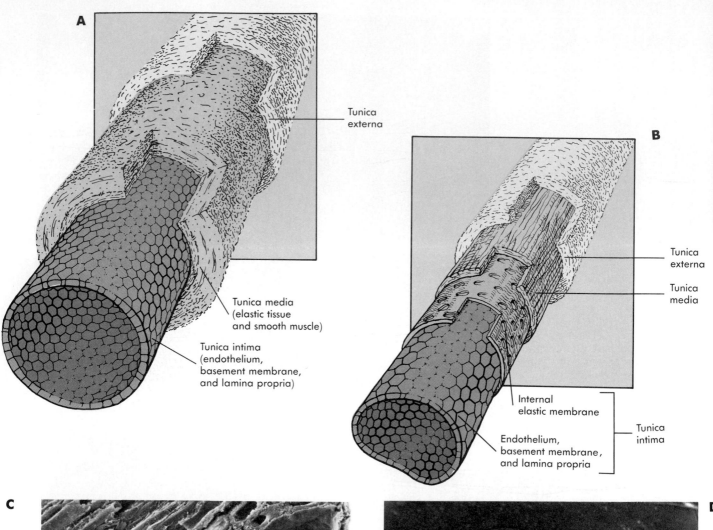

A

Tunica
externa

Tunica media
(elastic tissue
and smooth muscle)

Tunica intima
(endothelium,
basement membrane,
and lamina propria)

B

Tunica
externa

Tunica
media

Internal
elastic membrane

Tunica
intima

Endothelium,
basement membrane,
and lamina propria

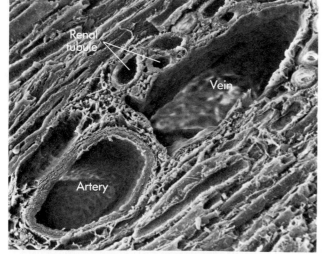

C

Renal
tubule

Vein

Artery

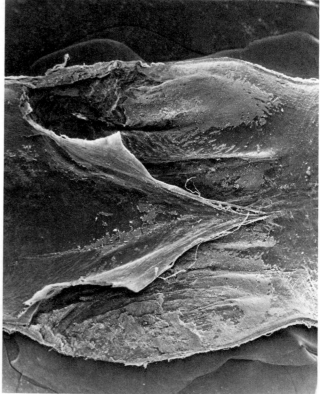

D

FIGURE 14-9

Structure of **A** arterial, and **B** venous walls. The tunica intima is composed of a basement membrane, a single thickness of endothelial cells reinforced with connective tissue. The tunica media is composed of elastic tissue and circular smooth muscle, and the tunica externa of longitudinal smooth muscle.

C A scanning electron micrograph comparing an artery and a vein.

D A scanning electron micrograph of the interior of a vein showing venous valve.

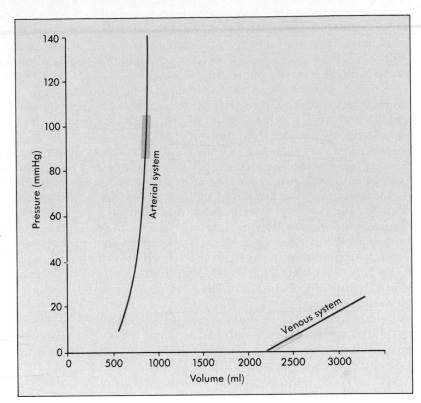

FIGURE 14-10
Pressure-volume curves for the arterial system and the venous system. The colored portions of the curves indicate approximate physiological ranges. Arterial pressure rises rapidly with increased volume, so that at the normal arterial volume of about 700 ml, the mean arterial pressure is almost 100 mm Hg. In the much more compliant venous system, the normal volume of more than 2 liters is held under a central venous pressure of only a few mm Hg.

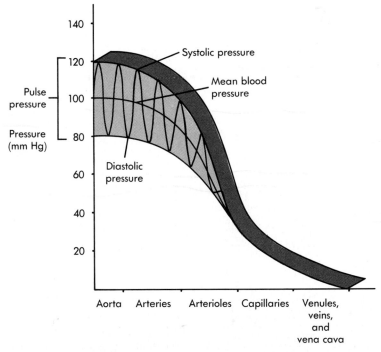

FIGURE 14-11
Blood pressure in each of the major blood vessel types.

object. For example, a force applied to the ends of a rubber band will stretch it. Stretch is one type of strain. Since the change in length is large, the rubber band is said to have a high compliance. In contrast, immense forces are needed to deform a steel girder, and steel has a very low compliance. Here we apply the concept of compliance to the cardiovascular system. Later you will learn that a similar concept is used to describe the mechanical properties of the lung.

The connective tissue of the walls of arteries and veins contains a mixture of the proteins, **collagen** and **elastin**. Elastin is easily stretched; collagen is stiffer. The proportions of the two strongly influence the **compliance,** or ease of stretch, of the vessel wall. The tone, or state of contraction, of the vascular smooth muscle is a second major determinant of a vessel's compliance. The compliance of the vessel wall determines the rate at which pressure inside a vessel increases as the volume of blood con-

Blood and the Vascular System   **403**

tained in it is increased. A vessel with high compliance can accept a large volume of blood with a relatively small increase in pressure; forcing the same volume of blood into a vessel with low compliance would produce a larger change in pressure.

Arterial walls have a relatively high collagen content, which makes the arteries considerably less compliant than the veins (Figure 14-10). The low compliance of arteries has an important role in maintaining blood flow through the circulation during diastole. During the ejection phase of the heart cycle, blood is entering the aorta much more rapidly than it is able to flow away into the rest of the vasculature. The aorta is stretched by the additional volume, but because its compliance is low, a large fraction of the energy applied by the heart to the blood during systole is stored in the stretched wall of the aorta and large arteries. During diastole, the **elastic recoil** of the aorta and large arteries slowly releases this stored energy, maintaining a relatively high mean arterial pressure and a relatively constant flow of blood through the capillaries. Thus the arteries and arterioles tend to smooth out the pulsatile flow of blood from the heart (Figure 14-11).

The arteries must be neither too compliant nor too rigid. The effect of too little compliance is illustrated by the effect of the increase in collagen content that occurs with age. The arteries of a 70-year-old person are about half as compliant as those of a young adult. As a result, the heart of a 70-year-old person does considerably more work to eject the same stroke volume than does the heart of a 20-

year-old person. This decreasing compliance of arteries is partly responsible for the increase in normal arterial blood pressure that occurs with age. The normal blood pressure for 20-year-old men is 123/76; that for 70-year-old men is 145/82. An even larger increase over this age interval is normal for women.

The volume of blood in the veins is about two thirds the total blood volume (Figure 14-12), but the venous pressure is only a few mm Hg. This is possible because the walls of veins contain a high proportion of elastin to collagen and thus are very compliant (Figure 14-10). Because they contain so much of the blood volume, small changes in venous compliance have dramatic effects on arterial blood pressure. The veins are often called the **capacitance elements** of the circulatory system because their capacity to hold blood is six times greater than that of the arteries.

Structure of Capillaries

Capillaries (Figure 14-13; Figure 14-7, B) are not miniature arteries or veins; they lack the smooth muscle coats of other vessels. Generally, capillaries are permeable to solutes and fluid. The structure of the capillary endothelium, and thus the permeability, differs significantly in different organs. The capillaries of most organs have closely joined endothelial cells whose intercellular junctions allow water and small hydrophilic solutes to pass readily between plasma and interstitial fluid. Typical capillaries have a low permeability to molecules the size of

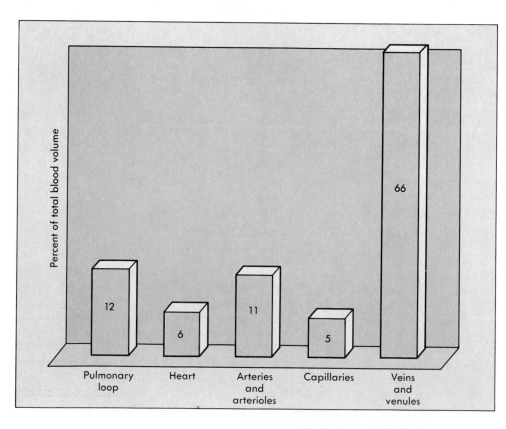

FIGURE 14-12
The relative volumes of blood in different parts of the circulatory system at any instant (values are typical of resting adults).

THE CARDIOVASCULAR SYSTEM

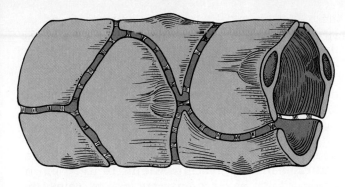

FIGURE 14-13
The structure of a generalized capillary. The spaces between endothelial cells are exaggerated in this diagram; they are normally about 4 nm wide, and the capillary wall is about 1 μm thick.

plasma proteins. The capillaries of most regions of the central nervous system have a low permeability even to small organic solutes. The capillary endothelium of these capillaries constitutes a **blood-brain barrier.** In the kidney, endocrine glands, intestine, and liver, capillaries may have large pores (**fenestrated capillaries**; see Figure 19-11) or even gaps in the endothelium (**discontinuous capillaries**).

MECHANICS OF BLOOD FLOW
Definition of Blood Flow

- *What are the three measurements of flow?*
- *What is the general relationship among driving force, flow rate, and resistance.*
- *What specific parameter of the systemic loop corresponds to driving force? Flow rate? Resistance?*
- *What three factors determine the flow resistance of a single vessel?*
- *Which vessels contribute most to the total peripheral resistance?*

Flow may be used in three different contexts: (1) as **flow,** (2) as **average velocity,** and (3) as the **linear velocity** of a small element of fluid. The flow rate is the total volume of fluid that passes a given point per time unit. In the cardiovascular system, cardiac output and the blood flow through the kidneys are examples of different flows. Flow can be determined by collecting and measuring the blood leaving a blood vessel per unit time.

Average velocity of flow is defined as the flow at a given point divided by the cross-sectional area. Average velocity indicates how much time blood spends in particular vessels of the circulatory system. The units of average velocity of flow are volume/(time) (area). The linear velocity is the distance travelled by small volume of blood per unit time. The units of linear velocity are distance/time.

The distinctions among the three terms can be illustrated by the flow of water in a river. If there are no tributaries and if evaporation and rainfall are neglected, the flow in the river must be the same at all points along its length. Where the banks are close together, the cross-sectional area is reduced, and rapids—areas of increased average velocity—are the result. Where the river is broad, the cross-sectional area through which flow is occurring is great, and water motion may be unnoticeable. At any given point along the river's course, not all of the water molecules are moving with the same velocity. The current, which represents the linear velocity of a small element of water in the river, is swiftest near the river's center and slowest near the bank. The difference results from the fact that frictional forces at the bank of the river hinder movement of water molecules close to the bank. These molecules in turn slow down others farther out, and so on. As the distance from the bank increases, the bank's influence diminishes and linear velocity increases.

Flow in blood vessels is almost always laminar—each element of the stream flows in a straight line without swirling or mixing across the axis of the vessel (Figure 14-14, A). A difference between the flow of blood in vessels and the flow of water in a river is that in rivers the flow is frequently turbulent. This is evidenced by the formation of eddies. The tendency to form eddies increases as the size of the stream and its velocity increase. This tendency result from irregularities in the river's banks. Even the largest blood vessels in the body do not experience turbulent flow at normal average flow rates. However, turbulent flow does occur in the vicinity of the pulmonary and aortic valves, and it may be increased by surface irregularities or by stenosis (narrowing) of the valves (Figure 14-14, B). Increased turbulence causes the aorta to vibrate audibly and is one cause of heart murmur.

If each of the elements of the vascular system were merged into one single vessel of corresponding total volume and net cross-sectional area, the result would look something like Figure 14-15. In this figure, arteries and arterioles have been lumped together for simplicity, as have venules and veins. The flow rate (Q) must remain constant throughout the system. Flow in each of the three elements of the system is the product of the total cross-sectional area of each class of vessels $(A_1, A_2, \text{ and } A_3)$ and the average velocity of flow in those vessels $(V_1, V_2, \text{ and } V_3)$. The constancy of the product of area and average velocity is called the **continuity equation of flow.** Continuity of flow means that the larger the total cross-sectional area, the lower the average flow velocity.

The average velocity of blood in different types of vessels depends on the total cross-sectional area presented by each vessel type. Average velocity through arteries is rapid, so blood spends little time in passing from the heart to the capillaries. Because the overall cross-sectional area of capillaries is large, blood in capillaries has the lowest velocity of all the vessels in the circulation. A typical systemic capil-

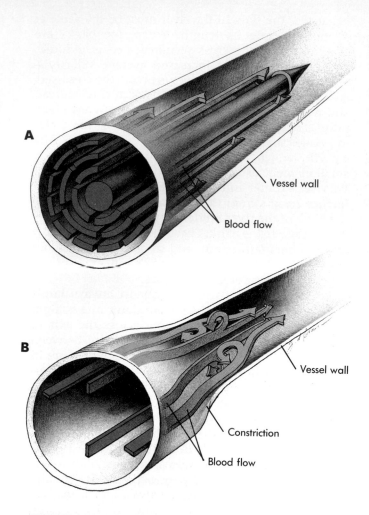

FIGURE 14-14

A Laminar flow. Fluid flows in long smooth-walled tubes as if it were composed of a number of concentric layers. Velocity is zero in the thin shell of fluid immediately next to the wall. It increases progressively in more central layers, rising to a maximum in the center.
B Turbulent flow. Turbulent flow is caused by numerous small currents flowing crosswise or oblique to the long axis of a vessel. It is more likely to occur at points where flow is high and there are abrupt changes in diameter, such as near the valves of the heart.

lary has a length of approximately 1 mm, and the flow velocity is about 1 mm/sec, so a red cell spends about 1 second in a capillary.

Capillary walls consist of a single layer of endothelial cells (Figure 14-13), the minimum possible barrier to diffusion. Diffusional equilibrium of the tissues with the blood in a capillary is achieved because the permeability of the capillary walls is extremely high, the blood is moving relatively slowly, and capillaries approach to within 10 to 30 μm of every cell. As blood enters venules and veins, the average velocity increases to a value similar to, but less than, that of arteries. The pattern of flow through the circulatory system is characterized by a rapid movement of blood to and from capillaries and by a slow movement of blood through capillaries (Figure 14-16).

Relationship of Blood Flow to Driving Force and Resistance

The laws of fluid mechanics apply to the flow of blood through the circulation and to the distribution of the cardiac output among the different organs. Blood flow through a blood vessel results from a driving force, in the form of a pressure gradient. The **perfusion pressure** along the length of any blood vessel is given by the difference between the pressure at the proximal end of the vessel and the distal end. The relationship of blood flow to perfusion pressure follows the general form of force-flow relationships described in Chapter 2:

Blood flow = Perfusion pressure/flow resistance

Using this equation, the flow resistance can be calculated from the pressure gradient between two points and the resulting blood flow between the same two points. The decrease in the pressure with distance in a blood vessel is therefore proportional to the vessel's flow resistance. Think of the resistance of a circulatory element as "using up" the

FIGURE 14-15

The principle of continuity of flow. Because there is no place where a net fluid gain or loss occurs, the total flow *(Q)* must remain constant. Flow is the product of average velocity *(V)* and area *(A)*. Therefore, as the cross-sectional area increases (for example, going from arterioles to capillaries), average linear velocity of flow must decrease *(V₁ to V₂)*.

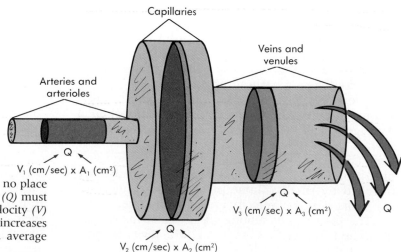

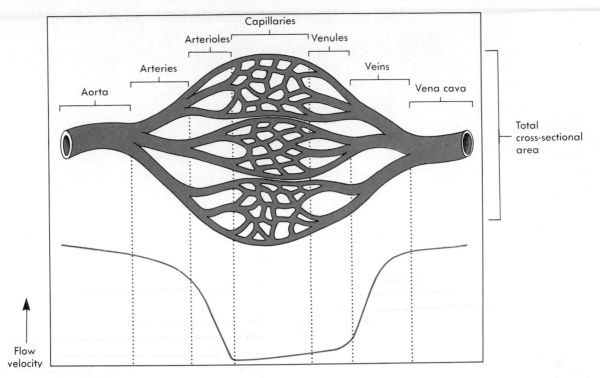

FIGURE 14-16
The relationship between net cross-sectional area and average velocity of flow in different vessel types.

driving force of the blood that travels through it. The greater the resistance of an element, the greater the fraction of the arteriovenous pressure gradient that is used up as blood passes through it.

The relationship between blood flow, perfusion pressure, and resistance can be restated in terms that apply to the entire vascular system acting as a unit. The mean arterial pressure (MAP) is a time-weighted average of the arterial pressure over the heart cycle. The driving force for blood flow through the systemic loop is the difference between the MAP and the central venous pressure (CVP) which is the same as right atrial pressure. (Because the venous pressure does not change significantly over the heart cycle, it need not be averaged.) The total peripheral resistance (TPR) is the net flow resistance of the systemic loop. The relationship between flow through the systemic loop (Q), the driving force (MAP - CVP), and total peripheral resistance (TPR) follows the form of the general relation between force, resistance, and flow stated above:

$$Q = (MAP - CVP)/TPR$$

According to the Frank-Starling Law, the heart pumps out all the blood that comes to it, and the cardiac output in a steady-state equals the venous return as averaged over times longer than a few heartbeats. Thus in this equation, **Q** can stand equally well for the cardiac output or the venous return.

Effects of Vessel Dimensions and Branching on Resistance and Flow

The flow resistance of individual segments of the vasculature depends on the geometry of the vessels involved. The physics of flow through a collection of rigid tubes of different diameters is well understood. Even though arteries and veins are not rigid tubes, but living tissue, the same principles can be applied because the changes in their diameters produced by the pressure changes within them are small compared to the diameter itself.

The simplest case is a vessel of uniform diameter that does not branch. The flow resistance of such a vessel is given by the equation:

$$R = L\eta/r^4$$

where **R** is flow resistance, **L** is the length of the vessel, is the viscosity, or thickness, of blood, and **r** is the radius of the vessel. Of the factors in the equation, the vessel radius is the most powerful determinant of flow resistance because it is raised to the fourth power, so small changes in radius cause large changes in resistance. For example, if the radius of a vessel is halved, its flow resistance is increased by 2^4, or 16 times. The vessel length is a less powerful factor in the equation, so it is possible for major arteries and veins to be quite long and still have relatively low resistance as compared with that of shorter but narrower vessel types. Radius changes are an important aspect of the control of blood flow and pressure; vessel length changes are

FOCUS ON PHYSIOLOGY

Measurement of Blood Pressure

Systemic arterial blood pressure can be measured by several techniques. In animal experiments and during cardiac catheterization or heart surgery in human patients, pressure at various points in the vasculature or even within the heart itself can be measured by inserting a tube, or cannula, into a blood vessel and connecting this tube to a blood pressure transducer, which generates an electrical signal that can be recorded using a polygraph (Figure 14-B, *1*). The pressure transducer responds to rapid changes in pressure such as those which occur during systole.

In the classical method of auscultation, a pressure cuff is inflated around a major peripheral artery, usually the brachial artery in the arm, and connected to a pressure-measuring device known as a sphygmomanometer (Figure 14-B, *2*). Initially the cuff is inflated so that cuff pressure exceeds arterial systolic pressure, collapsing the brachial artery and preventing flow at all stages of the heart cycle. The pressure in the cuff is then gradually decreased. As cuff pressure reaches the arterial systolic pressure, blood can spurt through only when the arterial pressure is at its highest value. Each spurt is heard as a tapping sound (called the first Korotkoff sound) in a stethoscope placed on the arm distal to the cuff. The first sound corresponds to the arterial systolic pressure, typically 120 mm Hg. As the cuff pressure is decreased further, more blood passes through at each beat. The resulting sound first changes from a tapping to a louder thud and then becomes muted as the cuff pressure continues to decrease. When the cuff pressure falls to arterial diastolic pressure, typically 70-80 mm Hg, the brachial artery is not closed by the cuff at any stage of the heart cycle. The sounds become inaudible at the diastolic pressure.

---

usually small and unimportant in control of the vasculature.

The viscosity of a fluid is a measure of the internal work necessary to make the fluid flow. A practical example of the effect of viscosity is the extra work that is needed to get ketchup or salad dressing to flow from a narrow-mouthed bottle, from which less viscous fluids such as water would flow freely. The viscosity of blood is determined by the composition of blood, the nature of the vessel in which it is flowing, and the mean flow velocity. Plasma has a viscosity about 3, twice that of water; whole blood has a viscosity approximately three times that of water. Thus red blood cells contribute significantly to blood viscosity, and an increase in the hematocrit level increases the work the heart must do to maintain the cardiac output.

When blood vessels branch, parallel pathways for blood flow result. Branching of major arteries causes the circulations of almost all organs to parallel each other (Figure 14-17). Within individual organs (with few exceptions), all arteries parallel each other, all capillaries parallel each other, and so on. The effect of branching on the net flow resistance depends on two factors: (1) the radii of the branches, and (2) the number of branches.

The key to understanding how vessels connected in parallel behave is the fact that the pressure gradient is the same across every vessel of the parallel combination. The flow through any single vessel (Q_1, Q_2, and Q_3 in Figure 14-18, *A*) is inversely proportional to its resistance. The total flow through the ensemble of parallel vessels is the sum of each of the individual flows. Each element of

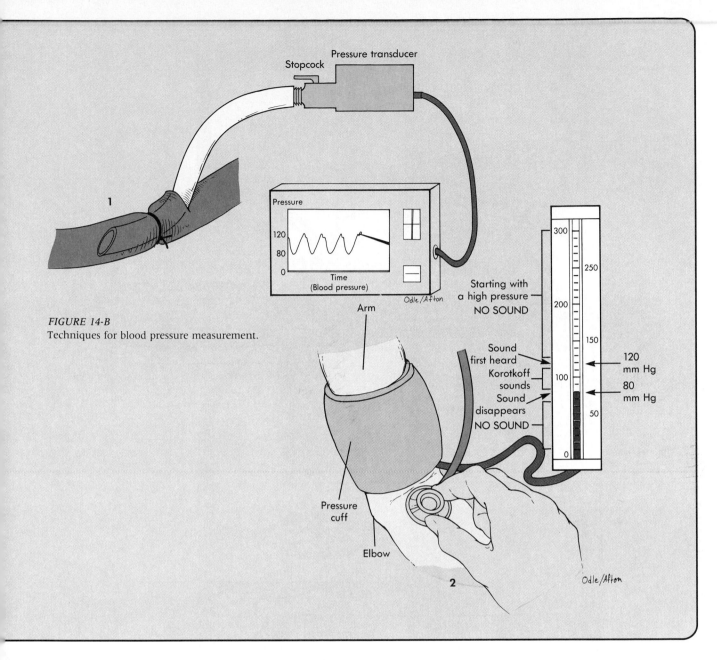

FIGURE 14-B
Techniques for blood pressure measurement.

fluid may take one of several possible pathways and effectively experiences only the resistance of the particular vessel it enters. The total flow resistance of a set of parallel vessels is always less than the flow resistance of a single branch vessel because fluid has "somewhere else to go"; that is, to another parallel vessel.

An analogy can be made with traffic patterns: if a large volume of freeway traffic passes through parallel toll booths, each individual driver will encounter the delay (resistance to flow) only of his or her own pathway in the parallel array. Whether the flow of traffic will be slowed or not depends on two factors: (1) how many booths there are, and (2) how rapidly the toll can be collected from vehicles passing each booth.

To apply this analogy to blood vessels, suppose

a stem vessel gives rise to a number, N, of identical branch vessels, each with a resistance, R' (Figure 14-18, B). A single drop of blood approaching the branch vessels may take any one of N identical paths. The total flow in the entire ensemble of branches is N times the flow through any single branch. But the pressure difference between the ends of the ensemble is the same as the pressure difference between the ends of each vessel. The flow resistance of the parallel combination of N vessels of resistance R' must therefore be N times less (R' divided by N) than the single vessel resistance R'. That is:

$$R(total) = R'/N$$

As in the analogy of toll booths on the freeway, when a stem blood vessel branches into vessels of

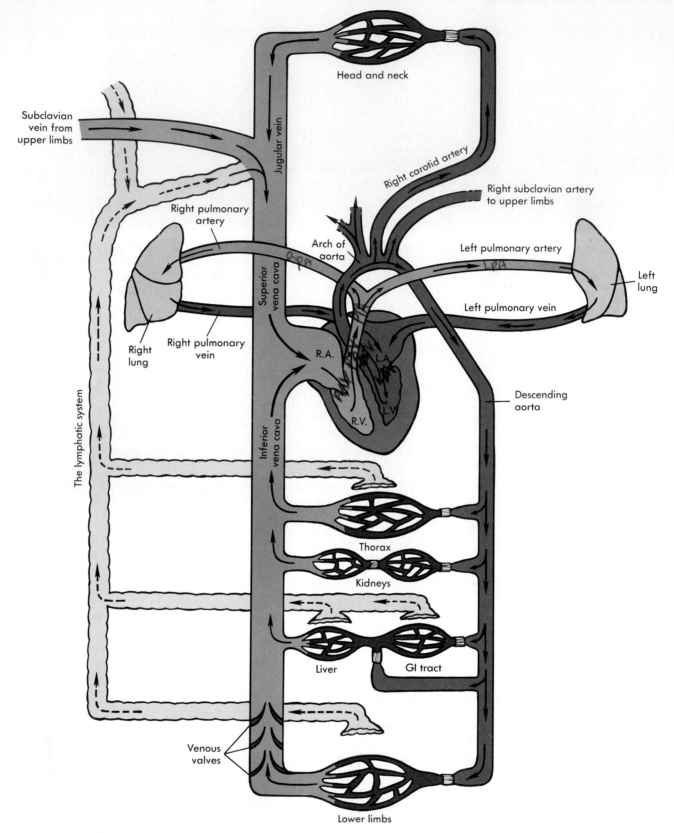

FIGURE 14-17

Diagram of the circulatory system and lymphatic system. The vessels of the lymphatic system begin as blind tubes in the tissues and return extracellular fluid to the heart via the superior vena cava and right atrium.

THE CARDIOVASCULAR SYSTEM

Blood Is Thicker than Water

As the hematocrit increases, so does the viscosity of blood. At a normal hematocrit of 45%, the viscosity is about 3 times higher than water, but blood may become 6 to 8 times as viscous as water if the hematocrit reaches 60%. High hematocrits are seen in disorders in which red cell production is elevated (polycythemia) and in severe water loss (dehydration). At the opposite extreme, hematocrits as low as 20% to 25% can be observed in disorders in which red cell production is low (anemia) and following severe blood loss. At a hematocrit of 25%, the viscosity of blood is only twice that of water.

Some athletes have used a technique called **blood doping** to improve their performance in endurance events, such as bicycling or distance running. In this technique, some of the athlete's blood is drawn 5 to 6 weeks in advance of an important meet, and the red cells are separated and frozen for storage. In the time before the event, erythropoiesis restores the hematocrit to normal. Immediately before the meet, the stored red cells are injected, raising the hematocrit by 8% to 20%. The increased hematocrit persists for at least 2 weeks. In principle, large increases in the hematocrit could dramatically increase maximal performance during exercise because the maximum sustained effort of a trained athlete is limited by the rate at which oxygen can be delivered to the working muscles. However, the usefulness of blood doping is limited because the increased viscosity of the blood increases the flow resistance of all the organs. With increases in the hematocrit to more than 65%, the advantage of higher oxygen-carrying capacity is outweighed by the resulting limitation of perfusion of working tissues and the increased load on the heart. At present, it appears that little or no increase in performance results from this technique.

smaller diameter, the resistance of the collection of branch vessels can be either greater than or less than the resistance of the stem. The outcome depends on the number of branches (the more branches, the lower the net resistance) and the resistance of each branch (the smaller the branch vessels as compared with the stem, the higher the net resistance). This is illustrated in two examples.

For the first example, consider a vessel that branches to yield 1,000,000 branch vessels, each of which has a tenth the radius of the parent. The r^4 relationship for resistance means that the resistance of each branch vessel will be increased by a factor of 10,000 over that of the stem. Because these vessels are all in parallel, the total collective resistance is 10,000 divided by 1,000,000, or 1% of the original value. Extensive branching has outweighed the fourth-power size dependence that would otherwise have increased the equivalent flow resistance. This example corresponds to the degree of branching occurring at the level of the capillaries and explains why, as a group, capillaries have a relatively low flow resistance, despite the small size of each individual capillary.

If branching is less extensive, the situation can be different. For the second example, suppose that the stem vessel gives rise to only 1000 branch vessels instead of 1,000,000. The resistance of each

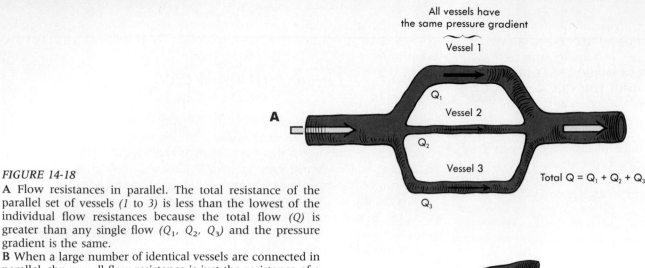

All vessels have
the same pressure gradient

Vessel 1

A

Q₁

Vessel 2

Q₂

Vessel 3

Q₃

Total Q = Q₁ + Q₂ + Q₃

FIGURE 14-18

A Flow resistances in parallel. The total resistance of the parallel set of vessels *(1 to 3)* is less than the lowest of the individual flow resistances because the total flow *(Q)* is greater than any single flow *(Q₁, Q₂, Q₃)* and the pressure gradient is the same.

B When a large number of identical vessels are connected in parallel, the overall flow resistance is just the resistance of a single vessel divided by the number of vessels in parallel with each other.

One vessel

B

N vessels
each of resistance R′

$$R_{total} = \frac{R'}{N}$$

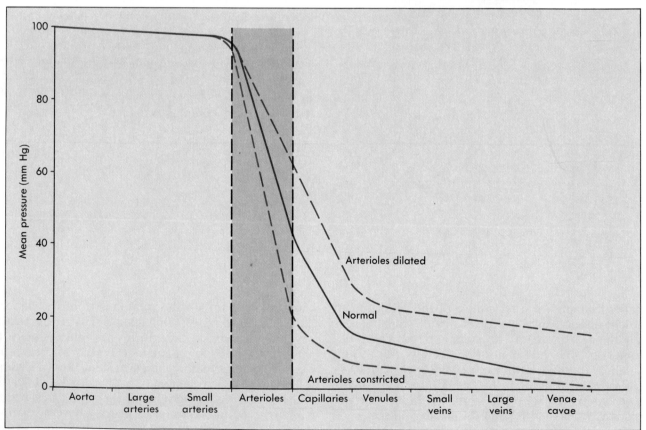

FIGURE 14-19

Pressure profile of the systemic loop. Pressure drops by only a few mm Hg between the heart and the arteriolar branch points. The pressure drop is much larger as blood flows through arterioles. A smaller pressure drop occurs in capillaries, and the last leg of the circuit through venules and veins is completed under very low pressure. Arteriolar dilation reduces the pressure drop in arterioles and proportionately increases the pressures in capillaries, venules, and veins. Arteriolar constriction increases the pressure drop in arterioles and reduces capillary and venous pressures. For simplicity, mean arterial pressure is assumed to remain constant.

branch vessel would still be increased by 10,000 over that of the stem. However, with only 1000 branches in parallel, the collective resistance of the branches would be 10 times greater than that of the stem. This example corresponds to the transition between arteries and arterioles. When blood flows from arteries into arterioles, the number of arterioles is not large enough to compensate for the large resistance of each single arteriole. As a result, the arterioles make the largest contribution to the total peripheral resistance of all the vessel types and are referred to as the **resistance vessels** of the circulation.

The different net resistances of different vessel types are reflected in the pressure profile of the systemic loop (Figure 14-19). More than half of the pressure drop occurs as blood passes through arterioles, with another drop about half as great in the passage through capillaries. In contrast, flow through the large arteries and veins is accomplished with very small pressure drops.

Importance of Arterioles for Control of Blood Flow Distribution and Peripheral Resistance

The dominating resistance of arterioles has two important consequences. First, the arterioles determine how the cardiac output is distributed among different tissues and organs. Contraction of the smooth muscle of arteriolar walls results in a decrease in radius, or **constriction**; relaxation of the muscle has the opposite effect and is called **dilation**. Dilation of the arterioles of a particular tissue decreases the flow resistance of that tissue and has the effect of diverting more of the cardiac output to the capillary beds of that tissue.

The sum of the resistances of the arterioles in the various branches of the systemic loop is the major determinant of the **total peripheral resistance**. Total peripheral resistance is increased by arteriolar constriction and decreased by arteriolar dilation. These changes also affect the pressure profile. Arteriolar dilation decreases the pressure drop in arterioles and increases capillary and venous pressure; arteriolar constriction increases the pressure drop in arterioles and reduces capillary and venous pressure (Figure 14-19). The importance of total peripheral resistance as a determinant of cardiac output and mean arterial pressure is discussed in the next chapter.

An important difference between the systemic loop and the pulmonary loop is that the pressure gradient between arteries and veins is only about 14 mm Hg in the pulmonary loop as compared with the 90 to 100 mm Hg of the systemic loop. Because the output of the right and left hearts must be the same over time, the resistance of the pulmonary loop must be about 14% of that of the systemic loop. One important factor in the low resistance of the pulmonary loop is the fact that the pulmonary arterioles are substantially larger than the systemic ones. Pulmonary circulation is discussed further in Section IV, Respiration.

TRANSFER OF FLUID AND SOLUTES BETWEEN CAPILLARIES AND TISSUES
Exchange of Nutrients and Wastes across Capillary Walls

> - *What is the Starling Hypothesis of Capillary Filtration?*
> - *What is the source of the pressure that favors movement of fluid from capillaries to interstitial spaces?*
> - *What feature of the capillary wall is responsible for the colloid osmotic gradient between interstitial fluid and plasma?*
> - *What is edema? What changes favor the development of edema?*

In almost all tissues, solute transfer between plasma and interstitial fluid is driven solely by concentration gradients rather than by the expenditure of metabolic energy. For example, glucose uptake by metabolizing cells tends to reduce the concentration of glucose in the interstitial fluid, resulting in a gradient of glucose concentration between plasma and interstitial fluid. Because glucose is poorly soluble in lipid, almost all of the glucose that diffuses between the plasma and interstitial fluid must pass through the intercellular pores of the capillary endothelium. At the same time, O_2 diffuses from blood to interstitial fluid to cytoplasm, and CO_2 produced by oxidative metabolism diffuses in the reverse direction along the same route. (Movement of both gases is driven by gradients of partial pressure, a measure of the abundance of gas molecules in solution.) Dissolved CO_2 and O_2 are quite lipid soluble, so their route passes mainly through the endothelial cells.

The brain is an important exception to the rule that solute exchange across capillary walls occurs by diffusion. Capillaries in most regions of the brain lack open intercellular pores, and hydrophilic solutes cannot readily pass across them, although lipid-soluble substances can diffuse through brain capillary walls quite easily. This feature of brain capillaries has been termed the **blood-brain barrier**. The glucose needed for energy and the amino acids needed for synthesis of proteins and peptide neurotransmitters are actively transported across the walls of brain capillaries. The blood-brain barrier gives the brain a large measure of control over the composition of its interstitial space.

Fluid Exchange Between Capillaries and Interstitial Spaces

A capillary wall can be thought of as a filter whose pores retain large plasma solutes, allowing the passage of water and small solutes. At each point along its length, a capillary has both an outward hydro-

The Blood-Brain and Blood-CSF Barriers

The brain's interstitial fluid, the cerebrospinal fluid (CSF), is produced by the choroid plexus, a membrane attached to the roofs of the four brain ventricles. The total mass of the choroid plexus tissues is only a few grams, but it secretes CSF at about 0.5 ml/min, sufficient to replace the 150 ml of the CSF five to six times a day. The CSF flows through the ventricles of the brain and around the brain and spinal cord. Just as a fetus is cushioned in the uterus by the amniotic fluid, the CSF supports the brain, reducing its weight and protecting it from sudden head movements and blows to the head.

Communication between nerve cells of the brain requires precise regulation of the composition of the CSF because the excitability of nerve cells depends on ionic gradients between the cytoplasm and interstitial fluid and because many small organic molecules normally found in the blood are also used for chemical communication between brain cells. The blood concentrations of ions (such as K^+) and amino acids can vary significantly as a result of absorption of a meal. If the brain were exposed to these variations, overall alterations in excitability and random activation of the pathways that use amino acid transmitters would undermine communication within the brain.

The brain is isolatd from the plasma in two ways. The endothelial cells lining the capillaries of the brain have a low permeability to ions and most small organic solutes, a property referred to as the blood-brain barrier in the text. The blood-brain barrier has facilitated diffusion transport pathways that allow substances that the brain needs in large amounts, such as glucose, to rapidly diffuse down their concentration gradients. The choroid plexus is a "reverse kidney," removing waste products from the CSF into the blood and pumping needed materials from the blood into the CSF. These activities make up the blood-CSF barrier. Like the kidney, the cells of the choroid plexus have different active transport systems confined to different parts of the cell. The choroid plexus also pumps HCO_3^- to maintain the pH of the CSF constant in the face of changes in plasma pH.

An important consequence of the presence of the blood-brain barrier is that antibiotics such as penicillin cannot be delivered to the brain by injection into the systemic vasculature. In contrast, caffeine, nicotine, cocaine, and alcohol are lipid soluble and readily cross the blood-brain barrier where they affect cerebral function. An important task of pharmacology is to identify or modify drugs to increase their lipid solubility, allowing them to cross the blood-brain barrier.

The blood-brain barrier is less tight in some specialized regions of the brain, such as the hypothalamus, pituitary, and pineal glands. In these regions peptide hormones and glucose can enter the blood from the brain. Recently, a method of temporarily opening the blood-brain barrier was discovered. An injection of hypertonic solute into the circulation of the brain disrupts the continuous tight junctions of the capillary endothelial cells, probably by shrinking the cells themselves. The effect lasts for up to several hours. This technique permits drugs that normally do not cross the blood-brain barrier to be delivered to brain cells.

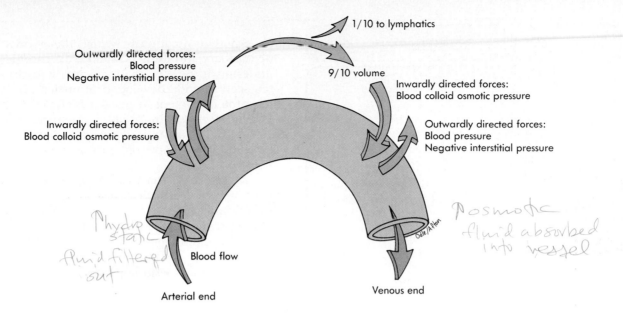

Outwardly directed forces:
Blood pressure
Negative interstitial pressure

1/10 to lymphatics

9/10 volume

Inwardly directed forces:
Blood colloid osmotic pressure

Inwardly directed forces:
Blood colloid osmotic pressure

Outwardly directed forces:
Blood pressure
Negative interstitial pressure

Blood flow

Arterial end

Venous end

FIGURE 14-20

Starling's Hypothesis of Capillary Filtration. The major pressures composing the total pressure are the hydrostatic pressure of the blood in the capillary and the colloid osmotic pressure of the blood and in the interstitial spaces. The total pressure is outwardly directed at the arteriolar end and inwardly directed at the venous end. About 90% of the filtered plasma is reabsorbed, with the remaining 10% entering the lymphatics.

static and an inward colloid osmotic pressure gradient across its wall. A net loss or gain in the fluid volume carried by the capillary is determined by which of these forces is stronger.

The movement of fluid between the blood and interstitial-fluid compartment is described quantitatively by Starling's Hypothesis of Capillary Filtration:

Net fluid filtration = K (Net hydrostatic forces − Net osmotic forces)

According to Starling's Hypothesis, the constant **K**, representing the permeability of the capillaries of a given tissue, is multiplied by the difference between the net **hydrostatic forces** and the net **colloid osmotic forces** exerted by the proteins in blood and the interstitial fluid. These colloid osmotic forces are often referred to as **oncotic** forces.

Fluid exit from the capillary is driven by the hydrostatic pressure gradient that drives blood through the capillary; this pressure is greatest at the arteriolar side and least at the venular side (Figure 14-20). The tissue, or interstitial, hydrostatic pressure is very small and may even be a negative value, so the blood pressure that drives fluid out of the capillary is relatively unopposed by tissue hydrostatic forces.

Fluid entry into the capillary is driven by the colloid osmotic force attributable to plasma proteins in the blood when the blood enters the capillary and that are not filtered out. This value increases only slightly as the blood moves through the capillary (Figure 14-20) and is opposed along the length of the capillary by a small but measurable tissue colloid osmotic force attributable to the small amount of protein that does leak out into the interstitial fluid. The smaller organic molecules and salts tend to be at the same concentration in blood and interstitial fluid and therefore do not contribute to the net osmotic force.

To summarize, if the (outward) hydrostatic pressure in the capillary is greater than the (inward) plasma osmotic pressure, fluid is pushed out of the capillary into the interstitial fluid. If the reverse is true, fluid is "pulled" into the capillary. In most peripheral capillary beds both occur. At the arterial end, the hydrostatic gradient dominates the osmotic gradient, causing fluid filtration. On the venous side the decreased hydrostatic pressure and increased plasma osmotic pressure cause fluid absorption.

In general, there is a slight excess of filtration over absorption (Figure 14-20). However, just where this occurs on the microscopic level has been difficult to determine. Starling believed the tissue hydrostatic pressure to be a small positive value (that is, slightly for fluid movement into the tissues). In this view, the tissues could be likened to a sponge saturated with water. Some investigators now believe that the tissue hydrostatic pressure may be negative by as much as 7 mm Hg. If so, the tissues would be like a sponge that is not saturated. If this is the case, both the capillary blood pressure and the suction of the negative tissue hydrostatic pressure favor filtration of fluid from the blood into the interstitial-fluid compartment at all points along the capillary.

Small, local variations in the capillary blood

TABLE 14-4 Causes of Edema

Factor altered	Possible causes
1. Increased capillary blood (hydrostatic) pressure	Arteriolar dilation Heart failure Prolonged standing
2. Decreased plasma osmotic pressure	Protein undernutrition Failure of plasma protein synthesis (liver disease) Excessive loss of plasma proteins in urine (kidney disease)
3. Increased capillary permeability	Inflammation Allergies Burns
4. Increased tissue osmotic pressure	Release of protein by damaged cells
5. Obstruction of lymphatic vessels	Injury Parasitism of lymphatic system (filariasis)

pressure, the water permeability of different parts of the capillary, or the tissue colloid osmotic and hydrostatic pressures could combine to favor filtration or reabsorption in other capillaries or in the same capillary at a different time. Some investigators now believe that for open capillaries (those whose precapillary sphincters are open), the net force favors filtration along their entire length, and that for closed capillaries, the net force favors reabsorption. In any case, the net effect for the body is that about 10% of the plasma filtered is not reabsorbed, resulting in net transfer of 2 to 4 liters of fluid per day from the systemic circulation to the interstitial fluid.

The balance between capillary filtration and reabsorption can be altered by changes in any of the elements in Starling's Hypothesis. One example of a disease process in which elevated mean capillary pressure alters the filtration-reabsorption balance is right ventricular failure. When this occurs, the venous pressure increases, increasing the mean pressure in the systemic capillaries and thus increasing the rate of filtration relative to reabsorption. Accumulation of fluid in the interstitial spaces causes swelling, **(edema)** (Table 14-4).

The lungs are particularly susceptible to edema. The total mass of lung tissue is small; therefore, relatively small shifts of fluid from the plasma to the lungs can have serious effects. The normal capillary blood pressure of the lungs is only about 7 mm Hg, consistent with the low pressures throughout the pulmonary loop. These low blood pressures reduce the probability of pulmonary edema under normal conditions. Failure of the left ventricle results in a backup of blood in the pulmonary circulation, with a consequent increase in the pulmonary capillary pressure, causing **pulmonary edema.**

Another situation in which the normal balance of capillary filtration is altered is chronic protein undernutrition (e.g., the marasmic Kwashiorkor often seen in underdeveloped countries). In this case, the protein content of plasma is depressed. The resulting decrease in osmotic pressure reduces absorption. The resulting general edema causes the bloated appearance of seriously undernourished people.

A shift in the balance of capillary filtration may have a beneficial effect. For example, if bleeding significantly decreases blood volume, capillary blood pressure falls. As a result, there is a net shift of fluid from the interstitial spaces to the plasma. As much as 1 L of volume can be transferred to the plasma over about an hour. The interstitial fluid is a reserve that can be drawn on to maintain plasma volume.

Lymphatic System

- *What is the role of the lymphatic system in regulating the volume of interstitial fluid?*
- *What changes would favor transfer of fluid from tissues to plasma across capillary walls?*

The lymphatic system (Figure 14-17) returns excess interstitial fluid formed in capillary filtration to the circulatory system and salvages the small amounts of protein that escape through the capillary walls. **Lymphatic capillaries** (Figure 14-21, *A*) terminate among blood capillary beds and have large pores that permit the entrance of proteins as well as smaller solutes. The **lymphatic capillaries** drain into **lymphatic veins**, which ultimately join to form two **lymphatic ducts**; these join with the subclavian veins near the heart. The duct on the left side, called the **thoracic duct,** serves the entire body except the right shoulder and the right side of the head, which are served by the **right lymphatic duct.**

Lymph is propelled in the direction of the ducts by contractions of surrounding skeletal muscle and, to a lesser extent, by contractions of smooth muscle surrounding the lymphatic vessels themselves. Like systemic veins, lymphatic veins have one-way valves (Figure 14-21, *B*). The lymphatic vessels are an ideal location for elements of the immune system, **(lymph nodes)** (see Chapter 24), because lymph coming from an infected region of the body carries invading bacteria. In infected or injured tissues, the proteins released from damaged cells increase the colloid osmotic pressure of the interstitial fluid, tipping the balance in favor of filtration. Consequently, a local edema develops in these areas (Table 14-4). In the case of infection, the extra interstitial fluid accumulation flushes bacteria into the lymphatic system, where they more readily trigger an immune response.

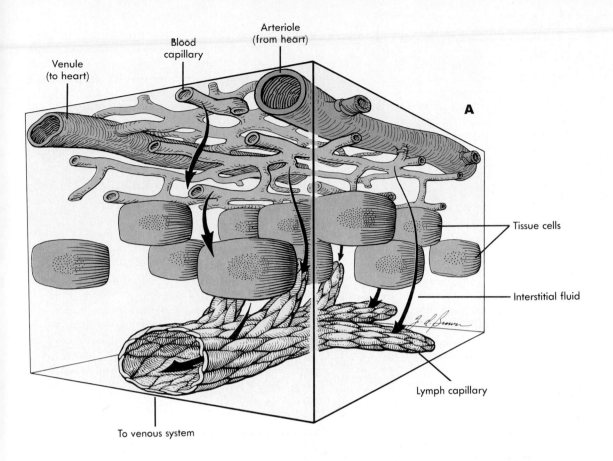

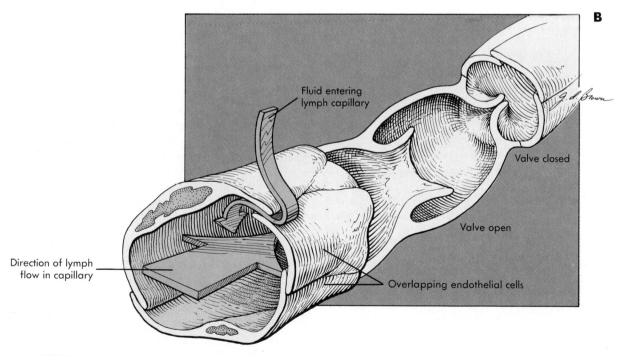

FIGURE 14-21

A Movement of fluid from blood capillaries into tissues and from tissues into lymph capillaries. Lymphatic capillaries are close to blood capillaries. Major features of their structure include valve-like pores between their endothelial cells and supporting fibers running from lymphatic walls into surrounding tissue to prevent collapse of the lymphatic capillaries. **B** One-way flow of lymph in lymph capillaries. The overlap of the lymph capillaries' endothelial cells allows easy entry of interstitial fluid but prevents movement back into the tissue. Valves in lymphatic veins, like those of blood veins, ensure one-way flow in the direction of the heart.

SUMMARY

1. **Blood** is a liquid tissue consisting of **red cells, white cells,** and **platelets** suspended in **plasma.** Red cells function in O_2 and CO_2 transport, white cells in immunity, and platelets in **hemostasis.** The **hematocrit** level is the ratio of red cell volume to total volume.

2. Three phases of hemostasis are (1) formation of a platelet plug at the site of vessel injury, (2) clot formation, and (3) clot dissolution. Clotting is the result of conversion of **fibrinogen** to **fibrin,** which is triggered by a factor cascade initiated either by factors present in blood (the **intrinsic pathway**) or by factors released by injured tissue (the **extrinsic pathway**). Clot dissolution is initiated by the factors that trigger clotting but becomes effective more slowly.

3. The ratio of volume to pressure in a vessel is determined by its **compliance.** Arteries are low-compliance elements containing a small volume of blood under high pressure. The arterial walls store energy during systole and recoil to maintain blood flow into the vasculature during diastole. Veins are the high-compliance, **capacitance** elements of the circulation.

4. Depending on context, "flow" indicates either the volume of blood passing a point over time (flow rate), the linear velocity of a single drop of blood (flow velocity), or the average velocity of all drops in a stream (average flow velocity).

5. The flow rate through a vessel is equal to the perfusion pressure divided by the vessel's **flow resistance.** In terms of the whole systemic loop, the **venous return,** or **cardiac output,** corresponds to flow rate; the **mean arterial pressure** minus the **central venous pressure** corresponds to perfusion pressure; and the **total peripheral resistance** corresponds to flow resistance.

6. Flow resistance is proportional to vessel length, fluid viscosity, and the inverse fourth power of vessel radius. Of the factors, vessel radius has the most important implications for flow through blood vessels.

7. Branching causes a decrease in net flow resistance. Decreases in radius increase net flow resistance. When both occur, there may be an increase or decrease in the net flow resistance, depending on the number of branches and the radius of each branch. In the case of arterioles, a net increase in resistance occurs; for capillaries, branching is sufficient for a decrease in resistance to occur. The net resistance of arterioles largely determines total peripheral resistance; dilation or constriction of individual arterioles controls the perfusion of individual tissues.

8. Extrinsic control of vascular smooth muscle is generally mediated by the sympathetic branch of the autonomic nervous system, which mainly causes vasoconstriction.

9. **Capillaries** are the sites of all exchange of materials between tissues and plasma. This exchange is favored by the slow average velocity of flow of blood in capillaries. Solutes cross capillary walls by diffusion; brain capillaries are an exception in that they actively transport glucose and amino acids.

10. The hydrostatic pressure gradient across capillary walls tends to drive filtration in the arteriolar ends of capillaries, and the colloid osmotic gradient tends to drive absorption in the venular ends. Excess interstitial fluid is returned to the circulation by **lymphatic** vessels. Pressure imbalances favoring filtration cause **edema;** decreased blood pressure following blood loss causes the colloid osmotic force to dominate and thus favors fluid movement into the plasma.

1. What advantage is conferred by the complexity of the factor cascade that triggers clotting? Which of the stages of hemostasis involve positive feedback? Which one represents negative feedback?

2. The diameter of a capillary is smaller than that of an arteriole, yet collectively the capillaries have a lower flow resistance than the arterioles. Explain.

3. What would be the effect of an increase in tissue colloid osmotic pressure on the volume of interstitial fluid? An increase in venous pressure?

4. Why does systolic pressure normally increase with age?

5. What effect would a decrease in venous compliance have on the volume of blood in veins? On the venous pressure? On the venous return? On the total peripheral resistance? What effect would arteriolar dilation have on the total peripheral resistance? On venous return?

6. What factors determine whether branching will increase or decrease the net flow resistance of a stem vessel and its branches?

7. Describe the role of each of the following in blood clotting.
   Hageman factor
   Thrombin
   Tissue thromboplastin

8. A sample of blood centrifuged in a glass capillary tube yields the following measurements: length occupied by plasma: 29 mm; length occupied by red blood cells: 21 mm. What is the hematocrit?

9. What is the effect on vascular resistance and blood flow of activating each of the following pathways:
   a. Sympathetic adrenergic input to fibers with alpha receptors.
   b. Adrenergic fibers on vascular muscle with beta receptors.
   c. Sympathetic cholinergic fibers.
   d. Parasympathetic fibers.
   In which tissue types is each of these pathways important?

10. An artery gives rise to two branches, one of which has half the radius of the other. If the blood flow rate in the artery is 10 ml/min, what would be the flow rate in each of the branches?

Choose the MOST CORRECT Answer

11. Plasma protein(s) that is (are) important in maintaining body fluid balance:
    a. Globulins   c. Fibrinogen
    b. Albumins   d. Thrombin

12. The ion is an essential co-factor in hemostasis:
    a. Na^+ c. K^+
    b. HCO_3^- d. Ca^{++}

13. The common endpoint in both the intrinsic and extrinsic pathways of hemostasis is:
    a. The release of tissue thromboplastin
    b. The production of prothrombin activator
    c. The activation of the Hageman factor
    d. The release of phospholipid PF3

14. In comparison to arteries, veins have:
    a. Thicker tunica media   c. Less compliance
    b. Smaller lumens   d. One-way valves

15. These vessels are often called the capacitance elements of the circulatory system:
    a. Arteries   c. Capillaries
    b. Veins   d. Arterioles

16. Which of these characterictics does NOT describe capillaries?
    a. Walls consist of a single layer of endothelial cells
    b. High diffusion rate
    c. High cross-sectional area
    d. High blood velocity

17. The capillaries that form the blood-brain barrier:
    a. Lack open intercellular pores
    b. Prevent passage of hydrophilic solutes across the walls
    c. Allow diffusion of lipid-soluble substances easily
    d. All of the above are true

● SUGGESTED READINGS

BANKHEAD CD: Combo may avert MI development, *Medical World News* 30 (15):51, August 14, 1989. Describes how thromboxane inhibitors combined with a serotonin antagonist averts myocardial infarction.

COMAROW A: One man, pummeled by positrons, *U.S. News & World Report* 109 (6):60, 1990. Describes how positron emission tomography is used to test condition of arteries.

EDWARDS DD: Searching for the better clot-buster, *Science News* April 1988, p. 230. Describes recent drugs that offer improved treatment of fibrolytic blood clots.

FEINBERG W: Antithrombotic therapy in stroke and transient ischemic attacks, *American Family Physician* 40 (5):53s Nov 15, 1989. Self-explanatory.

GOLDE DW: The stem cell, *Scientific American* 265(6), p. 86, December 1991. The stem cell of the bone marrow gives rise to red blood cells, platelets, and cells of the immune system.

SEGAL M: New hope for children with sickle cell disease, *FDA Consumer* 23 (2) p. 14, March 1989.

WEISS R: In a similar vein, *Science News* September, 1987, p. 201. Discusses fluosol-DA as a possible substitute for blood.

ZOLER ML: Calcium blockers may help stave off atherosclerosis, *Medical World News* 31:(12):10, June 25, 1990. Discusses the use of agents that block calcium movement into smooth muscle cells as anti atherosclerosis agents.

ZOLER ML: CAD payoff is finally seen in BP drug therapy efforts. *Medical World News* 31 (10):34, May 28, 1990. Discusses the relationship between coronary artery disease and blood pressure. A meta-analysis of more than a dozen hypertension treatment trials has shown what each trial, individually, failed to show: lowering blood pressure by drug treatment causes a statistically significant reduction in coronary artery disease events.

On completing this chapter you should be able to:

- Describe the autonomic innervation of blood vessels, distinguishing special features of blood vessels in skeletal muscle, the coronaries, and the external genitalia
- Diagram the baroreceptor reflex
- Describe how antidiuretic hormone, the atrial natriuretic hormone system, and the renin-angiotensin-aldosterone system regulate mean arterial blood pressure
- Appreciate the significance of the cardiovascular operating point in the regulation of central venous pressure, mean arterial pressure, and cardiac output
- Understand how standing up results in a drop in venous return and how mean arterial pressure is maintained by the baroreceptor reflex, venous valves, and the skeletal and respiratory pumps
- Outline the changes that follow moderate hemorrhage
- Describe the cardiovascular responses to exercise
- List factors that might be expected to lead to hypertension

*W*hen our primate ancestors began to walk upright, they imposed new stresses on many parts of the body, including the cardiovascular system. In most animals, the major axis of the circulatory system is horizontal, and most of the blood volume is close to heart level. In standing human beings, about 70% of the blood volume is in highly compliant veins below the heart. To respond to postural changes, the cardiovascular system requires receptors that sense relative hydrostatic pressure fluctuations and initiate appropriate reflex responses.

In addition to the problems associated with an upright posture, the cardiovascular system is also stressed by exercise, blood loss, heat, and disease. For example, strenuous exercise in elite athletes causes roughly a twentyfold increase in blood flow to skeletal muscle and a tenfold increase in cardiac output, with very little change in mean arterial pressure. This stability requires the integration of neurally and hormonally mediated cardiovascular reflexes with the intrinsic autoregulation that occurs in the heart and blood vessels.

Some of the mechanisms involved in cardiovascular integration are explained in this chapter; others are not yet fully understood. For example, how do the cardiovascular centers of the brain "know" how much sympathetic input to give the heart and vessels to meet the challenge of rapidly changing workloads? Some evidence suggests that the cardiovascular regulatory centers of the brain receive input from motor centers that provide "feed-forward" information on the demands that contracting skeletal muscle is about to place on the cardiovascular system.

CARDIOVASCULAR REFLEXES
Components of the Cardiovascular System

> - *What variables describe the overall function of the cardiovascular system?*
> - *What two variables best describe the function of the heart? What three variables describe the vascular system?*
> - *What property of the vasculature is affected by arteriolar constriction or dilation? Which one is affected by changes in venous compliance?*

The conceptual organization of the cardiovascular system is shown in Figure 15-1. The properties of the heart are myocardial contractility and heart rate. The properties of the peripheral vessels are the total peripheral resistance, blood volume, and venous compliance. However, the heart and blood vessels are not isolated. They form an integrated system. System variables include the cardiac output, mean arterial pressure, central venous pressure, and venous return.

Extrinsic Control of Vascular Smooth Muscle

Vascular smooth muscle is present in the walls of all blood vessels except capillaries. Changes in vascular smooth muscle tone (tension) control total peripheral resistance, the compliance of arteries and veins, and the distribution of blood flow. The two separate major control systems that affect vascular smooth muscle tension are (1) extrinsic control by hormones and the autonomic nervous system (Table 15-1), and (2) intrinsic control, or autoregulation, by chemical factors produced in the immediate area of the blood vessels (Table 15-2). The relative importance of these two systems depends on the type

of vessel and the organ or tissue served by that vessel.

The smooth muscle of most blood vessels is innervated by the sympathetic branch of the autonomic nervous system (ANS). The sympathetic nerve fibers that connect to blood vessels usually exhibit some degree of **resting tone** (a low level of nerve impulse frequency in the efferent fibers). Resting tone allows a single branch of the ANS to both constrict and dilate the target vessels. An increase in sympathetic adrenergic input to a vessel that possesses mainly α_1-adrenergic receptors increases smooth muscle tension above the resting level; the effect of smooth muscle contraction on the vessel is a decrease in the radius of the vessel, or **vasoconstriction**. A decrease in the sympathetic tone allows the smooth muscle to relax; the resulting increase in the radius of the vessel is **vasodilation.**

The arterioles and veins of most vascular beds contain α_1-adrenergic receptors, but the effect of sympathetic stimulation on the two types of vessels differs. The arterioles **(arteriolar vasoconstriction)** are the major site of flow resistance in the body. Like the faucets in a house, they act to redirect blood flow from one organ to another. Arteriolar vasoconstriction also increases the total peripheral resistance. Sympathetic stimulation of veins **(venoconstriction)** does not affect total peripheral resistance very much because venous resistance is low compared with arteriolar resistance. The major effect of venoconstriction is to decrease the compliance of vein walls (stiffen them).

Three exceptions to this pattern of response to sympathetic stimulation exist. The arterioles that supply blood to skeletal muscle possess a mixture of

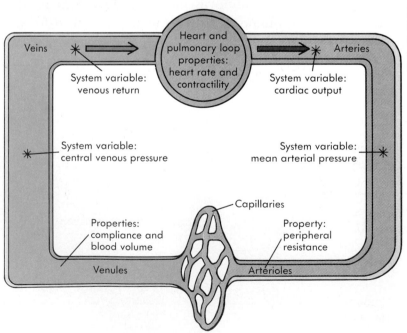

FIGURE 15-1
Schematic diagram of the heart and systemic circulation showing the properties associated with each part of the system and the variables that affect both the heart and the vasculature.

TABLE 15 1   Extrinsic Regulation of Blood Vessels

Agent	Effect	Comments
Sympathetic nerves		
α_1-Adrenergic	Constriction	The dominant effect of the sympathetic nervous system on the vasculature and the major cause of changes in the total peripheral resistance and venous compliance.
β_2-Adrenergic	Dilation	Skeletal muscle arterioles. Effect is minor compared with α_1-effect but may be important during moderate elevation of epinephrine.
Cholinergic	Dilation	Skeletal muscle arterioles. Important in preparation for activity, but effect is small compared to the autoregulatory increase in working muscle.
Parasympathetic nerves		
Cholinergic	Dilation	Important in genital sexual response. Organs that are activated by parasympathetic input, such as the salivary glands and GI tract, may vasodilate as indirect effect.
Hormones		
Antidiuretic hormone (vasopressin)	Constriction	Octapeptide released by hypothalamic neurons that project to posterior pituitary in response to signals, which include decreased atrial pressure.
Angiotensin I, II	Constriction	Results from renin secretion by kidneys in response to decreased renal perfusion, a consequence of decreased mean arterial pressure.

α_1- and β_2-adrenergic receptors (see Table 15-1). Low levels of circulating epinephrine preferentially activate the β_2-receptors and induce dilation. Higher plasma levels of epinephrine, or norepinephrine released directly from sympathetic fibers, activate α_1-receptors, causing arteriolar constriction.

The arterioles of the skin and skeletal muscle are an exception to the rule that vascular smooth muscle is innervated only by sympathetic adrenergic inputs. These vessels are innervated by both sympathetic adrenergic fibers that mediate constriction and sympathetic cholinergic fibers that mediate dilation. These inputs are important for diverting blood flow to the skin to increase heat loss and to skeletal muscle to increase perfusion before exercise. Another good example of vasodilation is blushing. The sweat glands in the skin are controlled by sympathetic cholinergic fibers. The associated vasodilation may occur indirectly through the release of bradykinin (see Table 15-2).

The arterioles of the external genitalia are an exception to the rule that vessels receive only sympathetic innervation because the external genitalia are also innervated by parasympathetic fibers that mediate dilation. These fibers have an important role in the vascular changes that occur with sexual arousal (see Table 15-1).

Intrinsic Control of Vascular Smooth Muscle

In **flow autoregulation,** the diameters of arterioles of individual vascular beds changes, adjusting the

TABLE 15-2   Intrinsic Control of Systemic Blood Vessels (Autoregulation)

Autoregulatory agents	Comments
Metabolic vasodilators	
Decreased O_2	Due to high rate of oxidative metabolism.
Decreased pH	Reflects increased levels of CO_2 and lactic acid.
Increased K^+	Due to activity of excitable cells.
Paracrine agents	
Prostaglandins	May dilate or constrict depending on the particular prostaglandin involved. Also important in vascular responses in inflammation and hemostasis.
Bradykinin	Dilator important in inflammation and in autoregulation in sweat glands.
Histamine	Dilator important for vascular responses in inflammation and allergic response.
Adenosine	Vasodilator important in some tissues such as the heart.

flow to meet the metabolic demands of the tissue, despite variations in perfusion pressure. This process was presented in Chapter 5 as an example of negative feedback. The feedback of tissue metabolism on arteriolar radius may be mediated by several substances referred to collectively as **local factors.**

Depending on the tissue involved, local factors are generally the end products of energy metabolism, such as lactic acid and carbon dioxide (CO_2); or substances released in the course of activity, such as potassium (K^+); or paracrine substances, such as **prostaglandins, adenosine, bradykinin,** and **nitric oxide** (see Table 15-2). Oxygen and other nutrients may also act as local factors when their concentrations are decreased by an increase in tissue activity. With an increase in metabolic rate, the changes in the concentrations of the local factors cause arterioles to dilate. Such changes in the local environment are particularly important for metarterioles because they receive little or no innervation. Similar factors stimulate opening and inhibit closing of precapillary sphincters.

Skeletal muscle provides an excellent example of how local factors affect autoregulation of flow. The fraction of the cardiac output that enters a resting skeletal muscle is relatively low because arterioles are constricted and precapillary sphincters are frequently closed. Much of the blood flow that does enter the muscle passes through shunts because the shunt resistance is lower than the resistance of the path through the constricted arterioles and the precapillary sphincters, most of which are closed.

If the muscle contracts repeatedly, end products of metabolism accumulate and nutrients and oxygen are depleted. In response, arterioles dilate and precapillary sphincters remain open longer. The decrease in the tissue flow resistance caused by arteriolar dilation delivers a larger share of the cardiac output to the working muscle. The open precapillary sphincters allow this increased flow to pass through capillaries. Thus the effect of a mismatch between production of local factors and their removal by blood flow is an increase in the delivery of blood to the affected area. This is termed **reactive hyperemia.**

Differences between Vascular Beds

The relative importance of extrinsic inputs and autoregulation in determining blood flow differs from tissue to tissue and even in the same tissue under different conditions. For example, the sympathetic nervous system has its largest vasoconstrictive effect on the vessels of the skin and abdominal organs. The dually innervated vessels of the external genitals are very responsive to changes in their autonomic inputs occurring during the sexual-response cycle (see Chapter 25). In contrast, autoregulation is believed to be much more important than extrinsic control in the heart and brain.

In skeletal muscle, the level of tissue activity alters the balance between extrinsic and intrinsic control. In resting skeletal muscle, sympathetic (norepinephrine) inputs dominate and exert mainly a vasoconstrictive effect. The blood flow to resting skeletal muscle can be increased somewhat by autonomic vasodilation (the result of cholinergic fibers and epinephrine); this occurs when exercise is anticipated. In working skeletal muscle, autoregulation can increase blood flow far above that attributed to autonomic input; thus autoregulation is the mechanism by which the skeletal muscles capture a large fraction of the cardiac output during exercise.

Baroreceptor Reflex

- *What is the functional role of the baroreceptor reflex?*
- *Both ADH and ANH release are affected by stretch of atrial walls. What variable is directly regulated by this input?*
- *Both ADH and angiotensins cause arteriolar constriction. What property of the vascular system is regulated by these hormones?*

Figure 15-2 illustrates the location of the cardiovascular **baroreceptors.** One important group of baroreceptors is located in the walls of the common carotid artery, at the point where it branches to form the internal and external carotid arteries. At this point the carotid bulges slightly and forms the **carotid sinus.** Other baroreceptors are scattered throughout the wall of the aortic arch. Afferent fibers from these baroreceptor cells enter cranial nerves IX and X and ascend to the cardiovascular center of the brainstem medulla. There they synapse on interneurons that determine the outflow of impulses in parasympathetic and sympathetic pathways to the heart, sympathetic pathways to the vessels, and sympathetic activation of the adrenal medulla (Figure 15-3).

The importance of carotid sinus receptors was discovered in the 1920s by a physician named Hering, who noticed that merely stroking the area of the neck above the carotid sinuses slowed the heart and caused fainting in some sensitive patients. Hering then showed with animal experiments that distending the carotid sinuses slows the heart and lowers pressure, whereas reducing pressure at the sinuses by clamping the carotid arteries below the carotid sinus increases heart rate and blood pressure.

Baroreceptors discharge action potentials at a rate proportional to the stretch of the arterial wall (Figure 15-4, *A*). The rate rises and falls in response to arterial pressure changes over the heart cycle. For the cardiovascular integrating center in the medulla, there is a "setpoint" firing pattern that corresponds to the normal mean arterial blood pressure. If mean arterial blood pressure in the carotid arteries begins to fall, the firing rates of the aortic and carotid

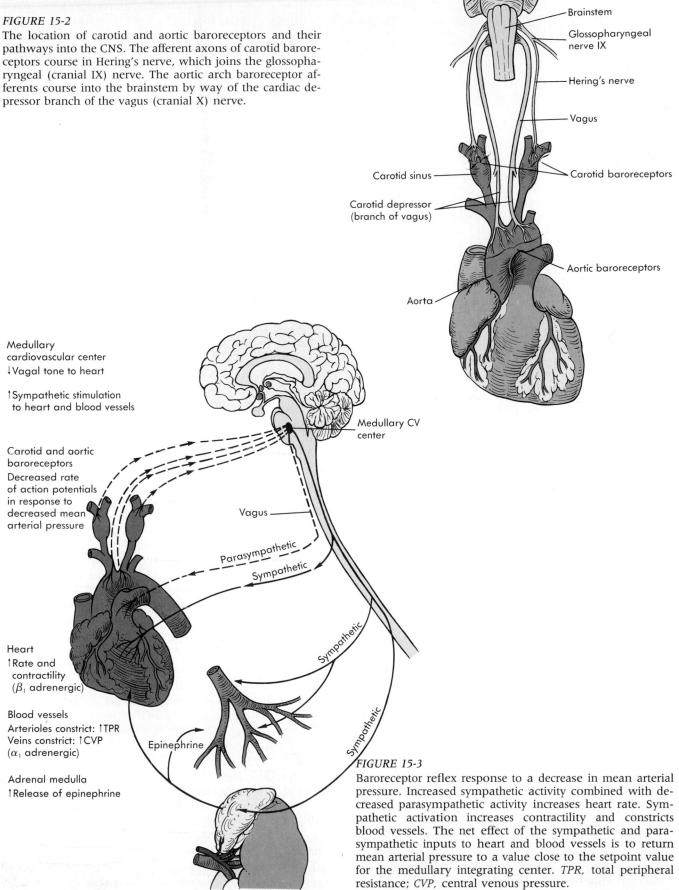

FIGURE 15-2
The location of carotid and aortic baroreceptors and their pathways into the CNS. The afferent axons of carotid baroreceptors course in Hering's nerve, which joins the glossopharyngeal (cranial IX) nerve. The aortic arch baroreceptor afferents course into the brainstem by way of the cardiac depressor branch of the vagus (cranial X) nerve.

Brainstem

Glossopharyngeal nerve IX

Hering's nerve

Vagus

Carotid sinus

Carotid depressor (branch of vagus)

Carotid baroreceptors

Aortic baroreceptors

Aorta

Medullary cardiovascular center
↓Vagal tone to heart

↑Sympathetic stimulation to heart and blood vessels

Medullary CV center

Carotid and aortic baroreceptors
Decreased rate of action potentials in response to decreased mean arterial pressure

Vagus

Parasympathetic

Sympathetic

Heart
↑Rate and contractility (β_1 adrenergic)

Blood vessels
Arterioles constrict: ↑TPR
Veins constrict: ↑CVP
(α_1 adrenergic)

Epinephrine

Sympathetic

Sympathetic

Adrenal medulla
↑Release of epinephrine

FIGURE 15-3
Baroreceptor reflex response to a decrease in mean arterial pressure. Increased sympathetic activity combined with decreased parasympathetic activity increases heart rate. Sympathetic activation increases contractility and constricts blood vessels. The net effect of the sympathetic and parasympathetic inputs to heart and blood vessels is to return mean arterial pressure to a value close to the setpoint value for the medullary integrating center. *TPR*, total peripheral resistance; *CVP*, central venous pressure.

Regulation of the Cardiovascular System

425

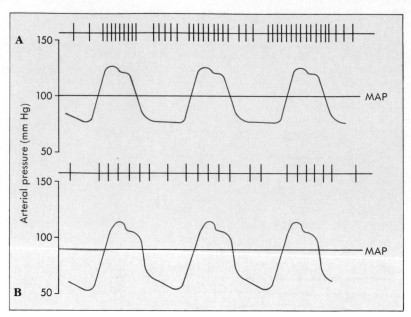

FIGURE 15-4
Idealized response of a carotid baroreceptor to two mean arterial pressures. The upper trace in each block shows the rate of action potentials; the lower trace, the record of arterial pressure. An increase in the rate of action potentials occurs with the pressure rise of each systole. The medullary cardiovascular center averages the rate of baroreceptor action potentials over time so that its response reflects the mean arterial pressure.

baroreceptors decrease (Figure 15-4, *B*). The medullary cardiovascular control center reacts to a drop in the frequency of action potentials from baroreceptors by increasing sympathetic activity to the heart and blood vessels and by reducing parasympathetic tone to the SA node (see Figure 15-3). These changes in autonomic inputs to the heart increase its rate and force of contraction. Increased sympathetic input to arterioles of the systemic loop causes a general arteriolar constriction and an increase in total peripheral resistance. Together, these changes rapidly restore carotid pressure to normal and maintain cardiac output.

The baroreceptors can also oppose an increase in blood pressure above the reflex set point. For example, if epinephrine is injected into a normal subject, the immediate effects are the same as those described for activation of the sympathetic nervous system: an increase in heart rate, arteriolar and venous constriction, and a dramatic increase in blood pressure. However, the heart slows within a few seconds. The slowing results from an increase in parasympathetic input to the heart, reflexively elicited by the pressure increase detected by the baroreceptors. At the same time, any sympathetic input to the cardiovascular system that may have been occurring at the time of the epinephrine injection is diminished. These adjustments soon restore the blood pressure normal.

Carotid baroreceptors play a more significant role in the baroreceptor reflex than do the baroreceptors in the aortic arch. However, a maximum response requires participation of both sets. The response of these receptors is not linearly related to the increase or decrease in blood pressure; instead, their response becomes more intense the farther the blood pressure departs from the setpoint. This nonlinearity increases the speed and therefore the effectiveness of the baroreceptor reflex.

Role of Blood Volume Regulation in Blood Pressure Regulation

The baroreceptor reflex can respond in seconds to blood pressure changes caused by changes in posture or moderate hemorrhage, but blood pressure regulation over periods of hours to days requires the cooperation of a family of neural and hormonal mechanisms that regulate both plasma volume and the baseline tone of vascular smooth muscle. The mechanisms of plasma volume regulation by the kidney are discussed more fully in Chapter 20. The volume regulatory systems are outlined here because of their importance in blood pressure homeostasis. The three main volume regulatory systems are the antidiuretic hormone system, the atrial natriuretic hormone system, and the renin-angiotensin-aldosterone system (Figure 15-5).

Antidiuretic hormone (ADH or **vasopressin)** has three major effects: (1) it regulates the fraction of water recycled into the body from urine in the kidney, thus affecting the rate of fluid loss from the plasma; (2) the ADH system affects the sensation of thirst, so it contributes to regulation of fluid intake as well as loss; and (3) it is a vasoconstrictor, so increases in plasma ADH tend to increase total peripheral resistance.

As described in Chapter 5, ADH is a small peptide neurohormone secreted by hypothalamic neurons whose axons pass into the posterior pituitary. The ADH neurons integrate two types of sensory input. One type of input comes from baroreceptors in the wall of the right atrium, which are responsive to changes in central venous pressure. Venous pressure is a sensitive indicator of blood volume. Decreases in central venous pressure signal abnormally low plasma volume and increase the rate of secretion of ADH. The increased ADH secretion causes increased thirst, reduction in the rate of fluid loss in urine, and increased vascular tone (see Figure 15-5). The effect on vascular tone requires especially high levels of ADH and may not be significant under most circumstances.

The ADH neurons also respond to changes in the osmotic pressure of plasma, increasing their secretion when plasma becomes more concentrated and decreasing it when the plasma is too dilute. As a result, factors that affect plasma osmolarity also affect blood pressure; the implications of this are treated further in Chapter 20.

Recently, the atrial walls have been shown to contain endocrine cells that secrete a peptide called **atrial natriuretic hormone (ANH)**. Secretion of ANH increases in proportion to atrial stretching, which reflects central venous pressure (see Figure 15-5). The detailed effects of ANH are still being explored, but at present it appears that a major effect of ANH is to decrease the reabsorption of Na^+ and water in the kidney, increasing the rate of loss of plasma solutes and water in the urine.

Secretion of the steroid hormone, **aldosterone**, by the adrenal cortex is controlled in part by a regulatory cascade that originates in the kidney, and also by the blood levels of Na^+ and K^+ as sensed by the adrenal cells themselves. Endocrine cells of the kidney increase their secretion of the enzyme **renin** in response to either decreased renal perfusion or decreased plasma Na^+ concentration. Decreased renal perfusion is a signal of decreased mean arterial pressure because, when arterial pressure falls, the baroreceptor reflex constricts the arterioles of the kidney as well as those of most of the other abdominal organs. Thus renin secretion is part of a feedback loop that regulates mean arterial pressure (see Figure 15-5).

In the plasma, renin catalyzes the conversion of a plasma protein, **angiotensinogen**, into **angiotensin I**. Angiotensin I is converted to **angiotensin II** by angiotension converting enzyme present in organs such as the lung and liver. Angiotensin II is converted to **angiotensin III** in the adrenal cortex; this form of angiotensin stimulates aldosterone secretion. Like ADH, angiotensin II is believed to stimulate thirst. Both angiotensin I and II stimulate contraction of vascular smooth muscle, increasing peripheral resistance.

Aldosterone decreases the rate of Na^+ loss in urine. Because Na^+ is ingested continually with food, increased aldosterone secretion increases the body's total content of Na^+. Because almost all Na^+ remains in the extracellular fluid, the ultimate effect of an increase in total Na^+ is an increase in the volume of tissue interstitial fluid and plasma, elevating the blood pressure.

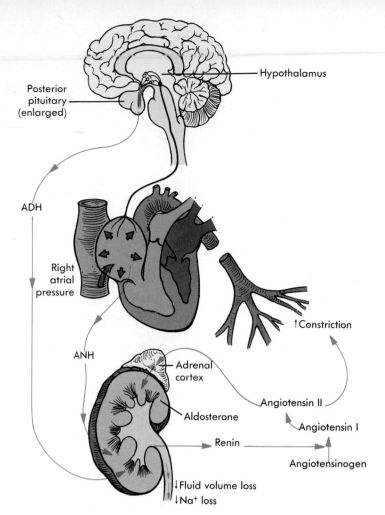

FIGURE 15-5
Summary of three hormone systems that regulate blood volume in response to changes in atrial pressure (ADH and ANH) or renal perfusion (renin-angiotensin-aldosterone). A significant loss of extracellular fluid results in increased secretion of ADH and aldosterone and inhibition of ANH; their net effect is to decrease renal excretion of Na^+ and water and to increase thirst.

Hypertension

- *What is hypertension? In what ways can it be reversed by therapy?*

Excessive secretion of renin is one cause of chronic elevation of arterial blood pressure (**hypertension**). The hypertension is caused both by the increased plasma volume (**hypervolemia**) that results from excess aldosterone secretion and by the increased total peripheral resistance resulting from the vasoconstriction caused by the angiotensins. Hypertension may also result from other factors that elevate plasma volume, from elevated sympathetic activity, or from disorders that increase the total peripheral resistance. Hypertension may be worsened by excessive consumption of NaCl because salt retention causes increases in plasma volume. In more than 95% of individuals with hypertension, the underlying cause of the disease is unknown (referred to as

Measurement of Cardiac Output

Cardiac output can be measured using a method based on a dilution principle, first introduced by the physiologist A.V. Fick. An indicator substance is added to the volume of venous blood returning to the heart, and its concentration is measured at an arterial site with the assumption that it has been completely mixed with the blood that passed with it through the heart.

The original version of the Fick Method used oxygen taken up by the lungs as the indicator. Oxygen is added to the blood in a steady stream at a known rate by respiration, and its subsequent dilution can be measured by collecting data on the rate of oxygen uptake and the difference between the oxygen concentration in systemic venous blood and in systemic arterial blood (the A-V O_2 difference). It then becomes possible to calculate cardiac output using the equation:

Cardiac output = O_2 consumption ÷ A-V O_2 difference

For example, suppose that venous blood oxygen is 14 ml/dl blood and arterial blood oxygen is 19 ml/dl blood. This means that each deciliter of blood that passes through the heart receives 5 ml of oxygen, the A-V O_2 difference. A liter (1000 ml of blood) would therefore receive 50 ml of oxygen. If the rate of oxygen consumption is 250 ml/min, the cardiac output is:

Cardiac output = 250 ml/min ÷ 50 ml O_2/L = 5 L/min

This method sounds simple, but it is complicated by the fact that venous blood draining from various organs differs in oxygen content. Differing venous blood becomes well mixed only after it passes through the right ventricle. Therefore the venous blood sample must be collected from the pulmonary artery. This requires cardiac catheterization—passing a cannula into a vein in the arm or leg and from there to the vena cava, through the right atrium, right ventricle, and into the pulmonary artery—entailing some risk to a cardiac patient.

Various dyes can also be used as indicators. The dye is injected rapidly into a vein, and the blood is repeatedly sampled from an artery. In its course through the heart, the dye becomes well mixed with the blood. Some seconds after the dye is injected, it starts to appear at the

---

essential hypertension). Hypertension usually causes no immediate symptoms but poses a long-term threat to health because of the stress it places on the heart and vessels. Whatever the cause of hypertension, therapy typically involves some combination of **diuretics** (drugs that increase urinary loss of fluid and solutes) and adrenergic antagonists (drugs that decrease cardiac contractility and dilate arterioles by blocking the effect of sympathetic stimulation).

INTERACTION OF THE HEART AND VESSELS
Effect of Cardiac Output on Central Venous Pressure

- *In the vascular function curve, which system variable is taken as the dependent variable? Which is the independent variable?*
- *How is the cardiovascular operating point determined graphically?*
- *What is the cardiac reserve? What factors affect the magnitude of the cardiac reserve?*

sampling site (Figure 15-A). It then passes through the rest of the systemic loop, returning to the sampling site more and more diluted as it mixes with the total blood volume and is taken into the tissues. The final dilution of the dye is an approximate measure of the total blood volume. A problem with dyes is that the second pass of dye starts to appear at the sampling point just as the first pass is ending; this pileup makes computing the mean concentration of the first pass more difficult. More recently thermodilution of a pulse of heat or cold has been used; the heat or cold dissipates into the body after passing the sampling site and does not return for a second pass.

FIGURE 15-A
The indicator dilution method for determining cardiac output. The indicator is injected into a major vein as a pulse. As it mixes with the rest of the venous blood in the heart and pulmonary loop, the indicator is diluted and the pulse spreads out because of the variations in linear flow velocity in different parts of the vessels. The pulse arrives at the sampling site in a major artery, where its mean concentration is measured by rapid sampling. The spreading effect allows the arrival of the second passage of the dye to catch up with the tail of the first pass. After two passes the indicator has mixed almost completely with the total blood volume.

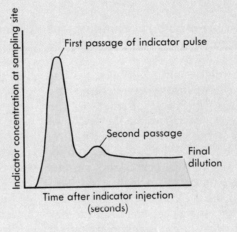

Central venous pressure both affects and is affected by the pumping action of the heart. Consider what would happen if the heart were stopped for a time. After heart pumping ceased, blood flow through the circulation would continue until pressure became equal throughout the system. Because the veins are the most compliant vessels, the blood pressure throughout the system after flow had ceased would be determined by the compliance of the veins and the volume of blood in them. A typical value for this pressure would be about 7 mm Hg.

If the heart began pumping again, it would move some blood from the venous side into the arterial side of the circulation. Because the arteries are far less compliant than the veins, the effect of shifting a small volume of blood from the venous side to the arterial side of the circuit would be a large increase in arterial pressure and a small decrease in venous pressure. The greater the cardiac output, the greater the volume shifted would be and the greater the drop in venous pressure. To restate the relationship in terms of venous return, the greater the difference

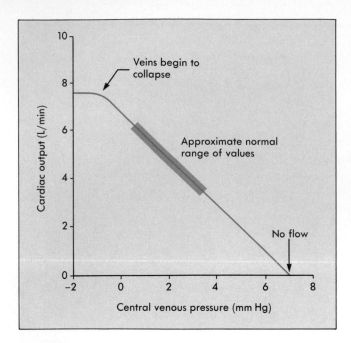

FIGURE 15-6
Vascular function curve. When flow through the vasculature is stopped, venous pressure is maximal; as cardiac output increases, venous pressure decreases. When venous pressures become lower than atmospheric pressure, veins begin to collapse, allowing no further increase in cardiac output.

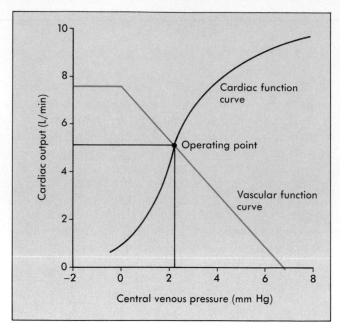

FIGURE 15-7
When the vascular function curve and the cardiac function curve are plotted on the same axes, they intersect at an operating point, whose coordinates give the cardiac output and central venous pressure determined by the interaction of the two systems.

between mean arterial pressure and central venous pressure generated by heart activity, the greater the venous return.

The graph of the relationship between cardiac output (venous return) and central venous pressure is the **vascular function curve** (Figure 15-6). The vascular function curve describes the behavior of the peripheral vessels, whereas the cardiac function curve describes the behavior of the heart as a pump. Both curves relate cardiac output and central venous pressure. Because the vascular function curve will later be superimposed on the same axes as the cardiac function curve, the central venous pressure has been plotted on the horizontal axis, even though it is now the dependent variable. Cardiac output is shown on the vertical axis. The upper limit of venous pressure is the value when flow is zero; the lower limit is set by the fact that veins begin to collapse and restrict flow at pressures less than the atmospheric pressure (indicated by the plateau in the curve). The vascular function curve can be measured in animal experiments in which the heart is replaced by a mechanical pump to eliminate effects of the vascular reflexes.

Cardiovascular Operating Point
Depending on its autonomic inputs and state of health, the heart can maintain a cardiac output of as little as 2 to 3 L/min or of more than 20 L/min. The vasculature could, if maximally dilated, permit a venous return even greater than the heart

could pump. The actual flow through the circuit at any instant depends both on the state of the vasculature, as specified by the properties of plasma volume, total peripheral resistance, and venous compliance, and on the performance of the heart, as specified by the heart rate and myocardial contractility. The vascular function curve describes the performance range of the vasculature; the cardiac function curve describes the performance range of the heart.

If the vascular function curve and the cardiac function curve are plotted on the same axes, the intersection of the two curves is the **cardiovascular operating point** (Figure 15-7). This intersection is the one point on the graph at which cardiac output equals venous return. Thus, of the wide range of blood flows possible for the heart and vasculature, the operating point specifies the actual flow that results when the heart and vasculature interact.

Vascular Determinants of Cardiac Output and Mean Arterial Pressure
Cardiac output and central venous pressure are affected by changes in the total peripheral resistance, the compliance of veins, and the blood volume. For simplicity, the effects of vascular changes on cardiac output and venous pressure are first described with the assumption that the heart rate and contractility do not change. In reality, changes in the status of the vessels are almost always accompanied by changes in the heart.

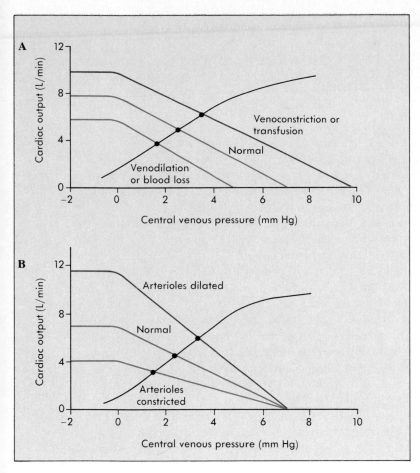

FIGURE 15-8

A The effect of changes in blood volume or venous tone on the vascular function curve and the cardio-vascular operating point. Venoconstriction and blood transfusion increase the cardiac output and central venous pressure; venodilation or blood loss decreases cardiac output and venous pressure.

B The effect of arteriolar dilation or constriction on the vascular function curve and the operating point. Arteriolar dilation (that is, a decrease in total peripheral resistance) increases cardiac output and central venous pressure; arteriolar constriction (that is, an increase in total peripheral resistance) decreases cardiac output and central venous pressure. The common feature of the venous tone, volume, and arteriolar diameter changes is that they affect central venous pressure and thus increase or decrease cardiac output according to the Frank-Starling Law.

An increase in blood volume, such as from a blood transfusion, appears primarily in the veins because veins are more distensible than arteries. The effect of increased blood volume is to increase the central venous pressure at any given cardiac output (Figure 15-8, A). Consequently, the cardiac output increases as the heart obeys the Frank-Starling Law. In other words, the curve representing cardiac output moves up and to the left along the vascular function curve. The new curve is identical to the control curve, but displaced to the right because of the increased blood volume and higher venous pressure. An opposite shift of the vascular function curve occurs when there is a decrease in blood volume, such as by hemorrhage (see Figure 15-8, A).

Decreases in venous compliance (venoconstriction), in the absence of flow (cardiac output = 0) also increases the central venous pressure, shifting the intercept of the vascular function curve to the right. When one then transfers blood from the venous to the arterial side by turning the pump on, one obtains a curve that is almost but not exactly parallel to the original.

When arterioles dilate, blood flows from the arterial side to the venous side, filling the veins and increasing the central venous pressure (Figure 15-8, B). Changes in arteriolar tone do not alter the inter-cept of the vascular function curve because there is no effect at greater flow. If there is arteriolar vaso-constriction, the transfer of blood to the venous side is decreased, and the central venous pressure decreases. With arteriolar vasodilation the opposite occurs. The greater the cardiac output, the greater the change in central venous pressure. In other words, as shown in Figure 15-8, the increased stretching of the ventricles during diastole causes the cardiac output to increase to the value (the new operating point) represented by the intersection of the new vascular function curve and the cardiac function curve (see Figure 15-8, B).

As metabolically active tissues increase their metabolic rate, the resulting decrease in total peripheral resistance causes the heart to increase the cardiac output. This is a very effective means of regulating cardiac output because it does not require any direct nervous or hormonal communication between the active tissues and the heart. It is the major means by which cardiac output is regulated in heart transplant recipients because a transplanted heart is not innervated. However, greater increases in cardiac output are possible in normal individuals because in them the heart is also stimulated by adrenergic input.

The difference between the actual cardiac output that matches venous return at any instant and

Causes and Consequences of Heart Failure

Acute damage to the myocardium can result from a sudden occlusion of a coronary blood vessel (a myocardial infarct, or heart attack) or from the effects of hemorrhagic shock. Coronary artery disease may slowly reduce coronary blood flow, resulting in progressive damage to the heart muscle. Decreases in myocardial contractility that result from disease or damage are generally referred to as **heart failure.**

In heart failure, right atrial pressure rises as unpumped blood piles up in the vena cava. The increase in right atrial pressure stretches the ventricles until cardiac output equals venous return. Thus the ventricles compensate for their decreased contractility at the expense of an increased end-diastolic volume. Some decrease in arterial blood pressure also occurs. As the ventricles continue to fail, right atrial pressure becomes higher and higher, so that the volume of blood left in the ventricles at the end of a systole may become several times greater than the stroke volume. Ultimately the cardiac muscle is stretched so much that further stretching does not elicit an increase in cardiac output. In fact, there is a decrease in the strength of contraction and a marked drop in systemic arterial blood pressure.

If heart failure takes place slowly, the decrease in arterial pressure that occurs leads to an increased release of aldosterone. Aldosterone acts to retain Na^+, so that during the course of the failure, plasma volume actually increases. This extra blood volume further increases the load on the failing heart. Some of the extra volume is transferred to the interstitial fluid because the increased venous pressure "backs up" into the capillaries. Edema in the lower parts of the body occurs when the patient is erect, and pulmonary edema, causing difficulty in breathing, may occur when the patient is reclining. These are the symptoms of congestive heart failure, *congestive* being a term used to describe the accumulation of fluid.

One frequent procedure used in the diagnosis of heart failure, the stress test, involves monitoring cardiac function in an individual during exercise on a treadmill (Figure 15-B).

The treatment of heart failure depends on the cause, and interventions appropriate for one cause may well result in additional damage for another. If the failure results from inadequate blood flow to the heart (cardiac ischemia), treatment is aimed at increasing the oxygen supply to the heart. In such cases, stimulating cardiac contractility with drugs such as epinephrine or digitalis may be a

the maximum output that could be maintained by the heart is called the **cardiac reserve** (Figure 15-9). The cardiac reserve of a healthy resting person is about 3 L/min, so that without sympathetic stimulation the cardiac output could rise no higher than about 8 L/min, assuming a resting cardiac output of about 5 L/min. The cardiac reserve sinks toward zero in those with progressive heart failure. Sympa- thetic stimulation can more than double the cardiac reserve; thus one advantage of extrinsic control of the heart is that it makes possible much greater increases in cardiac output than would be possible from the effect of the Frank-Starling Law alone.

Central control of the heart and blood vessels is also necessary because autoregulation of flow by individual tissues is a selfish process that can compro-

two-edged sword. These drugs increase the contractility of the undamaged portion of the heart, but they also increase the oxygen consumption of the heart, usually causing further damage to a region that is inadequately perfused. The internal work of the heart depends on the systolic pressure it must generate and on its end-diastolic volume. Any intervention that serves to decrease right atrial pressure or mean arterial pressure should be beneficial to a failing heart. Consequently, diuretics (drugs that increase urine production) are given to reduce right atrial pressure, and vasodilators to reduce peripheral resistance.

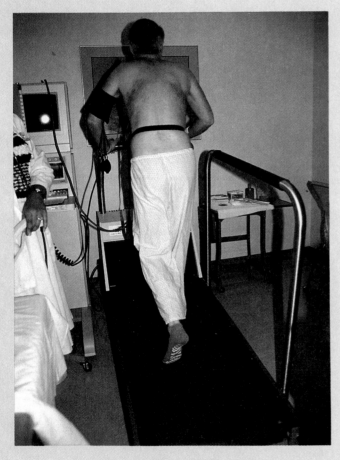

FIGURE 15-B
A patient undergoing a stress test on a treadmill. Cardiac function is monitored as the exercise level is progressively increased.

mise maintenance of mean arterial pressure. For example, consider blood loss, whose immediate effects on cardiac output and central venous pressure are illustrated in the bottom curve in Figure 15-8, *A*. The drop in cardiac output would cause all tissues to become underperfused. In the absence of any overriding central control, a general arteriolar dilation would occur because local autoregulation in the underperfused tissues acts to increase their blood flow. The mean arterial pressure would therefore fall further. In this case autoregulation, if not countered by central control mechanisms such as the baroreceptor reflex, would result in a dangerous lack of control over the driving force for perfusion of essential organs such as the brain and the heart itself.

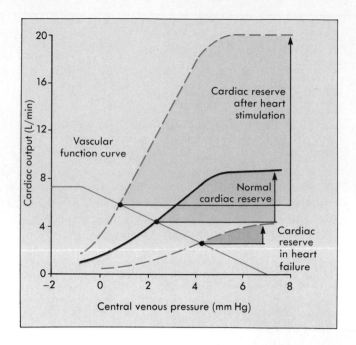

FIGURE 15-9
Cardiac reserve is the difference between the actual cardiac output that matches venous return at any instant and the maximum output that could be maintained by the heart.

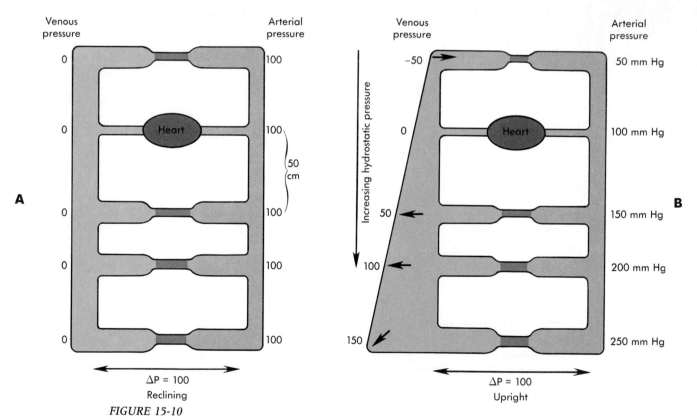

FIGURE 15-10
Pressures in the cardiovascular system for a person who is, **A**, lying down and, **B**, standing. The pressures in mm Hg are given at various levels in both the arteries and veins. In a recumbent person, the mean arterial and venous pressures are uniform because they depend only on the pumping action of the heart. In a standing individual, the hydrostatic pressure component must be added (below the heart) or subtracted (above the heart). However, the hydrostatic pressure affects only the transmural pressures across the blood vessels *(small solid arrows)*. The pressure difference between the arteries and veins (perfusion pressure) remains unchanged. As a result of the increase in transmural pressure, blood tends to pool in the veins of lower extremities (indicated by the changes in size).

THE CARDIOVASCULAR SYSTEM

THE CARDIOVASCULAR SYSTEM IN EXERCISE AND DISEASE
Standing Upright: The Baroreceptor Reflex and Venous Pumps

> - *What processes make it possible to stand upright without fainting?*
> - *What are the immediate reflexive responses to hemorrhage?*
> - *What role does capillary filtration play in homeostasis of plasma volume after hemorrhage?*

In a person who is lying down, the mean arterial blood pressure of the major arteries of the head, trunk, and legs is similar; the same is true for the pressure in great veins (Figure 15-10, *A*). Standing up adds a gravitational component to the arterial and venous pressures of the lower body and reduces the pressures in the upper body above the heart (Figure 15-10, *B*). As a consequence, arterial pressure in the lowermost parts of the body increases, whereas in the head it decreases. The venous pressure in the feet is increased from a few mm Hg to as much as 150 mm Hg, whereas in the head it becomes significantly negative. Prolonged standing or chronic elevated central venous pressure may distend the veins of the legs enough to weaken their walls, resulting in permanently distended **varicose veins.**

The pressure changes that occur on standing up have implications for the movement of blood through the circulation. The increased venous pressure in the lower body increases the volume of blood in lower veins by as much as 500 to 700 ml, reducing the central venous pressure that is responsible for returning blood to the heart. If these changes were unopposed, the drop in venous pressure and venous return would result in a substantial decrease in cardiac output. Arterial pressure would soon become insufficient to drive blood into the parts of the body above the heart, and fainting would result from inadequate perfusion of the brain. In its effect on arterial pressure, standing up is equivalent to immediately losing more than 500 ml of blood. Standing erect without fainting is possible because of the response of the baroreceptor reflex and because venous return is assisted by the contraction of limb muscles and the changes in the intrathoracic pressure caused by respiration.

The baroreceptor reflex has three effects: (1) a slight elevation of the heart rate, (2) an increase in the total peripheral resistance, and (3) a decrease in venous compliance. All of these changes start within a few seconds and are completed within a minute or so after standing up. The changes maintain arterial blood pressure at its normal value at the level of the baroreceptors. Blood pressure measured using a manometer and arm cuff will actually rise slightly after a person stands up because the arm cuff is positioned lower than the baroreceptors and includes a hydrostatic component.

In some people, the baroreceptor reflex is weak or absent, and standing frequently results in fainting. This condition is called **orthostatic hypotension.** Even if the baroreceptor reflex is functioning normally, people standing quietly still experience some venous pooling, causing the stroke volume to drop by about 40%, a drop that may decrease arterial blood pressure enough to cause fainting. However, when an erect person is physically active, pooling is reduced by the operation of two secondary pumps, the **skeletal muscle pump** and the **respiratory pump.**

The veins that pass between and next to the skeletal muscles of the limbs are squeezed and partly emptied with every contraction of the muscles (Figure 15-11). The one-way valves ensure that blood squeezed out of veins moves in the direction of the heart. When the skeletal muscles relax, the veins refill from their arterial ends. Soldiers standing for prolonged periods at attention are taught to rhythmically contract their leg muscles to reduce the pooling of blood and prevent fainting.

During the inspiratory phase of respiration the pressure within the chest cavity falls below atmospheric pressure; during expiration it rises above atmospheric pressure. These pressure changes are transmitted to the portion of the inferior vena cava

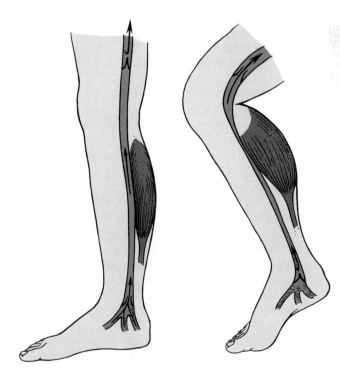

FIGURE 15-11
The skeletal muscle pump depends on the presence of valves in the veins and the fact that muscles adjacent to the veins squeeze these thin-walled vessels when the muscles contract.

FIGURE 15-12

The respiratory pump draws venous blood toward the heart as a result of the changes in pressure in the thoracic cavity during the respiratory cycle. In inspiration, the thoracic cavity enlarges, and, as it does so, the subatmospheric pressure pulls blood from the portion of the inferior vena cava in the abdominal cavity into the thoracic portion of that vein. When deep breaths are taken, the diaphragm moves downward during inspiration, pressing on the abdominal organs and further increasing the pressure difference experienced by the blood in the abdominal and thoracic portions of the inferior vena cava.

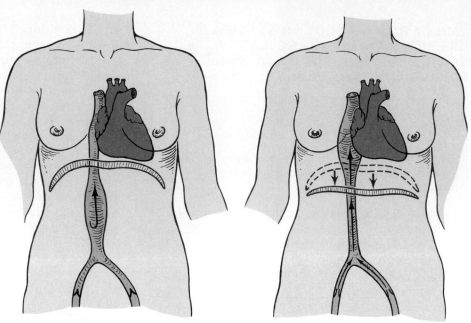

Expiration

Inspiration

that is located within the chest cavity (Figure 15-12). The result is that some blood is sucked into the vena cava from the abdominal veins during inspiration and then driven into the right atrium during expiration. Inspiration also increases abdominal pressure, pushing blood upward out of the abdominal veins. The respiratory pump is more effective during exercise because the thoracic pressure changes are greater. The skeletal and respiratory pumps are usually sufficient to overcome the gravitational effect on central venous pressure and maintain stroke volumes equal to or higher than those of people who are reclining quietly.

Hemorrhage and Hypovolemic Shock

The decreased blood volume (**hypovolemia**) caused by hemorrhage may reduce venous pressure to the point that end-diastolic volume is insufficient to maintain cardiac output and arterial blood pressure. Life is threatened when decreased arterial pressure (**hypotension**) causes critical organs such as the heart and brain to become underperfused (Figure 15-13). The baroreceptor reflex is the first line of defense against decreasing arterial blood pressure. The reflexive increase in heart rate and peripheral resistance partly compensates for decreased stroke volume and cardiac output. General arteriolar constriction caused by the reflex does not affect the arterioles of the heart and brain. The brain has few sympathetic fibers and both the brain and coronary arteries lack α-adrenergic receptors. Autoregulation by local factors is the major determinant of flow resistance in these organs, aided by the presence of β_2-adrenergic receptors in the coronary arteries mediating vasodilation. In effect, perfusion of the skin, skeletal muscle, and abdominal organs is sacrificed to maintain cardiac and cerebral blood flow.

The drop in mean arterial pressure and the reflexive increase in arteriolar constriction after hemorrhage cause a general decrease in capillary hydrostatic pressure. As noted in Chapter 14, decreases in capillary hydrostatic pressure favor transfer of fluid from interstitial spaces into the blood. Within a few hours after blood loss, as much as several hundred milliliters of plasma may be replaced by this process and by a shift of volume from the lymphatic circulation to the blood circulation. Hours to days may be required to replace the water, ionic constituents, and plasma proteins lost in severe bleeding. Replacement of lost blood cells may require several weeks.

Sweating, diarrhea, vomiting, or kidney diseases that cause excessive urine production can result in large losses of extracellular fluid volume without loss of blood cells. The threat of these fluid losses to circulatory function is similar to that seen in hemorrhage, and the reflexive and hormonal compensations are also similar.

Hemorrhage that results in rapid loss of more than 20% of blood volume may result in **hypovolemic shock**. In hypovolemic shock, underperfusion of the heart and nervous system compromises heart contractility and blocks cardiovascular reflexes. A vicious cycle is set in motion in which the arterioles of underperfused tissues dilate, further depriving the heart and brain of blood.

Performance in Sustained Exercise: The Cardiovascular Limit

- *What mechanisms are responsible for the large increase in cardiac output possible during exercise?*
- *What adjustments enable mean arterial pressure to remain essentially constant during exercise?*

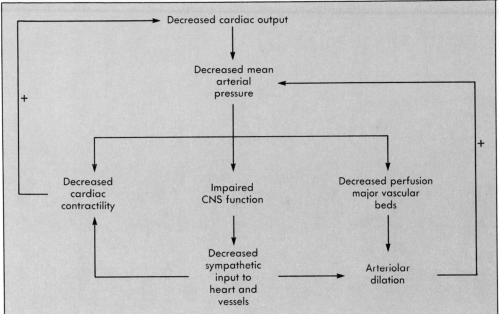

FIGURE 15-13
The positive feedback that can lead to death from hypovolemic shock.

Sustained exercise relies heavily on slowly fatiguing, high-oxidative skeletal muscle fibers. In principle, the workload these fibers can sustain could be limited by their own metabolism or contractility, by the ability of the lungs to exchange CO_2 for O_2, or by the rate at which the circulation can move blood from the pulmonary loop to working muscles. When the cardiac output of an exercising human reaches a maximum, no further increases in O_2 uptake or in workload can be sustained. Thus cardiovascular adjustments are crucially important for maximum performance in endurance exercise. In well-trained athletes, cardiovascular performance may be the limiting factor that separates champions from also-rans. World-class endurance athletes typically can generate maximum cardiac outputs in excess of 40 L/min, so much greater than those of average well-trained athletes as to suggest that truly superior performers in sports such as the marathon, cross-country skiing, and cycling racing owe something to genetics as well as to training.

Maximization of blood flow to working muscles necessitates that the cardiac output be increased and that the working muscles receive the largest possible share of the increased output. In nonathletes, the blood flow to skeletal muscle may increase by twentyfold; in elite endurance athletes, the increase can be as much as fortyfold. These increases are achieved by a combination of four processes: (1) a large vasodilation in the working muscle, (2) autoregulation of cardiac output according to the Frank-Starling Law, (3) an increase in heart rate and contractility resulting from increased adrenergic input, and (4) sympathetically mediated vasoconstriction in many vascular beds of the visceral circulation (Figure 15-14 and Table 15-3).

In resting muscle, arterioles are constricted and much of the blood flows through metarteriolar connections to venules; only about 1% of the capillaries are open at any given instant. When skeletal muscle fibers begin to contract, local factors dilate the muscle's arterioles, delivering a greatly increased share of the cardiac output to the muscles (see Table 15-3). Without any other cardiovascular changes, dilation of muscle arterioles would decrease the peripheral resistance. Venous return would increase, causing the cardiac output to increase as a consequence of the Frank-Starling Law.

The activation of sympathetic efferents to arterioles causes a general arteriolar constriction, especially in the abdominal viscera, which diverts some blood flow from the viscera into working muscles (see Figure 15-14). However, the increase in resistance of the abdominal organ circulation is far overbalanced by the decrease in resistance of the vascular beds of working muscles. In sum, total peripheral resistance typically falls during moderate-to-heavy exercise (see Table 15-3), and the greater the mass of working muscle, the greater the decrease in total peripheral resistance.

The large increases in cardiac output typical of exercise could not be achieved without changes in the properties of both the heart and vasculature. An increase in cardiac performance alone would cause only a fractional increase in cardiac output. Vascular changes by themselves are slightly more effective but would increase cardiac output only by the amount of the cardiac reserve (see Figure 15-9). To achieve even the several fold increase in cardiac output typical of submaximal exercise (see Table 15-3), cardiac performance and vascular changes must occur together (Figure 15-15). In terms of the cardiac function and venous return curves, increased

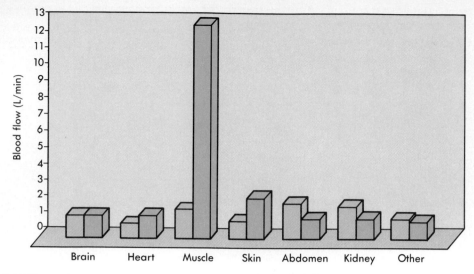

FIGURE 15-14
Typical changes in organ perfusion with exercise. The left bar in each pair shows the resting blood flow; the right bar shows the flow during exercise.

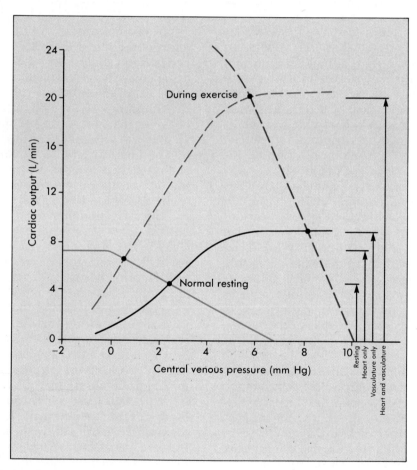

FIGURE 15-15
The importance of both cardiac and vascular responses in attaining high cardiac output in exercise. The two solid curves are the resting cardiac and vascular function curves. The broken curves are curves typical of submaximum exercise. Extrinsic stimulation of the heart alone would raise the cardiac output only slightly. The vascular responses (arteriolar dilation and venoconstriction) can increase cardiac output only to the extent of the cardiac reserve. Interaction of the vascular responses with the cardiac response raises cardiac output by a large factor: fourfold in this example. Considerably larger increases are possible for trained athletes engaging in maximum exercise.

THE CARDIOVASCULAR SYSTEM

TABLE 15 3   *Typical Circulatory Changes with Submaximal Exercise*

Parameter	Resting	Exercise
Cardiac output	5 L/min	20 L/min
Mean arterial pressure	90 mm Hg	105 mm Hg
Systolic pressure	120 mm Hg	160 mm Hg
Diastolic pressure	70 mm Hg	60 mm Hg
Stroke volume	70 ml	130 ml
Total peripheral resistance	18 mm Hg/L/min	5.25 mm Hg/L/min
Muscle blood flow	1 L/min	10 L/min
Flow to abdominal viscera	1.2 L/min	0.5 L/min
Cutaneous blood flow	0.5 L/min	1.5 L/min

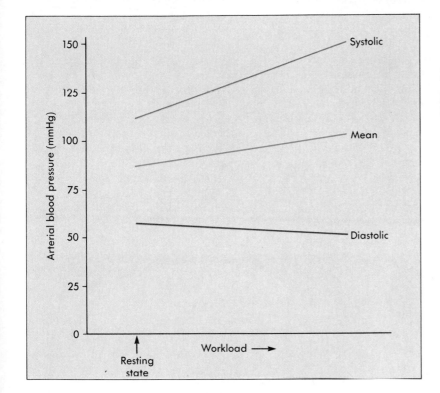

FIGURE 15-16
The effect of increasing workload on systolic pressure, diastolic pressure, and mean arterial pressure.

contractility shifts the former up and to the left, while the venous return curve shifts upward and to the right because of the combination of arteriolar vasodilation and venoconstriction. The combination shifts the operating point by more than either alone.

During moderate exercise, increased contractility of the heart and increased stroke volume (see Table 15-3) usually cause pulse pressure to increase (Figure 15-16). At the same time, the decrease in total peripheral resistance speeds blood runoff from the aorta into the arterial tree, allowing aortic pressure to fall rapidly between heartbeats. In well-muscled people the diastolic pressure may even decrease with increasing workload. The net effect of these changes is to leave mean arterial pressure almost unchanged. Thus changes in the driving force

for blood flow do not play an important part in increasing performance of the cardiovascular system during exercise. Blood pressure is regulated by the baroreceptor reflex during exercise as during rest. In this process the medullary cardiovascular center must balance the demand of working muscle for blood flow against the necessity of maintaining a mean arterial blood pressure adequate to perfuse the brain.

Vascular Changes in Regulation of Core Body Temperature

In cold environments, the arterioles and veins nearest the surface of the skin are highly constricted, diverting most of the blood flow through arteriolar-venular shunts (see Chapter 14) and reducing the rate of heat loss to the environment from the body

About 50% of all deaths from heart disease in developed nations are the result of atherosclerosis in the arteries of the heart or brain. Atherosclerosis is a disease process that results in the formation of abnormally thickened regions, called plaques, of the vascular wall (Figure 15-C). Plaques are characterized by an abnormal proliferation of modified smooth muscle cells and deposits of large quantities of cholesterol. Damage to the heart or brain can result from progressive blood deprivation as plaque development narrows arteries. Because the plaque is an abnormal surface, a clot (or thrombus—see Chapter 14) may form on it and block the artery, causing a heart attack or stroke.

A recent technique, called angioplasty, has been used to enlarge narrowed arteries. In this procedure, a deflated balloon is inserted into the diseased vessel and then briefly expanded several times (Figure 15-D). This procedure resembles cardiac catheterization and, because angioplasty does not require surgery, it may reduce the number of coronary bypass operations performed and thus decrease total health-care costs. In a third of patients treated by balloon angioplasty, the coronary arteries become blocked again within 6 months.

Lasers are a new tool to remove the fatty plaques of coronary artery disease. Lasers are introduced by threading a fiberoptic catheter into a vein in the arm and guiding the catheter into a partially blocked blood vessel. The laser is then used to burn away plaque.

Lasers are also used to treat clots in leg arteries. This condition, termed *phlebitis,* is not only quite painful, but if a clot in a peripheral vessel breaks free it can lodge in the lung, causing a potentially life-threatening embolism.

Considerable progress has occurred in understanding the development of plaques. The initial event in the formation of a plaque is believed to be an injury to the arterial endothelium that causes an inflammatory lesion. Of all aspects of the disease, the least is known about the cause of the initial lesions. Candidates include chemical agents from the environment, substances produced in the body itself, or the mechanical stress of elevated blood pressure.

Once the lesion has formed, the white blood cells and platelets that invade it release growth factors (see Chapter 14) that cause surrounding smooth muscle cells to proliferate, migrate toward the interior of the arterial wall, and accumulate cholesterol. The progress of atherosclerosis is greatly accelerated by eating foods excessively high in

FIGURE 15-C
A micrograph showing an atherosclerotic plaque in a coronary artery.

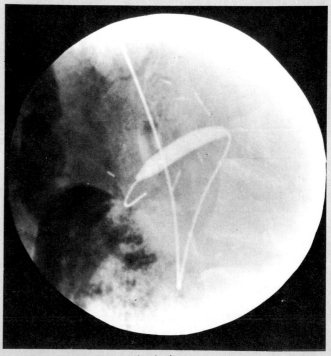

Angioplasty

FIGURE 15-D
In angioplasty, a catheter with a deflated balloon on the end is inserted into a peripheral vein and carefully manipulated into a narrowed coronary artery. It is then inflated several times to restore the artery to its normal diameter.

cholesterol content and by having high plasma levels of low-density lipoprotein, one form in which cholesterol is transported in the blood. High-density lipoproteins, on the other hand, tend to remove cholesterol from the circulation, and high levels of these lipoproteins are associated with a decreased incidence of atherosclerosis.

Although a large fraction of the population develops at least some plaques, not all people develop clinical disease as a result. The incidence of coronary artery disease, one manifestation of atherosclerosis, increases with age and is higher in men than in women in all age categories (Figure 15-E).

The Framingham study was a landmark investigation of 5000 inhabitants of a small Massachusetts town who were monitored more than 20 years. This study found that hypertension, elevated blood lipid levels, cigarette smoking, obesity, low levels of physical activity, and diabetes are all risk factors associated with a higher incidence of coronary artery disease.

In addition to reducing cholesterol intake, there is evidence that other dietary factors can reduce the risk of atherosclerosis. For example, fish oils contain omega-3 fatty acids, which seem to increase high-density lipoprotein levels. Exercise seems to decrease low-density lipoprotein levels while simultaneously increasing high-density lipoprotein levels.

Because the body can synthesize cholesterol from other substances, it is important to know the entire lipid profile of an individual to accurately assess his or her risk of coronary artery disease.

The first 1 to 3 hours following a heart attack are the most critical. Heart muscle is dying and there is only a narrow "window of opportunity" within which to save these vital cells by restoring normal blood flow. There has thus been an intensive effort to identify drugs that can dissolve blood clots, generally called clot-busters. In the body, clots are dissolved naturally by plasmin, an enzyme formed from the plasma protein plasminogen. The conversion of plasmin to plasminogen is triggered by the same factors that cause initial clot formation (Hagemann factor, see Figure 14-4), but the plasmin-to-plasminogen conversion is slow, allowing time for the vessel to repair itself. The substance responsible is tissue plasminogen activator. The three newest drugs are tissue plasminogen activator (TPA), and two other enzymes with relatively high specificity for blood clots, streptokinase and anistreplase (Eminase). All reduce the death rate by at least 25%, equivalent to 100,000 lives saved per year. Using the natural anticoagulant heparin in addition may increase survival rates further. The idea here is that vessels newly opened by the clot-busters are at the highest risk of reocclusion, which heparin would tend to prevent. Despite the excitement over these drugs, there are still hurdles to be surmounted, including increased risk of cerebral hemorrhage, a potentially fatal

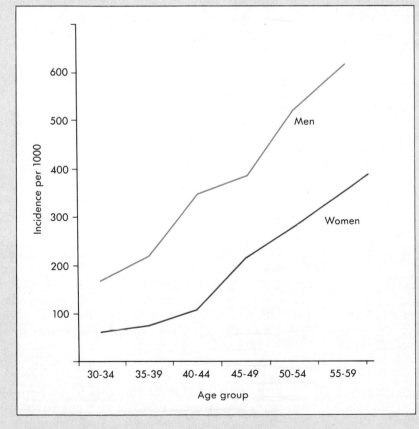

FIGURE 15-E
Incidence of coronary heart disease in the Framingham study population. Incidence is given as the cumulative number of people who developed disease over the 24 years of the study.

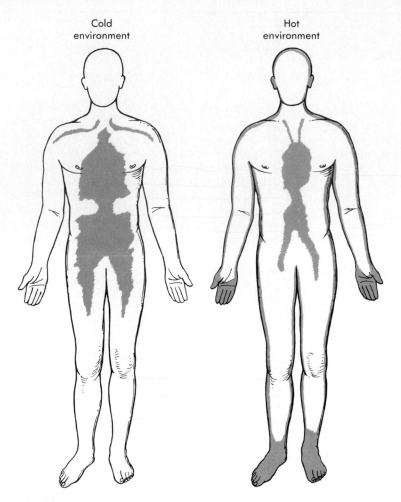

Cold
environment

Hot
environment

FIGURE 15-17
Major locations of venous volume in a cold environment *(left)* and in a hot environment. In the cold, cutaneous arterioles are highly constricted, and most flow takes the deeper arteriovenous shunts. In nonexercising people in hot environments, cutaneous arterioles are greatly dilated, and a significant fraction of the blood volume is near the skin surface.

core. This vasoconstriction causes the paleness seen in chilled hands and feet.

As environmental temperature increases, heat loss from the body core is increased by reflexive dilation of the cutaneous vessels. A shift of blood volume into the cutaneous circulation occurs (Figure 15-17). If the heat stress is great, the volume shift and the decrease in total peripheral resistance may be only partly compensated for by vasoconstriction in other vascular beds. As a result, venous return and arterial blood pressure may decrease. Standing quietly in an erect position is therefore more likely to cause fainting if the environment is warm than if it is cold.

A hot environment makes maintenance of arterial blood pressure during exercise a difficult balancing act for the cardiovascular center. The increased metabolism of exercising heart and muscles increases the heat load; perfusion of the skin for heat loss conflicts with diversion of cardiac output to muscles. Also, the fluid loss from the plasma caused by sweating further reduces venous return. This combination of stresses can be life-threatening for athletes performing heavy exercise in hot environments.

The effect of exercise on cutaneous perfusion varies depending on whether the environment is hot or cold. In cold environments, cutaneous blood flow is low at rest and increases during exercise; in hot environments, cutaneous blood flow is high at rest and decreases with increasing workload.

THE CARDIOVASCULAR SYSTEM

SUMMARY

1. Most blood vessels are innervated only by sympathetic adrenergic fibers. **Vasoconstriction,** the predominant adrenergic effect, is mediated by α_1-receptors. In some tissues such as skeletal muscle, **vasodilation** is mediated by β_2-adrenergic receptors and by cholinergic sympathetic fibers. Vessels of the external genitalia are dually innervated; parasympathetic inputs cause dilation, and sympathetic inputs cause constriction.

2. The **baroreceptor reflex** counters mean arterial pressure changes over a time scale of seconds to minutes. The medullary **cardiovascular center** integrates inputs from baroreceptors in the walls of the carotid sinus and aortic arch. Decreased mean arterial pressure, indicated by decreased baroreceptor firing, results in decreased vagal tone to the heart, increased sympathetic input to heart and vessels, and increased release of epinephrine from the adrenal medulla. The increased heart effectiveness and vasoconstriction return arterial blood pressure towards the setpoint value.

3. Regulation of mean arterial pressure over hours to days involves three hormonal systems that regulate extracellular fluid volume: the **antidiuretic hormone** system, the **atrial natriuretic hormone** system, and the **renin-angiotensin-aldosterone** system.

4. The properties of the cardiovascular system are **total peripheral resistance, blood volume, venous compliance, heart rate,** and **contractility.** Given a particular set of values for these properties, the **cardiovascular operating point** is the one flow value for which the cardiac output equals the venous return. A change in one or more of the system's properties changes the outcome of the interaction between heart and circulation and affects the system variables: central venous pressure, mean arterial pressure, and cardiac output.

5. Standing results in pooling of blood in compliant veins of the lower body and a drop in venous return and central venous pressure, with a consequent drop of mean arterial pressure in the arteries above the heart. Mean arterial pressure is maintained by the baroreceptor reflex, which is assisted by the pumping action applied to large veins by muscular movements of locomotion and respiration (the **skeletal** and **respiratory pumps**).

6. The responses to moderate **hemorrhage** resemble those involved in standing, except that lost fluid, plasma proteins, and blood cells must be replaced. Fluid transfer from interstitial spaces to plasma can make up for some lost volume; ultimately, fluid replacement involves responses of volume-regulatory hormonal control systems. Replacement of proteins and cells requires days. Loss of more than 20% of blood volume may cause **hypovolemic shock,** a positive feedback cycle in which underperfusion of the heart and brain reduces the ability of the system to compensate for the lost blood.

7. The cardiovascular responses to exercise include:
   a. Vasodilation and a large increase in perfusion of working muscle, mediated mainly by autoregulation of flow.
   b. An increase in the **cardiac reserve,** which allows the heart to match cardiac output to increased venous return from working muscle.
   c. Sympathetically mediated vasoconstriction of abdominal viscera, partly compensating for the decreased resistance of working muscle.

   The response of cutaneous vessels to exercise reflects a compromise between **thermoregulation** and regulation of mean arterial pressure.

8. In most cases, causes of hypertension are unknown; drug therapy for hypertension may involve some combination of reduction of total peripheral resistance, heart effectiveness, or blood volume.

1. How would venous return be altered by transfusion of blood? By arteriolar constriction? By decreases in venous compliance?

2. The cardiac output of a denervated heart still increases significantly during exercise. What mechanism makes this possible?

3. If myocardial contractility is increased by a drug, what will be the effect on central venous pressure? On cardiac output? On mean arterial pressure?

4. Define cardiac reserve. What factors increase the cardiac reserve? What factor might decrease it?

5. A healthy person lying on a tilt table is shifted into an upright position. What happens to the baroreceptor firing rate? To the action potential frequency in the nerve fibers innervating the SA node? To the mean capillary pressure in the foot? To the venous compliance? What other changes are initiated by the baroreceptor reflex?

6. A person's hematocrit is measured before and several hours after donation of a pint of blood. The second hematocrit is found to be lower than the first. Explain the changes that decreased the hematocrit during the hours following the donation.

7. A healthy person loses 1 liter of blood. How would the following factors change as an immediate consequence of the blood loss, before any compensation has had time to take place?
   Cardiac output
   Central venous pressure
   Mean arterial pressure
   Total peripheral resistance

8. What effect will the reflexive compensation for loss of the liter of blood in Question 7 have on:
   Total peripheral resistance
   Central venous pressure
   Cerebral blood flow
   Renal blood flow

9. Why is regulation of both mean arterial pressure and body temperature more difficult for people exercising in a hot environment?

10. Venoconstriction
    a. Decreases compliance of vein walls
    b. Increases compliance of vein walls
    c. Increases total peripheral resistance
    d. All of the above are true.

11. Which of the following is a cardiovascular system variable?
    a. Total peripheral resistance
    b. Cardiac output
    c. Blood volume
    d. Venous compliance

12. If mean arterial blood pressure in the carotid arteries begins to fall:
    a. Aortic baroceptor firing rates increase
    b. Carotic baroceptor firing rates decrease
    c. Sympathetic activity to heart decreases
    d. Parasympathetic activity to SA node increases

13. Which of the following autoregulatory agents ("local factors") does not cause vasodilation?
    a. Release of bradykinin
    b. Release of histamine
    c. Decrease in O_2
    d. Increase in pH

14. Antidiuretic hormone
    a. Is a vasodilator
    b. Increases rate of fluid loss in urine
    c. Causes increased thirst
    d. Decreases vascular tone

15. This hormone's major effect is to decrease the reabsorption of Na^+ and water in the kidney, increasing the rate of loss of plasma solutes and water in the urine:
    a. Antidiuretic hormone
    b. Atrial natriuretic hormone
    c. Aldosterone
    d. Renin

16. Which of these factors increases central venous pressure?
    a. Increase in blood volume
    b. Increase in venous compliance
    c. Arteriolar vasocontriction
    d. Decrease in cardiac output

● *SUGGESTED READINGS*

ALDER V: Beyond balloons, *American Health,* March 1988, p. 14. Describes how lasers and subminiature drill bits have been used to treat coronary artery disease.

BEARDSLEY TM: Exercising choice; case (almost) closed: fitness does seem to prolong life, *Scientific American* 260 (2):24, 1989. Discusses the cardiovascular benefits of exercise and the overall decreases in mortality exercise provides.

EDWARDS DD: Repairing blood pressure damage, *Science News,* May 1988, p. 292. Describes how new antihypertensive drugs can improve long-standing damage to the cardiovascular system.

EISENBERG S: Type A and coronary artery disease, *Science News,* November 7, 1987, p. 292. An examination of data linking personality characteristics to the incidence of heart disease. It also examines whether stress-control techniques are effective in preventing coronary artery disease.

EISENBERG S: Smoking raises female heart attack risk, *Science News,* November 22, 1987, p. 341. An examination of epidemiological data linking increased risk of heart attack in women with increased use of tobacco.

ERON C: Fatty acids cut heart artery narrowing, *Science News,* September 24, 1988, p. 197. Discusses studies suggesting that omega-3 fatty acids can reduce the risk of atherosclerosis.

GRADY D: Can heart disease be reversed? *Discover,* March 1987, p. 54. Discusses the ways diet modification, exercise, and drugs affect coronary artery disease.

GUNBY P: Laser may provide better channel, smoother lumen in future coronary artery occlusions, *Journal of the American Medical Association,* March 13, 1987, p. 1283. Self-explanatory, but points out that the technique needs further refinements before it can be generally applied.

MUKERJI B, ALPERT MA, MUKERJI V: Cardiovascular changes in athletes, *American Family Physician* 40(3):169, 1989. Describes how physical exertion such as long-distance running increases the capacity of the dilatation of the ventricles of the heart (diastolic volume) and increases the thickness of the walls of the left ventricle.

PERRY P: The pretended self, *Psychology Today* 23(5):60, 1989. Describes the connection between heart disease and social isolation.

YATANI A, OKABE K, CODINA J, BIRNBAUMER L, BROWN AM: Heart rate regulation by G proteins acting on the cardiac pacemaker channel, *Science* 249:1163, 1990. Explains how muscarinic and beta adrenergic receptors of the autonomic nervous system affect the behavior of the K^+ channel that determines the rate of pacemaker discharge in the heart.

ZOLER ML: Cardiologists clash on PET's forte, *Medical World News* 31(10):31, 1990. Discusses the pros and cons of positron emission tomography.

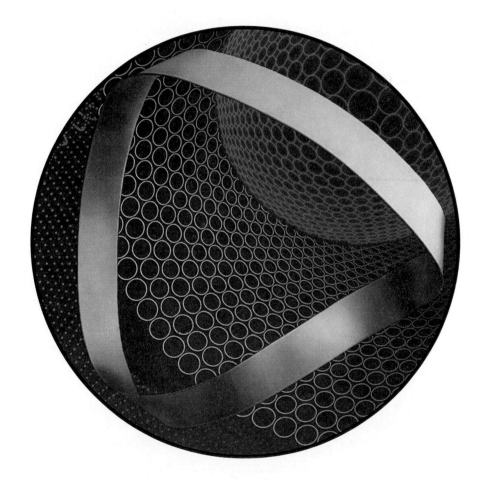

CHAPTER 16

Respiratory Anatomy and the Mechanics of Breathing

On completing this chapter you should be able to:

- Describe the components of the respiratory system, identifying the classes of airways, gas exchange elements, and the muscles responsible for ventilation
- Describe the mechanisms defending the lung from damage
- Understand how the terminal bronchioles are neurally and hormonally regulated
- Appreciate the factors that affect gas exchange in the lung
- Define the lung volumes and capacities and provide approximate normal values for them
- Understand the factors that affect respiratory minute volume
- Explain why intrapleural pressure is always less than alveolar pressure
- Distinguish between the elastic and flow-resistive components of the work of breathing
- Understand the role of pulmonary surfactant
- Summarize the basic characteristics of obstructive and restrictive lung disease
- Compare pulmonary and systemic blood flow
- Describe the conditions under which there may be an imbalance in the ventilation-perfusion ratio of regions of the lung, and show how this affects the O_2 and CO_2 content of alveolar gas

*T*he term **respiration** can be used in two senses. At the cellular level it refers to cellular oxidative energy metabolism. At the organismal level, respiration refers to the ventilation of gas exchange surfaces with atmospheric air. The respiratory system links the circulatory system with the atmosphere, an infinite source of oxygen and an infinite sink for the carbon dioxide generated by cellular oxidative metabolism. In some amphibians, gas exchange by diffusion across the skin (cutaneous respiration) is sufficient to support the animal's relatively low metabolic rate. If this were the mechanism of gas exchange in human beings, the high rate of metabolism would necessitate a body surface area so great that the entire body plan would have to be unrecognizably different. The lungs constitute only a few percent of total body mass but contain within their structure an enormous surface area that permits a rapid and complete gas exchange by diffusion. Much of respiratory physiology is concerned with the problems posed by the ventilation of such a complex structure.

In the mythology of the ancient Greeks, *pneuma*, or breath, was an invisible personal spirit that gave its possessor life. Healthy people take breathing for granted because it is almost effortless and goes on usually without conscious awareness. For those with respiratory disease, every breath may be hard won. Respiratory diseases often result from breathing contaminated air, such as cigarette smoke or polluted air. Poor air quality is sometimes regarded as an exclusively modern problem, but autopsies of mummified Egyptians and the frozen bodies of Inuits (prehistoric inhabitants of Alaska) suggest that inhalation of sand particles and smoke from cooking fires caused lung disease long before the Industrial Revolution.

The respiratory system has become adapted to produce sound. In speaking and singing, air flowing in the respiratory system vibrates the vocal cords, and the resonant properties of air-filled spaces associated with the airway favor some frequencies over others. This gives each individual his or her own unique voice.

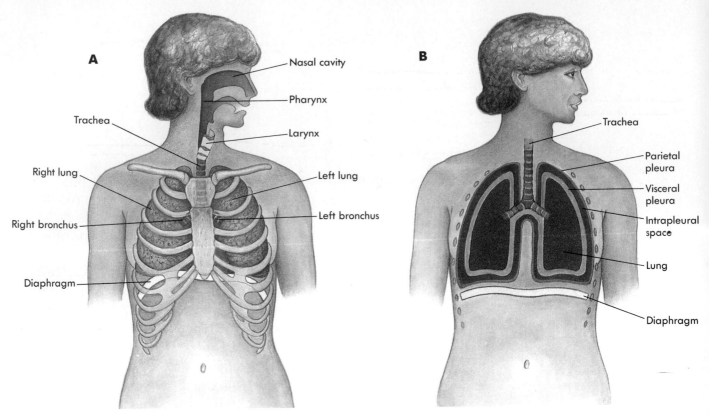

A

- Nasal cavity
- Pharynx
- Larynx
- Trachea
- Right lung
- Left lung
- Right bronchus
- Left bronchus
- Diaphragm

B

- Trachea
- Parietal pleura
- Visceral pleura
- Intrapleural space
- Lung
- Diaphragm

FIGURE 16-1

A   The anatomy of the thorax showing major airway components (pharynx, larynx, trachea, and left and right bronchi), the chest wall, and abdomen. The thoracic cavity is separated from the abdominal cavity by the diaphragm.
B   The lungs are separated from each other and from the chest wall by the parietal and visceral pleura. For clarity, the size of the intrapleural space is greatly exaggerated.

STRUCTURE AND FUNCTION IN THE RESPIRATORY SYSTEM

- *What is the intrapleural fluid?*
- *How are the airways between the trachea and the alveoli subdivided?*
- *How are sounds produced?*
- *What mechanisms protect the lung from air particulates?*
- *How does lung structure maximize its diffusing capacity?*

Lungs and Thorax

The **thoracic cavity,** or chest cavity, is a compartment bounded by the rib cage (made up of the ribs, **intercostal muscles** connecting one rib to another, and connective tissue) and closed by a sheet of skeletal muscle called the **diaphragm** (Figure 16-1, *A*). The **pleura** is a sheet of epithelial tissue that lines the interior of the thorax and surrounds each lung (Figure 16-1, *B*). Each lung is enclosed in a separate pleural cavity. The portion of the pleura lining the interior of the thoracic cavity is the **parietal pleura;** the **visceral pleura** covers each lung. The **intra-**

pleural space is enclosed by the parietal and visceral pleura. Within the intrapleural space there is a small amount (several milliliters) of intrapleural fluid that minimizes friction between the parietal and visceral pleural as the lungs inflate and deflate. The left lung is smaller than the right and consists of two lobes, whereas the right lung has three.

Structures of the Airway

The lungs are ventilated with atmospheric air by way of the treelike **airway.** The airway consists of all structures that conduct air between the atmosphere and the **alveoli,** the membranous lung structures across which gas exchange takes place. The parts of the airway that lie outside the lung are the nose, **pharynx** (throat), **larynx** (which houses the vocal cords), **trachea,** and right and left **primary bronchi.**

Within the lung, each primary bronchus branches extensively (Figure 16-2; Table 16-1). The first branching gives rise to secondary bronchi; the second branching gives rise to segmental (tertiary, quaternary, and so on) bronchi that supply discrete areas within the lobes. Subsequent branchings result ultimately in about 150,000 smaller bronchioles called **terminal bronchioles** (enlarged in Figure 16-2). The larger bronchi, up to the terminal bronchioles are surrounded by rings of cartilage that prevent their collapse. Terminal bronchioles are sur-

FIGURE 16-2

The anatomy of the respiratory airway. Air moves through the conducting zone of the lung (the trachea, left and right primary bronchi, secondary and tertiary bronchi, and bronchioles) by bulk flow. The figure shown is a simplification; there are believed to be 20 to 23 generations of branches, which cannot all be shown on the same scale. Terminal and respiratory bronchioles, alveolar ducts, alveolar sacs, and alveoli are shown greatly enlarged. The alveolar ducts and sacs form the exchange zone of the lung. Here diffusion replaces bulk flow because the amount of branching is so great that it reduces the average flow to nearly zero.

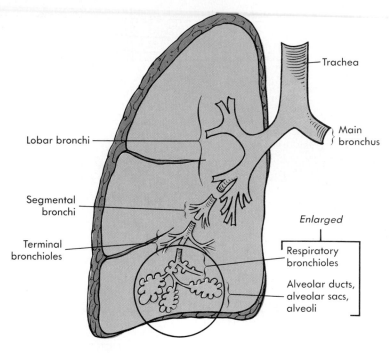

TABLE 16-1   Subdivisions of the Respiratory Tree

Name	Diameter	Number
Trachea	1.8 cm	1
Main bronchi	1.2 cm	2
Lobar bronchi	0.8 cm	3 on right, 2 on left
Segmental bronchi	0.6 cm	10 on right, 8 on left
Terminal bronchioles	1.0 mm	About 48,000
Respiratory bronchioles	0.5 mm	About 300,000
Alveolar ducts	400 μm	About 8×10^6
Alveoli	250 μm	About 6×10^8

rounded only by multiunit smooth muscle cells. Contraction of these muscles can adjust the diameter of the terminal bronchioles in response to autonomic innervation and circulating epinephrine. The absence of cartilage in the terminal bronchioles makes them susceptible to collapse if enough external pressure is applied across their walls.

Bronchioles with diameters smaller than 0.5 mm are referred to as **respiratory bronchioles.** Respiratory bronchioles differ slightly from terminal bronchioles in that they contain a few alveoli in their walls and are the beginning of the gas exchange region of the lung. The respiratory bronchioles eventually terminate in the gas exchange zone of the lung, the **alveolar ducts** and **alveolar sacs** (see Figure 16-2). Alveolar ducts contain smooth muscle cells, and thus their diameter is regulated by muscular contraction or relaxation. The alveolar ducts and alveolar sacs give rise to numerous spherical alveoli, so gas exchange can occur in both.

However, alveoli located in the alveolar sacs provide most of the surface area for gas exchange.

The bronchi and bronchioles are major sites of **airway resistance.** Airway resistance is affected by autonomic input and by paracrine agents (Table 16-2). Parasympathetic activity constricts bronchiolar smooth muscle. Sympathetic activation of adrenergic receptors via adrenal epinephrine leads to relaxation of bronchiolar smooth muscle, which dilates the airways and decreases their flow resistance. During exercise, increased sympathetic activity reduces the work of breathing. Histamine is a paracrine substance released in response to infection and, in susceptible people, to substances called **allergens** that trigger inappropriate responses of the immune system. The effects of histamine include constriction of bronchiolar smooth muscle, which increases the effort of ventilation and decreases alveolar ventilation. In severe allergic attacks, the airway resistance may rise by twentyfold, and the effect on ventilation may be life threatening.

The airway structures are specially designed for the bulk flow of air by forced convection, and no exchange of O_2 or CO_2 with the blood occurs across their surfaces. During passage across the warm, moist surface of the airway, the air is warmed to body temperature and saturated with water vapor. The water vapor protects the delicate surfaces of the lungs from desiccation. The airways are surrounded by smooth muscle that can respond to activation of the autonomic nervous system, inspired particles and chemicals, and substances released locally by cells along the respiratory tract. Except for the terminal bronchioles, the conducting airways are supported by incomplete rings of cartilage that make it quite difficult to collapse them.

Sound Production

Sound production is one of the functions of the human respiratory system (Figure 16-3, *A*). In normal breathing, the vocal cords (Figure 16-3, *B*) of the

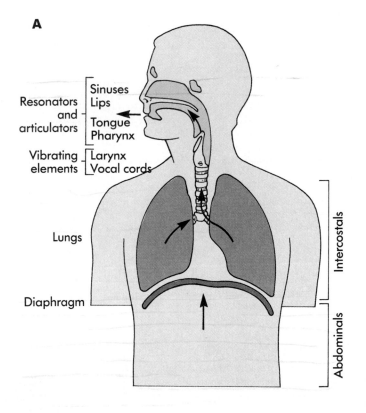

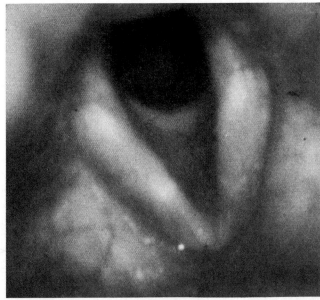

FIGURE 16-3
A   Mechanical elements that contribute to sound production.
B   The vocal cords as seen through a fiberoptic device known as an endoscope.

larynx are abducted toward the wall of the larynx and away from the airstream in normal breathing. For sound production, the vocal cords are moved into the airstream, which vibrates them as it passes. The intensity of the sound produced is controlled by the position of the vocal cords in the airstream. The pitch depends on the characteristic, or resonant, frequency at which the vocal cords vibrate and is controlled by muscles that regulate the tautness of the vocal cords (see Chapter 10 for a discussion of resonance). For high-pitched sounds the muscles contract, both tightening and thinning the cords. This raises the resonant frequency, just like tightening the strings on a guitar.

The range of sound frequencies produced by adult men is typically lower than for women because of sex-related differences in the dimensions of the vocal cords. The sounds produced by the vocal cords are altered by the articulators and resonators (the lips, tongue, mouth, nose and nasal sinuses, pharynx, and chest cavity—see Figure 16-3, *A*). Differences in these structures are responsible in large part for the different voice qualities of different individuals.

Speech requires the coordination of many sets of muscles that are innervated by motor neurons that are widely distributed in the spinal cord and the brain. Good vocalization requires exquisite control of the respiratory muscles: the diaphragm, the intercostal muscles, and the abdominal muscles. Holding a note, for example, requires that the pres-

sure in the airway be constant, so the pressure source must be operated in a different manner from the pattern of rising and falling pressure characteristic of quiet breathing (see Figure 16-3, *A*). Such command is developed only after many hours of training.

Protective Functions of the Airway

Along with inspired air comes an incredible array of environmental materials, including dust particles, pollen, bacteria, fungal spores, and viruses. The upper airway possesses mechanisms that protect against particulate contaminants and the entry of bacteria and viruses. First, large particles (over 5 to 10 microns in diameter) settle out or are caught on the hairs and sticky mucus in the nasal passages. Sensory endings in the nasal passages, trachea, and bronchi elicit the sneezing and cough reflexes, which reduce the intake of foreign particles. Breathing through the nose rather than the mouth improves the chances that larger particles will be stopped in the first levels of the respiratory tract, because air is drawn across a longer pathway before it enters the pharynx.

The second mechanism for cleansing the air is the ciliated epithelial lining of the airway (Figure 16-4). Mucus is continuously secreted by epithelial cells and glands. The cilia propel the mucus upward toward the trachea at a rate of as much as 1 cm/min. Trapped particles are thus carried from the airway into the pharynx to be swallowed or coughed up;

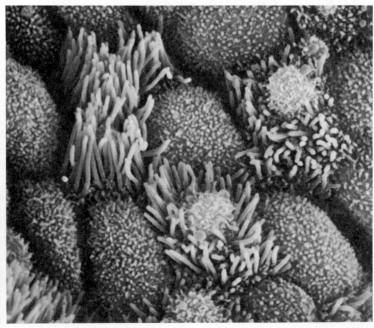

20 μm

FIGURE 16-4
The epithelium of a small bronchus is shown at low (**A**) and high (**B**) magnification in scanning electron micrographs. Note that the epithelial surface is provided with tufts of cilia.

Respiratory Anatomy and the Mechanics of Breathing

the process has been called the **mucociliary escalator.** The mucociliary escalator is inhibited by a variety of materials, including tobacco smoke, cold air, and many drugs. Heavy smokers, for example, typically cough repeatedly due to the accumulation of mucus and the irritation caused by the foreign particles.

A third level of protection consists of the white blood cells that leave nearby capillaries and move about on the surface of the airway and alveoli. The specific mechanisms whereby these cells inactivate infectious organisms and remove small particles will be described in Chapter 26.

In spite of the airway's defenses, very small particles (of the order of 5 microns diameter or less) reach the alveoli; some lodge there permanently. The lungs of all citizens of industrialized nations accumulate a burden of carbon particles that is clearly visible at autopsy. The burden of foreign material in the lungs is heavier in smokers than nonsmokers.

A family of disabling lung diseases results from exposure to specific particulates. The "black lung" of coal miners, the "brown lung" of textile mill workers, the silicosis of glass workers, and the asbestosis of steamfitters and insulation workers are a few examples of such diseases. A common feature of all of these diseases is a scarring, or **fibrosis,** of the alveolar membrane that affects both the length of the path for gas diffusion and the elasticity of the lung.

Acini and Alveoli

The basic functional unit of the lung is an **acinus,** a structure that resembles a bunch of grapes (Figure 16-5). Each acinus arises from a single terminal bronchiole, the stem of the bunch, which branches into about 100 alveolar ducts. Each alveolar duct terminates in about 20 alveoli, each 100 to 300 microns in diameter. There are about 300 million alveoli in each lung, giving it a sponge-like consistency. The alveoli are supported by a mesh of elastic connective tissue cells called **parenchymal cells.**

Movement of gas between the atmosphere and the blood occurs in two stages: bulk flow by forced convection and diffusion. In the **conduction zone** of the lung, consisting of the network of branching bronchi and bronchioles, there is bulk flow of air (convection). Bulk flow is a much more effective mechanism than diffusion for transporting substances over long distances. Between the respiratory bronchioles and the alveoli, referred to as the **diffusion zone** of the lung, there is no bulk flow (the average velocity is nearly zero, having decreased at each level of branching) and exchange occurs by diffusion. Gas exchange between alveoli and pulmonary capillary blood also occurs by diffusion.

Diffusion is favored by a large surface area and impaired by a long path length. Almost every aspect of the structure of the lung and pulmonary circula-

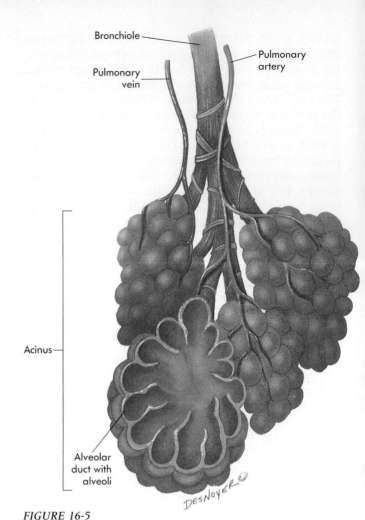

FIGURE 16-5
A pulmonary acinus consists of a terminal bronchiole, alveolar ducts and sacs, and associated circulation.

tion reflects several general adaptations that minimize the energy expenditure for forced convection, while maximizing the effectiveness of diffusional exchange in the lung. The effectiveness of diffusion requires minimizing the distance that gases must move by diffusion between alveolar air and capillary blood, and maximizing the surface available for diffusional gas exchange.

The barriers that lie in the diffusional path between alveolar gas and capillary blood (Figure 16-6) are (1) the thin film of moisture on the alveolar surface, (2) the single thin layer of epithelial cells with their basement membrane, (3) the parenchyma and interstitial fluid, and (4) the capillary wall. The total path length posed by these structures is only about 0.5 to 1.5 microns (Figure 16-7). The surface area for gas exchange of alveoli in the lung is 60 to 80 m², about the same as that of an entire tennis court! The large surface area of alveoli is matched by dense vascular beds—in all there are about 30 billion pulmonary capillaries, or about 100 capillaries per alveolus. Alveoli can be visualized as tiny air bubbles whose entire surface

Asbestosis and Lung Cancer: Consequences of Exposure to Smoke and Pollutants

Asbestos is a general term for several fibrous, heat-resistant minerals used in insulation, floor tiles, roofing, and other products. A large number of workers have been exposed to high levels of asbestos fibers; the entire population has received some level of exposure from asbestos products. The full potential of asbestos to cause disease is not readily apparent because, in many cases, decades pass between the first exposure to asbestos and the onset of illness. As certain types of asbestos fibers accumulate in the lungs, they are encapsulated, forming fibrotic lesions similar to those caused by other dusts. About 9% of asbestos insulation workers die of the consequences of fibrotic disease.

One recent survey shows that about 19% of insulation workers die of lung cancer, a rate about 7 times greater than the general population. The effect of cigarette smoking on the incidence of lung cancer in asbestos workers is an example of how risk factors may interact to multiply the incidence of a disease. Nonsmokers with asbestos exposure die of lung cancer at a rate about five times greater than nonsmokers who do not have documented asbestos exposure. Smokers die of lung cancer at a rate about 10 times greater than nonsmokers. Asbestos workers who are also smokers die of lung cancer at a rate that may be as great as 50 times that of nonsmokers without asbestos exposure. About 8% of insulation workers die of mesothelioma, a cancer of the pleura that is extremely rare in the general population. This cancer appears to be the result of migration of asbestos fibers from the lung to the pleura.

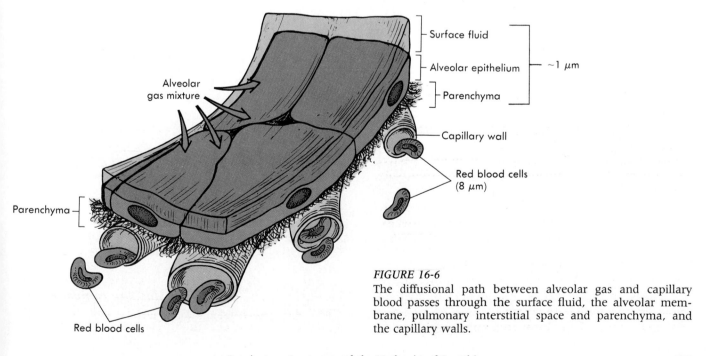

FIGURE 16-6
The diffusional path between alveolar gas and capillary blood passes through the surface fluid, the alveolar membrane, pulmonary interstitial space and parenchyma, and the capillary walls.

is bathed by the blood flowing through the lung. In all, the lungs may contain as much as 1500 miles (about 2400 km) of blood vessels, and each milliliter of blood may be spread out over about 70 miles (about 110 km) of capillaries.

RESPIRATORY VOLUMES AND FLOWS

- *What is the FRC? What forces are involved in passive expiration?*
- *What is the difference between a pulmonary volume and a pulmonary capacity? Name and define the four static volumes and four static capacities.*
- *Why is the alveolar Pco_2 much higher than that of the inspired air?*
- *What variables determine the alveolar minute volume?*

Lung Volumes and Capacities

The lung volume changes that occur during inspiration and expiration can be measured by **spirometry** (Figure 16-8). In the most common method, the subject breathes from a tube connected to an inverted reservoir that floats in water, rising as air enters it during expiration and sinking as air is withdrawn during inspiration. The volume changes in the reservoir correspond to fluctuations of the lung volume. The change in vertical position of the reservoir is thus a measure of change in lung volume. The volume of air contained by the lungs if they are maximally filled can be divided into four nonover-

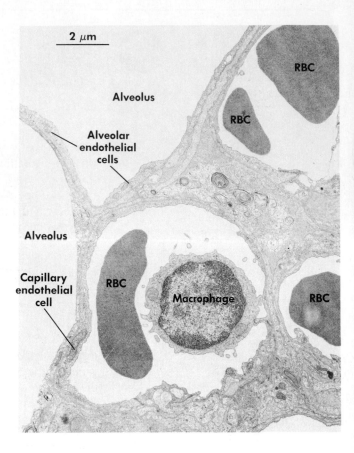

FIGURE 16-7
Pulmonary capillaries and adjacent alveoli in cross section in the electron microscope.

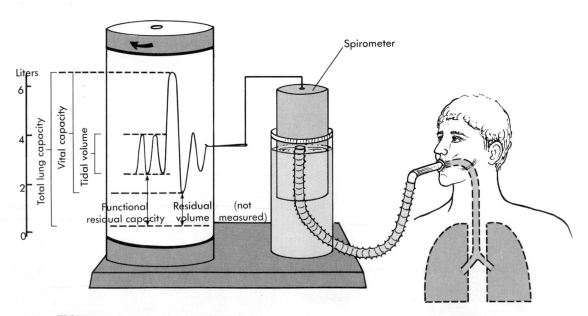

FIGURE 16-8
Static lung volumes can be measured in an intact subject by spirometry. Changes in lung volume are measured during normal quiet breathing and when a subject is asked to inhale maximally and then to exhale as much air as possible. The residual volume and functional residual capacity cannot be measured using a spirometer.

THE RESPIRATORY SYSTEM

FOCUS ON PHYSIOLOGY

Cystic Fibrosis: An Inherited Defect of Cl⁻ Transport

Secretion of saliva and sweat, maintenance of a moist surface on the airways of the respiratory system, and the digestive secretions of the intestine and exocrine pancreas all involve a process in which active transport of Cl^- is followed by Na^+ and water. The first step in this process is secondary active transport of Cl^- to the interior of the cell across one face of the cell membrane along with Na^+ and K^+. In the second step, the Cl^- leaks out of the cell through channels in the membrane on the opposite side of the cell.

Cystic fibrosis (CF) is a genetic disease that affects about 1 of every 2000 white newborns. It appears in children who received two copies of a defective allele, the recessive CF gene, one from each parent. About 5% of the population carries a copy of the CF gene paired with a dominant, nondefective allele that protects them from the symptoms of CF. A child whose parents are both carriers of the CF gene has a 25% chance of inheriting two copies of the recessive gene and thus the disease.

The major symptoms of CF are (1) chronic pulmonary disease in which the airways become clogged by thick mucus, (2) insufficient pancreatic secretion, and (3) an increase in the electrolyte concentration of sweat which reflects the fact that reabsorption of ions from sweat after its initial formation is compromised. Analysis of sweat has been a useful tool in diagnosis of CF. No cure has been found for CF, and few patients survive beyond young adulthood.

All of the symptoms of CF are traceable to failures of epithelial ion transport in various organs. Recently it has become clear that a general failure of Cl^- channels to respond to some of the neural and hormonal signals that normally regulate them could account for all of the diverse signs and symptoms of CF. The abnormal regulation of Cl^- channels has been confirmed in tissue samples from CF patients, using the method of patch clamping (see pp. 198).

Until it becomes possible to correct defective genes, hope for an effective treatment for CF depends on better understanding of the basic secretory and absorptive mechanisms of epithelia. The CF gene has been cloned, and it will soon become clear whether the protein product of this gene is the Cl^- channel itself or some component of an intracellular regulatory system.

The recent identification of at least one CF gene is a big first step toward a cure by gene therapy. In addition, genetic testing should allow couples considering having children to find out whether they are CF carriers. Prenatal diagnosis of CF by amniocentesis or chorionic villus biopsy (see boxed essay in chapter 25) will also be possible, enabling parents to terminate such a pregnancy, or to prepare to care for a child who will be ill from birth.

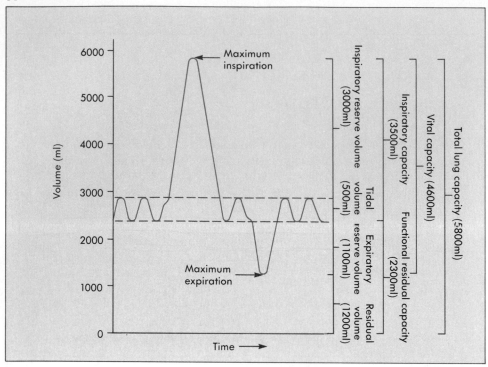

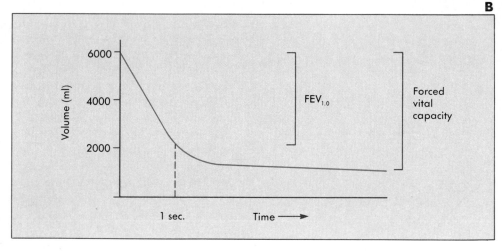

FIGURE 16-9
A The static lung volumes and capacities.
B The 1-second forced expiratory volume and the forced vital capacity.

lapping **volumes** (Figure 16-9, *A*; Table 16-3). The term **capacity** is used to refer to measures of lung function that encompass more than one volume. The volume of air that remains in the lung after a passive expiration is the **functional residual capacity** (FRC) and is about 2300 ml, usually 40% to 50% of the total volume of the lung. The volume of air alternately inspired and expired in a single breath is the **tidal volume** (V_T). A normal tidal volume is about 500 ml.

The volume of air that can be inspired after a normal inspiration is the **inspiratory reserve volume** (IRV) and may be in the range of 3000 ml. The volume that can be inspired following a normal expiration is called the **inspiratory capacity** (IC).

The inspiratory capacity is the sum of the tidal volume and the inspiratory reserve volume and is typically about 3500 ml. The volume of air that can be expelled from the lung after the end of a passive expiration is called the **expiratory reserve volume** (ERV) and is about 1100 ml. The sum of the tidal volume and the inspiratory and expiratory reserve volumes equals the maximum amount of air that can be moved into and out of the lungs by muscular effort. This is called the **vital capacity (VC).** The vital capacity is typically 4.6 L in young men and 3.1 L in young women and is correlated with height. The air remaining in the lung after a maximum expiration is the **residual volume** (RV), typically about 1200 ml, which cannot be determined

TABLE 16-3   *Respiratory Volumes and Capacities*

Volumes	Definition	Typical resting values (ml)
Tidal volume (V_T)	The volume of gas inspired or expired in a single respiratory cycle	500
Inspiratory reserve volume (IRV)	The maximum volume of gas that can be inspired starting at the end of a normal inspiration	3000
Expiratory reserve volume (ERV)	The maximum volume of gas that can be expired starting from the end of a normal expiration	1100
Residual volume (RV)	The volume of gas that remains in the lungs after a maximum expiration	1200
Forced expiratory volume ($FEV_{1.0}$)	The maximum amount of gas that can be expired in the first second of an FVC, following a maximum inspiration	4000
Anatomic dead space volume (V_D)	The volume of the conducting (nonexchanging) portion of the respiratory system	150
Capacities		
Total lung capacity (TLC)	The total amount of gas in the lungs at the end of a maximum inspiration (the sum of all four lung volumes)	5800
Vital capacity (VC)	The maximum volume of gas that can be inspired after a maximum expiration (the sum of the expiratory reserve volume, normal tidal volume, and inspiratory reserve volume)	4600
Inspiratory capacity (IC)	The maximum amount of gas that can be inspired starting from the FRC (the sum of the tidal and inspiratory reserve volumes)	3500
Functional residual capacity (FRC)	The amount of gas in the lungs at the end of a normal expiration (the sum of the expiratory reserve and residual volumes)	2300
Forced vital capacity (FVC)	The amount of gas that can be expelled from the lungs by expiring as forcibly as possible, after a maximum inspiration	5000

using a spirometer. The sum of the vital capacity and the residual volume is the **total lung capacity (TLC).** A normal value for the total lung capacity of an adult male is 5800 ml (see Figure 16-9, *A*).

The volumes described above are called **static volumes** because they are measured in the absence of air flow. In the case of tidal volume, flow is zero at the ends of inspiration and expiration. The other volumes are measured during breath-holding at various stages of inspiration. The rate at which air can be forced through the airways can also be an important measure of pulmonary function. The changes in lung volume measured during a forced expiration are called the **dynamic lung volumes.** The two most commonly measured dynamic lung volumes are the 1-second **forced expiratory volume** ($FEV_{1.0}$) and the **forced vital capacity (FVC).**

Static lung volumes and capacities are measured using slow inspirations and expirations, so as to avoid the possible collapse of airways, which would tend to trap gas within the lung. The forced expiratory volume and forced vital capacity are measured by requesting that a subject inhale to the total lung capacity and then exhale as forcibly as possible (Figure 16-9, *B*). The $FEV_{1.0}$ is the amount of air that can be expelled from the lung during the first second. The total amount of air eventually expelled from the lung is defined as the forced vital capacity. In young, healthy individuals, the $FEV_{1.0}$ is about 4 L, and the forced vital capacity is about 5 L (a ratio of 0.8). Normally the forced vital capacity is about the same as the vital capacity measured during a slow expiration because little or no gas is trapped in the lung. The maximum air flow is usually measured, and a normal value would be 8 to 12 L/sec.

Alveolar Ventilation and the Anatomical Dead Space

The **respiratory rate (RR)** is the number of breaths per minute. The total amount of air entering and leaving the respiratory system per minute, or **respiratory minute volume ($\dot{V}$)** (Table 16-4), is the product of the tidal volume and the respiratory rate in breaths/min. A normal respiratory minute volume is around 5 L/min, but $\dot{V}$ values of as much as 130 L/min can be attained by young adult males during strenuous exercise.

Not all the air inhaled and exhaled enters the al-

TABLE 16-4   Factors Involved in Ventilation

Factor	Definition	Typical resting value
Respiratory rate (RR)	Number of breaths/min	12
Respiratory minute volume ($\dot{V}$)	Tidal volume × Respiratory rate (V_T × RR)	6000 ml
Anatomical dead space volume (V_D)	Volume of airway from nose to terminal bronchioles	150 ml
Alveolar minute volume ($\dot{V}_A$)	(Tidal volume − Dead space volume) × Resp. rate $(V_T - V_D)$ × RR	4200 ml

veoli where it can be exposed to the blood. Air in the trachea, bronchi, and nonrespiratory bronchioles does not contact the pulmonary capillaries, and these areas are referred to as the **anatomical dead space (V_D)** of the lung. A portion of the respiratory minute volume ventilates the dead space; the remainder, the **alveolar minute volume ($\dot{V}_A$),** ventilates the alveoli and is subject to gas exchange. The amount of fresh air that reaches the alveoli with each normal breath is equal to the tidal volume minus the dead space volume. The anatomical dead space volume of a healthy 70 kg man is about 150 ml. Thus for a single breath of 500 ml, only 350 ml of new air reaches the alveoli. This is only 12.5% of the total volume of the lung, and accounts for the fact that the CO_2 content of the alveolar air is 40 mm Hg. The alveolar minute volume is equal to the fraction of each tidal volume that reaches the alveoli multiplied by the number of breaths/min [$(\dot{V}_A = (V_T-V_D)RR$]. For a person whose respiratory rate is 12 breaths/min, anatomical dead space volume is 150 ml, and tidal volume is 500 ml, the alveolar minute volume is 12 breaths/min × 350 ml/breath = 4200 ml/min.

The effect of anatomical dead space on the dynamics of gas movement in the respiratory tract is illustrated in Figure 16-10. The airway, which has no capillary contact area, is represented by a single tube leading to the cluster of alveoli. Just before air is exhaled from the lungs, the alveoli are full of air that has been accumulating CO_2 and giving up O_2 (see Figure 16-10, A). Although the actual amounts of these two gases in the alveoli have not changed very much, as Chapter 17 will show, the air in the alveoli is represented as "old" air, in contrast to the "fresh" air that will enter the respiratory system from the atmosphere upon inspiration. At expiration, some of the "old" air is left in the anatomical dead space (see Figure 16-10, B). Upon the next inspiration, this "old" air enters the lungs and mixes with the "fresh" air (see Figure 16-10, C).

Different combinations of V_T and RR can give quite different alveolar ventilations (Table 16-5). As the V_T approaches the VR, $\dot{V}_A$ approaches zero (see Figure 16-10, D). Thus alveolar ventilation (and

TABLE 16-5   Alveolar Ventilations Produced by Different Combinations of Respiratory Rate and Tidal Volume

Respiratory rate	Tidal volume (ml)	Dead space volume (ml)	Alveolar ventilation (ml/min)
15	500	150	5250
15	250	150	1500
15	150	150	0
52.5	250	150	5250
6	1025	150	5250

therefore gas exchange) is compromised if breathing is very shallow. Alveolar ventilation may be threatened by additions to the dead space. For example, when swimmers breathe through a snorkel tube, tidal volume must be increased by the volume of the additional dead space if alveolar ventilation is to remain at its normal value. Alveolar ventilation is regulated by respiratory control reflexes described in Chapter 18.

THE MECHANICS OF BREATHING
Lung–Chest Wall System

- *Why is intrapleural pressure always less than atmospheric pressure?*
- *What normally prevents two lungs from collapsing?*

The compliance of a container (see Chapter 2) is the change in its volume that results from imposing a pressure difference across its walls. The larger the volume change in response to a given pressure change, the greater the compliance. The lungs are elastic, highly compliant structures. When the lungs are removed from the chest, the alveoli collapse and the lung volume is reduced to the volume of the bronchi and bronchioles, which do not collapse because of their rings of cartilage. When the alveoli contain air, the lung exerts an elastic recoil like that

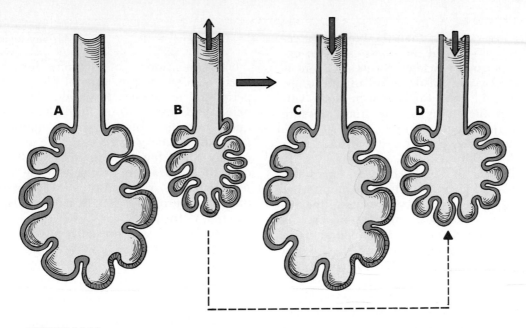

FIGURE 16-10
A conceptual illustration of the respiratory dead space. In **A** the lung and dead space are illustrated as they appear just before exhalation. In **B** exhalation forces some of the air out of the lung and dead space. During inhalation (**C**), alveolar air that entered the dead space at the end of the previous exhalation returns to the alveoli, followed by fresh air. In **D** tidal volume does not exceed dead space volume, and no fresh air enters the region of the lungs in which gas exchange occurs.

of an air-filled balloon, which will tend to collapse by expelling air if its opening is unobstructed. However, the lungs are contained within the chest cavity, which is a much less compliant structure. The two are linked together by the intrapleural fluid, which is neither expanded nor compressed by pressure changes.

The interplay of forces in the lung–chest wall–diaphragm system can be appreciated using a simple analogy. When the lung is placed inside the chest and "connected" to the chest wall by removing the air between them (leaving only a few ml of intrapleural fluid), the lung wants to collapse. However, this inward recoil is opposed by the tendency of the chest to expand. When the intercostal muscles are relaxed, the inward pull or recoil of the lung is matched exactly by the outward spring of the diaphragm and chest wall (Figure 16-11). This is the state of the system between the end of an expiration and the beginning of the next inspiration in a quietly breathing person. Because the trachea is open to the atmosphere, the alveolar pressure is equal to the atmospheric pressure, and no air flow is occurring. To maintain any volume other than the FRC requires active muscular effort, even if the air flow is zero.

As a result of the inward recoil of the lung and outward spring of the chest wall, the intrapleural pressure is several cm H_2O less than atmospheric pressure (1 cm H_2O = 0.74 mm Hg). During the late prenatal and postnatal period of a baby's life, differ-

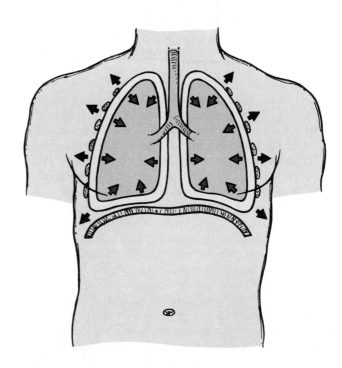

FIGURE 16-11
At the FRC, the elastic recoil of the lung *(inward-pointing arrows)* is balanced by the outward recoil of the chest wall. At this volume, intrapleural pressure is about -5 cm H_2O.

ential growth of the chest and lungs creates a permanent negative pressure in the intrapleural space. The chest cavity initially is very pliable, but it increases in size more rapidly than the lungs. The result is that the tendency of the chest cavity to expand to the dimensions dictated by its structure is always opposed by the elastic recoil of the lungs. The expansion of the chest cavity that produces volume changes in the lungs results from pressure changes in the intrapleural fluid that are superimposed upon the negative pressure that resulted from the differential growth.

The significance of the balance of forces between the lung and chest wall is evident in what happens during thoracic surgery or penetrating injury to the chest. When the intrapleural space is punctured, air is pulled into the intrapleural space by the negative pressure. The lung and chest cavity are now uncoupled, so their volumes can change independently of one another. The lung collapses to its minimum volume, while the chest expands slightly. This condition is called a **pneumothorax.** It is treated by resealing the intrapleural space and connecting a drain that can be used to create a negative intrapleural pressure during the healing process.

Respiratory Movements, Pressure Changes, and Air Flow

> • *Which muscle groups are involved in inspiration? Which groups are involved in forced expiration?*

Inspiration—flow of air into the lungs—results from an increase in thoracic volume and consequent decrease in pressure inside the lungs. Inspiration during quiet breathing results mainly from contraction of the diaphragm, which moves downward and increases the intrathoracic volume (Figure 16-12). Larger tidal volumes are achieved by contraction of the external intercostal muscles (Figure 16-13). Contraction of external intercostal muscles lifts the rib cage outward and upward, while the diaphragm moves downward (compare Figure 16-14, *A* and *B*). All of the volume increase of the thorax is transmitted to the lungs, because the volume of the intrapleural fluid remains constant. Because the bronchi and bronchioles are rigid structures compared with alveoli, almost all of the increase in lung volume is attributable to increased alveolar volume.

Boyle's law says that, at constant temperature, the product of the pressure and volume of a gas is constant (see Figure 16-16). Therefore when the volume of the lung increases, the pressure within the lung must decrease. When the pressure in the alveoli (P_{Alv}) falls below atmospheric pressure (P_{Atm}), as it does during inspiration, air enters the lungs. The rate of air flow depends on two factors: the magnitude of the pressure gradient and the resistance of the airway, according to the universal relationship between force, resistance and flow:

$$\text{Air flow} = (P_{Alv} - P_{Atm})/\text{Airway resistance}$$

This relation is of the same form as that between cardiac output, arterial blood pressure, and peripheral resistance. However, air is a much less dense and less viscous fluid than blood. Consequently, driving forces in the respiratory system are much smaller than those in the circulatory system—usually only a few mm Hg. Such small pressures are difficult to measure precisely using mercury manometers, so respiratory physiologists measure pressure in units of cm H_2O. One mm Hg is equal to about 1.36 cm H_2O.

The intrapleural pressure, always less than the pressure within the alveoli, becomes more negative on inspiration (Figure 16-15), because stretching the lungs increases their recoil force. At end-inspiration, when the inspiratory muscles relax, the recoil of the lung exceeds that of the chest wall, causing expiration until the system returns to the FRC (Figure 16-14, *C* and *D*). This is referred to as passive expiration because it occurs without muscular contraction. While the volume of the lung is dropping back to the end-expiratory value, the alveolar pressure rises above atmospheric pressure (see Figure 16-15), driving air from alveoli through the airway into the atmosphere. Intrapleural pressure rises toward the resting value. This is the mechanism of expiration in quiet breathing. It is called passive expiration because it is not driven by any muscle contractions. As noted earlier, the volume of the lung at the end of a passive expiration is referred to as the functional residual capacity (FRC).

The increases in tidal volume that occur during effort or excitement are the result of larger movements of the chest wall. The changes of intrapleural pressure and alveolar pressure during the respiratory cycle are correspondingly greater (illustrated by the broken lines in Figure 16-15). The forceful expiration of vigorous breathing is the result of contraction of the abdominal and internal intercostals (see Figure 16-13). Contracting these muscles increases the alveolar pressures attained during expiration and causes the volume of the thorax to decrease to a value lower than the FRC.

The Work of Breathing

> • *What are the two components of respiratory work?*

Work—force times distance (or pressure times volume)—is a measure of energy expenditure (see Chapter 2). The work that is done by respiratory muscles is the sum of two components: **elastic work** done against the elasticity of the lung-chest wall and **flow-resistive work** done against airway resistance. The alveolar pressure changes reflect flow-resistive work; the intrapleural pressure

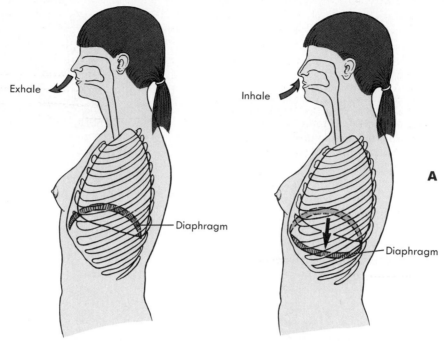

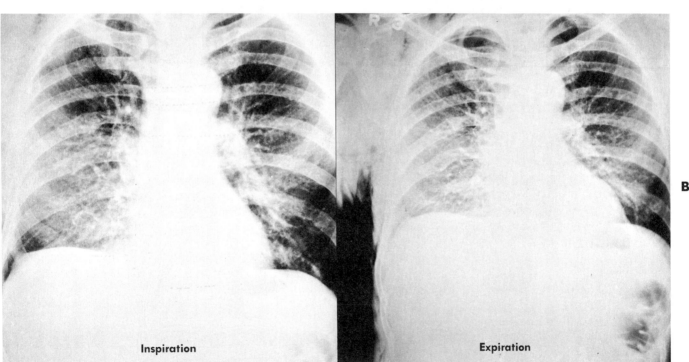

Inspiration   **Expiration**

FIGURE 16-12
A   The movement of the diaphragm. At rest, the negative intrathoracic pressure pulls the relaxed diaphragm upward toward the thoracic cavity. Contraction of the diaphragm pulls it downward toward the abdomen, increasing the thoracic volume.
B   Chest radiographs taken at the end of a normal inspiration *(left) and* at the end of a normal expiration *(right)*.

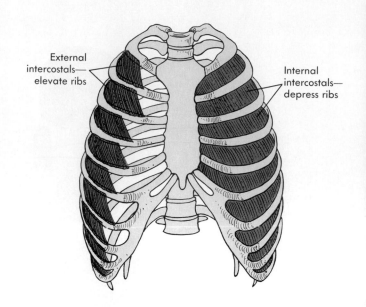

External intercostals— elevate ribs

Internal intercostals— depress ribs

FIGURE 16-13

The intercostals are the respiratory muscles that move the rib cage. External intercostals elevate the ribs, increasing thoracic volume. Internal intercostals depress the ribs during forced expiration, decreasing thoracic volume. The overlying muscles have been omitted for clarity.

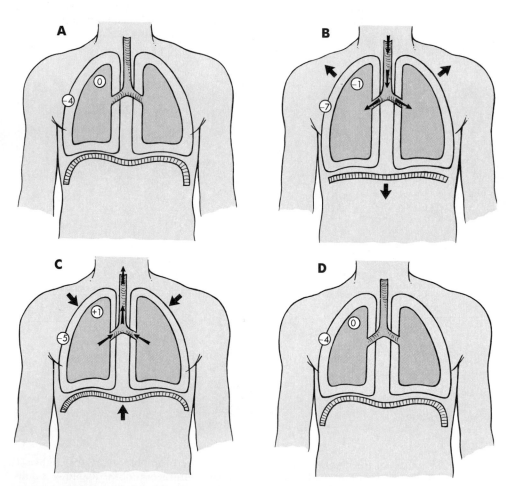

FIGURE 16-14

Figures of the chest during a cycle of inspiration (**A** to **B**) followed by expiration (**C** to **D**). The direction of air movement, the intrapleural pressure, and the alveolar pressure are shown.

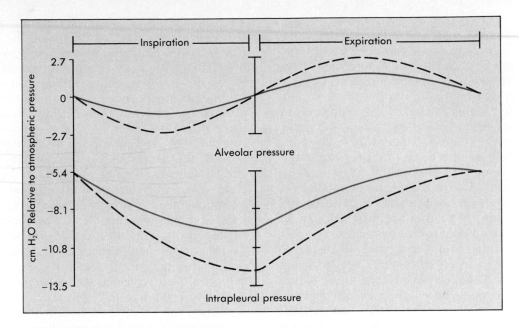

FIGURE 16-15

Changes in alveolar pressure and intrapleural pressure over the respiratory cycle. The solid traces show a cycle of 500 ml tidal volume (quiet breathing). The broken traces show a cycle in which a tidal volume of 1 L was moved in a cycle lasting the same amount of time.

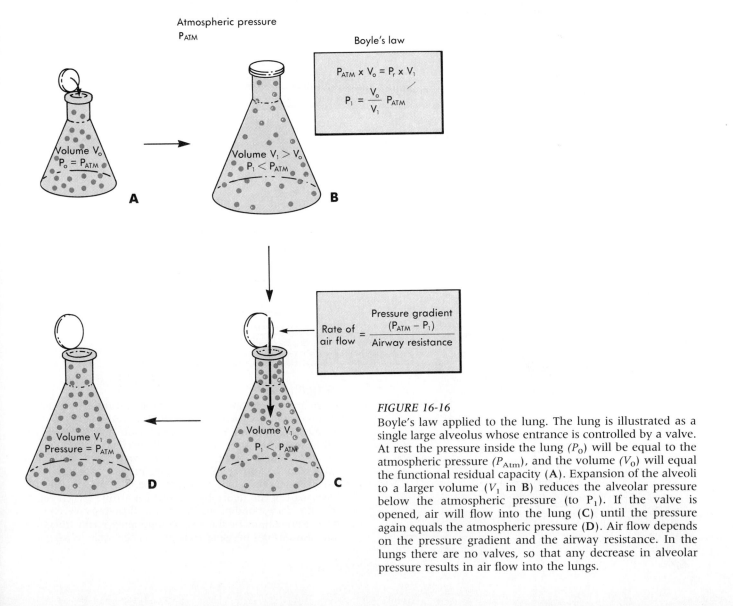

FIGURE 16-16

Boyle's law applied to the lung. The lung is illustrated as a single large alveolus whose entrance is controlled by a valve. At rest the pressure inside the lung (P_o) will be equal to the atmospheric pressure (P_{Atm}), and the volume (V_o) will equal the functional residual capacity (**A**). Expansion of the alveoli to a larger volume $(V_1$ in **B**) reduces the alveolar pressure below the atmospheric pressure (to P_1). If the valve is opened, air will flow into the lung (**C**) until the pressure again equals the atmospheric pressure (**D**). Air flow depends on the pressure gradient and the airway resistance. In the lungs there are no valves, so that any decrease in alveolar pressure results in air flow into the lungs.

changes reflect both elastic and flow-resistive work.

The sum of the two work components in intrapleural pressure is illustrated in Figure 16-17. The upper curve shows the alveolar pressure changes during one cycle of inspiration followed by passive expiration. The heights of the arrows represent the differences between alveolar pressure and atmospheric pressure at different times in the cycle. For a normal tidal volume of 500 ml, the pressure gradient between alveoli and atmosphere ($P_{Alv} - P_{Atm}$) varies from about +1.4 cm H_2O during expiration to about −1.4 cm H_2O on inspiration. The arrows would become larger if either the flow velocity increased or the airway resistance increased.

The farther the lung-chest wall is displaced from the FRC, the greater is the force needed to expand the thorax. The green line in the lower part of Figure 16-17 shows how intrapleural pressure would change over a respiratory cycle if only elastic work were considered. The elastic work for a single cycle is proportional to the volume change of the thorax and has no relationship to air flow. The algebraic summation of the elastic and flow-resistive components to give the actual intrapleural pressure change is shown in the lower part of Figure 16-17.

A given alveolar ventilation can be achieved by deep breaths occurring at a low rate, very shallow breaths at a high rate, or various combinations between these two extremes. Figure 16-18 shows how the elastic and flow-resistive work components are affected by changing the respiratory rate while adjusting the tidal volume to keep alveolar ventilation constant. For very slow, deep inspirations and expirations, the flow-resistive component is negligible and the elastic component accounts for almost the entire change in intrapleural pressure. As ventilation increases, the tidal volume drops and the chest excursion that accompanies each breath is smaller, so the elastic work decreases (red curve in Figure 16-18). At the same time, air flows more and more rapidly through the airway, and the work needed to overcome airway resistance becomes progressively greater (green curve in Figure 16-18).

When the elastic and flow-resistive work components are summed to show how the energy cost of breathing varies with respiratory rate (black curve in Figure 16-18), the work is seen to be greatest at very low and very high respiratory rates, with a minimum in a middle range. Tidal volume and respiratory rate tend to be set by the nervous system

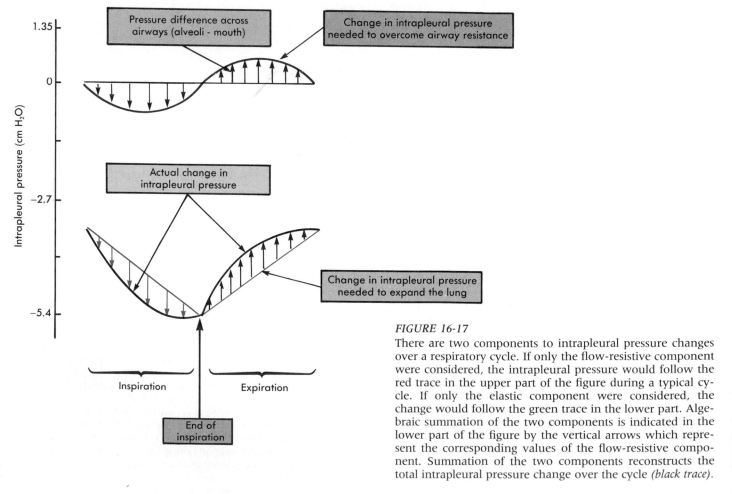

FIGURE 16-17

There are two components to intrapleural pressure changes over a respiratory cycle. If only the flow-resistive component were considered, the intrapleural pressure would follow the red trace in the upper part of the figure during a typical cycle. If only the elastic component were considered, the change would follow the green trace in the lower part. Algebraic summation of the two components is indicated in the lower part of the figure by the vertical arrows which represent the corresponding values of the flow-resistive component. Summation of the two components reconstructs the total intrapleural pressure change over the cycle (*black trace*).

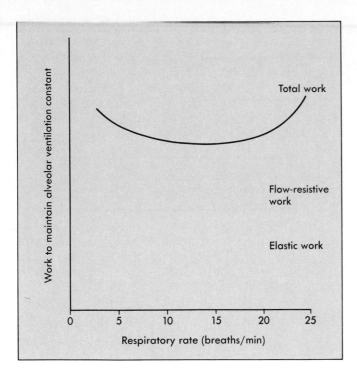

The elastic component and the flow-resistive component of the work needed to maintain a constant alveolar minute volume are affected by tidal volume. If tidal volume increases, the same alveolar ventilation requires fewer breaths per minute. The resistive work of breathing is low, but elastic work is high. If respiratory rate increases, the tidal volume is diminished accordingly. The elastic work decreases, but the resistive work increases. The sum of the two components is minimum at some intermediate point—about 12 breaths/min for a resting rate of alveolar ventilation.

at the levels that minimize the effort needed to maintain the necessary level of ventilation. For typical resting rates of ventilation, the energy cost of ventilation is minimal at a value very close to the actual resting respiratory rate of about 12 breaths/min.

Roles of Pulmonary Surfactant

- *What is the law of Laplace? How does it apply to alveoli?*
- *What are two important functions of pulmonary surfactant?*

The **surface tension** of water is the result of attractive forces between water molecules near an air-water interface (see Chapter 2). This is the force that makes water droplets bead on a nonabsorbent surface like the hood of a car. The alveolar membranes themselves are so compliant that the surface tension of the thin layer of water on the air side of the alveolar membrane contributes about 70% of the elastic recoil of the alveoli. Even so, the measured surface tension in the alveolar walls is much lower than would be expected from calculations of the surface tension of a layer of pure water. The reason is that the water on the surface of the alveoli contains phospholipid molecules in combination with several proteins, called **lung surfactant.** Surfactant is produced by a special category of epithelial cells called **type II alveolar cells.**

Like all phospholipids, lung surfactant molecules are amphipathic (see Chapter 3), with a hydrophilic end and a hydrophobic end. The surfac-

tant molecules disrupt the attractive forces between the water molecules that are responsible for surface tension. The reduction of surface tension in the thin water layer on the surface of the alveoli reduces the surface tension by 7 to 14 times and reduces the force required to expand the lungs by a factor of three to four, greatly reducing the work of breathing. The protein component of surfactant, on the other hand, increases the osmotic pressure of the film of liquid lining the alveoli to that of the plasma, thereby protecting the alveoli from desiccation.

The **law of Laplace** relates the pressure (P) inside a sphere to the tension (T) in its walls. If r is the radius of an alveolus, then the alveolar pressure is given by:

$$P = 2T/r$$

An implication of the law of Laplace is that, if surface tension were equal in alveoli of all sizes, the pressure within large alveoli should be less than in small ones (Figure 16-19, *A*). If so, pressure differences between larger and smaller alveoli would force air out of smaller alveoli into large ones, and all of the alveoli of an acinus would merge to form one large alveolus. The large alveolus thus formed would be less effective because its surface area would be smaller relative to the volume of air in it. However, surfactant makes the surface tension of small alveoli less than that of larger alveoli. This effect can be understood by visualizing the surfactant molecules as closer together in the smaller alveolus than in the larger alveolus, just as the color is more dense in a balloon that has less air in it. The more concentrated surfactant reduces the surface tension

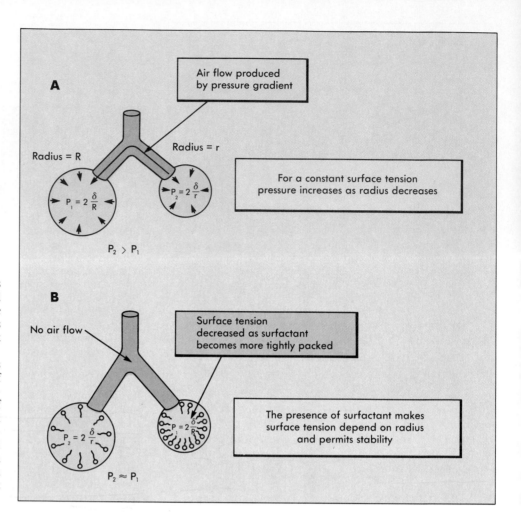

A Air flow produced by pressure gradient

Radius = R   Radius = r

$P_1 = 2\frac{\delta}{R}$ $P_2 = 2\frac{\delta}{r}$

For a constant surface tension pressure increases as radius decreases

$P_2 > P_1$

B No air flow

Surface tension decreased as surfactant becomes more tightly packed

$P_2 = 2\frac{\delta}{r}$ $P = 2\frac{\delta}{r}$

The presence of surfactant makes surface tension depend on radius and permits stability

$P_2 \approx P_1$

FIGURE 16-19
A The law of Laplace predicts that a smaller alveolus should have a higher internal pressure than a large alveolus, and thus air should flow into the largest alveolus.
B Collapse of smaller alveoli is prevented by the presence of lung surfactant because surfactant reduces the magnitude of the surface tension in the alveoli and compensates for the differences in alveolar size. As alveoli expand, surfactant concentration decreases, so the pressure in small alveoli is very close to that in large alveoli.

more in smaller alveoli than in larger ones. The volume-dependent effect of lung surfactant almost completely eliminates the pressure differences that would otherwise exist between alveoli of different sizes (Figure 16-19, *B*).

Infants born prematurely may suffer from low lung compliance in a condition called **respiratory distress syndrome** or **hyaline membrane disease.** This condition is caused by inadequate quantities of lung surfactant. Lung surfactant begins to be synthesized about the 32nd week of fetal development. Surfactant synthesis requires the hormone cortisol and is completed only a few weeks before normal delivery. In normal infants the first deep breath inflates the lungs, which then remain inflated due to the presence of surfactant. Infants with respiratory distress syndrome must make strenuous efforts to breathe because they must reinflate their lungs with each breath. In extreme cases the increased work of breathing and lung collapse can lead to exhaustion and death.

The importance of pulmonary surfactant can be appreciated by comparing surfactant-depleted lung tissue with normal lung tissue (Figure 16-20, *A* and *B*). The alveolar ducts are enlarged, but the alveoli

themselves tend to be collapsed. The decrease in lung compliance caused by a lack of surfactant greatly increases the work of breathing.

It is possible for women who are threatened with premature delivery to be given an injection of the adrenal hormone, cortisol, which will stimulate the early synthesis of surfactant in the fetus. Recently a number of artificial surfactant preparations have been developed and have proved effective in the management of respiratory distress syndrome. Artificial surfactant is composed of two phospholipids with no protein and has physical properties similar to natural surfactant. Surfactant preparations are given in liquid form through an endotracheal tube. Typical doses range from 60 to 200 mg/kg administered either before the infant's first breath as a preventive measure, or after the development of acute respiratory distress syndrome. In clinical trials with babies under 30 weeks, artificial surfactant reduced neonatal mortality by 50%, with no evidence of subsequent ill effects after natural surfactant synthesis reached normal levels. So far, artificial surfactant appears to be a safe therapy, though additional clinical trials with long-term follow-up are still needed.

A

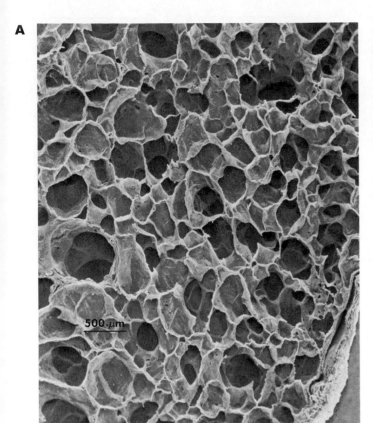

500 µm

B

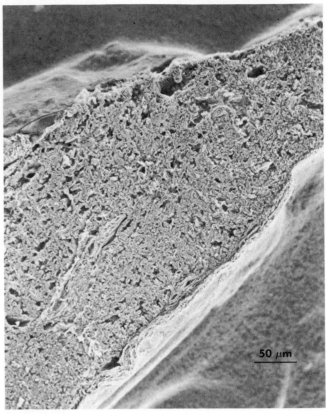

50 µm

C

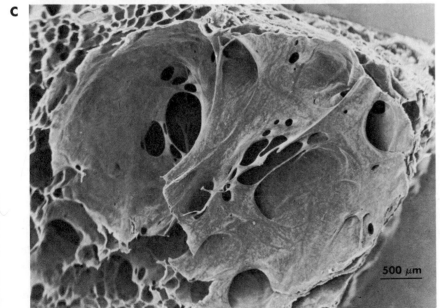

500 µm

FIGURE 16-20
Scanning electron micrographs of a normal lung (**A**), a surfactant-depleted lung (**B**), and a lung with emphysema (**C**). Note that the micrographs in **A** and **C** were taken at 10 times the magnification of that in **B**.

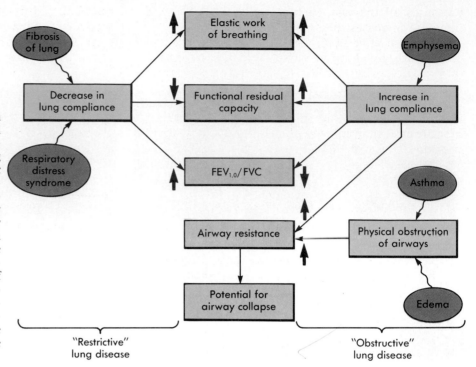

FIGURE 16-21
A diagram illustrating the changes in respiratory function in obstructive and restrictive pulmonary disease. Restrictive disease involves a decrease in lung compliance. It increases the elastic work of breathing, decreases the functional residual capacity (FRC), and increases the ratio of forced expiratory volume to the forced vital capacity ($FEV_{1.0}$/FVC). Obstructive diseases can involve an increase in lung compliance or a physical obstruction of the airways. In this case the elastic work of breathing is decreased, but the flow-resistive work increases. The FRC is increased, and the $FEV_{1.0}$/FVC ratio is reduced. In all obstructive diseases the small airways tend to collapse because airway resistance is increased.

Pulmonary Resistance and Compliance in Disease

> • *How do obstructive and restrictive pulmonary disease differ?*

Pulmonary diseases can be divided into two general families (Figure 16-21). **Obstructive diseases** increase the airway resistance and thus the flow-resistive work of breathing. Obstructive diseases are subdivided into two types: (1) diseases such as emphysema, in which a loss of lung tissue (see Figure 16-20, *C*) decreases the elastic recoil of the lung; and (2) diseases such as asthma, in which bronchiolar constriction and increased mucus secretion increase airway resistance, and pulmonary edema, in which the alveoli and airways are plugged by fluid. **Restrictive diseases**, in which lung compliance is decreased, include pulmonary fibrosis and respiratory distress syndrome. Restrictive diseases limit the volume changes of the lung and increase the elastic work of breathing.

Increased airway resistance causes the collapse of some smaller airways during expiration. This is because a forced expiration increases the intrapleural pressure and produces a significant pressure drop between the alveoli and the atmosphere. If airway resistance is increased by disease, the pressure drop is larger and may drop below intrapleural pressure in the terminal bronchioles. They will then be compressed, trapping alveolar gas.

Cigarette smoking is a major cause of emphy-

sema, a disease that results in loss of the parenchymal elastic elements of the lung. Loss of these components of lung tissue results in increased lung compliance. This has the same effect on alveoli as an increase in airway resistance because the intrapleural pressure is less negative, and negative intrapleural pressure normally helps distend the small airways. Even if terminal bronchioles are otherwise unoccluded, they tend to collapse during expiration and require the opposing force of the negative intrapleural pressure to reopen them during inspiration.

Obstructive diseases are diagnosed by measuring the $FEV_{1.0}$. The $FEV_{1.0}$ is normally about 4.0 L, compared with a forced vital capacity (FVC) of 5.0 L. This produces a ratio of $FEV_{1.0}$/FVC of about 0.80. In a person with obstructive disease, the FRC is increased. However, the main abnormality is a decreased expiratory flow rate and an increase in the amount of gas trapped behind collapsed airways. Both the $FEV_{1.0}$ and the FVC are decreased. Typical values in an emphysemic person would be 1.3 L and 3.1 L, respectively. The ratio is thus about 0.40. In contrast, the vital capacity measured during slow expiration may be normal.

People with obstructive disease tend to breathe slowly, maintaining a large functional residual capacity. When the FRC is large the lungs are more fully inflated at all stages of the respiratory cycle. Under these conditions the intrapleural pressure is more negative, helping to keep airways distended. Such people also obtain some relief by pursing their lips. The resulting external resistance increases the pressures all the way back along the airways and therefore increases the distending force.

The restrictive lung diseases cause a decrease in compliance, resulting in a more negative intrapleural pressure. Airway collapse during forced expiration does not occur. It is easy to expel air, and the ratio of vital capacity to functional residual capacity is usually closer to 1.0 in persons with restrictive lung disease than in normal subjects. However, the absolute values of $FEV_{1.0}$ and forced vital capacity are decreased because the lung volumes are generally decreased.

ADAPTATIONS OF THE PULMONARY CIRCULATION
Pressure and Flow in the Pulmonary Loop

- *What is the function of recruitment of pulmonary blood vessels?*
- *What are the advantages of the low flow resistance and low pressures of the pulmonary loop?*
- *What changes in pulmonary circulation and ventilation result from standing up?*

Because the flow resistance of the pulmonary vessels is low compared with that of the systemic circulation, the normal mean pulmonary arterial pressure of only about 18 mm Hg is sufficient to drive blood through the pulmonary loop at a rate equal to that of the systemic loop. Unlike the systemic loop, the flow resistance of the pulmonary circulation is almost completely independent of changes in cardiac output. There are two reasons for this independence. When cardiac output increases and pulmonary pressure begins to rise, blood vessels that were previously collapsed begin to open. This process is termed **recruitment.** The pulmonary arteries are also more compliant than the systemic arteries. They expand as the pressure within them increases, and thus their flow resistance falls. These features of the pulmonary loop make it possible for the large increases in output of the left heart that occur in exercise to be matched by the relatively weak right heart, without a substantial increase in pulmonary arterial pressure. Recruitment increases the overall functional lung surface area and reduces the physiological dead space (see below).

Capillary Filtration and Reabsorption in the Lung

The low pulmonary arterial pressure results in a mean pressure of about 7 mm Hg in the pulmonary capillaries, biasing the balance of filtration and absorption strongly in favor of absorption. This normally protects the lung from edema. Pulmonary edema can occur as a consequence of inflammation or left ventricular failure and significantly impairs lung function in two ways. First, expansion of the pulmonary interstitial fluid lengthens the diffusional path between capillaries and alveoli. Second, some of the excess interstitial fluid is forced into the alveoli, impairing alveolar ventilation.

Ventilation-Perfusion Matching

The ratio of alveolar ventilation to alveolar blood flow is the **ventilation-perfusion ratio (V/Q).** For the respiratory system as a whole, the ratio is the alveolar minute volume divided by the cardiac output ($\dot{V}_A/CO$); about 4.0 L/min air $\div$ 5.0 L/min blood, or 0.8. For most efficient gas exchange, V/Q should be uniform throughout all regions of the lung. Regional imbalances of V/Q result in **physiological dead space** and **physiological shunt.** The meaning of these terms can be illustrated by the most extreme cases of V/Q mismatch: an alveolus that is ventilated but receives no perfusion and an alveolus that is perfused but receives no ventilation. Ventilation of an alveolus that receives no perfusion has the same effect as ventilation of an equivalent volume of the anatomical dead space. Perfusion of an alveolus that is not ventilated is the same as allowing that amount of blood flow to bypass (shunt) the lung.

Ventilation-perfusion matching is a trivial problem in a person who is lying down, because both intrapleural pressure and blood pressure are constant throughout the lung. As explained in Chapter 14, the effect of gravity on hydrostatic pressure at different points within the body has implications for the vascular systems of standing humans; there are also effects on the lungs. In a standing person, the hydrostatic component of blood pressure decreases above the heart and increases below it. Thus blood pressure is greater at the base of the lung than at the apex. Pulmonary blood vessels are more compliant than those in the systemic circulation, so the effect of the greater blood pressure at the base of the lung is to expand the vessels of the base and substantially increase perfusion at the base compared with the apex.

In a standing person the intrapleural pressure (P_{pl}) is more negative at the apex of the lung (about -13 cm H_2O) than at its base (about -3 cm H_2O; Figure 16-22). Because the alveoli experience the intrapleural pressure, alveoli at the apex of the lung tend to be more fully expanded than alveoli near the base. If alveolar compliance were independent of lung volume then expansion would not matter, but such is not the case. As explained earlier in this chapter, larger alveoli have lower compliances. Alveoli at the base of the lung are smaller and thus expand more during inspiration than those at the apex. That is, in addition to greater perfusion, they receive a larger fraction of the inspired air than those at the apex and are thus better ventilated.

From the description just given, it might appear that the effects of gravity on ventilation and perfu-

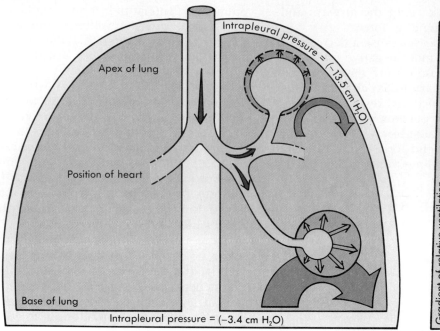
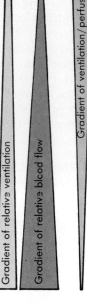

Intrapleural pressure = (−13.5 cm H_2O)

Apex of lung

Position of heart

Base of lung

Intrapleural pressure = (−3.4 cm H_2O)

Gradient of relative ventilation

Gradient of relative blood flow

Gradient of ventilation/perfusion

Alveoli at base
receive greater fraction
of total lung ventilation and perfusion

FIGURE 16-22

Ventilation-perfusion relationships in the lung. Because the intrapleural pressure is more negative at the apex than at the base, alveoli at the base of the lung are less expanded than those of the apex prior to inhalation and receive a larger fraction of the inspired air. The base of the lung also receives a larger blood flow, because the outward transmural pressure across the pulmonary vessels is greater, increasing their diameter and decreasing their flow resistance. However, these gradients do not balance, and the ratio of ventilation to perfusion (V/Q) decreases from the apex to the base. Arrows to the right of the figure indicate the magnitude of the three gradients of blood flow, ventilation, and the ventilation to perfusion ratio.

sion would cancel each other out. However, the apex-to-base ventilation gradient and perfusion gradient are not identical. With no compensation by blood vessels and bronchioles, the base of the lung would receive about twice as much ventilation as the apex, but about ten times more blood flow.

The lung compensates for this regional variation in alveolar ventilation and blood flow in two ways: (1) increases in the alveolar carbon dioxide and decreases in oxygen cause the nearby bronchioles to enlarge, redirecting the inspired air toward relatively underventilated alveoli at the apex of the lung; while (2) decreases in oxygen or increases in carbon dioxide cause a vasoconstriction and shift

TABLE 16-6	*Ventilation-Perfusion Matching*		
Imbalance	*Change in alveolar gas concentration*	*Response of pulmonary arterioles*	*Response of bronchioles*
Ventilation > perfusion	Increased O_2 Decreased CO_2	Dilate	Constrict
Perfusion > ventilation	Decreased O_2 Increased CO_2	Constrict	Dilate

blood away from the underventilated apical portions of the lung (Figure 16-23; Table 16-6). The response of pulmonary arterioles to alveolar O_2 and CO_2 is exactly opposite to the response that systemic arterioles make to the same factors in systemic flow autoregulation. As a result of these intrinsic compensations, the actual discrepancy between alveolar ventilation and perfusion is less severe than it would be in their absence. However, there is still a difference in V/Q between the apex and base of the lung in a standing person.

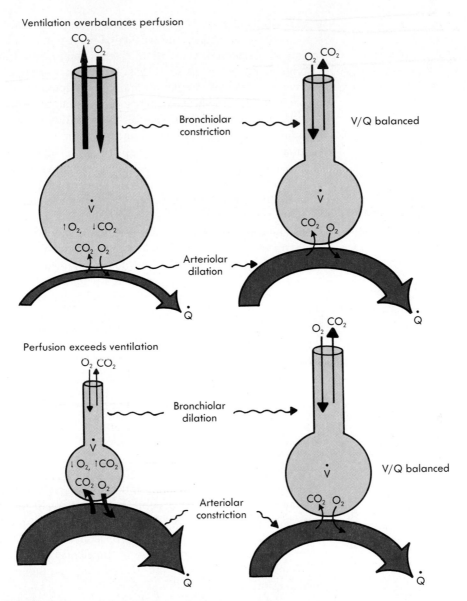

FIGURE 16-23
Compensation by bronchioles and pulmonary arterioles for ventilation-perfusion mismatch is driven by changes in alveolar gas composition. The arrows show relative rates of gas movement between alveoli and blood and alveoli and airway. The response of arteriolar smooth muscle is reflected in a change in local perfusion shown by the magnitude of the flow arrow. The response of the smooth muscle of terminal and respiratory bronchioles is reflected by a change in local alveolar ventilation, as indicated by the diameter of the bronchiole in the figure.

Cardiopulmonary Resuscitation and the Heimlich Maneuver

Respiratory failure impairs the function of the heart and brain within seconds. In heart attacks, the weakened heart pumping may lead rapidly to cessation of ventilation. If ventilation is interrupted (asphyxiation), the heart soon ceases to pump. Cardiopulmonary resuscitation restores ventilation and blood pumping until the victim recovers or other medical assistance can be obtained.

In cardiopulmonary resuscitation the rescuer(s) ventilate the victim with air from their own lungs (mouth-to-mouth resuscitation) and simultaneously apply rhythmic pressure to the victim's chest. When the chest is compressed, intrathoracic pressure rises, causing the equivalent of a forced expiration and at the same time expelling blood from the heart and major thoracic blood vessels. When the compression is released, thoracic pressure decreases to the resting value, filling the lungs and the heart. Enough ventilation and blood circulation can be maintained in this way to prevent brain and heart damage.

Cardiopulmonary resuscitation must be applied immediately and cannot be used effectively by untrained people. Substantial increases in the survival of victims of heart attacks and other accidents have been shown in cities in which a significant portion of the population has been trained in the method. The American Heart Association suggests that all Americans learn cardiopulmonary resuscitation, starting in the eighth grade.

Choking on food is a major cause of accidental death. It results from inspiration of food into the trachea or lodging of food in the pharynx, blocking air flow into the trachea. Powerful reflexive responses act to oppose blockage of the trachea or pharynx, but these are less effective if blunted by alcohol, a common accompaniment of meals. Sudden inability to speak or breathe is a distinguishing characteristic of choking.

The Heimlich maneuver is an effective means of relieving choking. To carry it out on an adult victim, the rescuer embraces the victim from behind, placing both arms around the victim's waist just above the beltline. The rescuer makes a fist with one hand and presses the thumb side of the fist upward into the victim's abdomen with the other hand, repeatedly making a sharp flexion of the elbows. The effect is to press the diaphragm upward into the thorax, elevating thoracic pressure. The increased pressure in the airway ejects the food, as a cork is ejected from a bottle. If the victim is lying down, the upward thrust can be applied by placing the heel of one hand on the abdomen above the navel and just below the sternum. Once the airway is clear, cardiopulmonary resuscitation can be applied if necessary.

SUMMARY

1. The respiratory system consists of the **airways:** the nasal passages, **pharynx, trachea, bronchi,** and **bronchioles;** the gas exchange elements: the **alveolar sacs** and **alveoli;** and the **diaphragm** and **intercostal muscles** that are responsible for ventilation.

2. The airway warms and humidifies inspired air. Its **mucous secretions** and **cilia** filter and remove particulate contaminants and microbes. The flow of air through the airway is the energy source for sound production in the larynx. The diameter of the terminal bronchioles of the airway is under neural and hormonal control, which affect air flow by changing the **airway resistance.**

3. The gas exchange function of the lung is maximized by the short diffusion distance between pulmonary capillary blood and alveolar gas and by the enormous surface areas of alveoli and capillaries.

4. The volume of gas in the lungs can be divided into the tidal, inspiratory reserve, expiratory reserve, and residual volumes. The lung capacities are the sum of two or more volumes. The **vital capacity** is a measure of the largest possible tidal volume; the **functional residual capacity** is the volume of air present in the lung at the end of a passive expiration.

5. The **respiratory minute volume** is the total volume of air moved by the respiratory system per minute. Of this volume, part ventilates the anatomical dead space and the rest, the alveolar ventilation, ventilates the alveoli.

6. Respiratory muscles cause volume changes of the chest wall, which result in changes in the intrapleural pressure according to Boyle's law. The lungs and chest wall are mechanically coupled by the intrapleural fluid. The **intrapleural pressure** is always less than alveolar pressure because of the inward **elastic recoil** of the lung.

7. The total work done in ventilation is the sum of work done in changing the volume of the lung-chest wall (elastic component) and the work done to move air through the system (flow-resistive component). For a given alveolar ventilation, elastic work increases with increasing tidal volume and decreasing frequency, while flow resistive work increases with increasing frequency and decreasing tidal volume.

8. **Pulmonary surfactant** reduces the surface tension of the air-liquid interface in the alveoli, with two major consequences for the lung. First, the work of expanding the alveoli is decreased. Second, the tendency of small alveoli to merge into larger ones is reduced, so that the large surface area of the alveoli is maintained.

9. Pulmonary diseases can be divided into **obstructive disease,** in which air flow is impeded, and **restrictive disease,** in which the elasticity of the alveoli is reduced. The two types of disease can be distinguished on the basis of their effects on the static and dynamic lung volumes and capacities.

10. Large increases in pulmonary blood flow can result from small pressure changes because of the high compliance of pulmonary vessels and because small increases open (recruit) additional pulmonary vessels. The low resistance of the pulmonary loop favors capillary absorption over filtration and thus protects the lung from edema.

11. In a standing person, gravity tends to cause the base of the lung to be better ventilated and better perfused than the apex. The resulting inequities of the **ventilation-perfusion** ratio affect the O_2 and CO_2 content of alveolar gas and initiate compensatory responses of bronchiolar and arteriolar smooth muscle.

1. Where are the following located and what are their main functions?
   Bronchi
   Respiratory bronchioles
   Terminal bronchioles
   Alveoli

2. What structures compose a pulmonary acinus?

3. What structures correspond to the anatomical dead space? If a subject who has an anatomical dead space volume of 100 ml is breathing through a snorkel tube that has a volume of 200 ml, what must her tidal volume be to maintain an alveolar minute volume of 3 L at 12 breaths/min?

4. Define the following lung volumes:
   Residual volume
   Tidal volume
   Functional residual capacity
   Vital capacity

5. Why is the intrapleural pressure in a normal individual negative in relation to the atmospheric pressure?

6. When the lung volume is at the functional residual capacity, in which direction is the chest recoiling? In which direction is the lung recoiling? How do these recoil forces compare with one another?

7. Why do intrapleural pressure changes not mirror alveolar pressure changes?

8. How does the intrapleural pressure compare with the alveolar pressure during inspiration and expiration?

9. How would the FRC differ between a premature infant lacking pulmonary surfactant and a normal full-term infant?

10. Subject A has a lung compliance that is lower than that of Subject B. All other physiological parameters are the same for the two subjects. How would the work of breathing differ between the two individuals?

11. Compare the apex of the lung with the base in terms of their:
    Alveolar ventilation
    Blood flow
    Ventilation/perfusion ratio
    Intrapleural pressure

12. Airway resistance in a subject is doubled. What effect would this have on the intrapleural pressure during inspiration? On the alveolar pressure? On the work of breathing?

Choose the MOST CORRECT Answer

13. The respiratory airways:
    a. Protect against particulate contaminants through sneezing and coughing reflexes
    b. Trap particles in mucus and propel upward to the pharynx by the mucociliary escalator
    c. Inactivate infectious particles and remove small particles through the action of white blood cells
    d. All of the above are true

14. When one blows up a balloon, increased tidal volume is typically drawn from:
    a. Inspiratory reserve volume
    b. Expiratory reserve volume
    c. Residual volume
    d. Both a and b.

15. _____ is a measure of dynamic lung capacity.
    a. Inspiratory capacity
    b. Forced vital capacity
    c. Total lung capacity
    d. Forced expiratory capacity

16. During inspiration:
    a. Intrathoracic volume decreases
    b. Lung volume decreases
    c. Lung pressure increases
    d. Intrathoracic volume increases

17. Restrictive pulmonary diseases are characterized by:
    a. Increase in airway resistance
    b. Increase in functional residual capacity
    c. Increased lung compliance
    d. None of these is true

● *SUGGESTED READINGS*

EDWARDS DD: Mending a torn screen in the lung, *Science News* 132:277, May 2, 1987. Describes new treatments for alpha-antitrypsin deficiency.

KAMBERG ML: Take a deep breath and prepare to meet your respiratory system, *Current Health* 16(3):4, 1989. Discusses the risks of smoking within the context of respiratory function.

LIPMAN MM: Emphysema takes smokers' breath away, *Consumer Reports Health Letter* 2(9):70, 1990. Discusses occurrence of chronic obstructive lung disease in smokers.

PEDERSEN M: Ciliary activity and pollution, *Lung* 168:368, 1990. A current, topical review.

RALOFF J: New clues to smog's effect on lungs, *Science News* 132:86, August 8, 1987. Describes how the lung's defense mechanisms can be damaged by environmental contamination.

WEIBEL ER: *The pathway for oxygen,* Harvard University Press, Cambridge, Mass., 1984.

CHAPTER 17

Gas Exchange and Gas Transport in the Blood

On completing this chapter you should be able to:

- Apply the gas laws to gas mixtures in the respiratory tract and gases dissolved in the blood and tissues
- Understand the concept of partial pressure
- Describe how O_2 is transported in the blood by hemoglobin
- Appreciate why the **S**-shaped oxyhemoglobin dissociation curve maximizes O_2 loading in the lungs and unloading in the tissues
- Describe the factors that shift the oxyhemoglobin dissociation curve to the right
- Compare the oxyhemoglobin dissociation curve for adult hemoglobin with those for fetal hemoglobin and myoglobin
- Describe how CO_2 in the blood is carried as HCO_3^-
- Apply the Henderson-Hasselbalch equation to bicarbonate transport in the blood

Respiration could not be understood until the birth of chemistry as a science. The English physician Joseph Black both discovered CO_2 and showed that it is produced by metabolism. One of his experiments was performed in 1754 in an air duct in the ceiling of a church in Glasgow, where 1500 people had gathered for a 10 hour service. Black dripped limewater—containing $Ca(OH)_2$—over rags and collected the precipitate of Ca_2CO_3 that resulted from its reaction with the CO_2 exhaled by the congregation below. The French chemist Antoine Lavoisier (1743-1794) found that, when placed in a sealed container, candles went out and animals died as soon as most of the enclosed O_2 had been replaced by CO_2.

The mechanisms by which O_2 and CO_2 are transported in the blood began to be understood in the second half of the nineteenth century. In 1886, Christian Bohr showed the quantitative relationship between the partial pressure of O_2 availability and O_2 binding by hemoglobin. In 1904, together with K.A. Hasselbalch and August Krogh, he showed that O_2 unloading in systemic capillaries increases, in the presence of CO_2, a process now referred to simply as the Bohr effect.

In the last decades of the nineteenth century, the mechanism whereby O_2 and CO_2 are exchanged across the alveolar membrane was the subject of a scientific controversy. One hypothesis held that the lung can actively secrete CO_2 and absorb O_2 into the blood against a concentration gradient. In 1904 August Krogh and his wife Marie showed that diffusion can account for alveolar gas exchange. The final defeat of the secretion theory was provided by Joseph Barcroft in 1919. Barcroft spent 6 days in a glass chamber maintained at subatmospheric pressure. During this exposure to simulated high altitude, Barcroft found that gas exchange always occurred down a concentration gradient.

A

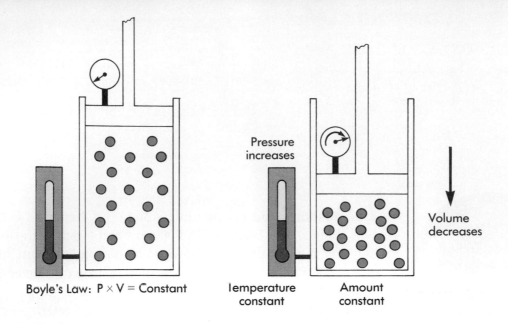

Boyle's Law: P × V = Constant

Pressure increases

Temperature constant

Amount constant

Volume decreases

B

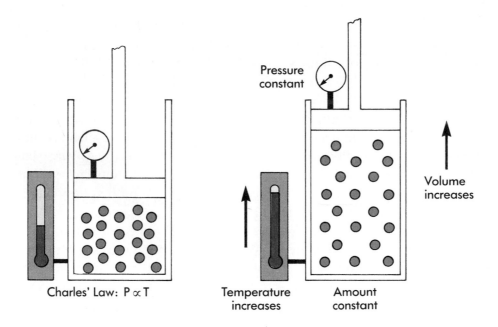

Charles' Law: P ∝ T

Pressure constant

Temperature increases

Amount constant

Volume increases

C

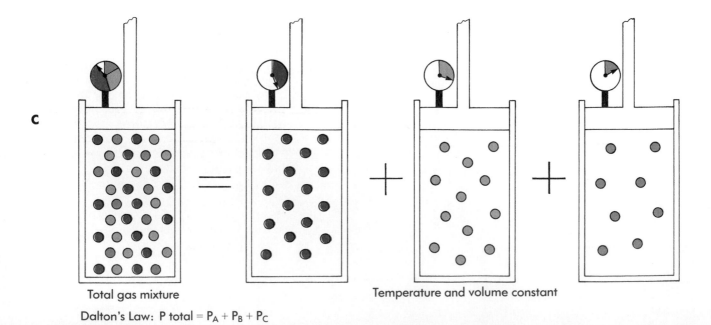

Total gas mixture

Temperature and volume constant

Dalton's Law: P total = P_A + P_B + P_C

PHYSICS AND CHEMISTRY OF GASES

- *What are the three variables whose relationships are described by the gas laws? State the equation that summarizes the relationship.*
- *What is meant by the partial pressure of a gas? What two factors determine the partial pressure?*
- *What two factors determine the number of gas molecules per unit volume of water for gases in solution?*

Molecular Properties of Liquids and Gases

Water and many other compounds can exist as solids, liquids, or gases. Both liquids and gases are fluids, forms of matter in which individual molecules can move freely in relation to one another. But, in some respects, liquids and gases behave differently. These differences have to do with the density of the constituent molecules. Liquids are dense enough that intermolecular effects have significant consequences; surface tension, density, and viscosity are examples of properties that arise from such interactions. For most gases (or gas mixtures) at physiological pressures, the density is so low that collisions between molecules very rarely occur; thus the gas molecules do not interact. Such gases are termed **ideal gases.** A consequence of the large distances and infrequent interactions between gas molecules is that gases are **compressible.** In contrast, the forces between molecules in liquids are so great that the volume of a liquid will not change significantly if the pressure on the liquid is increased.

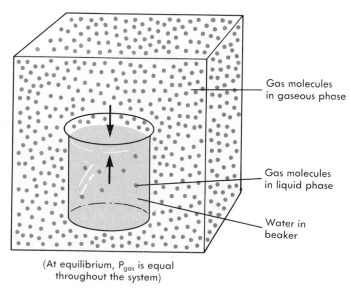

(At equilibrium, P_{gas} is equal throughout the system)

FIGURE 17-1
The gas laws. (See also p. 480.)

Gas Laws

Gas molecules continuously strike the walls of whatever container they are in and produce a **force per unit area,** or **pressure.** The pressure of a gas depends on two factors: (1) the average number of collisions per unit time; and (2) the average velocity of the molecules. Collision frequency is proportional to the number of molecules per unit volume, or concentration of the gas. The higher the temperature, the greater the average velocity of the gas molecules.

The gas laws describe the relationships among the volume, pressure, and temperature of an ideal gas. If one of the three variables is held constant, the relationships between the other two are given by Boyle's law and Charles' law. In **Boyle's law** (Figure 17-1, *A*), which has already been shown to apply to ventilation (Chapter 16), temperature is fixed, and increases in the volume of a gas decrease gas molecule density, thus decreasing the pressure (and vice versa). In **Charles' law** (Figure 17-1, *B*), volume is not allowed to change; under that condition, the average kinetic energy (pressure) of gas molecules increases or decreases with rising or falling temperature. **Gay-Lussac's law** allows volume to change with changes in temperature, and as a consequence, pressure also changes (Figure 17-1, *C*).

The three gas laws described in the previous paragraph can be combined into the **gas equation:**

$$PV = nRT$$

where **n** is the number of moles of a gas, **R** is the **universal gas constant,** and **T** is the temperature in degrees Kelvin. For any compound, a mole of a gas contains 6.023×10^{23} molecules (**Avogadro's number**). At a temperature of 0° C (273° Kelvin) and a pressure of 760 mm Hg (1 atmosphere), 1 mole of any gas occupies exactly 22.4 liters. At body temperature (37° C or 310° K) and 1 atmosphere (atm) of pressure, a mole of gas will occupy 25.5 liters. The gas equation applies to gas mixtures as well as a pure gas (Figure 17-1, *D*). Because gas molecules do not interact substantially an ideal gas of whatever type exerts the same pressure per number of molecules at 0° C and 1 atm.

Partial pressure of any single gas in a gas mixture is that pressure that would exist if a single gas was present by itself. **Air** is a mixture of gases whose major constituents are O_2 and N_2 (Table 17-1). The total pressure of a mixture of gases equals the sum of the individual partial pressures of its components. The partial pressure of one constituent in a mixture is determined by multiplying the percentage of the total pressure which that gas exerts by the total pressure of the mixture (see Figure 17-1, *D*).

The partial pressure of O_2 in dry atmospheric air at sea level can be computed from the following data: average atmospheric pressure of 760 mm Hg,

TABLE 17-1	Constituents of Clean, Dry Atmospheric Air at Sea Level	

Gas	Percent of total	Partial pressure (mm Hg)*
Nitrogen (N_2)	78.08	593.40
Oxygen (O_2)	20.95	159.22
Inert gases (argon, neon, helium, krypton, xenon)	0.9654	7.34
Carbon dioxide	0.0314	0.24

*Partial pressures total is slightly higher than 760 mm Hg because of rounding.

and O_2 content of the air of about 21%. The partial pressure of oxygen (Po_2) is:

$$P_{O_2} = 760 \text{ mm Hg} \times 0.21 = 160 \text{ mm Hg}$$

Thus the part of the total pressure that is attributable to O_2 is 160 mm Hg. Similarly at sea level the partial pressures of nitrogen and carbon dioxide are:

$$P_{N_2} = 760 \times 79\% = 600 \text{ mm Hg}$$
$$P_{CO_2} = 760 \times 0.03\% = 0.2 \text{ mm Hg}$$

Water Vapor

If a beaker of water is placed in a closed compartment containing dry air, some water molecules will leave the liquid phase and enter the gas phase (Figure 17-1, E). As this occurs, an oppositely directed flux of water molecules develops. When the system has come to equilibrium, the vapor-to-liquid and liquid-to-vapor water fluxes are equal, and the gas phase is saturated with water vapor. The partial pressure of water vapor depends on the average thermal energy of the water molecules (an application of Charles' law); the saturation point is higher, that is, more molecules of water will move into the vapor phase at a higher temperature. At the normal lung temperature of 37° C, the partial pressure of water vapor (or **vapor pressure**) is 47 mm Hg in air saturated with water.

Inspired air passes over the warm, moist surfaces of the airway and becomes saturated with water at body temperature. The addition of water vapor does not change the total pressure of the air, but its inclusion means that the partial pressures of other component gases must decrease. During inspiration, the partial pressure of O_2 in the tracheal air is 21% of (760 - 47) mm Hg, or about 150 mm Hg (Figure 17-2).

Gases in Solution

Not only do gases become saturated with water, but water will become saturated with dissolved mole-

cules of the gases with which it is in contact. As the gas molecules collide with the gas-water interface, some of them will dissolve in the water. The number of surface collisions per unit time made by molecules of each type of gas in the air is proportional to the partial pressure of each type. As the number of dissolved molecules of each gas type increases, the number of molecules that pass from water back to air also increases. At equilibrium, the partial pressure of each constituent in the gas phase is equal to the partial pressure of that constituent in solution, and there is no net movement of gas in either direction across the interface.

Henry's law says that at equilibrium the concentration of dissolved gas (number of moles of gas per unit volume of water) depends on the partial pressure of the gas, its **solubility**, and the temperature (see Figure 17-1, E). A liquid exposed to two gases that have the same partial pressure but different solubilities will have different numbers of molecules of the two gases dissolved per unit volume at equilibrium. Gas molecules of all ideal gases bombard the liquid with the same energy, but the more soluble a gas, the more likely that the collision will result in entry of the gas molecule into the liquid (or the less likely that a molecule colliding with the surface from the liquid phase will escape from the liquid). The solubility coefficient for CO_2 in plasma at 37° C is approximately 21 times greater than that of O_2.

The rate at which dissolved gas diffuses in solution is determined by the magnitude of the partial pressure gradient and the diffusion coefficient of the particular gas involved. The diffusion coefficient depends on temperature and the nature of the diffusion path. For living tissue at body temperature, the diffusion coefficient of CO_2 is about 20 times greater than that of O_2.

GAS EXCHANGE IN ALVEOLI

- *What is the respiratory quotient? What determines its value?*
- *What physical factors determine the rates of diffusion of O_2 and CO_2 between blood and alveoli?*
- *How is it possible for CO_2 to be exchanged at approximately the same rate as O_2 when the partial pressures of the two gases differ by about tenfold?*

Composition of Alveolar Gas

The molar ratio of CO_2 production to O_2 consumption is the **respiratory quotient** (RQ). The RQ varies with the specific metabolic pathways used for energy production. The overall equation for glucose metabolism is:

$$C_6H_{12}O_6 + 6O_2 \rightarrow 6CO_2 + 6H_2O$$

If glucose were the only fuel used by the body, the RQ would be 1.0. Fats and proteins contain less

O_2 than carbohydrate and therefore require more O_2 for their oxidation. Metabolism of fat to CO_2 results in an RQ of about 0.7; similarly, protein metabolism results in an RQ of slightly less than 0.8. Typically a mixture of all three fuels is being metabolized at any time, so that the RQ is in the range of 0.8 to 0.85. At rest a normal individual consumes about 250 ml of O_2 per minute. With an RQ of 0.8, this individual would produce about 200 ml of CO_2 per minute. That is, averaged over the entire body, slightly more than 4 molecules of CO_2 are produced by the tissues for every 5 molecules of O_2 consumed.

Even though the alveoli are continually ventilated, the partial pressure of O_2 in the alveoli is not as high as in the inspired air because the inspired air mixes with the volume of "old" air retained in the anatomical dead space (the volume not exchanged with inspired air) at the end of each expiration. The partial pressure of CO_2 in the alveoli is higher than in inspired air for the same reason. Composition of the alveolar air (for which typical values were given in Figure 17-2) depends on the ventilation of the alveoli (frequency and depth of breathing) and the rates of O_2 uptake and CO_2 release.

The relationship between alveolar Pco_2, alveolar Po_2, the Po_2 of inspired air, and RQ is described by the **alveolar gas equation**:

$$\text{Alveolar } Po_2 = \text{Inspired air } Po_2 - (\text{Alveolar } Pco_2/RQ)$$

At sea level, the inspired Po_2 is about 150 mm Hg (see Figure 17-2). A normal alveolar Pco_2 under these conditions is 40 mm Hg. For a respiratory quotient of 0.8, the alveolar gas equation predicts that alveolar Po_2 will be 150 mm Hg − (40 mm Hg/0.8), or 100 mm Hg.

The alveolar gas composition changes very little during each respiratory cycle. There are two reasons for this stability. First, as noted in Chapter 16, the linear velocity of air in the smallest airways is so slow that diffusion is the dominant mode of gas movement. There is relatively little bulk flow of air in and out of each alveolus with each breath. Also, the approximately 350 ml of alveolar ventilation of each breath is diluted into more than 2 L.

During exhalation, the alveolar air mixes with air present in the anatomical dead space that was not altered by gas exchange, so expired air has an average Po_2 of 120 mm Hg and Pco_2 of 32 mm Hg (see Figure 17-2).

Gas Composition of Pulmonary Blood

Mixed venous blood typically has a Po_2 of 40 mm Hg and a Pco_2 of 46 mm Hg (see Figure 17-2). The typical resting values for alveolar Po_2 and the alveolar Pco_2 are 100 mm Hg and 40 mm Hg, respectively. The driving force for O_2 diffusion at the arteriolar end of the pulmonary capillaries is the difference between the Po_2 of alveolar air (100 mm Hg) and that of the pulmonary arterial blood (40 mm Hg), or 60 mm Hg. The driving force for CO_2 diffusion into the alveoli, calculated similarly, is only 6 mm Hg. Nevertheless, approximately equal volumes of O_2 and CO_2 are exchanged across the alveolar membrane. This is possible because CO_2 has a much higher diffusion coefficient in water than O_2.

Blood gas partial pressures come into equilibrium with the partial pressures of alveolar air in the first third of the 1 second that each drop of blood spends in a pulmonary capillary. Figure 17-3 shows how Po_2 varies with distance along the capillary. The linear velocity of blood through the capillary may be twice as fast in exercise (*curve B* in Figure 17-3), but full equilibration of pulmonary capillary blood with alveolar air still occurs in healthy individuals, because the diffusion path between alveolar air and blood (Figure 17-4) is so short.

Because complete equilibration occurs between alveolar air and pulmonary blood, the partial pressures of dissolved gas in systemic arterial blood might be assumed to be identical to those of alveolar air. This is not usually the case because as shown in Chapter 16, a small fraction of the pulmonary blood flow bypasses the alveoli, and because the ventilation and perfusion are not perfectly matched in all alveoli. However, the differences are normally so slight that they can be neglected in a discussion of the basic mechanisms of gas exchange.

OXYGEN TRANSPORT IN THE BLOOD

- *What is measured by the percent saturation of hemoglobin?*
- *What percent saturation values for hemoglobin are characteristic of the pulmonary venous blood? Of the systemic venous blood?*
- *On average, about what percent of the O_2 content of systemic arterial blood is unloaded in the tissues in a resting individual?*
- *What portion of the oxyhemoglobin dissociation curve is most strongly affected by temperature, CO_2, pH, and 2,3 diphosphoglycerate?*
- *What is the functional significance of the right shift caused by these factors?*
- *How do the dissociation curves of myoglobin and hemoglobin differ? What is the functional significance of the difference?*

Hemoglobin and the Problem of Oxygen Solubility

At body temperature and an arterial Po_2 of 100 mm Hg, 0.3 ml of O_2 is dissolved per 100 ml of plasma (0.3 vol%). A normal cardiac output of 5 L/min could supply 15 ml/min of dissolved O_2 (5 L/min × 3.0 ml O_2/liter), but this is only 6% of the body's resting oxygen consumption of 250 ml O_2/min. Therefore if dissolved O_2 were the only form in

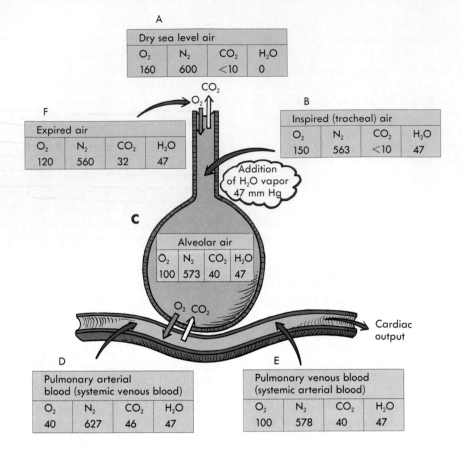

FIGURE 17-2

The partial pressures of oxygen, carbon dioxide, nitrogen, and water vapor in the atmosphere (**A**); the trachea (**B**); the alveoli (**C**); and the pulmonary arterial and venous blood (**D** and **E**) under normal physiological conditions. (All units are mm Hg.) The sum of all the partial pressures must equal the barometric pressure.

FIGURE 17-3

The upper portion of the figure schematically illustrates a red blood cell traversing a pulmonary capillary, while the lower portion shows how the partial pressure of oxygen varies from the arterial to the venous end of the capillary. Full equilibration of oxygen between the alveoli and pulmonary capillaries occurs in the first third of a pulmonary capillary under resting conditions *(curve A)* and occurs only slightly further along when cardiac output is increased *(curve B)*. Oxygen fails to equilibrate fully only in disease conditions in which the barrier to diffusion is increased, a condition referred to as diffusion limitation *(curve C)*.

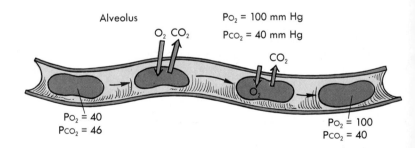

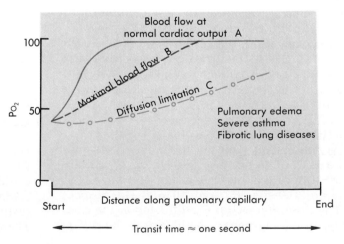

A

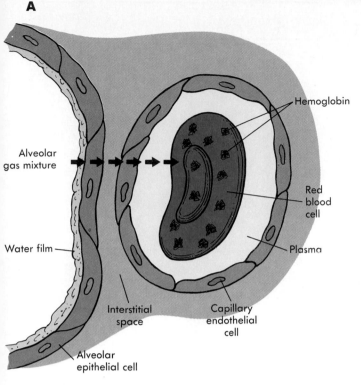

Alveolar gas mixture

Water film

Interstitial space

Alveolar epithelial cell

Hemoglobin

Red blood cell

Plasma

Capillary endothelial cell

B

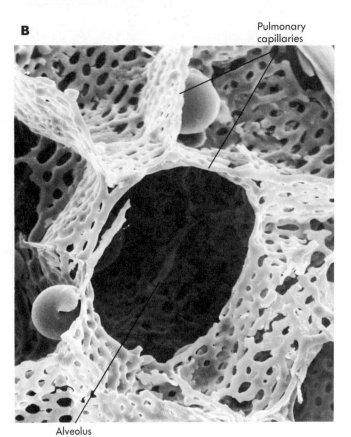

Pulmonary capillaries

Alveolus

FIGURE 17-4

A The diffusional path between alveolar air and hemoglobin.

B A scanning electron micrograph of a vascular cast of lung tissue. The capillaries were injected with a resin that was allowed to harden. The tissue was then digested away from the hardened resin, leaving the cast. The large space is an alveolus surrounded by a dense mesh of capillaries.

which O_2 were transported in the blood, the cardiac output would have to be 80 L/min just to meet the needs of tissues during rest. For the heart to pump this volume of blood is clearly impossible.

The actual O_2 content of systemic arterial blood is about 20 ml/100 ml of blood, or about 20 vol% (Table 17-2). About 99% of the O_2 is carried by **hemoglobin,** an oxygen-binding protein contained within the red blood cells (Figure 17-5; see also Chapter 14). Hemoglobin, abbreviated **Hb,** is a protein in which a **heme group** is attached to each of four subunit polypeptide chains, two **α chains** and two **β chains.** Each polypeptide chain folds to produce a cavity in which an iron-containing heme group binds O_2. Therefore four O_2 molecules can combine with one molecule of hemoglobin.

Hemoglobin fully saturated with oxygen (**oxyhemoglobin**) is a brilliant red. Hemoglobin that has lost one or more of the four O_2 molecules (**deoxyhemoglobin**) is dark red. Venous blood is darker than arterial blood because it contains more deoxyhemoglobin than oxyhemoglobin. The reaction of deoxyhemoglobin with O_2 to form oxyhemoglobin takes less than 0.01 second, so it is never the rate-limiting step in O_2 transfer.

The ease with which hemoglobin accepts an additional molecule of O_2 depends on how many of the sites are already occupied by O_2 molecules. If one of the four sites is occupied, binding of O_2 to the second and third sites becomes easier. This is an example of an **allosteric effect** (see Chapter 3). The result of this cooperativity between O_2 binding sites is that the amount of oxygen bound to hemoglobin increases in an S-shaped, or sigmoid, fashion as the P_{O_2} increases. The relationship between the P_{O_2} and the extent of O_2 binding to hemoglobin is given by the **oxyhemoglobin dissociation curve** (Figure 17-6). The oxyhemoglobin dissociation curve can be described either in terms of the **relative saturation** of hemoglobin (left-hand vertical scale in Figure 17-6) or by the **oxygen content** of the arterial blood (right-hand vertical scale). The P_{50} for hemoglobin is the value of P_{O_2} at which the hemoglobin is half saturated with oxygen (that is, on the average, each hemoglobin molecule has two O_2 molecules bound).

The O_2 **carrying capacity** is the volume of O_2 contained in a volume of O_2-saturated blood; the carrying capacity depends on the concentration of effective hemoglobin. The carrying capacity is re-

	Systemic arterial (= pulmonary venous)	Mixed systemic venous (= pulmonary arterial)	Arteriovenous difference
Oxygen			
P_{O_2}	100 mm Hg	40 mm Hg	60 mm Hg
In physical solution	0.3	0.1	
As HbO_2	19.5	14.4	
TOTAL	19.8 vol%	14.5 vol%	5.3 vol%
Carbon dioxide			
P_{CO_2}	40 mm Hg	46 mm Hg	6 mm Hg
In physical solution	2.7	3.1	
As bicarbonate	43.9	47.0	
As carbaminoHb	2.4	3.9	
TOTAL	49.0 vol%	54.0 vol%	5.0 vol%

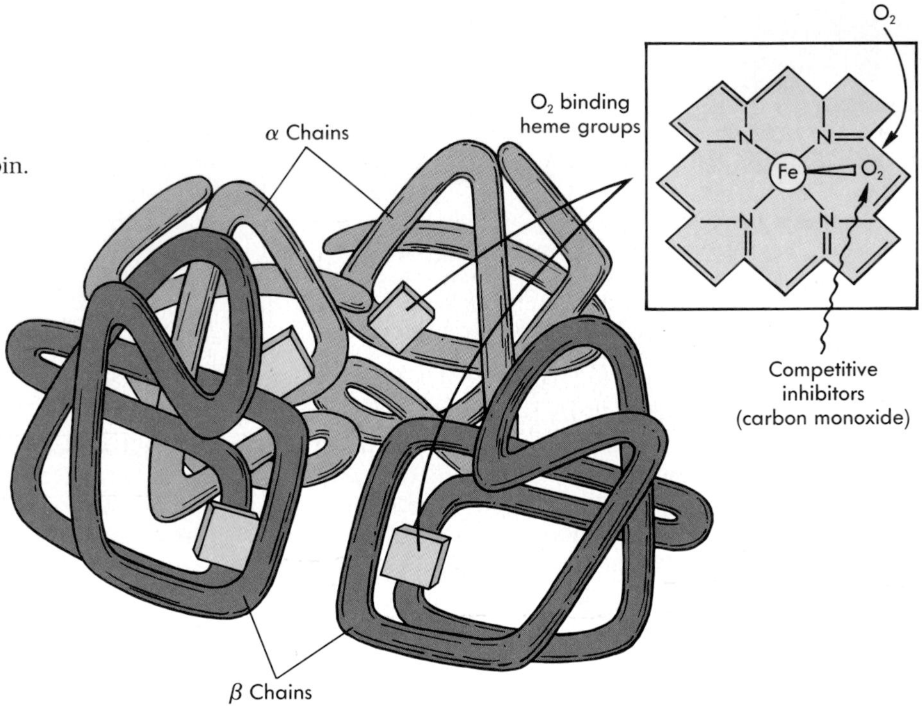

FIGURE 17-5
The molecular structure of hemoglobin.

α Chains

O_2

O_2 binding heme groups

Fe

Competitive inhibitors (carbon monoxide)

β Chains

duced in various forms of **anemia,** which may be caused by a reduction in the number of red blood cells, insufficient hemoglobin production, or the production of abnormal hemoglobin. An anemic condition can exist for a period of days to weeks after significant blood loss in otherwise normal individuals. Failure of hemoglobin production occurs in dietary **iron deficiency anemia,** because iron is needed for synthesis of heme groups. Dietary deficiency or failure of absorption of vitamin B_{12} causes **pernicious anemia,** in which formation of red blood cells is impaired.

In **sickle cell anemia,** substitution of a single amino acid in the β chains causes hemoglobin to aggregate into large polymers when P_{O_2} is low. These polymers distort the red blood cells into a characteristic sickle shape that prevents their passage through small blood vessels. People with various forms of **thalassemia,** an inherited defect in the DNA that codes for either the α or β chains, have higher than normal hematocrit levels, but the red blood cells contain less hemoglobin, and the O_2 binding characteristics of the hemoglobin are abnormal.

Oxygen Loading in the Lungs

The shape of the oxyhemoglobin dissociation curve has important implications for O_2 uptake in the

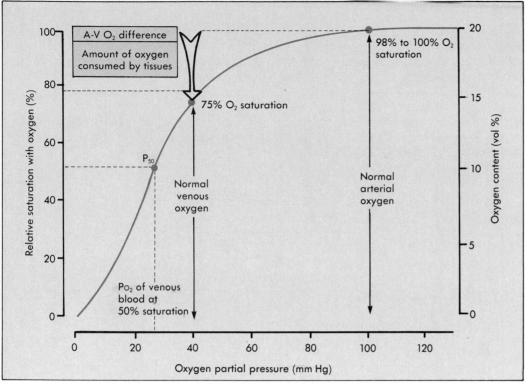

FIGURE 17-6

Oxyhemoglobin dissociation curve for arterial blood under systemic arterial conditions. The left-hand vertical axis gives percentage saturation; the right axis gives the corresponding values in vol% for blood with a typical O_2 carrying capacity. The P_{50} is the value of Po_2 at which the hemoglobin is one half saturated. At the Po_2 of mixed venous blood the hemoglobin is about 75% saturated; in other words, about 25% of the O_2 carried by systemic arterial blood has been unloaded.

lungs and delivery in the tissues. The normal alveolar Po_2 of 100 mm Hg is on the flat portion of the oxyhemoglobin dissociation curve, so that the hemoglobin remains almost fully saturated even in the face of relatively large changes in the alveolar Po_2. This means that the O_2 loading is virtually complete even when ventilation is not optimum.

The safety factor for loading is particularly important at high altitudes. For example, at an altitude of 3000 meters the atmospheric pressure is 520 mm Hg, as compared with the sea level value of 760 mm Hg, and the Po_2 of the inspired air is 110 mm Hg rather than 150 mm Hg. The alveolar gas equation can be applied to this situation by assuming an RQ of 0.8. The alveolar Pco_2 will be 40 mm Hg, and the alveolar Po_2 would be reduced by 50 mm Hg to 60 mm Hg (*point A* in Figure 17-7). At a Po_2 of 60 mm Hg, hemoglobin is still 88% saturated with O_2.

Oxygen Unloading in Systemic Capillaries

Dissolved O_2 and O_2 bound to hemoglobin are in equilibrium, so that the diffusion of free O_2 into the tissues results in the dissociation of O_2 from hemoglobin. The Po_2 in the extracellular fluid surrounding the tissues is about 40 mm Hg. When arterial blood enters capillaries and encounters this lower Po_2, it gives up, or **unloads,** some of its bound O_2 according to the dissociation curve. For values of Po_2 in this range, the oxyhemoglobin dissociation curve is steep, and small decreases in Po_2 can cause relatively large changes in the amount of bound O_2 unloaded. The significance of the steepness of this part of the curve is illustrated by the change in

O_2 delivery that occurs in exercise. At rest, the amount of O_2 unloaded is about 5 vol%, which corresponds to delivering, on the average, only one of the four O_2 molecules bound by the hemoglobin molecules. During strenuous exercise the O_2 content drops, on the average, from 20 vol% to 10 vol% between arteries and veins (Figure 17-8, *right side*), representing an unloading of 10 vol%. On the average, therefore, the hemoglobin molecules have delivered two rather than one of the four O_2 molecules that they carried. A decrease in average tissue Po_2 of only about 16 mm Hg can result in this doubling of the amount of O_2 delivered to the active tissues.

Effects of CO_2, pH, Temperature, and 2,3 DPG on Oxygen Unloading

The delivery of O_2 to the tissues is affected by four factors that alter the affinity of hemoglobin for O_2 (Figure 17-9; Table 17-3). The first two factors, CO_2 and pH, are closely related. Most of the CO_2 that enters the capillaries of the systemic loop reacts with water to form carbonic acid, H_2CO_3, which dissociates to give a bicarbonate ion, HCO_3^-, and a hydrogen ion, H^+. When the H^+ concentration increases (lowering the pH), some H^+ binds to hemoglobin and reduces the affinity of hemoglobin for O_2. The oxyhemoglobin dissociation curve is shifted to the right, in the direction of increased O_2 delivery (*dashed curve* in Figure 17-9). Binding of some of the CO_2 to hemoglobin results in formation of **carbaminohemoglobin,** which has the same effect as H^+ binding on O_2 delivery (a right shift).

FIGURE 17-7

The effects of a decrease in the inspired $Psco_2$ on the delivery of oxygen to the tissues viewed using the oxyhemoglobin dissociation curve. The normal arteriovenous O_2 difference of about 5 vol% is shown as the larger, downward-directed yellow arrow and results in a fall in Po_2 to 40 mm Hg *(yellow, left-directed arrow)* in the venous blood. At 3000 meters the inspired Po_2 is about 60 mm Hg, hemoglobin is 88% saturated with oxygen, and the arterial O_2 content is 18 vol% *(point A)*. The same arteriovenous O_2 difference of 5 vol% *(yellow arrow under point A)* causes the venous Po_2 to fall only slightly below the normal value of 40 mm Hg *(black, left-directed arrow and point B)*.

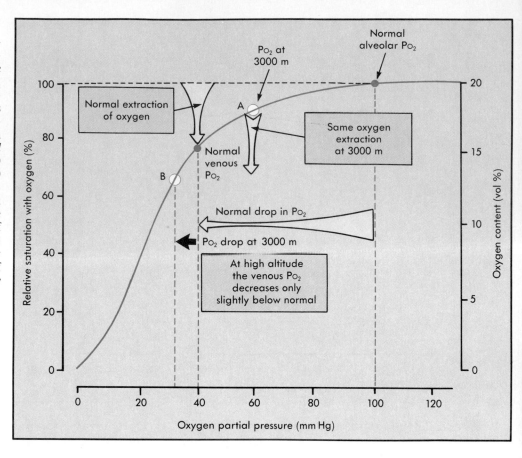

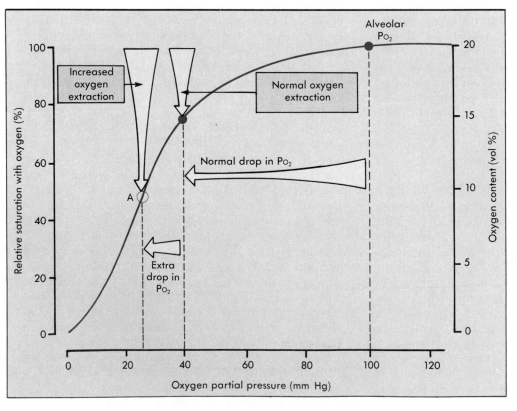

FIGURE 17-8

The effects of an increase in total body metabolism (assuming constant cardiac output) viewed using the oxyhemoglobin dissociation curve. The normal arteriovenous O_2 difference of 5 vol% is shown as the small, downward-directed arrow and results in a fall in venous Po_2 to 40 mm Hg *(large, left-directed arrow)*. If the arteriovenous O_2 difference is increased to 10 vol% *(large, downward-directed arrow)*, the venous Po_2 falls only slightly below the normal value of 40 mm Hg *(small, left-directed arrow and point A)*.

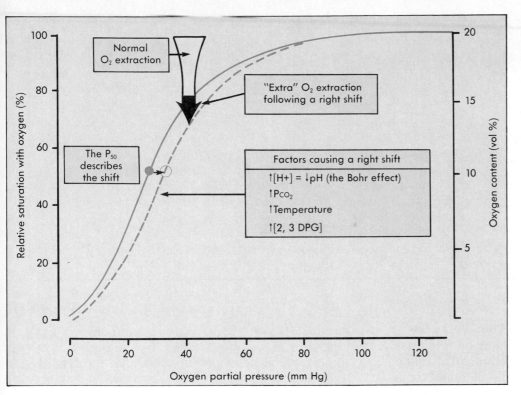

FIGURE 17-9

Factors shifting the oxyhemoglobin dissociation curve to the right (increase the P_{50}) correspond to a decrease in the affinity of hemoglobin for O_2 and increase the amount of O_2 that can be extracted from hemoglobin at a given P_{O_2} *(black portion of the downward-directed arrow).* Such factors include a decrease in the pH (Bohr effect), an increase in the P_{CO_2}, an increase in temperature, and an increase in the concentration of 2,3 DPG. Note that all these factors combine to favor combination of oxygen with hemoglobin in the lung while, at the same time, facilitating unloading of O_2 to metabolically active tissues.

TABLE 17-3	Factors that Bring About a Rightward Shift of the Oxyhemoglobin Dissociation Curve

Factor	Source
Decreased pH	Results from increased production of CO_2 and lactic acid in active tissues
Increased P_{CO_2}	Increased in venous blood from tissues with high rate of oxidative metabolism; acts by formation of carbaminohemoglobin
Increased temperature	Higher in fever and in active tissues, especially skeletal muscle
Increased 2,3 DPG (result of Hypoxia)	Produced by red blood cells; increased in anemia and high-altitude adaptation

This pH and CO_2-dependent right shift is referred to as the **Bohr effect.** The significance of the Bohr effect is that it is a response to the metabolites released by active tissues, which always have a lower pH and higher CO_2 release rate than inactive tissues. The P_{O_2} does not have to change for active tissues to be preferentially supplied with O_2 because the properties of hemoglobin are altered in response to these by-products of metabolism. In human beings, a decrease in plasma pH from 7.4 to 7.2 increases the P_{50} from 26 mm Hg to 30 mm Hg. The increase in P_{50} results in a 15% greater unloading of O_2.

The third factor is that active skeletal muscle may have a local temperature several degrees higher than resting muscle because heat is also a by-product of oxidative energy metabolism. The increased temperature favors additional O_2 unloading. For example, at 43° C the P_{50} increases from 26 mm Hg to 37 mm Hg. Conversely, at 30° C the P_{50} decreases to 18 mm Hg. The temperature effect is reversed when blood passes through the pulmonary circulation, because the temperature of the lungs is about 1° C lower than core body temperature.

The last of the four factors that affects the oxyhemoglobin dissociation curve is **2,3 diphosphoglycerate (2,3 DPG)**, a metabolite produced by red blood cells. Production of 2,3 DPG is increased at low P_{O_2}, so when red blood cells pass through inadequately oxygenated tissues they produce more 2,3 DPG. Doubling the 2,3 DPG concentration increases the P_{50} of hemoglobin from 26 mm Hg to 35 mm Hg. Thus the effects of pH, CO_2, elevated temperature, and 2,3 DPG production all complement one another in the sense that they favor oxygen unloading from hemoglobin in metabolically active tissues. The 2,3 DPG levels become depleted in stored blood, and the oxyhemoglobin dissociation curve is shifted far to the left even though the pH, P_{CO_2}, and temperature are all normal. If stored blood is transfused without any pretreatment (such as heating and oxygenation), the P_{O_2} in the tissues can fall to low levels.

Hypoxia and Oxygen Therapy

Hypoxia is a condition of inadequate O_2 delivery to a body region or to the whole body. Inadequate O_2 delivery may result from inadequate blood perfusion (ischemia); from a disorder in the function of hemoglobin, as in anemia or carbon monoxide poisoning; or from failure of ventilation or blood oxygenation in the lungs (asphyxiation). Breathing pure O_2 is effective therapy if ventilation is depressed or gas exchange is impaired. It is less effective when ventilation and pulmonary gas exchange are normal, because arterial hemoglobin will already be almost saturated with O_2, and the only effect of O_2 breathing will be to increase the dissolved O_2 content of plasma.

Oxygen Transfer from Mother to Fetus

The fetal O_2 carrying capacity is higher than that of adult blood because of two additional adaptations: the hematocrit of fetal blood is higher, and the fetal red blood cells contain a higher concentration of hemoglobin. However, these adaptations are not sufficient to ensure adequate O_2 delivery to the fetus. The placenta, the organ for exchange of gases and nutrients, does not normally allow maternal and fetal blood to mix. Instead, the fetal capillaries in the placenta are bathed in maternal blood. The Po_2 of the maternal blood entering the placenta is 32 mm Hg, similar to the Po_2 that would be seen in blood at an altitude of 10,600 m (35,000 feet), well above the normal limits of human habitation. If the fetus possessed adult hemoglobin, blood entering its circulation would be only 60% saturated, and O_2 extraction by the developing fetus would cause dangerously low O_2 levels in fetal tissue. Instead, fetal blood is able to load O_2 from maternal blood because of the synthesis of a different form of hemoglobin, **fetal hemoglobin,** during fetal life.

Fetal hemoglobin resembles adult hemoglobin in that it also binds oxygen cooperatively and therefore has an S-shaped dissociation curve (Figure 17-10). The difference is that polypeptide chains are substituted for the adult β chains (Table 17-4). This shifts the dissociation curve of fetal hemoglobin to the left along the Po_2 axis (Figure 17-10). A position to the left of the adult curve means that fetal hemoglobin binds O_2 more readily and can remove more of it from the maternal blood than adult hemoglobin. Consequently, fetal hemoglobin can be fully saturated with O_2 at a lower partial pressure than adult hemoglobin. It does not get a chance to be fully saturated, however, because the maternal blood arriving in the placenta has a Po_2 considerably lower than that of arterial blood.

Blood passing from the placenta to the fetus in the umbilical vein is about 80% saturated with O_2. The gain in loading provided by the higher O_2 affinity of fetal hemoglobin means that the fetal tissues must have a lower Po_2 to provide the necessary gradient for O_2 delivery. Fetal tissues, therefore, operate at a lower Po_2 than adult tissues. Fetal hemoglobin is relatively insensitive to 2,3 DPG, so this mechanism of enhancing unloading is not important before birth.

Myoglobin: An Intracellular Oxygen Binding Protein

Myoglobin is a heme-containing protein found in energetic tissues such as red skeletal muscle and cardiac muscle. Myoglobin differs from hemoglobin in that it consists of a single chain and can bind only a single O_2 molecule. Because each molecule has only one heme unit, binding of O_2 to myoglobin cannot involve cooperativity, and thus the dissociation curve for myoglobin is hyperbolic rather than S-shaped (see Figure 17-10). Myoglobin binds O_2 more strongly than hemoglobin (see Table 17-4). The oxymyoglobin dissociation curve lies to

TABLE 17-4	Subunit Composition and P_{50} Values for Hemoglobin, Fetal Hemoglobin, and Myoglobin	

Hemoprotein	Subunits	P_{50} (arterial conditions)
Adult hemoglobin	α, α, β, β (tetramer)	26 mm Hg O_2
Fetal hemoglobin	α, α, γ, γ (tetramer)	20 mm Hg O_2
Myoglobin	α (monomer)	8 mm Hg O_2

THE RESPIRATORY SYSTEM

Physiology of Carbon Monoxide Poisoning

Carbon monoxide (CO), a product of incomplete combustion of hydrocarbons, competes with O_2 for the hemoglobin binding site and thus blocks its O_2-carrying capability. Binding of CO results in formation of **carboxyhemoglobin** (COHb). The affinity of the Hb binding site for CO is more than 200 times greater than that for O_2, so once CO has bound to a heme group it may prevent that heme group from transporting O_2 for as much as several hours. Binding of CO to one of the four sites of hemoglobin also abolishes the cooperativity of O_2 binding at the other sites, shifting the oxyhemoglobin dissociation curve to the left. This shift makes unloading of O_2 more difficult.

Because of its very high affinity for hemoglobin, very low levels of CO in environmental air can result in formation of significant amounts of COHb and a decrease in the blood's O_2 carrying capacity. In nonsmokers breathing clean air, the COHb level in the blood is less than 1%. Heavy smokers, nonsmokers who breathe tobacco smoke in confined areas, and freeway drivers or those who live and work on heavily travelled streets may accumulate levels of COHb as high as 7%. The problem posed by the reduced carrying capacity and the left shift are most serious for organs such as the heart that extract a large fraction of the arterial O_2. Elevated levels of COHb are regarded as a potential contributing factor in heart attacks and premature births.

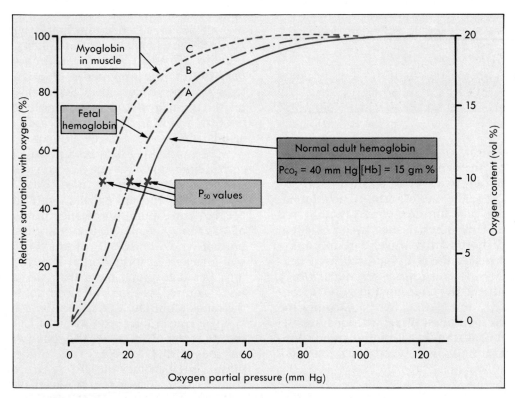

FIGURE 17-10
The dissociation curves for adult hemoglobin *(A)*, fetal hemoglobin *(B)*, and myoglobin *(C)*. These dissociation curves depend on the P_{O_2} and can be expressed either in terms of the percent saturation of hemoglobin *(left-hand vertical axis)* or, given the hemoglobin concentration, as the O_2 content of the blood *(right-hand vertical axis)*. Note the sigmoidal (S-shaped) nature of the dissociation curves for adult and fetal hemoglobin compared with the dissociation curve for myoglobin. The position of each dissociation curve can be described by stating its P_{50}. The further to the left a particular curve is, the more tightly the corresponding molecule binds O_2.

the left of the hemoglobin curves, so that the loading portion of the myoglobin curve lies within the range of partial pressures in which hemoglobin unloads.

Myoglobin has two roles. First, it represents an intracellular reserve of O_2. Second, when the P_{O_2} in the muscle cell has fallen very low, the presence of myoglobin increases the rate of O_2 diffusion from the cell surface to the interior. Recall that the rate of diffusion is related to the magnitude of the gradient of the diffusing substance. If myoglobin is present, O_2 can move through the cytoplasm in two forms: (1) physically dissolved, and (2) "riding" from one myoglobin to another in the direction of the lower O_2 concentration. This effect is particularly strong when the P_{O_2} has fallen to very low values, because many of the myoglobin molecules have given up their bound O_2 and move O_2 from the region just inside the muscle cell membrane toward the cell interior. Thus when O_2 is readily available, the cell can accept and store more O_2 because of the presence of myoglobin, and when O_2 is in short supply, myoglobin can help move it to the mitochondria where it is most needed by providing a second route for diffusion.

CARBON DIOXIDE TRANSPORT AND BLOOD BUFFERS

- *In what forms is CO_2 carried in the blood? What is the role of carbonic anhydrase in CO_2 transport?*
- *What is the Haldane effect? How does it differ from the Bohr effect?*
- *What is the Cl^- shift? How does it facilitate CO_2 transport?*

The CO_2-H_2CO_3-HCO_3^- Equilibrium

Carbon dioxide is the major waste product of oxidative metabolism. CO_2 reacts with water to form carbonic acid (H_2CO_3). The rate of carbonic acid formation is increased one hundred times within red blood cells and many other tissues by an essential enzyme called **carbonic anhydrase.** Carbonic anhydrase is present inside the red blood cells and catalyzes both the forward and reverse reactions. Which reaction is dominant depends on the relative concentrations of CO_2 and H_2CO_3. At equilibrium the rates of the forward and reverse reactions are the same. Once formed, carbonic acid rapidly and spontaneously dissociates to give a bicarbonate ion and a hydrogen ion:

$$CO_2 + H_2O \rightleftharpoons H_2CO_3 \rightleftharpoons H^+ + HCO_3^-$$

In solution, CO_2, H_2CO_3, H^+, and HCO_3^- coexist in a chemical equilibrium. The relative proportions of the reactants are constant, as predicted by the law of mass action (see Chapter 3). The relationship is described by the **Henderson-Hasselbalch equa-**tion, written in terms of pH, the negative logarithm of H^+ concentration:

$$pH = 6.1 + \log_{10} [HCO_3^-]/0.03 \, P_{CO_2}]$$

In this equation, the plasma pH is proportional to the plasma concentrations of HCO_3^- and CO_2. The constant 6.1 is the effective acid dissociation constant for the overall reaction of CO_2 and H_2O. The concentration of dissolved CO_2 in the plasma is equal to the P_{CO_2} multiplied by the solubility coefficient of CO_2 in plasma (0.03 mmol/L/mm Hg).

Blood Buffers and Control of Plasma pH

A **buffer** is a substance that opposes changes in H^+ concentration by binding some H^+ when H^+ becomes more abundant and releasing some H^+ when H^+ becomes less abundant. The CO_2-H_2CO_3-HCO_3^- equilibrium is a buffer system (usually referred to as "bicarbonate buffer"). Addition of H^+ results in formation of H_2CO_3 and CO_2; a decrease in H^+ is opposed by dissociation of some H_2CO_3 to keep the proportions of reactants constant. The CO_2-H_2CO_3-HCO_3^- system is an important buffer for extracellular fluids. Other buffer systems are also important.

For comparison, Table 17-5 includes all of the major buffer systems of the body. Hemoglobin is an important blood buffer because it is the most abundant protein of the red blood cell cytoplasm, and like all proteins it has ionizable -COOH and -NH_3 groups. The phosphate buffer system is of some importance in blood but is more important as a urinary buffer (the importance of urinary buffer systems will be discussed in Section V). Proteins are relatively insignificant as buffers in plasma and interstitial fluid but are the major buffers of the intracellular fluid.

The normal pH of arterial plasma is 7.4. This pH corresponds to a plasma H^+ concentration of 4 x 10^{-8} M, or 40 nanomolar (nM). Plasma pH values of less than 6.9 and more than 7.6 are life threatening, yet they correspond to a change in H^+ concentration of less than 1 μ mole/L from the normal value! The normal whole-body CO_2 production of 200 ml/min would result in formation of about 15 moles of H^+ ions per day, much more than the capacity of the body's buffer systems, if it were not for the fact that the lungs eliminate CO_2 as rapidly as it is produced.

The continuous production of CO_2 by oxidative energy metabolism potentially has a very large effect on the acidity of the extracellular fluid. If the plasma pH becomes either too acid or basic, small changes in P_{CO_2} can restore the plasma pH to normal. The CO_2-H_2CO_3-HCO_3^- system cannot oppose changes in pH that are caused by addition of one of the reactants of the system, such as CO_2. Precise balance of the rate of CO_2 loss by ventilation to the rate of CO_2 production by metabolism is thus essential. The respiratory system has a central role in the

TABLE 17-5   *Major Buffer Systems of the Body*

System	Reactants	Importance
Bicarbonate	$CO_2 + H_2O \leftrightharpoons H_2CO_3 \leftrightharpoons H^+ + HCO_3^-$	Most important buffer system of blood plasma
Hemoglobin	$Hb - H \leftrightharpoons Hb^- + H^+$	Buffers interior of red blood cells
Phosphate	$H_2PO_4^- \leftrightharpoons H^+ + HPO_4^=$	Most important buffer of urine
Protein	$Pr - H \leftrightharpoons Pr^- + H^+$	Most important buffer system of intracellular fluid; quantitatively the most important buffer of the whole body

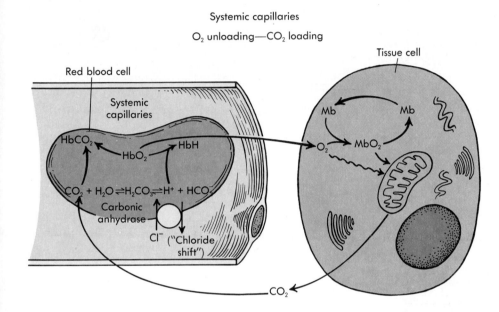

Systemic capillaries

O_2 unloading—CO_2 loading

FIGURE 17-11
Cooperation between CO_2 loading and O_2 unloading in systemic capillary blood. The increase in P_{CO_2} within red blood cells drives the CO_2 dissociation reaction to the right, liberating H^+ and HCO_3^-. The reaction is accelerated by carbonic anhydrase. Some of the H^+ is accepted by hemoglobin, which becomes a better buffer as it becomes deoxygenated (the Haldane effect). The HCO_3^- is exchanged across the red cell membrane for Cl^-, allowing the rightward reaction to continue. The oxyhemoglobin more readily becomes deoxygenated in the more acidic environment (the Bohr effect). Some of the oxyhemoglobin exchanges its O_2 for CO_2, forming carbaminohemoglobin (shown as Hb_{CO_2}). Within the cells, O_2 diffusion to mitochondria is assisted by myoglobin.

regulation of acid-base balance through its ability to control the plasma P_{CO_2}. This topic will be discussed in more detail in Chapter 20.

Carbon Dioxide Loading in Systemic Capillaries: The Haldane Effect

Dissolved carbon dioxide gas diffuses readily from cells into interstitial fluid (Figure 17-11). As red blood cells pass through systemic capillaries, CO_2 that diffuses into them from plasma is converted to H^+ and HCO_3^-, a reaction accelerated by the carbonic anhydrase inside red cells. Hemoglobin accepts H^+, driving the reaction to the right, and allowing the blood to accept additional CO_2. The HCO_3^- ions are returned to the plasma in exchange for Cl^- ions (Figure 17-11). This Cl^- movement to balance bicarbonate movement is called the **"chloride shift."** It prevents the build up of HCO_3^- inside the red cell that would otherwise oppose further dissociation of CO_2.

The more the hemoglobin is deoxygenated, the more CO_2 the blood can load. This phenomenon, called the **Haldane effect** (Figure 17-12), links CO_2

loading to O_2 unloading because it involves the same two factors as involved in the Bohr effect (see Figure 17-11). These two factors are binding of H^+ and formation of carbamino complexes. First, deoxyhemoglobin is a better buffer than oxyhemoglobin. As hemoglobin becomes deoxygenated, even more CO_2 can be accepted in the blood. Second, deoxyhemoglobin forms carbamino complexes more readily than does oxyhemoglobin. Of the two factors, 70% of the Haldane effect comes from increased carbaminohemoglobin formation and 30% from increased buffering.

The total CO_2 content of the blood increases from 49 vol% in the arteries to 54 vol% in the veins (see Table 17-2). Of this total the most, 44 to 47 vol%, is transported as bicarbonate, giving a plasma bicarbonate concentration of 24 mEq/L. The remainder of the CO_2 carried by the blood is nearly equally divided between carbaminohemoglobin and dissolved CO_2.

Changes of the RQ over its normal range of 0.7 to 1.0 can be accommodated by changes of about 2 vol% of the CO_2 content of mixed venous blood. For

FIGURE 17-12

CO_2 dissociation curve for systemic arterial blood *(lower curve)*, systemic venous blood *(middle curve)*, and completely deoxygenated blood *(upper curve)*. This plot demonstrates that the more deoxygenated blood is, the greater its CO_2 content is at any value of P_{CO_2}. In contrast to the oxyhemoglobin dissociation curve, the dissociation curve for CO_2 is nearly linear over the physiological range of CO_2 partial pressures. Normally, 4 vol% of CO_2 are added to the mixed venous blood by the tissues *(vertical arrow)* and would increase the venous P_{CO_2} from point A to point B. However, when O_2 dissociates from hemoglobin there is an increased formation of reduced hemoglobin and carbamino compounds, both of which increase the CO_2 carrying capacity of the blood. This is referred to as the Haldane effect, and as a result the venous P_{CO_2} only increases to 46 mm Hg (X *on the 75% saturation curve).*

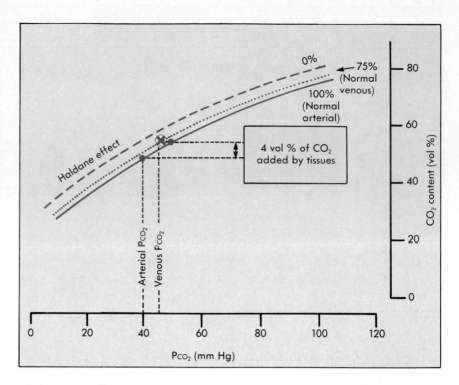

Pulmonary capillaries

O_2 loading—CO_2 unloading

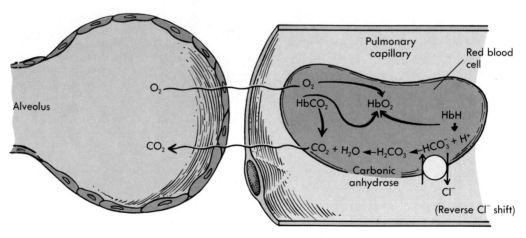

FIGURE 17-13

Cooperation between CO_2 unloading and O_2 loading in pulmonary capillary blood. The decreasing P_{CO_2} pulls the CO_2 dissociation reaction to the right; the reaction is accelerated by carbonic anhydrase. H^+ is released by deoxyhemoglobin as it becomes fully oxygenated (the Haldane effect in reverse). HCO_3^- is provided by the reverse chloride shift. The increase in pH slightly favors O_2 binding to hemoglobin (the reverse of the Bohr effect). Some CO_2 is released from carbaminohemoglobin, freeing the hemoglobin to accept O_2.

example, the RQ value corresponding to the blood gas values in Table 17-2 is equal to the ratio of AV differences for CO_2 and O_2, or 5 vol%/5.3 vol% = 0.94. An increase in venous CO_2 content of 0.3 vol% would result from elevating the RQ to 1.0; a decrease of RQ to 0.7 would result in a decrease of a little more than 1 vol% from the value of venous CO_2 given in Table 17-2.

Carbon Dioxide Unloading in the Lungs

The conditions that favored O_2 unloading and CO_2 loading in the tissues are reversed when blood arrives in the lungs (Figure 17-13). The pH of the blood in the lungs is normal, P_{CO_2} is low, the temperature may be cooler than in very active systemic tissues, and the high levels of O_2 reduce 2,3 DPG formation. Thus the same factors that shift the he-

Physiology of Diving

People trained in free diving can spend a minute or more under water without breathing; some aquatic mammals such as whales, seals, and porpoises can survive dives lasting tens of minutes. During these periods the needs of energy metabolism are provided by O_2 stored in lung air, bound to hemoglobin and myoglobin, and dissolved in body fluids. The central physiological problem of prolonged diving is that of maintaining heart and brain function. Almost all mammals show some cardiovascular response to immersion of the head; this has been called the **diving reflex.** In humans the diving reflex typically results in a slowing of the heart rate by several beats/min. The reflex is more profound if the water is cold. It is probably of little importance in recreational and sport diving, but when combined with the peripheral vasoconstriction and reduced metabolism induced by hypothermia, it may increase chances of survival in people threatened with drowning in cold water.

In aquatic mammals, diving is usually accompanied by dramatic reflexive bradycardia and peripheral vasoconstriction. The effect is to force skeletal muscle first to draw on its store of O_2 in myoglobin and then shift to anaerobic glycolysis. The O_2 that would otherwise have been used by muscle is spared for the heart and brain. During the dive, some of the lactic acid that is produced by the anaerobic muscle can enter the pathway of oxidative metabolism in heart muscle; some can enter the pathway of gluconeogenesis in the liver and be reconverted to glucose (see the discussion of **oxygen debt** in Chapter 11). Most of the lactic acid accumulates in the muscle and is washed out when blood flow to skeletal muscle is restored at the end of the dive.

Scuba or helmet divers who work at depths greater than 30 meters or so breathe air at higher than atmospheric pressure and accumulate dissolved gases in blood and tissues. If the diver's return to the surface is too rapid, some of the dissolved gas comes out of solution, forming bubbles in tissues and blood vessels. The bubbles are mainly N_2, since N_2 is the major constituent of air, and is less soluble than either O_2 or CO_2. The pain resulting from these bubbles has caused the condition to be named **the bends.** Permanent damage can result if blood flow to parts of the brain is blocked. The condition can be prevented by controlled decompression, which allows the partial pressures of the blood gases to return gradually to atmospheric values.

Diving animals typically exhale at the beginning of a dive. The increased pressure experienced by the animal if it dives to a considerable depth further reduces the volume of gas in the lungs (Boyle's law) so that most of the remaining gas is in the dead space and the lungs are collapsed. Reducing air volume in the lungs, where the gases are exposed to capillaries, minimizes the transfer of dissolved N_2 to the animal's blood so that bubble formation is not a problem when the animal returns to the surface.

Gas Exchange and Gas Transport in the Blood

SUMMARY

1. Gas mixtures in the respiratory tract and gases dissolved in the blood and tissues behave according to the gas laws.
2. Gases always diffuse from an area with a higher partial pressure to an area with a lower partial pressure.
3. The majority of the O_2 transported in the blood is carried by **hemoglobin** (19.7 vol%); the remainder (0.3 vol%) is dissolved in the plasma.
4. The plateau portion of the S-shaped **oxyhemoglobin dissociation curve** allows hemoglobin to load O_2 to near 100% saturation even if alveolar P_{O_2} falls below normal, and the steep portion of the curve allows a large amount of O_2 to be delivered to the tissues with a relatively small drop in P_{O_2}.
5. The oxyhemoglobin dissociation curve is shifted to the right by conditions prevailing in the tissues (increases in H^+, CO_2, 2,3 DPG, and temperature). This right shift decreases the affinity of hemoglobin for O_2; the greater the shift the greater the O_2 delivery in the tissues.
6. The leftward position of the fetal hemoglobin oxyhemoglobin dissociation curve in comparison with the adult hemoglobin curve reflects a greater affinity for O_2. Thus fetal tissues can selectively take up oxygen from the maternal circulation, despite the low P_{O_2} of the maternal blood.
7. The affinity of **myoglobin** for O_2 is higher than that of either form of hemoglobin, representing a reserve supply in muscle.
8. The majority of the CO_2 in the blood is carried as HCO_3^-. Smaller amounts are dissolved in the plasma or combined with hemoglobin. The presence of **carbonic anhydrase** in the red blood cells catalyzes the reaction of CO_2 with H_2O that leads to formation of H_2CO_3. Carbonic acid dissociates to form H^+ and HCO_3^-. The chemical equilibrium between the reactants and products is represented by the **Henderson-Hasselbalch equation.**

moglobin dissociation curve to the right in active tissues cause a leftward shift in the lung and facilitate the combination of oxygen with hemoglobin.

The higher P_{O_2} in the alveoli drives O_2 into the blood, loading the hemoglobin to near 100% saturation. As the deoxyhemoglobin is converted to oxyhemoglobin, it unloads some of the H^+ that it had been carrying (see Figure 17-13). The law of mass action dictates that the addition of H^+ to the red blood cell cytoplasm will drive the formation of H_2CO_3, which is rapidly converted into CO_2 and H_2O in the presence of carbonic anhydrase. The dissociation of H^+ from the hemoglobin and its reaction with available HCO_3^- creates a gradient favorable for entry of bicarbonate ions into the red cells. The movement of HCO_3^- into the red cell is electrically balanced by exit of Cl^-; this is the **reverse Cl^- shift.** Some CO_2 carried in the blood as carbaminohemoglobin is released, adding to the amount of CO_2 unloaded and freeing additional hemoglobin for binding with O_2.

● STUDY QUESTIONS

1. Why are the steady-state alveolar P_{CO_2} higher and the alveolar P_{O_2} lower than those of the inspired air? Suppose overall metabolic activity doubled without a corresponding change in alveolar ventilation. What would happen to the alveolar partial pressures of O_2 and CO_2? Why?
2. What effect does the respiratory quotient have on the composition of alveolar gas?
3. What determines the O_2 carrying capacity of blood?
4. An individual has a normal rate of O_2 consumption but a hemoglobin concentration 50% of normal. How would the magnitude of the arteriovenous O_2 difference in this individual be different from a normal individual?
5. How is the difference between the dissociation curves for myoglobin and hemoglobin related to the physiological roles of hemoglobin in gas transport in the blood and myoglobin as an oxygen reserve in muscle?
6. An individual voluntarily doubles his alveolar ventilation. What would happen to plasma pH? Why?

7. The normal arterial P_{O_2} is 95 mm Hg, slightly less than the normal alveolar P_{O_2} of 100 mm Hg. Why? The alveolar-arterial P_{O_2} difference can be very large in cases of emphysema. Why? Would there be a similar difference for CO_2?

8. Diagram the steps in the pathway for diffusion of CO_2 (A) from tissues into the blood and (B) from the blood into the alveolar gas mixture.

9. When an individual begins to breathe from an atmosphere whose total pressure is one-half that of sea level, what changes would occur in the following: arterial P_{O_2}, venous P_{O_2}, arterial O_2 content?

Choose the MOST CORRECT Answer

10. This law states that at equilibrium the concentration of a gas depends on its partial pressure, its solubility, and the temperature:
    a. Boyle's law
    b. Henry's law
    c. Charles' law
    d. Gay-Lussac's law

11. This anemia is caused by an inherited defect in the DNA that codes for beta chains in hemoglobin molecule that causes hemoglobin to aggregate in large polymers:
    a. Sickle cell anemia
    b. Pernicious anemia
    c. Thalassemia
    d. Iron deficiency anemia

12. Hypoxia may result from:
    a. Ischemia
    b. Malfunctioning hemoglobin
    c. Increasing altitude
    d. More than one answer is correct

13. Which of these factors does NOT favor the release of oxygen from hemoglobin into metabolically active tissues?
    a. Lowering the pH
    b. Production of 2,3 diphosphoglycerate by RBCs
    c. Decrease in body temperature
    d. Increase in CO_2

14. Choose the FALSE statement:
    a. The hematocrit of fetal blood is higher than that of adult blood.
    b. Fetal blood is highly sensitive to 2,3 DPG for enhancing O_2 unloading.
    c. Fetal hemoglobin has a higher O_2 affinity at a lower partial pressure than adult hemoglobin.
    d. Fetal red blood cells contain a higher concentration of hemoglobin.

15. The most important buffers of intracellular fluid are:
    a. Bicarbonate
    b. Hemoglobin
    c. Phosphates
    d. Proteins

16. The majority of the CO_2 in the blood is carried in this form:
    a. Dissolved in plasma
    b. Carbaminohemoglobin
    c. 2,3 DPG
    d. Bicarbonate ion

● *SUGGESTED READINGS*

AUBIER M: Pharmacological treatment of respiratory insufficiency, *Recent Prog Med* 81(3):193, 1990. Fatigue of respiratory muscles plays an important role in respiratory failure. This review deals essentially with respiratory muscle dysfunction and its treatment.

CARPENTER B: The newest health hazard: breathing, *U.S. News & World Report* 106(23):50, June 12, 1989.

COBURN RF, BARON CB: Coupling mechanisms in airway smooth muscle, *Am J Physiol* 258(4 Pt 1):119, 1990. This review documents available information about coupling mechanisms involved in airway smooth muscle force development and maintenance and relaxation of force.

DOUGLAS JS: Receptors on airway smooth muscle, *Am Rev Respir Dis* 141:123, 1990. Review of the regulation of airway smooth muscle response in health and disease.

CHAPTER 18

Regulation of the Respiratory System

On completing this chapter you should be able to:

- Understand how respiratory motor movements are affected by centers in the medulla and pons
- List the factors affecting the intensity of respiratory drive, rate of breathing, and tidal volume
- Explain how, by monitoring the pH of CSF, the central chemoreceptors serve to regulate the arterial P_{CO_2}
- Appreciate how inputs from the peripheral chemoreceptors become important during hypoxia and acid-base imbalance
- Describe the ventilatory responses to high-altitude exposure
- Understand the role of the respiratory system in responding to acute metabolic acidosis and lung disease
- Describe factors that may be involved in determining the ventilatory increase during exercise

Breathing, like the heartbeat, goes on automatically. Yet the generation and control of the ventilatory cycle are in many ways different from the heart cycle. The respiratory muscles are skeletal muscles. Unlike cardiac muscle, respiratory muscles depend on nervous stimulation for contraction. Each breath must be driven by a burst of action potentials in motor neurons of the respiratory muscles. The neuronal mechanisms by which the central pattern of respiratory activity is generated are still poorly understood.

Although substantial voluntary control over respiration is possible, internal stimuli have an overriding importance, evidenced by the fact that most people cannot hold their breath for much longer than 1 minute. The main internal stimuli for regulation of breathing are the P_{CO_2} and pH of the arterial blood and the cerebrospinal fluid. The respiratory control system protects the extracellular fluid from the threat of acidification posed by metabolic production of CO_2, and it compensates immediately for alterations in pH caused by ingestion or production of acids other than CO_2, and bases. The respiratory system is one of the two pillars of acid-base regulation described in Chapter 20, the kidneys being the other. Under most circumstances the "safety zone" of the oxyhemoglobin dissociation curve ensures that the hemoglobin of systemic arterial blood will be saturated with oxygen, no matter what changes in alveolar ventilation occur as part of acid-base regulation.

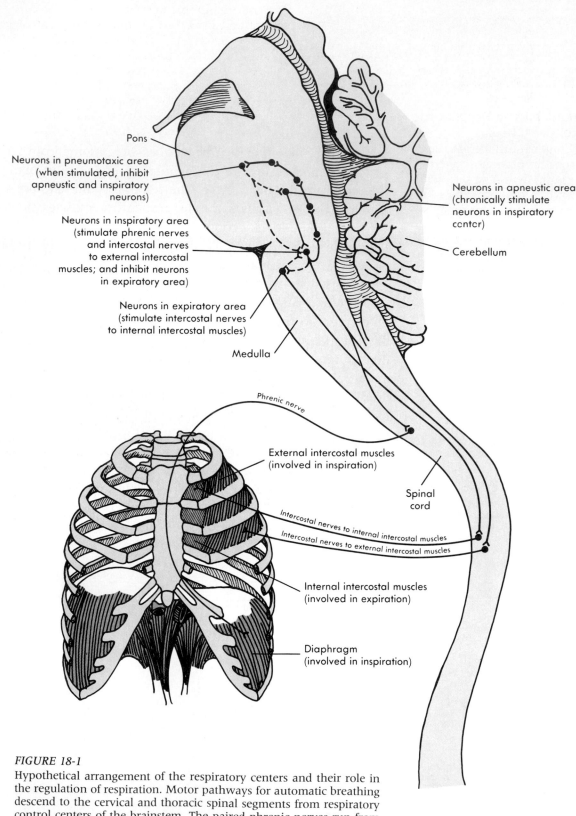

Pons

Neurons in pneumotaxic area
(when stimulated, inhibit
apneustic and inspiratory
neurons)

Neurons in apneustic area
(chronically stimulate
neurons in inspiratory
center)

Neurons in inspiratory area
(stimulate phrenic nerves
and intercostal nerves
to external intercostal
muscles; and inhibit neurons
in expiratory area)

Cerebellum

Neurons in expiratory area
(stimulate intercostal nerves
to internal intercostal muscles)

Medulla

Phrenic nerve

External intercostal muscles
(involved in inspiration)

Spinal
cord

Intercostal nerves to internal intercostal muscles

Intercostal nerves to external intercostal muscles

Internal intercostal muscles
(involved in expiration)

Diaphragm
(involved in inspiration)

FIGURE 18-1
Hypothetical arrangement of the respiratory centers and their role in
the regulation of respiration. Motor pathways for automatic breathing
descend to the cervical and thoracic spinal segments from respiratory
control centers of the brainstem. The paired phrenic nerves run from
the third, fourth, and fifth cervical segments through the thoracic cavity to the diaphragm. The intercostal muscles are innervated by the
spinal segmental nerves of the 1st through 11th thoracic spinal segments.

THE RHYTHM OF BREATHING

- *How is quiet breathing programmed?*
- *What is respiratory drive?*
- *How do stretch receptor afferent fibers function in determining the duration of a respiratory cycle?*

The Central Motor Program for the Respiratory Cycle

The breathing pattern is produced by interactions between specific brainstem neurons, stretch receptors in the lung, neural signals from higher levels of the CNS, and inputs from the chemoreceptors (Figure 18-1). Groups of neurons in the medulla and pons generate a central pattern for **respiration;** a series of action potentials that drives the spinal motor neurons innervating the diaphragm, the internal and external intercostal muscles of the rib cage, and the muscles of the abdominal wall. The diaphragm is innervated by the paired **phrenic nerves** from the cervical spinal cord, and the intercostal and abdominal muscles are innervated by thoracic spinal segmental nerves (see Figure 18-1). As a result, high thoracic spinal cord injuries may paralyze the accessory respiratory muscles of the rib cage and abdominal wall but spare neural control of the diaphragm. Inspiration and passive expiration are thus easy, but forced expiration and coughing are difficult or impossible. Inability to clear the airways by coughing causes the high incidence of serious respiratory infections in such individuals.

In quiet breathing, the diaphragm is responsible for about two thirds of the thoracic volume change. Figure 18-2 shows the pattern of action potentials in the phrenic nerve during quiet breathing.

At end-expiration (the FRC) few, if any, motor axons in the phrenic nerve are active. The onset of inspiration is marked by the recruitment of phrenic motor neurons. As inspiration progresses additional phrenic motor neurons are recruited, and the discharge frequency of active motor neurons increases. These changes are the result of a burst of activity in medullary respiratory neurons that synapse on the spinal motor neurons of the phrenic nerve. The burst of action potentials in the respiratory motor neurons comes to an end abruptly. At end-inspiration there is a sudden drop in discharge frequency and a rapid derecruitment. This relaxes the inspiratory muscles; as a result the respiratory system passively recoils to the functional residual capacity. The cycle is soon repeated.

At increased ventilation rates, motor neurons innervating the inspiratory muscles (primarily the external intercostals) are recruited during inspiration (see Figure 18-2). As the total motor drive to the inspiratory muscles increases, so does the total inspiratory force. Further increases in tidal volume are achieved by stronger activation of the inspiratory muscles. In addition, tidal volume may be increased by more forceful expirations due to the recruitment of the expiratory muscles. Active expiration may drive the end-expiratory volume below the functional residual capacity (Chapter 16). When expiration is forced in this manner, the central motor program consists of a burst of action potentials in the motor units of inspiratory muscles, followed by a burst of action potentials in the motor units of internal intercostal nerves (see Figure 18-2).

In summary, the tidal volume of a respiratory cycle is determined by the number of motor neurons recruited and their firing rate during inspira-

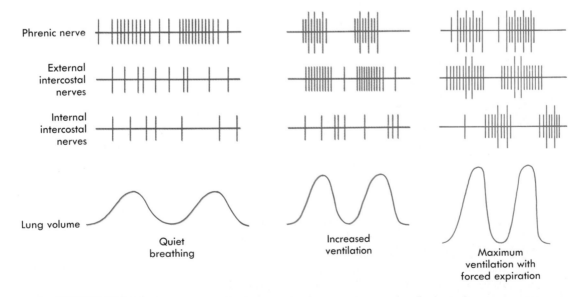

FIGURE 18-2

An idealized pattern of action potentials in the phrenic nerve, a nerve to external intercostals, and a nerve to internal intercostals, in three levels of pulmonary ventilation.

TABLE 18-1	Locations and Sensitivities of Receptors Involved in Respiratory Control	
Receptor	Location	Stimulus
Stretch receptors	Smooth muscle layer of bronchi and bronchioles	Stretch of lungs during inflation; mediate Hering-Breuer reflex
Irritant receptors	Among airway epithelial cells	Noxious gases, allergens, lung inflammation; stimulate rapid, shallow breathing
Central chemoreceptors	Ventrolateral surface of medulla, near roots of cranial nerves IX and X	pH changes in CSF-induced changes of arterial P_{CO_2}
Carotid bodies	In crotch of carotid bifurcation on top of carotid sinus	Mainly sensitive to decreased arterial P_{O_2}; also to pH and P_{CO_2}
Aortic bodies	Scattered over surface of aorta and major thoracic arteries	

tion and expiration; the respiratory rate is determined by the time interval between inspiratory bursts. The rate and depth of these components of the respiratory cycle are adjusted unconsciously by the respiratory centers of the brain to minimize the energy cost of breathing.

A separate pathway is used for conscious control of the respiratory muscles, for example, in speaking, singing, and voluntary breath-holding. In this corticospinal pathway (Chapter 12), the cortex shares control of the spinal respiratory motor neurons with the brainstem. Some axons in this pathway also project to motor neurons of other somatic muscle groups. Damage to the corticospinal pathway causes a general paralysis that abolishes voluntary control of respiration but preserves automatic respiration.

Damage to the pathway between the medulla and respiratory motor neurons causes a paradoxical condition called **Ondine's curse,** in which the capacity for automatic control is lost but voluntary control remains. The patient must therefore remember to breathe. The name comes from a legend in which the water nymph Ondine punished her faithless husband by taking away his automatic body functions. In the legend, the husband died when he fell asleep. Modern sufferers of the curse must sleep in respirators.

Important feedback about the degree of inflation of the lung is provided by stretch receptors in the bronchi and bronchioles that are activated by increasing lung volume (Table 18-1). Afferent fibers from the lung receptors run to the CNS in the vagus nerves; this is the **Hering-Breuer reflex.** After the vagus nerves of an experimental animal are cut, the tidal volume increases (and respiratory frequency decreases) as the inspiratory drive is freed from their inhibitory input. Stimulation of the central end of one of the cut vagus nerves can cause **apnea** (cessation of breathing) because the animal maintains an expiration as long as the stimulation con-

tinues. These results suggest that, in the intact system, the rising level of activity in lung stretch receptors as the lungs are expanded helps to terminate each inspiration.

Scattered among the epithelial cells of the airway are **irritant receptors** (see Table 18-1) responsible for the sensation of distress when noxious or irritating gases or particulates are inhaled. Irritant receptors also respond to chemical mediators of the immune system released upon exposure to allergens and during inflammation. These receptors tend to promote rapid, shallow breathing and to stimulate bronchiolar constriction.

How Does the CNS Generate the Basic Rhythm of Breathing?

Some early investigations suggested that the medulla contained separate inspiratory and expiratory centers (Table 18-2), each of which drove the appropriate population of spinal motor neurons. It was tempting to hypothesize that mutual inhibitory synaptic connections between the two kept the system oscillating, like two children playing on a seesaw. More recently, the organization of the respiratory areas in the medulla has been shown to be complex, with some cells in both areas active during inspiration and some active during expiration. In fact, some inspiratory cells are active only early in the cycle and others later in the cycle.

While not yet proved, it is possible that some of the respiratory neurons in the medulla possess intrinsic rhythmicity, analogous to heart pacemakers. Unlike heart pacemakers, the respiratory pacemakers probably generate bursts of action potentials rather than single action potentials. These neurons may set a basic respiratory rhythm, or their rhythmicity may help to stabilize a rhythm that is determined by synaptic connections.

A crude test of the importance of connections between respiratory centers in different parts of the pons and medulla was made by cutting the brain-

THE RESPIRATORY SYSTEM

TABLE 18-2		Respiratory Control Centers
Center	**Location**	**Effect**
Pneumotaxic	Upper pons	Tends to terminate inspiration
Apneustic	Lower pons	Promotes inspiration
Inspiratory	Dorsal medulla	Active during inspiration; stimulation promotes inspiration; generates basic respiratory rhythm
Expiratory	Ventral medulla	Some neurons active during all phases of the cycle, especially during forceful expiration

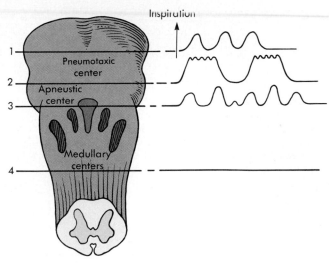

FIGURE 18-3

Identification of respiratory centers in the pons-medulla of cats by cutting it at three levels. The resulting respiratory patterns are shown to the right of each cut. See the text for further explanation.

stems of decerebrate anesthetized animals at different levels. These experiments demonstrated that the entire basic respiratory control system lies within the pons-medulla (between cuts 1 and 4 in Figure 18-3) because a cut above the pons has no effect on the cycle and a cut below the medulla abolishes ventilation. Separation of the upper pons from the lower pons and the rest of the brainstem (a cut at level 2 in Figure 18-3) produces a breathing pattern characterized by deep and very long inspirations, followed by rapid expiration. This pattern is particularly evident when the vagus nerves are also cut. Thus loss of the part of the brainstem between cut 1 and cut 2 biases the system in the direction of inspiration, and more so if the remaining part of the control system does not receive the inhibitory stretch receptor input. The area located somewhere between cut 1 and cut 2 has been called the **pneumotaxic center** (see Table 18-2; Figure 18-1). Its role is to terminate each inspiration.

If the medulla is separated from the pons (a cut at level 3 in Figure 18-3), the respiratory rate is increased and the system spends most of its time in expiration. The area between cut 2 and cut 3 was named the **apneustic center** (see Table 18-2; Figure 18-1). This is only a functional designation, but stimulation of this center can prolong inspiration. The pneumotaxic and apneustic centers can be thought of as providing a modulatory influence on the basic rhythm generated by the medulla (Figure 18-4). Neither the pneumotaxic center nor the apneustic center are necessary for the generation of a basic rhythm because breathing after cut 3 is still rhythmic although the breathing pattern is irregular.

While the applicability of these data to humans is uncertain, whatever the origin of the rhythm, the basic respiratory rhythm is strengthened and modified by tonic inputs from chemoreceptors that monitor arterial P_{CO_2}, P_{O_2}, and pH, and inputs from the reticular activating system that sets the brain's level of arousal. Other, less well-defined inputs are important in driving respiration at a high level during exercise. Somatosensory inputs can affect the respiratory centers. For example, people gasp reflexively when slapped or plunged into cold water. Emotions affect respiration; anxiety increases ventilation, and depression decreases it. These inputs are integrated at various levels in the control system (see Figure 18-4), providing a continuous **respiratory drive** whose magnitude is reflected in the intensity and frequency of respiratory cycles driven by the medullary center. Under resting conditions much of the respiratory drive is attributable to chemoreceptor inputs, which are discussed in the following sections.

THE CHEMICAL STIMULI FOR BREATHING
Arterial P_{CO_2}—The Central Chemoreceptors

- *Why does a change in arterial P_{CO_2} have a greater effect on CSF pH than on plasma pH? What is the significance of the sensitivity of CSF pH to arterial P_{CO_2}?*
- *What is ventilatory response?*
- *How do the central chemoreceptors function? The peripheral chemoreceptors?*

At normal or near-normal arterial P_{O_2}, changes in the arterial P_{CO_2} have a much greater effect on respiration than do variations in the arterial P_{O_2}. For example, an increase in the arterial P_{CO_2} (**hypercapnia**) of 2 to 3 mm Hg will approximately

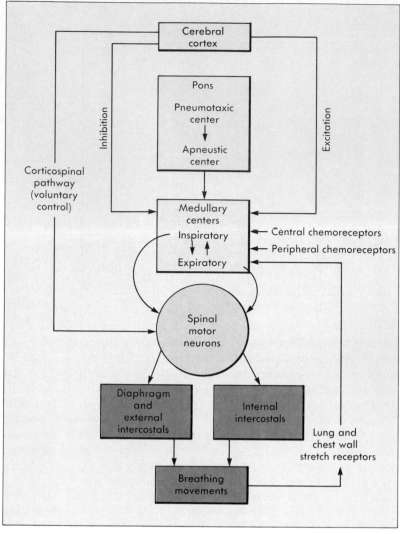

FIGURE 18-4
A hypothetical diagram of information flow in the respiratory control system, including some of the inputs that establish respiratory drive. The feedback loops based on the effect of ventilation on blood gas composition are not shown.

double the respiratory minute volume. In contrast, the arterial P_{O_2} would have to decrease by 30 mm Hg to stimulate respiration to the same degree. The greater sensitivity to CO_2 rather than O_2 provides effective regulation of breathing for two reasons: (1) the S-shape of the oxyhemoglobin dissociation curve protects against decreases in the O_2 content of blood that might occur as a result of small decreases in alveolar P_{O_2}; and (2) small changes in the arterial P_{CO_2} produce large shifts in the plasma pH, so that CO_2 production is a constant threat to total body acid-base balance. Since CO_2 reacts with water to form H_2CO_3, and then HCO_3^- and H^+, we will refer to CO_2 as an acid for convenience, though strictly speaking it is H_2CO_3 that is the acid.

The most important sensors of arterial P_{CO_2} are in the medulla of the brain, probably close to the subarachnoid space. These **central chemoreceptors** do not actually sense arterial P_{CO_2} directly, but instead respond to changes of pH in their immediate environment, the extracellular fluid. To understand how a change in pH of brain extracellular fluid translates into a sensitive measure of arterial P_{CO_2}, it is necessary to know something about the

two fluid systems of the brain: the extracellular fluid and the cerebrospinal fluid.

Cerebrospinal fluid (CSF) is a special secretion formed by the **choroid plexi**, highly vascularized structures on the walls of the ventricles (see the boxed essay in Chapter 14). The CSF surrounds the brain and spinal cord and fills the subarachnoid space and the ventricles. The CSF maintains a constant extracellular environment in the brain because the CSF and the interstitial fluid surrounding the brain cells are separated only by the thin, relatively leaky layer of ependymal cells (a type of epithelium) that line the ventricles of the brain. For convenience, the total extracellular fluid volume of the brain, consisting of the CSF and a smaller volume of interstitial fluid, will be referred to as CSF in subsequent discussion.

Because the walls of brain capillaries are not freely permeable to ions such as H^+ and HCO_3^- (see Chapter 14), they function as a **blood-brain barrier** that effectively isolates the brain interstitial fluid and CSF from the arterial blood. Ions such as H^+ and HCO_3^- cannot readily cross the blood-brain barrier without the help of specific transport mecha-

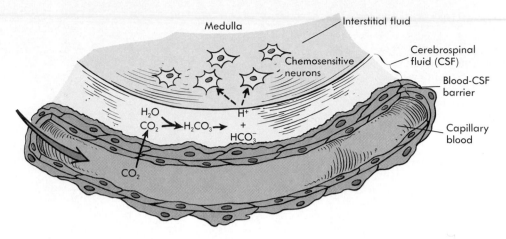

FIGURE 18-5

Activation of central chemoreceptors by changes in arterial P_{CO_2}. Although the receptors are sensitive to small changes in the P_{CO_2}, they are not directly stimulated by CO_2, but instead by the pH of the cerebrospinal fluid. CO_2 readily diffuses across the blood-brain barrier and forms carbonic acid, which dissociates in the CSF and interstitial fluid of the brain. By this mechanism, an increase in arterial P_{CO_2} acidifies the CSF and interstitial fluid.

nisms, but dissolved CO_2 can cross readily because it is uncharged and highly soluble in both water and the lipid membrane (Figure 18-5). As stated earlier, the central chemoreceptors do not detect plasma P_{CO_2} directly but respond to the pH of their immediate environment, the CSF. However, any increase in the arterial P_{CO_2} results in an immediate and equivalent increase in the P_{CO_2} of the CSF. Once in the CSF, the CO_2 reacts with water to form carbonic acid (H_2CO_3), which dissociates to produce H^+ and HCO_3^-:

$$CO_2 + H_2O \underset{\text{anhydrase}}{\overset{\text{carbonic}}{\rightleftharpoons}} H_2CO_3 \rightleftharpoons HCO_3^- + H^+$$

According to the law of mass action (see Chapter 3), elevating the P_{CO_2} of CSF drives the reaction to the right and thus increases both the H^+ and HCO_3^- concentrations. Hydrogen and HCO_3^- ions formed within the CSF are trapped there because they cannot cross the blood-brain barrier into the plasma (see Figure 18-5).

As compared with blood, CSF is weakly buffered because it contains little protein and a lower [HCO_3^-]. The low buffer capacity of the CSF makes the pH of CSF sensitive to changes in plasma P_{CO_2}. When the arterial P_{CO_2} increases, the P_{CO_2} of the CSF increases proportionately. The subsequent reaction of CO_2 with H_2O in the CSF decreases its pH. The response of the central chemoreceptors to this change in pH increases the respiratory drive and thus alveolar ventilation (see Figure 18-4). The **ventilatory response** is the change in alveolar ventilation that results from any change in the inspired gas or the plasma chemistry. In the case of CO_2, the ventilatory response is nearly linearly related to the change in arterial P_{CO_2} (Figure 18-6). In experi-

ments, decreasing the pH of the CSF in the subarachnoid space near the chemoreceptors resulted in an increase in ventilation within seconds.

The Choroid Plexi and the Ventilatory Setpoint

The changes in CSF pH that result from changes of arterial P_{CO_2} are only temporary because the choroid plexi oppose changes in CSF pH by increasing or decreasing their transport of HCO_3^- from CSF to blood (Figure 18-7). As the pH of the CSF is restored toward its normal value, the ventilatory drive of central chemoreceptors diminishes. In effect, in the process of protecting the brain from changes in blood pH, the choroid plexi reset the central chemoreceptors, a process that serves the same function in the respiratory system as receptor adaptation does in sensory systems. If the central chemoreceptors were not reset, they would become unresponsive to variations in the metabolic load of CO_2. Such changes in the setpoint of the ventilatory response are important in situations in which there is a chronic change in arterial P_{CO_2}, such as lung disease and adaptation to high altitude (discussed later).

Arterial P_{O_2} and pH—The Carotid and Aortic Chemoreceptors

Although arterial P_{CO_2}, as monitored by the central chemoreceptors, is responsible for most of the respiratory drive, the respiratory centers also respond to changes in arterial P_{O_2} and pH detected by **peripheral chemoreceptors**. The major peripheral chemoreceptors are located in the **carotid** and **aortic bodies** (Figure 18-8). Unlike the central chemoreceptors, the peripheral chemoreceptors are in direct contact with the arterial blood. These pe-

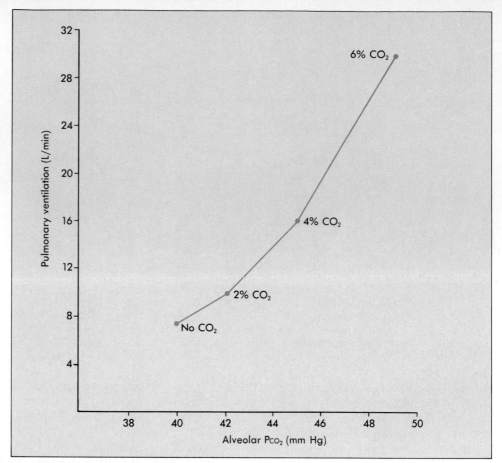

FIGURE 18-6
The relationship between alveolar P_{CO_2} and ventilation. The subjects breathed either air or air containing 2%, 4%, or 6% CO_2. The resulting values of alveolar P_{CO_2} can be expected equal the P_{CO_2} of arterial blood.

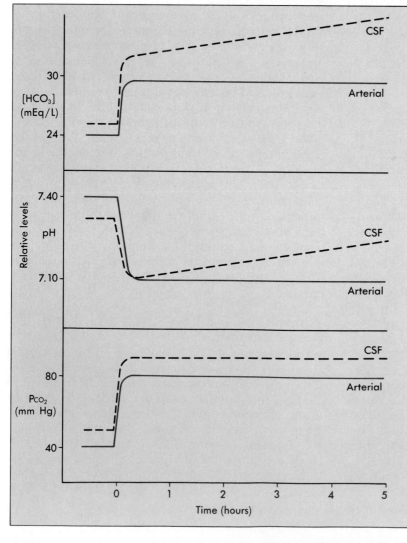

FIGURE 18-7
Relative changes in P_{CO_2}, pH, and HCO_3^- concentrations in CSF and arterial plasma after addition of CO_2 to inspired air at time 0. The P_{CO_2} of CSF tracks that of arterial blood, but addition of bicarbonate to the CSF by the choroid plexi causes the CSF pH to become less acid over the subsequent hours, a process that is continuing at the end of the time scale shown. In the case of a decrease in plasma P_{CO_2}, the pattern would be similar except that the restoration of CSF pH would involve net loss of HCO_3^- from the CSF.

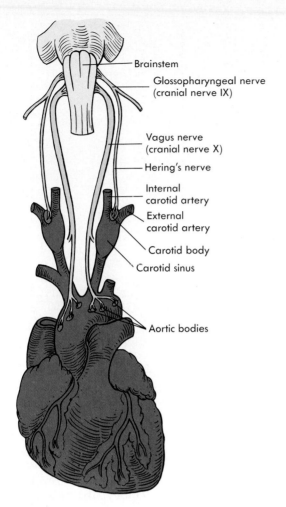

FIGURE 18-8
The locations of carotid and aortic arch chemoreceptors. The carotid bodies are near the carotid sinus baroreceptors discussed in Chapter 15, but are anatomically and functionally distinct from them.

ripheral chemoreceptors are most responsive to changes in arterial P_{O_2}. As arterial P_{O_2} falls, chemoreceptor activity increases slowly at first but rapidly as lower values are reached (Figure 18-9). In addition to the P_{O_2}, chemoreceptors are also responsive to arterial pH, especially in the case of the aortic chemoreceptors. Because they respond to pH, the peripheral chemoreceptors respond indirectly to P_{CO_2} because the reaction of CO_2 with water produces HCO_3^- and H^+.

Distinguishing the effects of peripheral chemoreceptors from those of the central receptors is difficult because it is hard to change just one variable. Normally O_2 and CO_2 levels change in opposite directions when alveolar ventilation changes, and changes in CO_2 also affect the plasma pH. Experimenters had to develop ways to vary O_2, CO_2, and pH independently in order to study the function of particular chemoreceptors. For instance, the central effects of peripheral chemoreceptor activity are best studied when the cerebrospinal fluid is artificially perfused so that its values do not change when arterial values of CO_2, O_2, and pH are experimentally altered.

When arterial P_{O_2} is decreased, the frequency of action potentials in the afferent nerves from both the aortic and carotid bodies increases. This increased neural activity is transmitted along the glossopharyngeal and vagus nerves to the brainstem respiratory centers. The respiratory centers respond by increasing respiratory frequency and tidal volume, therefore increasing alveolar ventilation. A similar effect is seen when the plasma P_{CO_2} is increased and when the plasma pH is decreased.

When arterial P_{O_2} is decreased and arterial P_{CO_2} and pH are held at normal values (40 mm Hg and 7.40, respectively), there is little increase in ventila-

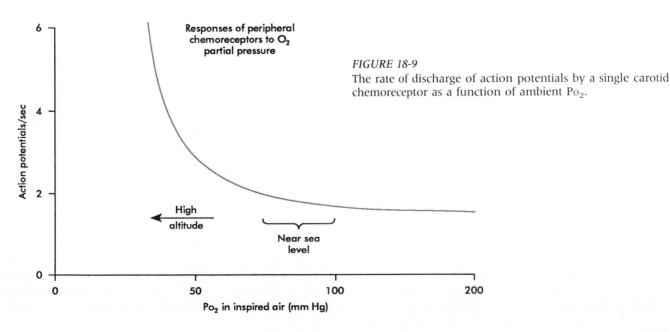

FIGURE 18-9
The rate of discharge of action potentials by a single carotid chemoreceptor as a function of ambient P_{O_2}.

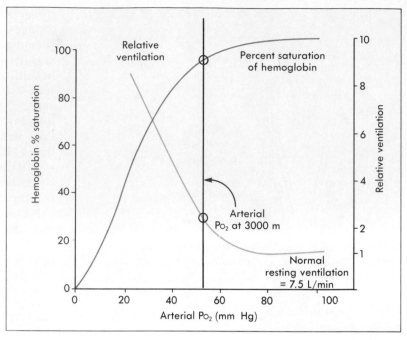

FIGURE 18-10
Ventilation (green) relative to control as a function of P_{O_2}. No increase occurs until the arterial P_{O_2} drops below 80 mm Hg. This corresponds to the shoulder of the oxyhemoglobin dissociation curve (red curve). The hemoglobin saturation at 3000 meters and the corresponding value of relative ventilation are indicated by the vertical line.

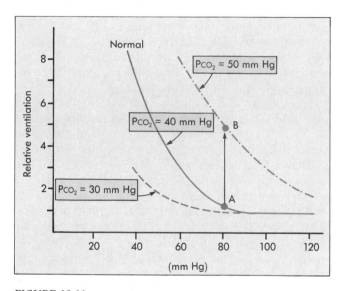

FIGURE 18-11

The effect of arterial P_{CO_2} on the ventilatory response to O_2. When the CO_2 level increases, the response to decreasing arterial P_{O_2} is augmented at all P_{O_2} levels. Conversely, respiratory minute volume is increased by increases in the P_{CO_2} at any given P_{O_2} (point A to point B).

tion until the arterial P_{O_2} falls below about 60 to 70 mm Hg (Figure 18-10). At a P_{O_2} of 60 to 70 mm Hg, hemoglobin is still 90% saturated with O_2. The peripheral chemoreceptors start to become very effective at generating respiratory drive at arterial P_{O_2} values at which hemoglobin is less than fully saturated with O_2. The respiratory minute volume rapidly increases as the arterial P_{O_2} decreases below 60 mm Hg. Such a condition might be encountered at an altitude of 3000 meters where arterial P_{O_2} is about 60 mm Hg, and alveolar ventilation is about

three times its normal resting value. The increased ventilation at altitude blows off CO_2, which would result in hypoventilation if it were not for the peripheral hypoxic drive to breathing.

Peripheral chemoreceptors respond to CO_2 as well as to O_2 (Figure 18-11). If P_{CO_2} is elevated, ventilation is increased at all partial pressures of O_2. For instance, if the arterial P_{CO_2} were elevated to 50 mm Hg (upper, dotted curve), the increase sums with the increase in lung ventilation attributable to decreased P_{O_2}. The lower, dashed curve shows that the responsiveness to O_2 decreases when the P_{CO_2} is lowered to 30 mm Hg. If arterial P_{O_2} is held constant while arterial P_{CO_2} is increased, the relative ventilation increases (vertical arrow from point A to point B in Figure 18-11).

These changes in the sensitivity of the peripheral chemoreceptors to oxygen are not produced directly by dissolved CO_2, but instead by the effect that P_{CO_2} has on plasma pH. The evidence that H^+ rather than CO_2 is the actual stimulus for the peripheral chemoreceptors was obtained by experimentally maintaining the pH constant while increasing the arterial P_{CO_2}. When pH is held constant, P_{CO_2} has no effect on the activity of peripheral chemoreceptors. The peripheral chemoreceptors are responsible for at most about 15% to 20% of the overall respiratory response to CO_2, so the response of central chemoreceptors to CO_2 is far more important for regulation of arterial P_{CO_2}.

The respiratory control centers are sensitive to very small changes in arterial pH detected by peripheral chemoreceptors. If P_{O_2} and P_{CO_2} are normal, a decrease of 0.2 pH units increases ventilation by a factor of five (Figure 18-12). The peripheral chemoreceptors are the only receptors that are capa-

THE RESPIRATORY SYSTEM

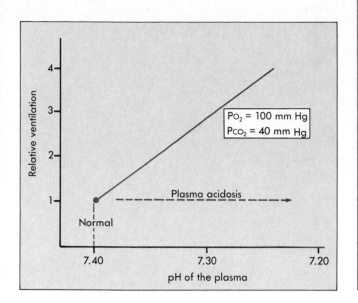

FIGURE 18-12

The ventilatory response to plasma pH driven by the peripheral chemoreceptors. In contrast to their O_2 response, the ventilatory response driven by the peripheral chemoreceptors is highly sensitive to small increases in plasma H^+ ion concentration (acidosis). The peripheral chemoreceptors are thus a major element in the control of total body acid-base balance.

ble of responding to acids or bases that cannot cross the blood-brain barrier.

INTEGRATED RESPONSES OF THE RESPIRATORY CONTROL SYSTEM

- *How do the choroid plexi function in adaptation to high altitudes?*
- *In what situations might adaptation of the central chemoreceptors have dangerous consequences?*
- *Is there a feed-forward mechanism of respiratory control in exercise? What reasons are there to believe that feedback mechanisms must also be involved?*

Interactions Between Peripheral and Central Chemoreceptors

The effects of central and peripheral chemoreceptors on ventilation are usually synergistic, as shown in Figure 18-13. Four situations in which the effects of central and peripheral chemoreceptors are *not* synergistic are breath-holding after hyperventilation, high-altitude acclimatization, chronic acidosis, and respiratory disease. These are discussed in the following sections.

Breath-Holding

During breath-holding, plasma P_{CO_2} rises and P_{O_2} falls. At some point the increase in ventilatory drive, resulting primarily from the increased P_{CO_2}, will overcome any psychological resistance, and

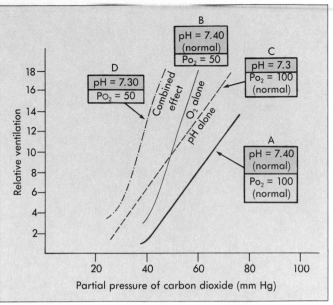

FIGURE 18-13

The overall ventilatory response of an intact individual to oxygen, carbon dioxide, and plasma pH. **Curve A** shows that under normal conditions (arterial pH = 7.40; arterial P_{O_2} = 100 mm Hg), the arterial P_{CO_2} represents the major factor controlling the respiratory minute volume. However, this response is augmented by decreases in the arterial P_{O_2} and pH. **Curve B** shows the effect of the P_{CO_2} when the P_{O_2} is reduced to 50 mm Hg. Ventilation increases and so does the sensitivity to additional changes in the P_{CO_2}. **Curve C** illustrates the effect of a decrease in arterial pH to 7.3. Ventilation increases, but the sensitivity to CO_2 is about the same as normal. **Curve D** shows the large increases in ventilation produced by CO_2 when both the P_{O_2} and pH are low.

breathing will begin again. It is possible to hold one's breath longer if breath-holding is preceded by inspiration rather than expiration, because the additional volume provides an additional O_2 reserve and CO_2 sink. Even though the drop in P_{O_2} has less of an effect in overriding the central control and initiating involuntary ventilation, the hypoxia that can result from voluntary breath-holding rarely threatens the heart and brain. However, when breath-holding is preceded by a period of hyperventilation, the duration of breath-holding increases, not because the hyperventilation increases P_{O_2}, but because it reduces arterial P_{CO_2} before breath-holding begins. Consequently, the P_{CO_2} may not rise high enough to restart ventilation before arterial P_{O_2} has fallen so low that brain hypoxia causes unconsciousness. Swimmers who hyperventilate to increase breath-holding time can become unconscious underwater.

Acclimatization to Altitude

The first example of opposing chemoreceptor effects is seen when a person ascends rapidly to high altitude. Many people live and vacation at altitudes

509

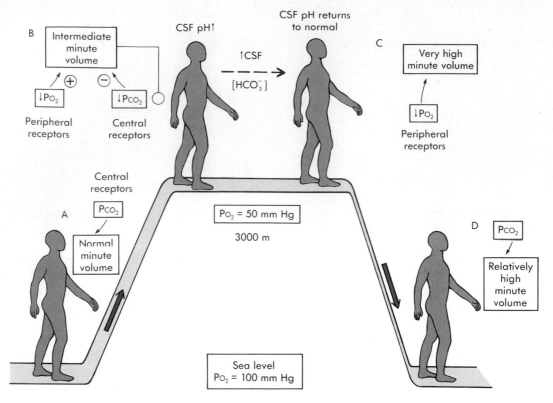

FIGURE 18-14
Changes in respiratory regulation during and after altitude acclimatization. At first, ventilation is less than it should be because the central chemoreceptors counteract the peripheral hypoxic drive *(box B)*. With time, the pH of the cerebrospinal fluid returns toward its normal value because the rate of HCO_3^- transport across the blood-brain barrier changes and ventilation increases *(box C)*. After an acclimatized individual returns to sea level, respiration remains elevated until the central chemoreceptors are reset *(box D)*.

sufficient to significantly reduce blood oxygenation. For example, at an altitude of 3000 meters (about the altitude of many mountain resort areas in Colorado and California, for example), the inspired Po_2 is only 100 mm Hg, causing arterial Po_2 to fall from its normal value of 95 mm Hg to about 50 mm Hg. At this arterial Po_2 hemoglobin is about 87% saturated. When people acclimated to sea level travel to such altitudes, the lower arterial Po_2 stimulates the peripheral chemoreceptors and increases alveolar ventilation (Figure 18-14). The increased ventilation decreases the arterial Pco_2 and reduces stimulation of the central chemoreceptors. Given these contradictory inputs, the output of the respiratory center may oscillate between hyperventilation and short periods of apnea; generally ventilation is maintained at a level below that needed to maintain tissue Po_2 at normal levels. The sensations experienced by the transition to high altitude are fatigue, dizziness, headache, and nausea; when they are severe enough to cause distress, these symptoms of generalized hypoxia are called **acute mountain sickness.** The symptoms tend to be worse the more rapid the ascent; travel by air from sea level to a region that is several thousand feet above sea level is

more likely to provoke these symptoms than hiking to a high altitude.

The choroid plexi correct the abnormally high CSF pH by reducing their secretion of HCO_3^- from blood to CSF. Within 24 to 48 hours the CSF pH returns almost to normal, even though the arterial Pco_2 is still low. As a result, the central chemoreceptors no longer oppose the response of the respiratory control center to the reduced arterial Po_2, and alveolar ventilation gradually increases. At this point the respiratory center is said to have adapted to the altitude. Adaptation is not something that needs to occur only at extreme altitude; some degree of adaptation of the respiratory center occurs in people who live at any altitude above sea level.

The respiratory changes described in the preceding paragraphs are part of a general process of adaptation to altitude that also involves an increase in the hematocrit and a right shift of the oxyhemoglobin dissociation curve (Table 18-3). The increase in the hematocrit is brought about by increased secretion of the hormone erythropoietin (Chapter 14) by the kidneys. The normal hemoglobin concentration at sea level is 15 g/100 ml, and in the Peruvian Andes (at about 5000 meters) it may be elevated

TABLE 18-3   *Physiological Adaptation to High Altitude*

Variable	Change	Time scale	Comments
Ventilation	Increased	7 to 10 days	Respiratory control shifts from CO_2 to O_2
Hematocrit	Increased	7 days	Results from decrease in plasma volume (early) and increased secretion of erythropoietin (later)
Blood viscosity	Increased	2 to 7 days	Due to increased hematocrit
Cardiac output	Increased	Immediate onset	Rises in proportion to decreased blood O_2
Hemoglobin P_{50}	Increased	24 hr	Due to increased 2, 3 DPG; opposes left shift due to lowered plasma pH

above 19 g/100 ml. With the resulting increase in O_2 capacity, the blood may actually carry a normal load of O_2 even though the hemoglobin is not saturated. The right shift is the result of increased formation of 2,3 DPG by red blood cells (Chapter 17). Although the right shift does not benefit O_2 loading at the reduced Po_2, it does increase the ability of the tissues to extract O_2 from the hemoglobin. It also opposes a left shift of the curve that might be expected as the plasma Pco_2 decreases (**hypocapnia**). The threshold for such adaptations is about 1600 meters; a number of major cities (for example, Mexico City, Guadalajara, Denver, and Albuquerque) are near or above the threshold.

When an adapted person descends to sea level, the process of adaptation is reversed. The arterial Po_2 increases to its sea-level value of 100 mm Hg, and the component of respiratory drive due to peripheral chemoreceptors is greatly diminished. As a result, the arterial Pco_2 rises above normal until the choroid plexi have had time to adjust the CSF pH and reset the central chemoreceptors (see Figure 18-14).

Acute Acidosis

Acids other than CO_2 are called **fixed acids** because, unlike volatile CO_2, they cannot be removed from the body by the respiratory system. Fixed acids dissociate in plasma to yield H^+ and an anion called the **conjugate base.** Any change in plasma pH caused by fixed acids is conventionally referred to as **metabolic acidosis,** even if the acid is not of metabolic origin. Generally, conjugate base ions do not readily cross the blood-brain barrier. Nevertheless, ventilation is increased in metabolic acidosis, because a decrease in plasma pH stimulates the peripheral receptors. The resulting decrease in plasma Pco_2 opposes the change in pH caused by the fixed acid.

In acute metabolic acidosis (just as with acute exposure to high altitude) a conflict develops between the peripheral and central chemoreceptors that prevents the respiratory compensation for acute acidosis from being completely effective. With the onset of the acidosis, the peripheral chemoreceptors stimulate increased ventilation. The resulting decrease in Pco_2 starts to restore the arterial pH to normal. But meanwhile, the pH of the CSF begins to become more alkaline. The conjugate base and H^+ could not penetrate the blood-brain barrier to make the CSF more acid, but the decrease in arterial Pco_2 causes the CSF Pco_2 to decrease and the pH of the CSF to increase. As this happens, the response of central chemoreceptors begins to oppose the excitatory input from the peripheral receptors, preventing full compensation for the acidosis.

If acidosis becomes chronic (continues for days or longer), there is time for the change in CSF pH to be adjusted by the choroid plexi. Then the peripheral chemoreceptors regain full control, and a more complete compensation can occur. A number of diseases exist in which acid end products of metabolism accumulate in the plasma; the respiratory response is an important factor in protecting the plasma pH in such conditions. The cooperation of the respiratory control system and the renal system in regulation of acid-base balance will be described in Chapter 20.

Respiratory Disease

Arterial Pco_2 may be chronically elevated in lung disease, injury to the brainstem, or drug-induced depression of the CNS. If alveolar ventilation decreases, the arterial Pco_2 is increased (Figure 18-15), and the arterial pH must thus decrease. The arterial Po_2 also decreases, and all of these changes stimulate respiration.

After a time, the central chemoreceptors adapt, and the peripheral chemoreceptors become the major factor that is driving respiration. During the same time, the kidneys secrete H^+ ions and restore the plasma pH to normal. At this point the person's hypoxia is the only factor driving respiration. If the person is now given 100% O_2 to relieve his hypoxia, respiratory drive may be insufficient to maintain ventilation, and he may well stop breathing and die. Such patients are also much more likely to stop breathing under anesthesia, since even moderate depression of the CNS by anesthetics can abolish their weakened respiratory drive.

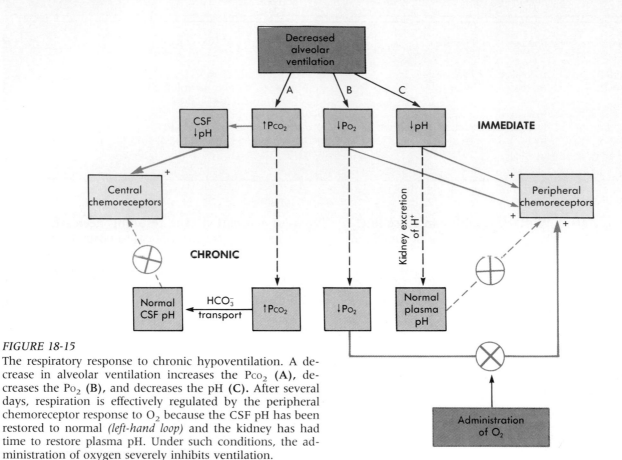

FIGURE 18-15

The respiratory response to chronic hypoventilation. A decrease in alveolar ventilation increases the P_{CO_2} **(A)**, decreases the P_{O_2} **(B)**, and decreases the pH **(C)**. After several days, respiration is effectively regulated by the peripheral chemoreceptor response to O_2 because the CSF pH has been restored to normal *(left-hand loop)* and the kidney has had time to restore plasma pH. Under such conditions, the administration of oxygen severely inhibits ventilation.

The Ventilatory Response to Exercise

Ventilation rises within a few seconds of the beginning of exercise. It may increase from its resting level of about 7 L/min to 70 L/min in heavy exercise. This increase can be divided into two phases, an initial abrupt increase and a later, slower rise to the steady state value (see Figure 18-17). The increase in ventilation supports the oxidative component of energy metabolism of skeletal muscle during exercise. The magnitude of the increase in respiration must be appropriate for the amount of effort that is being expended. Given what is known about regulation of ventilation during rest, it was natural to hypothesize that working muscles signal the change in respiratory demand through an increase in arterial P_{CO_2}, a decrease in arterial P_{O_2}, or a decrease in arterial plasma pH. In fact, arterial P_{CO_2}, P_{O_2}, and pH change slightly or not at all in exercise involving moderate effort (Figure 18-16). None is great enough to increase ventilation four to five times through known feedback mechanisms. What, then, drives the increased respiration and matches the magnitude of the increase to the metabolic need?

The increase in ventilation at the start of exercise is regarded by most exercise physiologists as an example of "feed-forward" regulation, in which the control system anticipates the needs of the body for gas exchange. Possibly the parts of the brain that

are concerned with carrying out motor programs learn how much stimulation to give the respiratory centers in the preparatory ("on your mark—get set") phase of exercise. This would account for the fact that the onset of elevated respiration occurs too soon to be accounted for by any of the changes in blood chemistry associated with exercise (Figure 18-17). Furthermore, if venous blood is prevented from returning to the heart of an experimental animal, the animal's ventilation still increases at the start of exercise.

The critical match between metabolism and respiration is achieved for the most part during the slower phase of the respiratory increase. It is difficult to believe that this match could be achieved by the respiratory control centers without some direct information about the rate of oxidative metabolism in the active muscles. This feedback has been hypothesized to take the form of blood-borne chemical signals from exercising muscles. However, the search for specific chemical stimulants of respiration released by exercising muscles has been frustrating. During exercise the amount of O_2 extracted by muscles increases, so the venous P_{O_2} decreases. One hypothesis suggested that this decrease was the signal for increased ventilation. However, there are no known chemoreceptors on the venous side of the circulation. Other workers have suggested that

Sleep Apnea, Sudden Infant Death Syndrome, and Respiratory Depression

Some adults and children stop breathing for a few minutes periodically during sleep; they have a condition called sleep apnea. In sleep, the major alteration in respiration is an increase in arterial P_{CO_2}, which results from a decrease in the sensitivity to CO_2. In sufferers from sleep apnea, this decrease in sensitivity to the major source of respiratory drive is apparently sufficient to stop breathing. Hypoxia develops after breathing stops, and the hypoxia is generally sufficient to awaken the patient and temporarily correct the problem. In infants, however, such an episode of apnea may end in death.

Infant sleep apnea is believed to be a major cause of sudden, unexplained infant death during sleep, or sudden infant death syndrome (SIDS). In adults, hypoxia enhances the response of the respiratory centers to CO_2.

In newborn preterm infants the reverse is true: hypoxia produces respiratory depression, especially in a cool environment. SIDS victims are outwardly healthy, but some postmortem evidence is consistent with the hypothesis that these infants have subtle metabolic or respiratory problems. Many SIDS victims have enlarged carotid bodies, and in some cases there is also an abnormal persistence of deposits of brown fat, a heat-generating tissue that is important for thermoregulation in young infants but normally disappears in older infants. Studies of other children in the families of victims suggest a familial decreased responsiveness to hypoxia and CO_2. This evidence suggests that in SIDS victims both the normal regulation of P_{CO_2} and the "backup" response to hypoxia may fail, allowing the infant to fall into a vicious cycle in which hypoxia and CO_2 acidosis depress the respiratory centers into unresponsiveness.

Drugs whose action depresses the CNS, including the general anesthetics used in surgery, opiate drugs such as heroin and morphine, barbiturates, and alcohol, also decrease responsiveness to CO_2. Respiratory depression is the most common cause of death in cases of overdose of such drugs and, in the operating room, patients may require artificial ventilation. Physicians should use such drugs with extra care on patients with diminished respiratory drive, such as those with chronic lung disease. The susceptibility of the respiratory centers to the action of opiate drugs indicates that endogenous opiates produced by the brain (see Chapter 8) may set the sensitivity of the respiratory system. If so, some cases of sleep apnea may be treated with drugs that antagonize the binding of opiates to their receptors.

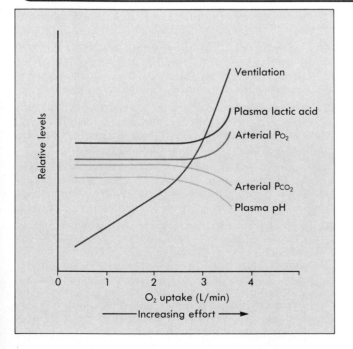

FIGURE 18-16
Changes in blood gas composition, plasma pH, and plasma lactic acid levels with increasing work load, as measured by the rate of O_2 uptake. The beginning of oxygen debt is indicated by the inflections in the curves.

FIGURE 18-17

Changes in ventilation during and after 8 minutes of exercise at a fixed work load. There is an initial, very rapid increase in ventilation and a slower secondary increase to a new higher steady-state level. These changes are almost mirrored during recovery, except that respiration may continue at a slightly higher rate for some time after exercise is over.

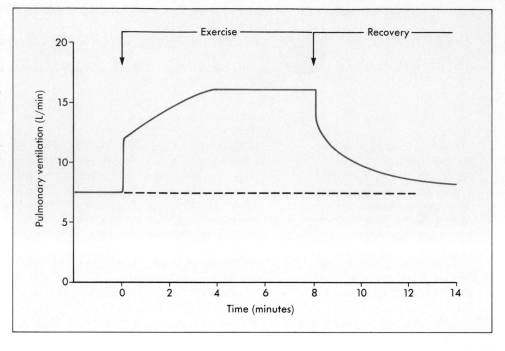

small changes in plasma K^+ concentration that occur when K^+ is released from excited muscles may increase respiratory drive during exercise. Tests of this hypothesis have so far been inconclusive.

The muscle signal does not have to be a chemical one. After muscular exercise has begun, sensory input from muscle and joint receptors could augment the respiratory drive. Also, the respiratory centers have been shown to be sensitive to temperature. A small increase in brain temperature due to exercise could be one of the signals that stimulates the respiratory centers.

At the highest levels of effort, the body's ability to deliver O_2 is not sufficient to support the work performed. At this point oxidative metabolism is replaced by anaerobic metabolism in some tissues, and an oxygen debt develops (Chapter 12). At these levels of effort, ventilation increases substantially as peripheral chemoreceptors are stimulated by increasing levels of lactic acid in the plasma (see Figure 18-16), and ventilation continues at a high level even though arterial P_{CO_2} is decreasing and arterial P_{O_2} is increasing.

● *STUDY QUESTIONS*

1. Describe the pattern of activity in the phrenic nerve during quiet breathing. How does the pattern in this nerve and the innervation of the internal and external intercostals change with increases in respiratory drive?

2. Where are the peripheral chemoreceptors located? To what factors do they respond? What is the magnitude of their contribution to respiratory drive under normal conditions?

3. What is known about the location of the central chemoreceptors? To what factor do they respond? What is the magnitude of their contribution to respiratory drive under normal conditions?

4. The ventilatory response to P_{CO_2} and to P_{O_2} is different. How does this difference relate to the way in which these two gases are carried by the blood?

5. How do the ventilatory responses to changes of CSF pH and arterial P_{O_2} differ?

6. On climbing a 14,000 foot peak in Colorado most individuals will experience fatigue. Account for this in terms of the blood gas composition at 14,000 feet and the responses of central and peripheral chemoreceptors.

7. A normal individual has an arterial P_{O_2} of 50 mm Hg, an arterial P_{CO_2} of 40 mm Hg, and an arterial pH of 7.4. Describe her ventilation. Which chemoreceptors are most important for her ventilatory response?

8. A normal individual has an arterial P_{O_2} of 100 mm Hg, a P_{CO_2} of 40 mm Hg, and an arterial pH of 7.2. Describe his ventilation. Which chemoreceptors are most important for his ventilatory response?

9. A normal individual has an arterial P_{O_2} of 100 mm Hg, an arterial P_{CO_2} of 45 mm Hg, and an arterial pH of 7.4. Describe her ventilation. Which chemoreceptors are most important for her ventilatory response?

10. In an experimental animal whose respiratory centers are intact, both vagus nerves are cut. What will be the immediate effect on tidal volume? A few minutes afterward, alveolar ventilation is measured. Do you expect it to be different from the alveolar ventilation measured just before the vagus nerves were cut? Why or why not? What respiratory feedback loop is interrupted by the vagotomy? Which ones remain intact after both vagus nerves are cut?

SUMMARY

1. During quiet breathing, inspiration is driven by bursts of action potentials passing along axons in the **phrenic nerve** to the **diaphragm**. Increases in ventilation are obtained by recruitment of the spinal motor neurons of **internal intercostal** muscles during inspiration and of **external intercostal** and abdominal muscles during expiration.

2. The respiratory motor program is initiated by centers in the medulla and modulated by inputs from the **pneumotaxic** and **apneustic** centers of the pons.

3. The intensity of respiratory drive determines the inspiratory frequency rate and the **tidal volume** of each cycle. Many inputs are integrated to determine the **respiratory drive.** These include peripheral and **central chemoreceptors,** somatosensory input, and input from regions of the brain involved in arousal and parts of the brain concerned with voluntary movement.

4. Under normal conditions, the input from the **central chemoreceptors** provides the major component of respiratory drive. These receptors monitor the pH of CSF, which is determined over the time scale of minutes to hours by the arterial P_{CO_2}.

5. Input from the **aortic** and **carotid** bodies, **peripheral chemoreceptors** sensitive to arterial P_{O_2}, pH, and (indirectly) P_{CO_2}, is less important in normal conditions but becomes more important during **hypoxia** and **acid-base imbalance.**

6. The peripheral and central chemoreceptors have a synergistic effect on the respiratory control system: input from either enhances the sensitivity of the system to input from the other. This enhancement is appropriate in most but not all circumstances.

7. In circumstances such as high-altitude exposure, acute **metabolic acidosis**, and lung disease, the peripheral and central chemoreceptors may give contradictory signals to the respiratory control system initially, preventing a fully effective ventilatory response. After some days, the contradictory signals from the central chemoreceptors diminish as the **choroid plexi** correct the CSF pH. This process resets the sensitivity of the central chemoreceptors.

8. Currently, a feed-forward mechanism is hypothesized for the ventilatory increase during exercise, since it cannot be accounted for by changes in any of the parameters monitored by the chemoreceptors, and it occurs too rapidly to be driven by metabolic signals.

Choose the MOST CORRECT Answer

11. The breathing pattern is produced by: a. Inputs from chemoreceptors; b. Stretch receptors in the lungs; c. Interaction between specific brainstem neurons; d. All of the above are true.

12. This respiratory control center trends to terminate inspiration: a. Pneumotaxic center; b. Apneustic center; c. Medullary inspiratory center; d. Medullary expiratory center.

13. These receptors are mainly sensitive to decreased arterial P_{O_2}: a. Stretch receptors; b. Irritant receptors; c. Carotid bodies; d. Central chemoreceptors

14. These receptors mediate the Hering-Breuer reflex: a. Stretch receptors; b. Irritant receptors; c. Carotid bodies; d. Aortic bodies

15. Choose the FALSE statement: a. There is a greater respiratory response to arterial P_{CO_2} changes than to P_{O_2}; b. When arterial P_{CO_2} increases, the P_{CO_2} of the CSF decreases; c. Decreasing the pH of the CSF increases ventilation; d. Respiratory minute volume rapidly increases as the arterial P_{O_2} decreases below 60 mmHg.

16. As an individual accustomed to living at sea level adapts to an area of increased altitude: a. The hematocrit will decrease; b. Arterial P_{O_2} decreases; c. Stimulation to peripheral chemoreceptors is reduced; d. Alveolar ventilation will decrease.

● SUGGESTED READINGS

BARNES PJ: Muscarinic receptors in airways: recent developments, *J Appl Physiol* 68(5):1777, 1990. Describes the advances in our understanding of muscarinic receptors in airways and the implications for understanding airway control and airway diseases.

BERGER AJ: Recent advances in respiratory neurobiology using in vitro methods, *Am J Physiol* 259:24, 1990. Describes recent advances in our understanding of the neurobiology of respiration including the neurons generating the respiratory rhythm and the effects of the chemoreceptors.

DAVSON HK, et al: *The physiology and pathophysiology of the cerebrospinal fluid,* Edinburgh, 1987, Churchill-Livingstone. Describes the role of the blood-brain barrier in the regulation of the partial pressure of carbon dioxide in the arterial blood.

DEMPSEY JA, JOHNSON BD, SAUPE KW: Adaptations and limitations in the pulmonary system during exercise, *Chest*, 97:81, 1990. Self-explanatory.

LOPEZ BJ, LOPEZ JR, URINA J, GONZALEZ C: Chemotransduction in the carotid body, *Science* 241:580, July 19, 1988. Describes patch-clamp (see Chapter 7) studies showing that oxygen tension may be sensed by K^+ channels in type I chemoreceptor cells.

SPECTOR R and JOHANSON C: The mammalian choroid plexus, *Scientific American* 261(5):68, 1989. Describes the transport processes of the brain's kidney.

Physiological Aspects of Exercise

WHAT IS EXERCISE PHYSIOLOGY?

Exercise is usually thought of as structured physical activity that increases fitness or improves ability in sports. It is important to realize that, from a physiological point of view, any voluntary activity that uses skeletal muscles is exercise—lifting a barbell is not fundamentally different from lifting a bag of groceries, nor is running to catch a bus different from competitive sprinting. From this point of view, light-to-moderate exercise is part of almost everyone's life. More extreme exercise tests the capabilities of almost all of the body's homeostatic systems.

Exercise physiology focuses on two areas:(1) how homeostatic mechanisms meet the short-term stresses of exercise and (2) how the long-term adaptive changes occur in the structure and biochemistry of the musculoskeletal, cardiovascular, and respiratory system during training. One purpose of this focus unit is to show how the musculoskeletal, cardiovascular, and respiratory systems (described separately in sections II, III, and IV) work together during exercise. Also, it is convenient to note some implications of exercise for the digestive, endocrine, and renal systems; these are treated in subsequent chapters. Table 18-A summarizes much of the following discussion.

ENERGY SOURCES FOR EXERCISING SKELETAL MUSCLE

As described in Chapter 4, ATP may be produced rapidly but inefficiently by the substrate-level phosphorylation steps of glycolysis, or it may be produced more efficiently by the oxidative reactions coupled with the Krebs cycle. As described more fully in Chapter 12, skeletal muscle fibers have evolved into fiber types that show varying degrees of specialization for either brief, powerful contractions (Type II) or sustained, less powerful contractions (Type I). The highest degree of specialization for brief contractions is seen in fast-twitch glycolytic fibers (Type IIB), which have few mitochondria and are committed to anaerobic metabolism. At the other extreme are Type I slow-twitch oxidative fibers, which deliver much slower, weaker contractions but are also resistant to fatigue. Fast oxidative fibers (Type IIA) are regarded as an intermediate between these two extremes.

Individual muscles are mixtures of glycolytic and oxidative fibers, but for each muscle the mix matches the functions normally assumed by that muscle. For example, the muscles of the neck and those attached to the spine are predominantly composed of oxidative fibers, whereas flexors of the arms are predominantly composed of glycolytic fibers. In each individual, the numbers of fast and slow fibers in each skeletal muscle are thought to be essentially constant after maturity, having been established by the muscles' innervations (see Chapter 12). Thus both the total number of fibers and the relative numbers of fast and slow fibers are probably more strongly affected by heredity than by sex or training. Some evidence provided by muscle samples taken from individual athletes suggests that exceptional performance in events requiring either strength or endurance is correlated with disproportionate numbers of either fast or slow fibers. The contractile capabilities and, to some extent, the metabolic properties of skeletal muscle fibers are affected by training, as discussed below.

WHOLE-BODY METABOLISM IN EXERCISE

The immediate energy supply for muscle contraction is the intracellular store of high-energy phosphates—ATP and creatine phosphate. By themselves, these stores could support only very brief efforts—a 6-second sprint, a 1-minute walk, or lifting a moderate weight. Additional work requires continuous regeneration of the stores of high-energy phosphate by anaerobic glycolysis or by oxidative phosphorylation.

During brief, heavy exercise, metabolism of glycolytic fibers is almost independent of that of the rest of the body because glycolytic fibers do not require oxygen and possess large internal stores of glycogen. Unlike glycolytic fibers, oxidative fibers depend on the circulatory system for a continuous supply of glucose, other metabolic intermediates, and oxygen. This difference in dependence on the circulatory system is reflected in the numbers of capillaries that serve each fiber type (see Table 12-2).

The magnitude of the glycogen stores of muscle poses an important limitation on the duration of maximum exercise because the circulatory system

TABLE 18-A   *Physiological Changes in Exercise*

System	Factors involved	Consequences
Cardiovascular		
Increased heart rate and contractility	Increased sympathetic input, decreased parasympathetic input	Increases cardiac output
Vasodilation in active muscle	Local vasodilatory substances and β-adrenergic effect of epinephrine	Increases blood flow to skeletal muscle, decreases peripheral resistance
Vasoconstriction in abdominal and renal beds	Increased sympathetic input to blood vessels (α-adrenergic)	Helps maintain arterial pressure, diverts blood to active muscles
Increased venous return	Decreased peripheral resistance, sympathetic venoconstriction, action of respiratory and skeletal muscle pumps	Supports high cardiac output
Respiratory		
Increased tidal volume and respiratory rate	Increased central respiratory drive	Increases alveolar ventilation
Increased diffusing capacity	Effect of increased lung perfusion	Increases gas exchange
Endocrine		
Increased levels of circulating epinephrine and norepinephrine	Increased sympathetic activity	Stimulates glycogen and fat breakdown
Increased plasma levels of cortisol	Increased release of ACTH-releasing hormone by hypothalamus	Stimulates glycogen and fat breakdown
Decreased insulin levels	Effect of sympathetic input to pancreas	Inhibits conversion of glucose to glycogen
Increased glucagon levels	Effect of sympathetic input to pancreas	Increases breakdown of glycogen and fat
Thermoregulatory		
Elevated core body temperature	Elevated setpoint of central thermostat	Increases rates of metabolic reactions, favors O_2 unloading
Increased cutaneous blood flow	Effect of cutaneous sympathetic afferents	Increases rate of heat loss to environment
Increased sweating	Effect of cutaneous sympathetic afferents	Increases rate of heat loss to environment

cannot deliver glucose to glycolytic fibers rapidly enough to sustain maximum force development. Depletion of muscle glycogen stores during maximum exercise is one of several factors highly correlated with both a sharp decline in performance and a subjective feeling of exhaustion. Marathon runners sometimes refer to this subjective effect as "hitting the wall."

Glycogen loading (carbohydrate loading) is a training regimen that results in a temporary but significant increase in the amount of glycogen stored in both fast and slow fibers. The regimen is based on the observation that carbohydrate deprivation increases the ability of muscles to take up and store glucose. Accordingly, athletes who are preparing for a major endurance event first deprive themselves of carbohydrates, and then switch to a high-carbohydrate diet 2 or 3 days before the event and stop intensive training. Glycogen levels in muscle are tem-

porarily driven to levels as much as two times higher than those obtained with muscle hypertrophy alone. Doubling of glycogen stores has been shown to increase the maximum duration of endurance exercise by as much as 50%.

Production of lactic acid rises with the intensity of exercise, but the rise steepens as the power developed by exercising muscles approaches about half the workload that could be supported by oxidative metabolism (Figure 18-A). The level of effort at which the steep rise begins is called the **anaerobic transition**. The maximum blood level of lactic acid that can be developed is a measure of an individual's potential for brief, intense muscle work. Athletes involved in sports that require such work (for example, a half-mile sprint) can achieve very high blood lactic acid levels, whereas untrained individuals and endurance athletes have low or moderate maximum lactic acid levels.

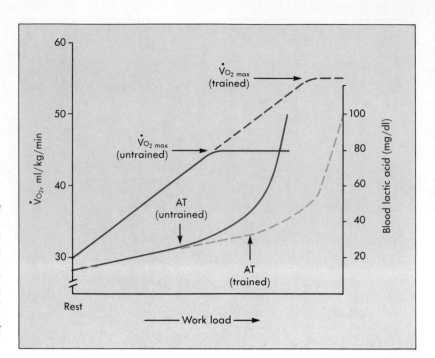

FIGURE 18-A
The relationship between oxygen uptake (red curves) and blood lactic acid levels (blue curves) as the intensity of exercise increases. At about 50% of the maximum oxygen uptake, the blood lactic acid levels begin to rise exponentially—this point is called the anaerobic transition (AT). In endurance-trained people (dashed curves), the maximum oxygen uptake ($VO_{2\ max}$) is increased and the AT occurs at a higher workload as compared with those of untrained people (solid curves).

Some of the lactic acid produced by glycolytic fibers may be reconverted to glucose by way of the Cori cycle (see Chapter 12); the rest is reconverted to pyruvate and further metabolized to CO_2 and water by oxidative skeletal muscle, the heart, and other organs that continuously rely on oxidative metabolism. Lactic acid is a metabolic intermediate rather than an end product of muscle contraction, and all exercise is supported by oxidative metabolism.

The energy cost of a particular exercise can be estimated by measuring the total amount of oxygen consumed in the process, including both that consumed during the exercise and that consumed after the exercise is over (the "oxygen debt"; see Chapter 11). The total energy expended by a person weighing 70 kg in performing various activities is shown in Table 18-B. Note that the overall energy utilization can increase by about thirteenfold during sustained, heavy exercise (marathon running) and by as much as fiftyfold during brief periods of very intense effort (sprinting).

Carbohydrates and fats are the main metabolic fuels of exercise; protein metabolism accounts for only about 10% of the total energy expended. The carbohydrate stores are within the muscle cells and in the liver. Carbohydrate metabolism accounts for about 40% of the total energy used during rest. During maximum exercise, the contribution of carbohydrate metabolism to total energy use increases to 75% to 80%. As exercise duration increases, fat metabolism accounts for a larger part of the total energy. Fat is stored in deposits adjacent to muscle cells within muscles and at other sites in the body. The protein that may be consumed during sustained

TABLE 18-B	Total Energy Expended in Various Activities By a Person Weighing 70 Kg

Activity	Kcal/min
Sleeping	1.21
Sitting	1.67
Standing	1.83
Light housework, office work	2.5
Bicycle riding (5.5 mph)	3.17
Walking with a 43-pound load	4.83
Running (5.7 mph)	12.00
Marathon running	16.5
Sprinting (15.8 mph)	65.17

heavy exercise is in the muscle cells. Replacement of all the energy sources occurs during a recovery period following the exercise. When glucose is the substrate, the uptake of 1 L of O_2 results in the release of about 5.1 kcal of energy. When fat is the substrate, the amount of energy released is 4.69 kcal/L O_2. For mixed substrates (the normal situation), an approximate value of 5.00 kcal/L O_2 can be used.

ENDOCRINE RESPONSES

The adrenal glands secrete epinephrine and cortisol, two hormones that stimulate breakdown of fat and glycogen. At the beginning of exercise, an immediate increase occurs in plasma catecholamine levels. This is the result of increased sympathetic activity, which includes elevated adrenal secretion. Both the increased levels of circulating catecholamine (primarily epinephrine) and the increased input of norepinephrine delivered by the sympathetic innervation of liver and adipose tissue promote mobilization of stored lipid and glycogen. Epinephrine also stimulates glycogen breakdown in skeletal muscle and inhibits glycogen formation (see Chapter 23). The rise in sympathetic activity is followed within a few minutes by an increase in plasma levels of adrenocorticotropic hormone (ACTH). One effect of the increased levels of ACTH is an increase in cortisol secretion by the adrenal cortex. Cortisol contributes to increased resistance to physical stress. It is important during exercise because of its permissive effect—it must be present for epinephrine and norepinephrine to be fully effective.

The pancreas secretes insulin and glucagon, two hormones that have opposing effects on fat and glycogen metabolism (see Chapter 23). Exercise inhibits insulin secretion and increases glucagon secretion, probably as the result of increased sympathetic input to the pancreas. Thus muscles must increase their glucose uptake during exercise despite a decrease in plasma insulin levels. Because the number of insulin receptors in muscle is up-regulated by exercise, regular exercise can reduce the amount of insulin needed by diabetics to control their plasma glucose levels. The increased glucagon secretion adds to the effects of epinephrine and cortisol in releasing energy substrates from storage.

OXYGEN UPTAKE AND THE RESPIRATORY SYSTEM

Metabolic needs at rest correspond to an oxygen uptake of about 250 ml/min. Oxygen uptake increases in proportion to the mass of muscle used; the highest rates of oxygen uptake are attained only when both arms and both legs are active. During maximum exercise with an energy consumption of 10 to 25 kcal/min, oxygen uptake rises to 2 to 5 L/min, an increase of 8 to 20 times. This increase is reflected in a similar increase in pulmonary ventilation, which rises from 5.6 L/min to as much as 120 to 140 L/min. This increase is due to increases in both tidal volume and respiratory rate (Figure 18-*B*). Tidal volume can be as much as 50% of the vital capacity and is taken mostly from the inspiratory reserve. The factors involved in the rapid increase in ventilation are poorly understood (see Chapter 18). However, feedback processes from receptors that monitor blood gas composition are not involved because plasma P_{CO_2}, P_{O_2}, and pH are essentially unchanged in moderate exercise (see Figure 18-16).

The increased ventilation increases the elastic and flow-resistive work of breathing, thereby increasing the total energy requirement for breathing (see Figure 16-8). At rest, O_2 consumption of the respiratory muscles is approximately 2 ml/min; it may increase to 100 ml/min during heavy exercise.

Exercise increases the **diffusing capacity** of the lungs; this is a measure of the amount of lung surface effectively exposed to gas exchange. The increase results from the greater expansion of the lungs and also from improvement of the ventilation/perfusion ratio, which results from increased blood flow through the lungs (see Chapter 16).

Exceptional athletes can consume O_2 at rates as high as 7 L/min, or 90 ml/kg/min. Maximum O_2 consumption ($VO_{2\ max}$) is relatively constant for each individual. It declines with inactivity and increases with training, particularly in elite athletes. In both men and women, $VO_{2\ max}$ is maximum at about the time growth is completed and declines by

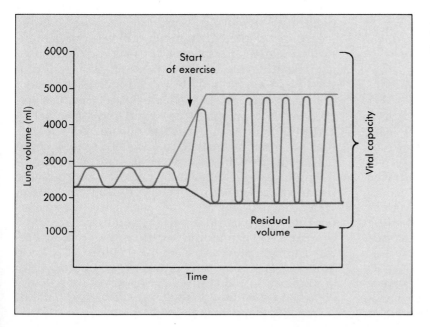

FIGURE 18-B
Typical changes in the pulmonary volumes with the onset of exercise.

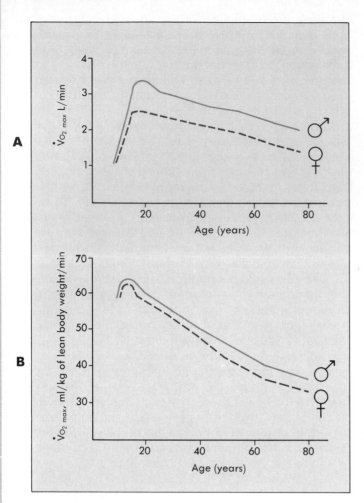

FIGURE 18-C

A Maximum O_2 uptake of men *(solid curve)* and women *(dashed curve)* reaches a maximum in late teens and declines thereafter. These values are on a per person basis and not scaled to body weight. There is some evidence suggesting the decline can be minimized by training.

B Maximum O_2 uptake of men *(solid curve)* and women *(dashed curve)* scaled on the basis of lean body weight. Note the much smaller difference between the male and female curves in this plot as compared with the unscaled curves shown in A.

about 20% by middle age (Figure 18-C). There is a difference in the mean $VO_{2\ max}$ of men and women as expressed on a total-body-weight basis (see Figure 18, *C, a*), but this difference almost disappears when the values are scaled to lean body weight (see Figure 18-C, *b*).

The variability of $VO_{2\ max}$ in the healthy population is greater than can be readily accounted for by the effects of training. The $VO_{2\ max}$ values of identical twins are much more similar than those of fraternal twins. These findings suggest that $VO_{2\ max}$ is primarily determined by heredity. The performance of athletes in events requiring sustained high levels of exertion is almost certainly limited by the rate of

O_2 delivery to the heart and working skeletal muscle because glycolytic fibers cannot support sustained effort. This suggests that athletes with the greatest potential for excellence in endurance events such as the marathon, cross-country skiing, and long-distance cycling could be identified by screening the population of prospective athletes for those with high $VO_{2\ max}$ values. However, many other factors are involved in athletic excellence, and measurement of $VO_{2\ max}$ can, at best, identify those who probably do not have the potential to develop into world-class endurance athletes.

THE CARDIOVASCULAR SYSTEM

Cardiovascular responses to exercise are described in Chapter 15. Cardiac output may be 5 to 6 L/min at rest and 30 to 40 L/min during heavy exercise, an increase of six to seven times. This increase is achieved by increases of both heart rate and stroke volume (see Figures 13-22 and 15-15). The amount of oxygen carried by each liter of blood (the **O_2 carrying capacity**; see Chapter 17) is essentially constant if small changes in the hematocrit (see below) are neglected. If oxygen delivery is to increase by as much as twentyfold, there must also be an increase in the **oxygen extraction**—the percentage of bound oxygen released from hemoglobin as the blood passes through the capillaries of working tissues. Similarly, each liter of blood that travels to the lungs must deliver a larger fraction of its burden of CO_2. Factors that affect oxygen extraction and CO_2 transport include the pH, P_{CO_2}, and P_{O_2} of the working tissues (see Chapter 17) and the fraction of the blood flow that passes through true capillaries instead of through arteriovenous shunts (see Chapter 15).

The hematocrit increases slightly during exercise because fluid is redistributed from plasma to interstitial space. This change adds at most a few percent to the oxygen carrying capacity. Various measures adopted by serious athletes can temporarily increase the oxygen carrying capacity more significantly. One of these measures is a period of training at a high altitude, which increases the rate of red cell synthesis physiologically (see Chapter 18). Some individuals, however, have also taken erythropoietin to artificially stimulate red cell production. Another method, blood doping (see Chapter 14) involves increasing the hematocrit by transfusing a person with his or her own concentrated red blood cells that had been removed and stored. However, it is not clear that increasing the hematocrit improves performance significantly.

Exercise changes the partitioning of cardiac output among different tissues (see Figure 15-14 and Table 15-3). These changes can be summarized as:
1. Increase in blood flow to exercising muscle, determined by local vasodilator substances and adrenal epinephrine.

2. Decrease in blood flow to nonexercising muscle and abdominal and renal vascular beds, caused by central sympathetic activity.
3. Increase in cutaneous perfusion, primarily determined by central thermoregulatory centers.

The cardiovascular changes are affected by the nature of the exercise. **Dynamic exercise** is characterized by movement of body parts so that muscle contractions are accompanied by shortening (that is, isotonic contractions). The total peripheral resistance typically decreases during dynamic exercise; the extent of the decrease is determined by the fraction of the total muscle mass that is involved. **Static exercise** is characterized by tension development without length change (that is, isometric contractions). Some bodybuilding exercises are purely static, but weightlifting and wrestling also have substantial static components. In static exercise at more than about 20% of maximum effort, blood flow through working muscles is impaired as the contracting fibers collapse blood vessels. There are two important consequences of flow impairment: (1) most of the energy used in static exercise must come from anaerobic glycolysis, and (2) the net effect of static exercise is a large increase in total peripheral resistance. This results in much larger increases in arterial blood pressure than are seen in dynamic exercise, making static exercise potentially harmful for people with heart or vascular disease.

THERMOREGULATION AND FLUID BALANCE

Because most of the energy released from substrate in energy metabolism appears as heat, the ten to twentyfold increases in metabolism that occur in exercise (Table 18-B) represent a significant change in the body's heat budget. In the first several minutes of exercise, muscle temperature rises rapidly while core body temperature rises more slowly (Figure 18-D). Skin temperature decreases initially because of sympathetic vasoconstriction, but, as core body temperature rises, the central thermoregulatory centers cause cutaneous vasodilation and sweating, which allow more rapid heat transfer to the environment. After some minutes of exercise, core body temperature attains a new steady state at a value that may be as high as 40° C to 41° C. This higher core body temperature reflects the resetting of the central thermostat to a higher setpoint. Probable beneficial effects of increased temperature include an increase in the velocity of all chemical reactions in the active tissues and an increase in O_2 delivery because of the right shift of the oxyhemoglobin dissociation curve (see Figure 17-9).

If heat loss to the environment is restricted by high environmental temperatures, high humidity, or heavy clothing, the heat produced by muscular activity is a serious threat to thermal homeostasis that may result in collapse (heat stroke) or even death.

At rest, about 15 ml of fluid per hour are lost by evaporation from the body surface. During exercise, fluid loss through evaporation from respiratory surfaces and from sweating increases. The net loss may be as much as 2 to 5 L/hr. Loss of as little as 3% of body weight begins to degrade performance, and loss of 8% to 10% may have serious consequences. Plasma electrolytes (Na^+, K^+, and Cl^-) are lost in sweat, but their rate of loss is lower than the rate of fluid loss because sweat is more dilute than plasma. Maintaining electrolyte balance is of much less concern for exercising people than is maintaining fluid

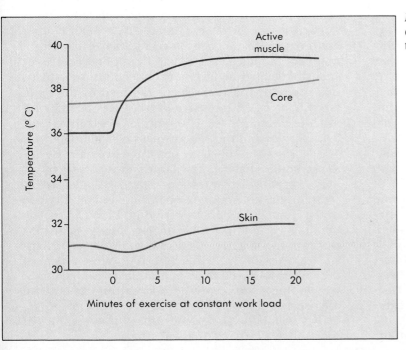

FIGURE 18-D
Changes in core body temperature, muscle temperature, and skin temperature during exercise.

balance, and using salt supplements or isotonic fluids during training is of questionable value and may be dangerous.

The American College of Sports Medicine recommends that athletes training and competing in hot weather should have free access to fluid that contains not more than 25 g of glucose, 10 mEq of Na^+, and 5 mEq of K^+ per liter. Such dilute solutions are physiologically appropriate and also rapidly emptied from the stomach into the intestine. In contrast, most commercial soft drinks, including some that are advertised as appropriate for use by athletes, contain 80 to 100 g of sucrose per liter, concentrations high enough to retard stomach emptying (see Chapter 22) and thus slow their absorption. However, such beverages may be appropriate for use in cold weather when rapid absorption is not as critical.

GENDER-RELATED DIFFERENCES IN ATHLETIC PERFORMANCE

There are general differences in the construction, composition, and capabilities of male and female bodies after puberty (see Chapter 25). In particular, the musculoskeletal system makes up a larger percentage of the average total body mass in men than in women, and the weight-adjusted cardiac output, vital capacity, hematocrit, and blood volume of men are greater than those of women. Some of these differences are attributable to the fact that, on average, fat makes up a larger fraction of total body weight in women than in men; adjustment of the values for lean body weight reduces many of the differences between men and women (see Figure 18-C). In general, athletic training minimizes the gender-related differences in performance that are seen in the general population.

The gender-related differences in the mass of muscle, skeleton, and fat are partly consequences of the different metabolic and developmental effects of male and female sex hormones, and partly consequences of differences in male and female activity patterns (see Chapter 25). In particular, the male sex hormone, testosterone and its metabolites stimulate the growth and functional state of skeletal muscle and bone. These effects are also obtained from synthetic derivatives of testosterone (anabolic steroids), which are used by many athletes, both male and female (see the essay in Chapter 12).

Hormonal levels in women rise and fall during the course of the menstrual cycle. Because sex steroids are known to affect metabolism and fluid balance, these hormonal changes may affect athletic performance. It is difficult to separate the purely physiological effects of the menstrual cycle on exercise performance from its psychological effects. Nevertheless, some effects of female hormones on exercise physiology are now known. For example, the rise in progesterone during the latter part of the menstrual cycle has been shown to increase the responsiveness of the respiratory centers, causing hyperventilation during exercise, which in turn increases the work of breathing and leads to respiratory alkalosis (see Chapter 20). This effect reduces performance, although the effect is less pronounced in female athletes than in untrained women.

Some evidence suggests that to have normal menstrual cycles, women must possess a certain amount of body fat (see Chapter 25). Some dedicated female athletes have very small amounts of body fat and also have irregular menstrual cycles or even fail to have them (a condition called **amenorrhea**). Amenorrhea seems to have little effect, either good or bad, on athletic performance. It may favor development of a higher hematocrit because it reduces the rate of iron loss from the body. However, if amenorrhea continues for long periods, the lack of estrogen normally provided by active ovaries might contribute to loss of bone calcium and to the development of osteoporosis (see the essay, Chapter 23).

Although it has been taken for granted that androgens (male sex hormones) favor athletic performance, very respectable performances have been given by women who were in the early stages of pregnancy. Hormones released by the placenta, the organ that nourishes the fetus before birth, cause changes in the mother's physiology that contribute to the success of the pregnancy (see Chapter 26). These changes may also favor athletic performance. Some concern has arisen that female athletes might exploit this effect to improve their performance in important events, such as Olympic competition.

EXERCISE AND THE IMMUNE SYSTEM

Moderate exercise has a beneficial effect on overall health and may reduce the incidence of minor infectious disease. However, exhaustive exercise does not promote disease resistance and may even reduce immune responses. The B and T lymphocytes are responsible for recognition of infectious organisms that have previously been encountered by the body (see Chapter 24). Intensive exercise has been found to reduce both the levels of antibody secretion by B lymphocytes and the effectiveness of T lymphocytes. The reasons are not clear, but intensive exercise induces some changes similar to those during immune response to infection. These include an increase in the core body temperature and an increase in the number of circulating white blood cells. The immune system, like other major body systems, includes some negative feedback controls (see Chapter 24). Possibly, the infection-like effect of severe exercise is followed by increased negative feedback, which blunts the response to a real infection. Also, intense exercise makes such severe demands on the body's metabolic stores and repair processes that it constitutes a form of physical stress.

TABLE 18-C	Effects of Training	
Factor	Strength	Endurance
Muscle mass	Increased	Little effect
Fiber diameter	Increased diameter of fast fibers	Little effect on either fiber type
Glycogen content	Increased	Increased
Number of myofibrils	Increased in fast fibers	Little effect
Glycolytic enzymes	Increased in fast fibers	Little effect
Oxidative enzymes	Little effect	Increased in oxidative fibers
Capillary density	Decreased	Increased
Mitochondrial density	Decreased	Increased
Cardiovascular system	Little effect	Maximum CO and $\dot{V}_{O_{2max}}$ increased

ADAPTIVE EFFECTS OF ACTIVITY AND TRAINING

Skeletal muscles respond to levels and types of activity; this maintains a level of muscle development corresponding to the demands placed on that muscle over time. A period of bed rest or the immobilization of a limb in a cast results in **disuse atrophy**, a decrease in muscle mass and contractility. Training, in contrast, can increase muscle capability (**hypertrophy**). Hypertrophy and atrophy are relative terms that refer to the adaptive responses of muscle to use. As shown in Table 18-C, the effects of training depend on which aspects of muscle performance, strength or endurance, the training regimen emphasizes.

Strength training, which leads to muscle hypertrophy, selectively affects fast muscle fibers. Such training involves contracting muscles against heavy resistance. It can lead to increased peak isometric tension and increased shortening speed. One of the first effects of a strength training regimen is an increased ability to activate motor units voluntarily. Trained weight lifters appear to be able to cause simultaneous tetanic contraction in almost all of their motor units. When all motor units can be voluntarily activated, additional increases in strength result from increases in contractility. The major effects of fast fiber hypertrophy are increases in:

1. The number of myofibrils.
2. The amount of stored glycogen.
3. The fiber diameter, reflecting changes 1 and 2. On the whole muscle level, this is seen as an increase in muscle mass.

Note that this type of training regimen does not increase capillary density, but the increase in muscle mass actually decreases the number of capillaries per gram of muscle. The same is true for the number of mitochondria per gram of muscle. Strength training is not currently believed to result in formation of additional skeletal muscle fibers in human athletes.

Endurance training (sometimes called aerobic training) selectively affects oxidative fibers. It can increase the ability to sustain submaximum exercise and the threshold level of effort at which an oxygen debt begins to be incurred (see Chapter 12). The changes seen include increased:

1. Size and number of mitochondria.
2. Capillary/muscle fiber ratio.
3. Capacity to oxidize fat, resulting in a greater reliance on fat as a fuel.
4. Levels of myoglobin.

As noted above, endurance training increases $VO_{2\ max}$. This is the result both of adaptive changes in the respiratory and cardiovascular systems that increase oxygen delivery and of the increases in the oxidative capacity of skeletal muscle described above. The cardiovascular effects include:

1. Increased mass and contractility of the myocardium.
2. A lower resting heart rate, which increases the range of the cardiac output, given the increased stroke volume that results from effect 1.
3. Increased capillary density in the myocardium.
4. Decreased peripheral resistance during rest.

All of these changes contribute to increased maximum cardiac output. In untrained but otherwise healthy people, such changes can typically be attained by an exercise regimen that involves 15 to 60 minutes of exercise 3 to 5 days a week in which the heart rate rises to 60% to 90% of the maximum value. The maximum heart rate can be estimated by subtracting the person's age from 220.

KIDNEY AND BODY FLUID HOMEOSTASIS

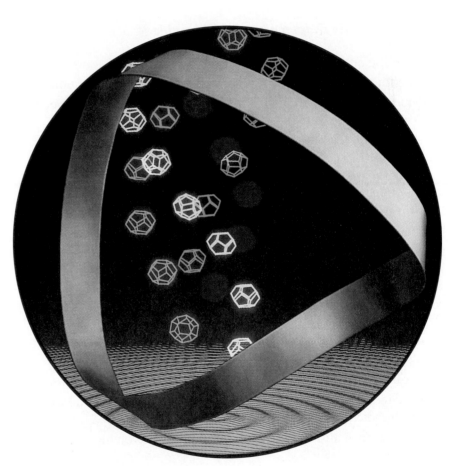

CHAPTER 19

The Kidney

On completing this chapter you should be able to:

- Describe the anatomy of the excretory system.
- Outline the reflex and voluntary pathways controlling urination.
- Understand the structure of the nephrons and distinguish between cortical and juxtamedullary nephrons.
- Describe the process of ultrafiltration across glomerular capillaries and describe the composition of the glomerular filtrate.
- Describe the mechanism for reabsorption of most of the filtered Na^+ and water in the proximal tubule and understand how this leads to the passive reabsorption of other small solutes.
- Understand the mechanisms of active reabsorption and secretion, defining the transport maximum.
- Appreciate how the loop of Henle acts to create an osmotic gradient in the kidney and understand the role of urea in urine concentration.
- Describe the transport processes occurring in the distal tubule and how these are affected by aldosterone.
- Understand how antidiuretic hormone affects the water permeability of the collecting duct.
- Appreciate the role of the vasa recta in preserving the medullary osmotic gradient.
- Define clearance and be able to solve problems involving clearance.
- Define tubuloglomerular feedback and trace the feedback loop.
- Explain how atrial natriuretic hormone modulates the rate of loss of Na^+ in urine.

The kidneys receive about 1.1 L/min of blood flow, a larger fraction of the cardiac output than the brain. It is 20% of the total cardiac output, about the same as that received by resting skeletal muscle. Urine is the end product of a process in which about 200 L of a protein-free filtrate of plasma generated within the kidney each day passes through tiny tubular structures that modify its composition and greatly reduce its volume. Typically, about 99% of the filtered plasma volume is reabsorbed before the stage of final urine is reached, and the output of urine is only 1 to 2 ml/min, some 0.2% of the cardiac output.

As filtered plasma is modified in the kidney, the changes in its composition reflect the kidney's three major functions. First, the kidney is the primary route by which metabolic end products such as urea, uric acid, creatinine, ammonia, phosphate, and sulfate leave the body. Second, the kidney maintains homeostasis of body fluid volume and solute composition. Dietary intake of water and mineral ions such as Cl^-, Na^+, K^+ and Ca^{++} must be closely balanced with urinary excretion. Third, the kidney controls the plasma pH. The kidney excretes acidic end products of metabolism and, along with the lungs (see Chapter 18) helps regulate the plasma bicarbonate concentration.

The importance of functioning kidneys is seen in the profound consequences of end-stage kidney disease: people whose kidneys have failed can live for only a few days while their bodies accumulate solutes, fluid, and acidic metabolic end products. To survive, these people must rely on periodic cleansing of the blood by a dialysis unit, or artificial kidney, or must receive a kidney transplant.

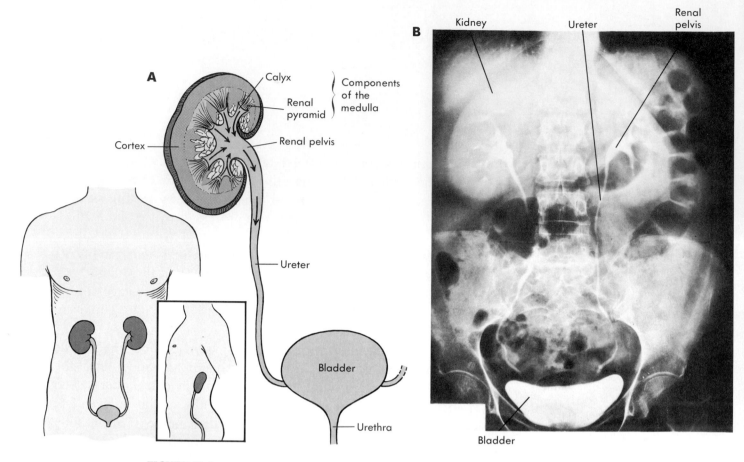

FIGURE 19-1
A The anatomy of the kidneys. Each kidney is composed of an outer cortex and an inner medulla. Urine flows into the renal pyramids, to the calyces and pelvis, and finally through the ureter to the bladder. The bladder empties to the exterior by way of the urethra.
B An X-ray pyelogram made by passing contrast medium from the bladder up the ureters to the pelvis. The contrast medium absorbs X-rays better than does tissue, so it outlines the hollow, fluid-filled parts of the excretory tract. The calyces are surrounded by contrast medium, giving them a cuplike appearance.

STRUCTURE OF THE EXCRETORY SYSTEM
Anatomy of the Kidney, Ureters, and Bladder

> - *What is the role of each of the following structures in transport and storage of urine?*
>   *Bladder*
>   *Ureter*
>   *Urethra*
> - *What muscular structures are involved in urination? Which of these are skeletal muscle, and which are smooth muscle?*
> - *How does voluntary initiation of urination differ from the reflexive urination of infants?*

Each human kidney weighs about 200 to 300 g and is about the size and shape of a fist (Figure 19-1, *A*). The bulk of the kidney is divided into an outer **cortex** and an inner **medulla.** The initial steps of urine

formation and modification occur in the cortex. The final step in the process of urine production takes place in tubes called **collecting ducts,** whose parallel arrangement gives the innermost parts of the medulla, the **renal pyramids,** a striped appearance. The collecting ducts of the renal pyramids drain into the **renal calyces** (singular, **calyx**), the **renal pelvis,** and eventually the **bladder** by way of the **ureter.** Figure 19-1, *B* shows an X-ray image of kidneys in which the calyces and pelvis of each kidney have been made visible by filling them with a contrast medium that absorbs X-rays.

The Urinary Bladder: Urine Storage and Urination

Urine passes from the renal pelvis through the ureters into the urinary bladder (see Figure 19-1), a hollow sac of smooth muscle lined with epithelial cells, that stretches as it is filled with urine. Urine is

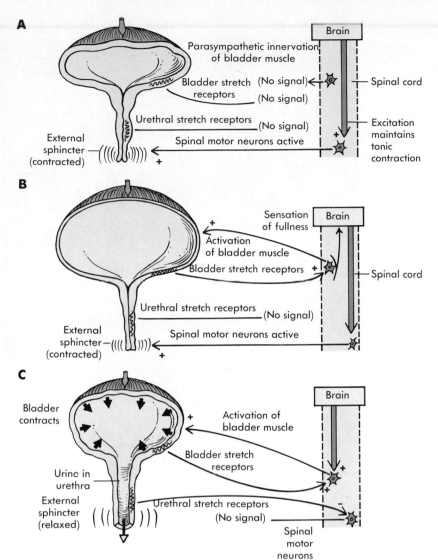

A

Parasympathetic innervation of bladder muscle

Bladder stretch receptors (No signal)

(No signal)

Urethral stretch receptors (No signal)

Spinal motor neurons active

External sphincter (contracted)

Brain

Spinal cord

Excitation maintains tonic contraction

B

Sensation of fullness

Activation of bladder muscle

Bladder stretch receptors +

Urethral stretch receptors (No signal)

External sphincter (contracted)

Spinal motor neurons active

Brain

Spinal cord

C

Bladder contracts

Activation of bladder muscle

Bladder stretch receptors

Urine in urethra

External sphincter (relaxed)

Urethral stretch receptors (No signal)

Brain

Spinal motor neurons

FIGURE 19-2
Micturition.
A The bladder contains little urine and therefore no stretch reflex is activated. The brain maintains the tonic drive on the motor neurons that innervate the skeletal muscle of the external sphincter.
B The volume of urine in the bladder has increased enough to activate the stretch receptors in the bladder wall. The stretch receptors excite the parasympathetic innervation of the bladder, and each contraction activates a positive feedback loop. The external sphincter continues to be closed in response to signals from the brain, but the sensation of the need to void has been felt.
C Voluntary control of bladder emptying occurs when the signals from the brain to the external sphincter motor neurons cease and stimulation of the bladder parasympathetic innervation occurs. This is assisted by the stretch reflex and the inhibition of external sphincter contraction while urine is in the urethra.

stored in the bladder until it is allowed to flow out through the **urethra** in the process of **urination (micturition)**.

The smooth muscle of the bladder forms the **internal sphincter** at the junction of the urethra with the bladder. A second, **external sphincter** located more distally in the urethra is composed of skeletal muscle (Figure 19-2, *A*). Stretch receptors are present in the bladder and also in the muscle of the internal sphincter. Filling of the bladder is detected by the stretch receptors of the bladder. The excitation of these receptors initiates a reflex contraction of the smooth muscle of the bladder, and each contraction leads to another contraction because the stretch receptors are strongly excited each time the bladder contracts but does not empty (Figure 19-2, *B*). After several bladder contractions, the reflex pathway becomes refractory and the stretch receptors do not cause bladder contraction for a period varying from several minutes to an hour, despite the

increasing distension of the bladder. After this period, however, the cycle is repeated.

The greater the volume of urine in the bladder, the stronger the bladder contractions become. At some point, the contractions are sufficient to stretch open the internal sphincter and force some urine into the urethra. This initiates a second stretch reflex that inhibits the spinal motor neurons that maintain tonic contraction of the external sphincter muscles. At this point, urination will occur if the urethral stretch receptor input is able to prevail over the brain's control of the external sphincter. Once urination has begun, positive feedback continues as long as urine is flowing through the urethra, allowing the bladder to be completely emptied (Figure 19-2, *C*).

This description of bladder control is simplified; many other reflex pathways also contribute. For instance, sudden, frightening events interrupt the brain's maintenance of external sphincter control

The Kidney 529

A

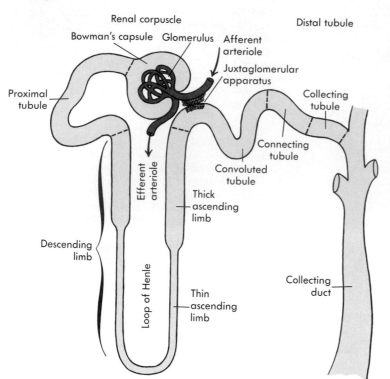

B

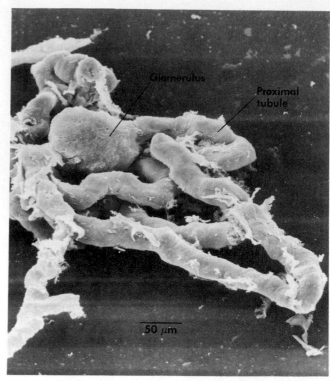

50 µm

FIGURE 19-3

A The functional anatomy of a nephron. All nephrons begin with the glomerulus, the site of plasma filtration. The hollow portion of the nephron surrounding the glomerulus is Bowman's capsule. From Bowman's capsule, the tubular fluid passes through the proximal tubule, loop of Henle, distal tubule, and collecting duct. The afferent arteriole contacts a spot on the initial portion of the distal tubule; the juxtaglomerular apparatus is located at this junction.

B A scanning electron micrograph of the renal corpuscle and proximal tubule of a single nephron dissected from the kidney.

and also activate bladder contraction by increased autonomic activity. Involuntary urination may result.

In infants, urination is purely reflexive. Voluntary control of urination develops about 2 to 3 years after birth. The series of bladder contractions initiated by bladder stretch receptors probably results in a signal to the brain that provides a sensation of the need to void the bladder. Voluntary initiation of urination is accomplished by descending pathways that inhibit the spinal motor neurons of the external sphincter and surrounding muscles of the pelvic floor. Once these pathways are inhibited, urine enters the urethra, initiating the positive feedback that promotes urination. The reflexive emptying of the bladder is assisted by contraction of muscles of the lower abdomen, which increases the pressure of the bladder and promotes reflexive contraction. Damage to pelvic nerves may abolish or weaken the reflex, causing the bladder to be chronically overfilled (**urinary retention**). Damage to nerves serving the external sphincter and skeletal muscles of the pelvic floor results in a loss of sphincter tone and a peri-

odic leakage of urine (**urinary incontinence**). Incontinence is also common in lower spinal cord injuries.

Nephrons: The Functional Units of the Kidney

- *What are the structural differences between cortical and juxtamedullary nephrons?*
- *The kidney contains two types of capillary beds in series. What are the names and locations of the two beds? How do the capillary beds of cortical and juxtamedullary nephrons differ?*
- *What structures compose the juxtaglomerular apparatus? Where in the nephron is this apparatus located? What are its two major functions?*

Each kidney contains about 1.2 million **nephrons**, or **renal tubules** (Figure 19-3). The term "tubule" is also used to refer to particular portions of the nephron. At the head of each nephron is a **glomerulus**, a tuftlike capillary bed enclosed in a funnel-shaped structure called **Bowman's capsule**. The glomerulus and its capsule form a **renal corpuscle**.

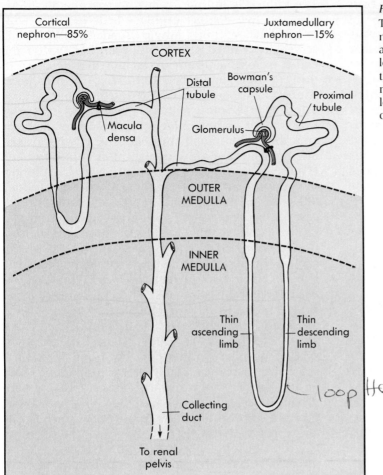

FIGURE 19-4

The structures and locations of the two classes of nephrons. The renal corpuscles of cortical nephrons are near the surface of the cortex, and their short loops of Henle penetrate only the outermost layer of the medulla. The renal corpuscles of juxtamedullary nephrons are typically deeper in the cortex, and their long loops of Henle penetrate into the innermost parts of the medulla, paralleling the collecting ducts.

All of the renal corpuscles are in the cortex of the kidney.

The first step in urine formation is glomerular filtration, in which a portion of the plasma water and small solutes flowing through the glomerular capillaries are forced through the walls of the capillaries into the renal corpuscles. The resulting **glomerular filtrate** is processed to produce final urine as the filtered tubular fluid flows from renal corpuscles through the intermediate regions of the nephron to the collecting duct. In the remainder of the discussion we will generally use the term urine to refer to the final end-product of renal processing, with intermediate stages designated as simply the tubular fluid.

Descriptions of the nephron employ the terms **proximal** for the region closer to the corpuscle and **distal** for the region farther away from the corpuscle. The **proximal convoluted tubule** is a twisted region that lies entirely within the cortex of the kidney. A hairpinlike loop in the tubule called the **loop of Henle,** connects the proximal convoluted tubule with the distal tubule. Because the loop of Henle dips downward into the medulla and then bends back upward to the cortex to join the distal tubule, it is described as having **descending** and **ascending limbs** (see Figure 19-3, A).

The distal tubule is divided into specialized regions. The initial region of the distal tubule immediately following the ascending limb of the loop of Henle is twisted and is referred to as the **distal convoluted tubule.** The next region of the distal tubule is relatively straight and is called the **distal connecting tubule.** The connecting region delivers the tubular fluid into the **distal collecting tubule** (see Figure 19-3, A).

The collecting regions of the distal tubules of 5 to 10 nephrons drain into each cortical collecting duct. Hundreds of cortical collecting ducts converge to form each of the final collecting ducts that descend into the medulla. Each of these ducts carries the urine from approximately 3000 nephrons into the hollow renal pelvis, from which it drains via the ureters into the bladder (see Figure 19-1).

Nephrons can be divided into two classes based on their structures and location in the kidney: about 85% are **cortical nephrons,** and the rest are **juxtamedullary nephrons** (Figure 19-4). The glomeruli of cortical nephrons are located in the outer regions of the cortex, and almost all of the length of cortical nephrons is within the renal cortex, with only a small portion of the loop of Henle descending into the outer zone of the medulla. The glomeruli of juxtamedullary nephrons are located deeper within the cortex, and their loops of Henle have long, **thin descending limbs** that plunge deeply into the renal medulla. The ascending limbs have thin walls within the medulla, but the wall thickness is greater in the cortical portion. Thus the ascending part of the loops of Henle of the juxtamedullary nephrons can be divided into **thin** (medullary) and **thick** (cortical) **ascending limbs.** In contrast, the loops of Henle of cortical nephrons have short, thin descending segments and remain thick throughout the rest of their length. Although the juxtamedullary nephrons are fewer in number, their total bulk is greater than that of the cortical nephrons because their loops of Henle are so much longer (the loops of nephrons in Figure 19-4 are not drawn to scale).

The epithelial cells of the nephron differ in structural details in different parts of the nephron, but some common features are recognizable. The cells are knit together by bands of tight junctions (Figure 19-5, *A* and *B;* see also Chapter 4) in much the same way that beverage cans are linked by flexible plastic to form a six-pack. The apical surfaces of the cells face the tubular fluid on the inside of the nephron and correspond to the tops of the cans.

In most parts of the nephron, the apical surfaces are covered with microvilli, projections of membrane that increase the surface area available for transmembrane transport activity. Microvilli are particularly evident in the proximal tubule (Figure 19-5, *C*). In contrast, the apical surface of cells of the collecting duct, in which active solute transport occurs at a lower rate, has only stubby microvilli and the cells themselves are flatter (Figures 19-5, *D* and 19-5, *E*). The **basolateral** surfaces of tubular cells face away from the interior of the tubule and correspond to the sides and bottoms of the cans in a six-pack.

The transport proteins of apical and basolateral membranes differ from one another, with the result that the cells can move a substance in across one surface and out the other side. For example, almost all parts of the nephron reabsorb Na^+ from the tubular fluid. The first step in this process is entry of Na^+ down its concentration gradient across the apical membrane. The entry of Na^+ is frequently coupled to entry or exit of other substances, a phenomenon called gradient-mediated cotransport or countertransport (see Chapter 6). Exit of Na^+ from the cytoplasm is driven by the Na^+-K^+ pump, which is confined to the basolateral membrane.

Renal Blood Vessels and the Juxtaglomerular Apparatus

The structural relationship between nephrons (Figure 19-6, *A*) and their vasculature (Figure 19-6, *B*) has important functional implications. In most regions of the systemic circulation there is but one set of capillaries between the arterioles and venules.

The kidney is unique in having two sets of arterioles and two distinct capillary beds in series with one another within a single organ. Each kidney is served by a renal artery that branches within the cortex to form **afferent arterioles** that give rise in turn to the glomerular capillaries of individual glomeruli. Glomerular capillaries then join to form **efferent arterioles,** which are recognizable as arterioles rather than venules by their narrow diameter and abundant smooth muscle. The efferent arterioles branch into diffuse **peritubular capillary beds** (Figure 19-6, *C*). In the case of cortical nephrons, the peritubular capillaries surround the short loops of Henle and the proximal and distal tubules. In the case of juxtamedullary nephrons, the peritubular capillary beds include capillaries that serve the cortical portions of the nephrons and also the **vasa recta,** which are capillaries that accompany the loops of Henle into the medulla (see Figure 19-6, *C*). Peritubular capillaries join to form venules, small veins, and ultimately the renal vein.

Recall from Chapter 14 that fluid can move across capillary walls from plasma to interstitial spaces (capillary filtration) or from interstitial spaces to plasma (capillary reabsorption) depending on the balance of hydrostatic and osmotic pressure driving forces across the wall, and that both filtration and reabsorption can typically take place in different parts of the same capillaries. The kidney's two sets of capillary beds cause the two processes of filtration and reabsorption to take place in different parts of the kidney. Filtration is the dominant process in glomerular capillaries, and this is the first step in urine formation. The subsequent parts of the nephron are involved in recovery of most of the filtered fluid; this recovered fluid enters the peritubular capillaries and so returns to the bloodstream.

The cortical portions of each nephron are intertwined like a tangled telephone cord, and the initial part of the distal tubule bends around to contact the afferent arteriole at the place where it enters the

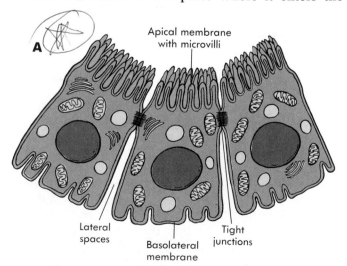

A

Apical membrane
with microvilli

Lateral
spaces

Basolateral
membrane

Tight
junctions

KIDNEY AND BODY FLUID HOMEOSTASIS

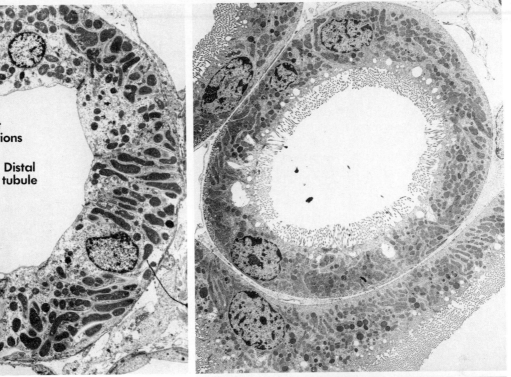

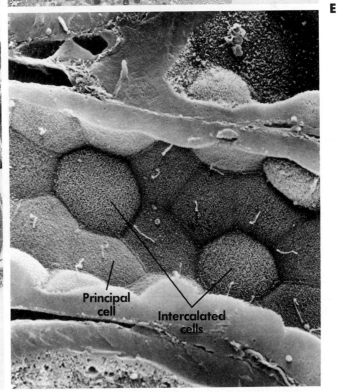

FIGURE 19-5

A *(left)* General features of renal tubular epithelial cells.

B Section through a portion of the wall of a distal tubule, showing parts of two tubular epithelial cells and a peritubular capillary. The tight junction connecting the adjacent cells is visible. Between the two cells, a lateral space separates the basolateral membranes of the two cells.

C Photomicrograph of section through a proximal tubule. Proximal tubular cells have extensive apical microvilli and contain many mitochondria.

D A transmission electron micrograph of the collecting duct showing the two main types of cells: principal and intercalated cells.

E A scanning electron micrograph of the interior of the collecting duct.

The Kidney

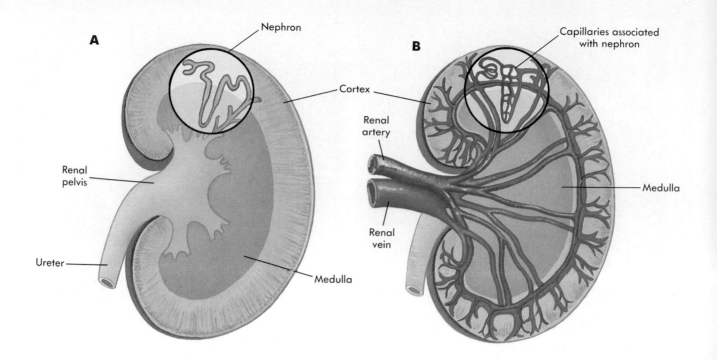

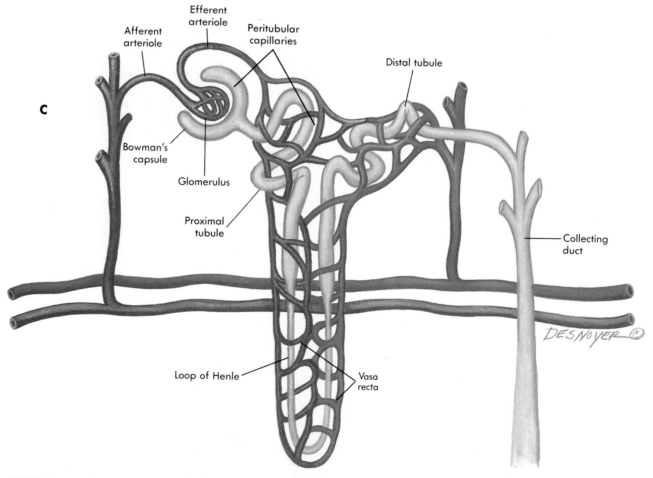

FIGURE 19-6

A   Location of a juxtamedullary nephron in the kidney.

B   Relationship of its glomerular and peritubular capillaries to the renal blood vessels.

C   An enlarged view of the nephron with its afferent and efferent arterioles, capillary beds, and venules. Branches of the peritubular capillaries, the vasa recta, follow the loop of Henle of juxtaglomerular nephrons in their plunge into the medulla.

When Kidneys Stop Working: Dialysis and Kidney Transplant

Chronic kidney failure can result from infection, diabetes mellitus, inherited renal degeneration, and damage by the body's own immune system. If the failure becomes complete, two treatment options exist: dialysis and kidney transplant.

Dialysis is a method of removing toxic wastes from the blood. This may be done in two ways. In hemodialysis, the original procedure, the patient receives surgical connections (usually located on the lower arm) that allow arterial blood to pass from the patient's circulation through the dialysis unit and back into a vein. Inside the dialysis unit (Figure 19-A, *top*), the blood passes through hollow fibers of semipermeable membrane that allow small solutes to pass down their concentration gradients from the blood into an isotonic dialyzing fluid in the machine. In this procedure the patient is connected to the dialysis unit three times a week. The patient must carefully manage salt and water intake because the dialysis machine, unlike the kidney, does not regulate blood volume and total body Na^+.

In the newer approach, the peritoneal membrane lining the abdominal cavity is used as the dialysis membrane. For this procedure, called continuous ambulatory peritoneal dialysis (CAPD), the patient receives an abdominal catheter that allows the abdominal cavity to be filled with dialysis fluid (see Figure 19-A, *bottom*). The dialysis fluid is changed several times daily. This procedure makes work and travel easier. Diet and fluid restrictions are eased, but great care is needed to prevent peritoneal infections. In some cases a combination of hemodialysis and peritoneal dialysis is used.

Dialysis is a short-term solution to renal failure. Not only is dialysis expensive, but it severely restricts the activity of an individual. Because a single healthy kidney can meet the excretory needs of the body, a more permanent solution is the transplantation of a kidney from a healthy donor.

A transplanted kidney from a living donor genetically related to the recipient has the best chance of success, but donors are frequently unrelated victims with healthy kidneys, who have died in accidents. A successful transplant means that the recipient is cured of renal disease, but there are some drawbacks. Even organs from closely related donors may be rejected by the recipient's immune system (the immune basis of transplant rejection is described in Chapter 26). To reduce chances of rejection, the recipient must be treated with immunosuppressive drugs, and such drugs not only increase susceptibility to infections, but may be directly toxic to the liver and the bone marrow.

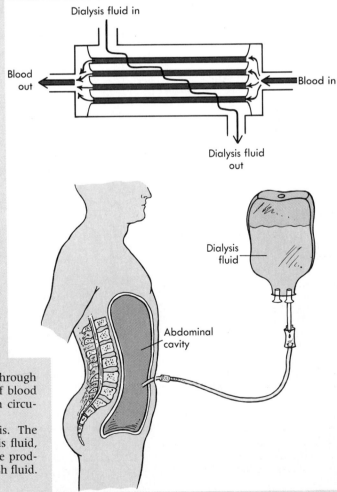

FIGURE 19-A

Top   Interior of a hemodialysis unit. Blood passes through hollow fibers of dialysis membrane. Small solutes of blood are exchanged into the isotonic dialysis fluid, which circulates around the fibers.

Bottom   Continuous ambulatory peritoneal dialysis. The abdominal cavity is being filled with isotonic dialysis fluid, which will accumulate small solutes, including waste products, for several hours before being replaced with fresh fluid.

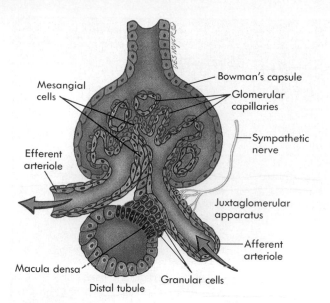

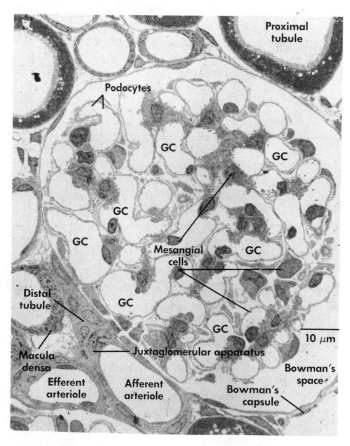

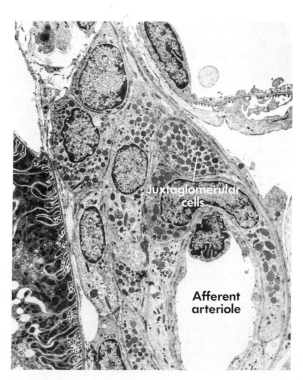

FIGURE 19-8
A   A photomicrograph of a section through a glomerulus, showing the juxtaglomerular apparatus *(lower left)* adjacent to the distal tubule and afferent and efferent arterioles.
B   An enlarged view of the juxtaglomerular cells.

glomerulus. At this junction is a structure called the **juxtaglomerular apparatus** (see Figure 19-3, *A*). The juxtaglomerular apparatus consists of a group of specialized tubular epithelial cells, the **macula densa** and the **granular,** or **juxtaglomerular cells** of the adjacent arteriolar wall (Figures 19-7 and 19-8). The juxtaglomerular apparatus probably has two major functions. First, it is known to be responsible for secretion into the blood of **renin.** Renin is an enzyme that initiates a regulatory cascade resulting ultimately in secretion of the steroid hormone **aldosterone** from the adrenal cortex. Aldosterone is one of two major hormones involved in control of total body Na$^+$ content. The role of aldosterone and other hormones in regulating the volume and composition of extracellular fluid will be discussed in detail in Chapter 20. The juxtaglomerular apparatus is believed also to be involved in the kidney's ability to maintain a nearly constant rate of blood perfusion and glomerular filtration in spite of changes in renal arterial pressure, and to match the glomerular filtration rate and the rate of fluid absorption from the nephron. This process, called **tubuloglomerular feedback,** will be discussed later in this chapter.

MECHANISMS OF URINE FORMATION AND MODIFICATION

- *Define the following terms.*
  Glomerular filtration rate (GFR)
  Filtered fraction (FF)
  Tubular maximum (T_m)
- *What is the standing gradient hypothesis? In this hypothesis, what is the role of lateral spaces between the cells of the proximal tubular epithelium?*
- *How does the protein content of the urine compare with that of the plasma? Why is there a difference?*
- *What tubular processes tend to make urine more dilute than plasma? More concentrated? What hormone regulates urine osmolarity?*
- *Describe the force that removes water from the urine in the descending limb of the loop of Henle.*
- *Describe the contribution of the vasa recta to the kidney's urinary concentrating mechanism.*

Ultrafiltration and the Composition of the Glomerular Filtrate

Bowman's capsule (Figure 19-9, *A*) and the glomerular capillaries are structurally adapted for a high rate of capillary filtration. The surface area of glomerular capillaries is extensive (Figure 19-9, *B*). The glomerular capillary wall (Figure 19-10) consists of three layers. Of these, the **fenestrated capillary endothelial layer** is a very coarse screen that probably only keeps blood cells from blocking the glomerular filter (Figure 19-11). The **basement membrane,** or **basal lamina,** is an intermediate filter that keeps back only larger plasma proteins, in part

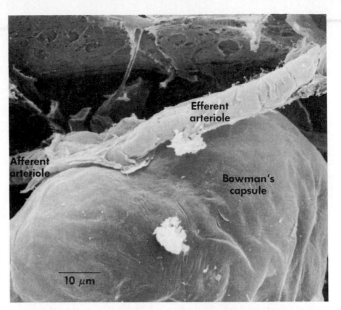

A

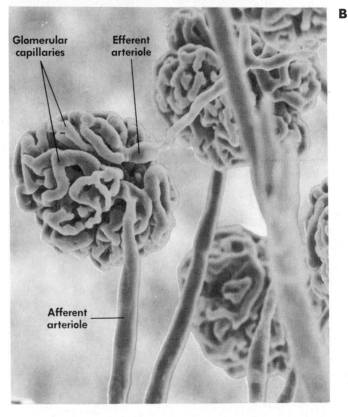

B

FIGURE 19-9

A Scanning electron micrograph of Bowman's capsule showing afferent and efferent arterioles.
B Scanning electron micrograph of casts of glomerular capillaries with their afferent and efferent arterioles, made by perfusing the renal circulation with resin. After the resin hardened, the kidney tissue was digested away.

because both plasma proteins and the basal lamina are negatively charged. The most selective filtration is believed to take place at the diaphragms of the **slit pores,** which are formed by footlike projections of supporting cells (**podocytes,** see Figure 19-10).

The glomerular filter is freely permeable to water, mineral ions such as Na^+, K^+, Ca^{++}, and Cl^-, and to small organic molecules such as glucose. A substantial fraction of small plasma proteins (those up to molecular weights of about 20,000 daltons) is filtered, but plasma proteins with molecular weights above 40,000 daltons are filtered only in trace amounts, so urine is normally almost protein-free.

Most of the small amount of plasma protein filtered is reabsorbed by specific transport pathways, but there is still a small but significant protein loss in urine, about 100 mg/d. Among these proteins are some that have medical significance, such as chorionic gonadotropin, the hormone detected by at-home pregnancy tests (see Chapter 25). In healthy individuals, urinary protein loss is trivial in comparison with protein intake. In some kidney diseases, however, the normal recovery of filtered proteins is impeded. In other kidney diseases, damage to the glomeruli increases the amount of plasma protein filtered. In both situations not all of the filtered protein can be recovered, and abnormally large quantities of protein appear in the final urine (**proteinuria**). In severe kidney disease the loss of protein can be so large as to cause muscle wasting unless dietary protein is supplemented. In the aftermath of strenuous exercise, hemoglobin and myoglobin released into the plasma from stressed muscle may also appear in urine (**hemoglobinuria**), giving it a brownish tinge.

The Glomerular Filtration Rate and Its Determinants

The **glomerular filtration rate (GFR)** is the net rate of formation of filtrate by the two kidneys. The GFR is equal to the **renal plasma flow rate (RPF)**, the rate of plasma flow through the renal arteries, multiplied by the fraction of this plasma flow that is filtered (the **filtered fraction, FF**), so

$$GFR = (RPF)(FF).$$

The kidneys receive a large part of the total cardiac output: for a typical cardiac output of 5.6 L/min, the kidneys might receive about 1.2 L/min, or about 20%. This rate of perfusion greatly exceeds what would be needed to satisfy the metabolic needs of the kidneys. Almost all of the renal blood flow passes through glomerular capillaries. Thus a much larger volume of capillary filtrate is formed in the glomeruli than in the rest of the body's capillary beds put together. Ordinarily, FF is about 20% of RPF. The kidneys receive about 1580 L/d of whole blood; the RPF is calculated by subtracting from the renal blood flow the fraction of the blood volume that is cells (the hematocrit). Assuming a hematocrit of 45%, the RPF is about 870 L/d. If the filtered fraction is 20%, the GFR is 125 ml/min or 174 L/d. Thus in a day the kidneys make a volume of filtrate that is about 60 times the total plasma volume of about 3 L; in other words, a volume of filtrate equal to the plasma volume is made about every 30 minutes.

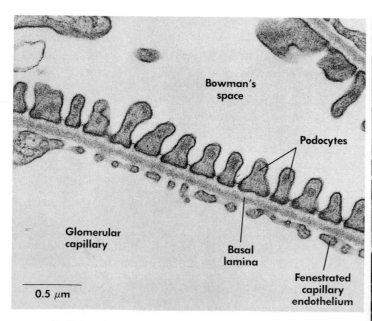

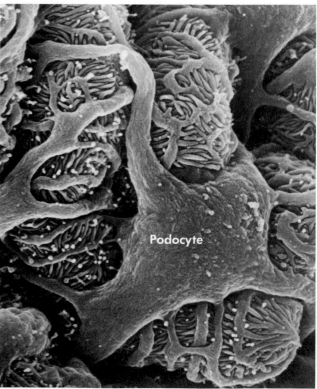

FIGURE 19-10

A Transmission electron micrograph of a portion of glomerular capillary wall showing structures involved in glomerular filtration.
B A scanning electron micrograph of a podocyte viewed from Bowman's space.

The filtered fraction is determined by three factors: (1) the net driving force for fluid movement, or **filtration pressure**, across the glomerular capillary walls, (2) the permeability of the renal filter to fluid, and (3) the total surface area available for filtration. Each of these factors will be considered in turn in following paragraphs.

The filtration pressure is the sum of hydrostatic and osmotic forces, just as in formation of interstitial fluid by capillary filtration in other systemic capillaries (see Chapter 14). The mean hydrostatic pressure in the glomerular capillaries is 50 mm Hg (Figure 19-12). Normally, the opposing hydrostatic pressure within Bowman's capsule is 10 mm Hg or

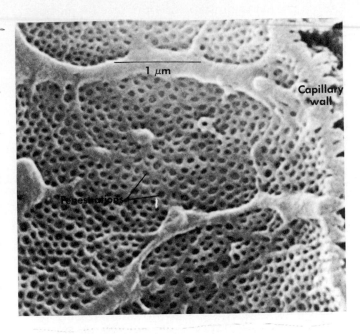

FIGURE 19-11
The surface of a glomerular capillary at much higher magnification. The first step in urine formation is filtration of water and solutes through numerous small openings (fenestrations).

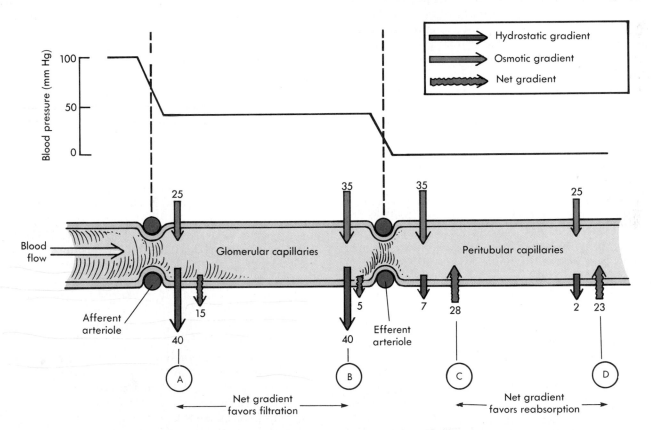

FIGURE 19-12
The graph shows the capillary hydrostatic pressure in the kidney. Note that because the afferent and efferent arterioles are connected in series, the pressure in the glomerular capillaries is 45 mm Hg, much higher than in the systemic capillaries. The hydrostatic pressure in the peritubular capillaries is about 10 mm Hg, only slightly higher than the venous pressure. The diagram shows the balance of hydrostatic and osmotic forces in the two renal capillary beds. The purple arrows (and the associated values) show the net hydrostatic pressure, the gray arrows show the net osmotic pressure, and the red arrows indicate the net force for filtration or reabsorption at various points along the nephron (**A** to **D**). Note that fluid is always filtered in the glomerular capillaries and reabsorbed in the peritubular capillaries.

less, so the net hydrostatic driving force for glomerular filtration is about 40 mm Hg. Recall from Chapter 6 that only impermeant solutes can sustain an effective osmotic gradient across a membrane. The unfiltered plasma proteins and the cations that remain with them to balance their negative charges generate an osmotic force that opposes filtration (indicated by open arrow in Figure 19-12, *A*). This force amounts to about 25 mm Hg (see also capillary filtration in Chapter 14). The net filtration force is the difference between the net hydrostatic force and the net osmotic force, or (in approximate figures):

40 mm Hg hydrostatic −
25 mm Hg protein osmotic =
15 mm Hg filtration

The relationship given above is for the head of a glomerular capillary. The filtration force decreases as blood flows along the length of the glomerular capillaries. This results mainly from an increase in the osmotic gradient. The hydrostatic pressure decrease is slight, because glomerular capillaries are short. The rapid loss of fluid by filtration into Bowman's capsule causes the osmotic gradient due to unfiltered plasma protein to increase significantly as blood flows along the length of the glomerular capillary, reaching a value of about 35 mm Hg near the end of the capillary bed.

If the balance of hydrostatic and osmotic forces changes, the GFR will change. GFR increases (1) if the mean glomerular capillary pressure rises as a result of either dilation of the afferent arterioles or constriction of the efferent arterioles, or (2) if the concentration of plasma proteins falls, because this reduces the force favoring reabsorption. GFR falls if (1) the hydrostatic pressure in Bowman's capsule rises (for example, if the ureters are occluded) or (2) plasma proteins escape into Bowman's capsule, because protein in Bowman's capsule makes the net osmotic force across the glomerular wall smaller.

Per unit of surface area, glomerular capillaries are 10 to 100 times more permeable than capillaries in most parts of the circulation, so the rate of formation of renal filtrate is higher than the rate of lymph formation throughout the rest of the body. The permeability of the renal filter probably does not change except in cases of renal disease. However, the surface area available for filtration is regulated by **mesangial cells** (see Figures 19-7 and 19-8), which surround the glomerular capillaries and affect the amount of glomerular surface that is available for filtration. The mesangial cells have contractile elements like smooth muscle. When they contract, they decrease the surface available for filtration. Contraction of mesangial cells is believed to be controlled by hormones and paracrine agents; this is one mechanism by which GFR can be regulated in response to changes in total plasma volume. The second way in which hormones can control GFR is by affecting the tone of the afferent arterioles; con-

striction of the afferent arterioles decreases the filtration pressure in glomerular capillaries and thus decreases GFR, whereas dilation of afferent arterioles increases GFR.

In summary, the first step in urine production is formation of a large volume of glomerular filtrate in the renal corpuscles. The glomerular filtrate has all of the characteristics of plasma except that the concentration of protein is much smaller. The rest of the nephron modifies the composition of this primary filtrate by reabsorbing some filtered solute molecules, and secreting others into the tubular fluid. As a consequence of net reabsorption of solute, the volume of fluid in the nephrons decreases, until only about 1% or 2% of the original filtrate volume reaches the ureters as final urine. The adaptive significance of formation of urine by filtration in the glomerulus followed by reabsorption and secretion in the later parts of the nephron is that the kidney tends to excrete every solute for which there are no specific tubular reabsorptive pathways. This "fail-safe" excretionary mechanism is an evolutionary advance that makes it possible for the kidney to excrete any soluble chemical that might enter the body, including drugs and bacterial toxins.

Functions of the Proximal Tubule: Reabsorption and Secretion

The proximal tubule carries out two basic activities: reabsorption of some filtered solutes and secretion of other solutes. Of these two processes, the dominant one in terms of rates of particle movement is reabsorption. In the proximal tubule, net solute reabsorption is closely followed by osmotic movement of water, so that of the 174 L/d of glomerular filtrate only about 60 L/d enters the loops of Henle from the proximal tubules.

The loss of many of the plasma constituents that are freely filtered across the glomerular capillary walls, such as glucose, amino acids and vitamins, must be avoided. The proximal tubule actively reabsorbs these substances. The energy for active reabsorption of many such substances (see Table 19-1), is provided by the Na^+ concentration gradient across the apical membrane of the renal tubule cells. In this process, called Na^+ gradient-coupled cotransport, the energy for active transport comes ultimately from the Na^+-K^+ pump located in the basolateral membrane of proximal tubular cells. This pump uses ATP to maintain the Na^+ gradient across the apical membrane.

Although many solutes are transported by the proximal tubule, the rates of reabsorption of Na^+ and Cl^- are higher than the rate of transport of any other solute. As in other epithelia, the first step of the mechanism of Na^+ transport (Figure 19-13) is Na^+ entry across the apical membrane. In the proximal tubule, much of the Na^+ that enters the cells is exchanged for H^+. The Na^+ taken up into the cells is actively transported from cytoplasm to interstitial

KIDNEY AND BODY FLUID HOMEOSTASIS

Reabsorbed substances (coupled to Na$^+$ gradient)

Sugars

D-Glucose
D-Galactose

Amino acids

Neutral (for example, L-alanine, L-glutamine)
Acidic (for example, L-glutamate, L-aspartate)
Basic (for example, L-arginine, L-ornithine)

Inorganic ions

Phosphate
Sulfate

Metabolites

L-Lactate
Succinate
Citrate

Actively secreted substances

Substances generated within the body

H$^+$ ion (coupled to Na$^+$ absorption)
Hydroxybenzoates
Hippurates (for example, paraaminohippuric acid)
Neurotransmitters (for example, dopamine, acetylcholine, epinephrine)
Bile pigments (responsible for yellow color of urine)
Uric acid

Drugs and toxins

Antibiotics (for example, penicillin G, cephalothin)
Atropine

Morphine
Saccharin
Paraquat (an herbicide)

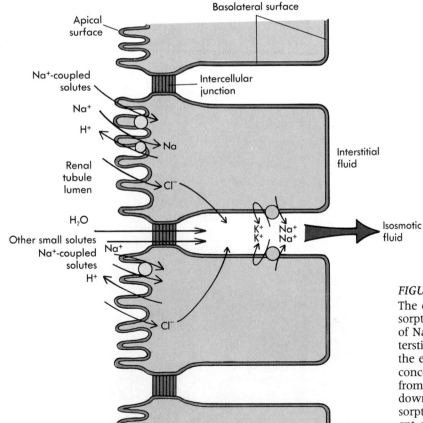

FIGURE 19-13

The cellular mechanisms of Na$^+$ and water reabsorption in the proximal tubule. Active transport of Na$^+$ from the renal epithelial cells into the interstitial fluid occurs on the basolateral surface of the epithelial cells, reducing the intracellular Na$^+$ concentration. Sodium enters the epithelial cells from the renal tubule by facilitated diffusion down its electrochemical gradient, driving the absorption of numerous other solutes by Na$^+$ gradient-driven cotransport. Electroneutrality is maintained in part by secretion of H$^+$ and in part by facilitated diffusion of Cl$^-$. The result is salt reabsorption, creating an osmotic gradient for water flow. Small, nonpolar solutes such as urea are reabsorbed passively as their concentrations in the tubular fluid rise above plasma concentrations.

fluid across the basolateral membrane by the Na^+-K^+ pump. Reabsorption of Na^+ is closely followed by movement of Cl^- and other anions which maintain electroneutrality. At first, it was difficult to understand how the active solute reabsorption of the proximal tubule drives reabsorption of water, because there is never a substantial osmotic gradient between proximal tubular fluid and interstitial fluid. Two factors are involved. First, the cell membranes and intercellular junctions of the proximal tubule are so permeable to water that only a small osmotic driving force (about 1-10 mOsmol/L) would be adequate to account for the large rate of fluid movement. According to the **standing gradient hypothesis**, some of this osmotic driving force is concealed within the epithelium itself. According to this hypothesis (Figure 19-13), the transported solutes enter the long narrow lateral spaces between cells of the tubular wall, generating an osmotic gradient between the tubular fluid and the spaces. Relatively permeable tight junctions (see Chapter 4) between the cells of the tubular wall allow water to flow into the spaces. As a result of equilibration of water with actively transported solute within the lateral spaces, the reabsorbed fluid becomes isosmotic with the tubular fluid and with the interstitial fluid of the kidney by the time it exits the lateral spaces. In summary, both water and solute are reabsorbed in the proximal tubule without affecting the overall ratio of solute to water in the tubular fluid (Figure 19-14).

Carrier-mediated solute transport processes are similar in many respects to enzyme-catalyzed reactions. The binding of the transported molecule to the carrier involves the same principles as binding of substrate to an enzyme's active site. Most carrier-mediated transport pathways specifically transport a small family of solutes with similar chemical structures. For example, five or six separate transport systems exist for corresponding families of the 20 common amino acids. Substances using the same carrier may compete with or block each other. The rate of reabsorption of a given substance from the tubular fluid depends on its concentration, according to the relationship for enzyme-catalyzed reactions shown in Figure 19-15, A. There is a finite number of carriers for any particular family of substances along the proximal tubule. When all carriers are occupied, further increases in the concentration of the solute in the tubular fluid do not result in increases in transport rate. The maximum rate of reabsorption from tubular fluid for any particular substance is called its **transport maximum (T_m)**. The units of T_m are amounts reabsorbed by the two kidneys per unit time.

The importance of the T_m can be evaluated by comparing it to the rate of delivery of solute to the tubules by the glomerulus. For any substance X, the **filtered load** L_x or the amount filtered per unit of time, is computed as:

$$L_x = P_x \times GFR$$

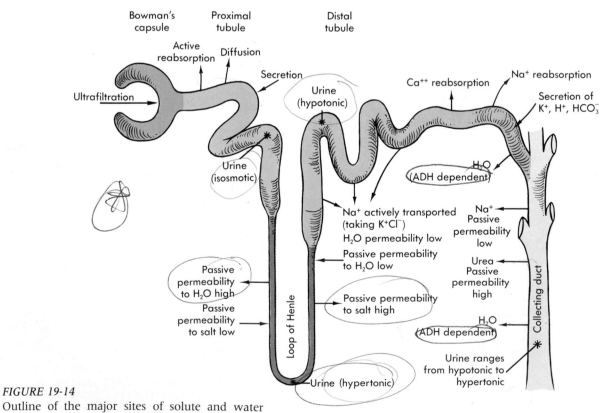

FIGURE 19-14
Outline of the major sites of solute and water movement across the nephron.

where P_x is the plasma concentration of the substance, usually expressed per deciliter as a percent. Clinically, P_x is often expressed either in units of g% or mg%. For example, the normal plasma glucose concentration is about 100 mg% (100 mg/100 ml). The filtered load of glucose is 1.00 g/L × 174 L/d = 174 g/d, or about 121 mg/min.

Curves *1*, *2*, and *3* in Figure 19-15 show how the rates of glucose filtered per unit time, that reabsorbed, and excreted glucose vary when plasma glucose concentration increases. The transport maximum for glucose is 375 mg/min. Because the normal filtered load of glucose is well below the T_m, glucose reabsorption is normally complete (see Figure 19-15, *curve 2*), and only trace amounts of glucose remain in the tubular fluid. However, the filtered load of glucose increases with increasing plasma concentration (Figure 19-15, *curve 1*). If the plasma glucose concentration increases above 300 mg%, the filtered load exceeds the transport maximum of 375 mg/min (see Figure 19-15, *point X*). When this occurs, glucose begins to be excreted in the final urine (see Figure 19-15, *curve 3*), and the amount of glucose excreted increases as the plasma glucose concentration increases. Such abnormally high levels of plasma glucose are characteristic of diseases that affect carbohydrate metabolism such as diabetes mellitus. High levels of plasma glucose may also occur briefly in healthy individuals under conditions of stress or after a meal high in carbohydrates.

Solutes that are not reabsorbed from tubular fluid by specific carrier-mediated processes may still be reabsorbed passively, if the tubular wall is permeable to them. Water reabsorption and active solute reabsorption from proximal tubule fluid tend to increase the concentrations of those solutes that are not actively reabsorbed. Some of these solutes are able to diffuse across the tubule wall down their developing concentration gradients.

In addition to its reabsorptive processes, the proximal tubule is the site at which secretion of numerous substances occurs (see Table 19-1 and Figure 19-14). Some secreted substances are generated by normal body metabolism; other secreted substances are drugs or toxins. The tubular cells actively transport these substances from the unfiltered blood within peritubular capillaries. Active secretion by the proximal tubule is particularly important in the case of metabolites or toxins that are able to bind to plasma proteins. Excretion of these sub-

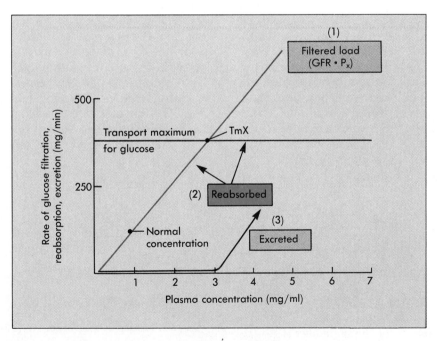

FIGURE 19-15

Curve 1 The filtered load of glucose as a function of plasma glucose concentration. The filtered load varies in direct proportion to the plasma concentration.
Curve 2 The rate of glucose reabsorption as a function of plasma glucose concentration. Reabsorption equals the filtration rate until the T_m is reached. At glucose filtration rates above the T_m, glucose reabsorption does not increase further.
Curve 3 The rate of glucose excretion as a function of plasma glucose concentration. The rate of glucose excretion equals the filtration rate (**curve 1**) minus the rate of reabsorption (**curve 2**). The rate remains negligible until the T_m is reached. The vertical axis of this plot is in units of mg glucose/min filtered (**curve 1**), reabsorbed (**curve 2**), and excreted (**curve 3**).

stances would be very slow if it depended on filtration alone. Secretion greatly reduces the lifetime of some drugs in the body, necessitating larger and more frequent doses than would otherwise be necessary. Penicillin is an example of such a rapidly secreted drug.

Secretion via specific transport pathways involves the same general principles as does absorption. The filtered load of a secreted substance increases linearly with its plasma concentration (Figure 19-16). The rate of secretion increases rapidly until the T_m for secretion is reached, and then the rate remains constant. The rate of excretion of such solutes is the sum of the filtered load and the secretion rate. Once the plasma concentration of the substance exceeds the T_m for secretion, the contribution of secretion is maximal, and further increases in the plasma concentration of the substance increase only the filtration rate.

Small proteins (of molecular weights of 20,000 daltons or less) are able to pass through the glomerular filter in significant amounts. However, as noted previously, most of the filtered proteins are reabsorbed directly, or digested by brush border enzymes (with the resulting amino acids being reabsorbed), in the proximal tubule. Direct reabsorption occurs by endocytosis (see Chapter 4), and subsequently these proteins are metabolized in the proximal tubule cells. Almost all protein-type hormones (see Chapter 5) are of a size that allows them to be filtered. Filtration and subsequent metabolism of protein hormones by the kidney significantly shortens their lifetime in the blood. A percentage of the filtered molecules of some protein hormones escape reabsorption and can be detected in the urine, making it possible to assess a person's hormonal status from an analysis of the urine. An example of such a test is the at home pregnancy test, which detects human chorionic gonadotropin, a specific protein hormone produced by an embryo after it implants in the uterus (discussed in Chapter 25).

Functions of the Distal Tubule: Formation of Dilute Urine

The proximal tubule greatly reduces the volume of the tubular fluid, but it has no effect on the overall ratio of water to solute, so tubular fluid entering the loop of Henle is isosmotic to plasma. The thick ascending limb of the loop of Henle and the segments of the distal tubule play an important role in producing urine with a lower osmotic pressure than that of plasma. Thus the thick ascending loop and distal tubule cause reabsorption of solute without proportionate reabsorption of water (see Figure 19-14). This is possible because these segments, unlike the proximal tubule, are almost impermeable to water (Table 19-2).

The major transport process of the thick ascending loop and the initial part of the distal tubule (convoluted region) is reabsorption of Na^+ and Cl^-. In the first step of salt reabsorption, an Na^+ ion and a K^+ ion are taken across the apical membrane of the tubule cells in company with two Cl^- ions (Figure 19-17). In this process, called **Na^+, K^+, $2Cl^-$ cotransport,** the Na^+ gradient provides the energy for uptake of both the K^+ and the Cl^- ions. Subsequently, the Na^+ that enters the cell by this route is actively pumped across the basolateral membrane, the Cl^- diffuses passively into the interstitial fluid, and the K^+ transported in at both membranes replaces K^+ that is constantly leaking out of the cells.

The middle part of the distal tubule (the **connecting tubule**) is believed to be responsible for reabsorption of filtered Ca^{++}. This segment of the nephron plays a key role in homeostasis of plasma Ca^{++}, described in the next chapter. The dominant

FIGURE 19-16
Curve 1 The filtered load of a filtered and secreted substance as a function of concentration of the substance in plasma.
Curve 2 The rate of tubular secretion as a function of plasma concentration.
Curve 3 The rate of excretion of the substance as a function of plasma concentration. The rate of excretion is the sum of the rate of filtration (curve 1) and the rate of secretion (curve 2). The vertical axis of this plot is in units of mg substance/min filtered (**curve 1**), secreted (**curve 2**), and excreted (**curve 3**).

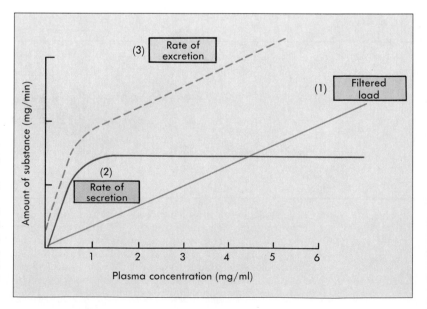

KIDNEY AND BODY FLUID HOMEOSTASIS

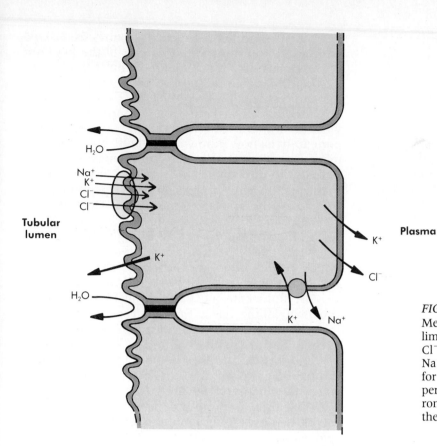

H₂O

Na⁺
K⁺
Cl⁻
Cl⁻

**Tubular
lumen**

K⁺

H₂O

K⁺   Na⁺

Plasma

K⁺

Cl⁻

FIGURE 19-17
Mechanisms of salt transport in the thick ascending limb and distal convoluted tubule. Entry of Na⁺, Cl⁻, and K⁺ are coupled, and all are driven by the Na⁺ gradient. Na⁺ and Cl⁻ leave the tubular cells for the interstitial fluid. The tight junctions are not permeable to water in these segments of the nephron, so relatively little water movement occurs and the tubular fluid becomes hypotonic.

TABLE 19-2   *Transport and Permeability Properties of Nephron Segments Related to Urine Concentration and Dilution*

Segment	Active salt transport	Permeability			Osmolarity	
		H₂O	NaCl	Urea	Beginning	End
Proximal tubule	+	+ + + +	+ + +	+ + +	Iso	Iso
Loop of Henle						
Descending limb	0	+ + + +	0	+	Iso	Hyper
Thin ascending limb	0	0	+ + +	+	Hyper	Hyper
Thick ascending limb	+ + + +	0	0	0	Hyper	Hypo
Distal tubule						
Convoluted tubule	+	0	0	0	Hypo	Hypo
Collecting tubule	+ to >+ + +*	0 to >+ + +*	0	0	Hypo	Hypo/iso*
Collecting duct	+	0 to + + +*	0	+ + +	Hypo/iso*	Hypo/hyper

*Varies in response to hormonal regulation.

transport process of the final part of the distal tubule (the collecting region) is reabsorption of NaCl, but it is also able to secrete H⁺ and to both reabsorb and secrete K⁺.

Depending on the concentrations of hormones that regulate salt reabsorption and tubular water permeability, solute reabsorption across the wall of the distal tubule may not be accompanied by proportionate movement of water. Therefore, the tubular fluid leaving the distal tubule may be more dilute than plasma. If this fluid were to pass relatively unaltered through the collecting duct, a dilute urine would result. Dilute urine is produced as a homeostatic response to a water load that dilutes the extra-

The Kidney

cellular fluid. The hormonal mechanisms that cause the kidney to produce dilute urine under such conditions are explained in Chapter 20.

The Loops of Henle and the Collecting Duct: Formation of Concentrated Urine

Formation of a final urine more concentrated than plasma allows the body to rid itself of toxic metabolic end products and excess ions with a minimum loss of water. The kidneys concentrate tubular fluid by maintaining the medullary interstitial fluid at a concentration typically as much as 4 to 5 times that of the plasma (that is, about 1400 mOsM versus the 300 mOsM of plasma). The osmotic gradient between the concentrated medullary interstitial fluid and the dilute tubular fluid entering the collecting duct provides a driving force for water to move from the tubular fluid in the collecting duct to the medullary interstitial fluid.

Physiologists began to understand the urine concentrating mechanism when they realized that only those animals with loops of Henle that penetrate the renal medulla can form concentrated urine, and that maximum urine tonicity depends on the overall length of these loops. Five features of the anatomy of the kidney are important for urine concentration: (1) the loop of Henle plunges deep into the renal medulla and then bends upward sharply so that the ascending limb stays near the descending limb, (2) the descending and ascending limbs have different water permeabilities, (3) NaCl transport occurs only in the ascending limb, (4) the capillaries of the vasa recta form loops that accompany the loops of Henle, and finally (5) the collecting duct passes from the cortex through the medulla, passing the loops of Henle in the medulla before it drains into the renal pelvis.

The permeability properties and solute transport activities (and thus energy requirements) of the loop of Henle and the collecting duct (see Table 19-2) are reflected in the structure of their cells. The descending limb has relatively flat cells with few mitochondria. Consequently, the descending tubule itself is relatively thin. The cells of the descending limb are not believed to transport Na$^+$. The changes in tubular fluid composition that occur as the tubular fluid passes through the descending limb are the result of passive exchange with the interstitial fluid (Figure 19-18). The descending limb is highly permeable to water but not permeable to salt. The thin part of the ascending limb of the loop also has relatively flat cells, and like the descending limb, this portion of the tubule does not actively transport ions. In contrast to the descending limb, the thin ascending limb has a low water permeability but a high salt permeability (see Figure 19-14).

Unlike the other parts of the loop, the cells of the thick ascending limb have many mitochondria and other intracellular structural features associated with active ion transport. This part of the nephron actively transports NaCl from the tubular fluid to interstitial fluid, but its passive permeability to both salt and water is low (see Table 19-2 and Figure 19-14). Historically there was controversy about whether Na$^+$ or Cl$^-$ was the ion actively transported. It is now known that both Na$^+$ and Cl$^-$ move from the tubular fluid in the thick ascending limb to the interstitial fluid against electrochemical gradients. In this process, ATP is spent to move Na$^+$ across the basolateral membrane. The uptake of Na$^+$, Cl$^-$ and K$^+$ across the apical membrane (Na$^+$, K$^+$, 2Cl$^-$ cotransport) is driven by the Na$^+$ electrochemical gradient across the apical membrane, which results from the action of the basolateral Na$^+$–K$^+$ pump.

The fluid entering the loop of Henle from the proximal tubule has a NaCl concentration and osmolarity similar to that of plasma, and, as in plasma, NaCl is its most abundant solute. As tubular fluid passes into the descending limb, it passes through regions of the kidney with increasingly concentrated interstitial fluid. The wall of the descending limb is permeable to water, so as water leaves the descending limb, the tubular fluid comes into osmotic equilibrium with more and more concentrated interstitial fluid. Osmotic withdrawal of water from the descending limb concentrates the solutes in the tubular fluid (see Figure 19-14), whereas the water that is removed joins the interstitial fluid, enters the outer medullary level of the vasa recta and ultimately is carried away to the renal vein.

As the tubular fluid rounds the bend of the loop and moves along the thin ascending limb, it passes through regions with more and more dilute interstitial fluid. The wall of the thin ascending limb is permeable to NaCl, but not permeable to water, so as the tubular fluid ascends, osmotic equilibrium is reached by movement of NaCl out of the tubular fluid. In this way, much of the Na$^+$ that was brought into the medulla by the descending limb of the loop of Henle is unloaded into interstitial fluid in the innermost parts of the medulla.

The thick ascending limb provides the energy for urine concentration. In the thick ascending limb, some of the NaCl that escaped unloading in the medulla is actively transported from the tubular fluid. In this part of the tubule water cannot follow, so a concentration gradient develops between the tubular fluid and the interstitial fluid. As the tubular fluid leaves the thick ascending limb and passes through the distal tubule, more NaCl is actively recovered. Active transport of NaCl to the interstitial fluid across the thick ascending limb and distal tubule makes those solutes that remain behind more important in determining the osmotic pressure of the tubular fluid that reaches the collecting duct.

KIDNEY AND BODY FLUID HOMEOSTASIS

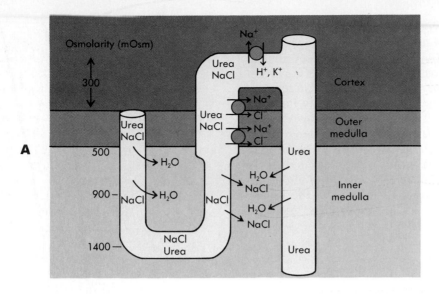

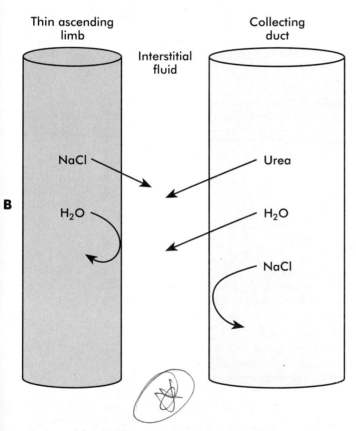

FIGURE 19-18
Mechanism of formation of concentrated urine according to the two-solute hypothesis.

A Overall view of the loop of Henle, distal tubule and collecting duct: The osmolarity of the interstitial fluid at different levels of the medulla is shown on the scale at the left. The tubular fluid leaving the proximal tubule is isotonic. As the tubular fluid travels through the descending limb of the loop of Henle, water leaves the descending limb, drawn by the increasing osmotic pressure of interstitial fluid in the medulla. As a result, the tubular fluid in the descending limb becomes progressively more concentrated. As the tubular fluid passes through the thin ascending limb, NaCl, but not water, diffuses out, so that the osmotic pressure of interstitial fluid decreases. In the thick ascending limb, more salt is removed by active reabsorption. The tubular fluid entering the distal tubule is more dilute than plasma with respect to NaCl, while urea has been concentrated by the reabsorption of water. Urea and water diffuse down their concentration gradients as tubular fluid passes through the collecting duct. The remaining solutes in the tubular fluid are concentrated further by the water reabsorption, and a urine as concentrated as the interstitial fluid at the innermost part of the medulla may be formed if ADH levels are high. If ADH levels are low, a final urine similar to the dilute urine in the distal tubule is excreted.

B The two driving forces that generate a high solute concentration in the medullary interstitial fluid are the NaCl gradient between ISF and thin ascending limb, and the urea gradient between collecting duct and ISF. Water cannot leave the thin ascending limb in response to the osmotic gradient, but can be reabsorbed from the collecting duct in the presence of antidiuretic hormone.

One such solute is urea. By the time the tubular fluid enters the collecting duct, removal of much of the remaining NaCl has made it hypotonic, and urea has become one of its major solutes.

As the tubular fluid descends the collecting duct, it passes once again through regions of increasing interstitial fluid osmotic pressure. The epithelium of the collecting duct actively absorbs NaCl at a low rate, but its passive permeability to NaCl is low, so little NaCl returns to the interstitial fluid from collecting duct urine. The key element in regulation of urine concentration is the permeability of the collecting duct epithelium to water, because it determines the extent to which water can follow the osmotic gradient from tubular fluid in the collecting duct to the medullary interstitial fluid. The water permeability of the final segment of the distal tubule and the collecting duct is regulated by **antidiuretic hormone (ADH),** a peptide hormone released by the posterior pituitary (see Chapter 5). ADH increases the water permeability by opening **water channels in the membranes** of the epithelial cells through which water, but not solute, can pass (Figure 19-19). When ADH is present, water can equilibrate across the epithelial cells of the final distal tubule and collecting duct, and the descending tubular fluid becomes concentrated as water is reabsorbed (Figure 19-20). Thus in the presence of high concentrations of ADH the final urine leaves the collecting duct greatly reduced in volume but almost as concentrated as the extracellular fluid in the innermost, highly concentrated regions of the medulla (see Figure 19-14).

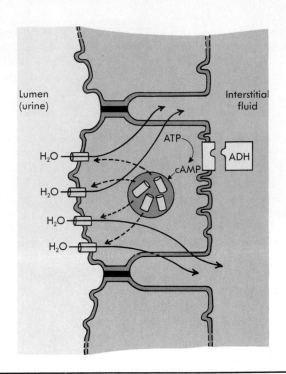

FIGURE 19-19

Current hypothesis explaining the action of ADH on water permeability of the collecting duct. The binding of hormone to receptor increases intracellular levels of cAMP, causing preformed water pores stored in the endoplasmic reticulum to be incorporated in the apical cell membrane.

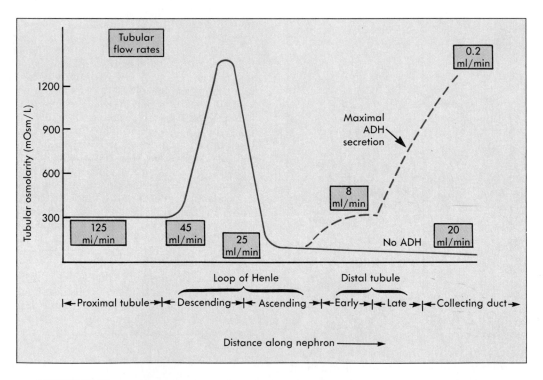

FIGURE 19-20

An overview of the osmotic pressure and volume flow changes along the length of the nephron in the presence and absence of ADH. The solid line on the graph indicates the osmolarity of the tubular fluid at various points along the nephron. The tubular fluid is initially isotonic, increases in osmolarity in the descending loop of Henle, and decreases in osmolarity in the ascending loop of Henle. The tubular fluid enters the distal tubule hypoosmotic to plasma. It then can become either more hypoosmotic (if ADH is absent—*solid line*) or hypertonic (if ADH is present—*dashed curve*). The numbers in the boxes are the tubular flow rates; they indicate the degree of water reabsorption that has occurred. Specific transport properties of the different regions are summarized in Table 19-2.

KIDNEY AND BODY FLUID HOMEOSTASIS

How the kidney generates the high concentration of solute in the interstitial fluid of the medulla is still not fully understood. There is no doubt that the parallel, opposite flows of tubular fluid in the two limbs of the loop of Henle and in the collecting duct are essential. This arrangement has been called a **countercurrent multiplier.** As the name suggests, the parallel flow of descending and ascending tubular fluid multiplies the power of ion transport processes of the thick ascending limb and distal tubule to create a concentration gradient.

Historically, several attempts have been made to explain how the countercurrent multiplier generates the osmotic gradient between the outermost and innermost parts of the medulla. In the **classical hypothesis,** a high concentration of solute at the tip of the loop of Henle was maintained by active transport of ions out of the ascending loop and water diffusion out of the descending loop. In this hypothesis, the loop of Henle could be likened to a pair of escalators running in parallel, one going up and the other going down. If riders wanting to go to the top were transferred from the ascending escalator to the descending escalator at various points (countercurrent exchange), there would be a crowd of frustrated people repeatedly getting on the up escalator only to be forced to ride back down.

Unfortunately, when the permeability and transport properties of the loop were measured in pieces of tubule dissected from the kidney, the results (see Table 19-2) did not entirely support this hypothesis. A crucial point against the classical hypothesis is the fact that the descending loop is permeable to water but impermeable to ions. In the escalator analogy, this impermeability to ions would prevent ions that are ejected from the up escalator (the thick ascending limb) from getting on the down escalator (the descending limb).

Currently the most plausible hypothesis is the two-solute hypothesis (see Figure 19-18). This hypothesis builds on the finding that, along with NaCl, urea makes up a large fraction of the total solute of the medullary interstitial fluid. The high concentrations of NaCl and urea in the medullary interstitial fluid result because (1) NaCl is the major solute of tubular fluid in the thin ascending limb, and urea is a major solute in the tubular fluid of the medullary collecting duct; and (2) the thin ascending limb is more permeable to NaCl than to urea, and the collecting duct is more permeable to urea than to NaCl.

In summary, two driving forces are at work in the two-solute hypothesis (Figure 19-18, *B*): the NaCl gradient across the thin ascending limb and the urea gradient across the collecting duct. Both of these gradients are created by the active reabsorption of NaCl by the thick ascending limb. Both gradients drive solute into the medullary interstitial fluid, concentrating both NaCl and urea in the interstitial fluid. The high osmotic concentration of solute in the medulla provides the driving force for water recovery from the medullary collecting duct.

Role of the Vasa Recta in Urine Concentration

Only about 10% of the filtered water is still present in the tubular fluid entering the collecting ducts. When additional reabsorption of water occurs as the collecting ducts travel through the medulla, the amount that moves into the interstitial fluid of the medulla is not great enough to overwhelm the ability of the medullary structures to form a concentrated interstitial fluid. However, if the composition of plasma flowing through the medulla were identical to the composition of the plasma in other parts of the circulation, the medullary interstitial fluid would rapidly equilibrate with the more dilute plasma. This does not happen because the structure of the vasa recta forms a **countercurrent exchanger** that keeps the solute that has been concentrated in the medulla from escaping into the general circulation.

The vasa recta supplies the needs of the medullary cells for nutrients and gas exchange without washing away the the high concentration of solute formed in the medullary interstital fluid. It does this because its loop shape allows countercurrent exchange of solute and water between ascending and descending blood. This is a passive process driven by the medullary solute gradient. The vasa recta capillaries follow the hairpin course of the loops of Henle, and these capillaries are permeable to both NaCl and water (Figure 19-21). The plasma within the capillaries becomes as concentrated as the adjacent interstitial fluid as the vasa recta follows the descending loop, developing a very high osmolarity deep within the medulla. As the blood reascends from the medulla to the cortex, the extra solute is lost and water is absorbed. In the end, the plasma leaves the vasa recta with an osmotic pressure only slightly greater than it had when it entered; thus the vasa recta removes solute from the medulla at a low rate. However, the rate of solute removal by the vasa recta depends on the rate of flow in these capillaries, so blood flow in the vasa recta must be low to preserve the medullary osmotic gradient.

The hypertonicity of the medullary interstitial fluid imposes osmotic stress on the cells that are exposed to it. Red blood cells crenate briefly while they pass through the vasa recta. In individuals with sickle cell anemia (see Chapter 17), shrinkage of red blood cells in the vasa recta greatly increases sickling, a characteristic deformation of the red cells. Sickled red cells tend to block renal capillaries, and kidney damage from this blockage is one of the effects of the disease. Tubular cells protect them-

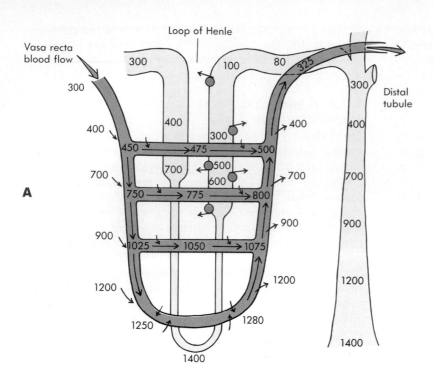

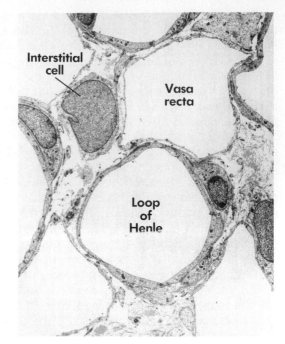

A

B

Loop of Henle

Vasa recta blood flow

Distal tubule

Interstitial cell

Vasa recta

Loop of Henle

FIGURE 19-21

A Overview of the concentrations of solute in the renal tubule and the vasa recta. A countercurrent system is one in which fluid flows in parallel tubes in opposite directions. Heat, solutes, and water diffuse from one tube to the other so that, in the absence of any energy-driven process, the fluid in both tubes has nearly the same composition. The renal tubule acts as a countercurrent multiplier system because salt is actively transported. The vasa recta accompany the loop of Henle and act as a countercurrent exchange system to prevent dissipation of the medullary concentration gradient by blood flow. The numbers give the osmolarity of the tubular fluid, plasma, and interstitial fluid in the renal tubule and vasa recta at several levels.

B An electron micrograph showing the relationship between a capillary of the vasa recta and the loop of Henle.

Endocrine Control of Urine Concentration

In the absence of ADH, the collecting portion of the distal tubule and the collecting duct are not water permeable, and water is not recovered from collecting duct fluid (see Figure 19-20). Recall that the initial filtered load of water was about 174 L/d. The 20% of the filtered load of water that remains unreabsorbed in the absence of ADH (10% in the distal tubule and 10% in the collecting duct) corresponds to a urine output of about 20 ml/min, or a little less than 30 L/d. This flow rate represents an absolute upper limit of the kidney's ability to handle a water load (though rarely does the urine flow exceed 18 L/d). A high urine flow rate might be maintained for a brief period by a normal individual after rapidly drinking several liters of water. Damage to the pituitary or hypothalamus can result in a disease called **diabetes insipidus,** in which little or no

ADH is secreted. A chronic high rate of production of dilute urine is characteristic of this disease.

Normal levels of ADH result in the production of about 1 L/d of slightly hyperosmotic urine. If water intake is restricted or if sweating is profuse, plasma levels of ADH rise. As a result, urine osmolarity may increase to 1200 to 1400 mOsM, four to five times plasma osmolarity (see Figure 19-20), and the urine flow rate can be as low as 0.2 ml/min, or 300 ml/d. The factors involved in control of ADH secretion are discussed in detail in the next chapter.

INTEGRATED FUNCTION OF THE KIDNEY

- *What are the units of renal clearance?*
- *What is the physiological basis for using PAH for measurement of RBF? For using inulin or creatinine for measurement of GFR?*
- *What are the effects of tubular secretion and tubular reabsorption on the clearance of a substance? How do the clearances of these substances compare with the GFR as measured by inulin clearance?*

TABLE 19-3   Renal Handling of Some Representative Substances

Substance	Plasma concentration	24-hour filtered load	Actively reabsorbed/ secreted	Typical percent of filtered load excreted	24-hour clearance (L)
Glucose	100 mg%	174 gm	R	0	0
Na⁺	142 mEq/L†	24.7 Eq	R	1	1.7
Urea	30 mg%	47.5 gm	‡	60	105
Phosphate	2 mEq/L	348 mEq	R	20	35
K⁺	5 mEq/L	518 mEq	R + S	10	17
H⁺	10^{-4} mEq/L	0.1 mEq*	S	800*	1394
Bicarbonate	24 mEq/L	4.2 Eq	‡	0	0
Cl⁻	103 mEq/L	18 Eq	R	1	1.7
Creatinine	1 mg%	1.4 m	Neither	100	171

*Essentially all H⁺ loss is the result of tubular secretion.
†Milliequivalents/L (mEq/L) is equal to the product of the concentration in millimoles/L times the valence of an ion.
‡See text.

Concept of Renal Clearance

The concentrations of filterable solutes that appear in the glomerular filtrate are the same as their plasma concentrations. However, excretion rates of different solutes may differ because either absorption or secretion of filtered substances may occur. The sequence of processes in which the kidney deals with a given substance is called **renal handling.** Handling of some representative substances is summarized in Table 19-3.

One way to quantify the handling of a particular substance by the kidney is to determine the **renal clearance** of that substance. The clearance is the number of liters of plasma that are completely cleared of the substance by the kidney per a unit of time (so the units are ml/min). The idea of completely clearing some liters of plasma of one of its solutes is hypothetical because very few substances are completely removed from the plasma during a single pass through the kidney. However, clearance measurements have very important clinical application.

The clearance (C_X) of a substance (X) is given by the equation:

$$C_X = (U_X)(V)/P_X$$

Where U_X and P_X are the concentrations of X in the urine and plasma respectively, and V is the rate of urine production. Because U_X and P_X have the same units, these units cancel and clearance has units of volume/time.

If the calculated clearance of a solute equals the GFR, the amount excreted equals the filtered load and the solute must not have undergone net secretion or reabsorption during its passage down the renal tubule (Figure 19-22). If the clearance is less than the GFR, net reabsorption has occurred be-

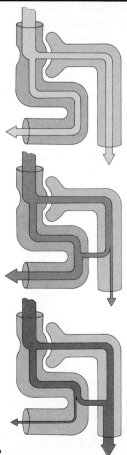

FIGURE 19-22

The three ways in which the kidney may handle materials appearing in the glomerular filtrate. Some substances *(aqua line)* are filtered into Bowman's capsule, but then are neither reabsorbed nor secreted. Others are filtered and reabsorbed into the peritubular capillaries *(purple line)*. Finally, some materials are both filtered and actively secreted from the peritubular capillaries into the renal tubule *(red line)*. The rate of loss of a substance in the urine can thus be equal to, less than, or greater than the filtered load *(inset)*. The thicknesses of the arrows indicate the amount of each of the substances flowing in the plasma and urine.

Experimental Methods of Renal Physiology

Several experimental techniques have been used to characterize the function of different parts of the nephron. These include the **micropuncture method,** the **perfused renal tubule** preparation, and methods in which pieces of renal tubules are suspended in an appropriate medium. Much of what is known about the human kidney

is the result of experiments using the kidneys of other mammals or even of amphibians, reptiles, and fish, on the assumption that many of the fundamental mechanisms of urine formation and modification are common to all vertebrates.

In the micropuncture method, a micropipette is introduced into nephrons at various points to

remove a sample of the tubular fluid (Figure 19-B, *A*). The rates at which fluid is reabsorbed by different parts of the nephron can be determined by measuring the concentration of a nonreabsorbed marker substance (usually inulin, see the text) at different points along the nephron. Similarly, the ratio between the inulin

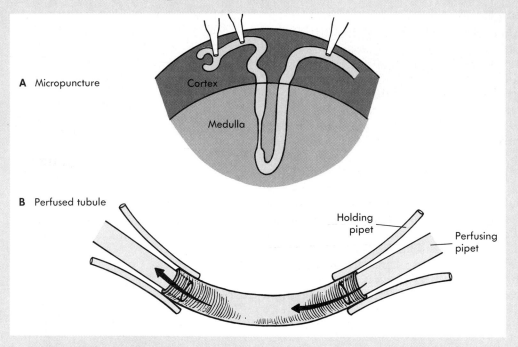

FIGURE 19-B

A   Micropuncture of cortical nephrons with fine glass pipettes can provide nanoliter samples of fluid from Bowman's capsule, the proximal tubule, and the distal convoluted tubule.

B   Pieces of tubule can be dissected from any region of the kidney and mounted between fine glass pipettes. The outer pipette holds the tubule in place; the inner pipette is for perfusion of the tubule. The isolated tubule is bathed in physiological saline solution. Changes in the composition of the perfusate as it passes through the tubule can be measured. The permeability of the tubule to water and salt are measured using chemical techniques or radioisotope markers.

concentration and that of other solutes can reveal solute absorption or secretion. The success of this method depended on accurate methods for handling, and chemical assay of, very small fluid samples. Also, micropuncture can be applied only to those parts of cortical nephrons which are accessible from the kidney surface. Thus it is not possible to apply this method to the loop of Henle or to the juxtamedullary nephrons.

Portions of nephrons can be dissected from the kidney and mounted between two perfusion pipettes (see Figure 19-B, *B*). A very slow flow of fluid through the tubule is set up, and the effect of the mounted piece of tubule on the composition of the fluid can be measured. This method permits study of the parts of the nephrons that are deeper in the kidney, but it has the disadvantage of disrupting the normal structural and functional relationships between the different parts of the nephron and between the nephron and its circulation.

If the kidney cortex is gently minced, pieces of proximal and distal tubules are released. The ends of these segments tend to seal themselves off. If a substance that is actively secreted by these parts of the nephron is dissolved in the medium, it becomes concentrated in the tubules. This is especially evident if the substance is colored because an optical assay of the tubular fluid is possible. Chemical or radioisotope methods may also be used to follow renal handling of a particular substance. In a further refinement of this basic approach, sonication of the tubule pieces with a generator of high-frequency sound waves breaks up the cells of the tubules, causing the formation of membrane vesicles derived from either apical or basolateral surfaces of the cell membrane. The vesicles from the different regions can be separated for measurement of their ability to transport specific solutes.

Most recently, the composition of tubular fluid in the different parts of nephrons has been measured using the **X-ray microanalysis** method. In this method, living kidney tissue is quickly frozen and sliced. A beam of electrons scans across the slice. Different chemical elements scatter the electrons in characteristic ways; thus, with appropriate calibration, the elemental composition of the tubular fluid and extracellular fluid of different kidney regions can be determined from the energy spectrum of the scattered electrons.

Because some other epithelial tissues share some of the transport mechanisms of the nephron, it is frequently possible to study these transport processes in a more convenient tissue. For example, the skin epithelium of the frog is similar in many ways to the distal tubule. This skin has been a valuable experimental preparation in which to study the mechanisms of diuretic drugs and the ways in which the distal tubule responds to the hormones, aldosterone and antidiuretic hormone.

cause more solute was filtered than appeared in the urine (see Figure 19-22, *inset,* line 2). Any solute that is fully reabsorbed by the kidney has a clearance of zero. Solutes that have clearances greater than the GFR have undergone net secretion as well as filtration; that is to say, more plasma was cleared of the solute than was filtered during the measurement interval.

The flow rate of final urine is much less than the GFR because much filtered fluid is reabsorbed. A substance that is filtered but not secreted or reabsorbed becomes concentrated in the urine as fluid is reabsorbed. Because the rate of filtration of such a substance is proportional to the GFR, its clearance is an accurate measure of the GFR. For example, substance F in Figure 19-22 was filtered into Bowman's capsule but not subsequently reabsorbed or secreted. What is the clearance of such a substance? The amount of F excreted in the urine must equal the original filtered load:

$$\text{Filtered load of F} = (\text{GFR})(P_F) = (V)(U_F)$$

The clearance of F must be equal to the volume of fluid filtered along with it. Solving for the glomerular filtration rate gives:

$$\text{GFR} = (U_F)\,(V)/P_F = C_F$$

Inulin, a plant polysaccharide, is one example of a solute that is filtered but neither secreted nor reabsorbed. The inulin clearance is determined by injecting inulin into a subject, waiting for the injected inulin to be distributed throughout the blood, and then measuring the plasma and urine concentrations of inulin and the rate of urine production. It is frequently more practical to use the **creatinine** clearance as a measure of GFR. Creatinine is naturally produced at a low rate as a metabolite of muscle creatine. It is neither reabsorbed nor metabolized by the kidney and is secreted at a negligible rate. Measurement of creatinine clearance only requires samples of blood and urine and measurement of the rate of urine production and gives results nearly the same as those obtained by measuring inulin clearance.

For those substances which are filtered and reabsorbed (*substance R* in Figure 19-22), some or all of the substance filtered from glomerular blood is returned to the peritubular blood, making the clearance less than the GFR. For substances such as glucose that are essentially totally reabsorbed, the clearance is zero. For solutes that are filtered and secreted (*substance S* in Figure 19-22), some solute enters the urine at the filtration step, an additional amount is transferred from peritubular blood to urine, and the clearance is greater than the GFR.

If the kidney secretes a solute so efficiently that the solute is completely eliminated by the time the blood leaves the kidney, its clearance will be equal to the total **renal plasma flow (RPF)** because the amount of plasma cleared of the solute is all the plasma that entered the kidney. This includes both the filtered fraction entering Bowman's capsule and the unfiltered plasma in the peritubular capillaries. One example of such a solute is **paraaminohippuric acid,** or **PAH.** PAH is produced in the body in small amounts as an end product of metabolism of aromatic amino acids. About 90% of the PAH entering a healthy kidney is removed. The rate of PAH excreted in the urine ($U_{PAH} \times V$) equals the plasma concentration of PAH (P_{PAH}) times the RPF; thus the clearance of PAH approximates the RPF. The total **renal blood flow (RBF)** can be calculated from the PAH clearance by dividing it by the fraction of whole blood that is plasma. If the hematocrit is 45%, 0.55 of the blood is plasma, and the renal blood flow RBF is

$$\text{RBF} = C_{PAH}/0.55$$

If both the RPF and the GFR are known, the filtered fraction (FF) can be computed:

$$\text{FF} = \text{GFR}/\text{RPF}$$

Clearance measurements are of enormous importance to physicians, because they tell what the kidney does in health and what it may be failing to do in disease. However, they are of little value in revealing the mechanisms of renal function. For example, reabsorption of a filtered substance from one part of the nephron might be accompanied by secretion into another. If reabsorption and secretion happened to balance exactly, the clearance of the substance would equal its filtration rate. It would be a mistake to conclude from this that the substance was filtered but neither secreted nor reabsorbed.

How Na$^+$ and K$^+$ Are Handled by the Kidney

> - *Describe the differences between the handling of Na$^+$ and K$^+$ by the kidney. What is meant by obligatory reabsorption?*
> - *What is the effect of aldosterone on the clearances of Na$^+$ and K$^+$?*
> - *Trace the feedback loop in tubuloglomerular feedback.*

About 90% of the filtered load of Na$^+$ is reabsorbed in the proximal tubule and thick ascending limb of the loop of Henle. This **obligatory Na$^+$ reabsorption** is not under hormonal control and occurs at about the same rate no matter what the state of Na$^+$ balance in the body. As discussed in more detail in the following chapter, reabsorption of Na$^+$ in the distal tubule and collecting duct, carried out by principal cells (see Figures 19-5,*D* and 19-5,*E*) occurs at a rate controlled by the adrenal cortical hormone, aldosterone. In the absence of aldosterone, the distal tubule reabsorbs about 8% of the filtered

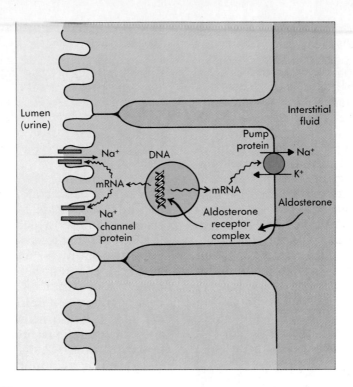

FIGURE 19-23

Mechanism of action of aldosterone. The hormone molecules enter the cytoplasm and bind to receptors; the hormone-receptor complexes enter the nucleus, where they affect the transcription of specific genes that code for proteins involved in Na^+ transport, including apical Na^+ channel proteins and Na^+-K^+ pump proteins. The increase in apical membrane channel proteins results in a more rapid leakage of Na^+ into the cells from distal tubular fluid. The increased rate of entry is handled by the increased number of basolateral Na^+-K^+ pumps, so the rate of Na^+ recovery from distal tubular fluid increases.

load, leaving 2% to be excreted. This level of Na^+ excretion corresponds to a daily loss of an amount of Na^+ equal to the total Na^+ content of the body. This condition is typical of Addison's disease, in which inadequate amounts of adrenal cortical hormones are secreted. In the presence of typical plasma levels of aldosterone, Na^+ excretion is much less than 1% of the filtered load.

Aldosterone, like other steroid hormones (see Chapter 5), acts by inducing the transcription of messenger RNAs that code for specific proteins involved in Na^+ transport (Figure 19-23). These include Na^+ channels in the luminal membrane that allow Na^+ to enter the principal cells from the tubular fluid and Na^+-K^+ pump proteins in the basolateral membrane (see Chapter 6). The response of distal tubular cells to changes in levels of aldosterone requires several hours to be completed, because new proteins must be synthesized and transported to the cell membrane.

A second hormone involved in Na^+ homeostasis is **atrial natriuretic hormone** (ANH), which increases Na^+ and water loss by several mechanisms. The major effect of the hormone is to increase the glomerular filtration rate. This is accomplished by two complementary effects: dilation of the afferent arterioles (which increases filtration pressure) and relaxation of the mesangial cells (which increases the effective glomerular filter area). In addition to its renal effects, ANH has a number of systemic effects that tend to decrease blood pressure and blood volume. The homeostatic roles of aldosterone and ANH are discussed more fully in the next chapter.

Potassium handling by the nephron provides an interesting example of interacting filtration, reabsorption, and secretion. Potassium ions are freely filtered. Between 85% and 90% of the filtered K^+ is obligatorily reabsorbed by the proximal tubule and the loop of Henle. Depending on the state of K^+ balance in the body, the rate of K^+ excretion in final urine may be as little as 1% of the filtration rate or as great as 200% of the filtration rate. The distal tubule, and to some degree the collecting duct, are able to either reabsorb or secrete K^+, depending on the need. Of the two major cell types of the distal tubule and cortical collecting duct (Figures 19-5, D and E), principal cells carry out K^+ secretion, and intercalated cells are believed to carry out K^+ reabsorption. It is believed that K^+ reabsorption goes on at a relatively constant rate, whereas K^+ secretion is altered by the hormone aldosterone to meet the needs of K^+ homeostasis. The feedback loops by which aldosterone regulates both Na$^+$ and K$^+$ balance are described in Chapter 20.

The Juxtaglomerular Apparatus and Tubuloglomerular Feedback

Each nephron adjusts its glomerular blood flow and filtration rate by a negative feedback process that tends to keep the rate of delivery of fluid and NaCl to the early distal tubule approximately constant in spite of changes in systemic arterial blood pressure that occur as a result of changes in posture and physical activity. This phenomenon, called **tubuloglomerular feedback**, is analogous to the autoregulation of ventilation and perfusion in the lung (see Chapter 16). It protects each nephron from large variations in filtration that would disrupt Na$^+$ reabsorption in the distal tubule. Each of the approximately 200 million human nephrons filters about 12 micromol of Na$^+$ per day and reabsorbs all but about 0.05 micromol. An error of only 1% in either filtration or reabsorption would cause a gain or loss of 240 mmol/day, which is about 10% of the total body Na$^+$ content.

The most likely hypothesis for the mechanism of tubuloglomerular feedback is based on the fact that the distal tubule of each nephron is physically joined to the glomerulus and afferent arteriole by the juxtaglomerular apparatus (see Figures 19-3, 19-7 and 19-8). The macula densa of the juxtaglomerular apparatus is a specialized group of epithelial cells located on the wall of the distal tubule. These cells are believed to sense the concentration of NaCl in tubular fluid passing them and to cause the juxtaglomerular apparatus to release a paracrine signal that travels the short distance to the afferent arteriole. For example, an increase in renal blood pressure would increase the filtered load of fluid and NaCl. This increase would be sensed by the macula densa, which would increase its rate of release of paracrine signal. The paracrine signal would constrict the afferent arteriole, reducing the pressure of blood entering the glomerulus and restoring the rate of filtration to normal. The identity of the supposed paracrine signal is still unknown, but is believed to be a prostaglandin.

In addition to its role in tubuloglomerular feedback, the juxtaglomerular apparatus is involved in hormonal control of blood volume and blood pressure by its release of an enzyme, **renin**. Renin initiates a regulatory cascade that causes production of a family of peptide hormones called **angiotensins**. Angiotensins affect thirst and salt appetite, cause vasoconstriction, and modulate the rate of release of aldosterone by the adrenal cortex. As noted above and described in Chapter 20, aldosterone plays a major role in regulation of Na$^+$ transport by the kidney and other organs such as the sweat glands, salivary glands and colon.

Renal Contributions to Acid-Base Regulation

- *What is accomplished by tubular acidification of urine?*
- *What are the two sources of the HCO$_3^-$ that is returned to the general circulation from the kidney?*
- *Describe the mechanism of pH trapping of NH$_3$ in urine.*

The kidney has a major role in regulating the content of "fixed" acids and bases in the body. Fixed acids are those which cannot be eliminated as gases; recall that the respiratory system has the major responsibility for regulating the level of the volatile acid, CO$_2$. For people who consume the high-protein diet typical of developed nations, catabolism of sulfur-containing amino acids, phospholipids and other substances containing phosphorus and sulfur results in production of 70 to 100 mM/d of fixed acids (HCl, H$_3$PO$_4$, H$_2$SO$_4$), that must be eliminated by the kidney. Catabolism of protein results in the formation of HCO$_3^-$ and NH$_4^+$ (ammonium ion; see Chapter 23). Some of the HCO$_3^-$ and NH$_4^+$ is converted to urea (CO[NH$_2$]$_2$). Retention of the remaining HCO$_3^-$ would help counter the effect of fixed-acid production, but the remaining NH$_4^+$ must be eliminated from the body.

As described in following paragraphs, acidification of the tubular fluid simultaneously accomplishes the goals of excreting fixed acids and NH$_4^+$ and recovering filtered HCO$_3^-$. Acidification of tubular fluid is the result of active secretion of H$^+$ by the tubular epithelium and takes place in the proximal tubule, the cortical collecting tubule, and the collecting duct. In the proximal tubule, acidification results in the conversion of filtered HCO$_3^-$ to CO$_2$. The carbonic anhydrase enzyme located on the apical surface of proximal tubular cells is essential for this process. Because CO$_2$ is a nonpolar molecule that freely diffuses through tissues, almost all of the resulting CO$_2$ returns to the blood. The pH of blood plasma is one or two orders of magnitude higher than that of the acidified tubular fluid. At the higher pH, the Law of Mass Action dictates that the CO$_2$ undergo the reverse reaction, combining with water to form carbonic acid that then dissociates into HCO$_3^-$ and H$^+$ (see Chapter 3). The rate of the reverse reaction in the blood is accelerated by the carbonic anhydrase enzyme of red blood cells, described in Chapter 17. Under normal conditions, almost all of the approximately 4 moles of HCO$_3^-$ filtered each day is recovered in this way. After the HCO$_3^-$ has been recovered, the bulk of additional H$^+$ that is secreted is accepted by other buffers in the urine. For example, the dibasic form of phosphate (HPO$_4^=$) can accept H$^+$, becoming the monobasic form (H$_2$PO$_4^-$)

(Figure 19-24). Phosphate is quantitatively the most important urinary buffer.

Recovery of filtered bicarbonate results in no net loss or gain of acid for the body because the H^+ and the HCO_3^- involved balance out. However, the tubular cells can also generate what is called "new" bicarbonate. Each tubular cell produces CO_2 in oxidative metabolism. Within the cells, the metabolic CO_2 reacts with water to form H^+ and HCO_3^-. The tubular cells separate the two ions, sending the H^+ into the tubular fluid and the HCO_3^- into the blood. In this way, the kidney can contribute to the body's store of bicarbonate buffer.

The process of NH_4^+ excretion starts in the liver, where the amine (NH_3) groups of amino acids are stripped off before conversion of the carbon skeletons to glucose or pyruvate. In a process called **transamination**, the NH_3 groups are transferred to the amino acid, glutamate, which becomes glutamine in the process. Glutamine passes into the blood and is taken up by tubular cells, where it is deaminated. For each molecule of glutamine, two NH_3 molecules are released within the cells. The carbon skeleton of glutamine is metabolized by the renal cells into CO_2; some of this CO_2 may reappear in the blood as new bicarbonate.

The NH_3 released in tubular cells is concentrated in the tubular fluid by a process called **pH trapping** (Figure 19-24). Like CO_2, NH_3 is a nonpolar, highly diffusible molecule, but it readily reacts with free H^+ to form NH_4^+. The relative concentrations of H^+ and NH_4^+ in tubular fluid are determined by the Law of Mass Action. Proton secretion by the tubular epithelium may make the concentration of H^+ in tubular fluid as much as two orders of magnitude greater than that in the interstitial fluid, or about pH 5.5. At this low pH, the equilibrium concentration of NH_4^+ in the tubular fluid is high. Once the NH_3 has diffused into the more acid tubular fluid, almost all of it is converted to polar NH_4^+, which is trapped there because its charge prevents it from recrossing the tubular epithelium.

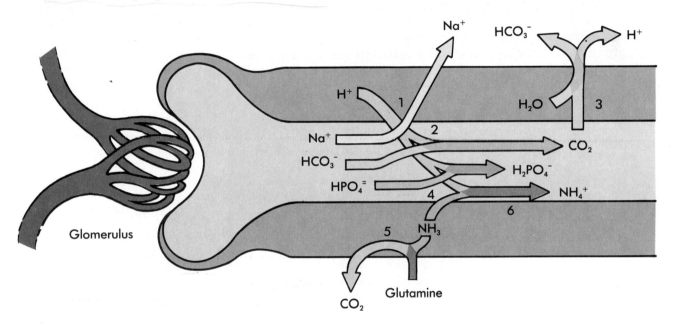

FIGURE 19-24
Interrelationships between H^+ secretion, HCO_3^- reabsorption, and ammonia trapping. H^+ is secreted, largely in exchange for Na^+ (1). The secreted H^+ converts the filtered HCO_3^- to CO_2 (2), which is passively reabsorbed (3), returning both the filtered HCO_3^- and the secreted H^+ to the body. Additional secreted H^+ is accepted by other urinary buffers, mainly phosphate (4). Glutamine is deaminated in the renal tubular cells (5), releasing NH_3, which diffuses into the tubule where it is converted to NH_4^+ (6). The charged NH_4^+ is trapped in the urine and is excreted.

Mechanisms of Action of Diuretic Drugs

Drugs that increase the rate of urine production over the normal rate of 2 to 3 ml/min are called diuretics. Diuretics are used to reduce extracellular fluid volume—an important therapy for hypertension and congestive heart failure. Because of the large market for diuretics to control hypertension, pharmaceutical firms have sponsored much research into their mechanisms. Diuresis can be the result of interference with several different specific renal mechanisms.

A simple factor that can cause diuresis is the filtration of large amounts of a substance that cannot be reabsorbed from the tubular fluid (Figure 19-C, *1*). One such substance is mannitol. When the proximal tubule reabsorbs Na^+, the tubular osmolarity decreases. Water begins to follow, but this increases the concentration of mannitol. Because mannitol cannot be reabsorbed, it limits water reabsorption. The failure of water to follow Na^+ results in a lower tubular Na^+ concentration and a larger concentration gradient between the tubular lumen and the lateral intracellular spaces. As a result, active Na^+ transport is decreased and the backwards leak of Na^+ through the tight junctions is increased. Eventually more Na^+ must therefore pass through the proximal tubule without being absorbed.

Substances that the kidney handles as it handles mannitol are called osmotic diuretics. Not all osmotic diuretics are drugs; some metabolites can have the same effect. One reason the proximal tubule must be permeable to urea is because urea would otherwise act as an osmotic diuretic. Osmotic diuresis occurs in diseases such as diabetes mellitus. Here the plasma levels of glucose are so high that the filtered load of glucose often exceeds its T_m, and the unreabsorbed glucose acts as an osmotic diuretic.

Some drugs inhibit Na^+ transport across the tubule, with consequences similar to those of osmotic diuretics. This is the mechanism of action of drugs containing mercury (mercurials), which were used before more modern diuretics were developed. Amiloride blocks Na^+ reabsorption and acidification by inhibiting gradient-mediated exchange of Na^+ for H^+ (Figure 19-C, *2*). The unreabsorbed Na^+ acts as an osmotic diuretic. Inhibitors of carbonic anhydrase such as acetazolamide (Diamox) selectively block bicarbonate reabsorption in the proximal tubule. This failure to reabsorb bicarbonate may have the side effect of increasing the plasma H^+ concentration.

Loop diuretics inhibit the Na^+, K^+, $2Cl^-$ gradient-mediated cotransporter in the thick ascending limb of the loop of Henle (Figure 19-C, *3*), preventing the absorption of solute that drives the renal concentrating mechanism. Furosemide and bumetanide are examples of loop diuretics. Loop diuretics also increase the fraction of the filtered Na^+ load reaching the distal tubule, resulting in the loss of both K^+ and H^+ into the urine. Because loop diuretics tend to cause depletion of K^+ as well as of Na^+, patients who take them are frequently given K^+ supplements.

Antagonists of aldosterone, such as spironolactone (Figure 19-C, *4*), allow Na$^+$ that would otherwise have been reabsorbed to pass into the collecting duct and oppose the two-solute mechanism, which works more effectively if urea rather than Na$^+$ is the major solute in the urine.

Inhibitors of antidiuretic hormone secretion include alcohol, caffeine, and water (Figure 19-C, *5*). However, because antidiuretic hormone acts on the collecting region of the distal tubule and collecting duct, the resulting diuresis does not usually affect plasma K$^+$ or H$^+$ levels.

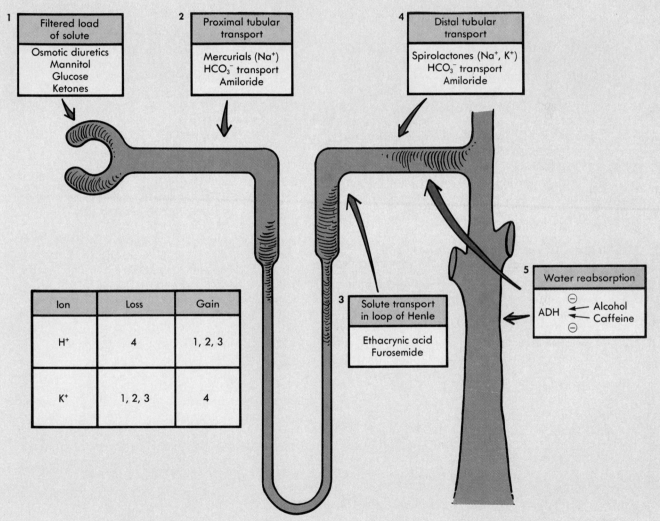

1
Filtered load of solute

Osmotic diuretics
Mannitol
Glucose
Ketones

2
Proximal tubular transport

Mercurials (Na$^+$)
HCO$_3^-$ transport
Amiloride

4
Distal tubular transport

Spirolactones (Na$^+$, K$^+$)
HCO$_3^-$ transport
Amiloride

3
Solute transport in loop of Henle

Ethacrynic acid
Furosemide

5
Water reabsorption

ADH ⊖← Alcohol
⊖← Caffeine

Ion	Loss	Gain
H$^+$	4	1, 2, 3
K$^+$	1, 2, 3	4

FIGURE 19-C
Sites of action of diuretics. The inset gives the numbers of sites at which diuretic action may affect handling of H$^+$ or K$^+$.

SUMMARY

1. The excretory system consists of the **kidneys,** the **ureters,** which connect the kidneys to the urinary bladder, and the **urethra,** which connects the bladder to the outside of the body. **Urination** is the result of reflexive contraction of the bladder smooth muscle accompanied by opening of the external sphincter (skeletal muscle). The reflex is modulated by pathways from the brain to give voluntary control.

2. **Nephrons,** or renal tubules, are the functional units of the kidney. Cortical nephrons carry out the same functions as juxtamedullary nephrons, except that they do not assist in the formation of the medullary osmotic gradient.

3. About 20% of the renal plasma flow is converted to the glomerular filtrate by **ultrafiltration** across glomerular capillaries. The glomerular filtrate is similar to plasma, except that it contains little of the larger plasma proteins.

4. The functions of the proximal tubule are as follows:
   a. Reabsorption of most of the filtered Na^+ and water.
   b. Passive reabsorption of other small solutes in proportion to their abundance in the filtrate.
   c. Active reabsorption of some essential solutes such as glucose (typically by a Na^+ gradient-coupled process).
   d. Active secretion of some organic metabolic end products.
   e. Active secretion of H^+, which rids the body of its daily production of fixed acid and recovers filtered bicarbonate by conversion to CO_2, which diffuses back into the peritubular capillaries.

5. The **loop of Henle** concentrates descending tubular fluid using salt removed from the ascending limb. The result is an osmotic gradient between the cortex and medulla of the kidney. The interstitial fluid in the medulla is further concentrated by entry of urea from the concentrated tubular fluid in the collecting ducts. The tubular fluid leaving the ascending limb of the loop of Henle is made hypoosmotic as a result of the reabsorption of salt from it without osmotic movement of water.

6. The distal tubule recovers filtered Na^+ by active transport, diluting the tubular fluid and causing other solutes, such as urea, to make up a larger percentage of the total solute content. The rate of Na^+ reabsorption (and K^+ secretion) is increased by the hormone, **aldosterone.**

7. In the presence of ADH, the collecting duct is permeable to water and the tubular fluid in the collecting duct is concentrated by osmotic water movement into the medullary interstitial spaces. Some urea is reabsorbed in this process, contributing to osmotic movement of water and to the solute content of the medullary interstitial fluid. Decreasing the ADH concentration increases the amount of water that is not reabsorbed, making the final urine more dilute.

8. The loop shape of the **vasa recta** mirrors that of the loops of Henle. Its function is to preserve the medullary osmotic gradient by **countercurrent exchange,** while also providing metabolic support to the medullary cells. Plasma descending the vasa recta equilibrates with the concentrated medullary interstitial fluid. As the plasma ascends the loop, it equilibrates with the less and less concentrated interstitial fluid and leaves the kidney at the same concentration as normal plasma. In this way, the fluid and solute reabsorbed in the medullary portions of the nephrons are returned to the general circulation and the osmotic gradient is not disrupted.

9. The excretion rate of any component of plasma is the sum of filtration and secretion minus reabsorption. The net of these processes can be expressed as the solute's **clearance,** defined as the volume of plasma cleared of the substance per a unit of time. The clearance of substances filtered but not secreted or reabsorbed (for example, inulin and creatinine) is equal to the **glomerular filtration rate** (GFR). The clearance of solutes that are filtered and reabsorbed (for example, Na^+, K^+, Cl^-, glucose, and HCO_3^-) is much less than the GFR. The clearance of secreted substances (for example, PAH and H^+) is greater than the GFR.

10. The kidney autoregulates its rate of formation of glomerular filtrate by **tubuloglomerular feedback.** The feedback loop runs from the macula densa of the juxtaglomerular apparatus, which senses the rate of delivery of fluid and NaCl to the tubule, and secretes a paracrine signal to the nearby afferent arteriole of the glomerulus.

1. Juxtamedullary and cortical nephrons differ anatomically and in their physiological roles. Describe the anatomical difference. Explain the functional difference.

2. Describe the blood supply to the kidney. What is the significance of the presence of two capillary beds in series?

3. In response to mild sympathetic activation, the total RBF decreases, but the GFR is almost unchanged. What protects the GFR from changes in perfusion pressure?

4. In what region of the nephron is the tubular fluid always hyperosmotic to plasma? In what region is it always isosmotic?

5. What is Na^+ gradient-mediated cotransport? Give some examples of substances reabsorbed by this process.

6. Account for the fact that inhibition of ADH secretion by alcohol ingestion results in the production of a large volume of dilute urine.

7. In a diabetic the plasma glucose concentration is 10 mg/ml; the inulin clearance is 125 ml/min; and the urine flow rate is 10 ml/min. What is the filtered load of glucose? Assuming a T_m for glucose of 400 mg/min, what is the rate of glucose reabsorption? What is the rate of glucose excretion? What is the concentration of glucose in the urine?

8. Describe the classical and two-solute hypotheses for generation of the medullary concentration gradient. What evidence argues against the classical hypothesis? What is the role of urea in the two-solute hypothesis?

9. What factors can disrupt the ability to produce concentrated urine?

10. What reabsorptive process is driven by tubular acidification? Why does ammonium excretion increase when H^+ secretion is stimulated?

11. A low-protein diet impairs urine concentrating ability. Explain. (Hint: the answer involves an end product of protein metabolism.)

Choose the MOST CORRECT Answer.

12. Filtration occurs in the:
    a. Peritubular capillaries
    b. Vasa recta
    c. Glomerular capillaries
    d. Renal arteries

13. The juxtaglomerular apparatus secretes:
    a. Aldosterone
    b. Renin
    c. Sodium
    d. Potassium

14. Which of these substances would NOT typically be present in quantity in glomerular filtrate?
    a. Water
    b. Glucose
    c. Proteins
    d. Mineral ions

15. This glomerular capillary wall layer functions as an intermediate filter to keep back larger plasma proteins:
    a. Fenestrated capillary endothelial layer
    b. Mesangial cell layer
    c. Slit pores
    d. Basal lamina

16. This substance is neither actively reabsorbed nor secreted by the renal tubules:
    a. Glucose
    b. Phosphate
    c. Creatinine
    d. Sodium

17. These nephron segments are almost impermeable to water:
    a. Thick ascending limb of Henle; distal convoluted tubule
    b. Thick ascending limb of Henle; proximal tubule
    c. Proximal tubule; descending limb of Henle
    d. Proximal tubule; collecting tubule

18. Angiotensins:
    a. Affect thirst and salt appetite
    b. Modulate release of aldosterone
    c. Cause vasoconstriction
    d. All of the above

19. Acidification of tubulogomerular fluid accomplishes all EXCEPT:
    a. Excretion of fixed acids
    b. Increase in urine volume
    c. Excretion of NH_{4+}
    d. Recovering filtered HCO_{3-}

● SUGGESTED READINGS

BEEUWKES R: Renal countercurrent mechanisms: how to get something for (almost) nothing. In TAYLOR CR, JOHANSEN K and BOLIS L: *A Companion to Animal Physiology*, Cambridge, Mass., 1982, Cambridge University Press. Describes the countercurrent multiplier mechanism and two-solute hypothesis for renal concentration.

FLIEGER K: Kidney disease: when those fabulous filters are foiled. *FDA Consumer* 24(2):26, 1990. Discusses kidney diseases and hemodialysis.

ROBINSON JR: *Reflections On Renal Function.* Oxford, 1988, Blackwell Scientific Publications. Introduction to renal physiology that is valuable for its readability and historical perspective.

SCHAFER JA: Fluid absorption in the kidney proximal tubule. *News in Physiological Sciences* 2:22, 1987. Explains the standing-gradient osmotic-flow hypothesis for proximal tubule salt and water absorption.

SKOTT O and BRIGGS JP: Direct Demonstration of Macula-Densa Mediated Renin Secretion, *Science,* September 1987. Discusses the renin-angiotensin-aldosterone system for blood pressure regulation.

STEPHENSON JL: Models of the urinary concentrating mechanism. *Kidney International* 31:648, 1987. Reviews and evaluates hypotheses of urinary concentration by the kidney.

CHAPTER 20

Control of Body Fluid, Electrolyte, and Acid-base Balance

On completing this chapter you should be able to:

- Describe how total body water is distributed between the intracellular, interstitial, and vascular compartments.
- Understand why cell volume is determined by cytoplasmic protein and extracellular fluid by total body Na^+.
- List the routes for daily water turnover.
- Understand the effects of atrial natriuretic hormone on glomerular filtration rate and Na^+ reabsorption.
- Describe antidiuretic hormone secretion response to plasma osmolarity and atrial stretch.
- Describe the factors affecting aldosterone secretion by the adrenal cortex, and explain how aldosterone increases the recovery of filtered Na^+ in the distal tubule.
- Understand the mechanisms for the reabsorption of filtered K^+ and how plasma K^+ directly stimulates K^+ secretion.
- Describe how plasma Ca^{++} and phosphate are regulated by calcitonin, parathyroid hormone, and 1,25 DOHCC.
- Appreciate how the extracellular fluid pH is buffered by the bicarbonate buffer system, and how the state of this buffer system is controlled by the combined action of lungs and kidneys.
- Describe how H^+ secretion first recovers filtered HCO_3^-, and then results in the formation of "new" HCO_3^- from CO_2, increasing plasma $[HCO_3^-]$ and pH.

The control systems that regulate extracellular fluid provide a constant chemical environment for the cells. It has been said that we are what we eat; in the case of body fluid composition we are also what we drink, and what we do and do not excrete. In fact, the pioneer renal physiologist Homer Smith said that body fluid composition depended not so much on what we ate and drank as on what the kidney kept in. If the body fluids are to remain in a steady state, gains of salts and water from the diet must exactly match evaporative losses of fluid from the body surface; obligatory losses of salts and water with feces, tears, and other secretions; and loss of salts and water in urine. The sheer number of routes of gain and loss make regulation of body fluids a web of interacting processes.

Where the regulatory problem is complex, complex regulatory systems evolve. For example, three known interlocking endocrine systems regulate the volume and Na^+ content of extracellular fluid. One of these, the atrial natriuretic peptide system, was only recently discovered, and its full complexity remains to be worked out. Other endocrine systems involved in Na^+ and volume regulation may yet be discovered. The K^+, Ca^{++}, phosphate, and acid-base regulatory systems face similarly complex problems of matching gains and losses.

Failure of body fluid homeostasis has varying consequences, depending on the particular regulatory system and substance involved. Changes in the concentrations of K^+, H^+, or Ca^{++} affect the excitability of muscles and nerves immediately. In contrast, the total body Na^+ content and extracellular fluid volume may increase for some time without outward symptoms, until the resulting hypertension has taken its toll on the heart and blood vessels.

THE FLUID COMPARTMENTS OF THE BODY

- *What impermeant solutes determine distribution of fluid between the intracellular and extracellular compartments?*
- *What impermeant solute exercises a key role in determining the distribution of fluid between the vascular and interstitial subcompartments?*
- *Why is regulation of total body Na⁺ essential to long-term regulation of blood volume and blood pressure?*

The Intracellular and Extracellular Compartments

To understand the topics treated in this chapter, it is necessary to visualize the fluid compartments of the body (see Chapter 1 and Chapter 14). Total body water averages 60% of an individual's weight in males and 50% in females. The sex difference is due to a greater average contribution of fat to body weight in women than men—fat cells contain less water than most other cell types. Body water exists in two main compartments: (1) the **intracellular compartment**, consisting of fluid inside the cells; and (2) the **extracellular compartment**, containing the fluid outside of the cells. The intracellular compartment contains two thirds of the total body water and includes water in red blood cells; the extracellular compartment contains the remaining one third.

Extracellular fluid is located in many separate spaces of the body. The two main subcompartments are interstitial fluid and blood plasma; other subcompartments are cerebrospinal fluid, the aqueous humor of the eyes, the fluid within the lumen of the gastrointestinal tract, urine in the excretory tract, etc. Neglecting these smaller subcompartments, interstitial fluid makes up approximately 75% of the extracellular fluid; the remaining 25% is blood plasma. Figure 20-1 shows a breakdown of the major body fluid compartments for a 75 kg man with total body water of 45 liters. About 30 liters of total body water is intracellular, 11.25 liters is interstitial, and 3.75 liters is plasma. As a rule of thumb, the intracellular compartment is about twice the volume of the extracellular compartment and two thirds of the total body fluid volume.

The dominant cation of extracellular fluid is Na⁺ (145 mEq/L), and the most abundant extracellular anions are Cl⁻ (104 mEq/L) and HCO₃⁻ (24 mEq/L) (Figure 20-2). The osmolarity of the extracellular fluid (both plasma and interstitial fluid) is approximated by dividing the total content of osmotically active solute particles by the volume; this gives a normal value of about 300 mOsm/L (Figure 20-1). Because all of the plasma solutes do not behave as perfect osmolytes (see Chapter 2), the measured osmolarity of normal plasma is slightly less than the value calculated in this way, but the difference is not critical for our purposes.

The main difference between the plasma and interstitial fluid is the presence of a higher concentration of proteins in the plasma. The plasma proteins are the only plasma solutes that cannot freely cross the capillary wall; their restriction to the plasma is responsible for the osmotic gradient across capillary walls which largely counteracts the hydrostatic pressure of capillary blood. The opposing effects of plasma hydrostatic pressure and plasma protein osmotic pressure determine the distribution of fluid between the intracellular and extracellular compartments (Chapter 14).

The major differences between intracellular and extracellular fluid are that K⁺ is the dominant cation instead of Na⁺, and protein and other organic

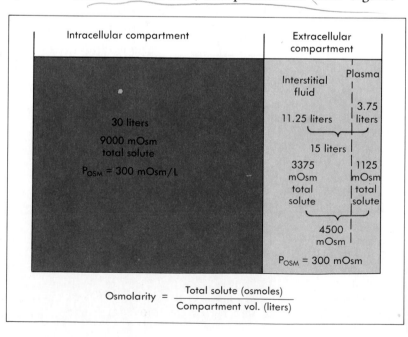

FIGURE 20-1
The fluid compartments of the body. Total body water is divided into intracellular (two thirds) and extracellular (one third) compartments. Three fourths of the extracellular compartment is interstitial fluid and one fourth is plasma. The osmolarity of the intracellular and extracellular compartments is always equal when the two systems are at equilibrium. The volume and total solute values given are for a 75 kg man. In this and similar figures, the area of each box indicates its volume.

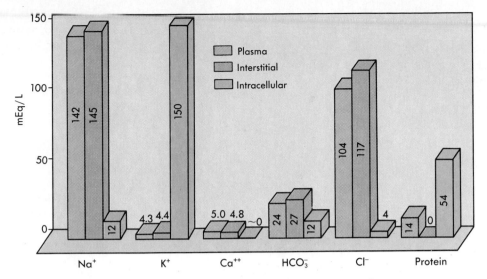

FIGURE 20-2
Electrolyte and protein concentrations of the body fluid compartments.

molecules are the major anions instead of Cl^- and HCO_3^-. Recall from Chapter 6 that the impermeant cytoplasmic protein anions would cause cells to swell uncontrollably except for the fact that the Na^+-K^+ pump makes Na^+ an effectively impermeant extracellular ion. Because the steady-state distribution of water between the intracellular and extracellular compartments is determined by the distribution of impermeant solutes, the cytoplasmic proteins largely determine the volume of the intracellular compartment and Na^+ largely determines the volume of the extracellular compartment. The system is in steady state when the osmolarity of the solutes in the cytoplasm balances that of extracellular fluid (Figure 20-3). All net water movement between the intracellular and extracellular compartments is the consequence of osmotic gradients. Addition of impermeant solute to the extracellular compartment would increase the osmolarity of that compartment initially, but water would move from the intracellular compartment until the osmolarities of the two compartments were once again equal. Similarly, transfer of solute from one compartment to another would cause a proportionate shift of fluid volume from the donor compartment to the recipient compartment.

Fluid and Salt Balance

A healthy person maintains a precise balance between water intake and water loss. A person may drink 800 to 1500 ml of water daily and take in an additional 500 to 700 ml of water in food (Figure 20-4). Oxidative energy metabolism produces 200 to 300 ml of water per day. The total daily water gain is therefore 1.5 to 2.5 liters. This water gain is balanced by **bulk fluid losses** of 800 to 1500 ml as urine and 100 to 150 ml in feces. In addition, 600 to 900 ml of water is lost by evaporation from the

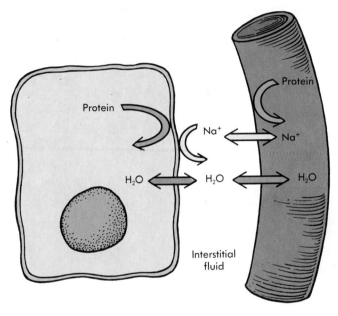

FIGURE 20-3
The permeability properties of the plasma—interstitial fluid barrier and the interstitial fluid—cytoplasm barrier determine which solutes may equilibrate throughout the system and which must remain confined in one or more compartments. Plasma proteins and cytoplasmic proteins do not freely enter the interstitial fluid, so its protein content is low. Na^+ can permeate the capillary wall but is effectively excluded from the cytoplasmic water. Water can freely equilibrate across all three barriers, so at equilibrium the osmolarity of fluids must be equal throughout the system.

lungs and the skin surface; water loss by these routes is called **insensible water loss**. Sweating is responsible for additional evaporative loss.

Depletion of body water (**dehydration**) occurs when losses consistently exceed gains. Some common causes of dehydration are excessive sweating and insensible loss in a hot environment, severe vomiting or diarrhea, and kidney diseases that re-

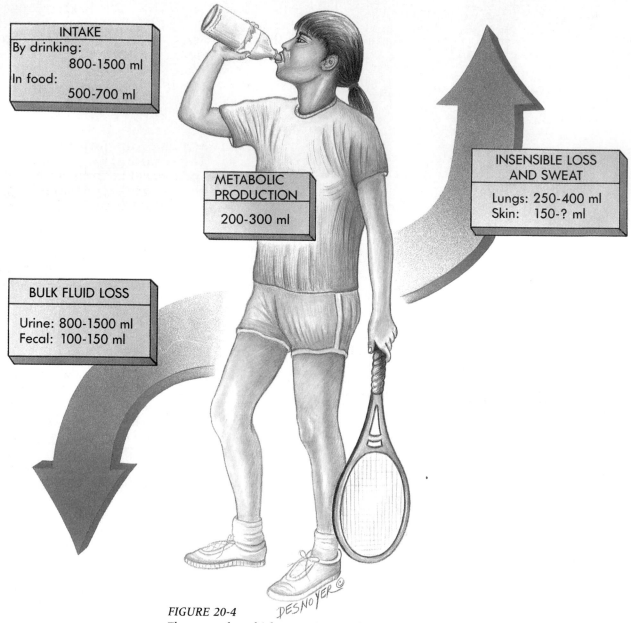

INTAKE
By drinking:
800-1500 ml
In food:
500-700 ml

METABOLIC PRODUCTION
200-300 ml

INSENSIBLE LOSS AND SWEAT
Lungs: 250-400 ml
Skin: 150-? ml

BULK FLUID LOSS
Urine: 800-1500 ml
Fecal: 100-150 ml

FIGURE 20-4
The routes by which water is gained or lost from the body.

sult in production of large quantities of dilute urine. Excessive ingestion of alcohol causes dehydration because alcohol inhibits secretion of antidiuretic hormone, causing production of a large volume of dilute urine.

Salt gains must balance losses to maintain body fluid homeostasis. In addition to the direct consumption of table salt, many foods contain "hidden" salt in the form of preservatives, such as sodium bisulfide, or additives. Some beverages contain a significant amount of salt. Salt is lost in the urine and in various body fluids, such as sweat, saliva, tears, and secretions of the gastrointestinal tract. Abnormal losses of body fluids occur in vomiting, diarrhea, and excessive sweating and can result in large salt losses.

Challenges to Salt and Water Homeostasis

The problems of extracellular fluid homeostasis can be reduced to three simple homeostatic challenges: (1) gain or loss of isotonic NaCl solution, (2) gain or loss of pure water, and (3) gain or loss of pure NaCl. Although these are presented as simple challenges, they could result from multiple imbalances of intake and loss. For example, if a person lost NaCl and water by sweating and subsequently made up the lost volume by drinking pure water, the net result would be the same as a loss of pure NaCl. The examples given are based on an individual who weighs 75 kg and has 45 L total body water (30 L intracellular and 15 L extracellular; see Figure 20-1). Round numbers are used in these examples to make the calculations easier to follow.

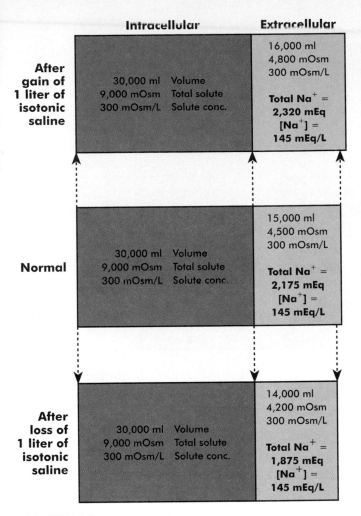

	Intracellular	Extracellular
After gain of 1 liter of isotonic saline	30,000 ml Volume 9,000 mOsm Total solute 300 mOsm/L Solute conc.	16,000 ml 4,800 mOsm 300 mOsm/L Total Na$^+$ = 2,320 mEq [Na$^+$] = 145 mEq/L
Normal	30,000 ml Volume 9,000 mOsm Total solute 300 mOsm/L Solute conc.	15,000 ml 4,500 mOsm 300 mOsm/L Total Na$^+$ = 2,175 mEq [Na$^+$] = 145 mEq/L
After loss of 1 liter of isotonic saline	30,000 ml Volume 9,000 mOsm Total solute 300 mOsm/L Solute conc.	14,000 ml 4,200 mOsm 300 mOsm/L Total Na$^+$ = 1,875 mEq [Na$^+$] = 145 mEq/L

FIGURE 20-5

The effects on intracellular and extracellular fluids of gain or loss of 1 L of isotonic NaCl solution. Since the added or lost Na$^+$ cannot cross into the intracellular compartment, the volume of the intracellular compartment does not change. Added fluid remains in the extracellular compartment; lost fluid comes only from the extracellular compartment.

Figure 20-5 shows the effects of gain or loss of 1 L of isotonic saline, containing 300 mOsmoles of solute, on the fluid volumes of the 75 kg man of Figure 20-1. Since cells are effectively impermeable to Na$^+$, additions or losses of isotonic NaCl affect only the volume of the extracellular compartment—the intracellular volume does not change. Seventy-five percent of the extracellular volume change occurs in the interstitial compartment and the other 25% in the plasma compartment. The resulting condition of abnormally increased plasma volume is called **hypervolemia** (hyper = high; vol = volume; emia = blood); the corresponding term for abnormally low blood volume is **hypovolemia**.

The second simple homeostatic challenge is gain or loss of pure water. The simplest cause of pure water gain is drinking water or other dilute fluids. The simplest case of pure water loss is caused by insensible evaporation from the skin and respiratory surfaces, or production of dilute urine. Dilute urine can be thought of as consisting of some volume of isotonic urine to which an additional volume of pure water has been added. The volume of isotonic solution represents a loss of isotonic solution; the volume of pure water represents a net loss of pure water. The example of net loss of 1 liter of water is about what would normally occur over 12 hours if losses were not replaced by drinking (Figure 20-4); less time would be required to lose this amount of water in a dry enviroment.

If there is a net loss of pure water from the extracellular compartment, the extracellular osmolarity increases. The resulting osmotic gradient between the intracellular and extracellular fluid causes water to move from cells to extracellular fluid until osmotic equilibrium is restored. The reverse occurs when there is a net gain of water. The volume of water gained or lost from each compartment is proportional to the original volume of that compartment, so two thirds of the change will occur in the intracellular compartment and one third in the extracellular compartment (Figure 20-6). For example, starting with a normal osmolarity of 300 mOsm/L, each liter of water gained or lost will change the osmolarity of both the intracellular and the extracellular fluids by about 6 mOsm/l. A fundamental difference between the challenge of gain or loss of pure water and the challenge of gain or loss of isotonic solution is that the water challenge affects both the intracellular and the extracellular volumes, while the isotonic challenge affects only the extracellular volume.

The third simple challenge is gain or loss of pure NaCl. The gain or loss of 150 mmoles of NaCl (or about 8.5 g) in the example in Figure 20-7 is equivalent to the amount of solute in 1 liter of isotonic saline and is within the range of daily intake of NaCl in a cool environment. If this gain is imagined as occuring all at once, the subject's extracellular fluid osmolarity will initially increase to about 320 mOsm/L. The increased osmolarity of the extracellular fluid will cause net movement of water from the intracellular compartment to the extracellular compartment until the osmolarities of the intracellular fluid and the extracellular fluid are equal (Figure 20-7). The final osmolarity of intracellular and extracellular fluid will be 300 mOsm/L + (300 mOsm/45 L) = 307 mOsm/L. The volume of the extracellular compartment will have increased by about 650 ml at the expense of the intracellular compartment. The plasma [Na$^+$] will be about 149 mEq/L; the condition of elevated plasma [Na$^+$] is called **hypernatremia**. Loss of the same amount of pure NaCl would cause a transfer of 650 ml from the

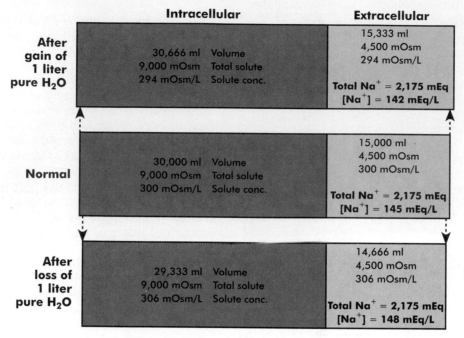

FIGURE 20-6

The effects on the volume and solute concentration of gain or loss of 1 L of pure water. In the case of water gain, the additional water equilibrates between the two compartments on the basis of the total solute in each compartment, so the final volume of both compartments is increased and the final solute concentrations are decreased. In the case of pure water loss, the volumes of both compartments are decreased and the solute concentrations of both rise.

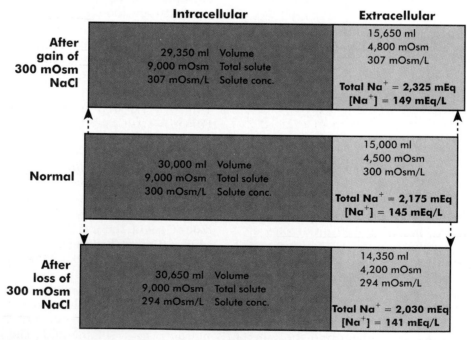

FIGURE 20-7

The effect of gain or loss of 8.5 grams (300 mOsm) of pure NaCl. The added salt is confined to the extracellular compartment. Water moves from the intracellular compartment to the extracellular compartment until the osmolarity of the two compartments is equal. The volume of the intracellular compartment decreases (the cells shrink) and that of the extracellular compartment increases by the same amount. When salt is lost, water moves from the extracellular to the intracellular compartment, and the cells swell.

extracellular compartment to the intracellular compartment and would result in **hyponatremia.**

These examples illustrate the importance of the total body content of Na$^+$ in determining the volume of the extracellular compartment. An increase in plasma volume increases the venous pressure and cardiac output. If the total peripheral resistance did not change, the mean arterial blood pressure would increase. The effect of salt ingestion on venous and arterial blood pressure is the reason that physicians advise individuals with high blood pressure or heart disease to limit their salt intake.

MECHANISMS OF REGULATION OF WATER AND NA$^+$ BALANCE

- *What process prevents urine production from slowing when sympathetic activation decreases renal blood flow? What effect does ANH have on glomerular filtration rate?*
- *Why is atrial stretch a good measure of plasma volume? Which regulatory systems use this measure?*
- *What two types of sensory input are integrated to regulate ADH secretion? Which of the two inputs dominates?*
- *What effectors are affected by the regulatory cascade set in motion by renin?*
- *What is the role of atrial natriuretic hormone in volume homeostasis and blood pressure regulation? At what sites in the body does it have its effects?*

Recall from Chapter 19 that the mechanism of urine formation in the kidney involves formation of an isosmotic glomerular filtrate followed by obligatory absorption of most of the filtrate in the proximal tubule, so that a much smaller volume of isosmotic tubular fluid reaches the distal nephron (consisting of loop of Henle, distal tubule, and collecting duct). Differential absorption of solutes and water by the distal nephron results in production of urine that may be several times as concentrated as plasma, or several times more dilute than plasma. It is important to keep in mind that the rates of loss of solute and water are determined by the rate of production of urine as well as its concentration. For example, an increase in the rate of urine flow represents an increase in the rate of loss of solute and water even if the concentration remains the same.

The role of the kidney in osmotic and volume homeostasis is to modulate the relative rates of loss of solute and water in urine. For example, in response to a net loss of pure water the rate of urine production decreases and the urine concentration increases; both changes are the result of increased reabsorption of water by the distal nephron. This response reduces the rate of net water loss from the body without reducing salt loss. If this were the only homeostatic response, the normal plasma osmolarity could ultimately be restored, but the lost volume can be restored only by drinking. As will be seen in subsequent sections, the behavioral responses of thirst and salt appetite are fully integrated with the renal control systems.

Neural and Endocrine Effects on Glomerular Filtration Rate

In the absence of any response by the distal parts of the nephron, increasing or decreasing glomerular filtration rate (GFR) would amount to increasing or decreasing loss of isotonic fluid. Mild sympathetic activation, such as occurs when a person stands up or begins to exercise, constricts renal arterioles, reducing the rate of blood flow to the kidney (see Chapter 15). However, the tubuloglomerular feedback described in Chapter 19 maintains GFR almost constant in the face of such changes in renal perfusion. Only when the sympathetic nervous system is intensely activated, as in severe bleeding or hemorrhage, does the renal blood flow fall low enough to significantly decrease GFR.

The kidneys normally respond to an increase in total body Na$^+$ and extracellular fluid volume with **natriuresis,** an increase in the rate of urinary loss of Na$^+$ and water. In 1980, muscle cells of the heart atria were discovered to contain a factor that significantly increased natriuresis. This 28 amino acid factor can now be called **atrial natriuretic hormone (ANH)** (or **atrial natriuretic peptide**). A significant positive correlation between right atrial pressure and ANH release has been demonstrated in human subjects. Increasing Na$^+$ intake in human volunteers also resulted in significant alterations in plasma ANH. It is now clear that ANH has multiple actions throughout the body that contribute to decreasing plasma volume and blood pressure. A major effect is increased GFR (Figure 20-8), believed to be the result of dilation of renal afferent arterioles and of relaxation of glomerular mesangial cells. Other effects of ANH will be described in subsequent sections.

Control of Water Reabsorption by the Antidiuretic Hormone System

The hypothalamus contains **osmoreceptor** neurons that are stimulated by increases in plasma osmolarity, as would occur in the examples of net loss of pure water or gain of pure NaCl. Increased activity of the hypothalamic osmoreceptors results in increased release of antidiuretic hormone (ADH) from the posterior pituitary (Chapter 5). ADH increases the water permeability of the collecting duct (see Chapter 19). The greater the plasma levels of ADH, the more water can be recovered in response to the osmotic gradient provided by the medullary interstitial fluid. With increased levels of ADH, the kidney recovers more water from glomerular filtrate, so less water is excreted and the urine is more concentrated.

For example, if extracellular fluid osmolarity

Control of Body Fluid, Electrolyte, and Acid-Base Balance

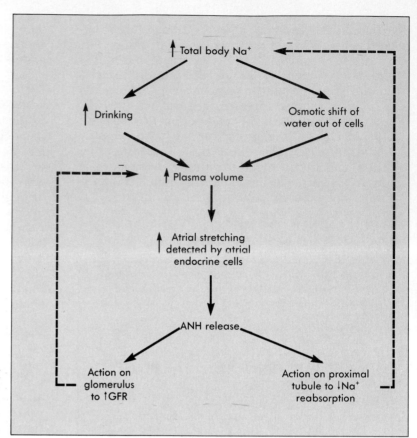

FIGURE 20-8

The sequence of events in the response of the ANH system to an increase in total body Na$^+$. The feedback loop is completed when the decreased reabsorption of Na$^+$ and volume triggered by the hormone restores extracellular fluid volume to normal.

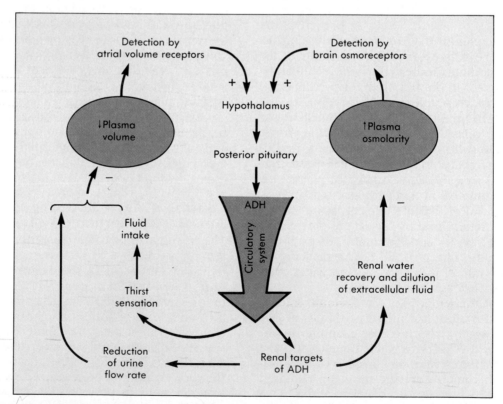

FIGURE 20-9

The response of the ADH system to a decrease in plasma volume or to an increase in plasma osmolarity is an increase in ADH secretion. The negative feedback pathways show the corrective effect of increased renal water reabsorption and increased water consumption on plasma volume and osmolarity.

KIDNEY AND BODY FLUID HOMEOSTASIS

were elevated (a pure salt challenge as in Figure 20-7), water would move from the intracellular compartment into the extracellular compartment. The loss of 250 ml of urine that had an osmolarity of 1200 mOsm/L would decrease total body water by 250 ml, but the excretion of the solute would cause plasma osmolarity to decrease toward normal, and 500 ml of water would shift back into the intracellular compartment. This water shift would make the intracellular volume and extracellular fluid osmolarity more nearly normal at the expense of the extracellular volume. The feedback loop that stimulates ADH secretion during increases in plasma osmolarity is shown in Figure 20-9.

Water loads that dilute the plasma decrease ADH secretion profoundly. When ADH levels fall, urine passing through the collecting duct gives up relatively little of its water, and a large volume of dilute urine is produced. This has the same effect as loss of a small amount of isotonic solution plus loss of a large volume of pure water. The net effect is now an intracellular-to-extracellular water shift.

The major responses of the ADH system are to changes in plasma osmolarity, but the system also receives input from stretch receptors in the atria. Recall from Chapter 15 that right atrial pressure is a sensitive indicator of plasma volume. When plasma volume is normal, the stretching of the atria during the cardiac cycle has an inhibitory effect on ADH secretion; when plasma volume is decreased, the atria are less stretched and the inhibition is removed. After hemorrhage, for example, total body water and venous pressure are decreased but the plasma osmolarity is normal. Since the osmolarity remains normal, ADH secretion is controlled entirely by the atrial volume receptors, which promote water retention (see Figure 20-9). In dehydration, ADH secretion is stimulated both by the decrease in venous pressure and by the increase in osmolarity. In this case, the net secretion of ADH produced by the combined volume receptor reflex and osmoreceptor reflex is greater than that which would be caused by either factor acting alone.

When both the blood volume and the blood osmolarity deviate from normal, ADH secretion is usually dominated by input from the hypothalamic osmoreceptors. It is adaptive for osmolarity regulation to receive a higher priority than volume regulation, because the volumes of cells and the activity of nerves and muscles depend from minute to minute on the correct osmolarity, whereas changes in volume can be tolerated and may be compensated for by the cardiovascular reflexes.

Control of Sodium Reabsorption and Vascular Tone by the Renin-Angiotensin-Aldosterone Hormone System

The endocrine cells of the juxtaglomerular apparatus are sensitive to three factors: (1) the filtered load

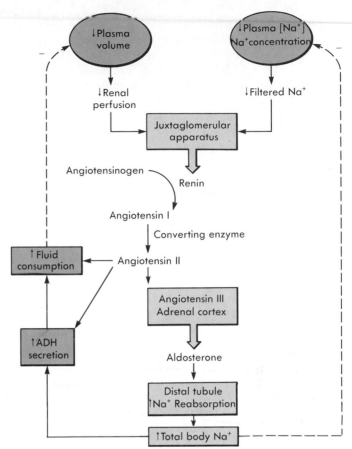

FIGURE 20-10
The response of the aldosterone system to a decrease in plasma volume or in plasma Na$^+$ concentration is an increase in renin secretion, which leads ultimately to an increase in aldosterone secretion. The increased recovery of filtered Na$^+$, along with continued Na$^+$ intake from diet, increases total body Na$^+$. The dashed lines show the corrective effect of aldosterone secretion. If the initial problem were a volume decrease, the osmoreceptors of the ADH system will ensure that increases in total body Na$^+$ result in restoration of volume.

of Na$^+$, (2) the blood pressure in afferent arterioles, and (3) inputs to the juxtaglomerular cells from sympathetic nerves. The first factor is a measure of plasma Na$^+$ concentration; the second and third factors are measures of plasma volume. If the Na$^+$ concentration or the blood pressure to the kidney drops or if sympathetic activity rises, the juxtaglomerular cells increase their secretion of renin (Figure 20-10).

Renin is a protease that catalyzes the first step of a regulatory cascade. Each renin cleaves a 10 amino acid fragment, **angiotensin I**, from each of a number of molecules of a blood protein **angiotensinogen.** Angiotensin I is converted to the 8 amino acid peptide **angiotensin II** by a converting enzyme **(ACE)** that resides primarily in the endothelial cells of blood vessels. Angiotensin II itself is a potent vasoconstrictor and also stimulates ADH. In the adrenal cortex it is converted to the 6 peptide **an-**

giotensin III, which stimulates secretion of the steroid hormone **aldosterone** by the adrenal cortex.

Aldosterone increases reabsorption of Na^+ by the distal tubule. Because so much of the filtered load of Na^+ is reabsorbed by more proximal parts of the nephron, aldosterone can affect only the potential reabsorption of the last 3% of the filtered load. Nevertheless, this 3% (equivalent to about 30 to 50 grams of NaCl per day) is an essential contribution to overall Na^+ balance in the body. Aldosterone also stimulates Na^+ reabsorption at several other sites, including the colon and the reabsorptive portion of sweat and salivary glands. At all of these sites, aldosterone decreases the rate of Na^+ loss. If the rate of Na^+ intake remains unchanged, increasing plasma levels of aldosterone will increase the total body Na^+ content. Because extracellular fluid volume is closely linked to total body Na^+, the renin-angiotensin-aldosterone system (RAA) exercises a strong influence on extracellular fluid volume and thus on blood pressure. Furthermore, angiotensins are potent vasoconstrictors, so they may also contribute to regulation of arterial blood pressure.

Under normal conditions, increases or decreases in plasma aldosterone change the extracellular fluid volume rather than extracellular fluid Na^+ concentration. The reason for this lies in the relative response times of the ADH and aldosterone systems. Aldosterone takes effect only after several hours and controls such a small part of the filtered load of Na^+ that it takes several days for aldosterone to affect the plasma Na^+ concentration. ADH acts on 20% of the filtered load of water, and its effect is rapid. Therefore water reabsorption keeps up with Na^+ reabsorption, and aldosterone-induced Na^+ retention is equivalent to gain of isotonic saline.

Disorders of the RAA system cause the mean arterial pressure to depart from its setpoint. For example, renin secretion is inappropriately high in some adults. The effects of the regulatory cascade cause a chronic **high-renin hypertension**. In several adrenal disorders there is hypersecretion of aldosterone, which results in a characteristic Na^+ retention and hypertension. In Addison's disease, a general failure of secretion of adrenal steroids, one symptom is negative Na^+ balance and hypotension.

Control of Na^+ and Water Loss by the Atrial Natriuretic Hormone System

The major role of the ANH system (see Figure 20-8) is to oppose accumulation of excess plasma. Plasma excess corresponds to gain of isotonic fluid; loss of excess plasma requires a decrease in total body Na^+ (i.e. natriuresis) as well as loss of some extracellular water. Effective natriuresis requires both increase in GFR and inhibition of Na^+ and water recovery in the distal nephron. In addition to its effect on GFR, the ANH system opposes the ADH system at almost every step. Increased stretch of the atria stimulates ANH secretion, while it inhibits ADH secretion. ANH is a potent inhibitor of ADH release, and it opposes the ADH effect in the collecting duct by activating an antagonistic second messenger system.

The ANH system acts as a brake on the RAA system as well as the ADH system. ANH inhibits reabsorption of filtered Na^+ in the distal parts of the nephron, antagonizing the stimulatory effects of aldosterone. The feedback loop is completed when the decreased reabsorption of Na^+ and volume triggered by the hormone restores extracellular fluid volume to normal. Another aspect of ANH-RAA antagonism is the fact that angiotensins are vasoconstrictors, whereas ANH is a vasodilator. Although the natriuretic effect of ANH is fully confirmed in human volunteers, much of what is known about the mechanisms of action of this recently discovered hormone is based on animal experiments.

Thirst and Salt Appetite

The reserve capacity of the kidneys to clear water from the body is so great and the ADH response so rapid that excessive voluntary fluid intake cannot normally cause more than a transient decrease in plasma osmolarity. However, thirst is the primary defense against decreases in plasma volume. The kidneys can, at best, only slow the loss of water through one of the several routes by which it leaves the body. The ADH, RAA, and ANH systems all seem to be involved in control of thirst. The stimuli that elicit thirst (hypovolemia, increased plasma osmolarity) are the same as those that increase ADH secretion. This fact and the close association of ADH neurons to thirst centers of the hypothalamus suggest that ADH secretion and thirst share a common pathway. Angiotensin II is the most powerful thirst-causing substance known, so the RAA system ultimately affects not only renal performance but also the rate of fluid intake (Figure 20-10). ANH appears to antagonize the thirst response to ADH as well as secretion of ADH by the hypothalamus.

When dietary salt is severely restricted, the combined effects of the ADH and RAA systems cause daily loss of Na^+ to fall from about 100 mEq to a few tens of mEq. Dietary intake must match this minimal loss. Trade and technology have made table salt abundant and cheap, but during most of human evolution salt was a scarce and valuable commodity for most people. Human beings have a powerful drive to consume salt; when salt is readily available most people voluntarily consume more of it than the minimum needed to maintain Na^+ balance. Therefore in modern societies body salt content almost never falls so low that it is regulated by feedback, although such regulation is important in other animals.

KIDNEY AND BODY FLUID HOMEOSTASIS

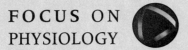

Hypertension

Hypertension, or high blood pressure, is defined as a sustained systolic pressure of 140 mm Hg or more and diastolic pressure of at least 90 mm Hg. Untreated hypertension is associated with an increased risk of coronary disease, kidney failure and stroke. Factors that are associated with an increased incidence of hypertension include stress, advancing age, lack of exercise, obesity and a high-salt diet. In about 1/10 of cases, hypertension can be accounted for as a consequence of some known disease process. One example of such a process is compromise of renal blood flow by atherosclerosis or as a result of a developmental defect in the renal artery. These conditions result in an inapropriate secretion of renin that increases aldosterone levels above those appropriate for homeostasis. Aldosterone levels also rise in Cushing's syndrome, an abnormally high secretion of some or all adrenal cortical steroids. Excessive levels of aldosterone cause total body Na^+ content to rise, increasing extracellular fluid volume. Emotional stress or tumors of the adrenal medulla can result in excessive blood levels of epinephrine, which causes hypertension by its vasoconstrictive and cardiac stimulatory effects.

The 9 out of 10 cases for which there is no demonstrable cause are termed essential hypertension. Essential hypertension may run in families, suggesting that there is a genetic predisposition. Several strains of experimental animals have been bred for a tendency to develop hypertension; these strains are potential models for inherited hypertension. The incidence of hypertension differs among cultures and between racial groups. In particular, Americans of black African descent have twice the risk of developing hypertension as Americans of European descent. Some evidence suggests that at least part of this difference is of genetic origin.

A key step in development of essential hypertension is resetting of the setpoint of the baroreceptor reflex that normally opposes increases in arterial blood pressure. Recently, endothelin, a paracrine agent released by the endothelium or lining of blood vessels, has been found to be capable of altering the responsiveness of carotid sinus baroreceptors. Investigations are underway to evaluate the possible role of endothelin in development of hypertension.

There are several families of antihypertensive drugs. These include beta-blockers such as propranolol, which antagonizes the effect of sympathetic inputs to the heart, reserpine, which depletes adrenal stores of epinephrine, and diuretics such as furosemide, which cause a salt-losing diuresis. Angiotensin-converting enzyme (ACE) blockers are a promising new family of drugs especially effective against high-renin hypertension. Renin initiates the cascade by converting angiotensinogen to angiotensin I. The hypertensive effects of renin depend on conversion of angiotensin I to angiotensin II by angiotensin-converting enzyme; angiotensin II both constricts systemic arterioles and is converted to angiotensin II which stimulates aldosterone release. ACE blockers inhibit the enzyme that converts angiotensin I to angiotensin II, thus preventing both the vascular and renal effects of the cascade.

POTASSIUM HOMEOSTASIS

- *Why is the regulation of K⁺ concentration so important?*
- *What factors affect the relative rates of H⁺ and K⁺ secretion by the distal tubule?*
- *Why do secretion of both K⁺ and H⁺ diminish in Addison's disease?*

Location of K⁺ in the Body

About 98% of total body K⁺ is in the intracellular compartment. Each cell is largely responsible for its own K⁺ homeostasis, and the cytoplasmic concentration of K⁺ differs among cell types, although K⁺ is the major cytoplasmic cation in all cells. Assuming that the plasma K⁺ concentration remains within normal limits, the total body K⁺ is regulated by the rate of K⁺ pumping of cells. The regulatory system for plasma K⁺ acts on less than 2% of the total body K⁺. Why is its job so important?

The resting potential of cells depends mainly on the K⁺ concentration gradient across their membranes. The resting membrane potential is largely determined by the ratio of the intracellular K⁺ concentration (a large value) to the extracellular K⁺ concentration (a small value). In this ratio, the small number exercises a kind of leverage. A change of 1 mEq/L in the intracellular concentration would have a trivial effect on cell membrane potentials. In contrast, because the extracellular K⁺ concentration is a relatively small number, increasing it or decreasing it by even as little as 1 mEq/L would have a significant effect on the resting potentials of all cells in the body. Precise regulation of the relatively small amount of K⁺ that is in the extracellular fluid is an important homeostatic function of the kidney.

Regulation of K⁺ Excretion by Aldosterone

The normal daily intake of K⁺ is 40 to 120 mEq. Most of this intake is balanced by renal excretion, with only small losses in feces and sweat. About 65% of the filtered load of K⁺ is reabsorbed in the proximal tubule as a consequence of fluid reabsorption. An additional 20% to 30% is reabsorbed in the ascending loop of Henle by the Na⁺, K⁺, 2 Cl⁻ cotransport system. The distal tubule receives the remaining 10% of the filtered load, and it is in the distal tubule and collecting duct that all K⁺ regulation occurs.

The rate of loss of filtered K⁺ in the urine depends on the relative rates of opposing processes of tubular secretion and absorption. Each process is carried out by a separate type of epithelial cell (Figure 20-11). Active reabsorption is carried out by intercalated cells that possess an active K⁺ pump in their lumen-facing membranes. The K⁺ that is actively transported into intercalated cells diffuses out

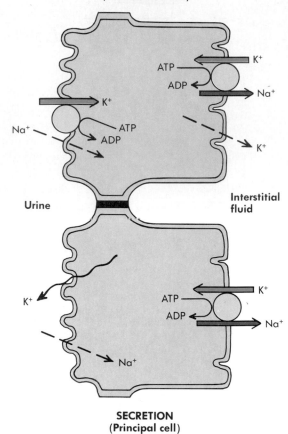

REABSORPTION (Intercalated cell)

SECRETION (Principal cell)

FIGURE 20-11

The distal tubular mechanisms for the renal handling of K⁺. In the reabsorptive cell (an intercalated cell), the apical K⁺ pump drives K⁺ into the cell and it leaks out into the extracellular fluid across the basolateral membrane. These cells also reabsorb Na⁺. Na⁺ leaks into the cell and is pumped out across the basolateral membrane in exchange for K⁺. The amount of K⁺ that passes through the basolateral membrane is the sum of that transported in across the apical membrane and that transported in by the Na⁺−K⁺ pump across the basolateral membrane. In the secretory cell (a principal cell), aldosterone induces a large leak of K⁺ across the apical membrane and stimulates the Na⁺−K⁺ pump. The K⁺ transported into the cell across the basolateral membrane leaks into the urine. At the same time, Na⁺ is leaking into the cell and being transported out by the Na⁺−K⁺ pump, so both cell types also reabsorb Na⁺. See Chapter 19 for photomicrographs of intercalated and principal cells.

of them into the cortical interstitial fluid. Intercalated cells have the ability to recover all of the K⁺ that is delivered to the distal tubule. In **hypokalemia** (low plasma K⁺), reabsorption predominates, and nearly all of the filtered load of K⁺ is recovered.

Active K⁺ secretion in the distal tubule and cortical collecting duct is carried out by principal cells, which are also responsible for most of the Na⁺ reabsorption of the distal nephron. The process is driven by the basolateral Na⁺=K⁺ pump (see Chapter 19), which concentrates K⁺ in the cytoplasm. There are K⁺ channels in the apical membrane of principal cells, so K⁺ transported into the cells from the blood

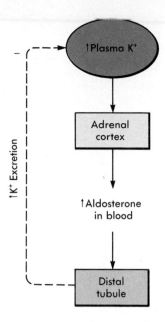

FIGURE 20-12
The direct response of the adrenal cortex to hyperkalemia. The effect of the increased aldosterone secretion is to increase K^+ secretion by the secretory cells of the distal tubule and thus to increase K^+ clearance by the kidney. The corrective effect is shown by the dashed line.

leaks out into the tubule lumen. When the plasma K^+ concentration rises above normal (**hyperkalemia**), the secretion of aldosterone by the adrenal cortex increases (Figure 20-12). Earlier in this chapter aldosterone was shown to increase in hyponatremia and hypovolemia as a result of activation of the renin-angiotensin cascade. The increase in aldosterone secretion in hyperkalemia is a direct effect of K^+ on the adrenal cortex and does not involve the renin-angiotensin pathway. Aldosterone acts on principal cells to increase the activity of the basolateral pump and the number and activity of apical channels, increasing K^+ secretion as well as Na^+ reabsorption. In hyperkalemia, the reabsorptive activity of intercalated cells continues but is partly canceled by the aldosterone-stimulated secretion, so urinary loss of K^+ increases.

In patients with adrenal gland tumors that secrete aldosterone at abnormally high rates, plasma Na^+ levels rise above normal and plasma K^+ falls below normal. The opposite condition results when the aldosterone-secreting cells of the adrenal cortex are damaged, as in Addison's disease. In this case, hypovolemia and low total body Na^+ are accompanied by hyperkalemia. Increased plasma K^+ can cause fatal cardiac arrhythmias and disorders of central nervous function and is the main reason that loss of adrenal cortical function is life threatening.

Interaction Between Na^+ and K^+ Regulation and Acid-Base Balance

All Na^+ reabsorbed from the urine must be matched either by absorption of an anion or by secretion of

another cation. In the proximal tubule, Cl^- and other anions can follow the reabsorbed Na^+, but in the distal tubule the passive permeability of the tubule wall to anions is low. Reabsorption of Na^+ must be matched by secretion of K^+ and H^+. Consequently, H^+ and K^+ behave as if they competed with one another for secretion (Figure 20-13, *A*; note, however, that H^+ and K^+ actually are secreted by different pathways). Therefore the plasma K^+ concentration and H^+ concentration tend to vary in opposite directions.

Because of the electrical link between tubular transport of Na^+, H^+, and K^+, a change in plasma H^+ concentration can cause a change in plasma K^+ concentration. Acidosis leads to increased H^+ secretion. This is associated with decreased K^+ secretion, resulting in increased plasma K^+ concentration. Alkalosis, in contrast, is often associated with decreased plasma K^+ concentration. When the plasma K^+ concentration decreases, H^+ secretion increases and the plasma pH becomes less acidic, causing alkalosis (Figure 20-13, *B*). Conversely, acidosis can be caused by increased plasma K^+ concentration.

A reciprocal relationship between H^+ and K^+ is seen in several pathological situations. In Addison's disease less Na^+ is reabsorbed so less H^+ can be secreted, causing acidosis. Aldosterone-secreting adrenal tumors have the opposite effect, resulting in alkalosis as well as reduced plasma K^+ (Figure 20-13, *C*). Osmotic diuretics increase the amount of Na^+ reaching the distal tubule, and the increase in Na^+ reabsorption increases secretion of K^+ and H^+, which tends to decrease plasma K^+ concentration and produce alkalosis. Finally, impairment of renal function by infection or decreased renal blood flow reduces both H^+ and K^+ secretion, causing both hyperkalemia and acidosis.

CALCIUM AND PHOSPHATE HOMEOSTASIS

- *Why is it necessary to distinguish plasma free Ca^{++} from total plasma Ca^{++}? What factors affect the ratio of free Ca^{++} to bound Ca^{++} in the plasma?*
- *Why must Ca^{++} regulation and phosphate regulation be closely linked?*
- *What is the one essential difference between the hormonal response to hypocalcemia and the hormonal response to hypophosphatemia?*

Transfer of Ca^{++} and Phosphate Between Bone, Intestine, and Kidney

Calcium's chemical activity in solution is significantly less than would be expected from its concentration, because calcium salts do not dissociate completely. The physiologically significant value is the concentration of "free" ionic Ca^{++} rather than the total plasma Ca^{++}. Plasma normally contains 2.5 mM Ca^{++}, of which about 46% is protein-bound

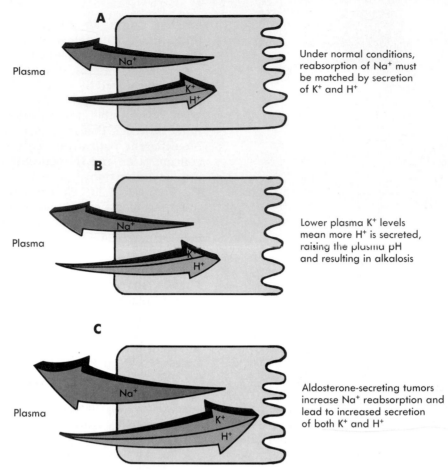

A

Plasma

Na⁺

K⁺
H⁺

Under normal conditions, reabsorption of Na⁺ must be matched by secretion of K⁺ and H⁺

B

Plasma

Na⁺

K⁺
H⁺

Lower plasma K⁺ levels mean more H⁺ is secreted, raising the plasma pH and resulting in alkalosis

C

Plasma

Na⁺

K⁺
H⁺

Aldosterone-secreting tumors increase Na⁺ reabsorption and lead to increased secretion of both K⁺ and H⁺

FIGURE 20-13

A summary of the interactions between the plasma concentrations of Na^+, K^+, and H^+.

and about 6% is bound by phosphate ($H_2PO_4^-$/ HPO_4^{2-}) and other small anions. This leaves about half of the total Ca^{++} as free Ca^{++}. The free and bound Ca^{++} exist in chemical equilibrium with each other according to the Law of Mass Action. Increases in the concentration of any agent that binds Ca^{++} (for example, plasma protein or phosphate) will decrease the free plasma Ca^{++}. Also, the ratio of bound to free Ca^{++} is very sensitive to plasma pH: acidosis increases the concentration of the ionized form, and alkalosis decreases it.

The plasma free Ca^{++} must be closely regulated because it is important for membrane excitability and for contractility in cardiac and smooth muscle. Elevated extracellular Ca^{++} (**hypercalcemia**) reduces excitability, causing lethargy, fatigue, and memory loss. Depressed plasma Ca^{++} (**hypocalcemia**) causes muscle cramps, convulsions, and other symptoms of increased neuromuscular excitability. Because plasma phosphate affects the ratio of free to bound Ca^{++} and because Ca^{++} is deposited in bone as a phosphate salt, Ca^{++} homeostasis and phosphate homeostasis must be closely linked.

Most of the Ca^{++} in the body is contained in bone. Although bone seems substantial and permanent, it undergoes continuous remodeling. Bone cells called **osteoblasts** deposit bone; a second population of cells called **osteoclasts** break it down. When deposition goes on at a greater rate than breakdown, the mass of bone increases and there is a net transfer of Ca^{++} and phosphate from the plasma to bone. When breakdown predominates, both Ca^{++} and phosphate are released into the plasma.

Calcium and phosphate can be transferred to and from bone to regulate plasma Ca^{++} and phosphate. The total amount of Ca^{++} in bone is so much greater than that in plasma that breakdown of some bone to maintain plasma Ca^{++} does not significantly weaken the skeleton. However, over time, Ca^{++} uptake from the diet must equal Ca^{++} loss in urine, or the strength of bone may be decreased. Active intestinal uptake of Ca^{++} from the diet ranges from 20% to 70% of the available Ca^{++}; the efficiency of the uptake is regulated in response to changes in plasma Ca^{++}. The kidneys are the major route of Ca^{++} and phosphate loss. Typically the renal clearances of both Ca^{++} and phosphate are less than the glomerular filtration rate because of tubular reabsorption. The rates of reabsorption of both Ca^{++} and phosphate are hormonally regulated, as well.

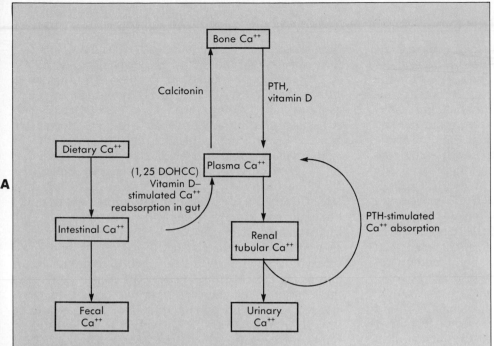

FIGURE 20-14
A Routes of gain and loss of Ca^{++} from the body. Some pathways are stimulated by hormones as indicated.
B A scanning electron micrograph of compact and spongy bone.

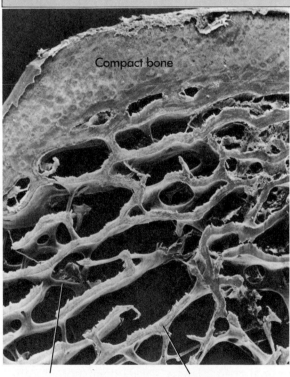

Compact bone

Spongy bone   Trabecula

Endocrine Regulation of Total Body Calcium and Phosphate

The hormones involved in Ca^{++} homeostasis are: (1) **calcitonin**, a protein produced and released by the parafollicular cells of the thyroid gland; (2) **parathyroid hormone** or **parathormone (PTH)**, a protein secreted by the parathyroid glands near the thyroid; (3) **1,25 dihydroxycholecalciferol (1,25 DOHCC)**, a steroid (see Chapter 3), and (4) gonadal steroids (estrogens or testosterone).

Of these hormones, calcitonin is perhaps the least significant in humans. Calcitonin stimulates bone deposition and decreases plasma Ca^{++} (Figure 20-14). Decreases in calcitonin from its basal level allow for Ca^{++} release from bone, but plasma Ca^{++} is regulated adequately after removal of the thyroid gland.

Secretion of PTH increases when the free Ca^{++} concentration of plasma flowing through the parathyroid glands decreases. Parathyroid hormone has three distinct effects (see Figure 20-14), two of which are on the kidney: (1) it causes breakdown of bone; (2) it increases renal reabsorption of Ca^{++}; and (3) it decreases renal reabsorption of phosphate, increasing the clearance of phosphate. The increase in phosphate clearance is an important part of the picture. If the phosphate released in bone mobilization remained in the plasma, it would bind Ca^{++} and oppose restoration of normal free Ca^{++} levels.

1,25 DOHCC is a derivative of **vitamin D$_3$** from meat and fish and **vitamin D$_2$** from plant sources. In the body, vitamin D$_3$ may be synthesized from cholesterol. A key event in the process is the opening of one of the rings of the steroid structure, which occurs in the skin as an effect of ultraviolet light. Similarly, irradiation of milk with ultraviolet light increases its content of vitamin D$_3$. Vitamin D$_2$ and D$_3$ are of identical potency in humans and are collectively referred to as **cholecalciferol**.

The liver adds a hydroxyl group to the 25th carbon of cholecalciferol, while the kidney performs a second hydroxylation, yielding 1,25 DOHCC. Both increased PTH and decreased plasma phosphate stimulate formation of 1,25 DOHCC, so either low plasma Ca^{++} or low plasma phosphate increases levels of 1,25 DOHCC.

In the intestine, 1,25 DOHCC stimulates Ca^{++} and phosphate absorption. It must be present for normal absorption of Ca^{++} and phosphate from the diet. In its absence, children develop **rickets**, a condition of impaired bone growth. In adults the corresponding condition is called **osteomalacia** ("bad bone"). Deficiency of the hormone became more common when industrialization forced people to spend the daylight hours indoors, diminishing the body's intrinsic production of 1,25 DOHCC. Given that the intestinal effect of 1,25 DOHCC is necessary for normal bone growth and maintenance, it may be surprising to find that 1,25 DOHCC, like PTH, causes bone breakdown. The intestinal effect and the bone effect of 1,25 DOHCC both tend to increase plasma Ca^{++} and plasma phosphate as well.

To understand how PTH and 1,25 DOHCC work together to regulate both plasma Ca^{++} and plasma phosphate, consider the hormonal responses to decreases of either Ca^{++} alone (**hypocalcemia**) or phosphate alone (**hypophosphatemia**) (Figure 20-15). If plasma Ca^{++} decreases, PTH levels increase, increasing 1,25 DOHCC levels. The 1,25 DOHCC and PTH release Ca^{++} and phosphate from bone, 1,25 DOHCC increases uptake of Ca^{++} and phosphate from the diet, and PTH causes the kidney to retain the Ca^{++} and clear the extra phosphate.

If plasma phosphate falls (**hypophosphatemia**), 1,25 DOHCC levels will increase because of the effect of phosphate on 1,25 DOHCC activation in the kidney. But PTH levels will decrease because lower levels of plasma phosphate automatically increase levels of plasma free Ca^{++} according to the Law of Mass Action as described above. In hypophosphatemia, as in hypocalcemia, 1,25 DOHCC releases both Ca^{++} and phosphate from bone, but the kidney, responding to decreased PTH, increases its clearance of Ca^{++} and decreases its clearance of phosphate. Thus in hypophosphatemia the kidney retains phosphate mobilized from bone and clears the calcium. The key difference between the responses to decreased plasma Ca^{++} and decreased plasma phosphate is that in hypocalcemia both PTH and 1,25 DOHCC levels rise, while in hypophosphatemia 1,25 DOHCC levels rise but PTH levels fall (see Figure 20-15).

Gonadal steroids are not involved in regulation of plasma Ca^{++} levels, but some estrogen or test-

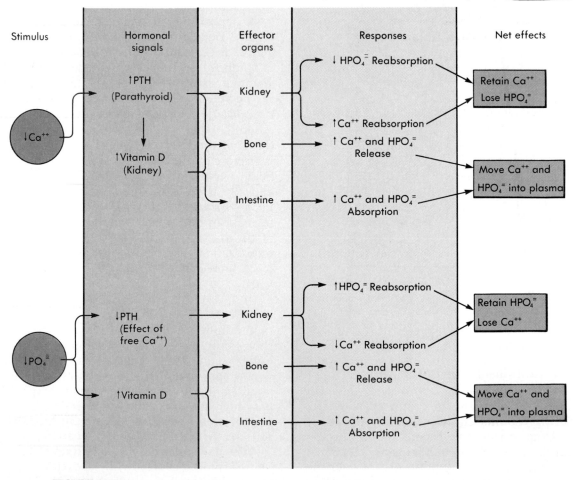

FIGURE 20-15
Comparison of the sequence of events in the homeostatic responses to hypocalcemia and hypophosphatemia.

osterone is necessary for long-term maintenance of normal bone mass after puberty. In the absence of estrogen or testosterone, Ca^{++} is steadily lost from bones; this loss ultimately results in a condition of spongy, fragile bone called **osteoporosis.** In men, levels of testosterone adequate to prevent osteoporosis are generally maintained throughout life, even though testosterone secretion declines in old age (see Chapter 25). In women, estrogen levels fall below the level that maintains bone mass after menopause (the cessation of ovarian function in middle age) or if reproductive cycles are interrupted by an eating disorder or intensive athletic training (see box essay on osteoporosis, Chapter 24).

ACID-BASE HOMEOSTASIS

- *What physiological processes make it possible for bicarbonate to be an effective buffer for extracellular fluid?*
- *What is alkalosis? Acidosis? Is it possible for a patient to have a plasma pH that is within normal limits and still be acidotic or alkalotic? Is it possible for a patient to be acidotic and alkalotic at the same time?*
- *What is meant by "anion gap"? Is the gap caused by a lack of anions?*

The Importance of pH Regulation

Small changes in the relative abundance of H^+ and OH^- have widespread effects in the body. Perhaps the most general effect of pH is on enzyme function. Typically enzymes are able to perform their catalytic functions optimally only when the pH is within a narrow range that is characteristic for each individual enzyme. This range is called the enzyme's **pH optimum.** The pH optimum of most enzymes that operate within the physiological interior is close to the normal pH of cytoplasm or extracellular fluid, depending on the location of the enzyme. Changes in pH also affect the excitability of nerve and muscle cells; decreases in pH reduce excitability, while increases increase it.

The pH of arterial plasma is regulated at 7.4. The pH of cytoplasm is usually slightly lower than that of plasma. The difference is due largely to the electrical potential across cell membranes; in many cells the H^+ and OH^- ions apparently distribute themselves passively across the cell membrane according to the inside-negative membrane potential. Some cells appear to be able to regulate their internal pH by active transport of H^+ or OH^-.

Buffer Systems

The remainder of this chapter requires an understanding of the chemical basis of buffer systems. A Focus Unit has therefore been provided, summarizing the chemistry of acids and bases with emphasis on buffer chemistry. The major buffer systems of the body are summarized in Table 20-1. For present purposes, keep the following facts in mind:

1. Acids are solutes that dissociate into H^+ and a **conjugate base.** The ratio between the undissociated acid and its products is determined by the law of mass action according to an equilibrium constant (K_a) that is characteristic for each acid. It is frequently convenient for computations to use the negative logarithm of the K_a, the pK_a. **Strong acids** dissociate relatively completely and thus have large K_a values. In the case of **weak acids,** a substantial concentration of undissociated acid is maintained at equilibrium, so the K_a is low.

2. Weak acids and their conjugate bases can serve as buffer systems. If H^+ is added to the systems, some of the added H^+ will react with conjugate base and be converted to the undissociated acid form, whereas if H^+ is removed from the system, some undissociated acid will dissociate and partly restore the H^+ concentration. Weak acids are most effective as buffers when the pH of the system is close to the pK_a of the acid, because under these conditions the concentrations of the undissociated acid and its conjugate base are nearly equal. Buffer effectiveness also depends on the total concentration of buffer present, referred to as the **buffer capacity.**

3. As strong acid is added to a buffer system in which the pH is near the pK_a, the pH decrease with each addition of strong acid is quite small until the concentration of conjugate base becomes small relative to the concentration of buffer acid. At this point further additions of acid cause the pH of the solution to drop more

TABLE 20-1	Buffer Systems of the Body
Buffer system	**Description/function**
Hemoglobin (protein)	Intracellular proteins and the major plasma protein, hemoglobin, act as buffers and compose about 75% of the total buffer capacity of the body
Bicarbonate	Although components of the bicarbonate buffer system are present at relatively low concentrations and the buffer system has a pK_a of 6.1, the fact that the components of this system are regulated causes it to play an important role in controlling the pH of the extracellular fluid
Phosphate	The concentration of phosphate in the extracellular fluid is low compared with the other buffer systems, but phosphate is an important intracellular buffer and is the major buffer in the urine

rapidly, and the conjugate base is said to be titrated. The amount of strong acid needed to titrate the conjugate base is a measure of the buffer capacity of the system. If the buffer system were titrated instead with strong base, the pH would rise slowly until almost all of the buffer acid had been converted to conjugate base, whereupon further addition of strong base would cause large increases in pH.

4. A major buffer system in the body is the bicarbonate buffer system (see Table 20-1):

$$CO_2 + H_2O \rightleftharpoons H_2CO_3 \rightleftharpoons H^+ + HCO_3^-$$

The first reaction, facilitated by the enzyme carbonic anhydrase, is the combination of water and carbon dioxide to form carbonic acid. Carbonic acid is a weak acid and dissociates to H^+ and HCO_3^-. The equilibrium constants for the two reactions can be summed together and treated as a single constant. The summed pKa of the bicarbonate buffer system is 6.1 at body temperature. The Henderson-Hasselbalch equation applies the Law of Mass Action to the bicarbonate buffer system (see the Focus Unit following this chapter):

$$pH = 6.1 + \log [HCO_3^-]/0.03 \, P_{CO_2}$$

Multiplying the plasma P_{CO_2} by 0.03 in the denominator converts it to carbon dioxide concentration.

Respiratory and Renal Contributions to Regulation of Plasma pH

The bulk of the buffer capacity of the whole body is due to protein (Table 20-1). Most of the protein is intracellular (including hemoglobin in red blood cells), but it is difficult to assess the state of intracellular buffer systems, so most of this section has to do with regulation of the extracellular pH.

The normal values for components of the bicarbonate buffer system of arterial plasma are: (1) pH = 7.4; (2) P_{CO_2} = 40 mm Hg; and (3) $[HCO_3^-]$ = 24 mEq/L. Acidosis and alkalosis can arise in two fundamentally different ways. One way is by an excess or deficit of CO_2, the other is by an excess or deficit of fixed acid. **Respiratory acidosis** is caused by excessive plasma P_{CO_2}. Carbon dioxide is referred to as a volatile acid to distinguish it from fixed acids that cannot be eliminated by the respiratory system. **Respiratory alkalosis** arises from a subnormal plasma P_{CO_2}. Respiratory acidosis can be caused by a failure of respiratory function (Table 20-2), either at the respiratory control system level or at the lung level. Respiratory alkalosis can be caused by deliberate hyperventilation, anxiety, or rapid ascent to high altitudes.

Metabolic acidosis and **metabolic alkalosis** are the terms for an excess of any fixed acid or base, whether of metabolic origin or not. Both metabolic acidosis and alkalosis can be caused by the exces-

TABLE 20-2	Typical Causes of Acid-Base Abnormalities
Condition	**Typical causes**
Metabolic acidosis	Diarrhea resulting in excess loss of bicarbonate from the intestine
	Renal failure resulting in an inability to secrete H^+ ions
	Untreated diabetes resulting in excess keto acid formation
	Hyperkalemia
Metabolic alkalosis	Vomiting resulting in the loss of gastric acid
	Abuse of antacids
	Aldosterone-secreting tumors
Respiratory acidosis	Lung diseases such as chronic bronchitis and emphysema that decrease alveolar ventilation
	Injury to respiratory control centers; barbiturate overdose; hypoventilation*
Respiratory alkalosis	Hyperventilation*: rapid ascent to high altitude

*NOTE: In patients who are artificially ventilated, failure to monitor blood gases can result in either respiratory acidosis or alkalosis.

sive dietary intake of acid or base (see Table 20-2), and can also arise from a wide variety of disease conditions and intoxicants. For example, in renal failure the kidney becomes unable to secrete H^+, resulting in acidosis.

The **Davenport diagram** (Figure 20-16) is a useful graphic representation of the Henderson-Hasselbalch equation. The vertical axis is the arterial plasma HCO_3^- concentration; the horizontal axis is the arterial pH. The circled point indicates the normal state of the bicarbonate buffer system. It is the point on the graph specified by the normal value of plasma pH (7.4) and the normal plasma HCO_3^- concentration (24 mEq/L). If P_{CO_2} is held constant at 40 mm Hg, addition of fixed acid or base will cause the pH and $[HCO_3^-]$ values to trace out the central dashed line. This is a titration curve for the bicarbonate buffer system at this P_{CO_2}.

Sometimes such titration curves are referred to as **CO_2 isobars** (isobar means "constant pressure"). A decrease in P_{CO_2} to 20 mm Hg (right-hand curve) results in an increase in the plasma pH to a value slightly above 7.6 and a decrease in bicarbonate to about 20 mEq/L (respiratory alkalosis). An increase in P_{CO_2} to 80 mm Hg (left-hand curve) results in a plasma pH less than 7.2 and a plasma HCO_3^- concentration of about 28 mEq/L (respiratory acidosis). The pKa of the bicarbonate buffer system is 6.1. At pH values around the physiological range, it might seem from the chemist's point of view that the bicarbonate buffer system is too far from its pKa to be

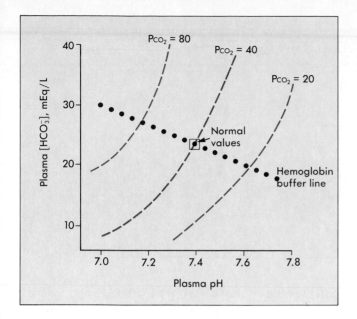

FIGURE 20-16

A Davenport diagram predicts the behavior of the bicarbonate buffer system in response to added fixed acid or base and to changes in P_{CO_2}. The squared point corresponds to the normal values of plasma pH, bicarbonate concentration, and P_{CO_2}. When any two of these values are normal, the Law of Mass Action dictates that the third value will be normal as well. A change in the P_{CO_2} changes the plasma bicarbonate and plasma pH values, tracing out the dotted line (the buffer line). The dashed lines are titration curves for the plasma system at the three different values of P_{CO_2} indicated.

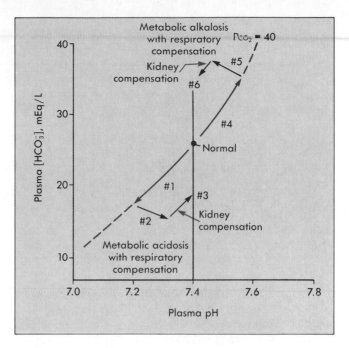

FIGURE 20-17

The effect on plasma pH of challenges to the buffer system and the regulatory systems. *Arrow 1* shows the change in the buffer system induced by injection of acid, followed by rapid but incomplete respiratory compensation *(arrow 2)*, and then by slower renal compensation *(arrow 3)*. The effect of injection of alkali *(arrow 4)* is similarly followed by rapid, incomplete respiratory compensation *(arrow 5)* and slower renal compensation *(arrow 6)*.

an effective buffer. This is not the case because the bicarbonate buffer system is not a "closed" system. During additions of fixed acid to the plasma, the P_{CO_2} remains constant unless there is a change in ventilation. The CO_2 that is formed during the titration of the bicarbonate buffer is lost to the atmosphere because the arterial blood is in equilibrium with alveolar gas. Thus the blood buffer system is referred to as an "open" system because formation of CO_2 in the buffer reaction does not increase the P_{CO_2}.

If the plasma P_{CO_2} is increased, the law of mass action dictates that the pH will decrease and the HCO_3^- concentration will increase. Similarly, if the P_{CO_2} is decreased, the pH rises and the HCO_3^- concentration decreases (see Figure 20-16). In response to changes of P_{CO_2}, the pH and HCO_3^- of the system will trace out the dotted line, which is called the **buffer line.** The slant of the buffer line results from the presence of other buffers (hemoglobin, protein, etc.) in blood. Each point on the buffer line lies on a titration curve for blood at that P_{CO_2}. Figure 20-16 shows just three of the many such curves possible, each corresponding to a different P_{CO_2}.

Most of the organic and inorganic acids produced by cellular metabolism are almost completely dissociated at the pH of body fluids (that is, they are strong acids). The rate of production of acids other

than carbonic acid in individuals who consume a typical high-protein diet is about 1 mmol/kg body weight/day. The bicarbonate buffer system minimizes the immediate impact of this acid on the pH of extracellular fluid. The daily production of metabolic acid would reduce bicarbonate stores only slightly, but this small reduction must be opposed by renal secretion of acid to prevent weakening of the bicarbonate reserve that protects the pH against the effects of abnormal acid loads.

The respiratory system and kidneys are both involved in minimizing the change in arterial pH that might be caused by the normal metabolic acid production (lactic acid, for example) and by abnormal loads or losses of fixed acid or base. This cooperation is illustrated by the normal response to a load of fixed acid, such as might be caused experimentally by injection of lactic acid. Arrow 1 in Figure 20-17 shows the immediate effect of injection of the acid; the injected acid is buffered by the bicarbonate, and both the pH and the plasma $[HCO_3^-]$ decrease along the titration curve. The amount of CO_2 produced by the buffer titration is relatively small compared with the body's own production of metabolic CO_2 and is easily lost by way of the lungs, so that P_{CO_2} remains constant.

Within a few seconds the decrease in arterial plasma pH is sensed by the peripheral chemorecep-

TABLE 20-3   *Renal and Respiratory Responses to Acid-Base Abnormalities*

Term	Compensation(before or after)	$[HCO_3^-]$	P_{CO_2}	pH
Metabolic acidosis	Before	Decreased	Normal	Decreased
	After	Decreased or normal	Decreased or normal	Decreased or normal
Metabolic alkalosis	Before	Increased	Normal	Increased
	After	Increased or normal	Increased or normal	Increased or normal
Respiratory acidosis	Before	Increased	Increased	Decreased
	After	Increased	Increased	Less decreased
Respiratory alkalosis	Before	Decreased	Decreased	Increased
	After	Decreased	Decreased	Increased or normal

tors, and the respiratory response described in Chapter 17 begins. Ventilation increases and arterial P_{CO_2} decreases. When P_{CO_2} decreases, the law of mass action mandates that both the H^+ concentration and the HCO_3^- concentration decrease. The decrease in H^+ concentration carries the arterial pH back in the direction of the normal value, but at the expense of a further decrease in $[HCO_3^-]$ (*arrow 2* in Figure 20-17).

In acid-base physiology, compensation is said to be complete if it carries the plasma pH back into the normal range. Complete compensation does not imply that all constituents of the bicarbonate buffer system are restored to their normal concentrations. The response of the respiratory system to acute changes in plasma pH is rapid but usually cannot be complete (Chapter 18). The injected lactic acid cannot easily penetrate the blood-brain barrier, so the pH of the CSF is not decreased by the injection and is in fact increased by the respiratory response to peripheral acidification. At some point in the response, the increase in CSF pH cancels out the respiratory response to the peripheral acid. At this point the condition of the buffer system can be described as **metabolic acidosis with respiratory compensation** (Table 20-3). In the Davenport diagram of metabolic acidosis (Figure 20-17), the extent of respiratory compensation is shown by how far the point is displaced to the right of the buffer titration curve.

Metabolic acidosis can become chronic in disease conditions such as diabetes mellitus (see Chapter 23) in which there is a sustained increase in the production of acidic end products. If acidosis continues for several days or longer, there is time for renal compensation to become effective. Normally the kidney recovers all of the filtered bicarbonate as a result of its acidification of the urine (Chapter 19). To solve the deficiency of bicarbonate in chronic metabolic acidosis, the kidney must not only save all of the filtered bicarbonate, it must also make some "new" bicarbonate to replace what was ti-

trated away by the acid (Figure 20-18, *A*).

The new bicarbonate is new only in the sense that it does not represent recovery of filtered bicarbonate. In acidosis the proximal tubular secretion of H^+ exceeds the amount necessary to recover all of the filtered bicarbonate. After all the filtered bicarbonate is recovered, the proximal tubular cells continue to secrete H^+ from their cytoplasm into the tubular lumen. The proximal tubular cells contain the enzyme carbonic anhydrase, which catalyzes conversion of CO_2 to H^+ and HCO_3^-. The H^+ is secreted into the urine and accepted by other urinary buffers such as phosphate, leaving HCO_3^- behind in the cells. The "new" HCO_3^- resulting from this process is returned to the blood to replace the HCO_3^- that reacted with the lactic acid. As the renal generation of new bicarbonate and loss of H^+ proceeds, the plasma HCO_3^- concentration and arterial pH increase toward normal (*arrow 3* in Figure 20-17). When this process is complete the respiratory system and kidneys have completely compensated for the acidosis, in that the plasma pH is again normal. However, note that the P_{CO_2} and the plasma $[HCO_3^-]$ are both displaced from their normal values by the compensation and will remain so until the injected acid is removed by metabolism or excretion.

The coordinated response to an alkaline load can be traced in the same way. In this case, the immediate response to injection of base (metabolic alkalosis) would be conversion of some H_2CO_3 to HCO_3^-. The plasma pH would increase along the titration curve (*arrow 4* in Figure 20-17). In response to the increase in plasma pH, respiration would be depressed, increasing the plasma P_{CO_2}. The result of the respiratory compensation would be a further increase in HCO_3^- concentration and restoration of the arterial pH almost to normal (*arrow 5* in Figure 20-17). At this point the condition would be described as **metabolic alkalosis with respiratory compensation** (see Table 20-3). The response of the respiratory control system to alkalosis is typically much weaker than that to acidosis.

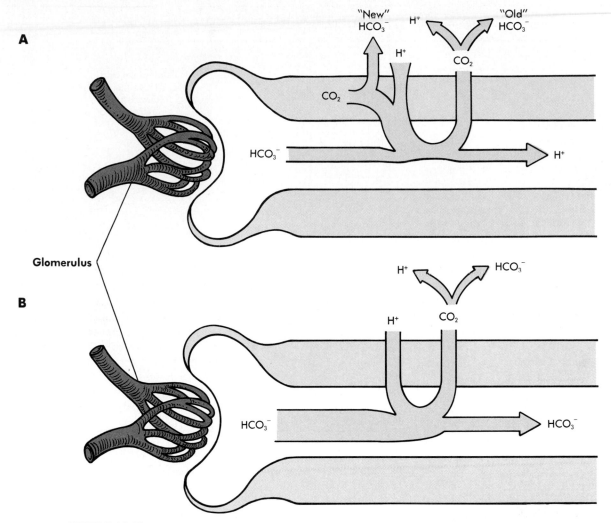

A

B

Glomerulus

FIGURE 20-18
Tubular responses to acidosis and alkalosis. The proximal tubule is shown, but similar processes go on in the distal tubule.
A The tubular response to acidosis. Secretion of H^+ increases. If the acidosis is due to fixed acid, the filtered load of HCO_3 decreases; if it is due to a respiratory problem, it increases. In either case, the secretion of H^+ is more than enough to recover all the filtered HCO_3^-. Extra secretion of H^+ has the effect of adding "new" HCO_3^- to the blood in addition to the recovery of filtered HCO_3^-; actually the "new" bicarbonate comes from CO_2 produced by oxidative metabolism in the tubular epithelium.
B The tubular response to alkalosis. In respiratory alkalosis the filtered load of HCO_3^- decreases; in metabolic alkalosis it increases. In either case, the secretion of H^+ is less than normal, and some of the filtered HCO_3^- is not recovered. The unreabsorbed HCO_3^- constitutes a net loss of alkali from the body, while all of the secreted H^+ is reabsorbed as CO_2.

Chronic metabolic alkalosis is frequently caused by habitual ingestion of antacid medication. In chronic metabolic alkalosis the task of the kidney is to dispose of excess HCO_3^- and conserve H^+. As HCO_3^- filtration is increased and H^+ secretion is simultaneously decreased, part of the filtered load of HCO_3^- is not reabsorbed, and HCO_3^- begins to spill into the final urine (Figure 20-18, *B*). For every HCO_3^- lost in the final urine, there is a net gain of one H^+ in the plasma, which will return the plasma pH to the normal range (*arrow 6* in Figure 20-17).

Renal Compensatory Responses in Respiratory Disease

Renal or pulmonary disease places a burden of compensation on the system that remains intact. For example, in acute lung disease with impaired gas exchange, the increased Pco_2 would decrease the plasma pH and increase the HCO_3^- concentration (*arrow 1* in Figure 20-19). The compensatory renal response would increase the HCO_3^- concentration further, but with time would probably be capable of restoring the arterial pH to the normal range (*arrow*

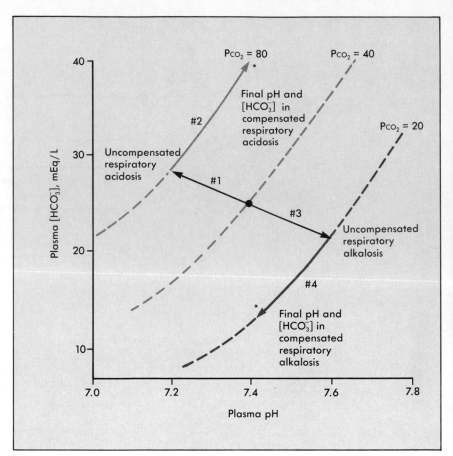

FIGURE 20-19

Respiratory acidosis *(arrow 1)* followed by slow but complete renal compensation *(arrow 2)*. Similarly, respiratory alkalosis *(arrow 3)* is followed by complete renal compensation *(arrow 4)*.

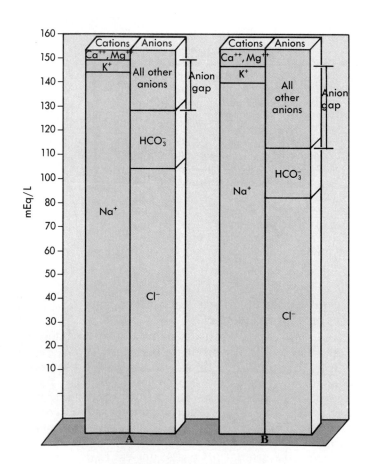

FIGURE 20-20

A A Gamblegram of plasma ionic constituents showing typical normal values.

B A Gamblegram of an untreated diabetic in ketoacidosis. Notice that the contributions of K^+, Ca^{++}, and Mg^{++} to the total cations have increased as a result of the acidosis. The contributions of Cl^- and HCO_3^- to the total anions have decreased, making room for the conjugate base ions of the ketoacids and increasing the anion gap.

Ethylene Glycol Intoxication

Acid-base disorders are usually not the direct result of consumption of acid or base, but some drugs and toxins are metabolized to acids or bases by the body, resulting in an acid-base problem. The toxicology of ethylene glycol is an example. Ethylene glycol is a major ingredient of antifreeze and windshield deicer. Children and pets may consume these liquids because of their sweet flavor, and alcoholics sometimes resort to them as substitutes for beverage alcohol. Ethylene glycol itself is relatively harmless, but it is metabolized by the liver to extremely toxic ketoaldehydes and ultimately to oxalic acid. Many of the consequences of ethylene glycol intoxication are due to the resulting metabolic acidosis. As with diabetic ketoacidosis, the anion gap increases at the expense of plasma [HCO_3^-].

The diagnosis of ethylene glycol intoxication is difficult because blood and urine of patients with suspected drug overdose are not routinely assayed for ethylene glycol. The intoxication without indications of ethanol ingestion, the metabolic acidosis, and the presence of crystals of oxalate in urine are important clues. The treatment consists of three basic approaches: removing the ethylene glycol from the blood, correcting the metabolic acidosis, and preventing further metabolism of the ingested ethylene glycol to oxalic acid. The last of these is frequently accomplished by giving ethanol, since ethanol competes with the ethylene glycol for metabolism by the liver, and the metabolites of ethanol are somewhat less toxic.

2). At this point, the existence of a **respiratory acidosis with renal compensation** could not be diagnosed from the arterial pH, but could be recognized readily from the elevated plasma [HCO_3^-] and P_{CO_2}. Furthermore, there could be no doubt that the acid-base disturbance was of respiratory and not metabolic origin, because in metabolic acidosis the [HCO_3^-] is depressed by the fixed acid and even further depressed by the respiratory response that reduces arterial P_{CO_2}.

Respiratory alkalosis occurs as an initial reaction to high altitude (Chapter 18), as a consequence of intoxication with poisons that stimulate the respiratory center, and in some neurotic anxiety states. Acute respiratory alkalosis is characterized by increased arterial pH and decreased plasma [HCO_3^-] (*arrow 3* in Figure 20-19). The compensatory renal response restores arterial pH at the expense of a further decrease in [HCO_3^-] (*arrow 4*).

Renal Compensatory Responses in Metabolic Acidosis and Alkalosis

Untreated diabetes mellitus is an example of disease-related metabolic acidosis. In diabetes mellitus, the lack of insulin, the hormone necessary for glucose entry into most cells, inhibits glucose metabolism and causes a high rate of fat catabolism. The end products of fat catabolism, the ketone bodies acetone, β-hydroxybutyric acid, and acetoacetic acid are released into the plasma in large quantities, causing **ketoacidosis.** The production of metabolic acid may exceed the rate at which H^+ can be secreted by the kidney, placing a large burden of compensation on the respiratory system.

In metabolic acidosis, blood chemistry may reveal the presence of unusual organic anions. The cation and anion composition of blood plasma can be shown in a type of bar graph called a **Gamblegram** (Figure 20-20). The Gamblegram is a rearrangement of the bar graph shown in Figure 20-2 that makes it clear that the total concentration of positive charges in the plasma exactly equals the total concentration of negative charges.

In standard clinical analyses of blood plasma, only the Na^+, K^+, Cl^-, HCO_3^- concentrations and pH are determined. The sum of [Na^+] + [K^+] accounts for the bulk of plasma cation (see Figure 20-20, *A*), but the sum of [Cl^-] + [HCO_3^-] leaves a substantial amount of plasma anion unaccounted for: lactate, phosphate, ketone bodies, and protein. The difference between the sum of [Na^+] + [K^+] and [Cl^-] + [HCO_3^-] is an approximate measure of the

Control of Body Fluid, Electrolyte, and Acid-Base Balance

SUMMARY

1. **Total body water** is distributed between the **intracellular compartment** (two thirds of the total) and the **interstitial** and **vascular** subcompartments of the **extracellular fluid**. The interstitial fluid comprises about 75% of the extracellular fluid.

2. The effective **osmotic pressures** of the intracellular and extracellular fluids are equal. The distribution of fluid between the two compartments is determined by the impermeant solute content of each compartment. Thus cell volume is determined by cytoplasmic protein and extracellular fluid by total body Na^+.

3. Total water turnover is 1.5 to 2.5 L/day. Roughly one-half of the volume lost is in the form of urine, about one-tenth as fecal water, and the rest by evaporation from skin and respiratory surfaces and secretions such as sweat and tears. Of these routes of loss, only the urine composition can effectively be regulated to balance water loss with gains from drinking and metabolism.

4. Secretion of the peptide hormone atrial natriuretic hormone (ANH) is stimulated by atrial stretch. This hormone increases **glomerular filtration rate** and antagonizes the Na^+-conserving tubular mechanisms. An increase in plasma levels of the hormone tends to reduce extracellular fluid volume.

5. The posterior pituitary secretes the peptide **antidiuretic hormone** (ADH) in response to increased osmolarity of plasma. Secretion of the hormone may be inhibited by increased atrial stretch. ADH increases water reabsorption across the wall of the collecting duct, stimulating water conservation and resulting in the production of a hyperosmotic urine. ADH could be said to be the hormone of water conservation.

6. The steroid hormone **aldosterone** is secreted by the **adrenal cortex** in response to a regulatory cascade that begins with secretion of **renin** from the **juxtaglomerular apparatus**. Renin catalyzes the conversion of plasma angiotensinogen to angiotensin I. Angiotensin II is produced by an endothelial converting enzyme. **Angiotensin III** is produced by the adrenal cortex and stimulates its secretion of **aldosterone**. Secretion of renin increases in response to decreases in glomerular filtration or the Na^+ concentration of the filtrate. Aldosterone increases the recovery of filtered Na^+ in the distal tubule, and could be characterized as the hormone of salt conservation. Under normal conditions increases in total body Na^+ are matched so rapidly by water retention that the main effect of the hormone is to increase extracellular fluid volume.

other plasma anions, including plasma proteins, and is traditionally called the **anion gap**. This is a deceptive name, since it implies that some anions are missing, whereas they simply are not measured in standard assays. The values in Figure 20-2 give an anion gap of about 18 mEq/L; if nonideal solute behavior is taken into account, the normal anion gap is about 15 mEq/L.

Changes in the anion gap indicate the cause of metabolic acidosis. In metabolic acidosis the conjugate base molecules of the acids that are present in abnormal abundance have replaced the bicarbonate ions that were titrated to CO_2 when the acids were added. Thus in metabolic acidosis the total anion concentration may well be normal, but the contribution of HCO_3^- to the total anion is decreased in favor of conjugate base anions that are not measured in the standard assay. Thus the anion gap increases (see Figure 20-20, *B*), and this increase indicates the presence of unmeasured anions. In the case of diabetic ketoacidosis, the unknown anions are acetoacetate and β-hydroxybutyrate. The HCO_3^- has been titrated out by the acids, which have partly taken its place in the anion bar.

● STUDY QUESTIONS

1. A 40-year-old female weighs 48 kg. Plasma Na^+ is 145 mEq/L. Assume that total body water is 50% of total body weight. What is the intracellular volume? What is the total intracellular solute? The intracellular osmolarity? The total body Na^+? If this individual were given 3 L of isotonic saline intravenously, what would be the new values for each of these variables?

2. In congestive heart failure, plasma levels of ANH increase. Explain why.

3. Explain why dietary Na^+ consumption usually does not lead to hypernatremia but may well contribute to hypertension.

4. Salt substitutes sold under such names as "light salt" usually consist in part of KCl, which has a salty taste that is not as pleasant as that of NaCl. Why would consumption of KCl as a substitute for NaCl be less likely to increase blood pressure?

5. Atherosclerosis of the renal artery restricting blood flow to one kidney results in hypertension. Explain why. Hint: The hypertension is not due to failure of urine formation by the flow-impaired kidney, since hypertension does not develop as a matter of course if one kidney is entirely removed.

6. Secretion of both ADH and ANH is affected by atrial stretching. Why is this part of the circulation monitored so closely by the systems that regulate extracellular volume?

7. A constituent of natural licorice, glycyrrhizic acid, causes hypokalemia, alkalosis, and Na^+ retention. What single action of the drug on the kidney could account for all these symptoms?

8. Hyperparathyroidism is characterized by lethargy, but frequently the disease is discovered when minor accidents result in broken bones. Explain the lethargy and bone fragility. What differences in plasma phosphate concentration would you expect in this condition?

9. From a strictly theoretical point of view, the bicarbonate buffer system is not a good one for extracellular fluid. Why not? What physiological processes make it possible for bicarbonate to be an effective buffer for extracellular fluid?

10. Respiratory compensation for acute metabolic alkalosis or acidosis is usually not complete. Explain.

11. An individual has a plasma $[HCO_3^-]$ of 12 mEq/L and a Pco_2 of 40 mm Hg. What is the plasma pH? What would Pco_2 need to be to restore the plasma pH to normal if plasma $[HCO_3^-]$ remained at 12 mEq/L?

12. An individual has a plasma pH of 7.1, a plasma $[HCO_3^-]$ of 24 mEq/L, and a Pco_2 of 80 mm Hg. Can a single acid-base abnormality explain all of these figures? How would you describe the acid-base problem in this case?

Choose the MOST CORRECT Answer

13. The dominant cation of extracellular fluid is:
   a. Na^+
   b. Ca^{++}
   c. K^+
   d. H^+

14. The "hormone of water conservation" is:
   a. Aldosterone
   b. Calcitonin
   c. ADH
   d. Atrial natriuretic hormone

15. Atrial natriuretic hormone is responsible for all of the following EXCEPT:
   a. Increase of water absorption across the collecting duct wall
   b. Increase in glomerular filtration rate
   c. Reduction in extracellular fluid volume
   d. Inhibition of Na^+-conserving tubular mechanisms

16. Of the following hormones, the least significant in humans is:
   a. Parathyroid hormone
   b. Calcitonin
   c. 1,25 dihydroxycholecalciferol
   d. Atrial natriuretic hormone

17. About 98% of the total body concentration of the cation is found in the intracellular fluid:
   a. Na^+
   b. Ca^{++}
   c. H^+
   d. K^+

18. The major effect of aldosterone is to:
   a. Increase recovery of filtered Na^+ in the distal tubule
   b. Decrease extracellular fluid volume
   c. Increase production of dilute urine
   d. Inhibit distal tubular K^+ secretion

19. Renal compensation of acidosis involves:
   a. Decreased secretion of H^+
   b. Decreased blood pH
   c. Inhibition of HCO_3^- reabsorption
   d. Increase in tubular H^+ secretion

● **SUGGESTED READINGS**

BEAUCHAMP GK: The Human Preference for Excess Salt, *American Scientist*, 75:27, 1987. Discusses salt appetite and the possible relationship of salt consumption to hypertension.

COWLEY AW: Jr Long-term control of arterial blood pressure. *Physiological Reviews* 72:231-300, January 1992. Reviews current state of understanding of the ADH, RAA and ANH control systems.

EDWARDS RM: Direct assessment of glomerular arteriole reactivity. *News in Physiological Sciences* 3:216-219, 1988. Gives additional evidence for effects of ANH on afferent and efferent arterioles.

FACKELMANN KA: The African gene? Searching through history for the roots of black hypertension. *Science News* 140:254-255, October 19, 1991.

KNOX FG, GRANGER JP: Control of Sodium Excretion: the Kidney Produces under Pressure, *News in Physiological Sciences*, 2:26, 1987.

ZEIDEL ML: Renal actions of atrial natriuretic peptide. *Annual Review of Physiology* 52:747-759, 1990. Recent review of evidence for multiple renal effects of ANH.

ACIDS, BASES, AND BUFFER SYSTEMS

The Law of Mass Action and Acid-Base Equilibrium

A reversible chemical reaction between two reactants (A and B) to form products (C and D) is described by the equation:

$$A + B \underset{k_2}{\overset{k_1}{\rightleftharpoons}} C + D \tag{1}$$

where k_1 and k_2 are called *unidirectional rate constants* and describe the probability for the reaction to occur in either direction. The rates for the left-to-right (forward) and right-to-left reactions are k_1 [A] [B] and k_2 [C] [D], respectively (the brackets denote concentrations). The larger the concentrations on one side of the reaction, the greater the unidirectional rate and the more the reaction is driven toward the opposite side. At equilibrium all concentrations must be constant, and this means that the unidirectional reaction rates are equal (k_1[A] [B] = k_2 [C] [D]). Solving for the ratio of k_1/k_2 gives the Law of Mass Action:

$$K_{eq} = k_1/k_2 = [C] [D]/[A] [B] \tag{2}$$

We call K_{eq} the *equilibrium constant* for the reaction (1). A large K_{eq} means that A and B react easily to form large amounts of C and D (using up A and B). A small K_{eq} means that A and B do not readily react and, at equilibrium, the concentrations of C and D are low when compared with those of A and B. Acids are defined as compounds that dissociate in water to yield H^+ ion, whereas bases combine with H^+ ions. If acids are designated by the form HA, reaction (1) becomes:

$$HA \rightleftharpoons H^+ + A^- \tag{3}$$

The product A^- is called the conjugate base of the acid A because it can combine (or conjugate) with H^+ to regenerate HA. For a general acid dissociation, the law of mass action (equation 2) is:

$$K_a = [H^+] [A^-]/[HA] \tag{4}$$

Solving this equation for $[H^+]$ gives:
$$[H^+] = K_a [HA]/[A^-] \tag{5}$$
In equations 4 and 5 the equilibrium constant for an acid dissociation is designated as K_a to distinguish it from equilibrium constants of other types of chemical reactions. Strong acids have large values of K_a, meaning that HA readily dissociates to H^+ and A^- and the concentration of undissociated HA is very small. Weak acids partly dissociate (some undissociated HA remains).

The fact that acids and bases are dissolved in an abundance of water cannot be neglected. Water exists in equilibrium with its ions H^+ and OH^-. Such a small fraction of the water is in the undissociated form that the concentration of water can be treated as a constant. If that is so, then the quantity $[H^+]$ $[OH^-]$ must also be constant. The latter constant is called K_w and is 10^{-14} moles/L at 25°C.

The pH Scale

The degree to which a solution is acidic is a measure of its $[H^+]$ (see Chapter 2). In an acidic solution with $[H^+] = 10^{-2}$ M, the K_w dictates that the $[OH^-]$ be 10^{-12} M. The K_w also dictates that an aqueous solution can have an $[H^+]$ no greater that 1 M and no less than 10^{-14} M. Thus the pH scale, a convenient means of indicating the $[H^+]$, arises from K_w. When $[H^+] = [OH^-]$ at 25°, the concentration of both ions is 10^{-7} M (pH = 7.0). The most acidic solutions have a pH of 0 ($[H^+] = 1$ M); the least acidic solution have a pH of 14 ($[H^+] = 10^{-14}$M). Since the pH scale is a logarithmic scale, its advantage is that it covers the enormous range of variation of $[H^+]$ in a small range of numbers. It is important to remember that a change in pH of one unit corresponds to a tenfold change in $[H^+]$.

The pK_a Is A Measure of the Strength of An Acid

Just as it is easy to handle a wide range of $[H^+]$ using the pH scale, so also is it easier to express and calculate with K_a values that are on a similar scale. The pK_a of an acid is the negative logarithm of its K_a. Strong acids thus have low pK_a values, and weak acids have higher ones. If negative logarithms are taken of both sides of equation 5, the result is in terms of pH and pK_a:

$$pH = pK_a + \log [A^-]/[HA] \tag{6}$$

This is the Henderson-Hasselbalch equation. The equation has three variables: the concentration of H^+, the concentration of the conjugate base A^-, and the concentration of the undissociated HA. If any two of the variables are known, the third can be calculated. If the acid in the system is a weak one, the equilibrium of this acid with its conjugate base constitutes a buffer system. The undissociated acid acts as a reserve of H^+; the conjugate base acts as a sink for extra H^+.

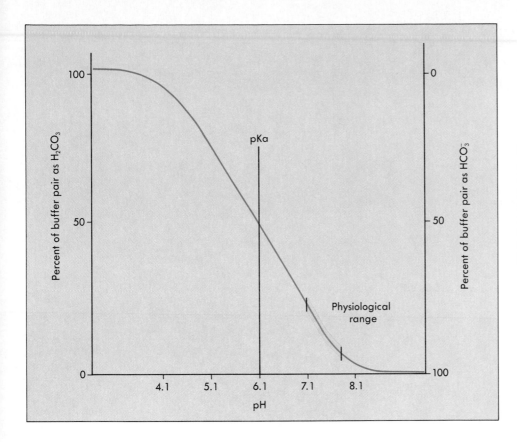

FIGURE 20-A
A titration curve for the HCO_3^- buffer system of plasma. At the pK_a for the system the concentrations of HCO_3^- and H_2CO_3 are equal. The physiological range occupies a small segment of the curve more than one pH unit from the pK_a.

Titration Curves Are Determinations of the Buffering Ability of Buffer Solutions

If a solution of weak acid is titrated with a strong base while the pH is measured, a *titration curve* (Figure 20-A) is generated. At first the pH rises rapidly, but when the pH approaches to within one unit of the pK_a of the weak acid, the curve flattens so that the pH changes relatively slowly as more base is added. The change in pH per amount of base added is least at the point where the pH of the system is equal to the pK_a of the acid. At this point the concentrations of conjugate base and undissociated acid are equal. As still more base is added, the pH begins to rise rapidly again, especially after the pH is more than one unit above the pK_a. Ultimately, essentially all of the weak acid is converted to conjugate base, and the buffering capacity of the buffer is exhausted.

The two reactions of the bicarbonate/carbonic acid/CO_2 buffer system (hereafter called the bicarbonate buffer system) are:

$$CO_2 + H_2O \rightleftharpoons H_2CO_3 \rightleftharpoons H^+ + HCO_3^- \qquad (7)$$

The equilibrium constants for the two reactions can be summed together and treated as a single constant. The summed pK_a of the bicarbonate buffer system is 6.1 at body temperature. The Henderson-Hasselbalch equation written for this system is:

$$pH = 6.1 + \log [HCO_3^-]/0.03 \; P_{CO_2} \qquad (8)$$

Multiplying the plasma P_{CO_2} by 0.03 in the denominator converts it to CO_2 concentration.

Figure 20-A shows a titration curve for the HCO_3^- buffer system of blood. The concentration of HCO_3^- and H_2CO_3 are equal at the pH of 6.1. The part of the curve that is of physiological importance is the very short segment between pH 7.0 and 7.6. This is the buffer line indicated in Figure 20-17; it looks like a straight line in Figure 20-17 because over the small range of physiological pH shown in Figure 20-17 the curvature is insignificant. It is important to keep in mind that during titrations of the plasma bicarbonate buffer system the P_{CO_2} stays constant.

The buffer capacity of a solution is a measure of the total number of H^+ acceptors and H^+ donors available. The greater the buffer capacity, the smaller the change in pH for any given quantity of added strong acid or base. The buffer capacity of whole blood is greater than that of the plasma alone because of the presence of hemoglobin in the red blood cells. The effect of the hemoglobin is to increase the buffer capacity of the blood and to reduce the slope of the buffer line. In effect, the hemoglobin substitutes for some of the HCO_3^- acceptor, and for some of the H_2CO_3 as an H^+ donor, when the blood is titrated.

THE GASTROINTESTINAL TRACT

CHAPTER 21

Gastrointestinal Organization, Secretion, and Motility

On completing this chapter you should be able to:

- Describe the anatomy of the gastrointestinal tract and assign general physiological functions to each of its components.
- Identify the location and role of secretory and absorptive epithelial cells.
- Describe the blood supply to the gastrointestinal tract, identifying systemic blood vessels and lymph lacteals.
- Appreciate how muscular contractions mix and move chyme from one region of the gastrointestinal tract to another.
- Distinguish between segmentation and peristalsis.
- Understand the functions of mucus on the surface of the GI tract.
- Describe the composition of secretions in the stomach and small intestine and the nature of bile secreted by the liver.
- Understand the transport mechanisms responsible for acid and bicarbonate secretion.
- Describe the circulation and physiological role of bile salts.
- Understand how intestinal motility mixes the chyme with digestive secretions, promotes absorption of nutrients, and propels the chyme from one region of the intestine to another.

*T*he gastrointestinal (GI) system is a tubular system the surface of which, like the skin and lung surfaces, forms one of the boundaries between the physiological interior and the outside world. There are four major operations of the gastrointestinal tract and its glands: (1) transport and mixing of food within the tube system, (2) secretion of fluid, salts and digestive enzymes, (3) digestion, and (4) absorption. Food transport and mixing are carried out by the smooth muscle that surrounds all parts of the gastrointestinal tract. The conditions created within the spaces by warmth, secretion, and mixing result in digestion, the process in which highly organized food structures containing complex macromolecules are reduced to simple molecular forms that can be absorbed by the intestine.

Much human food consists of relatively intact parts of other organisms. A salad of raw vegetables, for example, consists of large fragments of plant parts, containing cells with intact cell walls. The proteins and other large molecules that had been synthesized by the plant cells are potential nutrients locked within the cells. The first task of the gastrointestinal system is to break up the food structure to make these potential nutrients available.

After potential nutrients have been released from food, further steps of digestion chemically convert them into simpler molecules that can be absorbed by the intestine. For example, dietary proteins are, as proteins, nutritionally useless. They cannot be absorbed efficiently from the intestine, and even if they were absorbed they could not be incorporated directly into human cells. Their usefulness depends on the ability of the GI tract to convert them to the common currency of free amino acids.

The task of the GI tract is both helped and hindered by gastronomy—cooking and other cultural techniques of food preparation assist in the disruption of the food structure, but they also may result in loss of some valuable nutrients.

ORGANIZATION OF THE GASTROINTESTINAL SYSTEM

> - *What are the general functions of the GI tract?*
> - *What are the major anatomical components of the GI tract?*

The Anatomy of the Mouth and Esophagus

Figure 21-1, *A* illustrates the location of the major components of the digestive system in the body. Individual elements of the digestive system can be visualized by having a person drink a suspension of barium (a contrast medium that absorbs X-rays), and then taking X-rays at various times thereafter

FIGURE 21-1
A The digestive system depicted in place in the body. All organs except the mouth and esophagus are located in the abdomen beneath the diaphragm.
B An X-ray of the stomach and small intestine. A subject was given a barium milk shake (barium is opaque to X-rays) to drink shortly before the X-ray was taken.

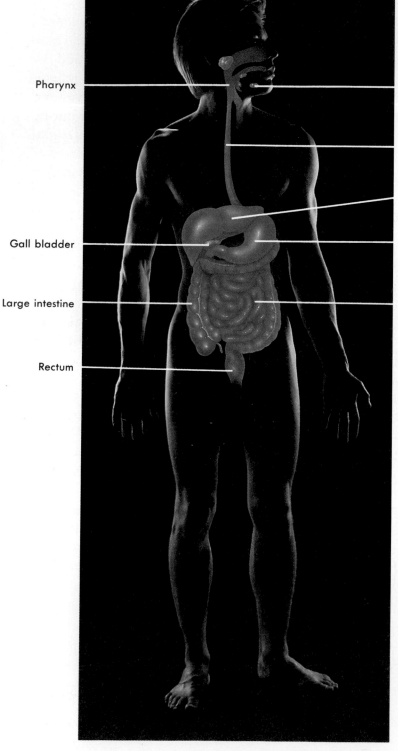

A

Pharynx

Gall bladder

Large intestine

Rectum

Salivary glands

Esophagus

Liver

Stomach

Small intestine

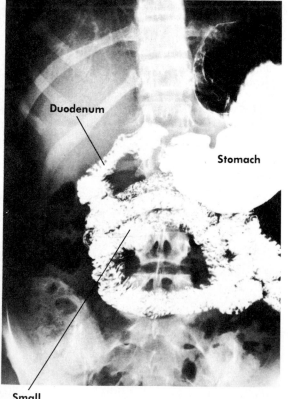

B

Duodenum

Stomach

Small intestine

THE GASTROINTESTINAL TRACT

TABLE 21-1   *Subdivisions of the Gastrointestinal Tract*

Major component	Subdivision	Functional role
Mouth		Gross mechanical breakdown; lubrication with saliva
Pharynx		Swallowing; junction with the esophagus is termed the upper esophageal sphincter
Esophagus		Movement of food from the pharynx to the stomach; the junction of the esophagus with the stomach is the lower esophageal sphincter
Stomach	Body	Initial storage of food
	Fundus	Storage/secretion
	Antrium	Vigorous mixing of food with secretions to form a semisolid chyme; the pyloric sphincter (or phlorus) separates the stomach from the duodenum
Small intestine	Duodenum	Segment receiving liver and pancreatic secretions; important site for the regulation and overall coordination of GI function
	Jejunum	Absorption of the majority of the end products of digestion
	Ileum	Fluid reabsorption; junction with the large intestine is termed the ileocecal sphincter
Large intestine	Ascending colon	Fluid reabsorption
	Transverse colon	Fluid reabsorption
	Descending colon	Fluid reabsorption
	Sigmoid colon	Storage of feces
	Rectum	Storage and elimination of feces
	Anus	Most distal opening of the GI tract; it is regulated by the internal and external anal sphincters

(Figure 21-1, *B*). Table 21-1 lists the subdivisions of each of the components in Figure 21-1, *A*.

In the mouth, mastication (chewing) reduces the size of the food particles and mixes them with secretions from the three sets of salivary glands (Figure 21-2) that open into the mouth. The enzyme α-amylase (sometimes called ptyalin) present in the saliva initiates digestion of starches. The **esophagus** connects the **pharynx** with the **stomach** (see Figure 21-1, *A*). The upper portion of the esophagus is formed from striated muscle, the middle portion contains a mixture of striated and smooth muscle,

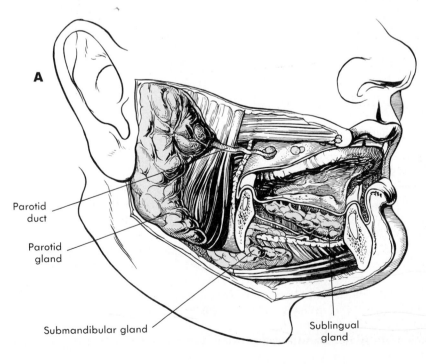

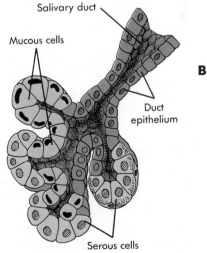

FIGURE 21-2
A   The locations of the salivary glands.
B   An idealized schematic drawing of the histology of a salivary gland and its duct. The salivary glands differ in the relative abundance of the cells shown.

FIGURE 21-3

A The anatomy of the stomach.
B The lining of the stomach seen through an endoscope, a tube about 1 cm in diameter containing fiber optics that is passed down the esophagus to allow physicians to examine the stomach lining. The rugae are clearly visible without magnification. **C** The pylorus, also seen through an endoscope.

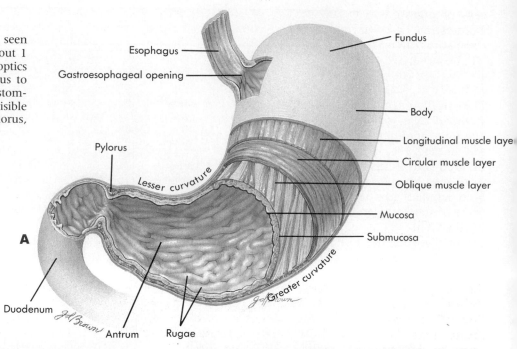

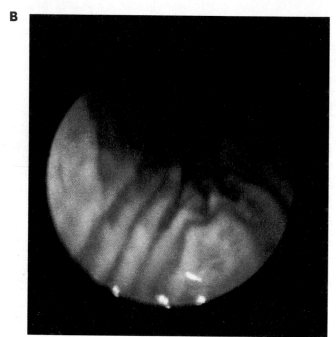

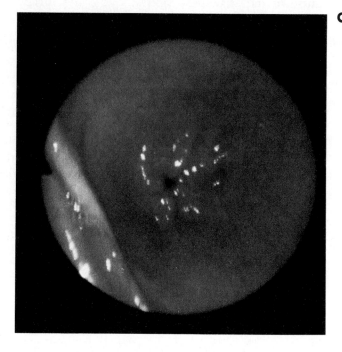

and the lower part contains only smooth muscle. No digestion occurs in the esophagus beyond that which begins in the mouth.

The Anatomy of the Stomach

The stomach (Figure 21-3, *A*) is surrounded by several layers of smooth muscle and is divided into two regions that serve different functions. The **body** and **fundus** are specialized for storage (see Table 21-1). Relaxation of the smooth muscle in the body and fundus allows the stomach to expand when a quantity of food enters it. The **antrum** is specialized for mixing food with gastric secretions to form a broth-

like material called **chyme.** The digestion of protein begins in the stomach, but the mixing and storage functions of the stomach are more significant than the protein digestion that occurs there.

The luminal surface of the stomach has numerous folds **(rugae),** that can be easily seen without the aid of a microscope (Figure 21-3, *B*). The antrum of the stomach and the small intestine are separated by a region of smooth muscle with a relatively high intrinsic tone, the **pylorus** (Figure 21-3, *C*; Table 21-1). The pylorus acts as a sphincter, regulating entry of chyme into the intestine from the stomach. The pylorus could be regarded as a gatekeeper that regulates access of ingested materials to the part of the gastrointestinal tract in which absorption mainly occurs; anything on the stomach side of the gate could still be evacuated by vomiting.

Intestinal Anatomy

The small intestine (see Figure 21-1, A), also surrounded by layers of smooth muscle, is the major site for both digestion and absorption of food. It is 6 to 7 m long, has an average diameter of 4 cm, and is divided into three parts on the basis of their function (see Table 21-1). The duodenum is the 20 cm long segment just distal to the pylorus. The secretions of the liver and pancreas enter the small intestine at the level of the duodenum, which also contains several hormone-secreting cells involved in the overall control of the GI tract (see Chapter 22). The next 250 cm of the small intestine is termed the jejunum, while the remaining 400 cm is called the ileum. The surface area for digestion and absorption in all regions of the small intestine is much greater than the length of the region indicates because of the structural modifications that increase surface area.

The large intestine (see Figure 21-1, A; Table 21-1) consists of the cecum, colon, and rectum. The cecum is a blind pouch at the beginning of the large intestine, joined to the ileum by the ileocecal valve. The vermiform appendix, a narrow, short tube opening into the cecum, has no digestive function and consists largely of lymphoid tissue. The colon is subdivided anatomically into the ascending, transverse, descending, and sigmoid regions and is one third the length of the small intestine. It stores feces and absorbs water. The anus, the opening of the most distal portion of the large intestine, or rectum, is regulated by two sphincters: the internal anal sphincter is a thickened region of the smooth muscle, and the external anal sphincter is made up of striated muscle.

HISTOLOGY OF THE GASTROINTESTINAL TRACT

- *What modifications increase membrane surface area in the stomach and small intestine?*
- *What are the functional categories of epithelial cells of the GI tract?*
- *How do the secretions of the GI tract differ in the stomach and small intestine?*

The serosa, the covering of the GI tract facing the body cavity, is visceral peritoneum that is continuous with the parietal peritoneum that lines the body cavity. A mesentery is a double layer of peritoneum that forms a sheet connecting the digestive organs to the wall of the body cavity. Mesenteries hold the organs in place and form a route for nerves and blood vessels to reach the GI tract.

The wall of the GI tract is composed of four types of tissues: connective, smooth muscle, neural, and epithelial, organized into concentric layers. An example of this organization is the cross section of the small intestine shown in Figure 21-4. The thickness of each layer and the size of the lumen vary in different parts of the GI tract, but the general arrangement of the layers is the same throughout. Proceeding from the body cavity side of the tract to the lumen, the layers are: the serosa, muscularis externa, submucosa, and mucosa (Table 21-2). Between and among the layers are blood and lymphatic vessels and nerve processes.

The muscularis is subdivided into layers of longitudinal and circular smooth muscle fibers. A network of autonomic axons and ganglia (neuron cell bodies and synaptic connections) called the myenteric plexus (Auerbach's plexus) is located between the muscle layers of the muscularis. Nerve fibers from the plexus pass among the muscle fibers, making diffuse synaptic connections with them (see Figure 21-4; Table 21-2).

The submucosa is made up of connective tissue, blood and lymphatic vessels, and a variety of exocrine glands. Recall, also, that exocrine glands secrete their products via ducts, in contrast to endocrine glands that secrete hormones directly into the blood (see Chapter 5). Surrounding the blind ends of exocrine glands are acini, the secretory cells. Ducts connect the secretory portion of the glands with the lumen of the GI tract.

Between the submucosa and the circular muscle layer of the muscularis is a second network of autonomic nerve fibers (see Figure 21-4; Table 21-2), the submucosal plexus (or Meissner's plexus). The myenteric and submucosal plexuses and their interconnections make up the enteric nervous system, which is responsible for coordination of muscular and glandular functions. The enteric nervous system is capable of considerable integration of information; some reflexive pathways involve only the plexuses and not the central nervous system—these are called "short" or enteric reflexes. In some cases, hormonal signals secreted by endocrine cells in the mucosa are involved (see below). Conventional or "long" reflexes originating in the GI tract are mediated by afferents that course mainly in the vagus (X cranial) nerves, and by autonomic efferents. The sympathetic innervation of the GI tract passes through the prevertebral ganglia (celiac ganglion and superior and inferior mesenteric ganglia). The parasympathetic innervation of the GI tract is supplied by the vagus nerves, except for the most distal parts of the colon and rectum, which receive parasympathetic input from sacral spinal segments (see Figure 12-4).

The innermost layer of the wall, the mucosa, contains three subdivisions: (1) the epithelium, (2) the lamina propria, the basement membrane the

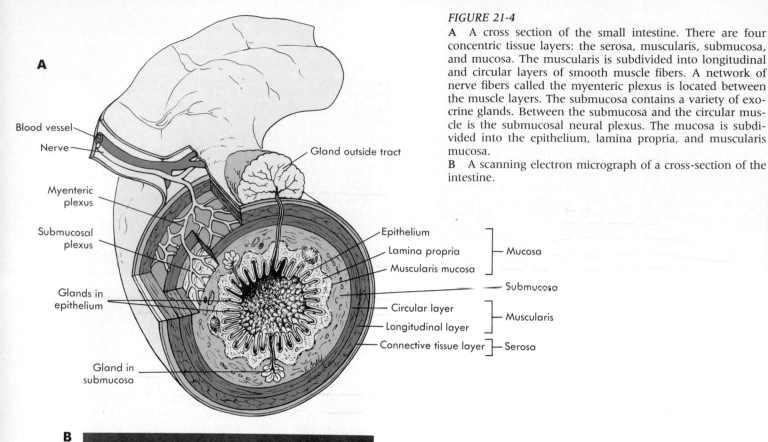

A A cross section of the small intestine. There are four concentric tissue layers: the serosa, muscularis, submucosa, and mucosa. The muscularis is subdivided into longitudinal and circular layers of smooth muscle fibers. A network of nerve fibers called the myenteric plexus is located between the muscle layers. The submucosa contains a variety of exocrine glands. Between the submucosa and the circular muscle is the submucosal neural plexus. The mucosa is subdivided into the epithelium, lamina propria, and muscularis mucosa.

B A scanning electron micrograph of a cross-section of the intestine.

Labels on Figure A

- Blood vessel
- Nerve
- Myenteric plexus
- Submucosal plexus
- Glands in epithelium
- Gland in submucosa
- Gland outside tract
- Epithelium
- Lamina propria — Mucosa
- Muscularis mucosa
- Submucosa
- Circular layer
- Longitudinal layer — Muscularis
- Connective tissue layer — Serosa

TABLE 21-2	Layers of the Gastrointestinal Tract	
Layer	**Sublayer**	**Type of cell present**
Serosa		Connective tissue
Muscularis	Longitudinal	Longitudinally oriented muscle
	Myenteric plexus	Enteric nerve fibers; autonomic fibers
	Circular	Circularly oriented muscle
	Submucosal plexus	Enteric nerve fibers; autonomic fibers
Submucosa		Many glands
Mucosa	Muscularis mucosa	Smooth muscle
	Lamina propria	Lymph nodules
	Epithelium	Absorptive cells; glands; receptor cells

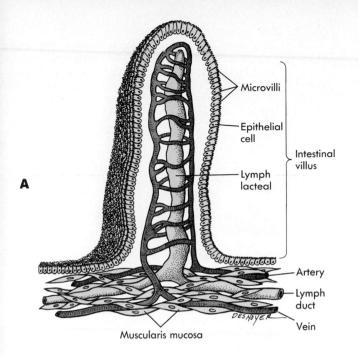

A

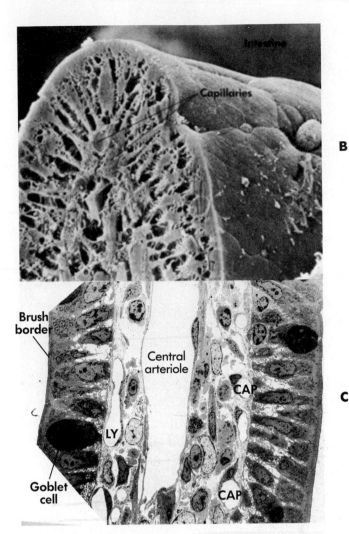

B

C

FIGURE 21-5
A   Diagram of the organization of intestinal villi.
B   A scanning electron micrograph showing the cross section of a single villus.
C   An electron micrograph of a villus showing the location of the central arteriole.

epithelial cells rest on, and (3) a layer of circular muscles called the **muscularis mucosa** (Figure 21-4; Table 21-2). The mucosa forms an enormous number of **villi,** fingerlike projections about a millimeter in length (Figure 21-5) that give the mucosal surface a furry appearance. At the bases of the villi are the openings of the tubelike **crypts of Lieberkuhn** (Figure 21-6). Inside each villus there are blood capillaries and lymphatic capillaries called **lacteals** (Figure 21-5). The lacteals and blood capillaries provide the two routes by which nutrients move from the vicinity of the intestine into the general circulation.

Beneath the epithelial cells lies the **lamina propria,** which is made up of connective tissue, elastin, and collagen fibers and contains blood vessels, some glands, and lymph nodes. The muscularis mucosa is a thin layer of smooth muscle that thrusts the epithelium into permanent, visible ridges called the **plicae circulares** (Kerckring's folds; Figure 21-7).

Surface Area of the Gastrointestinal Tract
The small intestine has the largest internal surface area in relation to its external surface of any part of the GI tract. The plicae circularis and villi combine to increase its luminal area 10 to 30 times over that of a smooth cylinder of the same external diameter. The large intestine is 50% larger in diameter and 33% of the length of the small intestine but, because it lacks villi, the large intestine accounts for only 3% to 4% of the total GI surface area.

Additional enlargement of the surface area is attributable to membrane specializations of the epithelial cells. The luminal membranes of the epithelial cells of the stomach and intestines consist of a **brush border** of minute projections called **microvilli,** each about 1 μm long and 0.1 μm in diameter (Figure 21-8). There are about 2 billion microvilli per square centimeter of intestinal epithelium, and they further increase the surface area of the small intestine by a factor of 20 to 40. When the increase in surface area attributable to the villi and plicae circulares folds is added to that attributable to the microvilli, the total absorptive area of the small intestine is about 200 m².

The Life Cycle of Cells of the Mucosal Epithelium
In the small intestine, new cells are produced by mitosis in the crypts of Lieberkuhn (see Figure 21-6). Immature cells lack digestive or absorptive abil-

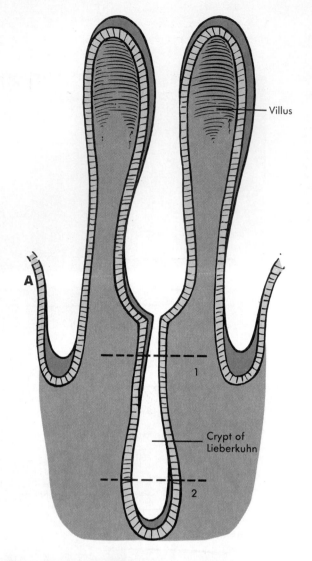

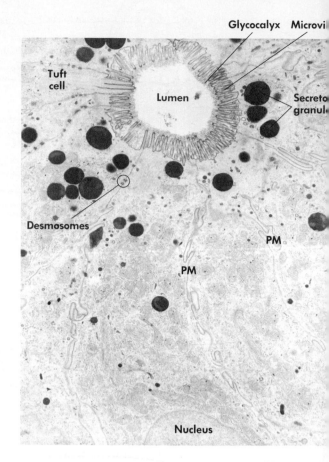

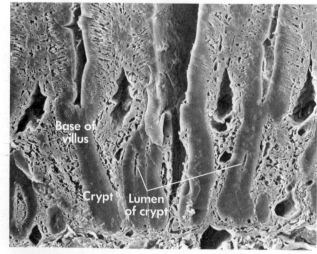

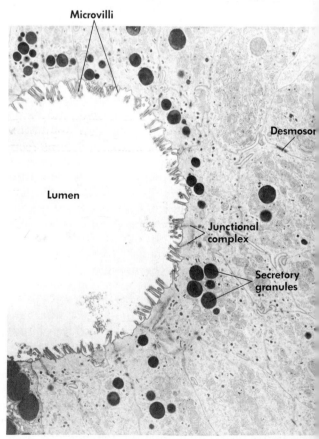

FIGURE 21-6

A Structure of a crypt of Lieberkuhn.

B A light micrograph of the base of a villus showing the crypts of Lieberkuhn.

C, D Transverse sections showing cells surrounding the crypt lumen in the upper *(line 1 in A)* and lower *(line 2)* portion of the crypt.

Most exocrine cells in the GI tract are located in glands in the mucosa and submucosa and are connected to the lumen via ducts (see Figure 21-4), while virtually all absorption is carried out at the tips of intestinal villi. In addition to its secretory and absorptive functions, the GI tract epithelium contains some cells that are modified to serve as endocrine or paracrine cells.

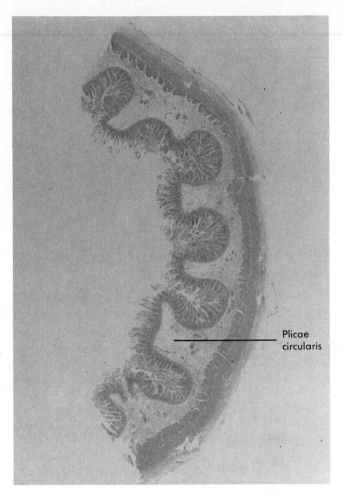

FIGURE 21-7
Photomicrograph of a section of intestine showing villi and plicae circulares.

ities. Over a period of 2 to 5 days, they migrate up the sides of the villi. During their migration, digestive enzymes and the transport proteins responsible for absorption of organic materials and ions are synthesized and inserted into the cell membranes. Migration of the new cells pushes the older cells to the tips of the villi, where they slough off into the intestinal lumen. The enzymes and mucus present in secretory cells are released when they are sloughed, and this supplements the release from intact cells.

The entire lining of the small intestine is replaced in 5 to 6 days. Continuous replacement counteracts the damage normally caused by abrasion and digestive enzymes and allows the system to recover rapidly from acute intestinal infections. A similar replacement process occurs in the stomach.

Functions of the Mucosa

Epithelial cells of the GI tract perform two general functions: secretion of enzymes, salts, and fluid into the lumen, and absorption of nutrients, salts, and water from the chyme to the blood (Table 21-3).

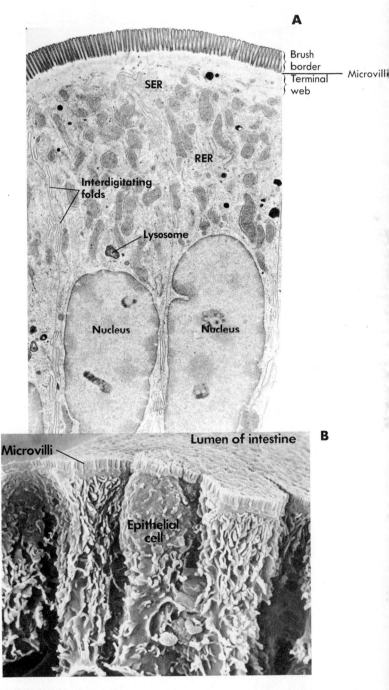

FIGURE 21-8
A   An electron micrograph of two adjacent epithelial cells showing the details of the microvilli or brush border.
B   A scanning electron micrograph of several epithelial cells.

TABLE 21-3   The Functions of Intestinal Epithelial Cells

Function	Cell type	Characteristics	Physiological role
Absorption	Columnar	Microvilli	Increase surface area
		Membrane enzymes	Digestion
		Facilitated diffusion and co-transport	Absorb salt; absorb products of digestion
		Basolateral Na^+-K^+ pump	Absorb water and salt
		Tight junctions	Seal lumen
Exocrine secretion	Goblet	Mucin-filled vesicles	Lubrication; protection
	Crypt and gland	Located in mucosa and submucosa; connected via ducts	Fluid secretion; secretion of enzymes
Endocrine secretion	Paracrine	Sensory microvilli	Sensory receptors
		Vesicles at base	Hormone secretion

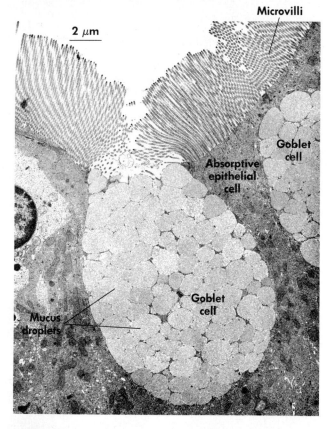

FIGURE 21-9
An electron micrograph of the apical portion of a goblet cell, full of vesicles of mucin ready to be secreted.

The mucosa is so named because it is always coated with a protective mucus secreted by **goblet cells** (Figure 21-9; Table 21-3), the most abundant secretory cells of the GI tract. The lumen-facing pole of a goblet cell has large vesicles filled with **glycoprotein mucins**. When mucin-filled vesicles undergo exocytosis (see Chapter 4), the released mucin combines with water to form **mucus.** The composition of the mucins and the pH of the watery fluid that accompanies them is different in each region of the GI tract. In general, the fluid is similar to plasma but higher in HCO_3^-, so that the mucus layer that lines the lumen is slightly alkaline. The protection provided by alkaline mucus is most important to the cells in the stomach and the initial part of the small intestine, especially the duodenum, because the acidified chyme from the stomach enters this region and its pH is only gradually neutralized by pancreatic secretions. The rate of mucus and fluid secretion increases with food intake. Mucus also serves to clump indigestible materials together in the large intestine.

Absorptive epithelial cells contain numerous transport systems and digestive enzymes in their **apical** (lumen-facing) surfaces and have Na^+-K^+ pumps on their **basolateral** surfaces (Figure 21-10). They transport sugars, amino acids, and water from the lumen to the interstitial fluid and then to the capillary blood. Adjacent epithelial cells have tight junctions at their apical surfaces that allow small solutes to pass but prevent the movement of large molecules (proteins, for example) from the intestinal lumen to the interstitial space.

Scattered in the mucosal epithelium of the GI tract, and concentrated particularly in the stomach and duodenum, are cells called **enteroendocrine** or **argentaffin** (silver-staining) cells (see Figure 21-10). These cells act as chemoreceptors or mechanoreceptors. Sensory microvilli project from their apical surfaces into the lumen, where they monitor the chemical and physical properties of the chyme. Many enteroendocrine cells secrete substances into the blood that satisfy the definition of a hormone. However, some enteroendocrine cells release substances that act locally and do not enter the general circulation (paracrine agents; Chapter 5). In some instances it is difficult to draw a distinction between a hormone and a paracrine agent because it is hard to determine how widespread the agent's ac-

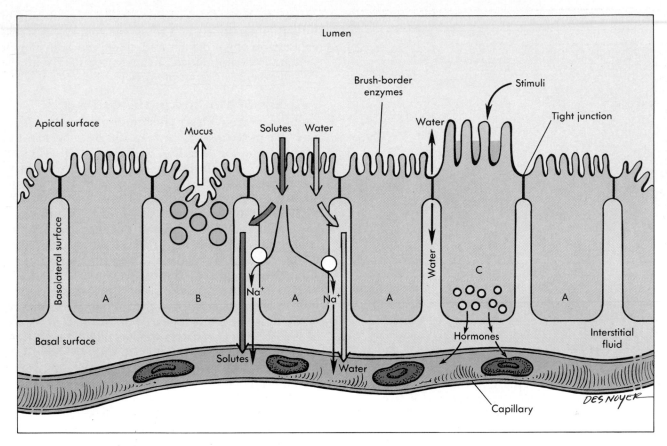

FIGURE 21-10

Types of epithelial cells in the GI tract. In the intestine, absorptive cells **(A)** have transport systems and brush-border digestive enzymes in their lumen-facing (apical) surfaces and Na^+ pumps on their basolateral surfaces. They transport sugars and amino acids *(purple arrow)* and water *(blue arrow)* from the lumen to the interstitial fluid. The solutes then enter the capillary blood. Adjacent epithelial cells have tight junctions that prevent large molecules from reaching the interstitial space but that can allow water and small ions to move in response to osmotic forces. Goblet cells **(B)** secrete mucins. Endocrine cells **(C)** act as chemoreceptors or mechanoreceptors and release hormones from their basal surfaces.

tions are and where the receptors are located. Two well-known paracrine agents, histamine and prostaglandin E_2, have opposing effects on gastric acid secretion (see p. 612-613).

GASTROINTESTINAL SMOOTH MUSCLE

- *What are the four layers of the wall of the tube system?*
- *What is the electrical mechanism of the rhythmic contraction of GI smooth muscle?*
- *What structures make up the "enteric brain"?*

The smooth muscle cells of the GI tract are typically of the single-unit type (see Chapter 12). GI smooth muscle cells are 50 to 100 μm long and 2 to 5 μm wide. Smooth muscle action potentials have a smaller amplitude and last longer (10 to 20 milliseconds) than those in neurons but are briefer than

cardiac action potentials. As in cardiac muscle, an inward current of Ca^{++} is an important component of the action potential in smooth muscle. The conduction velocity of action potentials along smooth muscle fibers is low, because activation of the Ca^{++} channels is slow.

Smooth muscle cells are usually arranged in sheets, with the sheets oriented longitudinally, obliquely (as in the stomach) or circularly (see Figures 21-3, *A,* and 21-4). Smooth muscle cells can be excited by action potentials from other cells or can exhibit an intrinsic pattern of periodic depolarizations called **pacemaker activity,** reminiscent of the spontaneous activity of cardiac pacemakers (see Chapter 12). In most parts of the GI tract, pacemaker activity takes the form of **slow waves** of depolarization followed by repolarization that comprise the **basic electrical rhythm** (Figure 21-11, *A*). Each slow wave lasts 3 to 20 seconds, depend-

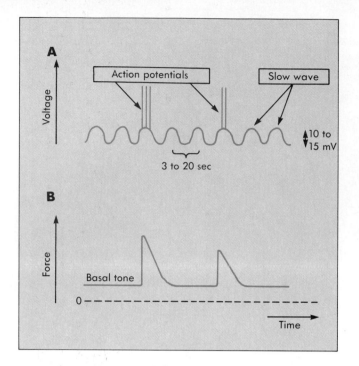

FIGURE 21-11

A Smooth muscle cells exhibit pacemaker activity called slow waves that arise in the longitudinal muscle layer. The membrane voltage varies by 10 to 15 mV, and the slow waves last 3 to 20 seconds. Bursts of action potentials generated at the top of the slow wave can initiate smooth muscle contraction. The force depends on the number of action potentials.

B Most GI muscle exhibits resting tension, or basal tone.

ing on the location. Groups of cells in the longitudinal muscle layer usually act as the slow wave pacemakers. The basic electrical rhythm causes isolated strips of GI tract smooth muscle to undergo rhythmic contraction and relaxation, even in the absence of any neural or hormonal input.

Slow wave depolarizations maintain a level of tonic tension in smooth muscle, but these depolarizations by themselves typically cause only minimal waves of contraction. In the presence of excitatory neural or hormonal input, however, a slow wave may exceed the threshold for action potential generation, in which case one or more action potentials occur during the wave. These action potentials initiate more vigorous contraction (Figure 21-11, *B*). Smooth muscle contracts slowly, and the force generated depends on the number of action potentials occurring at the peak of the slow wave (compare *1* and *2* in Figure 21-11, *B*).

In the interval between slow wave-initiated bursts of action potentials, smooth muscle tension is usually above zero, termed **basal tone** (see Figure 21-11, *B*). Most extrinsic control of motility occurs by modifying the basal tone of the GI smooth muscle. Stretch elicits two responses (Figure 21-12): (1) in **stress activation** a brief vigorous stretch can in-

crease the force of contraction and is one means of initiating peristalsis and segmentation; (2) in **stress relaxation** the stomach and colon can adjust their volume to accommodate an increased load with little increase in internal pressure.

Patterns of Motility in the GI Tract

Motility in the GI tract falls into three major categories: (1) **peristalsis,** which results in movement of chyme along the length of the tube system, (2) **segmentation,** which results in mixing with digestive secretions and increased contact with the mucosal surface (3) mixing motions of villi and microvilli. Peristalsis involves waves of contraction and relaxation that travel for variable distances, depending on the location and the stage of digestion (Figure 21-13, *A*). The first phase of peristalsis is contraction of the longitudinal muscle layer and relaxation of circular muscle. In the second phase, the circular muscle layer contracts and the longitudinal layer relaxes. This pattern is repeated in an adjacent region, as the wave of peristalsis spreads along the GI tract, causing the contents of the GI tract to move in a proximal-to-distal direction. Effective peristalsis cannot be generated by the spontaneous activity of smooth muscle pacemakers alone, but requires the coordination of the enteric nervous system.

The strength of a peristaltic wave and the total distance it travels before it dies out is determined in part by the effectiveness of the electrical connections between muscle cells in the region of the GI system under consideration. Propagation of the wave is assisted by distension of the tract wall by chyme. Although peristalsis occurs throughout the tract, it is strongest in the swallowing pattern of the esophagus (so one can swallow while doing a handstand), moderately strong in the stomach, and relatively weak in the intestines.

In segmentation (Figure 21-13, *B*) alternating rings of contracted and relaxed circular muscle mix the intestinal contents without moving them very far in either direction. As is true for peristalsis, segmentation is coordinated by the enteric nervous system and is stimulated by distension.

Individual villi are capable of motion due to the presence of strands of smooth muscle within the submucosa. The motion of villi is greatly increased when food is present; the increase in motion is attributed to an as yet unidentified hormone believed to be released by endocrine cells in the GI tract. The structure of microvilli suggests that they may also be capable of motion. Each microvillus contains 20 to 30 actin microfilaments (Figure 21-14) that extend some distance into the apical cytoplasm of the epithelial cells, a region referred to as the **terminal web** (see Figure 21-8). Actin is involved in muscle contraction (see Chapter 12) and may make movement and extension of the microvilli possible. The movement of villi and microvilli helps reduce the

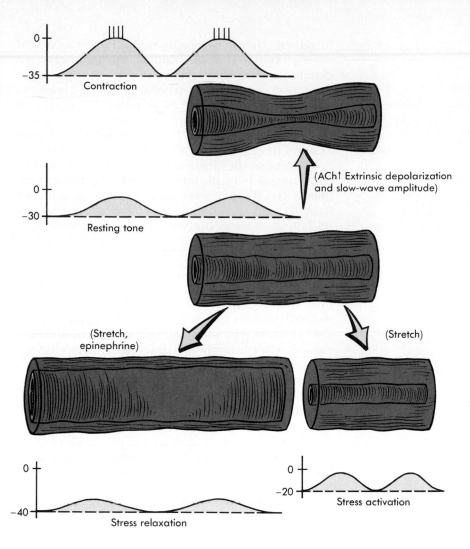

0
−35
Contraction

0
−30
Resting tone

(ACh↑ Extrinsic depolarization and slow-wave amplitude)

(Stretch, epinephrine)

(Stretch)

0
−40
Stress relaxation

0
−20
Stress activation

FIGURE 21-12
Factors that regulate intestinal contractility and the underlying electrical events. Extrinsic control is exerted by the parasympathetic branch of the autonomic nervous system. Intrinsic responses include stress activation and stress relaxation.

A Peristalsis

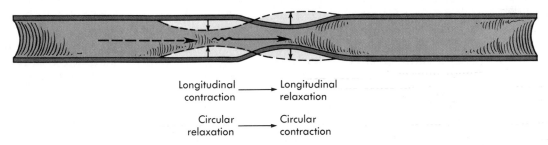

Longitudinal contraction ⟶ Longitudinal relaxation

Circular relaxation ⟶ Circular contraction

B Segmentation

2-4 cm

Circular contraction ⟷ Circular relaxation

FIGURE 21-13
Forms of gastrointestinal motility.
A Peristalsis involves coordinated waves of contraction and relaxation that travel for variable distances and cause propulsion.
B Segmentation involves alternating contraction and relaxation of the circular muscle layer that mixes the chyme.

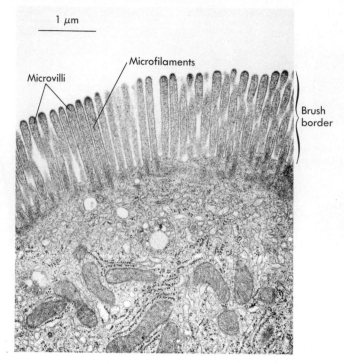

FIGURE 21-14
High-magnification electron micrograph showing the micro-filament bundles of intestinal microvilli.

effect of the **unstirred layer** of fluid immediately adjacent to the apical surface of the cells. Within the unstirred layer, solute movement is slow because it occurs primarily by diffusion rather than bulk flow.

Neural and Hormonal Control of Motility

Coordination of the circular and longitudinal muscle layers in peristalsis and segmentation is one of the functions of the enteric nervous system (Figure 21-15). The mucosal layer contains the endings of sensory neurons specialized for chemoreception and mechanoreception. The cell bodies of these receptor cells are in the submucosal plexus. Enteric interneurons transmit information between the submucosal plexus and the myenteric plexus. Effector neurons in the myenteric plexus respond to the sensory signals by initiating local contractile or secretory responses. Some myenteric motor neurons innervate the longitudinal and circular muscle layers of the muscularis externa and the muscularis mucosa, and other myenteric neurons synapse with the neurons in the submucosal plexus that control secretory cells.

The enteric nervous system exercises a remarkable degree of autonomy, but is also influenced by inputs from the autonomic nervous system (see Figure 21-15). Postganglionic parasympathetic fibers enter the myenteric plexus and activate motor neurons or interneurons to influence contraction or secretion. Autonomic control of the circulatory supply

to the GI system is coordinated within the central nervous system to match the inputs on muscles and glands, so that when digestion and absorption are promoted, an appropriately large blood supply to the stomach and intestines provides for uptake of the nutrients. In addition, the enteric nervous system possesses receptors for a number of hormones that can modify its function (see Figure 21-15).

Recent studies have shown that the enteric nervous system is much more complex than the autonomic ganglia that serve other viscera. In addition to the classical autonomic transmitter chemicals acetylcholine and norepinephrine, the enteric neurons utilize perhaps as many as twenty other neurotransmitters and neuromodulators. These include serotonin, substance P, vasoactive intestinal peptide (VIP), gamma aminobutyric acid (GABA), enkephalins, somatostatin and histamine. The complex webs of neurons that make up the two gut plexi are surpassed only by the brain and spinal cord in their capability for information processing, justifying the term *enteric brain* for this network.

THE ROLE OF THE MOUTH AND ESOPHAGUS

- *What basic mechanisms are involved in production of saliva? Why is the composition of saliva more like that of plasma when the flow rate is high?*
- *What are the functions of salivary amylase?*
- *What structures are involved in swallowing?*

Function and Control of Salivary Secretion

Saliva has three components: (1) a slightly alkaline electrolyte solution that moistens food; (2) mucus, which serves as a lubricant; and (3) the enzyme α-amylase, which initiates starch digestion. Although the similar pancreatic enzymes can digest the entire carbohydrate load, the α-amylase present in the saliva alters the taste of certain foods by breaking down starches to sugars. The digestive action of the salivary amylase assists in cleansing the teeth by reducing starches to more soluble compounds.

Of the three pairs of salivary glands (see Figure 21-2), the submandibular glands produce 70% of the normal daily secretion of saliva, the parotid glands another 25%, and the sublingual glands the remaining 5%. The rate of flow of saliva into the mouth varies from about 0.1 ml/min at rest to as much as 4 ml/min during maximum secretion. There are three types of cells in salivary glands: (1) **mucous cells** secrete mucins; (2) **serous cells** secrete the primary salivary fluid into the **acini;** and (3) **duct cells** line salivary ducts and modify the primary fluid (Table 21-4; Figure 21-16). Submandibular glands have both serous and mucous cells, parotid glands contain only serous cells, and sublingual glands mainly have mucous cells. As a result,

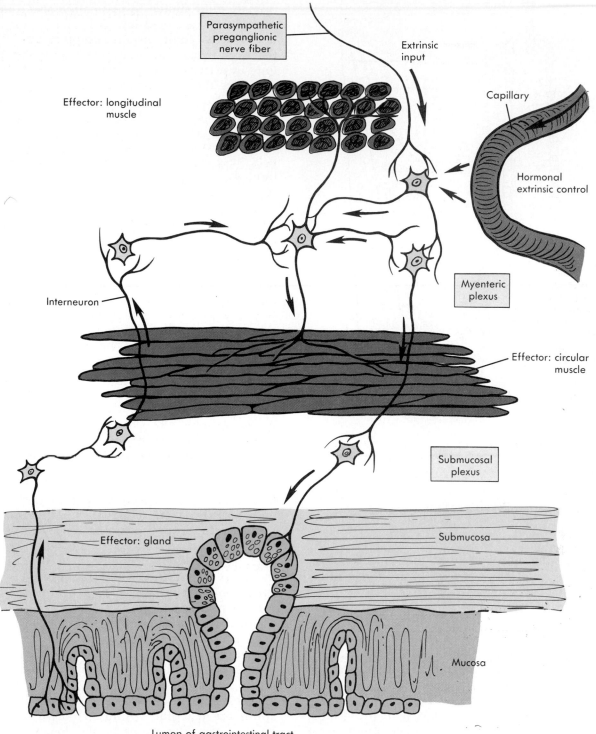

Parasympathetic preganglionic nerve fiber

Extrinsic input

Effector: longitudinal muscle

Capillary

Hormonal extrinsic control

Interneuron

Myenteric plexus

Effector: circular muscle

Submucosal plexus

Effector: gland

Submucosa

Mucosa

Lumen of gastrointestinal tract

FIGURE 21-15
The organization of the enteric nervous system. Neurons in the myenteric plexus primarily control muscle contraction, whereas the submucosal plexus controls secretion. However, the two networks are extensively interconnected. Extrinsic parasympathetic input is exerted on the myenteric plexus; hormones can also exert extrinsic control.

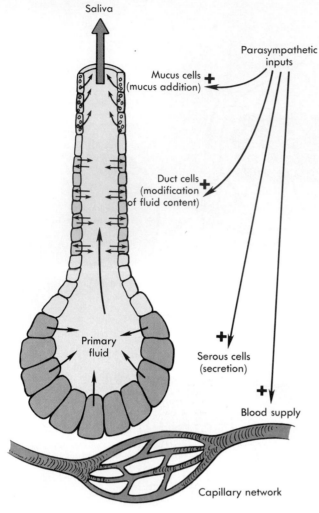

FIGURE 21-16
Summary of parasympathetic control of salivary secretion.

ing glands in the GI system. Secretion is produced in two stages (see Figure 21-16). In the first stage, a primary fluid similar to plasma but also containing α-amylase is secreted into the acini by a combination of active ion transport and plasma filtration. In the second stage, the duct cells modify the primary fluid by active reabsorption of Na^+ and HCO_3^-. This decreases the osmolarity of the luminal fluid and the osmotic gradient created drives water absorption from the lumen. The saliva secreted by the unstimulated glands is relatively dilute, but the concentration rises with higher rates of secretion. This is because at high rates of secretion the saliva flows through the duct too rapidly to incur much reabsorption of solute.

Swallowing

Swallowing is a stereotyped motor pattern that involves coordination of striated muscle and smooth muscle. There are three phases of swallowing, corresponding to the anatomical location of the material being swallowed. After a voluntary **oral phase** (Figure 21-17, *A*), in which the readiness of the material to be swallowed is assessed and the material is pushed to the back of the mouth by the tongue, the **pharyngeal phase** and **esophageal phase** follow reflexively.

Both of these phases are controlled by a **swallowing center** in the brainstem. In the pharyngeal phase (lasting about 1 second), the entry of a **bolus** of food into the pharynx initiates an orderly sequence of contractions of pharyngeal muscles that propel the bolus of food toward the esophagus (Figure 21-17, *B*). In this process, the soft palate is reflected backward and upward, closing off the nasal passsages from the pharynx. As the bolus approaches the esophagus, the **upper esophageal sphincter** relaxes and the epiglottis covers the opening of the larynx (Figure 21-17, *C* and *D*). Once the bolus enters the esophagus, the upper esophageal sphincter constricts to prevent reflux of food into the pharynx (Figure 21-17, *E*).

In the esophageal phase of swallowing, a primary peristaltic wave of contraction sweeps along the entire esophagus in the direction of the stomach, propelling the bolus of food down the esophagus in 5 to 10 seconds (Figure 21-17, *F*). As the bolus nears the stomach, the **lower esophageal sphincter (cardiac sphincter)** relaxes. After the bolus enters the stomach, the lower esophageal sphincter contracts, preventing reflux. The esophageal phase also involves a **receptive relaxation** of the stomach. The primary peristaltic wave usually empties the esophagus, but any remaining food triggers secondary peristaltic waves.

The pharyngeal phase of swallowing is depressed by anesthetics and is affected by brainstem damage. Caffeine and alcohol inhibit the contraction of the smooth muscle in the lower esophageal sphincter, and if the strength of contraction drops

the composition of saliva in the mouth depends on the relative secretion from each of the three different pairs of glands.

Salivary glands are innervated by the autonomic nervous system. Although the two branches typically have contrasting actions on other parts of the GI tract, in the case of the salivary glands, both parasympathetic and sympathetic stimulation promote salivary secretion. The sympathetic effect is to temporarily increase secretion by causing contraction of smooth muscle associated with the glands, whereas the parasympathetic effect produces a longer lasting increase in the blood flow to the glands that is important for promoting secretion (see Figure 21-16). Salivary secretion is enhanced by the physical presence of food in the mouth (an intrinsic reflex loop) and by visual, olfactory, or mental stimuli (extrinsic controls).

Mechanism of Salivary Secretion

The mechanism of secretion in salivary glands has elements that are common to all electrolyte-secret-

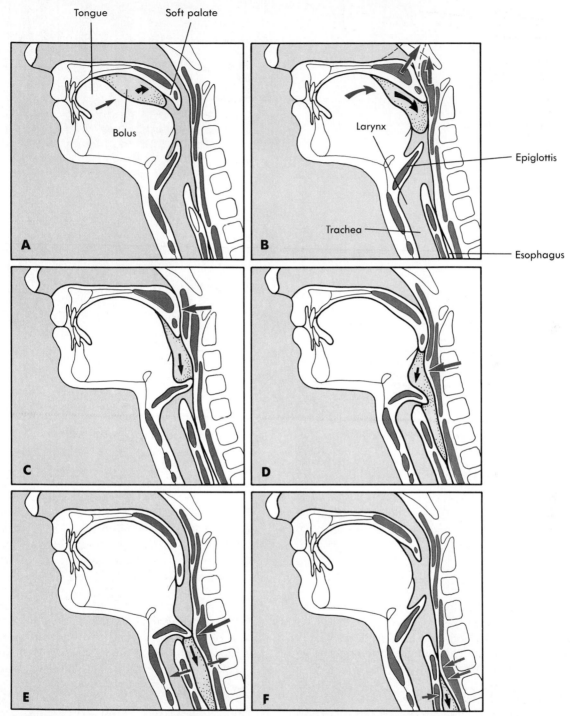

FIGURE 21-17
Swallowing.
A During the voluntary phase a bolus of food *(yellow)* is pushed toward the pharynx.
B to **E** During the pharyngeal phase the soft palate closes off the entrance to the nasopharynx, and the epiglottis closes off the entrance to the trachea with contractions forcing the bolus of food into the esophagus.
F In the esophageal phase the bolus of food is moved by successive contractions toward the stomach.

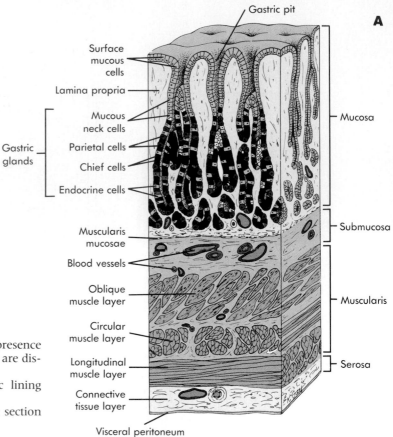

A

Gastric pit

Surface mucous cells

Lamina propria

Gastric glands
- Mucous neck cells
- Parietal cells
- Chief cells
- Endocrine cells

Mucosa

Muscularis mucosae

Blood vessels

Oblique muscle layer

Circular muscle layer

Longitudinal muscle layer

Connective tissue layer

Visceral peritoneum

Submucosa

Muscularis

Serosa

FIGURE 21-18

A The organization of the gastric lining. Note the presence of several types of gastric glands in gastric pits. These are discussed later in the text.

B A scanning electron micrograph of the gastric lining showing the location of numerous gastric pits.

C A scanning electron micrograph of a transverse section through a gastric pit.

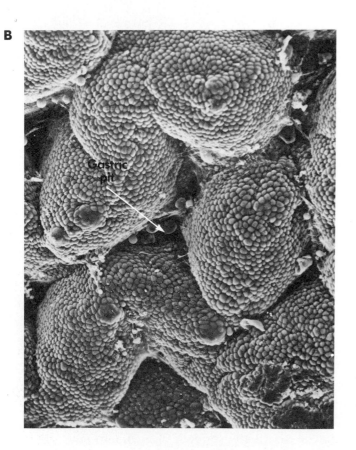

B

Gastric pit

C

Gastric pit

Gland

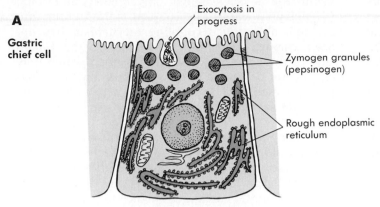

A
Gastric chief cell

Exocytosis in progress

Zymogen granules (pepsinogen)

Rough endoplasmic reticulum

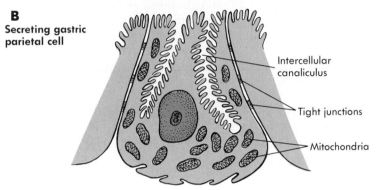

B
Secreting gastric parietal cell

Intercellular canaliculus

Tight junctions

Mitochondria

FIGURE 21-19
Gastric gland cell types.
A   Chief cell, showing rough endoplasmic reticulum.
B   Parietal cell, showing large numbers of mitochondria.
C   Gastric G cell, showing secretory granules filled with the hormone gastrin.
D   Mucus neck cells, showing large mucus granules.

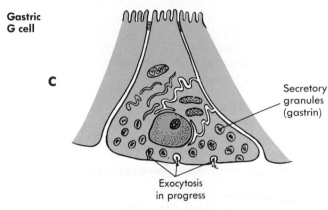

Gastric G cell

C

Secretory granules (gastrin)

Exocytosis in progress

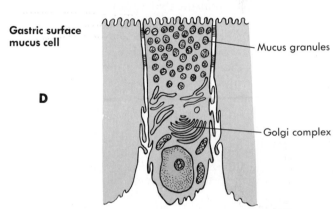

Gastric surface mucus cell

D

Mucus granules

Golgi complex

below a certain level, **gastric reflux** may occur. The resulting irritation of the esophageal lining by HCl causes heartburn. If the lower esophageal sphincter fails to relax, food may be unable to enter the stomach. This condition is called **achalasia**; it is usually the result of damage to the myenteric plexus of the esophagus. The esophagus enters the abdominal cavity through an opening in the diaphragm called the **esophageal hiatus.** If the lower esophageal sphincter or a portion of the stomach protrudes into the thoracic cavity (a condition called a **hiatal hernia**), gastric reflux and heartburn often result. It is not unusual for a woman to experience this condition late in pregnancy due to the pressure of the uterus on the abdominal contents.

GASTRIC SECRETION AND MOTILITY

- *What cellular mechanisms contribute to gastric acid secretion?*
- *What aspects of gastric motility are involved in vigorous mixing of food with gastric secretions?*
- *What is the function of the pylorus?*

Gastric Secretion

The acid secreted by the stomach has four primary roles: (1) it facilitates the reduction of food particles into chyme; (2) it denatures proteins and nucleic acids (see Chapter 3); (3) it transforms the inactive form of gastric proteolytic enzymes, termed pepsinogens, to pepsins (which then can activate more pepsinogens); and (4) it destroys ingested bacteria.

Each square centimeter of gastric mucosa contains about 20,000 depressions called **gastric pits** (see Figure 21-18). Each pit receives the secretion of 3 to 7 gastric glands. The glands include several types of cells (see Table 21-3). **Chief** (peptic) cells are located at the base of the glands and secrete an inactive precursor of pepsin called pepsinogen (Figure 21-19, *A*). **Parietal** (oxyntic) cells are distributed along the length of the gland (Figure 21-19, *B*). They secrete HCl and an **intrinsic factor** necessary for vitamin B_{12} absorption by the intestine. Gastric glands in the antrum lack parietal cells and cannot secrete HCl, but by the time chyme reaches the antrum it has already been acidified. Glands in the lower part of the stomach also contain enteroendocrine cells called **G cells** that secrete the hormone gastrin (Figure 21-19, *C*). The function of this hormone is described below and in Chapter 22.

Surface mucus cells (Figure 21-19, *D*) secrete the thick mucin that protects the epithelium of the stomach and duodenum from the harsh, acidic conditions in the lumen. Mucin-secreting cells are stimulated by mechanical or chemical irritation and by parasympathetic inputs. The protective mucus barrier can be damaged by bacterial or viral inflammation, by certain foods, and by drugs such as aspirin. Since aspirin also increases acid secretion by inhibiting prostaglandin synthesis, it can be particularly harmful when taken in large amounts or consumed by sensitive individuals.

The Cellular Mechanism of Acid Secretion

Figure 21-20 shows the cellular mechanisms presently thought to be involved in HCl secretion by parietal cells. The H^+ that is secreted results from the combination of H_2O with CO_2 produced in the metabolic activity of the cell:

$$(CO_2 + H_2O \rightarrow H_2CO_3 \rightarrow HCO_3^- + H^+)$$

Parietal cells have a $K^+ - H^+$ **ATPase** active transport system on their luminal surfaces that actively secretes H^+ in exchange for K^+. The HCO_3^- that is left behind exits the basolateral side of the cell and enters the circulation. The increase in H^+ secretion that accompanies a meal results in an increase in plasma $[HCO_3^-]$ called the **alkaline tide**. Some of the HCO_3^- movement across the basolateral

membrane is in exchange for Cl^- entry, and some Cl^- entry appears to be coupled to Na^+ entry into the cell, although a coupling mechanism has not been characterized. Cl^- passes across the lumen side of the cell by facilitated transport, which may or may not be linked to the K^+ efflux that accompanies elevated HCl secretion. The net result of the transport systems in the parietal cells is secretion of HCl.

Since the plasma pH is 7.4 and the pH of the secreted fluid is 0.84, H^+ ions are secreted against a concentration gradient of about 2 million to 1. This requires large amounts of ATP and accounts for nearly all the O_2 consumed by the stomach. The energy demands of this task are attested to by the large number of mitochondria present in the cytoplasm of this type of cell (see Figure 21-19, *B*). In addition to their special transport systems, parietal cells have an $Na^+ - K^+$ pump on the basolateral surface that maintains intracellular concentration gradients for these two ions.

Control of Gastric Acid Secretion by the Enteric Nervous System

Control of acid secretion by the parietal cells of the gastric mucosa involves both neurons of the plexuses and nonneuronal enteroendocrine cells. The parietal cells are stimulated to secrete acid by three substances: acetylcholine released by synapses from enteric cholinergic fibers, gastrin from G cells, and histamine from H cells in the stomach wall. H cells

FIGURE 21-20
Diagram illustrating the processes thought to be involved in the secretion of HCl in the stomach.

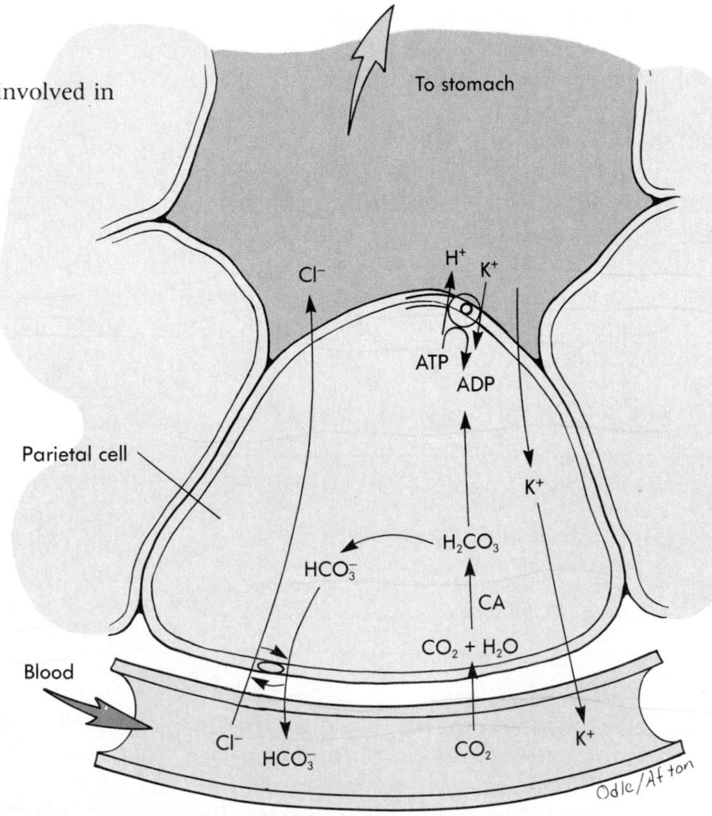

THE GASTROINTESTINAL TRACT

are closely related to mast cells of the immune system. Maximum rates of acid secretion are attained only when all three stimulants are present. Input from parasympathetic neurons running in the vagus nerve is a potent stimulus for gastrin release. When the stomach is empty, acid secretion is inhibited by a decline in the levels of acetylcholine, gastrin and histamine, and by an increase in secretion of somatostatin by S endocrine cells.

Histamine is well-known as a stimulant of fluid secretion in the respiratory system. Antihistamines, drugs that block histamine receptors, are useful in reducing the runny noses of colds and allergic responses. Recognition that histamine is also involved in control of gastric acid secretion led to the development of the drug cimetidine, a blocker of the distinct class of histamine receptors characteristic of parietal cells. Cimetidine is now prescribed for patients with gastric ulcers. In the future it may be possible to use analogues of somatostatin for this purpose.

Any factor that increases acid secretion above normal levels can overcome the protective mucus barrier and damage the lining of the stomach, causing a **peptic ulcer**. More commonly, excessive gastric acid secretion ulcerates the duodenum or esophagus (Figure 21-21), because these parts of the tube system are not as well protected against low pH as the stomach lining. There are several possible underlying causes of excessive acid secretion. Anxiety can cause increased parasympathetic activity, which indirectly brings about HCl secretion, and alcohol directly stimulates parietal cells. Prostaglandins normally present in the stomach wall inhibit acid secre-

tion. Since aspirin inhibits prostaglandin synthesis, it can increase acid secretion.

Gastric Motility

Three functions of the stomach are related to its motility: (1) it stores meals of varying size, (2) mixes food with gastric secretions to form chyme, and (3) releases chyme into the small intestine at a rate the remainder of the GI tract can handle. In the first few hours after a meal, slow, mild contractions of the body of the stomach force the gastric contents toward the antrum. Pyloric and antral motility is highest during and just after meals. Vigorous mixing of food with HCl, mucus, and enzymes to form chyme occurs in the lower portion of the body and the antrum. The motility of the body and antrum are controlled by a pacemaker region located near the middle of the body. This pacemaker initiates slow waves in the longitudinal muscle (about three per second) that spread into the circular muscle layer and sweep into the antrum toward the pylorus. The result is a wave of peristalsis, or **propulsion** (Figure 21-22, *A*). As this peristaltic wave nears the pylorus, it overtakes and passes the contents that are being propelled along. When the wave reaches the pylorus, the antrum and pylorus contract, closing the pyloric sphincter. A few milliliters of material may escape through the sphincter, but most of the gastric contents are forced back towards the antrum (**retropulsion**; Figure 21-22, *B*).

When food first enters the stomach, it is a dense mass. The contents become more fluid over the next few hours with antral mixing and addition of secretions, and both propulsion and retropulsion in-

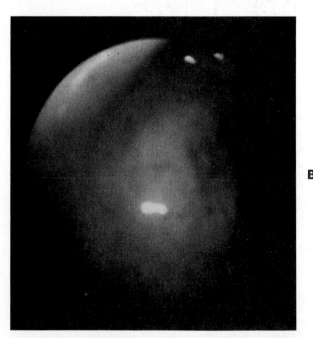

A

B

FIGURE 21-21
The normal duodenum (**A**) and a duodenal ulcer (**B**) as seen through the endoscope.

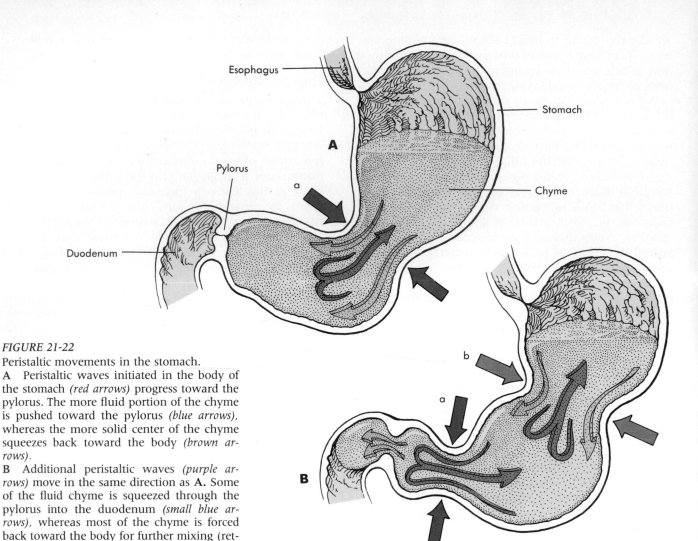

Esophagus

Stomach

A

Chyme

Pylorus

a

Duodenum

b

a

B

FIGURE 21-22

Peristaltic movements in the stomach.

A Peristaltic waves initiated in the body of the stomach *(red arrows)* progress toward the pylorus. The more fluid portion of the chyme is pushed toward the pylorus *(blue arrows)*, whereas the more solid center of the chyme squeezes back toward the body *(brown arrows)*.

B Additional peristaltic waves *(purple arrows)* move in the same direction as **A**. Some of the fluid chyme is squeezed through the pylorus into the duodenum *(small blue arrows)*, whereas most of the chyme is forced back toward the body for further mixing (retropulsion).

crease. As liquid chyme is produced, the pyloric sphincter remains open for longer periods and gastric emptying gradually accelerates. Complete emptying of the gastric contents after a meal may require several hours because gastric motility and emptying are subject to negative feedback by signals that arise in the intestine and reflect the volume and composition of the chyme entering the duodenum. These negative feedback mechanisms are described in Chapter 22.

SECRETION AND MOTILITY IN THE INTESTINE

- *What categories of enzymes are secreted by the pancreas?*
- *What are zymogens?*
- *How are pancreatic enzymes activated in the intestine?*
- *What are the major secretions of the liver? What is the role of the gallbladder? What are the possible fates of bile acids that reach the intestine in bile?*

Anatomy of the Duodenum and Associated Structures

In the duodenum, the acidic chyme from the stomach is mixed with the alkaline secretions of the intestine itself and those of two digestive glands, the pancreas and liver. The secretion of the pancreas is called **pancreatic juice**; that of the liver is called **bile.** In many animals, including humans, bile is stored in a structure called the **gall bladder** during times when no chyme is entering the duodenum. The anatomical relationship of the pancreas, liver and duodenum are shown in Figure 21-23. Note that the pancreatic duct and the bile duct join to form the **hepatopancreatic ampulla** (or **ampulla of Vater**)which opens into the duodenum. Entry of bile and pancreatic juice into the ampulla is separately controlled by sphincters indicated in the Figure; the final entry of mixed bile and pancreatic juice into the duodenum is controlled by the **sphincter of Oddi.**

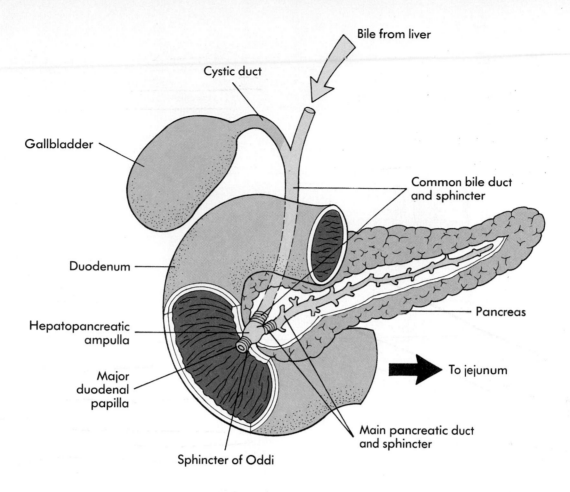

Bile from liver

Cystic duct

Gallbladder

Common bile duct
and sphincter

Duodenum

Pancreas

Hepatopancreatic
ampulla

Major
duodenal
papilla

To jejunum

Main pancreatic duct
and sphincter

Sphincter of Oddi

FIGURE 21-23
The duodenum and routes for entry of digestive secretions of the pancreas and liver.

Secretions of the Intestine and Pancreas

Intestinal epithelial cells secrete an alkaline solution similar in its electrolyte composition to that of the pancreas. The majority of intestinal secretion is carried out by cells of the crypts of Lieberkuhn (see Figure 21-6). In the duodenum, a significant amount of fluid is secreted by submucosal glands that empty into the crypts. Intestinal fluid secretion is increased by mechanical irritation and by factors that stimulate pancreatic secretion. It also can be stimulated by bacterial toxins, such as those produced by the organism that causes cholera. Excess intestinal fluid secretion results in **secretory diarrhea.** The resulting loss of alkaline intestinal contents can lead to dehydration and metabolic acidosis; this is generally the actual cause of death in diarrheal diseases such as cholera.

The pancreas arose from the duodenum in evolution, and like the duodenum it incorporates both digestive and endocrine functions. However, these functions are served by distinct parts of the pancreas. The endocrine portion of the pancreas consists of numerous small islets of Langerhans scattered among the much larger mass of exocrine tissue. The islets of Langerhans secrete several hormones important in regulating metabolism, including **insulin, glucagon** and **somatostatin.** These will be discussed in Chapter 23.

The exocrine pancreas consists of bulbous acini that lead into ducts that ultimately merge to form the main pancreatic duct. The **acinar cells** secrete the enzymes, and the **duct cells** secrete the primary electrolyte solution (Table 21-3; Figure 21-24). Pancreatic secretion of fluid averages 2 L/day, with most secretion occurring after meals. Much of the increase in secretion is due to parasympathetic input. Secretin and cholecystokinin, two hormones released by enteroendocrine cells of the duodenum during entry of chyme into the intestine, are also important stimulants of pancreatic secretion. These factors will be discussed further in Chapter 22.

The cellular mechanism that forms the primary pancreatic fluid in pancreatic ducts is similar to that of gastric acid secretion, but opposite in its direction (Figure 21-25). Within the cells, CO_2 and H_2O react to form H_2CO_3, which dissociates to H^+ and HCO_3^-. Na^+ leaks down its electrochemical gradient from the ducts into the cells, driving the coupled

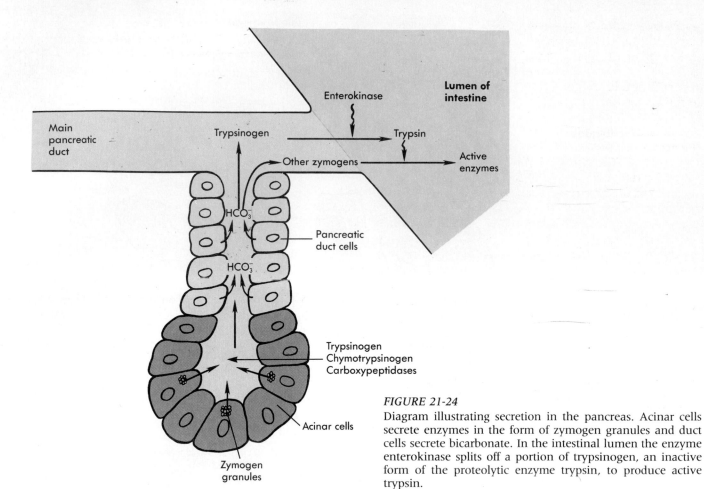

FIGURE 21-24
Diagram illustrating secretion in the pancreas. Acinar cells secrete enzymes in the form of zymogen granules and duct cells secrete bicarbonate. In the intestinal lumen the enzyme enterokinase splits off a portion of trypsinogen, an inactive form of the proteolytic enzyme trypsin, to produce active trypsin.

FIGURE 21-25
Bicarbonate production in the pancreas.

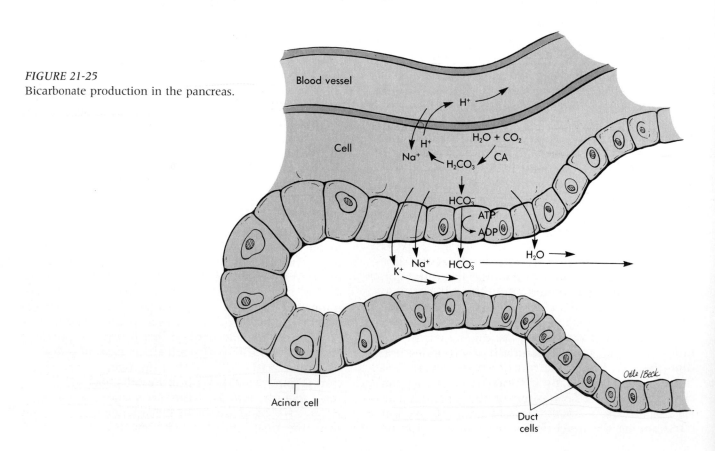

movement of H^+ ions out the basal surface of the duct cell. The HCO_3^- moves into the duct lumen in exchange for Cl^-, and a basolateral Na^+-K^+ pump maintains the intracellular Na^+ concentration of the duct cells at a level lower than that of the extracellular fluid, providing the energy for the entire process.

Pancreatic Enzymes

The acinar cells located at the ends of the pancreatic ducts secrete digestive, or hydrolytic enzymes in the form of zymogen granules (see Figure 21-24). Zymogens are inactive precursors of digestive enzymes that are subsequently activated in the intestine. Intracellular production and storage of the enzymes in an inactive form prevents the enzymes from digesting the secretory cells themselves. The membrane-bound vesicles containing the zymogens are released by exocytosis. The six categories of enzymes released from the pancreas are: (1) proteases (peptidases, including trypsin, chymotrypsin and carboxypeptidase) specific for particular kinds of peptide bonds, (2) nucleases that hydrolyze RNA and DNA to their component nucleotides, (3) elastases that digest the collagen molecules of connective tissue (4) phospholipases that reduce phospholipids to fatty acids and a polar compound, (5) lipases that degrade triglycerides into fatty acids and glycerol, (6) a pancreatic α-amylase similar to that in saliva.

How are inactive pancreatic zymogens converted to their active forms? It was noted earlier that pepsins in the stomach are activated by acid. There is very little HCl in the intestine, so a different mechanism is required from that which activates pepsins in the stomach. Intestinal epithelial cells have a membrane-bound enzyme called enterokinase (see Figure 21-24). Enterokinase converts the inactive proteolytic enzyme trypsinogen to trypsin, which then activates other pancreatic zymogens by removing specific small peptides from them. Trypsin is also capable of activating trypsinogen, so once a small amount of trypsin is produced, the process of zymogen activation is rapid. The pancreas possesses a trypsin inhibitor that normally protects it from premature activation of its digestive enzymes.

The Liver and Bile

As will be explained in Chapter 22, fat digestion requires bile, a mixture of substances synthesized by the liver. Bile has six components: (1) bile acids; (2) the phospholipid lecithin; (3) cholesterol; (4) the end product of hemoglobin degradation, bilirubin, also referred to as a bile pigment; (5) toxic chemicals that have been converted into a harmless form, or detoxified, by the liver; and (6) a $NaHCO_3$ solution similar to that produced by the pancreas (see Table 21-4).

Bile acids are synthesized from cholesterol at a rate of about 0.5 g/day and comprise about half of the solute content of bile. Before they are secreted, bile acids are combined, or conjugated, with the

| TABLE 21-4 | Secretions Released into the Lumen of the Gastrointestinal Tract |

Gland	Cell type	Secretion	Functional role
Salivary	Duct	Fluid ($NaHCO_3$)	Suspension; lubrication; starch digestion
	Mucus	Mucus	
	Serous	Amylase	
Gastric	Parietal	Acid (HCl)	Solubilization; destruction of bacteria;
		Intrinsic factor	vitamin B_{12} absorption
	Chief	Pepsin	Protein digestion
	Surface mucus	Thick mucus	Protection
	Mucus neck	Thin mucus	Lubrication
Pancreas	Acinar	Proteases	Protein digestion
		Elastases	Elastin digestion
		Collagenases	Collagen digestion
		Lipases	Fat digestion
		Amylase (as zymogen granules)	Starch digestion
	Duct	Fluid ($NaHCO_3$)	Buffer acid from stomach
Liver		Fluid ($NaHCO_3$)	Buffer acid; emulsification of fats; elimination of waste materials
		Bile acids	
		Cholesterol	
		Bile pigments	
		Varied organic compounds	
Intestine	Most cells	Fluid ($NaHCO_3$)	Buffer acid Lubrication
	Goblet	Thin mucus	

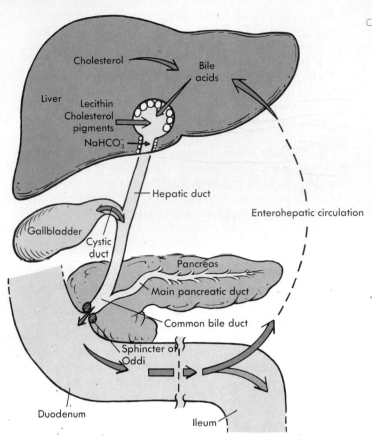

FIGURE 21-26

Bile secretion into the duodenum and the enterohepatic cycle of bile salts. Bile is manufactured in the liver and stored in the gallbladder. The sphincter of Oddi regulates the passage of bile and pancreatic secretion into the duodenum (see also Figure 21-23). The majority of the bile salts are transformed to bile acids in the ileum and recirculated to the liver. An individual molecule may complete this cycle several times during digestion of a single meal.

amino acids glycine or taurine to make them more water soluble. Conjugated bile acids are called **bile salts.** The total bile acid and bile salt content of the body is only 3 to 4 g. Bile salts are reused, and the entire bile acid content of the body recirculates about 10 times a day (Figure 21-26). The 0.5 g/day of bile acid that is lost in the intestines (open arrow) is about equal to the normal production rate by the liver. Recycling of secreted bile salts occurs in stages. Bacteria in the intestine deconjugate bile salts into bile acids, and these bile acids are absorbed in the small intestine. They travel to the liver in the hepatic portal circulation. The liver converts the reabsorbed bile acids into bile salts and secretes them along with newly synthesized bile acids.

Between meals, bile is stored in the gallbladder (Figures 21-23 and 21-26). The **sphincter of Oddi,** located where the **common bile duct** and the **pancreatic duct** enter the duodenum, is normally closed, so bile that is continuously secreted by the liver is forced through the **cystic duct** into the gall-

bladder. The gallbladder actively absorbs Na$^+$ and water and concentrates the bile five- to twenty-fold. Since its capacity is about 30 to 40 ml, the gallbladder can store the equivalent of 200 to 800 ml of bile. Upon appropriate stimulation, the sphincter of Oddi relaxes and the smooth muscle of the gallbladder contracts, releasing concentrated bile. One major stimulant of relaxation of the sphincter of Oddi and gall bladder contraction is the duodenal hormone **cholecystokinin;** the role of this and other duodenal hormones in controlling digestion will be described in Chapter 22.

Normally, the cholesterol secreted in bile is kept in solution by the detergent action of the lecithin. Excessive secretion of cholesterol by the liver or over-concentration of the bile in the gallbladder can result in supersaturation of bile with cholesterol. If this condition is chronic, the cholesterol that precipitates from solution may form sandlike or pebble-like aggregates called **bile duct stones** or **gallstones,** depending on their location (Figure 21-27, *A*). These stones cause severe discomfort and impair digestion if they migrate into the hepatopancreatic ampulla and block it. If this happens, the normally low concentration of endogenous trypsin in the pancreas can rise high enough to overcome the protective effects of the trypsin inhibitor that is also present in the pancreas. If this occurs, the pancreas begins to digest itself (**pancreatolysis.**)

An estimated 25 million Americans have gallstones, most between 3 and 18 mm in diameter. Some can be removed by sphincterotomy, a procedure in which an endoscope is passed down through the mouth to the intestine. The opening to the bile duct is widened mechanically and the lower end is snipped to remove the stone. The digestive system usually works almost as well without a gallbladder because bile can simply flow directly from the liver to the small intestines, but some dietary adjustments are necessary.

In lithotripsy (Figure 21-27, *B*) a patient is immersed in a tub of water (water transmits sound much more effectively than air) and focused ultrasound waves are used to disrupt gallstones and kidney stones. These mineral deposits absorb the ultrasound energy more readily than the surrounding tissues, literally "blasting" them into small fragments. In the few days following lithotripsy, the natural flow of bile (or urine) carries the stone fragments out of the gallbladder (or kidney) into the intestine (or bladder) where they are eliminated from the body. In the original design there was a need to sedate patients and admit them for follow-up for several days. Newer lithotripters have been developed in which less focused ultrasound waves travel through a larger area of the body. There is little or no discomfort and anesthesia is not required, so lithotripsy can be done on an outpatient basis in some 90% of all patients.

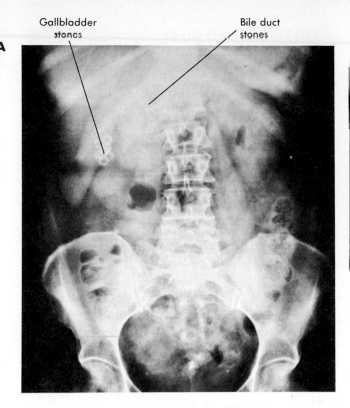

Gallbladder stones

Bile duct stones

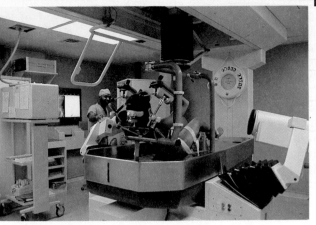

A

B

FIGURE 21-27
A X-ray of a patient who had both gallstones and bile duct stones.
B A new technique called lithotripsy, illustrated here, involves immersing a patient with gallstones in water and focusing high-frequency sound waves on the gallstones. These sound waves often disrupt gallstones, and the patient is able to avoid surgery.

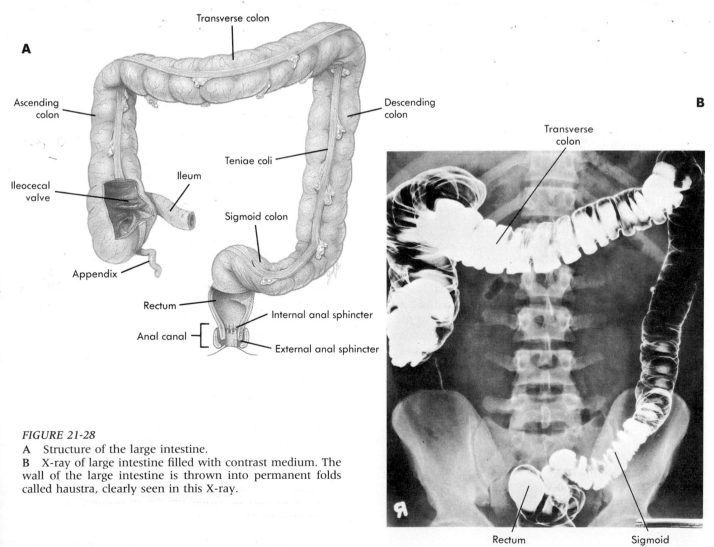

A

B

Transverse colon

Ascending colon

Descending colon

Teniae coli

Ileocecal valve

Ileum

Sigmoid colon

Appendix

Rectum

Internal anal sphincter

Anal canal

External anal sphincter

Transverse colon

Rectum

Sigmoid colon

FIGURE 21-28
A Structure of the large intestine.
B X-ray of large intestine filled with contrast medium. The wall of the large intestine is thrown into permanent folds called haustra, clearly seen in this X-ray.

Ursodiol (Actigall), a naturally occurring bile acid, has recently been approved by the FDA for treatment of patients with gallstones composed of cholesterol. Ursodiol acts like a detergent, gradually dissolving cholesterol-based gallstones. Complete dissolution of gallstones by what has been called "bile acid therapy" is a slow process, requiring several months or more and ursodiol cannot dissolve gallstones containing mainly calcium salts or bile pigments. However, in addition to its direct effects, urdodiol inhibits intestinal absorption of cholesterol from food and decreases the synthesis of cholesterol in the liver. Not only should this decrease gallstone formation, but it makes the drug a potentially useful candidate for lowering plasma cholesterol levels, and thus potentially reducing the incidence of coronary artery disease.

INTESTINAL MOTILITY

- *What are the modes and functions of intestinal motility?*
- *What is a mass movement?*
- *In what part of the intestine are haustra found?*

Intestinal motility has three functions: (1) mixing the chyme with the digestive enzymes, fluids, and bile secreted by the pancreas and gallbladder; (2) circulation of chyme within the lumen to facilitate contact with the absorptive epithelial cells; and (3) propulsion of the chyme toward successive regions of the GI tract. When the rate of segmentation is high, chyme moves more slowly and the mixing promotes digestion and absorption.

Peristaltic contractions are superimposed on segmentation, and their relative importance varies along the small intestine. In the duodenum and jejunum, peristaltic waves are infrequent and they propagate for only about 10 cm, while segmentation activity is high. This limits chyme movement to about 1 cm/min (0.5 m/hour). In the later portions of the jejunum and in the ileum, peristaltic waves increase in amplitude and duration and begin to

FIGURE 21-29

Reflexes in the colon and rectum. In the large intestine waves of peristalsis, called mass movements, occur several times each day. Entry of material into the large intestine is regulated by the ileocecal sphincter. As a result of each mass movement, feces enter the rectum through the internal anal sphincter, stimulating the urge to defecate. Defecation involves voluntary relaxation of the external anal sphincter.

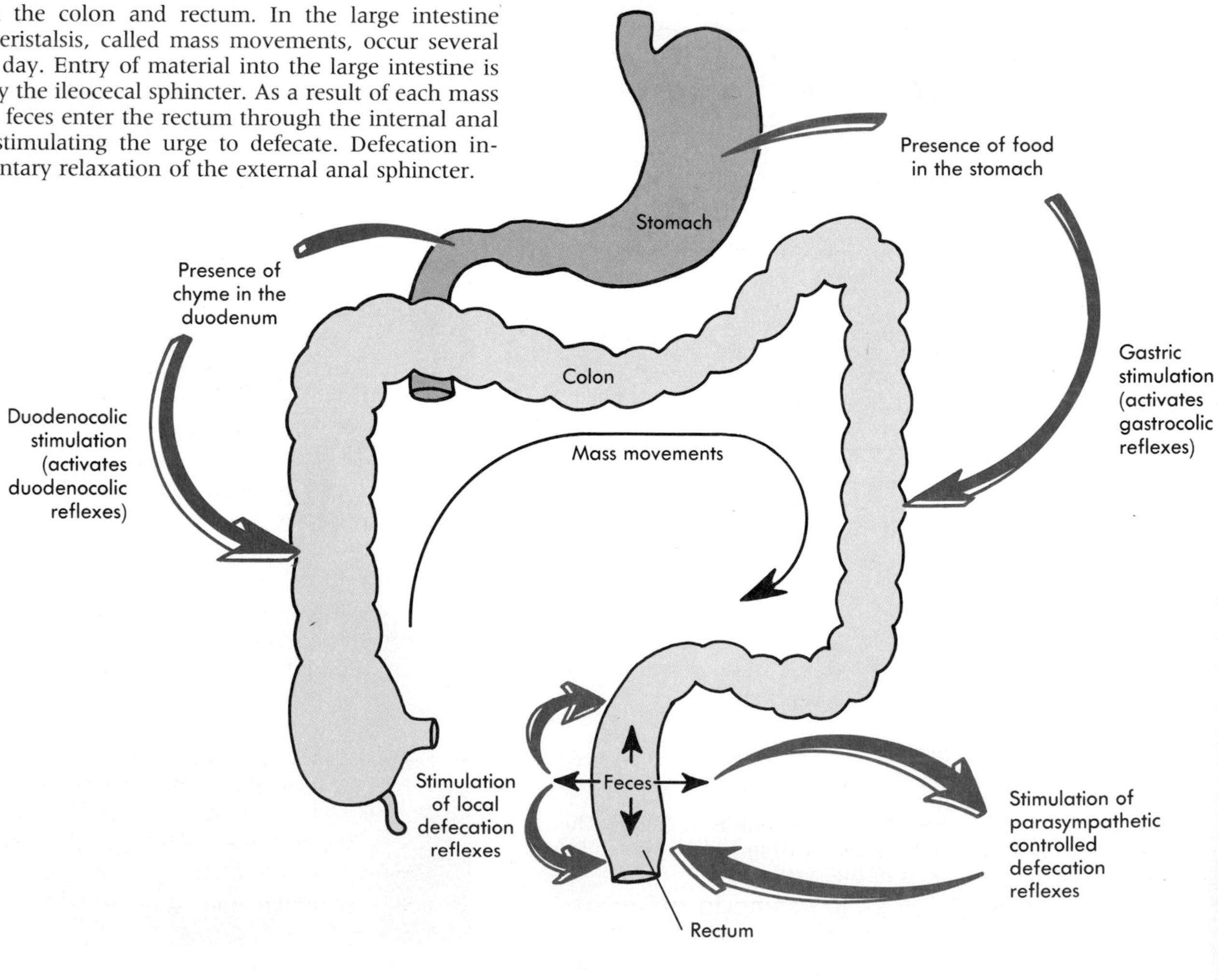

Presence of food in the stomach

Stomach

Presence of chyme in the duodenum

Gastric stimulation (activates gastrocolic reflexes)

Colon

Duodenocolic stimulation (activates duodenocolic reflexes)

Mass movements

Stimulation of local defecation reflexes

Feces

Stimulation of parasympathetic controlled defecation reflexes

Rectum

THE GASTROINTESTINAL TRACT

dominate segmentation, increasing the propulsion of material toward the large intestine. Overall, material travels the 7 m from the pylorus to the large intestine in about 5 hours.

The large intestine has little longitudinal smooth muscle, and what it has is concentrated in three tonically contracted bands called the **teniae coli** (Figure 21-28) which cause the large intestine to be drawn up into a series of baggy folds called **haustra.** The haustra slowly move the 1 to 2 L of chyme that enters the large intestine each day back and forth, allowing ample time for absorption of most of the remaining fluid from the chyme. The rate of movement of the contents of the colon is about 5 cm/hour, except during **mass movements** (Figure 21-29). In a mass movement, segmentation ceases and a strong peristaltic wave sweeps along the colon, moving some material into the rectum. Mass movements take place several times a day, generally occurring upon awakening in the morning and after meals, and can transport chyme all the way from the cecum to the sigmoid colon.

The end of the ileum and the colon are separated by the **ileocecal valve,** which governs the rate of emptying of the small intestine (see Figure 21-28, *A*). The ileocecal valve is closed most of the time, preventing reflux of material from the large intestine into the ileum. The ileocecal valve opens when the end of the ileum is distended by chyme, and it closes more tightly when the colon is distended. Factors that increase gastric motility tend to open the ileocecal valve and permit an increase in the rate of entry of material into the colon.

Defecation

The anal canal has two sphincters (see Figure 21-28, *A*): the **internal anal sphincter** which is composed of smooth muscle and the **external anal sphincter** which is composed of skeletal muscle. Distension of the normally empty rectum by a mass movement produces a reflexive relaxation of the internal anal sphincter and contraction of the external anal sphincter. This reflex is carried over sacral spinal nerves that include both somatic and parasympathetic autonomic components. In infants the sensation of rectal fullness results in reflexive defecation. In older children and adults the reflex is brought under voluntary control and the sensation of fullness is recognized as an urge to defecate. If conditions allow, the urge is followed by voluntary relaxation of the external anal sphincter. A series of reflex contractions in the descending and sigmoid colon and rectum then permits the evacuation of feces. These contractions are assisted by an increase in pressure in the abdominal cavity brought about by the contraction of abdominal skeletal muscles.

If conditions do not allow defecation, the contraction of the external sphincter is sufficient to halt the rectal contractions, and the urge to defecate subsides. This sequence of stimulation followed by voluntary postponement of defecation can be repeated many times because the rectum can accommodate a fairly large quantity of material through stress relaxation.

SUMMARY

1. The organization of the GI system allows sequential processing of the food. In the mouth, food particles are reduced in size and mixed with saliva. In the stomach, the food is acidified, protein digestion begins, and food is stored until the chyme produced by mixing food with secretions is delivered to the small intestine. In the small intestine, the contents are neutralized by bicarbonate secretions, and digestion and absorption of proteins, carbohydrates, and lipids occur. The role of the large intestine is to store unabsorbed material.

2. The wall of the GI tract includes structures that perform many functions. The epithelium has both **secretory** and **absorptive** cells. Smooth muscles in the gastrointestinal tract wall are responsible for movement of the contents. Blood vessels and **lymph lacteals** are available for supplying the tissues' needs and for absorbing nutrients.

3. The muscular contractions that are responsible for mixing and moving the food from one region to another are the result of smooth muscle membrane depolarizations and reflex pathways of the enteric neural plexuses. The patterns include **segmentation** and **peristaltic waves.**

4. The secretions of the GI tract perform several functions. **Mucus** lubricates and protects all regions of the GI tract. It is present in saliva, along with fluid and α-amylase. In the stomach HCl and **pepsin** are secreted. In the small intestine, HCO_3^- and enzymes for the digestion of carbohydrates, proteins, and lipids are added from the pancreas. The **bile** secreted by the liver enters the duodenum along with the pancreatic secretions.

5. Acid and bicarbonate secretions in the GI tract are the result of active transport and facilitated transport processes. Blood leaving the stomach is more alkaline than blood in the rest of the circulation, especially after a meal. Blood leaving the intestine and pancreas, the sites of bicarbonate secretion, is more acidic than arterial blood. Over time, the secretion of acid and base in stomach and intestine tend to cancel one another.

6. **Bile salts** produced in the liver and released with bile into the duodenum are essential for fat digestion and absorption because of their **emulsifying** effect. After they have served their function in the proximal part of the small intestine, they are acted on by bacteria and reabsorbed in the form of **bile acids.** The bile acids return to the liver in the form of bile salts.

7. Intestinal motility mixes the chyme with digestive secretions, promotes absorption of nutrients, and propels the chyme from one region of the intestine to another. Segmentation (mixing) predominates in the early portions of the small intestine; peristalsis in the later portions; **mass movements** are the major propulsive activity of the large intestine.

1. Describe the roles of the mouth, esophagus, stomach, small intestine, and large intestine in the sequential processing of food.

2. What are the two steps in saliva formation?

3. The tonic tension of smooth muscles in the GI tract can be increased or decreased. How are these changes accomplished, and how does the altered tension contribute to GI system functioning?

4. Explain why only a small amount of chyme enters the small intestine from the stomach with each peristaltic wave.

5. Which cells in the stomach show the greatest evidence of oxidative metabolism? What is the function of these cells?

6. List the categories of enzymes secreted by the pancreas.

7. What is enterokinase, and what role does it play in digestion?

8. What voluntary and involuntary processes contribute to defecation?

Choose the MOST CORRECT Answer

9. Food is mixed with gastric juices to form chyme in this area of the stomach:
   a. Fundus
   b. Body
   c. Pylorus
   d. Antrum

10. The major site for both digestion and absorption of food in the gastrointestinal tract is the:
   a. Stomach
   b. Small intestine
   c. Esophagus
   d. Large intestine

11. The enteric nervous system is found in this layer of the gastrointestinal tract:
   a. Serosa
   b. Muscularis
   c. Submucosa
   d. Mucosa

12. Motility in the GI tract is accomplished by:
   a. Peristalsis
   b. Segmentation
   c. Mixing motions of villi and microvilli
   d. All of the above

13. The ___ cells found in gastric glands secrete HCl and an intrinsic factor necessary for intestinal absorption of vitamin B_{12}:
   a. Chief
   b. Parietal
   c. G
   d. Mucus

14. The most abundant secretory cells of the GI tract are:
   a. Goblet cells
   b. Enteroendocrine cells
   c. Paracrine cells
   d. Crypt cells

15. Which of the following is NOT released by pancreatic acinar cells?
   a. Elastases
   b. Phospholipases
   c. Proteases
   d. Enterokinases

16. The function of the basolateral Na^+-K^+ pump in the intestinal epithelial cells is to:
   a. Absorb products of digestion
   b. Secrete digestive enzymes
   c. Absorb water and salt
   d. Function as sensory receptors

● SUGGESTED READING

COHEN LA: Diet and cancer, *Scientific American* 257(5):42, 1987. Recommendations aimed at reducing the incidence of cancers associated with nutrition are based on epidemological studies and animal experiments.

FURNESS JB, COSTA M: *The Enteric Nervous System*, Edinburgh, 1987, Churchill-Livingstone. Complete description of neural control of gastrointestinal function.

MARANTO S: Bugged by an ulcer? You could have a bug, *Discover* 8:10, 1987. Considers whether some ulcers are not primarily related to stress, but are caused by specific infections.

OHNING G, SOLL A: Medical treatment of peptic ulcer disease, *American Family Physician* 39(4):257, 1989. Self-explanatory.

SANDERS KM: Colonic electrical activity: concerto for two pacemakers. News in Physiological Sciences 4:176, 1989. Motility of the proximal colon is a product of the interaction of two pacemakers.

UVNAS-MOBERG K: The gastrointestinal tract in growth and reproduction. *Scientific American* July 1989, 261(1):78. Shows how a mother's digestive tract adapts to the increased demands of pregnancy.

Will fiber save your life? *Consumer Reports Health Letter*, 2(3):17, 1990. Discusses the relationship between soluble and insoluble fiber and cholesterol, low density lipoproteins, coronary disease and colorectal cancer.

WOLFE MM, SOLL AH: The physiology of gastric acid secretion. *N Eng J Med* 319(26):1707, 1988. Describes how acid and pepsin are secreted, how secretion is regulated, and how gastric secretion may injure the gastroduodenal mucosa.

CHAPTER 22

Digestion and Absorption

On completing this chapter you should be able to:

- Describe how water, amino acids, and sugars are absorbed in the gastrointestinal tract.
- Appreciate the role of the transmembrane Na^+ gradient in carrier-mediated co-transport of most amino acids and sugars into the intestinal epithelial cells.
- Describe carbohydrate digestion in the mouth, stomach, and small intestine, noting the role of brush-border enzymes in the small intestine.
- Understand the steps in lipid digestion, noting the role of the bile in emulsifying fats.
- Trace the pathway of lipids from the intestinal lumen to the bloodstream.
- Describe the processes involved in iron absorption.
- Understand the mechanisms for absorption of water-soluble and fat-soluble vitamins, noting the specialized transport mechanisms for vitamins.
- Characterize the cephalic, gastric, and intestinal phases of digestion.
- Compare long-loop and short-loop reflexes in the control of digestion.
- Identify the major known gastrointestinal tract hormones—gastrin, motilin, gastric inhibitory peptide, secretin, and cholecystokinin— and describe how each regulates gastrointestinal tract function.

*I*n digestion, the GI tract exposes chyme to mechanical disruption, gastric acid, enzymes and detergent. This combined assault breaks down food structure, making nutrients, vitamins and minerals locked up in the structure available. It converts many of the complex molecules of food, such as proteins, nucleic acids, polysaccharides and triacylglycerols to simpler ones: monosaccharides, free amino acids and free fatty acids, which can be absorbed. Absorption of nutrients, together with water and electrolytes, is carried out almost entirely in the small intestine. The maximum absorptive capacity of the intestine is much greater than the normal intake of most nutrients. Even when the amount of food a person consumes greatly exceeds his or her nutritional needs, the intestine still absorbs nearly 100% of the fat, protein, and carbohydrate ingested. Caloric intake must therefore be regulated at the level of ingestion because it is not regulated at the level of absorption. In contrast, a few substances, such as calcium and iron, are usually not completely absorbed by the intestine, so adjustments in the rate of absorption of these substances from the gastrointestinal (GI) tract are important in regulating blood levels.

The GI system is intrinsically regulated to an extent unmatched among organ systems. Its single-unit smooth muscle is capable of spontaneous rhythmic contraction even in the absence of nervous or hormonal inputs. The enteric brain of the GI tract, consisting of the two plexuses with their web of interconnections, contains about 10^8 neurons, a number similar to that in the spinal cord. Although the plexuses are technically parasympathetic ganglia, most of the neurons are neither adrenergic nor cholinergic, but instead use small peptides as neurotransmitters and neuromodulators. The GI tract also has its own endocrine systems, which help coordinate motility and secretion in different parts of the tube system. The underlying kinship between GI endocrine cells, the neurons of the plexuses, and the central nervous system is seen in their chemical signals. Peptides that were first known as gut hormones are now known to also serve as neurotransmitters in the plexuses and in the central nervous system.

AN OVERVIEW OF DIGESTION AND ABSORPTION

> - *What barriers do nutrients cross in the course of absorption?*
> - *How does the route differ for lipids as compared to carbohydrates and amino acids?*
> - *What is the role of the hepatic portal system?*
> - *What role does the blood that enters the liver from the hepatic artery play in hepatic function?*

Digestion: The Task at Hand

The forms in which nutrient molecules can be absorbed into the body are usually quite different from the more complex molecules provided by the diet. The digestion of lipids, carbohydrates, and proteins has one factor in common: they all involve **hydrolytic reactions** (see Chapter 3). In hydrolytic reactions, a bond that links subunits of the food molecule is split, and a hydrogen and hydroxyl ion from a water molecule are added to the two ends of the subdivided molecule. Particular types of digestive enzymes catalyze the hydrolysis of specific categories of bonds.

As proteins, enzymes are affected by the pH, solute composition and temperature of their environments. In the process of evolution, each digestive enzyme has been adapted to function optimally under the conditions in which it normally operates. Thus pepsin, a protease secreted by the stomach, is most effective in the pH range of 1.5 to 2.5 characteristic of gastric chyme, whereas pancreatic proteases are most effective in the pH range of 7 to 8 characteristic of intestinal chyme. In the case of polysaccharides and proteins, digestion involves hydrolysis of several distinct types of bonds and the cooperation of several types of enzymes is generally necessary to complete the job. In the case of triacylglycerols and phospholipids, the job is even more complex, because before enzymatic hydrolysis can proceed effectively, the water-insoluble lipid must be **emulsified**—broken up into small droplets—by detergent bile salts secreted by the liver.

Pathways for Absorption from the Gastrointestinal Tract

After food is reduced to molecules that can be taken into the body, the process of absorption begins. Subsequent sections of this chapter will follow the digestion and absorption of particular nutrient types individually. The four structures that lie between the lumen of the intestine and the blood (Figure 22-1) are (1) the layer of mucus covering the microvilli, (2) the mucosal epithelium, (3) the interstitial space, and (4) the capillary walls. The pathways taken through these barriers differ for different substances.

Most sugars and amino acids are transported into epithelial cells of the small intestine by carrier-mediated co-transport systems that are driven by Na^+ moving down its electrochemical gradient into the cells (for an example of gradient-mediated cotransport, see Chapter 6). The energy cost of this transport system is ultimately paid by the ATP expended by the Na^+-K^+ ATPase to pump out the Na^+ ions entering in this co-transport. Lipids follow a different route, moving first into the intestinal lymphatic capillaries (lacteals) and then entering the systemic circulation along with the lymph.

The Liver and Hepatic Portal Circulation

The **liver** plays important roles in digestion, energy metabolism, biosynthesis, and detoxification (Table 22-1). The role of the liver in energy metabolism is

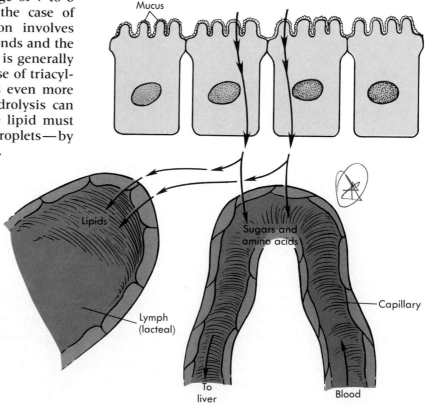

FIGURE 22-1

To enter the blood or lymph, materials absorbed from the intestinal lumen must traverse an external mucous layer, the epithelial cell membrane, the epithelial cell cytoplasm (which includes the core of the microvilli), the basolateral membrane, the interstitial fluid, and the membranes and cytoplasm of the capillary or lymph endothelial cells. The route through the mucosal barrier and into the blood is different for different chemical classes of nutrients.

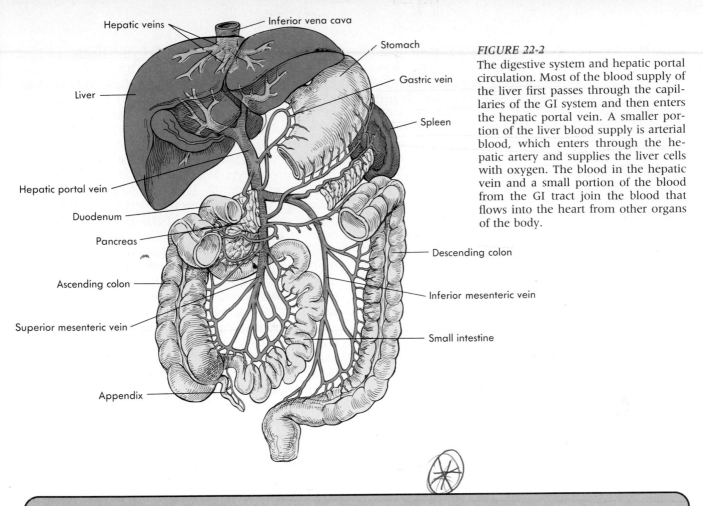

Hepatic veins
Inferior vena cava
Stomach
Gastric vein
Liver
Spleen
Hepatic portal vein
Duodenum
Pancreas
Descending colon
Ascending colon
Inferior mesenteric vein
Superior mesenteric vein
Small intestine
Appendix

FIGURE 22-2
The digestive system and hepatic portal circulation. Most of the blood supply of the liver first passes through the capillaries of the GI system and then enters the hepatic portal vein. A smaller portion of the liver blood supply is arterial blood, which enters through the hepatic artery and supplies the liver cells with oxygen. The blood in the hepatic vein and a small portion of the blood from the GI tract join the blood that flows into the heart from other organs of the body.

TABLE 22-1   Functions of the Liver

Physiological process	Product(s)	Action
Digestion	Bile salts; bile acids; lecithin	Emulsifies fats so they can be absorbed.
	Cholesterol	Recirculates bile salts via the entero hepatic circulation.
Detoxification	Urea; detoxified drugs; inactivated hormones	Adds polar groups to drugs, some hormones, and several metabolites so that they can be excreted by the kidney. Converts ammonia to the much less toxic urea.
Biosynthesis	Plasma proteins	Synthesizes all plasma proteins except the immunoglobulins, including clotting factors and complement (discussed in Chapter 26).
	Lipoproteins	Makes protein carriers for cholesterol and the triglycerides.
Energy metabolism	Glucose; fatty acids, ketones	Stores glucose as glycogen; converts amino acids to glucose; synthesizes and degrades lipids.
Other functions	Iron; heme; transferrin	Destroys aging or damaged red blood cells.

described in Chapter 23. Its digestive function is the secretion of **bile,** which is essential for lipid digestion in the intestine. Bile is secreted continuously. During the interdigestive period it passes from the liver into the gallbladder where it is stored and concentrated (see Chapter 21). During digestion, contractions of the gallbladder expel bile into the duodenum.

Blood that has passed through the capillaries of the stomach, large and small intestines, spleen, and pancreas is carried by the **hepatic portal vein** to the liver (Figure 22-2). This blood passes through a second set of capillaries in the liver before returning to the heart. About 75% of the blood that enters the liver comes from the GI system by way of the hepatic portal vein, and the other 25% (bringing vital oxygen to the liver cells) comes from the systemic circulation by way of the hepatic artery. Blood from both origins leaves the liver in the **hepatic vein.** The adaptive significance of the hepatic portal circulation, which allows the liver first access to amino acids and glucose absorbed by the intestine, arises from the central role of the liver in regulating blood levels of glucose, amino acid and fatty acids. During absorption of nutrients from a meal, the liver stores some of the absorbed glucose as glycogen and converts some amino acids to fatty acids which can be stored in adipose tissue. The importance of the liver in metabolism during feeding and fasting will be explained in detail in Chapter 23.

The structure of the liver reflects both its secretion of bile and the fact that much of its blood flow comes from the GI tract. Liver cells, or **hepatocytes,** are arranged in a series of sheets only one or two cells thick, separated by the liver capillary network (Figure 22-3). The **lobules,** or functional units of the liver, are sheets of hepatocytes organized around a central vein. Large capillaries, or **sinusoids,** run between the sheets of hepatocytes. The sinusoids receive blood enriched by absorbed nutrients from a branch of the hepatic portal vein. This blood has a low O_2 content because it has already delivered O_2 to the GI tract. However, the sinusoids also receive blood with a high O_2 content from the hepatic artery, and this mixes with the portal blood. The hepatic artery blood supply to the liver meets the liver's needs for gas exchange and also allows the liver to regulate the amounts of sugars and amino acids in the general circulation (see Chapter 23). The bile is secreted by the hepatocytes into a network of microscopic tubules called **bile canaliculi,** located within each lobule of the liver. The bile canaliculi converge into larger and larger tubules that eventually join to form the bile duct.

The liver plays an important part in the metabolism and eventual excretion of numerous substances found in the blood, including drugs, hormones, and metabolic end products (Table 22-1). The liver converts some substances to more water-soluble forms that can be eliminated by the kidney. Generally this is accomplished by **conjugation**—adding to the substance a polar group such as glucuronic acid, taurine, or glycine. An example of a substance handled by conjugation is **bilirubin,** the product of toxic porphyrins released in the degradation of hemoglobin. Conjugated substances may be excreted in the bile, ultimately leaving the body with the feces, or they may reenter the blood and be excreted by the kidneys through a secretory pathway specific for organic anions (see Chapter 19). The liver also converts some substances to less toxic forms. For example, ammonia is converted to the less toxic urea. Enzymes in the liver degrade some hormones and convert certain drugs and toxins to inactive forms. Although the general name for this process is **detoxification,** some drugs (and some toxins) are actually made more toxic by liver metabolism. The ability of the liver to metabolize drugs decreases with age, so drug dosages appropriate for young adults may produce excessively high plasma

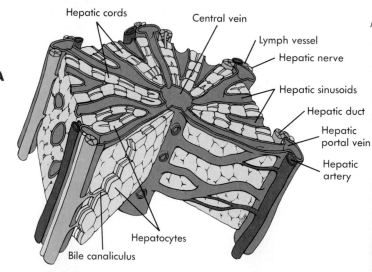

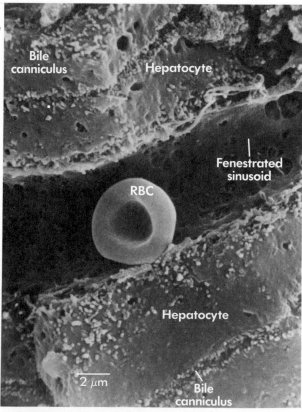

FIGURE 22-3
A The liver is composed of lobules, which consist of stacks of sheets of cells through which blood drains in large spaces, called sinusoids, into a central vein. In the portion of a lobule shown here, the mixing of blood from the hepatic artery and hepatic portal vein is illustrated, as are the canaliculi through which bile drains in the opposite direction to the bile duct.
B A scanning electron micrograph of a hepatocyte showing a bile canaliculus.

THE GASTROINTESTINAL TRACT

levels, and thus toxic side effects, in the elderly.

Another role of the liver is to synthesize all plasma proteins except the globulins. The proteins secreted by the liver include albumin (important in maintaining the plasma osmotic pressure), the protein carriers for plasma transport of cholesterol and triglycerides, clotting factors, angiotensinogen, and the complement proteins involved in the immune response (see Chapter 26).

CARBOHYDRATE DIGESTION AND ABSORPTION

- *What is the difference between a monosaccharide, a disaccharide, and a polysaccharide? In which of these forms is carbohydrate absorbed?*
- *What types of enzymes are involved in carbohydrate digestion? Explain why cellulose cannot be digested.*
- *What are the mechanisms of intestinal sugar absorption? What driving force results in absorption of glucose and galactose?*

Dietary Sources of Carbohydrate

Dietary carbohydrates fall into three major categories: **monosaccharides, disaccharides,** and **polysaccharides** (see Chapter 3 and Figures 3-5 and 3-6). Monosaccharides include **glucose** and **fructose.** Dietary disaccharides include **sucrose** (table sugar), a disaccharide of glucose and fructose that constitutes on average 30% of the carbohydrate intake; and **lactose** (milk sugar), a disaccharide of glucose and **galactose** that accounts for about 6% of the dietary carbohydrate. Ribose and deoxyribose (present in RNA and DNA) and the disaccharide **maltose** are in the diet in small amounts. **Starch** is the major dietary polysaccharide. Plant starch (**amylopectin**) and animal starch (**glycogen;** see Figure 3-7) together typically constitute over 50% of the total daily carbohydrate intake in the average Western diet. Starch contains both branched-chain and straight-chain polymers of glucose.

Starch Digestion by Amylases and Brush-Border Enzymes

The salivary glands of most people secrete an enzyme, **α-amylase,** that attacks the linkages between adjacent glucose molecules in the straight-chain regions of starch. Salivary amylase is inactivated by stomach acid, leaving a considerable fraction of dietary starch to be digested by a similar pancreatic amylase (Table 22-2). The salivary and pancreatic amylases cannot break all of the bond configurations between monosaccharides and are particularly ineffective at dealing with the monosaccharides near branch points. The result of digestion of starch by the amylases is not pure glucose, but a mixture of oligosaccharides of 2-3 monosaccharide units each, including maltose, sucrose and maltotriose. Digestion of the oligosaccharides to monosaccharides is accomplished by **brush-border enzymes,** which are bound to the mucosal membranes of intestinal cells (Figure 22-4). The activity of these enzymes releases monosaccharides in close proximity to the surface of the microvilli, facilitating absorption.

Cellulose, the main structural element in plants, is a polysaccharide of glucose abundant in whole grains, fruits and vegetables. However, few of the glucose units of cellulose become available in the human digestive system, because the bonds between the units are resistant to attack by the amylase enzymes secreted by animals. Cellulose is a major component of dietary **fiber,** the class of dietary carbohydrates that cannot be digested and absorbed in the small intestine (see Frontiers box, Chapter 21). The undigested cellulose contributes to the volume of material in the alimentary canal and stimulates motility of the large intestine. In some animals, such as cattle, microbes that live in specialized parts of the stomach provide enzymes that can digest cellulose. The human large intestine contains bacteria that can partly degrade cellulose. The glucose so released is not absorbed by the colon in that

---

TABLE 22-2   *Pancreatic Enzymes*

Enzyme	Substrate	Products
Amylase	Plant starch, glycogen	Maltose, short chains of glucose molecules
Trypsin, chymotrypsin	Internal peptide bonds	Free amino acids, small peptides
Carboxypeptidases	Peptide bonds at carboxy end of peptides	Free amino acids
Aminopeptidases	Peptide bonds at amino end of peptides	Free amino acids
Elastase, collagenase	Internal peptide bonds of collagen	Small peptides
Lipase	Triacylglycerols	Free fatty acids, 2-monoacylglycerols, glycerol
Phospholipase	Phospholipid	Lysophosphatides, free fatty acids
Hydrolase	Cholesterol esters	Cholesterol, free fatty acids
RNAase	RNA	Short chains of ribonucleic acid
DNAase	DNA	Short chains of deoxyribonucleic acid

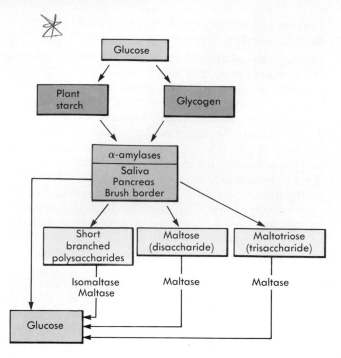

FIGURE 22-4

The two dietary sources of starch are both composed of glucose molecules and can be hydrolyzed by α-amylases. The four major products of starch digestion by the amylases are listed, along with the brush-border enzymes which complete their digestion to glucose.

form, but significant amounts of short-chain fatty acids and acetate that are produced from the glucose by the bacteria are absorbed.

Monosaccharide Absorption

Although starch digestion produces only glucose, the dietary carbohydrates, sucrose and lactose, are digested to glucose, galactose, and fructose. All three of these monosaccharides are then taken up by the epithelial cells. There is an Na^+-dependent co-transport system for glucose and galactose in intestinal epithelial cell membranes (Figure 22-5) that is similar to that found in the proximal tubular cells of the kidney (see Chapter 19). Because glucose and galactose compete for the same carriers, their carriers can be saturated by a lot of either sugar or by some of both. The energy that indirectly supports this transport is expended by the Na^+-K^+ pump, which bails out Na^+ that enters with the sugar molecules. This expenditure of energy ensures that the Na^+ gradient driving rapid absorption is maintained.

In addition to the Na^+-dependent system, there are Na^+-independent carrier systems for glucose and fructose. Monosaccharides in the cytoplasm of the intestinal epithelial cells reach the interstitial fluid by facilitated diffusion across the basolateral membranes (see Figure 22-5). The monosaccharides subsequently diffuse into the blood.

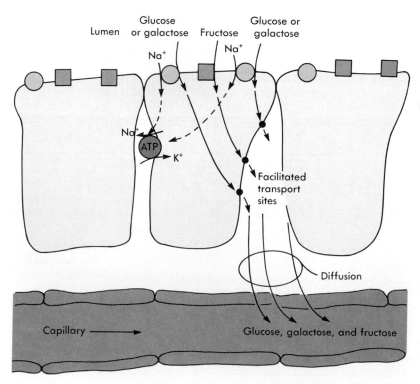

FIGURE 22-5

Two transport systems for sugars are shown: galactose and glucose share one Na^+-dependent co-transport system, and fructose uses a second, Na^+-independent transport system. The sugars are probably moved out of the cells by specific carriers *(black dots).* Movement into the blood is by diffusion.

Lactose Intolerance

Normally, carbohydrate digestion and absorption is completed midway through the jejunum, and little monosaccharide reaches the ileum. The ileum has a much lower capacity for absorbing monosaccharides and can be overloaded if the small intestine is diseased. Most disorders of digestion involve deficiencies of the carbohydrate-specific enzymes present on the microvilli of the intestinal epithelial cells. One such condition is called lactose intolerance.

Lactose is the major carbohydrate component of milk and is degraded by lactase, a brush-border enzyme, to glucose and galactose. In most of the world's population this enzyme is present in infants but disappears from the intestine after the first few years of life. However, about 90% of Caucasian Europeans and Americans continue to express the enzyme as adults. This is regarded as an evolutionary adaptation for cultural use of dairy products from domesticated animals. When lactase is absent, lactose cannot be digested or absorbed and instead remains in the lumen, where it is osmotically active, like the Mg^{++} present in laxatives. The result is diarrhea and fluid loss. In addition, bacteria in the large intestine metabolize lactose to produce large quantities of CO_2 gas, which causes distension and pain. A temporary intolerance of lactose can occur even in children who possess lactase if an intestinal infection causes the cells lining the intestine to be shed more rapidly than usual. Because the immature cells do not have as many functioning enzymes, digestion of lactose is less efficient. Individuals with lactose intolerance can consume milk products such as yogurt and cheese without distress because the action of the bacteria used to make these products breaks down the lactose in them before they are consumed. Lactose-intolerant individuals may consume fresh milk and milk products if they also take daily doses of lactase, or may purchase milk containing cultures of lactose-digesting bacteria (acidophilus milk).

PROTEIN DIGESTION AND ABSORPTION

- *How are the conditions for protein digestion different in the stomach and small intestine?*
- *What are the most common end products of protein digestion in the stomach and small intestine?*
- *How is the mechanism for amino acid absorption similar to the mechanism for sugar absorption?*

Enzymes of Protein Digestion

There are more types of amino acids in proteins than there are simple sugars in carbohydrates (see Chapter 3, Figures 3-8, 3-10, and 3-11). Proteins are formed in a dehydration reaction that unites the **carboxy terminal** of one amino acid with the **amino terminal** of another in a **peptide bond** (see Figure 3-10). Enzymes (**peptidases** or **proteases**) that digest proteins recognize particular amino acids or classes of amino acids and hydrolyze, attacking the peptide bonds between them (Figure 22-6, *A*). Peptidases fall into two classes, **endopeptidases** that hydrolyze bonds in the center of polypeptide chains, and **exopeptidases** that chop away at the free ends of polypeptide chains. Of the latter, those that attack the free carboxy end are termed **carboxypeptidases**; those attacking the amino end are called **aminopeptidases** (Figure 22-6, *B*).

Pepsin is an endopeptidase secreted by the chief cells of the stomach. The low pH of the stomach is important for the action of pepsin for two reasons. First, the enzyme is most active when the pH is between 1.5 and 2.5. Second, the low pH **denatures** dietary proteins, disrupting some of the internal bonds that determine the tertiary structures and causing the proteins to assume more linear configurations. Denaturation exposes peptide bonds that

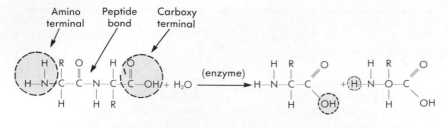

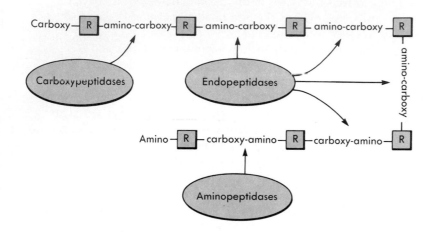

FIGURE 22-6

A A dipeptide (its amino acids are labeled *R*) is hydrolyzed to yield two amino acids. The peptide bond and the amino and carboxy terminals of the dipeptide are indicated, and the sites where the H^+ and OH^- are added when the peptide bond is broken are shown on the right.

B A polypeptide (its amino acids are labeled *R*) is digested by three classes of peptidases: the carboxypeptidases act from the carboxy end, the aminopeptidases act from the amino end, and the endopeptidases are a diverse family of enzymes that can attack specific classes of peptide linkages within the chain.

would otherwise be deep inside the protein's tertiary structure to the action of the digestive endopeptidases. Pepsin is able to hydrolyze some but not all classes of peptide bonds, so the main products of pepsin digestion of proteins are shorter peptide chains called **peptide fragments**. Pepsin is inactivated when the acidity of the chyme entering the duodenum is neutralized by pancreatic and duodenal fluid secretions. In sum, pepsin begins the job of protein digestion, but 85% of the peptide bonds remain to be hydrolyzed in the small intestine by pancreatic and intestinal brush-border enzymes.

Both classes of peptidases (endopeptidases and exopeptidases) are secreted by the pancreas to mix with the chyme as free enzymes. Examples of pancreatic endopeptidases are **trypsin** and **chymotrypsin** (see Table 22-2). Brush-border peptidase activity accounts for the remaining 35% of protein digestion, which occurs in the small intestine. The brush border of intestinal epithelial cell membranes has several types of endopeptidases and exopeptidases.

Mechanisms of Amino Acid Absorption

The products of GI protein digestion are amino acids and a few small peptides. There is a transport system for each of the three major classes of amino acids, acidic, basic, and neutral (see Figure 3-8). There is also a separate transport system that can accept small peptides, which are subsequently degraded into their constituent amino acids by en-

zymes in the epithelial cell cytoplasm. As for monosaccharides, the most amino acids are absorbed across the apical membrane of intestinal cells by gradient-mediated co-transport with Na^+. Amino acids probably exit the epithelial cells into the interstitial fluid by a Na^+-independent mechanism. Once in the interstitial fluid, the amino acids enter the plasma and travel along with the products of carbohydrate digestion through the hepatic portal vein to the liver.

FAT DIGESTION AND ABSORPTION

- *What features of the structure of bile acids are responsible for their ability to emulsify fats?*
- *Why is the action of lipases so dependent on the increase in surface area that results from emulsification?*
- *How are digestive products of lipids absorbed into the blood?*

Fat Emulsification in the Intestine

Most dietary fat consists of **triacylglycerols** (triglycerides or neutral fats), **phospholipids, cholesterol** or **cholesterol esters** (cholesterol joined to a single fatty acid chain), and some **fatty acids** (see Chapter 3). Neutral fats are insoluble in water and form large droplets that coalesce and separate completely from the water phase. Fatty acids and phospholipids are slightly more water soluble, but they still tend to aggregate into relatively large sheets, or

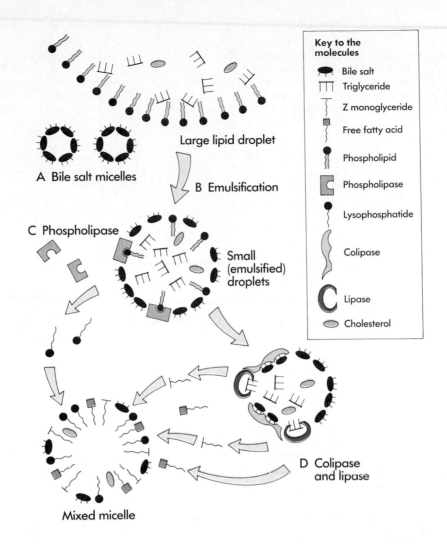

FIGURE 22-7
The steps in lipid solubilization and digestion.
A Bile salts enter the lumen of the GI tract in the form of spherical micelles, with polar groups extending toward the surrounding water and hydrophobic regions oriented toward the center of the micelle. Mixing of intestinal contents brings the bile salt micelles into contact with large aggregates of lipids.
B Some of the lipids are separated from the large masses when their surfaces are covered with bile salts. This process, called emulsification, increases the surface area of the lipids.
C Pancreatic lipases, with the help of colipase, attack the lipid bonds exposed on the surface of the emulsified lipid droplet.

bilayers, with the polar groups facing outward and the hydrocarbon chains facing inward. The fats tend to float on top of the other ingested material, and the churning of the stomach reduces the size of the lipid droplets only slightly.

The first problem that must be overcome to digest lipids is to break up large lipid droplets into smaller ones. This increases the surface area for attack by enzymes that are in the aqueous phase. The bile secretions contain a class of molecules that reduce the size of the lipid aggregations by acting very much the way detergents act to cut grease. The

process, called **emulsification**, is carried out by bile salts derived from the steroid molecule, cholesterol.

Bile salts have polar groups attached to one side, causing them to be **amphipathic** (see Chapter 3)—able to associate with polar molecules such as water and with nonpolar molecules like lipids. Bile salts are detergents that reduce the surface tension of fat droplets and make them more susceptible to disruption into smaller droplets by segmentation of the intestine. In the hepatic ducts, bile salts aggregate to form **micelles** (Figure 22-7, *A*). In the intestine, bile salts are transferred from biliary micelles to the

FIGURE 22-8

A triacylglycerol (for example, stearic acid) is hydrolyzed to glycerol and three fatty acids (long chain represented by R) by the addition of three molecules of H_2O. Alternatively, and more commonly, the triacylglycerol is hydrolyzed with two H_2O molecules to yield two free fatty acids and a 2-monoacylglycerol.

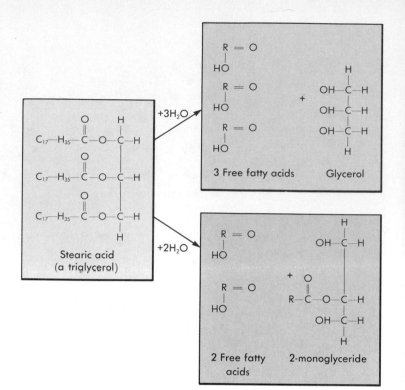

Fat Digestion by Pancreatic Lipases

The digestion of fat **(lipolysis)** is accomplished by lipid-digesting enzymes called **lipases,** which are secreted by the pancreas (see Table 22-2; Figure 22-7, *C*). Unlike the proteases, the lipases are secreted in their active forms. Pancreatic lipase action on triacylglycerols produces **free fatty acids, 2-mono-acylglycerols,** and small amounts of **glycerol** (Figure 22-8). The reason that fairly large amounts of 2-monoacylglycerols are produced is that lipase tends to attack the fatty acid chains on the end carbons of the glycerol backbone, leaving the one on the center carbon intact.

In addition to lipase, the pancreas secretes an enzyme called **hydrolase** (see Table 22-2), which breaks down cholesterol esters into fatty acids and cholesterol. Another enzyme that digests lipids is **phospholipase** (see Table 22-2), whose end products are **lysophosphatides** (phospholipids from which the second fatty acid has been cleaved) and fatty acids. Pancreatic lipases act only at the boundary between a lipid droplet and water (see Figure 22-7, *C*). Phospholipases also require bile salts for their activity. The lipid-digesting enzymes reduce ingested lipids to fatty acids, cholesterol, lysophosphatides, 2-monoacylglycerols, and glycerol.

Pancreatic lipases cannot effectively attack lipids in emulsified droplets, because the attached bile salts and the projecting heads of the phospholipids create such a clutter on the surface of the droplet that lipase molecules cannot reach the triacylglycerol. Pancreatic phospholipase must first remove some of the phospholipids from the surface of the droplet. The action of lipase is greatly enhanced by **colipase,** a coenzyme secreted by the pancreas. Colipase attaches itself to the surface of the droplet by binding to bile salts; then it binds to lipase, anchoring it next to the surface (Figure 22-7). The lipase can then begin to convert triacylglycerol into monoacylglycerol and free fatty acid.

Micelle Formation and Fat Absorption

The end products of lipase and phospholipase activity (free fatty acids, cholesterol, lysophosphatides, 2-monoacylglycerols, and glycerol) combine with the bile phospholipid lecithin, bile salts, and cholesterol to form structures 50 to 100 nm in diameter called **mixed micelles** (Figures 22-7 and 22-9). In the mixed micelles, the hydrophobic molecules and the parts of amphipathic molecules that are hydrophobic are located in the center of the sphere, whereas the surface of the sphere is studded with the polar projections, consisting primarily of the charged groups on the bile salts.

Mixed micelles move around within the chyme and eventually come into contact with the microvilli of the intestinal epithelial cells (Figure 22-9). The products of lipolysis are slightly soluble in water, so an equilibrium exists between the micelles and the

surfaces of fat droplets, coating them and preventing small droplets from recoalescing to form larger ones (Figure 22-7, *B*). The result of mixing bile salts with the intestinal contents is the formation of **emulsified droplets** about 1 μ in diameter. This greatly increases the surface area available for the actions of digestive enzymes and therefore reduces the time required for lipid digestion.

THE GASTROINTESTINAL TRACT

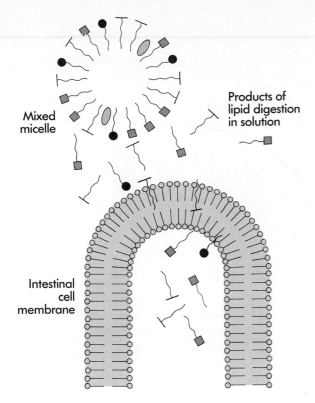

Mixed
micelle

Products of
lipid digestion
in solution

Intestinal
cell
membrane

FIGURE 22-9

Absorption of lipids is facilitated by the accumulation of fatty acids, glycerol, and other products of fat digestion in mixed micelles. This occurs near the site of lipase activity. The mixed micelles give up their by-products of lipid digestion in the vicinity of the microvilli, where they diffuse across the lipid bilayer membrane. The micelles may be carried back to the site of lipid digestion and then pick up products of lipid digestion.

aqueous solution surrounding them. When the concentration of lipid in the aqueous phase becomes very high, as it is near the site of digestion, the micelles accumulate dissolved lipids. As mixing of the intestinal contents carries these loaded micelles into the vicinity of the microvilli, the local concentration of fatty acids and 2-monoacylglycerols in the aqueous phase is much lower because they are dissolving in and disappearing across the lipid bilayer of the epithelial cell membranes. A new equilibrium is established under these circumstances, and the products of lipolysis move out of the mixed micelle down their concentration gradient into the aqueous environment, where they are free to diffuse into the epithelial cells. Most of the bile salts do not enter the cells of the jejunum, and bile salts can continue the process of micelle formation when the bile salts are removed from the intestine, reprocessed, and returned to the site where lipids are being digested.

When bile salts reach the ileum, they are actively transported into the epithelial cells and returned to the liver in the portal circulation (see Chapter 21). Their uptake by the liver stimulates

bile salt release, an action that is effective in delivering them to the duodenum as long as the sphincter of Oddi is relaxed (see actions of cholecystokinin, below). This rapid cycling of the bile salts allows a limited supply of bile salts to be used repeatedly in digestion of a meal high in lipids.

Chylomicron Formation in Intestinal Epithelial Cells

Once inside epithelial cells, fatty acids, 2-monoacylglycerols, cholesterol, and lysolecithin do not simply cross the basolateral membrane and directly enter the interstitial fluid. Instead, triacylglycerols are resynthesized from fatty acids and 2-monoacylglycerols by enzymes located in the smooth endoplasmic reticulum (Figure 22-10). Other enzymes in the smooth endoplasmic reticulum reesterify cholesterol or synthesize phospholipids. Intracellular triacylglycerols, phospholipids and cholesterol esters are then packaged into tiny vesicles in the Golgi complex. Finally, a lipoprotein synthesized by the intestinal cells is incorporated into the surface of the vesicles, surrounding each lipid droplet with polar groups. The droplets formed within the epithelial cell are called **chylomicrons.**

Chylomicrons leave the epithelial cells by exocytosis: the vesicles containing chylomicrons are transported to the basolateral surface of the epithelial cell where they fuse with the basolateral cell membrane, releasing the chylomicrons into the extracellular fluid. Once released, the chylomicrons do not enter the capillaries because they are too large to pass through the pores between the capillary endothelial cells. Instead they move into the lacteals, the lymphatic capillaries that extend into each of the villi (see Figures 22-1 and 22-10). The intercellular junctions of the lacteals are large enough to allow the entry of chylomicrons. The chylomicrons are carried in the lymph to the systemic venous circulation where the main lymphatic duct (the thoracic duct) empties into the venous system at the junction of the left subclavian and internal jugular veins (see Chapter 14). During digestion of a meal rich in fat, the lacteals appear white because they are filled with chylomicrons. This appearance gave them their name, which means "milky."

ABSORPTION OF IRON AND VITAMINS

- *How do the actions of HCl and the presence of transferrin in the lumen both contribute to iron absorption?*
- *When hemorrhage reduces the amount of hemoglobin in the circulatory system, what sources of iron are available in the body for synthesis of new hemoglobin?*
- *What factors are involved in absorption of Vitamin B12?*
- *Why is absorption of fat-soluble vitamin supplements more effective when the vitamins are taken after a meal rather than on an empty stomach?*

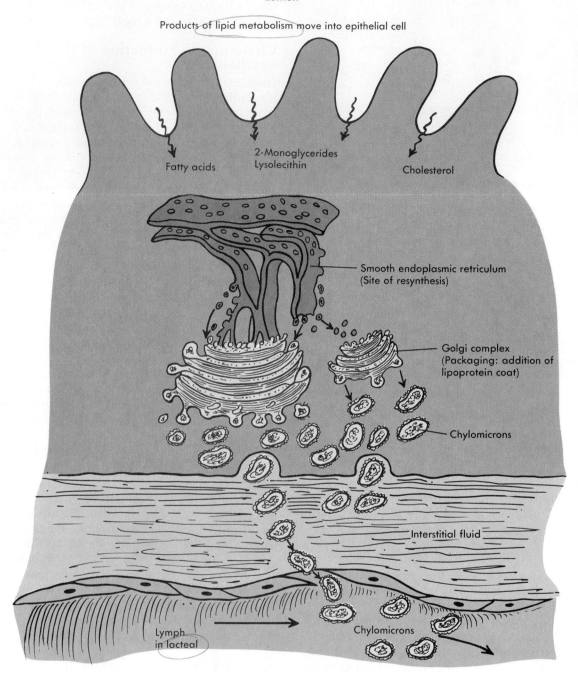

Lumen

Products of lipid metabolism move into epithelial cell

Fatty acids

2-Monoglycerides
Lysolecithin

Cholesterol

Smooth endoplasmic reticulum
(Site of resynthesis)

Golgi complex
(Packaging: addition of
lipoprotein coat)

Chylomicrons

Interstitial fluid

Lymph
in lacteal

Chylomicrons

FIGURE 22-10

Once the products of lipid metabolism move into the epithelial cell, they enter the smooth endoplasmic reticulum and are resynthesized into larger molecules (cholesterol esters and triacylglycerols). They are then packaged in the Golgi complex and rendered hydrophilic by the addition of β-lipoprotein. The resulting droplets of lipids, called chylomicrons, leave the cell by exocytosis, travel through the interstitial fluid, and enter the lymph lacteals.

Fake Fat: New Hope for the Obese?

"Fake fats" are commercial products with the flavor and physical properties of triacylglycerol, but which are chemically mostly protein or carbohydrate, and which contain little or no cholesterol. Use of fat substitutes can reduce the caloric content of high-fat foods by 60% to 70%. The first fake fat to be approved by the FDA was Simplesse, a blend of egg whites and milk, low in calories but not entirely fat-free. A 4 oz serving of the ice cream Simple Pleasures has 120 calories and just under a gram of fat, compared to 250 calories and 15 grams of fat in regular ice cream. Simplesse may be put into processed foods such as mayonnaise, salad dressings, and sour cream; however, it has the disadvantage that it cannot be heated. Another product, Olestra, awaits FDA approval. It is a synthetic product made up of a sucrose polymer and is fat-free. Because it can be heated Olestra promises to have more widespread use.

Will fake fats be the newest diet fad? It's too early to tell. There are subtle differences between the taste and texture of these substitutes and triacylglycerol. Also, nutrition experts question whether the use of fake fats will translate into leaner bodies and healthier hearts. For example, artificial sweeteners such as aspartame have made diet soft drinks, candies and other foods increasingly popular, but total per capita sugar consumption in the United States has remained more or less constant.

Iron Absorption

Our diet contains materials other than carbohydrates, protein, and fats. On the positive side, there are the vitamins and trace elements necessary for health, and on the negative side, there are substances that are potentially toxic. Regulation of the absorption and metabolism of Na^+, K^+, Ca^{++}, and phosphate were discussed in relation to kidney function (see Chapter 20). Iron is another mineral present in relatively large quantities in the body, particularly in the myoglobin of muscle, the hemoglobin of red blood cells and in storage in the liver, spleen and intestine. Iron is very efficiently recycled by the body (Figure 22-11). The need for dietary iron in adults arises mainly from loss by bleeding and the death of intestinal cells.

Red blood cells have an average life span of 120 days. About 10% die in the circulation. The rest are removed by phagocytes in the liver and spleen when they near the end of their effective life span. The products of red blood cell disintegration and the digestion products of the cells trapped by the phagocytes are broken down by spleen and liver cells. The iron from the hemoglobin is either returned to the plasma, where it travels bound to **transferrin,** an iron-carrying protein, or it is stored in an accessible form. The iron stored in the liver and spleen can be released in the bile and absorbed in the intestine when blood levels of iron drop.

Transferrin-bound iron in the plasma travels to the site of red blood cell formation, the bone marrow. There it can either be used immediately in the synthesis of more hemoglobin in the developing red blood cells or it can be stored. Because the iron present in the body is reused so efficiently, little iron needs to be accumulated from the diet unless blood is lost. When blood loss occurs, the bone marrow cells call on their own stored iron and take up the circulating iron at an increased rate. The level of transferrin-bound iron can be replenished in part by release of the iron stored in the liver and spleen, but uptake is still regulated by the absorption mechanisms, which are not increased for several days after blood loss.

The amount of iron absorbed daily is in the range of 0.5 to 1.0 mg for adult men and about twice that for women in their reproductive years, who must compensate for the loss of blood in menstruation. Children and pregnant women need still greater quantities of iron because the increasing volume of the circulatory systems demands a net increase in body iron. Also, individuals who move to a higher altitude or suffer from chronic anoxia respond by increasing the number of red blood cells

FIGURE 22-11
The iron cycle in the body.

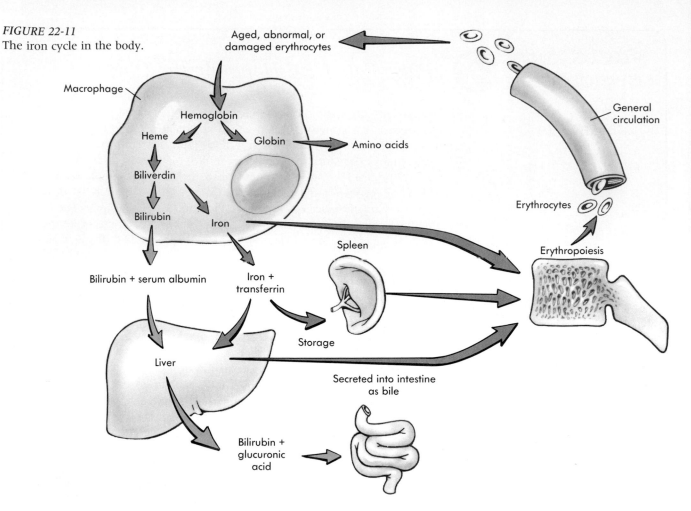

present in the circulatory system, thus increasing the O_2 carrying capacity of the blood.

Ionized iron can exist in two oxidation states: Fe^{++} (ferrous) and Fe^{+++} (ferric). Dietary sources of iron include animal muscle myoglobin, vegetables, and fruits. Gastric HCl releases the iron from plant sources and also reduces it to Fe^{++}, which forms insoluble compounds with other food constituents less readily than does Fe^{+++} (Figure 22-12). Transferrin secreted by the epithelial cells of the intestine binds Fe^{++}, allowing it to be absorbed by the epithelial uptake system. The complex of two Fe^{++} per transferrin is recognized by a receptor on the mucosal surface of intestinal epithelial cells. Binding of Fe^{++}-transferrin complexes to the receptors stimulates endocytosis, which moves the transferrin carrier and iron into the cell (see Figure 22-12). The entire heme group of myoglobin or hemoglobin can also enter the intestinal epithelial cells, where an enzyme splits the Fe^{++} from heme.

Iron derived either from heme groups or transferrin has one of two fates in the epithelial cell: (1) Some of it is bound to a carrier molecule and is subsequently actively transported across the basolateral surface into the interstitial fluid, where it binds to transferrin molecules produced by vascular endothelial cells. Transferrin-bound iron in plasma is available to the bone marrow stem cells that form erythrocytes (Figure 22-11). (2) Some of it is complexed with an intracellular iron-binding protein, **ferritin,** where it is stored for later use. If the supply of iron in the diet drops, stored iron is released from the ferritin, binds to the carrier, and moves into the blood. When an intestinal cell dies, the ferritin-bound iron it contains is lost in the feces (see Figure 22-12).

Free iron is toxic—one role of binding by ferritin and transferrin is to protect the body from free iron ions. The regulation of iron absorption by transferrin secretion in the stomach and intestine and the further restriction imposed by the limited number of intracellular transferrin molecules prevents excess iron from entering the circulation in adults. However, in young children, the mechanisms regulating iron uptake are geared to admit much more iron, and there is no homeostatic mechanism for removing iron from the body once it has been absorbed. Children can be poisoned by ingesting iron supplements intended for adults because excess iron can form insoluble deposits in the brain.

Vitamin Absorption

Vitamins are organic substances essential for proper body function that cannot be synthesized in

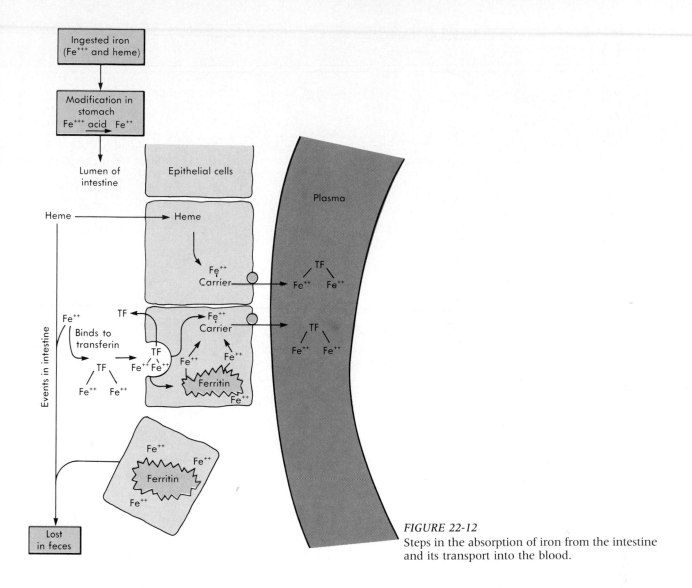

FIGURE 22-12

Steps in the absorption of iron from the intestine and its transport into the blood.

adequate amounts by the body, and so must be obtained from dietary sources or bacterial fermentation in the large intestine. As shown in Table 22-3, the vitamins fall into two major groups: the **water-soluble vitamins** and the **fat-soluble vitamins.** Water-soluble vitamins are readily lost during food preparation because of their solubility and heat lability. Specific recognition molecules have been identified for most water-soluble vitamins, and they may move into the intestinal epithelial cells by simple passive transport, facilitated transport, or active transport. Once absorbed by intestinal epithelial cells, the water-soluble vitamins follow the same pathway into the circulation traveled by amino acids and sugars.

Handling of **Vitamin B$_{12}$ (cobalamin)** is somewhat different from that of the other water-soluble vitamins, and resembles that of iron. The intestinal uptake system for B$_{12}$ recognizes the vitamin only when it is bound to **intrinsic factor,** a glycoprotein synthesized and released by parietal cells of the stomach. When vitamin B$_{12}$ enters the epithelial cells, intrinsic factor remains behind in the lumen. At the basolateral surface of the epithelial cells, vitamin B$_{12}$ attaches to a specific plasma protein that serves as a carrier, rapidly delivering the vitamin to the bone marrow, where it is important in synthesis of red blood cells, and to storage sites in the liver and kidney.

Fat-soluble vitamins are absorbed in the same way as by-products of fat digestion, partitioning into micelles and passing into the lymph before entering the general circulation. In general, these vitamins are present in foods containing lipids, and absorption of these vitamins from dietary supplements is facilitated by consuming them with foods that contain lipids. Excess consumption of some vitamins, particularly vitamins A and D, can lead to **hypervitaminosis,** a disease condition that occurs when these vitamins, normally stored by the liver, accumulate to toxic levels. Vitamin E is stored in many different sites, particularly in adipose fat and membrane lipids, and higher levels of consumption can be tolerated.

TABLE 22-3 *Major Vitamins*

Vitamin name	Dietary sources	Function	Effects of deficiency
Water-soluble vitamins			
B_1 (thiamine)	Whole grains; yeast; nuts; eggs; liver	Coenzyme for carbohydrate metabolism	Beriberi; impaired neural function
B_2 (riboflavin)	Legumes; fish; root crops; meat; whole grains	Coenzyme for protein and carbohydrate metabolism	Blurred vision; dermatitis; nausea
Niacin	Yeast; meats; whole grains; nuts	Coenzyme in Krebs cycle	Pellagra
B_6 (pyrodoxine)	Meats; salmon; tomatoes; corn; spinach; yogurt; whole grains	Coenzyme in amino acid and lipid metabolism	Dermatitis; nausea
B_{12} (cyanocobalamin)	Meat; liver; kidney; milk products; eggs	Erythrocyte formation; amino acid metabolism	Pernicious anemia; neural malfunction
Pantothenic acid	Green vegetables; cereals; liver; kidney; yeast	Coenzyme in steroid hormone synthesis and Krebs' cycle	Neuromuscular degeneration; adrenal hormone deficiency
Folic acid	Green leafy vegetables; liver	Enzyme in purine and pyrimidine synthesis and blood cell production	Anemia; abnormally large blood cells
Biotin	Yeast; liver; eggs; kidney; GI bacteria	Coenzyme in fatty acid synthesis; pyruvate conversion to glucose in liver	Mental depression; muscular fatigue; dermatitis
C (ascorbic acid)	Citrus fruit; tomatoes; green vegetables	Promoter of metabolic reactions; wound healing, antibody production, and collagen formation	Scurvy; anemia; connective tissue degeneration
Fat-soluble vitamins			
A (retinol)	Fish liver oils; green and yellow vegetables; cheese	Formation of rhodopsin; regulates bone and tooth formation	Night blindness; epithelial and neural disorders
D (1,25 dihydroxy-calciferol)	Fish liver oils; egg yolk; milk; ultraviolet light on skin	Calcium and phosphorus absorption	Rickets in children; softened or deformed bones
E (tocopherol)	Nuts; wheat germ; vegetable oils	Antioxidant; RNA, DNA, and red blood cell formation	Abnormal membrane function
K	GI bacteria; green leafy vegetables; liver	Coenzyme for prothrombin and clotting factors synthesis	Spontaneous bleeding; delayed clotting

ELECTROLYTE AND WATER ABSORPTION

- *What are the sources of the fluid content of intestinal chyme?*
- *What mechanisms contribute to net absorption of fluid in the intestine? Why is the absorbed fluid isotonic?*
- *What disorders can result in diarrhea or constipation?*

The Scale of Daily Movement of Material in the GI Tract

Food and water intake varies enormously from one person to another and from day to day. Table 22-4 summarizes the typical daily intake of an adult in the United States. Solid food (upper portion of the table) includes 300 to 800 g of carbohydrate, 75 g of protein, and 60 to 100 g of fat (mostly as neutral fats, or triacylglycerols). Fats have twice the energy content of carbohydrates and proteins per gram, so 30% to 40% of the calories come from fats, half of the calories come from carbohydrates, and the remaining 10% to 20% come from protein. Of the carbohydrates, typically 60% is starch, 30% sucrose (table sugar), and 10% lactose (milk sugar). The GI tract secretes about 20 g of protein in the form of digestive enzymes each day. These proteins are broken down along with the ingested proteins, and the amino acids are recycled.

Material	Source	Amount
Carbohydrate	Food	300 to 800 g
Protein	Food	70 to 80 g
	Secreted enzymes	20 g
Fat	Food	60 to 100 g
Salt (NaCl)	Food	20 to 30 g
	Secretions	30 g
Water	Pure water and water in foods	1 to 1.5 L
Secretions	Salivary glands (NaHCO$_3$)	1.5 L
	Stomach (HCl)	3.0 L
	Pancreas (NaHCO$_3$)	2.0 L
	Liver (bile)	0.5 L

TABLE 22-4   Daily Gastrointestinal Inputs

It takes about 4 hours to digest an average meal, and little of the available nutrients escapes. Indeed, the GI system could theoretically absorb as much as 20 lbs of sucrose per day! Fat absorption is more limited but still exceeds the body's requirement for lipids. The digestive system is so efficient that feces consist mainly of indigestible materials such as plant cellulose, inorganic matter, cellular debris shed from the intestinal lining, and bacterial debris. A normal diet can include daily ingestion of 20 to 30 g of NaCl (see Table 22-4), and in addition to this, 30 to 40 g of NaCl is secreted into the mouth and stomach along with watery secretions. Most of this ingested and secreted NaCl is absorbed. Other electrolytes consumed in gram amounts include K^+, Mg^{++}, Ca^{++}, and phosphate. The K^+ content of food and the K^+ secretion are 10 to 15 times lower than the values for Na^+. In contrast to the high degree of Na^+ conservation, about 25% of the K^+ that enters the intestine is normally lost with feces. A number of vital trace elements, including Ca^{++}, Fe^{++}, Zn^{++}, Cu^{++}, I^-, and F^-, are also ingested and are absorbed to varying extents.

Inputs of fluid into the GI tract are shown in the bottom two sections of Table 22-4. An average of 1.0 to 1.5 liters of water are consumed each day through drinking and in association with food. GI secretions average 7 L/day, including 1.5 L of saliva, 3 L of gastric secretion, 2 L of pancreatic juice and 0.5 L of bile. This makes a total fluid load of 8.5 L/day. Most of this load is absorbed in the small intestine and some of the remaining water is absorbed in the large intestine, so only around 100 ml/day of water is lost with the feces.

Mechanisms of Solute and Water Transfer Across the GI Tract Wall
The solute content of food usually is high enough to make the gastric chyme more concentrated than plasma, but the stomach wall is relatively tight to water movement, so there is little osmotic movement of fluid between extracellular fluid and gastric chyme. In contrast, the "tight junctions" between epithelial cells in the duodenum and small intestine are very leaky to water. As hyperosmotic chyme flows from the stomach into the duodenum, water flows osmotically into the chyme from the extracellular fluid, rapidly making the chyme isosmotic.

Following duodenal equilibration, water, electrolytes, and digestive end products are absorbed as the chyme passes through the rest of the GI tract. Most water absorption in the GI tract occurs by an isotonic mechanism similar to that found in the proximal tubules of the kidney. That is, water reabsorption is secondary to the active absorption of Na^+, amino acids and monosaccharide described earlier in this chapter. As the absorbed solutes emerge into the lateral interstitial spaces between the intestinal epithelial cells, the fluid in the lateral spaces is made slightly hypertonic. Water rapidly flows down its osmotic gradient across the leaky "tight junctions", so that by the time the absorbed fluid emerges from the lateral spaces of the epithelium it has become isotonic (Figure 22-13). As a result of this hidden osmotic equilibration, there is net movement of solute and water across the small intestinal epithelium without any measurable osmotic gradient developing between chyme and the blood plasma.

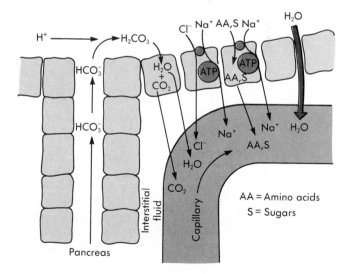

FIGURE 22-13
In the small intestine, reaction of H^+ with HCO_3^- produces H_2CO_3, which enters the bloodstream as CO_2 and H_2O. Na^+ absorption involves passive Cl^- movement along with the Na^+ which is actively pumped out the basolateral surface of the cells. Water follows passively. Sugars and amino acids are moved from the lumen into the blood by coupled transport systems, and their absorption also drives water absorption.

Digestion and Absorption

Effect of Gastrointestinal Tract Disorders on Body Fluid Homeostasis

Both the composition and volume of body fluid are threatened by disorders in which large amounts of the secretions of the GI tract are lost. Vomiting is a response to irritation of the stomach and intestine by chemical agents, bacterial or viral infection, or overdistension. For healthy people, vomiting has no lasting consequences, but in some illnesses and eating disorders, vomiting occurs repeatedly and constitutes a significant loss of fluid and acid from the body. Because H^+ and K^+ act as if they compete for a common carrier (see Chapter 20), the alkalosis resulting from vomiting causes an increase in K^+ secretion by the distal tubule, so that hypokalemia results. The hypokalemia caused by chronic vomiting may result in death from cardiac arrhythmia.

Diarrhea is a symptom of irritation or infection of the intestine. It may result from increased motility or from interference with the normal control of secretion and reabsorption of fluid and solutes, or both. For the majority of dietary fluid and the secretions of the stomach, pancreas, liver, and initial portion of the small intestine to be absorbed, the chyme must pass through the small intestine and large intestine slowly. A large number of bacteria and viruses can infect the intestinal mucosa, releasing substances that inhibit segmentation and enhance peristalsis. The increase in motility moves the chyme through the large intestine before water and solutes can be absorbed, resulting in diarrhea.

The most serious varieties of diarrheal disease are the result of bacterial toxins that greatly increase intestinal secretion. Cholera toxin is one example. Cholera toxin is believed to act on a receptor that normally recognizes a peptide hormone that regulates secretion in the intestine. Binding of the toxin with the receptor increases intracellular levels of cyclic AMP. The second messenger stimulates secretion and inhibits solute absorption. The secretion contains quantities of Na^+, K^+, and HCO_3^-. The loss of this secretion, together with dietary fluid and solutes that would otherwise have been absorbed, can rapidly result in a condition of dehydration and acidosis if not treated.

These symptoms, and not the infection itself, are the aspect of the disease that threatens the patient's life. The course of the disease is particularly rapid in infants and small children and is a major cause of infant mortality in developing nations. Death can usually be prevented if the patient is rehydrated orally with a simple solution that contains Na^+, K^+, HCO_3^-, and glucose; only in severe cases is intravenous rehydration or other medication necessary. Mothers in third-world countries can be supplied with this solution very cheaply, and they can be taught to use this treatment even if trained medical assistance is unavailable. As a sad postscript, it is very likely that diarrhea-producing toxins have been purified for possible use in chemical warfare, since their effects are both immediate and very disorganizing.

Mechanisms of Diarrhea and Constipation

Diarrhea, a condition characterized by excessive volume and fluidity of feces, can occur as a result of excessive secretion of electrolytes and water in the intestine (secretory diarrhea—see Focus box), as a result of failure to absorb electrolytes or nutrients normally (these would be termed malabsorption syndromes), or as a result of excessive gut motility that moves chyme through the intestine too rapidly for efficient absorption of solutes and water. Both excessive motility and excessive secretion can result from irritation or inflammation of the intestinal mucosa—these are the mechanisms of the diarrhea that accompanies some intestinal diseases.

Conversely, reduced gut motility or voluntary retention of feces allows more time for water reabsorption in the colon and causes **constipation**, a condition in which feces are so dry and compact that they are difficult to expel. The presence in intestinal chyme of ions that cannot be absorbed, such as Mg^{++} and SO_4^{--} (ingredients in several laxatives and antacids), causes an osmotic effect that retains water with the chyme and ultimately increases fecal volume and fluidity. This osmotic laxative effect is similar in principle to the action of osmotic diuretics in the kidney (see Focus box, Chapter 19).

REGULATION OF GASTROINTESTINAL FUNCTION

> - *What are the three phases of digestion? To what extent do they overlap in time?*
> - *What and where are the dominant sensory inputs during each phase?*
> - *What physical and chemical changes occur to food in the mouth? The stomach? The duodenum? The small intestine? The large intestine?*
> - *What is the dominant hormone of the gastric phase, and what are its effects?*
> - *Name the major duodenal hormones and describe the regulatory role of each one.*
> - *What factors affect the rate of stomach emptying?*
> - *What are the major coordinating reflexes of the GI tract?*

Gastrointestinal Control Systems

The digestion of a meal begins even before the first mouthful is taken, and continues for several hours after eating is completed. Three stages of the process can be distinguished. The first is the **cephalic phase,** a period in which the bulk of the food is yet to be eaten and the responses of the digestive system are triggered by sensory inputs and mental processes. During the second or **gastric phase,** most or all of the meal is in the stomach, and the responses of the system are triggered mainly by the composition and bulk of what has been eaten. In the last or **intestinal phase,** food (now converted to chyme) is entering the intestine for further digestion and absorption; in this phase sensory inputs to the duodenum and more distal parts of the intestine dominate the regulatory picture. The three phases are not sequential, but overlap one another in time as shown in Figure 22-14.

The task of regulating and coordinating digestion is complex. The rate of secretion in each part of the GI tract must be matched to the need for digestive enzymes, bile, acid and buffer, as determined by the volume and chemical composition of the meal. The movement of chyme from stomach to intestine and from point to point within the intestine must be organized so that the small intestine is not congested at any point with unmanageable volumes of chyme. Orderly movement of material through the GI tract and the secretion of fluid and enzymes required for digestion are coordinated by reflex pathways. These pathways are activated by the chemical composition of the food and the fact that its presence is detected by stretch receptors. A large portion of the cardiac output is directed to the digestive tract during digestion and absorption as a result of release of local autoregulatory factors. Many of the inputs that affect motility, digestion, and absorption are mediated by the parasympathetic branch of the autonomic nervous system by way of the vagus nerves and the sympathetic branch by way of abdominal plexuses (Table 22-5; see also Figure 12-4). These are "long" reflexes. The pathways of a second class of reflexes ("short") lie entirely within the enteric nervous system.

Hormones released by enteroendocrine cells of the GI tract are important in coordinating secretion and motility in the different parts of the tube system. These cells are sensitive to the presence of food and secretions in the canal and can also be activated by neural inputs. Table 22-6 summarizes the stimuli that cause the release of the best understood hormones, their effects on GI secretion, GI motility, and the secretion of other hormones, and their roles in GI control.

Cephalic Phase of Digestion

Smelling, tasting or simply thinking about food initiates the **cephalic phase** of digestion, which triggers responses that prepare the mouth and stomach for reception of food (Figure 22-15). In the cephalic phase, secretion of saliva in the mouth, blood flow to the stomach, and gastric secretion of acid, mucus, and pepsin (as pepsinogen) all increase. These responses continue and are enhanced as long as food is being chewed and swallowed.

The cephalic phase is mediated by reflexes that involve the parasympathetic branch of the autonomic nervous system and by a hormonal reflex pathway. Parasympathetic inputs stimulate the salivary glands and also stimulate the parietal cells to release HCl, the chief cells to release pepsinogen, and the G cells to release the hormone **gastrin.** As shown in Chapter 21, acetylcholine is not the final

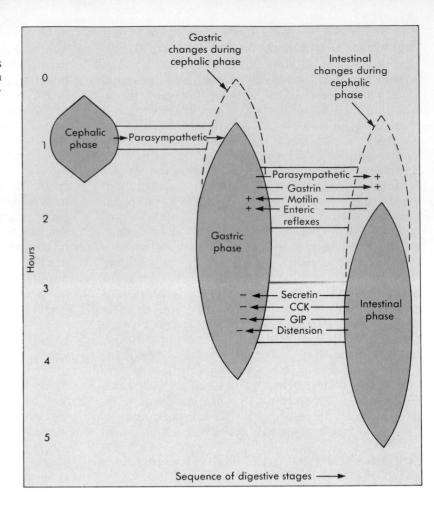

FIGURE 22-14
The temporal overlap between the three phases of digestion and the signals that travel between the different regions in the coordination of digestion.

TABLE 22-5   Some Reflexes of the Gastrointestinal Tract

Reflex name	Stimulus and effect
Reflexes involving the autonomic nervous system	
Receptive relaxation	Swallowing food relaxes the fundus of the stomach.
Enterogastric reflex	Acid and hypertonic solutions in the duodenum inhibit gastric emptying.
Gastrocolic reflex	Distension of the stomach and duodenum increase the motility of the ileum, cecum, and colon.
Gastroileal reflex	Ileocecal sphincter opens when gastric emptying begins.
Reflexes involving only the enteric nervous system	
Myenteric reflex	Local stimulation of the intestine causes contraction of the muscle immediately above the point of stimulation and relaxation of the muscle below the point of stimulation.

---

transmitter in all of these pathways; several peptide neurotransmitters and neuromodulators are believed to play important roles in the plexuses of the enteric nervous system, so the transmitter chemistry is rather complex. Gastrin moves into the blood and travels through the circulation to the stomach epi-thelium, where it potentiates parasympathetic stimulation of the parietal cells and chief cells.

Gastric Phase of Digestion

When the stomach is empty, its wall is collapsed in folds. As soon as food is swallowed and enters the

644   THE GASTROINTESTINAL TRACT

TABLE 22-6   *The Gastrointestinal Hormones*

Hormone	Stimuli	Effects	Role
Gastrin	Amino acids Distension pH > 3 Parasympathetic activity	+Acid secretion +Antral motility +Pepsinogen secretion	Facilitates gastric digestion
Gastric inhibitory peptide	Glucose Distension Hypertonicity	+Intestinal motility −Gastric secretion and motility	Limits gastric emptying; prepares the intestine for substrate
Motilin	pH > 4.5	+Gastric motility +Intestinal motility	Facilitates digestion in intestine
Secretin	Hypertonicity pH < 4.5 Parasympathetic activity	+Pepsinogen secretion +Pancreatic bicarbonate and zymogen secretion −Gastric motility and secretion	Limits gastric emptying; neutralizes the chyme leaving the stomach
Cholecystokinin	Fats Amino acids	+Pancreatic zymogen secretion +Intestinal secretion and motility −Gastric secretion and motility +Gallbladder contraction	Limits gastric emptying; promotes digestion in the intestine

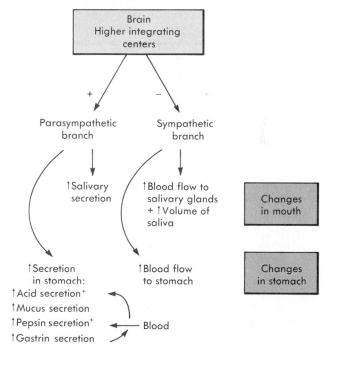

FIGURE 22-15

Activation of the cephalic phase of digestion and the responses it evokes. The consequences of parasympathetic activation are increases in secretion; inhibition of sympathetic activity allows an increase in blood flow to the salivary glands and stomach.

stomach, the **gastric phase** of digestion is initiated, continuing as long as there is chyme in the stomach. The stimuli that trigger this phase (Figure 22-16) relate to the presence of the food. The two major factors are distension of the stomach wall and the chemical content of the food. Distension of the an-

trum promotes gastrin release by both local enteric reflexes and by long-loop parasympathetically mediated reflexes.

Even in the presence of favorable stimuli from the parasympathetic nervous system and the presence of food, relatively little gastrin would be re-

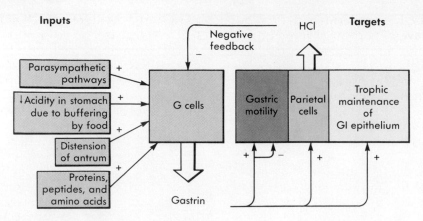

FIGURE 22-16
The control of gastrin secretion and its targets.

leased if the pH of the stomach lumen remained at the low values characteristic of the empty stomach. At the beginning of the gastric phase, food arriving in the stomach dilutes and buffers the gastric acid, and the pH of the stomach contents rises. The rise in pH permits gastrin secretion to occur in response to chemical and neural stimuli.

Dietary protein is a powerful stimulant of gastric secretion (Figure 22-16). The proteins arriving in the stomach are met by the HCl and pepsins that were released in the cephalic phase and that continue to be released in response to cephalic stimuli, which last as long as food intake continues. The acid denatures the ingested proteins, attacking their secondary and tertiary structure (see Figure 3-11) and opening folds in the amino acid chains so that the peptide bonds are more accessible to digestion by the gastric pepsins. As digestion proceeds, concentration of free amino acids and peptide fragments in the gastric contents rises. These products of protein digestion are **secretagogues**—substances that stimulate gastric secretion by promoting release of gastrin and pepsin.

Certain beverages and foods often served as first courses of meals contain secretagogues and therefore promote gastrin release. For example, using clear soups as the first course of a meal provides amino acids that are detected by the stomach's chemoreceptors and speed preparation for protein digestion by promoting gastrin release. Caffeine also acts as a secretagogue and can aggravate an ulcer by promoting acid secretion when food may not be present to absorb the acid.

The stimuli that promote gastrin release provide positive feedback as long as protein is present to be digested. As acid secretion continues in the gastric phase, the buffering capacity of the proteins is exceeded. Gastrin secretion is inhibited when the chyme reaches a pH of 3.0 (see Figure 22-16). This inhibition terminates acid secretion by the stomach as the end of the gastric phase approaches.

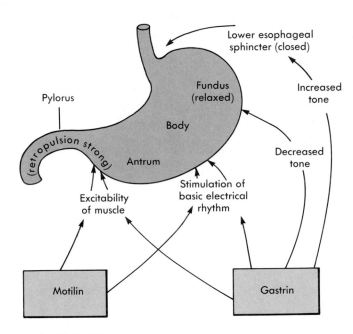

FIGURE 22-17
Control of the stomach by inputs from gastrin and motilin.

Control of Gastric Motility and Stomach Emptying

At the beginning of the gastric phase, the smooth muscle of the stomach relaxes as food enters. This **receptive relaxation**, a vagally-mediated long reflex (Table 22-5), allows a substantial amount of food—perhaps a liter or more of volume—to be stored in the stomach before vigorous contractions begin.

The dominant hormone of the gastric phase is gastrin, released by G cells of the gastric mucosa (Figures 22-16 and 22-17; see also Chapter 21). As long as gastrin levels are high, the fundus and body of the stomach are relaxed and serve mainly to store chyme. Also, the excitability of smooth muscles in

THE GASTROINTESTINAL TRACT

the pyloric region is high, so pyloric contraction occurring during the wave of peristalsis severely limits delivery of chyme to the intestine. Gastrin also increases lower esophageal sphincter tone, preventing stomach contractions from causing a reflux of food into the esophagus. As food continues to enter the stomach, the fundus and upper body store it, while the lower body and antrum mix food with the mucus, buffer solutions, acid, and enzymes to form chyme.

During the gastric phase, pancreatic secretion is stimulated, preparing the intestine to receive the food that will soon be entering it (see Figure 22-14). Distension of the stomach directly stimulates pancreatic fluid and enzyme secretion via parasympathetic reflexes. Gastrin also stimulates pancreatic enzyme secretion because gastrin is chemically very similar to the hormone, **cholecystokinin (CCK),** which regulates pancreatic secretion in response to food in the intestine.

Intestinal Phase of Digestion

As soon as food begins to enter the duodenum, the **intestinal phase** of digestion is triggered. The hormones secreted and the effects they have are summarized in Table 22-6. As was true for the gastric phase, the stimuli for the intestinal phase relate to the stretching of the wall of the intestine and the chemical content of the food. The intestine is the site for digestion of all categories of nutrients, and the quantities of enzymes that are secreted and the time allotted for digestion are determined by the amounts of protein, carbohydrate and fat in the chyme. Both reflexes and hormones are mediators of these effects of meal composition on intestinal activity (see Figure 22-14).

As the antrum becomes filled with fluid chyme, the process of **gastric emptying** begins. During gastric emptying, the pyloric sphincter remains slightly open so that each wave of peristalsis in the antrum expels a small amount of chyme into the duodenum. This process is repeated until the stomach is entirely empty. The rate of gastric emptying is regulated by reflexive and hormonal signals from the duodenum to the stomach. Signals from the duodenum also coordinate pancreatic and liver secretion, intestinal motility, and ultimately contribute to the feeling of satisfaction and fullness that follows a meal. The sensory inputs that initiate these duodenal signals are the volume, acidity and osmolarity of chyme in the duodenum, and the carbohydrate, protein and fat composition of the chyme.

The volume of chyme in the duodenum is detected by duodenal mechanoreceptors, which reflexively inhibit gastric motility and increase pyloric tone by a vagally-mediated feedback pathway called the **enterogastric reflex** (see Table 22-5). Surgical removal of the parasympathetic innervation of the GI tract is sometimes used to control ulcers. But be-cause this procedure also eliminates the enterogastric reflex, it is one cause of a syndrome of rapid gastric emptying called **gastric dumping.** Gastric dumping is intensely unpleasant while it is occurring, and also causes diarrhea by overwhelming the absorptive capacity of the intestine.

The acidity of duodenal chyme is detected by a dual control mechanism involving the duodenal hormones **secretin** and **motilin,** that slows stomach emptying when the pH of duodenal chyme is lower than about 4.5 and accelerates it when the pH of the chyme is above this value. Secretin is released by duodenal **S cells** (Figure 22-18) when the pH of the chyme in the duodenum falls below 4.5. Hypertonic conditions and fat content of the chyme also stimulate secretin release. Secretin also tends to inhibit the motility of the stomach but stimulates secretion of pepsin. Secretin is the single most important inhibitor of gastrin release and therefore of acid secretion. Secretin also stimulates pancreatic bicarbonate secretion, which has the effect of buffering the HCl. Thus secretin protects the intestine from excessive acid by two negative feedback mechanisms: (1) reducing the acidity of the incoming chyme through inhibition of gastric acid secretion, and (2) reducing the acidity through the buffering action of the pancreatic bicarbonate secretion.

A second set of duodenal enteroendocrine cells releases the hormone **motilin** when the pH of the duodenal chyme is greater than 4.5. Motilin increases the strength of gastric contractions and the tone of the pyloric sphincter (Table 22-6; see Figure 22-17). Motilin secretion is high early in the period following a meal, when little acidic chyme has entered the small intestine, and its role is to stimulate the mixing action of the stomach. Motilin secretion may also stimulate pepsin secretion.

Duodenal osmoreceptors respond to the osmotic strength of the chyme, allowing the system to differentiate between stretching caused by water ingestion and that related to food ingestion. If food is present, stretching, elevated osmotic pressure, and nutrient molecules are all detected. These stimuli cause **gastric inhibitory peptide (GIP)** to be released from **K cells** in the mucosa of the intestine. This hormone acts with other signals from the intestine to slow the digestive activity of the stomach and induce the stomach to assume a storage role (see Figure 22-14). GIP accomplishes these effects by reducing pepsin and HCl secretion and gastric motility.

Carbohydrates and proteins are digested more readily than are lipids, so a meal high in lipids requires a much longer time for digestion and absorption. Early in the intestinal phase, when the chyme is buffered by the intestinal mucus and bicarbonate secretions that were stimulated as part of the gastric phase, amino acids arriving in the duodenum stimulate still more gastrin release via reflexes in the enteric nervous system. Thus, during this early period,

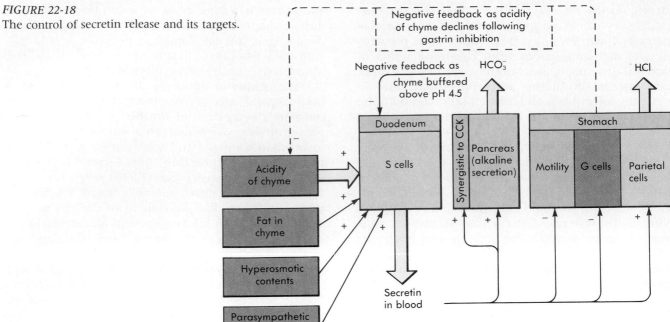

FIGURE 22-18
The control of secretin release and its targets.

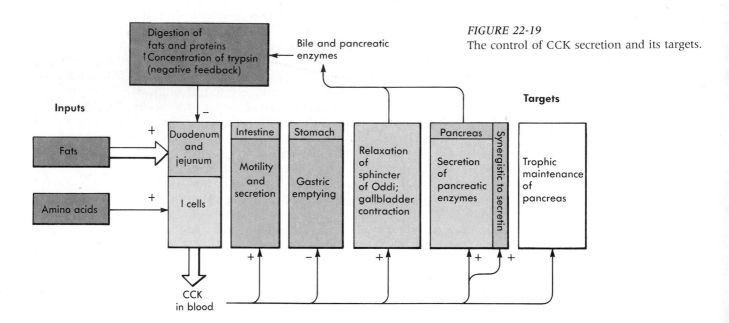

FIGURE 22-19
The control of CCK secretion and its targets.

the feedback from the intestinal receptors may have a positive effect on stomach activities, and HCl secretion in the stomach remains high. This facilitates the preparation of proteins for intestinal digestion. Later in the intestinal phase, the feedback to the stomach becomes mainly inhibitory (see Figure 22-14).

Fats and fatty acids present in the chyme are detected by duodenal chemoreceptors. These, and to a lesser extent, amino acids, polypeptides, and proteins, stimulate the release of cholecystokinin (Fig-

ure 22-19). This remarkable hormone has seven roles in the digestive system:

1. It increases pancreatic enzyme secretion.
2. It relaxes the sphincter of Oddi.
3. It causes contraction of muscle cells of the gall-bladder.
4. It inhibits gastric emptying.
5. It potentiates the action of secretin on the pancreatic secretion of enzymes.
6. It promotes secretion and motility in the small intestine.

If body mass is to be in a steady state, total caloric expenditure must closely match caloric intake. The closeness of this match is critical because on a time scale of months or years the values of intake and expenditure to be matched are large compared to the amount of energy stored as fat. Even a slight error would result in significant weight gain if continued over time. From this point of view, most people regulate their body weights quite well. Unfortunately, the value at which weight is regulated is frequently somewhat higher than the value that is optimal for health, and cultural ideals of best weight may differ markedly from what is healthy. Roughly a third of adults in developed nations are overweight; an exact figure is impossible to determine because ideal weights are difficult to define precisely. The steady-state body weight for any given individual is determined by a complex interaction of metabolic regulatory systems involving appetite, food availability, and the regulation of energy flow into storage (fat and glycogen) and heat production. Some of the aspects of metabolic regulation are treated in Chapter 23.

Appetite regulation is believed to involve interactions among several regulatory pathways in the brain, some of which receive information from the stomach and intestine. Some of the pathways involve endogenous opiates (enkephalins and endorphins) as modulators (see Chapter 8). These neuromodulators are generally involved in pathways that involve stress and reward, and they stimulate eating.

A hypothetical cycle of overeating develops when emotional stress is balanced by the temporary increases in opiate levels that follow eating. The resulting weight gain increases the emotional stress, thus increasing the drive to eat. Ultimately an addiction to the high opiate levels develops, and the behavior persists because returning to normal eating patterns would involve discomfort similar to the withdrawal syndrome experienced by drug addicts.

The incidence of the eating disorders anorexia nervosa and bulimia has increased dramatically in recent years. Approximately 1% of adolescent females have anorexia and up to 19% of college women may have bulimia. The incidence in males is considerably lower. The sex difference may be related to the fact that in modern western culture weight gain generates higher levels of anxiety for females than for males. Anorexia nervosa is best described as self-inflicted starvation; its victims become dangerously thin while denying hunger. Bulimia is at the opposite extreme; people with bulimia alternate between binges of eating and daily fasting, self-induced vomiting (purging), and abuse of laxatives and diuretics to control their weight. These diseases probably represent aspects of a single disorder because some patients show features of both. In the view of some therapists, this disorder could also be described as a food addiction. People with anorexia are successfully denying their addiction; those with bulimia are yielding to it.

Eating disorders are signs of more generalized disturbances of mental functioning, social adjustment, and body image. In addition to the stress such disorders place on the energy balance of the body, they are potentially fatal threats to homeostasis of body fluids. For example, repeated purging may lead to alkalosis and hypokalemia and ultimately to heart arrhythmia. Abuse of diuretics and laxatives causes hypovolemia, electrolyte imbalances, and acid-base disorders.

Because those with anorexia and bulimia characteristically deny that they have a health problem, their conditions may not receive timely medical attention. Repeated purging causes a distinctive pattern of erosion of dental enamel and also an easily recognized enlargement of the salivary glands. Dentists have an important role in diagnosing eating disorders and bringing them to the attention of the patient's physician.

7. It is a trophic hormone for the pancreas.

The negative feedback that limits CCK release acts through the reduction of fats and proteins present in the intestine as these are digested and absorbed. A specific inhibition is provided by the presence of trypsin in the duodenum because this inhibits the release of CCK and thus decreases enzyme secretion (see Figure 22-19). Fat in the duodenum also stimulates the secretion of GIP. Secretin, GIP, and CCK all directly decrease gastric motility (see Table 22-6). A meal with high lipid content will greatly extend the gastric and intestinal phases of digestion, as the slow processing of the fats by the intestine suppresses gastric emptying. The CCK released during such a meal is also believed to act on the hypothalamus, increasing the sensation of

satiety, or feeling of fullness and absence of hunger.

At the beginning of the intestinal phase, the jejunum is empty and the ileum contains chyme composed mainly of fiber, dead intestinal cells and unabsorbed fluid. Gastric emptying triggers reflexive signals that clear the remains of the last meal from the ileum into the cecum and prepare the intestine to accept the latest meal. The **gastroileal reflex** (see Table 22-5) initiates motility in the jejeunum and ileum and relaxes the sphincter located between the ileum and the large intestine. As the chyme from the previous meal is pushed into the terminal ileum, the ileocecal valve opens and with each contraction of the ileum some chyme passes into the cecum. At the same time, the colon becomes more active and one or more mass movements usually result. This stimulation of large intestine motility during stomach emptying is called the **gastrocolic reflex** (see Table 22-5), and is probably mediated by gastrin and by long neural reflexes.

SUMMARY

1. Digestion of food macromolecules is accomplished by enzymes that catalyze hydrolytic reactions. **Amylase** begins carbohydrate digestion in the mouth, and **pepsins** begin protein digestion in the stomach, but most digestion is accomplished in the small intestine by enzymes from the pancreas or attached to the **brush-border** membranes of the intestine.

2. Digestion of lipids requires the emulsifying action of **bile acids** to increase the lipid surface area so that pancreatic **lipases** can hydrolyze the lipids. **Colipase** facilitates the actions of lipases. Bile acids also facilitate absorption by collecting the products of lipid digestion into **mixed micelles.** Some resynthesis of lipids occurs in the epithelial cells, and these lipids are released into the lymph in small droplets called **chylomicrons.**

3. Iron absorption is regulated by the secretion of the carrier molecule, **transferrin,** into the intestinal lumen. Most iron in the body is recycled between old red blood cells and the cells in the bone marrow that produce new red blood cells. The liver removes iron from **heme** and the iron is released into the blood, where it is carried by transferrin to the bone marrow.

4. Vitamins fall into two categories depending on their solubilities: **water-soluble** and **fat-soluble.** Most water-soluble vitamins are absorbed by transport mechanisms, but **vitamin B$_{12}$** must be bound to the carrier, **intrinsic factor,** which is secreted by the stomach, to be taken into intestinal cells. Fat-soluble vitamins partition with the fats into mixed micelles and chylomicrons.

5. The movement of water, amino acids, and sugars into the body from the alimentary canal depends largely on the movement of Na$^+$. Sodium ions diffuse from the lumen into the epithelial cells in response to a gradient created by active transport of sodium out of the cells into the interstitial fluid. As Na$^+$ moves out of the lumen, water follows by osmosis. The gradient for Na$^+$ also drives the carrier-mediated transport of most amino acids and sugars into the epithelial cells. The carriers recognize particular classes of sugars or amino acids.

6. Digestion can be divided into three phases, depending on where food stimuli are acting. The **cephalic phase** begins even before food ingestion, when the idea or stimuli arising from the presence of food are detected. The **gastric phase** of digestion begins as soon as food enters the stomach, and the **intestinal phase** begins as soon as food enters the intestine. The three phases therefore have considerable overlap in time.

7. Stimuli arising from the physical and chemical presence of food in the mouth, the stomach, and the intestine regulate the movement of material through the GI tract and the rate of secretion of enzymes. Regulation of digestion involves reflexes that use long parasympathetic pathways, short reflex loops, and hormonal secretions.

8. The best-understood GI-tract hormones are **gastrin, motilin, GIP, secretin,** and **CCK.** Gastrin stimulates gastric acid and pepsinogen secretion and gastric motility. Motilin stimulates gastric and intestinal motility and pepsinogen release. GIP stimulates intestinal and pancreatic secretion and inhibits gastric secretion and motility. Secretin stimulates pancreatic secretion and inhibits gastric secretion and motility. CCK stimulates gallbladder contraction and inhibits gastric secretion.

THE GASTROINTESTINAL TRACT

1. How are each of the following regions of the GI tract functionally specialized for digestion and reabsorption?

Mouth	Duodenum
Esophagus	Jejunum and Ileum
Stomach	Large intestine

2. What enzymes digest carbohydrates? Where does most carbohydrate digestion occur? How are carbohydrates absorbed?

3. How are proteins digested? What are the two sites of protein digestion? How are amino acids absorbed? Can peptides be absorbed?

4. What are the roles of the bile salts in fat digestion and absorption?

5. In what ways is absorption of fat different from absorption of amino acids and carbohydrates?

6. For each of the following GI hormones, state (a) its source, (b) the most important physiological stimulus for its release, (c) the most important inhibitor of its release, and (d) its target organs.

Gastrin	Cholecystokinin
Secretin	Gastric inhibitory peptide

7. What events occur during the cephalic phase of digestion? The gastric phase? The intestinal phase?

8. When gastric motility is elevated the tone of the pyloric sphincter is also high. What is the effect of sphincter excitability on gastric emptying?

9. How does acidic chyme in the duodenum result in increased pancreatic secretion?

10. What factors determine the rate of gastric emptying? Will a meal with a high fat content empty more or less rapidly than a meal with a high carbohydrate content? What feedback control is involved?

Choose the MOST CORRECT ANSWER.

11. Which of the following are functions of the liver?
    a. Production of bile
    b. Detoxification of drugs, metabolites, etc.
    c. Glycogen storage
    d. All of the above

12. This water-soluble vitamin requires a carrier, the intrinsic factor, to be transported into intestinal cells:
    a. Thiamine (B_1)
    b. Riboflavin (B_2)
    c. Cyanocobalamin (B_{12})
    d. Folic acid

13. A deficiency of this vitamin can result in night blindness, epithelial, and neural disorders:
    a. Vitamin A
    b. Vitamin E
    c. Vitamin K
    d. Vitamin B_{12}

14. This reflex is stimulated by acid and hypertonic solutions in the duodenum, which inhibit gastric emptying:
    a. Myenteric reflex
    b. Enterogastric reflex
    c. Gastrocolic reflex
    d. Gastroileal reflex

15. This hormone limits gastric emptying and neutralizes chyme leaving the stomach:
    a. Gastrin
    b. Gastric inhibitory substance
    c. Motilin
    d. Secretin

16. Which of the following is NOT an effect of cholecystokinin?
    a. Promotes pancreatic zymogen secretion
    b. Initiates emptying of gallbladder
    c. Promotes antral motility
    d. Inhibits gastric secretion

17. This phase of digestion begins before food is ingested:
    a. Cephalic phase
    b. Gastric phase
    c. Intestinal phase
    d. Colic phase

18. This carbohydrate cannot be digested and absorbed in the small intestine:
    a. Glycogen
    b. Lactose
    c. Cellulose
    d. Plant starch

● *SUGGESTED READINGS*

AVERY ME, SNYDER JD: Oral therapy for acute diarrhea—the underused simple solution, *New Eng J Med* 323 (13):891, 1990. Describes composition and use of oral rehydration solutions that can be used in underdeveloped countries.

DAVENPORT HW: *A Digest of Digestion*, Chicago, 1975, Yearbook Medical Publishers. A delightfully written summary of gastrointestinal physiology, now dated but still valuable for its organ-level physiology and its down-to-earth details, such as what causes intestinal gas and why students may have diarrhea.

FIELD M, RAO MC, CHANG EB: Intestinal electrolyte transport and diarrheal disease, *New Eng J Med* 321(12):800, 1989. Reviews mechanisms and regulation of intestinal electrolyte transport and how they are altered in diarrheal disease.

UVNAS MK: The gastrointestinal tract in growth and reproduction, *Scientific American* 261(1):78, 1989. Current review of hormonal regulation of metabolism.

WHAT OUR ANCESTORS ATE, *The New York Times Magazine*, June 5, 1988, p. 54. Compares the modern diet with archaeological findings regarding past eating habits.

WININCK M: Control of Appetite. In *Current Concepts in Nutrition*, New York, 1988 John Wiley & Sons, Inc. Describes appetite regulation by control centers in the brain and receptors in the GI tract. Discusses the fact that the CNS pathways involved in satiety involve endogenous opiates and mediate responses to stress and reward.

CHAPTER 23

Endocrine Control of Organic Metabolism and Growth

On completing this chapter you should be able to:

- Understand how insulin and glucagon regulate blood glucose.
- Distinguish between the absorptive and postabsorptive states.
- Understand the differences between type I diabetes and type II diabetes.
- Describe the adaptive responses of the body to periods of starvation.
- Describe how increases in blood levels of glucose, bulk aspects of ingested food, and neural and hormonal signals produce satiety.
- Understand the factors involved in long-term regulation of body weight.
- Describe how the relative proportions of fat, muscle, and ossified skeleton vary over the life cycle.
- Describe the processes by which thyroid hormones set the body's metabolic rate.
- Describe the adaptive responses of the body to cold.

*O*rganic metabolism is regulated on several time scales. Blood levels of nutrients such as glucose are controlled over a time scale of minutes. Appetite for food is matched with requirements for energy, repair, and growth over a time scale of hours to days. The longest regulatory time scale reflects the differing metabolic requirements of different life stages. The overall metabolic picture is different for young children, adolescents, young adults, mature adults, and older adults. It differs between men and women, and is greatly affected in women by pregnancy and milk production.

In adults maintaining a stable weight, overall organic metabolism is balanced, although most tissue components are continually being degraded and replaced. A slight excess of carbohydrate and fat intake over metabolism to CO_2, water, and energy leaves a small extra amount of fat in fat cells each day. Continued over time, even small imbalances would lead to large changes in the mass of fat. A minor imbalance in favor of carbohydrate and fat breakdown over intake would have the opposite result: a steady erosion of first the fat reserve and ultimately the protein structure of the body.

In many less well-developed societies, periodic famine is a fact of life for most people. It is also true that in most highly developed societies, some people are sometimes unable to meet their daily nutritional needs. In undernutrition, the metabolic state of the body is an exaggeration of the state occurring during the short fasts that intervene between the last meal of the day and breakfast on the next day. Stored nutrients move into the blood and are converted to forms that can support energy metabolism. In undernutrition, metabolism is fundamentally altered, resulting in a decrease in the amount of energy spent by cells on turnover of their constituents, so that the cells become more thrifty in their use of energy. This pattern is the opposite of the one seen during growth in children and adolescents. Undernutrition, particularly when it involves protein intake, is fundamentally incompatible with normal growth of organ systems such as the central nervous system and skeleton, which take on the adult pattern gradually during growth.

CHARACTERISTICS OF THE ABSORPTIVE STATE

- *Define anabolism and catabolism. Give examples for the treatment of protein, carbohydrates, and fats by the body.*
- *How does the Law of Mass Action combine with the actions of insulin to direct metabolic pathways during the absorptive phase?*
- *In what ways does insulin stimulate anabolism in the absorptive phase?*
- *How is the number of glucose receptors on the beta cells of the pancreas adjusted by diet?*
- *Why does protein anabolism depend on consumption of proteins containing all the essential amino acids in the same meal?*
- *What is the role of apoproteins in lipid absorption, and why is it good to have a high proportion of the HDL particles to LDL particles in the blood?*
- *What are the characteristic symptoms of diabetes mellitus? Why does each occur?*
- *What are the important differences between Type I and Type II diabetes mellitus? How is the regulation of insulin receptors believed to be different in the two types?*

Role of Insulin

In the bodies of adults, some periods of the day are dominated by the process of **anabolism** (synthesis), whereas other periods are dominated by the process of **catabolism** (degradation) (Table 23-1; see Chapter 4, pp. 76-82). Anabolic processes include glycogen formation from glucose (**glycogenesis**), triacylglycerol synthesis from free fatty acids (**lipogenesis**), protein synthesis from free amino acids, and glucose formation from amino acids (**gluconeogenesis**). Catabolic processes include glycogen breakdown (**glycogenolysis**), lipid breakdown (**lipolysis**), and protein breakdown (**proteolysis**). Because many processes must go on continuously (for example, the simultaneous loss and replacement of the epithelial cell lining of the GI tract), the difference between net anabolism and net catabolism is usually determined by what the body is doing with nutrients that enter during absorption and what it is doing when absorption is not occurring.

Glycogen, the storage form of glucose, is a polysaccharide consisting of many sugar residues that is poorly soluble in water and is found in cells in the form of granules. Triacylglycerols, the storage form of free fatty acids, are water-insoluble and exist in adipose (fat) cells as droplets. There are several advantages of using these insoluble molecules for energy storage. Insoluble molecules do not cause an osmotic problem for cells as large amounts of soluble molecules would, they do not diffuse away from their sites of deposit, and they do not participate in cellular metabolism unless acted upon by specific enzymes. Strictly speaking, there are no proteins that serve exclusively to store amino acids; all proteins have primary roles as structural elements or enzymes. However, protein catabolism continuously frees amino acids that may reenter protein synthetic pathways or be catabolized for energy, depending on the metabolic state of the body.

The period during which nutrients are being absorbed from the intestine is called the **absorptive state** of metabolism. During the **postabsorptive state,** nutrients are not entering the body and must be made available from internal stores. The change in disposition of nutrients that occurs at the end of the absorptive phase is the result of changes in levels of hormones and in the pattern of activity of the autonomic nervous system. Storage of nutrients as glycogen and fat dominates the absorptive state. This temporary storage is reversed in the postabsorptive stage, when stored glycogen and fat molecules are broken down and released to supply the body's needs for energy and synthesis.

During the absorptive period, the hormone **insulin** regulates the entry of glucose and amino acids into cells. Insulin is released from **beta cells** of the pancreatic **islets of Langerhans** (Table 23-2 and Figure 23-1). Insulin is synthesized as an inactive precursor called **proinsulin** and subsequently packaged into secretory vesicles in the Golgi apparatus. Proinsulin is hydrolyzed to insulin inside these secretory vesicles by a proteolytic enzyme. Insulin is then released by exocytosis, like a neurotransmitter. A normal plasma glucose concentration is in the range of 80 to 110 mg/dl of plasma, and at this level,

---

TABLE 23-1   *Anabolic and Catabolic Processes*

Category	Process	Description
Anabolic	Glycogenesis	Formation of glycogen from glucose
	Lipogenesis	Formation of fatty acids and then triglycerides (esterification)
	Ketogenesis	Synthesis of ketones
	Gluconeogenesis	Synthesis of glucose from amino acids
Catabolic	Glycogenolysis	Breakdown of glycogen to glucose
	Lipolysis	Breakdown of triacylglycerols to fatty acids and glycerol, and then to acetyl-CoA
	Proteolysis	Breakdown of protein to amino acids

TABLE 23-2   *Hormones that Regulate Metabolism*

Hormone	Chemical characteristics	Secreted by
Insulin	Protein	Beta cells of pancreatic islets of Langerhans
Glucagon	Protein	Alpha cells of pancreatic islets of Langerhans
Epinephrine	Catecholamine	Adrenal medulla
Cortisol	Steroid (glucocorticoid)	Adrenal cortex
Thyroid hormones (T_3 and T_4)	Iodinated thyronines (tyrosine derivatives)	Thyroid gland
Growth hormone (somatotropin)	Protein	Anterior pituitary
Somatostatin	Small protein (14 amino acids)	Delta cells of pancreatic islets of Langerhans

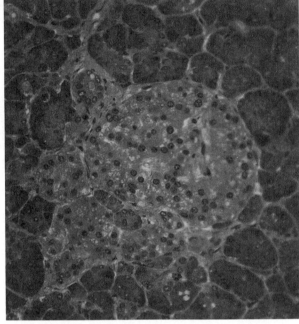

A

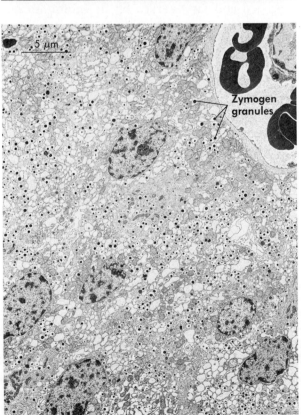

B

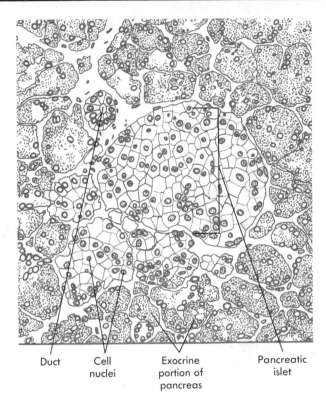

Duct   Cell nuclei   Exocrine portion of pancreas   Pancreatic islet

FIGURE 23-1

A Islet of Langerhans surrounded by exocrine tissue (acini and ducts) in the pancreas. The photomicrograph (left) is interpreted in the labeled diagram (right). Beta cells are the most common cell type in the islets and are usually found in the center of the islet. Alpha cells are almost always located near the outside of the islet. Delta cells are the least common type.

B Electron micrograph of islet cells, showing cytoplasm filled with vesicles of hormone. A capillary with red blood cells is seen in the upper right corner.

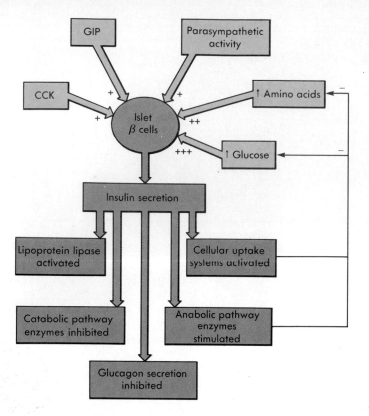

FIGURE 23-2
Inputs to beta cells and effects of insulin, including negative feedback on glucose and amino-acid levels.

| TABLE 23-3 | Factors Determining the Plasma Levels of Insulin and Glucagon |

Factor	Effect on insulin secretion	Effect on glucagon secretion
Hyperglycemia	Increase	Decrease
Hypoglycemia	Decrease	Increase
Increased plasma amino acids	Increase	Increase
Parasympathetic input to pancreas	Increase	Decrease
Sympathetic input to pancreas	Decrease	Increase
Somatostatin	Decrease	Decrease

plasma insulin is about 10 microunits/ml (U/ml—an arbitrary measure of concentration).

The most potent stimulus for insulin secretion is elevated glucose levels in the blood (Figure 23-2 and Table 23-3). Insulin secretion is blocked when the plasma glucose level drops below 50 mg/dl, and maximum insulin secretion occurs at a plasma glucose level of 300 mg/dl. The release of insulin from beta cells also depends on how many glucose receptors are present on the beta cells. High-carbohydrate diets increase the density of glucose receptors on

beta cells, making this feedback system more sensitive. That is, after adaptation to high levels of glucose, a given level of plasma glucose causes a greater rate of insulin secretion. Conversely, low-carbohydrate diets decrease glucose receptor density and thus decrease insulin secretion at a given level of plasma glucose.

In addition to elevated levels of glucose, the beta cells are also stimulated to release insulin in response to elevated levels of amino acids in the blood, to the hormones, gastric inhibitory peptide (GIP), and cholecystokinin (CCK), and to increased parasympathetic activity (see Figure 23-2). Other hormones can also raise plasma glucose levels, and these will often have the secondary effect of stimulating insulin release.

Insulin is called the hormone of abundance because it promotes the use of glucose, amino acids, and fats as they come into the blood during the absorptive period (Table 23-4). Insulin is required for glucose transport into nearly all cells. The primary exceptions are liver cells, brain tissue, erythrocytes, and cells in the renal medulla. Insulin is an anabolic hormone that promotes repair of structures composed of proteins and lipids. This stimulation of synthesis is partly due to the fact that increased uptake of glucose, amino acids, and fats leads to high levels of these reactants in the cells, driving reactions in the direction of net synthesis by the Law of Mass Action. Insulin further promotes the storage of extra nutrients as (1) glycogen in muscle and liver cells, and (2) fat in liver and adipose cells by stimulating specific enzymes. While promoting substrate storage, insulin exerts a strong inhibition on catabolic pathways. It prevents the breakdown of glycogen and fat reserves by its inhibition of hormone-sensitive enzymes in several key catabolic pathways. Figure 23-2 summarizes the general effects of insulin on the body.

Diabetes Mellitus:
Cellular Famine in the Midst of Plenty

Diabetes mellitus is a condition that results when inadequate uptake of glucose from the blood by the body's cells causes high levels of glucose in the blood (**hyperglycemia**). The high levels of glucose saturate the glucose transport system in the kidney so that glucose "spills over" into the urine (**glycosuria**), making it sweet, a property that could be diagnosed by physicians in ancient times. In addition to hyperglycemia and glucosuria, the characteristic symptoms of severe, untreated diabetes mellitus include **ketosis**, a metabolic acidosis caused by elevated levels of the **ketone bodies** acetoacetate, β-hydroxybutyrate and acetone. These are products of fatty acid breakdown by β-oxidation (see Chapter 4). Protein catabolism also occurs, increasing urea production. The presence of glucose, ketone bodies, and urea in urine causes osmotic diuresis

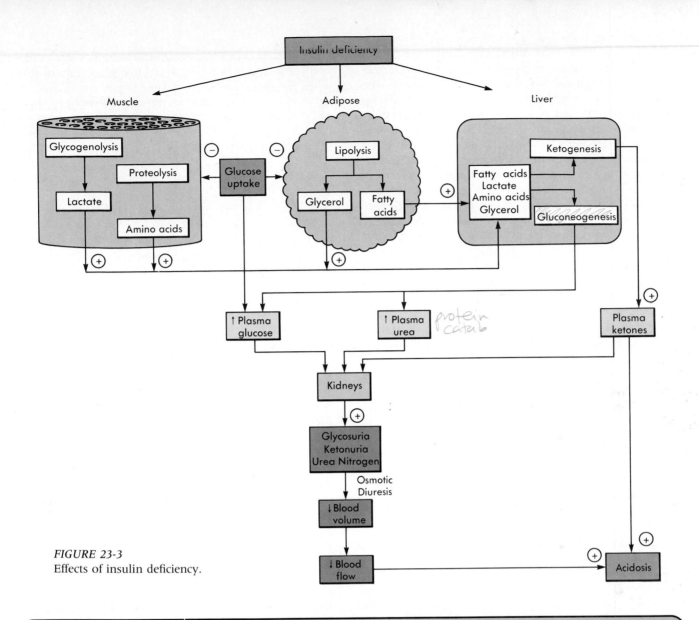

FIGURE 23-3
Effects of insulin deficiency.

TABLE 23-4	*Major Effect of Insulin during the Absorptive States*

Target tissues	Insulin effects	Metabolic consequences
Muscle	Increases glucose uptake	Increased muscle glycogen
	Increases uptake of some amino acids	Increased protein synthesis
Liver	Stimulates enzymes of glycogen synthesis; inhibits enzymes that break glycogen down	Increased liver glycogen
	Stimulates enzymes of pathways leading to synthesis of fat from fatty acids and keto acids	Ingested amino acids and fatty acids are converted into lipids for shipment to fat cells
Fat	Increases glucose uptake	Plasma glucose is converted to alpha-glycerophosphate and used in fat synthesis
	Stimulates lipoprotein lipase	Plasma lipids are taken up for fat synthesis
	Inhibits hormone-sensitive lipase	Decreased fat breakdown in fat cells

(Figure 23-3), the symptom that gave the disease its name, which means "excessive, sweet urine".

There are two types of diabetes mellitus: **Type I** or **juvenile onset diabetes,** which accounts for about 20% of cases, results from inadequate secretion of insulin from the beta cells. Type I diabetes is also called **insulin-dependent** diabetes because it can be treated by insulin replacement therapy. If untreated, the failure to use glucose and inhibit the catabolic pathways that contribute to gluconeogenesis by the liver result in several metabolic derangements (see Figure 23-3). Most threatening is the production of acidic ketones from fatty acids in the liver. Ketone production in a nondiabetic, fasting individual cannot reach levels that result in **metabolic acidosis** because elevated ketone levels elicit secretion of insulin, which opposes ketone production. In diabetics, this negative feedback pathway does not work, and the resulting **ketoacidosis** may be so severe that it causes **diabetic coma** or death.

Diabetics who take insulin without also eating carbohydrate are in danger of a rapid drop in blood glucose. Decreases of plasma glucose to the range of 20 to 30 mg/dl may cause **insulin shock,** which can be fatal because of its effects on the brain. Rapid administration of glucose reverses insulin shock. Newer methods of supplying insulin (such as minipumps) have greatly improved the patient's ability to match the supply of insulin to the body's needs.

The remaining 80% of diabetic patients have **type II** or **adult-onset diabetes** in which release of insulin appears to be in the normal range, but insulin receptors either do not respond normally to insulin, or their density is abnormally low. Although in principle insulin injection might be used to overcome the low receptor density, insulin may lead to further reduction of receptor number by the target cells **(down regulation)**; thus most patients with type II diabetes are treated by controlling their diet and by increasing exercise. Beneficial effects of exercise include stimulation of glucose uptake by the muscle cells and the production of more insulin receptors.

In many cases, type II diabetes is associated with obesity, and the scarcity of insulin receptors may be a response in these sensitive individuals to the elevated levels of insulin secreted in response to consumption of excessive quantities of food. Reduced caloric intake can result in **up regulation,** an increase in the number of insulin receptors in the membranes of target cells. The change can be seen in a few days, even before significant weight loss has occurred.

A number of disease conditions characteristically accompany diabetes mellitus, including cataracts, retinal degeneration that may result in blindness, peripheral nerve disease, and vascular disease. It is not unusual for Type I diabetics to become blind and to lose limbs to gangrene. Until recently, it was not clear how these conditions were caused by the primary disorder of glucose metabolism. It is now clear that elevated blood glucose levels cause glucose residues to be attached to many important proteins that do not normally possess them. This abnormal glycosylation of proteins affects their function, and a high incidence of glycosylated protein ultimately causes tissue deterioration.

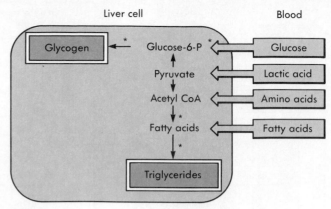

FIGURE 23-4

Reactions in liver cells in the absorptive state. Steps stimulated by insulin are indicated by an asterisk.

Roles of the Liver in the Absorptive State

The hepatic portal system transfers nutrient-rich blood from the GI tract directly to the liver, where approximately 75% of the glucose and amino acids absorbed from the intestine can be taken up. The role of the liver during the absorptive period is to transform incoming nutrients into more complex forms suitable for storage. Depending on their structure, amino acids can be converted to pyruvate for use in glycogen formation, or they can be converted directly into keto acids, intermediates in fatty acid synthesis (Figure 23-4).

Liver cells do not need insulin for uptake of amino acids and glucose. However, insulin exerts a powerful effect on carbohydrate metabolism in the liver by its effects on enzymatic pathways within liver cells. In the first step of glycolysis (see Chapter 4), glucose is converted to glucose-6-phosphate, a molecule that cannot diffuse back out across the liver cell membrane. This conversion is stimulated by insulin (see Figure 23-4), effectively trapping glucose inside the liver cells. Thus this insulin-sensitive enzymatic reaction favors storage during the absorptive period when insulin levels are high. The synthesis of glycogen from glucose-6-phosphate is simultaneously stimulated by another rate-limiting insulin-sensitive enzyme, **glycogen synthetase.** Although the liver forms glycogen mainly from glucose, some of the glycogen stored by the liver during the absorptive state is synthesized from lactate carried by the blood to the liver from white (fast glycolytic) muscle cells (Figure 23-5; see also Figure 23-4 and Chapter 11).

Transport and Storage of Lipids

Although an average of about 200 grams of glycogen can be stored in the liver, most of the nutrients entering the liver follow metabolic pathways to lipids. Glucose metabolized to acetyl-CoA is used to form the fatty acids of triacylglycerols. Triacylglycerols can be stored in the liver as triacylglycerols, or they can be released into the plasma and travel to

GROWTH, METABOLISM, REPRODUCTION AND IMMUNE DEFENSE

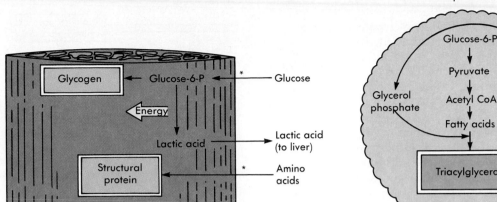

FIGURE 23-5
Reactions in white muscle cells in the absorptive state. Steps stimulated by insulin are indicated by an asterisk.

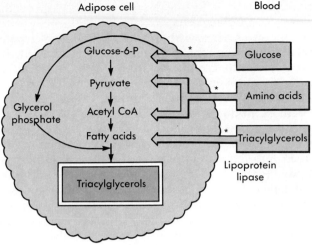

FIGURE 23-6
Reactions in adipose cells in the absorptive state. Steps stimulated by insulin are indicated by an asterisk.

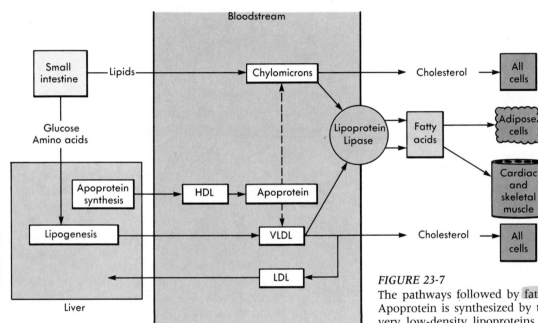

FIGURE 23-7
The pathways followed by fats during the absorptive state. Apoprotein is synthesized by the liver and delivered to the very low-density lipoproteins by the high-density lipoproteins. The apoproteins facilitate uptake of lipids. Low-density lipoproteins and chylomicron remnants are recycled.

adipose cells (Figure 23-6). When excess acetyl-CoA is formed in carbohydrate or amino acid metabolism, liver cells also produce **ketones,** small organic acids that can be metabolized by most tissues (see Table 23-1). In this way amino acids also participate in lipogenesis.

Lipids travel in the blood in the form of **lipoproteins,** particles consisting of complexes of triacylglycerol, phospholipid and cholesterol with specific carrier proteins. Plasma lipoproteins are divided into categories based on their density as determined by centrifugation. The greater the content of lipid compared to protein, the lower the density of the lipoprotein. The protein components of li-

poprotein are recognized and bound by specific receptors on cell membranes, so that the particular proteins associated with the different lipoprotein classes serve as shipping labels that target the lipids for different sites in the body.

The liver releases lipids in the form of **very low density lipoproteins (VLDLs),** which spend an average of 3 hours in the circulation. When the VLDLs deliver their lighter elements to cells, they become **low-density lipoproteins (LDLs)** (see Figure 23-7). Low-density lipoproteins have an important health implication. Although they ultimately return to the liver, the cholesterol they contain may be taken up by a variety of cells, including those of the arterial walls.

Blood is rich with lipids following a typical meal, both from the chylomicrons that entered the blood from the lymph and from the lipids manufactured in the liver (Figure 23-7). Chylomicrons have a relatively short average lifetime in the plasma—about 8 minutes. Both VLDLs and chylomicrons contain triacylglycerols that must be broken down before they can enter cells. Two factors are of particular importance in fat uptake. The first, **lipoprotein lipase (capillary lipase),** is present in the endothelial cells of the capillaries that supply adipose tissue (see Figure 23-7) and, to a lesser extent, cardiac and skeletal muscle. Lipoprotein lipase enzymatically liberates free fatty acids from the triacylglycerols, so the presence of this enzyme determines which tissues will receive the lipids. Insulin is one of the hormones that stimulates lipoprotein lipase activity (see Figure 23-6 and Table 23-4). The second factor that is important for lipid uptake is the **high-density lipoproteins** (HDLs) which are synthesized mainly by the liver (and, to a lesser extent by the intestine). HDL particles circulating in the plasma facilitate uptake of lipids by transferring to VLDLs and chylomicrons a class of molecules called **apoproteins,** which are important both for the attachment of HDL particles to membranes and for the activation of lipoprotein lipase.

Elevated levels of plasma cholesterol and triacylglycerols are statistically associated with an increased risk of atherosclerosis and coronary disease, but the risk is strongly affected by the relative amounts of HDL and LDL. In addition to their delivery of apoproteins to VLDLs and chylomicrons, HDL particles collect cholesterol that is liberated from cell membranes into the plasma and carry it back to the liver for excretion as bile acids. In contrast, LDLs can distribute cholesterol to the blood vessel wall. A low ratio of HDL to LDL is likely to result in disease, while a high ratio may carry a reduced risk of disease even if total plasma lipids are above normal. Factors that have been identified as promoting the amount of HDLs in the blood include exercise and consumption of so-called omega oils from cold water fish such as tuna.

After fatty acids enter adipose cells, they are joined to a glycerol backbone (see Chapter 3, p. 42) to form triacylglycerols (see Figure 23-6 and Chapter 4). This process requires an intermediate phosphorylated form of the glycerol molecule called **glycerol phosphate.** The rate of synthesis of glycerol phosphate is a critical determinant of the rate of fat storage. Adipose cells do not take up glycerol along with the fatty acids, but instead must synthesize glycerol phosphate from glucose. High rates of fat storage are attained only when blood levels of both glucose and insulin are elevated after a meal (see Figure 23-6).

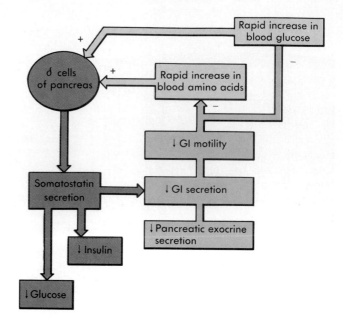

FIGURE 23-8
Inputs to delta cells and effects of somatostatin, including negative feedback, which reduces entry of glucose and amino acids into the circulation.

Protein Synthesis

The increase in plasma amino acids after a protein meal is one factor that stimulates insulin secretion (Figure 23-2). Insulin stimulates amino acid uptake, driving protein synthesis by the Law of Mass Action. Protein synthesis requires not only that many amino acids be present, but that all the **essential amino acids** (those the body cannot synthesize; see Chapter 3, p. 47) be present in the proper proportions at the same time. Otherwise, synthesis of some proteins will not occur. The requirement for a balance of essential amino acids means that dietary protein sources must either be **complete proteins,** in that they contain all the essential amino acids (an example is meat), or, alternatively, be **complementary incomplete proteins** consumed in the same meal. Vegetarian diets typically are built around the combination of one of the whole grains with legumes or nuts and dairy products, and such diets provide all of the essential amino acids.

Somatostatin

Somatostatin (see Table 23-2) is secreted by pancreatic **delta cells** (see Figure 23-1) and by similar enteroendocrine cells in the duodenal mucosa (see Figure 21-8). This peripheral somatostatin must be distinguished from the somatostatin secreted by the hypothalamus into the hypothalamic-anterior pituitary portal circulation in control of growth hormone secretion (Chapter 5), which does not reach the general circulation. Somatostatin decreases GI motility and secretion, pancreatic secretion, and the release of insulin and glucagon (discussed below).

The release of somatostatin is increased by high plasma levels of both glucose and amino acids. Plasma somatostatin levels normally rise only slightly during the absorptive state, but they increase more dramatically if consumption of large amounts of food causes a rapid rise in blood levels of sugars and amino acids. Under these circumstances, somatostatin is capable of slowing all aspects of digestion and absorption (Figure 23-8). It thus prolongs the absorptive state and prevents nutrients from entering the body more rapidly than processing and storage mechanisms can handle.

CHARACTERISTICS OF THE POSTABSORPTIVE STATE

- *Why is the decrease in insulin the single most important aspect of the hormonal pattern of the postabsorptive state?*
- *What is the role of glucagon in the postabsorptive state? Under what circumstances is its secretion needed in the absorptive state?*
- *What effect does insulin secretion have on glucagon secretion? Does glucagon have the same effect on insulin secretion?*
- *What are the roles of cortisol and epinephrine in the postabsorptive state?*

Onset of the Postabsorptive State

In the typical pattern of three meals a day, the full effects of the postabsorptive state are rarely encountered during the day, so the only period in which the body enters the postabsorptive state to any significant extent is during the night, before "breaking the fast" at breakfast. In the first portion of the 8 to 12 hours between the evening meal and breakfast, insulin promotes glycogen and fat storage in the liver and adipose tissue and net synthesis in all body cells. As the levels of glucose and amino acids begin to drop, the stimulation of beta cells diminishes. This automatically lowers blood insulin levels, marking the beginning of the postabsorptive state (see Table 23-3).

By itself, the drop in insulin levels adapts the body for a period in which nutrients must be supplied from the body's stores. Withdrawal of insulin removes the facilitation of insulin-dependent pathways and entry steps and also removes the inhibition exerted by insulin on catabolic pathways in carbohydrate and lipid metabolism. The rate of glycogenolysis in the liver and of lipolysis in both liver and adipose tissue increases as soon as blood insulin reaches its low fasting level (see Table 23-1). Liver glycogen stores by themselves are usually sufficient to supply the body's need for glucose in the period between the end of the absorptive period following dinner and breakfast the next day. However, the postabsorptive state elicits its own hormones, which oppose the actions of insulin and participate in the transformation of glycogen, fats, and protein into molecules that can be metabolized for energy. The extent to which any particular hormone is active in the postabsorptive phase is determined by how long the postabsorptive period lasts and by other factors such as exercise or emotional stress.

Role of Glucagon

Glucagon (see Table 23-2), the major hormone of the postabsorptive state, is released from the alpha cells of the islets of Langerhans (see Figure 23-1). Glucagon is a catabolic hormone (Figure 23-9 and Table 23-5) that promotes glycogenolysis in the liver. At the same time that the liver is stimulated to release glucose derived from stored glycogen, it is also stimulated to convert other molecules (chiefly pyruvate, lactic acid, amino acids, and glycerol) into glucose (gluconeogenesis, see Table 23-1). The major source of starting material for gluconeogenesis is the muscles, which have both a store of glycogen and a large amount of protein. Muscle cell reactions in the postabsorptive state are shown in Figure 23-10, *B*. (Note that in this and the corresponding illustrations for liver and adipose cells, the reactions of the absorptive state have been duplicated as part *A* for ease of comparison.)

Glucagon stimulates lipolysis in adipose tissue by stimulating the hormone-sensitive lipase (Figure 23-11, *B*; see Table 23-1). The fatty acids that are released in lipolysis travel in the blood bound to serum albumin until they are taken up by cells. Through its alteration of metabolic pathways, glucagon increases the levels of glucose, fatty acids, and glycerol in the blood. When insulin levels are low, glucose does not easily enter most cells. Cells preferentially metabolize fatty acids, reserving glucose for those cells which do not have insulin-dependent uptake mechanisms. Neurons are the major cell type in this category.

Glucagon also promotes the use of fatty acids as an alternative energy source. The metabolism of fatty acids for energy increases the levels of acetyl-CoA in the liver, and this favors release of **ketones (ketone bodies)** from liver cells (Figure 23-12, *B*). These ketones, which include acetone, acetoacetic acid, and β-hydroxyacetoacetic acid (β-hydroxybutyric acid), are alternative energy sources. This promotion of fatty acids and ketones as energy sources is called **glucose sparing**. In addition to stimulating the catabolic and gluconeogenic pathways, glucagon inhibits the enzymes stimulated by insulin in glycogenic and lipogenic metabolic pathways (see Figure 23-9 and Table 23-5).

Control of Glucagon Secretion

Glucagon secretion is inhibited by the blood levels of glucose typical of the absorptive state, as well as

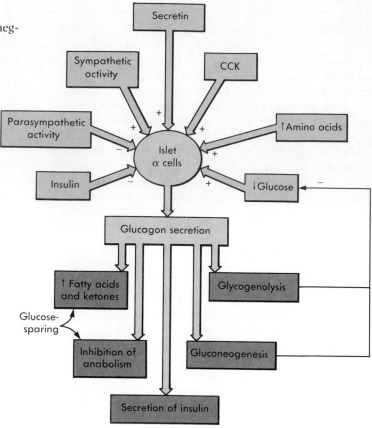

FIGURE 23-9
Inputs to alpha cells and effects of glucagon, including negative feedback, which increases plasma glucose levels.

A Absorptive state   White muscle cell   Blood

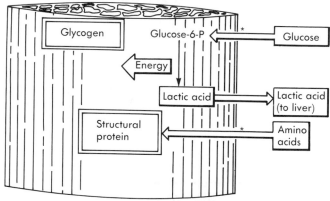

FIGURE 23-10
Reactions in white muscle cells in the absorptive state **(A)** and the postabsorptive state **(B).** Steps stimulated by insulin are indicated by an asterisk. The absence of insulin in the postabsorptive state inhibits glucose uptake.

B Postabsorptive state

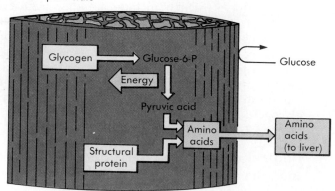

TABLE 23-5	Effects of Glucagon and Catecholamines during the Postabsorptive State	

Target tissue	Hormone effects	Metabolic consequences
Liver	Stimulates enzymes of glycogen breakdown; inhibits enzymes of glycogen synthesis	Conversion of glycogen to glucose, which is released into blood
	Stimulates enzymes of gluconeogenesis	Conversion of lactate made by muscle into glucose for use by CNS
	Stimulates enzymes of fatty acid oxidation; inhibits enzymes of lipogenesis	Conversion of fatty acids released by fat into ketone bodies, which can be used for energy by most cells, sparing glucose for CNS
Fat	Stimulates hormone sensitive lipase	Increased lipolysis in fat, with release of fatty acids and glycerol for conversion to ketones by liver

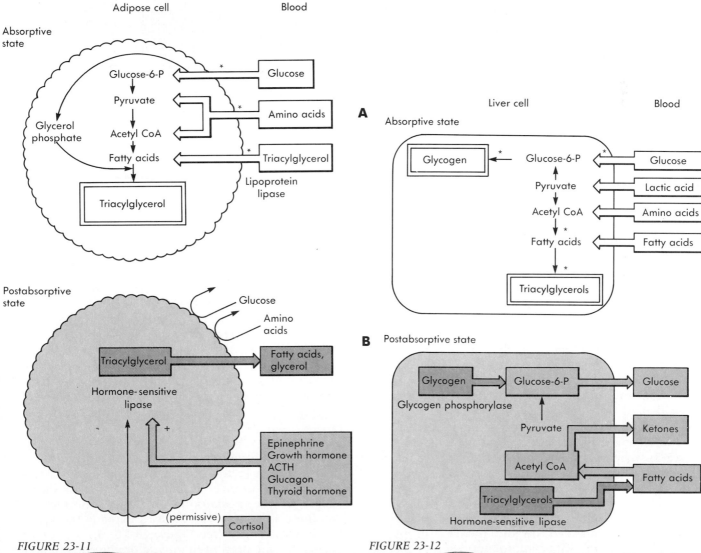

FIGURE 23-11
Reactions in adipose cells in the absorptive state **(A)** and the postabsorptive state **(B)**. Steps stimulated by insulin are indicated by an asterisk. The absence of insulin blocks the uptake of glucose and amino acids during the postabsorptive state.

FIGURE 23-12
Reactions in liver cells in the absorptive state **(A)** and the postabsorptive state **(B)**. Steps stimulated by insulin are indicated by an asterisk.

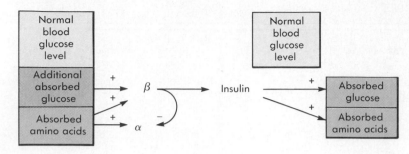

A Both carbohydrates and proteins consumed:

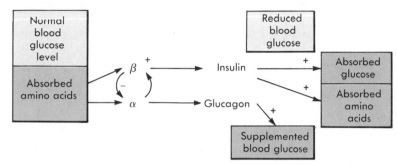

B High protein meal consumed:

FIGURE 23-13

Comparison of the hormonal responses of the pancreatic α and β cells to a meal with a typical mixture of carbohydrates and proteins (**A**) with the responses to a meal with high proteins in the absence of carbohydrates (**B**). When no glucose is coming from the intestine, insulin still increases the transport of glucose into cells, and preservation of the normal level of blood glucose is accomplished by glycogenolysis, which is stimulated by glucagon.

by the high level of parasympathetic activity that exists during the early phases of digestion. Through interactions within the pancreas, the glucagon-producing alpha cells are inhibited when the insulin-producing beta cells are stimulated and vice versa (see Table 23-3). In the postabsorptive state, plasma glucose levels, insulin secretion, and parasympathetic activity decrease and glucagon secretion increases. Thus intervention by glucagon prevents a further fall in plasma glucose.

In certain circumstances, glucagon plays an important role in the absorptive state. Plasma amino acids and the hormones, CCK and secretin, stimulate glucagon release at the same time that they stimulate insulin secretion (Figure 23-13; see Table 23-3). When a carbohydrate-rich meal is consumed, the elevation of blood glucose overrides amino acid, CCK, and secretin stimulation and keeps glucagon levels low. If a protein or fat-rich meal is consumed, little glucose enters the bloodstream, but insulin secretion is stimulated by the increase in plasma amino acids and other stimulatory factors that accompany a meal (see Figure 23-2). After such a meal, glucagon secretion is also stimulated, mainly by the increase in amino acids and decrease in glucose (see Figure 23-9). If glucagon were not secreted to mobilize glucose stores there would be an uncontrolled drop in plasma glucose after a protein meal, because insulin stimulates cellular uptake of glucose as well as amino acids (see Figure 23-13).

Glucagon secretion increases (and insulin secretion decreases) whenever the sympathetic branch of the autonomic nervous system is activated (see Figure 23-9 and Table 23-3). This occurs during exercise and in stressful situations. The combined effect of the increase in the ratio of glucagon to insulin and the direct effect of catecholamines causes the liver to release glucose at a rate much higher than it does during overnight fasting. This effect prevents hypoglycemia during strenuous exercise.

Conversion of Protein to Glucose

Turnover of protein can take place in all tissues, but most protein in the body is in skeletal muscles, so the protein turnover in muscle is the most significant. In the postabsorptive state, movement of amino acids out of the muscles exceeds the rate of amino-acid entry (see Figure 23-10, *B*). The amino acids that are present in muscle protein do not enter the blood in the same proportions at which they are found in the muscle. This is because amino acids are not all treated the same way when proteins are broken down. The essential amino acids are likely to be reused in protein synthesis, whereas nones-

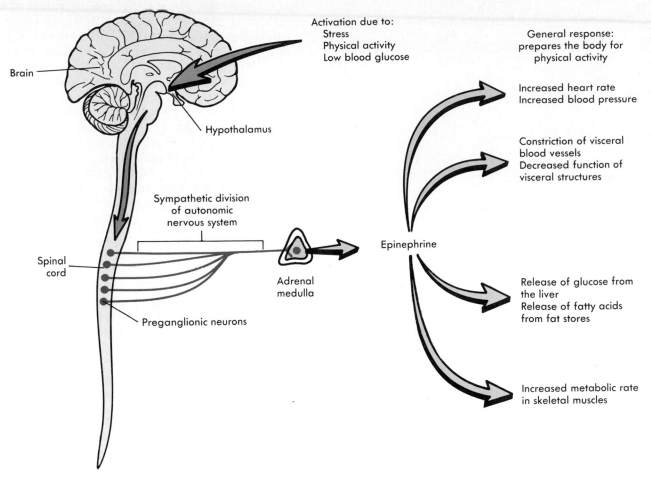

Activation due to:
Stress
Physical activity
Low blood glucose

General response:
prepares the body for
physical activity

Increased heart rate
Increased blood pressure

Constriction of visceral
blood vessels
Decreased function of
visceral structures

Epinephrine

Release of glucose from
the liver
Release of fatty acids
from fat stores

Increased metabolic rate
in skeletal muscles

Brain

Hypothalamus

Sympathetic division
of autonomic
nervous system

Spinal
cord

Adrenal
medulla

Preganglionic neurons

FIGURE 23-14

Regulation of the secretion of epinephrine from the adrenal medulla. Inputs that cause increased secretion of epinephrine include stress, physical activity, and low blood glucose. Epinephrine elevates glucose, blocks glucose use in cells that have insulin-dependent uptake systems, induces glucose sparing, and promotes lipid availability.

sential amino acids tend to be released into the blood to be metabolized for energy.

Some nonessential amino acids are synthesized from carbohydrates in the muscle and released to supply the energy needs of other tissues. For instance, pyruvate is not able to leave the muscle cells, but it can be aminated (with amino groups taken from other amino acids) to form alanine or glutamate. The alanine concentration of blood increases during the postabsorptive state (see Figure 23-10, *B*).

Adrenal Hormones

The hormone **cortisol,** a steroid hormone produced by the adrenal cortex (see Table 23-2), is one of a family of steroids called **glucocorticoids** because of their importance for maintaining blood glucose levels. Cortisol is not typically elevated in the postabsorptive state, but it must be present at normal levels for glucagon to stimulate catabolic pathways—this is an example of a permissive hormonal effect.

Epinephrine (see Table 23-2) may also contribute to the postabsorptive state (Figure 23-14), espe-

cially if stress occurs during this period. The degree of stress associated with the postabsorptive state is determined in part by how fast plasma glucose levels drop. When the plasma glucose concentration falls rapidly, the sympathetic nervous system is activated and epinephrine is released from the adrenal medulla. The liver and adipose tissue are innervated by sympathetic endings, and norepinephrine released from those endings, as well as blood-borne epinephrine, stimulates liver gluconeogenesis and glycogenolysis and hormone-sensitive-lipase activation in adipose cells. Epinephrine also stimulates glucagon secretion and inhibits insulin secretion.

Metabolic Homeostasis in Starvation

During prolonged fasting or the low nutrient intake levels that produce starvation, the body enters a new state that is somewhat different from the postabsorptive state. One of the changes seen in fasting and starvation is a progressive decline in basal metabolic rate. This change increases the chances of survival by slowing the rate at which the body's energy stores are consumed. The stores of nutrients

TABLE 23-6 Caloric Content of the Three Major Categories of Nutrients

Category	Kcal/gram*	Kcal/1 O_2 consumed
Carbohydrate	4.2	5.0
Protein	4.3	4.5
Fat	9.4	4.7

*A kilocalorie (kcal) is the amount of heat required to raise the temperature of 1000 grams of water by 1° C. The caloric content is measured by measuring the heat liberated by complete combustion of a known weight of the nutrient, using an instrument called a bomb calorimeter.

that form the basis of survival consist mostly of fats and muscle protein. Fats contain about twice as much energy per gram as do carbohydrates or protein (Table 23-6). The 10 kg of fat available in a typical individual is sufficient for survival of a 30- to 40-day fast. A smaller amount of fat is also available from the liver. Fats are used for energy during starvation just as in the postabsorptive state because insulin secretion is depressed and glucagon levels are elevated. The proteins present in the muscles and other tissues can be metabolized to extend the survival period up to two months (Table 23-7).

The stress of fasting and starvation activates the sympathetic nervous system. The resulting secretion

TABLE 23-7 Substrate Use and Storage

Substance	Daily turnover (grams)	Storage form	Organ	Amount stored (grams)	Functional reserve
Glucose	250	Glycogen	Liver	200	<1 day
			Muscle	200	
Amino acids	150	Protein	Muscle	6000	1 to 2 weeks
Fatty acids	100	Triacylglycerols	Adipose tissue	10,000	4 to 6 weeks
			Liver	400	1 to 2 days

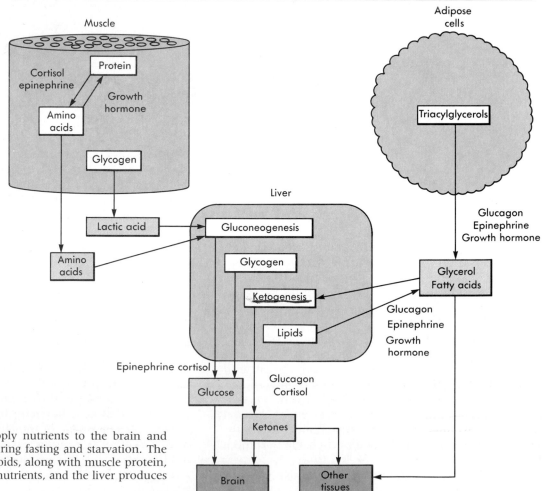

FIGURE 23-15
Pathways used to supply nutrients to the brain and body cells (left side) during fasting and starvation. The stored glycogen and lipids, along with muscle protein, provide the source of nutrients, and the liver produces glucose and ketones.

A Possible New Pancreatic Hormone, Amylin

A proposed new pancreatic peptide hormone, amylin, may have a key role in Type II diabetes. Amylin was originally identified in the islets of Langerhans of patients with Type II diabetes. Since then amylin secretion has been shown to be present in normal pancreatic tissue where amylin is synthesized at about 20% the rate of baseline beta cell insulin production. Amylin and insulin contain regions with similar amino acid sequences. This sequence homology suggests that the two molecules have a common evolutionary ancestor. Amylin is stored with insulin in the vesicles of beta cells and secreted along with insulin.

Amylin stimulates glycogenolysis and inhibits glycogen formation by preventing the initial phosphorylation of glucose to glucose-6-phosphate. As a result of these effects, amylin increases plasma glucose levels. As far as its effects on liver and muscle are concerned, amylin is a kind of anti-insulin. In Type I diabetes both insulin and amylin secretion decrease. In Type II diabetes the secretion of insulin is normal, but amylin secretion is elevated, presumably due to an increased rate of amylin synthesis. The excess amylin is deposited in the islet cells as amyloid, an abnormal protein-carbohydrate complex. If overproduction of amylin proves to be an important factor in Type II diabetes, it may be possible to directly inhibit amylin synthesis within the beta cells. In any case, measuring amylin levels is a promising new tool in diagnosing early Type II diabetes.

Amylin may also be an important paracrine agent within the islets of Langerhans. In animals, amylin augmented glucose-stimulated insulin release from pancreatic beta cells, and inhibited amino-acid dependent glucagon release from the alpha cells. Amylin also reduced somatostatin release. Amylin does not affect the exocrine pancreatic secretion of HCO_3^- or enzymes. While the details are still to be worked out, this would make a total of four pancreatic hormones that have powerful regulatory effects on glucose metabolism: insulin, glucagon, somatostatin, and amylin.

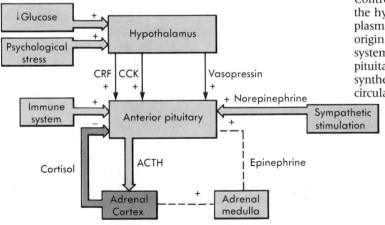

FIGURE 23-16
Control of glucocorticoids (cortisol) by releasing factors from the hypothalamus; these factors are stimulated by a drop in plasma glucose levels and stressful stimuli of psychological origin. The immune system and the sympathetic nervous system and epinephrine can act at the level of the anterior pituitary to stimulate cortisol release, and cortisol stimulates synthesis of epinephrine. Negative feedback is provided by circulating levels of glucocorticoids on the anterior pituitary.

of epinephrine by the adrenal medulla promotes glucose release from the liver. Epinephrine also inhibits insulin release, so uptake of glucose by insulin-dependent uptake systems is severely restricted, and potentiates glucagon release. Epinephrine stimulates the enzymes that mobilize fat stores and promote gluconeogenesis, increases lactic acid uptake by the liver, and promotes glycogenolysis that maintains plasma glucose levels. Epinephrine also spares glucose by stimulating lipolysis in adipose tissue and the liver and the uptake of fatty acids into muscle. The overall result of epinephrine is to produce a metabolic pattern similar to, but more exaggerated than, the postabsorptive state (Figure 23-15).

Many of the actions of epinephrine require that cortisol also be present. Epinephrine stimulates cortisol secretion by acting on the anterior pituitary cells that secrete **adrenocorticotropic hormone (ACTH)**. Control of glucocorticoid secretion (Figure 23-16) is also exerted on the hypothalamus, which releases three factors that initiate ACTH secretion: **corticotropin-releasing hormone (CRH), vasopressin,** and **CCK.** Stressful situations activate antidiuretic responses through peripheral actions of vasopressin.

Cortisol accentuates the effects of glucagon and norepinephrine. Like epinephrine secretion, cortisol secretion is increased whenever plasma glucose levels fall rapidly. Other stimuli causing cortisol release include tissue damage and infections. Cortisol is strictly a catabolic hormone that antagonizes the anabolic effects of insulin and mobilizes stored nutrients, and so its presence during starvation is required for the body to significantly tap the muscle proteins to provide amino acids for energy, lipogenesis, and gluconeogenesis (Figure 23-15). In the liver, cortisol increases the rate of conversion of fatty acids to ketone bodies (see Figure 23-15).

Muscle protein could maintain total body metabolism for several weeks, but continued protein breakdown ultimately leads to serious protein deficiencies. Growth hormone plays an important role in prolonged fasting because it prevents the extensive protein breakdown that would occur if only epinephrine and cortisol levels increased.

The liver's role in conversion of amino acids to glucose is crucial during starvation because the blood level of glucose must be maintained for normal brain function to continue, and fats cannot be converted to glucose. Most cells use fats for energy, but the brain, which is an obligate consumer of glucose when food intake levels are normal, is able to adjust its enzymes to allow it to use ketones for up to half of its energy needs after a week of fasting (see Figure 23-15). The two parts of the body that are protected from net catabolism the longest during starvation are the heart and brain.

REGULATION OF FOOD INTAKE
Satiety Signals and Short-Term Regulation of Food Intake

Food is not consumed continually, but is taken in meals with intervening periods of nonconsumption. Proteins, carbohydrates, and fats also differ in their energy content per gram (Table 23-6). Factors that influence total intake are how long meals last, the quantity of food consumed in each meal, the nutrient content of a meal, and how frequently meals occur (Figure 23-17). What signals initiate and terminate food consumption? Hunger does not correspond particularly well with the nutritional state of the body as measured by the circulating levels of glucose or stored fats. The desire to eat is determined by many factors associated with social patterns of food intake or the aromatic and visual appeal of food. Yet there appear to be some signals that regulate the amount of food consumed and the total caloric intake.

Changes in metabolic rate occur even before nu-

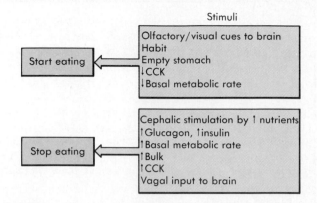

FIGURE 23-17
Regulation of food consumption is determined by the amount of food consumed in each meal and the length of the interval between meals. Food consumption is regulated by signals. Some of the best-studied stimuli for the onset and termination of feeding are listed on the right.

trients begin to be consumed. These changes are caused by hormonal stimulation in the cephalic stage of digestion. Estimates of the richness of food can be made on the basis of past experience, and this influences the level of circulating hormones before ingestion. Thus the association cortex is involved, along with areas of the brain involved with reward systems.

Both insulin and glucagon levels are elevated in the cephalic stage of digestion before food absorption; thus stored nutrients are mobilized, and their uptake into cells is stimulated. A neural connection between sugar receptors in the mouth and the hypothalamus is one pathway that promotes these hormonal changes. Early signals related to incoming food lead to metabolic changes that provide a feeling of **satiety,** of having consumed enough, even before absorption of the meal is complete (Figure 23-17).

Termination of a meal is also influenced by signals arising from the GI tract that indicate both bulk and nutrient levels of ingested food (see Figure 23-17). In the stomach, stretch receptors detect the presence of food, and the hunger pangs resulting from contraction of the empty stomach are alleviated by the bulk aspects of food. In experiments in which nutrient-poor bulk is ingested, the first few meals are terminated on the basis of bulk before adequate caloric intake occurs. However, this control is overridden within days by mechanisms that respond to the body's overall nutritional state so that such experimental subjects will increase intake of the nutrient-poor food over time.

In spacing meals, signals of satiety that arise from the pancreas and duodenum and from nutrients flowing to the liver in the hepatic portal system are probably more important than are gastric satiety signals. Insulin is regarded as a likely satiety signal, since its levels accurately track the amount of glucose and amino acid brought in by a meal. In animal experiments, doses of insulin similar to those that would result from a meal suppressed feeding within a few minutes. Another probable satiety signal from the duodenum is the duodenal hormone cholecystokinin (CCK; see Chapter 22). Entry of chyme into the duodenum increases the level of CCK in the blood, and in experiments injection of CCK into the blood or peritoneal fluid reduced food consumption. How do these hormonal signals reach the brain? There are areas of the ventricles where the blood-brain barrier is relatively leaky; these are near areas of the hypothalamus that are involved in control of feeding. Alternatively, the hormones may act on sensory projections from the intestine and abdominal organs. There is some evidence for such a pathway from experiments in which cutting the abdominal vagus nerves was found to diminish the antifeeding effect of injected CCK. Finally, the elevation of temperature that accompanies digestion continues all during the absorptive phase, and the tendency to eat again is correlated with the drop in metabolic rate as the body approaches the postabsorptive state (see Figure 23-17).

Long-Term Control of Food Intake by Nutritional State

The mass of the fat stores of adults is remarkably stable over long periods. For example, the average woman in the United States gains 11 kg between the ages of 25 and 65 years, while she consumes about 1.8×10^7 kg of food during the same period. The 11 kg increase corresponds to an average daily overconsumption equivalent to about 1/4 of a slice of bread. It has been difficult for investigators to determine whether this stability of fat stores is achieved by direct feedback regulation of the mass of fat, or is just the result of very good short-term regulation of meal size, energy expenditure and metabolic rate. Historically, three simple hypotheses about regulation of body weight and appetite have been advanced; these are the glucostatic, lipostatic and thermostatic hypotheses.

The **glucostatic** hypothesis of long-term regulation of eating is based on feedback relating to the availability of glucose to body cells. Although the level of glucose in the blood is not well correlated with the daily pattern of eating, long-term carbohydrate deprivation does produce hunger. However, this hypothesis has been partly supplanted by the **lipostatic** hypothesis, which asserts that feeding frequency and duration are regulated by a signal that arises from the fat deposits in the body. Surgical removal of body fat results in an increase in food consumption, and increases in body fat induced by various experimental procedures cause a temporary decrease in food intake. The identity of the signal

from adipose tissue to brain is not yet known. It appears to be a blood-borne factor, because exchange of blood between normal and obese animals suppressed feeding in the normal animals.

The general nutritional state of the body is related to the intake of all categories of food. Carbohydrates, fats, and amino acids all contribute to the anabolic reactions of the absorptive state, and the heat that is produced in these reactions is another signal related to total nutrient intake. There is some evidence that heat from nutrient anabolism may serve as a satiety signal. As body fat accumulates, metabolic rate increases, with an accompanying elevation of heat production. If heat production during the anabolic phase is a factor that tends to limit consumption, then the elevation of body temperature associated with accumulation of excess body fat would be an automatic mechanism that tends to return the body to a leaner weight. This is the **thermostatic** hypothesis of body weight regulation.

Each of these three simple hypotheses accounts only partly for long-term regulation of body weight, so it is probable that the feeding centers of the hypothalamus integrate a number of factors to control the complex behavior of eating.

GROWTH AND METABOLISM OVER THE LIFE CYCLE

- *What are the mechanisms by which growth hormone affects growth and metabolism? What is the role of growth hormone in adults?*
- *What changes in muscle mass, adipose mass and skeletal mass occur with aging? What are some possible causes for these changes?*

Growth and Growth Hormone

Human growth hormone (hGH, also called **somatotropin)** is secreted from the anterior pituitary in an oscillatory pattern that usually peaks during the early hours of the sleep cycle (Figure 23-18). The release is stimulated by **growth hormone-releasing hormone (GHRH)** from the hypothalamus and inhibited by **somatostatin (growth hormone-inhibiting hormone, GHIH)** from the hypothalamus.

The principal effect of hGH is seen in children—fetal and early postnatal development does not require it. The rapid metabolic rate of children and adolescents is largely attributable to the stimulation of net anabolism by hGH in these ages. hGH is needed for cartilage formation **(chondrogenesis)** and for the subsequent calcification **(ossification)** of bone by osteoblasts. These processes are promoted by hGH's stimulation of mitosis and protein synthesis in cells in the **epiphyseal plates,** the growing regions of long bones (Figure 23-19). (At sexual maturation or puberty, described in Chapter 24, gonadal steroids begin to stimulate both bone lengthening and epiphyseal closure, as part of the

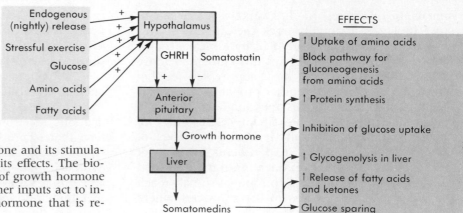

FIGURE 23-18

Control of the secretion of growth hormone and its stimulation of the somatomedins that mediate its effects. The biological rhythm that controls the release of growth hormone produces a nightly peak of secretion; other inputs act to increase the overall amount of growth hormone that is released.

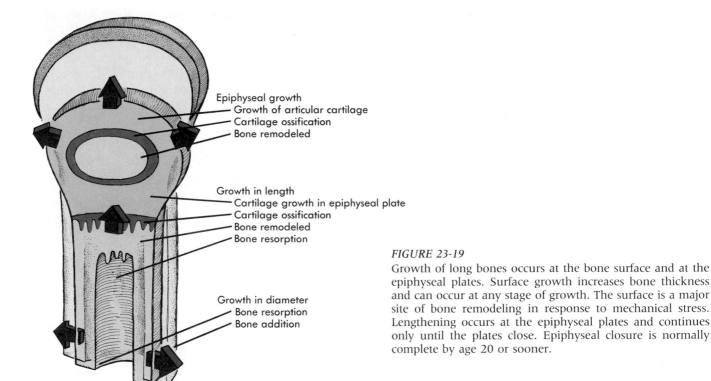

Epiphyseal growth
 Growth of articular cartilage
 Cartilage ossification
 Bone remodeled

Growth in length
 Cartilage growth in epiphyseal plate
 Cartilage ossification
 Bone remodeled
 Bone resorption

Growth in diameter
 Bone resorption
 Bone addition

FIGURE 23-19

Growth of long bones occurs at the bone surface and at the epiphyseal plates. Surface growth increases bone thickness and can occur at any stage of growth. The surface is a major site of bone remodeling in response to mechanical stress. Lengthening occurs at the epiphyseal plates and continues only until the plates close. Epiphyseal closure is normally complete by age 20 or sooner.

growth spurt of puberty.) Protein synthesis is seen in soft tissues such as muscle as well as at the epiphyseal plates, and protein synthesis is favored over all other metabolic pathways by hGH. Rapidly growing children and adolescents, who typically have little body fat, exemplify the effects of this hormone on metabolism. Body fat is mobilized as an energy source by hGH at the same time that amino acids, another potential energy source, are being incorporated into protein.

Growth hormone exerts its effects through induction of **somatomedins,** a family of hormones secreted into the blood by the liver and other tissues. The somatomedins contain significant amino acid sequence homology to insulin and share some of its effects, so they are also called **insulin-like growth factors.** Somatomedins act on target tissues to promote mitosis and protein synthesis (see Figure 23-18). The induction of new enzymes and an increase in ribosomes, which are needed to synthesize protein in the target cells, requires some time, so the effects of hGH occur more slowly than those of hormones which act by changing the effectiveness of enzymes already present in the cell. Somatomedins promote protein synthesis by increasing the rate of uptake of amino acids into muscle (an insulin-like effect), and they increase the availability of amino acids to the muscle cells by opposing the use of amino acids in gluconeogenesis. There is evidence that somatomedins also inhibit hGH secretion, providing negative feedback.

Growth hormone opposes fat deposition by de-

Synthetic Growth Hormone

Failure of the anterior pituitary to secrete growth hormone (hGH) during childhood causes one form of dwarfism. The growth hormone proteins produced in other animals, although similar to the human form, are not sufficiently similar to be active in humans. (In this, growth hormone differs from some other hormones. For example, insulin from animal sources is essentially as effective in humans as the human form.) Until recently, inadequate growth hormone secretion could be treated only with hormone extracted from the pituitaries of human cadavers, making the therapy very expensive. With the introduction of techniques that allow human genes to be cloned and expressed in bacteria (see box, pp. 70), biosynthesized hGH became available at a much lower cost. This hormone is identical to the form made by the anterior pituitary. As a result of this advance, other uses of the hormone became economically feasible. It is now possible, at least in theory, for children who have normal levels of hGH secretion to be given extra hormone to ensure that they will be exceptionally tall. Given the respect our culture has for tall people, and the importance of height in basketball, this might seem a desirable option.

Growth in stature is ultimately a self-terminating process. Until recently, hGH was not believed to have any role in long-term homeostasis in adults whose skeletal growth was complete. However, new studies suggest that decreasing levels of hGH with age are partly responsible for the loss of muscle mass and skin elasticity seen in elderly men, and that administration of biosynthesized hGH might slow or reverse these effects. The effects of hGH in promoting muscle growth and inhibiting fat deposition have led to abuse of this hormone by athletes. At present, hGH abuse is much more difficult to detect by drug testing than abuse of steroid hormones.

creasing glucose uptake and utilization in adipose tissue. Conversely, hGH increases synthesis of the capillary lipases needed for dietary triacylglycerol uptake into adipose tissue and promotes the breakdown of the triacylglycerols allowing fatty acids to be released. In this way, hGH maintains blood glucose levels, decreases storage of fat, and promotes the use of fatty acids for energy. This effect of hGH on fat metabolism is seen only if the plasma insulin levels are low because insulin normally inhibits the catabolism of fat through its inhibition of intracellular hormone-sensitive lipases.

Deficiency of hGH or a lack of hGH receptors in early childhood results in **dwarfism.** The stature of pygmies apparently results from lack of one of the major somatomedins. An abnormally high level of hGH secretion produces excessive growth of the long bones, leading to **gigantism.** Administering hGH during childhood and puberty can likewise result in additional growth of long bones. This treatment, which allows normal development in children with deficient pituitary function, can also in-

crease the rate of growth in normal children. Biotechnology has now reduced the cost of producing hGH. Some parents ask for the use of this hormone in children with normal growth potential in an attempt to promote growth and increase heights. In adults, hGH secretion increases in response to the stress experienced during sustained exercise and in starvation. In both cases, hGH acts to conserve the protein stores of the body. Deficiency of hGH in adults does not have dramatic effects, but there is evidence that some of the muscle loss and reduced exercise capacity associated with aging are the result of decreased secretion of hGH.

Acromegaly is a condition caused by grossly excessive secretion of hGH during adulthood. Because the epiphyseal plates are closed, hGH has no effect on long bone growth in adults, but it may potentially affect bone width. Acromegaly is characterized by thickening of the bones of the fingers and toes, some muscle hypertrophy, disfiguring growth of the flat bones of the back and face, and coarse skin. Anterior pituitary tumors are a major cause of acrome-

The loss of muscle mass with advancing age is accompanied by net decalcification of the skeleton, called osteoporosis. In osteoporosis, the loss of calcium is accompanied by loss of the cells that deposit the calcium salts and is therefore irreversible. Osteoporosis is a real threat to women in their postmenopausal years because maintenance of bone mass is dependent on gonadal steroids (estrogen and testosterone), and the level of estrogen drops with the cessation of menstruation. Osteoporosis is less frequent in men because male gonadal steroid (testosterone) secretion continues throughout life.

Concern about the increased likelihood of bone fractures that accompanies osteoporosis has led to recommendations for increased intake of calcium. This has fostered the addition of calcium to all types of foods.

However, much of the added calcium is not in a form that the body can use; even the calcium in spinach cannot be absorbed. Dairy products remain the best and safest sources of calcium, but it is now recognized that, although ingestion of adequate amounts of calcium is necessary, dietary calcium is unable to maintain bone mass in the absence of estrogen and other factors that promote bone deposition.

Recent studies compared groups of women who were matched with respect to known or suspected risk factors for development of osteoporosis (that is, low calcium intake, cigarette smoking, moderate alcohol consumption, and skeletal type) and who differed only in their amount of daily exercise.

The conclusion of the study was that a lifelong pattern of regular exercise confers significant protection against the ravages of osteoporosis. The exercise must be of the weight-bearing sort, such as jogging, because some otherwise beneficial exercises, such as swimming, consume energy but do not seem to help maintain the skeleton. It is important to note that the protective effect of exercise on bone mass requires gonadal steroid hormones. Female athletes who train intensively frequently decrease their adipose tissue below the threshold needed to maintain menstrual cycles (see Chapter 24). These women lose skeletal mass and can develop osteoporosis.

galy, but tumors in other tissues may also produce excessive quantities of hGH-like protein.

Effect of Aging on Body Composition

The relative contributions of the skeleton, muscle mass, and adipose tissue to total body weight change over the life cycle. Children tend to be plump in early childhood and then become thin in the period dominated by hGH when rapid long-bone growth is occurring. In adulthood, a gradual shift in the proportions of weight is attributable to changes in body protein, fat, and skeleton, even when body weight is maintained at a constant level. Considering the population at large, muscle and skeletal mass reach a peak about age 35 and almost all people lose muscle tissue as they get older. This loss is attributable both to a reduction of muscle mass through loss of filament proteins and to a reduction in the number of muscle cells. Loss of muscle mass is much less evident in individuals involved in daily physical labor and is largely a function of disuse. As is apparent from the data presented in Figure 23-20, the percentage of total body weight represented by fat more than doubles for men ranging in age from 20 to 55 years. Taken together with the loss of muscle mass, it could be argued that the ideal weight of physically inactive individuals decreases with increasing age rather than remaining constant.

Because muscle has a much higher metabolic rate than adipose tissue, the shift in weight from muscle mass to fat reduces the overall metabolic rate of the body. Ideally, the changes in body composition should automatically result in reduced food consumption. Unfortunately, eating habits formed over a lifetime are difficult to change, and lack of exercise exacerbates the problem just at a time in life when weight gain and the threat of atherosclerosis become significant. Because the beneficial effects of exercise are well known, it is wise to develop and maintain a pattern of exercise that can be continued throughout adulthood.

THYROID HORMONES AND THE BASAL METABOLIC RATE

- *What factors determine the basal metabolic rate?*
- *What factors affect thyroid activity? What are the sources of T_3?*
- *How do T_3 and T_4 increase the rate of metabolic heat production?*

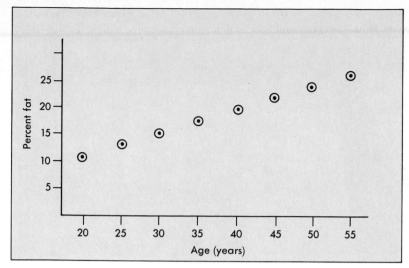

FIGURE 23-20
Average increase in fats as a proportion of total body weight with increasing age in men.

In adults, thyroid hormones regulate the overall rate of metabolic turnover. Directly or indirectly, they affect the intermediary metabolism of fats, carbohydrates and proteins, and the rate of terminal oxidative metabolism. These effects have consequences for regulation of body weight and ability to maintain core body temperature by production of metabolic heat.

A rough measure of the general rate of energy transformation in the body is the **basal metabolic rate**, the rate at which the body utilizes stored energy when the person is in an alert but resting condition and has fasted for 12 hours. The direct measurement of the basal metabolic rate is made with the person in a small chamber that has sensors in the walls that can measure heat liberated from the body. An indirect method is based on the measurement of the rate of O_2 uptake in respiration because the rate of release of energy from food is quantitatively related to O_2 consumption (see Table 23-6). Typically, a mixture of fuels is oxidized during the postabsorptive state, so that about 4.8 kcal of energy are released for every liter of O_2 consumed in respiration. The basal metabolic rate of adults is about 20 kcal/kg body weight/day, or roughly 1 kcal/min for a typical adult. The total metabolic rate of an individual is the sum of basal metabolism and activity-specific metabolism and varies according to the sex of the individual, physical activity, and genetic factors. Average rates are about 39 kcal/kg/day for men and about 34 kcal/kg/day for women.

Many hormones affect the overall rate of conversion of energy sources to CO_2, water, ATP, and heat. These include metabolic hormones that respond to short-term changes in the nutritional state (see Table 23-2); for example, the effect of insulin may increase the basal metabolic rate by as much as 20% during the absorptive state. Sex hormones have effects on metabolism (as suggested by the sex difference in metabolic rate) but are not involved in regulation of metabolic rate. The **thyroid hormones** (see Table 23-2) are of particular importance in regulating the basal metabolic rate and in long-term adaptation to stress, including prolonged exposure to cold. The major form of thyroid hormone secreted by the thyroid gland is **thyroxine (T_4)**. Much of the secreted T_4 is converted by the liver and kidney to a more active form, **triiodothyronine (T_3)**, by removal of one iodine atom (I). Most of the metabolic effect of the thyroid hormones is actually caused by T_3 rather than by the secreted form of T_4.

The thyroid contains spherical follicles into which a large storage protein, thyroglobulin, is secreted (Figure 23-21). Thyroid hormones are formed from the amino acid, tyrosine, in a two-stage process (Figure 23-22). First, tyrosine residues on thyroglobulin are singly or doubly iodinated to form monoiodotyrosine (MIT) or diiodotyrosine (DIT). Next, an MIT and a DIT are joined to form T_3, or two DIT are joined to form T_4. Then the hormone molecule is cleaved from the thyroglobulin and secreted into the blood (Figure 23-23). About 80% of the total hormone secreted by the thyroid is T_4. The requirement for iodine to synthesize the hormone is met by active accumulation of iodide (I^-) by the thyroid from the blood.

The receptors for thyroid hormones are in the nuclei of cells. The affinity of the receptors for T_3 is much higher than for T_4. The major effect of thyroid hormones is to alter enzyme systems that affect metabolic rate. The thyroid hormones increase metabolic rate through induction of specific proteins, such as receptors that are incorporated in the cell membrane, the Na^+-K^+ ATPase, and enzymes involved in specific anabolic pathways. The result of these actions is an increase in O_2 consumption (basal metabolic rate) and protein synthesis within

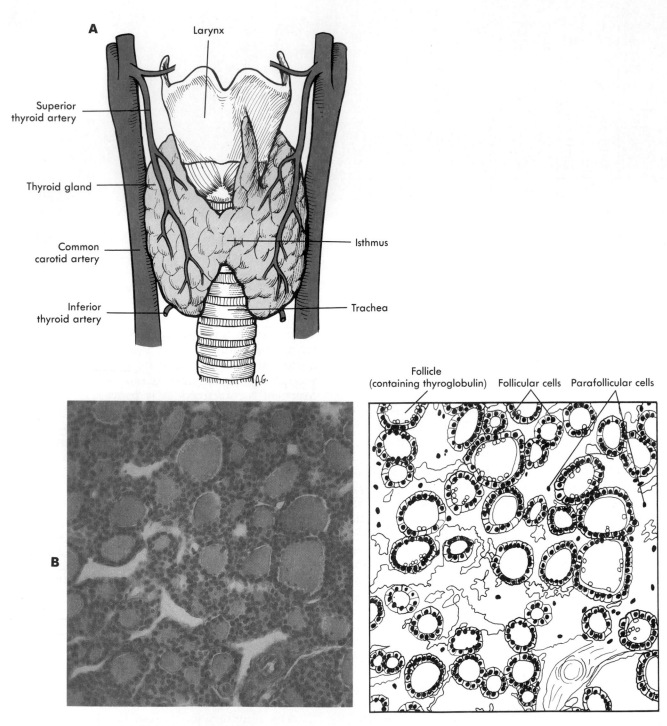

FIGURE 23-21
A Location of the thyroid.
B Micrograph of thyroid follicles, interpreted in drawing at right. Each follicle consists of a central core of thyroglobulin, which represents a reserve of iodine and hormone precursor. The surrounding follicle cells have receptors for TSH and are responsible for I-uptake from the blood, thyroglobulin synthesis, and synthesis and release of T_3 and thyroxine.

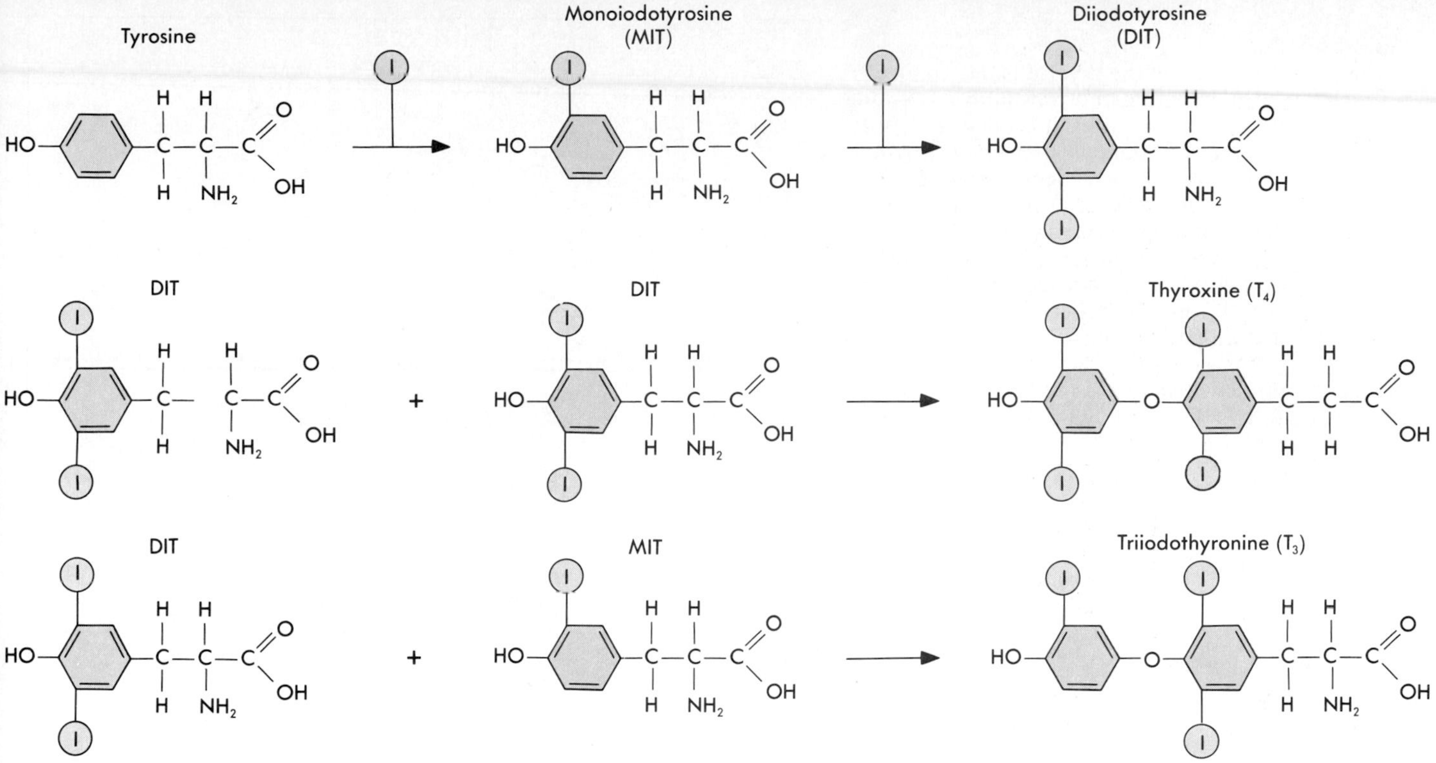

Figure 23-22 (top)

Tyrosine

Monoiodotyrosine (MIT)

Diiodotyrosine (DIT)

DIT + DIT → Thyroxine (T$_4$)

DIT + MIT → Triiodothyronine (T$_3$)

FIGURE 23-22

The two steps in synthesis of T$_3$ and thyroxine from tyrosine in the thyroid follicular cells. The first step is formation of either monoiodotyrosine (MIT) or further iodination to form diiodotyrosine (DIT). In the second step, either the diiodinated benzene ring from one molecule of DIT is joined to a second DIT to form thyroxine, or the monoiodinated iodinated ring from an MIT is joined to a DIT to form T$_3$. Formation of additional T$_3$ in target tissues is the result of removal of an I from T$_4$ (pathway not shown).

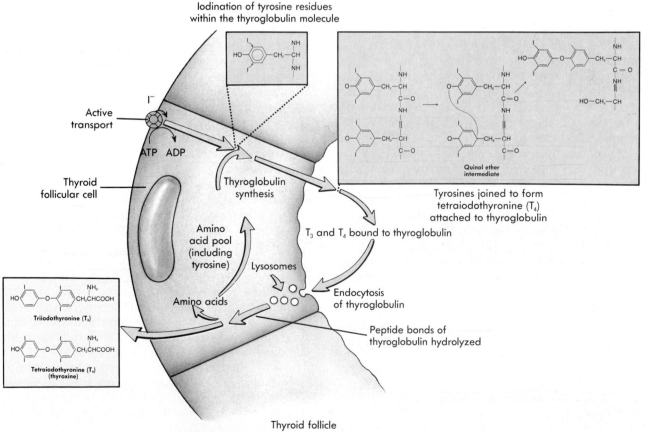

Iodination of tyrosine residues within the thyroglobulin molecule

I⁻

Active transport

ATP   ADP

Thyroid follicular cell

Thyroglobulin synthesis

Amino acid pool (including tyrosine)

Lysosomes

Amino acids

Quinol ether intermediate

Tyrosines joined to form tetraiodothyronine (T$_4$) attached to thyroglobulin

T$_3$ and T$_4$ bound to thyroglobulin

Endocytosis of thyroglobulin

Peptide bonds of thyroglobulin hydrolyzed

Triiodothyronine (T$_3$)

Tetraiodothyronine (T$_4$) (thyroxine)

Thyroid follicle

FIGURE 23-23

Cellular events in formation and secretion of T$_3$ and T$_4$.

FIGURE 23-24

Control of secretion of the thyroid gland hormones, T_3 and T_4. The hypothalamus has both a releasing factor, thyrotropin-releasing hormone (TSHRH), and an inhibiting hormone, somatostatin (thyroid-inhibiting hormone), which regulate the release of thyroid-stimulating hormone (TSH) by the anterior pituitary. Some of the factors that affect TSHRH secretion are shown. See Figure 23-17 for additional factors that affect somatostatin secretion. Blood levels of T_3 and T_4 are affected by rates of secretion from the thyroid and by rates of conversion of T_3 to T_4 in target tissues. Feedback control of thyroid function is mediated in part by effects of plasma levels of T_3 and T_4 on pituitary TSH secretion and on hypothalamic secretion of TSHRH and somatostatin (feedback loops shown in red).

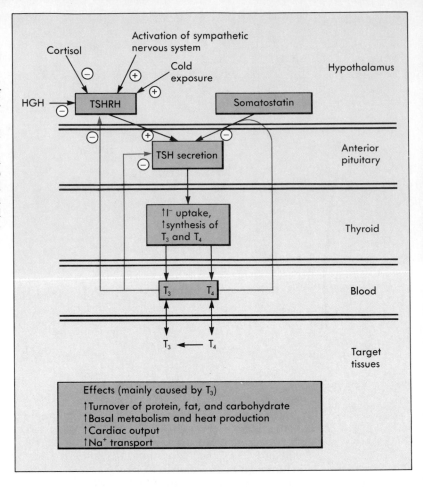

Regulation of Thyroid Function

The major stimulus for thyroid hormone secretion (Figure 23-24) is anterior pituitary **thyroid-stimulating hormone (TSH)** or **thyrotropin**. Secretion of TSH is in turn controlled by the hypothalamus, which produces **thyroid-stimulating hormone releasing hormone (TSHRH)** and somatostatin, which inhibits TSH release as well as hGH release. Normally, TSH secretion is very nearly constant, as is thyroid hormone release. Activation of the sympathetic branch of the autonomic nervous system directly stimulates the thyroid gland. hGH and cortisol inhibit TSH secretion, whereas I⁻, which is required for synthesis of thyroid hormone, directly inhibits growth of the thyroid gland. Thyroid hormone has a negative feedback effect on TSHRH release and TSH secretion.

An example of the effect of interrupting this feedback loop is seen in dietary iodine deficiency. When there is not enough I⁻ to make normal amounts of T_4, the plasma levels of T_4 decline, causing an increase in TSH secretion. The additional TSH stimulates thyroid enlargement. The enlarged gland causes a characteristic swelling of the neck called a **goiter.**

cells. Thyroid hormone effects on lipid and carbohydrate metabolism include increasing glucose uptake, glycogenolysis, and fatty acid use.

Thyroid Disorders

Thyroid hormones are required for normal growth and development and could be viewed as tissue growth factors, especially for the central nervous system. Absence of thyroid hormones in early infancy causes a severe mental and growth retardation termed **cretinism.** Bone growth is retarded, especially that of long bones, and developmental milestones such as sitting, standing and walking are delayed.

In adulthood, deficits of thyroid hormone secretion **(hypothyroidism)** reduce the basal metabolic rate, cause fatigue and lethargy, and ultimately diminish physical and mental activity. Ability to maintain core body temperature in cold environments is impaired. Cardiac output is decreased, and a characteristic edema (myxedema) develops.

Hypothyroidism may be caused by failure of TSH secretion **(hypopituitary hypothyroidism)** or by damage to the thyroid itself **(primary hypothyroidism).** The two forms may be distinguished by measurement of plasma TSH levels, which will be depressed in hypopituitary hypothyroidism, but elevated above normal in primary hypothyroidism because of the lack of negative feedback of T_4 and T_3 on the anterior pituitary and hypothalamus. One possible cause of primary hypothyroidism is iodine deficiency, which impairs hormone secretion even though the gland itself is functional. The high level

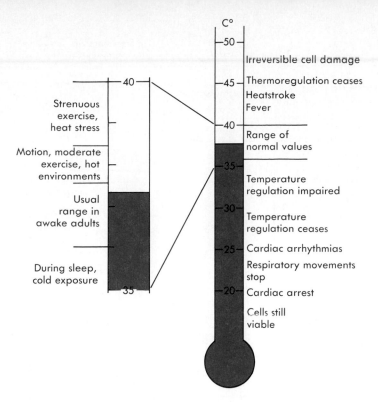

C°

50

Irreversible cell damage

45 — Thermoregulation ceases
Heatstroke
Fever

40 — Range of
normal values

35 — Temperature
regulation impaired

30 — Temperature
regulation ceases

25 — Cardiac arrhythmias

Respiratory movements
stop

20 — Cardiac arrest

Cells still
viable

40

Strenuous
exercise,
heat stress

Motion, moderate
exercise, hot
environments

Usual
range in
awake adults

During sleep,
cold exposure

35

FIGURE 23-25
The normal range of core body temperature, and effects of departures from this range.

of TSH in this condition causes visible enlargement of the thyroid, or **goiter. Pseudohypothyroidism,** in which hypothyroid symptoms occur but thyroid secretion is within normal limits, is the result of impaired conversion of T_4 to the more potent T_3 by target tissues. This condition occurs in a large percentage of hospital patients and can be treated by increasing dietary carbohydrate intake.

Excess thyroid hormone secretion **(hyperthyroidism)** pathologically increases the metabolic rate and makes patients hot, irritable and nervous. Weight loss occurs in spite of increased food consumption. Tachycardia and atrial arrhythmias may occur. There is a characteristic protruding condition of the eyeballs (exopthalmos). As with hypothyroidism, hyperthyroidism may be caused by a pituitary disorder or by a primary disorder of the thyroid. A common cause of hyperthyroidism is **Grave's disease**, in which an error in the immune system causes it to generate antibodies to the TSH receptor of thyroid tissue. These antibodies have a TSH-like effect on the thyroid.

Thermoregulation and Metabolic Adaptation to Cold

Body heat is produced as a consequence of metabolic reactions because some of the energy that is released when bonds are broken is liberated as heat. The more metabolically active a tissue is, the more heat it produces. Major heat-producing organs in the body are muscle and the visceral organs, such as the liver and kidney. The skin and respiratory surfaces are the major sites of heat exchange with the environment. When body heat production is greater than necessary to maintain a normal body temperature, blood flow to the skin increases transfer of heat to the external environment. Evaporative heat loss is the only mechanism for cooling the body when the ambient temperature is above body temperature.

The usual range (set point; see Chapter 5, p. 86) of values of the **core body temperature**—the temperature of the heart, brain, thoracic and abdominal organs, and blood in major central vessels—is between 36° C and 37.6° C in adults (Figure 23-25). Departures from this narrow range have life-threatening implications, especially when body temperature rises above 45° C. Body temperature typically falls to the low end of the normal range in sleep and cold exposure. It rises to the high end of the normal range in exercise and heat stress. Body temperature may rise as high as 43° C when the normal set point is elevated in fever. Control of body temperature involves two basic types of responses: (1) modulation of heat loss, and (2) modulation of heat production (Figure 23-26 and Table 23-8).

When the external temperature is well below body temperature, blood is shunted toward internal organs by constriction of arterioles of the skin and appendages **(peripheral vasoconstriction)**, and behaviors that decrease the area of skin available for heat exchange, such as folding the arms or hunching the shoulders, are elicited by the sensation of cold. **Shivering thermogenesis** results from activa-

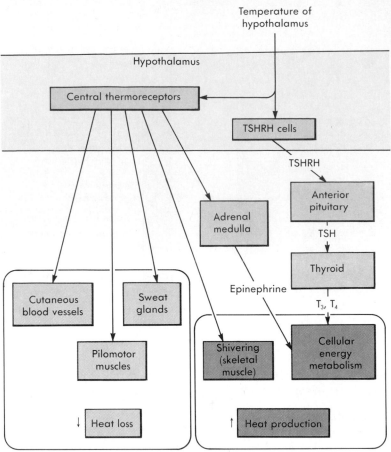

FIGURE 23-26

Pathways in central control of core body temperature by hypothalamic centers. The responses of the effectors to departures from the thermostatic set point are given in Table 23-8.

TABLE 23-8   *Summary of Centrally Mediated Thermoregulatory Responses*

	Hypothalamic temperature	
Effectors	*Above the set point*	*Below the set point*
Cutaneous arterioles	Dilated	Constricted
Sweat glands	Secreting sweat	No sweat
Pilomotor muscles	Relaxed	Contracted ("goose bumps")
Skeletal muscles		Shivering thermogenesis
Adrenal medulla		Epinephrine secretion increased
Thyroid		Thyroid hormone secretion increased

tion of a special pattern of repetitive, asynchronous contractions of skeletal muscle fibers, which requires synthesis and hydrolysis of ATP with its accompanying heat production. Although peripheral temperature receptors play a role in these responses, their input is coordinated with the receptors in the hypothalamus that are sensitive to the temperature of the blood, as evidenced by the fact that localized heating or cooling of specific regions of the hypothalamus elicits changes in blood distribution, sweating, or shivering.

The increased basal metabolic rate and heat production that result from long-term exposure to cold (**nonshivering thermogenesis**) are controlled by responses of the thyroid gland acting in conjunction with the sympathetic branch of the autonomic nervous system and epinephrine released from the adrenal medulla. Exposure to cold in many types of mammals stimulates TSHRH secretion via a neural input to the hypothalamus and this increases TSH and thyroid hormone release. In man, the thyroid is necessary for adaptive changes made in response to cold, but the circulating levels of the hormone do not become significantly elevated, which may simply reflect increased rates of uptake. Although thyroid hormone actions can increase the metabolic rate of cells, a primary contribution of thyroid hormones to nonshivering thermogenesis is to stimulate synthesis of increased numbers of receptors for epinephrine on the target tissue cells (especially of the liver, pancreas, and muscle). These tissues then increase their metabolic rate under the control of epinephrine.

T_4 and T_3 may also increase the basal metabolism of cells by increasing the permeability of the cell membrane to Na^+, which imposes increased demands on ATP production to drive the Na^+-K^+ pump, and by increasing the rates of **futile cycles**, in which storage molecules are repeatedly synthesized and degraded in energetically wasteful but heat-producing reactions. The result is that the basal metabolic rate of the body is elevated, and individuals who have adapted in this way are able to survive cold exposure that might kill unadapted individuals.

Human infants have a very limited capacity to regulate body temperature by either sweating or shivering, but they do have a mechanism for producing heat that is lost after the first year or so of life (although this mechanism is present throughout life in many mammals). **Brown fat** is a special, metabolically active tissue located in the neck, chest, and between the scapular bones in the human infant. The location of brown fat allows infants to channel heat to the thorax and to blood in the arteries that serve the brain. The color of brown fat results from the many mitochondria and numerous blood vessels present in the fat cells, which reflect the high metabolic capabilities of this tissue as compared with white adipose cells. Activation of brown fat by sympathetic inputs elicits heat production through pathways that allow the process of oxidative metabolism to be largely uncoupled from phosphorylation. The result is that stored lipids are catabolized, but very little useful work (ATP production) is performed, so the sole purpose served by these fat deposits is to rapidly generate heat.

SUMMARY

1. Hormones secreted by the **pancreas** regulate blood glucose; **insulin** promotes uptake of glucose when the supply in the blood is high; **glucagon** increases the level of glucose in the blood by promoting **glycogenolysis, gluconeogenesis,** and **glucose sparing.**

2. **Diabetes mellitus** produces the postabsorptive state in the presence of excess glucose because either insulin secretion is insufficient or absent **(type I diabetes),** or the body does not have enough insulin receptors to allow insulin in the blood to have its normal effects on uptake and synthesis **(type II diabetes).**

3. The **postabsorptive state** is defined by the drop in glucose and therefore in the insulin concentration in the blood. This drop in insulin removes catabolic pathway inhibition, and glucagon secretion stimulates catabolic pathways.

4. In starvation, **epinephrine** promotes the postabsorptive pattern of energy use, **cortisol** (a **glucocorticoid** from the adrenal cortex) is essential for the action of epinephrine and glucagon, and hGH protects muscle protein by mobilizing fatty acids and promoting their use as fuel.

5. Stimulation of catabolic pathways begins even before food is ingested or absorbed. The resulting increase in blood levels of glucose, bulk aspects of the ingested food, and neural and hormonal signals (CCK and insulin) all produce **satiety.**

6. Long-term regulation of body weight requires that intake be matched to nutrient use. The amount of glucose available to body cells, signals arising from fat cells, and heat production associated with metabolism of fats, carbohydrates, and proteins may all contribute to regulation of intake.

7. The relative proportions of fat, muscle, and ossified skeleton that make up the body weight vary over the life cycle, with muscle and skeletal weight peaking around age 35 and fat increasing as a proportion of body weight as the muscle and skeletal weight decline. Substitution of less metabolically active fats for muscle decreases overall metabolic rate.

8. The **thyroid hormones T_4 and T_3** set the body's **basal metabolic rate,** which can be increased in response to cold. Thyroid hormones and activation of the sympathetic nervous system promote **non-shivering thermogenesis** and adaptation to cold. In infants, **brown fat** is a major site of metabolic heat production. Other thermoregulatory responses include peripheral vasoconstriction and **shivering thermogenesis** by involuntary contraction of skeletal muscles.

1. Where are lipids stored in the body? Carbohydrates? Where is most of the protein in the body located?

2. What molecules supply most of the energy for body cells during the absorptive state? During the postabsorptive state?

3. Carbohydrates and amino acids can be transformed into lipids by liver cells. What happens to the lipids then?

4. What are the effects of insulin on carbohydrate, protein, and lipid metabolism in (a) muscle, (b) adipose cells, and (c) liver cells?

5. What factors increase insulin secretion?

6. What hormones oppose the effects of insulin? What signals lead to elevation of both insulin and glucagon levels after a protein-rich meal?

7. What factors are thought to contribute to the sensation of satiety? Where are the feeding and satiety centers located in the brain?

8. Why is a low ratio of HDL to LDL associated with elevated risk of developing atherosclerosis?

9. Why are women more likely to suffer from osteoporosis than are men? What factors are known to retard the progress of osteoporosis?

10. Describe the major effects of (a) hGH and (b) epinephrine.

11. What hormones promote cold adaptation, and how do they interact to effect changes? How is the response of an infant different from that of an adult? What metabolic responses can be made to chronic cold exposure, and what hormones are involved?

12. A starving person uses up the body's supply of nutrients in what order? What role is played by (a) epinephrine, (b) cortisol, and (c) hGH during starvation?

Choose the MOST CORRECT Answer.

13. Select the catabolic process.
    a. Glyconeogenesis
    b. Glycogenolysis
    c. Glycogenesis
    d. Lipogenesis

14. Insulin is secreted by:
    a. Adrenal cortex
    b. Alpha cells of pancreatic islets of Langerhans
    c. Beta cells of pancreatic islets of Langerhans
    d. Delta cells of pancreatic islets of Langerhans

15. Which of these factors results in a decrease of insulin secretion?
    a. Hyperglycemia
    b. Increase in plasma amino acids
    c. Parasympathetic stimulation of pancreas
    d. Presence of somatostatin

16. The major hormone of the postabsorptive state is:
    a. Glucagon
    b. Cholecystokinin
    c. Epinephrine
    d. Cortisol

17. The hormone needed for cartilage formation and for ossification and bone growth in children is.
    a. ACTH
    b. Thyroxine
    c. Somatostatin
    d. hGH

18. If the core body temperature drops below the set point, all of the following responses will occur EXCEPT:
    a. Constriction of cutaneous blood vessels
    b. Epinephrine secretion decreases
    c. Thyroid hormone secretion increases
    d. Contraction of pilomotor muscles

19. Which of the following is NOT a function of epinephrine?
    a. Stimulates lipolysis in adipose tissue
    b. Increases lactic acid uptake by the liver
    c. Promotes glycogenolysis to elevate plasma glucose levels
    d. Increases insulin release

20. This disorder is an error in the immune system causing it to generate antibodies to the TSH receptor of thyroid tissue:
    a. Grave's disease
    b. Cretinism
    c. Goiter
    d. Acromegaly

• SUGGESTED READINGS

Adiposity, fat distribution and cardiovascular risk, *American Family Physician*, 41(3):962, 1990. Shows how distribution of body fat is an independent predictor of cardiovascular morbidity and mortality.

ALTSCHUL AM: *Weight Control, A Guide For Counselors and Therapists*, New York, 1987, Praeger Publishers. Practical manual for an integrated program of weight reduction and control.

BATES SR, GANGLOFF EC: *Atherogenesis and Aging*, New York, 1987, Springer-Verlag New York, Inc. Discusses the changes in the handling of dietary fat during aging and the implications for the formation of plaques in blood vessels.

BLOCH GB: Fat master, *Discover*, 11(4):30, 1990. Describes how a fat-regulating protein links body fat to the operation of the immune system.

COOPER GR, MYERS GL, SMITH SJ, SCHLANT RC: Blood lipid measurements, *Journal of the American Medical Association* 267(12):1652, 1992. Describes the causes of variation in plasma lipid and lipoprotein levels and their effects on the risk of coronary disease.

Effect of fatty acids on serum cholesterol, *American Family Physician*, 1990 41(5):1600, 1990. Describes the physiological aspects of the presence of unsaturated fatty acids in human nutrition.

FORBES JM, HERVEY GR: The Control Of Body Fat Content: International Monographs in Nutrition, Metabolism and Obesity 1, London, 1990, Smith-Gordon & Co.

FRISCH RE: Fatness and Fertility, *Scientific American*, 258(3):88, 1988. Argues that a woman must store a threshold minimum of fat to reproduce and how gaining or losing just a few pounds can dramatically affect fertility.

LAWN RM: Lipoprotein(a) in heart disease, *Scientific American* 266(6):54, 1992. Lipoprotein(a) transports cholesterol, binds with blood clots and can increase the risk of a heart attack.

LE MAHO Y, ROBIN J-P, CHEREL Y: Starvation as a treatment for obesity: The need to conserve body protein, *News in Physiological Sciences* 3:21-24, 1988. Deals with some hazards of starvation as a means of weight loss.

YATANI A, BIRNBAUMER L, BROWN AM: Direct coupling of the somatostatin receptor to potassium channels by a G protein, *Metabolism*, 39(9): 91, 1990. Describes G protein-coupled second messenger systems in cells where somatostatin inhibits and the general mechanism of metabolic regulation by this hormone.

CHAPTER 24

Reproduction and Its Endocrine Control

On completing this chapter you should be able to:

- Describe the anatomy of male and female reproductive systems.
- Describe how gonadal sex is determined by the sex chromosome complement, and how the presence of male or female gonads determines the course of development of male or female internal accessory structures and genital sex.
- Understand the genetic and endocrine causes of some representative disorders of sexual development.
- Trace the formation of male and female gametes, and know the similarities and differences between this process in male and female gonads.
- Describe changes characteristic of male and female puberty and their endocrine basis.
- Describe the endocrine feedback loops that control gonadal function in men and women.
- Trace the events of the menstrual cycle and explain the roles of the hypothalamus, anterior pituitary, ovary, and corpus luteum in control of the cycle.

*F*rom the moment of conception to maturity as a man or woman, sexual development is a road with many branches. The first branch occurs at conception—one of the father's contributions to his children are genes that determine their sex. A specific signal from the genes causes the reproductive organs of male fetuses to develop along the male pattern before birth. Normally this signal is not present in female fetuses, and female reproductive organs develop automatically in its absence.

As early as 6 months after birth, the child begins to develop a sense of identity as a girl or boy. The development of an appropriate sexual identity is highly dependent upon interactions with the adults who care for the child. Ten to fourteen years after birth, hormones secreted by the brain cause reproductive capacity to develop, and the physical characteristics that distinguish mature men and women begin to emerge. Before birth, during childhood and at puberty, the individual has the potential to develop in either the female or male direction.

The ultimate consequence of sexual development is production of reproductive cells that will pass genes to the next generation. Each individual produces an enormous number of reproductive cells. Of the stock of several million reproductive cells each female possesses at birth, only several hundred will develop to the point where they have a chance to be fertilized—the rest will degenerate at various stages. In males, reproductive cells are formed continuously after sexual maturity is reached, and each ejaculation releases perhaps a hundred million. Each reproductive cell produced by an individual carries a unique assortment of his or her genes. As a result, each child possesses one of a very large number of different possible combinations of the genes of the parents.

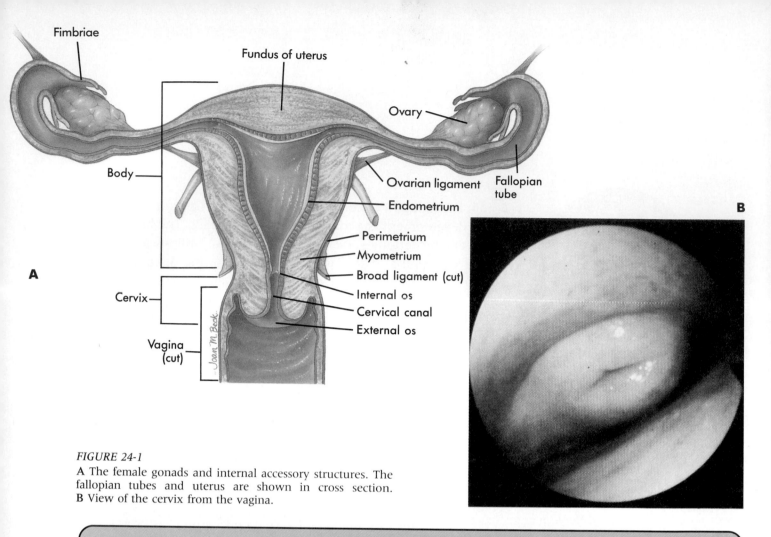

A **B**

FIGURE 24-1
A The female gonads and internal accessory structures. The fallopian tubes and uterus are shown in cross section. B View of the cervix from the vagina.

	Male	Female
TABLE 24-1	**The Male and Female Reproductive System**	
Gonads	Testes	Ovaries
Gonadal steroids	Testosterone	Estrogen (mainly estradiol), proges- terone
Accessory structures	Penis	Fallopian tubes
	Scrotum	Uterus
	Duct system composed of	Vagina
	Rete testis	Labia majora
	Efferent ductules	Labia minora
	Epididymis (2)	Clitoris
	Vas deferens (2)	
	Ejaculatory duct	
	Penile urethra	
	Glands include	
	Prostate	
	Seminal vesicles (2)	
	Cowper's glands (2)	
Secondary sex characteristics	Male pattern of body, pubic, and facial hair; larger larynx (deeper voice); greater bone mass and growth of long bones; narrow pel- vis; greater ratio of muscle mass to total body weight; higher he- matocrit level; male behavior (?)	Female pattern of body and pubic hair; wider pelvis; greater ratio of fat mass to total body weight; female pattern of fat deposition; female behavior (?)

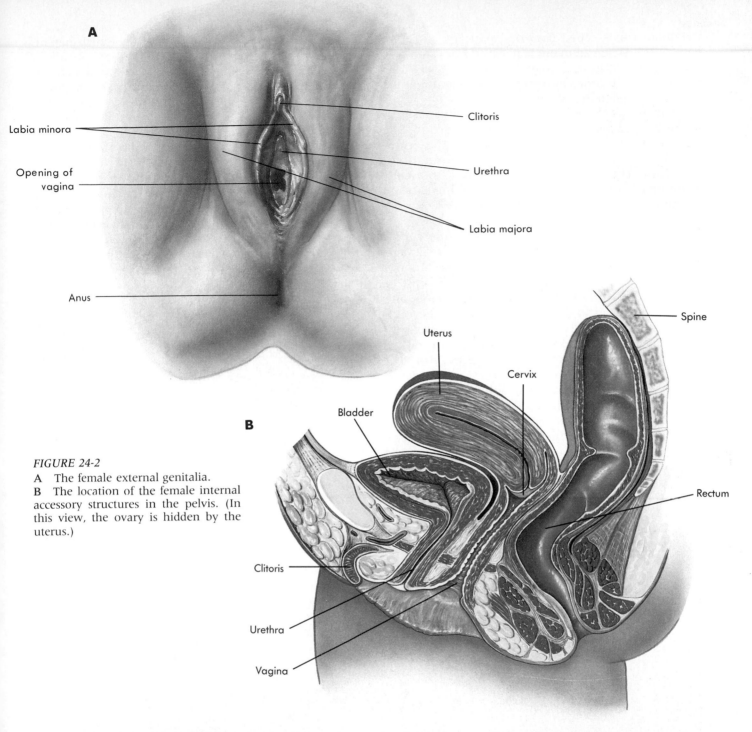

A Labia minora

A Opening of vagina

A Anus

Clitoris

Urethra

Labia majora

Uterus

Cervix

Spine

Bladder

B

Rectum

Clitoris

Urethra

Vagina

FIGURE 24-2
A   The female external genitalia.
B   The location of the female internal accessory structures in the pelvis. (In this view, the ovary is hidden by the uterus.)

THE ANATOMY OF MALE AND FEMALE REPRODUCTIVE ORGANS
Female Reproductive System

- *What are the female gonads? Where are they located?*
- *What is the function of each of the female accessory organs?*

In both sexes, the reproductive systems, or **genitalia,** consist of the primary reproductive organs (**gonads**) that produce reproductive cells and secrete sex hormones and **accessory structures** that support the function of the gonads.

In females, the gonads are **ovaries** (Figure 24-1; Table 24-1). Accessory structures are the **Fallopian tubes** or **oviducts**, the **uterus**, the **vagina**, the labia majora and minora, and the clitoris (Table 24-1). Ovaries are located within the pelvic cavity, where they are protected by the bony pelvic girdle. Each ovary weighs about 15 g and is attached to the uterus by ligaments. A Fallopian tube, or oviduct, 10 to 12 cm long, is associated with each ovary and provides a route for the female reproductive cells (**ova**; one is an **ovum**) to travel to the uterus. Although the Fallopian tubes open into the uterus, they do not enclose the ovaries. Instead, finger-like processes of the

Fallopian tubes called **fimbriae** partially surround each ovary. When an ovum is released from an ovary, it enters the peritoneal cavity and must be caught by the fimbriae in order to begin its passage down the Fallopian tube to the uterus.

The uterus (Figure 24-1, *A*) is a highly muscular organ located between the bladder and rectum. The wall of the uterus consists of a lining, the **endometrium,** and a coat of smooth muscle, the **myometrium.** The space within the uterus is normally quite small, but the uterus enlarges and increases in strength during the development of a baby, or **fetus.** A narrow passage, the cervical **canal,** leads from the interior of the uterus through the **cervix** into the vagina (Figure 24-1, *B*). The external vaginal opening is posterior to the urethra and anterior to the rectum. It is enclosed by two pairs of "lips" or labia which surround and protect the vaginal and urethral openings and functionally extend the vaginal canal, which leads to the uterus. The inner mucus-membrane covered folds, the **labia minora,** are surrounded by the fleshy outer **labia majora.** The region circumscribed by the labia minora is called the **vestibule of the vagina,** although it also includes the opening of the urethra (Figure 24-2).

Access from the exterior of the body into the pelvic cavity is gained by way of the vagina, the cervical canal, the uterus and the Fallopian tubes. The body cavity is normally protected from the entry of foreign material by the narrowness of the cervical canal, the mucus which is present in the canal, and the acidity of the female reproductive tract. However, this access to the body interior, which has no parallel in the anatomy of the male, can result in dangerous pelvic inflammatory infections which can be life-threatening or produce sterility in females.

The structures that surround the opening of the vagina are called the **external genitalia.** These include the clitoris, the labia minora and the labia majora. The **clitoris** serves as a sensory focus in the sexual response. It is positioned anteriorly within the vestibule, beneath the anterior hood-like extension of the labia minora (Figure 24-2, *A*). The breasts technically do not belong to the reproductive system, but are sexually dimorphic structures supported by the female sex hormones which play a reproductive role in nourishment of the young.

Male Reproductive System

- *What are the functions of the three major categories of testicular cells?*
- *What and where are the erectile tissues in the male system?*
- *What are the components of the male gland and duct system?*

The male gonads, the **testes** (Figure 24-3) produce male reproductive cells **(spermatozoa** or **sperm cells).** Within the testes, spermatozoa are produced in **seminiferous tubules** (Figure 24-4), which make up 80% of the testicular mass. In the seminiferous tubules, cells called spermatogonia divide and

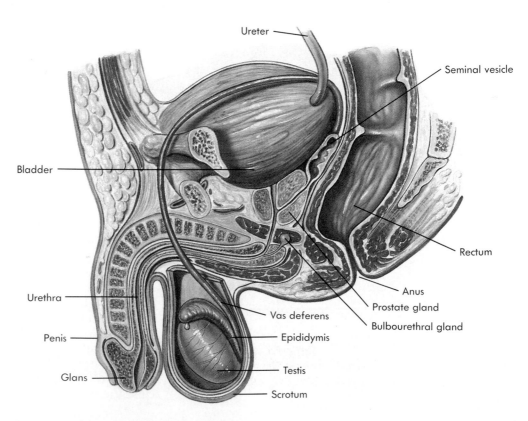

FIGURE 24-3
The testes and male accessory reproductive structures.

GROWTH, METABOLISM, REPRODUCTION AND IMMUNE DEFENSE

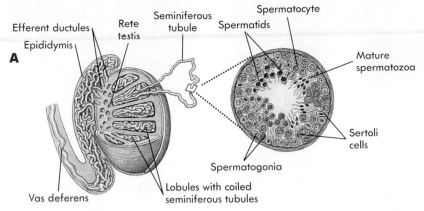

A Efferent ductules — Epididymis — Rete testis — Seminiferous tubule — Spermatids — Spermatocyte — Mature spermatozoa — Sertoli cells — Spermatogonia — Lobules with coiled seminiferous tubules — Vas deferens

FIGURE 24-4

A *(Left)* Testis with epididymis and vas deferens, showing arrangement of the rete testis and seminiferous tubules.

(Right) Cross section of a single seminiferous tubule showing spermatogonia, Sertoli cells, and cells in various stages of meiosis and spermiogenesis.

B Electron micrograph of a portion of seminiferous tubule showing two spermatogonia.

C Electron micrograph of seminiferous tubule showing four spermatids surrounded by Sertoli cells. The four spermatids outlined in red probably arose from a single spermatogonium. Secondary spermatocytes are adjacent.

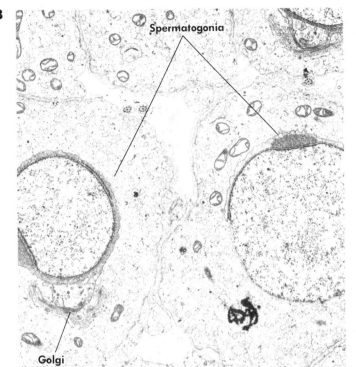

B Spermatogonia — Golgi

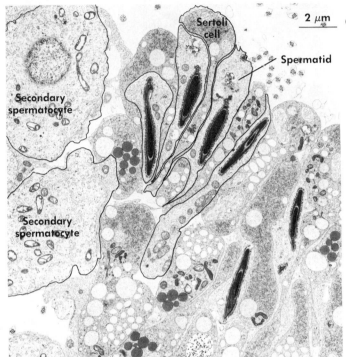

C Sertoli cell — Spermatid — Secondary spermatocyte — Secondary spermatocyte — 2 μm

give rise to **spermatozoa. Sertoli cells** nurture the developing spermatozoa. **Leydig cells** (interstitial cells) located between the seminiferous tubules secrete the male sex hormone **testosterone.**

The male accessory structures (Table 24-1 and Figures 24-3 and 24-4, *A*) include a system of ducts and glands and the **penis,** which introduces spermatozoa into the vagina of the female. Starting from the seminiferous tubules, the ducts are: the **rete testis, efferent ductules** (both of these structures are within the testes as shown in Figure 24-4), the **epididymis** (containing the convoluted duct of the epididymis), the **vas deferens,** and the **ejaculatory duct,** which leads to the **urethra.** Of these, all but the last are paired. The penis contains three vascular sinuses (Figure 24-5): the two **corpora cavernosa** and the **corpus spongiosum** through which the urethra passes. Together these are referred to as the **erectile tissues** because they become filled with blood during erection of the penis.

The glands of the male system (Figure 24-3) include the two **seminal vesicles,** the **prostate gland** and **Cowper's glands** (also called **bulbourethral glands**). The secretions of these glands combine with the spermatozoa to produce a mixture called **semen.** The spermatozoa make up less than one-tenth of the total volume of semen. The bulk of the semen is contributed by the seminal vesicles, with smaller contributions from the prostate and bulbourethral glands.

In the semen the sugar fructose supplies energy for motility, bicarbonate buffers vaginal acidity, prostaglandins stimulate contractions of the female reproductive tract, and clotting factors promote temporary retention of the semen within the vagina. These constituents contribute to the viability of spermatozoa in the vagina and increase the probability of fertilization.

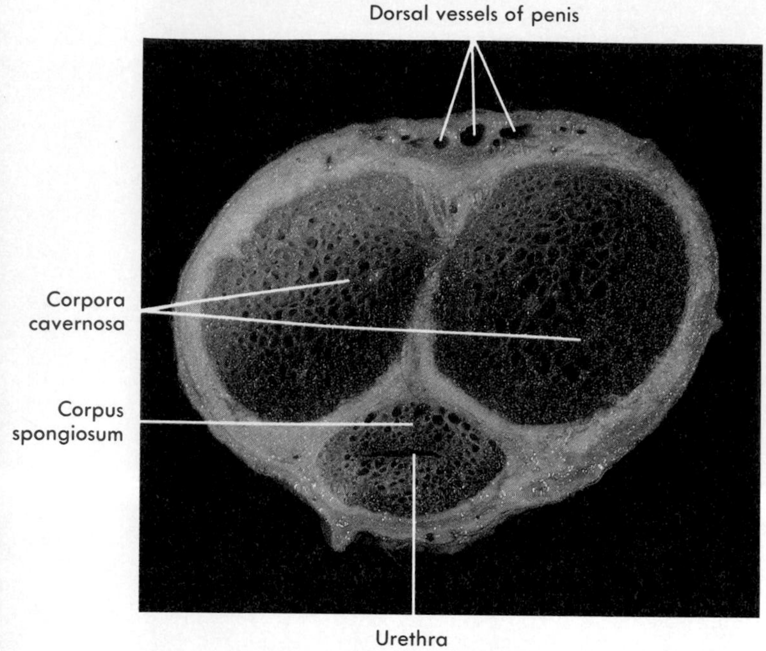

Dorsal vessels of penis

Corpora cavernosa

Corpus spongiosum

Urethra

FIGURE 24-5

Cross section of penis showing erectile tissue (corpora cavernosa and corpus spongiosum), urethra, and dorsal vein and arteries.

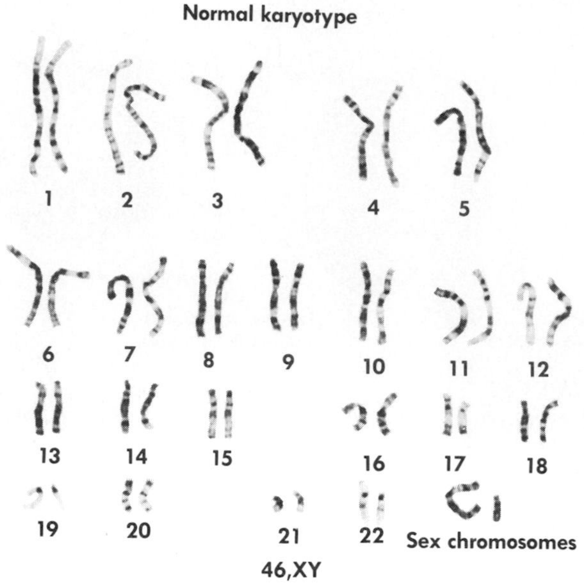

Normal karyotype

46,XY

Sex chromosomes

FIGURE 24-6

Chromosomes of a cell as they appear during metaphase of mitosis. The chromosomes are cut out of a photomicrograph of the cell nucleus and arranged in homologous pairs to form a karyotype. The sex chromosome complement is XY.

THE GENETIC BASIS OF REPRODUCTION AND SEX DETERMINATION
Meiosis: Formation of Haploid Gametes

- *What is meant by the diploid chromosomal complement? What are homologous chromosomes? What is a bivalent?*
- *What are differences and similarities between spermatogenesis and oogenesis?*

Human cell nuclei contain a total of 46 chromosomes, including 22 **homologous pairs** of **autosomal** (or **somatic**) **chromosomes** and a homologous pair of **sex chromosomes** (Figure 24-6). One chromosome of each homologous pair is contributed by each parent. Except for the sex chromosomes, the members of a pair of homologous chromosomes are of identical size and appearance under the microscope and carry corresponding genes. Each pair of sex chromosomes is normally composed of either an **X chromosome** and a **Y chromosome** (the normal situation in males) or two X chromosomes (the normal situation in females).

The total of 46 chromosomes represents the combination of 23 chromosomes contributed by each parent (N) and is referred to as the **diploid** or **2N** complement. In mitotic cell division that occurs in growth and repair (see Chapter 3), each chromosome replicates itself, and each daughter cell receives a set of chromosomes that, except in occa-

sional accidents, is identical to that of the mother cell.

Meiosis is the form of cell division that produces **gametes** (reproductive cells) in the ovary and testis. Meiosis results in cells with the **haploid** chromosomal complement (N). Meiosis involves two successive cell divisions, called the **first** and **second meiotic divisions** and only one chromosome duplication event. At the completion of meiosis, each of the four resulting gametes has one representative of each of the homologous pairs of autosomal chromosomes, and one of the sex chromosomes. Combination of the nuclei of two gametes in fertilization restores the diploid complement of chromosomes in the fertilized ovum (a **zygote**).

On the chromosomal level, the events of meiosis are similar in the ovary and testis. To simplify the explanation, the process of **spermatogenesis** in the testis will be followed first (Figure 24-7), and then the important differences between spermatogenesis and oogenesis will be noted.

In the testes of sexually mature males, spermatogonia continuously undergo mitotic divisions. Of the two cells produced by each mitotic division, one will remain a spermatogonium, and the other will begin a cell line that will give rise to 16 spermatozoa. In this way, the population of spermatogonia is maintained so that production of spermatozoa can continue throughout adulthood.

The cell line that is ultimately to produce sper-

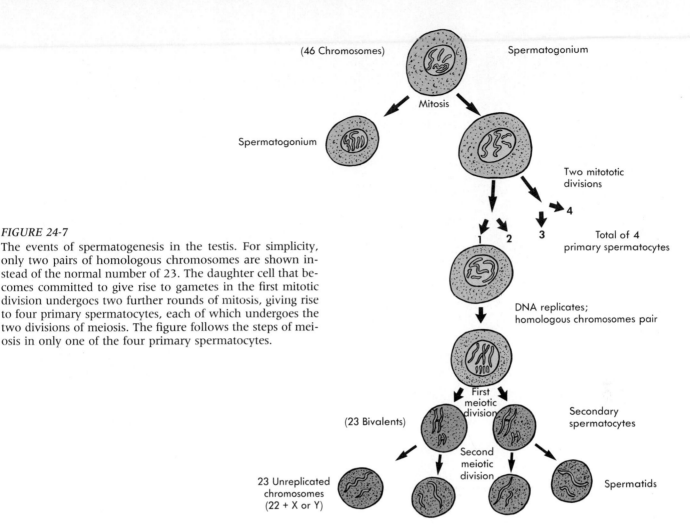

(46 Chromosomes)   Spermatogonium

Mitosis

Spermatogonium

Two mitototic
divisions

4

1   2   3   Total of 4
primary spermatocytes

DNA replicates;
homologous chromosomes pair

First
meiotic
division

(23 Bivalents)   Secondary
spermatocytes

Second
meiotic
division

23 Unreplicated
chromosomes
(22 + X or Y)   Spermatids

FIGURE 24-7
The events of spermatogenesis in the testis. For simplicity, only two pairs of homologous chromosomes are shown instead of the normal number of 23. The daughter cell that becomes committed to give rise to gametes in the first mitotic division undergoes two further rounds of mitosis, giving rise to four primary spermatocytes, each of which undergoes the two divisions of meiosis. The figure follows the steps of meiosis in only one of the four primary spermatocytes.

matozoa first goes through two cycles of mitosis (Figure 24-7), resulting in four **primary spermatocytes,** each of which will undergo meiosis. The primary spermatocytes, like spermatogonia, are diploid cells. Prior to the **first meiotic division,** the chromosomes of each primary spermatocyte replicate. Each original chromosome and its copy remain attached to one another, forming a **bivalent.** At this point each primary spermatocyte contains 23 homologous pairs of bivalents. As primary spermatocytes undergo the first meiotic division, each daughter cell (a **secondary spermatocyte**) receives one bivalent from each of the homologous pairs. Each of the eight secondary spermatocytes derived from a single spermatogonium possesses two copies of one of each of the original 23 pairs of homologous chromosomes.

The two secondary spermatocytes resulting from the first meiotic division are not genetically identical. Even though homologous chromosomes look identical under the microscope and carry the same genes, they are not identical because they carry different **alleles,** alternative forms of the gene. These alternative states of a gene provide for the enormous diversity of structure and function displayed by the

members of a species. The chromosomes from the individual's two parents come together to produce a unique blend of features and then are segregated in the gametes to provide different contributions to the formation of a new individual.

Because bivalents are distributed randomly between the two daughter cells, the chromosomes derived from one's parents may be assorted into a large number of new combinations. The primary spermatocytes of an individual can give rise to more than 4 million different combinations of bivalents in the secondary spermatocytes. Still more variation in the possible genetic makeup of secondary spermatocytes is introduced by a process called **crossing over,** in which homologous bivalents trade pieces of each other while they are aligned during the first meiotic division (Figure 24-7).

Each of the two secondary spermatocytes resulting from the first meiotic division of a primary spermatocyte undergoes a **second meiotic division** (Figure 24-7). In the second meiotic division, each bivalent splits, and each of the two daughter cells **(spermatids)** receives one of the replicates. Thus each spermatid has one copy of one chromosome from each of the pairs of homologous chromosomes

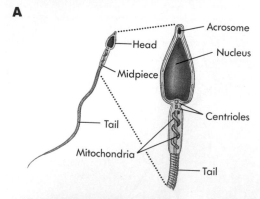

A

Acrosome — Head — Midpiece — Tail — Mitochondria — Nucleus — Centrioles — Tail

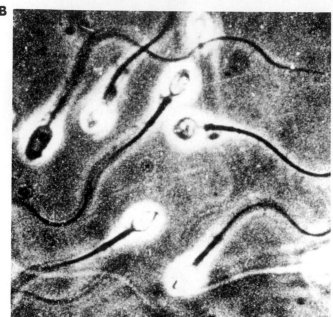

B

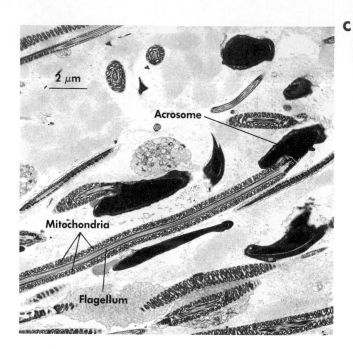

C

2 μm

Acrosome

Mitochondria

Flagellum

FIGURE 24-8
A The structure of a spermatozoan.
B Light micrograph of human spermatozoa after ejaculation. The heads are glowing because they have been labeled with a fluorescent antibody.
C Electron micrograph of sectioned rat spermatozoa. Parts of heads, midpieces with mitochondria, and flagella are visible.

that were present in the secondary spermatocyte. Each spermatid has 23 chromosomes, the haploid number. Because there were two cycles of meiotic division, each of the four primary spermatocytes was the ancestor of two secondary spermatocytes and four spermatids. Thus the cell line established by the original mitotic division of a single spermatogonium terminates in 16 spermatids.

Spermiogenesis

The spermatids cannot function as reproductive cells because they lack the cellular equipment needed to reach and fertilize an ovum. In a process called **spermiogenesis** or **spermiation,** spermatids acquire the specialized structures of spermatozoa (Figure 24-8, *A* and *B*).

In spermiogenesis, most of the cytoplasm is discarded, leaving a sperm head 2 to 3 μm in diameter and about 5 μm long, which consists of a nucleus surrounded closely by plasma membrane. Each spermatid develops a **flagellum** (or "tail") for locomotion, a **midpiece** containing mitochondria to provide the ATP to drive the flagellum, and an **acrosome** on the head (Figure 24-8, *A* and *C*). The acrosome is located beneath the plasma membrane

and contains enzymes needed for penetration of an ovum, including hyaluronidase, protease, and corona-penetrating enzyme.

Most cells depend on glucose for their energy source but spermatozoa depend on the fructose secreted by the seminal vesicles. The average velocity of locomotion of spermatozoa is about 50 microns/second (4 mm/min). Ejaculated spermatozoa cannot fertilize an ovum until they have been **capacitated,** a process involving undefined chemical factors in the female reproductive tract, which endow the sperm with the ability to fertilize an ovum (see below). The need for spermatozoa to be capacitated was a problem in the development of in vitro fertilization (see p. 748).

Oogenesis and Follicle Development

Oogenesis and spermatogenesis differ in the timing of the processes in the life cycle, in the fates of daughter cells, and in the differentiation of the cytoplasmic parts of the cells (Table 24-2). Spermatogenesis begins at puberty and continues until old age. In the ovaries all oogonia are converted to primary oocytes before birth. The first meiotic division of primary oocytes begins during embryonic develop-

TABLE 24-2   *Comparison of Oogenesis and Spermatogenesis*

	Oogenesis	*Spermatogenesis*
Stem cells	Oogonia	Spermatogonia
Timing of meiosis	Arrested at first meiotic metaphase before birth; resumed at ovulation after puberty; second meiotic division occurs after fertilization; process ceases at menopause	Begins at puberty and continues until death
Fates of daughter cells	Only primary oocytes remain at birth; each primary oocyte gives rise to one ovum; the other daughter cells (polar bodies) degenerate	Each spermatogonium gives rise to 16 spermatids that all become spermatozoa
Cytoplasmic division	Each oocyte receives most of the cytoplasm; polar bodies get chromosomes but almost no cytoplasm	Cytoplasm is divided equally between daughter cells; almost all cytoplasm is shed during spermiogenesis

ment but is arrested at metaphase (Figure 24-9; see Chapter 3 for stages of cell division). At birth of a female each ovary contains about 2 million **primordial follicles,** each consisting of a primary oocyte surrounded by a single layer of supporting granulosa cells. Each primary oocyte contains 46 homologous pairs of bivalents.

Throughout life, follicles are lost by a degenerative process called **atresia.** About 400,000 follicles remain at the time of puberty. During puberty, cycles of **ovulation** begin. In each cycle, many follicles begin to mature, and some of them become **primary follicles,** contributing hormones that promote ovulation. Usually only one follicle completes maturation and releases its oocyte. By the time of **menopause,** the cessation of ovarian function that typically occurs in the 5th decade of life, the stock of oocytes is so reduced that the cycles of ovulation cannot be maintained. In a woman's reproductive life, only a few hundred of the original number of follicles are actually ovulated; the rest degenerate at one stage or another.

The first meiotic division of the primary oocyte (Figure 24-9) is stimulated to occur around the time of ovulation. The division of cytoplasm is not equal. One of the two daughter cells retains almost all of the cytoplasm, becoming a **secondary oocyte.** The other daughter cell, a tiny **first polar body** with one set of chromosomes, has no future but may divide before degenerating. If the secondary oocyte is

fertilized, it undergoes the second meiotic division (Figure 24-9). As in the first meiotic division, the chromosomes are divided equally but the cytoplasm is divided unequally, leaving most cytoplasm with

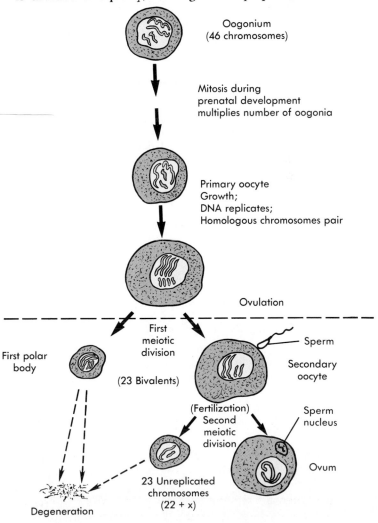

FIGURE 24-9
The events of oogenesis in the ovary. Because of the unequal divisions of cytoplasm in ovarian meiosis, each primary oocyte gives rise to a single ovum.

the large **ovum** and very little with the set of chromosomes in the **second polar body.** The relatively large cell that results from the unequal division of cytoplasm in ovarian meiosis is an adaptation that provides the ovum with a supply of nutrients and cellular organelles to distribute among the cells of the early embryo.

Determination of Genetic Sex

> • *What is the developmental basis for the claim that the basic human body plan is female?*

In spermatogenesis, the reduction of the chromosomes to the haploid number during the first meiotic division results in spermatozoa that have either an X or a Y chromosome. The autosomal chromosomes and the X chromosome carry genes that specify basic structural and functional aspects of the body common to both sexes—they do not normally carry genes that determine the course of sexual development. The Y chromosome is much smaller than the X chromosome (Figure 24-6) and carries only a few genes. The determination of maleness depends on only one-tenth of one percent of the DNA of the Y chromosome.

All of the gametes produced in oogenesis carry an X chromosome because the oogonia, like all the woman's cells, do not have Y chromosomes. The genetic sex of the zygote is determined by the spermatozoan. If it carries a Y chromosome, the sex chromosome complement of the zygote will be XY (Figure 24-6); if it carries an X, the zygote's complement will be XX. Classically, male genetic sex was defined as possession of a Y chromosome and absence of a Y chromosome was regarded as conferring female genetic sex. Recent studies showed that a small part of the Y chromosome, called the SRY (sex-determining region Y) gene, is all that is needed to induce the differentiation of testes and male sexual behavior, and this small piece of DNA can (rarely) become attached to an X chromosome in XX embryos. The SRY gene consists of only about 14,000 base pairs. Its product acts as a switch that controls the action of many other genes on autosomal chromosomes.

FIGURE 24-10
Inheritance of sex chromosomes begins a sequence in which prenatal development differentiates appropriate gonads and genitalia. Rearing helps the child acquire an appropriate gender identity and sexual orientation. Disorders of sexual development can arise at any step of this sequence.

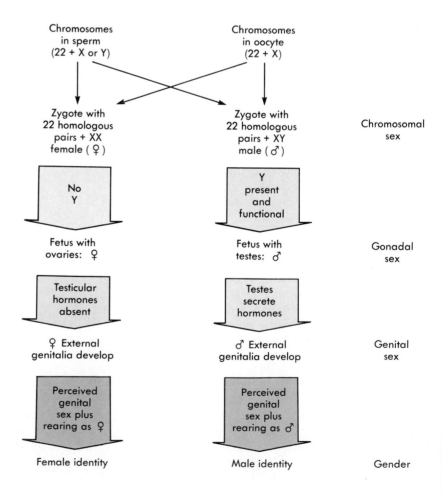

GROWTH, METABOLISM, REPRODUCTION AND IMMUNE DEFENSE

Embryonic Development of Reproductive Organs

- *What are the factors that determine a person's*
- *Genetic sex?*
- *Gonadal sex?*
- *Genital sex?*

The development of reproductive organs before birth can be divided into two steps (Figure 24-10). In the first step, the genetic sex of the fetus, determined by the presence or absence of the SRY gene, results in the development of the male or female gonads, establishing the **gonadal sex.** The second step is formation of appropriate male or female accessory organs, including both internal and external genitalia. Development of male structures depends upon the responses of the tissues to the hormones released by the testes in male fetuses. In the absence of hormones released by testes, the female structures develop. The development of appropriate external genitals establishes the **genital sex** of the fetus, the basis for the assignment of a child as male or female at birth.

An embryo has **indifferent gonads** that can become either testes or ovaries and a **primitive urogenital tract** that can give rise to either male or female accessory organs. Indifferent gonads contain three types of cells: (1) cells that will give rise to gametes (oogonia or spermatogonia), (2) precursors of cells that will nourish the developing gametes (granulosa cells in the ovary; Sertoli cells in the testis), and (3) precursors of cells that will produce sex hormones (thecal cells in the ovary; Leydig cells in the testis).

The primitive urogenital tract consists of two duct systems: the **Wolffian ducts,** which can become the male gland and duct system; and the **Müllerian ducts** (Figure 24-11), which can develop into female internal accessory structures. Male or female external genitalia develop from an external **genital tubercle, urogenital groove,** and **genital swellings** (Figure 24-12). Thus the gonads and accessory reproductive structures of the early embryo are **bipotential,** but unless the SRY gene product from the Y chromosome is expressed, the female potential is automatically seen in development.

The SRY gene causes the indifferent gonads to differentiate into testes at 6 to 8 weeks. The presence of testes establishes **male gonadal sex.** Shortly after the testes differentiate, the Leydig cells begin to secrete testosterone. Early production of testosterone stimulates the Wolffian ducts to differentiate into the male gland and duct system: epididymis, vas deferens, and seminal vesicle (Figures 24-3 and 24-11). The Sertoli cells of the testis differentiate between 9 and 10 weeks and secrete a protein called **Müllerian regression factor** (MRF) that suppresses development of the (female) structures from Müllerian ducts. Testosterone also inhibits differentiation of the external genitalia into female structures and stimulates the primitive urogenital tract to develop into the prostate gland and male external genitalia (Figures 24-11 and 24-12), establishing **male genital sex.**

If the cells of the indifferent gonads lack a Y chromosome, they differentiate into ovaries at 9 to 10 weeks after conception (Figure 24-11), establishing female gonadal sex. Little or no hormone secretion occurs from the ovaries, and the mother's estrogen and progesterone production is so high that a hormonal signal from the fetal ovaries would be overwhelmed in any case. The significant factor in female development is the absence of testosterone. At about 18 to 20 weeks the Müllerian ducts develop into Fallopian tubes, uterus, and the innermost one-fourth of the vagina. The Wolffian ducts spontaneously degenerate at 10 to 11 weeks, so that male accessory sex organs do not appear. The external genital tubercle differentiates into the clitoris, the genital swellings form the labia majora, and the the genital fold develops into the outer portion of the vagina and the labia minora, establishing **female genital sex** (Figure 24-12)

HORMONAL CONTROL OF SEXUAL MATURATION
Role of the Hypothalamus and Anterior Pituitary in Puberty

- *What causes the gonads to begin to function at puberty?*
- *What are the male and female secondary sexual characteristics? What causes them to develop in males? In females?*
- *What are the sources and roles of androgen in women?*

During the 10 to 12 years between infancy and the onset of sexual maturation, or **puberty,** the gonads produce no gametes and make almost no steroid hormones. The accessory sexual structures grow slowly during this period, although cases in which meat or milk contaminated with hormone-like substances have caused pre-pubescent children to develop precociously show that children can exhibit premature secondary sexual development.

The hypothalamus initiates the changes that lead to sexual maturity by increasing activity of neurons that release **gonadotropin releasing hormone** (GnRH). This GnRH causes the anterior pituitary to release the gonadotropins **luteinizing hormone** (LH) and **follicle-stimulating hormone** (FSH). These hormones were named for their effects in women, but they also control reproduction in men (Table 24-3).

The combined effect of the two gonadotrophic

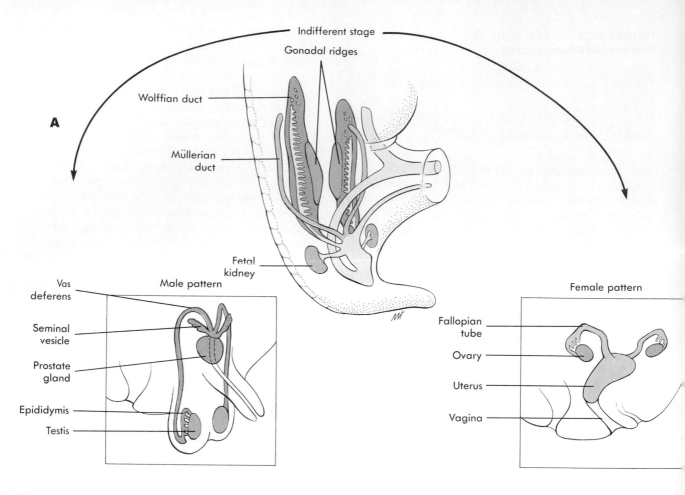

A

Indifferent stage

Gonadal ridges

Wolffian duct

Müllerian duct

Fetal kidney

Male pattern

Vas deferens

Seminal vesicle

Prostate gland

Epididymis

Testis

Female pattern

Fallopian tube

Ovary

Uterus

Vagina

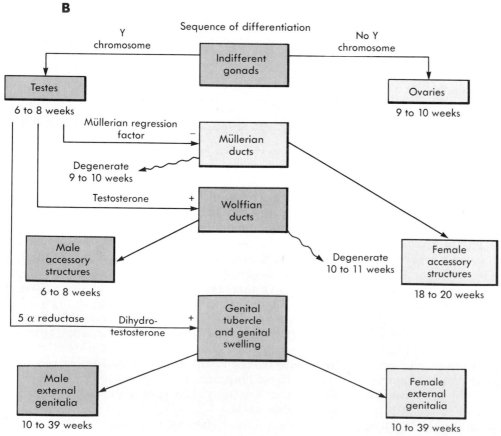

B

Sequence of differentiation

Y chromosome

No Y chromosome

Indifferent gonads

Testes
6 to 8 weeks

Ovaries
9 to 10 weeks

Müllerian regression factor

− Müllerian ducts

Degenerate 9 to 10 weeks

Testosterone + Wolffian ducts

Male accessory structures
6 to 8 weeks

Degenerate 10 to 11 weeks

Female accessory structures
18 to 20 weeks

5 α reductase

Dihydro-testosterone + Genital tubercle and genital swelling

Male external genitalia
10 to 39 weeks

Female external genitalia
10 to 39 weeks

FIGURE 24-11
A   The early embryo *(center)* is sexually bipotential. The embryonic gonads can become either ovaries (right) or testes *(left)*, depending on the presence or absence of a Y chromosome. In the case of male gonadal sex, secretion of testosterone causes the Wolffian ducts *(brown)* to develop into the male internal accessory structures *(left)*; Müllerian inhibiting factor secreted by the testes causes the Müllerian ducts *(blue)* to deteriorate. Testosterone also causes the external genitalia to assume the male form. In the absence of testes, the Wolffian ducts degenerate, the Müllerian ducts develop into the female internal accessory structures, and the external genitalia assume the female form.
B   The sequence and timing of the events of sexual differentiation.

GROWTH, METABOLISM, REPRODUCTION AND IMMUNE DEFENSE

FIGURE 24-12
The anatomical differentiation of external genitalia. The alternative fates of the bipotential embryonic structures are color-coded.

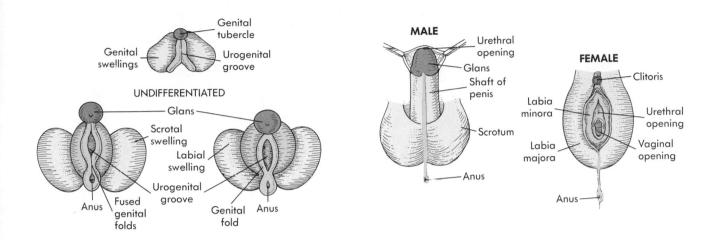

TABLE 24-3	*Effects of Gonadotropins*	
	Males	*Females*
Follicle-stimulating hormone (FSH)	Stimulates spermatogenesis and speriogenesis by stimulating Sertoli cells	Stimulates follicle maturation; is required for follicles to progress to the secondary stage and begin to secrete estradiol
Luteinizing hormone (LH)	Increases testosterone secretion by stimulating Leydig cells; LH is required for spermatogenesis because spermatogenesis requires testosterone	A surge of LH causes the mature Graafian follicle to ovulate and transform into a corpus luteum (luteinization)

hormones is to stimulate growth of the gonads to their adult size, to stimulate gametogenesis, and to initiate secretion of sex steroids. The gonadal steroid secretion initiates sex-specific changes in body growth and development of **secondary sexual characteristics,** the sex-specific body features that develop after puberty (see Table 24-1).

In women, release of gonadotropins occurs on a cyclical basis, driving the female **menstrual cycle.** In men, the average release is relatively steady over periods as long as several days, but there are a number of pulses of gonadotropin release during each day, and cycles of hormone fluctuation lasting days to weeks can also be detected in men.

The factors that affect the brain's timing of puberty are poorly understood. In girls the age at first menstruation **(menarche)** is an easily dated event related to pubertal development. Nutritional factors probably have some part in initiating puberty, because evidence suggests that a threshold body weight or a certain proportion of body fat must be reached for puberty to occur. The mean age of menarche has decreased continuously since the beginning of the last century in populations of European origin (Figure 24-13). The causes of this trend are not clear, but changes in nutrition and other cultural influences, such as artificial lighting may be involved.

Gonadal Steroids and Sexual Maturation

Testosterone is the major sex steroid secreted in males. It masculinizes the secondary sexual characteristics at puberty and maintains the male accessory structures (Table 24-4). Testosterone causes en-

FIGURE 24-13
Historical decline in the mean age of menarche in a number of populations of European origin.

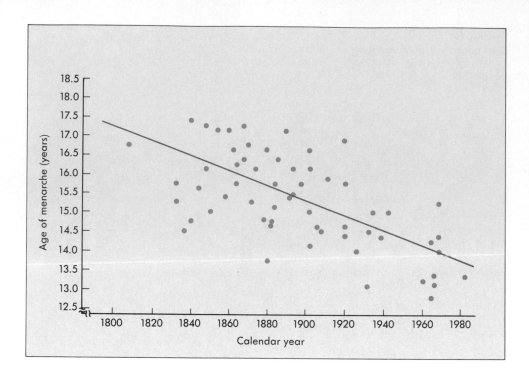

TABLE 24-4   *Effects of Testosterone and Its Metabolites*

Prenatal

1. Inhibits development of Müllerian structures (female internal accessory organs)
2. Masculinizes external genitalia
3. Establishes male pattern of steady release of gonadotropin after puberty

At Puberty

1. Stimulates protein synthesis, increasing growth and mature muscle mass
2. Increases basal metabolic rate
3. Stimulates long bone growth (and also closure of epiphyses)
4. Stimulates development of male accessory sexual organs and maintains them in functional state
5. Establishes male secondary sexual characteristics
6. Necessary for spermatogenesis
7. Stimulates formation of red blood cells
8. Stimulates libido (and may have other behavioral effects)

largement of the larynx and growth of the penis and seminal vesicles. It is mainly responsible for initiating the growth spurt of male puberty, and it ultimately brings growth in stature to an end by stimulating closure of epiphyses at the ends of long bones. Testosterone stimulates growth of muscles and is probably responsible for the higher hemat-

ocrit of men. It is required for spermatogenesis and plays an important role in establishing and maintaining sexual interest, or **libido.**

Some target tissues convert testosterone to **estradiol, dihydrotestosterone,** or **5 α-androstenediol,** which is the form that binds with their receptors. These conversions are catalyzed by enzymes the target tissues possess: aromatase in the case of those that require estradiol; 5 α-reductase in the case of dihydrotestosterone; and both 3 α-reductase and 5 α-reductase in the case of 5 α-androstenediol (Figure 24-14).

Dihydrotestosterone stimulates the growth of the scrotum and prostate and maintains prostatic secretion. It is the form of testosterone necessary for masculinizing the external genitalia in fetal development (Figure 24-10). An inborn deficiency of reductase causes a disorder of sexual development in which the Wolffian structures (which are activated by testosterone) develop normally but the external genitalia fail to masculinize completely (Table 24-5). Both dihydrotestosterone and 5 α-androstenediol stimulate the growth of facial, body, and pubic hair and are a contributing factor in male pattern baldness.

In females there are two major reproductive steroids, **estradiol** (an **estrogen**) and **progesterone** (a **progestin;** Figure 24-14). Estradiol is associated with almost all female-specific changes that occur after puberty (Table 24-6). Both the **mammary glands,** which secrete milk during lactation, and the adipose tissue of the breasts increase in response to estradiol. Estradiol supports the subcutaneous fat deposits of women, which are, on the av-

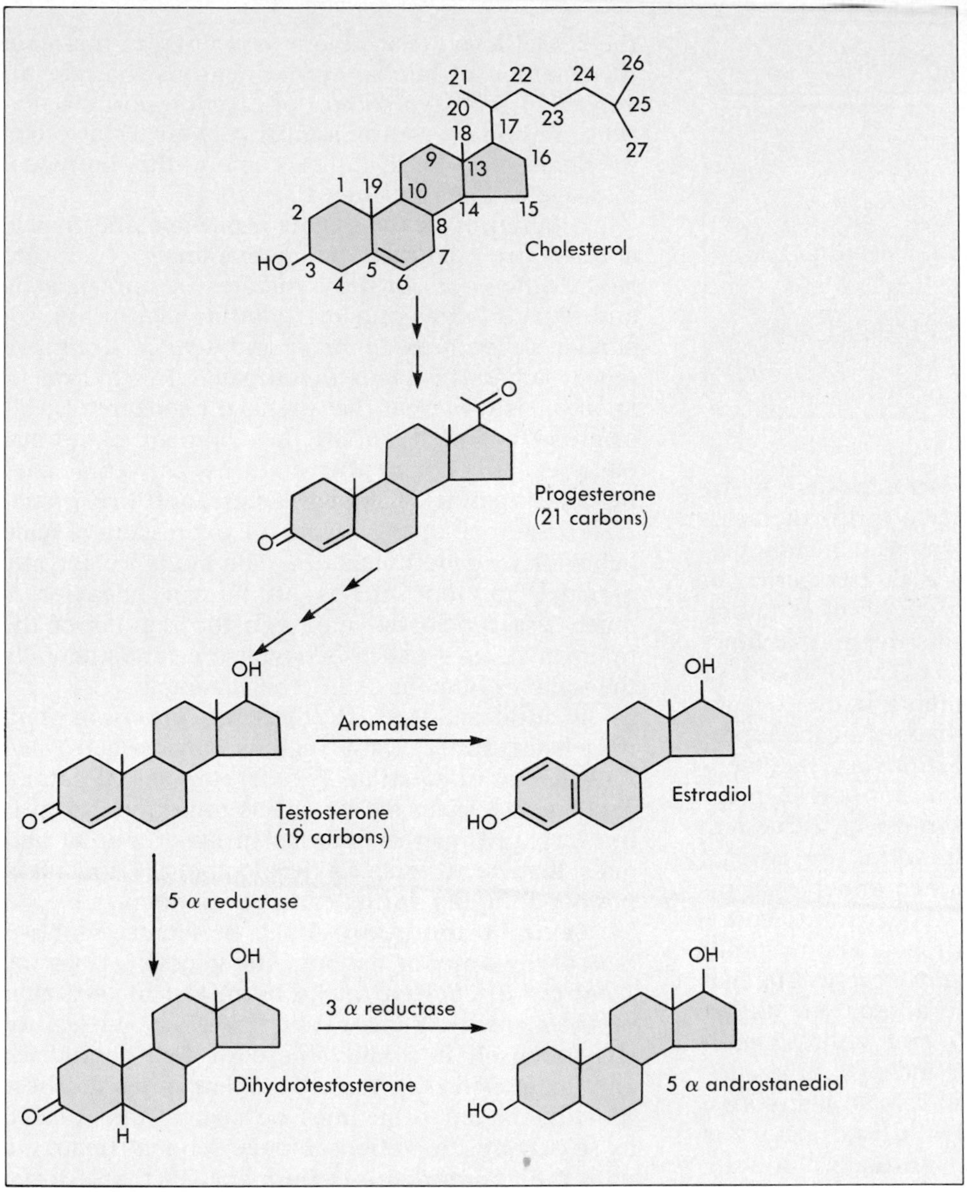

FIGURE 24-14
Pathways of steroid synthesis and metabolism. In the testis, testosterone is synthesized from cholesterol by a pathway that includes progesterone. Multiple arrows indicate places where several intermediate steps in the pathway are not shown in the figure. In some of its target tissues, testosterone may be metabolized to estradiol, dihydrotestosterone, or 5 α-androstenediol. (The ovary makes estradiol by a similar route.) The individual carbons of the steroid backbone are numbered so that reactions involving them can be more easily identified—so, for example, the enzyme 5 α-reductase breaks a double bond joining carbons 4 and 5.

TABLE 24-5   *Some Endocrine Disorders that Affect Genital Sex*

Name	Defect	Effects
Adrenogenital syndrome	Adrenals produce excess of DHEA	Masculinized female external genitals
Androgen-insensitivity (testicular feminization) syndrome	Defective androgen receptor	XY genotype has undescended testes, female body plan, and external genitalia; no internal accessory structures
5 α-reductase deficiency	Testosterone not converted to dihydrotestosterone	XY genotype has testes and male internal accessory organs but incompletely masculinized external genitalia
Hypopituitary hypogonadism	Secretion of FSH and/or LH inadequate	Weak puberty, reduced fertility (both sexes); decreased libido in males

Reproduction and Its Endocrine Control

TABLE 24-6   Effects of Estrogen at Puberty

1. Establishes female pattern of body fat deposition, bone growth, and body hair
2. Stimulates growth of external genitalia
3. Stimulates growth of internal accessory sexual organs and maintains them in functional state
4. Stimulates growth of breasts, particularly the ductal system
5. Stimulates growth of uterine endometrium during follicular phase of female cycle

erage, twice as great as the male fat deposits. On the other hand, women have only two-thirds the average male muscle and skeletal mass. The broader pelvis of the female skeleton and a shorter period of growth of long bones are also effects of estradiol. After puberty, estradiol maintains the normal function and structure of the accessory sex organs. Progesterone also plays a role each month in the changes in the breasts and associated changes in the reproductive tract. Progesterone's actions in the menstrual cycle and its major role in support of pregnancy are described below and in the next chapter.

Shortly before sexual maturation, the adrenal glands of adolescents of both sexes undergo an increase in growth and activity. The adrenal cortex produces small amounts of androgen as a by-product of synthesis of cortisol and aldosterone. The major androgens produced by the adrenal are **dehydroepiandrosterone (DHEA)** and androstenedione. Although these are weaker androgens than testosterone, they are important in female maturation because in women the adrenal is a major source of the relatively small amount of androgen produced. Adrenal androgen is responsible for the growth of pubic and underarm hair in women, although it is the simultaneous presence of estrogen which tends to inhibit the male pattern of beard and chest hair. Adrenal androgen may contribute to the growth spurt that occurs at the time of female puberty. The sexual interest of women also appears to depend on adequate levels of androgen, rather than estradiol and progesterone. At menopause, secretion of steroids by the ovaries drops, but androgen produced by the adrenals maintains sexual interest.

Sex Hormones and Development of the Nervous System

- *What factors are believed to be involved in the development of gender identity?*

There are anatomical and functional differences between the nervous systems of males and females. At

the spinal level, testosterone is known to maintain the function of lumbar spinal neurons that are involved in reflexive control of erection and ejaculation in animals. Neurons must convert testosterone to estradiol (Figure 24-14) before the hormonal message can be transduced.

Differences in the behavior of male and female animals are characteristic in mammals. There are slight differences in the structures of human male and female brains and the possible significance of similar differences in male and female brains of other mammals is being investigated. In experimental animals it is clear that prenatal exposure to testosterone is important for development of normal male sexual behavior after maturity, as well as having an influence on aggressiveness, and that prenatal testosterone can stimulate the expression of male behavior in genetic females. The evidence for any prenatal hormone effects on human behavior is much less conclusive, although the regions of the brain involved in such behavior are fundamentally the same in humans as in other mammals.

In addition to sexual differences in certain brain nuclei, the size of these regions may be correlated with sexual orientation. The developmental pattern for these nuclei in the human is unknown but it is likely, based on the findings in experimental animals, that the differential development is a result of hormonal inputs during early development.

Sexual orientation—one's preference for partners of the same or the opposite gender—does not seem to be reflected in the hormonal picture after sexual maturity. Some people are **gender-dysphoric**—intensely uncomfortable with their genital sex and the gender that normally accompanies it. These individuals can sometimes be successfully treated by sex-change operations and subsequent treatment with the appropriate sex hormone. For male-to-female transsexuals it is relatively easy to construct a functional vagina. Penises constructed by plastic surgeons are much less satisfactory.

Some children are born with external genitals that are intermediate in structure between the male and female forms, making it difficult to assign them to their correct gonadal sex. Studies of children who have been raised with a gender at odds with their gonadal sex suggest that gender identification is more dependent on early social conditioning than on gonadal sex or genetic sex.

Disorders of Sexual Differentiation

- *Why is the adrenogenital syndrome a problem for female and not male fetuses?*
- *Compare the site of the defect in the masculinization of individuals with 5 α-reductase syndrome and testicular feminization syndrome.*

TABLE 24-7   Disorders of Genetic Sex

Name	Sex chromosomes	Body plan	Effects
Turner's syndrome	XO	Female	Infertility, no puberty
Klinefelter's syndrome	XXY	Male	Underdeveloped testes, infertility, weak puberty
Supermale	XYY	Male	Effects (if any) are not certain
Superfemale	XXX	Female	Reduced fertility

The homologous sex chromosomes may fail to separate properly in meiosis, resulting in spermatozoa or oocytes that carry either no sex chromosome or two sex chromosomes. If these abnormal gametes are involved in production of a zygote, chromosomal complements of XO (the O means that one of the gametes contributed no sex chromosome), XXY, XYY, or XXX may result (Table 24-7).

The first possibility (XO), called **Turner's syndrome,** results in female gonadal sex and the female body plan, as would be expected from the principle that the body develops as female in the absence of a Y chromosome. However, two X chromosomes are apparently needed for normal ovarian function. The ovaries of females with Turner's syndrome do not develop beyond a primitive stage, and do not assume either a reproductive or an endocrine function at puberty. These women are thus infertile and must be given estrogens to initiate a feminizing puberty and maintain feminine secondary sexual characteristics.

The second possibility (XXY) results in **Klinefelter's syndrome.** The gonads and body plan are male, but the function of the gonads is impaired. Males with this symptom undergo what is at best a weakly masculinizing puberty. They are typically tall, because the growth of long bones continues for longer than it would if testosterone were present. The third possibility, **supermale** (XYY), results in men who appear normal. The fourth possibility (XXX) has been called the **superfemale** complement; the women who have this complement are essentially normal, but their fertility is reduced. All possible sex chromosomal abnormalities may be accompanied to a greater or lesser extent by mental retardation or problems of personality development.

Excessive androgen production by the adrenals can masculinize the external genitalia of a fetus whose genetic and gonadal sex is female; this is the **adrenogenital syndrome** (see Table 24-5). Typically this occurs as a result of a genetic defect in the fetus or its mother that blocks a step in the pathways of steroid metabolism leading to cortisol and aldosterone. The lack of cortisol causes an increase in secretion of ACTH, which powerfully stimulates steroid metabolism in the adrenal. Excessive production of androgen results when steroid molecules that normally are precursors for cortisol and aldosterone accumulate upstream from the blocked steps in the pathway. These precursors then enter the alternative pathway leading to DHEA and androstenedione, raising blood levels of the androgens high enough to masculinize the developing external genitals. The degree of masculinization is variable but may be so complete that an error of sex assignment is made at birth. This error may become apparent only when a feminizing puberty occurs.

In some genetic males, there is a defect in the gene for 5 α-reductase enzyme. Some parts of the primitive urogenital groove depend on dihydrotestosterone rather than testosterone, so the lack of this enzyme results in failure of the external genitalia to masculinize before birth. The urogenital groove may fail to close, and the penis remains small.

An extreme example of failure to masculinize is provided by the **testicular feminization syndrome,** in which the genetic and gonadal sex are male, but androgen receptors are defective. The Müllerian structures degenerate because MRF is secreted, but the Wolffian structures also fail to develop because their receptors cannot respond to the testicular androgen. Thus there is neither a female nor a male internal accessory system, and the testes remain in the abdomen. The insensitivity of the genital tubercle to androgen causes the external genitalia to take the female pattern even though testosterone is present. Consequently, the individuals with this syndrome appear female at birth. They undergo a feminizing puberty under the influence of the estrogen produced by the testes in the course of their much larger production of androgen. The individuals with this condition are genetic and gonadal males but their gender and genital sex are female. The condition typically comes to medical attention because menstruation does not begin at puberty, or because of a complaint of infertility.

ENDOCRINE REGULATION OF MALE REPRODUCTIVE FUNCTION
Feedback Regulation of Spermatogenesis and Testosterone Secretion

- How do the following hormones function in men?
  FSH
  LH
  Testosterone
  Inhibin
- How do hormonal feedback loops regulate gonadal function in men?

The hypothalamus secretes GnRH in bursts throughout the day, but the total amount of GnRH secreted remains fairly constant from day to day, maintaining secretion of the gonadotropins FSH and LH by the pituitary.

FSH controls spermatogenesis by stimulating Sertoli cells (Figure 24-15; Table 24-3), and the Sertoli cells exert negative feedback on pituitary release of FSH by secreting the hormone **inhibin.** Inhibin concentration reflects sperm production. At one time, this feedback loop was viewed as a prime target for development of a male contraceptive. Elevating the level of inhibin would reduce gamete production without interfering with the production of testosterone, leaving the associated hormone-dependent male secondary sex characteristics and libido intact. However, the fact that inhibin is a pro-

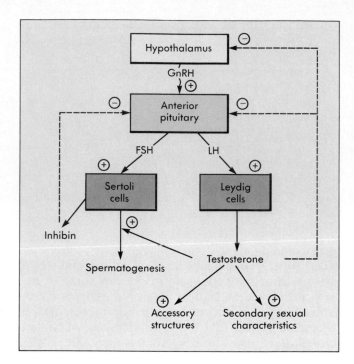

FIGURE 24-15
Negative feedback effects of testosterone and inhibin from the testes regulate the anterior pituitary's secretion of the gonadotropins LH and FSH. Testosterone is necessary for spermatogenesis; it also stimulates the growth and maintains the function of accessory structures and secondary sexual characteristics.

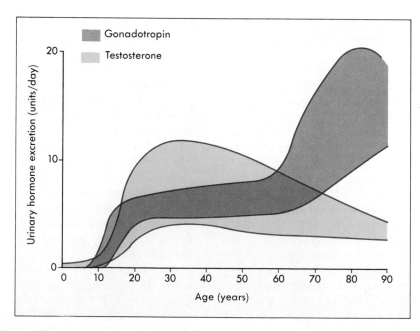

FIGURE 24-16
Rates of urinary excretion of androgens and gonadotropin over the male life span. These rates are rough indications of the plasma levels of the corresponding hormones. Most of the androgen in male plasma is testosterone. As plasma testosterone levels fall in later life, the decrease in negative feedback causes gonadotropin levels to rise. (From data of Pedersen-Bjergaard and Tonneson, *Acta Med Scand* 131[suppl 213]:284, 1948.)

tein rather than a steroid has complicated pharmacological delivery, since proteins are relatively unstable molecules, oral administration would inactivate a protein, and they do not cross membranes and move into the blood readily from a skin patch or implant.

Production and secretion of testosterone is carried out by the Leydig (interstitial) cells. Testosterone exerts feedback control on LH secretion at two levels (Figure 24-15). Grossly elevated testosterone levels inhibit secretion of both LH and FSH by inhibiting secretion of GnRH. This is the reason that abuse of androgens in body-building can cause infertility. A more subtle effect occurs at the level of the anterior pituitary, where elevated testosterone levels specifically inhibit secretion of LH. This completes a feedback loop by which the endocrine function of the testis can be regulated somewhat independently of spermatogenesis.

Changes in Testicular Function with Age

Levels of testosterone rise after puberty and stabilize at their maximum value by about age 20. Thereafter, testosterone levels remain relatively steady until about the fourth decade of life, when a decrease becomes apparent (Figure 24-16). The rate of decrease with age is highly variable from one individual to another.

Coincidental with a decrease in testosterone secretion is an increase in gonadotropin levels (Figure 24-16). The brains of older men are just as competent to release GnRH, and their anterior pituitaries just as able to release LH and FSH. It is the testes themselves that must be affected by age. The elevated output of GnRH from the hypothalamus and LH and FSH from the anterior pituitary following the decrease in hormone secretion by the testes is just what would be expected from the negative feedback loops shown in Figure 24-15.

Target tissues up-regulate their testosterone receptors as testosterone levels fall, somewhat cushioning the physiological effects of the change in testosterone level. In spite of the decrease, testosterone levels are adequate to support spermatogenesis throughout the male lifespan after puberty, to maintain the function of accessory structures, and to maintain sexual interest.

ENDOCRINE REGULATION OF FEMALE REPRODUCTIVE FUNCTION
The Ovarian Cycle

- What are the roles of the granulosa cells and the thecal cells in estradiol production in the ovary?
- Under what conditions does estradiol inhibit gonadotropin secretion? Under what conditions does it stimulate it? What is the function of positive feedback of estradiol secretion on gonadotropin secretion in the menstrual cycle?
- What is the function of the corpus luteum? From what structure is it derived?

The ovaries of sexually mature women undergo a regular cycle driven by hormonal oscillations that involve the hypothalamus, the anterior pituitary and the ovaries themselves. The timing of these events is summarized in Table 24-8. The first step in the sexually mature stage of maturation of follicles involves division of the **granulosa** cells that surround the ovum. This corresponds to the transition between the primary oocyte and a primary follicle (Figure 24-17). The number of primary follicles in each ovary declines throughout child-

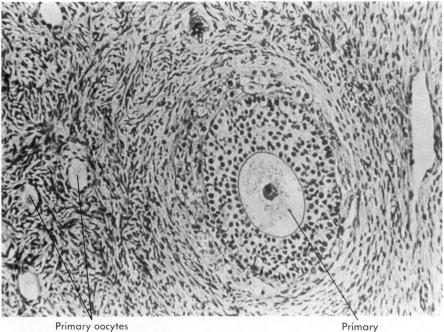

FIGURE 24-17
A photomicrograph of ovarian tissue showing a primary follicle containing a maturing primary oocyte *(right)* and several primordial follicles containing non-maturing primary oocytes *(left)*.

Primary oocytes

Primary follicle

Reproduction and Its Endocrine Control

703

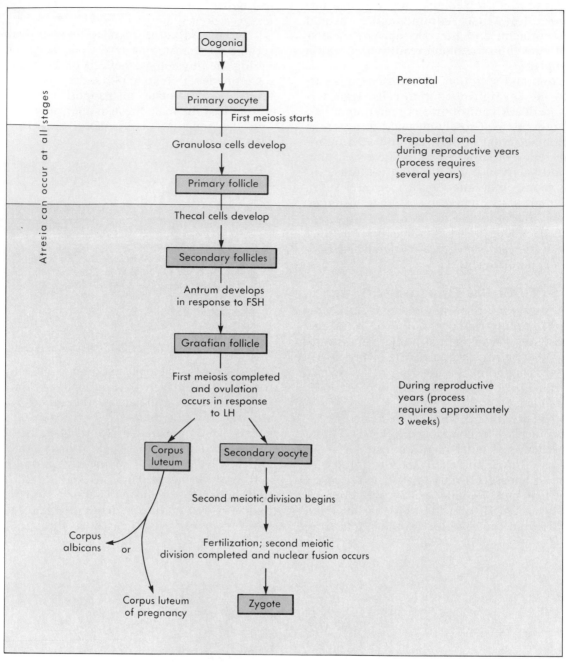

FIGURE 24-18
Flow-chart of oogenesis and follicle development.

hood, but at puberty, the potential for some of the approximately 400,000 remaining oocytes to develop into mature ova is first seen (Figure 24-18). Each month approximately 1000 follicles will begin developing, and of these, some will start development coincident with the most favorable stage in the hormonal cycle. The hormonal conditions will promote their development to more advanced levels of maturation than those which happen to begin development at a different part of the cycle. Of the 6 to 10 follicles that are promoted to the antral stage each cycle, typically only one follicle completes the cycle of maturation and undergoes ovulation (Figure 24-19).

The first half of the ovarian cycle is called the **follicular stage** because it is dominated by development of follicles, and particularly, the one follicle which progresses to the stage of ovum release, or **ovulation** (Figure 24-19). The granulosa cells secrete a viscous substance to form the **zona pellucida,** a layer around the ovum that sperm must penetrate in order to fertilize the ovum. Granulosa cell proliferation produces additional layers of cells; the outer layer of cells differentiates into **theca cells.**

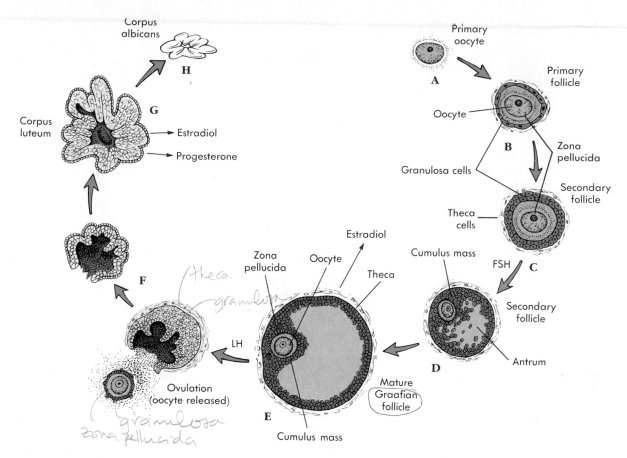

FIGURE 24-19

The life cycle of an ovarian follicle. A single primary oocyte (**A**) serves as the basis for a primary follicle (**B**). Several primary follicles advance to the secondary follicle stage (**C**) over several menstrual cycles. Usually only one follicle per cycle continues to mature (**D**) under the influence of FSH and reaches the Graafian stage (**E**). A surge of LH at midcycle causes the Graafian follicle to ovulate (**F**). After ovulation, the remnant of the follicle becomes a corpus luteum (**G**). If pregnancy does not occur, the corpus luteum degenerates after about 12 to 14 days, leaving a scar-like corpus albicans (**H**).

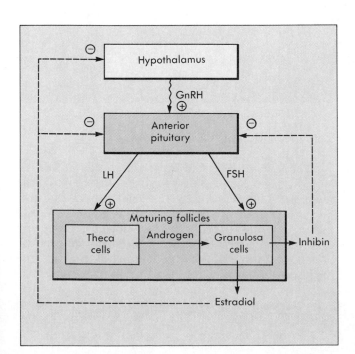

FIGURE 24-20

Negative feedback loops by which estradiol and inhibin secreted by the ovaries regulate gonadotropin secretion by the anterior pituitary (compare with the male pattern shown in Figure 24-15). Negative feedback is the dominant factor that regulates gonadotropin secretion during the luteal phase of the menstrual cycle.

Surrounding the ovum there are these concentric structures: (1) the zona pellucida next to the ovum, (2) a layer of granulosa cells, and (3) an outer layer of theca cells (Figure 24-19). A follicle that has reached this stage is called a **secondary follicle.**

Continued maturation of the secondary follicles leads to development of secretory capabilities. Theca cells produce androgen, which is converted to estradiol by the granulosa cells (Figure 24-20). In addition to estradiol, the follicle also secretes the hormone inhibin. Just as in the male reproductive

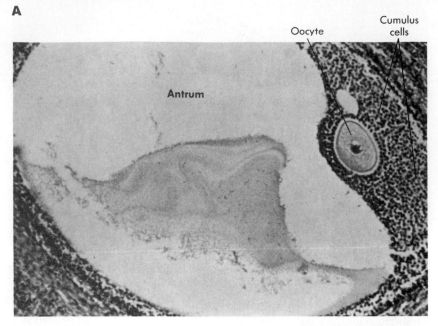

A

Oocyte

Cumulus
cells

Antrum

FIGURE 24-21

A   A photomicrograph of a section through a Graafian follicle, showing the fluid-filled antrum and the oocyte surrounded by cumulus cells.

B   A photograph of an ovary showing a Graafian follicle protruding from its surface. This photograph was made through a laparoscope.

B

Ovary

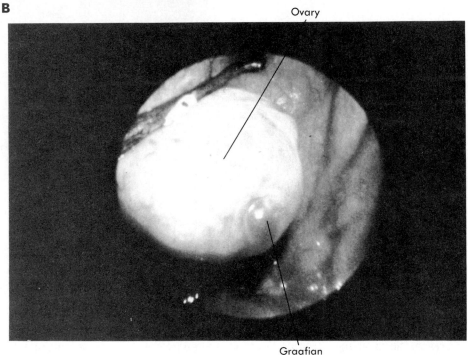

Graafian
follicle

system, inhibin levels reflect gamete formation, allowing for feedback regulation of FSH secretion independently of LH secretion.

FSH is required for continued maturation of secondary follicles (Figures 24-18 and 24-20). Proliferation of granulosa and thecal cells continues, but the defining feature of this stage is development of a fluid-filled cavity called the **antrum** (Figure 24-19 and 24-21). The antral fluid is secreted by the granulosa cells. The ovum is supported on a mound of granulosa cells (the **cumulus mass** or **cumulus oophorus**).

The level of estradiol increases slowly at first, as more and more of the developing follicles reach the secretory stage (Figure 24-22). The amount of estradiol then begins a steeper climb, because estradiol itself promotes mitosis of the hormone-secreting cells and the enzymes required for hormone production (an example of positive feedback). By the time the estradiol production is climbing steeply, it is largely produced by the "leading" follicle, and development of other follicles is inhibited by unkown factors.

In the final stage of follicular maturation, the

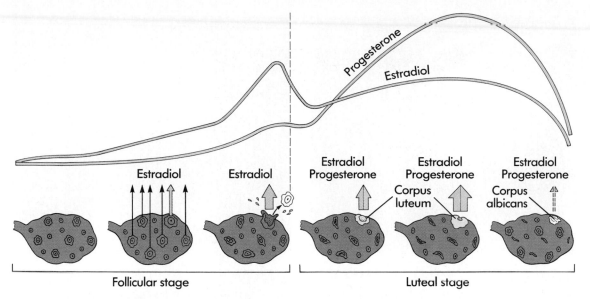

Progesterone

Estradiol

| Estradiol | Estradiol | Estradiol Progesterone | Estradiol Progesterone Corpus luteum | Estradiol Progesterone Corpus albicans |

Follicular stage   Luteal stage

FIGURE 24-22
Events in the ovarian cycle relating the follicle development to hormone production.

ovum is almost completely surrounded by antral fluid, which causes the follicle to visibly bulge from the ovary surface. The overlying tissue on the surface of the ovary becomes thin following local reduction in blood flow and enzyme release, and the internal pressure of the fluid increases. At this stage the follicle is called a **Graafian follicle** (Figure 24-21, *A* and *B*). A Graafian follicle enlarges from 5 mm to 10 to 20 mm in diameter in the last 48 hours before rupture.

The stage is now set for ovulation. The ovum is released from the ovary into the pelvic cavity in a rather explosive fashion, accompanied by antral fluid and a little blood (Figure 24-23). A few women report that they can feel this event. The ovum, (actually a secondary oocyte) surrounded by its layer of granulosa cells, is typically sucked into the nearby Fallopian tube, which responds to chemicals released at the time of ovulation with muscular contractions which draw it closer to the ovary and increase the beating of cilia. Once in the oviduct, the ovum begins a trip that will last 4 to 5 days.

The **luteal stage** of the ovarian cycle follows ovulation (Figure 24-22). The remnants of the follicle collapse and undergo a period of repair and alteration that prepare the cells to secrete both estradiol and progesterone. The follicle is now called a **corpus luteum,** literally "yellow body", because of how it looks on the surface of the ovary. It undergoes cell proliferation and its hormone production steadily increases until degeneration of the corpus luteum late in the cycle leads to a decline in hormone production. In the absence of a hormonal signal that reflects the occurrence of pregnancy, the corpus luteum will spontaneously degenerate into a corpus albicans, a white scar tissue region on the

ovary which disappears in a matter of months.

The ovarian cycle can be summarized as follows: development of follicles leads to production of estradiol by a number of follicles and the eventual suppression of the competing follicles by the one which becomes the Graafian follicle and releases an ovum. These events are occurring in both ovaries each cycle, but typically only one ovary develops a Graafian follicle. The period prior to ovulation is characterized by a gradual and then a steep increase in production of estradiol by follicle cells. The event of ovulation temporarily disrupts follicle cell hormone production, but repair and modification results in a corpus luteum which secretes both estradiol and progesterone. The levels of both these hormones rise steadily and then drop off as the corpus luteum begins to degenerate (Figure 24-22).

Hypothalamic/Anterior Pituitary Stimulation of the Ovaries

The hypothalamus releases gonadotropin releasing hormone into the portal circulation in a pulsatile fashion in both males and females. The relatively steady supply of this factor is necessary to support the cyclic female hormonal pattern that involves feedback between hormones produced in the ovary and the anterior pituitary.

The gonadotropic hormones FSH and LH from the anterior pituitary are both needed to stimulate ovarian cycles of follicle development. High levels of estradiol and progesterone inhibit LH and FSH release, but in the period after the corpus luteum degenerates, the level of these hormones rises. Thus, in the early portion of the follicular phase of the ovarian cycle, follicle stimulating hormone, as its name suggests, stimulates granulosa cell mitosis. As

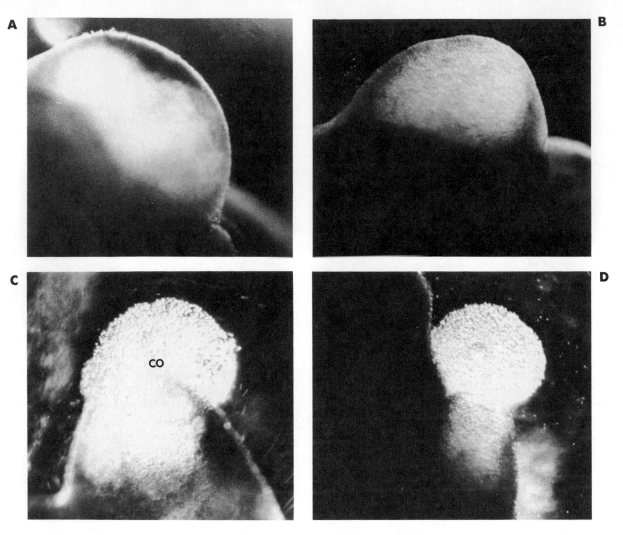

FIGURE 24-23
This series of photographs shows the events of ovulation in a hamster.
A and **B**   The follicle is swelling while its surface is being weakened by enzymatic action.
C and **D**   The fluid and cells are emerging as the follicle begins to collapse.

the number of granulosa cells rises and theca cells differentiate, the capability of secretion by the follicles develops. Hormone production depends upon responses of the granulosa and theca cells to both FSH and LH. The LH stimulates thecal cells to synthesize androgen and FSH stimulates the granulosa cells to make the enzymes necessary for converting androgen to estradiol (Figure 24-20). FSH also stimulates production of the hormone inhibin by the granulosa cells.

During the early part of the follicular phase, when estradiol levels are rising slowly, the estradiol provides feedback that follicle development is underway. The negative feedback from estradiol reduces the release of gonadotropins from the pituitary and the release of GnRH from the hypothalamus. In addition, the protein hormone inhibin is synthesized by the follicle cells and specifically in-

hibits the cells that release FSH (Figure 24-20). This is just what would be expected of negative feedback hormonal regulation: release of the hormones that stimulate a process is suppressed by signals that reflect accomplishment of the process.

However, when the levels of estradiol begin to rise sharply, a strange reversal occurs: the negative feedback is converted to a positive feedback. Cells of the anterior pituitary no longer respond to estradiol as a negative feedback signal. They increase their output of gonadotropins even more steeply than the rate of rise of the estradiol. The flip-flop in response pattern (compare Figure 24-24 with Figure 24-20) is seen in elevated output of both LH and FSH. The increase of LH in response to rapidly rising estradiol is called the **LH surge.** This is the hormonal trigger for ovulation (Figure 24-25). Estradiol stimulates the production of gonadotropin

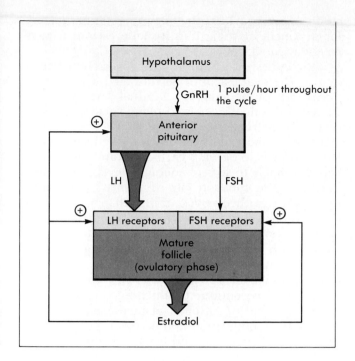

FIGURE 24-24
Pathways that provide positive feedback to LH secretion during the follicular stage of the menstrual cycle. The positive feedback culminates in a surge of LH that causes a Graafian follicle to ovulate (compare with Figure 24-22). Negative feedback from inhibin is thought to diminish the output of FSH.

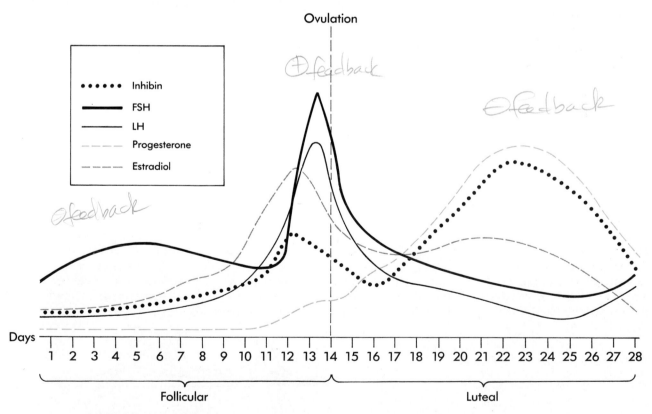

FIGURE 24-25
Relationship between the cyclic changes in the gonadotropins and the major hormones produced by the ovary. (From data of McLachlan et al. 1987.)

receptors in follicle cells (Figure 24-24). Granulosa cell responses to LH just before ovulation include the rapid increase in antral size of the Graafian follicle and the release of hydrolytic enzymes which break down the cell adhesion molecules and connective tissue in the ovarian capsule overlying the follicle.

The significance of the name luteinizing hormone is now apparent: the LH surge results in ovulation and production of the corpus luteum. It is also LH which stimulates synthesis of progesterone by the corpus luteum. The luteal phase is characterized by rising levels of estradiol, progesterone and inhibin. The positive feedback to the anterior pituitary by rising estradiol levels is not seen, and this may be attributable to inhibition by progesterone and/or inhibin. Late in the cycle, when the deterioration of the corpus luteum leads to a decrease in estradiol, progesterone and inhibin production, the level of gonadotropins begins to climb (Figure 24-25).

Recent studies in primates indicate that the entire cycle of positive feedback that leads to ovulation can be carried out by the ovaries and pituitary. A constant pattern of pulsatile release but no coordinating signal from the hypothalamus is required to drive the female cycle, and the pattern of GnRH secretion of males and females is not significantly different. If the pathway between the hypothalamus and the pituitary is experimentally interrupted and pulses of GnRH are given at the rate of about one every hour, ovulation occurs normally. The cycle cannot be carried out if GnRH is totally absent, or if the timing of the pulses is abnormal. These results show that a timing signal from the hypothalamus is not necessary for triggering the LH surge. Such studies have provided the experimental framework for treatment of some fertility failures which have an endocrine basis.

The cycles of estradiol and progesterone secretion by the ovaries continue throughout the reproductive life of the female. A gradual decrease in cycle regularity finally leads to menopause, the cessation of menstruation. In the period leading up to menopause, the supply of follicles is reduced, fewer are stimulated to develop each cycle, and the ovaries become unable to secrete adequate levels of steroids in response to gonadotropin. Evidence for a change in the ovaries themselves, rather than in the hormonal control system, is reflected in a rise in gonadotropin levels after menopause (Figure 24-26). Just as with the testis, the effect of aging is attributable to changes in the gonad, ultimately making the ovary incompetent to perform, while the negative feedback loops continue to function.

The decline of estrogen at menopause reduces some of the feminization that occurs at puberty. The sex-specific fat deposits and the size of the accessory reproductive structures diminish. Some vaginal atrophy and reduction in vaginal lubrication may be experienced and decrease in bone calcification may increase the risk of osteoporosis (see Chapter 23). Many women experience changes in the circulatory system known as "hot flashes." These and other consequences of estrogen decline can be effectively treated with estrogen replacement therapy, which has the added advantage of protecting women from the increased susceptibility to heart attacks that otherwise follows menopause. However, there is an increased risk of cervical cancer with estrogen replacement therapy; therefore, the relative risks must be carefully weighed.

FIGURE 24-26
Normal ranges of rates of urinary excretion of gonadotropin and estrogens in a population of women of various ages. These rates are rough indicators of the plasma levels of the corresponding hormone. During the reproductive years, much of the variation in the range of normal values is due to changes over the menstrual cycle. (From data of Pedersen-Bjergaard and Tonneson, *Acta Endocrinol* 1:38, 1948.)

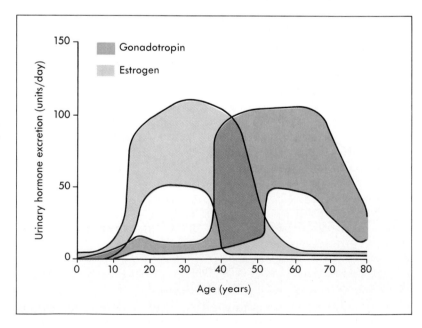

GROWTH, METABOLISM, REPRODUCTION AND IMMUNE DEFENSE

The Uterine or Endometrial Cycle

- *The uterine cycle can be divided into three phases. What events and processes are characteristic of each phase? How do the uterine phases correspond to what is going on in the ovaries?*
- *What changes are seen in the endometrium over the menstrual cycle? What are the effects of estradiol and progesterone on the endometrium? What processes are responsible for the onset of menstruation?*

The uterus is the major, but not the only target of the changing levels of steroids in women. The monthly shedding of the tissue in the uterus gives the cycle its name, because menstrual refers to a monthly process. Most mammals do not shed the uterine lining, and the cycles of these animals may be characterized by other features, such as the effects of estrogen on behavior of female dogs and cats, which exhibit **estrous** cycles or "heats."

The menstrual cycle is charted from the first day of menstruation, the flow of blood from the uterus (Table 24-8). This discharge of blood and associated tissue is the result of the reduction in the levels of estradiol and progesterone at the end of the luteal phase of the ovarian cycle, when the corpus luteum is degenerating. The duration of menstruation is typically 3 to 5 days, although, as with the duration of the entire cycle, there is much individual variation.

This **menstrual phase** is followed by the **proliferative phase** of the uterine cycle. This roughly corresponds to the portion of the ovarian follicular phase in which the follicles are secreting estradiol (Figure 24-27). In response to estradiol, mitosis in the basal layer of the endometrium greatly increases the number of cells in the uterine lining. At the start of the proliferative phase the endometrium is about 1 mm thick. As estradiol levels rise, the cells of the outer layers of the endometrium multiply, increasing endometrial thickness to 2 to 3 mm. More blood vessels (**spiral arteries**) are added, and exocrine glands (**spiral glands**) develop (Figure 24-22 and Table 24-8). A thick layer of epithelial cells, glands, and blood vessels develops, producing a cushiony lining. This type of development continues up to the time of ovulation, when the ruptured follicle transforms into a corpus luteum and begins to secrete progesterone as well as estradiol. The **secretory phase** of the uterine cycle corresponds to the luteal phase of the ovarian cycle, and derives its name from the fact that the glands in the uterine lining begin secreting a rich mixture of glycoproteins that has been called "uterine milk." Progesterone also stimulates the spiral arteries to develop tight coils, a feature not seen in other arteries of the body. The thickness (4 to 6 mm), metabolic rate, and glycogen content of the endometrium are great-

est when estrogen and progesterone levels peak at days 22 to 23 of the menstrual cycle, about 7 to 8 days after ovulation. All of these changes make the uterus a receptive environment for implantation, if fertilization of the ovum occurs.

Toward the end of the luteal phase, when the corpus luteum is beginning to degenerate, hormone production drops. Withdrawal of estradiol and progesterone leads to degenerative changes in the hormone-dependent uterine tissue. Much of this degeneration is caused by the fact that estradiol and progesterone inhibit the synthesis of uterine prostaglandins that act as local vasoconstrictors. When hormone levels decrease, prostaglandin synthesis increases and causes vasoconstriction of the spiral arteries. Periodically, these arteries contract so strongly that they close and patches of the endometrial lining supplied by that circulation die from ischemia. Lysosomes released from damaged tissue contribute to digestion of the extracellular matrix. Relaxation of the spiral arteries after constriction allows the blood to rush into the weakened tissue, pushing the deteriorating layers of endometrial tissue off the basal layer. The result is that the outer portion of the endometrial lining is sloughed off into the uterine cavity.

Although the amount of blood typically lost averages 50 ml (2.5 tablespoons), the quantity can vary greatly, and the loss accounts for the greater dietary iron requirement of women, compared with men. Prostaglandin-mediated cramping of uterine muscle and excessive bleeding can be partially alleviated by prostaglandin blockers such as aspirin and acetaminophen.

In addition to the endometrium, other tissues in the female body respond to the hormonal cycle. The breasts undergo changes each cycle which parallel the uterine changes. The breasts are composed of the overlying skin and specialized nipple and areola externally; internally is the smooth muscle, supportive ligaments, fatty tissue, circulatory and lymphatic elements and the mammary glands (Figure 24-28). A duct system unites the glandular alveoli into lobes, and the ducts that drain each lobe converge as separate lactiferous (milk-conducting) ducts that open in the nipple. In the proliferative phase, estradiol promotes an increase in the fat deposits of the breasts and an increase in duct proliferation. Nipple growth is stimulated and nipple pigmentation enhanced. In the secretory phase, both the duct and glandular development is promoted and there is an increase in fluid accumulation in the breast tissue. The glands receive additional hormonal stimulation during pregnancy, but milk production requires the hormonal pattern after delivery (see Chapter 25). At the time of menstruation, cellular changes and accelerated cell death reduce the hormone-dependent buildup until the next cycle stimulates proliferation once again.

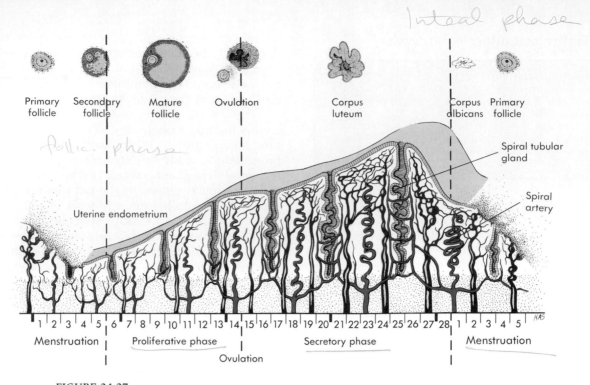

luteal phase

follic. phase

FIGURE 24-27
Changes in the lining of the endometrium over the menstrual cycle, illustrated in relation to the corresponding ovarian events.

TABLE 24-8	*Phases of the Menstrual Cycle*		
Days	**Ovary**	**Uterus**	**Hormonal picture**
1 to 4	*Follicular phase begins:* corpus luteum degenerates; several follicles begin to mature	*Menstrual phase:* outer layer of endometrium sheds after estradiol and progesterone levels fall	Estradiol, progesterone, FSH, and LH levels all low
5 to 13	*Follicular phase continues:* FSH stimulates follicle maturation	*Proliferative phase:* endometrium regrows	Estradiol: rising Progesterone: low FSH and LH: low
14	*Ovulatory phase:* LH surge causes ovulation	Cervical mucus becomes thin and watery	LH and FSH: sharp rise Estradiol: falls after follicle ovulates
15 to 28	*Luteal phase:* corpus luteum secretes estradiol and progesterone	*Secretory phase:* endometrium secretes uterine milk	Estradiol and progesterone: high FSH and LH: low

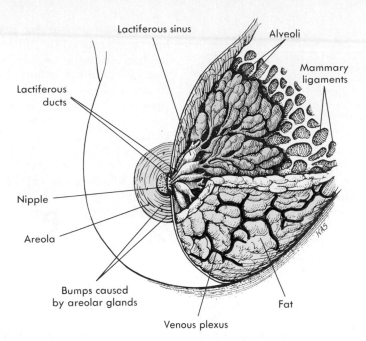

FIGURE 24-28
The structure of the breast. The mass of the breast is largely fat; the shape is maintained by a network of supporting ligaments. The mammary glands consist of alveoli that empty into a system of sinuses emptied to the nipple through ducts.

Timeline for the Menstrual Cycle

On the average, the menstrual cycle lasts 28 days; the normal range is 21 to 35 days. The first day of menstrual flow is numbered as day 1. As soon as menstruation is over, the rising estradiol promotes rebuilding of the endometrial lining. Ovulation occurs, on the average, on day 14. Thus the first 14 days include menstruation and the proliferative phase of uterine development and correspond to the follicular phase of the ovarian cycle. After ovulation, the secretory phase of uterine development continues under the stimulation of both progesterone and estradiol for approximately 14 days, leading up to the next period of menstrual flow. This second half of the endometrial cycle corresponds to the luteal stage of the ovarian cycle (Table 24-8).

These generalities are useful for textbook discussions of the events in menstruation, but they sometimes obscure the variability in the pattern and the sources of the variability. For one thing, the events leading up to ovulation are a greater source of individual differences in the duration of the cycle than the fairly constant second portion of the cycle, which is determined by the lifespan of the corpus luteum. In women with relatively shorter cycles, the estrogen levels may rise faster or the pituitary may be more sensitive, driving ovulation sooner. On the other hand, the longer cycles may result from more gradual or delayed development of follicular hormone production or a less easily triggered LH surge. In any case, the ovary communicates with the anterior pituitary when a follicle is ready to be ovulated. However, it isn't as simple as that, because there is evidence that some women ovulate in response to sexual intercourse, and ovulation may readily occur as early as the tenth day of the cycle. Thus the rising levels of estrogen that force the LH surge may be the backup system for stimulating ovulation in the absence of stimulation during sexual activity. Ovulation in response to the stimulation of mating is the typical pattern in many mammals, including rabbits and cats, and it assures optimal timing for the meeting of egg and sperm. Hypothalamic influence in initiating or delaying induced ovulation is not understood, (although oxytocin is one of the triggers), but this part of the brain is involved in emotional responses and is the neural link between experience and the timing of events in the menstrual cycle.

CONTRACEPTION

- *What are the currently available methods of hormonal birth control? What alternatives to hormonal contraception are available?*
- *What physiological mechanism accounts for the action of estrogen/progesterone birth control pills?*

Nonhormonal Methods of Contraception

The variability in the length of the menstrual cycle sometimes makes it difficult for human couples to determine the most fertile part of the cycle. It is possible to document the time of ovulation by recording body temperature and features of cervical mucus (described below) and the degree of dilation of the cervix itself. There are small alterations in body temperature, which are best detected if measured upon awakening: temperature increases slightly following the LH surge (presumably due to elevated progesterone) and remains elevated for the rest of the cycle. If a woman's cycle is very regular, such records can help her predict when ovulation is likely to occur, but the LH predictor kits available at pharmacies allow women to determine the time of ovulation more precisely.

Other methods of charting the menstrual cycle include detecting changes in the **cervical mucus,** a secretion that protects the opening of the uterus. It becomes thin and fluid in the first half of the cycle under the influence of estrogen, and sperm can move through it most easily at the time of ovulation. After ovulation, the mucus becomes thick and blocks the cervical canal. The opening of the cervical canal changes visibly near the time of ovulation: it appears closed during most of the cycle (as observed using a speculum, see Figure 24-1, *B*), but the smooth muscle relaxes, allowing the canal to remain open around the time of ovulation. With experience, a woman can detect the gaping of the cervical canal by finger contact with the cervix.

The **rhythm method** of birth control depends on alteration of sexual practices during a woman's fertile period to avoid the meeting of ovum and sperm. Because of the many factors that can affect the time of ovulation, this method may be effective only in women with very regular menstrual cycles whose lives allow careful record keeping and whose sexual partners are supportive of this choice. The high failure rate argues against using this method except in cases where the possible pregnancy is acceptable. Much easier indicators of the fertile period might be developed to greatly improve the potential of the rhythm method. The estradiol and/or LH surge associated with the fertile period might be detected by dip sticks that measure levels in saliva, with a similarly detected rise in progesterone signaling the end of the fertile period.

The use of **barrier methods** of contraception has been spurred by the associated protection these methods provide against sexually-transmitted diseases, including acquired immunodeficiency syndrome (AIDS). The condom is effective because it prevents the transfer of body fluids (semen) to the female's body. A new disposable product that is currently being evaluated for market is the female condom. It is a loose condom with an outer ring that fits over the external genitalia and an inner ring which fits over the cervix. Both products are more effective if used with a spermicide. The **diaphragm,** the **cervical cap,** and the **vaginal sponge** act by blocking access of spermatozoa to the cervix. These physical barriers may also be combined with a chemical barrier of spermicide.

Surgical methods of fertility control are primarily sought by those who wish to terminate their reproductive potential. A relatively permanent block to gamete transport is attained by **tubal ligation,** a surgical procedure in which the Fallopian tubes are ligated and severed, closing the connection between the abdominal cavity and the uterus. The corresponding surgical sterilization operation in men is called a **vasectomy;** the two vasa deferentia are exposed with small incisions, ligated, and cut.

The **intrauterine device,** or **IUD,** is not a barrier method of contraception. The mechanisms of action of IUDs are not well understood. The presence of the device in the uterus apparently alters the uterine environment sufficiently to interfere with the process by which an embryo implants in the uterine wall. Many of the IUDs have generated unacceptable side-effects in a significant fraction of their users and their use has been associated with complications such as tubal pregnancies and infections that can reduce fertility. However, on a world wide perspective IUDs offer a simple and cheap method of reducing population growth, and so IUDs continue to be improved. A major advance has been the combination of slow release of hormones with the device, which improves effectiveness and puts the hormones where they are most effective, rather than elevating whole-body levels in order to reach the same targets.

The practical effectiveness of different means of contraception (Table 25-1) depends on intrinsic physiological and anatomical factors, which may vary from individual to individual. The effectiveness of methods that require attention and action on the part of the user, such as abstinence, the rhythm method, and the barrier methods that are applied before sexual intercourse, obviously depends on conscientious application. The low but finite failure rate of the surgical methods is due to the occasional failure of the vas deferens and the Fallopian tubes to close fully after they are ligated.

GROWTH, METABOLISM, REPRODUCTION AND IMMUNE DEFENSE

Hormonal Birth Control

The first hormonal birth control agents were developed on the basis of observations that steady levels of estrogen in the presence of progesterone inhibited gonadotropin secretion and thus interfered with follicle maturation and ovulation. The goal of the "pill" was to mask the rise in estrogen which triggers the LH surge by providing relatively high levels of estrogen for the first half of the cycle. Ovulation was prevented, and therefore no pregnancy could occur. In most cases the regimen followed in hormonal birth control simulates the menstrual cycle; estrogen is taken for a number of days, then estrogen in combination with progesterone, followed by several days in which both steroids are withdrawn and the endometrial lining is shed. If periodic shedding of the endometrium is not promoted, the endometrium becomes overdeveloped and small episodes of bleeding occur. (Over the long run, the chances of endometrial cancer increase when hormone-stimulated buildup of the endometrium is not periodically interrupted.)

As the potential hazards of large doses of estrogen and progesterone were recognized, lower doses were developed. Possible stimulation of breast cancer by estrogen has remained a concern, but this risk must be evaluated in the light of evidence that the combination pill provides some protection from endometrial and ovarian cancers, tubal pregnancies and ovarian cysts. However, the major danger from use of estrogen is the complications resulting from blood clots that can result in stroke and heart attacks. This risk greatly increases in women who smoke, and is another important reason for not smoking.

In contrast to the formulations used in the early period after oral contraceptives were introduced, the doses of estrogen used for contraception today fall into the medium dose (50 mg) and low dose (ranging from 40 to 20 mg) categories. In many cases the low-dose formulations do not reliably prevent ovulation and yet are still effective contraceptives (Table 25-1). This is because other processes that contribute to fertility, such as the transport of gametes in the fallopian tubes, the hospitability of the uterine environment for implantation, and the ease of cervical mucus penetration by spermatozoa are adversely affected by the presence of the exogenous steroids. The levels of progesterone also vary among the different formulations, and because progesterone may be associated with some of the unpleasant side-effects of the pill, the minimum that is effective is usually sought. The selection of the appropriate formulation for a given woman is primarily accomplished on a trial-and-error basis, and any exogenous hormones will require a period of adjustment by the body, so this method of birth control does require medical supervision and patience on the part of the woman.

The mini-pill is a birth control pill which contains only progesterone. This has the advantage of eliminating the major risks of hormonal contraceptives, which are associated with estrogen rather than progesterone, but it has a higher pregnancy rate (Table 24-10) and the side effects include irregular spotting or bleeding.

TABLE 24-9   *Failure Rates of Contraceptive Methods (The units are in pregnancies/100 women/year)*

Method	Theoretical failure rate	Failure rate in actual use
Tubal ligation	0.004	0.006
Vasectomy	0.15	0.15
Estrogen/progesterone pill	0.34	4 to 10
Progestin-only pill	1 to 15	5 to 10
Norplant	—	0.04
Condom with spermicide	Less than 1	5
IUD	1 to 3	0.5 to 4
Condom	3	10
Diaphragm with spermicide	3	2 to 20
Cervical cap	3	2 to 20
Spermicide alone	3	2 to 30
Vaginal sponge	9 to 11	13 to 16
Coitus interruptus	0 to 9	20 to 25
Rhythm	2.5 to 13	21 to 40
Unprotected intercourse	90	90
Abstinence	0	?

Data from Vessey et al.: Efficacy of different contraceptive methods, *Lancet,* April 10, 1982; and Sloane: *Biology of Women* (2nd edition), John Wiley & Sons, Inc., New York, 1985; Franklin: The birth control bind, *Health* July-August, 1992.
*The theoretical effectiveness is an estimate of the failure rate based on the intrinsic characteristics of the methods; the use rate reflects additional failures resulting from variation in anatomy, physiology, behavior, and motivation in the user population. Use failure rates may vary greatly in different populations.

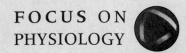

RU 486: A Menstruation Inducer

The number of illegal abortions worldwide is staggering, as is the loss of women's lives. For instance, in Brazil more than four million illegal abortions are performed annually, and tens of thousands of women die or are injured in the process. Apparently, women are not deterred even by the threat of death. The need for a safer alternative to surgical abortions stimulated the research that led to RU 486, a drug that has been called an abortion pill. The goal was to produce an antagonist to progesterone, the hormone that maintains the uterine environment of pregnancy. RU 486 is a steroid structurally similar to progesterone. It crosses the cell membrane and binds to the progesterone receptor in the cytoplasm of target cells, forming an inactive complex (see p. 111 for a description of the intracellular mechanism of action of the lipid-soluble steroid hormones).

Administration of RU 486 causes shedding of the endometrial lining either before implantation has occurred or within the first 49 days after conception. It is thus a contragestin, counteracting the uterine conditions necessary for gestation. In practice, RU 486 is given in a single dose, followed 36 to 48 hours later by an oral prostaglandin that increases the frequency and strength of uterine muscle contractions and spiral artery constriction, leading to detachment and expulsion of the endometrial lining. The success rate is greater than 95% and in most cases the embryo and all endometrial fragments are expelled within 24 hours. Heavy bleeding may follow any abortion procedure, so the treatment is typically given in a medical facility where women can be monitored and treated if necessary. The complications have generally been attributable to the prostaglandin rather than the RU 486, and experiments to select the prostaglandin with the least side effects and the lowest effective dosages are continuing. In France, 30% of women who decide to terminate a pregnancy choose this chemical approach. The advantages of this method are that surgery is not required and emotional trauma is minimized.

In addition to its use as a contragestin, RU 486 may aid in induction of difficult labors and may be used to treat uterine fibroid tumors and forms of breast cancer that are promoted by the trophic effects of progesterone. Other conditions that have responded to RU 486 include Cushing's syndrome, meningioma and endometriosis. However, the FDA has thus far declined to license RU 486 for any medical use in the United States.

The newest method of contraceptive steroid delivery is Norplant, a flexible tube the size of a matchstick that releases its content of progestin over a period of 5 years. Six Norplant tubes are inserted under the skin of a woman's upper arm. The advantage of this method is that it substitutes a single contraceptive decision for a daily regimen of pill-taking. The effects and side-effects are the same as for the progestin-only pill, although some women also experience bruising from the implant. Norplant is most effective in women who weigh less than 110 lbs; it is not effective in women who are overweight because the amount of estrogen produced by fat cells may promote fertility despite the inhibiting effects of the progesterone. A woman who decides to resume fertility after receiving the implants can have them removed and fertility may return within 48 hours.

Another approach to birth control that is available in many countries is the progesterone antagonist, RU 486 (see box). RU 486 is an abortion inducer because it brings about the shedding of the

TABLE 24-10   *Phases of the Sexual Response*

Phase	Characteristic changes		
	Male	*Female*	*Both sexes*
1. Excitement	Erection	Vagina lubricates; clitoris swells	Blood pressure and skeletal muscle tone increase; blood flow to pelvic area increases; hyperventilation
2. Plateau	Scrotum contracts, elevating testes	Orgasmic platform forms; uterus is elevated, causing vaginal tenting	Sex flush appears; hyperventilation; heart rate 100 to 160 beats/minute
3. Orgasmic	Ejaculation	Orgasmic platform contracts rhythmically	Heart rate 110 to 180 beats/minute; blood pressure and respiration maximum
4. Resolution	Erection subsides; refractory period	Uterus returns to normal position	Respiration and blood pressure return to normal; pelvic vasodilation reversed; skeletal muscle tone decreases

endometrial lining, either removing the possibility of implantation or including an already-implanted embryo with the shed tissue. There is no significant difference between this and the action of the low-dose pill and the IUD, both of which may allow fertilization but interfere with implantation. Similarly, some prostaglandin-based contraceptives now reaching the public act by interfering with implantation.

THE HUMAN SEXUAL RESPONSE
The Sexual Experience: A Whole-Body Response

- *What are the stages in the female and male sexual response cycles? What changes characterize each phase?*
- *What autonomic inputs are responsible for erection and ejaculation?*
- *What aspects of the female genital responses favor impregnation?*

Beginning in the late 1950s William Masters and Virginia Johnson began to measure physiological changes in the bodies of men and women during the sexual response. Until then, physiologists and physicians had virtually no information about the physiological changes accompanying sexual function. Masters and Johnson demonstrated that the emotional and genital components of the response are accompanied by changes in blood pressure, respiration, and regional blood flow. These changes involve both branches of the autonomic nervous system.

Considering both the genital and the extragenital changes, the sexual response pattern of both men and women can be divided into four phases (Table 24-10): an initial **excitement phase** during which the genital and extragenital changes occur, a **plateau phase** during which the changes are sustained, an **orgasmic phase** during which both the emotional and physiological responses are most intense, and a **resolution phase** in which excitement subsides and the physiological parameters return to their baseline values. In men there is typically a refractory period of varying length between orgasms during which at least some resolution occurs. In women the intensity and number of orgasmic peaks are variable, and occasionally women may return to the orgasmic phase without having to pass through the resolution phase (Figure 24-29).

In both sexes there are elevations of blood pressure and skeletal muscle tone during the excitement phase, and even greater increases during the orgasmic phase. As the orgasmic phase approaches, both men and women may display a rash-like vasodilation, the **sex flush,** over the face and chest. Specific vasodilation in the genital area is driven by parasympathetic inputs that run to the blood vessels of the genitalia. These efferents constitute an exception to the rule that blood vessels receive only sympathetic input. As described in the preceding chapter, the male and female external genitalia arise from the same embryonic structures, and in both cases vasodilation is important for the genital component of sexual response.

One might expect that, since the male and female genitalia are different, the subjective feelings of men and women during the sexual response would be different also. However, in one study in which male and female subjects described their feelings during sexual response, readers found it

Endometriosis: A Challenge for Diagnostic and Preventive Medicine

Endometrial tissue that escapes its normal location in the lining of the uterus complicates the lives of millions of women. Until recently, problems resulting from endometriosis could rarely be diagnosed. Women were sometimes told that their reluctance to accept the female role was the cause of painful, heavy, or irregular menstruation or infertility.

The way endometrial tissue moves from the uterus to other sites in the body has not been documented, but the most common ectopic site is on the ovary, suggesting that the tissue moves up the oviducts and out into the peritoneal cavity. The displaced tissue can range from microscopic implants to extensive adhesions, nodules, or cysts. The endometrial tissue responds to the cyclic changes in ovarian hormone levels with cell proliferation, increases in vascularization and then degeneration. Tissue that swells and then bleeds into the peritoneal cavity stimulates inflammatory responses that result in formation of scar tissue and adhesions. The normal independence of the organs in the pelvis is compromised by adhesions, which tie the organs together, often kinking the oviduct or binding the ovary to the uterus. Although half of the cases of endometriosis discovered during abdominal surgery for other causes are asymptomatic, the adhesions can cause pain during intercourse, defecation or urination and are a major cause of infertility in women over age 24. Although endometriosis is very rarely fatal, similarities to benign cancer include the capacity of the tissue to invade and penetrate organs and grow to compress the abdominal contents.

If infertility, menstrual complications or a range of other symptoms suggest the presence of endometriosis, diagnosis can be accomplished by laparoscopy. This procedure, a minor operation in which a light-containing telescope is inserted through a small incision into the abdomen, has greatly improved the ability of physicians to diagnose

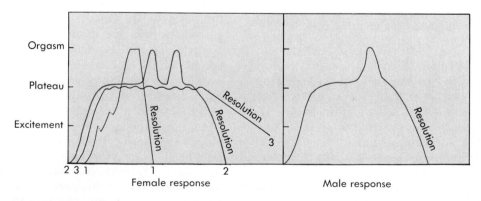

FIGURE 24-29
Comparison of the phases of sexual response in males and females. The female response may follow one of three basic patterns, including a shorter or longer plateau phase and from none to several orgasmic phases. The male pattern is typically more uniform.

endometriosis as well as pelvic inflammatory disease, ectopic pregnancy or other abnormal conditions involving abdominal organs.

Endometriosis can assume a variety of forms, from clear spots and fibers to the classical spots called "chocolate cysts" (brown due to clotted blood enclosed in tissue sacs); a biopsy will be taken to identify the cell types. Some implanted tissue and adhesions may be removed by scalpel, electric cautery or laser under laporoscopic observation but larger growths may require more extensive surgery. The decision about how to deal with the endometriosis depends on the severity, the symptoms and whether it prevents a desired pregnancy. Often if infertility results from a blocked oviduct or scars covering the ovary, surgery can correct the problem.

An alternative to surgical treatment is to pharmacologically suppress ovarian cycling, because the troubling symptoms are alleviated in women who become pregnant. One of the drugs used is Danazol, a synthetic androgen which can relieve symptoms and cause the endometrial tissue to regress; it operates on several levels, but is effective primarily because it interferes with feedback from ovarian hormones to the pituitary, decreasing secretion of estrogen and progesterone. The pseudomenopause that results is a temporary alternative to surgical menopause (removal of ovaries and uterus) and can be used by women who need relief while maintaining their options for future pregnancies. Side effects of this expensive drug may include androgenic effects and the troubling symptoms of menopause. In many cases, the knowledge that the painful symptoms of endometriosis have an organic basis that is not life-threatening can help a woman to live with the condition. Advice on prevention and development of a cause-related therapy will depend upon a better understanding of this disease, whereas current treatments can only repress the condition or provide temporary relief.

impossible to reliably distinguish the male and female responses if all direct references to the gender of the subjects were deleted from the accounts.

Male Genital Responses

The penis must become stiff in order to penetrate the female vagina; this process is called **erection**. The penis has an extensive network of venous sinuses, the corpus spongiosum and corpora cavernosa (Figure 24-5); entry of blood into them is under neural control. In the absence of sexual stimulation, the arterioles leading to the sinuses are constricted and the resistance of the veins draining the sinuses is low. Consequently there is little blood in the venous sinuses, and the penis is flaccid. During the excitement phase, tactile, visual, olfactory, or imaginary stimulation increases parasympathetic activity and decreases sympathetic activity. The penile smooth muscle surrounding the blood sinuses relaxes and the arterioles leading to the corpus spongiosum and corpora cavernosa dilate. Both aspects of smooth muscle relaxation are mediated by the newly described neurotransmitter nitric oxide. Blood flows into the sinuses, and as the venous sinuses fill with blood, they expand against the outer tunic of the penis, compressing the veins that drain the erectile tissue. The blood pressure within the sinuses rises to match the arterial presssure. The process is reversed by sympathetic-mediated contraction of the smooth muscles around the sinuses and constriction of the arterial smooth muscle, reducing the inflow of blood.

Failure to achieve erection in response to appropriate stimulation (**impotence**) can be caused by worry or anxiety and by drugs that affect the autonomic nervous system. Blockage of the arterial system that provides inflow to the penis can result from physical injury to the region or factors that increase atherosclerotic changes, including hypertension, hyperlipidemia, cigarette smoking and diabetes mellitus.

Semen is expelled through the penis during **ejaculation,** the major genital event of the male orgasmic phase. Ejaculation is a spinal reflex mediated by sympathetic nerves that stimulate contraction of the smooth muscles of the tubular system, of the accessory glands, and of muscles at the base of the penis. Ejaculation is normally a two-stage process. In the first stage **(emission)** the smooth muscle lining the internal gland and duct system contracts, forcing semen from the epididymis into the ductus deferens. On each side, the ductus deferens joins with the duct of the seminal vesicle to form the ejaculatory duct, which runs within the prostate gland and empties into the urethra. (The semen cannot enter the bladder because the sphincter at the base of the bladder constricts during ejaculation.) In the second stage of ejaculation **(expulsion)** the smooth muscles of the ducts and glands and skeletal muscle at the base of the penis contract rhythmically, expelling the semen from the penile urethra. These contractions are accompanied by sensations of intense pleasure.

Female Genital Responses

Like the penis, the clitoris (Figure 24-2) is the sensory focus for stimulation during sexual activity. The vasodilation that occurs during the excitement phase causes erection of the clitoris and also engorgement of the labia. The clitoris may be stimulated directly or indirectly through movement of the labia, and clitoral stimulation is almost always necessary for orgasm. Although not easily observed, the female genital response is as dramatic as erection of the penis. In the excitement phase, the vagina is readied for intromission by an increase in the secretion of fluid across the vaginal wall **(vaginal lubrication).** The vaginal smooth muscles relax and the length of the vagina is extended by distension of the engorged labia minora anteriorly and elevation of the uterus posteriorly. The **orgasmic platform,** a constriction resulting from increased vasocongestion in the outer third of the vagina, forms near the end of the plateau phase. During the orgasm, rhythmic contractions of the orgasmic platform occur with timing similar to the contractions seen in the male orgasm (intervals of .8 sec for 10 to 12 contractions or more). The uterus also contracts during orgasm, but with a different rhythm, and with contractions that are patterned like those of labor, beginning at the upper portion and spreading downward. During the resolution phase the uterus returns to its usual position in the pelvis, potentially dipping with opened cervical os into a pool of semen deposited in the vagina. For a short period, the orgasmic platform can function to restrict loss of the semen from the vagina, before the vasocongestion in the pelvic area diminishes and the muscular and vascular characteristics return to normal. Resolution is slower if a woman fails to achieve orgasm.

GROWTH, METABOLISM, REPRODUCTION AND IMMUNE DEFENSE

SUMMARY

1. The primary reproductive organs **(ovaries and testes)** arise from the bipotential indifferent gonads in fetal development. Ovaries develop automatically unless a protein coded for on the Y chromosome (the SRY gene) is present.

2. The female internal accessory organs **(Fallopian tubes, uterus,** and **inner vagina)** arise from the embryonic **Müllerian duct** system automatically unless **Müllerian regression factor,** secreted by the testes, is present. The male internal gland and duct system **(rete testis, efferent ducts, epididymis, vas deferens, seminal vesicle, prostate gland,** and **ejaculatory duct)** arise from the embryonic **Wolffian duct** system in response to testosterone secreted by the fetal testis.

3. The female **external genitalia** (labia majora, labia minora, clitoris, and vagina) develop from the embryonic urogenital groove and genital tubercle automatically unless **dihydrotestosterone,** a derivative of testicular testosterone, is present. In the presence of dihydrotestosterone, the urogenital groove closes completely, forming the penile urethra, and the genital swellings fuse to form the **scrotum** into which the testes descend. The genital tubercle becomes the **penis.**

4. The stem cells for gamete formation **(spermatogonia** in the testis and **oogonia** in the ovary) undergo **mitosis,** producing cells that become committed to gamete formation **(primary oocytes** or **primary spermatocytes).** The latter undergo **meiosis** which reduces the chromosomal complement to the haploid number (23 chromosomes). Each primary spermatocyte gives rise to four **spermatids,** which subsequently differentiate into **spermatozoa.** In oogenesis, the end products are one **ovum** and **polar bodies.**

5. Puberty is initiated by an increase in hypothalamic release of GnRH, which stimulates the pituitary's secretion of gonadotropins, causing the gonads to secrete sex steroids and produce gametes. The testes secrete the steroid **testosterone;** the ovaries secrete **estrogen** (mainly estradiol) and **progesterone.** Testosterone secretion is relatively steady; secretions of estrogen and progesterone follow an ovarian cycle of maturation of **follicles** followed by ovulation and conversion of the follicle to a **corpus luteum.**

6. At puberty, the sex hormones stimulate sex-specific developmental changes, including effects on fat deposition, muscle, and bone growth; growth of the accessory reproductive structures to their adult size; and effects on behavior and the intensity of sexual interest.

7. **Sertoli cells** of the testis support spermatogenesis and spermiogenesis; **Leydig cells** secrete testosterone. Spermatogenesis is controlled by a feedback loop in which **inhibin** secreted by Sertoli cells inhibits FSH secretion. Testosterone secretion is controlled by feedback loops in which high testosterone levels inhibit secretion of GnRH by the hypothalamus and the release of LH by the pituitary.

8. The ovarian cycle consists of **follicular** and **luteal** phases. Each maturing follicle includes an oocyte, a layer of **granulosa cells,** and **theca cells.** Theca cells synthesize androgen, which is converted to estradiol by granulosa cells. Ultimately one follicle becomes a **Graafian follicle.** The secretion of estrogen rises rapidly as the Graafian follicle grows, triggering a surge of LH secretion which results in ovulation. After ovulation, FSH transforms the remnant of the follicle into a corpus luteum which secretes both estradiol and progesterone during the subsequent luteal phase of the cycle, which is terminated by degeneration of the corpus luteum.

9. The uterine cycle begins with **menstruation,** loss of the lining of the uterine endometrium. During the follicular phase of the ovary, estradiol stimulates regrowth of the endometrium (the proliferative phase). After ovulation, estradiol and progesterone secreted by the corpus luteum cause the endometrium to secrete **uterine milk** (the secretory phase), creating an environment favorable for implantation of an embryo.

10. The **sexual response** is a whole-body response involving both branches of the autonomic nervous system as well as the central nervous system. The response can be divided into **excitement, plateau, orgasmic,** and **resolution** stages.

1. List the primary reproductive organs and the accessory reproductive structures for males and females.

2. Describe the factors that determine whether internal sexual organs will differentiate into male or female structures. What structures arise from the Müllerian duct? The Wolffian duct?

3. What factors normally determine whether the external genitalia develop as male or female? What are some ways in which disorders of development can result in a person whose genital sex is at odds with his or her gonadal and genetic sex?

4. Describe the function of each of the following:
   Graafian follicle
   Theca cell
   Granulosa cell
   Corpus luteum

5. Trace the changes in LH, FSH, estradiol, inhibin and progesterone over the menstrual cycle. What is the physiological role of each of the four hormones?

6. What is the major physiological role of each of the following structures?
   Sertoli cells
   Leydig cells
   Vas deferens
   Seminal vesicle

7. What is the function of the midpiece of a spermatozoan? The acrosome? What is the importance of capacitation for spermatozoa? What is the function of the cortical reaction of oocytes?

8. What are the sites of secretion and physiological effects of each of the following hormones? For the ones that are abbreviated, supply full names.
   GnRH
   FSH
   LH
   Testosterone
   Estradiol
   Progesterone
   Inhibin

9. What consequences would you expect for male children who have the genetic defect that causes adrenogenital syndrome in female children?

Choose the MOST CORRECT Answer.

10. Select the FALSE statement.
    a. Only primary oocytes remain at birth.
    b. Each spermatogonium gives rise to 16 spermatids that all develop into spermatozoa.
    c. Oogenesis begins at puberty and continues until death.
    d. During oogenesis, each oocyte receives most of the cytoplasm, while polar bodies get chromosomes but almost no cytoplasm

11. Which of the following is NOT an effect of luteinizing hormone (LH)?
    a. Increases testosterone secretion by stimulating Leydig cells
    b. Stimulates secretion of estradiol
    c. Causes mature Graafian follicle to ovulate
    d. Promotes production of a corpus luteum

12. This genetic disorder results from the absence of a Y chromosome (the genotype is XO):
    a. Klinefelter's syndrome
    b. Testicular feminization syndrome
    c. Adrenogenital syndrome
    d. Turner's syndrome

13. Testosterone is responsible for all of the following EXCEPT:
    a. Decreases basal metabolic rate at puberty
    b. Inhibits development of Müllerian structures during prenatal development
    c. Establishes male secondary sex characteristics at puberty
    d. Is necessary for spermatogenesis

14. This endocrine disorder results in masculinized female external genitals due to excessive production of DHEA by the adrenals:
    a. Adrenogenital syndrome
    b. Androgen-insensitivity syndrome
    c. Turner's syndrome
    d. Hypopituitary hypogonadism

15. The ovulatory phase of the menstrual cycle occurs during:
    a. Days 1 to 4
    b. Days 5 to 13
    c. Day 14
    d. Days 15 to 28

16. A surge in this hormone signals ovulation:
    a. Estrogen
    b. Progesterone
    c. Estradiol
    d. Luteinizing hormone

17. Select the correct order of the phases of sexual response:
    a. Plateau, orgasmic, excitement, resolution
    b. Excitement, plateau, orgasmic, resolution
    c. Resolution, orgasmic, plateau, excitement
    d. Excitement, orgasmic, plateau, resolution

● SUGGESTED READINGS

BURNETT AL, LOWENSTEIN CJ, BREDT DS, CHANG TSK, SNYDER SH: Nitric oxide: a physiologic mediator of penile erection. *Science* 257:401, 1992. Describes the immunohistochemical localization of nitric oxide synthase in innervation of the penis and experiments that established the role of nitric oxide in erection.

CHERFAS J: Sex and the single gene, *Science* 252:782, 1991. Describes how the SRY gene that determines maleness was found.

EDWARDS DD: Keeping sex under control, *Science News* 133:88, 1988. Reports on discovery of FSH antagonists in women treated with a GnRH analog. These substances have been given the name sex "antihormones." A more detailed paper by Dahl et al. appeared in *Science* 239:72, 1988.

DJERASSI C: Fertility awareness: jet age rhythm method? *Science* 248:1061, 1990. The inventor of the first birth control pill suggests that modern methods for keeping track of the menstrual cycle may supersede chemical contraceptives.

FRISCH RE: Fatness and fertility, *Scientific American* 258(3):88, 1988. Discusses the possible reversible effect of fat tissue on human female reproductivity.

KRANE RJ, I GOLDSTEIN, SI DE TEJADA: Impotence, *New Engl J Med* 321(24):1648, 1989. Summary of the mechamism of erection and the causes of impotence and its treatment.

LEVAY S: A difference in hypothalamic structure between heterosexual and homosexual men. *Science* 253:1034, 1992. Postmortem examination of the brain of women, men presumed to be heterosexual, and homosexual men revealed a size similarity in the brains of women and homosexual men in one region of the anterior hypothalamus. Possible relevance to sexual behavior is discussed with reference to animal studies and the need for additional research on the causation and significance of differences in brain structure.

MCLACHLAN RI, ROBERTSON DM, HEALY DL, BURGER HG, DE KRETSER DM: Circulating immunoreactive inhibin levels during the normal human menstrual cycle, *J Clin Endocrinol Metabolism* 65:954, 1987. Levels of pituitary and ovarian hormones were measured in women throughout the normal menstrual cycle. Inhibin was found to closely follow the pattern of progesterone.

MILLAR RP and KING JA: Evolution of gonadotropin-releasing hormone: multiple usage of a peptide, *News in Physiological Sciences* 3:49, 1988. There are several forms of GnRH that have different regulatory functions.

REINISCH JM, ROSENBLUM LA, and SANDERS SA: *Femininity, basic perspectives*, Oxford University Press, New York, 1987. Presents discussion of possible innate differences between men and women.

WASSARMAN PM: Fertilization in mammals, *Scientific American*, 258(6):78, 1988. Presents recent studies of the mechanism of recognition of oocytes by spermatozoa.

WRIGHT K: Playing demigod: biologists find limits to tinkering with reproduction, *Scientific American* 260(5):30, 1989. Balanced discussion of the potential risks and rewards in gene therapy.

CHAPTER 25

Pregnancy, Birth, and Lactation

On completing this chapter you should be able to:

- Describe the early development of the human zygote, its implantation in the uterus, and differentiation of the embryo and chorion.
- Understand the endocrine functions of the corpus luteum and placenta.
- Understand the architecture of the fetal circulation, how it functions in oxygen transport between maternal blood and fetal tissues, and the changes in the circulation that must occur at birth.
- Know the timing of key events of embryonic and fetal development and describe the changes in maternal physiology caused by pregnancy.
- Describe the stages of labor and the endocrine changes that are hypothesized to initiate labor and increase the strength of uterine contractions as labor progresses.
- Describe the control of breast development and milk production by prolactin, placental lactogen, progesterone, and estrogen.
- Trace the hormonal reflex that results in oxytocin secretion and milk letdown when an infant suckles.
- Describe the treatment for infertility.

*T*he preceding chapter described the structure, development, control and functioning of the reproductive system—the equipment of reproduction. It remains to be shown in this chapter how the reproductive systems of men and women bring together gametes, how the mother's system supports the developing child until it can live outside her body, and how it continues the child's nourishment after birth.

Prenatal development depends on the nutritive and endocrine functions of the placenta, an organ that links the developing infant to its mother's circulatory system and maintains a favorable environment for the fetus. After approximately 39 weeks of development, strong contractions of the muscular wall of the uterus expel the infant into life as a physically separate individual.

To survive, the newborn infant must rapidly make dramatic adjustments in its circulatory system, respiratory system, and metabolism. After birth the child and mother can continue their interdependent relationship through breast-feeding, which supplies both nutrients and temporary immunity. The child's suckling in turn has a potent effect on the mother's endocrine system, providing for continued milk production, suppressing ovulation and favoring return of the uterus to a nonpregnant condition.

PREGNANCY
Meeting of the Egg and Sperm

> • *What anatomical path is followed by the ovum after ovulation? Where does fertilization occur? What are the events of fertilization?*

Upon ovulation, the ovum is swept into the Fallopian tube by the cilia of the fimbriae (Figure 25-1). The ovum is carried toward the uterus by continued ciliary activity assisted by contractions of the smooth muscle cells of the Fallopian tube. An ovum can be fertilized for only about 10 to 12 hours after ovulation, during which period it is still in the Fallopian tube. Spermatozoa can survive in the female reproductive tract for perhaps 3 days, so intercourse (or artificial insemination) is likely to cause pregnancy only during a relatively brief window of time in the menstrual cycle. Fertilization normally occurs at the ovarian end of the Fallopian tube, so spermatozoa deposited into the vagina in sexual intercourse must travel through the uterus and much of the length of the Fallopian tube to meet the ovum.

Some spermatozoa may appear in the Fallopian tubes within minutes after being deposited in the vagina. The spermatozoa could not have made the trip so rapidly if they depended only on their own motility. This suggests that the Fallopian tubes perform a rather tricky task of assisting the movement of both spermatozoa and ovum toward one another. During its passage through the female tract, the spermatozoa undergo a process called *capacitation,* which removes glycoproteins from the membrane surface. Then the spermatozoa that reach the area immediately around the ovum undergo *activation,* the final anatomical and behavioral changes that allow the spermatozoan to penetrate the ovum.

The ovum or its associated follicular cells attracts the spermatozoa by release of a peptide. The response to this chemical is lacking in some sperm, and its presence or absence is enough to make the difference between successful and unsucessful attempts at in vitro fertilizations (those performed outside the female's body). Even with the assistance of Fallopian contractions and the chemical attractant, so many spermatozoa are lost along the way that only a small fraction of the ejaculated spermatozoa (50 to 100 out of 250 million to 500 million) succeed in reaching the oocyte. Consequently, men whose semen contains fewer than 20 million spermatozoa per milliliter are regarded as infertile.

To achieve fertilization (Figure 25-2), a sperm must recognize its target and then penetrate the coat of granulosa cells, the zona pellucida, and the oocyte's cell membrane. This task seems forbidding when the enormous difference in size between oocytes and spermatozoa is considered. However, there is evidence that the ovum releases a chemical that attracts sperm, greatly increasing the chances that enough sperm will reach the surface to contribute to the breakdown of the zona pellucida and allowing one sperm to penetrate and fertilize the ovum.

The recognition that occurs when the sperm head touches the outlying granulosa cells (Figure 25-3) can be compared to receptor binding. The acrosome of the sperm head releases enzymes that allow the head to dissolve its way through the layers surrounding the oocyte. It is the binding of recognition molecules on the acrosome with receptors located on the corona radiata and zona pellucida that causes it to release its enzymes; this **acrosomal reaction** does not readily occur in spermatozoa that have not been capacitated. When the sperm head comes into contact with the oocyte (Figure 25-2), the cell membranes of the two cells fuse and the sperm nucleus enters the oocyte cytoplasm.

Within seconds after the entry of the sperm nucleus, the oocyte's cell membrane undergoes a **cortical reaction.** Vesicles that lie just under the cell membrane fuse with it, forming a **fertilization membrane** (Figures 25-4, *A* and 25-5). This reac-

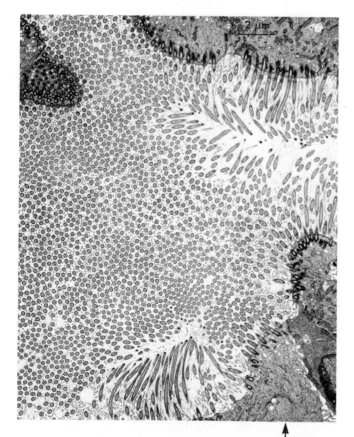

FIGURE 25-1

Fallopian tube

Electron micrograph of a section of Fallopian tube showing the coat of cilia. Many cilia are seen in cross section *(center);* in a few cases the plane of the section passes for some distance along the length of the cilium *(upper right, bottom).*

FIGURE 25-2

Two spermatozoa penetrating the corona radiata of an ovum. The heads of the spermatozoa are labeled with fluorescent antibody to make them more visible under the microscope.

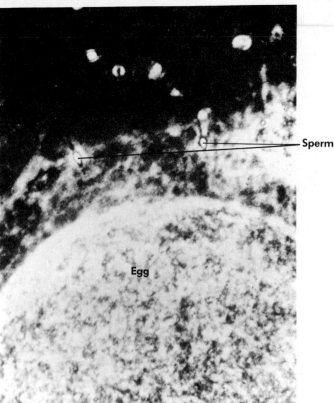

FIGURE 25-3

A scanning electron micrograph of spermatozoa attaching themselves to the surface of an ovum. This picture was made under laboratory conditions; the density of spermatozoa is much greater than would normally occur in the Fallopian tube.

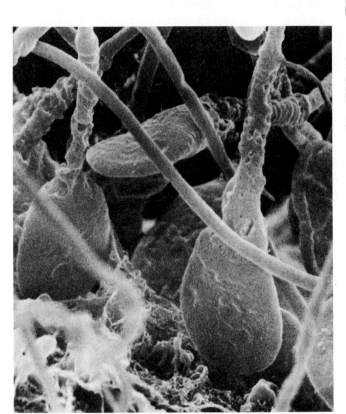

tion involves entry of Ca^{++}, and the process may be similar to that involved in release of synaptic vesicles. The cortical vesicles contain enzymes which alter the sperm receptor molecules on the zona pellucida, making them unable to bind with the sperm. Simultaneous fertilization of an oocyte by more than one spermatozoan is called **polyspermy**. This pre-

vents normal development of the zygote, but occurs rarely because the fertilization membrane typically prevents the binding of additional sperm.

From Fertilization to Implantation

- *What maternal hormones are important for maintaining the pregnancy? What hormones does the fetus produce, and what is the function of each?*
- *What is a blastocyst? What part of the blastocyst gives rise to the fetal portion of the placenta? What part gives rise to the embryo?*
- *What is meant by the terms luteal phase and placental phase of gestation? Do they overlap in time?*
- *What are the endocrine functions of the corpus luteum and the placenta?*

Entry of the head of the spermatozoan into the ovum cytoplasm stimulates the egg to undergo the second meiotic division, producing the second polar body (Figure 25-5). The zygote divides once in 30 hours, then again and again, and progresses from a solid ball of cells to a fluid-filled ball, a **morula**, (Figure 25-4) which continues to travel down the oviduct toward the uterus. The mass remains constant from the single fertilized cell to the morula, so each cycle of cell divisions results in smaller and smaller cells. The egg completes its trip to the uterus in about 4 days. In the meantime, the uterus is continuing to develop into a suitable environment for

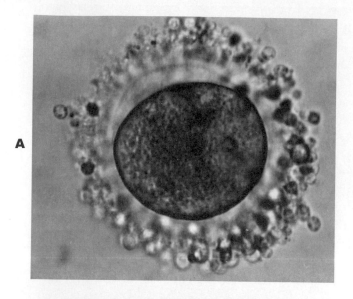

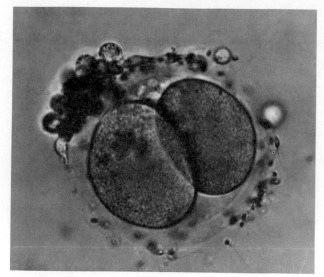

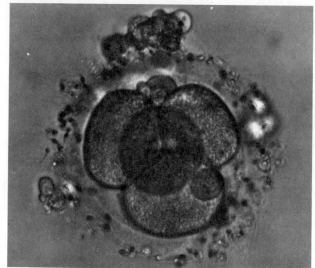

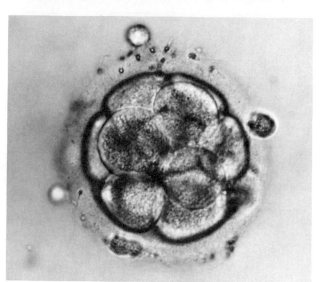

FIGURE 25-4
Early stages of human development.
A The zygote within its fertilization membrane, surrounded by the corona radiata of follicular cells.
B The first mitotic division after fertilization.
C The eight-cell stage after three rounds of mitosis.
D Morula.

nourishing the developing embryo. While the morula is traveling in the Fallopian tube, the post-ovulatory follicle becomes a corpus luteum. In addition to secreting estradiol, the corpus luteum begins secreting progesterone, and the progesterone induces development of spiral arteries and the onset of secretion by uterine glands. By the time the morula arrives in the uterus, it usually consists of 48 cells. The morula floats freely in the lumen of the uterus and is nourished by uterine secretions as it undergoes further development to a form called a **blastocyst** (Figure 25-6).

The blastocyst consists of a fluid-filled sphere of cells, the **trophoblast,** which encloses an **inner cell mass.** The fates of these two parts of the blastocyst are very different. The inner cell mass will become the **embryo** and ultimately develops into a baby's body; the trophoblast will become the **chorion,** which forms an outer covering of the embryo. The chorion gives rise to the fetal component of the **placenta,** the organ that mediates transfer of nutrients, gases, and wastes between maternal and fetal blood and secretes hormones that support the pregnancy.

About the seventh or eighth day after ovulation, the blastocyst lodges against the wall of the uterus. Secretions of the trophoblast dissolve the endometrium, and the blastocyst literally eats its way into the endometrium, nourished by the breakdown

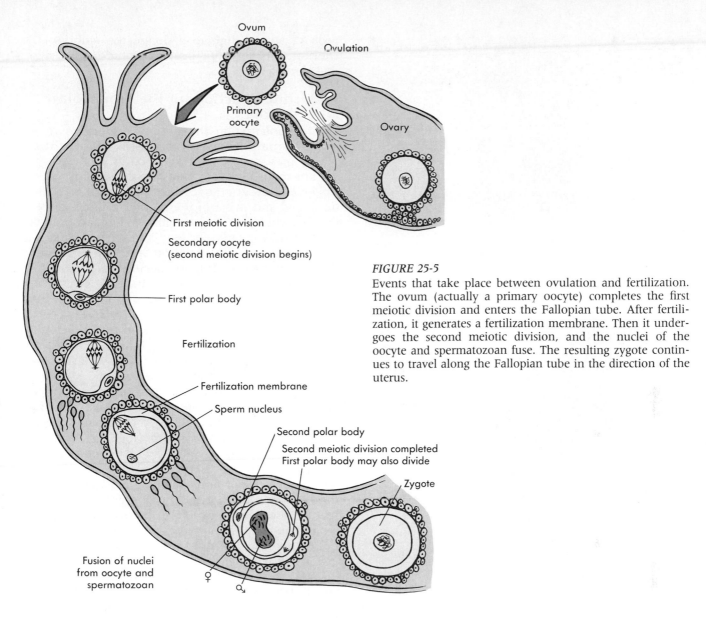

Ovum

Ovulation

Primary oocyte

Ovary

First meiotic division

Secondary oocyte
(second meiotic division begins)

First polar body

Fertilization

Fertilization membrane

Sperm nucleus

Second polar body

Second meiotic division completed
First polar body may also divide

Zygote

Fusion of nuclei
from oocyte and
spermatozoan

♀

♂

FIGURE 25-5
Events that take place between ovulation and fertilization. The ovum (actually a primary oocyte) completes the first meiotic division and enters the Fallopian tube. After fertilization, it generates a fertilization membrane. Then it undergoes the second meiotic division, and the nuclei of the oocyte and spermatozoan fuse. The resulting zygote continues to travel along the Fallopian tube in the direction of the uterus.

of molecules released by enzymatic action (Figure 25-7, *A*). This process is called **implantation**. Finally the blastocyst is completely embedded in the endometrium (Figure 25-7, *B*).

From the beginning of the luteal phase, progesterone secreted by the corpus luteum maintains the uterus in a state receptive for implantation. Each luteal phase of the menstrual cycle is somewhat like a pseudopregnancy, the development of a pregnancy-like hormonal pattern. Even without contraception, about half of all zygotes are believed to die after failing to implant. In such cases the woman never becomes aware that fertilization has occurred.

The situation of the embryo that succeeds in implanting is still far from secure. It has already completed a journey that in proportion to its size, was immense, but it is still tiny in comparison to the size of its mother, consisting of only several hundred cells. From the point of view of its mother's immune system, the embryo is an invader, because

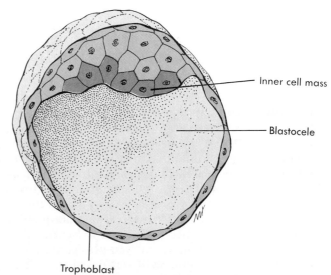

Inner cell mass

Blastocele

Trophoblast

FIGURE 25-6
Preimplantation blastocyst.

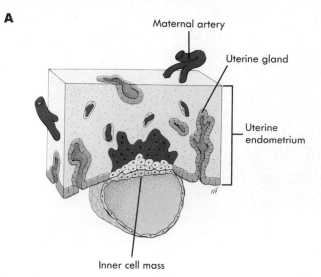

A

Maternal artery

Uterine gland

Uterine endometrium

Inner cell mass

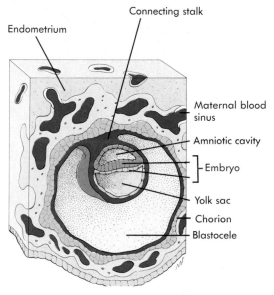

Connecting stalk

Endometrium

Maternal blood sinus

Amniotic cavity

Embryo

Yolk sac

Chorion

Blastocele

B

FIGURE 25-7
A The beginning of implantation. The implanting blastocyst is shown in green.
B After implantation. The inner cell mass is differentiating into an embryo, amnion, and yolk sac. The chorion *(shown in green)* is stimulating the growth of surrounding maternal blood vessels.

the difference in its genetic makeup is reflected in molecules on the surface of the cells which are different from those of the both parents. This would seem to make it vulnerable to attack by the mother's immune system, which is designed to detect and destroy foreign cell types, but the embryo is immunologically privileged (see below). Finally, the embryo is dependent for its survival on continued secretion of estrogen and progesterone by the corpus luteum. Without some signal that pregnancy has begun, the corpus luteum will self-destruct in only a few days. The resulting loss of the uterine endometrium would terminate the pregnancy. Even normal spontaneous contractions of the uterus would

probably be enough to dislodge the embryo. The chorion and placenta provide solutions to all of these problems.

Endocrine Functions of the Trophoblast and Placenta

Before implantation the trophoblast begins to secrete **human chorionic gonadotropin (hCG)**, which prevents the self-destruction of the corpus luteum and causes it to maintain the secretion of estrogen and progesterone. The appearance of hCG in the blood and urine of the mother (Figure 25-8) is the first chemical indicator that pregnancy has begun. hCG is chemically very similar to LH, so the identification of it is based on antibodies generated against the portion of the molecule that differs from LH. An assay for hCG is the basis of early pregnancy tests, both those used by physicians and those available for home use (Figure 25-6). The secretion of chorionic gonadotropin by the trophoblast continues during the first 3 months **(first trimester)** of pregnancy and later a lower level of secretion is maintained by the placenta.

There is a gradual transfer of hormonal support of the pregnancy from the corpus luteum to the placenta. By the beginning of the second trimester, the placenta assumes responsibility for secretion of estrogen, progesterone, and hCG. Although the placenta can synthesize progesterone from cholesterol, it is unable to perform the complete chemical synthesis of estrogens from cholesterol. This requires that the intermediates be transferred to the fetal adrenal gland or through the mother's circulation to her own adrenal glands, where progestins are converted to dehydroepiandrosterone (DHEA) (see Figure 24-14). The placenta then completes the synthesis of estrogens from DHEA. The fetal adrenal also depends on the placenta for supplying progesterone for conversion to cortisol. These pathways of steroid metabolism are an excellent example of interdependence within the maternal-fetal-placental unit (Figure 25-9).

As the placenta begins to secrete estrogen and progesterone, its production gradually overshadows that of the corpus luteum until it is sufficient to maintain the pregnancy. Meanwhile, secretion of HCG decreases and the corpus luteum is regressing. As the contribution of the corpus luteum becomes negligible, the **luteal phase of the pregnancy** is over, and the ovary bearing the corpus luteum could be removed without endangering the pregnancy. The **placental phase of the pregnancy** refers to roughly the second and third trimesters, and refers to the fact that hormonal support of the pregnancy is now provided by the placenta. During the second and third trimesters the plasma levels of estrogen and progesterone rise continuously, reaching a peak just prior to the time of labor (see Figure 25-8).

In addition to hCG, estrogen, and progesterone,

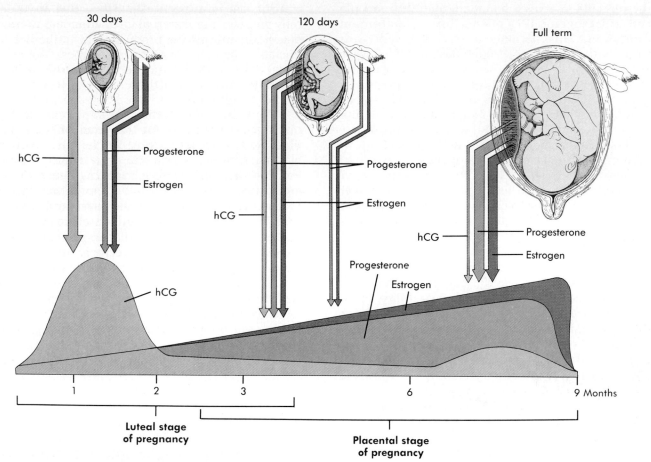

FIGURE 25-8

Changes in the concentrations of some reproductive hormones during pregnancy. During the first trimester *(left)*, hCG secreted by the chorion and placenta maintains secretion of estrogen and progesterone by the corpus luteum in the ovary. During the second trimester *(middle)*, the placenta begins to take over secretion of estrogen and progesterone from the corpus luteum, and secretion of hCG decreases. By the last trimester *(right)*, the placenta is the major source of estrogen and progesterone. Approximate periods of the partially overlapping fetal and placental stages of the pregnancy are indicated. Levels of estrogen and progesterone drop shortly before term.

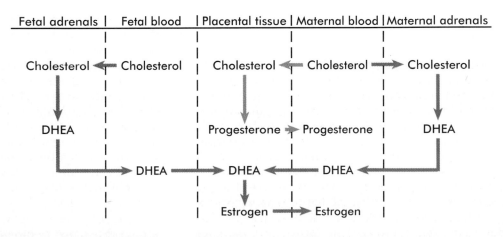

FIGURE 25-9

Metabolism of cholesterol to progesterone and estrogen in the placenta requires the cooperation of the adrenals of mother and fetus. The contribution of the adrenals is conversion of progestins to DHEA, which the placenta can convert to estrogens.

Pregnancy, Birth, and Lactation

the placenta secretes a hormone called **human placental lactogen (HPL)** or **chorionic somatomammotropin.** This hormone is secreted very early, along with HCG, by the trophoblast, and the level of HPL rises throughout the first and second trimesters and reaches a peak in the last month of pregnancy. Placental lactogen is structurally similar to both growth hormone (Chapter 23) and prolactin (a hormone involved in milk production, discussed later in this chapter), and HPL has similar effects on the mother. Placental lactogen stimulates breast development in preparation for postnatal lactation (milk production), supports fetal bone growth, and alters the way the mother uses metabolic substrates by substituting lipids for glucose as the cellular energy source. Lipid utilization by the mother makes more glucose available to the fetus. The placenta also secretes antidiuretic hormone, aldosterone, and renin. All these act to increase the maternal extracellular fluid volume, which enhances the ability of the mother's circulatory system to meet the demands of the fetus and placenta.

One of the most recently recognized hormones of pregnancy is **relaxin.** Like estrogen and progesterone, relaxin is initially secreted by the corpus luteum, but later in pregnancy most relaxin probably comes from the uterine muscle and placenta. Relaxin softens the cervix and uterus, providing one of the earliest anatomical changes associated with pregnancy. Later, it softens connective tissue between the bones of the pelvic girdle, so that the pelvic aperture is enlarged and the baby passes through more easily at the time of delivery. Relaxin withdrawal just before delivery may contribute to the uterine smooth muscle responsiveness to contraction-producing agents present in labor.

The Structure and Functions of the Placenta

As the embryo continues to grow, simple diffusion is inadequate to supply nutrients and oxygen and to carry away carbon dioxide and wastes. This problem is partly solved as the fetus develops a circulatory system and the chorion gives rise to the fetal part of the placenta. The fetal blood must be close enough to the maternal blood for exchange to take place, but actual mixing of the two blood supplies is prevented, reducing the chance that the mother's immune system will mount an attack against the fetus. The structure of the placenta is such that placental villi containing fetal capillaries are immersed in **maternal blood** sinuses (Figures 25-10 and 25-11). Diffusional exchange of gases and some nutrients occurs across the capillary walls, and other nutrients are actively transported. The fetal derivatives of the chorion constitute the **fetal-maternal barrier** that usually protects the fetus. Its remarkable ability to act as an immunoabsorbant barrier to antifetal antibodies is under investigation, especially

since it has now been established that some fetal cells do enter the maternal circulation and therefore potentially initiate the production of antibodies (see Prenatal Diagnosis of Genetic Diseases). Some maternal antibodies are transferred directly to the fetus, protecting it from those diseases to which the mother is immune.

The remarkable ability of the placenta to organize maternal support for the needs of the fetus is illustrated by **ectopic** pregnancies, in which the blastocyst implants at a site other than the uterus. Possible sites are the abdominal cavity and the Fallopian tube. Even in ectopic sites the placenta can protect the fetus from immune attack and stimulate sufficient new maternal blood vessels to support fetal development. Embryos that implant in a Fallopian tube, the most common ectopic site, will eventually rupture the tube, causing abdominal bleeding that threatens the life of the mother. These pregnancies must be terminated. Implantation at other ectopic sites in the body cavity is not necessarily life threatening for mother or child and may go undetected until the pregnancy is quite advanced. In these very ununusual pregnancies it has been possible to deliver the baby by Caesarean section.

The Fetal-Placental Circulation

- *What is a vascular shunt? What are the three vascular shunt pathways of the fetal circulation? What happens to these shunts at birth?*
- *Where in the fetal circulation is the blood most oxygenated? Least oxygenated? Why is blood in the fetal dorsal aorta less saturated with oxygen than blood in the umbilical vein?*
- *Why is the fetal pulmonary loop a high-resistance pathway? What effect does this have on blood flow through the loop?*

The fetal circulatory system is adapted to include the placental circulation (Figure 25-12). Two **umbilical arteries** branch from the fetal descending aorta and pass into the umbilical cord that connects the fetus with the placenta. A single **umbilical vein** returns blood from the placenta. The umbilical vein joins the venous circulation of the fetal liver, which is connected to the inferior vena cava by the **ductus venosus.** This shunt allows blood returning from the placenta to pass directly to the vena cava by a low-resistance route. The blood that enters the fetal vena cava is a mixture of relatively well-oxygenated blood from the umbilical vein and placenta and less well-oxygenated blood flowing in the hepatic portal vein. As this blood flows from the ductus venosus into the vena cava, there is a further dilution with fetal venous blood.

The lungs of the fetus are collapsed and present a high resistance to blood flow. Since they serve no respiratory function before birth, it is unnecessary that all of the cardiac output pass through them.

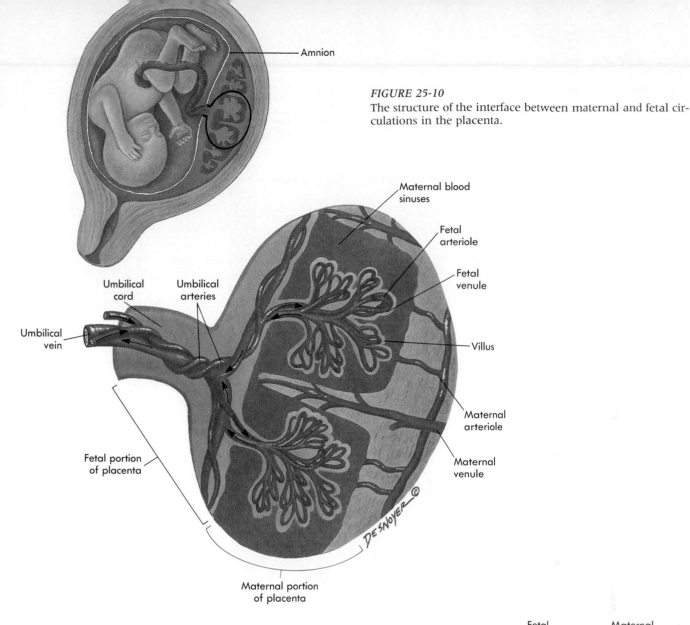

- Amnion

- Maternal blood sinuses
- Fetal arteriole
- Fetal venule
- Villus
- Maternal arteriole
- Maternal venule
- Umbilical cord
- Umbilical arteries
- Umbilical vein
- Fetal portion of placenta
- Maternal portion of placenta

DESNOYER ©

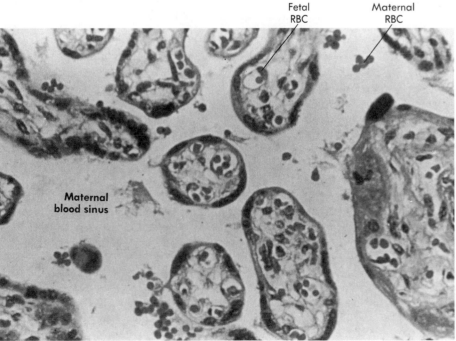

- Fetal RBC
- Maternal RBC
- Maternal blood sinus

FIGURE 25-11
Photomicrograph of a section of placenta showing fetal capillaries in a maternal blood sinus.

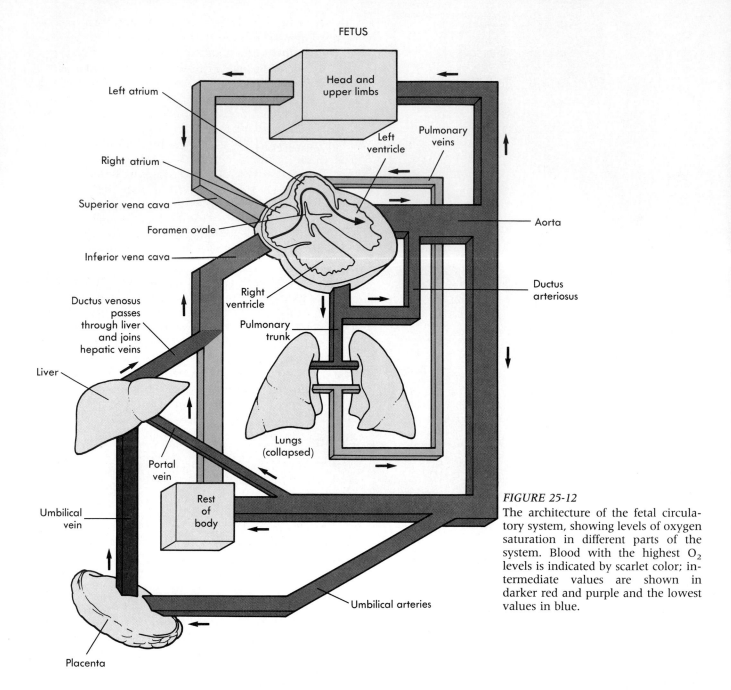

FIGURE 25-12
The architecture of the fetal circulatory system, showing levels of oxygen saturation in different parts of the system. Blood with the highest O₂ levels is indicated by scarlet color; intermediate values are shown in darker red and purple and the lowest values in blue.

Venous blood returning to the right heart from the body mixes with placental blood and enters the right heart. Most of this blood is not pumped to the lungs, but instead takes one of two shunts to the systemic circulation. One shunt pathway is provided by an opening, the **foramen ovale,** in the septum between the atria. The other is the **ductus arteriosus,** a connection between the aorta and the pulmonary artery. This arrangement of the placental circulation has some important consequences. The placental blood equilibrates with partly deoxygenated maternal blood, returning to the fetus only about 80% saturated with oxygen. The placental blood mixes with fetal venous blood that is about

27% saturated, so that the arterial blood is only about 58% saturated with oxygen. Consequently, fetal tissues must be adapted to function at lower O₂ partial pressures than those normally seen by the corresponding adult tissues.

Fetal hemoglobin is adapted to function at lower partial pressures than those that will be typical after birth. Fetal hemoglobin, like adult hemoglobin, is a tetramer, (see Chapter 17) but it contains two gamma subunits in place of the two beta subunits of adult hemoglobin. The substitution shifts the O₂ dissociation curve leftward compared to that of adult hemoglobin (Figure 25-13). The shift allows fetal hemoglobin to load from partly

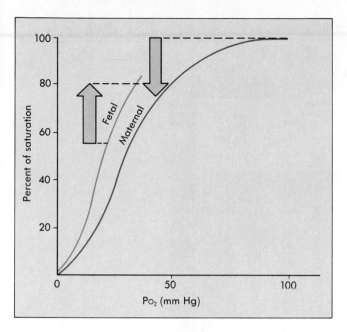

FIGURE 25-13
The oxygen dissociation curves of maternal and fetal hemoglobin. The leftward position of the curve for fetal hemoglobin indicates a higher oxygen affinity. As maternal blood enters the placental blood sinuses, the P_{O_2} drops and O_2 is unloaded from maternal hemoglobin *(indicated by arrow pointing down)*. The oxygen given up by the maternal hemoglobin is accepted by the fetal hemoglobin *(arrow pointing up)*. Even with the leftward shift of its dissociation curve, the fetal hemoglobin is still not saturated when it leaves the placenta.

deoxygenated maternal blood and increases its oxygen carrying capacity by 20% to 30% at the P_{O_2} of placental venous blood.

Oxygen delivery and CO_2 exchange are homeostatic problems for the mammalian fetal circulation. The rate of gas exchange across the placenta is affected by the rate of blood flow through the placental loop. This is determined in turn by the relative resistances of the placental circuit and the systemic circulation of the fetal lower body. Fetal-placental blood flow is affected by two hormone systems best known in adults for their role in extracellular fluid volume regulation: the ADH system and the renin-angiotensin system. A drop in the P_{O_2} of fetal arterial blood stimulates release of ADH whereas an increase in the P_{CO_2} or a decrease in the pH stimulates renin release and a rise in angiotensin II. Both ADH and angiotensin II increase the resistance of the fetal systemic loop, increasing the fraction of the cardiac output that takes the placental loop. The effect is to increase oxygen delivery to the fetus, particularly to the vital organs of the head and upper body. The participation of these two hormones in regulation of the fetal-placental circulation is not entirely surprising, since both ADH (vasopressin) and angiotensin have vasoconstrictor effects in the adult circulation.

At birth, the fetal circulatory and respiratory systems must undergo dramatic changes to assume their function in the extrauterine environment (Figure 25-14). When the umbilical circulation is interrupted, the ductus venosus and umbilical arteries close off and eventually atrophy. The ductus arteriosus constricts and later becomes permanently closed. The inflation of the lungs with the first breath lowers the resistance of the pulmonary loop, and all of the blood pressures in this loop fall below their respective values in the systemic loop. The foramen ovale is provided with a flap that permits blood to flow from the right atrium to the left atrium, but not in the reverse direction. The development of a pressure gradient between the right and left atria causes the flap to be pushed shut. Eventually the flap more or less fuses to the rest of the interatrial septum, and separation, of the pulmonary and systemic loops is accomplished.

Occasionally the shunts between the pulmonary and systemic loops persist after birth and may even be present throughout life. For example, the ductus arteriosus may fail to close **(patent ductus)**. If such a persistent anatomical connection between the two loops results in blood leakage from the systemic loop into the pulmonary loop, it can put a strain on the pulmonary loop, which is normally at lower pressure than in the systemic loop.

From Implantation to Birth: The Developmental Timetable

> • *What major events are characteristic of the embryonic period?*

The human **gestation period,** or duration of prenatal development, is around 39 weeks, counting from the first day of the last menstrual period. The gestation period is divided into three 3-month parts—the first, second, and third **trimesters.** Most of the organ systems begin to differentiate within the **embryonic phase** of development—the first 8 weeks of gestation (Figure 25-15, *A* and *B*). This is the part of gestation in which sensitivity to drugs and other chemicals is greatest. The subsequent 31 weeks of development are the **fetal phase.**

By about a week after implantation, the beginnings of the heart and nervous system are recognizable. By 2 to 3 weeks the heart has begun to beat, but the mother may not yet suspect that she is pregnant. By 3 to 4 weeks, most of the major organ systems have begun to differentiate. The embryonic thyroid, parathyroids, liver, and pancreas are present; the genital tubercle is recognizable; and limbs, eyes, and nose have begun to form. At this point the body of the embryo is about 0.6 cm long (see Figure 25-15, *A*). It is enclosed in a double-walled chamber attached to the uterine endometrium by the placenta. The outer wall of the

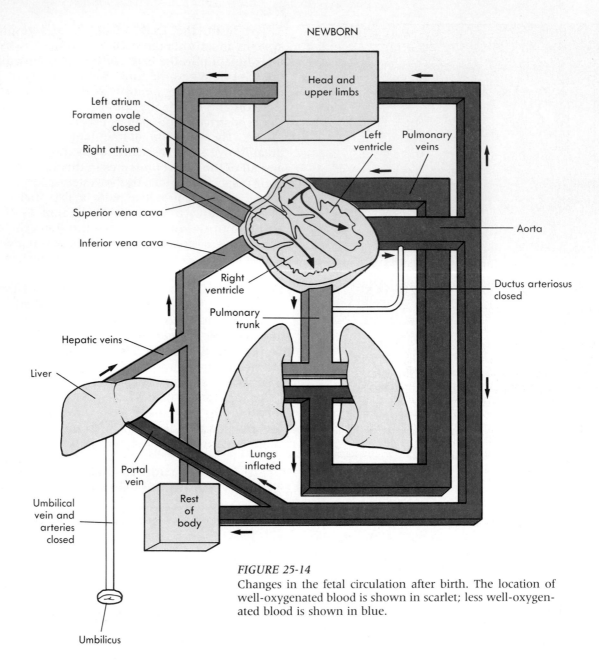

FIGURE 25-14
Changes in the fetal circulation after birth. The location of well-oxygenated blood is shown in scarlet; less well-oxygenated blood is shown in blue.

chamber is the chorion, which is continuous with the placenta; the inner wall is the **amnion** (Figure 25-16). Within the amnion, the embryo, attached to the placenta by its umbilical cord, floats in a miniature sea of **amniotic fluid.** A **yolk sac** attached to the abdomen of the embryo is one site of formation of blood cells (Figure 25-16). By the sixth week of development, the embryonic central nervous system, at first a simple tube of neural precursor cells, shows subdivisions characteristic of the adult CNS and has begun to innervate the developing appendages; sensory structures such as eyes and ears have begun to differentiate.

By the end of the first trimester, the embryo, now a **fetus,** is 7 to 8 cm long (Figure 25-15, *C* and *D*). Its heartbeat can be detected. Its face is recogniz-

ably human, its appendages are complete, and its genital sex can be determined. The development of each individual organ system can be thought of as a thread that is woven together with other threads to generate an organism that, as development goes forward, is increasingly capable of independent life. By the end of the second trimester, the fetus is 25 to 37 cm long, weighs 550 to 700 g, and differentiation of all of its organ systems is well advanced. It is approaching the stage at which, with medical assistance, it might survive if born prematurely. During the third trimester, the fetus grows rapidly and also accumulates a store of fat important for survival in the first few days after delivery. Some organs, such as the heart and kidney, are already functioning. At about the eighth month the lung begins to prepare

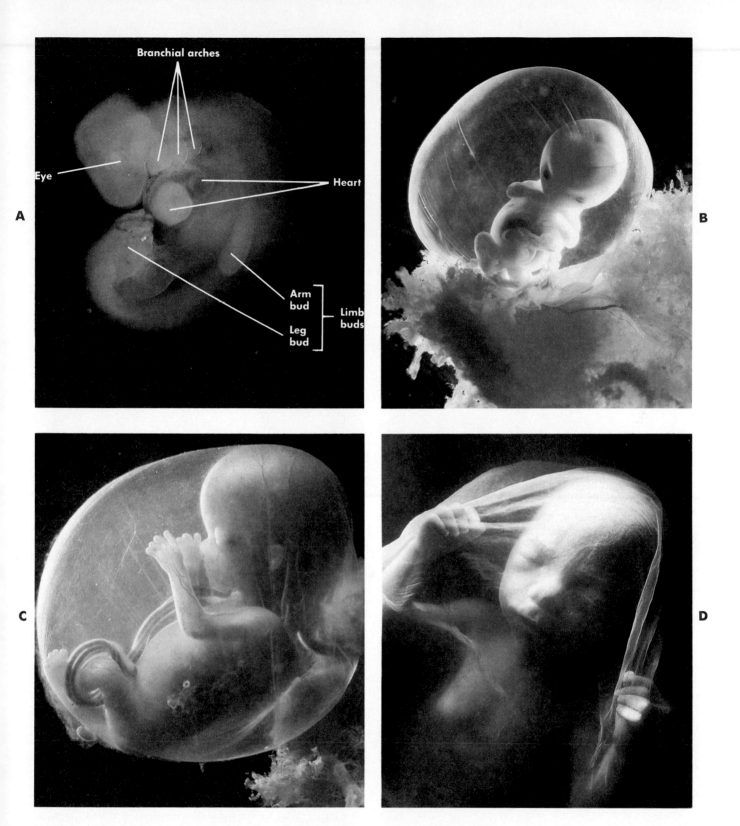

FIGURE 25-15
Human embryos and fetuses.
A   At 35 days.
B   At 49 days.
C   At the end of the first trimester.
D   At 4 months.

FIGURE 25-16
The embryo and surrounding structures after about 1 month of development (see also Figure 25-15, *A*).

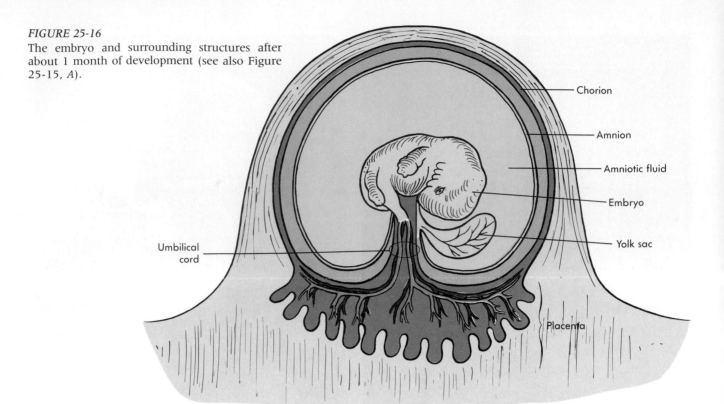

Chorion

Amnion

Amniotic fluid

Embryo

Yolk sac

Umbilical cord

Placenta

for function by secreting **pulmonary surfactant.** The presence of surfactant greatly increases the chance of survival by reducing the alveolar surface tension (see Chapter 16). Fetal nervous system development continues throughout the intrauterine period and is not complete even at birth. This is the reason that inadequate nutrition, substance abuse by the mother, and exposure to **teratogens** (agents that cause birth defects) can have such a devastating effect on the nervous system at any period of the gestation.

At birth the baby typically weighs between 3 and 3.5 kg and is about 50 cm long. Birth is a milestone in the developmental process but does not represent the end of it, since many organ systems, including the central nervous system, the reproductive system, and the muscular-skeletal system, do not arrive at their final form until puberty or later.

Alterations of Maternal Physiology by Pregnancy

> • *What are some of the effects of pregnancy on the mother's physiology?*

Pregnancy has dramatic effects on the physiology of the mother. Most of these are related to the hormones produced by the placenta (Table 25-1). Progesterone's major function has been associated with maintaining a pregnancy, and progesterone acts on smooth muscle, particularly uterine smooth muscle,

to reduce the tendency of the normally excitable muscle to contract, even in response to fetal movements. As a consequence of the large amounts of progesterone in the mother's blood, the smooth muscle of the GI tract is also relaxed, which slows digestive processing and can lead to constipation if it is not counteracted with additional fiber intake in the diet and adequate exercise.

Only about 30% of the weight gain of pregnancy is accounted for by the fetus and placenta (Figure 25-17); the rest is due to changes in the mother's body that support the pregnancy and subsequent lactation (discussed below; see Table 25-2). By the end of pregnancy the mother's blood volume may have increased by about 30% as the result of the placental steroids and increased secretion of aldosterone by the mother's adrenals. The increase in blood volume causes her cardiac output to increase by a similar percentage. Respiratory output must increase proportionally. Several athletes have turned in excellent Olympic performances while in the early stages of pregnancy.

The early gain in weight is often referred to as "water weight." There is a tendency for women to have a lowered threshold for thirst, and the fluid intake results in increases in the blood volume as a result of coordinated alterations in renal function. However, under conditions of adequate food intake, the early weight gain also represents true nutrient storage in the form of fat. This weight gain results from two factors, the alteration of metabolism to fa-

TABLE 25-1   *The Hormones of Pregnancy, Labor, and Lactation*

Hormone	Secreted by	Effects
Estrogen	Corpus luteum, placenta	Maintains endometrium; contributes to breast development
Progesterone	Corpus luteum, placenta	Maintains endometrium; inhibits contraction of myometrium; stimulates breast development
Placental somatomammotropin (fetal lactogen)	Placenta	Increases fat and carbohydrate catabolism in mother, making stored nutrients available to fetus; contributes to breast development
Oxytocin	Posterior pituitary	Stimulates uterine contraction; causes milk letdown
Relaxin	Corpus luteum, placenta, uterus	Softens cervix; loosens ligaments of pelvis; inhibits uterine contractions; potentiates effect of oxytocin
Prolactin	Anterior pituitary	Maintains milk production after delivery

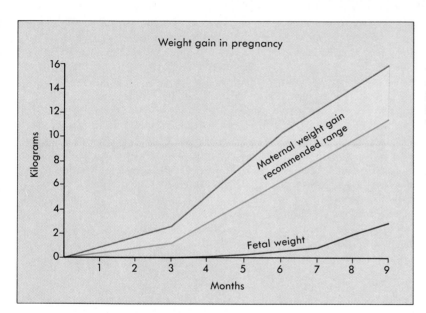

FIGURE 25-17
Maternal weight gain during pregnancy (in kilograms). The portion of the weight gain due to the growth of the fetus itself is shown in the lower curve.

vor insulin-mediated transfer of sugar into the fat cells and an alteration of behavior to reduce activity, reducing the demands for glucose utilization to supply energy demands. Most women experience fatigue and a greater need for sleep early in pregnancy. At least part of this effect is attributable to the hormone CCK, which acts on the brain to increase sleepiness and increases the release of insulin.

The quality of the mother's nutrition during the first trimester is important because during this time her intake of protein, vitamins such as folic acid, and minerals such as calcium, phosphorus, and iron should exceed the immediate demands of the embryo so that the extra nutrients and minerals stored during the first trimester can be drawn on during

TABLE 25-2	*Typical Components of Maternal Weight Gain during Pregnancy*

Component	Weight (kg)
Baby	3.2
Placenta	0.5
Amniotic fluid	0.9
Breasts	0.9
Blood	1.4
Uterus	0.9
Other	8.1
TOTAL	15.9

A genetic defect occurs in approximately 2% of all pregnancies recognized, and defects probably account for a larger percentage that are terminated spontaneously so early that they are never detected. The likelihood of some genetic defects is higher in parents belonging to particular ethnic groups (Table 25-A). The incidence of pregnancies involving fetuses with chromosomal abnormalities rises as a function of maternal age, presumably due to the fact that the follicles of older women have had longer to develop chemical alterations that result in chromosomal nondisjuction. For example, Down's syndrome is a form of mental retardation that results from possession of an extra chromosome number 21. The incidence of abnormalities involving the sex chromosomes also rises with maternal age. Parents whose age or genetic makeup puts them at risk for conceiving a child with a genetic abnormality may elect to undergo a process of prenatal diagnosis.

The method of prenatal testing which has been in use the longest is called **amniocentesis** (Figure 25-A). This process is performed relatively far into the pregnancy, in the 14th to 16th week of gestation. It involves removing a sample of fluid from the amniotic sac, which contains some live fetal cells which have been sloughed off the membrane. A needle is passed through the mother's abdomen and into this sac, using ultrasonography to locate the relevant structures and direct the needle away from the fetus (Figure 25-A). The viable cells must be cultured for 2 to 3 weeks to increase the number and allow enough to be observed in metaphase so that the chromosomes of the dividing cells can be examined microscopically for abnormalities. The fetal cells can also be analyzed at an earlier stage for markers that indicate genetic abnormalities, such as neural tube defects (see below) or

the abnormal enzyme responsible for Tay-Sachs disease, in which myelination of neurons is impaired.

The risk of miscarriage following amniocentesis is 0.5%. This must be balanced against the incidence of birth defects, which is 1% to 2% in women over 35. The advanced state of the pregnancy before the sample is taken and the delay before the results are known is a major drawback.

The wait can be a very harrowing experience for the parents, and if the results lead them to choose an abortion, the pregnancy is so advanced that relatively safer abortion procedures limited to the first trimester cannot be used. Some facilities are able to offer early amniocentesis procedures in the 9th to 14th week, but techniques offering even earlier results are also becoming routine.

One of the newer techniques, chorionic villus biopsy, involves inserting a catheter through the cervix to collect a small piece of the placenta. This method can provide diagnostic information in the 7th to 12th week of gestation. No culture of the cells is necessary, so results are available within a week. Experience with this method is reducing the risks to a level similar to the risks associated with amniocentesis.

Percutaneous umbilical blood sampling (PUBS), also called cordocentesis, is used late in a pregnancy (18 to 36 weeks, even up to the end of the pregnancy) to check for chromosome abnormalities if amniocentesis is inconclusive, or because problems are suspected. In addition to genetic abnormalities, Rh-factor incompatibility, and infections such as German measles and herpes can be detected. Under continuous ultrasound localization, a sample of fetal blood is taken from the umbilical cord. The results are available within 3 days and the risk of miscarriage is 1% to 2%. If Rh-factor incompatibility is detected, the fetus can be saved by a blood transfusion.

Ideally, prenatal tests will entail no risk to the pregnancy and the results should be available early in the pregnancy and with little delay. One such test is currently available. Physicians can offer a test called maternal serum alpha fetoprotein screening, which tests for neural tube defects, which include spina bifida and anencephaly, malformations of the spine or brain.

The incidence of false positive results is high, and many perfectly normal conditions, such as twins or a more advanced gestational age than expected, can give false positives. Parents who have experienced loss of a pregnancy involving a neural tube defect have a higher than average chance of encountering the same condition in a second pregnancy, so the peace of mind that can come from a normal reading can be a great relief. Alternatively, the opportunity to prepare for Cesarean delivery of infants with known defects can greatly reduce the severity of the child's injuries. Encouraging news is the fact that folic acid supplements begun prior to or early in pregnancy can greatly reduce the incidence of neural tube defects.

The newest technique for obtaining fetal cells for genetic diagnosis is still under investigation. It is based on the discovery that the blood of pregnant women contains some live fetal cells. The relatively rare fetal cells in the sample of mother's blood can be concentrated by a process called flow cytometry, which uses laser light to separate cells on the basis of defined features. Two cell surface markers characteristic of fetal cells and other factors are used to obtain a sample greatly enriched in the fetal cells, which can be fixed on a glass slide and sent for analysis. DNA probes that bind to segments of DNA that are responsible for 90% to 95% of birth defects are available, and probes for the rarer defects will also be developed. The detection procedure, called fluorescent in situ hibridization (FISH), is expected to offer rapid screening that can be verified, if necessary, by standard chromosomal techniques.

Ultrasound imaging of the fetus is acheived by imaging the echoes from pulses of sound waves, which are reflected and scattered to a different extent by the various maternal and fetal tissues. Sometimes anatomical defects are detected by this method.

All of these procedures give parents information that may cause them to terminate the pregnancy, or may allow them to prepare to deal with the additional problems posed by an abnormal condition or disease, if they elect to continue the pregnancy. Ultimately prenatal diagnoses like these may give physicians the opportunity to provide therapy to fetuses with genetic and developmental diseases.

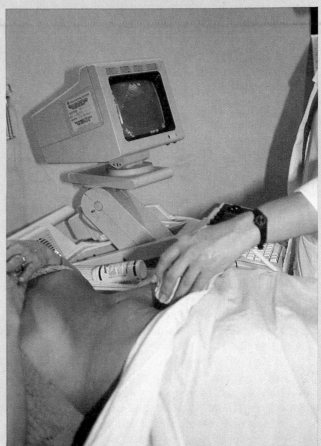

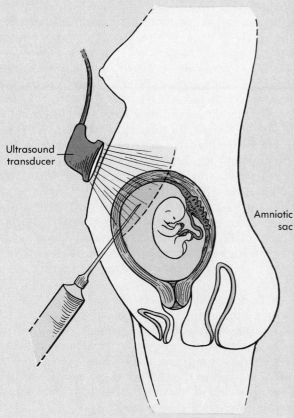

Ultrasound transducer

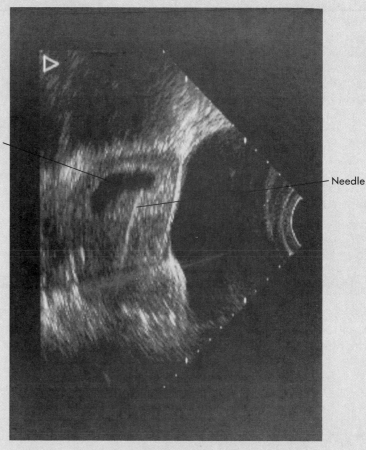

Amniotic sac

Needle

FIGURE 25-A
Ultrasonogram of amniotic sac with needle in place for amniocentesis. As shown in the inset, the ultrasound imager reconstructs an image of a two-dimensional section through the uterus. In this case the section passed through the uterus along a plane that included the amniotic sac and part of the placenta, but not the fetus itself. Passing the needle into the amnion along this plane ensures that it will miss the fetus.

TABLE 25-A *Some Genetic Diseases Detectable before Birth*

Condition	Deficit	Incidence
Cystic fibrosis	Single gene	1:2500
Hemophilia	Single gene (X chromosome)	1:3000
Sickle cell anemia	Single gene	1:500 (among Blacks)
Tay-Sachs disease	Single gene	1:3000 (among Jews)
Down's syndrome	Chromosomal	1:600 to 1:100 (for women age 20s to 40s)

the second and third trimesters, when fetal growth may impose greater demands than can be met by intake. This fact underscores the value of beginning a program of good prenatal nutrition as early as possible, perhaps even before pregnancy.

In the second and especially the third trimester, the hormones that support pregnancy create a bias that favors fetal growth rather than maternal nutrient storage. The same levels of nutrients are available to both the fetal and the mother's body cells, but they are not equally able to use them. One of the ways this is accomplished is through interactions between insulin and human placental lactogen. HPL blocks the effects of insulin on maternal cells, reducing their ability to take up glucose and leaving the blood levels of glucose high so that this nutrient is available for the fetal cells. Some women develop a mild diabetic condition during the latter portion of a pregnancy, and the exaggeration of already existent diabetes mellitus by the hormonal pattern of pregnancy makes it difficult for a diabetic women to undergo a pregnancy.

BIRTH AND LACTATION
What Determines the Length of Gestation?

- *What factors are hypothesized to be responsible for the onset of labor? What are the roles of oxytocin and prostaglandins in labor?*
- *What are the three stages of labor?*

During most of the normal gestation period, the uterine smooth muscle is relatively quiet, but as the fetus grows, the muscle is stretched. Stretching increases the contractility of the muscle (**stress activation**), and some uncoordinated contractions may occur throughout pregnancy (**Braxton-Hicks contractions**).

What factors initiate and maintain human labor? One factor that has been recognized for many years

and which has been utilized in induction of labor is the rising sensitivity of the uterine muscle to oxytocin. Early in pregnancy there are few uterine oxytocin receptors, but oxytocin receptors appear in increasing numbers as pregnancy progresses. Athough labor involves a response to oxytocin secretion, the cervix does not dilate properly in the absence of uterine prostaglandins, and therefore induced labor does not progress normally if the prostaglandins are not released.

Binding of oxytocin to its receptors has two effects: (1) it causes uterine smooth muscle contraction, and (2) it stimulates the uterus to make prostaglandins ($PGF_2\alpha$ and PGE_2) that also activate uterine smooth muscle. Some researchers believe that the consistent rise in prostaglandin production close to term occurs because fetal inhibition of prostaglandin production is withdrawn. Prostaglandins are also used to artificially induce labor or assist in the delivery. The cells in the endometrial-derived portion of the placenta that are responsible for the large increase in prostaglandin release are macrophages. These cells are highly responsive to a range of stimuli, including steroid hormones, oxytocin, and factors associated with infections and inflammation. The rise in prostaglandins which is known to bring about preterm labor may result from the exposure of these cells to infections that have spread to the fetal membranes.

Despite increasing levels of oxytocin and prostaglandins, progesterone directly inhibits uterine smooth muscle and blocks its response to oxytocin and prostaglandin. Used as a drug, progesterone is frequently effective in preventing threatened premature labor. However, maternal progesterone levels do not fall abruptly before labor. Therefore, a drop in progesterone does not precipitate labor. Another possible contributing factor in the timing of labor is relaxin, which promotes an increase in the number of oxytocin receptors on uterine muscle. The amount of relaxin in the maternal blood peaks just before delivery and then drops off rapidly, so that the relaxing effect it exerts on uterine muscle is removed.

An understanding of the factors that normally prevent premature activation of labor would have medical applications, because premature babies require much medical attention and have a much poorer chance of survival than do full-term infants. Factors involved in the timing of labor differ among different species of mammals that have been studied. Attention has focused on the secretion of cortisol by the fetal adrenal cortex. Cortisol stimulates secretion of lung surfactant and increases the ability to mobilize metabolic stores between meals and to thermoregulate. In some animals there is a dramatic increase in cortisol secretion by the fetal cortex several weeks before birth, which could serve as a signal to the placenta or maternal endocrine system

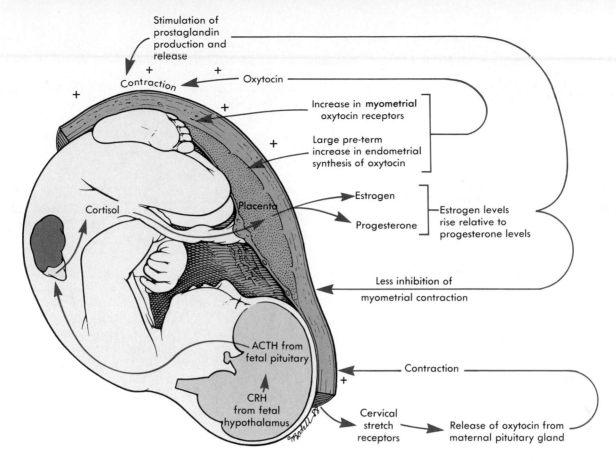

FIGURE 25-18

Labels within the figure:

- Stimulation of prostaglandin production and release
- Contraction
- Oxytocin
- Increase in myometrial oxytocin receptors
- Large pre-term increase in endometrial synthesis of oxytocin
- Cortisol
- Placenta
- Estrogen
- Progesterone
- Estrogen levels rise relative to progesterone levels
- Less inhibition of myometrial contraction
- ACTH from fetal pituitary
- CRH from fetal hypothalamus
- Contraction
- Cervical stretch receptors
- Release of oxytocin from maternal pituitary gland

FIGURE 25-18
Factors that may be involved in initiation of labor. Secretion of ACTH by the fetal pituitary increases near term, causing an increase in cortisol secretion by the fetal adrenals. This in turn is believed to inhibit progesterone synthesis by the placenta, causing estrogen levels to rise relative to progesterone levels, and increasing synthesis of a prostaglandin that increases the excitability of the myometrium. Once contractions of the uterus become strong enough to trigger oxytocin secretion, a positive feedback loop is set in motion.

that labor should begin (Figure 25-18).

At the onset of labor, the pressure of the head of the fetus against the cervix initiates a hormonal reflex (Figure 25-18) that increases secretion of **oxytocin.** Oxytocin greatly potentiates uterine contractions, setting up a positive feedback loop of strong, coordinated waves of contraction that continues until delivery is complete.

The Stages of Labor

During the last trimester the fetus usually takes a head-downward position so that its head is against the cervix. Normally, it will be delivered headfirst. In **breech birth,** the buttocks or legs enter the cervix first. Other presentations include *transverse.* The approach of labor is indicated by **effacement,** or thinning of the cervix in response to relaxin, and promoted by uterine contractions.

There are three stages of labor (Figure 25-19). Stage one is marked by the onset of regular uterine contractions that open or **dilate** the cervix to 7 to 10 cm from its prelabor diameter of 0 to 2 cm. During this stage, uterine contractions occur in waves in which the intervals between contractions are about 3 minutes. The duration of stage one is variable.

Stage two begins when the baby begins to move through the cervix and ends when it is delivered. Stage two may last as long as several hours in first

deliveries, but may be as brief as a few minutes in women who have previously delivered. In stage three, the placenta (afterbirth) is delivered; this stage may last only a few minutes.

Prolactin and Lactation

- *What factors are responsible for priming of the breasts for milk production during gestation?*
- *What are the roles of prolactin and oxytocin in lactation?*
- *What is the effect of prolactin secretion on gonadotropin secretion and ovulatory cycling?*

The evolution of mammary glands from sweat glands was a defining step in mammalian evolution. In nonlactating women, breast size is determined almost entirely by the distribution of adipose tissue, a secondary sexual feature that does not contribute to milk production. At puberty, breast development

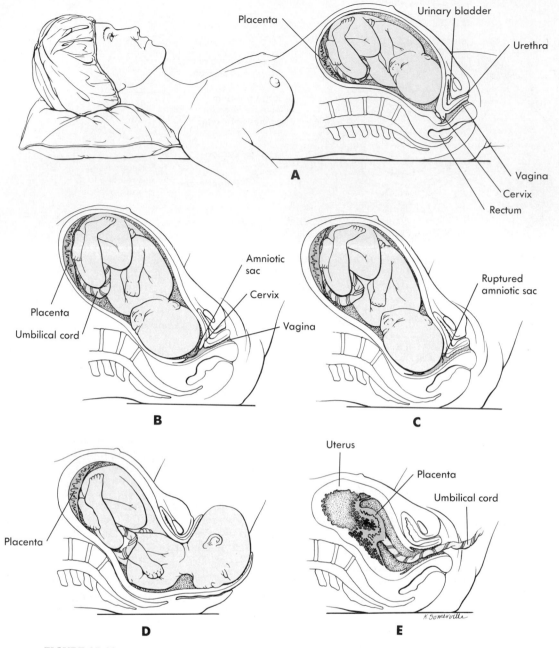

FIGURE 25-19
Stages of labor.
A Fetal position before the start of labor.
B and C First stage of labor. In **C**, contractions have ruptured the amniotic sac.
D Second stage of labor begins.
E The third stage of labor begins with separation of the placenta from the wall of the uterus.

is stimulated mainly by estradiol (Figure 25-20), although insulin, thyroid hormones, cortisol, and growth hormone have permissive effects. Progesterone results in proliferation of the sinuses and ducts and in generalized breast enlargement (so that the size of the breast varies during the menstrual cycle).

During pregnancy, estrogen and progesterone levels are elevated well beyond those of the luteal phase of the menstrual cycle (Figure 25-8), and

therefore the alveoli and alveolar ducts proliferate considerably. Two other hormones, **prolactin** and **placental lactogen,** are also important for preparing the breasts for milk production.

During lactation, milk production occurs in the alveoli of mammary glands (Figure 25-20). Milk from the alveoli is secreted into a series of **lactiferous sinuses,** surrounded by smooth muscle, that eventually lead by way of ducts to the nipple.

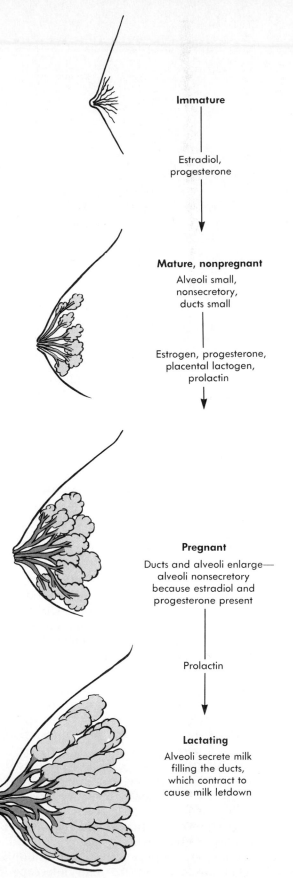

Immature

↓ Estradiol, progesterone

Mature, nonpregnant
Alveoli small, nonsecretory, ducts small

↓ Estrogen, progesterone, placental lactogen, prolactin

Pregnant
Ducts and alveoli enlarge—alveoli nonsecretory because estradiol and progesterone present

↓ Prolactin

Lactating
Alveoli secrete milk filling the ducts, which contract to cause milk letdown

FIGURE 25-20

Hormonal influences on the breast at puberty, during pregnancy, and during lactation.

Prolactin, an anterior pituitary hormone, stimulates alveolar development and milk production. How the hypothalamus controls release of prolactin is not completely understood. Current evidence suggests that prolactin release is inhibited by the neurotransmitter dopamine and it may be stimulated by a small polypeptide from the posterior pituitary.

Prolactin is secreted in small amounts beginning at puberty, but the secretion is greatly increased during pregnancy. Further increases in prolactin secretion occur during lactation. Prolactin stimulates growth and secretory activity of the mammary alveoli. Placental lactogen acts in the same way as pituitary prolactin, augmenting alveolar growth. Although prolactin and placental lactogen reach high levels in the plasma toward the end of pregnancy, milk production does not begin then because of the inhibitory effects of the high levels of estrogen and progesterone in the later stages of pregnancy. In effect, the breasts are primed for milk production during pregnancy but not switched on. Levels of estrogen and progesterone fall rapidly after delivery, because the placenta is no longer present and follicles are not maturing in the ovaries. The drop in estrogen and progesterone allows milk production to begin.

Milk production is sustained by continued prolactin secretion as long as suckling continues. According to a current hypothesis, when an infant nurses, sensory signals from the nipples (Figure 25-21) cause the hypothalamus to increase the secretion of prolactin releasing factor, maintaining plasma prolactin levels and milk production. Prolactin release could also be sustained by decreases in dopamine release or by other controlling factors as yet undiscovered. Increases in the frequency or duration of suckling stimulate prolactin release, with a time lag of about a day, so that the rate of milk production increases to match the needs of the growing baby. If a mother does not nurse her infant, prolactin levels fall and milk production stops after a few days.

Oxytocin and Milk Letdown

Milk production and the movement of milk from the alveoli into the ducts, called **milk letdown,** are distinct processes. Milk letdown involves a reflex in which stimulation of the nipple by an infant's suckling causes oxytocin release from the posterior pituitary (see Figure 25-21). Oxytocin causes contraction of the smooth muscle cells around the alveoli and delivers milk to the nipples. The reflex becomes conditioned after a short time, so that the sound or sight of the infant can cause milk letdown to start even before suckling begins. Oxytocin released in response to suckling also stimulates uterine contraction. This side-effect of breast-feeding is important after delivery because it reduces blood loss and

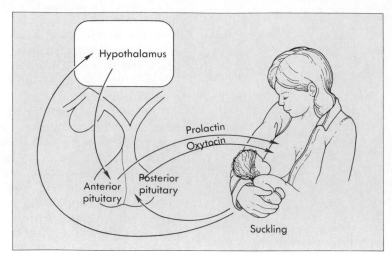

FIGURE 25-21
The hormonal reflexes caused by suckling. Both prolactin release from the anterior pituitary and oxytocin release from the posterior pituitary are stimulated by afferents from the nipple. Prolactin sustains milk production, and oxytocin causes milk letdown.

speeds the recovery of the stretched uterus.

Milk production does not begin for a day or so after birth. During this time the breasts secrete **colostrum,** a yellow watery fluid that contains large quantities of maternal antibodies. Antibody secretion continues after milk production begins, and by the fourth month of lactation it may amount to as much as 0.5 g/day. This **passive immunization** tends to protect the infant from infections, particularly those that result from entry of infectious agents across mucous membranes (see Chapter 26).

The Contraceptive Effect of Lactation

Pregnancy and lactation both place significant burdens on the mother's metabolism and nutrient balance. For most of human evolution, children were breast-fed for several years after birth, and this is still the practice in some societies. Evidence suggests that lactation, although not a reliable means of contraception under the conditions of child-rearing that prevail in developed countries, can have a significant effect on birth-spacing in developing countries. The contraceptive effect results from reflexive inhibition of gonadotropin secretion by the stimulus of suckling.

FERTILITY ENHANCEMENT

Infertility is a medical problem for one of every twelve couples in the United States. Approximately 40% of the cases are attributable to male factors, most of which are easy to detect. Blocked seminal ducts sometimes occur and reduce sperm count. Varicose veins of the scrotum can reduce fertility by increasing the temperature of the testes. Such structural defects can frequently be treated by surgery. If the problem is inadequate numbers of sperm, collection and concentration of the sperm followed by artificial insemination is a possibility.

Investigation of female infertility begins with an evaluation to determine if the woman ovulates. Failure to ovulate may be overcome in some cases by treatment with gonadotropins to stimulate follicle development and ovulation. These drugs are sometimes able to deliver too much of a good thing, because multiple ovulation and the conception of as many as six or seven fetuses can occur. Multiple births make exciting news stories but represent injudicious medicine. Such pregnancies require a disproportionate amount of medical attention, expense, and risk to the mother. Using ultrasonography and other methods, physicians can estimate the number of follicles that will ovulate in a given cycle in a patient being treated with gonadotropin; conception can be avoided if it appears that multiple ovulations will occur.

If the woman appears to ovulate and the hormonal system is adequate to support pregnancy (enough progesterone produced to support the endometrium) then the possibility of a physical obstruction is evaluated. Anatomical defects or the blockage of Fallopian tubes can be examined by introducing an X-ray-opaque dye into the uterus. Approximately 25% of female infertility can be attributed to blocked tubes, scar tissue from infections in the peritoneal cavity and endometriosis. By using what may be the smallest telescope around, it is now possible to inspect and repair Fallopian tubes. In falloposcopy, an 0.5 mm telescope is threaded through the cervix into the Fallopian tubes and connected to a video camera. This provides a high-resolution color image of sites of scar tissue or

The Birth Experience Contributes to the Survival of the Newborn

During birth, infants are intermittently deprived of oxygen by compression of the umbilical cord. After birth, they must quickly adapt to an environment in which they must breathe air, thermoregulate, and be fed at intervals rather than continuously. Recent evidence suggests that the stress of normal vaginal birth stimulates secretion of catecholamine hormones (primarily norepinephrine) that improve infant survival during vaginal delivery and help initiate physiological adaptations to extrauterine life. The first evidence for an important role of catecholamines during birth was the finding that catecholamine levels in the blood of healthy infants at birth were more than five times higher than the highest levels achieved by adults during exercise. Removal of the adrenals from new born animals reduced their ability to survive episodes of hypoxia such as those caused by uterine contractions during labor.

In fetal animals, injection of epinephrine stops secretion of fluid into the lungs and causes fluid to be reabsorbed. This effect would contribute to clearing fluid from the lungs so that air breathing could be effective. Catecholamines also stimulate secretion of pulmonary surfactant, which reduces the work of breathing (Chapter 16).

Mobilization of nutrient stores and production of heat are stimulated by catecholamine in infants just as in adults (Chapter 23). This effect may help the infant survive the first day of postnatal life, because milk production typically does not start until the second day. Cooling after birth is a significant threat, because it is estimated that a wet newborn can lose heat at a rate as great as 200 cal/kg/min. In contrast the maximum rate of heat production in a resting adult is only about 90 cal/kg/min. Newborn infants possess a special heat-producing organ: a deposit of brown fat (Chapter 23) that forms a mantle beneath the skin of the upper thorax, accounting for 2% to 6% of body weight. Brown fat is brown because of its abundant mitochondria, which color the otherwise unpigmented tissue because of their cytochromes. Under the influence of catecholamines, brown fat begins to release large amounts of heat. This is possible because, in the mitochondria of brown fat, oxidative metabolism is not coupled to phosphorylation, so all of the potential energy of the lipid appears as heat. The brown fat disappears a few weeks after birth.

Secretion of epinephrine and norepinephrine from the adrenal medulla in children and adults in response to a stressful situation is stimulated by activity in sympathetic preganglionic nerves leading to the adrenal glands. However, in infants the autonomic nervous system is not fully developed at the time of birth, and the adrenals are not yet connected to the nervous system; secretion of adrenal catecholamines in infants must be stimulated by an indirect mechanism. The nature of the indirect pathway is not clear, but two factors that are involved are hypoxia and pressure on the head. These stimuli are experienced by infants during vaginal delivery, but not during surgical delivery (Caesarean section). The use of surgical delivery as a means of circumventing the hazards of vaginal delivery has increased in developed countries, as improvements in surgical practice have reduced the risk of the surgery to the mother. Yet, surgical delivery would seem to deprive infants of a hormonal response that may contribute significantly to their survival in the hours immediately following birth.

---

blockage, allowing video-aided Fallopian surgery.

Transcervical balloon tuboplasty (TBT) applies a technique developed to open partially-blocked coronary blood vessels to the opening of obstructed Fallopian tubes. In TBT a catheter with a deflated balloon is inserted through the vagina into the uterine cavity and up into the Fallopian tube while a fluoroscope projects an X-ray image onto a television screen. The balloon is inflated and deflated several times to reopen the tube. By providing relief for one or more ovarian cycles, this procedure may be effective for thousands of infertile women whose Fallopian tubes are blocked, even if the Fallopian tubes eventually close down again.

Some cases of infertility result from male-female incompatibility. In some women, the mucous secretions of the cervix never become penetrable to spermatozoa. In others, the vaginal secretions are too

SUMMARY

1. After fertilization in the upper Fallopian tube, the zygote divides to form first a **morula** of undifferentiated cells and then a **blastocyst** consisting of a **trophoblast** and **embryoblast**. After implantation the trophoblast becomes the **chorion**, which gives rise to the fetal component of the **placenta**. The embryoblast becomes an **embryo**, which after the eighth week of development is termed a **fetus**.

2. Implantation is followed quickly by secretion of **human chorionic gonadotropin**, which sustains the corpus luteum for about the first 4 weeks of pregnancy.

3. The placenta exchanges nutrients and gases with maternal blood and protects the fetus from attack by the mother's immune system. The placenta also secretes estrogen and progesterone to maintain a hospitable uterine environment and **placental lactogen** to prepare the breasts for lactation and to mobilize nutrients from maternal stores.

4. The fetal circulatory system has three shunts: the **ductus venosus** that allows blood to reach the vena cava from the umbilical vein without passing through the hepatic circulation; the **ductus arteriosus,** a route between the pulmonary arteries and the aorta; and the **foramen ovale** between the right and left atria. The last two shunts allow blood to bypass the pulmonary loop while the lungs are collapsed and nonfunctional. The shunts normally close shortly after birth. The fetal blood must load and transport adequate oxygen at a lower partial pressure than adult blood; the oxygen dissociation curve of fetal hemoglobin is left-shifted compared to adult hemoglobin.

5. Development of organ systems is initiated during the embryonic period—the first 8 weeks of development. During this period the developmental process is most sensitive to drugs and toxins. By the end of the second trimester, organ systems are almost completely differentiated. The third trimester is a period of growth and nutrient storage that prepares the fetus for independent life.

6. **Labor** is normally initiated after approximately 39 weeks of gestation and consists of three stages: in stage one, uterine contractions dilate the cervix; in stage two the baby passes through the cervix; and in stage three the placenta is delivered. Initiation of labor is not fully understood. The progress of labor is assisted by the hormone **oxytocin,** which stimulates uterine contraction. As term approaches, there is an increase in the number of oxytocin receptors. The first contractions of labor initiate a positive feedback reflex that increases oxytocin secretion from the posterior pituitary. Oxytocin also stimulates secretion of a prostaglandin that is important in dilation of the cervix.

7. The breasts are primed for milk production by high levels of estrogen and progesterone secreted by the placenta during pregnancy. Milk production is hypothesized to be initiated after delivery by release of a small polypeptide from the posterior pituitary that stimulates secretion of prolactin from the anterior pituitary.

8. **Milk letdown** is a hormonal reflex in which stimulation of the nipples causes oxytocin secretion, which in turn causes the smooth muscle of the mammary ducts to contract. Lactation can suppress ovulation for some time after pregnancy, especially if the infant is fed at frequent intervals during the day and receives no other food.

9. Methods to improve the chances of parenthood for infertile couples include initial testing procedures that can identify the problem in 80% to 90% of the cases. Male infertility can be treated by artificial insemination, sperm washing and treatments that make it easier for the sperm to enter the ovum. Female infertility can be corrected with hormone treatments, by surgically correcting anatomical blockage of the female tract or by bypassing portions of the female tract to place gametes together in the oviduct or uterus.

acidic or contain antibodies that lead to the destruction of the spermatozoa. In these cases, artificial insemination, which deposits a fresh sample of the husband's semen directly into the uterus, can bypass the problem. Alternatively, the process known as sperm washing removes the surface molecules that the female antibodies recognize on the sperm. Finally, the outer surface of the zona pellucida can be removed (zona drilling) to permit easier access for a limited number of sperm, or an individual sperm can be injected into an ovum, following procedures outlined below.

A last resort for infertile couples is assisted fertilization that involves placing gametes or a zygote into the Fallopian tube. Gonadotropin treatments are used to induce the maturation of several follicles in a single cycle. The maturation of the follicles is followed with ultrasonography. When several follicles are judged to be close to maturity, a small incision is made in the woman's abdomen for the insertion of a laparoscope, a fiber-optic device that allows the physician to see the ovaries, and also to puncture mature follicles and to suck up the oocytes that are released. Several oocytes are collected in this way. The in vitro fertilization technique calls for fertilizing the ovum in a dish containing sperm and introducing the zygote into the uterus. Failures of implantation are more common than with the newer techniques of gamete intrafallopian transfer (GIFT) or zygote intrafallopian transfer (ZIFT), which allow the reproductive tract time to recover from the mechanical manipulations before the zygote enters the uterus. Improved success with implantation is achieved by the thinning of the egg's outer surface using the techniques developed for zona drilling and sperm microinjection: the abraded surface of the ovum is the "sticky" site at which initial attachment to the endometrium occurs.

● *STUDY QUESTIONS*

1. What factors stimulate oxytocin secretion? What is the role of oxytocin in labor? In lactation? Are the target tissues involved similar in any way?

2. How is prolactin secretion controlled? What is the role of prolactin in lactation?

3. What is an ectopic pregnancy? What are the possible outcomes of such a pregnancy?

4. Name and describe the effects of placental hormones on maternal physiology.

5. Why is it important to augment the mother's nutrition early in pregnancy?

6. If premature delivery is anticipated, the fetus may be treated with cortisol. What is the benefit of this?

7. During the first 8 weeks of development, the development of a child is most threatened by drugs and toxins. What developmental processes are going on during this time?

8. If the ductus arteriosus fails to close after birth, what effect would this probably have on:
   a. Blood flow through the systemic loop
   b. The oxygenation of systemic arterial blood
   c. The pressure of blood in the pulmonary loop

9. What are the roles of progesterone in pregnancy? Progesterone is sometimes used medically to prevent a threatened premature delivery. What is the reasoning behind this use?

10. How do the structure and oxygen affinity of fetal hemoglobin compare to those of adult hemoglobin? What is the adaptive value of the difference in oxygen affinity?

Choose the MOST CORRECT Answer.

11. Fertilization occurs in the:
    a. Vagina
    b. Cervix
    c. Uterus
    d. Fallopian tube

12. The appearance of the hormone ___ in the blood and urine of the mother is the first chemical indicator that pregnancy has begun. This is the basis of early pregnancy tests.
    a. Progesterone
    b. Human chorionic gonadotropin
    c. Estrogen
    d. Human placental lactogen

13. In fetal circulation, there is a connection between the pulmonary artery and the aorta called the:
    a. Ductus venosus
    b. Ductus arteriosus
    c. Foramen ovale
    d. Umbilical artery

14. ___, the major hormone associated with maintaining a pregnancy, acts upon uterine smooth muscle to inhibit contraction.
    a. Relaxin
    b. Oxytocin
    c. Prostaglandins
    d. Progesterone

15. This hormone stimulates uterine contractions during labor and causes milk letdown:
    a. Relaxin
    b. Prolactin
    c. Oxytocin
    d. Fetal lactogen

● *SUGGESTED READINGS*

BAUMAN DE and NEVILLE MC: Nutritional and physiological factors affecting lactation, *Federation Proceedings* 43:2430, 1984. A symposium covering the impact of malnutrition, prematurity, diet, and duration of breast feeding on milk composition.

HANSEN JT and SLADEK JR: Fetal research, *Science* 246:775, 1989. Describes ethical issues and potential benefits of research using fetuses and fetal tissues.

MODAHL C: The love hormone, *Mademoiselle* 96(11):112, 1990. Discusses the role of oxytocin in the sexual response, labor, and lactation.

NORWITZ ER, STARKEY PM, BERNAL AL, and TURNBULL AC: Identification by flow cytometry of the prostaglandin-producing cell populations of term human decidua, *J Endocrinol* 131:327, 1991. Describes the technique for sorting the cell types in the human decidua and the finding that macrophages represent the major source of prostaglandins.

PEDERSEN CA, ET AL: Oxytocin and maternal, sexual, and social behaviors, *Anals of the New York Academy of Science* 652:1-492, 1992. This volume resulted from a conference on the regulation of oxytocin and its many functions in reproduction.

ROBEN-JONATHAN N, HYDE JF, and MURAI I: Suckling-induced rise in prolactin: mediation by prolactin-releasing factor from posterior pituitary, *News in Physiological Sciences* 3:172, 1988. Short review of evidence for control of prolactin secretion by dopamine and an unknown peptide from posterior pituitary.

RODIN J and COLLINS A, editors: *Women and reproductive technologies: medical, psychosocial, legal, and ethical dilemmas.* Hillsdale, NJ, 1991, Lawrence Earlbaum Associates, Publishers. Discussion of all sides of the new technologies and their impacts on women and the family.

SMALL MF: Sperm wars, *Discover* 12(7):48, 1991. The role that sperm competition may play in natural selection.

TREVATHAN WR: *Human birth: an evolutionary perspective,* Aldine de Gruyter, New York, 1987. A comparison of structural and behavioral determinants of birth in primates.

CHAPTER 26

The Immune System

On completing this chapter you should be able to:

- Describe the nonspecific elements of immune system response.
- Understand the activation and function of complement.
- Know how inflammation develops and promotes immune function.
- Describe the benefits derived from systemic responses to infection, such as fever and altered blood constituents.
- Understand the role of the major histocompatibility complex antigens in immune system recognition of self and nonself (antigen presentation complex).
- Describe the structure of antibodies and how the structure is related to antigen binding and the biological consequences of binding.
- Understand the role of antibody-antigen interactions in agglutination, lysis, neutralization, and opsonization.
- Know the roles in specific immune responses played by helper T cells, cytotoxic T cells, suppressor T cells, and B cells.
- Be able to describe the development of humoral immunity and the roles of plasma and memory cells.
- Understand how malfunctions of the immune system lead to hypersensitivity and autoimmune diseases.

*I*mmunology is the science of the body's resistance to invasion by pathogenic organisms. The immune system is able to distinguish friend from foe and to "remember" old enemies. The molecular basis of recognition and memory is similar in principle to the binding of hormones or neurotransmitters with their receptors, but it involves different proteins.

Without immune responses, the body would be rapidly overcome by opportunistic bacteria, fungi, viruses, protozoa, and parasitic worms. Without the vigilance of the immune system, the continual transformation of small numbers of normal cells into precancerous cells would cause many more cancers. Consequences of loss of immune function are seen in acquired immunodeficiency syndrome (AIDS), in which a viral infection alters the immune system. Death from cancer or infection is one possible consequence of infection with the human immunodeficiency virus (HIV).

Autoimmune diseases result from misdirection of the immune system against specific tissues of its host. Inappropriate immune system responses also cause allergies. Positive feedback that normally accelerates initial stages of an immune response can, under some circumstances, generate responses totally out of proportion to the threat, causing tissue damage or death.

The techniques of molecular biology have revealed much about the workings of the immune system, and the immune system has provided biologists with potent tools for molecular recognition. Immunology is a rapidly advancing science; this chapter can provide only an overview of the immune system and recent developments in immunology. Our aim is to highlight aspects of immune system function that convey an appreciation of its role in the health of the individual.

NONSPECIFIC COMPONENTS OF THE IMMUNE RESPONSE

- *What distinguishes specific and nonspecific immune responses?*
- *What cell types are involved in the nonspecific defense mechanisms?*
- *How does an inflammatory response develop?*
- *How is complement activated and how do membrane attack complexes result in cell lysis?*
- *Describe the sequence of events that lead to fever and benefits fever and other systemic responses provide in regaining health.*

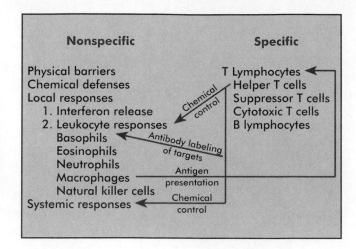

FIGURE 26-1

Overview of the nonspecific and specific elements in the immune system. Some of the many interactions between the two categories of response are indicated.

Barriers and Local Cellular Defenses

Two approaches are utilized in the body's resistance to invasion by pathogens. The first is called the **nonspecific** or **innate** response, and does not depend upon prior exposure to the infectious agent. The **specific** or **acquired** immune response, which evolved along with the vertebrates, provides a defense specifically targeted against a particular infectious or foreign agent to which the system has been sensitized (Figure 26-1). Both specific and nonspecific systems are assisted by responses to injury or infection that involve local alterations in circulatory system function and chemical cascades involving blood proteins.

Nonspecific defense mechanisms include (1) physical and chemical deterrents to invasion across body surfaces, (2) local responses to invasion, and (3) whole-body responses. The physical barriers to body invasion are the body surface and the mucous membranes of the respiratory, digestive, excretory, and reproductive tracts. Chemical deterrents include secretion of sweat and oils and the acid secretions in the digestive tract and in the female reproductive tract, which kill potentially pathogenic microorganisms. The secretion of mucus and the cleansing action of cilia, which transport the mucus toward the body's exterior surfaces, prevent the accumulation of potentially infectious agents within the respiratory system.

If penetration of the physical barriers occurs, the invading agent or foreign material is met by a local response. In the case of viral invasion, one such response comes directly from the infected cells. Individual cells that have become infected by viruses release an antiviral agent called **interferon** that binds to interferon receptors on other cells, inducing the synthesis of protective proteins in those healthy cells that inhibit the replication of viruses. This is a non-specific protection. The name interferon comes from the fact that infection with one virus interferes with infection by other, unrelated viruses. If the viral infection becomes systemic rather than local, the levels of interferon in the blood rise detectably.

Other aspects of the nonspecific cellular defense are mediated by specialized cells within the tissues and phagocytic cells that arrive from the bloodstream. **Mast** cells are located in perivascular tissue in all tissues, but especially in the lungs and GI tract. These large cells respond to exposure to components of bacterial cell walls and certain other stimuli by releasing histamine, heparin (an anticoagulant) serotonin, bradykinin, lysosomal enzymes, and other active substances. Many of these products elicit dilation of blood vessels and development of gaps between the capillary endothelial cells. These responses favor accumulation of plasma and movement of **white blood cells**, or **leukocytes**, into the region.

All the blood cells, including the erythrocytes, are derived from the same stem cell (Figure 26-2). The leukocytes constitute the cellular circulating agents of the immune system. **Granulocytes** are the commonest type of leukocyte. Categories of granulocytes are often distinguished on the basis of the affinity of their granules for different stains. These are **neutrophils, eosinophils,** and **basophils.** Neutrophils make up 50% to 75% of the total leukocyte population. They can engulf and digest bacteria, which are destroyed by the hydrolytic enzymes from lysozymes. Eosinophils constitute 1% to 6% of the leukocytes and are especially effective against parasitic infections. Basophils, less than 1% of the leukocyte population, release **histamine, heparin** (an anticoagulant), and other local chemical media-

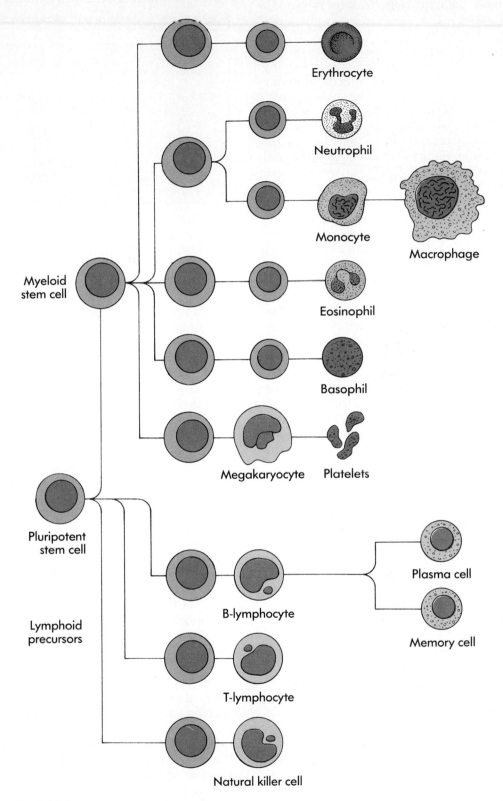

Erythrocyte

Neutrophil

Monocyte

Macrophage

Myeloid
stem cell

Eosinophil

Basophil

Megakaryocyte   Platelets

Plasma cell

Pluripotent
stem cell

B-lymphocyte

Memory cell

Lymphoid
precursors

T-lymphocyte

Natural killer cell

FIGURE 26-2
Differentiation of blood cell types from a single type of stem cell found in the bone
marrow.

tors and resemble the mast cells in staining properties and responses.

Large granular leukocytes, which give rise to **natural killer cells**, belong to the lymphocyte lineage and their proliferation can be controlled by signals from cells involved in the specific immune response (Figure 26-2). They make up approximately 5% to 10% of the lymphocyte population and exhibit two categories of cytotoxic action. The action identified with natural killer cells is to spontaneously kill tumor and virus-infected cells, and they probably represent a first defense against these abnormal body cells. This is a nonspecific action in the sense that prior exposure and recognition of a previously encountered chemical structure is not involved. The other action exhibited by an overlapping population of large granular leukocytes is called **antibody-dependent cellular toxicity**, which directs them to kill cells marked as part of the specific immune system responses.

Monocytes comprise 2% to 10% of the circulating leukocyte population. The monocytes in the blood are on their way to a destination within the body tissues. They exit the bloodstream and enter interstitial spaces, where they transform into **macrophages,** cells specialized for phagocytosis. Macrophages are strategically positioned to respond locally to invasion (for instance, in the lungs, liver, kidney, lymph nodes, connective tissue, and GI tract). In these tissues, they may be designated by special names, such as the mesangial cells of the kidney and the microglia of the brain. Macrophages also assist in initiating and controlling specific immune responses, as described below, and they can be stimulated to greater activity by signals released during a specific immune response (Figure 26-1).

Inflammation and the Complement and Kinin-Kallikrein Systems

Inflammation is a local response that reflects the body's attempt to rid itself of injured cells or infectious material. When a tissue is damaged or infected, the surrounding area displays the redness, swelling, warmth, and pain characteristic of inflammation (Table 26-1). Three processes contribute to this response: (1) blood flow to the inflamed area increases (causing redness); (2) capillary permeability increases, allowing humoral agents of the immune system to get to the affected area and resulting in fluid accumulation; and (3) leukocytes migrate from blood vessels into the affected tissue, attracted by substances released in the damaged tissue. The promotion of this response by activation of mast cells has already been discussed. Two cascade systems of proteolytic enzymes present in body fluids contribute to the inflammatory response. These are the **kinin-kallikrein system** and the **complement system.** A third class of substances, prostaglandins, regulate the intensity of inflammation.

TABLE 26-1	Mediators of Inflammation

Substance(s)	Effect(s)
Bradykinin (histamine)	Vasodilation; increased capillary permeability; activation of pain fibers
Complement	Bacterial cell lysis; attracts leukocytes; stimulates phagocytes
Prostaglandins	Potentiate bradykinin; increase set point for body temperature

The kinin-kallikrein system is activated by damaged tissue surfaces. The end products of the cascade are **kinins,** the most important of which is **bradykinin.** Bradykinin causes vasodilation, which produces the redness and warmth of inflamed areas. It also increases the permeability of capillaries, resulting in local edema and increased lymph flow from the damaged area. Bradykinin activates neural afferents, causing the burning pain characteristic of skin abrasions and burns. Histamine released from mast cells and basophils has similar effects.

The complement system organizes a local attack on bacterial invaders. Complement activation leads to **lysis** (killing by membrane or cell wall disruption) of bacteria and foreign cells, increased vascular permeability, attraction of neutrophils, promotion of phagocytosis, and stimulation of histamine release by mast cells. The vascular changes and chemical responses promote inflammation. The complement system can be activated by elements of the nonspecific immune system (the **alternative pathway**) or by elements of the specific immune system (the **classical pathway,** see below).

The complement system (Figure 26-3) is composed of plasma proteins designated C1 through C9 and fragments that result from cleavage, which are designated with lower-case letters. Additional elements are currently being described. Activation of the complement cascade can occur by two pathways, beginning with antigen-antibody interaction (the classical pathway) or by bacterial invasion (the nonspecific pathway). The nonspecific part of the complement system is activated at a minimal level unless exposure to cell walls of some bacteria, yeasts, or cobra venom triggers an elevated response. These produce an active enzyme called C3 convertase, which cleaves C3 into C3a and C3b (Figure 26-3), producing the first of several fragments with biological activity and promoting the cascade of events that leads to cell lysis. There is an additional alternative route leading to the formation of C5b, not shown on Figure 26-3. When the cascade produces the protein called C5b, this hydrophobic molecule can insert into a membrane along with serum proteins C6, C7, and C8. These are joined by six molecules of C9, and together they assemble a remarkable structure: a hollow cylinder that induces a leak in the cell wall or membrane (Figure 26-3, Figure 26-4). This **membrane attack complex** leads to lysis. Many of these destructive holes can quickly be assembled in the membrane of the target cell or bacterium.

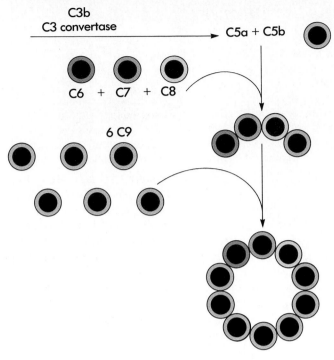

FIGURE 26-3
Activation of complement by two pathways leads to the lytic membrane attack complex and other biological consequences of the cascade. Amplification that results from production of the activated enzymes results in many active molecules in each category and thousands of membrane attack complexes.

Roles of Prostaglandins in Inflammation and Fever

Fever is a systemic response to infection. It is generally agreed that moderate elevations of body temperature improve the body's disease-fighting capacity and should not be routinely blocked by **antipyretics** such as aspirin and acetaminophen. Fever can be activated by foreign molecules, the **exogenous pyrogens**, which include bacterial by-products called **endotoxins**. The endotoxins, in turn, cause phagocytic cells to release **endogenous pyrogens** that can set body temperature at a higher level by resetting the hypothalamic regulatory set-point.

The **endogenous pyrogens** belong to a family of chemical messengers called **cytokines**, which are released by the body's cells. These proteins can act locally or be distributed through the blood. The cytokines released by leukocytes are also called **lymphokines** or **interleukins**, and represent a large family of proteins (see Frontiers Box). The interleukin most associated with the promotion of fever is **interleukin-1, (IL-1)** which is released by stimulated macrophages. Among other actions, interleukin-1 promotes the conversion of the membrane component arachidonic acid to prostaglandins (Figure 26-5). The macrophages primarily produce PGE_2, the prostaglandin associated with production of fever. **Antipyretic** (fever-opposing) agents such as aspirin and acetaminophen act by blocking prostaglandin synthesis.

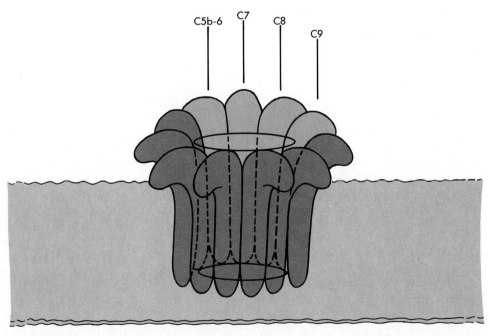

FIGURE 26-4
A membrane attack complex is assembled in the membrane of the target cell from one of each of the complement proteins C5 through C8 and a variable number of C9.

The immune system is regulated by a family of cytokines, many of which are called **interleukins** (Table 26-A). The large number of interleukins and the requirement for receptors on their target cells resembles communication in the endocrine and nervous systems, with their great diversity of hormones, neurotransmitters and neuromodulators. Interleukins may act at short range like neuromodulators or like hormones by circulating in the blood to distant targets. Some of the messenger molecules released by lymphocytes became known by other names before the lymphokine naming system was adopted in 1986. Lymphokines are interleukins that are released by lymphocytes. The communication between the immune system and the nervous system involves both interleukins and molecules that also serve as classical neurotransmitters, such as histamine. New members of the interleukin family are currently being identified by their effects and characterized with regard to structure, but many elements in this communication system are still only suspected on the basis of effects that cannot be explained, and understanding of the role of particular second messenger pathways in the target cells is still in its infancy.

Interleukins were first suspected when researchers attempted to duplicate the antigen activation of immune system cells in vitro, and found that the range of activities was increased when supernatants that had contained macrophages or activated lymphocytes were added. The first interleukin to be characterized was interleukin-2 (IL-2), which was originally called T-cell growth factor. IL-2 has been produced by recombinant DNA technology in commercial quantities, and it, along with

interferon gamma (INF-γ) is available for antitumor treatment. On the other hand, many of the drugs used to suppress the immune system in transplant patients act by inhibiting IL-2 production and secretion, but have many adverse side-effects. A better understanding of the regulation of synthesis and release of IL-2 might lead to a more specific means of blocking its production or release or a means of blocking the IL-2 receptor. Increasingly, the synergistic and antagonistic interactions between different interleukins have been shown to be crucial in regulating the range of their biological effects, so an optimal mechanism of regulating IL-2 probably already exists and only needs to be discovered.

One way that interleukins interact is by regulating the expression of receptors for interleukins on potential target cells. After a macrophage ingests and presents an antigen to a helper T cell, it completes the process of helper T cell activation by releasing IL-1. The activated helper T cells then express IL-2 receptors and begin to make IL-2. IL-2 acts as a switch to cause T cells to multiply rapidly and provide more IL-2, which activates natural killer cells and promotes mitosis and antibody secretion by B cell clones. The IL-1 has many other effects, including induction of fever, lowering of plasma levels of iron and zinc and elevation of copper, which is important as a cofactor for immune response enzymes. One mode of action is to induce the production of acute phase proteins by the liver, which radically alters the characteristic mix of blood proteins, providing proteins that promote antibacterial actions, such as lactoferrin, complement proteins and activators of complement as

well as clotting factors and many other proteins of unknown function.

IL-1 acts at the same synapses as the endogenous opiates, endorphins and enkephalins, but causes the opposite effect. Instead of suppressing pain perception, IL-1 accentuates it. This causes the muscle and joint aches and throbbing cycs characteristic of many febrile illnesses. IL-1 also seems to induce fatigue and sleep by acting directly on the central nervous system centers regulating the sleep cycle, thus conserving the body's energy to battle infection. IL-1 also causes the hypothalamus to release corticotropin releasing hormone, stimulating adrenal cortisol release. Cortisol mobilizes the body's glycogen and fat stores in response to the stress of infection and blocks anabolic (tissue-building) pathways, providing amino acids for the synthesis of the acute phase proteins. Another cytokine called tumor necrosis factor (TNF) also causes production of acute phase proteins and fever in addition to its toxic and disruptive effects on the tumor cells. Combinations of several interleukins eventually may be the best approach to promoting tumor regression. The ability to selectively block some of the generalized debilitating effects of the interleukins on nontumor cells would aid recovery in patients with tumors and parasitic infections. Some other interleukins are listed in Table 26-A. The interleukins and their functions listed in this table represent only a sketchy outline of a field that is rapidly expanding and which has great medical importance. Suggested readings at the end of this chapter provide additional information on this topic.

TABLE 26-A	Immune System Mediators
Gamma-interferon	Antiviral agent
Tumor necrosis factor	Tumor remission, inflammation
IL-1	T cell activation by macrophages Fever induction, inflammation promotion
IL-2	T and B cell differentiation and mitosis
IL-3	Bone marrow stem cell mitosis, differentiation of blood cell types
IL-4	Growth factor for B and T cells and mast cells
IL-5	B cell mitosis and differentiation
IL-6	Inflammation promotion, B cell activation
IL-7	Differentiation of B lymphocytes

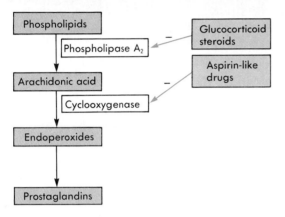

FIGURE 26-5
An abbreviated pathway of prostaglandin synthesis showing the steps at which anti-inflammatory steroids and aspirin-like drugs act.

After pyrogens have reset the hypothalamus to a higher temperature setpoint, the body temperature is regulated around the new setpoint as closely as normal body temperature is regulated in a healthy individual. The difference is that the sum of heat-loss and heat-gain mechanisms add up to a higher temperature. Depending on the conditions, an individual may even begin to shiver to increase body heat production. This response is accompanied by the sensation of being chilled, even as body temperature is rising. Fever enhances the immune system's defenses, rendering phagocytosis and lysis of invading organisms more effective. Associated with the production of fever are other systemic responses, including reduction of plasma levels of free iron, which is required for growth of many bacteria. The iron is removed from the plasma by being bound to **lactoferrin** released from leukocytes.

In summary, the nonspecific defenses against bacteria or other foreign matter include physical and chemical blockers of invasion by infectious agents, the release of interferon from virally-infected cells, and the responses of mast cells, fixed tissue macrophages, and the phagocytes and natural killer cells attracted from the bloodstream in the inflammatory reaction. In addition, chemical mediators in the bloodstream and systemic responses such as fever and the alteration in metal ion concentrations tend to reduce the survival and spread of the infectious agent.

SPECIFIC IMMUNE RESPONSES

- *What is an antigen and how is it distinguished from similar molecules produced by the body?*
- *What roles do the major histocompatibility complex antigens play in the immune system cell-to-cell communication that leads to activation of an immune response?*
- *How are the Fab and Fc regions of antibodies related to their fucntion?*
- *What happens in the primary and secondary stages of immune responses?*

Distinguishing Between Self and Nonself

In contrast to nonspecific immune responses, which do not involve recognition based upon prior exposure, specific immune responses involve activation of that small portion of the lymphocyte population that has receptors preadapted to bind specifically with the chemical structure of the foreign material. Immunity is typically acquired after an exposure, or **immunization**, with an agent that possesses chemical structures that are foreign to the body, such as molecules on the surface of a bacterium. In the first exposure, the specific response takes days to develop and typically plays a part in the individual's recovery from an infection. A memory of the foreign invader is preserved so that subsequent exposure to the same structure provokes such a rapid and efficient response that the threat is eliminated before symptoms develop. For this to happen, an individual's immune system must have a way to distinguish normal components in the blood and extracellular fluid and normal cells (**self**) from foreign substances and body cells that are abnormal (**nonself**).

The ability to distinguish self from nonself is based upon the chemical fit of molecules with the lock-and-key detection system of the specific portion of the immune system. The diversity of the detection system, which is able to find a fit with an enormous number of possible molecular configurations, will be described below. Any molecule that can be detected by such a system is called an **antigen** or **immunogen**. To function as an antigen, a molecule must have certain features of size and complexity. Examples of materials that cannot act as

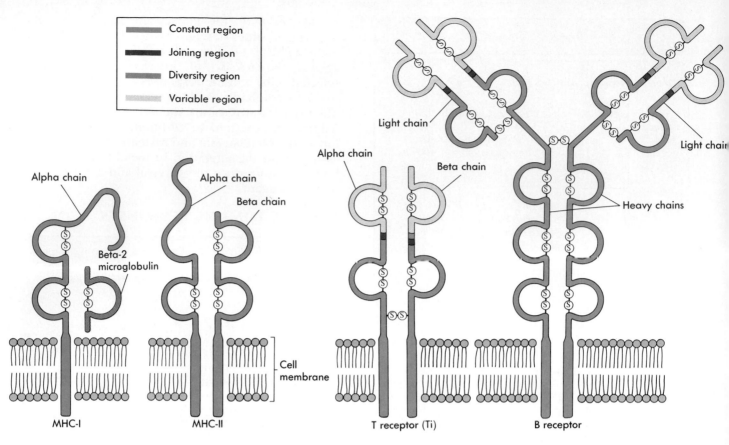

Constant region
Joining region
Diversity region
Variable region

Alpha chain

Alpha chain
Beta chain

Light chain
Light chain

Alpha chain
Beta chain

Heavy chains

Beta-2 microglobulin

Cell membrane

MHC-I
MHC-II
T receptor (Ti)
B receptor

FIGURE 26-6
Similarity of structures of MHC antigens, T-cell receptor, and antibodies (shown in the membrane-associated form of B receptors). All four classes of molecules have similar molecular structures and share similar amino acid sequences. The loops are formed by connecting disulfide bonds.

antigens are the cellulose in a splinter and the carbon particles that may be present in the lungs after exposure to smoke. Many small molecules (**haptens**) may be antigenic if they are chemically bound to larger molecules. The molecules produced by the body's own cells and displayed on its membranes are potentially antigenic. These molecules are **self antigens**. These include membrane receptors for transmitters and hormones and other molecules such as insulin that circulate in the blood. The **major histocompatibility complex (MHC)** is a chromosomal region that codes for the synthesis of self antigens that are used by the specific immune system for recognition of self. An **antigen** not native to the body is a **nonself antigen**. Generally, immune responses are mounted only against nonself antigens, but when the discrimination fails, an **autoimmune disease** can result.

In human beings, the gene cluster that contains the major histocompatibility complex is located on chromosome 6 and is called the **HLA**, for **human leukocyte antigen** complex. The MHC antigens are coded for by the most highly variable human genes

known. Each gene may be represented by one of at least 50 variant forms called **alleles**.

Two families of histocompatibility antigens, **Class 1** and **Class 2**, are present on human cells (Figure 26-6). Class 1 antigens are found on most cell membranes. (Exceptions are the membranes of erythrocytes and neurons.) Immune system interactions are based upon recognition of the MHC. For instance, some immune system cells can fight cells that are infected with viruses, because viral markers and gene products that normally are not present on differentiated cells appear like chicken pox on the membranes of the infected cells. The immune system cells recognize the infected cells by a combination receptor that requires the class I MHC and the nonself antigen. The differences between an individual's class I MHC and the class I MHC on tissue transplanted from another individual is the basis of graft rejection.

Class 2 antigens are present primarily on the surfaces of macrophages and some lymphocytes, as described below. These antigens are important in the **antigen presentation** aspect of the specific im-

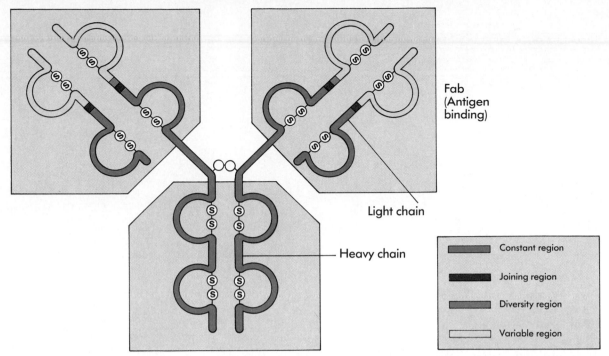

FIGURE 26-7
Structure of an antibody. Antibodies consist of two heavy and two light chains joined by disulfide bonds. The variable region binds to the antigen. The constant region of the antibody can activate the complement pathway or bind to receptors on the membrane of cells such as macrophages, basophils, or mast cells. The disruption of the disulfide bonds results in the formation of two Fab (for antigen-binding) fragments and an Fc (for crystallizable) fragment.

mune response. The structure of MHC antigens is compared with other recognition molecules in Figure 26-6; see also Figure 26-7.

Humoral Immunity

An important step in understanding immunity was taken with the discovery that individuals who had recovered from certain diseases had substances in their blood that could make another susceptible individual temporarily resistant to the disease. The specific disease resistance conferred by substances in the plasma of immune individuals is called **humoral immunity.**

Antibodies are the substances that confer humoral immunity. The term antibody was coined by the German scientist E. A. von Behring late in the last century before anything was known about the structure or origin of antibodies. Antibodies compose a family of plasma proteins called **immunoglobulins,** or **gamma globulins. Antiserum** is the serum (the fluid portion of plasma left after blood clots) of an immunized individual or animal that contains antibodies.

Antibodies are produced and released into the blood or other body fluids by lymphocytes called B cells. When B cells are activated by exposure to their specific antigen and chemical signals, they proliferate to produce two types of progeny: **plasma cells**

and **memory cells.** This is the antibody-dependent phase of B cell differentiation. Plasma cells begin to secrete antibodies into the plasma. The antibodies specifically bind to antigen, marking any organism bearing such molecules for destruction, as described below. The memory cells become resting, mature B cells exposing on their membranes the particular antibody coded by their particular gene rearrangement. Memory cells provide the expanded population that will be quickly activated upon a subsequent exposure to the antigen, as described in the section on activation and proliferation of B cells (see p. 767).

Immunity following an initial exposure to a source of nonself antigens is illustrated by measles, mumps, and chicken pox. If a child contracts any of these diseases, the memory mechanisms usually render the child **immune** to illness upon subsequent exposure. Immunity can also be produced by injecting a weakened bacterium or virus, a process called **vaccination. Active immunity** results from the production of antibodies by cells of the body's own immune system (Table 26-2).

In contrast to active immunity, the transfer of immunoglobulins produced by another individual results in **passive immunity** because it does not involve antibodies produced by the body's own reaction to antigen. Passive immunity provides only

TABLE 26-2   Comparison of Active and Passive Immunity

Type	Characteristics	Examples	
		Natural	Artificial
Active immunity	The person produces his or her own immune agents, either antibodies or T cells	Infection by bacteria or viruses	Vaccination with weakened virus or bacterial antigens
Passive immunity	The person is given antibodies produced by an actively immune person (or animal)	Transfer of IgG from the mother to the child across the placenta and IgA in breast milk	Immunoglobulin administration; monoclonal antibody therapy

temporary protection, lasting for a few days or weeks because there are no memory cells to respond to subsequent challenges (Table 26-2). Passive immunization is accomplished by injecting an antiserum or purified immunoglobulins. The transfer of some types of antibodies from mother to child across the placenta during prenatal development and in breast milk after birth is an example of naturally occurring passive immunization.

Antibody Structure

All types of antibodies contain a basic unit which is made up of two identical heavy amino acid chains joined by disulfide bonds to two identical light chains. These produce a Y-shaped protein called the **antibody monomer** (Figure 26-7). The antigen binding sites are located on the forked, or **Fab**, ends of the antibody molecule (Fab stands for antigen-binding fragment, which can be separated from the rest of the molecule by proteolytic digestion). The amino acid sequences on the heavy and light chains in the Fab portion of the molecule are the **variable region**, which is different for each antigen. Within the variable region, three short **hypervariable segments** form a cleft that acts as a binding site for some part of the antigen. The portion of the antigen molecule recognized and bound by the antibody is called the **determinant** portion. The determinant portion of the antigen fits within the cleft the way a key fits into a lock.

Antigenic determinants, or **epitopes**, must meet certain requirements of size and chemical reactivity; molecules may fail to generate antibody-mediated immune responses if they cannot be presented to the lymphocyte in such a way that they interact with a binding site. A large antigen molecule may possess more than one epitope, so several antibodies may be able to recognize a single antigen, each responding to a different epitope (Figure 26-8).

The remaining fragment, the stem of the Y (see Figure 26-7), has no antigen reactivity. This frag-

TABLE 26-3   Classes of Immunoglobulins

Class	Property	Functional role
IgG	Stimulates phagocytosis and complement reactions (can cross placenta)	Identifies microorganisms for engulfment or lysis and provides passive immunity in newborn infants
IgE	Binds to mast cells and basophils Stimulates histamine release	Inhibits parasite invasion; involved in allergic reactions
IgD	Surface Ig on many B cells	May be important in activation
IgA	Present in saliva, tears, breast milk, and intestinal secretions	Agglutinates infectious agents in secretions outside the body, antiviral
IgM	Stimulates phagocytosis and complement reactions	Identifies microorganisms for engulfment or lysis

ment can be crystallized, so it is called the **Fc** portion of the antibody molecule. In contrast to the very large number of possible Fab regions, there are only five categories of Fc structures, designated IgG, IgM, IgE, IgA, and IgD (Table 26-3 and Figure 26-8; *Ig* stands for *immunoglobulin,* so IgG is immunoglobulin G).

Relationship Between Immunoglobulin Structures and Actions

The class to which an immunoglobulin belongs is determined by its Fc portion. Once antibody is bound to the reactive site, the biological consequences of the binding depend on the Fc portion of the molecule. The biological consequences of immunoglobulin-antigen binding fall into four general

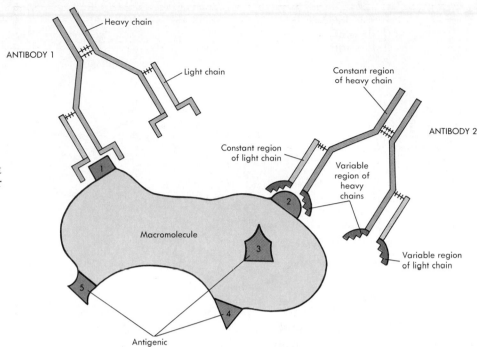

FIGURE 26-8
Antibodies may recognize different regions, called determinants or epitopes, of an antigenic molecule.

categories determined by the structure of each Fc class:

1. **Agglutination,** the binding of antibody-bearing particles or microorganisms together, is an action that all antibody classes can perform. The basic antibody unit, the monomer, is the structure of IgG, IgE, and IgD (Figure 26-9). Because this monomeric structure provides two identical binding sites, each antibody unit can link two antigens together (Figure 26-10). If the two antigens happen to be on different bacteria, the two invaders are linked together like chain-gang prisoners, and their ability to disperse in the body is reduced. They are also more likely to be phagocytosed. The antibodies of the dimeric IgA class of antibodies have a greater potential for linking antibodies to produce large clumps, and the pentameric IgM antibodies are even more effective in agglutination (Figure 26-10).

2. **Neutralization** of toxins and viruses and the immobilization of bacteria with cilia and flagella are accomplished by binding of Ig and/or release of lytic enzymes from lymphocytes and eosinophils attracted to IgG-coated surfaces. Parasitic worms may also be attacked by the eosinophils after they are tagged by IgG.

3. **Opsonization** is the binding of antibody to the surface of an invading microorganism, and it promotes phagocytosis by macrophages. The IgG antibodies are effective in this, because phagocytic cells have receptors for the Fc region of IgG, and linking to the antibody helps these cells to engulf the bacterium or foreign cell (Figure 26-11). The Fc region of some members of the IgG antibody class is responsible for their ability to pass through the placenta.

4. **Complement system** activation results from antibody-antigen binding to C1 in the classical pathway (Figure 26-3). IgM is the most efficient activator of complement, because a single molecule of the pentameric IgM can activate complement. When C1 binds to the Fe region of two IgG molecules that have bound antigen, activated C1 results. These activation steps involving cleavage of other plasma proteins into fragments that unite to form activated C3, where the alternative pathway joins the classical pathway at C3. IgG can also activate complement, but the antibody level must be high enough to allow cooperation between antibody molecules to occur. Not indicated on the complement pathway is the enormous amplification that results from production of the activated enzymes: a single activated C1 molecule can result in thousands of membrane attack complexes.

Different classes of antibodies are associated with different actions on the basis of their different structures and concentration at different sites within the body. The IgM and IgG antibodies are the major categories of antibodies produced and released into the plasma in response to systemic infection, as described next. The IgA immunoglobulins are found in exocrine secretions such as breast milk, respiratory and GI-tract mucus, saliva, and tears. The IgA in breast milk provides breast-fed infants with passive resistance to GI tract diseases. IgA protects against entry of foreign antigens across

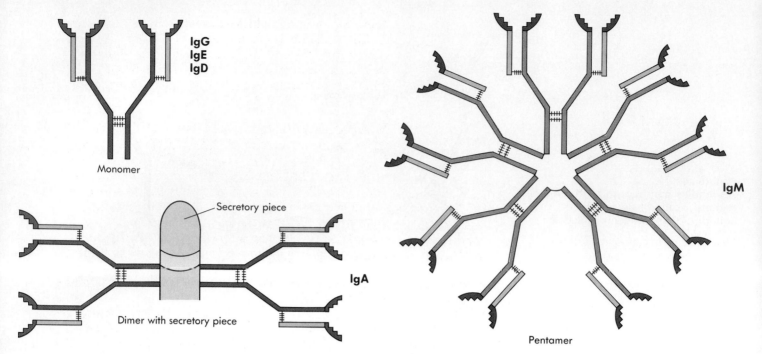

FIGURE 26-9
The basic antibody unit is a monomer, seen in classes IgG, IgE and IgD. IgA is a dimer, and IgM is a pentamer.

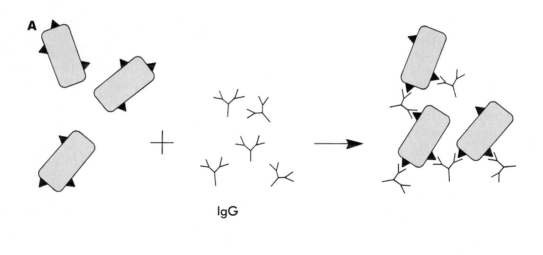

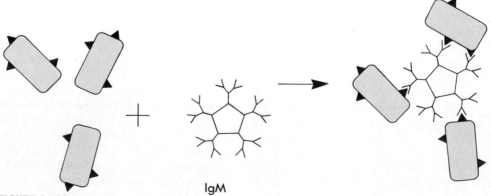

FIGURE 26-10
(A) Agglutination by IgG, a monomer. (B) IgM, a pentamer, is even more effective in agglutination reactions.

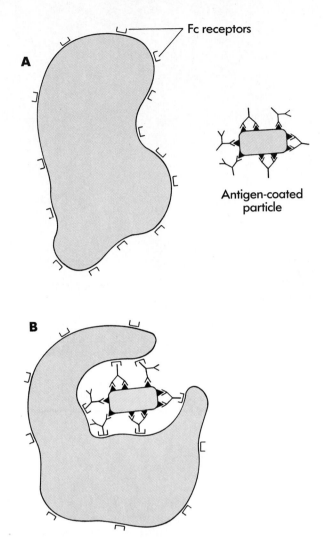

A

Fc receptors

Antigen-coated
particle

B

FIGURE 26-11
Opsonization facilitates phagocytosis.
A Binding of antibody to epitopes on the foreign parti-
cle leaves the Fc portion of the antibody free to dock on
Fc receptors on the surface of a phagocytic cell, pulling
the membrane around the foreign particle like a zipper **B**.

body surfaces by inhibiting bacterial adherence and
opposing formation of bacterial colonies on the
body surface. It is unable to damage bacteria by activating complement because the proteins of the
complement system are not present in exocrine secretions.

IgE is not significantly involved in agglutiniz-
ing or complement activating responses, and is not
directed against bacterial invaders. Its most obvious natural target is parasitic organisms, which
cause the normally very low titer of IgE to rise
many-fold, causing mast cells and basophils to
release histamine, which stimulates some pathways
of the inflammatory response. IgE also plays a
significant part in allergic reactions, as described
below. IgD is located on the surface of B cells, and
binding of antigen to IgD is thought to play a role
in activation of the B cell.

The Genetic Basis of Antibody Diversity

The studies of Austrian scientist Karl Landsteiner in
the first decades of the twentieth century demonstrated that the immune system is capable of making antibodies to a very large number of different
antigens. Estimates of the potential number of
unique antibodies range from 10^6 to as high as 10^{20}.
The variability in the amino acid sequences of the
hypervariable segments accounts for this diversity.
If only a few hundred genes code for immune proteins, then what is the source of this prodigious
variability?

The major source of variability is a process
called **somatic rearrangement.** Immune receptor
genes in antibody producing cells are assembled by
joining several separate segments of DNA that are
initially in different locations on the chromosome.
Assembling these fragments of genetic information
is a key event in the process of forming a clone, a
small number of identical cells that, in the case of
lymphocytes, can synthesize one antibody. Within
the founder cell of each clone, the gene specifying
the single variable region of the antibody that the
clone will make is assembled from a library of possible sequences (Figure 26-12). A great many more
B cell clones are produced than survive, so it is proposed that clones that happen to make antibodies to
self antigens are eliminated during B cell differentiation before birth.

Specific Immune System Cell Types

Lymphocytes are the cells that have the chemical
detection systems that allow them to recognize nonself antigens. Lymphocytes make up 20% to 40% of
the leukocyte population and are divided into two
general types: **T cells** and **B cells** (Figure 26-2).
These two types have distinct functions. B cells are
primarily located in the lymph node tissues but also
may move through the circulatory and lymphatic
systems. During development, individual B cells
are committed to production of a single **antibody**
specificity. An antibody is defined as a molecule
produced by a lymphocyte that has the ability to
combine specifically with an antigen. The antibodies produced by the B cells are expressed on their
surface as membrane bound receptors, but in
activated B cells, they are produced in large numbers and released and travel in the blood to sites
where they may encounter and bind their specific
antigen. T cells are the agents of cell-mediated specific immune reactions. They also are committed to
respond to particular antigens during development

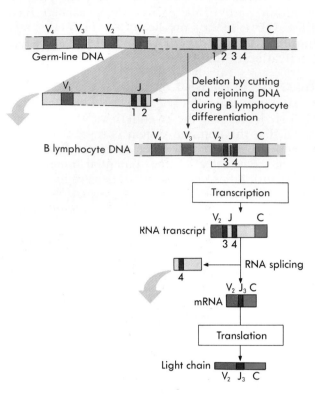

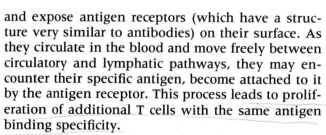

FIGURE 26-12
The gene for the variable region of a heavy chain is assembled from segments chosen at random from a large library of DNA segments. The choice is different in each clone, and a large number of different variable regions is possible.

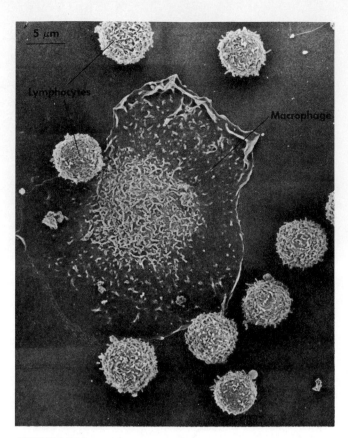

FIGURE 26-13
Scanning electron micrograph showing a macrophage (large central cell) contacting three T lymphocytes. Such an association would occur during antigen presentation.

and expose antigen receptors (which have a structure very similar to antibodies) on their surface. As they circulate in the blood and move freely between circulatory and lymphatic pathways, they may encounter their specific antigen, become attached to it by the antigen receptor. This process leads to proliferation of additional T cells with the same antigen binding specificity.

In addition to the B and T cells, macrophages are the third type of cell that participates in specific immunity (Figure 26-13). They are not able to show specificity in their actions, but during a nonspecific response to invasion of the body, macrophages engulf foreign material. They may carry antigens associated with the foreign invader to T cells. When the information-bearing macrophage locates a lymphocyte with the ability to bind that antigen, events that lead to activation of that cell are initiated. Some B cells also express class II MHC and can also present antigen. The events in activation will be considered after the characteristics of B and T cells and the nature of antigens and antibodies have been described.

Differentiation of the B and T Cells

The cells of the specific portion of the immune system, the B and T cells, arise from the same bone marrow stem cells that give rise to red blood cells, granulocytes and monocytes (Figure 26-2). Lymphoid progenitor cells produce precursors of both B cells and T cells. In the early development of birds, the cells that produce antibodies mature and differentiate in a lymphoid tissue called the **bursa of Fabricius**, an outpocketing of the cloaca. They were therefore named bursa or B cells to dintinguish them from the T cells, which undergo their differentiation in the **thymus gland**. Although the thymus is involved in differentiation of T cells in all vertebrates, differentation of B cells occurs early in embryonic development in the liver of mammals and subsequently is taken over by the bone marrow and possibly other locations (Figure 26-14).

During the differentiation of B cells, rearrangements of genetic material in individual cells produce genes with the information necessary to code for unique antigen binding properties. A founder cell of a B cell clone is eliminated if the antibody ex-

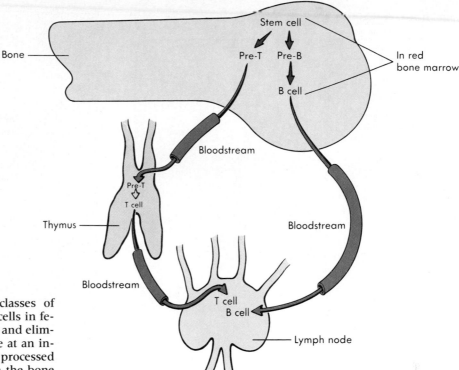

FIGURE 26-14
Differentiation of the two different classes of lymphocytes, B and T cells, from stem cells in fetal bone marrow. Clonal differentiation and elimination of self-specific clones take place at an intermediate site in the body. T cells are processed in the thymus. B cells are processed in the bone marrow.

posed on its surface binds to an antigen (a self antigen). The founder cells that fail to encounter an antigen in this trial stage, which occurs both prenatally and later in life, survive and form the vast population of individually specialized cells capable of recognizing nonself antigens.

In the thymus, T lymphocytes undergo a similar differentiation to produce a large number of antigen-specific clones. Self-directed clones are typically eliminated. The surviving clones migrate to blood, to become a major component of the circulating lymphocytes, moving easily between the blood and the lymphoid tissues. The thymus is large in infants, but begins to atrophy during childhood and has almost vanished by adulthood. In addition to its role as a site of differentiation of T cells, the thymus secretes several hormones important for stimulating blood cell production. Atrophy of the thymus occurs after puberty, but secondary lymphoid organs have already been populated by cells that differentiated in the thymus, so the effect of age-induced atrophy of the thymus on immune function is minor, compared with the effect of removing the thymus in infancy.

There are three subpopulations of T cells: **helper T** cells, **suppressor** T cells and **cytotoxic T** cells (Table 26-4). Helper T cells are involved in antigen presentation and secrete chemicals that promote the responses of cytotoxic T cells and B cells. Cytotoxic T cells participate in cell-mediated killing. Suppressor T cells are generated with a longer delay than the other two categories and play a role in damping the immune response.

COORDINATION IN THE IMMUNE SYSTEM
Activation of Helper T Cells

After macrophages engulf a microorganism or foreign particle, the phagosome containing the foreign material fuses lysosomelike with vesicles that are not lysosomes but which contain enzymes that alter or cleave the molecules into pieces (epitopes) that can be attached to class II MHC. These epitopes are displayed on the surface of the macrophages in association with the class II MHC molecules (Figure 26-15). The display of such a combination of antigen and class II MHC is called **antigen presentation.** Both cell-mediated immunity and humoral immunity are initiated by antigen-presenting cells. Macrophages, some B cells, and a few other cell types express Class II MHC molecules, so only these cells can present antigens in an **antigen presentation complex.**

Passing through lymph nodes, the antigen-presenting macrophages come into contact with helper T cells. Of the large populations of helper T cells, a very small number will have a compound receptor that can bind both the class II MHC and the epitope. The structure of the antigen-detecting portion of the **T-cell receptor** (Figure 26-6) is like one arm of an antibody. It consists of an **alpha** and a **beta** chain and is called T_cR. The portion of the re-

| TABLE 26-4 | *Subclasses of T Lymphocytes* |

Designation	Functional role
Cytotoxic T cells	Recognition/destruction of cells that have been infected with a virus or have been transformed in some way (for example, cancer cells).
Natural killer cells	Actions similar to cytotoxic T cells, but their activity does not require previous exposure to antigen.
Helper T cells	Regulators of the immune response; they increase the numbers of cytotoxic T cells, activate B cells, and secrete various lymphokines.
Suppressor T cells	Inhibit both cell-mediated (T-cell activity) and humoral (B-cell mediated) immunity; limit what might otherwise be an uncontrolled response.

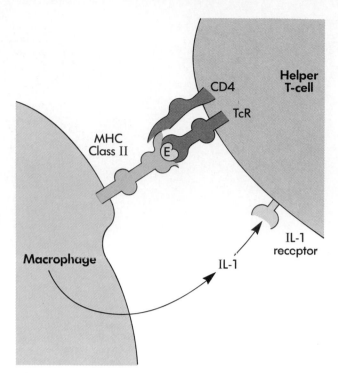

FIGURE 26-15

Activation of a helper T cell involves two signals: 1) The presentation of epitope in association with the MHC class II must bind specifically to the T$_c$R receptor. This association is thought to be stabilized by the CD4 on the helper cell. The second signal is the helper cell detection of interleukin-1 that is released by the macrophage.

ceptor that is able to recognize the MHC class II and stabilize the binding of T$_c$R with the epitope is called **CD4** or **T4** (Figure 26-15).

The first step in T cell activation is interaction between the presenting macrophage and a helper T cell competent to bind MHC II and the presented antigen. The second step is release of the lymphokine called Interleukin-1 (IL-1) from the macrophage and its detection by receptors on the helper T cell. These two signals, antigen presentation and IL-1, activate the helper T cell. It then releases additional lymphokines, a collection of proteins that includes Interleukin-2 (IL-2). IL-2 serves as a mitogen and growth factor for helper T cells and cytotoxic T cells. Activated helper T cells secrete other lymphokines that stimulate the activities of macrophages, converting them to "angry" macrophages. Several lymphokines, including an interferon, inhibit the migration of macrophages away from a site of infection.

Cytotoxic T Cell Activation and Function

Cytotoxic T cells are an important component of **cell-mediated immunity** (Table 26-4). The activation of the cytotoxic T cells is similar to the process for helper T cells, except that the antigen presenting cell in this case is an abnormal body cell, such as a tumor cell or a virally infected cell (Figure 26-16). The combination of MHC class I antigens, which are found on all body cells, and a nonself antigen stimulates the cytotoxic cell to produce receptors for IL-2, and reception of this signal (from helper T

cells) initiates proliferation of activated cytotoxic cells. The cytotoxic cells then target body cells displaying the activating antigen and destroy them by secretion of toxins. One such toxin is a protein called **perforin.** Several perforin molecules join into a ring to form a pore in the membrane of the target cell. The pore allows all small ions to cross the membrane, accompanied by water, and the target cell swells and eventually lyses.

Two conditions elicit destruction of body cells by cytotoxic T cells: tumors and viral infections. Transformation of body cells into cells that are not subject to normal growth control occurs fairly frequently. If unchecked, these "mistakes" result in a colony of cells, a tumor, which may become cancerous if the growth invades adjacent tissues or spreads to other regions of the body. Many transformed cells reexpress surface antigens characteristic of early stages of development or antigens that reflect their altered genes. These abnormal antigens can be recognized as nonself and the transformed cells destroyed by cytotoxic T-cells. The actions of natural killer cells and cytotoxic T cells constitute the body's **immune surveillance** that protects against cancer. The development of cancer constitutes a failure of this system. Such a failure is more

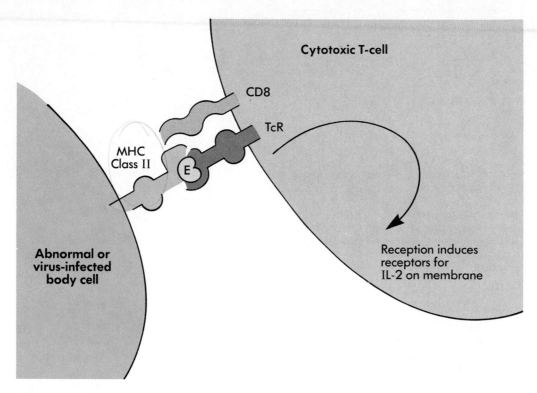

FIGURE 26-16
Activation of cytotoxic cells results from binding to the T_cR receptor of an appropriate epitope in association with the Class I MHC, which is found on any of the body's cells. Cytotoxic cells are distinguished from helper cells by possession of a CD8 antigen, which may assist in the binding. Interleukin receptors are induced on the cytotoxic cell's membrane after antigen presentation, and the IL-2 signal from helper cells stimulates differentiation and mitosis.

likely to occur when age reduces the effectiveness of the immune system.

Viruses are small packets of genetic material enclosed in a protein coat that invade and use the intracellular environment of a host cell to replicate their genes. Viruses take over the protein-synthesizing apparatus of a cell and force it to synthesize viral proteins. The membrane of such a cell can reflect the presence of the virus either by display of viral proteins or proteins from the cell's own genome not normally expressed by healthy cells (Figure 26-16). Either way, the cytotoxic T cell can find the combination of class I antigens and the abnormal antigen, bind to them, and destroy the cell. Chemical mediators released by helper T cells are important in stimulating such actions. This action supplements the action of interferons in viral resistance (see nonspecific immunity).

When tissue from one individual is transplanted into another individual, the recipient's immune system quickly reacts to destroy the foreign tissue in a **transplant rejection.** This is the classic example of cell-mediated immunity and is mediated by cytotoxic cells. Clinical intervention can prevent the rejection of organ transplants, at least temporarily.

Activation and Proliferation of B Cells

While a foreign antigen is activating a select population of helper T cells, the corresponding population of B cells is also encountering and binding the antigens (Figure 26-17). However, an isolated binding event is not sufficient to activate the B cells. The B cell receptors can move around on the membrane, and the antigen-antibody receptor complex can become crosslinked in a process ("capping") that resembles agglutination. This is followed by internalization of the linked antigen-antibody complexes (Figure 26-18). Lymphokines released by helper T cells are necessary to complete the activation of the B cells.

Activation of a B lymphocyte leads to the **primary response,** production of a clone of cells with identical antigen specificity but two possible fates. The progeny of activated B cells include a population of memory cells, which enter a resting state, and a population of relatively short-lived plasma cells, which begin to release antibodies (Figure 26-18). The transformation to antibody-releasing cells involves cell enlargement and development of a prominent endoplasmic reticulum, a characteristic of cells that manufacture large quantities of protein

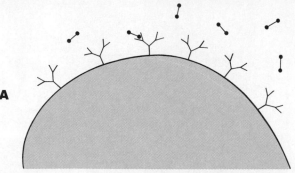

A

B-cell receptors free to move about in plasma membrane

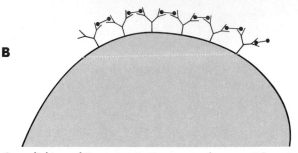

B

Cross-linking of Receptor-Antigen complexes or "Capping"

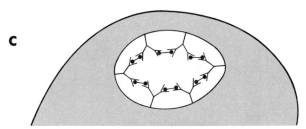

C

Internalized Receptor-Antigen complex

FIGURE 26-17
The interactions between B cells and antigens require that a complex of the antigens be internalized. Steps in the process include
A antigen binds to antibodies in the membrane;
B the antibody molecules drift together to facilitate binding, resulting in crosslinking of antigen-antibody complexes into a single clump, called "capping".
C the antigen-antibody complex is internalized.

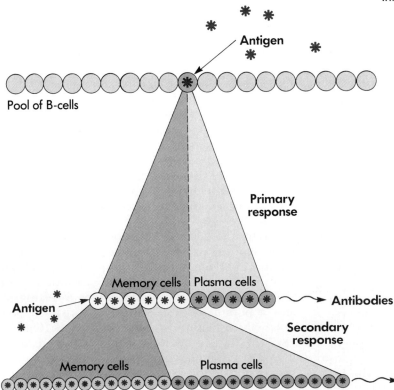

FIGURE 26-18
Antigen presentation results in activation of a small number of lymphocytes from the total pool. The primary immune response involves some antibody production but primarily lays the foundation for the rapid and vigorous secondary immune response to subsequent exposure to the same antigen, in which the memory cells divide rapidly to produce many antibody-releasing plasma cells and a larger population of memory cells.

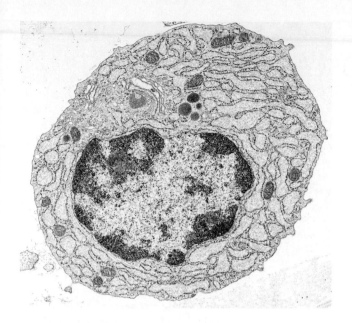

FIGURE 26-19
The structure of plasma cells. The cytoplasm is filled with rough endoplasmic reticulum for the synthesis of antibody protein.

(Figure 26-19). The role of the memory cells is described below.

At the same time that helper T cells stimulate an immune response, they also activate the subpopulation of suppressor T cells that bind the same antigen. Suppressor T cells release signals that act on T and B cells, inhibiting the cell-mediated and the humoral components of the specific immune response (Table 26-4). This action might at first appear self-defeating, but the suppressor T cell population is activated slowly, and gradually opposes the proliferation of immune system responses that might otherwise be self-perpetuating. In addition to their role in damping the immune response to a specific antigen, suppressor T cells probably play an important role in preventing autoimmune reactions by limiting the activity of those T and B cells that are able to bind self antigens but escaped deletion. A failure of this regulatory system could be one cause of autoimmune disease (see p. 777).

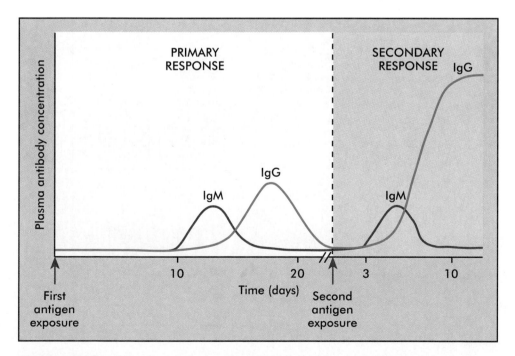

FIGURE 26-20
The timescale for the primary and secondary responses by B cells. A typical lag between antibody exposure and onset of antibody release is 10 days for the primary response and 3 days for the secondary response, which may be elicited years after the first response. Note the difference in the timing and quantities of the IgM and IgG immunoglobulins involved in the response.

Using Monoclonal Antibodies in Physiology and Medicine

Since the discovery of antibodies, their ability to target specific molecules with great precision has offered the possibility of therapies aimed at a very select population of the body's cells. Antibodies can be raised in the laboratory by injecting animals with antigen and collecting blood plasma after the animals have mobilized a secondary immune response. Because every antigen has at least several potential determinants, the population of antibodies raised in this way is a mixture of antibodies to different determinants. Such a mix of antibodies is said to be polyclonal.

Polyclonal antibodies not only recognize different determinants on the same antigen molecule, but also have a relatively high probability of including some antigen-binding sites that cross-react with other molecules. The precision of the recognition and targeting process could be refined if there were some way to ensure that the antibodies raised to an antigen were monoclonal; that is, derived from a single B-cell clone and specific for a single determinant.

This could be done if a single plasma cell of the appropriate clone were able to multiply indefinitely in cell culture, but plasma cells do not multiply well or live long either in culture or in the body.

However, in 1984, British scientist C. Milstein and Swiss scientist G. Kohler received a Nobel Prize for discovering a method for making large quantities of monoclonal antibodies. They found a way to make plasma cells multiply in culture by hybridizing them with myeloma cells, a variety of cancer cell that divides ceaselessly in culture when provided with the proper nutrients. These hybrids (called hybridoma cells) are obtained by cell fusion; the plasma cells are cultured together with myeloma cells and

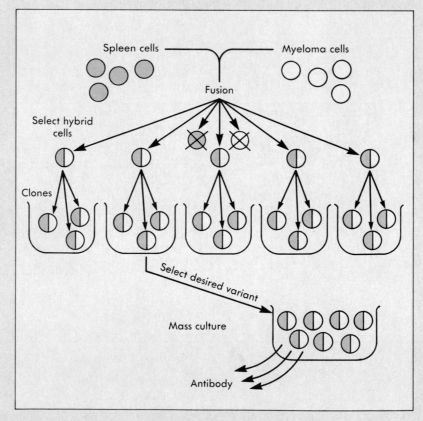

FIGURE 26-A

Production of hybridoma clones by cell fusion followed by selection of antibody-secreting clones and further selection and mass culture of a clone that secretes the desired antibody.

some of the cells fuse, resulting in cells that multiply like the cancer cells but that also continue to make antibodies (Figure 26-A).

The cells that fail to fuse are eliminated; the cancer cells die when deprived of a necessary nutrient, and the unfused plasma cells do not survive long in culture. Single hybridoma cells are isolated and allowed to multiply into a population of genetically identical cells that constitute a clone and can be cultured indefinitely. The antibody produced by each clone established in this way is screened for its specificity.

Monoclonal antibodies can be coupled (conjugated) to fluorescent molecules that make the sites at which they bind visible under the light microscope or to molecules containing a heavy metal such as gold that makes the location of the antigen visible with the electron microscope. Monoclonal antibodies to specific neurotransmitters can be used to trace neural pathways in the nervous system. Monoclonal antibodies can also be used to purify large amounts of their antigens by a process called affinity chromatography, in which tissue extracts are passed through a bed of particles to which the antibodies have been attached. The antibodies retain the desired antigen, allowing other substances to pass. Interferon was first commercially purified with this technique.

Monoclonal antibodies are being employed in medical diagnosis and therapy. Tumor cells can be identified with monoclonal antibodies specific for a particular characteristic antigen. The location of the tumor in the body can then be determined by conjugating the antibody to a radioactive label that can be detected by scanning the body. Monoclonal antibodies might direct the patient's immune system to a tumor, or they could be made to deliver toxic drugs specifically to tumor cells but not to normal cells, serving as a type of "magic bullet."

The Kinetics of Immune System Response

The first response to an antigen is called a **primary response** (Figure 26-18). Not all antibody classes are produced with the same rapidity or remain elevated for the same timespan after challenge by a foreign antibody (Figure 26-20). The primary immune response involves expression of IgM; if IgG is seen later in the primary response, the amount of IgM usually drops off at that time. The transition is thought to be accomplished when the cells that produced the IgM switch over to producing the same antibody specificity associated with a different Fc class. The complex IgM can activate complement and agglutinate particulate antigens but it does not activate macrophages. Continued presence of the antigen elicits the opsinization and neutralization reactions promoted by IgG.

The memory effect arises because **memory cells** can quickly give rise to a **secondary response** if the antigen is experienced again (Figure 26-18). The primary immune response requires 15 to 20 days to reach peak antibody production. This response is usually too slow and too weak to prevent development of the disease. In a secondary exposure, the time required for maximum production of antibody is much shorter, and the total amount of antibody produced is much greater because the number of cells in responding clones is greater from the beginning (Figure 26-20). For many diseases, the secondary response is usually rapid enough to prevent development of the disease. In other cases, the secondary response simply makes subsequent episodes of a particular disease less severe.

UNDESIRABLE IMMUNE SYSTEM RESPONSES

- *How are superantigens thought to elicit secretion of large amounts of IL-2?*
- *What cell type mediates organ transplant rejection and how does this occur?*
- *What are the types of hypersensitivity and which antibodies or cell-mediated responses are involved?*
- *How might autoimmune diseases arise?*

Superantigens Bypass Immune Cell Specificity

The antigen presentation complex is designed to allow the antigen to bind with the very selective region of the T cell Ti receptor. Certain proteins that elicit a massive overstimulation of the immune system act by binding to a different region of the T receptor (Figure 26-21; compare with Figure 26-15). These same molecules, known as **superantigens**, bypass processing by the macrophage and bind directly to the MHC class II antigen. Such superantigens may be able to activate a large subpopulation of the T cells. Examples of these molecules are proteins called enterotoxins produced by *Staphylococcus aureus* that cause food poisoning and toxic shock

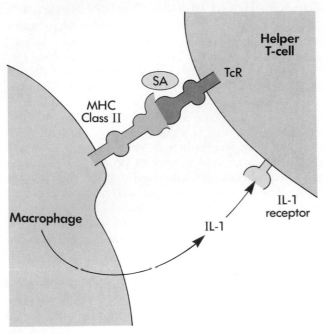

FIGURE 26-21

Superantigens such as some enterotoxins bind to a different portion of the MHC Class II molecule than antigens that have been processed by the macrophage for presentation, and they interact with a different portion of the T_cR receptor. This portion of the Ti molecule is not so variable as the antigen-binding site, so a given superantigen can bind with a relatively large portion of the helper T cells and activate them. The result is production of many lymphokines, including interleukin-2, which accounts for soem of the symptoms of food poisoning or toxic shock syndrome.

syndrome. The overstimulation of the helper T cells leads to release of interferon and other lymphokines, including high levels of IL-2. The consequences of this can include diarrhea, vomiting, fever, and shock. It is probable that because bacteria can grow in the body, toxic shock complications are more severe than food poisoning. In addition to massive nonspecific activation of T cell populations, the superantigens may in some cases kill off the overactivated cells, reducing the total T cell population, or cause autoimmune disease by activating resting populations of cells that produce antibodies against self (see below).

Transplantation and Transfusion Reactions

Transplantation of tissues from one individual to another has become a spectacular mechanism for restoring function. Corneal transplants and kidney transplants are common, and even heart and liver transplants are becoming ordinary. The major drawback to this approach, apart from the difficulty of obtaining healthy donor tissue, is the complication of transplant rejection. If the MHC class 1 antigens in donor and recipient cells are different, the difference, however small, forms the basis of cytotoxic T cell rejection of the implant.

Measures that suppress the rejection of the graft include corticosteroid antiinflammatory agents, which stabilize lysosomal membranes and thus suppress destruction of the cells. Antimetabolites suppress responses of the immune system by antagonizing RNA or DNA synthesis. Cytotoxic agents suppress any population of proliferating cells, and thus tend to inhibit T cell activation. Cyclosporine is a compound derived from a fungus that reduces the number of helper T cells relative to suppressor T cells. These are nonspecific effects that leave the immune system of the transplant recipient less able to fight off infections of all kinds.

Blood was the first tissue to be successfully transplanted. Red blood cells do not carry MHC antigens but do possess other surface antigens that can lead to **transfusion reactions** from the recipient's immune system. At least 15 different families of blood-group antigens have been recognized. In most cases, an individual whose own red blood cells do not possess the antigen in question can be successfully transfused at least once with blood that includes it. Subsequent transfusions incite an immune response that destroys the transfused cells and makes the recipient ill.

The most familiar family of red blood cell antigens is the basis for the **ABO blood group system.** Each individual possesses two homologous genes for antigens of this family, and there are three different alleles. People with different alleles make some of each antigen; for example, those with the blood type AB make both antigen A and antigen B. People with the O blood type do not have either the A or the B allele and produce only the precursor glycoprotein H.

The unusual feature of the ABO system is that immunity to nonself antigen is almost always developed even if there is no exposure. Those with type A blood possess antibodies to the B antigen even if they were never transfused, and vice versa. Those who lack both the A and B alleles have type O blood and produce antibodies to both antigens. Individuals with type AB blood do not produce antibodies to either antigen. How anti-A and anti-B antibodies arise is not understood.

Incompatibilities in ABO system are shown in Table 26-5. If a person with type A blood receives type B blood, two antibody-antigen reactions occur: (1) the donor's anti-A reacts with the A antigen of the recipient's red blood cells, and (2) the recipient's anti-B reacts with the B antigen of the donated red blood cells. The first reaction is of little consequence because donated antibodies are greatly diluted in the recipient's plasma. The second reaction is severe because the recipient's antibodies lead to destruction of the donated cells, releasing hemoglobin and cell fragments into the plasma that may interfere with renal filtration. The complement reaction makes the recipient quite ill. People with type AB blood can receive either types A or B blood with

TABLE 26-5   Summary of Transfusion Reaction

Recipient blood group	Blood group of donor			
	Group A	Group B	Group AB	Group O
A	No*	Yes†	Yes	No
B	Yes	No	Yes	No
AB	No	No	No	No
O	Yes	Yes	Yes	No

*No: No transfusion reaction will occur.
†Yes: Transfusion reaction will occur.

only a minor reaction because the donor's antibodies are greatly diluted. Similarly, although type O blood contains both anti-A and anti-B, people with types A, B, or AB blood can receive type O blood because the dilution effect protects the recipient's cells from a severe reaction.

Another example of erythrocyte antigens is the **Rh system.** Those who possess the antigen are **Rh-positive; Rh-negative** individuals lack the antigen. This antigen can cause Rh disease, a condition in which the placenta fails to protect the fetus from an immune attack by its mother's body. Fetal red blood cells normally do not enter the mother's circulation, but may do so late in gestation or at birth when placental blood vessels are disrupted. If the mother is Rh-negative but the fetus is Rh-positive (having inherited the gene for the antigen from its father), the fetal red blood cells trigger a primary immune response in the mother. The antibodies to the Rh antigen can cross the placenta, so in a subsequent pregnancy, the anti-Rh antibodies will attack fetal blood cells, causing **erythroblastosis fetalis,** or **Rh disease,** in an Rh-positive infant (Figure 26-22). This can be prevented by flooding the maternal bloodstream with anti-Rh antibodies just after each child is delivered. The injected anti-Rh destroys any fetal cells that may have entered the mother's blood before the antigen they bear can initiate a primary response.

Hypersensitivity

An exaggerated or inappropriately directed immune response can lead to tissue damage. Such deleterious reactions involve a runaway form of the same processes that provide protection when correctly targeted. Such damaging responses are collectively called hypersensitivity reactions, and they may act over different timescales, depending on the particular immune system elements involved. The first three types are mediated by antibody reactions and the third, cell-mediated responses, develop more slowly. **Type I,** or **anaphylactic reactions,** are mediated by IgE signals to mast cells, which respond within minutes. **Type II** hypersensitivity results

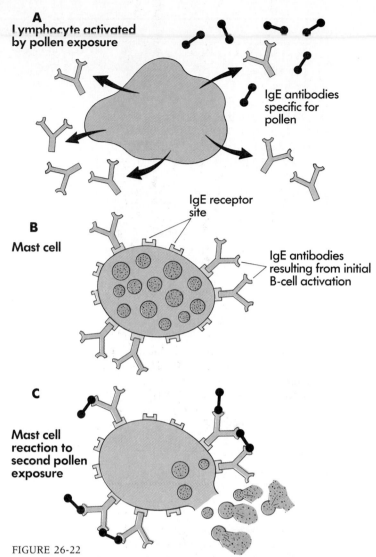

A Lymphocyte activated by pollen exposure

IgE antibodies specific for pollen

B Mast cell

IgE receptor site

IgE antibodies resulting from initial B-cell activation

C Mast cell reaction to second pollen exposure

FIGURE 26-22
The sequence of events that lead to an allergic attack.
A In the first exposure to an innocuous antigen to which the immune system can respond, B lymphocytes are activated and produce IgE antibodies to the antigen, which might be pollen grains.
B These antibodies then attach by their Fc regions to receptors on mast cells and basophils. The stage is now set for a vigorous response to the next encounter with the pollen.
C When the pollen is in high enough concentration to provide crosslinking of the IgE antibodies on the mast cell surfaces, the cells release their granules, liberating histamine and other active chemicals that cause the reactions associated with hay fever or asthma.

from self-directed IgM and IgG antibodies that mediate phagocytosis, cytotoxic activity and complement activation directed against the body cells. **Type III** hypersensitivity results from large aggregates of antigen and antibody, which form clumps that stimulate complement and attract granulocytes. Damage to local tissue results from release of lytic enzymes. **Type IV** or **delayed hypersensitivity** involves the activation of T cells and their release of lymphokines.

Acquired Immunodeficiency Syndrome (AIDS)

The acquired immunodeficiency syndrome, AIDS, results from infection by the human immunodeficiency virus, HIV-1. The HIV belongs to a class of viruses called retroviruses, whose genetic material is RNA rather than DNA (Figure 26-B). To be replicated within a cell, the retrovirus utilizes an enzyme called reverse transcriptase which reverses the DNA-to-RNA readout pattern to produce DNA that codes for the viral RNA. This DNA is spliced into the host cell DNA, where it can direct the production of more viral RNA and corresponding viral proteins and also be passed along to all the daughter cells.

Initial symptoms of HIV infection may include a sore throat and fever, which appear a few weeks to months after exposure, but often go unnoticed because they are common to so many mild illnesses. The initial phase involves rapid increases in the numbers of viruses in the blood followed by a rapid response by the immune system. The elevation in antibodies is correlated with the virtual disappearance of the virus from the bloodstream. The antibodies against HIV that appear in this initial response are maintained at high levels and form the basis of HIV testing. Half of the men who test positive for HIV remain asymptomatic or do not have an AIDS-defining illness for an average of 10 years. Unfortunately, the immune suppression does not last indefinitely, and intact viruses in resting cells or viral genetic material incorporated into the DNA of host cells serve as a reservoir for later bursts of virus production that the immune system fails to suppress (Figure 26-C).

Retroviruses spontaneously and rapidly mutate into new forms that allow them to escape the antibody recognition system developed for the original strain.

The HIV can infect several cell types, including epithelial cells,

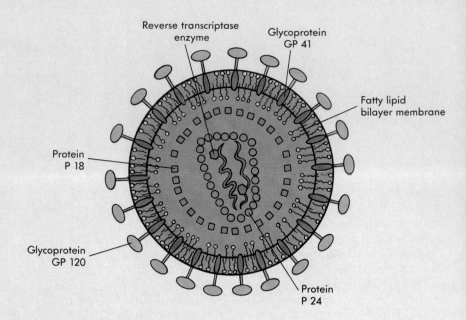

FIGURE 26-B
The structure of the HIV virus. The virus is surrounded by a protein coat that is studded with receptors for corresponding molecules on the lymphocyte cell membrane. The interior of the coat contains viral RNA and reverse transcriptase.

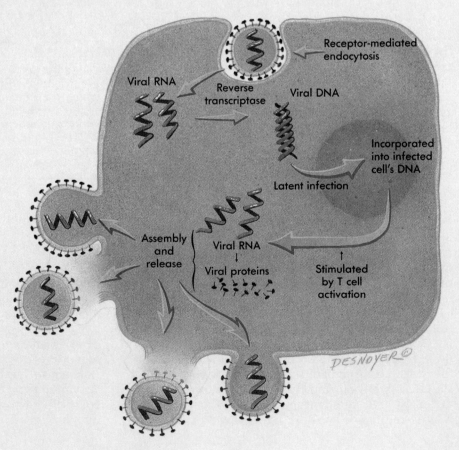

nervous system cells, and macrophages, but the most devastating aspect of HIV is that it attacks the helper T lymphocytes, the very cells that are crucial to mounting an immune response. The CD4 receptor on helper T cells that recognizes Class II MHC antigens on the antigen-presenting cells is the site where the virus attaches to the cell. Binding of the virus with the receptor stimulates endocytosis of the virus. Helper T cells comprise the majority of the cells with the CD4 antibody, so the terms CD4 (T4) cells and helper T cells are often used interchangeably. The presence of the viral genetic material in the cells is innocuous until the infected cell is activated during an immune response. Then, the infected cells will give rise to a clone of infected cells that produce lymphokines and so many viruses that the cells are lysed, and the liberated viruses from each infected cell spread out among the remaining CD4 cells.

The CD4 cell population is monitored as one indicator of whether the infection with HIV has progressed to the condition called AIDS. In healthy individuals, the ratio of helper T cells to cytotoxic T cells is 2:1. AIDS is diagnosed when the ratio drops as low as 0.5:1.0. As the number of helper T cells drops, infected individuals typically fall prey to opportunistic diseases that rarely develop if an individual's immune system is functioning properly. Early symptoms associated with AIDS include neurological disorders, fever, night sweats, fatigue, and diarrhea. Among the early signs of infection in women are abnormal Pap smears and chronic vaginal bacterial and yeast infections. On

the average, the interval between diagnosis of HIV and death resulting from complications of AIDS is much shorter in women than in men. The difference may be attributed to failure to detect the condition as early in women as in men. Physicians are not trained to detect the symptoms that appear in women and, historically, women have been in the minority among the HIV-infected population in the United States, so HIV infection is not always suspected. However, that is changing because women are the fastest-growing group of AIDS patients and account for half of the new cases worldwide.

Transmission of HIV occurs during sexual contact with an infected partner. Semen contains the virus, which can be transmitted to the bloodstream of the recipient of penetrating intercourse, either heterosexual or homosexual.

Vaginal secretions, as well as tears and saliva, can contain the virus, but the rate of transmission from an infected female to uninfected male partner is low, compared with the reverse situation. The sharing of contaminated needles and syringes by drug users continues to be a major mode of transmission. In contrast, transmission of the virus in transfusion of blood or blood products represents a vanishing risk in the United States because of donor screening, blood testing, and heat or chemical treatments of clotting factor products. Pregnant women infected by the virus may transmit the virus to a fetus, or a baby may be infected by breastfeeding. However, there is evidence that at least some of the babies born to infected mothers contract the virus during exposure

to fluids in the birth canal, so cleansing the canal of vaginal secretions and minimizing exposure to maternal blood may reduce the tragically large number of infected babies.

By March of 1992, 211,000 cases of AIDS had been confirmed in the United States by the Center for Disease Control; 139,000 deaths had already occurred. The number of people in the United States infected with HIV is estimated to be in the range of 1,000,000 to 1,500,000. Worldwide, the number of infected adults is estimated by the 1992 Conference on AIDS in Amsterdam to be 12 million, with an additional 1 million children infected. There is currently no cure for this disease, and the treatments that are available alleviate some of the symptoms present during progression of the disease without altering the course of the disease. Approaches include treating the opportunistic diseases that the immune system is incompetent to resist, such as pneumonia caused by the protozoan *Pneumocystis carinii*, cancers such as Kaposi's sarcoma, viral infections such as cytomegalovirus and *Herpes simplex*, and fungal infections such as candidiasis. Because the lymphokines normally released by helper T cells in response to infections are greatly reduced after the helper T cell population drops, this limits the response of the normal B and cytotoxic T cells, so one possible treatment may be to supply these factors. The other treatments currently in use are aimed at inhibiting replication of the virus. They include zidovudine (AZT, retrovir, HIVID), dideoxyinosine (ddI, which is converted in the body into

FIGURE 26-C
The sequence of events that leads from infection of a T cell to a latent infection followed at some later date by ac- tivation of viral replication and destruction of the infected cell, releasing the new viruses.

Continued.

dideoxyadenosine) and dideoxycytidine (ddC). These analogs of nucleotides are incorporated into DNA but they differ from the normal nucleotides in that the next nucleotide in the sequence cannot be linked to them, so the process of DNA production from the viral RNA is blocked. However, interfering with DNA replication affects all the populations of cells that normally divide at a rapid rate, such as the lining of the GI tract and bone marrow stem cells, so the treatment is inherently toxic. The complications that result from these drugs may be reduced by using two of them together at lower doses.

Attempts to induce immunity against HIV have thus far been thwarted by the mutability of the virus and the fact that some of the immune responses that it elicits in infected lymphocytes promote the viral infection. This is presumably the reason the antibody and cytotoxic T cells that are induced in the initial infection are insufficient to protect the patient.

One of the aims of designers of a vaccine is to use antigens derived from the virus that are structural and may serve as epitopes for the induction of immunity without introducing a possible threat. Although a primate animal model for studies of AIDS vaccines exists, human clinical studies will eventually be necessary. All such tests pose hazards, but the dangers associated with AIDS make such tests particularly threatening.

Although research on the induction of immunity through the use of vaccines is being actively pursued, recent interest has also focused on the rare individuals in the population who exhibit unusual resistence to infection or whose immune system is resisting the replication of HIV. Studies of some resistant individuals suggest that cytotoxic T cells may play a crucial role in the prolonged immunity, so clinical studies have been designed to better understand the basis of long-term survival in these naturally more resistant individuals. Understanding of the

mode of infection by HIV has accumulated rapidly, and hope for a way to prevent as well as cure this dread disease motivates many dedicated researchers. The fact remains that until AIDS can be prevented, education about risk behaviors is vital and should be a top priority for all nations. High risk behaviors are intravenous drug use that could involve sharing of blood-contaminated instruments and unprotected intercourse, both between heterosexual and homosexual partners. The risk increases with the number of partners and can be reduced by the use of latex condoms with spermicide. Probability of infection in a single exposure increases if other venereal diseases are present. Because of the lack of symptoms in early stages of the disease, many infected individuals are unaware that they may infect others, so unprotected sex must be avoided when there is even the slightest chance of exposure.

Allergic reactions fall under the type I responses involving IgE. An allergy is a hypersensitive reaction to an otherwise harmless substance. An **allergen** is an antigen capable of eliciting an allergic reaction in susceptible people, and the most common allergens are pollen, pet dander, feces of dust mites and cockroaches, house molds, stinging insects, medicines and foods. In the first exposure to a given allergen, T lymphocytes specific for the allergen bind it, starting the process that results in production of activated plasma cells that produce IgE antibodies. These attach to IgE receptor sites on body cells, particularly mast cells (Figure 26-22). Thus the initial response to the allergen is unnoticed, but sensitized mast cells, bristling with antibodies, are ready to bind the antigen the next time it is encountered. Binding of the allergen to the antibody receptors on the mast cells in such a way that antibodies

are cross-linked (Figure 26-22) releases granules containing histamine and other substances from the mast cells. These substances cause the symptoms of an allergic reaction. The susceptibility to particular IgE allergies appears to be determined largely by heredity and probably involves inheritance of particular MHC genes.

In **hay fever**, the response is mainly attributable to histamine release and is concentrated in the mucous membranes of the nose, although the eyes also may be affected. Antihistamines are used to block these reactions. In **asthma**, the response is concentrated in the bronchioles, and is not so much a reaction to histamine as to slower-acting substances, **leukotrienes**, which produce prolonged constriction of smooth muscle, the bronchial spasms that are life-threatening. These can be relieved by adrenalin-like agents called beta-agonists, which open blocked airways. The most threatening Type I hypersensitivity is called **anaphylactic shock**. This reaction involves not only the mast cells in one part of the body but spread of the antigens to react with sensitized mast cells close to the capillaries and basophils in the bloodstream. The histamine causes movement of fluid from the circulatory system into the tissues, producing a precipitous drop in blood pressure; if this threat is counteracted by norepinephrine, an individual may still die as a result of asthmalike bronchiolar constriction brought on by the leukotrienes.

The commonest example of type II hypersensitivity reactions is the transfusion reaction. Transfused blood cells are destroyed by activation of complement or responses to opsonization. Some cases of drug allergies result from the fact that the drugs, such as phenacetin, an analgesic, binds to leukocytes, or chlorpromazine, a tranquilizer, binds to erythrocytes, causing the abnormally labeled cells to be targeted for destruction.

Type III hypersensitivity is characterized by antigen-antibody deposits in localized regions, where the healthy cells are destroyed by lytic enzymes and membrane attack complexes and other actions related to complement activation. The antigens that lead to this destruction may be the result of persistent microbial infection, as in kidney infections, environmental antigens that can elicit a great variety of lung diseases brought about by occupational exposure, and self antigens. The responses to self antigens are described under autoimmune diseases in the following section.

Type IV hypersensitivity is seen in individuals with chronic diseases such as tuberculosis, leprosy, and leishmaniasis, in which the immune response is mediated entirely by T cell responses, which are reflected in the delayed skin response seen in the tuberculin test of exposure to the tubercle bacillus. Helminthic parasites represent another threat that is combated by T cells, and the worms may be so resistant to elimination that the attempt of the immune system to destroy them does the host more damage than the parasite alone would have caused. Another example of T cell mediated hypersensitivity is poison ivy dermatitis. The oil produced by poison ivy will elicit a response in virtually every individual, and can be used to test for T cell reactivity.

Autoimmune Diseases

Several diseases are recognized as attacks by the immune system on some part of the body or on systemically distributed targets (Table 26-6). When a specific organ is affected, as is the case for the thyroid destruction in Hashimoto's thyroiditis, the targeted antigen is very restricted in its distribution. On the other end of the spectrum, the antibodies may be directed against nonorgan-specific antigens

TABLE 26-6	Autoimmune Diseases
Disease	**Affected organ systems**
Juvenile-onset diabetes	Pancreatic beta cells
Rheumatoid arthritis	Joints
Ankylosing spondylitis	Spine
Multiple sclerosis	Myelin in the central nervous system
Thyrotoxicosis	Thyroglobulin
Rheumatic fever	Valves in the heart
Myasthenia gravis	Acetylcholine receptors at the neuromuscular junction
Ulcerative colitis	Intestine
Male infertility (some)	Spermatozoa
Systemic lupus erythematosus	Most organs
Amyotrophic lateral sclerosis	Motor neurons in the spinal cord

such as the cell nucleus, as is the case for systemic lupus erythematosis.

Autoimmune disease can, in principle, arise in the following ways:

1. T cells specific for antigens not seen during T cell maturation could become activated as a result of an accident that exposes self-proteins. Sperm and lens proteins are normally sequestered from immune cell contact, but immune reactions can readily occur if exposure in injury or operations allows the antigens to come into contact with the blood. Thus a man who has a vasectomy and later wishes to reverse the operation may find that he is still infertile because his sperm are destroyed by his immune system.

2. An antibody induced against an exogenous antigen may cross-react with a self antigen. An example of this is the induction of damage to heart muscle and valves and kidney glomerular basement membrane because these tissues bear antigens that closely resemble the epitopes presented by the streptococcus that causes rheumatic fever.

3. The MHC self-identification system may fail to protect healthy cells. Rheumatoid arthritis, multiple sclerosis, and myasthenia gravis are examples of immune attacks directed against the joints, the myelin coats of nerves, and the acetylcholine receptors, respectively. The incidence of particular MHC antigens in patients who have these diseases is approximately double the incidence in the population at large. These relationships suggest that a failure of the self-recognition property, which should be me-

SUMMARY

1. The immune system responds to invasion of the body by foreign material with nonspecific and specific defense mechanisms. Interactions between the two systems are mediated by cell-to-cell contact and chemical messengers, which include interleukins.

2. **Antibodies** are the agents of humoral immunity. The basic antibody structure is composed of two heavy protein chains linked together with two light chains to form a Y, whose arms, the **Fab regions**, contain highly variable amino acid sequences. The consequence of the variability is that a large variety of binding sites are created that can recognize various **antigen determinants or epitopes**. The extreme variability of the Fab regions is obtained by rearrangement in which DNA segments coding for this region are joined together randomly during clonal differentiation of B cells. The biological effects of antibody binding depend on the antibody class, as determined by the **Fc region** of the antibody. Antibodies are classed as IgG, IgM, IgE, IgA, or IgD. Biological effects include agglutination, neutralization, opsonization, or activation of the complement system.

3. **Histocompatibility antigens** are identity tags for the cell-mediated immune system. They are unique to each individual, arising from the most polymorphic gene loci known. **MHC-1** antigen is found on all nucleated body cells and directs cytotoxic T lymphocytes to their targets; **MHC-2** antigen is found on B cells, macrophages, and some types of T cells, and it channels interactions between T cells and B cells that occur in the course of a specific immune response.

4. **Cytotoxic** T cells recognize and attack cells that are simultaneously displaying MHC-1 antigen and nonself antigen. **Natural killer** cells are not specific but are directed to major classes of nonself antigens.

5. **Helper T cells** are activated by antigen presented by macrophages, stimulating helper T cells to deliver the antigen to activate B-cell clones to proliferate and give rise to memory cells and plasma cells. At the same time, helper T cells stimulate the activity of **suppressor T cells,** which limit the duration and scope of the immune response.

6. **Inflammation** is characterized by vascular dilation and nervous activity that results in local redness, swelling, warmth, and pain. All these characteristics are attributable to bradykinin released by damaged tissues and to **histamine** released by mast cells and basophils. The responses are nonspecific but are greatly augmented by the specific immune system.

diated by the MHC system, is at least a contributing factor in these diseases.

4. The suppressor T cells may fail to suppress self-directed immune reactivity. For example, in **systemic lupus erythematosus,** antibodies are generated by B cells against almost all tissues of the body. Sufferers of this disease may possess a genetic defect in the generation of a particular subclass of suppressor T cells, which appears to put them at risk for the disease.

● *STUDY QUESTIONS*

1. What accounts for the swelling and pain at the site of an infection?

2. Cells of the immune system fall into overlapping classes. Which cell types are phagocytes? Which are granulocytes? Which are leukocytes? Which are lymphocytes? What is the functional similarity between basophils and mast cells?

3. How does the cascade of complement reactions facilitate specific immune system responses?

4. Why are organ transplants are more likely to succeed if the donors are closely related? Why might even closely related donors not be acceptably matched to a recipient?

5. What mechanisms involving antibody binding result in damage to infecting organisms?

6. Which cell types express the MHC class I antigens? The class II antigens?

7. People with juvenile-onset diabetes who have been treated with insulin may develop antibodies to insulin; those without diabetes do not make such antibodies. Explain.

8. What is the role of the immune system in preventing cancer?

9. Trace the events of a primary immune response. How would a secondary response be different?

10. What failures of immune differentiation and control can cause autoimmune disease?

11. What is the role of suppressor T cells in an immune response?

Choose the MOST CORRECT Answer.

12. A nonspecific antiviral substance released by virally infected cells is:
    a. Complement
    b. Interleukin
    c. Interferon
    d. Opsonin

13. Bradykinin is responsible for all of the following EXCEPT:
    a. Increased permeability of capillaries resulting in edema
    b. Activation of neural afferents causing burning sensation
    c. Causes vasodilation producing redness of inflamed areas
    d. Promotes phagocytosis

14. The type of immunity conferred by transfer of antibodies across the placenta from mother to fetus and later acquired through breast milk is called:
    a. Naturally acquired active immunity
    b. Naturally acquired passive immunity
    c. Artificially acquired active immunity
    d. Artificially acquired passive immunity

15. Which portion(s) of the antibody molecule shows no antigen reactivity?
    a. Fab fragment
    b. Fc fragment
    c. Hypervariable segments
    d. All of the above reactive

16. ___ involves the binding of antibody to the surface of an invading microorganism thereby promoting phagocytosis by macrophages.
    a. Agglutination
    b. Neutralization
    c. Opsonization
    d. Complement activation

17. This immunoglobulin concentration increases during parasitic infections and allergic reactions:
    a. IgG
    b. IgM
    c. IgA
    d. IgD
    e. IgE

18. These lymphocytes participate in cell-mediated killing of antigen-bearing cells:
    a. Cytotoxic T cells
    b. Helper T cells
    c. Suppressor T cells
    d. B cells

19. Which of the following have been used to counteract transplant rejection?
    a. Antimetabolites
    b. Corticosteroid antiinflammatory agents
    c. Cyclosporine
    d. All of the above

20. An individual with type O blood can receive blood from which of the following?
    a. Type A
    b. Type B
    c. Type AB
    d. Type O
    e. All of the above

21. Select the type of hypersensitivity observed in individuals suffering from chronic diseases or parasitic diseases in which the immune response is mediated entirely by T cells.
    a. Type I, anaphylactic reaction
    b. Type II
    c. Type III
    d. Type IV, delayed hypersensitivity

● SUGGESTED READINGS

BRIGHT S, ADAIR J, SECHER D: From laboratory to clinic: the development of an immunological reagent, *Immunology Today* 12(4):130, 1990. Step-by-step description of the processing and testing of antibodies for medical applications.

COHEN J: AIDS research shifts to immunity, *Science* 257:152, July 10, 1992. The responses of the immune systems of those HIV-exposed individuals who are apparently able to fight off the infection are being studied to learn how the immune responses of more susceptible individuals can be boosted.

COLDITZ IG: The induction of plasma leakage in skin by histamine, bradykinin, activated complement, platelet-activating factor and serotonin, *Immunology and Cell Biology* 69:215, 1991. The relative potencies of these mediators of the inflammatory response was determined experimentally; prostaglandins enhances the effects of histamine and bradykinin.

DAWSON AA: *Lymphokines and interleukins,* Open University Press, Buckingham, 1992. A fascinating introduction to the chemical mediators of immune function, this book begins with an overview of the immune system and then introduces and discusses in very readable fashion the production and functions of a selection of lymphokines and interleukins. Emphasis is on basic biology and relevance to health issues.

EGLEZOS A, ANDREWS PV, BOYD RL, HELME RD: Modulation of the immune response by tachykinins, *Immunology and Cell Biology* 69:285, 1991. The intimate relationship between the nervous system and the immune system is reviewed in this article on neuroimmunomodulation. The relationship between emotional health and the ability of the body to overcome infection and cancer is dependent upon the links between neural and immunological control systems.

GOLDE DW: The stem cell, *Scientific American* 265(6):86, Dec., 1991. The main ingredient in bone marrow transplants is the stem cells, which can give rise to all the blood and immune system cells. Umbilical cord blood is another source of stem cells and these survive freezing for long periods; therefore, if they were taken at the time of birth and saved, they could serve as a source for totally compatible supply of healthy blood cells for that individual in the case of a medical crisis.

HAMILTON JD, ET AL: A controlled trial of early versus late treatment with zidovudine in symptomatic human immunodeficiency virus infection, *The New England Journal of Medicine* 326 (7):437, 1992. Early intervention with zidovudine is associated with certain side effects but in general, it delays the progression of HIV infection to AIDS. However, early treatment does not improve survival, compared with treatment initiated after the symptoms of AIDS are apparent.

JANEWAY CA JR: The immune system evolved to discriminate infectious self from nonself, *Immunology Today* 13(1):11, 1992. Review of the mechanism of discrimination.

JOHNSON HM, RUSSELL JK, PONTZER CH: Superantigens in human disease, *Scientific American* 266(4): 1992. Overstimulation of the immune system by certain proteins results in the massive, even life-threatening response seen in food poisoning and toxic shock syndrome. The superantigens, enterotoxins, activate helper T cells at a site distinct from the site that usually binds antigens, and thus bypass the specificity of the system. The massive activation of T cells leads to secretion of excess IL-2, which causes fever, nausea, vomiting, and shock.

MAK TW, WIGZELL H: AIDS: 10 years later, *The FASEB Journal* 5(10):264, 1991. The theme of this issue is AIDS, and it contains 12 review articles by experts in the field on various aspects of AIDS biology, including epidemiology, neural symptoms, vaccine production, and mother-to-child transmission.

REITSCHEL ET, BRADE H: Bacterial endotoxins, *Scientific American* 267(2):54, 1992. The outer membrane of many bacteria can elicit powerful responses from the immune system, but research may uncover ways to use the less damaging of these responses to boost immune responses in combating tumors.

RENNIE J: Tolerating self; experience teaches the immune system to recognize the self, *Scientific American* 263(3):50, 1990. Discusses the mechanisms of B and T cell processing in fetal and neonatal development.

ROSENBERG SA: Adoptive immunotherapy for cancer, *Scientific American* 262(5):62, 1990. Describes how cells of the immune system can be reprogrammed to selectively target host tissues.

WOOD JD: Communication between minibrain in gut and enteric immune system. *Federation Proceedings* 6:64, April 1991. Studies of the communication between neural and immune system cells in the enteric system are of clinical significance in the understanding of sudden-onset diarrhea and also as a model for how the less-accessible brain cells may communicate with the immune system cells.

Answers to Study Questions

Chapter 1

1. The internal environment of multicellular animals can be regulated; the presence of many cells allows for functional specialization of different cell types.

2. Diffusion is rapid enough to be physiologically significant for distances of cellular dimensions (1 to 10 microns); it is too slow to be of physiological significance for distances that are of the dimensions of most major organs or body parts.

3. Environmental stress is the sum of processes that tend to bring the organism into physical and chemical equilibrium with the environment. Depending on the environment, these processes might include heating, cooling, drying, dilution, and so forth.

4. Natural selection is the term for the process by which environmental stress, through its differential effects on survival and reproduction of genetically different individuals in a population, results in predominance of genes that favor reproduction and survival.

5. Student 1's answer is partly correct—the human appendix is much smaller than the corresponding part of the intestine in some other animals. However, because it is small does not mean that it has no function—the appendix contains some immune tissue. Also, not all functioning organs are vital organs essential to life; consider the reproductive organs, for example. Student 2 assumes that the operation of natural selection is rapid and perfect; in fact, natural selection is an ongoing process. The instructor assumes that present-day species can be arranged in an evolutionary chain. In fact, all the present-day primate lines are branches from an extinct common ancestor that can be assumed to have had the anatomical correlate of an appendix; in the course of evolutionary divergence of the different lines, the appendix was treated differently in each line, and at some point monkeys lost their appendices while "higher" primates retained them.

6. Negative feedback is a control process that tends to stabilize regulated variables; regulated variables of the internal environment include arterial blood pressure, blood composition, blood volume, and core body temperature.

Multiple Choice Questions:

7. c
8. d
9. b
10. b
11. d

Chapter 2

1. See the Glossary.

2. Examples of the bond types are given in order of stability. Examples of covalent bonding: the bond between oxygen atoms to form O_2; bonds between C and H in carbohydrates.
   Examples of ionic bonding: between Na^+ and Cl^- to form NaCl; between Na^+ and HCO_3^- to form $NaHCO_3$.
   Examples of hydrogen bonding: between the $-H$ and $-O-$ atoms of adjacent water molecules; between -H and -OH or -NH groups of proteins.
   An example of van der Waals interactions: attraction between adjacent nonpolar side groups of proteins.

3. The number of unpaired electrons.

4. Polar molecules have unequal distributions of Charge. The bond angle between the H atoms of water is determined by the positions of the orbitals of the unpaired electrons of the O atom. See Table 2-2.

5. When each molecule of the solute dissociates into more than one particle in solution, since osmotic pressure is determined by the concentration of particles rather than by the molar concentration.

6. They constitute one of the forces between different regions of protein molecules that cause the amino acid chains of proteins to assume their normal folded shape.

7. A molecule represents an ordered structure—a kind of negative entropy. To make one, first the constituent elements must first be brought together. Then they must be assembled into a definite arrangement of chemical bonds, many of which may require overcoming activation energies and net free energy changes. Synthesis of complex molecules by organisms is more than matched increases in entropy in the universe as a whole, so for each molecule synthesized in an organism, the net entrophy of the universe increases.

8. Resistance is a force that arises from the interaction of masses, fluids, or particles with their surroundings and opposes their movement. The universal relationship is: rate of movement equals driving force divided by resistance force.

Multiple Choice Questions:

9. d
10. b
11. c
12. a

Chapter 3

1. See the Overview and relevant subheads in the chapter.

2. Ally enzyme-mediated reactions involve binding of the reactants to an active site of the enzyme, which reduces the barrier of activation energy that must be overcome to allow the reaction to occur. The reaction rate depends on the concentration of enzyme and of reactants and products; when the concentrations of reactants are such that the active size of all enzyme molecules are continuously occupied, the enzyme is said to be saturated and the reaction rate is maximal.

3. DNA replication—S period
   chromosome condensation—prophase of mitosis
   chromosome separation—anaphase of mitosis

4. See "DNA Replication" under Nucleic Acids.

5. See "DNA Transcription into mRNA" under **DNA Transcription and Protein SYnthesis.**

6. In posttranscriptional processing, intron portions of the primary transcript are removed and the remaining exons may be spliced together in different combinations to form several different mRNA sequences.

7. The genetic code is *redundant* in that a given amino acid may be specified by more than one base triplet, or codon.

Multiple Choice Questions:

8. c
9. d
10. b
11. b
12. d

Chapter 4

1. See relevant subheadings in the chapter.
2. See Table 4-2.
3. The cells of tight epithelia are joined by one or more continuous

bands of tight junctions that prevent larger solutes from passing between the cells. In leaky epithelia the "tight" junctions are discontinuous, allowing large solutes and water to pass through the epithelium by paracellular pathways.

4. In the absence of a nucleus, the cell can neither reduce by mitosis nor direct the synthesis of new messenger RNA for protein synthesis. Without mitochondria, the red cells must rely on anaerobic glycolysis for energy production.

5. The two major pathways of energy metabolism are glycolysis, and the Krebs cycle. Glycolysis starts with carbohydrate and mediates substrate-level phosphorylation. In cells with mitochondria, glycolysis is a source of pyruvate for the Krebs cycle. The Krebs cycle also is the major pathway for energy production from catabolism of protein and fat. Glycolysis can be carried out without oxygen and can respond rapidly to changes in demand, appropriate characteristics for some skeletal muscle cells. Glycolysis is far less efficient than the Krebs cycle, which is used by almost all cells that must sustain a steady, high rate of energy production. Organs that depend primarily on the Krebs cycle include the heart, brain, and skeletal muscle cells that fatigue slowly.

6. Aerobic energy metabolism is roughly 10 to 20 times more efficient than anaerobic glycolysis at converting the energy of substrates into ATP.

7. As electrons move from NADH and FADH along the series of iron-containing molecules called cytochromes, a small amount of energy is released at each step. Some of this energy is used to drive the transport of H^+ across the mitochondrion to the outside. As the H^+ moves out, energy is transferred to protein complexes that can phosphorlyate ADP.

8. See Figure 4-23. The oxygens in the carbon dioxide come from the substrate. The oxygens in the water come from the atmosphere.

9. The amino acids produced by protein degradation and the fatty acids and glycerol produced by fat metabolism are converted to acetyl-CoA, which enters the Krebs Cycle. On a gram for gram basis, the energy yield of protein metabolism is similar to that of carbohydrate metabolism, but fats yield 2.3 times as much energy per gram as glucose.

Multiple Choice Questions:

10. c
11. b
12. d
13. a
14. d
15. a
16. b

Chapter 5

1. A simple feedback loop comprises one or more *sensors,* and *integrator,* and one or more *effectors.*

2. The gain affects how far the regulated variable will depart from the set point in response to perturbing factors. Given an error signal of constant magnitude, the higher the gain, the greater the effectors' response and the sooner the regulated variable returns to the set point.

3. Time delays in the feedback loop tend to cause the effectors' response to lag behind the changes in the error signal. If the error signal is continuously changing, the effectors' response never quite catches up with the changes; the result is termed a phase lag.

4. See "Levels of Physiological Regulation."

5. Some examples are given in the section "Mechanisms of Intrinsic Regulation"; others are given under **Extrinsic Regulations: Reflex Categories.** Perhaps you can think of others.

6. See "Chemical Classes of Hormones" under **Endocrine Reflexes.**

7. Whether the target organ possesses receptors for the hormone.

8. Generally, hormones may affect the activity of enzyme systems already present in the cell, or they may influence the rate of synthesis of enzymes or even the presence or absence of particular enzymes, by affecting the relevant genes. For specifics, see

"Hormones with Membrane-Bound Receptors" under **Endocrine Reflexes.**

9. Prostaglandins are synthesized from cell membrane lipids and typically act at the local level.

10. See "Functional Classes of Hormones" under **Endocrine Reflexes**

11. The posterior pituitary contains secretory nerve terminals that release small peptide neurohormones. The anterior pituitary cells are not neurons; their hormones are proteins.

12. In many cases hormone secretion does not occur at a steady rate, but rather occurs in pulses throughout the day or at particular times of day, or in response to particular temporary challenges such as exercise, eating, or stress or excitement.

13. d
14. b
15. b
16. a
17. d
18. d
19. d
20. b
21. d
22. a

Chapter 6

1. In the fluid mosaic model, the lipids of the cell membrane are fluid enough to allow membrane proteins to be mobile within the plane of the membrane but not to rotate through the plane of the membrane or to migrate from one side to the other. Thus proteins that initially are inserted into one or the other of the membrane surfaces usually reamin there, and proteins that span the membrane maintain a consistent orientation with respect to the two membrane surfaces.

2. Some factors that might be involved, depending on the solute:
   a. The lipid solubility of the solute
   b. The size of the solute
   c. For ionic solutes, whether the solute could exist in an uncharged (and thus more lipid-soluble) form
   d. The existence in the membrane of proteins that formed specific permeability pathways or acted as transport proteins for that solute

3. a. For a nonelectrolyte, 300 mOsm = 300 mM
   b. For a solute that dissociated into two univalent ions (for example, NaCl), 150 mM = 300 mOsm (neglecting the isotonic coefficient)
   c. For a solute that dissociated into one divalent and two univalent ions (for example, $CaCl_2$), 100 mM = 300 mOsm (again neglecting the isotonic coefficient)

4. The answer depends on which solutes the cell membrane is permeable to. There cannot be an osmotic gradient across the cell membrane, but only impermeant (or effectively impermeant) solutes can contribute to osmotic movement of water across the membrane.

5. The Na^+-K^+ pump makes the membrane effectively impermeable to Na^+, allowing extracellular Na^+ to balance the osmotic effect of intracellular proteins.

6. Two facts explain the inside-negativity of the resting potential: the membrane is mainly permeable to K^+ and the K^+ gradient is outwardly directed. See "Determinants of the Resting Potential of a Cell" under **The Membrane Potential.**

7. Increasing the extracellular K^+ decreases the magnitude of the transmembrane K^+ gradient and shifts the membrane potential nearer to the Na^+ equilibrium potential. The result is depolarization.

8. Mediated transport systems involve membrane transport proteins ("carriers") that are specific for particular molecular types and that can become saturated as the concentrations of the transported molecules are increased.

Multiple Choice Questions:

9. b
10. b
11. c

12. b
13. a
14. d

Chapter 7

1. These venoms prolong the open time of Na$^+$ channels, increasing the duration of the spike portion of the action potential. Since the channels remain open longer, the peak of the action potential has more time to approach the Na$^+$ equilibrium potential, increasing its amplitude slightly. Since the spike part of the action potential lasts longer, the absolute refractory period, which corresponds to this part of the action potential, also is longer.

2. See "Spread of Excitation" under **Propagation of Action Potentials.**

3. See Figure 7-12.

4. The height or amplitude of the potential is determined by the number of Na$^+$ channels that open (in other words, the magnitude of the inward Na$^+$ current relative to the constant outward K$^+$ leak) and the length of time the channels remain open. A stimulus that exceeds the threshold level of depolarization sets in motion a positive feedback cycle that causes all the Na$^+$ channels to open. Thus action potentials are not graded to the stimulus magnitude.

5. Inactivation of Na$^+$ channels and opening of the voltage-sensitive K$^+$ channels.

6. Inactivated channels are latched in the closed state and cannot open again until the membrane repolarizes. Removing inactivation leaves the channels in the closed state; in this state they can open again in response to depolarization.

7. See "Effect of Diameter and Myelination on Propagation Velocity" under **Propagation of Action Potentials.**

8. Within a nerve trunk, each axon individually follows the all-or-none law; the compound action potential is the sum of electrical currents caused by the population of axons in the trunk that are excited. The threshold stimulus intensity for different classes of axons varies. Some are activated at low stimulus intensity; additional classes can be activated at higher stimulus intensities.

9. A reduction in the density of active channels raises the threshold - the initial depolarization must be much greater to get the positive feedback cycle started. If enough channels are blocked, propagation of an action potential becomes impossible and sensations cannot be transmitted from the anesthetized area to the central nervous system.

Multiple Choice Questions:
10. d
11. a
12. c
13. d
14. c
15. c

Chapter 8

1. See "Input and Output Segments of Neurons" under **Neuron Structure and Information Transfer.**

2. The receptive field of a cutaneous receptor could be mapped by recording from the receptor's axon while stimulating the skin with single touches of a mechanical stimulator such as a bristle or fine needle. Summated responses could be elicited by rapidly repeated touches in the same spot or by simultaneously touching more than one spot in the receptive field.

3. Temporal summation refers to repeated stimulation of the same restricted part of the input segment of a neuron. Spatial summation refers to simultaneous stimulation at different spots on the cell's input segment.

4. Depolarization is the result of an increase in Na$^+$ permeability, although some channels that mediate depolarizing receptor potentials admit other ions as well. Hyperpolarization typically is the result of increased K$^+$ permeability. Photoreceptors are an exception to this rule; their hyperpolarization in response to light is the result of a decrease in Na$^+$ permeability (some Na$^+$ channels are constantly open in the dark and are closed by light).

5. See **Generation of the Neural Code.** The key to the process is an initial graded response (the receptor potential) that gives rise to a proportionate frequency of action potentials in the output segment of the receptor.

6. See "Intensity Coding By The Receptor Potential." The key to the process is an initial graded response (the receptor potential) that gives rise to a proportionate frequency of action potentials in the output region of the receptor.

7. The factors involved are:
   a. The pattern of depolarization
   b. The previous history of electrical activity
   c. The action of autoreceptors on the presynaptic terminal
   d. Neuromodulators
   All these interact in some way with Ca^{++} entry into the presynaptic terminal, which is the final determinant of transmitter release.

Multiple Choice Questions:
8. b
9. d
10. b
11. c
12. a

Chapter 9

1. Typically, areas with fine two-point discrimination have a disproportionately large representation in the primary somatosensory cortex (see Figure 9-10). The receptors in such areas typically have small receptive fields and the fields overlap extensively, so the receptor density is high.

2. See "The Reticular Activating System" and also "Sleep and the Wakeful State".

3. Once laid down in long-term memory, memories usually are not completely abolished by damage to any discrete part of the brain.

4. Such deficits might be seen in people who have sustained damage to the language areas of the dominant hemisphere, particularly the Broca area or connections between it and other language areas. Musical skills typically are localized in the nondominant hemisphere, particularly so in individuals who have not received extensive music training.

5. Mixed nerves include both afferent and efferent axons. This obtains for all spinal nerves and some but not all cranial nerves.

6. See Figure 9-13.

7. White matter contains myelinated axons; gray matter contains neuronal cell bodies. In the brain the outermost layers are gray and the innermost layers are white; in the spinal cord the reverse is true.

8. The dorsal column pathway (see Figure 9-15, *A*) carries modality specific fine touch, proprioception and sharp pain information to the primary sensory cortex by way of the thalamus; the nonspecific pathway (see Figure 9-15, *B*) projects mainly to parts of the limbic system and carries information that results in sensations such as heat, cold, and dull or burning pain.

9. See "The Somatosensory System" and Figure 9-14. Rapidly adapting touch receptors include hair follicle receptors, Meissner's corpuscles and Pacinian corpuscles; slowly adapting receptors include touch dome endings and Ruffini's endings.

Multiple Choice Questions:
10. d
11. d
12. a
13. b
14. c
15. c
16. a
17. b
18. a

Chapter 10

1. Sudden loud noises can damage the ear during the several tens of milliseconds of delay of the reflex loop.

2. See "Sympathetic Resonance and Frequency Coding in the Co-chlea" under **The Auditory System.**

3. Lateral inhibition between afferents from the cochlea is important in resolving pure tones of similar frequencies.

4. The polarity of the receptor potential of hair cells is determined by the direction of bending of the stereocilia. Rotations of the head cause the endolymph of the ampulla to move relative to the cupula; which way the stereocilia are bent depends on the direction of rotation of the head.

5. When the head begins to rotate, inertia causes the endolymph to lag behind the wall of the ampulla, bending the cupula. Stimulation of the hair cells drives compensatory eye movements (nystagmus). After a time, friction between the ampulla and the endolymph accelerates the endolymph so that both endolymph and ampulla are rotating at the same velocity, and the cupula is no longer bent. When the head stops rotating, the endolymph's momentum causes it to surge ahead, bending the cupula in the opposite direction. The resulting nystagmus is thus in the opposite direction.

6. See Figure 10-25.

7. See Figure 10-24.

8. As one reasonable hypothesis, a simple cell might receive input from some on-center, off-surround ganglion cells whose receptive fields form a row on the retinal surface (see Figure 10-29). When a stimulus bar or border lies along the excitatory centers of the receptive fields, the simple cell is maximally excited. If the bar lies immediately on either side of the row of on-centers, it stimulates the off-surrounds of the ganglion cells, and the simple cell is maximally inhibited. The connections needed for a complex cell are clearly much more complicated. A complex cell would have to receive input from several parallel rows of ganglion cells, with neural connections between the rows to mediate the motion sensitivity. Currently such connections are thought to be made by amacrine cells, but the details of the circuitry are not clear.

9. Taste receptors are modified epithelial cells, whereas olfactory receptors are neurons. In both cases discrimination is not the product of receptors labeled lines, because individual receptors can respond to more than one flavorant or odorant. Thus to perceive a given odor or flavor, the brain must integrate input from a large number of receptors with different response spectra.

10. The number of distinct odors is so large that if there were primary odors, the number of primary odors would also have to be quite large. In practice almost all odor sensations arise from complex mixtures of odorant molecules.

Multiple Choice Questions:

11. b
12. a
13. c
14. d
15. a
16. c
17. d

Chapter 11

1. The events can be divided into three groups: presynaptic events, postsynaptic events, and events that involve the contractile machinery. The presynaptic events culminate in transmitter release; the postsynaptic events begin with binding of ACh to the nicotinic receptors of the postsynaptic membrane and lead to initiation of an action potential in the muscle cell. The muscle action potential causes Ca^{++} release from the SR; the released Ca^{++} activates the contractile machinery. The details of each step in the sequence are found under appropriate headings in the text.

2. *Transverse tubules* conduct action potentials into the interior of muscle cells and mediate release of *inositol triphosphate,* causing the terminal cisternae of the *sarcoplasmic reticulum* to release Ca^{++} that it had sequestered during rest. The Ca^{++} diffuses to regulatory subunits of *troponin,* causing it to undergo a conformational change that moves *tropomyosin* away from actin-

myosin binding sites on the thin filaments. Myosin heads bind to actin and contraction begins.

3. In the absence of ATP, myosin heads are irreversibly bound to actin in the "uncocked" state and are unable to undergo power strokes—the resulting condition of stiffness is rigor mortis. In the absence of Ca^{++}, myosin heads are "cocked," but tropomyosin prevents them from binding to actin; this is the condition characteristic of resting muscle.

4. In an isotonic contraction the muscle shortens against a constant load; in an isometric contraction the muscle is prevented from contracting. A person lifting a load is performing what is approximately an isotonic contraction. The contraction is not perfectly isotonic because the effect of leverage changes the load on individual muscles as the elements of a joint move relative to one another. Isometric contractions occur when a person attempts to lift a load that is too heavy or when a load is held suspended in one position.

5. In the flexors of the left arm, which has a trivial load, relatively few motor neurons would be active at any instant and most of the active ones would serve slow muscle units. In the flexors of the right arm, many slow motor units would be active and some would show rapid bursts of action potentials. Some motor units of fast muscle fibers probably will have been recruited also, but since 1 kilogram is considerably less than the maximum load that most people can support with one hand, many fast units probably will not be recruited at all; others will be active for only brief periods.

6. See "Skeletal Muscle Fiber Types and Energy Metabolism" under **Muscle Energetics and Metabolism.**

7. See "Muscle Fatigue" under **Muscle Energetics and Metabolism.**

8. Important differences include:
contractile machinery: compare Figure 11-9 to Figure 11-30.
organization of myofibrils within the cell: compare Figures 11-2 and 11-27.
contractile properties: see Figures 11-15 and 11-29.
mechanism of excitation-contraction coupling: see Table 11-5.
innervation: see Figure 11-31.

9. The contractile machinery of smooth muscle must frequently resist steady forces. For example, the smooth muscles of arteries must constantly resist blood pressure that otherwise would distend the arterial system. Latch bridges are an economical way of maintaining such resistance to stretch over long periods of time.

10. Single-unit smooth muscles are those in which activation of a few fibers causes the entire population of fibers to contract, as a result of electrical coupling between the fibers. Cardiac fibers contract synchronously during a heartbeat because they are coupled by gap junctions. For the heart, as for single-unit smooth muscle, the stimulus to contract originates in a small population of pacemaker fibers. Each fiber is innervated in multiunit smooth muscle.

11. The major issues are: *Where does the Ca^{++} come from? How does electrical activity cause Ca^{++} release in the cell?* In smooth muscle, much of the Ca^{++} crosses the cell membrane through voltage-sensitive Ca^{++} channels, although there may be some contribution from sources inside the cell that is released by second messengers. In skeletal muscle, essentially all of the Ca^{++} is stored in the sarcoplasmic reticulum. Depolarization of the T tubules causes voltage-sensor channels in the T-tubular membrane to activate Ca^{++} release channels in the adjacent lateral cisternae of the SR. In cardiac muscle, the mechanism is similar to skeletal muscle except that some Ca^{++} that enters through voltage-sensitive channels in the cell membrane is important in promoting Ca^{++} release from the SR. In skeletal muscle, excitation causes the release of enough Ca^{++} to activate all crossbridges; in smooth and cardiac muscle, the levels of Ca^{++} attained in the cell during activation are typically not sufficient to activate all cross-bridges and thus Ca^+ can modulate contractile force as well as initiate contraction.

Multiple Choice Questions:

12. c
13. d

14. a
15. a
16. c

Chapter 12

1. Proximal muscles of the trunk, which are involved in posture and gross movement, are innervated by motor neurons whose cell bodies are in the medial regions of the ventral horn of the spinal cord. Distal muscles located in the appendages and involved in manipulation are innervated by neurons whose cell bodies are in lateral regions of the ventral horn. The direct corticospinal tract or pyramidal tract forms direct connections between the motor cortex and the spinal motor neurons involved in finely controlled movements of arms, hands, and fingers.

2. See "Structure of the Muscle Spindle," "Muscle Spindles as Length and Length Change Detectors," and "Control of the Stretch Reflex by Gamma Motor Neurons" under **Somatic Motor Control at the Spinal Level.**

3. As described in the section "Coactivation of α and γ Motor Neurons" under **Somatic Motor Control at the Spinal Level,** stretch reflexes are activated during voluntary movements if actual muscle length deviates from a new set-point length set by γ activation. The additional muscle fibers recruited by the reflex increase the gain of the feedback loop. In α-γ coactivation, it is the activation of γ motor neurons that resets the length set point of the reflex. If the reflex set point were not reset, the reflex would oppose the voluntary length change.

4. See "Golgi Tendon Organs: Monitors of Muscle Tension" under **Spinal Reflexes and Postural Stability.**

5. Muscle groups that move a joint in one direction are opposed by antagonist muscles that move the joint in the opposite direction. Typically the spinal circuits are arranged so that when the motor neurons serving agonists are activated, those serving antagonists are inhibited; this is reciprocal innervation. This innervation is important both for reflexes that mediate limb extension (stretch reflex and crossed extension) and those that mediate rapid flexion (withdrawal reflex), because it prevents the tone of antagonist muscles for acting as a drag on agonists when the reflexes are activated.

6. Facilitation is a use-related increase in the responsiveness of a motor neuron to excitatory synaptic input from afferents. It frequently is the result of spatial summation of increased excitatory input to the motor neurons from central interneurons. The sensitivity of spinal stretch and postural reflex pathways is modulated by input from the vestibular system of the brainstem along the vestibulospinal pathway.

7. See "The Cerebellum and Coordination of Rapid Movements" under **Motor Control Centers of the Brain.**

8. The caudate nucleus, putamen, and globus pallidus. The evidence that these subcortical nuclei possess motor programs is twofold: in primates, damage to the basal ganglia causes execution of inappropriate motor programs; in nonprimates, a fairly wide range of activities can be carried out by decorticate animals.

9. See "Motor Programs in the Central Nervous System" under **Motor Control Centers of the Brain.**

10. The cholinergic receptor types are nicotinic and muscarinic; the adrenergic receptor types are α_1, α_2, β_1, and β_2. See Tables 12-5 and 12-6 for examples of their locations, and Table 12-4 for examples of their effects.

11. See Summary items 1, 2, and 3.

12. Two important examples are vascular smooth muscle, which in most parts of the body receives only sympathetic (vasoconstrictor) input, and cardiac ventricular muscle fibers, which receive only sympathetic (excitatory) input. See Table 12-4. The effect of autonomic input is determined not only by which receptor type(s) the effector possesses, but also by the nature of the second message(s) set in motion by the binding of transmitter to receptor in that particular tissue. Thus activation of the same receptor type may have either excitatory or inhibitory effects, depending on the tissue involved.

Multiple Choice Questions:
13. c
14. a

15. c
16. c
17. a
18. d

Chapter 13

1. Since all cardiac fibers are activated during each heartbeat, the only means available to increase the total force is an increase in the force produced by each myocardial fiber. Such increases may result from increased ventricular filling, increased adrenergic input, or both. See "Intrinsic and Extrinsic Regulation of Cardiac Performance." In skeletal muscle more than enough Ca^{++} is released during a twitch to saturate all of the troponin molecules and thus activate all of the myosin heads. This is believed not to be the case in cardiac muscle, so that in cardiac muscle, unlike skeletal muscle, intracellular Ca^{++} levels can modulate force generation.

2. See Figure 13-14.

3. See Figure 13-14.

4. During ventricular contraction, the atria are repolarizing; the electrical signal of atrial repolarization is not normally seen in the ECG because it is overwhelmed by the much larger signal of ventricular depolarization. During this phase the pressure in the left ventricle is greater than atrial pressure but less than aortic pressure.

5. The plateau reflects the duration of the slow inward current of Na^+ and Ca^{++}; another factor that contributes to the length of the plateau is the the slow onset of the outward K^+ current, compared to nerve and skeletal muscle.

6. See Figures 13-12 and 13-14.

7. The fibers with the most rapid rate of diastolic depolarization have the highest intrinsic rate and drive the rest of the heart at their pace.

8. See "Abnormalities of Cardiac Rhythm" under **The Electrical Activity of the Heart.**

9. See Figure 13-4. Reentry is described in "Abnormalities of Cardiac Rhythm."

10. See "Excitation-Contraction Coupling in Cardiac Muscle" under **The Electrical Activity of the Heart,** and Table 13-4. The effect of adrenergic input on myocardial cells is an increase in the force of contraction and a disease in the duration of the active state, making systoles both more forceful and briefer.

11. See "Intrinsic Regulation of Stroke Volume" under **Intrinsic and Extrinsic Regulation of Cardiac Performance.**

12. See Figure 13-23.

Multiple Choice Questions:
13. b
14. a
15. c
16. c
17. c
18. a
19. d
20. a

Chapter 14

1. As in other regulatory cascades, multiple steps provide for amplification of the signal and allow several processes to be initiated or controlled by a single event. Since each step in the cascade typically is regulated by allosteric effectors, multiple steps also allow for finer control than a single step. Review the discussion of regulatory cascades in Chapter 5. In the early stages of clot formation, platelets that have been made sticky by contact with an abnormal surface release ADP, a signal that causes nearby platelets also to become sticky. As fibrin is formed, more and more platelets are stimulated to release ADP. Negative feedback is provided by formation of the clot-dissolving substance plasmin as one product of the clotting factor cascade.

2. Because there are many more capillaries than arterioles, the net cross-sectional area of all capillaries taken together is greater than that of all arterioles taken together. Thus the total resistance of capillaries is much less than the total resistance of arterioles.

3. An increase in tissue colloid osmotic pressure would decrease the transcapillary osmotic gradient. The resulting increase in interstitial fluid volume would tend to cause edema. An increase in venous pressure would increase the capillary hydrostatic pressure and cause an increase in interstitial fluid volume. See Table 14-4.

4. The elasticity of the walls of the arterial tree normally decreases with age. The resulting decrease in vessel compliance (see Figure 14-10) causes systolic pressure to increase.

5. A decrease in venous compliance would decrease the venous volume and increase the venous pressure. Venous return would increase, but total peripheral resistance would hardly be affected because the contribution of venous resistance to total peripheral resistance is very small in any case. Arteriolar dilation decreases total peripheral resistance, making blood flow from arteries to veins easier and increasing venous return.

6. The outcome depends on two factors, the number of branches and their dimensions. Generally the diameter of branch vessels is less than that of the stem. Because a vessel's resistance is very sensitive to its diameter, each branch vessel has a much higher resistance than the stem. The decrease in diameter with branching tends to increase the net resistance, but this effect may be overcome if the number of branch vessels is very large.

7. See Figures 14-4 and 14-6.

8. The total volume of blood in the tube corresponds to 29 + 21 mm = 50 mm. The hematocrit is (21 mm/50 mm) × 100 = 42%.

9. a. Vasoconstriction: decrease in blood flow; the dominant effect in most elements of the peripheral circulation
   b. Vasodilation: increase in blood flow; important in skeletal muscle in the presence of moderate levels of epinephrine
   c. Vasodilation: increase in blood flow; occurs in skeletal muscle before exercise
   d. Vasodilation: increase in blood flow; responsible for increased blood flow to the genital tissues during sexual response

10. The ratio of flow resistances is $1:2^4$, reflecting that resistance is proportional to the fourth power of the radius. This means that flow through the smaller vessel is 1/16 of that through the larger vessel. The total flow through both vessels is 10 ml/min; the flow through the smaller vessel is 10 ml/min × 1/16 = 0.625 ml/min, and the flow through the larger vessel is 9.375 ml/min.

Multiple Choice Questions:

11. b
12. d
13. b
14. d
15. b
16. d
17. d

Chapter 15

1. Effects on venous return of:
   transfusion: increase
   arteriolar constriction: decrease
   decreased venous compliance: decrease
   (See Figure 15-8)

2. Arteriolar dilation in exercising muscles increases venous return. The Frank-Starling Law results in an increase in cardiac output.

3. Assuming there is no reflexive compensation, increases in myocardial contractility increase mean arterial pressure. Cardiac output increases, but central venous pressure decreases (see Figure 15-9).

4. See Figure 15-9.

5. In response to sudden transition to upright posture:
   a. Baroreceptor firing rate decreases
   b. Parasympathetic input to the SA node decreases and sympathetic input increases, resulting in an increased heart rate
   c. Capillary hydrostatic pressure throughout the lower body increases

   d. Increase sympathetic input to veins decreases venous compliance
   e. Other reflexive responses: increased sympathetic activity constricts arterioles and increases myocardial contractility

6. The decrease in blood volume would decrease mean arterial pressure, except that arterial pressure is protected by the baroreceptor reflex. The reflexive increase in arteriolar constriction decreases the mean capillary hydrostatic pressure, shifting the balance of capillary filtration to favor net movement of fluid from interstitial spaces into the capillaries. The resulting transfer of fluid from interstitial spaces to the plasma replaces the lost plasma volume within a short time, but the lost erythrocytes cannot be replaced so rapidly. Thus the hematocrit decreases.

7. Before compensation for blood loss:
   cardiac output: decreased
   central venous pressure: decreased
   mean arterial pressure: decreased
   total peripheral resistance: unchanged

8. After reflexive compensation for blood loss:
   total peripheral resistance: increased
   central venous pressure: decreased
   cerebral blood flow: unchanged, because the reflex regulates mean carotid arterial pressure
   renal blood flow: reduced by reflexive arteriolar constriction

9. A heat load causes reflexive cutaneous vasodilation. The decrease in peripheral resistance caused by dilated skin arterioles is added to the decrease caused by dilated arterioles in exercising skeletal muscle. Under these conditions, it may not be possible to maintain mean arterial pressure without constriction of cutaneous arterioles, but such constriction would increase the heat load. Generally the blood flow through cutaneous vessels is managed reflexively to meet the needs of thermoregulation, as long as mean arterial pressure is normal. However, if mean arterial pressure drops significantly, the cardiovascular center causes cutaneous vasoconstriction, even if there is also a heat load.

Multiple Choice Questions:

10. a
11. b
12. b
13. d
14. c
15. b
16. a

Chapter 16

1. See Figure 16-2 and Table 16-1.

2. See Figure 16-5.

3. Normally, anatomical dead space is the volume of the airway structures, including nasal passages, trachea, bronchi, and bronchioles down to the level of respiratory bronchioles. For the diver breathing through a snorkel, the total dead space is 300 ml. The dead space ventilation at 12 breaths/min is 12 breaths/min × 300 ml/breath = 3600 ml/min. To maintain an alveolar minute volume of 3000 ml/min, the total ventilation must be 3600 ml/min + 3000 ml/min = 6600 ml/min. The tidal volume required at 12 breaths/min is (6600 ml/min) / (12 breaths/min) = 550 ml/breath.

4. See Table 16-3.

5. The "why" can pose at least three different questions. First, "What physical properties of the lung-chest wall cause a negative intrapleural pressure?" The answer is that the negative pressure is the result of the opposing outward springs of the chest wall and inward recoil of the lungs. Second, "What developmental processes led to this situation?" Since the volume of intrapleural fluid remains quite small, growth of the thorax relative to that of the lung during early development must increase the intrapulmonary volume and stretch the lung wall. Third, "What advantage could there be in having the lungs always partly inflated?" To answer this teleological question, the alternative must be considered. If the lungs were fully deflated at the end of a breath, the alveolar surfaces would become attached to one another by surface tension. Reinflating the lung for another

breath would require considerable energy. Furthermore, gas exchange would come to a halt during the part of the respiratory cycle when the alveoli were completely deflated.

6. See "Lungs and Thorax" under **Structure and Function in the Respiratory System.**

7. See Figure 16-17 and accompanying text. Some of the energy expended in a respiratory cycle is expended against the elasticity of the lung-chest wall system; the rest is expended against the flow resistance of the airway. The intrapleural pressure change reflects both the elastic and flow-resistive components; the alveolar pressure reflects only the flow-resistive component.

8. See Figure 16-15.

9. In the absence of pulmonary surfactant the alveolar surface tension is greater; therefore lung recoil is greater, and the functional residual capacity is decreased.

10. If lung compliance is decreased, the flow-resistive component of the work of inflating the lungs increases and ventilation requires extra effort.

11. See Figure 16-22.

12. If airway resistance increases, maintenance of normal alveolar ventilation requires a greater pressure gradient between the atmosphere and the alveoli. Thus both the alveolar pressure and the intrapleural pressure must be more negative during inspiration. The resulting increase in the flow-resistive component of respiratory work increases the total work of breathing. Other consequences of increased airway resistance are noted in "Pulmonary Resistance and Compliance in Disease" under the heading **The Mechanics of Breathing.**

Multiple Choice Questions:

13. d
14. d
15. b
16. d
17. d

Chapter 17

1. The effect of dead space (see Chapter 16) makes exchange of alveolar air with atmospheric air less than perfect (see "Composition of Alveolar Gas" under **Gas Exchange in Alveoli**). Increases in the rate of oxidative metabolism, if not matched by changes in ventilation, would increase the Pco_2 and decrease the Po_2 of alveolar gas. Since the blood flowing through pulmonary capillaries equilibrates with alveolar gas, the Pco_2 of systemic arterial blood must increase and the Po_2 must decrease.

2. The relationship is described by the alveolar gas equation (see p. 483). This equation can be solved for either alveolar Pco_2 or alveolar Po_2 as the dependent variable.

3. The hemoglobin content. Usually the hematocrit is a good indicator of hemoglobin content. This assumes that essentially all of the hemoglobin is functional—if some of it is nonfunctional (for example, converted to carboxyhemoglobin by carbon monoxide), the oxygen-carrying capacity would be reduced proportionately.

4. If the oxygen-carrying capacity of blood were halved, as for example in a severe case of anemia, the Po_2 of systemic arterial blood would remain normal, but the O_2 content of systemic arterial blood would be halved. This would result initially in a decrease in O_2 delivery and a decrease in tissue Po_2. A new steady state would be established when tissue Po_2 reached a value sufficiently low to extract O_2 from the blood at the normal rate. The Po_2 of mixed venous blood would reflect this new lower tissue Po_2, and the arteriovenous Po_2 difference would be increased.

5. The equilibrium concentrations of myoglobin and oxymyglobin for different Po_2 values are determined by the Law of Mass Action. The shape of the resulting function (the dissociation curve) is hyperbolic. The **S**-shaped dissociation curve of hemoglobin is the result of cooperation between the four subunits of the hemoglobin molecule—an allosteric effect that modifies the rate constant for binding of O_2 in response to the number of O_2 molecules already bound to the hemoglobin molecule.

The O_2 affinity of myoglobin is greater than that of hemoglobin, especially in the range of Po_2 values characteristic of systemic

capillaries. Thus O_2 can be efficiently transferred from hemoglobin to myoglobin for delivery to mitochondria.

6. Doubling of alveolar ventilation, without a proportional increase in oxidative metabolic rate, would decrease the Pco_2 of alveolar air and thus the Pco_2 of systemic arterial blood. Such a decrease would shift the plasma bicarbonate buffer equilibrium to the left, decreasing both plasma $[HCO_3^-]$ and plasma $[H^+]$. Thus plasma pH would increase. This is an example of respiratory alkalosis.

7. Small differences between the Po_2 of alveolar gas and arterial blood may arise as the result of ventilation-perfusion mismatch in the lung. This may cause physiological shunt, which is a net overperfusion of alveoli, or physiological dead space, a net underperfusion (see Chapter 16). In emphysema and other fibrotic lung diseases, the diffusional path for gas exchange between alveoli and pulmonary capillaries is increased, and pulmonary blood may not fully equilibrate with alveolar gas (see Figure 17-3). Such diffusional limitations apply both to O_2 and CO_2.

8. A: See Figure 17-11. B: See Figures 16-6 and 17-4, A.

9. See Figure 17-7.

Multiple Choice Questions:

10. b
11. a
12. d
13. c
14. b
15. d
16. d

Chapter 18

1. See Figure 18-2.

2. See Figure 18-8. Peripheral chemoreceptors respond primarily to Po_2, but also to pH and indirectly to Pco_2. Under normal conditions that contribute relatively little to respiratory drive. Halving the atmospheric pressure would halve the Po_2 of inspired air. Alveolar Po_2 would decrease; the input of peripheral chemoreceptors becomes important when arterial Po_2 falls below 60 to 70 mm Hg or when the pH of plasma is decreased.

3. See Figure 18-5. The exact location of the chemosensitive cells is not known. The central chemoreceptors respond to changes in the pH of cerebrospinal fluid; thus they respond indirectly to the Pco_2 of arterial blood. They make the largest single contribution to respiratory drive under normal conditions.

4. See "Arterial Pco_2—the Central Chemoreceptors" under **The Chemical Stimuli for Breathing.**

5. Compare Figures 18-6 and 18-9.

6. See Figure 18-10 and the left side of Figure 18-14. Fourteen thousand feet corresponds to about 4300 meters, so at this altitude arterial hemoglobin is considerably less than fully saturated (which would tend to increase respiratory drive by the peripheral receptors), but plasma Pco_2 is also decreased (which depresses the central component of respiratory drive in unacclimated people). Because of the conflicting central and peripheral signals, full ventilatory compensation for the decreased Po_2 of inspired air is not possible until altitude acclimation has occurred. The reduced O_2 delivery to working muscles results in more rapid fatigue.

7. This person is hypoxic, but plasma pH and Pco_2 are normal. See Figure 18-11 and curve B in Figure 18-13. Ventilation is driven to three or four times the normal value by the peripheral chemoreceptors; however, the input from these receptors also makes the respiratory control center more sensitive to any change in Pco_2.

8. See Figure 18-12, which predicts a ventilatory increase of fourfold to fivefold. Since plasma Pco_2 is normal, the ventilatory response probably is driven by the peripheral chemoreceptors, especially if the change in plasma pH is very recent.

9. Because arterial pH is normal, the increased Pco_2 cannot affect peripheral chemoreceptors; therefore any effect must be on central chemoreceptors. If the increased arterial Pco_2 is recent, ventilation will be increased by severalfold (see Figure 18-6 and curve A of Figure 18-13).

10. Vagotomy abolishes inhibitory input from lung and chest wall stretch receptors that normally mounts during inspiration (see Figure 18-4). The afferents from aortic arch chemoreceptors also are severed, abolishing part of the input from central chemoreceptors. The immediate effect of cutting stretch receptor afferents is an increase in tidal volume without a change in respiratory rate. The resulting increase in alveolar ventilation causes a transient respiratory alkalosis. The feedback loop that runs from central chemoreceptors to the respiratory center is still intact, and the effect of the alkalosis on this loop decreases the respiratory rate to a level that restores plasma P_{CO_2} and pH to their set-point values. A few minutes after vagotomy the tidal volume is still increased, but the respiratory rate is reflexively decreased, and alveolar ventilation is close to its value before vagotomy.

Multiple Choice Questions:
11. d
12. a
13. c
14. a
15. b
16. b

Chapter 19

1. For the anatomical comparison, see Figure 19-4. The main known functional difference between the two is that juxtamedullary nephrons are involved in formation of concentrated urine, whereas cortical nephrons are not.

2. The anatomy of the renal circulation is shown in Figure 19-6. As shown in Figure 19-12, the net pressure gradient in glomerular capillaries favors filtration resulting in urine formation. In the peritubular capillaries the net pressure gradient is reversed, allowing fluid reabsorbed from primary urine to reenter the circulation.

3. Differential regulation of the tone of afferent and efferent arterioles allows the GFR to be maintained despite changes in RBF. In this case constriction of the afferent arterioles by sympathetic input is matched by constriction of the efferent arterioles, keeping the net pressure gradient between glomerular capillaries and the Bowman capsule constant. This form of autoregulation is called glomerulotubular balance and is mediated by the juxtaglomerular apparatus.

4. See Figure 19-18.

5. In Na^+ gradient-mediated cotransport, the energy stored in the transmembrane Na^+ gradient is used to drive simultaneous "downhill" movement of Na^+, which is coupled with "uphill" movement of solute into the cell. Both monosaccharides and amino acids may be transported by this process. See Chapter 6.

6. Inhibition of ADH secretion results in a decrease in the water permeability of the wall of the medullary collecting ducts, reducing the rate of water reabsorption from collecting duct urine. The consequence is that the composition of final urine becomes more like that of distal tubular urine (see Figure 19-20), and the additional unreabsorbed water adds to urine volume.

7. The filtered load of glucose is the plasma concentration (10 mg/ml) multiplied by the GFR (125 ml/min), which equals 1250 mg/min. The filtered load exceeds the Tm of 400 mg/min, so the reabsorption rate is 400 mg/min and the excretion rate is the filtered load (1250 mg/min) minus the reabsorption rate (400 mg/min), which equals 850 mg/min. The urine concentration of glucose is the excretion rate divided by the urine flow rate: (850 mg/min)/(10 ml/min) = 85 mg/ml.

8. See "Loops of Henle and the Collecting Duct—Formation of Concentrated Urine" under **Mechanisms of Urine Formation and Modification.**

9. Some possible factors:
   a. Lack of ADH (diabetes insipidus)
   b. Inhibition of NaCl transport in the thick ascending limb of the loop of Henle; this is an effect of "loop" diuretics
   c. See the answer to question 11

10. Tubular acidification drives reabsorption of filtered bicarbonate. Since the chemical equilibrium between NH_3^+, H, and NH_4^+ de-

termines the concentration of NH_4^+ in the urine, the more acid the urine, the more ammonium is excreted.

11. In the two-solute hypothesis, a concentration gradient of urea provides part of the driving force for water recovery from the collecting duct. When a diet contains adequate calories but no excess protein, little urea is produced. Reduction of the concentrating ability of the kidney by such diets is evidence in favor of the two-solute hypothesis.

Multiple Choice Questions:
12. c
13. b
14. c
15. d
16. c
17. a
18. d
19. b

Chapter 20

1. Total body water constitutes half of body weight, or 24 liters (1 kg = 1 L). ECF constitutes one third of total body water, or 8 liters, so the remainder (the ICF volume) is 16 liters. If intracellular solute concentration is 300 mOsm/L (see Figure 20-1), the total intracellular solute is 16 L × 300 mOsm/L = 4.8 Osm. Assuming that all of the total body Na^+ is in the ECF, the total body Na^+ is 8 L × 145 mEq/L = 1.16 Eq. After administration of 3 liters of isotonic saline the ECF volume is 11 liters. The ICF volume does not change because Na^+ is effectively impermeant to cells. The ECF osmolarity does not change because the saline is isotonic. Isotonic saline (300 mOsm/L) contains 150 mEq/L of Na^+. The body Na^+ increases by 3 L × 150 mEq/L = 0.45 Eq to a total of 1.16 Eq + 0.45 Eq = 1.61 Eq.

2. In congestive heart failure, right atrial pressure (that is, central venous pressure) increases (see Figure 15-9), stretching the atria. This stimulates release of ANH.

3. The ECF Na^+ concentration is stabilized by osmotic transfer of fluid from the ICF and by potent thirst induction and renal fluid conservation induced by the ADH system. All of these tend to make increased ECF volume the major consequence of increased total body Na^+. Since arterial blood pressure is also strongly regulated by the cardiovascular reflexes, volume increases the most likely to cause hypertension for individuals whose ability to regulate blood pressure is impaired in some respect.

4. An increase in total body K^+ would affect mainly ICF volume, since ICF contains almost all the total body K^+.

5. Reduction of blood flow to the kidney is a potent stimulus for renin secretion. The affected kidney would release large amounts of renin, raising aldosterone levels inappropriately. The excess aldosterone may well cause an increase in total body Na^+ and ECF sufficient to overwhelm reflexive compensation and increase arterial blood pressure.

6. As shown in Figure 15-8, the right atrial pressure (that is, central venous pressure) is affected directly by changes in plasma volume. Furthermore, the reflexive responses that regulate mean arterial blood pressure in response to changes in plasma volume exaggerate the changes in right atrial pressure. Thus atrial pressure is a far more accurate indicator of plasma volume than mean arterial pressure.

7. Glycyrrhizic acid is an aldosterone agonist.

8. Hyperparathyroidism causes bone breakdown to dominate bone formation, reducing the total mass of bone and increasing both ECF levels of Ca^{++} and phosphate. The increased Ca^{++} depresses excitability of nerves and muscles (see "Extracellular Ca^{++} and Excitability" under **Refractory Period and Firing Frequency,** Chapter 7).

9. See "Buffer Systems" under **Acid-Base Homeostasis.**

10. See "Respiratory and Renal Contributions to Regulation of Plasma pH" under **Acid-Base Homeostasis.**

11. Plasma pH would be about 7.1, and a P_{CO_2} of about 20 mmHg would be needed for full respiratory compensation, as estimated from Figure 20-16.

12. The P_{CO_2} of 80 indicates respiratory acidosis. However, the pH and bicarbonate concentrations are somewhat lower than expected for pure respiratory acidosis, especially if sufficient time had elapsed for renal compensation; this suggests a combined respiratory and metabolic acidosis (see Figure 20-17).

Multiple Choice Questions:

13. a
14. c
15. a
16. b
17. d
18. a
19. d

Chapter 21

1. See Table 21-1. In answering this question, consider the following:
   Physical changes
   Food transport within the tube system
   Storage
   Secretion
   Digestion
   Absorption

2. See "Function and Control of Salivary Secretion" under **The Role of the Mouth and Esophagus.**

3. A complete answer to this question should take into account the intracellular mechanisms of control of contraction in smooth muscle (review Chapter 12), the patterns of motility in the gastrointestinal tract (see "Patterns of Motility in the Gastrointestinal Tract" under **Gastrointestinal Smooth Muscle**), and the effects of autonomic innervation (see "Neural and Hormonal Control of Motility" under the same heading).

4. See "Gastric Secretion and Motility."

5. Parietal cells are densely populated with mitochondria (Figure 21-19), reflecting the large amounts of energy needed to transport H^+ into the gastric lumen against a pH gradient of several orders of magnitude.

6. See "Pancreatic Enzymes" under **Secretion and Motility in the Intestine,** and Table 21-4.

7. Enterokinase is a protease secreted by intestinal cells that activates trypsinogen; the trypsin formed activates the rest of the pancreatic enzymes.

8. See "Defecation" under **Secretion and Motility in the Intestine.**

Multiple Choice Questions:

9. d
10. b
11. b
12. d
13. b
14. a
15. d
16. c

Chapter 22

1. In answering this question, you might consider the following aspects:
   a. The role in digestion of each of the anatomical regions in the list (as described in Chapter 21)
   b. The particular anatomical specializations of each region, such as crypts, gastric glands, salivary glands, and so forth
   c. The chemical environment of each region resulting from secretions of digestive glands and the gastrointestinal mucosa
   d. Cellular specializations such as microvilli and membrane transport systems

2. See Carbohydrate Digestion and Absorption.

3. See Protein Digestion and Absorption.

4. a. Emulsification of lipids by bile salts greatly accelerates lipase and phospholipase activity.
   b. Bile salts stabilize free fatty acids, monoglycerides, and diglycerides in the form of micelles, the form in which digested lipid is carried to the mucosal surface of intestinal cells.

5. Three major differences are:
   a. Lipid absorption into intestinal cells is a passive rather than an active process.
   b. Unlike monosaccharides and amino acids, the products of fat digestion (free fatty acids, mono- and diacylglycerols, glycerol) undergo resynthesis (into triacylglycerol) in intestinal cells.
   c. The reassembled lipids traverse the intestinal cells in vesicles rather than as solutes. They enter the blood via the lymphatic system rather than passing directly into intestinal capillaries. Consequently, they enter the general circulation rather than passing directly to the liver via the enterohepatic portal circulation. See Figures 22-7 through 22-10).

6. See Table 22-6.

7. See Figures 22-14 and 22-15.

8. The increased tone of the pyloric sphincter severely restricts the amount of chyme that enters the duodenum with each gastric contraction, allowing mixing to occur freely without it resulting in uncontrolled stomach emptying.

9. Acidification of the duodenum results in an increased secretion of the hormone secretin which stimulates pancreatic secretion.

10. Gastric emptying is affected by reflexes and hormones, both of which are responsive to the volume and acidity of chyme and to the presence of specific nutrients. See Figures 22-16 through 22-19 and Tables 22-4 and 22-5.

Multiple Choice Questions:

11. d
12. c
13. a
14. b
15. d
16. c
17. a
18. c

Chapter 23

1. Lipids: adipose tissue
   Carbohydrates: liver and skeletal muscle
   Protein: muscle typically is the site of most intracellular protein

2. During the absorptive state, most energy comes from glucose and amino acids absorbed from the intestine. During the postabsorptive state, the drop in insulin levels and the rise in glucagon and epinephrine levels causes the body to shift to stored glycogen and fat as energy sources.

3. See "Transport and Storage of Lipids" under **Characteristics of the Absorptive State.**

4. a. In muscle, insulin promotes storage of glucose as glycogen and favors protein synthesis because it allows an influx of amino acids. Some of the glucose is metabolized to lactate. The lactate enters the blood, and some of it undergoes gluconeogenesis in the liver. See Figure 23-9.
   b. In adipose cells, insulin favors fat deposition by stimulating glucose entry. After entry, the glucose is converted into the glycerol backbone of triglyceride. See Figure 23-11.
   c. Uptake of glucose and amino acids by the liver is insulin independent. However, insulin stimulates the enzymes of glycogen synthesis, indirectly favoring glucose uptake and conversion of absorbed glucose to liver glycogen. See Figure 23-10.

5. See Table 23-3.

6. Glucagon, epinephrine, and cortisol. Generally insulin secretion and glucagon secretion are reciprocal, but secretion of both hormones is stimulated by an increase in plasma amino acid levels.

7. See "Satiety Signals and Short-term Regulation of Food Intake" under **Regulation of Food Intake.**

8. See "Transport and Storage of Lipids" under **Growth and Metabolism Over the Life Cycle,** and the essay "Control of Blood Cholesterol Levels by the Liver."

9. Both testosterone and estrogen stimulate calcification and favor maintenance of bone mass. Because menopause, the cessation of female reproductive cycles, causes levels of estrogen to decrease, postmenopausal women are more likely to develop osteoporosis than men of the same age. Replacement of estrogen

can prevent osteoporosis. Dietary calcium intake must be adequate, but it is not clear whether excessive calcium intake has any beneficial effect. Exercise that loads bone tends to protect bone mass, as does ingestion of small amounts of fluoride.

10. For hGH, see "Growth and Growth Hormone" under **Changes in Metabolism Over the Life Cycle.** For epinephrine, see Table 23-5.

11. See "Thermoregulation and Metabolic Adaptation to Cold" under **The Thyroid Hormones and Basal Metabolic Rate.**

12. See "Metabolic Homeostasis in Starvation" under **Characteristics of the Postabsorptive State.**

Multiple Choice Questions:

13. b
14. c
15. d
16. a
17. d
18. b
19. d
20. a

Chapter 24

1. See Table 24-1 and Summary items 1 and 2.

2. See Figure 24-10 and 24-11.

3. For factors that control differentiation of external genitalia, see Figures 24-10 and 24-11. For disorders, see "Disorders of Sexual Differentiation" under **Hormonal Control of Sexual maturation.**

4. A Graafian follicle is the final stage of follicular development; it contains and nourishes a maturing oocyte and secretes estrogen. The thecal cells of the follicle synthesize androgen and export it to granulosa cells, which convert it into estrogen, the hormone mainly responsible for female secondary sex characteristics. After ovulation, the remnant of follicle transforms into a corpus luteum, which secretes estrogen and progesterone. The latter hormones maintain the secretory state of the uterine endometrium until the end of the cycle.

5. See Figure 24-22. LH (lutenizing hormone) and FSH (follicle-stimulating hormone) are trophic hormones responsible for maintaining ovarian and testicular function; in females, cycles of FSH and LH release drive maturation and ovulation of follicles and their subsequent transformation into corpora lutea. Estrogen maintains female secondary characteristics; its rise during the follicular stage of the cycle causes the uterine endometrium to proliferate. Progesterone converts the endometrium to the secretory phase in preparation for implantation of the zygote that will result if the ovulated ovum is fertilized.

6. Sertoli cells: the testicular cells that nourish maturing spermatozoa

Leydig cells: the testicular cells that secrete testosterone.

Vas deferens: the two vasa deferentia connect the testes to the rest of the male gland and duct system (see Figure 24-3)

The two seminal vesicles are glands that together with the prostate and Cowper glands secrete the fluid component of semen.

7. GnRH is gonadotropin-releasing hormone, a hypothalamic hormone that controls release of the anterior pituitary hormones LH and FSH. The functions of LH and FSH are given in Table 24-3. The functions of the gonadal steroid hormones testosterone, estradiol, and progesterone are given in Tables 24-4 and 24-6.

8. The male external genitalia would in any case be masculinized by testicular androgen during early development, but if excessive secretion of adrenal androgen continues after birth, a precocious (early) puberty results. Depending on the location of the defect in the pathways of steroid metabolism in the adrenals, excessive androgen secretion may be accompanied by excessive cortisol secretion, which diminishes immune response, or by excessive aldosterone secretion, which causes hypertension.

Multiple Choice Questions:

9. c
10. b
11. d

12. a
13. a
14. c
15. d
16. b

Chapter 25

1. Oxytocin secretion is stimulated by cervical stretch during labor and by suckling after labor. During labor its role is to stimulate uterine contractions. During lactation it stimulates milk letdown. Both of these processes involve increasing the activity of smooth muscle.

2. Prolactin maintains milk production while suckling continues. The factors that control its release are discussed in "Prolactin and Lactation" under **Birth and Lactation.**

3. Ectopic means "in the wrong place." The outcome depends on the site of implantation. Blastocysts that implant in the Fallopian tube do not succeed; when they rupture the fallopian tube, a surgical emergency results. Implantation at other sites within the abdominal cavity can result in a pregnancy that sometimes can be carried to term.

4. See "Endocrine Functions of the Trophoblast and Placenta" under **Pregnancy.**

5. See "Alterations of Maternal Physiology by Pregnancy" under **Pregnancy.** During the early part of pregnancy the mother can accumulate stores of some nutrients in preparation for the much higher demands of the fetus during the later stages of pregnancy. Folic acid, a vitamin, is an excellent example. The importance of this vitamin in development is such that it is recommended that women who anticipate becoming pregnant to begin taking folic acid supplements.

6. Cortisol stimulates the fetal lung to begin secreting pulmonary surfactant. The importance of this secretion for breathing is explained in Chapter 16 (p. 467). Lack of pulmonary surfactant causes a form of respiratory distress called hyaline membrane disease.

7. Because the events that determine the fates of antire organ systems occur during the embryonic phase of development, damage during this time typically has a widespread impact.

8. The effects of patient ductus arteriosus:
After birth the resistance of the pulmonary loop falls below that of the systemic loop, therefore some of the output of the left heart can pass through the ductus and reenter the pulmonary circulation, diminishing the effective cardiac reserve available for the systemic circulation. This left-to-right shunt raises the pressure of the pulmonary loop. Since some of the blood makes more than one pass through the pulmonary loop for each pass through the systemic loop, the diverted blood actually has an additional opportunity to become oxygenated, and the oxygen saturation of systemic arterial blood is normal.

9. Progesterone maintains the uterine endometrium during pregnancy and inhibits contraction of uterine smooth muscle, a factor that tends to prevent premature labor. Possibly, in women prone to premature labor the amount of progesterone secreted by the corpus luteum is inadequate to prevent the onset of stretch-induced contractions; this can be corrected by supplying additional progesterone.

10. See "The Fetal-Placental Circulation" under **Pregnancy.**

Multiple Choice Questions:

11. a
12. c
13. d
14. b
15. b
16. d
17. c

Chapter 26

1. See "Inflammation and the Complement and Kallikrein Systems" under **Nonspecific Components of the Immune Responses.**

2. The two major types of phagocytes are macrophages (transformed monocytes) and neutrophils (one class of granulocytes).

The three classes of granulocytes are neutrophils, basophils, and eosinophils.

"Leukocyte" is the general name for white blood cells, including granulocytes, monocytes, and lymphocytes.

Lymphocytes are the class of leukocytes that are responsible for specific immune responses; these include two general types, B cells and T cells.

Both basophils and mast cells secrete histamine, a mediator of inflammation.

3. The complement cascade, like other regulatory cascades (see Chapter 5), *amplifies the initial signal* set in motion by binding of antibody to antigen, and *branches the signal* to activate several different elements of immune response.

4. The more closely the donor and recipient and related, the more likely they are to share similar major histocompatibility antigens. However, there are a number of different MHC antigens, and differences could trigger rejection.

5. See "Inflammation and the Complement and Kallikrein Systems" under **Nonspecific Components of the Immune Responses** and "Relationship Between Immunoglobulin Structures and Actions" under **Specific Immune Responses.**

6. See "Distinguishing Between Self and Nonself" under **Specific Immune Responses.**

7. In juvenile-onset diabetes, the pancreas may perhaps never secrete insulin. According to the clonal selection hypothesis, if insulin is not present in the blood at the time of clonal selection of B cells, clones that would make antiinsulin are not eliminated. Consequently, these clones would survive and generate a primary immune response when insulin is given. On the other hand, the pancreas could be normal initially. If for some reason clones that could direct an immune response against pancreatic cells were not eliminated, this error could result in an autoimmune attack on the pancreas. This attack would stop insulin secretion by damaging β-cells and also probably would involve formation of antibodies to insulin as well as other proteins synthesized by pancreatic cells. Finally, before the availability of genetically engineered human insulin, diabetics used beef or pork insulins. Although these hormones can interact with the human insulin receptor, their amino acid sequences are not identical to that of human insulin, leaving open the possibility that these molecules could be recognized as "nonself" antigens.

8. The immune system usually succeeds in eliminating cells that express novel antigens, including cancer cells. However, the vigilance of the immune system declines with age.

9. See Figures 26-18 and 26-20 and related text.

10. See "Autoimmune Diseases" under **Undesirable Immune System Responses**

11. Suppressor T cells moderate the vigor of the cell-mediated specific immune response. They form part of a negative feedback system that usually prevents excessive damage to the host tissues in an immune response. See also Table 26-4.

Multiple Choice Questions:

12. c
13. d
14. b
15. b
16. c
17. e
18. a
19. d
20. d
21. d

Glossary

A

A band The region of a muscle sarcomere that contains myosin thick filaments. p. 292

abscissa The horizontal, or x-axis, of a graph.

absolute refractory period The interval following an action potential during which a second action potential cannot be generated. p. 173

absorptive state The period after a meal when excess substrates are stored for future use. p. 656

acceleration The rate of change of velocity with time. p. 33

accommodation The change in the radius of the lens that allows the eye to focus on objects at different distances. p. 270

acetylcholine (ACh) A chemical transmitter released at the junction of a nerve and a skeletal muscle, in the autonomic ganglia, and from postganglionic parasympathetic fibers. p. 204

acetylcholinesterase An enzyme that degrades acetylcholine. p. 300

acetyl-CoA A coenzyme that enables the end products of glycolysis, amino acids, and fatty acids to enter the Krebs cycle. p. 89

achalasia A condition in which food cannot enter the stomach from the esophagus. p. 610

acid A compound that dissociates in solution to produce H^+ ions. p. 27

acid-base balance Maintenance of the plasma pH at a normal value of 7.4. p. 579

acidity The H^+ content of a solution The lower the pH, the more acid the solution. p. 27

acidosis A condition in which the plasma pH is less than 7.4. p. 580

acini The blind ends of secretory (exocrine) glands. p. 597

acromegaly A clinical condition in adults with elevated levels of growth hormone. p. 673

acrosome The apical portion of male sperm; it permits penetration of the ovum. p. 726

F- and G-actin The muscle protein in thin filaments. F-actin filaments are composed of G-actin subunits. Actin and myosin interact when Ca^{++} is present. p. 295

action potential An all-or-none membrane depolarization that is propagated along nerve fibers without any decrease in amplitude. p. 162

activation energy The energy "hill" that must be surmounted for a chemical reaction to proceed in a given direction. p. 57

active force In a muscle fiber, the component of total tension generated by the interaction between the thick and thin filaments. p. 304

active site The region of an enzyme that specifically binds a substrate, reducing the overall activation energy for a reaction. p. 59

active state The state of a muscle fiber when Ca^{++} and ATP are present and the fiber is generating force. p. 297

active transport Transport of a substance against its concentration gradient by direct use of cellular energy (ATP). p. 146, 147

actomyosin The low-energy complex formed between actin and myosin in the absence of ATP; it is also known as a "rigor complex." p. 297

adaptation In sensory receptors, the decrease in amplitude of the generator potential despite the continued presence of a stimulus. p. 197

adenohypophysis See anterior pituitary.

adenosine A nucleotide that vasodilates the coronary arteries. p. 106, 251

adenosine triphosphate (ATP) The "energy currency" of the cell. It is synthesized from adenosine diphosphate (ADP) in glycolysis, the Krebs cycle, and through oxidative phosphorylation. The breakdown of ATP by hydrolysis is coupled with reactions that require energy, rendering the overall reaction favorable. p. 82

adenylate cyclase An enzyme that converts ATP to cyclic AMP. It often is coupled with membrane receptors through the activation of intermediate G-proteins. p. 111

adipose tissue Fat cells; the site of lipid storage and a major factor in endocrine regulation of organic metabolism. p. 10

adrenal cortex That portion of the adrenal gland that secretes the corticosteroid hormones. p. 110

adrenal medulla That portion of the adrenal that secretes epinephrine. It is an extension of the autonomic nervous system. p. 329

adrenergic receptors Receptors for the neurotransmitter norepinephrine. The four subclasses are designated α_1, α_2, β_1, and β_2. p. 349

adrenocorticotropic hormone (ACTH) A hormone secreted by the anterior pituitary that controls the release of the adrenal corticosteroids. p. 116, 669

aversion learning A form of conditioning specifically involving taste.

aerobic metabolism The complete degradation of glucose to carbon dioxide and water by the combined action of glycolysis, the Krebs cycle, and oxidative phosphorylation. p. 89

afferent In the nervous system, a nerve fiber leading from a receptor to the spinal cord or brainstem. p. 106

afferent arterioles The renal arterioles proximal to the glomerular capillaries. p. 532

affinity A measure of the relative binding of various substrates to an enzyme, or ligands to a receptor. p. 59

afterload The load against which an activated muscle must try to shorten Greater afterloads result in lower velocities. p. 306, 376

afterpotential The membrane potential following an action potential. p. 173

agglutination The clumping together of cells. p. 761

agonist Any compound, such as a drug, that can activate a receptor; to be distinguished from compounds that activate receptors under normal conditions. p. 116

agonist muscle A muscle that moves a particular joint in the same direction as another muscle.

airways The bronchi and bronchioles of the lungs. p. 450

albumin One of the plasma proteins, and the major determinant of plasma osmotic pressure. p. 392

aldosterone A hormone secreted by the adrenal that controls Na^+ reabsorption in the distal tubules of the kidneys. p. 110, 427

alkaline tide A term referring to the fact that plasma leaving the stomach has an excess of HCO_3^- as a result of acid secretion. p. 612

alkalosis A condition in which the plasma pH is greater than 7.4. p. 580

allosteric effect The modification of enzyme activity by the binding of a modulator to an allosteric site separate from the active site. p. 60

allosteric site See allosteric effect.

alpha helix One of the secondary structures of proteins. p. 55

alpha motor neuron A spinal cord neuron innervating skeletal muscle fibers. p. 334

alpha receptors See adrenergic receptors.

alpha wave One of the inherent electrical rhythms of the brain seen in the electroencephalogram of a relaxed, awake individual with eyes closed. The frequency is about 10 per second. p. 243

alveolar ducts In the lungs, the smallest airways leading to the alveoli. p. 451

alveolar gas equation A formula for calculating the alveolar partial pressure of oxygen if the inspired partial pressure of oxygen, the alveolar partial pressure of carbon dioxide, and the alveolar ventilation are known. p. 483

alveolar ventilation The volume of air moving into and out of the alveoli during each respiratory cycle. p. 459

alveoli (alveolar sacs) Thin, air-filled sacs in the lungs across whose walls oxygen and carbon dioxide are exchanged. p. 450

amacrine cell A cell in the retina that connects adjacent photoreceptors and is primarily sensitive to the movement of an object. p. 272

amiloride A drug that inhibits active Na^+ transport in the kidneys. p. 558

amino acids The building blocks of proteins; there are 20 amino acids. p. 54

aminopeptidase An enzyme that degrades proteins one amino acid at a time, beginning at the free amino end. p. 631

ammonia (NH_3) A lipid-soluble compound. In the kidneys, conversion of ammonia to ammonium ions permits the secretion of large amounts of H^+ ions. p. 527

amnesia Loss of memory.

ampere (A) The unit of electrical current. p. 37

amphipathic molecules Molecules such as phospholipids that have polar and nonpolar regions. They aggregate as micelles or bilayers. p. 45

ampulla The region of the semicircular canals that contains vestibular hair cells. p. 265

amygdala A component of the limbic system in the brain. p. 241

amylase An enzyme that degrades starches to monosaccharides and disaccharides. p. 629

amylopectin Plant starch. p. 629

anabolism Processes that build body tissues. p. 86, 656

anaerobic metabolism Metabolic processes that occur without oxygen. p. 87

analgesia Loss of the sensation of pain. p. 214

anastomoses Direct connections between arteries and veins.

anatomic dead space The region of the lung that cannot exchange gas. p. 460

androgens The male sex steroid hormones, principally testosterone. p. 129

anemia A condition in which the number of red blood cells is low. p. 392

anesthesia Loss of all sensation. p. 214

angiotensin I The hormone produced in response to renin release by the kidneys. p. 427

angiotensin II A vasoconstrictor produced from angiotensin I that stimulates secretion of ADH and thirst. p. 427

angiotensin III A substance produced from angiotensin II that stimulates secretion of aldosterone from the adrenal gland. p. 427

angiotensinogen A plasma protein that is cleaved by a converting enzyme in the presence of renin to produce angiotensin I. p. 427

angstrom(Å) An older unit of distance equal to 10^{-10} meters.

anion A negatively charged ion.

anion gap A measure of plasma anions other than Cl^- and HCO_3^-. p. 586

anorexia nervosa A condition in which an individual abnormally restricts food intake. p. 649

antagonist In receptor activation, a compound (drug) that will block a receptor. p. 116, 331

antagonist muscle A muscle that opposes the movement produced by activation of another muscle at a joint. p. 108

anterior chamber In the eye, the space in front of the lens. p. 268

anterior pituitary A multifunctional endocrine gland that releases adrenocorticotropic hormone (ACTH), thyroid-stimulating hormone (TSH), follicle-stimulating hormone (FSH), luteinizing hormone (LH), prolactin, and growth hormone (GH). p. 116

anterolateral tract One of the major ascending somatosensory tracts; it mediates pain, temperature, and "crude" touch. p. 236

antibody An immunoglobulin molecule. Released by lymphocytes, it specifically recognizes "foreign" macro-

molecules and activates other components of the immune system. p. 68, 763

anticodon The region of the transfer RNA molecule that contains base pairs complementary to base pairs (codons) of messenger RNA. p. 67

antidiuretic hormone (ADH) A hormone released from the posterior pituitary that controls water permeability in the kidneys. Its release is influenced by plasma osmolarity and venous pressure (vasopressin or vasopressin-ADH). p. 116, 426, 548

antigen Any macromolecule that produces a specific immune response. p. 757

antigravity muscles Postural muscles. p. 340

antiport An active transport process in which two substances move in opposite directions across a cell membrane by means of a carrier.

antipyretic A substance that reduces fever. p. 755

antrum The region of the stomach closest to the duodenum. p. 596

anus The most distal sphincter in the gastrointestinal tract. p. 597

aorta The large blood vessel that receives the output of the left ventricle. p. 358

aortic baroreceptors Stretch-sensitive receptors in the wall of the aorta that monitor arterial blood pressure. p. 424

aortic arch The proximal region of the main artery leading from the heart. p. 424

aortic bodies Oxygen- and pH-sensitive receptors near the aortic arch. p. 505

aortic valve The valve between the left ventricle and the aorta. p. 362

aperture The area of the pupil of the eye. p. 268

apex The portion of the cochlea farthest from the oval window; it is activated primarily by low-frequency sounds. p. 258

aphasia A deficit in expressing or understanding language. p. 242

apneustic center A brainstem region whose activity inhibits respiration. p. 503

apoproteins Proteins needed to activate lipoprotein lipase and to permit chylomicrons and very low density-lipoproteins (VLDLs) to enter the cells lining the intestine.

aqueous humor The thin, watery fluid in the anterior and posterior chambers of the eye. p. 268

arachidonic acid A phospholipid whose synthesis is activated by many hormones and neurotransmitters. p. 49, 394

argentaffin cells Gastrin-secreting cells in the stomach. p. 602

arrhythmias Irregularities in the heart rate. p. 368

arterial pressure pulse The periodic variation in pressure caused by the heart's pumping action. p. 363

arteries Blood vessels that branch out from the aorta and lead into arterioles. p. 358

arterioles Vessels of the circulation that have the highest flow resistance and therefore control blood distribution. p. 358

ascorbic acid Vitamin C. p. 640

association cortex A term denoting the areas of the brain that are not primarily sensory or motor. p. 227

association neuron An interneuron in the spinal cord or brain. p. 221

associative learning A "higher" form of learning that requires correlations between two different stimuli.

asthma A disease characterized by the plugging of airways. p. 776

ataxia Unsteadiness of gait or motor coordination. p. 345

atherosclerosis The formation of fatty deposits on the interior surface of arteries; these deposits eventually can obstruct blood flow. p. 369

atmospheric pressure The hydrostatic pressure exerted on objects on the earth's surface by the air. At sea level it normally is 760 mm Hg.

atom The basic constituent of all matter. p. 20

atomic number The number of protons in the nucleus; it defines an element. p. 20

atomic weight The sum of the protons and neutrons in the nucleus. Elements with the same atomic number but different atomic weights are called isotopes. p. 20

ATPase Any enzyme that hydrolyzes ATP to ADP. p. 140

atresia The degeneration of unfertilized follicles. p. 693

atria The chambers of the heart that receive blood from the systemic and pulmonary veins. p. 361

atrial contraction (atrial systole) The phase of the cardiac cycle during which the atria are contracting. p. 362

atrial receptors Stretch receptors in the wall of the right atrium that measure venous blood pressure.

atrial septum The tissue separating the left and right atria. p. 361

atriopeptin (atrial natriuretic hormone [ANH]) A hormone released in response to increased atrial pressure (blood volume) that increases glomerular filtration in the kidneys. p. 569

atrioventricular (A-V) node A group of cells between the atria and ventricles that relay electrical activity to the ventricles. The delay provided by these cells gives the atria time to contract before the ventricles are activated. p. 361

atropine A drug that blocks parasympathetic postganglionic muscarinic receptors for acetylcholine. p. 350

auditory canal The canal of the outer ear. p. 256

auditory cortex The region of the cerebral cortex that receives direct sensory input from the auditory system. It is located in the temporal lobe. p. 262

auditory system One of the "special senses." It is composed of the ear and the neural pathways leading to the auditory cortex. p. 256

Auerbach's plexus A component of the enteric (intrinsic) nervous system of the gut between the circular and longitudinal muscle layers. It is also known as the myenteric plexus. p. 597

autoimmune diseases Diseases in which normal regulation of the immune system breaks down, and tissues in the body are attacked by cells of the immune system. p. 758

automaticity A term denoting the heart's ability to generate its own intrinsic electrical rhythm. See **pacemaker**.

autonomic ganglia Groups of neurons in the autonomic nervous system that are situated outside the central nervous system. p. 223

autonomic nervous system (ANS) The component of the central nervous system that controls visceral smooth muscle and glands. p. 109

autonomic reflex An "involuntary" reflex involving visceral receptors and smooth muscle or glands. p. 108

autoregulation The ability of a tissue or organ to internally regulate its function internally without extrinsic neural or hormonal input. p. 103

autotransfusion The effective increase in blood pressure caused by a sympathetically mediated decrease in venous compliance.

A V block The failure of action potentials to pass through the AV node from the atria to the ventricles. p. 369

A V delay The time required for action potentials to pass through the AV node from the atria to the ventricles. p. 363

average flow velocity In the circulation, the linear velocity of the movement of an element of blood past a given point. p. 405

Avogadro's number The number of molecules of a compound contained in 1 mole It is a universal constant equal to 6.023×10^{23}. p. 30, 429

axon The component of a neuron that carries efferent signals. It usually is the longest component of the neuron. p. 190

axon hillock The region of the nerve cell body from which the axon arises; it is the site of origin of action potentials. p. 175

axoplasmic transport An ATP-dependent process involving neurofilaments and neurotubules that moves materials away from (anterograde) or toward (retrograde) the soma of a neuron. p. 192

B

B lymphocytes A subclass of lymphocytes that can be activated to differentiate into antibodysecreting plasma cells that mediate humoral immune responses. p. 763

baroreceptors Sensory receptors that detect pressure. They are present in the aorta and the carotid arteries. p. 424

basal body A submembranous structure in auditory and vestibular hair cells that determines their polarity. p. 260

basal ganglia A group of subcortical nuclei that are mainly involved in controlling refined muscle movements. p. 224, 341

basal tone Resting tone; a condition in which a tissue has some intrinsic level of activity in the absence of neural or hormonal control. The term applies to smooth muscle tension, spontaneous firing of nerve fibers, and so forth. p. 604

base A compound that will combine with a H^+ ion. In the cochlea, the region nearest the oval window, where high-frequency sounds are localized. p. 27, 258

basic electrical rhythm Periodic waves of depolarization in visceral smooth muscle; this pulsing serves as the pacemaker for gastrointestinal motility. p. 316, 603

basilar membrane In the inner ear (cochlea), the structure that localizes sound frequencies to spatially segregated areas. p. 259

basolateral The surface of renal cells that faces adjacent cells. p. 532

basophils A class of white bood cells that secrete histamine. p. 752

behavioral homeostasis The form of homeostasis that involves modification of the body's external environment. p. 12

beta cells Insulin-secreting cells in the pancreas. p. 118, 656

beta receptors See adrenergic receptors.

beta waves One of the intrinsic electrical rhythms of the brain seen in the electroencephalogram of an alert subject with eyes open. p. 243

bicarbonate (HCO$_3^-$) The base formed when carbon dioxide dissolves in plasma. p. 580

bile A fluid produced by the liver that contains lecithin (a phospholipid), bile salts, and bile pigments. It is required for lipid digestion. p. 617

bile pigment Bilirubin. p. 617

bile salts One of the ingredients of bile that aids fat digestion. p. 617

bilirubin One of the bile pigments derived from hemoglobin. p. 394

bipolar cells In the retina, the cells that relay activity from the rod and cone photoreceptors to the ganglion cells. p. 272

bipolar neuron A class of nerve cells that are characterized by two branching processes at opposite ends of the cell body.

bladder The organ in which urine is stored. See **micturition**. p. 528

blastocyst A hollow sphere of cells in the early stages of embryogenesis. See **trophoblast** and **embryoblast**. p. 728

bleaching Splitting of the visual pigment into opsin and retinal. p. 276

blood-brain barrier A term referring to the fact that the cerebrospinal fluid is functionally isolated from the plasma. p. 405

blood The circulating contents of the cardiovascular system. Blood has a fluid component (plasma) and a cellular component (the red and white blood cells). p. 10

blood clot A structure formed from platelets and fibrin that seals ruptures in the walls of blood vessels. p. 394

blood doping A process in which an individual's blood is removed and the erythrocytes later retransfused in an effort to increase oxygen-carrying capacity. p. 411

blood pressure The pressure of the blood; used alone, the term usually refers to the average arterial pressure. p. 363, 408

blood reservoirs The veins; they contain about 80 of the circulating volume.

blood type A classification of individuals based on the antigen present on the surface of their red blood cells. The blood types are A, B, AB, and O. p. 772

blue-sensitive cones Photoreceptors with maximum sensitivity at 420 namometers. p. 272

body fluid compartments A term denoting the division of the total body water into intracellular, interstitial, and plasma components. p. 564

body of the stomach The middle region of the stomach. p. 596

Bohr effect The shift in the oxyhemoglobin dissociation curve produced by changes in pH, CO$_2$, or temperature. p. 489

bolus The amount of food normally swallowed. p. 608

botulinum toxin A poison released by a bacterium sometimes found in improperly canned food; it can block neuromuscular transmission.

Bowman's capsule The proximal part of the nephron into which plasma is filtered. p. 530

Boyle's Law One of the Gas Laws. p. 462

bradycardia An abnormally low heart rate. p. 377

bradykinin A potent local vasodilator released during an inflammatory response. p. 211, 424

brainstem The most primitive region of the brain; it includes the medulla, pons, and midbrain and contains many control centers for the autonomic nervous system. p. 223

Broca's area One of the two "language areas" of the cortex; it is responsible for translating concepts and commands into a verbal format. p. 242

bronchi Large airways in the lungs that contain cartilage. p. 450

bronchioles The small airways of the lungs that are surrounded by smooth muscle. p. 450

brown fat Specialized adipose tissue in infants that is important in thermoregulation. p. 681

brownian motion The random motion of large particles or molecules in a solution, caused by the thermal energy of water molecules.

brush border The luminal, microvillous surface of intestinal epithelial cells. p. 599

buffer A mixture of an acid and its conjugate base that can minimize the pH changes caused by addition of strong acids or bases. p. 28, 579

buffer line The line on a Davenport diagram that describes hemoglobin buffering. p. 580

bulimia An eating disorder in which an individual cycles between food binges and exaggerated dieting or voluntarily induced vomiting. p. 649

bundle branches The component of the heart's conducting system that relays impulses from the A-V node and bundle of His to ventricles. p. 361

bundle of His The component of the conducting system that is activated immediately after the A-V node. See **bundle branches**. p. 361

bursa of Fabricius A lymphoid structure in birds where lymphocyte differentiation occurs. p. 764

C

C fibers Unmyelinated nerve fibers usually less than 1μm in diameter.

calcitonin Thyrocalcitonin; one of the hormones involved in regulating Ca^{++} and phosphate balance. p. 577

calcium (Ca^{++}) A divalent cation; it is involved in muscle contraction and synaptic transmission and may be an intracellular messenger. p. 111

calcium channel A voltage-dependent "pore" in the membrane that is responsible for the plateau of the cardiac action potential, for neurotransmission, and for stimulus-secretion coupling. p. 198, 205

caldesmon A protein that can inhibit crossbridge formation and thus muscle contraction by binding to the actin-tropomyosin complex. p. 319

calmodulin A calcium-binding protein; it mediates smooth muscle contraction and may be a general mechanism for intracellular regulation. p. 113

calorie (cal) A unit of energy; 1 kcal = 1000 cal. p. 33

capacitance The electrical "inertia" of cells that prevents their membrane potential from changing instantaneously. In the cardiovascular system, the change in volume produced by a given change in pressure. p. 404

capacitation In reproduction, the ability of sperm to fertilize an ovum. p. 692

capillary The smallest blood vessel; it is surrounded by a single layer of epithelial cells. p. 358

capillary filtration The movement of water from the interior of a capillary into the interstitial space. p. 415

carbamino compounds (carbamino-hemoglobin) The compounds formed when CO$_2$ reacts with hemoglobin. p. 487

carbohydrate Organic molecules composed of carbon, hydrogen, and oxygen in the ratio 1:2:1. p. 49

carbon An element; carbon has a valence of +4 and can combine with itself and with many other atoms to form complex organic compounds. p. 20

carbon dioxide (CO$_2$) One of the respiratory gases; CO$_2$ is produced by the cells while they consume oxygen. p. 487

carbon monoxide (CO) A gaseous by-product of hydrocarbons that competes with oxygen for hemoglobin and thus inhibits oxygen transport. p. 491

carbonic acid (H$_2$CO$_3$) The acid produced when CO$_2$ reacts with water. p. 492

carbonic anhydrase An enzyme that

accelerates the reaction of CO_2 with water. p. 59

carboxypeptidase An enzyme that degrades proteins one amino acid at a time, beginning at the C-terminal end. p. 617

cardiac cycle The stereotyped pattern of electrical and mechanical activity that occurs during each heartbeat. p. 362

cardiac failure The inability of the heart to pump an adequate amount of blood to supply tissue needs. p. 432

cardiac function curve The relationship between cardiac output and venous pressure. p. 430

cardiac glycosides Drugs that increase the heart's contractile strength.

cardiac muscle The form of striated muscle found in the heart. p. 10

cardiac output The amount of blood pumped per minute by each ventricle of the heart. p. 376

cardiac reserve The amount that the cardiac output can increase above the normal resting level as a result of changes in venous return. p. 432

carotid baroreceptor A stretch receptor that monitors arterial blood pressure. p. 424

carotid bodies O_2- and pH-sensitive sensory chemoreceptors. p. 505

carotid sinus The region of the carotid artery that contains the carotid baroreceptors. p. 424

carriers Lipid-soluble molecules that bind ions or polar molecules and transport them across membranes. p. 148

catabolism The breaking down of tissues for energy production. p. 86, 656

catalyst A substance that accelerates a chemical reaction but is not consumed in the process. See **enzymes.** p. 57

catecholamines A generic term for norepinephrine and epinepherine. p. 330

caudate nucleus One of the basal ganglia of the brain. p. 341

cecum The most proximal end of the large intestine, separated from the small intestine by the ileocecal sphincter. p. 597

cell The basic element of all living organisms. p. 10

cell membrane The structure composed of lipids and proteins that separates the cytoplasm of a cell from the surroundings. p. 74

cellular immunity A component of the immune response; it is mediated by T lymphocytes. p. 766

cellulose Plant starch; it is a rigid polymer of glucose. p. 52, 629

central chemoreceptors Regions of the medulla (brainstem) that are functionally sensitive to the partial pressure

of CO_2 in the arterial blood and that regulate respiration. p. 504

central nervous system (CNS) The brain and spinal cord. p. 190

central venous pressure The blood pressure in the right atrium. p. 418

cephalic phase In digestion, a term referring to the fact that olfactory, taste, and visual stimuli produce gastrointestinal activity. p. 643

cerebellum A region of the brain behind the pons that coordinates and refines motor movements. The cerebellum is important for postural stability. p. 224

cerebral cortex The outer layer of the cerebrum; the most sophisticated and phylogenetically recent area of the brain. p. 224

cerebral dominance The specialization of one hemisphere for language functions. p. 242

cerebrum The most phylogenetically recent area of the central nervous system, composed of the cortex and the corpus striatum (basal ganglia). p. 224

cerebrospinal fluid The extracellular fluid in the central nervous system. p. 221, 504

cervix The distal opening of the uterus facing the vaginal canal. p. 688

channel A permanent or semipermanent "pore" in a membrane through which ions or polar molecules can move at a high rate. p. 147

Charles Law One of the Gas Laws; at constant pressure, the volume of a gas is proportional to temperature. p. 480

chemical bonding The process by which atoms combine to form compounds. The term also refers to the forces that cause molecules to interact with one another. See **covalent bonds, hydrogen bonds, ionic bonds,** and **Van der Waals forces.** p. 22

chemical senses Olfaction and taste. p. 283

chemical synapses Synapses at which a molecule (neurotransmitter) released from a nerve cell causes permeability changes in another nerve or muscle cell. p. 202

chemiosmotic hypothesis The process by which energy is transferred from a reduced coenzyme to form ATP in the mitochondria of a cell. p. 92

chemoreceptors Sensory receptors sensitive to chemical compounds. p. 505

chief cells The cells in the stomach that secrete pepsinogens. p. 611

chloride (Cl^-) The principal extracellular anion. p. 149

cholecystokinin (CCK) One of the gastrointestinal hormones; it causes the gallbladder to contract and inhibits secretion of gastric acid. p. 618

cholesterol A steroid used to synthesize the steroid hormones. p. 48, 617

cholinergic A neuron that releases acetylcholine as its neurotransmitter. p. 349

chondrogenesis The initial stage of bone formation. p. 671

chorea The uncoordinated limb movements seen in Huntington's disease.

chromaffin cells Modified postganglionic neurons in the adrenal medulla that release norepinephrine. p. 329

chromophore Light-sensitive pigment in the photoreceptors. p. 276

chromosome The form of the genetic material (DNA) in the nucleus of a cell. p. 62

chronotropic Affecting the heart rate. p. 379

chylomicron A mixed micelle of fatty acids, phospholipids, cholesterol, and protein that can be absorbed by the epithelial cells of the small intestine during fat digestion. p. 635

chyme The semisolid contents of the gastointestinal tract. p. 596

chymotrypsin An enzyme that degrades amino acids at internal peptide bonds. p. 617

cilia Tiny hair-like projections on the surface of a cell.

ciliary body The source of aqueous humor in the anterior chamber of the eye. p. 268, 270

cimetidine A drug that inhibits gastric secretion by blocking histamine receptors. p. 613

circular muscle One of the muscle layers in the gastrointestinal tract. p. 596

cis-retinal The form of retinal before it is activated by photons of light. p. 276

clearance In the kidneys, the minimum amount of plasma that would contain the amount of some substance that is excreted. p. 551

clonal selection In immunity, the proliferative response of lymphocytes that recognize a specific foreign antigen. p. 770

coactivation The activation of both alpha and gamma (spindle) motor neurons are activated during normal muscle movements. p. 335

cochlea The inner ear; it contains the auditory receptors. p. 256

cochlear duct An alternative term for the scala media. p. 258

cochlear partition An alternative term for the basilar membrane. p. 259

codon The sequence of three nucleotides that "codes" for each of the 20 unique amino acids in DNA and RNA. p. 66

coenzymes Nonprotein compounds that are not enzymes themselves but that transfer small molecules or H$^+$ between metabolic pathways. They are catalysts, because they are not used up in the process. p. 59

cold vasodilation A condition in which prolonged cooling leads to a breakdown in the reflex vasoconstriction of peripheral blood vessels, causing "rosy cheeks."

colipase A substance that increases the effectiveness of pancreatic lipase in fat digestion. p. 634

collagen The connective tissue that makes up tendons. p. 292

collateral ventilation In the lungs a term referring to the fact that adjacent alveoli are connected by small pores.

collecting duct In the kidneys, the most distal portion of the nephronIts water permeability is regulated by ADH. p. 528

colloid osmotic pressure The osmotic pressure of the plasma. It is higher than expected because the plasma proteins are charged. p. 415

colon The part of the large bowel between the cecum and the rectum. p. 597

competitive inhibition Blocking of enzymes or receptors caused by the binding of a molecule at the site normally occupied by a substrate or ligand. p. 60

complement A collection of plasma proteins that are activated during an immune response and that mediate lysis of "foreign" cells. p. 754

complete protein A protein that contains all essential amino acids. p. 662

complex cells Neurons in the visual cortex that are sensitive to the movements of lines and edges. p. 282

compliance The ratio of the change in length (volume) of an object to the applied force (pressure). It is a measure of "stiffness." p. 36, 399, 403

compound action potential The extracellular signal recorded from a nerve trunk. It reflects the number of active fibers in a particular size category. p. 181

compressibility The ratio of the change in volume of a gas to the change in pressure. Liquids are incompressible.

compression The phase of a sound wave in which pressure increases. p. 256

concentration The amount of solute in a solution; it is usually expressed in moles.

condensation A chemical reaction involving removal of water.

conducting fibers In the heart, the bundle of His, right and left bundle branches, and Purkinje fibers. p. 361

conductance The reciprocal of resistance; it is equivalent to permeability. p. 37

conduction aphasia A condition in which an individual cannot repeat what has been said because of a lesion between the language areas of the brain and the auditory cortex. p. 242

cones Color-sensitive photoreceptors in the retina. p. 272

congestive failure In the heart, a condition in which the central venous pressure (and thus end-diastolic volume) are high. p. 377

conjugate base The base produced by the dissociation of an acid. p. 27, 511

connective tissue One of the four basic tissue types; it makes up the tendons and a large part of the body's "loose" tissue. p. 10

connexon The structure in a gap junction that forms a "pore" between adjacent cells. p. 84

constriction The narrowing of a blood vessel or airway in the lungs. p. 348

continuity equation In the absence of sources or sinks, the volume flow of fluids or gases, expressed as the product of the linear velocity and crosssectional area, remains constant.

contractility In muscle, the ability of the crossbridges to generate force. It is related to internal Ca^{++} levels and myosin ATPase activity. p. 299

contraction The generation of force in a muscle. p. 362

contralateral The opposite side of the body.

controller integrator The element in a negative feedback loop that evaluates the error signal and regulates the effector(s). p. 100

convergence The arrival of input from many different neurons at a single target neuron. p. 209, 237

Cori cycle The cycle in which lactic acid produced in muscle by glycolysis is reconverted to glucose by the liver and returned to the circulation. p. 310

cornea The transparent curved surface in the front of the eye. p. 268

coronaries The arteries that supply blood to the heart. p. 358

corpus callosum The array of nerve fibers that connects the two hemispheres of the brain. p. 224

corpus luteum In the menstrual cycle, the structure that is formed from a follicle after ovulation. It secretes progesterone and estrogen. p. 707

corpus striatum The area of the central nervous system "beneath" the cortex but "above" the thalamus and hypothalamus. p. 224

corresponding points The points on the retina of each eye upon which an image is formed by light rays from a single point in the visual field. p. 280

cortex See cerebral cortex or renal cortex. p. 528

cortical column A term referring to the fact that cortical neurons at different depths share similar receptive fields. p. 227

cortical nephron The nephrons in the kidney whose loops of Henle do not penetrate very deeply into the renal medulla. p. 531

cortical reaction In fertilization, the initial response to sperm penetration. p. 726

corticorubrospinal tract A descending motor tract that arises from the red nucleus. p. 346

corticospinal tract A descending motor tract that arises from the cortex. p. 346

cortisol The major steroid secreted by the adrenal cortex. p. 110, 118, 666

cotransport Transport of a substance against its concentration gradient, which occurs because the substance is coupled to the "downhill" movement of another substance, usually Na$^+$. p. 149

countercurrent exchanger In the kidneys, a term referring to the fact that the blood vessels of the vasa recta act to conserve the osmotic gradient in the medulla. p. 549

countercurrent multiplication In the kidneys, the processes that increase the osmolarity of the renal medulla and allow production of hypertonic urine. p. 549

counter-transport See antiport.

covalent bond A chemical bond formed when electrons are shared among the constituent atoms. Covalent bonds can be polar or nonpolar. p. 23

Cowper's glands In the male reproductive system glands that secrete fluids into the urethra. p. 689

cranial nerves Peripheral nerves that arise from neurons in the brainstem. p. 221

creatine phosphate A high-energy compound that "buffers" ATP in muscle. p. 308

cretinism Mental retardation caused by thyroid hormone deficiency during early development. p. 678

crista The portion of the vestibular system that contains hair cell receptors. p. 265

cristae of mitochondria Foldings of the inner mitochondrial membrane. p. 82

critical period The period during the development of the visual system when visual signals seem to govern formation of neural connections. p. 285

crossbridge The "heads" of the myosin molecules that interact with the actin thin filaments and generate force. p. 292

crossbridge cycling The rate of formation and hydrolysis of the actinmyosin interactions that generate force in a muscle fiber. p. 297

crossed extension The reflex in which a noxious stimulus to one limb leads to oppositely directed movements of limbs on the two sides of the body. p. 338

crypt cells In the intestine, cells located at the base of the villous folds. They secrete water and ions. p. 599

cupula A gelatinous structure that occludes the semicircular canals. When displaced by rotations, it activates the hair cells. p. 265

curare A potent blocker of nicotinic cholinergic receptors such as are found at the neuromuscular junction. p. 350

current The flow of charge; current = voltage/resistance (Ohm Law). p. 36

current of injury The current that flows between a damaged and a normally polarized region of heart muscle.

cutaneous Pertaining to the skin. p. 231

cyclic AMP One of the cyclic nucleotides; a major intracellular "second messenger." p. 111

cycling rate See crossbridge cycling.

cytochromes Components of the mitochondrial electron transport chain. p. 7

cytoplasm All the material inside a cell other than the nucleus. p. 74

cytoplasmic receptor In endocrinology, a hormone-binding site in the cytoplasm rather than in either the cell membrane or nucleus. p. 102

cytoskeleton In cells, a network of tubules and filaments that acts as an internal structural "framework." p. 82

cytotoxins A subclass of T lymphocytes that are involved in cell-mediated immunity. p. 765

D

D cells Pancreatic endocrine cells that secrete somatostatin.

Dalton Law The Avogadro number expressed in terms of partial pressures. p. 480

Davenport diagram A graphical representation of the Henderson-Hasselbalch equation that relates pH, bicarbonate concentration, and CO_2 partial pressure to one another. It is used to evaluate acid-base balance. p. 580

dead space See anatomical dead space.

deamination A chemical reaction that involves removal of an NH_2 group. p. 94, 557

decarboxylation A chemical reaction that involves removal of a carboxyl group. p. 89

decerebration Section of the brainstem above the vestibular nucleus. p. 340

decerebrate rigidity Exaggeration of stretch reflexes characteristically seen in a decerebrate animal.

decibel A logarithmic unit of measure for sound intensity. p. 256

decomposition A condition in which movements are fragmented into their component parts as a result of injury to the basal ganglia. p. 341

decorticate Lacking a cortex. p. 342

decremental conduction The persistence of voltage changes for a short distance along a nerve or muscle fiber in the absence of propagated action potentials. p. 163

decussate To cross. p. 235

defibrillation A procedure for suddenly depolarizing the entire heart in an effort to restore a normal pattern of electrical activation. p. 369

dehydration Loss of body water with an increase in osmolarity. p. 565

dehydrogenation A chemical reaction that involves removal of a hydrogen atom. p. 58, 177

delta waves One of the intrinsic electrical rhythms of the brain seen during the deeper stages of sleep. p. 243

demyelination Loss of the myelin sheath that surrounds myelinated nerves. p. 183

denaturation The breaking of weak chemical bonds with a concomitant inactivation of proteins and nucleic acids. p. 57

dendrite Processes on a nerve cell that receive input from other neurons. p. 190

dense bodies Structures in smooth muscle that are functionally equivalent to the Z lines of striated muscle; sites where actin filaments attach. p. 316

deoxyribonucleic acid (DNA) One of the nucleic acids; it stores genetic information in its nucleotide sequence and can replicate itself. Regions of DNA can be transcribed into messenger RNA and then into specific proteins. p. 43

dependent variable In algebra, the variable whose value changes when other, independent variables are altered. It usually is seen on the left side of an algebraic equation.

dephosphorylation A chemical reaction that involves removal of a phosphate group. It often is a way second messengers affect enzyme activity. p. 58

depolarization A condition in which the membrane potential becomes internally less negative. p. 154, 162

dermatome The region of the body innervated by one spinal nerve. p. 231

desensitization The process by which permeability changes in a receptor disappear with prolonged exposure to chemical compounds such as neurotransmitters.

desmosome A "loose" junction that mechanically couples cells. p. 84

detoxification The conversion of drugs and other toxic compounds in the liver into organic anions that can be excreted by the kidneys. p. 617

deuteranopia A condition in which an individual lacks green-sensitive photoreceptors. p. 275

diabetes insipidus A clinical condition in which large amounts of relatively dilute urine are produced because of a lack of ADH regulation. p. 550

diabetes mellitus A clinical condition in which plasma and urine glucose levels are abnormally high, usually as a result of insulin deficiency. p. 121, 658

diaphragm The muscle that separates the thorax and the abdomen. Its contraction is the major factor in resting respiration. p. 450

diastole The phase of the cardiac cycle during which the ventricles are relaxed. It includes rapid and reduced filling. p. 362

diastolic depolarization In pacemaker cells, a term referring to the fact that between action potentials, the membrane potential gradually becomes internally less negative. The slope of diastolic depolarization determines the heart rate. p. 379

diastolic pressure The minimum pressure in the arteries reached just before blood is ejected from the ventricles. p. 363

dichromat An individual who has only two types of color-sensitive cones. p. 275

dicrotic notch The displacement seen in the aortic pressure pulse when the aortic valve closes; the incisura. p. 375

diencephalon The thalamus and hypothalamus. p. 223

diffusion Net movement of substances from areas of high to low concentration as a result of their inherent thermal energy. p. 13, 141

diffusion coefficient A measure of the rapidity with which a substance diffuses; a kind of molecular mobility. p. 142

digestion The degradation of proteins, fats, and carbohydrates into their constituent amino acids, fatty acids, and simple sugars. p. 625

digestive enzymes Enzymes capable of breaking down proteins, carbohydrates, and fats to amino acids, simple sugars, and fatty acids. p. 626

digestive system The organ system composed of the stomach, intestines, and associated glands. p. 12

digestive vesicle The result of fusion between a lysozyme and an endocytotic vesicle; the site of intracellular breakdown of foreign materials.

dihydroxyphenylalanine (DOPA) One of the major central neurotransmitters.

dilation An increase in the cross-sectional area of blood vessels or airways. p. 348

disaccharide A compound formed from two monosaccharides; table sugar. p. 51, 567

distal muscles Muscles involved in fine movements of the extremities. p. 343

distal tubule In the kidneys, the region of the nephron where ADH and aldosterone regulate water and Na^+ reabsorption. It is located between the ascending loop of Henle and the collecting duct. p. 531

disulfide bond A chemical bond between cysteine residues in proteins.

diuresis The production of large amounts of hypotonic urine. p. 659

diuretic A substance that greatly increases urine flow. p. 428

divergence The sending of processes by a neuron to many target neurons. p. 209

DNA See deoxyribonucleic acid.

DNA-dependent RNA polymerase The enzyme responsible for "reading" a sequence of bases on DNA to form complementary RNA. p. 66

Donnan equilibrium A term referring to the fact that impermeable, negatively charged proteins inside a cell tend to produce voltage and osmotic pressure differences across the cell membrane.

dopamine One of the major central neurotransmitters. p. 110, 204

dorsal column pathway The ascending somatosensory tract that carries signals from morphologically specialized cutaneous receptors and proprioceptors. It is the most "direct" sensory pathway. p. 235

dorsal horn The region of the spinal gray matter where the primary sensory afferents first synapse. p. 327

dorsal root ganglion The structure outside the central nervous system that contains the cell bodies of the peripheral spinal nerves. p. 190

double bond A covalent bond formed when each atom donates two electrons. p. 23

double Donnan equilibrium A condition in which the activity of the Na^+ pump effectively makes the cell membrane impermeable to Na^+. p. 124

down-regulation A decrease in receptor density caused by the binding of a hormone, neurotransmitter, or neuromodulator to a cell. p. 115, 660

dual innervation In the autonomic nervous system, the reception by many organs of both sympathetic and parasympathetic input. p. 329

duodenum The most proximal region of the small intestine. p. 597

dwarfism Stunted growth caused by a deficiency of growth hormone. p. 673

dye dilution A procedure in which a dye injected into the circulation is used to determine cardiac output.

dynamic (gamma) motor neuron Motor neurons that innervate the central region of the muscle spindle stretch receptor. p. 334

E

echocardiography A procedure in which high frequency sound waves are used to examine the heart's pumping action.

ectopic pacemaker In the heart, usually a region that spontaneously discharges, even though it is not the normal or a reserve pacemaker; this often occurs in the ventricular muscle. p. 368

edema Accumulation of fluid in the interstitial compartment. p. 416

effectors A general term referring to striated muscle, visceral smooth muscle, or glands; the "motor" element of feedback systems. p. 98

efferent A term referring to nerve fibers leading from the central nervous system to muscles or glands. p. 106

efferent arterioles In the kidneys, the arterioles located between the glomerular and peritubular capillary beds. p. 532

efficiency In thermodynamics, the ratio of work done to heat consumed.

Einthoven triangle In cardiology, the geometrical arrangement of leads I, II, and III of an electrocardiogram. p. 371

ejaculation A coordinated series of contractions of the male reproductive system that expels semen. p. 720

elastic recoil The tendency of the lung (or any object) to shrink to lower volumes and of the chest wall to expand. p. 404

elastic restoring forces See elastic recoil; to be contrasted with externally applied forces. p. 40

elastic work The component of the work of breathing needed to overcome the elastic recoil of the lungs; to be contrasted to flow-resistive work. p. 462

electrical field The imaginary field set up by charged particles that determines the force a small test charge would experience.

electrical potential The nondirectional energy "well" or "hill" set up by charged particles.

electrical synapses Junctions between cells that involve direct connections between the two cytoplasmic compartments. p. 200

electrically excitable A term referring to cells capable of generating propagated, all-or-nothing action potentials. p. 165

electrically inexcitable A term referring to cells that undergo no voltage changes or only local and generally graded ones. p. 165

electrocardiogram (ECG) The extracellular signal produced by electrical activation of the heart. p. 37, 369

electrochemical gradient A term referring to the fact that movement of charged substances across cell membranes depends on both the concentration gradient and the electrical field.

electroencephalogram (EEG) The extracellular signal produced by electrical activation of the cerebral cortex. p. 38, 243

electrogenic pump A term referring to the fact that the Na^+-K^+ ATPase (sodium pump) transports more Na^+ ions than K^+ ions.

electrolyte A solution composed of ions. p. 27

electron The negatively charged particle that orbits the nucleus. p. 20

electron acceptor An atom that tends to acquire extra electrons. p. 23

electron donor An atom that tends to give up electrons. p. 23

electron shell A term referring to the fact that electrons occupy a set of fixed orbits at varying distances from the nucleus. p. 40

embryoblast The portion of the blastocyst that develops into a fetus.

emphysema A disease characterized by a loss of lung tissue, resulting in an increase in lung compliance and the collapse of airways during expiration.

emulsification The process by which water-insoluble fats form microscopic micelles in solution. p. 633

end-diastolic volume The maximum volume of a ventricle just before contraction. p. 363

end-systolic volume The minimum volume of a ventricle just after contraction. p. 363

endocrine reflex A reflex that involves glandular secretion rather than contraction or relaxation of muscles. p. 108

endocrine system The collection of ductless glands that secrete chemical messengers, called hormones, directly into the blood. p. 12

endocytosis The process by which a cell takes up extracellular materials by "pinching off" a vesicle from its membrane. p. 76

endogenous pyrogen A fever-causing substance produced by the body. p. 755

endolymph The fluid in the semicircular canals and a large part of the cochleats composition resembles that of extracellular fluid. p. 259

endometrium The internal layer of cells in the uterus. p. 688

endoplasmic reticulum (ER) An intracellular organelle consisting of large, flattened sheets. Rough ER has ribosomes embedded in its surface. p. 78

endorphins A neuroactive peptide (endogenous opiate) that suppresses pain. p. 204

endotoxin A substance released by bacteria that increases capillary permeability. p. 755

endplate potential (EPP) The voltage change produced at the neuromuscular junction as a result of transmitter release. p. 141

enkephalin One of the neuroactive peptides used as an inhibitory presynaptic transmitter in afferent pain fibers in the dorsal horn of the spinal cord. p. 204, 214

enteric nervous system The network of nerve cells and fibers intrinsic to the gastrointestinal tract. p. 597

enterogastric reflex A gastrointestinal reflex in which increased duodenal tension decreases gastric motility and increases pyloric tone. p. 647

enterohepatic circulation The continual recirculation of the bile between the liver and the gastrointestinal tract.

enterokinase An intestinal enzyme that activates pancreatic trypsinogen. p. 617

enthalpy In thermodynamics, a measure of heat content.

entropy In thermodynamics, a measure of "randomness." The lower the entropy, the more ordered a system. In all spontaneous processes, the total entropy of the universe increases. p. 38

enzymes The substratespecific protein catalysts for biochemical reactions. Enzymes reduce activation energy. p. 59, 139

eosinophils White blood cells that protect against parasites. p. 752

epididymis The male reproductive tract between the sites of sperm production (seminiferous tubules) and the vas deferens. p. 689

epimysium The connective tissue surrounding a muscle. p. 292

epinephrine The catecholamine released by the adrenal medulla. Epinepherine preferentially activates β_2-adrenergic receptors. p. 110, 118, 204

epiphyseal plate The growing region of the bones. p. 617

epithelial cells One of the four basic cell types. They line all body cavities, form the skin, and comprise most glands. p. 10

equilibrium A state in which all a system's variables are constant. p. 38, 141

equilibrium constant In the Law of Mass Action, a measure of the degree to which a chemical reaction takes place. It relates equilibrium concentrations of reactants and products to one another. p. 588

equilibrium potential The voltage that would exist across a membrane if it were exclusively permeable to one ion. p. 152

erection Swelling of the penis caused by its engorgement with blood. p. 719

error signal In negative feedback systems, the difference between the actual value of a variable and its ideal "set-point." p. 98

erythrocyte A red blood cell. p. 392

erythropoeitin A hormone that regulates production of red blood cells. p. 393

esophageal hiatus The opening from the esophagus to the stomach. p. 611

esophageal phase The phase of swallowing during which esophageal distension causes secondary waves of peristalsis. p. 608

esophagus The long tube that connects the pharynx with the stomach. p. 595

essential amino acid Amino acids that the body cannot synthesize. p. 54, 662

estrogen A female sex steroid secreted by the ovaries. p. 110

evoked potential Electrical changes recorded from the sensory and motor areas of the cortex that correspond to sensations and movements. p. 343

excitability The ease with which a nerve or muscle cell can be made to generate an action potential. The lower the threshold, the more excitable the cell. p. 173

excitation-contraction coupling The process by which an action potential in a muscle activates the muscle's contractile "machinery." p. 300

excitatory phase The initial stage of the human sexual response. p. 717

excitatory postsynaptic potential (EPSP) The depolarization observed at a synapse when a neurotransmitter increases the Na^+ permeability. p. 209

exocrine gland A gland that secretes substances through ducts; to be contrasted to endocrine gland. p. 10, 98

exocytosis The process by which a cell extrudes some of its contents enclosed within a secretory vesicle. p. 76

exogenous pyrogen A fever-causing substance released by bacteria. p. 755

exons Regions of DNA used as codes for protein synthesis; to be contrasted to noncoding regions that must be removed before transcription to RNA. p. 67

expiration The phase of respiration during which air leaves the lungs. p. 501

expiratory muscles Muscles of the chest wall that normally are not activated during passive expiration but that are recruited in forced expiration.

expiratory reserve volume (ERV) The difference between the volume of air in the lungs at the end of a passive expiration and the residual volume of air that cannot be expelled by active effort. p. 458

expressive aphasia A condition in which an individual cannot form gramatically correct sentences or phrases because of damage in the Broca area. p. 242

extensor A muscle that serves to extend a joint. p. 292

extracellular fluid (compartment) Fluid outside the cells. p. 13, 564

extrafusal fiber The "normal" fibers of a muscle. See intrafusal fibers. p. 334

extrapyramidal An older term referring to the descending motor tracts that do not pass through the medullary pyramids. p. 346

extrasystole A condition in which ventricular contraction is not preceded by atrial activation. It is felt as a "skipped beat" because the next heartbeat usually is delayed.

external anal sphincter The most distal anal sphincter, composed of both voluntary and smooth muscle. p. 597

extrinsic pathway Formation of a blood clot in the presence of damaged tissue by a process involving factor III rather than activation of Hageman factor. p. 395

extrinsic protein A protein associated with the internal or external surface of a membrane through weak electrostatic forces. p. 138

extrinsic regulation Control of muscles and glands by the nervous or endocrine systems. p. 103

F

facilitated diffusion Carrier-mediated diffusion; the translocation of substances down their concentration gradients via specialized pathways. p. 146

facilitation In synaptic transmission, an increase in the quantal content either during a train of nerve impulses or as a result of some other presynaptic process. p. 338

fallopian tubes The tubes that connect the ovaries with the uterus. p. 687

fascicles Small groups of muscle fibers; an anatomical rather than functional subunit of a whole muscle. p. 292

fast pain, first pain, or "pricking" pain The component of pain that is relayed via the neospinothalamic tract and that can be well localized. p. 234

fast glycolytic muscle fibers One of the three types of muscle fibers characterized by rapid contraction and susceptibility to fatigue. p. 308

fast oxidative muscle fibers Muscle fibers characterized by moderately rapid contraction, abundant mitochondria, and moderate resistance to fatigue. p. 308

fat-soluble vitamins The water-insoluble vitamins (A, D, E, and K). p. 639

fatigue Tiredness. In muscle, the decrease in contractile force with prolonged use. p. 312

fats Triglycerides. p. 44

fatty acids A hydrocarbon chain with a carboxyl group at one end If the hydrocarbon chain has one or more double bonds, the fatty acid is referred to as unsaturated. p. 44, 632

feature extraction The way in which the visual system resolves the outlines and movements of an object. p. 227

feed-forward regulation In control theory, changes that anticipate fluctuations in the controlled (regulated) variable.

fenestrations Large pores in the walls of capillaries in vascular beds such as the kidneys and some endocrine glands. p. 537

ferritin An iron-binding protein in cells lining the intestine. p. 638

fertilization The union of a sperm and an ovum. p. 726

fetal hemoglobin A form of hemoglobin characterized by an increased ability to bind oxygen at low partial pressure. p. 490

fiber types In muscle, the difference between the speed of contraction and

metabolic pathways used by a fiber. p. 308

fibrillation In the heart, disorganized electrical activity in the atria or ventricles that cannot produce a normal contraction. p. 369

fibrin The fibrous components of a blood clot. p. 394, 395

fibrinogen The plasma protein that polymerizes to form fibrin. p. 392

fibrosis In the lungs, a condition involving a decrease in lung compliance. p. 454

Fick equation The relationship between the net flux of a substance and its concentration gradient; the diffusion equation. p. 428

filtered load In the kidneys, the amount of a substance that is filtered into the glomerulus. It is expressed in milligrams per minute. p. 542

filtration See capillary filtration.

filtration fraction In the kidneys, the fraction of the renal plasma flow that is filtered into the glomerulus. p. 538

finger proteins Proteins coded by the Y chromosome that control transcription of specific genes.

first heart sound The sound associated with closing of the A-V valves. p. 363

First Law of Thermodynamics A statement of the principle of conservation of energy. p. 38

fixed receptor In endocrinology, a plasma membrane—associated receptor. p. 111

flavin adenine dinucleotide (FAD) One of the electron-carrying coenzymes. p. 87

flexion reflex The withdrawal of a limb from a noxious stimulus.

flexor A muscle that bends a joint. p. 292

flow Movement of a fluid, either liquid or gas. p. 33, 405

flow autoregulation See autoregulation.

flow resistance The result of internal frictional forces in a liquid. Flow is equal to a pressure gradient divided by the resistance. p. 406, 407

flow-resistive work In the lungs, the component of the work of breathing needed to overcome frictional forces in the airways. p. 462

flower-spray endings The afferent fibers that innervate the peripheral regions of the muscle spindle.

fluid mosaic model A conceptualization of a membrane in which intrinsic proteins move freely in a two-dimensional sea of lipid. p. 138

flux The rate of movement of materials across an interface. See **unidirectional flux** and **net flux.** p. 141

folic acid Vitamin B_{12}, required for iron absorption in the gastrointestinal tract. Its absence causes pernicious anemia. p. 640

follicle In the ovaries, the structure containing the ovum, granulosa cells, follicular cells, and thecal cells. One follicle matures and is ovulated during each menstrual cycle. p. 693

follicle-stimulating hormone (FSH) An anterior pituitary hormone required for germ cell (sperm and ovum) maturation. p. 116, 695

follicular phase The phase of the menstrual cycle that precedes ovulation. p. 704

fovea The area of the retina that contains the cone photoreceptors. p. 277

Frank-Starling Law The principle that the length-tension relation of cardiac muscle causes cardiac output to be a function of venous return. p. 326

free energy The energy of a system that is available to do "useful" work. p. 39

free nerve endings Sensory nerve terminals not characterized by any overt structural specialization which, nevertheless, are sensitive to particular modalities such as touch, temperature, or pain. p. 231

frontal lobe The most anterior area of the cerebral cortex; it is involved in motor control and emotional behavior. p. 224

fructose A 5-carbon monosaccharide bound to glucose in table sugar. p. 50

functional residual capacity (FRC) the volume of air normally remaining in the lungs after a passive rather than forced expiration. p. 458

functional residual volume (FRV) The difference between the functional residual capacity and the residual volume attatined in a maximum expiratory effort. p. 462

functional syncytium A group of cells that function as a unit because of electrical connections between them. For example the heart and unitary smooth muscle.

G

G cells Gastrin-secreting cells in the stomach. p. 611

G-protein A membrane-bound protein that serves as an intermediate between hormone or neurotransmitter binding to receptors and activation of adenyl cyclase. p. 111

gain In control theory, a measure of the sensitivity of a negative feedback system. More generally, it is equivalent to amplification. p. 100

galactose A monosaccharide linked to glucose to form lactose (milk sugar). p. 50, 629

gallbladder The organ that stores the bile produced by the liver. p. 614

gamete Reproductive cells; an ovum or sperm. p. 690

γ-aminobutyric acid (GABA) One of the central neurotransmitters. p. 204

gamma motor neurons Motor neurons that activate muscle spindle intrafusal fibers; to be contrasted to alpha motor neurons. p. 334

ganglion A tightly packed group of neurons outside the central nervous system. p. 222

ganglion cells In the retina, the neurons whose axons form the optic nerve. p. 272

gap junction A junction between cells that couples them electrically. p. 84

gas A phase of matter in which molecules do not interact. p. 481

gastric inhibitory peptide One of the intrinsic gastrointestinal hormones. p. 647

gastric juice The acid and enzyme secretions of the stomach.

gastric motility Contractions of the smooth muscle of the stomach, including segmentation and peristalsis. p. 643

gastric phase In digestion, the phase in which chyme in the stomach initiates a series of local neural and hormonal reflexes. p. 643

gastrin A gastrointestinal hormone that increases gastric secretion. p. 643

gastrocolic reflex A reflex that involves an increase in the large intestine's motility during the gastric phase of gastric secretion. p. 650

gastroileal reflex An increase in activity in the small intestine (ileum) that occurs during gastric emptying. p. 650

gastrointestinal tract The esophagus, stomach, and intestines. p. 597

gated ion channel Water-filled pores in the cell membrane selected for certain ions. Opening and closing of such channels may be regulated by the electrical field, by molecules such as hormones and neurotransmitters, or by physical stimuli. p. 148, 170

generator potential A receptor potential; the change in voltage that occurs when a sensory receptor is activated by a stimulus. p. 195

Gibbs-Donnan equilibrium See Donnan equilibrium.

gigantism An abnormal increase in body size caused by excess growth hormone. p. 673

gland A collection of cells specialized for secretion. Glands may have ducts (exocrine) or may secrete into the blood (endocrine). p. 125

glaucoma A condition in which intraocular pressure increases. p. 269

glia In the nervous system, the non-excitable satellite cells that surround axons. They may or may not make myelin. See **Schwann cells** and **oligodendrocytes**. p. 179

globus pallidus One of the basal ganglia. p. 341

glomerular filtration In the kidneys, the process by which plasma is filtered into the proximal region of the nephrons. p. 531

glomerular filtration rate The rate at which plasma is filtered into the glomerulus. It usually is expressed in milliliters per minute. p. 538

glomerulotubular balance In the kidney, the tendency of the proximal tubule to reabsorb a constant fraction of its filtered load.

glomerulus In the kidney, the most proximal portion of the nephron, composed of the glomerular capillaries and the Bowman capsule. p. 530

glucagon A hormone secreted by the pancreas that promotes mobilization of glucose from body stores. p. 118, 663

glucocorticoids Steroids secreted by the adrenal cortex that affect organic metabolism; includes cortisol. p. 667

gluconeogenesis The formation of glucose from amino acids. p. 656

glucoprivation Long-term depletion of carbohydrates. p. 612

glucose A monosaccharide; it is degraded in glycolysis to produce ATP. p. 71, 629

glucose-sparing A term denoting the use of fatty acids and ketones in metabolism. p. 663

glutamate One of the central neurotransmitters. p. 204

glycine One of the central neurotransmitters. p. 204

glycogen The storage form of glucose; a highly branched polymer. p. 52, 309

glycogen synthetase The enzyme that converts glucose phosphate to glycogen. p. 119, 660

glycogenolysis The breakdown of glycogen into glucose. p. 656

glycolipid A molecule composed of a lipid and a sugar residue; a component of many cell membranes. p. 44, 45

glycolysis A metabolic pathway in which glucose is degraded to pyruvate. p. 87

glycoprotein The combination of a protein with carbohydrate residues. p. 602

glycosuria The presence of glucose in the urine. p. 658

goblet cell A secretory cell in the intestinal epithelium. p. 602

Goldman equation An equation that relates the membrane potential to the concentration gradients and relative permeabilities of all ions capable of crossing the cell membrane. p. 153

Golgi apparatus The intracellular organelle that "packages" proteins synthesized in the rough endoplasmic reticulum into secretory vesicles. p. 81

Golgi tendon organs Stretch receptors in tendons that measure muscle tension. p. 337

gonadotropic hormone A generic term that includes the anterior pituitary hormones FSH and LH.

gonadotropin releasing hormone (GnRH) The hypothalamic releasing factor that controls the secretion of FSH and LH from the anterior pituitary. p. 695

gonads The ovaries and testes. p. 687

graded In receptor physiology, a term denoting that generator and synaptic potentials vary continuously in amplitude and do not have a threshold. p. 163

gram (g) The unit of mass.

gramicidin An antibiotic that forms a transmembrane channel in lipid bilayers. p. 150

granular neuron A neuron with an extensive set of highly branched dendrites.

granulosa cells Cells in the ovary that nurture the ovum. p. 703

gray matter The areas in the central nervous system that contain nerve cell bodies. p. 221

gray rami Afferent connectives between the sympathetic chain and the main part of the spinal cord. p. 329

green-sensitive cones Photoreceptors with maximum absorbance at 530 nanometers. p. 272

growth hormone (GH) An anterior pituitary hormone that has widespread effects on growth, development, and organic metabolism. p. 105, 110, 114, 116

guanine One of the five nucleotides in RNA and DNA. p. 62

H

H zone The region of the thick filament that is free of crossbridges. It corresponds to the "tail" regions of the myosin molecules. p. 292

habituation A form of nonassociative learning present in invertebrates that involves reactions to repetition of a single stimulus.

Hageman factor A plasma protein; its activation is the initial event in the formation of a blood clot. p. 394

hair cells Ciliated sensory receptors in the cochlea and vestibular apparatus. p. 258

hair follicle receptors One of the somatosensory touch receptors. p. 233

Haldane effect The effect of oxygen on the dissociation curve for carbon dioxide. p. 493

hapten A small molecule that can serve as an antigen if it becomes chemically bound to a larger molecule. p. 758

haustra Segmentation movements in the large intestine. p. 621

heart The organ that pumps blood in the circulatory system. p. 357

heart failure The inability of the heart to maintain an adequate cardiac output. p. 432

heart rate The number of times the heart contracts per minute. p. 362

heart sounds Sounds produced during the cardiac cycle by the closing of valves or by vibrations caused by rapid blood flow. p. 363

heat The transfer of energy resulting from differences in temperature. p. 40

heat stroke A condition in which the core body temperature is abnormally high. p. 442

helicotrema The tiny opening at the distal end of the cochlea; it connects the scala vestibuli and the scala tympani. p. 259

helix A three-dimensional, coiled structure that resembles a spring. p. 48

helper T cells A class of T lymphocytes that increase the activity of B cells and other T-lymphocyte subpopulations. p. 765

hematocrit The fraction of the blood occupied by red blood cells. p. 392

heme The component of hemoglobin to which oxygen molecules bind. p. 393

hemicolinium A drug that blocks choline uptake into cholinergic nerve terminals. p. 351

hemodynamic Having to do with the flow of blood. p. 359

hemoglobin (Hb) The oxygen carrying protein in the blood. p. 485

hemorrhage Loss of blood volume through bleeding. p. 569

hemostasis The process of blood clot formation. p. 394

Henderson-Hasselbalch equation The Law of Mass Action applied to the reaction of CO_2 with water to form bicarbonate and H^+. It describes the interrelationships between plasma pH, bicarbonate concentration, and the partial pressure of CO_2. p. 588

Henle loop In the kidney, the nephron between the proximal and distal tubules responsible for the concentration of the medullary interstitium. p. 531

heparin A substance produced in the lungs that inhibits blood clotting. p. 395

hepatic circulation The flow of blood in the liver. p. 358

hepatocytes Liver cells arranged in a series of sheets. p. 628

hexamethonium A drug that blocks nicotinic cholinergic receptors. p. 350

hiatal hernia Bulging of the wall of the stomach into the thorax. p. 611

high-density lipoprotein (HDL) A lipoprotein that facilitates lipid uptake into chylomicrons and very-low-density lipoproteins (VLDLs). p. 662

high-renin hypertension Hypertension caused by renal arteriolar atherosclerosis. p. 572

hippocampus An "older" region of the cortex involved in initial processing of memories and in emotional behavior. p. 224

histamine A substance released by white blood cells that increases capillary permeability. p. 105

histocompatibility antigens Markers on the surface of cells that distinguish "self" from "nonself." p. 778

homeostasis The constancy of the "internal environment." p. 12

homologous chromosome One of the 23 pairs of nonsex chromosomes. p. 690

homunculus The often exaggerated representation of the body surface onto the somatosensory or motor cortex. p. 341

horizontal cells Neurons that couple adjacent photoreceptors in the retina. p. 272

hormone A chemical "messenger" secreted by endocrine glands. p. 10, 98, 103

human chorionic gonadotropin (HCG) A placental hormone that maintains the corpus luteum during pregnancy. p. 730

humoral immunity The immune system response in which B lymphocytes secrete specific antibodies. p. 759

Huntington disease A disease of the central nervous system characterized by uncontrolled limb movements.

hyaline membrane disease See respiratory distress syndrome.

hydration (shell) The "surrounding" of a dissolved ion by water molecules. p. 27, 143

hydraulic filter In the cardiovascular system, the way in which the arteries convert the pulsatile output of the heart into steady flow.

hydrocarbon A molecule composed only of carbon and hydrogen. p. 40

hydrochloric acid (HCl) A strong acid secreted by the stomach. p. 611

hydrogen The lightest element. p. 23

hydrogen bond A chemical bond created between two molecules or different parts of the same molecule by sharing of an electron from hydrogen with an acceptor. p. 24

hydrogen ion (H^+) A hydrogen atom stripped of its electron. p. 27

hydrogenation A chemical reaction that involves addition of a hydrogen atom.

hydrolysis The breaking of a chemical bond by the addition of water. p. 626

hydrophilic "Water loving"; refers to the attraction between polar groups of compounds and water molecules. p. 24, 29

hydrophobic "Water hating"; refers to the tendency for nonpolar compounds to interact with each other in such a way that water is excluded. p. 24, 30

hydrostatic force The force exerted by a column of liquid or gas. p. 415

hypercalcemia A condition in which the plasma Ca^{++} is abnormally high. p. 576

hypereffective (hypoeffective) In cardiovascular physiology, hypereffective indicates an increase in cardiac contractility; hypoeffective, a decrease.

hyperplasia An increase in the number of cells in a tissue.

hyperpolarization A condition in which the membrane potential of a cell becomes internally more negative. p. 154, 162

hypertension Abnormally high arterial blood pressure. p. 427

hypertonic solution A solution in which a cell will shrink. p. 144

hypertrophy An increase in the size of a cell such as muscle. p. 312

hyperventilation An increase in alveolar ventilation that causes the alveolar Pco_2 to fall below 40 mm Hg. p. 410

hypervolemia A condition in which total body water is abnormally high. p. 427, 567

hypocalcemia A condition in which plasma Ca^{++} is abnormally low. p. 174, 576

hypoglycemia Condition in which plasma glucose is abnormally low. p. 658

hyposmotic A solution with an osmotic pressure lower than that inside cells. p. 143

hypotension Abnormally low arterial blood pressure. p. 436

hypothalamic-releasing factors Peptides released from terminals of the hypothalamic neurons that control secretion of anterior pituitary hormones.

hypothalamohypophysial portal system The network of blood vessels connecting the hypothalamus and the anterior pituitary gland that carries the releasing factors responsible for pituitary secretion. p. 358

hypothalamus The region of the brain that controls the release of hormones from the pituitary gland. p. 223

hypotonic solution A solution in which cells will swell. p. 144

hypovolemia A condition in which total body water is abnormally low. p. 436, 567

hypoxemia A condition in which the arterial P_{O_2} is abnormally low.

hypoxia Lack of oxygen. p. 490

I

I band The region of a muscle sarcomere that contains thin filaments but not thick filaments. In the sliding filament model, the region of nonoverlap. p. 292

ileocecal value The sphincter between the small and large intestines. p. 597, 621

ileogastric reflex The decrease in gastric motility caused by distension of the ileum.

ileum The most distal region of the small intestine; it empties into the cecum. p. 597

immune system The organ system that "recognizes" and attacks "foreign" materials, including bacteria and viruses. p. 12

immunoglobulins Antibodies Protein molecules in which one end is specialized for recognition of "foreign" molecules (antigens) and the other end activates components of the immune response. p. 392

implantation Insertion of a fertilized ovum into the uterine wall. p. 729

incisura See dicrotic notch.

incomplete heart block A condition in which not all atrial contractions lead to ventricular activation because the AV node has a limited ability to pass impulses at high frequency.

incus One of the bones of the middle ear. p. 257

inert A term referring to elements whose outer electron shell contains its full complement of electrons and which cannot combine with other atoms to form compounds.

infarct A damaged region of the heart muscle. p. 369

inflammation Swelling; the term also refers to nonspecific immune responses. p. 105

inhibitory postsynaptic potential (IPSP) At a synapse, the change in membrane potential produced by a neurotransmitter that increases the K^+ permeability. p. 209

inner ear The cochlea. p. 256

inner segment The part of rod and cone photoreceptors that contains the nucleus of the cell and synaptic vesicles. p. 272

innervation ratio The number of muscle fibers innervated by a motor neuron.

inotropic effect Causing an increase in cardiac contractility. p. 379

inorganic Compounds that do not contain carbon.

insensible water loss Water lost through normal respiration and evaporation. p. 565

insertion One of the connections of a muscle to the skeleton. Contraction pulls the insertion toward the origin. p. 292

inspiration The phase of respiration during which air enters the lungs. p. 501

inspiratory capacity (IC) The volume that can be inhaled after passive expiration. The difference between total lung capacity and functional residual capacity. p. 458

inspiratory muscles The muscles of the chest wall that can be activated to increase the tidal volume. p. 501

inspiratory reserve volume (IRV) The volume of air that can be inhaled beyond the resting tidal volume. p. 458

insufficiency A term denoting the "leakiness" of a heart valve. p. 381

insulin A hormone secreted by the pancreas that is necessary for glucose utilization in most tissues. p. 118, 659

insulin-dependent diabetes Excess blood sugar caused by decreased secretion of insulin. p. 659

insulin-independent diabetes Excess blood sugar caused by a decrease in the density of insulin receptors. p. 660

integrator (integration center) The collection of neurons in the spinal cord between afferent sensory fibers and motor fibers. p. 98, 106

intention tremor Shaking seen when an individual tries to reach for an object.

intercalated disks Structures in cardiac muscle that electrically couple adjacent cells. p. 313

interleukins Molecules released during the immune response that activate other cells of the immune system and act as endogenous pyrogens. p. 755

intermediate filaments The most stable elements of the cytoskeleton, thought to provide the basic anatomical structure for a cell. p. 83

intermediate zone In the spinal cord, the gray matter between the dorsal and ventral horns; it contains spinal interneurons.

internal anal sphincter The more proximal anal sphincter, composed of involuntary smooth muscle. p. 597

internal energy In thermodynamics, the energy "within" a system.

interneuron The neurons in the spinal cord that are not motor neurons. p. 106, 108, 221

interstitial fluid The fluid in the spaces between cells. p. 13

intestinal phase In digestion, the neural and hormonal reflexes initiated by entry of chyme into the duodenum. p. 643

intestine A generic term for the small and large bowel. p. 597

intracellular compartment The fluid inside the cells. p. 564

intrafusal fibers The muscle fibers inside the spindle controlled by spinal gamma motor neurons. These fibers regulate spindle sensitivity. p. 334

intrapleural pressure The pressure in the intrapleural space.

intrapleural space The fluid-filled space between the lungs and chest wall. p. 450

intrinsic factor A glycoprotein required for vitamin B_{12} absorbtion. p. 611

intrinsic pathway Formation of a blood clot by activation of Hageman factor. p. 395

intrinsic proteins Proteins tightly embedded in cell membranes, often extending entirely across them to form receptors or channels. p. 138

intron A noncoding region of DNA. See exons. p. 67

inulin A plant polysaccharide. p. 554

inulin clearance A measure of the rate of glomerular filtration. p. 554

ion A charged atom. p. 24

ionic bond A chemical bond formed by donation and acceptance of electrons rather than by sharing. p. 24

ionophore A molecule that can carry ions across membranes.

ipsilateral On the same side of the body.

iris The pigmented area surrounding the pupil of the eye. p. 268

iron deficiency anemia A decrease in the hemoglobin content of the blood, caused by a lack of iron in the diet. p. 393

ischemia Inadequate blood supply to a tissue.

islets of Langerhans The regions of the pancreas that contain the cells that secrete insulin and glucagon. p. 118

isometric contraction Generation of force by a muscle under conditions that prevent the muscle's length from changing. p. 304

isosmotic solution A solution with an osmotic pressure equal to that of a cell. p. 143

isotonic contraction Muscle shortening under conditions of constant load. p. 304

isotonic solution A solution in which the volume of a cell will not change. p. 144

isotopes Elements in which the number of protons in the nucleus is constant, but the number of neutrons varies. p. 20

isovolumetric contraction In the cardiac cycle, contraction of the ventricle under conditions of constant volume; an isometric contraction. p. 363

isovolumetric relaxation In the cardiac cycle, relaxation of the ventricle under conditions of constant volume. p. 363

J

jejunum The "middle" region of the small intestine between the duodenum and the ileum. p. 597

joint receptors Sensory receptors in the joint capsules that provide information about limb position and movement.

junctions See desmosome, gap junction, and tight junction.

juxtaglomerular apparatus A specialized set of cells in the kidneys that release renin when renal arterial pressure decreases. p. 537

juxtamedullary nephrons Nephrons whose glomeruli are near the medulla and whose Henle loops penetrate the inner medulla. p. 531

K

K cells Cells in the intestine that secrete gastric inhibitory peptide (GIP).

kallikrein The enzyme that generates bradykinin from its precursor. p. 754

Kerckring folds See plicae circulares.

ketoacidosis Acidosis caused by accumulation of ketones. p. 585

ketone bodies Small molecules that are produced during the breakdown of fatty acids. p. 658

kidneys The principal excretory organs. p. 528

kinetic energy The energy an object has as a result of mass and velocity.

kinocilium The cilium in hair cells that determines directional sensitivity. p. 264

Korotkoff sounds The sounds heard through a stethoscope placed on a peripheral vein; they are caused by the pulsations of the blood through a partially inflated pressure cuff. p. 408

Krebs cycle A metabolic pathway in which 2-carbon acetyl groups are degraded to CO_2 and water. It occurs in the cental region of mitochondria. p. 89

L

labia minora (majora) Skin folds covering the outer opening of the vagina. p. 688

labelled line In sensory physiology, the ordered way in which afferent fibers enter the central nervous system. p. 193

labyrinth The inner ear; the cochlea plus the vestibular apparatus.

lacteals Lymph vessels in the intestinal villi. p. 599

lactic acid The end result of glucose metabolism in the absence of oxygen.

lactose Disaccharide of glucose and galactose; milk sugar. p. 629

lactose intolerance A condition in which the enzyme needed to degrade lactose is absent. p. 51

lamina propria The component of the gastrointestinal mucosa that contains glands. p. 597

laminar flow Nonturbulent or steady flow. p. 405

language areas The regions of the cerebral cortex that are specialized for the decoding and encoding of verbal content. See Broca's area and Wernicke's area. p. 242

LaPlace Law The relationship between the tension in the wall of a sphere or cylinder and the transmural pressure: $(P = T/2r)$. p. 461

large intestine The most distal region of the gastrointestinal tract. p. 597

larynx The region of the airway that contains the vocal cords. p. 450

latency In muscle, the time between electrical activation and contraction. In spinal reflexes the time between a stimulus and a motor response. p. 335

lateral geniculate One of the brainstem relay stations in the visual system. p. 283

lateral inhibition In sensory physiology, the tendency of input to adjacent regions of a sensory surface to be mutually inhibitive. p. 235

lateral reticular facilitory area The region in the brainstem that contains neurons that enhance spinal reflexes.

lateral vestibulospinal tract One of the descending motor pathways.

Law of Mass Action The principle that increasing the concentration of a reactant or product proportionately increases the unidirectional reaction rate. p. 39

leaky epithelia Epithelia characterized by a high water permeability. They are found where osmotically driven water flow is important physiologically. p. 85

learning The ability to modify behavior because of previous experience. p. 237

lecithin A phospholipid; a component of membranes and the bile. p. 617

lens The transparent portion of the eye that focuses light rays on the retina. p. 268

lemniscal system The major ascending sensory pathway, which carries proprioceptive information and fine touch. p. 235

length constant The distance over which currents applied to a nerve fiber will dissipate by a given amount, usually a few millimeters.

leukocytes White blood cells. p. 392, 752

luteinizing hormone (LH) The anterior pituitary hormone needed for secretion of estrogen and testosterone by the ovaries and testes respectively. LH also controls ovulation. p. 708

Leydig cells Cells in the male reproductive system that secrete testosterone. p. 689

Lieberkuhn crypts Depressions in the gastrointestinal tract in which new epithelial cells are formed by division. p. 599

limb leads The electrode connections to the arms and legs that are used to record an electrocardiogram. p. 371

limbic system An older region of the cerebrum that is involved in emotional behavior. p. 223

linear velocity of flow The rate of movement of an element of fluid past a given point; to be contrasted to volume flow and average flow velocity. p. 405

lipases Enzymes that degrade triglycerides into fatty acids. p. 634

lipid bilayer The basic "framework" of cell membranes. Bilayers consist of aggregated phospholipids whose hydrocarbon tails interact with one another. p. 138

lipids Compounds composed largely of hydrocarbons that usually are not water soluble. See fatty acids, phospholipids, steroids, and triglyceride. p. 44

lipogenesis The synthesis of fatty acids. p. 656

lipolysis The breakdown of fatty acids into 2-carbon acetyl groups. p. 634

lipoprotein A combination of proteins with lipid residues. p. 44, 661

lipoprotein lipase An enzyme on endothelial cell membranes that liberates free fatty acids from triglycerides. p. 662

liquid crystal A term denoting the structures (bilayers and multilamellar vesicles) typical of phospholipid and other amphipathic molecules in solution.

liter (L) The unit of volume.

liver The organ that acts as a chemical "reprocessing plant."It can interconvert carbohydrates, amino acids, and lipids. p. 597

load See preload or afterload.

lobules Functional subunits of the liver. p. 628

logarithm The power to which the base (usually 10) must be raised to obtain a given number. Logarithms are never negative. p. 196

long-term memory Memories that cannot be electrically disrupted and are presumably the result of some permanent alteration in neuronal function. p. 240

longitudinal muscle One of the muscle layers in the gastrointestinal tract.

lower esophageal sphincter The physiological sphincter that separates the esophagus from the stomach. p. 608

lumen The space within a tissue or organ; also, the space between membranes of the endoplasmic reticulum.

lungs The site of blood-atmospheric gas exchange. p. 450

luteal phase In the menstrual cycle, the phase immediately after ovulation and before menstruation. p. 707

lymph nodes Areas along the lymphatic vessels that contain lymphocytes. p. 416

lymphatic duct The lymph vessel that receives fluid from the right shoulder and the right side of the head. p. 416

lymphatic system The set of vessels that transports the lymph. p. 12

lymphocyte A type of white blood cell that is involved in specific immunity. p. 763

lysis The swelling and subsequent disruption of a cell. p. 754

lysosome An intracellular organelle that contains hydrolytic (digestive) enzymes. p. 77

M

macrocolumns One of the functional ways in which sensory areas of the cortex are organized. p. 282

macrophage One of the white blood cells derived from monocytes. Macrophages can engulf and digest bacteria and "foreign" substances. p. 754

macula The portion of the vestibular system (utricle and saccule) that contains linear acceleration-sensitive hair cells. p. 264

macula densa In the kidney, a region of the early distal tubule adjacent to the glomerular efferent arterioles. It plays a major role in intrarenal regulation. p. 537

malleus One of the bones of the middle ear. p. 257

maltose A disaccharide of glucose and galactose.

mammary gland The milk-producing gland in the female breast. p. 744

MAO inhibitors Drugs that inhibit the enzyme monoamine oxidase; a class of antidepressants.

mass movement In the gastrointestinal tract, an organized peristaitic wave that sweeps chyme along the small and large intestines several times a day. p. 621

mean blood pressure Average arterial blood pressure; it is equal to the product of cardiac output and total peripheral resistance. p. 430

mean circulatory filling pressure The pressure that would exist everywhere in the circulatory system, if the heart were "off," as a consequence of blood volume.

mechanoreceptors Sensory receptors sensitive to mechanical force.

medial reticular inhibitory area A region of the brainstem that decreases spinal reflex tone.

mediated transport See facilitated diffusion.

medulla (brain) One of the three regions of the brainstem. The site of many of the visceral control centers. p. 347

medulla (kidney) The innermost region of the kidney; it contains the loops of Henle. p. 528

megakaryocytes Bone marrow cells that give rise to platelets. p. 394

meiosis Cell division without duplication of chromosomes. p. 63, 690

Meissner corpuscles A type of touch receptor in the skin. p. 233

Meissner plexus One of the two neural networks intrinsic to the gut.

membrane The selectively permeable structure of phospholipid molecules and proteins that surrounds all cells. p. 162

membrane permeability A measure of the ease with which different substances cross cell membranes. p. 162

membrane potential The voltage difference across a cell membrane. p. 162

memory Retention of experiences Long term memories are difficult to disrupt, whereas short-term memory can be affected by trauma. p. 237

memory cells A class of B lymphocytes that can quickly react to a second exposure to a particular antigen. p. 759

memory consolidation The process by which transient shortterm memories are converted to relatively stable longterm memories. p. 240

menarche The onset of menstruation. p. 697

menstrual cycle A monthly cycle of ovulation and menstruation in the female. p. 697

menstruation The monthly shedding of the uterine lining. p. 697

mesencephalic locomotor area A region of the brainstem that causes nearly normal walking movements when stimulated. p. 342

mesencephalon The region of the brain composed of the medulla, midbrain, and pons.

messenger RNA (mRNA) The ribonucleic acid that represents a "copy" of a region of a cell's DNA (genome), used as a template for protein synthesis. p. 63

metabolic acidosis A decrease in plasma pH below 7.4 caused by addition of acid other than carbonic acid. p. 511, 580

metabolic alkalosis An increase in plasma pH above 7.4 caused by the loss of acid other than carbonic acid or the addition of base. p. 580

metabolism Intracellular chemical reactions. p. 86

metarterioles Vessels that connect arterioles directly to venules; they sometimes are referred to as "shunt capillaries" to distinguish them from true capillaries. p. 399

meter (m) The unit of length.

micelles The structures into which amphipathic molecules aggregate in solution; they can be planar or spherical. p. 45, 138

microcolumns In the cortex, an organizational feature of many sensory areas.

microelectrode A glass capillary pulled to have a tip less than 1 μm, used to record voltages from cells. p. 151

microfilaments Thin filaments in the cytoplasm that are composed of actin. p. 82

micropuncture A technique for sampling regions of the renal tubule.

microtubules Hollow tubular structures in the cytoplasm made up of a protein known as tubulin. Both microtubules and microfilaments seem to be involved in axoplasmic flow, cell division, and motility. p. 83

microvilli Microscopic projections on the luminal surface of epithelial cells. They vastly increase the surface area for transport. p. 599

micturition Urination. p. 529

midbrain The brainstem just below the thalamus and above the spinal cord. p. 347

middle ear The region between the eardrum and the cochlea; it normally contains three small bones that transmit sounds. p. 256

milligram % (mg%) A unit of weight used in clinical settings. It equals the number of milligrams of a substance contained in 100 milliliters of water.

millimeters of mercury (mm Hg) A unit of measure for pressure.

mineralocorticoids Steroid hormones secreted by the adrenal cortex that affect salt balance; aldosterone is one example.

miniature endplate potential (mEPP) The endplate potential that results when one synaptic vesicle releases its neurotransmitter.

mitochondria The intracellular organelles in which the Kreb cycle and oxidative phosphorylation occur. The "powerhouses" of a cell. p. 82

mitochondrial matrix The central area of the mitochondria; it contains soluble mitochondrial enzymes. p. 82

mitral valve The valve between the left atrium and left ventricle. p. 362

mixed receptor A sensory receptor whose output depends on both the absolute magnitude of a stimulus and its rate of change with time. p. 200

mobile receptor hypothesis In endocrinology, a cytoplasmic hormone receptor.

modality The nature of a sensory stimulus—touch, smell, sound, visual, kinesthetic, hot, cold, pain, and so forth. p. 193

molarity The amount of a compound in moles in 1 liter of a solution. p. 30

mole A weight of a compound in grams equal to the compound's molecular weight; 1 mole contains the same number of molecules (the Avogadro number). p. 30

molecular weight The sum of the atomic weights of the atoms in a compound. p. 30

molecule A combination of atoms via covalent or ionic bonds. p. 22

monoamine oxidase The enzyme that degrades norepinephrine in adrenergic nerve terminals. p. 352

monochromat An individual who lacks all three cone photopigments and thus color vision. p. 275

monocyte A type of white blood cell. p. 753

monosynaptic Having only one synapse.

morula A solid ball of cells produced by the initial divisions of a fertilized ovum. p. 727

motility See gastric motility.

motor area The region of the frontal lobe of the cortex that initiates voluntary muscle movement. p. 341

motor endplate The neuromuscular junction; the region in which a motor nerve functionally contacts a muscle. p. 141

motor nerve The nerve fiber leading from the spinal cord to a muscle (skeletal or smooth) or a gland.

motor neuron A neuron in the central nervous system that controls skeletal or smooth muscle contraction or glandular secretion. p. 106, 108, 221

motor programs Characteristic patterns of muscular activation seemingly stored in certain regions of the central nervous system. p. 341

motor tract Groups of fibers in the spinal cord that carry impulses from motor areas of the brain. p. 221

motor unit The set of muscle fibers innervated by a motor neuron. p. 326

mucin The protein component of mucus. p. 602

mucosa (mucosal) The internal lining (surface) of the gastrointestinal tract. p. 597

mucus A thick secretion that protects epithelial cells such as those found in the gastrointestinal tract. p. 602

Müllerian duct The structure in the embryonic gonad that gives rise to the fallopian tubes, uterus, and proximal vagina in the female. p. 695

multiunit smooth muscle Smooth muscle in which cells are not electrically coupled. p. 316

multivesicular bodies Membrane-bound remnants of empty synaptic vesicles.

muscarinic receptors A subclass of cholinergic (ACh) receptors present on the parasympathetic effector organs. See **nicotinic receptors.** p. 349

muscle fatigue See fatigue.

muscle fiber The basic functional unit of a muscle; a muscle cell. See also **motor unit.**

muscle hypertrophy An increase in the size of muscle fibers without any change in the number of cells.

muscle spindle A specialized receptor within muscles that provides information about muscle length and its rate of change. p. 331, 334

muscle tension The force exerted by a muscle between its ends.

muscle tone The force generated in a muscle in the absence of any conscious contraction. p. 335

muscle unit See motor unit.

muscularis mucosa The innermost smooth muscle layer of the gut. p. 597

musculoskeletal system The skeletal muscles and bones. p. 12

myelin sheath An insulator around some nerve fibers that consists of several layers of cell membrane. p. 179

myenteric plexus One of the intrinsic neural networks in the gut It is located between the circular and the longitudinal muscle layers. p. 597

myesthenia gravis A disease characterized by muscle weakness involving the block of acetylcholine receptors at the neuromuscular junction by antibody.

myesthenic syndrome Eaton-Lambert syndrome. A disease characterized by muscle weakness, it involves a decrease in the presynaptic release of acetylcholine at the neuromuscular junction.

myocardium Muscle cells of the heart.

myofiber A heart muscle cell.

myofibrils The elements of a muscle that contain the thick and thin filaments organized into sacomeres. p. 292

myofilaments Thick (myosin) and thin (actin) filaments.

myogenicity The spontaneous activity of smooth muscle cells despite the absence of neural or hormonal input.

myoglobin An oxygen-carrying protein found in muscle. p. 309, 490

myopia Nearsightedness. p. 272

myosin One of the contractile proteins in muscle. p. 292

myosin heavy chain The high-molecular-weight component of myosin that contains the actin binding site and ATPase activity. p. 297

myosin isozymes Variants of myosin characterized by differences in ATPase activity and thus crossbridge cycling rate; fast and slow myosin. p. 299

myosin light chain The lower-molecular-weight component of myosin, located in the crossbridges, which regulates myosin ATPase activity. p. 297

myosin light chain kinase An enzyme that phosphorylates myosin light chains. p. 297

myotatic reflex See stretch reflex.

myxedema Decreased metabolic rate and mental activity in adults caused by a decrease in thyroid hormone. p. 678

Na⁺ activation In nerve and muscle tissue, the voltage-gated opening of Na⁺ selective ion channels. p. 171

Na⁺ inactivation In nerve and muscle tissue, a term referring to the fact that voltage-gated Na⁺ channels spontaneously close after briefly opening. p. 171

Na⁺-K⁺ ATPase The sodium "pump." An enzyme that hydrolyzes ATP to transport Na⁺ against its concentration gradient. p. 146

natural killer cells In immunity, cells resembling cytotoxic lymphocytes that may have a role in resistance to cancer. p. 754

near point The minimum distance over which the eye can focus on an object. p. 270

negative feedback The principle that deviation of a variable from its desired value results in actions that correct the deviation. p. 12, 98

neocortex The most recent region of the cortex not including the hippocampus and the limbic system. p. 224

neospinothalamic tract One of the major ascending sensory tracts. It relays fast pain and crude touch and is somatotopically organized.

nephron The basic functional unit of the kidney; in essence, it is a long tube. See **collecting duct, distal tubule, glomerulus, loop of Henle,** and **proximal tubule.** p. 530

Nernst equation An equation for the equilibrium potential of an ion. p. 152

net flux The difference between two unidirectional fluxes. For any process at equilibrium, the net flux must be zero. p. 141

neural code A sequence of action potentials.

neuroendocrine A term referring to the control of pituitary hormone release by the central nervous system. p. 108

neurohypophysis The posterior pituitary. p. 116

neuromodulator A substance that acts to alter the effect of a neurotransmitter. Neuromodulators usually activate some type of intracellular second messenger, which, in turn, modifies an ion channel. p. 205, 329

neuromuscular junction The synapse between a motor nerve and a muscle.

neuron A nerve cell, including the cell body and all its processes. p. 10, 221

neuropeptides Peptides released from nerve fibers.

neurotransmitter A chemical released by a neuron that activates specific receptors on a second neuron, muscle cell, or gland. p. 78, 202

neutron The neutral particle in the nucleus of an atom that has approximately the same mass as a proton. p. 20

neutrophil A white blood cell that can ingest and destroy bacteria. p. 752

nicotinamide adenine dinucleotide (NAD) A coenzyme that carries H⁺ ions from one metabolic pathway to another. p. 87

nicotine An alkaloid from tobacco that activates nicotinic parasympathetic receptors. p. 250

nicotinic receptors A subclass of cholinergic receptors; They are present at the neuromuscular junction and in autonomic ganglia. p. 349

nociception Pain. p. 338

nodes of Ranvier The "gaps" along a myelinated nerve fiber where Na⁺ channels produce action potentials separated by an internode wrapped in myelin. p. 179

nonassociative learning A primitive form of learning present in invertebrates that does not require correlations of diverse stimuli.

noncompetitive inhibition A condition in which enzyme activity is decreased or receptors are blocked when molecules bind to sites other than those at which the normal substrates or ligands bind. p. 60

nonpolar A state in which the constituent atoms forming covalent bonds "share" their electrons equally so that there is no time-averaged local charge. p. 24

nonshivering thermogenesis An increase in heat production caused by increases in metabolic rate and muscle tone. p. 680

norepinephrine The neurotransmitter at sympathetic postganglionic terminals and at many central synapses. It is a catecholamine and one of the biogenic amines. p. 110, 204

nuclear bag fibers Muscle spindle fibers with enlarged central regions. p. 334

nuclear chain fibers Uniform muscle spindle fibers. p. 334

nuclear pore The "hole" in the nuclear double membrane. p. 67

nucleic acid A polymer of nucleotides; DNA or RNA. p. 62

nucleolus A structure in the nucleus of some cells that contains the genetic material. p. 78

nucleotides Compounds formed from a sugar (ribose or deoxyribose), a phosphate group, and a purine or pyrimidine base. p. 62

nucleus The center of an atom; it contains protons and neutrons. It is also the intracellular organelle that contains most of a cell's DNA. p. 20

nystagmus Periodic back-and-forth movements of the eye, usually elicited by stimulation of the vestibular system. p. 267

occipital lobe The most posterior region of the cortex; it receives the primary visual input. p. 224

off pathway A way in which cells in the retina are organized to process visual signals. The term "off" is used because light decreases the activity of afferent fibers in the optic nerve. p. 278

ohm The unit of measure for electrical resistance. p. 37

Ohm's Law An equation that relates current flow to the applied voltage: $I = V/R$. p. 37

olfaction Smell. p. 286

olfactory lobe The region of the cortex that contains the neurons that respond to smell. p. 224

oligodendrocytes Glial cells in the central nervous system that produce myelin. p. 179

on pathway A way in which cells in the retina are organized to process visual signals. The term "on" is used because light increases the activity of afferent fibers in the optic nerve. p. 278

oocytes (oogonia) Stem cells in the female reproductive system. p. 693

open loop system A system that is not controlled. p. 102

operating point In the cardiovascular system, the point at which the venous pressure and cardiac function curves intersect. This defines the cardiac output and central venous pressure. p. 430

operon A regulatory unit of DNA that determines what sequences of base pairs will be used as a template for messenger RNA synthesis.

opiates Pain-killing drugs related to morphine. See **endorphins.** p. 252

opsin A component of the visual photopigment. p. 276

opsonization The presentation of antigens by macrophages. p. 761

optic chiasm The point where the optic nerve partially divides. p. 247

optic disc The region of the retina where afferent fibers from ganglion cells leave the eye to form the optic nerve. p. 279

optic nerve A cranial nerve formed by axons from the retinal ganglion cells. p. 278

oral phase The phase of swallowing that is initiated voluntarily. p. 608

organ A functional unit of the body composed of one or more tissues. p. 10

organ of Corti The structure in the cochlea that contains auditory hair cells. p. 258

organ system A collection of organs with a common physiological function. p. 10

organelles Identifiable structures within a cell. p. 74

organic A term referring to compounds that contain carbon.

origin A site where a muscle tendon attaches to the skeleton. p. 292

osmolarity A measure of the number of dissolved particles in a solution. Osmolarity is inversely related to the concentration of water. p. 30, 143

osmoreceptor A sensory receptor sensitive to osmolarity. p. 569

osmosis The process of water diffusion down its concentration gradient. p. 143

osmotically active (or inactive) A term referring to an impermeable substance that can or cannot produce a gradient for osmosis. p. 123

osmotic diuretics Substances that increase urine production because they are not reabsorbed by the renal tubules. p. 558

osmotic pressure The pressure between a solution containing a solute and water. p. 36, 143

ossicles The bones of the middle ear. p. 257

ossification Calcification of bone. p. 671

osteoblasts The cells that produce bone. p. 576

osteoclasts The cells that degrade bone. p. 576

otoliths Dense structures in the vestibular system that contain $CaCO_3$; they produce a shearing force of the hair cells in response to linear acceleration. p. 264

ouabain An herb that increases heart muscle contractility.

outer ear The external auditory canal, pinna, and eardrum. p. 256

outer segment The region of a photoreceptor that contains photopigments in an array of membranes. p. 272

oval window The point on the cochlea where the last of the middle ear bones transmits its vibrations. p. 257

ovary The female gonad; it contains follicles and secretes the hormones estrogen and progesterone. p. 687

ovulation The release of a mature ovum from its follicle. p. 693

oxidation The gaining of an electron.

oxidative phosphorylation The metabolic process in which H^+ ions carried by coenzymes are transferred to molecular oxygen while ATP is synthesized. p. 77, 87

oxygen One of the respiratory gases essential for oxidative phosphorylation. p. 429

oxygen content The amount of oxygen carried by the blood. It depends on oxygen partial pressure and hemoglobin concentration. p. 485

oxygen debt A term referring to the period after exercise when accumulated lactic acid is metabolized by the liver. p. 310

oxyhemoglobin dissociation curve A graphic representation of the relationship between the partial pressure of oxygen and the relative saturation of hemoglobin with O_2. p. 485

oxyntic glands Glands in the stomach that secrete hydrochloric acid and intrinsic factor. p. 551

oxytocin A posterior pituitary hormone that causes uterine contractions and milk let-down. p. 116

P

P_{50} The partial pressure of oxygen needed for half-maximum saturation of hemoglobin.

P wave The component of the electrocardiogram that corresponds to the electrical activation of the atrium. p. 370

pacemaker A cell that spontaneously depolarizes and therefore can initiate activity in adjoining cells. Pacemakers are found in the heart and gastrointestinal tract. See **sinoatrial (S-A) node** and **atrioventricular (A-V) node**. p. 316, 361

pacemaker potential Spontaneous depolarization in a pacemaker. p. 310

pacinian corpuscle A cutaneous mechanoreceptor sensitive to vibration. p. 200, 233

paleospinothalamic tract One of the ascending sensory pathways. It carries slow pain and temperature information but does not reach the cortex.

pancreatic islets See islets of Langerhans.

pancreatic juice A general term referring to the secretion of bicarbonate and enzymes by the pancreas.

pancreatic lipases Enzymes that degrade triglycerides to fatty acids. p. 634

pancreatolysis Self-digestion of the pancreas, caused by a lack of trypsin inhibitor. p. 618

papillary muscles In the heart, the muscles that restrain the A-V valves. p. 363

paraaminohippurate (PAH) An end product of aromatic amino acid metabolism that is totally secreted by the kidneys and thus is a measure of renal plasma flow. p. 554

paracellular route A term referring to the movement of water through the "tight junctions" that connect adjacent epithelial cells in the gastrointestinal tract, kidneys, and some glands. p. 85

paracrine agent (gland) A hormone that has only local effects, and the gland from which it is secreted. p. 98, 103

parasympathetic nervous system One of the two principal divisions of the autonomic nervous system. Characterized by long preganglionic fibers and short postganglionic fibers. Uses ACh as a transmitter at both its synapses. p. 329

parathyroid glands Glands near the thyroid that secrete parathyroid hormone. p. 676

parathyroid hormone A hormone that affects calcium and phosphate balance by altering renal transport. p. 577

paravertebral chain The series of sympathetic ganglia that lie adjacent to, but outside of, the spinal cord. p. 329

parietal cells Cells in the stomach that secrete hydrochloric acid. p. 611

parietal lobe The region of the cortex located between the frontal and occipital lobes that contains the somatosensory areas. p. 224

parietal pleura The tissue lining the thoracic cavity. p. 450

Parkinson's disease A disease of the central nervous system characterized by tremor caused by a deficit in the transmitter dihydroxyphenylalanine (DOPA). p. 345

parotid gland One of the salivary glands.

partial pressure The pressure that one gas present in a mixture of gases would exert if it were present alone in the same volume.

passive transport Diffusion of a substance down its concentration gradient via a selective channel or carrier. See **facilitated diffusion**. p. 146

patch clamp A technique for examining the movement of ions across small areas of a cell membrane. p. 318

pepsin The enzyme that breaks specific internal peptide bonds in a protein. p. 611

pepsinogen The inactive precursor of the enzyme pepsin. p. 611

peptides Short chains of amino acids joined by peptide bonds. p. 204

peptide bond A chemical bond between amino acids that is formed by removal of water. p. 55

peptide hormones Hormones composed of peptides. p. 625

perforin A toxin secreted after an activated T cell binds to its target. p. 766

perfusion pressure Pressure actively generated by the heart. p. 406

perilymph Fluid in the scala media of the ear that resembles intracellular fluid. p. 259

perimysium The connective tissue surrounding a group of muscle fibers referred to as a muscle fascicle. p. 292

peripheral chemoreceptors The oxygen- and pH-sensitive sensory receptors in the cardiovascular system. See **aortic bodies** and **carotid bodies**. p. 505

peripheral nervous system The region of the nervous system that lies outside the brain and spinal cord. p. 326

peristalsis Waves of contraction in the gut that persist for some distance along its length; to be contrasted to the process of segmentation. p. 604

peristaltic rush Coordinated peristalis that occurs two or three times a day.

peritubular capillaries In the kidneys, the capillaries that surround the proximal and distal convoluted tubules. p. 358

permeability (coefficient) The ability of a substance to pass across an interface such as a cell membrane or epithelial cell. p. 142

permissive In endocrinology, a term referring to the fact that the presence of one hormone is necessary for another to be fully effective.

pernicious anemia A decrease in hemoglobin content caused by a dietary deficiency of vitamin B_{12}. p. 393

perspiration Sweating. p. 569

perturbation A disturbance. p. 98

pH A measure of the concentration of H^+ ions: $pH = -\log[H^+]$. p. 27

pH trapping The accumulation of NH_4^+ in the renal tubules. p. 557

phagocytosis Endocytosis characterized by the intake of extracellular material, including cellular debris, foreign material, and bacteria. p. 77

pharyngeal phase The phase of swallowing that follows the voluntary oral phase. p. 608

pharynx The throat. p. 450, 595

phase (lag) A time measure of the relative position of sinusoidal signals such as sound waves.

phasic receptor A rapidly adapting sensory receptor in which the generator potential disappears even though the stimulus persists. p. 200

phosphodiesterase An enzyme that degrades cyclic GMP, often activated by a G-protein. p. 111

phosphoinositides Lipid compounds that often serve as internal second messengers. p. 111

phospholipid An amphipathic lipid molecule composed of two fatty acid chains and a polar group esterified on a glycerol "backbone." p. 44, 45, 632

phosphorylation The addition of a phosphate molecule. p. 58

photon A light wave. p. 277

photopigment The light-sensitive component of the visual photoreceptors. p. 272

photoreceptors Rods and cones in the retina; they contain photopigments. p. 272

photopic system Vision mediated by cone photoreceptors; color-sensitive vision. p. 272

phrenic nerve The motor nerve fiber that innervates the diaphram. p. 501

physiological shunt In the lungs, a condition in which some regions are not adequately ventilated compared to local blood flow. p. 471

piloerection "Goose bumps." p. 316

pinna The external portion of the ear. p. 256

pinocytosis The uptake of materials into a cell by means of a pinching off of small membrane-bound vesicles. p. 77

pituitary gland An endocrine gland located below the hypothalamus. The anterior portion secretes six hormones (ACTH, TSH, LH, FSH, prolactin, and GH) and is controlled by releasing factors secreted by the hypothalamus. The posterior portion is an extension of the hypothalamus and secretes ADH and oxytocin. p. 116

placenta The organ that develops during pregnancy from maternal and fetal tissue and serves as an interface between the maternal and fetal circulations. p. 728

plasma The fluid component of the blood. p. 392, 408

plasma cell Cells derived from B lymphocytes that secrete antibodies. p. 759

plasma proteins Proteins in the fluid component of the blood. p. 392

plasma membrane See **membrane**.

plasticity In neurophysiology, a term referring to a change in synaptic efficacy. p. 285

plateau In the cardiac action potential, the component during which the membrane potential remains in the vicinity of 0 millivolts; also called phase 2. It results from an increase in Ca^{++} permeability. p. 366

platelets Small cell fragments in the blood that contribute to the formation of blood clots (hemostasis). p. 392

pleated sheet One of the two secondary structures of proteins characterized by an association of adjacent amino acid chains. p. 55

plethysmograph A device for measuring lung volumes; it can measure the residual volume of the lung. p. 409

plicae circulares Visible folds in the lining of the gastrointestinal tract. p. 599

pneumotaxic center The region of the brainstem that can stimulate respiration. p. 503

pneumothorax A condition in which air enters the intrapleural space. p. 462

poise The unit of viscosity.

Poiseuille Law The equation that describes the flow resistance of a blood vessel in terms of its radius and the blood viscosity.

polar A molecule that has an unequal charge distibution, or dipole moment. p. 24

polycythemia An abnormally high concentration of red blood cells. p. 392

polymer A long chain made up of identical molecular subunits. p. 44

polypeptide A long chain of amino acids linked by peptide bonds.

polysaccharides A long chain of monosaccharides that may be unbranched (cellulose) or branched (glycogen and the starches). p. 51, 567

polysynaptic Having more than two synapses.

pons The region of the brainstem between the medulla and the midbrain. p. 347

portal circulation A term denoting vessels that link two organ systems. p. 116

positive feedback The principle that deviation of a variable from its desired value results in further deviation. p. 98

postabsorptive state A condition in which substrates are not being absorbed from the gastrointestinal tract and must be provided from body stores. p. 656

posterior chamber One of the spaces in the eye outside the lens; it contains a watery fluid, the aqueous humor. p. 268

posterior pituitary The region of the pituitary gland that secretes ADH and oxytocin; it is an extension of the central nervous system. p. 116

postganglionic In the autonomic nervous system, a term referring to the neurons or nerve fibers that lead from the autonomic ganglia to target tissues. p. 327

postsynaptic inhibition In synaptic transmission, a condition in which a terminal on a cell body or dendrite releases a transmitter that increases the K^+ permeability, decreasing the excitability of the postsynaptic neuron. p. 207

postsynaptic potential The voltage change in a neuron produced by the release of a neurotransmitter. See **excitatory postsynoptic potential** and **inhibitory postsynoptic potential**. p. 206

posttranscriptional processing Modification of the messenger RNA produced from DNA templates by removal of bases corresponding to noncoding regions. p. 67

postural reflexes Reflexes that maintain balance.

potassium (K⁺) The major intracellular positively charged ion. p. 153

potassium equilibrium potential (E_K) The voltage that would exist across a membrane if it were exclusively permeable to K^+. p. 153

potential Voltage. p. 194

potential energy The energy an object has because of its position; to be contrasted to kinetic energy and internal energy.

P-R interval The period in the electrocardiogram between the P wave and the ORS complex. p. 371

precapillary sphincter A small group of smooth muscle cells that controls the entry of blood to a capillary. p. 399

precordial leads A set of electrodes placed on the chest to record an electrocardiogram. p. 371

preganglionic neurons In the autonomic nervous system, the neurons or nerve fibers that lead from the central nervous system to the autonomic ganglia. p. 327

preload In muscle, the amount of force that must be exerted on a resting muscle to stretch it to a given initial length. In the heart, the preload corresponds to the venous pressure and determines the end-diastolic volume. p. 306, 376

presbyopia A loss in the normal ability of the lens of the eye to focus on objects at various distances. p. 270

pressure Force per unit area.

pressure overload A condition such as aortic stenosis in which the ventricles are forced to generate abnormally high pressure to eject blood. p. 381

pressure-volume loop See work loop.

presynaptic facilitation An increase in transmitter release at a synapse. p. 206

presynaptic inhibition In synaptic transmission, a condition in which an axon makes contact with the terminal of another axon (an axoaxonic synapse) and releases a transmitter that decreases the quantal content. p. 206

presystolic murmur An abnormal heart sound that precedes ventricular activation; it is associated with mitral stenosis. p. 381

prevertebral ganglion A sympathetic ganglion adjacent to the aorta.

primary afferent Nerve fibers whose cell bodies are located in the dorsal root ganglia and that relay sensory information into the spinal cord.

primary structure The amino acid sequence of a protein. p. 55

primary transcript The immediate result of DNA transcription; it contains exons and introns. Primary messenger RNA is modified to final messenger RNA in the process of posttranscriptional processing. p. 67

priming In endocrinology, the stimulation by one hormone of synthesis of the receptor for another. p. 115

proenzyme An inactive enzyme precursor that, when cleaved by another enzyme, produces an active enzyme. p. 68

progesterone A female sex steroid; it is secreted by the corpus luteum (and placenta) and affects uterine development. p. 110, 698

proinsulin The inactive precursor to the hormone insulin. p. 656

prolactin One of the anterior pituitary hormones; it stimulates milk secretion by the mammary glands. p. 116, 745

proliferative phase In the menstrual cycle, the phase before ovulation during which the endometrium of the uterus is increasing in thickness but is not secretory. p. 711

promoter The region of a gene that acts as a switch to turn on transcription of a sequence of base pairs (an operon). p. 66

propagation The ability of an action potential to travel long distances along a nerve fiber. p. 165

proprioceptors Receptors in joints and muscles that provide information about limb position and movement and muscle length and tension. p. 194

propriospinal neurons Spinal interneurons whose processes extend over several spinal cord segments and thus coordinate groups of muscles. p. 311

propulsion Contractions of the stomach that move chyme toward the pylorus. p. 613

prostaglandins A group of modified fatty acids that act as chemical messengers. p. 49, 424

prostate A gland in the male that secretes fluid into the urethra. p. 689

protanopia A condition in which red-sensitive photopigment is absent. Individuals with this condition cannot distinguish between red and green. p. 275

protease An enzyme that degrades proteins to smaller peptides. p. 59, 617

protein A large molecule composed of one or more polypeptide chains. p. 55

protein kinase A generic term for an enzyme that phosphorylates proteins. p. 61, 114

prothrombin A plasma protein that, when activated by one of the clotting factors, catalyzes the conversion of fibrinogen to fibrin. p. 395

proton The positively charged particle present in the nucleus of an atom The number of protons defines an element. p. 20

proximal muscle Muscle near the trunk that is primarily involved in maintaining postural stability. p. 343

proximal tubule The region of the nephron immediately after the glomerulus. p. 531

ptyalin A salivary enzyme that breaks down starch (amylase). p. 595

pulmonary circulation Blood flow through the lungs. p. 358

pulmonary edema Accumulation of fluid in the alveoli of the lungs. p. 381

pulmonary surfactant See surfactant.

pulse pressure The difference between arterial systolic and diastolic pressure. p. 363

punctate In sensory physiology, a term referring to the fact that primary afferent fibers innervate a limited area of the skin.

pupil The dark opening in the eye whose diameter is controlled by the iris. p. 268

pure tone The sound made by a single frequency. p. 256

Purkinje fibers The most distal component of the conducting system of the heart; it activates the interior of the ventricles. p. 362

putamen One of the basal ganglia of the brain. p. 341

pylorus The physiological sphincter between the stomach and the duodenum. p. 596

pyrimidines Cytosine, thiamine, and uracil. Three of the nucleotide bases. p. 62

pyruvate The end product of glycolysis. p. 88

Q

QRS complex The component of the electrocardiogram that corresponds to the electrical activation of the ventricles. p. 370

quantal content The number of vesicles released from a synaptic terminal in response to a single presynaptic action potential. p. 206

quantum The amount of neurotransmitter contained in a single synaptic vesicle. p. 206

quaternary structure The subunit structure of proteins. p. 57

random coil One form of the secondary structure of proteins.

rapid ejection The phase of the cardiac cycle during which the heart is ejecting blood into the aorta and aortic pressure is increasing.

rapid eye movement (REM) sleep The phase of the sleep cycle characterized by desynchronized brain activity and dreaming; also called paradoxical sleep. p. 243

rapid filling The phase of the cardiac cycle during which the ventriclar volume is increasing but ventricular pressure is still decreasing.

rarefaction The phase of a sound wave during which pressure decreases. p. 256

readiness potential An electrical potential that can be recorded from the cortex before the onset of a movement. p. 343

receptive aphasia A condition in which an individual can form sentences gramatically but cannot comprehend. p. 242

receptive field The region of a sensory surface that, when stimulated, changes the response of a neuron or nerve fiber. p. 193

receptive relaxation Relaxation of the stomach during the entry of food. p. 646

receptors Membrane-bound molecules with specific binding sites for molecules such as neurotransmitters or hormones. p. 98, 109, 138

receptor potential See generator potential.

reciprocal inhibition A term referring to the fact that activation of the motor neurons to one muscle usually inhibits motor neurons of antagonists at that joint. p. 331

recombinant DNA Foreign DNA that has been incorporated into a cell's own DNA. p. 70

recruitment The process by which additional motor units are activated during the contraction of a skeletal muscle. p. 302

rectum The most distal region of the gastrointestinal tract. p. 597

red blood cells The cellular elements of the blood that contain hemoglobin. p. 392

red nucleus A collection of neurons in the brain that are involved in control of refined movements. p. 223

red-sensitive cones Photoreceptors with maximum sensitivity at 560 nanometers. p. 272

reduced ejection The phase of the cardiac cycle during which blood is being ejected from the heart into the aorta, but aortic pressure is decreasing.

reduced filling The phase of the cardiac cycle during which the ventricular volume and pressure are both increasing. It follows rapid filling.

reduction A reaction involving the loss of an electron.

reentry A condition in which electrical activity in the heart is altered so that different regions excite one another in a vicious cycle. p. 369

referred pain Visceral pain associated with the sensory nerve fibers leading from corresponding areas of the body surface. p. 234

reflex A negative feedback system in which a sensory stimulus leads to a motor response that diminishes the magnitude of the eliciting stimulus. p. 10

refractory period The interval after an action potential when a second action potential either cannot be produced (absolute refractory period) or has an increased threshold (relative refractory period). p. 173

Reissner's membrane In the cochlea, a layer of cells separating the scala vestibuli and the scala media.

relative refractory period A period of time after an action potential when a second action potential can be obtained, but the threshold is increased. p. 173

relative saturation The percent saturation of hemoglobin at a given partial pressure of oxygen. p. 485

relaxin A hormone used to relax the birth canal during labor. p. 732

releasing (or release-inhibiting) hormone A hypothalamically released substance that elicits (or inhibits) secretion of hormones by the anterior pituitary. p. 116

renal calyx The interior potion of the kidney; it empties into the ureter. p. 528

renal cortex The outermost layer of the kidney; it contains glomeruli and the proximal and distal tubules. p. 528

renal medulla A structure located below the cortex that contains the loops of Henle. p. 528

renal pyramids A structure located beneath the medulla that has numerous collecting ducts, giving it a striped appearance. p. 528

renin A hormone released by the juxtaglomerular cells of the kidneys that catalyzes production of angiotensin and secretion of aldosterone. p. 427

repetitive firing The train of action potentials elicited by an applied stimulus. p. 174

replication The process by which DNA generates a copy of itself. p. 63

residual volume The volume of air in the lungs that cannot be expelled. p. 458

resolution phase The phase of the human sexual response after climax. p. 717

resonant frequency The frequency at which external forces applied to harmonic oscillators (objects such as pendulums or springs) cause a progressive increase in the amplitude of the resulting oscillations. p. 259

respiratory acidosis A decrease in plasma pH below 7.4 caused by the addition of carbonic acid; that is, by decreased alveolar ventilation. p. 580

respiratory alkalosis An increase in plasma pH above 7.4 caused by a deficit in carbonic acid; that is, by increased alveolar ventilation. p. 580

respiratory distress syndrome A condition in newborns that involves decreased pulmonary surfactant. p. 468

respiratory drive The combined input to the brainstem centers controlling respiration. p. 503

respiratory pump A term referring to the fact that venous return increases during inspiration as a result of a decrease in intrapleural pressure. p. 435

respiratory quotient The ratio of CO_2 produced to O_2 consumed. p. 482

resting potential The voltage difference across a cell that is not producing an action potential or being otherwise excited. p. 151, 162

resting tone See basal tone. p. 422

restriction endonucleases Enzymes that cleave DNA at specified base pair sequences; they are used in genetic engineering to restructure genes. p. 70

reticular formation A region of the brainstem consisting of a diffuse network of neurons. It is thought to be involved in monitoring the flow of information between the brain and spinal cord and in setting a level of cortical "awareness." p. 223

retina The interior surface of the eye; it contains photoreceptors and other visual system neurons. p. 268

retinal A modified form of vitamin A that serves as the immediate receptor for photons of light. p. 276

retrograde amnesia Loss of memory for preceding events.

retropulsion Contractions of the stomach that move chyme away from the pylorus. p. 613

reverse transcriptase An enzyme capable of synthesizing DNA from RNA. p. 70

reversible reaction A reaction that can proceed in either direction. p. 38

Rh factor Antigens on red blood cells that are different from A, B, and O antigens. If Rh antigens are present, an

individual is termed Rh positive (Rh^+). p. 773

rhodopsin The rod photopigment (retinal plus opsin). p. 276

ribosomal RNA The RNA component of the ribosome. p. 64

ribosome A large cytoplasmic particle made up of proteins and RNA to which messenger RNA molecules attach during their translation into protein. p. 64

rickets Bone weakness and underdevelopment in children caused by Ca^{++} deficiency. p. 578

right axis deviation A condition in the heart in which certain waves of the electrocardiogram are increased because of an increase in the muscle mass of the right ventricle.

righting reflex A postural reflex that maintains the position of the head in space; it is mediated by stretch receptors in the neck. p. 340

rigor mortis A condition after death in which the depletion of ATP results in a permanent actin-myosin complex. p. 297

rod One of the photoreceptors; it is not color sensitive. p. 272

rough endoplasmic reticulum An intracellular membrane system characterized by the presence of ribosomes. p. 75

round window The opening in the cochlea that faces the middle ear but is not attached to the ossicles. p. 259

Ruffini endings Sensory receptors in the skin that are sensitive to touch. p. 233

S

saccule A receptor in the vestibular system that is sensitive to linear acceleration and head position. p. 264

safety factor The ratio of the current flowing along a nerve fiber to the minimum amount needed to sustain conduction. p. 180

saliva The fluid secreted by salivary glands. p. 606

salivary glands Glands in the mouth that secrete the enzyme amylase, some electrolytes, and mucus. p. 606

salt A crystalline compound made up of positively and negatively charged ions that fully dissociate in water.

saltatory conduction The conduction of nerve impulses in myelinated nerve fibers. p. 179

sarcolemma The plasma membrane surrounding a muscle cell. p. 292

sarcomere Subunits of myofibrils that contain the thick and thin filaments. p. 292

sarcoplasmic reticulum The smooth endoplasmic reticulum of a muscle fi-

ber. It releases and resequesters Ca^{++} during contraction. p. 79, 299

satiety center The region of the brainstem where stimulation inhibits feeding. p. 670

saturated Having no double or triple bonds, as in saturated fatty acids. p. 44

saturation A condition in which increasing the concentration of a reactant (or transported substance) fails to increase the reaction (or transport) rate. p. 485

scala media The fluid-filled region of the cochlea that contains the organs of Corti. p. 258

scala tympani The fluid-filled region of the cochlea nearest the round window. p. 258

scala vestibuli The fluid-filled region of the cochlea nearest the oval window. p. 258

Schlemm canal A small duct connecting the anterior and posterior chambers of the eye; if it becomes plugged, glaucoma results. p. 264

Schwann cells The myelin-forming glial cells in the peripheral nervous system. p. 179

scientific notation A system in which numbers are expressed as the product of a number between 1 and 10 and a power of 10. For example, in scientific notation, $1,559,000 = 1\ 559 \times 10^6$ and $0.00034 = 3.4 \times 10^{-4}$.

sclera The outer layer of the eye; it appears to be white in color. p. 268

scopalamine A drug that blocks muscarinic cholinergic synapses. p. 350

scotopic system Vision mediated by rods; not color sensitive. p. 272

scrotum The external pouch containing the male testes. p. 688

Second Law of Thermodynamics The principle that heat can flow only from higher to lower temperatures; also, a definition of entropy. p. 38

second heart sound The sound associated with the closing of the aortic and semilunar valves. p. 363

second messenger A molecule formed by the interaction of a ligand with a receptor that subsequently acts to modify intracellular metabolic processes. p. 110

secondary peristalsis Waves of contraction in the esophagus triggered by the presence of residual food.

secondary structure The α helix or random coil conformation of a polypeptide chain. p. 55

secretin A hormone secreted by duodenal cells that stimulates pancreatic bicarbonate secretion. p. 647

secretory phase In the menstrual cycle, the period between ovulation and menstruation during which the uterine lining develops a secretory characteristic. p. 711

segmental reflexes Reflexes mediated by interneurons in one segment of the spinal cord. p. 235

segmentation A series of more or less stationary contractions of the gut that act to mix its contents. p. 604

semen The combination of sperm and fluids secreted by the accessory glands of the male reproductive system. p. 689

semicircular canals The rotation-sensitive components of the vestibular system. p. 264

semilunar valves Valves between the right ventricle and the pulmonary artery. p. 362

seminal vesicles In the male, glands that secrete fluid into the vas deferens. p. 689

seminiferous tubules In the male, the sites of sperm production. p. 688

sensitization A primitive, nonassociative form of learning in which the repeated presentation of a stimulus enhances an avoidance response.

sensory receptor See receptors.

sensory tract A bundle of fibers in the spinal cord and brainstem that carry information from one of the special senses or from nonspecific sensory nerve endings. p. 221

series elastic element In muscle physiology, a term referring to the fact that muscles act as if an elastic structure must be stretched before the force generated by the crossbridges can be "felt" externally. p. 304

serotonin One of the central neurotransmitters. p. 211

Sertoli cells In the male, the cells that line the seminiferous tubules and "nurture" sperm maturation. p. 721

serum Plasma devoid of the plasma proteins that are used to form blood clots. p. 394

set point In control theory, the value at which a negative feedback system will regulate a controlled variable. p. 12

sex chromosome Chromosome 24; in males, **Y**, and females, **X**. p. 690

sex steroids Estrogen, progesterone, and testosterone. p. 697

shivering thermogenesis Increased heat production caused by consciously sensed contractions of the skeletal muscles. p. 679

shock A condition in which the blood pressure is abnormally low. It can be caused by loss of blood volume, a decrease in the heart's pumping ability, or loss of normal venous tone. p. 777

short-term memory Memory easily disrupted by trauma. p. 240

signal sequence A particular sequence of base pairs that serves to "direct" messenger RNA to ribosomes in the rough endoplasmic reticulum. p. 79

simple cell In the visual cortex, a cell sensitive to lines and edges. p. 282

single-unit smooth muscle A condition in which smooth muscle cells are electrically coupled to one another. p. 316

sinoatrial (S-A) node The usual pacemaker of the heart; it is in the right atrium. p. 361

sinus arrest A condition in which the S-A node ceases to generate periodic action potentials.

sinusoids Large capillaries in the liver between sheets of hepatocytes. p. 628

size principle The principle that in the activation of a muscle, the smaller motor units are recruited first.

skeletal muscle A muscle that attaches to the skeleton. p. 10, 304

skeletal muscle pump A term referring to the fact that muscle contraction combined with unidirectional valves in the veins helps propel blood to the heart. p. 435

slow muscle fibers Fibers that contain a myosin with low ATPase activity.

slow Na$^+$/Ca^{++} channel An ion channel in heart muscle that allows both Na$^+$ and Ca^{++} to pass through it and whose rate of activation is slower than the "fast" Na$^+$ channel of nerve and skeletal muscle fibers.

slow oxidative muscle fibers One of the three classes of skeletal muscle characterized by abundant mitochondria, fatigue resistance, and myosin with low ATPase activity. p. 308

slow pain, second pain, or burning pain The component of pain that is not easily localized and that persists for some time after removal of a noxious stimulus. p. 234

slow waves Slow changes in the membrane potential of smooth muscle cells in the gut. Action potentials tend to occur at the peaks of these slow waves. p. 603

smooth endoplasmic reticulum An intracellular structure involved in the synthesis of membranes. See **rough endoplasmic reticulum**. p. 75

smooth muscle See **unitary smooth muscle** and **multiunit smooth muscle**.

sodium (Na$^+$) The main extracellular positively charged ion. p. 145

sodium equilibrium potential (E$_{Na}$) The voltage that would exist across a cell membrane if it were solely permeable to Na$^+$ ions. p. 153

sodium pump See **Na$^+$-K$^+$ ATPase**.

solubility The amount (weight) of a solute that can be dissoved in a given amount of solvent. p. 24, 27

solute Molecules dissolved in a solution. p. 13, 27

solvent The liquid in which molecules are dissolved, usually water. p. 27

soma The cell body of a neuron. p. 190

somatic motor system The collection of neurons that activates the skeletal muscles. p. 331

somatic motor reflex A reflex involving the skeletal muscles. p. 108

somatomedin An insulinlike hormone secreted by the liver in response to growth hormone. p. 105, 672

somatosensory system The collection of neurons that receives information from sensory receptors on the skin and from internal proprioceptors. p. 236, 284

somatostatin A hormone released by the pancreas that acts to prevent an overload of the gastrointestinal system. p. 662

somatotopic organization In sensory physiology, a term referring to the fact that afferent fibers enter the central nervous system in an orderly way, thus preserving a "map" of the sensory surface in ascending fiber tracts, relay nuclei, and the cortex.

somatotropin Growth hormone. p. 114, 116, 671

sound waves Sinusoidal variations in the pressure in the air.

spatial summation A process by which two stimuli close together in time produce a response greater than would have been seen if either were present alone. p. 195

spermatogenesis The production of sperm. p. 690

spermatozoa Sperm. The male gamete. p. 688

sphincter A muscular ring that, when contracted, closes off a blood vessel or a portion of the gastrointestinal tract. p. 399, 529

spike An action potential. p. 162

spinal cord The most primitive region of the central nervous system; it contains the "circuitry" for basic reflexes. p. 327

spinal nerve One of the peripheral nerve trunks that enter the spinal cord; to be contrasted to cranial nerves. p. 222

spinal reflexes Reflexes that are "hard wired" into the spinal cord.

spindle A length-sensitive stretch receptor in muscles. p. 334

spinothalamic tract The set of sensory nerve fibers that carries kinesthetic information and fine touch.

spirometer A device for measuring some lung volumes; it cannot be used to determine the residual volume. p. 456

splanchnic circulation Blood flow through the abdominal viscera.

spleen The largest lymph tissue; it is located between the stomach and diaphragm. p. 627

split brain A condition in which the fibers normally connecting the two hemispheres of the brain have been severed or damaged. p. 242

S-T segment The interval in the electrocardiogram that corresponds to the plateau phase of the ventricular action potential. p. 370

staircase phenomenon In the heart, a term referring to the fact that at high heart rates, the force of contraction increases; also called treppe. p. 330

standing-gradient osmotic flow A process in the kidney whereby Na$^+$ transport across the basolateral surface creates an osmotic gradient for water. p. 542

stapes One of the bones of the middle ear. p. 257

starch A glucose polymer characterized by a degree of branching intermediate between that of glycogen and cellulose. p. 629

Starling Law See **Frank-Starling Law**.

static gamma motor neurons Motor neurons innervating the peripheral ends of muscle spindles and thus selectively modifying the muscle's tonic properties. p. 334

static lung volumes Steady-state lung volumes measured with a spirometer. p. 459

steady state A condition in which some variables of a system are constant with time. p. 39

stenosis Narrowing; it usually refers to partial occlusion of a blood vessel. p. 381

stereocilia The cilia on the auditory and vestibular hair cells. p. 260

steroids A class of lipids in which polar groups are attached to a skeleton of four carbon rings. p. 44

stimulus (stimulation) A chemical or physical change in the environment. p. 162

stimulus-gated ion channel A ion channel that is opened or closed by a physical stimulus such as light, mechanical movement, odors, tastes, and so forth. p. 171

stimulus-secretion coupling In secretory cells, the process by which a stimulus elicits exocytosis. p. 175

stomach The component of the gastrointestinal tract that receives food from the esophagus. p. 595

stress activation (or relaxation) In smooth muscle, a process in which increased tension leads to contraction (or relaxation) of the smooth muscle cells. p. 742, 604

stretch receptors Sensory receptors sensitive to stretch. p. 529

stretch reflex The only spinal reflex with a monosynaptic component. In it stretching a muscle leads to its reflex contraction. p. 108, 331

striated muscle   Muscle characterized by a pattern of light and dark bands that arises from the ordering of its contractile proteins. p. 299

stroke volume   The amount of blood ejected from a ventricle in one heartbeat; the difference between end-diastolic and end-systolic volume. p. 363

sublingual gland   One of the salivary glands. p. 595, 606

submaxillary gland   One of the salivary glands. p. 595

submucosa   The connective tissue beneath the gut mucosa. p. 597

substance P   One of the neuroactive peptides; it is the transmitter released from primary afferent pain fibers. p. 211

substantia gelatinosa   A region of the dorsal horn of the spinal cord where the primary afferent pain fibers synapse. p. 341

substantia nigra   A region of the brainstem that has deeply pigmented neurons. p. 345

substrate   A molecule that binds to the active site of an enzyme. p. 59

substrate-level   phosphorylation Formation of ATP from ADP by transferring a phosphate group from another higher-energy compound and without involving oxygen metabolism. p. 87

subthreshold stimuli   Stimuli too low to produce an action potential. p. 167

sulci   The grooves in the cortex. p. 224

summation   See **spatial summation** and **temporal summation.** The term also refers to the additive contractile effect seen during repetitive stimulation of a skeletal muscle. p. 163, 302

suppressor cell   A subclass of T lymphocytes. p. 765

surfactant   In the lungs, an amphipathic lipoprotein that reduces the alveolar surface tension and thus the intrapulmonary pressure. p. 467

swallowing center   The region of the medulla that coordinates the pharyngeal and esophageal phases of swallowing. p. 608

sweating   The secretion of a hypotonic electrolyte solution by glands in the skin As sweat evaporates, it cools the body surface. p. 569

sympathetic nervous system   A division of the autonomic nervous system. p. 329

symport   Cotransport in which two substances move across the membrane in the same direction by means of a common carrier.

synapse   A junction between neurons or between a nerve and a muscle. p. 190, 200

synaptic cleft   The small space between the presynaptic and postsynaptic elements at a synapse. p. 202

synaptic delay   The time between release of transmitter from the presynaptic terminal and the beginning of the postsynaptic potential. p. 202

synaptic efficacy   Refers to the amount of neurotransmitter released as a result of a single presynaptic action potential. p. 206

synaptic potential   See **excitatory postsynaptic potential** and **inhibitory postsynaptic potential.** p. 162

synergist   A muscle that moves a joint in the same direction as another muscle. p. 116, 331

systemic circulation   The blood flow through all the organs except the lungs. p. 358

systole   The period during which the ventricles are contracting. p. 362

systolic murmur   A sound that occurs during rapid ejection in cases of backflow of blood into the left atrium caused by a leaky mitrial valve. p. 381

systolic pressure   The maximum arterial pressure reached during systole. p. 363

T

T lymphocytes   A subclass of lymphocytes. p. 763

T tubule   See transverse tubules.

T wave   The component of the electrocardiogram that corresponds to the repolarization of the ventricle. p. 371

tachycardia   An abnormally rapid heart rate. p. 377

target cell   In endocrinology, the cells that contain receptors for a particular hormone. p. 109

taste buds   The sensory receptors for taste. p. 283

tectorial membrane   The gelatinous structure on top of the auditory hair cells that exerts a shearing force when the basilar membrane moves up or down. p. 262

temporal lobe   An area of the cortex; it contains the auditory areas and one of the language centers. p. 224

temporal summation   The addition of stimuli close together in time to produce a response greater than that seen if either were present alone. p. 195

tendon   The band composed of collagen fibers that attaches a muscle to the skeleton. p. 292

tendon organs   Stretch-sensitive receptors in the tendons that provide information concerning muscle tension. p. 337

teniae coli   The longitudinal muscle in the large intestine. p. 621

tension   Equivalent to force. p. 304

tensor tympani   A muscle that inserts on one of the middle ear bones and regulates the transmission of sound. p. 257

terminal cisternae   In muscle, the lateral portions of the sarcoplasmic reticulum from which Ca^{++} is released. p. 299

tertiary structure   The threedimensional conformation of a polypeptide chain. p. 55

testes   The male gonads. p. 688

testosterone   The main male sex hormone secreted by the testes. p. 110

tetanus   Sustained contraction in a skeletal muscle at high frequencies. p. 304

thecal cells   A type of cell found in the ovaries. p. 704

thermodilution   A technique that uses a pulse of heat to measure cardiac output.

thermoregulation   The process by which body temperature is regulated. p. 679

theta waves   Low-frequency components of the electroencephalogram seen in stage I sleep. p. 243

thick ascending limb   The cortical portion of the Henle loop; it transports Cl^-. p. 532

thick filaments   The contractile filament in muscle that contains myosin. p. 292

thin ascending limb   The medullary, salt-permeable portion of the loop of Henle. p. 532

thin filaments   The contractile filament in muscle that contains actin. p. 292

thoracic cavity   The interior portion of the body that contains the heart and lungs. p. 450

thoracic duct   The major vessel into which the lymphatic veins from most parts of the body empty and return the lymph to the heart. p. 416

threshold   The voltage that a nerve or muscle cell needs to be depolarized to to produce an action potential. p. 167

thrombin   An enzyme needed for conversion of fibrinogin to fibrin in clotting. p. 395

thrombocytes   Platelets. p. 392

thromboplastin   Factor III; a clotting factor released by damaged cells that accelerates formation of a blood clot. p. 395

thymine   One of the pyrimidine bases in nucleotides. p. 62

thymus   An organ of the lymphatic system and the site of T-lymphocyte production. p. 764

thyroglobulin   The glycoprotein in thyroid follicles that binds thyroid hormone. p. 675

thyroid gland A gland in the neck that secretes thyroid hormone. p. 676

thyroid hormone Thyroxin (T_3); a hormone that affects organic metabolism. p. 675

tidal volume (TV) The amount of air entering and leaving the lungs during the respiratory cycle. p. 458

tight epithelia Epithelia in which the water permeability of the tight junctions coupling adjacent cells is low; it is found in regions where osmotic water flow is undesirable. p. 85

tight junction A junction between cells that prevents substances from leaking between them; it is seen in most epithelial tissues. p. 85

tissue A collection of cells The four basic classes are nerve, muscle, connective, and epithelial. p. 10

titratable acidity In the kidneys, the amount of H^+ contained in the urine that has been buffered by phosphate or has acted to decrease its pH.

tone The sound made by a single frequency. p. 256

tonic receptors Sensory receptors that respond to sustained stimuli with an equally sustained generator potential. p. 197

tonotopic A term referring to an orderly mapping of sound frequencies on neurons and/or fibers within the central nervous system. p. 263

torr (t) A unit of pressure.

total body water The total amount of water the body contains.

total lung capacity (TLC) Air in the lungs after a maximum inspiration. p. 459

total peripheral resistance (TPR) The equivalent flow resistance of all the systemic organs "seen" by the heart. p. 407

touch dome endings Specialized touch receptors in the skin. p. 233

trachea The main airway that connects the lungs and pharynx. p. 450

transamination The transfer of amide groups from one compound to another. p. 557

transcellular route A term referring to movement of substances across the membranes of epithelial cells rather than between adjacent cells (paracellular). p. 85

transcription The process by which a portion of a DNA molecule is "copied" to form messenger RNA. p. 63

transduction The process by which a sensory receptor converts a stimulus into a series of nerve impulses.

transfer RNA (tRNA) The species of RNA that binds an amino acid at one end while "reading" triplet nucleotide sequences on messenger RNA at the other end. Each of the 20 amino acids has a specific transfer RNA. p. 63

translation The process by which a messenger RNA molecule guides the synthesis of a protein. p. 63

transferrin An iron-carrying plasma protein. p. 637

transmitters Chemicals released by neurons that alter the membrane permeability of target cells through the activation of specific receptors. p. 202

transmural pressure Pressure across the walls of an organ or blood vessel.

transport maximum (T_m) In the kidneys, the maximum rate of tubular reabsorption of some substance. p. 542

trans-retinal The form of retinal after activation by light that cannot bind opsin.

transverse tubules In muscle, the invaginations of the surface membrane that allow an action potential to "penetrate" the interior of a muscle fiber and eventually cause the release of Ca^{++}. p. 299

tremor Shaking seen in a resting individual.

treppe See staircase phenomenon.

triacylglycerol See triglyceride.

trichromat An individual with normal color vision; that is, having three types of color-sensitive cones. p. 275

tricuspid valve The valve between the right atrium and right ventricle. p. 362

tricyclic antidepressants A class of antidepressant drugs characterized by a chemical structure consisting of three heterocyclic rings. Imipramine is an example.

triglyceride Neutral fat composed of three fatty acid chains esterified on a glycerol backbone. p. 44, 635

tritanope An individual who lacks blue-sensitive conesand thus cannot distinguish between blue and green. p. 275

trophoblast The component of the blastocyst that develops into the placenta after implantation in the uterine wall. p. 728

tropomyosin The muscle protein that winds along the thin filament and covers the actin-binding sites in the absence of Ca^{++}. p. 295

troponin The muscle protein that binds actin (subunit I), tropomyosin (subunit T) and Ca^{++} (subunit C). When Ca^{++} is present, it "pulls" tropomyosin to uncover binding sites for myosin. p. 295

trypsin An enzyme that degrades proteins to peptides by cleaving internal bonds. p. 617

trypsin inhibitor A substance that inhibits trypsin activation in the pancreas. p. 617

turnover The rate at which elements of a cell are replaced.

twitch The contraction of a muscle after a single action potential. p. 302

two-solute model The combined role of active salt transport and differential urea permeability in creating the medullary osmotic gradient. p. 549

two-point discrimination The ability to perceive whether a single or two closely spaced stimuli are present on the skin. p. 237

tympanic membrane The eardrum. p. 256

type IA nerve fibers Sensory nerve fibers with diameters of 13 to 20 μm.

type IB nerve fibers Sensory nerve fibers with diameters of 6 to 12 μm.

type II nerve fibers Sensory nerve fibers with diameters of 1 to 5 μm, which carry fast pain and touch.

U

ulcer An erosion of the wall of the gut, usually in the stomach or duodenum. p. 613

ultrafiltration In the kidneys, the process by which the plasma is filtered free of protein into the glomerulus as a result of arterial blood pressure. p. 143

unidirectional flux The rate of movement of a substance across an interface in a particular direction. The difference between two unidirectional fluxes is the net flux. The unidirectional flux is proportional to concentration.

unitary smooth muscle Smooth muscle in which all the cells are electrically coupled via gap junctions. p. 316

upper esophageal sphincter Smooth muscle between the esophagus and the larynx. p. 608

unsaturated Having one or more double or triple bonds. p. 44

up-regulation An increase in receptor density caused by the presence of a hormone. p. 115, 660

upstroke The rising phase of an action potential.

uracil One of the pyrimidine bases in nucleotides. p. 62

urea The major waste product produced by the breakdown of proteins; it contains two ammonia molecules linked with carbon dioxide. p. 527

ureters The tubes that connect the kidneys with the bladder. p. 528

urethra The tube that connects the bladder with the exterior. p. 529, 689

uric acid A waste product produced by the breakdown of nucleic acids. p. 527

urine The final fluid excreted by the kidneys. p. 537

utricle A receptor in the vestibular system that is sensitive to linear accelerations and head position. p. 264

V

vagina The muscular canal connecting the female uterus with the exterior of the abdomen. p. 687

vagus nerve One of the cranial nerves. p. 347

valance A measure of an atom's ability to donate or accept electrons. Electron donors have positive valences.

van der Waals forces The attraction between molecules that occurs as a result of induced dipole-induced dipole interactions; a weak form of chemical bonding. p. 24

varicose veins A condition in which peripheral veins swell because their compliance has increased. p. 435

varicosities Regions in autonomic nerve fibers at which neurotransmitter is released. p. 329

vas deferens The portion of the male reproductive system between the epididymus and the urethra. p. 689

vasa recta The capillary network in the kidneys that surrounds the Henle loops; part of the peritubular capillary network. p. 532

vasoconstriction Narrowing of a blood vessel. p. 422

vasomotion The random blood flow in a capillary bed, caused by the periodic opening and closing of various precapillary sphincters.

vasopressin Antidiuretic hormone (ADH). p. 116, 670

vein A thin-walled blood vessel that returns blood to the heart. p. 358

vena cava The largest vein; it leads into the right atrium. p. 358

venodilation An increase in the diameter of the veins.

venous pooling The increase in blood volume in the veins of the lower extremities that occurs with prolonged standing.

venous return The return of blood to the heart, measured in liters per minute. p. 376

ventilation Movement of air into and out of the lungs, measured in liters per minute. p. 501

ventilation-perfusion ratio The ratio between the amount of air and blood exchanged in a certain region of the lung. p. 471

ventral horn The portion of the gray matter of the spinal cord that contains the motor neurons. p. 327

ventral root The portion of a spinal nerve that enters the ventral horn of the spinal cord and contains efferent motor fibers. p. 327

ventricles The two chambers of the heart that expel blood into the systemic or pulmonary circulations. p. 361

ventricular fibrillation A disorganized pattern of electrical activity in the ventricle that cannot cause a coordinated contraction.

ventricular septum The tissue separating the two ventricles. p. 361

venule A small vein. p. 358

very low density lipoprotein (VLDL) A lipoprotein manufactured in the liver that carries low-molecular-weight lipids to the cells After these are removed, VLDL becomes low-density lipoprotein. p. 661

vesicle A small, membrane-bound sphere. p. 205

vestibular membrane Reissner membrane. p. 259

vestibular nucleus A collection of neurons in the brainstem that receives input from the vestibular hair cell receptors. p. 338

vestibulospinal tract Descending motor fibers that originate in the vestibular nucleus; the vestibulospinal tract is part of the medial motor system.

villi Large foldings of the intestinal mucosa that increase its surface area. p. 604

viscera Internal organs in the thorax and abdomen. p. 325

visceral pleura The tissue found on the exterior of the lungs. p. 450

viscosity A measure of the internal friction in a liquid.

visual cortex The area of the cortex that receives the primary visual projection; it is located at the rear of the brain.

visual purple Rhodopsin. p. 276

vital capacity (VC) The maximum amount of air that can be moved into and out of the lungs. Total lung capacity minus residual volume. p. 458

vitreous humor A thick fluid found in the portion of the eye between the lens and retina. p. 268

volume flow The amount of fluid in liters or milliliters per minute that passes a given point.

volume overload A condition in which the left ventricle must pump abnormally large amounts of blood because of a leaky mitral valve.

volume receptors Stretch receptors in the right atrium that measure central venous pressure.

vols% A clinical unit of measure for gases; the number of milliliters of a gas contained in 100 milliliters of blood.

volt (V) The unit of measure for electrical potentials. p. 37

voltage The difference in electrical potential between two points. p. 36, 143

voltage-clamp A technique for controlling the voltage of a cell membrane and measuring the movement of ions across it. p. 168

voltage-gated ion channel An ion channel that opens or closes when the electrical field across the membrane changes. p. 170

voluntary nerve A somatic motor nerve.

W

water The fundamental solvent for all biochemical processes. A strongly polar molecule in which other polar molecules and ions are soluble.

weak acid An acid that incompletely dissociates in water. p. 27

Wernicke's area One of the language areas of the brain; it encodes and decodes concepts and commands. p. 242

white matter Areas of the central nervous system made up of nerve fibers rather than cell bodies. p. 221

white rami Connections between the sympathetic ganglia and the spinal cord. p. 329

withdrawal reflex A spinal reflex involving removal of a limb from a potentially harmful stimulus. p. 108

Wolffian duct The part of the embryonic gonad that develops into the male reproductive ducts; it disappears in the female. p. 695

word deafness A condition in which comprehension is impaired only for verbal communication.

work Force times distance. p. 31, 38

work loop In cardiovascular physiology, the relationship between ventricular volume and ventricular pressure. p. 338

Z

Z structure In muscle, the structure into which the thin filaments insert in a regular hexagonal array. The Z lines define the limits of a sarcomere. p. 292

zona compacta The portion of the uterine lining nearest to the lumen.

zona pellucida One of the layers of a developing follicle. p. 704

zona spongiosa The portion of the uterine lining rich in blood vessels.

zonular fibers In the eye, fibers that act to change the curvature of the lens. p. 268

zygote The result of the combination of an egg and sperm. p. 690

Illustration Credits

Chapter 1

1-1: Jack Tandy after Harvey: *DeMotu Cordis*
1-4 Pagecrafters/Nadine Sokol
1-8, 1-9 Barbara Stackhouse
1-10 Cardiovascular Unit, University of Alabama Medical Center

Chapter 2

2-3, 2-15, 2-17: Jack Tandy
2-6, C: Michael Godomski/Tom Stack & Associates
2-7: William Ober
2-19, B: Courtesy of O.N. Markand, M.D.
2-19, C: Rebecca S. Montgomery

Chapter 3

3-3, D, 3-25: Jack Tandy
3-8, B: Brenda R. Eisenberg, Ph.D., University of Illinois at Chicago
3-8, C, 3-13: Dr. Andrew P. Evan, Indiana University
3-14, 3-20, 3-23: Ronald J. Ervin
3-26: William Ober

Chapter 4

4-1, 4-6, 4-7, A, 4-9: Ronald J. Ervin
4-2, 4-12, B, 4-18: Brenda R. Eisenberg, Ph.D., University of Illinois at Chicago
4-3, 4-4, 4-5, 4-11, 4-16: Dr. Andrew P. Evan, Indiana University
4-7, B: Birgit H. Satir, Release of content of secretory vesicle via the fusion of the vesicle membrane and the cell membrane. The content expands and empties to the outside of the cell.
4-8 A, 4-10, 4-12, A, 4-14, A, 4-17, B: William Ober
4-8, B: T.W. Tillack
4-8, C, 4-10: Charles Flickinger
4-9, B, 4-17, D, 4-25, 4-27: Jack Tandy
4-14, B: Ellen Chernoff
4-15: Michael P. Schenk
4-17, A: Carmillo Peracchia
4-17, C: Elliot Hertzberg

Chapter 5

5-2, 5-3, B, 5-21, 5-22, 5-C, 5, D, 5-E, 5-F, 5-G

5-3, A, 5-6,: Trent Stephens
5-7, 5-19: Barbara Stackhouse
5-11, 5-12, 5-13, 5-17: Andrew Grivas
5-9: Scott Bodell
5-18: Patricia Kane, Indiana University Medical Center, Radiology Department
5-A: Cynthia Turner Alexander/Terry Cockerham, Synapse Media Production
5-B: William Ober

Chapter 6

6-1, 6-2, 6-10: William Ober
6-6, 6-7: Ronald J. Ervin
6-9: Barbara Stackhouse
6-16: Joan M. Beck
6-17: Jack Tandy

Chapter 7

7-2, A, 7-13, 7-14, 7-19, 7-21, 7-23: Barbara Stackhouse
7-2, B: Alan Peters
7-9: Roger T. Hanlon
7-20, A: Scott Bodell
7-20, B: Brenda Eisenberg
7-22: Dr. Andrew P. Evan, Indiana University

Chapter 8

8-1, B, 8-2, 8-3, 8-17, 8-18, 8-19, 8-21, 8-24, 8-26
8-1, C, 8-1, D, 8-13, B, 8-13, C, 8-13, D: Dr. Andrew P. Evan, Indiana University
8-1, A, 8-23: Scott Bodell
8-5, 8-8, 8-9, 8-10, 8-12, 8-13, 8-16, 8-19, 8-20, 8-A: Barbara Stackhouse
8-13, A: Michael V.L. Bennett
8-14, A, 8-16: Brenda R. Eisenberg, Ph.D., University of Illinois at Chicago
8-15: Joan M. Beck/Andrew Grivas
8-22: M. Klymkowsky; In Stroud, R.M.: Acetylcholine Receptor Structure, Neuroscience Commentaries 1, 124-138, 1983. Reprinted by permission of the Society for Neuroscience

Chapter 9

9-1, 9-17, B: Jack Tandy
9-3, A, 9-5, 9-9: William Ober
9-3, B: Brainslav Vidic
9-4, 9-15, 9-19: Michael P. Schenk
9-6, 9-8, 9-9: Scott Bodell
9-10, A, 9-12: Barbara Stackhouse
9-10, B: A.W. Campbell: Histological Studies on the Localization of Cerebral Function, Cambridge University Press, England, 1905
9-12: G. David Brown
9-14: Janis K. Atlee/Michael P. Schenk
9-18: Terry Cockerham, Synapse Media Production
9-A, 9-B: Patricia Kane, Indiana University Medical Center, Radiology Department

Chapter 10

10-2, 10-3, 10-4, 10-8, 10-11, A, 10-12, A, 10-15, 10-16, 10-21, A, 10-28, A, 10-18, B, 10-32, 10-33: Marsha J. Dohrmann
10-6, A: Dr. Tilney
10-6, B: Dr. Andrew Evan, Indiana University
10-7, 10-9, 10-10, 10-14, 10-24, 10-30: Barbara Stackhouse
10-11, B, 10-12, B: Kathy Mitchell Grey
10-13, 10-18, 10-29: G. David Brown
10-17, 10-20, B, C: Kenneth G. Julian, C.R.A., F.O.P.S.
10-20, A: William Ober
10-21, B: E.R. Lewis/F.S. Werblin/Y.Y. Feevi, University of California, Berkeley
10-23: Donna Odle
10-28, C: Patricia Kane, Indiana University Medical Center, Radiology Department

Chapter 11

11-1, 11-16, 11-17, 11-18, 11-19, 11-26, 11-27, 11-30, 11-31: Barbara Stackhouse
11-2, 11-4, 11-5, 11-6, A, 11-7, 11-11, 11-13, B: Joan M. Beck
11-3, 11-8, 11-12, 11-21, 11-24, 11-28: Brenda R. Eisenberg, Ph.D., University of Illinois at Chicago

Index

Acylglycerols—cont'd
structure of, 44, *46*
Addison's disease
adrenal cortical hyposecretion in, 129
hyperkalemia and, 575
sodium reabsorption in, 575
symptoms of, 572
Adenine
base pairing of, 64
in DNA, 62
structure of, *62*
Adenohypophysis, 116
Adenosine
coronary blood flow and, 383
function of, 106
in intrinsic control of systemic blood vessels, 423t, 424
Adenosine biphosphate, 39
in hemostasis, 394
Adenosine diphosphate, 87
in enzyme control, 94
in metabolic control, 94
Adenosine monophosphate, cyclic; *see* Cyclic AMP
Adenosine triphosphate, 39, 87
as energy for muscle contraction, 516
in enzyme control, 94
function of, 82
hydrolysis of, in smooth muscle, 320-321
passage between cells, 84
production of, 86-87
from complete glucose oxidation, 92
in skeletal muscle, 308
structure of, *82*
Adenylate cyclase, 111
ADH; *see* Antidiuretic hormone
Adipose tissue, 10
metabolic rate of, 674
ADP; *see* Adenosine biphosphate
Adrenal cortex, hormones produced by, 110
defects in, 128-129
Adrenal gland, 125
during exercise, 518-519
hormones of
in postabsorptive state of metabolism, 667
target and actions of, 133
during puberty, 700
tumors of
aldosterone-secreting, 575
electrolyte levels in, 575
Adrenal medulla, 329
function of, 330
Adrenergic receptors, 349-350
of arterioles and veins, 422-423
of cardiac muscle, 379
locations of, 351t
work of breathing and, 451
Adrenergic synapses, transmitter synthesis in, 351-352
Adrenocorticotropic hormone
cortisol production and, 121
in Cushing's syndrome, 130
in diagnosis of cortisol deficiency, 129
during exercise, 519
function of, 116
in initiation of labor, *743*
releasing hormones and target/effect, 118t
secretion of, 128, 669
source, target cells, and action of, 131
Adrenogenital syndrome, 701
defect and effects of, 699t
Afterload of cardiac muscle, 376
Agglutination
defined, 761
by immunoglobulin G, *762*
Aging
body composition and, 674
and increase in fats as proportion of body weight, *675*
Air
constituents of, 482t
defined, 481
Airway(s), 475
anatomy of, *451*

Airway(s)—cont'd
autonomic constriction of, 348
components of, *450*
protective functions of, 453-454
structure of, 450-452
sympathetic dilation of, 348
Airway resistance, 475
allergic reactions and, 451-452
sites of, 451
Alanine, 54t
nucleotide code for, 66t
structure of, *53*
Albumin
functions of, 392
liver secretion of, 629
normal range of, 393t
Alcohol, 249t
blood-brain barrier and, 414
dependence on, 250
diuretic effects of, 559
effects of, 250
on esophageal sphincter, 608, 610
on gastric acid secretion, 613
Alcohol abuse, defined, 250
Alcohol dehydrogenase in ethylene glycol poisoning, 60
Alcoholism, nerve demyelination in, 183
Aldosterone, 49, 110
action of, 555
in blood pressure regulation, 427, *427*
in Cushing's syndrome, 130
function of, 125, 129
in heart failure, 432
in hyperkalemia, 575
placental, 732
precursors of, 129
in regulation of potassium excretion, 574-575, 575
in renal handling of sodium and potassium, 554-555, *555*
secretion of, 586
in sodium reabsorption, 572
in sodium regulation, 537
source, target, and actions of, 133
Alkaline tide, 612
Alkalosis
metabolic
cause of, 580, 580t
chronic, 583
compensation for, 582t
renal compensatory responses in, 585-586
with respiratory compensation, 582
respiratory
causes of, 580, 580t, 585
compensation for, 582t
renal compensation after, *584*
tubular responses to, *583*
from vomiting, 642
Alleles, 758
defined, 691
Allergens, effects of, 451
Allergic reactions, 451-452, 776
Allosteric activators, 60
Allosteric inhibitors, 60
Allosteric sites, 60-62, *61*
Alpha 1,4 glycosidic bonds, 51
Alpha fetoprotein, maternal serum, testing for, 740
Alpha particles, 21
Alpha waves, 243, *245*
Alpha-coactivation, 335-336, *337*
Altitude
chemoreceptor response to, 509-510
respiratory alkalosis from, 585
Alveolar cells, type II, 467
Alveolar duct, 454, *454*
diameter and functions of, 451t
Alveolar gas equation, 483, 487
Alveolar minute volume, *467*
defined, 460
definition and resting value of, 460t
Alveolar pressure, *465*, 467
Alveolar sacs, 475

Alveolar ventilation, 459-460
with different combinations of respiratory rate and tidal volume, 460t
ratio to alveolar blood flow, 471
Alveoli, 475
contaminated particles in, 454
diameter and functions of, 451t
diffusion in, *455*
function of, 450
membrane compliance of, 467
Alzheimer's disease
acetylcholine in, 204
growth factors in treatment of, 122
Amiloride, diuretic effect of, 558
Amino acid(s)
absorption of, mechanisms of, 632
carrier-mediated co-transport of, 626
daily turnover, storage, and functional reserve of, 668t
derivatives of, as hormones, 123
essential, 54, 54t, 662
defined, 54
in protein synthesis, 667
glucagon secretion and, 666
of hormones, 110
ionizable, 54
lysozyme sequence of, *54*
metabolism of, 92-94
as neurotransmitters, 204
nonessential, 54t
in energy metabolism, 667
nucleotide codes for, 66t
peptide bond formation by, 631
plasma, effect on insulin and glucagon secretion, 658t
in polypeptide chains, *55*
protein sequence of, 56, 71
sequence of, 43
in sickle cell anemia, 57
structure of, *53*, 54
synthesis of, 43
Aminopeptidases
in protein digestion, 631
substrate and products of, 629t
Amniocentesis
miscarriage risk from, 740
process of, 740
Amnion, 736, *738*
Amphetamine, 249t
action of, 251
addiction to, 248
as stimulants, 249
structure of, 251
uses of, 251
Amphipathic molecules, 45
Ampulla of Vater, 614
Amygdala, long-term memory and, 241
Amylase
digestive function of, 650
starch digestion by, 629-630
substrate and products of, 629t
α-Amylase
function of, 595
intestinal, 622
pancreatic, 617
in saliva, 606
Amylin, function of, 669
Amylopectin, 629
Anabolic steroids, 321
Anabolism
defined, 86
processes of, 656, 656t
Anaerobic transition, 517
Anal sphincter, 597, 621-622
Anaphylactic reactions, characteristics of, 773
Anaphylactic shock, 777
Anatomical dead space, 459-460, 483
definition of, 460, 460t
resting value of, 460t
Androgen(s)
adrenal production of, 700
in body-building, 703
in differentiation of external genitalia, 129
effects of, 700

Index

Cell(s)—cont'd
  type II alveolar, 467
  volume of, determinants of, 144-146
  white blood; *see* White blood cell
Cell cycle
  interphase period of, 64
  of liver cells, *65*
  mitosis in, 64-65, *65*
Cellulose
  degradation of, 629-630
  structure of, *52*, 52
Central nervous system, 247
  afferent input to, 108
  components of, 219, *221*, 221-222
  depressants of, 249-250
  drugs affecting, 249
  functions of, 219
  hearing pathways of, *263*
  integration in, 189
  motor programs in, 342-343
  neurons in, 190, *191*, 209
  regions of, *221*
  in respiratory control, 502-503
  stimulants of, 249
  subdivisions of, 223-224
    functions of, 225t
  synapses of, 203
Central venous pressure, 418
  cardiac output and, 428-430
  defined, 407
  typical value for, 429
  vascular determinants of, 430-432
  vascular function curve of, *430*
Centrifugation, 392, *392*
Cerebellum
  body maps in, *238*
  components and function of, 225t
  function of, 224
  interactions of, *343*
  in motor programs, 353
  in rapid movement coordination, 341
Cerebral cortex, 224; *see also* Brain; specific lobes
  alpha waves of, 243, *245*
  beta waves of, 243, *245*
  delta waves of, 243, *245*
  dominance of, 242
  fissures of, 224
  interactions of, *343*
  left, 242
  in memory, emotion, and self-image, 243
  organization of, 224, 227
  regions of, *227*
  right, 242-243
  structure of, *230*
  theta waves of, 243, *245*
Cerebral dominance, 242
Cerebral hemispheres, 223
  components and function of, 225t
Cerebral ventricles, 221
Cerebrospinal fluid, 221, 564
  barrier between blood and, 414
  buffering of, 505
  PCO_2 of, 499, *506*
  pH of, 499
  production of, 414
  in respiratory control, 504-505
Cervical cancer, estrogen replacement therapy and, 710
Cervical cap, 714
  failure rate of, 715t
Cervical mucus, evaluation of, for birth control, 714
Cervical plexus, 327
Cervix, *686*
  anatomy of, 688
  dilation of, 742, 743
Cesarean delivery, effect on newborn, 747
cGMP; *see* Cyclic guanosine monophosphate
Channels
  calcium; *see* Calcium channels
  gated, 148
  ligand gated, 207
  potassium; *see* Potassium channels
  protein, *147*, 147-148, 155

Channels—cont'd
  sodium; *see* Sodium channels
  toxins that block, 176-177
  voltage-gated, 170-173
Charge, defined, 20
Charles' law, *480*, 481, 482
Chemical reactions
  activation energy of, 57
  coenzymes in, 59
  defined, 22
  intermediate transitional states in, *58*
  oxidation-reduction, 58, 87
  rate of, enzymes and, 57-58
Chemical synaptic transmission, 202
Chemiosmotic hypothesis of oxidative phosphorylation, 92, 95
Chemistry
  of cells, 42-71
  organic, 43
Chemoreceptors, 109, 194
  in breath-holding, 509
  central, 504, 515
    activation of, 505, *505*
  peripheral, 505, 507, *507*, 515
    and acclimatization to altitude, 509-510
    interaction with central chemoreceptors, 509-513
    and respiratory response to CO_2, 508
  in respiratory control, 502t
Chest
  during inspiration and expiration, *464*
  wall of, 460-462, *461*
Chief cells, 611, *611*
Chimpanzees, genetic code of, versus human genetic code, 8-9
Chlordiazepoxide; *see* Librium
Chloride ions, normal range in blood, 393t
Chloride shift, 493
Chlorine ions, renal handling of, 551t
Chlorpromazine, 249t, 250
Chocolate cysts, 719
Choking, Heimlich maneuver for, 474
Cholecalciferol
  in calcium ion homeostasis, 577
  source, target, and actions of, 133
Cholecystokinin, 618, 647, 650, 670, 681
  effect on gastric motility, 649
  function of, 615
    digestive, 648-649
  glucagon secretion and, 666
  during pregnancy, 739
  as satiety signal, 671
  secretion of, control of, *648*
  source, target, and actions of, 135
  stimuli, effects, and role of, 645t
  targets of, *648*
Cholera, cause of death in, 615
Cholera toxin, action of, 642
Cholesterol, 44, 617, 632, 635
  atherosclerosis and, 441
  bile acid synthesis from, 617
  elevated plasma levels of, 662
  excessive secretion of, 618
  function of, 48-49
  in low-density lipoproteins, 661
  metabolism to progesterone and estrogen, *731*
  plasma, dietary fiber and, 621
  steroid synthesis from, 110
  structure of, *48*
  synthesis of, 441
Cholesterol esters, 632, 635, *636*
Cholesterol ions, normal range in blood, 393t
Cholinergic fibers, sympathetic, acetylcholine release by, 330
Cholinergic receptors, 349
  location of, 350t
Cholinergic synapses, transmitter synthesis in, 351-352
Chondrogenesis, 671
Chordae tendinae, 363
Chorion, 728, 736, 749
Chorionic somatomamotropin, characteristics and effects of, 732
Chorionic villus biopsy, 740

Choroid layer, 268, *268*
Choroid plexus
  cerebrospinal fluid production by, 414, 504
  function of, 414
  response to high CSF pH, 510
Chromaffin cells, 329
  substances released by, 329-330
Chromophore, 276
Chromosome(s)
  autosomal (somatic), 690, 694
  diploid number of, 63, 64, 690
  DNA deletions in, 128
  haploid number of, 64, 690, 694
  homologous pairs of, 63
  human, 690
  during meiosis, 64
  during metaphase of mitosis, *690*
  mutations of, 128
  sex, 690
  structure of, 62-63
  X, 690, 694
  Y, 690
    male determination by, 694
    sex-determining region of, 694
Chromosome 17, 128
  growth retardation and, 128
Chronotropic effect
  defined, 379
  positive versus negative, 379
Chylomicrons
  formation of, in intestinal epithelial cells, 635
  plasma lifetime of, 662
Chyme, 596, 611, 647
  breakdown of, 625
  circulation of, 620
  duodenal, 614, 647
  flow from stomach to duodenum, 641
  formation of, 613
  mixing of, 620
  transport of, 621
Chymotrypsin
  pancreatic, 617
  in protein digestion, 632
  substrate and products of, 629t
Cigarette smoking
  coughing from, 454
  deaths from, 250
  economic costs of, 250
  emphysema and, 470
  lung cancer and, 455
  passive, 250
Ciguatoxin, toxic effects of, 177
Cilia
  of airway passages, *453*, 453-454
  function of, 475
Ciliary body, 268, *269*, 270
Ciliary muscles in near vision, *270*
Cimetidine, action of, 613
Circadian rhythms, 240
Circulation; *see also* Cardiac entries; Cardiovascular system; Heart
  fetal, 749
    changes after birth, 736
  fetal-placental, 732, *732*, 734-735
  Harvey's research on, 4
  hepatic portal, 358, 626-629
  hypothalamic-anterior pituitary portal, 116
  overview of, 358, 360-361
  portal, 116
  pulmonary, adaptations of, 471
  systemic, variables affecting, *422*
Circulatory system
  capacitance elements of, 404, 418
  diagram of, *410*
  gas exchange by, *14*
  pulmonary and systemic loops of, 358, *358*, 388
  vessels of, pressure, length, diameter, area, and relative resistance of, 399t
Cisternae, terminal, 299, 322
Clitoris, 687, 720
Clotting, defects of, 395
  in hemophilia, 396-397

Drugs—cont'd
  cardiac muscle and, 315
  CNS and, 249t
  defined, 248
  dependence on, 248
  diuretic, mechanisms of, 558-559, 559
  extent of use of, 248
  illegal, 251-253
  legal, 250-251
  nervous system and, 248-253
  over-the-counter, 248
  and responsiveness to CO2, 513
  tolerance to, 248
  withdrawal from, 248
Duct cells
  salivary, 606
  secretions of, 615
Ductus arteriosus, 749
  in fetus, 734
  patent, 735
Ductus venosus, 749
Duodenum, 597
  anatomy of, 614
  entry routes for pancreatic and liver secretions, 615
  normal versus ulcerated, 613
  x-ray of, 594
Dwarfism, cause of, 673
Dynes, 32
Dyskinesias, defined, 341
Dysrhythmias, 368-369; see also Arrhythmias

E

Ear(s)
  frequency range of, 194
  inner, 256, 257
    components of, 258
  middle, 257, 257
  outer, 256-257, 257
Eardrum, 256
Eccles, John C., 170
ECG; see Electrocardiogram
Edema, 418
  causes of, 416t
  pulmonary, 381, 471
    mechanism of, 416
Edinger-Westphal nuclei, 347
Eicosanoids, 44, 49, 111
  characteristics of, 105
  subunits and function of, 45t
Einthoven's triangle, 371, 372
Ejaculation, 720
Ejaculatory duct, 689, 721
Elastase
  pancreatic, 617
  substrate and products of, 629t
Elastic recoil of aorta, 404
Elastic restoring force, 35, 35-36, 40
Elasticity, muscle, 304
Elastin in vessel wall compliance, 403-404
Electrical events
  comparison of, 162t
  molecular basis of, 166-168
Electrical force, 20
Electricity, volume conductor of, 37
Electrocardiography, 37, 369-371
  of cardiac muscle cell, 369, 370
  of heartbeat, 370, 370
  leads in, 371
  normal, 37
Electrochemical equilibrium, 152, 152
  Nernst equation and, 157-158
Electroencephalogram, 37, 38, 243
  of cortex activity, 245
  in waking and sleep states, 244
Electrolytes
  absorption of, 640-641, 643
  defined, 27
Electromagnetic spectrum, 274
  visible, 272
Electromyographic biofeedback, 332
Electron shells, 40
Electron transport chain, 91

Electrons, 20
  in bond formation, 22
  carriers of, 87
  orbitals of, 22
Elements
  atomic numbers of, 20
  in human body, 20t
  isotopes of, 20, 21
Embolism, 440
  defined, 395
Embolus(i), pulmonary, 395
Embryo, 278, 737, 749
  after 1 month's development, 738
  development of, 728, 735-736
  implantation of, 729-730, 730
  reproductive organs of, 695
  sexual bipotential of, 696
Embryoblast, 749
Emotion, gastric secretions and, 4
Emphysema, 469
  cigarette smoking and, 470
Emulsification of fats, 632-634
End-diastolic volume, 363, 388
  in Frank-Starling curve, 377
  stroke volume and, 384
Endocardium, 361
Endocrine cells, 601
Endocrine disease
  genital sex effects of, 699t
  from hormone deficiency, 128-130
  from hormone excess, 130
  mechanisms of, 128-130
  from transduction failure, 130
Endocrine organs, 10, 125-135; see also specific organs
  expanded list of, 125
  functions of, 108
  location of, 126
Endocrine system
  in blood glucose control, 118-119, 121
  in calcium regulation, 577-579
  components and functions of, 12t
  control of, 94, 219
  in control of urine concentration, 550
  during exercise, 517t, 518-519
  feedback in, 118
  glomerular filtration rate and, 569
  in metabolic control, 654-683
  in phosphate regulation, 577-579
  in reproductive system regulation
    female, 703-713
    male, 702-703
  sodium regulation by, 563
Endocytosis, 76-77, 81
  defined, 76
  receptor-mediated, 77
Endolymph, 259, 265
  during head rotation, 267, 267
  in rotational nystagmus, 267-268
Endometrial cycle; see Menstrual cycle; Uterine cycle
Endometriosis
  characteristics of, 718
  treatment of, 718-719
Endometrium, 688, 730
  during menstrual cycle, 711, 712
Endomysium, 292
Endonucleases, restriction, 70
Endopeptidases in protein digestion, 631
Endoplasmic reticulum, 78-79, 80, 81
  function of, 95
  rough, 75, 78
    characteristics and functions of, 75t
    ribosomes of, 78-79, 81
  smooth
    characteristics and functions of, 75t
    function of, 78-79
    of muscle, 78
Endorphins, 205
  function of, 128
  in pain perception, 214
  secretion of, 128
Endothelial cells, anticoagulant mechanism of, 395, 398

Endothelins
  function of, 105-106
  hypertension and, 573
Endothermy, 13
Endotoxins in inflammation and fever, 755
End-systolic volume, 363
Endurance training, 523
Energy
  activation, 71
    in biochemical reactions, 57
  chemical, 19
  conservation of, in first law of thermodynamics, 38
  defined, 31
  for exercising skeletal muscle, 516
  expenditure during exercise, 518, 518t
  free, 39
  levels of, 22, 22
  metabolism of, 86-94
    liver's role in, 626-627
    skeletal muscle fiber types and, 308-310
  photosynthetic source of, 19
  sources of, 518
  thermodynamics of, 38
  transformation of, 38
Enkephalins, 205, 250
  in enteric nervous system, 606
  in pain perception, 214
  in sympathetic pathways, 329
Enteric brain, 606, 625
Enteric nervous system; see Nervous system, enteric
Enteric reflexes, 597
Enteroendocrine cells, 602
Enterogastric reflex, 647
  stimulus and effect of, 644t
Enterokinase, function of, 617
Enterotoxins, 772
  Staphylococcus aureus, effects of, 771
Entropy, 38
Enzyme(s)
  activation of, 68
  active sites of, 59
  allosteric regulation of, 60-62
  allosteric sites of, 61
  angiotensin converting, in sodium reabsorption, 571
  ATPase, 140
  brush-border
    digestive function of, 650
    starch digestion by, 629-630
  catalysis by, 57-62
    active sites in, 59-60
    competitive inhibition in, 60
    concentration dependence in, 60
    Michaelis-Menten plot in, 60, 61
  control of, 94
  defined, 71
  digestive, 626
  effect on activation energy, 57, 58
  function of, pH and, 579
  functional classes of, 140t
  hydrolytic, 617
  membrane-bound, 139-140
  names of, 58-59
  pancreatic, 617
    substrate and products of, 629t
  pH and, 60
  pH optimum of, 60, 579
  in protein digestion, 631-632
  protein subunits in, 57
  and rate of biochemical reactions, 57-58
  substrate affinity of, 59
  temperature and, 60
Eosinophils, function of, 752
Ephedra, 251
Ephedrine, 249t
Epicardium, 361
Epidermal growth factor
  function of, 122
  origin and effect of, 105t
Epididymis, 721
Epilepsy, brain surgery for, 243
Epimysium, 292, 293

Epinephrine, 110
  binding to pacemaker cells, 379
  during birth, 747
  blood glucose levels and, 118, 121
  in cardiovascular response to exercise, 125
  chemical characteristics and secretion site of,
    657t
  chromaffin cell release of, 329-330
  cortisol and, 116, 669-670
  during exercise, 518-519
  function of, 204, 330
  in glycogen breakdown, 119, *120*
  half-life of, 115
  for heart failure, 432
  insulin release and, 669
  plasma glucose levels and, 667
  receptors, cell type, membrane effect, and effect
    on heart, 378t
  secretion of, regulation of, *667*
  source, target, and actions of, 133
  in starvation, 681
  structure of, *330*
Epiphyseal plates, 671
Epithelial cells
  intestinal
    absorptive function of, 602
    chylomicron formation in, 635
    enzymes of, 617
    in fat absorption, 634-635, *635*
    functions of, *601*, 601-603, 602t
    secretions of, 615
    types of, *603*
  junctions between, 84, *84*
Epithelial tissue, 10
  leaky, 85-86
  tight, 85
  tight junctions between, 85-86
Epitopes, 760, 766, 778
EPSPs; *see* Excitatory postsynaptic potentials
Equilibrium, 40
  defined, 141, *142*
  electrochemical, 152, *152*
    Nernst equation and, 157-158
  and law of mass action, 38-39
  sense of, 255
Equilibrium constant, 588
  in bicarbonate buffer system, 589
  defined, 38
  example of, 38-39
Equilibrium potential
  for K^+ and Na^+ ions, 166
  relationship to concentration ratio, 152-153
Equivalency, 30-31
  defined, 31
Erectile tissues, 689
Erection, 719
Error signals, 98, 123
ERV; *see* Expiratory reserve volume
Erythroblastosis fetalis, 773, *773*
Erythrocytes, 392; *see also* Red blood cells
  derivation of, *392*
Erythropoiesis
  defined, 393
  regulation of, 393
Erythropoietin
  function of, 122, 393
  source, target, and actions of, 133
Erythroxylon coca, 251
Esophageal hiatus, 611
Esophageal sphincter, 608
Esophagus
  anatomy of, 594-596
  function of, 595, 595t
  muscles of, 595-596
Estradiol, 721
  in breast development, 744
  effects of, 698, 700
  follicle secretion of, 705
  in follicular development, 708
  testosterone conversion to, 698
  in uterine cycle, 711
Estrogen, 49, 110, 721
  after delivery, 745
  after menopause, 710

Estrogen—cont'd
  bone maintenance and, 578-579
  in contraception, risks of, 715
  osteoporosis and, 674
  in positive feedback system, 102
  during pregnancy, 744
  during pregnancy, labor, and lactation, 739t
  during puberty, 700t
  synthesis from cholesterol, *731*
  target and actions of, 134
  urinary excretion of, in women, *710*
  in uterine cycle, 711
Estrogen pill, failure rate of, 715t
Estrogen replacement therapy, cervical cancer
    and, 710
Estrous cycle, 711
Ethanol; *see also* Alcohol
  effects of, 250
Ethics, experimentation and, 4, 6
Ethylene glycol
  intoxication from, 585
  toxicity of, 60
Eucaryotic cells, function of, 7
Eustachian tube, 257, *257*
Evoked potentials in motor program initiation,
    343
Evolution
  Darwin-Wallace theory of, 7
  Gould's theory of, 7
Excitability, defined, 161
Excitation, 299
Excitation-contraction coupling, 300-302, *301*
  in cardiac muscle, 315, 366, *367*, 368
  in smooth muscle, 316, 318-321
Excitatory postsynaptic potentials, 209
  integration of, 210
Excretory system, 560
Exercise
  adaptive effects of, 523
  anaerobic transition during, 517
  body weight and, 674
  cardiac output during, 384, *385*, *438*
  cardiovascular effects of, 125, 421, 435-439,
    442, 520-521
  cardiovascular limit in, 436-437, 439
  coronary artery disease and, 441
  cutaneous perfusion and, 442
  dynamic, 521
  endocrine responses to, 518-519
  energy cost of, 518, 518t
  environmental factors in, 521
  fluid balance during, 521-522
  gender-related differences and, 522
  glycogen loading and, 312
  hematocrit during, 411
  during hot weather, 521-522
  immune system and, 522
  lactic acid production during, *310*
  lung volumes during, *519*, 519-520, *520*
  organ perfusion during, *438*
  osteoporosis and, 674
  oxygen consumption during, 310, 519-520
  physiological changes during, 517t
  physiology of, 516-523
  respiration during, 515
  static, 521
  stroke volume during, 384, *385*
  submaximal, circulatory changes during, 439t
  sustained, 437
  thermoregulation during, 521-522
  training and, 523
  ventilatory response to, 512, 514
Exocrine glands, 597
  in negative feedback loop, 98
Exocytosis, 76-77, *78*
  defined, 76
Exons, 67
  defined, 67
Exopeptidases in protein digestion, 631
Exophthalmos, 679
Experiments
  with animals, controversy over, 6
  ethics in, 5, 6

Expiration, *464*
  control of, 503t
  passive, 462
Expiratory reserve volume
  defined, 458, 459t
  resting value of, 459t
Extracellular fluid, 586; *see also* Body fluids
  components of, 13
  control systems for, 563
  dominant cation of, 564
  homeostasis of, 566-567, 569
  ionic composition of, 152t
  normal osmolarity of, 145
  osmotic pressure of, *145*
  water-based solutions of, 19
Extrafusal fibers, 334, 335t
Eye(s); *see also* Visual system
  accommodation of, 270
  autonomic innervation of, 349t
  as image-generating device, 268, 270
  structure of, 268, *268*, *269*, 270
Eye, electromagnetic sensitivity of, 194

F

Fab, 760
Fabricius, bursa of, 764
Facial nerve, *222*
  pathway and function of, 223t
Facilitation, defined, 338
F-actin, 295
Factors, clotting; *see* Clotting factors
FAD; *see* Flavine adenine dinucleotide
FADH; *see* Flavine adenine dinucleotide, reduced
Fallopian tubes, *686*, 687-688, 721
  blocked, 746
    treatment of, 746-747
  with cilia, *726*
  contractions of, 726
  embryonic development of, 695
Falloscopy, 746
Famine, 655
Faraday's number, 157-158
Farsightedness, *271*, 272
Fascicles, 292, *293*
Fasting, basal metabolic rate in, 667-668
Fat(s)
  absorption of, 632-635
    micelle formation and, *632*, 634-635, *635*
  in absorptive state, *661*
    versus postabsorptive state, *665*
  brown, 681
    in infant temperature regulation, 681
    in newborn, 747
  caloric content of, 668t
  daily inputs of, 641t
  daily turnover, storage, and functional reserve
    of, 668t
  deposition of, human growth hormone and,
    672-673
  dietary
    cancer incidence and, 49
    composition of, 632
  digestion of, 617, 632-635
    by pancreatic lipases, 634, *634*
    in duodenal chyme, 648
  effects of insulin on, 659t
  as energy source, 518
  fake, 637
  female versus male deposits of, 698, 700
  intestinal emulsification of, 632-634
  as proportion of total body weight with age,
    675
  storage of, 671
Fatty acids, 44, 632
  as alternative energy source, 663
  beta oxidation of, 93, *93*
  in duodenal chyme, 648
  function of, 44
  metabolism of, 92-94
  omega-3, coronary artery disease and, 441
  saturated, 44
    versus unsaturated, *46*
  structure of, 44
  unsaturated, 44

Proline, 54t
nucleotide code for, 66t
structure of, *53*
Proopiomelanocortin, studies of, 128
Proprioception, axons mediating, 235
Proprioceptors, function of, 194, 231
Prostacyclin, 105
effects of, 104
function of, 395
Prostaglandin(s)
abbreviations for, 49
aspirin inhibition of, 234
effects of, 104t, 754t
on gastric acid secretion, 613
functions of, 49
in initiation of labor, *743*
in intrinsic control of systemic blood vessels,
423t, 424
and length of gestation, 742
in modulation of sensory neuron response, 211
production of, 105
source, target, and actions of, 135
structure of, 49
synthesis of, drug action and, *757*
in uterine cycle, 711
Prostaglandin E2, gastric acid secretion and, 603
Prostate, 689, 721
development of, 698
Proteases, 59
pancreatic, 617
pH range of, 626
in protein digestion, 631
Protein(s), 52, 54-55, *55*, *57*
absorption of, 631-632
alpha helix structure of, 55
amino acid sequence in, 71
biochemistry of, 43
breakdown of, 625
as buffer, 492, 493t
buffer systems and, 580
caloric content of, 668t
carriers, 155
catabolism of, 656
in diabetes mellitus, 658-659
channels, 155
codes for, 63
complementary incomplete, 662
complete, 662
conversion to glucose, 666-667
daily inputs of, 641t
defined, 55, 71
denaturation of, 611, 631-632
dietary, in stimulation of gastric secretion, 646
digestion of, 626, 631-632
in energy metabolism, 94
as energy source, 518
filtration of, 544
fluid mosaic model of, 138
formation of, 631
function of, 55, 57
G; *see* G protein
gastrointestinal conversion of, 593
in Golgi apparatus, 81
histidine groups of, 54
as hormones, 110, 123
junctional, 139
functions and examples of, 140t
membrane
extrinsic, 138
functional classes of, 140t
functions of, 138-141
intrinsic, 138
metabolism of, 86
molecules of, types of, 4
mRNA translation into, 67-68
olfactory binding, 286
as percentage of body composition, 55
plasma, 392, 393t
pleated sheet structure of, 55
posttranslational modifications of, 68
receptor, 138-139, 194-195, *195*
recognition, functions and examples of, 140t
structure of, 55, *55*, *57*
primary, 55, *56*

Protein(s)—cont'd
structure of—cont'd
quaternary, *56*, *57*
secondary, 55, *56*
tertiary, *56*, *57*
subclass, subunits, and functions of, 45t
synthesis of, 672
in absorptive phase of metabolism, 662
from messenger RNA, *69*
transport, 76, 139, 146-149
in artificial membranes, 150
channels, *147*, 147-148, *148-149*
functions and examples of, 140t
undernutrition and, 416
urinary excretion of, 538
Protein buffer system, constituents and role of,
29t
Protein C, anticoagulant function of, 398
Protein kinase, second messenger regulation of,
114
Protein kinase C, 113
Protein phosphatase, 114
Proteinuria, characteristics of, 538
Proteolysis, function of, 656, 656t
Prothrombin in hemostasis, 395
Proton-motive force, 92, 95
Protons, 20
Proximal tubule, 531
functions of, 540, 542-544, 560
potassium reabsorption by, 574
sodium and water reabsorption by, mechanisms
of, *541*
substances reabsorbed or excreted by, 541t
Pseudohypothyroidism, characteristics of, 679
Psilocybe, 252
PSPs; *see* Postsynaptic potentials
PTH; *see* Parathormone
Ptyalin, function of, 595
Puberty, 721
hypothalamus in, 695, 697
timing of, 697
Puffer fish, tetrodotoxin in, 176
Pulmonary arteries, 358
Pulmonary disease; *see* Lung disease; specific
conditions
Pulmonary edema, 381, 471
mechanism of, 416
Pulmonary emboli, 395
Pulmonary surfactant, 467-468, *468*, 475
characteristics of, 467
in fetus, 738
function of, 468
in respiratory distress syndrome, 468
Pulmonary veins, 358
Pulse, 363
Purine base, 62
Purkinje cells during ventricular excitation, 374
Purkinje fibers, 362, 368
electrical characteristics and innervation of,
361t
Putamen, 224, 341
PVC; *see* Premature ventricular contraction
Pyloric sphincter, 614
Pylorus, function of, 596-597
Pyramidal tract, *346*, 346-347
Pyrimidine base, 62
Pyrodoxine, dietary sources, functions, and
deficiency of, 640t
Pyrogens
endogenous, 755
exogenous, 755
Pyruvate
decarboxylation of, *89*
fates of, *88*
in Krebs cycle, 95
production of, *88*

Q

QRS complex, 370-371, *371*
Quadraplegia, 347
Quantum, defined, 205

R

R wave, cardiac vector for, *372*
Radiation in computed tomography, 229

Radioisotopes, 21
in physiology studies, 11
Rami
gray, *328*, 329
white, *328*, 329
Ranvier, node of, 191, *192*
Rate of movement, defined, 31-32
Reactions; *see* Chemical reactions
Reactive hyperemia, 424
Readiness potentials, motor program initiation by,
343
Reasoning, spatial, 243
Receptive field, 193-194, *194*
defined, 193
Receptor potentials, 162, 162t, 184, 194-195, *195*
characteristics of, 163, 165
intensity coding by, 195
linear and logarithmic coding of, 195-196, *197*
Receptor proteins, 194-195, *195*
Receptors, 215
for acetylcholine, 349, *350*
adaptation of, 215
adrenergic, 349-350
of arterioles and veins, **422-423**
of cardiac muscle, 379
locations of, 351t
work of breathing and, 451
agonists of, 350
autonomic
antagonists of, 350
distribution of, 350-351
carotid sinus, 424
cholinergic, 349
location of, 350t
coupling to ion channels, 194-195, *195*
dihydropyridine, 300
function of, 98
muscarinic, 349, 353, 378
muscle, 335t
neurotransmitter-sensitive, 207
nicotinic, 349, 353, 378
for norepinephrine, 349-350, *351*
olfactory, 283, 286-288, *288*
pain, 211
in respiratory control, 502t
ryanodine, 300
in sense of taste, 283
sensory; *see* Sensory receptors
temperature, 233-234
touch, 231, 233, *233*
graded responses of, 163, *164*
vestibular, *264*, 264-267
Recombinant DNA, ethical questions about, 70
Recoverin, function of, 277
Rectum, 597
reflexes in, *620*
Red blood cells, 392, 418
adult production of, 393
life span of, 637
morphology of, 392
normal range of, 393t
scanning electron micrograph of, *393*
Red nucleus, 223, 346
function of, 225t
5-Reductase, deficiency of, 110
5 α-Reductase enzyme
defect in, 699t, 701
effects of, 699t
Reductases, 58
Reentry, 369
Reflex(es)
antigravity, 340
autonomic, 108, 109, 123
baroreceptor; *see* Baroreceptor reflex
cardiovascular, 421, 422-428
in cephalic phase of digestion, 643-644
in colon, *620*
crossed extension, function of, 353
diving, 495
endocrine, 108, 109-121
enteric, 597
enterogastric, 647
stimulus and effect of, 644t
gastrocolic
function of, 650

Salt(s)—cont'd
  homeostasis of, 565-566
    challenges to, 566-567, 569
Sarcolemma, 292, *293*, 316
  excitation of, 299
Sarcoma, Kaposi's, 775
Sarcomeres, 292, 322
  of cardiac muscle, 376-377
  cross sections of, *295*
  crossbridge formation in, *291*
  H zone of, 292
  shortening of, in crossbridge formation, *294*
Sarcoplasmic reticulum, 79, *299*, 299, 299-300
  calcium in, 300
  calcium ion release from, 366
  of cardiac muscle, 366
  of smooth muscle, 318
  terminal cisternae of, 322
Satiety
  defined, 670
  signals of, 670-671
Saxitoxin, toxic effects of, 176-177
Scala media, *258*
Scala tympani, *258*
Scala vestibuli, *258*
Schizophrenia, dopamine and, 204
Schlemm, canals of, in glaucoma, 269
Sciatic nerve, 327
Sclera, 268, *268*
Scopolamine, effects of, 350
Scrotum
  development of, 698
  varicose veins of, 746
Seasonal affective disorder, melatonin secretion
    and, 241
Secobarbital, 249, 249t
Seconol; *see* Secobarbital
Secretagogues, defined, 646
Secretin
  effect on gastric motility, 649
  function of, 125, 615, 647
  glucagon secretion and, 666
  in intestinal phase of digestion, 647
  secretion of, control of, *648*
  source, target, and actions of, 135
  stimuli, effects, and role of, 645t
  targets of, *648*
Secretions, daily inputs of, 641t
Secretory diarrhea, 615
Sedative, defined, 249
Sedative-hypnotics, 249-250
  action of, 249
  examples of, 249t
Segmentation, intestinal, 604, *605*, 622
Selectivity filter, 148
Semen, 689
  ejaculation of, 720
Semicircular canals, 264, *265*, *266*, 289
Seminal ducts, blocked, 746
Seminal vesicles, 689, 721
Seminiferous tubules, 688, *689*
Senses
  chemical, 283-288; *see also* Olfaction; Taste,
    sense of
  special; *see* Vestibular system
Sensors, function of, 98
Sensory cortex, function of, 227
Sensory fibers, classification, diameter,
    conduction velocity, myelination, and
    function of, 233t
Sensory inputs
  divergent pathways of, *231*
  projection of, 227, 231
Sensory modalities, 193
Sensory nerve cell, electrically excitable regions
    of, 163, *164*
Sensory pathways, 219
Sensory processing, 227, 231
Sensory receptors
  adaptation by, 197, *197*, 200
  integration with brain processes, 189
  internal, 194
  mixed, 200
  nonadapting, 197

Sensory receptors—cont'd
  physiological properties of, 194-197, 200
  rapidly adapting, 200
  sensory coding by, 195-196, *197*
  stimulus intensity and, 195
  tonic, 197
  as transducers, 194
Sensory system; *see also* Neuron(s), sensory;
    Vestibular system
  action potential activation in, 192, *193*
  organization of, 192-194
Sensory tracts, 221
Septum, cardiac, 361
Series elastic elements, 304
Serine, 54t
  nucleotide code for, 66t
  structure of, *53*
Serotonin
  in enteric nervous system, 606
  function of, 204
  in modulation of sensory neuron response, 211
Serous cells, salivary, 606
Sertoli cells, 721
  embryonic differentiation of, 695
Setpoint, 98, *99*, 100, 123
  deviations from, 100
  temperature, 757
Sex
  genetic
    determination of, 690-695, *694*, *694*
    disorders of, 701t
  genital
    determination of, 695
    endocrine disorders affecting, 699t
  gonadal, determination of, 695
Sex hormones, 721
  basal metabolic rate and, 675
Sex-change operations, 700
Sexual development
  gonadal steroids and, 697-698, 700
  hormonal control of, 695, 697-698, 700-701
Sexual identity, development of, 685
Sexual orientation, 700
Sexual response, 717-720, 721
  emotional and physical changes in, 717
  female genital, 720
  male genital, 719-720
  male versus female, 717, *718*, 719
  phases of, 717
Shift work, mental alertness and, 240-241
Shivering thermogenesis, 679-680, 681, 757
Shock
  anaphylactic, 777
  hypovolemic, 436, *437*, 443
  insulin, 660
Shunt
  arteriolar-venular, 399
  physiological, 471
Sickle cell anemia, 393, 486
  amino acid substitution in, 57
  vasa recta in, 549
SIDS; *see* Sudden infant death syndrome
Siemens, 37
Silicosis, 454
Simplesse, 637
Single photon emission computed tomography,
    229
Sinoatrial node, 361
  cellular mechanism of, 364-366
  electrical characteristics and innervation of,
    361t
  function of, 362
Sinus(es)
  lactiferous, 744
  maternal blood, 732, *733*
Skeletal muscle, 292-295
  anatomy of, *292*
  antagonist, 108-109
  ATP in, 308
  attachment of, 292
  autoregulation of blood flow in, 424
  versus cardiac and smooth muscle, *320*, 320t
  versus cardiac muscle, 322, 366
  connective tissue sheath of, 292

Skeletal muscle—cont'd
  contractile force of, 326-327
  contraction of, 35, 325, 516
    at molecular level, 295-297, 299
    sliding filament model of, 292-293, *295*
  distal, 343
  diving reflex and, 495
  energy metabolism of, 308-310
  energy sources for exercising, 516
  extrinsic control of blood flow in, 424
  fiber types of, 308-310, 309t
  function of, 10
  hypertrophy of, 312
  innervation of, 327
  intrinsic control of blood flow in, 424
  maximum force of, 304
  motor unit activation in, 331
  negative feedback in, 98
  nerve fiber innervation of, *326*
  neurons in, 209
  origin and insertion of, 292
  preload and, 376
  proximal, 343
  and skeletal level co-adaptation, 304, 306, *306*
  somatic motor system control of, 325
  stretch reflex of, *108*
  striations of, 291
  structure of, 322
  during sustained exercise, 437
  synapses of, 203
Skeletal muscle pump, 435, *435*
Skeletal pump, 443
Skeletal system, components and functions of,
    12t
Skin
  microcirculation in, *401*
  temperature of, during exercise, 521, *521*
  touch receptors in, 231, 233, *233*
Sleep
  as active function, 243
  age and, 246
  cycle of, 246
  EEG during, 243, *244*
  rapid eye movement, 243, 246, 247
  slow-wave, 243, 246, 247
  stages of, 243, 247
Sleep apnea, 513
Slit pores, 537
Smell, sense of; *see* Olfaction; Olfactory entries
Smith, Homer, 563
Smoking; *see* Cigarette smoking
Smooth endoplasmic reticulum, characteristics
    and functions of, 75t
Smooth muscle
  action potentials of, 603
  basic electrical rhythm of, 316, *318*
  bronchiolar, factors affecting, 452t
  versus cardiac and skeletal muscle, *320*, 320t
  cells of, 603
  characteristics of, 291
  contractile states of, 318t
  contractile units of, *317*
  contraction of, 316, 322
  excitation-contraction coupling in, 316,
    318-321
  gastrointestinal, 603-606, 625
    cell pacemaker activity in, 603-604, *604*
  in large intestine, 621
  length-tension relationships in, 316
  longitudinal section of, *317*
  multi-unit, 316
  pacemaker activity of, 316
  physiology of, 316, 318-321
  sarcoplasmic reticulum of, 318
  single-unit, 316
  thick filaments of, 319, 322
  thin filaments of, 319-320, 322
  types of, 316
  vascular
    extrinsic control of, 422-423, 423t
    intrinsic control of, 423t, 423-424
Sodium; *see also* Salt(s)
  aldosterone in regulation of, 537
  in chemical reactions, 23

Thymosin, source, target, and actions of, 135
Thymus, 125
  aging and, 765
  B cell differentiation in, 764
  hormones of, target and actions of, 135
  T cell differentiation in, 764, 765
Thyroglobulin, 675
Thyroid gland, 125
  enlargement of, 678
  follicles of, 676
  location of, 676
Thyroid hormone, 110, 111, 681
  basal metabolic rate and, 674-675, 675, 678-681
  chemical characteristics and secretion site of, 657t
  disorders involving, 678-679
  formation of, 675, 677
  half-life of, 115
  hyposecretion of, 130
  regulation of, 678
  source, target, and action of, 132
  target and actions of, 132
Thyroid-stimulating hormone, 116, 678
  depressed levels of, 130
  function of, 116
  in hypothyroidism, 678
  releasing hormones and target/effect, 118t
  source, target cells, and action of, 131
Thyroid-stimulating hormone releasing hormone, 678
  secretion of, with cold exposure, 680-681
Thyrotropin, 116, 678; see also Thyroid-stimulating hormone
Thyrotropin-releasing hormone, 110
  source, target, and actions of, 132
Thyroxine, 110, 681
  basal metabolic rate and, 675, 681
  formation and secretion of, 677
  secretion of, control of, 678
  source, target, and actions of, 132
  synthesis of, 677
Tidal volume, 460, 515
  alveolar minute volume and, 467
  defined, 458
  definition and resting value of, 459t
  during exercise, 519
  factors determining, 501-502
Tight junctions, 85-86, 86
Tissue(s); see also specific tissue types
  adipose, 10
  autonomic innervation of, 349t
  cell types in, 68
  chemical regulation of, 104-106
  connective, function of, 10
  coordination of, by gap junctions, 104
  defined, 10
  epithelial, 10
  hormones of, target and actions of, 135
  intrinsic versus extrinsic regulation of, 103-104, 104
  muscle, 10
  nervous, function of, 10
  replacement of, 65
Tissue plasminogen activator after heart attack, 441
Tissue thromboplastin, 395
Titration curve of bicarbonate buffer system, 580-581, 581, 589
TLC; see Total lung capacity
TNF; see Tumor necrosis factor
Tobacco, use of, 250
Tocopherol, dietary sources, functions, and deficiency of, 640t
Tone, pure, 256
Tongue, regions sensitive to primary flavors, 285
Tonicity
  defined, 144
  versus osmolarity, 144t
Tonotopic map, 262-263
Total lung capacity
  defined, 459
  definition and resting value of, 459t
Total peripheral resistance, 418, 443
  arteriolar resistance and, 413

Total peripheral resistance—cont'd
  defined, 407
  during submaximal exercise, 439t
Touch, sense of, 193
Touch receptors, 231, 233, 233
  graded responses of, 163, 164
Toxic shock syndrome, 771-772
Toxicity, antibody-dependent cellular, 754
Toxins
  diarrhea-producing, 642
  ion channel specificity of, 176-177
  neutralization of, 761
TPA; see Tissue plasminogen activator
TPR; see Total peripheral resistance
Trace elements in enzyme-catalyzed reactions, 59
Tracer experiments, uses of, 21
Trachea, 450, 475
  diameter and functions of, 451t
Training
  endurance, 523
  strength, 523
Tranquilizers
  classification of, 249
  examples of, 249t
  uses of, 249
Transamination, process of, 557
Transcervical balloon tuboplasty, 747
Transducers, sensory receptors as, 194
Transducin, 277
Transfer RNA, 78
  function of, 63
  structure of, 67, 68
Transferrin, 637
  in iron absorption, 650
Transforming growth factors, function of, 122
Transfusion reactions, 772-773, 773t
Transplantation, renal, 535
Transplantation reactions, 767, 772-773
Transport
  active, 155
    versus passive, 146-147
  anterograde, 192
  axoplasmic, 192
  counter, 149, 155
  mediated, 146
  proteins in, 146-149
Transport maximum, defined, 542
Transport systems, characteristics and examples of, 146t
Transsexuality, 700
Transverse tubules, 299, 299
Triacylglycerols, 44, 632
  breakdown of, 625
  characteristics of, 656
  digestion of, 626
  storage and release of, 660-661
  structure of, 46
  synthesis of, 635
Triazolam, 249, 249t
Trichromats, 275
Tricuspid valve, 362, 362
Tricylglycerols, structure of, 634
Triglycerides, 636
  structure of, 46
  subunits and function of, 45t
Triiodothyronine, 110, 681
  basal metabolic rate and, 675, 681
  formation of, 677
  secretion of, control of, 677, 678
  synthesis of, 677
Tritanopes, 275
Tritium, 20
Trochlear nerve, 222
  pathway and function of, 223t
Trophoblast, 728, 749
  endocrine functions of, 730, 732
Tropomyosin, 295, 322
  in crossbridge cycle, 298
  in resting and contracting striated muscle, 302t
  with troponin, 302
Troponin, 322
  binding of, 295

Troponin—cont'd
  calcium binding to, 300
  in resting and contracting striated muscle, 302t
  with tropomyosin, 302
Trypsin, 649
  function of, 617
  pancreatic, 617, 618
  in protein digestion, 632
  substrate and products of, 629t
Trypsin inhibitor, 617
Trypsinogen, conversion to trypsin, 617
Tryptophan, 54t
  nucleotide code for, 66t
  structure of, 53
TSH; see Thyroid-stimulating hormone
TSHRH; see Thyroid-stimulating hormone releasing hormone
T-tubules, 299, 299
TTX; see Tetrodotoxin
Tubal ligation, 714
  failure rate of, 715t
Tubules
  distal; see Distal tubule
  proximal; see Proximal tubule
Tubuloglomerular feedback, 537, 560
  juxtaglomerular apparatus and, 556
Tufted cells of olfactory bulb, 288
Tumor(s)
  cytotoxic T cells and, 766-767
  growth factors and, 122
  pituitary, acromegaly from, 673-674
Tumor necrosis factor
  function of, 756
  immune function of, 757t
Tunica externa, 399, 402
Tunica intima, 399, 402
Tunica media, 399, 402
Turner's syndrome, 701
  sex chromosomes, body plan, effects of, 701t
Twitch, muscle fiber, 302, 303, 308
Two-point discrimination
  demonstration of, 239
  thresholds for, 239
Two-point discrimination test, 237
TX$_{A2}$; see Thromboxane
Tympanic membrane, 256-257, 257, 289
Tyrosine, 54t
  derivatives of, 110
  nucleotide code for, 66t
  structure of, 53

U

Ulcer(s)
  duodenal, 613
  gastric, cimetidine for, 613
  peptic, causes of, 613
  stress and, 5
Ultrafiltration, 560
  defined, 143
Ultrasound in prenatal diagnosis of birth defects, 741, 741
Umbilical arteries, 732
Undernutrition, 655
Uracil
  in RNA, 62
  structure of, 62
Uracil triphosphate, 87
Uranium, nucleus of, 20
Urea
  normal range in blood, 393t
  renal handling of, 551t
Ureter(s)
  anatomy of, 528
  function of, 560
Urethra, 529
  function of, 560
  male, 689, 690
Uric acid, normal range in blood, 393t
Urinary bladder; see Bladder
Urinary incontinence, 530
Urinary retention, 530
Urinary system, components and functions of, 12t
Urination, 529, 560
  control of, 348

Minerals

Mineral	Minimal daily requirement	Function
Important minerals in the body		
Calcium	0.8 to 1.2 gm	Bone and tooth formation; blood clotting; muscle contraction; nerve activity
Chlorine	1.7 to 5.1 gm	Blood acid-base balance; acid production (stomach)
Chromium	0.05 to 0.20 mg	Associated with enzymes in glucose metabolism
Cobalt	Unknown	Component of vitamin B_{12}; erythrocyte production
Copper	2.0 to 3.0 mg	Hemoglobin and melanin production; component of electron transport chain
Fluorine	1.5 to 4.0 mg	Extra strength to teeth
Iodine	100 to 150 µg	Thyroid hormone production; maintenance of normal metabolic rate
Iron	10 to 18 mg	Component of hemoglobin; ATP production in the electron transport chain
Magnesium	300 to 350 mg	Coenzyme constituent; bone formation; muscle and nerve function
Manganese	2.5 to 5.0 mg	Hemoglobin synthesis; growth; activation of several enzymes
Molybdenum	0.15 to 0.5 mg	Enzyme component
Phosphorus	800 to 1200 mg	Bone and tooth formation; important in energy (ATP) transfer; component of nucleic acids
Potassium	1.8 to 5.6 gm	Muscle and nerve function; heart rate
Selenium	0.05 to 0.2 mg	Component of many enzymes
Sodium	1.1 to 3.3 gm	Osmotic pressure regulation; nerve and muscle function
Sulfur	Unknown	Component of hormones, vitamins, and proteins
Zinc	10 to 15 mg	Component of several enzymes; carbon dioxide transport and metabolism; requirement for protein metabolism